B

Directory of Research Workers in Agriculture and Allied Sciences

R. Vernon, Editor

C·A·B INTERNATIONAL

C·A·B International
Wallingford
Oxon OX10 8DE
UK

Tel: Wallingford (0491) 32111
Telex: 847964 (COMAGG G)
Telecom Gold/Dialcom: 84: CAU001
Fax: (0491) 33508

British Library Cataloguing in Publication Data

Vernon, R.
  Directory of research workers in
  agriculture and allied sciences.
  1. Agriculture. Research organisations – Directories
  I. Title
  630'.72

  ISBN 0-85198-623-4

Printed in the UK by BPCC Wheatons Ltd, Exeter

CONTENTS

CONTENTS

INTRODUCTION

This Directory is both the last of a distinguished line and the first of a new one, being published at a turning point in the history of CAB International. From 1936 the erstwhile Commonwealth Agricultural Bureaux has been publishing successive editions of its 'List of Research Workers in Agricultural Sciences in the Commonwealth'. The List has been highly valued as a unique treasury of information on the research pursuits of agriculturalists in the many countries covered. The present publication continues this honourable tradition, identifying researchers at Government and state-aided establishments in 29 countries and three Dependent Territories, members of CAB until 1986.

But 1986 marked the launch of CAB International, an organization encompassing all of CAB's previous world services to agriculture, while able to admit non-Commonwealth countries into membership. In keeping with this truly international remit, CAB International hopes in future to extend its directories to cover progressively all countries of the world.

As a first step, the present publication makes a valuable contribution to communications worldwide by listing the names, positions, studies and locations of approximately 29,000 scientists whose research work contributes to the world of agriculture. In this context 'agriculture' is taken to cover agricultural economics, food science and technology, forestry, fisheries, veterinary science and agriculturally relevant work in allied disciplines such as mycology, entomology, botany, zoology etc. Museums, universities, training establishments and extension services are included among other research institutions.

For this edition, the bulk of the data was collected in 1986. Publication was postponed, however, in view of the sweeping reorganization of governmental departments in several of the larger countries. Amendments to reflect the changes were collected in early 1988, bringing the data up-to-date for most countries. One exception is Australia, where CSIRO has undergone substantial reorganization, still not finalized. Similarly, for the UK's AFRC, the structural changes of 1985 and 1986 are reflected but not any subsequent developments. Hungary was welcomed into full membership of CABI in December 1988, but with the need for timeliness in mind we have been unable to collect details of her numerous agricultural research scientists. It was decided not to delay publication further, but to print the available information and ask you, the reader, to be cautious in its use.

In such a large undertaking, spread over so many countries, there are bound to be some omissions and inexactitudes. But forthcoming editions will afford the opportunity to rectify, expand and update. Any changes and suggestions will be gratefully received if brought to the notice of the Editor, Directory of Research Workers, at CABI, Wallingford, Oxon, OX10 8DE, UK.

ACKNOWLEDGEMENTS

The task of collecting information from so many different and distant locations is a huge endeavour, and grateful thanks must be extended to the CABI Liaison Officers and others who contributed the majority of entries. The list below shows those in post during the initial data collection period, but where an appointment has been changed more recently, the name of the current officer is given below his predecessor.

CABI Liaison Officers

Australia
Dr J.B. Allen
CSIRO
P.O. Box 225
Dickson, ACT 2602
*(now Dr E. F. Henzell,
at the same address)*

The Bahamas
Director of Agriculture
Ministry of Agriculture, Fisheries
and Local Government
P.O Box 3028
Nassau

Bangladesh
The Executive Vice Chairman
Bangladesh Agricultural
Research Council
G.P.O Box 3041
Dhaka

Botswana
Permanent Secretary
Ministry of Agriculture
Private Bag 0028
Gabarone

Brunei Darussalam
Dr Morni bin Othman
Dept of Agriculture
Ministry of Development
Negara

Canada
Dr R. Trottier
Research Branch
Agriculture Canada
Sir John Carling Building
Ottawa
Ontario K1A OC5
*(now Dr G Neish, at the
same address)*

Cyprus
Dr A Louca
Ministry of Agriculture
and Nature Resources
Nicosia
*(now Mr L. Shacallis, at the
same address)*

Fiji
Permanent Secretary
Ministry of Agriculture, Fisheries
and Forests
Rodwell Road
Suva

The Gambia
Permanent Secretary
Ministry of Agriculture and
Natural Resources
The Quadrangle
Banjul

Ghana
Executive Chairman
CSIR
P.O. Box M32
Accra

Guyana
Dep Chief Agricultural
Officer (Research)
Ministry of Agriculture
P.O. Box 1001
Georgetown
*(now Dr Michael Granger,
Director, at the same address)*

Hungary
Mr S. Szabo
Agroinform
Agrarinformacios Vallalat
1 Attila ut 93
1012 Budapest
*(name given for completeness,
although Hungary joined CABI in
December 1988 and data are
not included in this book)*

India
Dr N.S. Randhawa
Indian Council of
Agricultural Research
Krishi Bhavan
New Delhi 110001
*(the help of Mr P.C. Bhose
must also be gratefully
acknowledged)*

Jamaica
Mr A.C. MacDonald
Ministry of Agriculture
P.O. Box 480
Kingston 6
*(now Dr R.J. Baker, at
the same address)*

Kenya
Mr W. W. Wapakala
Ministry of Agriculture
P.O. Box 30028
Nairobi

Malawi
Chief Agricultural Research
 Officer
Ministry of Agriculture and
Natural Resources
P.O. Box 30134
Capital City
Lilongwe 3

Malaysia
Kamaruddin bin Siaraf
Ministry of Agriculture
Jalan Mahameru
Kuala Lumpur
(now Mr Mohamed Yasin Saleh,
Principal Assistant Secretary,
at the same address)

Mauritius
Mr K. Lutchmeenaraidoo
Agricultural Service
Reduit

New Zealand
Dr J.H. Troughton
DSIR
Private Bag
Wellington
(now Dr R.T.J. Clarke,
at the same address)

Nigeria
The Director
Agricultural Sciences
Research Department
Moor Plantation
Pvt Mail Bag No. 5328
Ibadan

Papua New Guinea
The Secretary
Department of Primary Industry
P.O. Box 417
Konedobu

Sierra Leone
Senior Lecturer
Njala University College
University of Sierra Leone
PMB
Freetown

Solomon Islands
Mr B.G.C. Smith
Ministry of Agriculture and Lands
P.O. Box G13
Honiara
(now Mr James Roni, Under
Secretary (Agriculture),
at the same address)

Sri Lanka
Dr I Goonewardena
Department of Agriculture
Peradeniya

Tanzania
Director of Agriculture
Ministry of Agriculture
P.O. Box 9071
Dar−es−Salaam

Trinidad and Tobago
Chief Technical Officer
(Crop Research)
Ministry of Agriculture,
Lands and Food Production
Central Experimental Station
Centeno
P.O., Arima

Uganda
Permanent Secretary
Ministry of Agriculture and
Forestry
P.O. Box 102
Entebbe

United Kingdom
Under Secretary
Agricultural and Food
Research Council
160 Great Portland Street
London W1N 6DT
(now Mr F.G. Raymond,
Ministry of Agriculture, Fisheries and Food,
Great Westminster House
Horseferry Road
London SW1P 2AE)

Zambia
Permanent Secretary
Ministry of Agriculture and
Water Development
P.O. Box 50291, Ridgeway
Lusaka
(now Mr E. Lumande,
Mount Makulu Central Research Station,
P.O. Box 7, Chilango)

Zimbabwe
Mr P.R.N. Chigaru
Dept of Research and
Specialist Services
Ministry of Agriculture
P.O. Box 8108
Causeway
Harare
(now Mr R.J. Fenner,
at the same address)

HOW TO USE THIS DIRECTORY

In keeping with the aim of assisting communications worldwide, this Directory allows you to locate agriculturalists in at least four ways:

a) by name:

If you know a researcher's name, look up the **Index of Persons** starting on page 395, where you will find a reference to the country abbreviation (in italics), the institution entry number (in bold type) and the person number (in small type). A key to the country abbreviations is given below.

Example
Entry in the Index of Persons   Waage JK *UK* **4002** 4
Entry at institution 4002   4 Waage JK – AB,PhD,DIC *Chief Research Officer* applied ecology

b) by country:

To contact someone in a particular country, go straight to the main text, where countries are given in alphabetical order. The usual order of entries within a country places international organizations first, then government departments, then state–aided institutions and universities; but this order may be varied to reflect the circumstances of some countries with a complex infrastructure.

c) by institution:

The **Index of Institutions** starting on page 485 includes entries for all the organizations listed, including some divisions and offices of larger institutions. Each entry gives the abbreviation of the country concerned and a reference in bold type to the entry number of the appropriate institution or office, etc.

d) by subject:

Entries in the **Subject Index** starting on page 325 give a reference in bold type to the appropriate institution entry number, with person number or numbers in small light type.

Example
     abattoirs
       hygiene AUS **851** 2

Person no. 2 of institution 851 (which happens to be in Australia) works on abbatoir hygiene

The references include country abbreviations (e.g. UK, CAN) to help you judge the relevance of the research in question, but in practice you can find the required institution by its number, without needing to know the country. A full list of the abbreviations appears below. Please note that very broad subject areas, such as 'agronomy' are not listed in the index without a qualifier because the list of relevant researchers would be unmanageably large.

Country Abbreviations

AUS	Australia	GUY	Guyana	NZ	New Zealand
BGD	Bangladesh	HK	Hong Kong	PNG	Papua New Guinea
BHS	The Bahamas	IN	India	SLB	Solomon Islands
BMU	Bermuda	JAM	Jamaica	SLE	Sierra Leone
BRN	Brunei	KE	Kenya	TT	Trindad & Tobago
BWA	Botswana	LK	Sri Lanka	TZ	Tanzania
CAN	Canada	MSR	Montserrat	UG	Uganda
CY	Cyprus	MU	Mauritius	UK	United Kingdom
FJ	Fiji	MW	Malawi	ZM	Zambia
GH	Ghana	MY	Malaysia	ZW	Zimbabwe
GMB	Gambia	NGA	Nigeria		

Organizations and Research Workers, by Country

COMMONWEALTH GOVERNMENT

1 Australian Biological Resources Study

Bureau of Flora and Fauna, GPO Box 1383, Canberra City, 2601
Telex: AA62960.

Part of the Department of Arts, Heritage and Environment

1 Bridgewater P - BSc,PhD *Director* vegetation ecology

2 Distribution Studies Section

1 Busby JR - MSc,PhD *Head of Section* biogeoclimatology

3 Fauna Section

1 Richardson BJ - BSc,PhD *Head of Section* biochemical systematics
2 Dyne GR - BSc,PhD,FLS systematics; evolutionary biology
3 Longmore RC - BSc herpetology
4 Wallace CC - BSc,PhD systematics and ecology of corals
5 Walton DW - BSc,PhD *Executive Editor* systematics, medical zoology

4 Flora Section

1 Hnatiuk RJ - MSc,BSc,PhD *Head of Section* plant ecology
2 Chapman AD bibliography; nomenclature
3 George A - BA *Executive Editor* plant taxonomy
4 Hewson HJ - BSc,PhD taxonomy
5 Purdie R - BSc,PhD environmental biology

5 The Australian Wine Research Institute

Private Bag, Glen Osmond, South Australia, 5064
Tel: (08) 79 6817. Telex: AA88657.

1 Lee TH - BSc,PhD,FAIFST *Director*

6 Microbiology Group

1 Monk PR - PhD,DipTech *Group Leader* yeast growth; lactic acid bacteria
2 Gockowiak H - BSc immunofluorescence techniques, image analysis

7 Chemistry Group

1 Pocock KF - BAppSc,FAIFST phenolics of grapes and wine
2 Simpson RF - BSc,PhD,DipEd,JP taints, volatiles of wine
3 Somers TC - BSc,MSc,DSc,FAIFST,FRACI phenolics of grapes and wine
4 Strauss CR - BSc,PhD grape and wine flavour
5 Williams PJ - BSc,PhD,FRACI *Group Leader* grape and wine flavour

8 Extension Group

1 Baldwin GE - BAppSc *Group Leader* winemaking
2 Miller GC - BSc analytical chemistry
3 Steele WJ - BA librarian

9 Australian Wool Corporation

Research and Development Department, P.O. Box H274 Australia Square, Sydney, New South Wales, 2000
Tel: 022693911. Telex: AA20633. Cables: NWOOL. FAX: 02-2902169.

1 Booth PC - BScAgr,PhD *Manager*
2 Bennett Miss JM - BSc,PhD *Controller Textile Research*
3 Charlton D - MSc *Controller Distribution Research*
4 Graham MA - BScAgr,M.Com *Research Officer, Planning and Evaluation*
5 Hart RJ - BSc,PhD *Controller Wool Production Research*
6 Hudson PRW - BSc,PhD *Controller Wool Harvesting Research*

10 Commonwealth Scientific and Industrial Research Organisation (CSIRO)

Limestone Avenue, Campbell, A.C.T.
[225, Dickson, Campbell, A.C.T., 2602]
Tel: (062) 48 4211. Telex: AA62003.

1 Boardman NK - ScD,FAA,FRS *Chief Executive*

11 Institute of Animal Production and Processing

Limestone Avenue, Campbell, A.C.T.
[225, Dickson, Campbell, A.C.T., 2602]
Tel: (062) 48 4211. Telex: AA62003. Cables: NWOOL.

1 Donald AD - BVSc,MACVSc,PhD *Director*

12 Australian Animal Health Laboratory

P.O. Bag 24, Geelong, Victoria, 3220
Tel: (052) 265222. Telex: 38923.

The functions of this microbiologically secure laboratory are to provide diagnostic, research, training and vaccine development and testing services in support of control and eradication programs for exotic livestock diseases that enter Australia; to ensure that livestock imported through animal quarantine stations are free from exotic diseases; and to certify that livestock imported from Australia are free from specific diseases.

1 Snowdon WA - BVSc,PhD,MACVSc *Chief*
2 Dennett D - BSc,PhD *Scientific Services Officer* information services
3 Wood HE - DipMedRad,DipLib *Librarian*

13 Epidemiology

1 Parsonson IM - BVSc,PhD,MACVSc *Assistant Chief* virology
2 Forman AJ - BVSc,DipTropVetMed,PhD virology
3 Westbury HA - BVSc,MVSc,PhD virology

14 Molecular Virology Group

1 Della-Porta AJ - BSc,PhD
2 Eaton BT - BSc,MSc,PhD nucleic acid chemistry
3 Gorman JJ - BSc,PhD protein chemistry
4 Gould AR - BSc,PhD recombinant DNA analysis
5 McPhee DA - BSc,PhD cell biology

15 Pathogenesis and Immunity Group

1 Boyle DB - MSc,PhD immunochemistry

16 Microbiological Security Group

1 Ludford CG - BSc,PhD,MASM microbiological security

17 Division of Animal Health

Cnr. Flemington Road and Park Drive, Private Bag No. 1, Parkville, Victoria, 3052
Tel: (03) 342 9700. Telex: 21376 32677.

1 Bagust TJ - BVSc,PhD *Chief of Division*

18 Animal Health Research Laboratory

Cnr Flemington Road and Park Drive, Parkville, Victoria
[Private Bag No 1, Parkville, Victoria, 3052]
Tel: (03) 347-2311. Telex: 32677.

1 Anderson N - BVSc,PhD,MACVSc parasitic infections
2 Cottew GS - MSc bacterial infections
3 Culvenor CCJ - PhD,DPhil,DSc plant-associated
4 Dufty JH - BVSc,RDA bacterial infections
5 Edgar JH - BSC,PhD plant-associated toxins
6 Emery DL - BScVetSci,BVSc,PhD bacterial infections
7 Fahey KJ - MSc,PhD avian diseases
8 Ignjatovic Miss J - BSc,MSC avian diseases
9 Jago Mrs MV - BSc,MSc plant-associated toxins
10 Lepper AWD - BVetMed,MRCVS,PhD,MACVSc bacterial infections
11 Lloyd LC - BVSc,PhD,MACVSc bacterial infections
12 Peterson JE - BVSc,MACVSc plant-associated toxins
13 Plackett P - BA bacterial infections
14 Sonza S - BSc,PhD avian diseases
15 Stewart DJ - BVSc,PhD bacterial infections
16 Wood PR - BSc,PhD bacterial infections

19 McMaster Laboratory
Parramatta Road, Private Bag No 1, Glebe, N.S.W., 2037
Tel: (02) 660 4411. Telex: 21376.

1 Waller PJ - MVSc,PhD *OIC* mucosal immunity, immunopotentiation and immunogenetics
2 Beh KJ - BVSc,PhD mucosal immunity, immunopotentiation and immunogenetics
3 Burrell DH - MVSc bacterial infections
4 Grant W - PhD mucosal immunity, immunopotentiation and immunogenetics
5 Hennessy DR - MSc,PhD mucosal immunity, immunopotentiation and immunogenetics
6 Lacey E - BPharm,MSc,PhD parasitic infections
7 Lascelles AK - MVSc,PhD,FACVSc mucosal immunity, immunopotentiation and immunogenetics
8 Mukkur TK - MVSc,PhD mucosal immunity, immunopotentiation and immunogenetics
9 Outteridge PM - BSc,PhD parasitic infections
10 Steel JW - BSc,PhD mucosal immunity, immunopotentiation and immunogenetics
11 Wagland BM - BSc,PhD mucosal immunity, immunopotentiation and immunogenetics
12 Waller PJ - MVSc,PhD parasitic infections
13 Windon RG - BSc,PhD mucosal immunity, immunopotentiation and immunogenetics

20 Division of Animal Production
Great Western Highway, Prospect
[239, Blacktown, New South Wales, 2148]
Tel: Sydney (02) 631-8022. Telex: 27450.

21 Ian Clunies Ross Animal Research Laboratory

1 Scott TW - BScAgr,PhD,FTS *Chief of Division*
2 Alexander G - DAgrSc foetal and neonatal physiology, animal behaviour
3 Ashes JR - BSc,PhD ruminant physiology
4 Bennett JW - BSc animal husbandry aspects of biological wool harvesting
5 Black JL - BAgrSc,DipEd,PhD control of nutrient utilization in sheep; computer simulation analyses
6 Brown BW - MSc,DipTech(Sc) physiology of reproduction
7 Brownlee AG - BSc,PhD ruminant microbiology
8 Byrne Ms CR - MSc molecular biology
9 Ch'ang TS - MAgrSc,PhD selection for meat production and decreased adiposity in sheep
10 Chapman RE - BScApp,DSc ultrastructure of the wool follicle
11 Colebrook WF - BScAgr control of nutrient utilization in sheep
12 Connell PJ - BSc amino acids analysis
13 Cox RI - BSc,PhD endocrinology, immunophysiology, oestrogens, phyto-oestrogens, prostaglandins
14 Davis P - MIBiol digestion in ruminants
15 Dollin Mrs AE - BA cytogenetics
16 Evans MK - BA statistical analyses of sheep data
17 Evans R - BSc,DipEd statistical analysis of sheep data
18 Faichney GJ - MScAgr,PhD,DAgrSc rumen function, physiology of digestion
19 Fleming JF - BSc computer programming
20 Foldes A - BSc,DipEd,PhD function of the pineal gland in sheep
21 Franklin IR - BSc,PhD *Assistant Chief of Division*
22 Gordon GLR - BSc,PhD rumen microbiology
23 Graham N - McC,BAgr,PhD,DSc ruminant energy metabolism
24 Hales JRS - MSc,PhD thermoregulatory and cardiorespiratory function
25 Hazelton IG - BRurSc *Farm Manager* embryo manipulation
26 Hollis DE - FAIMLS,FACBS electron microscopy of cutaneous and reproductive tissues in sheep
27 Hoskinson RM - BSc,ASTC,PhD organic synthesis, immunophysiology
28 Hunt Ms CL - BSc microbiology, molecular biology
29 Jackson N - BSc,PhD genetic statistics in sheep breeding for wool production
30 Jones MR - BAgrSc immunophysiology in ruminant production
31 Lax J - MAgrSc,PhD genetic and environmental factors influencing meat production in sheep
32 Leche TF - MScAgr,PhD liaison with animal production industries concerned with nutrient requirements of livestock
33 Leish Mrs Z - RNDr,PhD metabolism and biochemical actions of agents used for biological wool harvesting
34 Margan DE - MSc whole animal calorimetry (sheep); ruminant digestive function
35 Mattner PE - BAgrSc,MVSc,PhD,MACVSc physiology of reproduction
36 Maxwell CA - DipTech pineal involvement and seasonality involvement of adrenergic systems in physiological process
37 McConnell SJ - BSc,PhD hormonal regulation of appetite, growth and energy metabolism
38 McGuirk BJ - MSc,PhD sheep genetics and wool production
39 Moore GPM - BSc,PhD regulation of follicle development and wool growth; biological wool harvesting
40 Murray JD - MSc,PhD somatic cell and cytogenetics
41 Nancarrow CD - BAgrSc,PhD foetal endocrinology and reproductive physiology of cattle and sheep

42 O'Keefe JH - DICQ,MSc,BA,ASTC,ARACI applications of chemical instrumentation (mass spectrometry, nuclear magnetic resonance and infrared) to biological problems
43 Ostrowski-Meissner HT - MSc,PhD *Director, Australian Feeds Information Centre*
44 Pan YS - MScAgr,PhD heat stress and skin and hair in cattle and pigs
45 Panaretto BA - BVSc,PhD effects of hormones on wool growth, biological wool harvesting
46 Phillips MW - BSc,PhD rumen microbiology
47 Pisansarakit P - MRurSc culture of hair follicle cells
48 Podger RN - MScAg poultry breeding
49 Radford HM - BSc neuro-endocrinology of reproduction
50 Raphael Mrs KA - BA,PhD embryology and biochemistry of skin development
51 Reis PJ - BScAgr,DScAgr nitrogen metabolism and wool growth
52 Rigby Mrs NW - BSc molecular biology
53 Rigby RDG - BSc,PhD synthesis of biologically active peptides
54 Scaramuzzi RJ - BScAgr,PhD physiology of reproduction
55 Searle TW - BSc whole body composition in sheep comparative slaughter trials
56 Sharry LF - MSc,ARMTC application of analytical and isotope chemistry to biological problems
57 Sheldon BL - BScAgr,PhD selection genetics - poultry and Drosophila
58 Stacy BD - BSc,PhD *Assistant Chief of Division*
59 Stevens Mrs D - MSc foetal and neonatal physiology
60 Sutton Mrs R - MSc,PhD molecular biology
61 Tunks DA - BSc nitrogen metabolism and wool growth, biological wool harvesting
62 Turnbull KE - BA,MSc physiology of reproduction
63 Van Dooren PH - BSc,ASTC isolation of biologically active peptides
64 Walton JR - BA,PhD ovine embryology
65 Ward KA - MSc,PhD molecular biology;gene transfer
66 Weston RH - BScAgr,Ms ruminant feed intake and digestion
67 Westwood NH - Dipl (Computing) statistical programming;data analysis
68 Wilson BW - BSc molecular biology
69 Wilson PA - BSc endocrinology, immunophysiology, phyto-oestrogens
70 Wong M - BSc,PhD endocrinology, immunophysiology
71 Wynn PC - MRurSc,DipEd,PhD skin biology, ruminant metabolic endocrinology
72 Yoo BH - MSc,PhD population and quantitative genetics of poultry

22 Pastoral Research Laboratory
New England Highway, Private Mail Bag, Armidale, 2350
Tel: (067) 78 4000. Telex: 166916.

1 Hutchinson KJ - MScAgr,PhD *Assistant Chief of Division*
2 Adams DB - MVSc,DPhil mucosal immunity, immunopotentiation and immunogenetics
3 Barger IA - BAgrSc,MScAgr parasitic infections
4 Bindon BM - BSc,MRurSc,PhD physiology and genetics of reproduction in sheep and cattle
5 Dash KM - BVSc,PhD parasitic infections and communications
6 Donegan Ms SM - BRurSc responses of dairy cattle to heat stress and climate room testing of bulls
7 Ellis KJ - BSc,PhD application of controlled release systems in mineral nutrition of grazing animals and their biochemical mode of action
8 Furnival EP - BSc,MRurSc quantitative studies on nutrition of grazing animals
9 Hedges DA - BScAgr application of pattern and multi-variate analysis to animal production
10 King Miss KL - BSc,BA management effects of invertebrate and microbial decomposers involved in nutrient cycling
11 Langlands JP - BScAgr,NDA,PhD,DSc mineral nutrition of grazing sheep and cattle
12 Lynch JJ - BVSc,PhD animal behavioural studies in sheep
13 Marjoram AR - Grad IREE,BSc remote sensing and assessment of fertilizer needs of pasture
14 Mottershead BE - BSc animal behavioural studies in sheep
15 Pickering FS - MScAgr,ARCI quantitative studies on nutrition of grazing animals
16 Piper LR - BRurSc,PhD genetics of reproduction and parasite resistance in sheep and cattle
17 Roberts JA - BVSc,BA,PhD parasitic infections
18 Rocks RL - BSc radiation safety and radioisotope measurements; salt block characteristics and licking behaviour by sheep
19 Till AR - MSc,PhD radioisotope studies of nutrient cycling in animal production systems
20 Vickery PJ - BScAgr,MRurSc,PhD nutrient status and production in grazed pastures; applications of remote sensing to pasture agronomy
21 Watson DL - BScAgr,PhD bacterial infections
22 Wheeler JL - BScAgr,PhD nutritive value of forages; role of fodder crops in grazing systems

23 Ryde Laboratory
Delhi Road, P.O. Box 184, North Ryde, 2113
Tel: (02) 886 4888. Telex: 22970.

1 Collins RK - BSc cytogenetics - its implications for poultry breeding
2 Thorne MH - BA cytogenetics - its implications for poultry breeding

24 Minerals Unit
Underwood Avenue, Floreat Park, Private Bag, PO, Wembley, Western Australia, 6014
Tel: (09) 387 4233. Telex: 92178.

1 Purser DB - BSc,PhD *Assistant Chief of Division*
2 Adams NR - BVSc,PhD reproductive efficiency; clover infertility
3 Anderson GW - BSc agroforestry and livestock production
4 Barrow NJ - MAgrSc,PhD soil characteristics and efficiency of fertilizer use
5 Bennett D - BScAgr,PhD systems analysis
6 Chandler ES - BSc,PhD trace elements in ruminant nutrition
7 Hemsley JA - BScAgr,DipEnvStud,PhD nutrition and wool growth
8 Klein L - BAppChem nutritive value of forages
9 Lindsay JR - MSc trace elements in ruminant nutrition
10 Peter DW - BAgrSc,PhD trace elements in ruminant nutrition
11 Taylor GB - MScAgr nutritive value and agronomic characteristics of clovers
12 Whelan BR - BAgrSc soil characteristics and efficiency
13 White CL - BScAgr,PhD trace elements in ruminant nutrition

25 At Bio-Engineering Section
Private Bag 10,, Clayton, Victoria, 3168
Tel: (03) 542 2326. Telex: 35675.

1 Kautzner B - ARMII physical chemistry of the rumen and development of controlled release systems
2 Laby RH - MBE,PhD physical chemistry of the rumen and development of controlled release systems

26 Division of Tropical Animal Production
Headquarters, Long Pocket Laboratories, Meiers Road, Private Bag No. 3, P.O., Indooroopilly, Queensland, 4068
Tel: 377 0711 (STD 07). Telex: 41615.

1 Mahoney DF - DVSc,PhD,MACVSc *Chief of Division*
2 Aylward JH - MSc,PhD bovine babesiosis (immunology)
3 Bremner KC - MSc,PhD *Officer-in-Charge* helminth diseases of cattle
4 Cybinski DH - BSc,BA,MSc virology (arboviruses)
5 East IJ - BSc,PhD helminth diseases of cattle
6 Eisemann CH - MSc blowfly immunology
7 Goodger BV - DipMedLabTech,FAIMT,BAppSc,MSc,PhD immunity and pathophysiology of bovine babesiosis
8 Johnston LAY - BVSc,MACVSc cattle tick immunology, blowfly
9 Kemp DH - BSc,PhD cattle tick immunology
10 Leatch G - BVSc,MSc bovine anaplasmosis
11 McKenna RV - DipMedLabTech cattle tick immunology
12 Muller MJ - BAgrSc entomology (arboviruses)
13 Nolan J - BAgrSc,PhD acaricide resistance
14 Riddles PW - BSc,PhD molecular biology
15 Schnitzerling HJ - DipIndChem acaricide resistance
16 Schuntner CA - BSc bovine babesiosis (serology)
17 St George TD - MVSc,DVSc virology (arboviruses)
18 Standfast HA - BSc entomology (arboviruses)
19 Stone BF - DSc,PhD,FRACI tick toxinology
20 Tracey-Patte PD - DipMedLabTech cattle tick immunology
21 Uren MF - DipAnHusb,MVSc virology (arboviruses)
22 Utech KBW - BVSc cattle tick immunology
23 Willadsen P - BSc,PhD cattle tick immunology
24 Wright IG - DVSc,PhD immunity and pathophysiology of bovine babesiosis

27 Tropical Cattle Research Centre
Bruce Highway and Ibis Avenue, North Rockhampton, Queensland, 4702
[PO Box 5545, Rockhampton North, Queensland, 4702]
Tel: (079) 36 1288. Telex: 49017.

1 Vercoe JE - MAgrSc,PhD *Assistant Chief and Officer-in-Charge* genotype x environment interactions; evaluation of genotypes
2 Bean KG - AssDipAppSc genetics (tropically adapted cattle); evaluation of genotypes
3 Bryan RP - BSc genetics (tropically adapted cattle)
4 D'Occhio MJ - PhD growth regulation; cattle infertility
5 Frisch JE - BAgrSc,PhD genotype x environment interactions; genetics (tropically adapted cattle)
6 Hetzel DJS - BAgrSc,PhD genetics (tropically adapted cattle); genotype evaluation
7 Hunter RA - BAgrSc,PhD regulation of food intake and metabolism
8 Lindsay DB - MA,BSc,PhD regulation of metabolism; manipulation of growth
9 McKinnon M - DipSci(Agr),DipSci(Agr) genetics (tropically adapted cattle); genotype evaluation
10 Munro RK - BVSc,PhD cattle infertility; embryo manipulation
11 O'Kelly JC - BSc,DipBiochem,PhD lipid metabolism
12 Seifert GW - BScAgr,MSc,PhD genetics (tropically adapted cattle); genotype evaluation
13 Williams TJ - BSc,PhD embryo manipulation; cattle infertility

28 Davies Laboratory
University Road, Townsville, Queensland, 4814
[Private Mail Bag, PO Aitkenvale, Queensland, 4814]
Tel: (077) 71 9511. Telex: 47181.

1 Hogan JP - BScAgr,PhD *Assistant Chief* protein and mineral metabolism
2 Kennedy PM - BSc,PhD ruminant intake and digestion
3 Lowry JB - BSc,PhD ruminant intake and digestion
4 McSweeney CS - BVSc,PhD ruminant intake and digestion; protein and mineral metabolism
5 Schlink AC - DipEconStats,BAgric,PhD ruminant intake and digestion; protein and mineral metabolism
6 Wadsworth JC - BSc,PhD protein and mineral metabolism

29 at Division of Biotechnology, Institute of Industrial Technologies
103 Delhi Road, North Ryde, New South Wales
[PO Box 184, North Ryde, New South Wales, Queensland, 2113]
Tel: (02) 886 4888. Telex: 22970.

1 Murray MD - BScVetSci,FRCVS entomology (arboviruses)

30 Division of Wool Technology
343 Royal Parade, Parkville, Victoria, 3052
Tel: (03) 342 4200. Telex: AA33983. FAX: (03) 347 5481.

The division carries out research on the structure, properties and biological activity of proteins and applies this knowledge to aid (1) industries based on animal protein products, such as wool and hides and skins, (2) industries based on plant protein products, such as feedstuffs, and (3) industries based on biotechnicological processes such as vaccine production, diagnostic probe development and industrial microbiology

31 Melbourne Laboratory
343 Royal Parade, Parkville, Victoria, 3052
Tel: (03) 342 4200. Telex: AA33983. FAX: (03) 347 5481.

1 Azad Dr AA - PhD gene cloning, characterization and expression
2 Baker Dr AT - PhD influenza virus structure
3 Blagrove Dr RJ - PhD studies on plant proteins
4 Brahimi-Horn Dr MC - PhD enzyme technology in wool textile processing
5 Colman Dr PM - PhD influenza virus structure, protein crystallography
6 Dopheide Dr TAA - PhD gene structure and function, protein structure and function
7 Dowling Mr LM - BSc wool chemistry
8 Elleman Dr TC - PhD gene structure and function, genetic engineering
9 Evans Dr NA - BSc,PhD *OIC* molecular structure of fibrous proteins in wool and hides
10 Evans Dr NA - PhD leather chemistry
11 Gordon Dr PG - PhD leather and skin technology
12 Gruen Dr LC - PhD studies on seed storage proteins
13 Hewish Dr DR - PhD monoclonal antibodies
14 Hudson Dr PJ - PhD gene characterization and diagnostic probes
15 Inglis Mr AS - MSc protein sequencing and analysis
16 Jagadish Dr MN - PhD expression vectors and site specific mutogenesis
17 Kortt Dr AA - PhD studies on plant proteins, ovine footrot vaccines
18 Leaver Mr IH - MSc wool chemistry and photostabilization of wool
19 Maclaren Dr JA - PhD protein modification and peptide synthesis
20 MacRae Mr TP - MSc molecular structure of fibrous proteins in wool and hides
21 Macreadie Dr IG - PhD gene characterization and antigen expression in yeast
22 Marshall Dr RC - PhD wool chemistry
23 Milligan Dr B - PhD wool and leather chemistry
24 Montgomery Dr KC - PhD studies of hides, skins and leather
25 O'Donnell Dr IJ - DSc protein structure, antigens, vaccines
26 Peters Dr DE - PhD biomedical use of collagen
27 Ramshaw Dr JAM - PhD collagen chemistry
28 Reddie Dr RN - PhD leather chemistry
29 Rivett Dr DE - PhD wool chemistry
30 Savin Dr KW - PhD membrane proteins, gene structure and function
31 Scroggie Dr JG - PhD leather chemistry
32 Shukla Dr DD - DrAgr identification of viruses in plants
33 Sparrow Dr LG - PhD wool chemistry
34 Stapleton Dr IW - PhD wool chemistry
35 Stewart Dr FHC - DSc amino acid and peptide chemistry
36 Tulloch Dr PA - PhD collagen structure, influenza virus structure
37 Varghese Dr JN - PhD influenza virus structure
38 Ward Dr CW - PhD *Assistant Chief of Division* protein structure, immunochemistry and gene structure and function
39 Werkmeister Dr JA - PhD monoclonal antibodies
40 Wilshire Dr JFK - PhD protein sequencing

32 Geelong Laboratory
P.O. Box 21, Belmont, Victoria, 3216
Tel: (052) 472611. Telex: 32143. FAX: (052) 472657.

1 Garnsworthy RK - MSc,PhD,MAIP *OIC*
2 Abbott GM - PhD composite yarn, cotton
3 Anderson CA - BSc,PhD *Assistant Chief of Division*
4 Angliss IB - MAppSc yarn dyeing, colour measurement
5 Bateup BO - PhD surface chemistry; shrinkproofing; wool scouring
6 Boston WS - BScAMIST *Manager Planning and Communication*

7 Brady PR - PhD dyeing
8 Evans DJ - PhD surface chemistry
9 Feldtman HD - AGInstTech mechanical properties
10 Fincher KW - AGInstTech dyeing
11 Fleischfresser BE - PRD chemical finishing
12 Freeland GN - AGInstTech shrink resistance
13 Garnsworthy BE - PhD new products
14 Guise GB - PhD shrink resistance
15 Harrigan FJ - ATI dyeing
16 Leeder JD - DAppSci fibre histology
17 Mickleson CA - PhD biological control of effluents
18 Nason DO - PhD sizing, weaving
19 Naylor GR - PhD pilling
20 O'Loughlin GJ - PhD yarn sizing
21 Pearson AJ - BE yarn colouration
22 Phillips DG - PhD new products
23 Plate DE - PhD spinning, sirospun, worsted processing
24 Rippon JA - PhD dyeing
25 Smith GC - PhD pesticides
26 Sokolav NN - BE yarn sizing
27 Treloar IM - AGInstTech physical testing
28 Troy PR - BScBE yarn structure, weaving
29 Walls GW - BScDSc wool processing, packs; self-twist spinning
30 Warner JJ - PhD wool grease purification, scouring
31 White MA - PhD tailoring

33 Sydney Laboratory
1-11 Anzac Avenue, Ryde, New South Wales, 2112
Tel: (02) 80 0211. Telex: 70827.

1 Hoschke BN - MSc,PhD *OIC*

34 Raw Wool
Tel: (02) 80 0211. Telex: 70827.

1 Andrews MW - BSc,PhD raw wool characteristics, marketing and processing
2 Chudleigh PW - MPhil measurement of dark fibre contamination in wool
3 Foulds RA - MPhil detection, measurement and prevention of dark fibre contamination in wool
4 Lunney HW - BSc,BE,MA measurement of mean fibre diameter and its variability
5 Rottenbury RA - MSc raw wool characteristics, marketing and processing
6 Thompson RL - BSc instrumentation for the measurement of staple length and staple strength
7 Winston CR - BSc measurement of the residual characteristics of raw wool

35 Textile Research

1 Brooks JH - MSc,PhD surface chemistry in improving the properties of wool products
2 Smith LJ - ASTC mini-processing methods from raw wool to finished fabric
3 Stuart IM - MSc heat and mass transfer in wool fabrics and clothing
4 Watt IC - MSc,PhD surface properties of wool fibres and fabrics

36 High Technology Systems

1 Caffin RN - MSc,PhD image analysis, robotics and advanced instrumentation in the preparation of raw wool for objective measurement
2 Higgerson GJ - BSc image processing and analysis applied to raw wool, fabrics and skin
3 Whitely KJ - BSc,PhD application of high technology systems
4 Wills LJ - BSc advanced instrumentation and image analysis applied to yarns, textiles, and various other materials

37 Division of Food Processing
Divisional Headquarters, Delhi Road, PO Box 52, North Ryde, New South Wales, 2113
Tel: (02) 887-8333. Telex: 23407.

1 Walker D - PhD *Chief of Division*

38 At Food Research Laboratory
Delhi Road, PO Box 52, North Ryde, New South Wales, 2113
Tel: Sydney (02) 887-8333. Telex: 123407.

1 Graham D - BSc,PhD,CB,FAIBiol *Officer-in-Charge* chemistry and biochemistry of food lipids, biotin and stress, and Sudden Infant Death Syndrome
2 Adams RF - AAIFST effects of dietary components on the microflora of the gut
3 Algie JE - BE,ASTC,MSc,DU postharvest physiology of fruit
4 Back Miss JF - BSc,DipEd thermochemistry of egg and other proteins and lipids
5 Barnett D - BSc allergens in foods
6 Bell GA - MA,PhD neuro-anatomical and neuro-chemical aspects of olfaction

7 Board PW - BSc,FAIFST technology of processed fruit, vegetable and meat products
8 Burley RW - MSc,PhD physical biochemistry of proteins in egg yolk
9 Casimir DJ - MSc,DipEd,PhD,FAIFST food engineering
10 Chandler BV - BSc,MScBiotech,PhD,FAIFST chemistry of processed foods, citrus fruits and citrus products
11 Chaplin GR - BScAgr,MSc,PhD storage of tropical fruits
12 Chau-Ngoc N - BE,MEng food engineering
13 Cole SP - BSc storage of tropical fruits
14 Cornell BA - BSc,PhD physical properties of membranes in foods
15 Evans AJ - MSc,PhD fatty acid composition of fish
16 Eyles MJ - BSc,PhD food viruses
17 Fisher LR - MSc,PhD state of water in foods, physical forces between colloid particles
18 Fogerty AC - MSc chemistry and biochemistry of food lipids
19 Ford GL - BA,MSc chemistry and biochemistry of food lipids
20 Francis GW - MSc state of water in foods, physical forces between colloid particles
21 Graham D - BSc,PhD chilling injury in plants, enzymology and metabolic regulation
22 Hocking Miss AD - BSc food spoilage fungi, action of food preservatives
23 Hockley DG - BSc enzymology of chilling injury in plants
24 Holland RV - MSc,PhD singlet oxygen in food processing and packaging
25 Hood RL - BSc,PhD fat metabolism, biotin and stress, and Sudden Infant Death Syndrome
26 Irving AR - BSc heat and mass transfer in storage, transport and processing of foods
27 Johnson RL - MSc,DipEd,PhD adsorptive processes in food manufacture, chemistry of citrus products
28 Kavanagh Mrs EE - BSc postharvest physiology of fruits and vegetables
29 Kennett BH - ASTC physical properties of membranes in foods
30 Laing DG - BSc,PhD psychological and physiological assessment of odour interactions in flavours
31 Lane AG - MSc,PhD disposal, fermentation and utilization of food effluents and wastes
32 Last JH - ASTC chemistry of food flavours
33 McBride RL - BSc,PhD sensory evaluation of foods
34 McGlasson WB - BAgrSc,PhD postharvest physiology of fruits and vegetables
35 Middlehurst J - MSc physical properties of membranes in foods
36 Nguyen THL - BSc chemistry of food flavours
37 O'Connell PBH - MSc plant messenger RNA studies
38 Oakenfull DG - MSc,PhD dietary fibre, bile salt micelles, interaction of bile salts and lipids, colloids stability
39 Panhuber HHN - BA,MSc pyschophysical and physiological assessment of odour interactions in flavours
40 Parker NS - BSc,PhD rheology of foods
41 Patterson BD - BSc,PhD chilling injury of fruit and vegetables
42 Pitt JI - MSc,ASTC,PhD food spoilage fungi, action of food preservatives
43 Richardson KC - BSc food technology liaison
44 Rooney ML - MSc singlet oxygen in food processing and packaging
45 Rutledge PJ - AAIFST food technology liaison
46 Satyan Mrs SH - MSc,PhD transport and storage of fresh fruit and vegetables
47 Separovic Miss FE - BA physical properties of membranes in foods
48 Sharp AK - BE,MEngSci,PhD heat and mass transfer in the storage, transport and processing of foods
49 Shaw KJ - BA mass spectrometry of food components
50 Shenstone FS - ASTC studies on avian erythrocyte membranes
51 Shorter AJ - BSc postharvest physiology of fruits and vegetables
52 Sidhu GS - BScAgr,PhD protected protein supplements for ruminants, carotenoids, vitamin E
53 Sleigh RW - MSc,PhD chemistry of proteins in eggs and other systems
54 Smillie RM - MSc,PhD,DSc chilling injury in plants, chloroplast development
55 Stafford IA - BE,ASTC controlled atmosphere and cool stores for fruit and vegetables
56 Stanley G - BSc,ASTC chemistry of food flavours
57 Steele RJ - BSc,PhD,MBA corrosion of metal packaging materials and trace metals in foods
58 Tindale CR - BApSc physiology of active amines and amino acids in foods
59 Tobin NFB - BSc production and assay of mycotoxins, ochratoxin and aflatoxins
60 Walker GJ - MSc publications
61 Warth AD - MSc,PhD biochemistry of bacterial spores, action of food preservatives
62 Wheeler Miss KA - BSc spoilage of fresh, cured and dried tropical fish
63 Whitfield FB - MSc,ASTC,PhD chemistry of food flavours

39 At School of Biological Sciences, Macquarie University
Macquarie University, North Ryde, New South Wales, 2113
Tel: Sydney (02) 88 9473.

1 Bishop DG - MSc,PhD structure and function of membranes
2 Brady CJ - MScAgr,PhD regulation of protein synthesis
3 Brown MA - MSc,DipTechSci,PhD structure and function of membranes
4 Pearson Mrs JA - MSc chilling injury of fruit and vegetables
5 Raison JK - BSc,PhD structure and function of membranes

6 Speirs J - MSc,PhD regulation of protein synthesis

40 At Meat Research Laboratory
Cnr Creek and Wynnum Roads, P.O. Box 12, Cannon Hill, Queensland, 4170
Tel: Brisbane (07) 399-3122. Telex: 40150.

1 Walker DJ - DSc,PhD,FAIFST *Officer-in Charge and Assistant Chief*
2 Chua HM - BEng,MBA materials handling
3 Davey KR - BChemEng,MEngSc,PhD chemical and mechanical aspects of meatworks operations
4 Egan AF - MSc,PhD mechanisms of meat spoilage
5 Eustace IJ - BAgSc microbiological aspects of meat processing
6 Grau FH - MSc,PhD meat spoilage organisms
7 Hamilton RG - BAppSc meat protein recovery
8 Harris PV - BSc,PhD mechanical properties of meat
9 Horgan DJ - BSc,BEcon,PhD muscle biochemistry
10 Johnson BY - BAgSc *Information Officer*
11 Kerr DT - DipMechEng implementation of electrical stimulation in abattoirs
12 King NL - MSc,PhD muscle and plant protein interactions
13 Kurth LB - BAppSci,MPhil meat science and technology
14 Leppik RA - MSc,PhD recovery of steroid compounds
15 Lovett DA - BSc meat physics and process engineering
16 Macfarlane JJ - MSc,FRACI,FAIFST meat science and technology
17 McKenzie IJ - MAppSci meat science and technology
18 Morton DJ - BSc,PhD fine structure of muscle
19 Park RJ - MSc,PhD,FRACI recovery of steroids
20 Powell VH - MSc,PhD,FRACI *Leader, Industry Section*
21 Rankin RJ - BEngMech alternative slaughter technology
22 Rowe RWD - BSc,MVSc,PhD fine structure and biology of muscle
23 Shaw FD - BSc,MVSc growth and differentiation of cells in culture
24 Shorthose WR - BSc,PhD effects of pre-slaughter treatment on meat quality
25 Smith MG - BSc microbial genetics
26 Thornton RF - BSc,PhD biology of ruminant adipose tissue
27 Tume RK - BSc,PhD biology of ruminant adipose tissue
28 Weidemann JF - BSc fine structure of muscle

41 At Dairy Research Laboratory
Graham Road, P.O. Box 20, Highett, Victoria, 3190
Tel: Melbourne (03) 556 2211. Telex: 33766.

1 Claughton Mrs SM - BAppSc sunflower protein isolate studies
2 Daley Miss J - BSc flavours and off-flavours
3 Dornom Miss H - BAgrSc *Information and Liaison Officer*
4 Dunkerley JA - DipAppChem whey utilization, protein products
5 Eddy DW - BSc cheese starters
6 Freeman NH food engineering
7 Hardham JF - AAIFST recombined products
8 Hayes JF - BSc whey utilization, protein products
9 Hillier AJ - BSc,PhD enzymology of lactic bacteria
10 Jameson GW - BSc,PhD cheese chemistry and technology
11 Kieseker FG - BAgrSc recombined products, new foods
12 Lloyd GT - BSc,PhD fermentation studies, concentrated starter cultures
13 Marshall SC - DipAppChem,DipChemEng whey utilization
14 Mitchell IR - BSc process development
15 Orsini A - BSc cheese starters
16 Pearce RJ - BSc milk proteins
17 Ramshaw EH - MA,PhD flavour chemistry
18 Stark W - ARACI flavour chemistry
19 Sutherland BJ - ARACI cheese chemistry and technology
20 Taylor DT - BSc cheese technology
21 Trerice BJ - BEngMech,MEngSc food engineering
22 Urbach Mrs GE - MSc flavour chemistry
23 Zadow JG - MSc,DAppSc *Officer-in-Charge and Assistant Chief*
24 Zadow JG - MSc protein products and new foods

42 Institute of Industrial Technologies
Normanby Road, Clayton, Victoria, 3168
Tel: (03) 542 2787. Telex: 32945.

Also at Limestone Avenue, Campbell, A.C.T. (PO Box 225, Dickson, A.C.T. 2602) Telephone: (062) 484510 Telex: AA62003

1 Adam CM - FIM,FIEAust *Director*

43 Division of Biotechnology
103 Delhi Road, P.O. Box 184, North Ryde, New South Wales, 2113
Tel: (02) 886 4888. Telex: AA22970. FAX 888 9271.

1 Grigg GW - MSc,PhD,ScD *Chief of Division* a new method for sequencing nucleic acids
2 Bender VJ - BSc,PhD enzymic peptide synthesis
3 Both GW - BSc,PhD studies on rotavirus
4 Dalrymple BP - BSc,PhD a new vaccine for footrot, using the techniques of recombinant DNA and genetic engineering technology
5 Finney K - BSc,PhD a new vaccine for footrot, using the techniques of recombinant DNA and genetic engineering technology
6 Hannan GN - BABiolSc,PhD cell aging
7 Hill RJ - MSc,PhD control of gene expression

8 Jennings PA - BSc,BVSc,PhD modification of gene expression
9 Lockett TJ - BSc,PhD developmental regulation of mammalian genes
10 Mattick JS - BSc,PhD a new vaccine for footrot, using the techniques of genetic engineering and recombinant DNA technology
11 McAuslan BR - BSc,PhD cell aging
12 Molloy PL - BSc,PhD control of gene expression
13 Moss BA - BSc,PhD enzymic peptide synthesis
14 Podger DM - BSc,PhD mutagenesis and antimutagenesis
15 Reisner AH - AB,PhD sequencing of nucleic acids
16 Rockwell GA - MSc,PhD cell aging
17 Sleigh MJ - BSc,PhD developmental regulation of mammalian genes
18 Underwood PA - BSc,PhD monoclonal antibodies
19 Whitfeld PL - BSc,PhD studies on rotavirus
20 Whittaker RG - BSc,MSc,PhD enzymic peptide synthesis

44 Division of Chemicals and Polymers
506 Lorimer Street, Fishermen's Bend
[PO Box 4331, Melbourne, Victoria, 3001]
Tel: (03) 647 7222. Telex: 33951. FAX: (03) 646 1984.

1 Solomon DH - PhD,DSc,FAA,FTS,FRACI *Chief of Division*
2 Anderson-Mackay JA - BSc,PhD herbicide synthesis
3 Holan G - FRMIT,FTS insecticides, acaricides, anthelmintics
4 Johnson WMP - BSc,PhD fungicide synthesis
5 Liepa AJ - BSc,PhD fungicide, herbicide synthesis
6 Loder JW - MSc,PhD,FRACI herbicide synthesis
7 Loft BC - BSc *Scientific Services Officer* patents, publications, enquiries
8 Middleton EJ - BSc,PhD,FRACI insecticide synthesis
9 Morton TC - BSc,PhD fungicide synthesis
10 Nearn RH - LRICDF fungicide synthesis
11 O'Keefe DF - MAppSc insecticide synthesis
12 Sasse WHF - BSc,PhD,FRACI insecticides, fungicides, growth regulators
13 Virgona CTF - BSc insecticide biology
14 Winzenberg KN - BSc,PhD herbicide synthesis

45 Institute of Plant Production and Processing
Limestone Avenue, Campbell, New South Wales, 2113
[225, Dickson, 2602]
Tel: (02) 888 9600. Telex: 120003. Cables: Bresearch, Sydney.

1 Henzell EF - AO,BAgrSc,DPhil,FAIAS,FTS *Director*

46 Division of Entomology
HQ, Black Mountain Laboratories, CSIRO, Clunies Ross Street
[PO Box 1700, Canberra, 2601]
Tel: (062) 46 4911. Telex: 62309. FAX 470217.

1 Whitten MJ - BSc,BA,PhD,FTS *Chief of Division* molecular genetics of insects
2 Akhurst RJ - MSc,PhD entomopathogenetic nematodes; nematode/bacteria interactions
3 Annis PC - BSc grain storage in modified atmospheres
4 Balderson J - BA,BSc taxonomy of Blattodea and Mantodea
5 Banks HJ - BA,PhD stored grain insects; grain storage in modified atmospheres
6 Barrer PM - BSc,PhD behaviour of stored product Lepidoptera and Coleoptera
7 Barton Browne LB - BSc,PhD sheep blowfly reproduction; insect behaviour
8 Beaton CA - BSc electron microscopy of insects and viruses
9 Bedo DG - BSc,DipEd,PhD cytogenetics of sheep blowfly
10 Bellas TE - BSc,PhD mating and oviposition behaviour, chemistry of insect pheromones, chemical analysis
11 Binnington KC - DipAIMLT,BApplSc,PhD insect physiology, insecticide action
12 Briese DT - BSc,PhD insect viruses, biological control of weeds
13 Britton EB - DSc *Hon Research Fellow* taxonomy of Coleoptera
14 Brown WV - BSc,PhD insect secretions
15 Calder AA - BSc,PhD taxonomy and biology of Curculionidae
16 Cardale Ms J - MSc,BA taxonomy of Hymenoptera
17 Carne PB - BAgrSc,PhD,DIC *Assistant Chief* ecology and taxonomy (Cleoptera: Scarabaeidae); forest entomology
18 Carver Ms M - BSc,PhD taxonomy and biological control of aphids
19 Cavanaugh Ms JA - BSc biological control of sitona weevil
20 Clarke G - BSc genetic control of sheep blowfly
21 Colless DH - BScAgr,PhD taxonomy of Diptera
22 Dallwitz MJ - BA,BSc,PhD computer applications in taxonomy
23 Daly Ms J - BSc,PhD genetics of insecticide resistance in *Heliothis*
24 Davidson WJ - MSc molecular biology of insect biology
25 Delfosse ES - PhD biological control of weeds; ecology
26 Desmarchelier JM - BSc,PhD alternative grain protectants; insecticide chemistry
27 Drake VA - MA,DPhil radar studies of insect migration
28 Edwards ED - BScAgr taxonomy of lepidoptera
29 Edwards Ms PB - BSc,PhD biological control of dung, dung beetle biology
30 Evans DE - BSc,PhD,DIC temperature regulation for grain pest control
31 Farrow RA - BSc,PhD,DIC insect migration, insect pests of *Eucalyptus*
32 Foster GG - BSc,PhD genetic control of sheep blowfly

33 Frodsham Ms AC - BScAgr *Scientific Liaison Officer* information transfer
34 Gilbert N - BA genetical ecology; insect behaviour
35 Graver JE - BSc grain storage in the tropics
36 Hackman RH - MSc,PhD insect physiology, cuticle chemistry
37 Halliday RB - BSc,PhD taxonomy and ecology of macrochelid mites
38 Horak Ms M - DiplNatETH,PhD taxonomy and biology of tortricid moths
39 Hughes RD - BSc,PhD,DSc,DIC,ARCS biological control of lucerne aphids
40 Kalionis B - BSc,PhD molecular biology of insect development
41 Key KHL - PhD,DSc,DIC,FAA *Hon Research Fellow* taxonomy of orthopteroids
42 Konovalov Ms CA - MSc sheep blowfly mass rearing and release studies
43 Lacey MJ - MSc,PhD mass spectroscopy
44 Lawrence JF - BA,PhD taxonomy of Coleoptera
45 Lenz M - PhD biology of termites and resistance of materials to termite attack
46 Lohe AR - BSc,PhD molecular biology of *Drosophila*
47 Longstaff BC - BA,PhD ecology of grain insects
48 Loughnan Ms ML - MSc dung quality, biological control of dung
49 Mahon RJ - BSc,PhD ecology; control of sheep blowfly
50 McInnes RS *Services Co-ordinator, Australian National Insect Collection*
51 Milner RJ - BSc,PhD microbial pathogens of insects
52 Moore AD - BAgrSc,GradDipSci biological control of weeds
53 Moore BP - BSc,PhD,DPhil *Hon Research Fellow* insect secretions; taxonomy of Coleoptera
54 Nagorcka BN - BSc,PhD mathematical biology, pattern formation in insect development
55 Naumann ID - BSc,PhD taxonomy of Hymenoptera
56 Nielsen ES - MSc,PhD taxonomy and biology of Lepidoptera
57 Norris KR - DSc *Hon Research Fellow* taxonomy of blowflies
58 Paine Ms TA - BSc computer applications in taxonomy
59 Readshaw JL - BSc,PhD ecology of orchard pests
60 Rentz DCF - MSc,PhD taxonomy of Orthoptera
61 Roberts RJ - BA,MSc,PhD,DIC ecology of insect pests of pastures and Eucalyptus
62 Rothschild GHL - BSc,PhD,DIC *Assistant Chief* insect pheromones, ecology and behaviour
63 Rumbo ER - BA,PhD electrophysiology of chemoreceptors, pheromones
64 Saint R - BSc,PhD molecular biology of insect development
65 Smith Ms C - BSc,GradDipSc biological control of weeds
66 Smith PH - BSc,PhD behavioural studies on sheep blowfly
67 Spradbery JP - BSc,PhD biology and ecology of screw-worm fly and wasps
68 Taylor RW - MSc,PhD taxonomy of Formicidae
69 Traynier RMM - BA,PhD oviposition behaviour of insects
70 Tyndale-Biscoe Ms M - BSc,PhD biological control of dung; bush fly studies
71 van Gerwen ACM - BAppSc biology and ecology of sheep blowfly
72 Vickers RA - WDA field studies of insect pheromones
73 Vogt WG - BSc,PhD biology and ecology of sheep blowfly and bush fly
74 Wapshere AJ - BSc,PhD biological control of weeds
75 Waterford CJ - BAppSc pesticide resistance in grain insects
76 Waterhouse DF - AO,CMG,DSc,FAA,FRS *Hon Research Fellow* insect ecology and physiology, biological control
77 Watson JAL - BSc,PhD taxonomy of Odonata, ecology and physiology of termites
78 Weir TA - BSc taxonomy of Coleoptera
79 Whittle CP - BSc,PhD mass spectrometry, insect pheromes
80 Winks RG - MSc,PhD,DIC pesticide resistance in grain insects
81 Wiseman JR - BSc grain storage in modified atmospheres
82 Woodburn TL - MSc biological control of weeds
83 Zimmerman EC - BSc,PhD,DIC *Hon Research Fellow* taxonomy of Curculionidae

47   At CSIRO Long Pocket Laboratories
Long Pocket Laboratories, Brisbane
[Private Mail Bag 3, Indooroopilly, Queensland, 4068]
Tel: (07) 377 0711. Telex: 41615.

1 Room PM - BSc,ARES,PhD *Officer-in-Charge* biological control of weeds
2 Common IFB - MA,DAgrSc *Hon Research Fellow*
3 Floyd R - BSc,PhD ecological modelling
4 Forno Ms IW - MSc,PhD biological control of aquatic weeds
5 Julien MH - BSc,MSc biological control of aquatic weeds
6 Maywald GF - BSc computer programming and mathematical modelling
7 Room PM - BSc,PhD,ARCS biological control of aquatic weeds
8 Sands DPA - DipAIMLT,MSc,PhD biological control of weeds; agent physiology; ecology of fruit-piercing moths
9 Sutherst RW - BSc,PhD ecological management and tick resistance studies
10 Wright AD - BSc biological control of aquatic weeds and terrestrial weeds

48   At CSIRO Rural Laboratories
Rural Laboratories, Perth
[Private Bag, PO Wembley, Western Australia, 6014]
Tel: (09) 387 4233. Telex: 92178.

1 Ridsdill-Smith TJ - BA,MSc,PhD *Officer-in-Charge* biological control of dung, bush fly studies
2 Matthiessen JN - BScAgr bush fly studies

3 Mazanec Z - BSc,DipFor ecology of jarrah leafminer

49   At CSIRO Tasmanian Regional Laboratory
"Stowell", Stowell Avenue, Hobart, Tasmania, 7000
Tel: (002) 201 444. Telex: 58300.

1 Bedding RA - BSc,PhD,DIC,ARCS parasitic nematodes for insect control
2 Bornemissza GF - BSc,PhD *Hon Research Fellow* dung beetle taxonomy
3 Taylor KL - BScAgr *Hon Research Fellow* taxonomy of Psyllidae

50   At Adelaide
CSIRO Division of Soils, Private Bag No.2, Glen Osmond, South Australia, 5064
Tel: (08) 274 9311. Telex: 82406.

1 Baker GH - BSc,PhD ecology and control of snails and Portuguese millipedes

51   At CSIRO Tropical Cattle Research Centre, Rockhampton North
PO Box 5545, Rockhampton, Queensland, 4701
Tel: (079) 36 1288. Telex: 49017.

1 Vercoe JE - MAgrSc,PhD dung beetle ecology, buffalo fly studies

52   At CSIRO Entomology Research Station, Warrawee, NSW
"Kooyong", 55 Hastings Road, Warrawee, New South Wales, 2074
Tel: (02) 487 1756. Telex: 70701.

1 Fletcher BS - BSc,PhD,DIC *Officer-in-Charge* ecology and control of fruit fly
2 Lukins RG - BSc biological control of white wax scale
3 Milne Ms WM - BSc,PhD,DIC biological control of aphids
4 Wellings PW - MSc,PhD insect/parasitoid/plant interactions, ecology

53   At CSIRO Darwin Laboratories, Northern Territory
Private Bag 44, Winnellie, Northern Territory, 5789
Tel: (089) 22 1711. Telex: 85294.

1 Lonsdale WM - BSc,PhD biological control of weeds

54   At Port Moresby, Papua New Guinea
Screw-Worm Fly Investigations, Papua New Guinea
[PO Box 6645, Boroko, Port Moresby, Papua New Guinea]
Tel: Port Moresby 259896.

1 Clarke GM - BSc *Officer-in-Charge* control of screw-worm fly

55   At Montpellier, France
CSIRO Biological Control Unit (France), 335 Avenue Abbe Paul Parguel, Montpellier, 34100, France
Tel: Montpellier 676 33881. Telex: 490304.

1 Briese DT - BSc,PhD *Officer in charge* biological control of weeds
2 Aeschlimann JP - BSc,PhD biological control of weeds
3 Hasan S - DSc biological control of weeds by pathogens

56   At Acapulco, Mexico
CSIRO Field Station (Mexico), Av. Costa de Oro No. 220, Fracc. Costa de Oro, PO Box 2049, Veracruz, Veracruz, Mexico
Tel: Mexico (52) (293) 3714 78.

1 Ponce de Leon R biological control of weeds

57   At Pretoria, South Africa
CSIRO Dung Beetle Research Unit, Private Bag X5, Lynn East, Pretoria, 0039, South Africa
Tel: South Africa 70 3226/7. Telex: 322-043.

1 Wright EJ - BSc,MSc,PhD *Officer-in-Charge* biological control of dung
2 Doube BM - BSc,PhD biological control of dung

58   At Grahamstown, South Africa
CSIRO Biological Control Unit, University of Cape Town, Rondesbosch, Cape Town, 7700, South Africa
Tel: South Africa 021 650 3649. Telex: 5722208.

1 Scott JK - BSc,PhD biological control of weeds

59   Division of Forestry and Forest Products
Bayview Avenue, Private Bag 10, Clayton, Victoria
[PO Box 4008, Q. Victoria Terr., A.C.T., 2601]
Tel: (03) 542 2244. Telex: 35675. FAX: (03) 543 6613.

1 Hewetson W - BSc,PhD,CHEN,FASC,FRSC *Chief*
2 Benson MLW - BScForDipFor forest biomass
3 Benyon P - BSc,BE modelling
4 Boland DJ - BScFor,MSc utilization of eucalypt gene resources
5 Borough CJ - BScFor breeding and physiology of *Pinus radiata*
6 Brown AGJ - BSc,MSc,DipFor *Assistant Chief and OIC* genetics and breeding

7 Burgess P - BScFor,DipFor reproductive biology and hybridisation in *Eucalyptus*
8 Byrne GF - BSc water movement in trees
9 Cheney NP - BScFor,DipFor forest fire behaviour
10 Crane WJB - BScFor,DipFor,PhD forest nutrition
11 Cremer KW - BScFor,DipFor,MSc forest reproduction and regeneration
12 Cromer RN - BScFor,DipFor,MFor forest nutrition
13 Doran JC - BScFor utilization of indigenous gene resources
14 Eldridge KG - BSc,MSc,DipFor,PhD eucalyptus and pine genetics
15 Griffin AR - BScFor,PhD forest genetics
16 Hartney VJ - BScAg,MSc vegetative propagation
17 Hoare JRL - BSc fire ecology
18 Kerruish CM - BScFor,DipFor small wood harvesting
19 Khanna PK - BSc,MSc,PhD mineral cycling, soil plant relationships
20 Kriedemann PE - BAgrSc,PhD physiology
21 Lacey CJ - BSc,MSc asexual reproductive strategies in trees
22 Leuning R - BSc,PhD environmental physics, microhabitat of trees
23 Matheson AC - BSc,BA,PhD eucalyptus and pine genetics
24 McMurtrie RE - BSc,PhD modelling and statistical analysis
25 Moran GF - BSc,PhD eucalypt and pine genetics
26 Myers BJ - BScF,MSc water relations
27 Ohmart CP - BScF,PhD insect tree interactions
28 Old KM - BSc,PhD forest pathology
29 Raison RJ - BRurSc,PhD mineral cycling and fire effects
30 Sands PJ - BSc,PhD modelling and statistical analysis
31 Snowdon P - BSc(Agr),MSc forest nutrition
32 Talsma T - IrAgr,DAgrSc movement of nutrients in the forest hydrological cycle
33 Thompson WA - PhD modelling and statistical analysis
34 Turnbull JW - BScFor,PhD utilization of eucalyptus gene resource
35 Wingate-Hill R - BSc,MSc,MA Debarking and removal of moisture from small wood

60 South Queensland Forest Research Group
Cunningham Laboratory, 306 Carmody Road, St Lucia, Queensland, 4067
Tel: (07) 377 0307. Telex: 42159.

1 Cameron DM - BSc,DipFor,MF *Officer-in-Charge* soil tree and grass interactions
2 Cotterill PP - BSc,PhD genetics and breeding of *Pinus radiata*

61 Tasmanian Forest Research Group
"Stowell", Stowell Ave, Hobart, Tasmania, 7000
Tel: (002) 20 1444. Telex: 58300.

1 Kile GA - BAgrSc,PhD *Officer-in-Charge* forest pathology dieback diseases
2 Beadle CL - MSc,PhD comparative physiology of eucalyptus
3 Ellis RC - BScFor,PhD ecology and silviculture of eucalyptus species
4 Podger FD - BScFor,DipFor,MSc,PhD forest pathology, dieback
5 West PW - BScFor,PhD forest mensuration, computer systems

62 Plantation Forest Research Centre
PO Box 946, Mt Gambier, South Australia, 5290
Tel: (087) 255555. Telex: 80024.

1 Fife DN - BSc *Officer-in-Charge* nutritional physiology, fertilizer, root growth
2 Mitchell BA - BScFor water relations, nutrition second rotation
3 Sheriff DW - BSc,PhD physiology, nutrition and photosynthesis

63 Western Australian Forest Research Group
Floreat Park, Private Bag, PO Wembley, Western Australia, 6014
Tel: (09) 3874233. Telex: 92178.

1 Hingston FJ - BSc,MSc,PhD *Officer-in-Charge* fertilizer and pollution effects on native forest production
2 Brooker MIH - BAgSc,MSc,BA eucalyptus taxonomy
3 Malajczuk N - BSc,PhD microbial ecology, forest nutrition
4 Murray DIL - BSc,PhD forest pathology
5 O'Connell AM - BSc,PhD forest nutrition

64 Division of Plant Industry
Clunies Ross Street and Barry Drive, Black Mountain
[PO Box 1600, Canberra, A.C.T., 2601]
Tel: (062) 464911. Telex: 62351. FAX: (062) 473785.

1 Peacock WJ - BSc,PhD,FAA,FRS *Chief of Division* chromosome structure and function, molecular biology and cytogenetics
2 Barlow BA - BSc *Assistant Chief*
3 Begg JE - BScAgr,PhD *Assistant Chief* physiology of water stress and efficiency of water use; growth, development and yield in the context of environment x genotype interactions

65 Research Scientists

1 Anderson JM - MSc,PhD molecular organisation and regulation of chloroplast membranes
2 Appels R - BSc,PhD cellular controls operating on chromosome structure and function

3 Appleby CA - Bsc,PhD,FAA respiration and haemoprotein function in symbiotic nitrogen fixation
4 Ashton AR - BSc,PhD molecular biology of C_4 photosynthesis
5 Aston AR - BScAgr,MSc,PhD systems and experimental analysis of mass and energy transfer in crops
6 Atwell BJ - BScAgric,PhD influence of soil properties on the exploration of soil by roots and the effect of root activity on crop growth and yield
7 Barlow BA - BSc,PhD,DSc genetic systems and taxonomy of Loranthaceae, Viscaceae, *Melaleuca* and *Heliotropium*
8 Bergersen FJ - DSc,FRS biochemistry and physiology of nitrogen fixation in legume nodules ^{15}N techniques
9 Bremner PM - BSc,PhD crop genotype x environment interactions, with particular reference to the water economies of wheat and oilseeds in drought conditions
10 Brettell RIS - BA,PhD somatic cell genetics in wheat and maize; somaclonal variation and disease resistance
11 Brock RD - MAgrSc,PhD irrigated crop management systems (SIRAGCROP); plant gene vector construction
12 Brockwell J - DDA field ecology of *Rhizobium* spp.; Legume-*Rhizobium* interactions; inoculation studies
13 Bromfield SM - MAgrSc,PhD problems of acid soils, manganese and aluminium toxicities in plants
14 Brown AHD - BScAgr,PhD plant population, quantitative and conservation genetics; isozyme polymorphisms in wild barley; evolutionary relations in *Glycine*
15 Burdon JJ - BSc,PhD population biology of weeds; interactions between fungal pathogens and plant communities; variation in rust pathogens of cereal crops
16 Burnell JN - BSc,DPhil mechanisms and regulation of C_4 photosynthesis
17 Chandler PM - BSc,PhD regulation of expression of plant genes
18 Chow WS - BSc,PhD The relationship between the molecular organisation of chlorophyll-protein and electron transport complexes, thylakoid membrane structure and photosynthetic capacities in plants adapted to different light climates or stressed conditions
19 Christian KR - BSc,PhD simulation of plant growth and agricultural systems
20 Chu PW - BScAgr,PhD the identification and characterization of the pathogens of two virus diseases of subterranean clover and development of diagnostic probes for their detection
21 Coombe JB - MAgrSc,PhD integrating crop and animal production
22 Cruikshank IAM - DSc phytoalexins biological and physiological studies on their induction in plant studies
23 Daday HV - MSc,PhD genetics and synthesis of ribulose 1,5-bisphosphate carboxylase in *Medicago*
24 Davidson JL - MAgrSc,PhD crop genotype x environment interactions
25 Demezas DH - BSc,PhD application of molecular biology to ecological studies of *Rhizobium*
26 Dennis ES - BSc,PhD molecular organisations and expression of plant genes; gene transfer in plants
27 Dickinson MJ - BSc,PhD mycoviruses in rusts
28 Dove H - BAgrSc,DipEd,PhD integration of animal, pasture and crop production systems; nutrition of young ruminants
29 Downes RW - MAgrSc,PhD development of agronomic, ecological and genetic strategies to improve farm productivity
30 Downie JA - BSc,PhD molecular biology of the symbiotic association between legumes and their root nodule bacteria
31 Dudman WF - BSc,PhD serology and immunochemistry of *Rhizobium* spp. and the chemistry and function of their polysaccharides
32 Dunin FX - BAgrSc hydrology of natural vegetation and the hydrological consequences of changes in land use; evapotranspiration of eucalypt forests
33 Ellis JG - BAgSc,PhD analysis of T-DNA genes in *Agrobacterium*; T-DNA mediated gene transfer
34 Evans LT - AO,MAgrSc,DPhil,DSc,FAA,FRS crop physiology; environmental control of flower initiation, photosynthesis and translocation
35 Fillery IRP - MAgrSc,PhD the availability of water and nutrients in uniformly coarse textured and texture-contrast soils in semi-acid areas
36 Finnegan EJ - BSc,PhD analysis of gene structure and function in plants
37 Fischer RA - MAgrSc,PhD field and controlled environment studies of yield potential determination in wheat; field studies of tillage systems
38 Fitt GP - BSc,PhD abundance and behaviour of adult *Heliotis* spp.
39 Frankel OH - Kt.,DSc,DAgrSc,FRS,FAA,FRSNZ principles and methods of the exploration and conservation of plant genetic resources; evolutionary considerations in nature conservation; molecular evolution of wheat ribosomal gene
40 Freer M - BAgrSc,PhD grazing management and nutrition of sheep for meat production
41 Freney JR - MSc,PhD efficiency of fertilizer nitrogen use by plants; nitrogen losses from soils; transformation of nitrogen in soil; development of methods for estimating nitrogen loss from the soil and nitrogen fixation in the field
42 Gerlach WL - BSc,PhD plant gene structure and function; molecular cytogenetics; genome organisation and evolution of plant viruses
43 Gibson AH - BScAgr,PhD physiology and genetics of nodule development and nitrogen fixation in legumes; ecology of *Rhizobium* spp.; microbiology of nitrogen-fixing bacteria associated with straw decomposition
44 Gifford RM - BSc,PhD physiological and environmental control of yield; biological response to rising atmospheric CO_2

45 Gill AM - BAgrSc,MSc,PhD ecological effects of fire; management of native vegetation; weedol ecology (especially dispersal)

46 Goodchild DJ - BScAgr,MA,PhD ultrastructure of chloroplasts in relation to biochemistry of photosynthesis, and of legume root nodules in relation to nitrogen fixation; ultrastructure of storage protein deposition in plants

47 Groves RH - BAgrSc,PhD factors affecting the distribution, productivity, and conservation of natural grasslands; biological control of weeds, especially *Chondrilla juncea* and thistles

48 Halsall DM - BSc,DipEd,PhD the microbiology of nitrogen-fixing bacteria associated with wheat stubble decomposition; genetics of nitrogen fixation; disease assessment procedures in Indian mustard

49 Harris RLN - BSc,PhD rational design and synthesis of herbicides, and plant growth regulators. Inhibitors of C_4 specific enzymes. New synthetic methods in heterocyclic chemistry

50 Hartley TG - MSc,PhD taxonomic revisions in tropical Australian Rutaceae

51 Hatch MD - BSc,PhD,FRS,FAA C_4 photosynthesis and C_4 specific weedicides

52 Hearn AB - BSc,DipAgric,DipTropAgron cotton production strategies as affected by water, nitrogen, crop sequence and climate

53 Helms K - MSc,PhD survey of virus diseases of pastures and crops

54 Henson IE - BSc,PhD,NDH research on the water relations of wheat and lupins under semi-acid conditions

55 Higgins TJV - MAgSc,PhD protein and nucleic acid synthesis in developing seeds

56 Hocking PJ - BHortSci,PhD mineral nutrition of oilseed rape and sunflower with emphasis on the effects of nutrition on components of yield

57 Hoggart RM - BSc,PhD optimising DNA uptake by plants initially concentrating on uptake via pollen

58 Hudson GS - BSc,PhD research on the characterisation of genes coding for proteins of photosynthetic complexes

59 Huppatz JL - BSc,PhD systemic fungicides; chemistry of nitrogen heterocycles

60 Hyland BPM - BSc,PhD

61 Jacobsen JV - BAgrSc,PhD mechanism of action of gibberellin and abscisic acid in cereal aleurone

62 Jarosz AM - BSc,PhD research on ecological, genetical and pathological aspects of the interaction between native and introduced plant species and various pathogens of ecological and agricultural importance

63 Jenkins CLD - BSc,PhD development of C_4-specific weedicide

64 Kanis A - PhD taxonomic revisions of Australian Amaranthaceae (excluding *Ptilotus*) and Mimosaceae (excluding *Acacia*)

65 Katekar GF - BSc,LLB,PhD mode of action of natural and synthetic plant growth regulators and herbicides

66 King RW - MSc,PhD environmental control of flower initiation; control of cereal grain development and premature sprouting

67 Kirk JTO - MA,PhD,ScD primary productivity of aquatic ecosystems; biochemistry of cruciferous oilseeds

68 Langridge JB - MSc,PhD synthesis of gene vector systems for plant cell transformation

69 Larkin PJ - BSc,PhD somatic cell genetics and somaclonal variation; gene transformation; protoplast biology

70 Lawrence GJ - BAgSc,PhD genetics of cereal seed proteins; host-pathogen interactions with plant fungi

71 Lazarides M - MSc taxonomic research in Poaceae of tropical and arid Australia

72 Leigh JH - BSc,PhD effects of grazing on contrasting natural vegetation types in areas of high and low rainfall in relation to management, especially fire; ecology of rare and threatened Australian flora; weed ecology

73 Lipsett J - MAgrSc boron deficiency in plants; influence of soil fertility on crop growth; effect of liming treatments on plant growth; nutrition of rape

74 Llewellyn DJ - BSc,PhD development of gene transfer systems in plants with *Agrobacterium*

75 Luckett DJ - BSc,MPhil,PhD breeding of high quality cottons; cotton genetics

76 Macnicol PK - MSc,DrRERNAT amino acid metabolism and transport

77 McInnes KJ - BSc,MSc,PhD collaborative research programs on (i) simultaneous water and solute movement in soils; (ii) nutrient availability and cycling; (iii) water availability in relation to root exploration of the soil profile, and (iv) the role of tillage in water and nutrient availability

78 Miller WA - BSc,PhD determination of the nucleotide sequence of the RNA of barley yellow dwarf virus

79 Munns RE - BSc,PhD physiology of salt stress

80 Myers LF - MAgrSc crop genotype x environment interactions; crop nutrition; pasture breeding

81 O'Brien PA - BSc,PhD *In Vitro* mutagenesis of seed protein genes and transformation studies in higher plants

82 Oram RN - BAgrSc,PhD plant improvement of Indian mustard, phalaris, lupins and barley, including reduced toxicity, tolerance to acid soils, improved seed yield and quality, disease and drought resistance

83 Passioura JB - BAgrSc,PhD influences of water and salt stress on physiology and yield of wheat and barley

84 Peoples MB - BAgrSc,PhD development and evaluation of methods to measure biological nitrogen fixation

85 Phillips JN - BA,MSc,PhD,FRACI design and synthesis of novel herbicides, fungicides and plant growth regulators. Structure-activity relationship and mode of action studies

86 Porra RJ - BSc,PhD biochemical genetics and gene regulation in maize and maize rust

87 Pryor AJ - BSc,PhD biochemical genetics and gene regulation in maize and maize rust

88 Randall PJ - BSc,PhD mineral nutrition of oilseed crops; influence of nutrition on the quantity and quality of protein in seeds of cereal, legume and oilseed crops

89 Rawson HM - BSc,PhD physiology of seed yield in wheat and sunflower; relationship between leaf development and photosynthetic rates

90 Richards RA - BSc,PhD wheat breeding and physiology in drought prone and saline environments

91 Roper MM - BSc,PhD nitrogen fixation in association with wheat stubble and other crop residues, with particular regard to ecological constraints and management practices

92 Ryan SA - BScAgr,PhD genetics of cell culture induced variation; analysis of somaclonal variation in wheat

93 Simpson JR - MSc,PhD improvement of nitrogen fertiliser efficiency for rice and wheat; nitrogen cycling in agricultural ecosystems; maintenence of soil fertility under intensive cropping; root growth and nitrogen fixation by legumes in adverse soil environments

94 Spencer D - BSc,PhD protein and nucleic acid synthesis in developing seeds

95 Surin BP - BSc(Hons) application of molecular biology to ecological studies of *Rhizobium*

96 Tanner GJ - BSc,DipAgricChem,PhD cell culture of lucerne for bloat safety and improved nutrition

97 Teitei T - MSc,PhD synthesis of herbicides and plant growth regulators

98 Thomson NJP - MScAgric,PhD breeding of more productive cotton varieties with insect-resistant characteristics

99 Trinick MJ - MSc,PhD Physiology of *Parasponia Rhizobium association*; ecology of rhizobia

100 Turner NC - BSc,PhD morphological and physiological mechanisms of adaption to drought, particularly sunflower

101 Walker JC - BSc,PhD optimising DNA uptake by plants, initially concentrating on uptake via pollen

102 Wardlaw IF - BAgrSc,PhD role of metabolism, photosynthesis and translocation in regulating growth responses to the environment

103 Waterhouse PM - BSc,PhD virus diseases of subterranean clover and crop plants and development of diagnostic probes

104 Watson JM - DipAppChem,DipAppBiol,Masters Prel Exam,PhD molecular biology of the symbiotic association between legumes and their root nodule bacteria

105 West JG - BSc,PhD taxonomic revisions and breeding systems of Australian Portulacaceae and Caryophyllaceae; cladastic studies

106 Whitfeld PR - BSc,PhD structure, organisation and function of chloroplast DNA

107 Wilson AGL - BSc insecticidal and application efficiency; assessment of *Heliothis* moth abundance; cotton pest ecology

108 Zwar JA - MAgrSc,PhD cellular mechanisms of gibberellin action

66 Experimental Scientists

1 Armstrong JS - MSc,FBSC,FACS development of simulation models to compare tillage strategies

2 Ashburner NA - BAgrSc coding new modules for incorporation in the SIRATAC program

3 Axelsen A - Q.D.A.H. meat and crop production systems

4 Bagnall DJ - MSc plant response to low temperature; environmental control of fruiting

5 Bhati R - BSc analysis of somaclonal population of wheat for alcohol dehydrogenase variants; maintain and regenerate wheat plants to provide additional somaclonal populations for analysis

6 Brook KD - BAgrSc development of SIRATAC on-farm operations

7 Craig S - MSc,PhD ultrastructure of storage protein deposition in plants; freeze-etch techniques; immunocytochemistry

8 Craven LA - MSc basic taxonomic research in three projects (i) biosystematic study of *Melaleuca* in Australia; (ii) biosystematic study of *Heliotropium*; (iii) a revision of the genus *Syzygium* in Papuasia

9 Davies P - BAgSc analysis of somaclonal variation in wheat

10 Dawson IA - BSc control of biological services; high temperature control of yield in wheat

11 DeMarco DG - BScAgr phosphorus uptake by plants

12 Donnelly JR - MAgrSc pasture and animal production in grazing animals

13 Dunstone RL - BA,MSc relationship between leaf development and photosynthetic rate in dicotyledons; physiology of jojoba

14 Gault RR - WDA field ecology of *Rhizobium* spp.; inoculation studies

15 Geissler AE - DipHort,MSc mode of action of natural and synthetic plant growth regulators and herbicides

16 Grant JE - BSc interspecific hybridisation of wild *Glycine* with soyabean

17 Gray M Australian alpine and sub-alpine flora; naturalised weedy aliens; taxonomy of *Celmisia*

18 Green AG - MScAgr,PhD genetics of oil quantity and quality in linseed; interspecific hybridisation in *Linum*

19 Irving GCJ - BSc biochemical aspects of nitrogen and phosphorus nutrition in plants and application to diagnostic methods

20 Jones DB - BScAgr development of cropping systems

21 Lodder MS - BScAg,DipLscapeTech investigation of seed yield in native grasses in relation to their use in landscaping

22 McKinney GT - MRurSc,BApplSc simulation of agricultural systems; mini and micro computer use

23 Nicolas ME - DAA, INA,PhD studies on the root growth; assimilate transfer and water relations of a range of wheat genotypes
24 Pinkerton A - BSc factors affecting root growth and development; nutrition of oil crop plants; application of X-ray fluorescence techniques to soil and plant analyses; nutrition flow in lambs
25 Pook EW - MSc tree growth dynamics in eucalypt forests
26 Reid PE - BAgrSc breeding for disease resistance and earliness in cotton
27 Schroeder HE - BSc genetics of storage proteins in legume seeds
28 Stafford BG - DipAppScAgr testing of recently developed components of SIRATAC programme with experiments on commercial farms
29 Tonnet ML - MSc,PhD chemical aspects of plant and animal production
30 Turner GL - BSc,MSc biochemistry and physiology of nitrogen fixation; stable isotope analysis
31 Waldron JC - MSc isolation of gene systems suitable for genetic transformation in plants
32 Williams JD - D.D.A. weed ecology, especially competition and the prediction of weed potential
33 Wilson LJ - BSc investigation of mite abundance, economic injury levels and pest management in cotton
34 Wimbush DJ - MSc effects of grazing and fire on natural vegetation in the Snowy Mountains region

67 At CSIRO, Wembley
Private Bag, Wembley, Western Australia

1 Hamblin AP - BA,MAgrSc,PhD soil/plant root interactions as they affect wheat and lupin production on sand plains and texture soils in the south west region of Western Australia
2 Turner NC - BSc,PhD morphological and physiological mechanisms of adaption to drought, particularly sunflower

68 Wheat Research Unit
C/o Bread Research Inst. of Australia, 51 Delhi Road, P.O. Box 7, North Ryde, New South Wales, 2113
Tel: (02) 888 9600. Telex: 120003. Cables: Bresearch, Sydney.

1 Wrigley CW - MSc,PhD *Officer-in-Charge* protein composition of cereal grains
2 MacRitchie F - BSc,PhD *Deputy Officer-in-Charge* lipid-protein associations in flour
3 Bason ML - BSc storage of tropical grains
4 Batey IL - BSc,PhD modification of wheat gluten for novel food uses
5 Campbell WP - BSc,PhD protein composition and quality in wheat grain
6 Castle SL - MSc,PhD molecular biology of wheat grain proteins
7 Cheung AB - MA,PhD design of electrophoretic equipment
8 Donovan GR - BSc,PhD cloning of genes for wheat grain proteins
9 duCros Ms DL - BA protein composition of durum wheat grain
10 Gras PW - BSc,PhD grain storage; cereal starch composition
11 Ronalds JA - MSc cereal analysis by near infrared reflectance spectroscopy
12 Ross AS - BScAg objective testing of sprouted grain
13 Skerritt JH - BSc,PhD characterization of cereal proteins with monoclonal antibodies

69 Division of Horticulture

70 Headquarters
Hartley Grove, (Off Waite Road), Glen Osmond, South Australia, 5064
[GPO Box 350, Adelaide, South Australia, 5001]
Tel: (08) 274 9244. Telex: 88000. Cables: Hortres, Adelaide.

1 Possingham JV - BAgrSc,MSc,DPhil,DSc,FST,FAIAS *Chief of Division* growth and replication of chloroplasts
2 Cheung WY - BSc,PhD genome homology in organelles
3 Downtown WJS - BSc,PhD environmental control of photosynthesis
4 Loveys BR - BSc,PhD hormonal control of gas exchange
5 Rezaian MA - BAgrSc,PhD,MSc plant virology
6 Robinson SP - BSc,PhD biochemistry of photosynthesis
7 Scott NS - BAgrSc,PhD genome homology in organelles
8 Skene KGM - BAgrSc,PhD tissue culture
9 Turnbull CGN - BSc,PhD hormonal control of plant development
10 Whisson DL - BSc,PhD genome homology in organelles
11 Wren RA - BA,ALAA *Librarian*

71 Merbein, Victoria
Private Mail Bag, Victoria, 3505
Tel: (050) 256 201. Telex: 55581. Cables: Coresearch, Merbein.

1 Walker RR - BAgSc,PhD *Officer-in-Charge* plant responses to salinity
2 Alexander - DMcE,BSc,MS tropical tree crops
3 Barlass M - BSc,MSc,PhD tissue culture
4 Clingeleffer P - BAgSc viticulture
5 Douglas TJ - BSc,PhD lipid biochemistry
6 Ruhl EH - MSc,PhD grapevine nutrition
7 Storey R - BSc,PhD root structure and function
8 Sykes SR - BSc,MSc,PhD genetics of salt tolerance
9 Vithanage HIMV - BSc,PhD reproductive biology

72 At Darwin
CSIRO Darwin Laboratories, Private Bag 44, Winnellie, Northern Territory, 5789
Tel: (089) 22 1711. Telex: 85294.

1 Chacko EK - BagSc,MSc,PhD tropical tree crops

73 Division of Soils

74 Headquarters
Waite Road, Urrbrae, South Australia
[Private Bag 2, Glen Osmond, South Australia, 5064]
Tel: (08) 274 9311. Telex: 82406. Cables: Coresearch, Adelaide.

1 Lee KE - DSc *Assistant Chief (Adelaide)* soil fauna; ecology; energy flow and litter composition
2 Amato M - BAppSc nitrogen cycling in cereal/pasture rotations
3 Beckmann GG - PhD pedology; soil landscape relations
4 Beech TA - BSc soil, plant, water analysis; methods development
5 Billing NB - BSc pedology
6 Bird AF - DSc ecology of parasitic nematodes
7 Bowen GD - MSc,DSc dynamics of colonisation of roots by micro-organisms and their effects on mycorrhizal and pathogen fungi; *Casuarina/actinomycetes* interactions
8 Bowman GM - MSc,PhD pedology; geomorphology
9 Brisbane PG - MAgrSc biological control of cereal root diseases
10 Buckerfield JC - BSc soil fauna
11 Butler JHA - BSc,PhD nitrogen cycling and nitrogen fixation in legume pasture-wheat rotations
12 Cartwright B - BSc,PhD nutrition; wheat on calcareous soils; boron toxicity
13 Clayton PM - BSc environmental pollution; micronutrients in soils, plants and waters
14 Emerson WW - BA,PhD soil structure; treatments to control physical properties of soils
15 Fitzpatrick RW - PhD mineralogical transformations; pedology
16 Fordham AW - PhD soil development; translocation of clay
17 Foster RC - BSc,PhD ultrastructure of rhizosphere in wheat and forest trees
18 Gurr CG - BSc soil physics instrumentation; soil water relations; gamma ray techniques for measuring water
19 Handreck KA - BSc,MAgrSc *Scientific Liaison Officer* potting mixes
20 Highnett CT - BAppSc uptake of water by plants, hydrologic properties of soils
21 Hughes MW measurement of soil water movement; hydrology using environmental isotopes
22 Hutson BR - BSc,PhD role of soil fauna in litter decomposition
23 Jackson RB - MSc dynamics of nitrogen transformations in soils
24 Janik LJ - DipIndChem infrared spectroscopy; soil chemistry
25 Kimber RWL - BSc,PhD amino acid racemisation dating of calcretes and calcareous soils
26 Ladd JN - MSc,PhD nitrogen cycling in legume-cereal rotations
27 Leaney FW - BSc radiocarbon dating
28 Martin JK - MSc,PhD biochemistry of the rhizosphere; plant nutrients and organic matter turnover in the rhizosphere; exudation of organic compounds
29 McLeod S - ARACI chemical analysis; instumentation
30 Merry RH - BAgrSc environmental pollution; micronutrients in soils and plants; acid soils
31 Milnes AR - PhD accumulation of secondary silica in acid Australian landscapes; dating of sediments
32 Neate SM - PhD biological control of cereal root diseases
33 Norrish K - MSc,PhD,FAA mineralogy of soil trace elements; techniques of x-ray and micro-probe analysis
34 Peter P - BEng mechanisms of volume change in expansive clays; measurement of suction in soils
35 Raupach M - DSc infrared spectroscopy; clay-organic complexes; soil chemistry
36 Rosbrook PA - MAgrSc *Casuarina-Frankia* symbioses
37 Rosser H - DipAppPhy mineralogy of soils; X-ray micro-analysis; scanning electron microscopy; electronics
38 Rovira AD - BScAgr,PhD ecology of soil bacteria; control of cereal root and rots and cereal nematodes
39 Self PS - PhD studies of soils; scanning electron microscope
40 Slade PG - BSc,MA,PhD structure of clay-organic complexes
41 Smettem KRJ - PhD water and solute movement; soil structure
42 Taylor RM - MSc pedogenesis of soil iron oxides; geochemistry of trace elements
43 Theodorou C - MAgrSc root morphology of *Pinus radiata*; mycorrhizal relations of *Pinus radiata* and *Casuarina*
44 Tiller KG - MSc,PhD soil chemistry and plant availability of micronutrients; environmental pollution
45 Walker C - PhD transpiration; partitioning of isotopes
46 Walker GR - PhD modelling of soil hydrology
47 Wells CB - MAgrSc hydrology and salinity; carbonates
48 Wright MJ - RDA pedology; hardpans in arid soils
49 Zarcinas BA - BAppSc inductively coupled plasma spectroscopy

75 Canberra Laboratories
Black Mountain, via Canberra City, A.C.T.
[GPO Box 639, Canberra, 2601]
Tel: (062) 46 4911. Telex: 61452. Cables: Coresearch, Canberra.

1 Smiles DE - DScAgr *Chief of Division*
2 Walker PH - MScAgr,PhD *Assistant Chief* sedimentary processes; pedology
3 Beatty HJ - ABTC soil and plant analysis; automatic analysis
4 Blackwell P - PhD development of macroporosity
5 Bond WJ - PhD solute transport in clay soils; hydrology of duplex soils
6 Chartres CJ - PhD pedology; fabric analysis of surface soils
7 Colwell JD - BScAgr,PhD,BA statistical quantification of soil fertility; optimal use of fertilizers; soil phosphorus
8 Greenslade PJM - PhD,DIC ant ecology and their role in soil processes
9 Kinnell PIA - MSc analysis and characterisation of natural rainfall in relation to runoff and sediment loss
10 Kirby JM - PhD mechanical properties of swelling clay soils
11 Moss AJ - PhD mechanisms of entrainment, transport and sorting of soils and sediments
12 Phillips IR - PhD solute transport in clay soils; hydrology of duplex soils
13 Ringrose-Voase AJ - PhD structural analysis of clay soils
14 Watson CL - BAgrSc,PhD soil physical analysis
15 Willett IR - BSc,PhD chemistry of flooded soils; land disposal of sewage sludge

76 Cunningham Laboratory
306 Carmody Road, St Lucia, Queensland, 4067
Tel: (07) 377 0209 (Brisbane). Telex: 42159. Cables: Coresearch, Brisbane.

1 Probert ME - PhD *Officer-in-Charge* soil fertility; phosphorus; sulphur
2 Bridge BJ - MSc,PhD landscape dynamics: water infiltration
3 Drinnan JC - DipIndChem X-ray spectrographic analysis
4 Fergus IF - MAgrSc soil fertility; potassium
5 Little IP - BScAgr geochemistry; mineralogy
6 McBratney AB - PhD soil variability; geostatistics
7 McGarry D - PhD pedology; soil properties affecting management of clay soils under irrigation
8 Moore AW - MSc,PhD data base management
9 Prebble RE - BSc soil water; instrumentation

77 Davies Laboratory, Townsville
CSIRO, Private Mail Bag, P.O. Aitkenvale, Queensland, 4814
Tel: (077) 71 9511. Telex: 47181. Cables: Coresearch.

1 Isbell RF - MSc *Officer-in-Charge* pedology and land use
2 Bristow KL - PhD soil physical properties; seedbed conditions; soil temperature
3 Coventry RJ - BSc,PhD pedology
4 Gillman GP - MSc soil analysis; chemistry of variable charge soils
5 Hicks WS - BSc chemical analysis
6 Holt JA - MSc rainforest soil fauna; termite ecology
7 Murtha GG - QDA soil survey and land use
8 Ross PJ - BSc soil physical properties
9 Spain AV - MAgrSc,PhD soil biology, nutrient cycling in rainforest; termite ecology
10 Williams J - BScAgr,PhD soil physics; hydrology

78 Division of Tropical Crops and Pastures

Research on tropical pastures and some tropical field crops

79 Headquarters
Cunningham Laboratory, 306 Carmody Road, St Lucia, Queensland, 4064
Tel: Brisbane (07) 3770209. Telex: 42159.

1 Jones RJ - BSc(Agric),PhD,FAIAS,DTA *Chief of the Division*
2 Adams GT - BAppSc *Senior Information Officer*
3 Evans TR - BSc,MAgrSc *Divisional Liaison Officer, North Queensland*
4 Eyles AG - BScAgr,MSc *Divisional Secretary Research*
5 Pridanikoff Ms V - BA,ALAA *Technical Service Librarian*
6 Smith FWN - BScAgr,PhD *Assistant Chief of Division*
7 Tuppack A - BA,DipLib,ALAA *Divisional Librarian*

80 At Davies Laboratory
Private Mail Bag, PO Aitkenvale, Queensland, 4814
Tel: (077) 71 9511.

1 Gillard P - BSc,PhD *Liaison Officer, North Queensland*
2 Penny Mrs N - BSc,ALAA *Librarian*

81 Analytical Chemistry

1 Hansen RW - BAppScChem,GradDipMgt development of methods for analysis of carbohydrates using high performance liquid chromatography; techniques for analysis using atomic absorption spectrophotometry
2 Johnson AD - BSc,MSc *Program Leader* development of analytical methods; investigation of mechanisms by which some plants influence nutrient uptake by chemical modification of their substrate micro-environment

82 At Davies Laboratory
Private Mail Bag, PO Aitkenvale, Queensland, 4814

1 Megarrity RG - BSc development of analytical methods; elucidation of the metabolic fate of mimosine in the ruminant; Leucaena detoxification by ruminants with augmented rumen microflora

83 Animal Nutrition

1 McLean RW - BScAgr,MRurSc beef production from pastures; diet selection by cattle grazing grass/legume pastures, and their nutrient intake, status and cycling
2 McLeod MN - BSc,MAgrSc,BEcon physical and biological factors controlling the resistance of feed particles to reduction in size within the ruminant; predicting feed intake and digestibility by laboratory techniques
3 Minson DJ - BSc,PhD,DSc,FRSC,FIBiol,FASAP *Program Leader* chemical and physical characters of tropical pasture plants controlling intake, digestion, and productivity of ruminant animals

84 Crop Adaptation

1 Beech DF - MAIAS agronomy and adaption of tropical grain legumes and oilseeds crops new to northern Australia
2 Ferraris R - BagrSc,PhD factors affecting growth and soluble carbohydrate content of sweet and forage sorghums: guayule (*Parthenium argentatum*) as a potential natural rubber source; establishment of summer crops on cracking clay soils
3 Leach GJ - BScAgrBot,PhD cropping systems in the sub-tropics; competition for light and soil water between crops of sorghum and sunflowers, and a leguminous intercrop chickpea growth, phenology and water use
4 Rees MC - BAgr,MAgrSc,PhD potential of a range of plant species as ground cover under sunflowers to protect clay soils from erosion; interactions between these plants and sunflowers affecting yield, soil water and fertility
5 Wood IM - BSc,BAgrSc *Joint Program Leader* role of soil, climatic and plant factors in crop establishment in northern Australia; crop growth and water balance modelling; assessment of *Sesbania* spp. for production of forage, biomass and galactomannan

85 Cropping Systems

86 At ACIAR/CSIRO Dryland Project
Ihmadu Bello Uni, Zaria, Nigeria

1 Cogle AL - BAgSc,MSc evaluation of the potential of legumes for supplying nitrogen in African cereal systems
2 Moore AW - BAgSc,MSc,PhD soil and climatic constraints to crop production in the semi-arid tropics

87 AT ACIAR/CSIRO Dryland Project
P.O. Box 41567, Nairobi, Kenya
Tel: Kenya 335725.

1 Jones RK - BAgrSc,PhD soil fertility management in food crop systems in the semi-arid tropics
2 Keating BA - BAgrSc,PhD assessment and management of climatic risk to crop production in the semi-arid tropics
3 Ockwell AP - BAgrSc,PhD the causes of adoption/non-adoption of technological innovations in African smallholder agriculture

88 At Tropical Ecosystem Research Centre
Private Bag No. 44, Winnellie, Northern Territory, 5879
Tel: (089) 22 1711.

1 Abrecht D - BRurSc,PhD crop establishment in a no-tillage system on sesquioxidic soils in the semi-arid tropics
2 Chapman AL - BAgrSc,MS improvement of crop yield modelling in relation to climatic risk description; nitrogen contribution by leucaena in cereal systems

89 At Davies Laboratory
Private Mail Bag, P.O. Aitkenvale, Queensland, 4814

1 Carberry PS - BScAgr evaluation and modification of crop production models for use in the semi-arid tropics
2 McCown RL - BSc,MSc,PhD *Joint Program Leader* ecological adaption of farming systems and technological innovations in the semi-arid tropics

90 Katherine Research Station
P.O. Box 21, Katherine, Northern Territory, 5780
Tel: (089) 72 1488.

1 Dimes JP - BAgrSc nitrogen availability in cereal-pasture legume rotations on sesquioxide soils in the semi-arid tropics

91 Information Technology

1 Dale MB - BSc,PhD,MBCS FLS BA *Computer Pattern Analyst* development of computer methods for expert systems on crop production, (SATISFY and REFINE) and automated inference procedures

2 Russell JS - BAgrSc,PhD,FTS,FAIAFS *Program Leader* integration of available climatic, soil, plant operational and economic information into expert systems. Two systems being studied are SATISFY (inland subtropical grain areas) and REFINE (sugar cane)

92 Legume Bacteriology

1 Date RA - BScAgr,PhD *Program Leader* selection and evaluation of *Rhizobium* for nodulation and nitrogen fixation in tropical legumes specificity and genotype interactions between legumes and *Rhizobium*; effect of soil temperature, moisture, pH and pH related factors on *Rhizobium*; techniques for assessing, preserving and utilizing *Rhizobium* germplasm

2 Bushby HVA - BAgrSc,Phd relations between rhizosphere colonisation and nodulation by inoculum strains of field grown legumes; relations between surface charge characteristics of rhizobia and soils of isolation; phosphorus nutrition of rhizobia

93 Nitrogen Economy and Agricultural Systems

1 Myers RJK - BRurSc,PhD *Program Leader* nitrogen (N) requirements of dryland cereal crops; modelling of N process in soils; modelling crop response to N; N and carbon dynamics of grass pastures

2 Catchpoole VR - BAgrSc,MAgrSc soil fertility/crop production studies on clay soils under permanent grass pasture, permanent legume, summer crop-winter fallow, and winter crop-summer fallow; nitrogen and carbon dynamics in grass pastures

3 Vallis I - BAgrSc,PhD nitrogen fixation, nitrogen losses, and soil nitrogen changes in *Stylosanthes* pastures; measurement of nitrogen fixation in grain legumes by ^{15}N dilution methods; competition for nitrogen between legume intercrops and maize

4 Weier KL - BAppScChem,MSc nitrate transformations in clay soils under different management systems; developing and applying special equipment and techniques for denitrification and C/N research

94 Pasture Agronomy and Ecology

Sub Tropical Eastern Australia

1 Jones RM - BScAgr,MSc *Program Leader, Pasture Agronomy and Management (Sub-Tropics)* population ecology of some tropical legumes (primarily Siratro, *Desmodium intortum* and *D. uncinatum*); development of management strategies to improve legume persistence; evaluation of legume introductions; utilization of leucaena

2 Clements RJ - BRurSc,PhD stability of legume-based pastures; agro-ecological significance of genetic diversity in pasture legumes; multi-disciplinary research on *Centrosema*

3 Cook SJ - BRurSc,PhD competitive interactions in pastures - the nature and role of competitive interference in botanical composition changes and in the growth and survival of seedlings; proceedings of meteorological data

4 Mott JJ - BSc,PhD ecology of both sown and native pastures, with specific emphasis on the autecological and synecological changes caused by grazing

5 Taylor JA - BAppSc,PhD plant-animal interactions in native and oversown native pastures; grazing behaviour, the grazing process and the response of grazed plants.

95 At Davies Laboratory
Private Mail Bag, P.O. Aitkenvale, Queensland, 4814

Tropical North Eastern Australia

1 Jones RJ - BScAgric,DTA,PhD,FAIAS *Program Leader* value of sown grasses with tropical legumes for growth of steers and pasture stability; relation between pasture characteristics and liveweight change; overcoming leucaena toxicity by transfer of rumen bacteria

2 Coates DB - BRurSc(Hons) nutrition of beef cattle grazing *Stylosanthes* (Verano and Seca) based pastures in semi-arid tropics; effects of phosphorus fertilizer and supplement on animal production; diet selection and intake studies

3 Edye LA - BAgrSc,MAgrSc *Stylosanthes* species: classification and evaluation (particularly *S. scabra* and *S. hamata*); adaptation to marginal environments; selection of improved cultivars from naturally occurring genotypes; selection of anthracnose resistant genotypes from hybrid populations

4 Gardener CJ - BSA,PhD effect of agronomic and ecological factors on sown legumes in grazed pastures; seed input and survival; legume and seedling competition; longevity; selection; spread through cattle; nitrogen input; species interaction

5 Gillard Dr P - BSc,PhD application of decision support systems to the adoption of improved pastures in north Queensland

6 McIvor Dr JG - BAgrSc,PhD agronomy and ecology of pastures in the seasonally dry tropics; autecology of native and introduced grasses with emphasis on adaption of introduced grasses to soil factors

96 Darwin Laboratories
Private Bag No 44, Winnellie, Northern Territory, 5789

Tropical North Western Australia

1 Andrew Dr MH - BScHons,PhD ecology and genetic resources of native *Sorghum* spp.; effects of defoliation and fire on native perennial grasses; stability of a legume-ley pasture; grazing management of native pastures

2 Winter Dr WH - BAgrSc,PhD *Program Leader* evaluation of pastures comprising native and introduced grasses and legumes; the role of phosphorus and sulphur in pasture ecology and animal nutrition; integration of legume ley and native pastures

97 Plant Breeding and Adaptation

1 Imrie Dr BC - BScAgr,PhD *Program Leader* mungbean, cowpea and soyabean; inheritance of characters contributing to adaptation and yields; breeding of better adapted varieties; evaluation of breeding strategies and selection criteria

2 Bray Dr RA - BAgrSc,PhD breeding and genetics of the genus *Leucaena*; genetics of disease resistance of lucerne; quantitative genetics; genetics of rust resistance of *Macroptilium atropurpureum*

3 Cameron Dr DF - BAgrSc,PhD genetics of anthracnose resistance and winter survival in *Stylosanthes*; breeding *Stylosanthes guianensis* for winter survival, anthracnose resistance and edaphic adaption; breeding *Stylosanthes scabra* for anthracnose resistance

4 Chakraborty Dr S - BSc,MSc,PhD epidemiology of anthracnose disease(*Colletotrichum gloeosporioides*) of *Stylosanthes* species; use of genetic diversity for control of anthracnose disease; development of field and glasshouse techniques for assessment of partial resistance in breeding populations

5 Hacker Dr JB - BScAgrBot,PhD genetic improvement of forage quality; *Setaria* seed production; *Digitaria smutsii* seed production and spring yield; *Digitaria milanjiana* dormancy and soil/climate adaption; native herbaceous legume distribution and taxonomy

6 Lawn Dr RJ - BAgrSc,MAgrSc,PhD implications of climatic factors for adaptation and breeding in soyabean and *Vigna* spp.; wet-soil culture of soyabean; weathering resistance in mungbean; fixation in *Vigna* spp

7 Manners Dr JM - BSc,PhD disease physiology of interactions between genotypes of *Stylosanthes* and races of *Colletotrichum gloeosporioides* development of a transformation system for *Stylosanthes* species; application of molecular biology techniques to studies of the physiology and biochemistry of *Stylosanthes/Colletotrichum* interactions

98 At Davies Laboratory
Private Mail Bag, P.O. Aitkenvale, Queensland, 4814

1 Stace Dr HM - BSc,MSc,PhD genetic structures of populations and species relationships in *Stylosanthes*

99 At Darwin Laboratories
Private Bag No 44, Winnellie, Northern Territory, 5789

1 Williams Mr RJ - BSc,MSc *Program Leader* plant introduction for the subtropics; plant geography and adaption; legume taxonomy; information retrieval systems

2 Done Dr AA - BSc,PhD quantitative genetics of variation in phenological and physiological characters in grain sorghum relevant to improving cultivars for the wet season of tropical Australia; grain sorghum breeding and drought evaluation of lodging resistance

3 Strickland RW - BScAgric,MScAgric growth and persistence of native and introduced forage plants in western Queensland in relation to soil type, temperature and soil moisture; taxonomic, agronomic and toxicity studies in *Cassia, Crotalaria*, and *Tephrosia*

100 At Samford Research Station
Old Cedar Creek Road, Samford, Queensland, 4520

1 Pengelly BC - Q.D.A.,BSc maintenance of the collections of tropical grasses and legumes; seed exchange including activities of CSIRO/ADAB Pasture Seeds Project; classification and description of the collections of *Macroptilium* and *Desmanthus* spp.

101 At Davies Laboratory
Private Bag, P.O. Aitkenvale, Queensland, 4814

1 Burt Dr RL - BScAgricBot,PhD classification of *Stylosanthes* and *Desmanthus*; relationships between environment at place of origin and success in Australian environments; plant adaptation and relationships with *Rhizobium*

102 Plant Nutrition

1 Smith Dr FWN - BScAgr,PhD *Program Leader* regulation of nutrient transport; mineral nutrition of summer cropping species; diagnosis of nutrient disorders; availability of soil phosphate

2 Kerridge Dr PC - BAgrSc,PhD maintenance fertilization of grass-legume pastures; interaction of phosphorus nutrition and moisture stress in tropical legumes

103 Plant Physiology

1 Ludlow Dr MM - BAgrSc,PhD,FIBiol *Program Leader* tropical crop and pasture plants; effects of environmental and biological factors on growth, photosynthesis and survival; adaptation to hot, dry environments; response of tropical pasture plants to defoliation

2 Davis R - BAComputing design of microcomputer systems and development of software; instrumentation and control systems for agricultural research

3 Foale Mr MA - BAgrSc,MAgrSc influence of genetic and soil factors, and of sowing pattern on root growth and water use efficiency of dryland grain sorghum; influence of canopy structure and solar altitude on grain sorghum photosynthesis

4 Wilson JR - BScAgr,PhD influence of growth, environment and plant anatomical structure on the chemical composition, digestibility, rate of tissue breakdown and intake of tropical forages; nitrogen use efficiency of grasses; leaf water relations

104 At Darwin Laboratories
Private Bag No 44, Winnellie, Northern Territory, 5789
Tel: (07) 3770209.

1 Muchow Dr RC - BAgrSc,MAgrSc,PhD yield physiology of tropical grain crops; comparative adaption of sorghum, maize and bulrush millet; competition for radiation and water in an intercropping system

105 Ruminal Biotechnology
Private Bag No 44, Winnellie, Northern Territory, 5789
Tel: (07) 3770209.

1 Hegarty Dr MP - BSc,PhD,FRACI *Program Leader* chemical and biological properties of deleterious nitrogenous compounds in some tropical pasture legumes and grain legumes and determination of their modes of action and development of ways of overcoming their effects

2 Ford Dr CW - BSc,PhD,FRSC enzymic degradation of plant cell carbohydrates; structural studies on cell wall carbohydrate complexes; photochemical changes in crops and pastures during water stress; methods development using gas-liquid chromatography and high performance liquid chromatography

3 Lambourne Dr LJ - BSc,MSc,PhD chemical composition, quality and possible toxicity of forages and feed and animal nutrition

106 Special Equipment Development
Private Bag No 44, Winnellie, Northern Territory, 5789
Tel: (089) 22 1711.

1 O'Neill Mr BM - DipAppElecEng,ME,MIEAust,S.M.I.R.E.E. developing data acquisition, instrumentation and control systems; automating laboratory instrumentation; developing improved methods for sensing and logging environmental parameters, particularly radiation, evaporation and soil moisture

2 Wall Mr BH - BE field data logging with micro-computers; field data acquisition and simultaneous reduction by micro-computers; water balance modelling; optimization of soil water flux models; portable instrumentation for crop micro-environmental measurements

107 Institute of Natural Resources and Environment
Limestone Avenue, Campbell, A.C.T.
[225, Dickson, A.C.T.]
Tel: (062) 484211. Telex: 62003.

1 Green RM - BSc,PhD,FTS *Director*

108 Division of Fisheries
Headquarters, CSIRO Marine Laboratories, GPO Box 1538, Hobart, Tasmania, 7001
Tel: (002) 20 6222. Telex: 57182. FAX: 23 7125.

109 Administration

1 Harden Jones FR - BSc,PhD *Chief* fish guidance mechanisms
2 Abbott DJ - BA,DipLib,ALAA *Librarian*
3 Dickson VM - MA,BLS,PhD *Scientific Editor*
4 Elliott NG - BSc,PhD *Scientific Assistant to the Chief* pollution, mollusc ecology

110 Algal Physiology and Ecology

1 Jeffrey SW - MSc,PhD *Head of Section* algal physiology; phytoplankton ecology; algal culture
2 Blackburn SI - BSc,PhD algal culture
3 Hallegraeff GM - BSc,PhD phytoplankton taxonomy and ecology
4 Kelly GJ - BScAgr,PhD photosynthesis; fatty acids

111 Fish population dynamics and stock assessment

1 Kirkwood GP - BSc,PhD *Head of Section* fish population dynamics
2 Eckert Miss GJ - BSc population dynamics; computer programming
3 Hampton JW - BSc computer programming; population modelling
4 Hearn WS - BSc population dynamics
5 Majkowski J - MSc,PhD population dynamics and modelling
6 Wayte Ms SE - BSc computer programming

112 Fisheries resources

1 Davenport Ms SR - BSc fish biology
2 Davis TLO - BSc,PhD fish biology

3 Lindholm RY - Filkand fish biology
4 Martin RB - MSc reproductive biology of fish; mollusc ecology
5 McLoughlin KJ - BSc fish biology
6 McLoughlin RJ - BSc marine sedimentology; mollusc ecology
7 Sainsbury KJ - BSc,PhD tropical fish population dynamics and management
8 Stevens JD - BSc,PhD shark biology and ecology
9 West GJ - BSc fish biology
10 Whitelaw AW - BAppSc,BSc fish biology
11 Young PC - BSc,ARCS,PhD *Head of Section* reproductive physiology of fish; mollusc ecology

113 Southern Temperate

1 Blaber SJM - MSc,PhD,DSc *Head of Section* tropical fish ecology; community ecology; ichthyology
2 Bulman Ms CM - BSc fish biology
3 Furlani Miss D - BSc fish biology
4 Gunn JS - BSc fish biology
5 Kenchington TJ - MSc,PhD morphometrics; population dynamics
6 Last PR - BSc,PhD morphometrics; population dynamics
7 May Ms JL - MSc fish biology
8 Stanley CA - BSc,PhD reproductive ecology and behaviour of fish
9 Young JW - BSc fish biology

114 Traditional fishing

1 Johannes RE - MSc,PhD artisanal fisheries

115 Biological oceanography

1 Harris GP - BSc,PhD *Head of Section* biological oceanography
2 Clementson Ms LA - BAppSc,MSc determination of nutrients
3 Griffiths FB - BSc euphausiid biology
4 Heron AC - MSc,PhD zooplankton biology; population dynamics
5 Rimmer DW - BSc larval fish
6 Tranter DJ - MSc,DSc biological oceanography

116 North and Northeast Regional Laboratory
233 Middle Street, P.O. Box 120, Cleveland, Queensland, 4163
Tel: (07) 286 2022. Telex: 42240.

1 Dall W - MSc,PhD,DSc *Officer-in-Charge* crustacean physiology
2 Crocos PJ - MSc,DipBioch prawn biology
3 Harris ANM - BSc fish biology
4 Hill BJ - MSc,PhD crustacean behaviour and ecology
5 Jackson CJ - BSc prawn larval ecology
6 Lavery SD - BSc biochemical genetics
7 Milton DA - BSc biochemical genetics
8 Moriarty DJW - BAgSc,PhD microbial ecology
9 Morris GB - BSc computer programming
10 Poiner IR - BSc,PhD fish and seagrass biology
11 Pollard PC - BSc microbial ecology
12 Rothlisberg PC - MSc,PhD prawn larval taxonomy, biology and ecology
13 Salini JP - BAgSc biochemical genetics
14 Smith DM - BAppSc prawn physiology
15 Somers IF - MSc,DipCompSci population dynamics
16 Staples DJ - BSc,PhD prawn ecology
17 Vance DJ - BSc prawn ecology
18 Wassenberg TJ - BAppSc prawn behaviour

117 West and North West Regional Laboratory
Leech Street, Marmion, Western Australia
[PO Box 20, North Beach, Western Australia, 6020]
Tel: (09) 246 8288. Telex: 93366.

1 Phillips BF - BSc,PhD *Officer-in-Charge* rock lobster ecology
2 Crossland CJ - MSc,PhD plant physiology
3 Edgar GJ - BSc,PhD seagrass epifauna
4 Howard RK - BSc,PhD rock lobster ecology
5 Jernakoff P - MSc,PhD rock lobster ecology
6 Kirkman H - MAgSc,PhD seaweed, seagrass biology
7 Major GA - BSc,DipEd,MAppSc seawater chemistry
8 Rainer SF - MSc,PhD macrobenthos ecology
9 Smith DF - BSc,MSc,PhD primary and secondary production
10 Wadley Ms VA - MSc community ecology

118 at Cairns - Northern Fisheries Research Centre
Cnr Aumuller and Tingira Streets, Portsmith, PO Bungalow, Cairns, Queensland, 4870
Tel: (070) 51 5588. Telex: 48261.

1 Trendall JT - BSc,PhD rock lobster ecology

119 Division of Water Resources
GPO Box 1666, Canberra, A.C.T., 2601
Tel: 464911. Telex: AA62337. Cables: Researchlabs, Canberra.

1 Allison GB - BSc,PhD *Chief of Division*

120 Griffith Laboratory
Private Mail Bag, Griffith, New South Wales, 2680
Tel: (069) 621700. Telex: 169990.

1 Sale PJ - BSc,PhD *Officer-in-Charge* aquatic plant ecology, limnology

121 Water Resources Management

1 Bowner KH - BSc,PhD *Program leader* residue chemistry, agricultural chemicals, eutrophication
2 Breen P - MAppSc effluent treatment by aquatic plant systems
3 Cary PR - NSc,PhD nutrition of aquatic plants
4 Dunbabin J - BScAgr mining effluent treatment by aquatic plant systems
5 Gunn AH - PhD suspended particle characterisation
6 Korth W - BSc analytical chemistry
7 Oliver RL - BSc,PhD algal physiology,ecology
8 Orr PT - BSc aquatic plant physiology
9 Sale PJM - BSc,PhD aquatic plant productivity, physiology
10 Weerts PJG - BAGeol,DipTropAgr on-farm water recycling

122 Irrigation Cropping Management

1 Smith RCG - BSc,DipAgrEc,PhD *Programme Leader* irrigation cropping systems
2 Barrs HD - BSc,PhD root measurement and physiology
3 Blackwell J - CDA,NDA,ND,AgrE irrigation engineering
4 Ceresa A - MIEE instrumentation
5 Heritage AD - BSc,PhD plant pathology
6 Humphreys E - MSc soil nitrogen dynamics
7 Jayawardane NS - BSc,PhD hydraulic behaviour of soils
8 Melhuish FM - BSc,PhD soil nitrogen dynamics
9 Meyer WS - BAgrSc,PhD plant soil water relations
10 Moss GI - BSc,PhD root physiology
11 Muirhead WA - BScAgr,MAgrSc,PhD agronomy of irrigated crops
12 Schaefer NL - BSc,PhD root nutrient relationships
13 Shell GSG - BSc,MSc agricultural physics and instrumentation
14 Stapper M - BSc,PhD crop modelling - expert systems

123 Greenhouse Technology Project

1 Fuller RJ - DipMechEng greenhouse covering materials
2 Garzoli KV - BMechE,ME,PhD *Project Leader* greenhouse design
3 Meyer CP - BSc,PhD modelling of greenhouse climate

124 Catchment Hydrology Program

1 Barnes CJ - BSc,PhD mathematical modelling; isotype hydrology
2 Braunack MV - BAgrSc,PhD soil physics; soil mechanics
3 Chin A - DipMath computer programming; modelling
4 Clark RL - BA,BSc,PhD vegetation and environment history
5 Crapper PF - BE,PhD *Mathematician* fluid dynamics
6 Crockford RH - ARMTC chemistry; ecology
7 Fleming PM - BE *Program Leader*
8 Galloway RW - MA,PhD *Geomorphologist* palaeoclimatology
9 Goodspeed MJ - BSc *Physicist* hydrology
10 Jakobsen P - BScAgr,ARMIT soil chemistry
11 Kalma JD - IngAgr,PhD applied climatology
12 Moore ID - BE,MEngSc,PhD *Hydrologist* water movement in porous media
13 Morland RT - BSc forest catchment hydrology
14 O'Loughlin EM - BE,MEngSc,PhD *Engineer* hydrology
15 Speight JG - MSc *Geomorphologist* terrain classification
16 Srikanthan R - BSc,MEng,PhD hydrology; modelling
17 Wasson RJ - BA,PhD *Geomorphologist* quaternary history
18 Williams BG - BAgrSc,PhD *Soil Scientist*

125 Resource Management Program

1 Booth TH - BSc,PhD *Systems analyst* land evaluation for forestry
2 Cocks KD - MAgrSc,PhD *Agricultural Economist* land use planning
3 Compangnoni PT - DipArch,DipT&CP,GradEKistics *Architect and Planner* land use planning
4 Davis JR - BSc,BEc,PhD *Physicist and Economist* land use planning
5 Graham AW - BSc rain forest ecology
6 Grant IW - BSc *Computer Programming* mapping programs and planning aids
7 Hopkins MS - BSc,PhD rainforest ecology
8 Hutchinson MF - MSc,PhD *Mathematician* computer modelling of climate, terrain
9 Ive JR - BAgrSc,LDA,BEc *Agricultural Economist* computer modelling
10 Mackenzie DH - BAgrSc *Agronomist* crop data base development
11 Mackey BG - BAppSc,MSc *Applied Biologist* environmental data base
12 Martin Ms CVA - BA,GradDipComp *Computer Programmer* computerized geographic data systems
13 Moncur MW *Agronomist* crop phenology, physiology
14 Nix HA - BAgrSc,QDA,FAIS *Program Leader*
15 Parvey CA - BAppSc *Geographer* computerized geographic data systems
16 Penridge LK - BSc,GradDipComp *Zoologist* modelling plant communities
17 Rattigan K - BAppSc *Agronomist* land evaluation for horticulture
18 Reece PH - BSc *Applied Biologist*
19 Walker J - BSc,PhD *Plant Ecologist* management of natural vegetation

20 Walker PA - MA *Transport Geographer* information systems
21 Wells KF - BSc,MSc,PhD *Plant Ecologist*

126 Land Resources Program

1 Austin MP - BSc,PhD *Plant Ecologist* inventory analysis and numerical classification
2 Belbin DL - BSc,GradDipComp *Geologist* computer programming and numerical taxonomy
3 Faith DP - BA,PhD *Evolutionary Ecologist* numerical classification
4 Gillison AN - QDA,BAgrSc,MSc,PhD *Plant Ecologist* vegetation classification, rapid resource survey methods, and plant biogeography
5 Harrison Mrs BA - BSc,GradDipComp *Computer Programmer and Forester* remote sensing
6 Heyligers PC - PhD *Plant Ecologist* coastal vegetation studies
7 Jupp DLB - BSc,PhD *Applied Mathematician* remote sensing, spatial analysis
8 Laut P - BA,PhD *Program Leader*
9 Margules CR - BAppSc,PhD *Ecologist* resource inventory analysis and conservation evaluation
10 Mayo KK - BA *Computer Programmer* remote sensing, digital image analysis
11 Nanninga PM - BSc,GradDipComp *Mathematician* computer geographic data systems
12 Nicholls AO - BSc,PhD *Plant Ecologist* plant competition
13 Tunstall BR - BSc,PhD *Ecophysiologist* hydrology and remote sensing
14 Williams OB - BAgrSc,MSc *Ecologist* pasture ecology and plant demography
15 Yapp GA - BA,PhD *Recreation Geographer* coastal zone studies

127 International Research Program

1 Angus JF - BAgrSc,PhD *Agronomist* agricultural land evaluation crop modelling
2 Bellamy Mrs JA - BA *Geographer* geomorphology and land inventory
3 Bleeker P - GeolDrs,PhD *Soil Scientist* soil taxonomy
4 Cuddy Ms SM - BA *Computer Programmer* geographic data systems
5 Fazekas Ms CA - BSc,MSc,DipD'Etudes Approfondies biometry; simulation-model optimization
6 Hackett C - BSc,PhD *Crop Physiologist* horticultural land evaluation; crop data base development
7 Hide RL - BA,PhD *Anthropologist* rural sociology, farming systems
8 Keig Mrs G - BSc,MSc *Biochemist/Medical Demographer* computer geographic data systems
9 McAlpine JR - BA *Geographer* land use studies, water balance studies
10 Saunders JC - BScAgr *Forester* land evaluation for forestry
11 Wetselaar R - IngAgr *Soil Scientist* nitrogen cycling in crop systems

128 Division of Wildlife and Ecology
"Gungahlin", Barton Highway, Canberra, A.C.T.
[PO Box 84, Lyneham, A.C.T., 2602]
Tel: (062) 42 1600. Telex: 62284.

1 Fennessy BV - BAgrSc,FAIAS *Scientific Liaison Officer*
2 Staples BA - BADipLib,ALAA *Divisional Librarian*
3 Walker BH - BScAgric,MSc,PhD *Chief of Division* dynamics of ecosystems; plant ecology; soil-plant-herbivore interaction

129 Fauna Surveys and Taxonomy

1 Calaby JH - DSchc *Assistant chief of Division* biology and taxonomy of marsupials and indigenous rodents; surveys of native mammals
2 Schodde R - BSc,PhD systematics, zoogeography and phylogeny of Australian land and freshwater birds and plants (Compositae); surveys of vertebrate fauna
3 van Tets GF - MA,PhD comparative behaviour and ecology of cormorants and related birds; paleoecology

130 Physiological Ecology

1 Brown GD - BSc,PhD environmental physiology of marsupials
2 Cork SJ - BSc,PhD nutritional and digestive physiology of herbivorous marsupials
3 Green BF - BSc,PhD water physiology; lactational physiology and bio-energetics of marsupials and other terrestrial vertebrates
4 Hinds LA - BSc,PhD endocrinology and reproductive physiology in marsupials
5 Merchant JC - BApplSc marsupial reproduction, lactation and growth; age determination
6 Nicholas KR - BSc,PhD biochemistry; endocrinology of lactation; cell biology and protein structure-function relationships
7 Poole WE - BSc reproduction, distribution and specification of grey kangaroos
8 Taylor LS - BSc electronics and biotelemetry
9 Tyndale-Biscoe CH - MSc,PhD *Program Leader* reproductive physiology and endocrinology of marsupials

131 Rangeland Management

1 Graetz RD - BSc,PhD application of remote sensing to rangeland management; assess productivity of Acacia woodlands

132 Systems Ecology

1 Braithwaite LW - MSc,PhD ecology of forest fauna and waterfowl
2 Caughley GJ - PhD,DSc *Program Leader*
3 Fullagar JC - MAgrSc,PhD population dynamics of birds and mammals; ecology of water birds and petrels
4 McIlroy JC - MAgrSc,PhD effect of poisoning on wildlife; ecology of feral pigs; vertebrate pest control
5 Short JC - MSc diet, density and movements of macropodids and other mammals
6 Spratt DM - MSc,PhD arthropod, helminth and haematozan parasite fauna of native mammals; pathogenesis of parasitic infections

133 Vertebrate Pest Research Centre

134 Rabbit Biology - Ecology

1 Newsome AE - MSc,PhD,DSc *Program Leader* ecology of mammals; predator/prey; eruptions, fire ecology; arid, temperate and tropical ecosystems
2 Parer IP - BAgrSc ecology of rabbits; epidemiology of myxomatosis
3 Wood DH - MSc,PhD ecology of rabbits (arid areas) - grazing/vegetation interactions; control methods

135 Rabbits - Biological Control

1 Robbins SJ - BSc,PhD *Leader* ecology and epidemiology of myxomatosis
2 Russell RJ - BSc,PhD molecular biology of myxoma virus

136 Rabbits - Behaviour

1 Hesterman ER - BSc rabbit behaviour; function of olfaction in behaviour
2 Mykytowycz R - DVM *Leader* rabbit behaviour with emphasis on role of smell and odoriferous glands

137 Rabbits - Physiology

1 Williams CK - BSc,PhD processes of adaption in the wild rabbit; population ecology of small mammals

138 Rodent Control Research Laboratory

See also Goodrich, BS and Pennycuik, P "At Sydney" below

1 Redhead TD - MSc,PhD *Laboratory Leader* population biology of rodents, theory and practice of pest control
2 Singleton GR - BSc,PhD population biology of rodents; parasites and their effect on house mice

139 Predators - Ecology

1 Catling PC - BSc,MACS ecology of predators and small mammals, fire ecology

140 Computing - Modelling

1 Parker BS - BSc,MACS distribution of rabbit populations; vertebrate fauna distribution; use of microcomputers for ecological studies
2 Pech RP - BSc,PhD modelling; remote sensing

141 At Alice Springs, Northern Territory

Alice Springs Laboratory, Heath Road (via South Stuart Highway), Alice Springs, Northern Territory
[PO Box 2111, Alice Springs, Northern Territory, 5750]
Tel: (089) 52 4255. Telex: 81272.

1 Pickup G - BA,PhD *Officer-in-Charge and Program Leader* soil erosion; sedimentation modelling; image processing
2 Foran BD - BAgSc,MScAgric management of arid rangelands; plant ecology; remote sensing
3 Freidel BD - BSc,PhD nutrient cycling, population dynamics of trees and shrubs, vegetation assessment
4 Griffin GF fire ecology, conservation management, natural functioning of arid ecosystems
5 Morton SR - BSc,PhD wildlife ecology
6 Stafford-Smith DM - BACantab,PhD ecosystem modelling, distributions of plants and animals; grazing impacts

142 At Atherton, Queensland

Atherton Laboratory, Maunds Road, Atherton, Queensland, 4883
Tel: (070) 911755. Telex: 48840.

1 Crome FHJ - BSc rainforest ecology; community ecology of vertebrates
2 Harrington GW - BSc,PhD OIC

143 At Darwin, Northern Territory

McMillans Road, Berrimah, Northern Territory
[Private Bag 44, Winnellie, Northern Territory, 5789]
Tel: (089) 22 1711.

1 Werner PA - BSc,MS,PhD *Officer-in-Charge and Program Leader* vegetation dynamics; plant ecology; tropical ecosystem ecology

2 Braithwaite RW - MSc,PhD ecology, behaviour and genetics of rodents and small marsupials; faunal use of environs of unconfined aquifers
3 Corbett LK - MSc,PhD ecology and behaviour of predators, especially carnivores; small mammal ecology; population dynamics of waterbirds
4 Ridpath MG - BSc,PhD habitat alterations and their effect on wildlife with emphasis on large mammal fauna

144 At Deniliquin, New South Wales

Charlotte Street, Deniliquin, New South Wales
[Private Bag PO, Deniliquin, New South Wales, 2710]
Tel: (058) 811133. Telex: 55457.

1 Hodgkinson KC - BAgrSc,PhD *Officer-in-Charge and Program Leader* arid zone plant population dynamics
2 Greene RSB - BSc,PhD edaphic studies
3 Hodgkinson KC - BAgrSc,PhD fire ecology; ecophysiology of grazed plants
4 Ludwig JA - BSc,MSc,PhD computer simulation modelling
5 Noble JC - HDA,BA,LittB,PhD fire in mallee ecosystems; plant population ecology
6 Tongway DJ - DipApplChem,ARACI soil chemistry changes in soil surface features
7 Wilson AD - BAgrSc,PhD grazing management of natural pastures
8 Young MD - BEc,MAgrSc economic and social influences on rangelands management

145 At Perth, Western Australia

Helena Valley Laboratory, CSIRO, Cnr Clayton and Fyfe Roads, Helena Valley via Midland, Western Australia
[Locked Bag 4, PO Midland, Western Australia, 6056]
Tel: (09) 252 0111. Telex: 92178.

1 Saunders DA - BSc,PhD *Officer-in-Charge and Program Leader* ecology of avian populations on remnants of native vegetation
2 Arnold GW - MScAgric,PhD ecology and behaviour of mammals
3 Brooker MG - MScAgr social dynamics of animal populations
4 Hobbs RJ - BSc,MA,PhD dynamics of plant communities
5 Rowley ICR - BAgrSc social dynamics of animal populations
6 Smith GT - BSc,PhD community ecology of lizards on small reserves

146 At Sydney, New South Wales

147 School of Chemistry

Macquarie University, North Ryde, New South Wales, 2113
Tel: (02) 889463.

1 Goodrich BS - BSc,PhD chemistry of attractants for enhancement of rodent baits

148 Ryde Laboratory

Delhi Road, North Ryde, New South Wales
[PO Box 184, North Ryde, New South Wales, 2112]
Tel: (02) 8870888. Telex: 22970.

1 Pennycuik P - MSc,PhD biology of attractants for enhancement of rodent baits

149 Tropical Forests Research Centre

P.O. Box 780, Atherton, Queensland, 4883
Tel: (070) 911755. Telex: 48840.

1 Harrington GW - BSc,PhD *Officer-in-Charge*
2 Stocker GC - BScFor,DipFor,MSc ecology of tropical rainforests
3 Tracey JG - DipAgr floristics and mapping of tropical rainforests
4 Unwin GL - BScFor forest physiology
5 Webb IS - BSc,PhD nutrient cycling in tropical rainforests

150 Centre for Environmental Mechanics

GPO Box 821, Canberra, A.C.T., 2601
Tel: (062) 46 4911. Telex: 62861. Cables: Enmech, Canberra.

Physical investigations of energy exchange and the movement of natural and introduced substances (e.g. water, carbon dioxide, salts, fertilizers) in the enivironment, with special reference to plants, soils and the lower layers of the atmosphere

1 Philip JR - BCE,DSc,FAA,FRS *Chief of Division* simultaneous transfer of heat, water, and soluble salts; water movement and volume change in porous media; moisture equilibrium in swelling soils; theoretical aspects of micrometeorology; mathematics of diffusion; thermodynamics of the soil-plant-atmosphere continuum; population dynamics; interactions between convection and molecular diffusion; fluid mechanics; free convection in porous media and Newtonian fluids
2 Bradley EF - BSc,PhD turbulent transfer of heat, moisture and momentum in the lower atmosphere in various situations - changes in surface roughness, advection, nonuniform surface topography, shelterbelts, forest micrometeorology
3 Broadbridge P - BSc,DipEd,PhD exactly solvable models of unsaturated flow, numerical models of unsteady, unsaturated flow from fractionally wetted surfaces; mathematical models of turbulence

4 Coppin PAW - BSc,PhD turbulent fluxes of momentum, heat and moisture in atmospheric boundary layers; wind tunnel studies of turbulent diffusion; advanced field instrument developed

5 Denmead OT - BAgrSc,PhD micrometeorology and transfer processes in and above plant canopies; water movement in plants and soils; photosynthesis and evaporation in the field; exchange of gaseous nitrogen compounds between soils, plants, and the atmosphere; gas transfer across air-water interfaces

6 Finnigan JJ - BSc,PhD wind tunnel instrumentation and research, airflow in plant canopies; mathematical models of airflow; boundary-layer turbulence; wave-turbulence interactions

7 Huang CH - BSc,MSc,PhD mechanics of erosion; flow in fractured media

8 Katen PC - MSc,PhD air pollution and surface layer turbulent transport processes including the dispersion, dry deposition and resuspension of gases and aerosol particles; instrument design and analysis; wind turbine siting; modelling of dispersion in wakes

9 Knight JH - BA,PhD soil water movement; mathematics of diffusion; numerical algorithms

10 Lang ARG - BSc,ASwTC,PhD plant optics - structure and growth of plant communities in relation to absorption and transmission of solar energy; advection; mechanisms of transfer processes in plant canopies

11 Millar BD - BScAgr,PhD plant water - retention and movement of water in plants and uptake from soil; influence of water status on plant growth; developing valid methods for measuring plant water status

12 Perroux KM transport in naturally occurring porous materials; measurement of physical properties of porous and field media in laboratory

13 Raupach MR - BSc,PhD wind tunnel studies and theoretical models of turbulent dispersion; boundary-layer turbulence; turbulent transport in and above plant canopies; erosion by wind

14 Sully MJ - BA,MSc,PhD variability of physical properties of field soils; preferential flow in porous materials

15 White I - BSc,PhD transport in heterogeneous porous materials; variability of field soil hydraulic properties; unstable flows in porous materials; evaporation from porous materials; marangoni effects in porous media

16 Zegelin SJ - MSc flow in layered materials; nondestructive methods of determining liquid and solute distributions in porous materials; rainfall infiltration

151 Institute of Information and Communications Technologies

Headquarters, GPO Box 1965, Canberra, A.C.T., 2601
Tel: (062) 818555. Telex: 62620.

152 Division of Mathematics and Statistics
Headquarters, GPO Box 1965, Canberra, A.C.T., 2601
Tel: (062) 82-2011. Telex: 62620.

1 Diggle PJ - MSc,PhD *Chief of Division*

153 A.C.T. Region
GPO Box 1965, Canberra, A.C.T., 2601
Tel: (062) 81555. Telex: 62620.

1 Carpenter Mrs SM - BSc,DipEd,MStats biometrics
2 Diggle PJ - BSc,MSc,PhD biometrics
3 Forrester RI - BSc,DipEd biometrics
4 Gates DJ - BSc,MSc,PhD modelling
5 Morton R - BSc,MSc,PhD biometrics
6 Muller WJ - BSc,CertEd,DipStats,MSc biometrics
7 Westcott M - BSc,PhD modelling
8 Williams ER - BSc,PhD experimental design
9 Wood JT - MA,MSc,PhD biometrics

154 New South Wales Region
P.O. Box 218, Lindfield, New South Wales, 2070
Tel: (02) 467 6572. Telex: 26296.

1 Sandland RL - BSc,AIA,PhD
2 Best DJ - MSc,DipMet biometrics
3 Brown GH - BSc,DipEd,MSc,PhD biometrics
4 Carter Mrs NB - BSc biometrics
5 Donnelly JB - BScAgr,BSc biometrics
6 Eagleson GK - BSc,PhD biometrics
7 Willcox Miss ME - BSc biometrics

155 Queensland Region
Cunningham Laboratory, 306 Carmody Road, St Lucia, Queensland, 4067
Tel: (07) 377 0209. Telex: 42159.

1 Bourne Mrs AS - BSc,MScSt biometrics
2 Haydock KP - BSc biometrics
3 Jones PN - BSc,MSc biometrics
4 Kerr JD - BSc biometrics
5 Ratcliff D - BSc,PhD biometrics
6 Sinclair DF - BSc,MStat,MS,PhD biometrics

156 South Australian Region
Private Bag No. 2, Glen Osmond, South Australia, 5064
Tel: (08) 274 9350. Telex: 82406.

1 Cellier KM - BSc biometrics
2 Constantine AG - BSc,PhD biometrics
3 Correll RL - BSc,MSc,PhD biometrics
4 Field JBF - BSc biometrics
5 Veitch LG - BSc biometrics
6 Wilkinson GN - BSc,MSc,DSc biometrics

157 Victoria/Tasmanian Region
Private Bag No. 10, Clayton, Victoria, 3168
Tel: (03) 542 2253. Telex: 35675.

1 Anderson Miss DA - BSc,MSc,PhD biometrics
2 Jarrett RG - BSc,PhD biometrician
3 Miller AJ - MSc,PhD biometrics
4 Ratkowsky DA - BChEng,PhD biometrics
5 Trajstman AC - BSc,PhD biometrics

158 Western Australian Region
Private Bag P.O., Wembley, Western Australia, 6014
Tel: (09) 387 0320. Telex: 92178.

1 Campbell NA - BSc,PhD biometrics
2 Maller RA - BSc,PhD biometrics
3 Palmer MJ - BSc biometrics

159 Institute of Minerals, Energy and Construction

Delhi Road, North Ryde, New South Wales
[PO Box 136, North Ryde, New South Wales, 2113]
Tel: (02) 887 8666. Telex: AA25817.

Also at Limestone Avenue, Campbell, A.C.T. (PO Box 225, Dickson, A.C.T. 2602) Telephone: (062) 48 4211 Telex: AA62003

1 Reid AF - PhD,DSc,FAA *Director*

160 Commonwealth Serum Laboratories Commission

45 Poplar Road, Parkville, Victoria, 3052
Tel: (03) 389-1911. Telex: AA32789. Cables: Serums, Melbourne.

1 McCarthy NJ - MA,MAdmin,MB,BS,FRACGP,FRACMA,FAIM *Managing Director*
2 Schiff P - MB,BS,BScMed,PhD,FRACP,FRCPA *Research and Development Director*
3 Cheyne I - PhD *Senior Research and Development Virologist*
4 Cox J - MSc *Senior Research and Development Bacteriologist*
5 Eggleton DG - MSc *Bacteriology Research*
6 Hampson AW - MSc *Scientific Manager, Research and Development*
7 Middleton HD - BVSC,DipAgr,MATA *Principal Veterinary Officer*

161 Department of Primary Industry

Edmund Barton Building, Broughton Street, Barton, Australian Capital Territory
[Department of Primary Industry, Barton, Australian Capital Territory, 2600]
Tel: (062) 72 3933, 61 62 72 3933. Telex: AA62188. FAX (062) 72 5161.

1 Fitzpatrick EN - BSc,MAgrSc *Deputy Secretary*
2 Gibson JL - BComm,MBusAd,FASA,CPA *Deputy Secretary*
3 Miller GL - BAgrEc,MA *Secretary*

162 Australian Agricultural Health & Quarantine Service

NFF House, Brisbane Avenue, Barton, Australian Capital Territory
[Barton, Australian Capital Territory, 2600]
Tel: (062) 72 3933. Telex: AA 62189. FAX (062) 72 3933.

The service develops, co-ordinates and administers national programs for animal health, welfare and disease control, for plant protection and for animal and plant quarantine

1 Gee RW - BVSC,FACVSc *Director*
2 Lane L - AASA *Deputy Director*

163 Animal Health Programs Branch
NFF House, Brisbane Avenue, Barton, Australian Capital Territory
Tel: (062) 72 3933. Telex: AA 62189.

1 Andrews LG - BVSC,MSc,CDAH,MACVSc *Principal Veterinary Officer*

2 Broadbent R - BVSc *Principal Veterinary Officer*
3 Davis I - BVSc,MVSt,QDAH *Principal Veterinary Officer*
4 Digby JG - BVSc,MBA *Assistant Director*
5 Geering WA - BVSc,MRCVS,MACVSc *Senior Principal Veterinary Officer*

164 Animal Quarantine and Exports Branch
NFF House,Brisbane Avenue, Barton, Australian Capital Territory
Tel: (062) 72 3933. Telex: AA 62189.

1 Doyle KA - BVSc,MACVSc *Assistant Director*
2 Dunn KJ - BVSc *Senior Principal Veterinary Officer: Animal Quarantine and Exports*
3 Robinson BA - BVSc *Principal Veterinary Officer: Animal Quarantine and Exports*

165 Australian Plague Locust Commission
NFF House,Brisbane Avenue, Barton, Australian Capital Territory
Tel: (062) 72 3933. Telex: AA 62189.

1 Hooper GH - BSc,DSc,PhD *Director*
2 Bryceson KP - BSc,PhD *Forecaster: Locust Plague Forecasting*
3 Hunter DM - MSc,PhD *Entomologist: Locust biology*
4 Matic T - BSc,PhD *Insecticide Officer: Locust Control*
5 McCulloch L - BA,MSc *Deputy Director*
6 Wright DE - BSc,MSc,DIC,PhD *Senior Forecaster: Local Plague Forecasting*

166 Development and Laboratories Branch
NFF House,Brisbane Avenue, Barton, Australian Capital Territory
Tel: (062) 72 3933. Telex: AA 62189.

1 Auty JH - MVSc,BA *Assistant Director*
2 Beckett R - BVSc,PhD *Senior Veterinary Officer*
3 Brennan RG - BVSc *Principal Veterinary Officer*
4 Hooper GN - BSc *Senior Principal Scientist*
5 Moore BL - BVSc,HDA *Principal Veterinary Officer*

167 Operations Branch
NFF House,Brisbane Avenue, Barton, Australian Capital Territory
Tel: (062) 72 3933. Telex: AA 62189.

1 Lehane L - BVSc,MSc,PhD *Principal Veterinary Officer*
2 Pryor WJ - BVSc,MVSc,FACVSc,PhD *Assistant Directors*
3 Tuckey KW - BVSc,MBA *Principal Veterinary Officer*

168 Plant Health and Quarantine Branch
NFF House,Brisbane Avenue, Barton, Australian Capital Territory
Tel: (062) 72 3933. Telex: AA 62189.

1 Catley A - BScAgr *Assistant Director*
2 Evans G - BScAgr,PhD *Officer in Charge, Plant Quarantine Research Station*
3 Khair GT - MSc *Nematologist: Nematology and Quarantine*
4 McLean GD - BSc,PhD *Plant Virologist: Plant Virus Diseases and Quarantine*
5 Navaratnam SJ - MSc *Principal Plant Quarantine Officer*
6 Paton R - BSc *Principal Plant Quarantine Officer*

169 Australian Fisheries Service

Northern Fisheries Unit, P.O. Box 1089, Cairns, Queensland, 4870
Telex: 48061.

The Australian Fisheries Service advises the Australian Commonwealth Government on the utilization and conservation of living marine resources, as well as on international matters concerned with fisheries. It administers Commonwealth legislation regulating commercial fishing beyond the three-mile zone of coastal waters. The sevice liaises with State and Territory Governments and agencies on the formulation and co-ordination of fisheries management, development and research.

1 Bain Dr R *Director*

170 Fisheries Information and Publications
Northern Fisheries Unit, P.O. Box 1089, Cairns, Queensland, 4870
Tel: (062) 72 3933. Telex: AA 62189.

1 Young C fisheries information and publications

171 Northern Fisheries Unit
Northern Fisheries Unit, P.O. Box 1089, Cairns, Queensland, 4870
Tel: (062) 72 3933. Telex: AA 62189.

1 Burnage G *Senior Investigation Officer*

172 Dairy and Intensive Livestock Division

See Field Crops Division for functions

1 Pajic B *Assistant Secretary* policies fo dairy marketing
2 Pettit G *Assistant Secretary* policies for dairy production and intensive livestock
3 Thorn E *First Assistant Secretary*

173 Development and Coordination Division

1 Allan PB - BScAgric *Assistant Secretary*
2 Core P - BRurSc *First Assistant Secretary*
3 Hyde KW - BAgrSc,DipAgrEc *Principal Executive Officer*

174 Export Inspection Service

The service operates Australia's export inspection system for meat and other products, develops and implements appropriate strandards of inspection and develops manpower policy and plans for current and future inspection requirements.

1 Langhorne P *Director*

175 Meat Operations & Technical Services Group

1 Corrigan PH - MVB,MVSt *Principal Veterinary Officer*

176 Non Meat Branch

177 Horticultural Crops

1 Jefferies M - BSc *Section Head, Entomology*

178 Australian Wool Measurements Standards Laboratory

1 Stutter A - BSc *Section Head, Textiles: Wool Measurement Standards*

179 Dairy

1 Abhayaratna N - BA,MSc *Section Head, Dairy Standards*

180 Fish

1 Walter D - BSc *Section Head, Fish Standards*

181 Field Crops

1 Horrigan W - BSc *Chief Entomologist, Stored Products Entomology*

182 Field Crops Division

The function of this division, the Meat and Wool Division, the Dairy and Intensive Livestock Division, and the Forestry and Horticultural Crops Division, is to provide policy advice and administration for (a) Commonwealth interest and responsibilities in the production and marketing of rural commodities and forest products; (b) activities of commodity marketing boards; (c) production control and production development, financial assistance, stabilisation, price support, equalisation; (d) international agreements relating to rural commodities

1 Carter W *Assistant Secretary (acting)* wheat and coarse grains
2 Honan N *First Assistant Secretary*
3 Messer R *Assistant Secretary (acting)* sugar and industrial crops

183 Forestry and Horticultural Crops Division

See Field Crops Division above for functions

1 Alderson R *Assistant Secretary* wine, dried fruits and vegetables
2 Bryant R *Assistant Secretary* forestry
3 Jenkins E *Assistant Secretary* canned and fresh fruits
4 Mackey G *First Assistant Secretary*

184 Meat and Wool Division

See Field Crops Division for functions

1 Blamey M *First Assistant Secretary*
2 Jenkins J *Assistant Secretary* meat
3 Sainsbury J *Assistant Secretary* wool
4 Waring E *Assistant Secretary* pastoral production

185 Bureau of Agricultural Economics

Macarthur House, Cnr Macarthur & Northbourne Avenues, Lyneham,
Australian Capital Territory
[1563, Canberra, Australian Capital Territory, 2601]
Tel: (062) 46 9111. Telex: AA 61667.

The Bureau is responsible for economic research and the economic
evaluation of policy options for Australia's agricultural, pastoral, fisheries
and forest industries

1 Stoeckel AB - BAgrSc,BEc,PhD *Director*
2 Rae JR - BEc *Deputy Director*
3 Wonder BS - BCom,MResEnvStud *Chief Research Economist*

186 Quantitative and Econometric Services Section

Macarthur House, Cnr Macarthur & Northbourne Avenues, Lyneham,
Australian Capital Territory
Tel: (062) 46 9111. Telex: AA 61667.

1 Ball Ms KM - BEc,GradDipEc *Senior Research Officer: Quantitative and*
econometric services
2 Brewer K - BSc,BA,MSc,PhD *Officer in Charge: Quantitative and*
Econometric Services
3 Hinchy MD - BEc,MSc *Senior Economist* quantitative and econometric
services
4 Lee MS - BSc,MSc,PhD *Principal Research Officer: Quantitative and*
Economic Services

187 Economic And Policy Analysis Branch

Macarthur House, Cnr Macarthur & Northbourne Avenues, Lyneham,
Australian Capital Territory
Tel: (062) 46 9111. Telex: AA 61667.

1 Lewis P - BSc,MSocSc *Senior Economist*
2 Urban PJ - BSc,MEc *Assistant Director*

188 International Economic Analysis Section

Macarthur House, Cnr Macarthur & Northbourne Avenues, Lyneham,
Australian Capital Territory
Tel: (062) 46 9111. Telex: AA 61667.

1 Chapman JN - BAgrSc *Principal Research Officer: International And*
Economic Analysis
2 Field Ms H - BSc *Senior Research Officer:International Economic Analysis*
3 Roberts IM - BEc *Officer in charge* international economic analysis
4 Tie GH - BEc,BA *Principal Research Officer:International Economic*
Analysis

189 Macroeconomic Analysis Section

Macarthur House, Cnr Macarthur & Northbourne Avenues, Lyneham,
Australian Capital Territory
Tel: (062) 46 9111. Telex: AA 61667.

1 Hogan Ms LI - BA *Senior Research Officer* macroeconomic analysis
2 Kirby MG - BEc,MEc,PhD *Officer in Charge* macroeconomic analysis

190 Policy and Project Evaluation Section

1 Curran WR - DipHort,BAgrEc,MEc *Officer in Charge* policy and project
evaluation
2 Minnis PW - BEc *Principal Research Officer* policy and project evaluation

191 Rural Production Economics Branch

1 Bardsley P - BSc,GradDipEc,PhD *Senior Economist* rural production
economics
2 Johnston BG - BAgrSc,MEc,PhD *Assistant Director* rural production
economics

192 Rural Resources Economics Section

1 Clark Ms J - BAgrEc *Senior Research Officer* rural resources economics
2 Flavel NJ - BAgrEc *Senior Research Officer* rural resources economics
3 Menz KM - BScAgr,MS,PhD *Officer in Charge* rural resources economics
4 Purtill Ms AM - BA,BEc *Principal Research Officer* rural resources
economics

193 Broadacre Economics Research Section

1 Farquharson RJ - BAgrEc *Officer in Charge* economics research

2 Moir BG - BEc,BA,DipEd,MA *Principal Research Officer* economics
research
3 Stoneham GC - DipBFME,MAgrEc *Principal Research Officer*

194 Horticultural Production Research Section

1 Chandra S - BSc,MSc,PhD,DipTropAgron *Principal Research Officer*
economics of horticultural production
2 Samuel SN - BSc,MSc,PhD *Officer in Charge* economics of horticultural
production
3 Williams KL - BAgrEc *Senior Research Officer* economics of horticultural
production

195 Intensive Livestock Production Economics Section

1 Backhouse MH - BEc,DipAgr *Principal Research Officer* intensive livestock
production economics
2 Beck AC - MEc,BScAgr,PhD *Officer in Charge* intensive livestock
production economics
3 Bhati UN - BScAgr,MSc,PhD *Principal Research Officer* intensive
livestock production economics
4 Hall NH - BSc,DipAgrEc,PhD *Officer in Charge* intensive livestock
production economics
5 Hunter RD - BScAgr,MAgrEc,DipAgrEcon *Principal Research Officer*
intensive livestock production economics
6 Knopke P - BEc *Senior Research Officer* intensive livestock production
economics
7 Tucker JC - BEc *Senior Research Officer* intensive livestock production
economics

196 Livestock Marketing and Outlook Branch

1 Dewbre JH - BSAgrEc,MSAgr,PhD *Senior Economist* livestock market
outlook
2 Martin WJ - BAgrSc,BEc,MAgrEc,PhD *Assistant Director* livestock
market outlook

197 Meat Marketing and Outlook Section

1 Brown CG - BAgrSc *Senior Research Officer* meat market outlook
2 Hill Ms DJ - BEc *Senior Research Officer* meat market outlook
3 McLeish RA - BAgrEc,MEc *Principal Research Officer* meat market
outlook
4 Porter DJ - BAgrEc *Senior Research Officer* meat market outlook
5 Sheales TC - BAgrEc,MScAgr,PhD *Officer in Charge* meat market outlook
6 Weeks PD - BEc *Principal Research Officer* meat market outlook

198 Wool and Other Fibres Outlook Section

1 Bramma KM - BAgrEc *Senior Research Officer* economics of wool and
other fibres
2 Flint PE - BAgrSc *Principal Research Officer* economics of wool and
other fibres
3 Haszler HC - BScAgr *Officer in Charge* economics of wool and other
fibres
4 Shaw IH - BEc *Principal Research Officer* economics of wool and other
fibres

199 Wool Market Analysis Section

1 Corra GS - BAgrEc *Principal Research Officer* wool market analysis
2 Ebbott TV - BEc *Senior Research Officer* wool market analysis
3 Foster MJ - BEc *Senior Research Officer* wool market analysis
4 Simmons PR - BScAgr,MAgr *Principal Research Officer* wool market
analysis
5 Trewin RC - BSc,BEc,PhD *Officer in Charge* wool market analysis

200 Crop Marketing and Outlook Branch

1 Blyth MJ - BAgrSc,PhD *Senior Economist* crop market outlook
2 Rose RN - BEc,MEc,PhD *Assistant Director* crop market outlook

201 Grain Marketing and Outlook Section

1 Geldard Ms JM - BAgrEc *Senior Research Officer* grain market outlook
2 Lawrence MJ - BEc,DipAgrEc *Officer in Charge* grain marketing outlook
3 Love GC - BEc *Principal Research Officer* grain market outlook
4 Sniekers PP - BAgrEc *Principal Research Officer* grain market outlook

202 Industrial Crops Marketing and Outlook Section

1 Borrell BW - BAgrCom,MAgrCom *Officer in Charge* industrial crops
market outlook
2 Connell PJ - BAgrSc,BEc *Principal Research Officer* industrial crops
market outlook
3 Jolly LO - BAgrEc *Senior Research Officer* industrial crops market
outlook

203 Fruit and Wine Marketing and Outlook Section

1 Dalziell IL - BAgrEc,PhD *Officer in Charge* fruit and wine market outlook
2 Davenport SV - BAgrEc *Senior Research Officer* fruit and wine market outlook
3 Muir DA - BAgrSc,MSA,PhD *Principal Research Officer* fruit and wine market outlook

204 Chief Commodity Analyst

1 Johnson CB - BEc,BAgrEc *Senior Research Officer* commodity analysis
2 O'Mara LP - BEc,PhD *Senior Economist* commodity analysis
3 Young R - MA,MSc,PhD *Chief Commodity Analyst* commodity analysis

205 Rural Statistics Section

1 Pietsch LR - BEc *Officer in Charge* rural statistics

206 Dairy, Forestry and Fisheries Branch

1 Dalton ME - BSc,MAgrEc,PhD *Senior Economist*
2 Kingma OT - BAgrSc,MAgrSc,PhD *Assistant Director*

207 Dairy Marketing and Outlook Section

1 Barrett DL - BSc,DipAgrEc *Senior Research Officer* dairy market outlook
2 Hogan JC - BAgrEc *Officer in Charge* dairy market outlook
3 Martin JF - BScAgr,MADE *Senior Research Officer* dairy market outlook
4 Skinner SJ - BAgrEc *Principal Research Officer* dairy market outlook

208 Forestry Economic Research Section

1 Hossain M - BScAgrEc,MScAgrEc,MADE,PhD *Senior Research Officer* forestry economic research
2 Paul PB - BEc,MScAgr *Principal Research Officer* forestry economic research
3 Wallace Ms IA - BEc *Principal Research Officer* forestry economic research
4 Watson WD - BAgrEc,MSc *Officer in Charge* forestry economic research

209 Fisheries Economic Research Section

1 Brown Ms D - BEc,DipAgrEc *Senior Research Officer* fisheries economic research
2 Collins DJ - BAgrEc *Senior Research Officer* fisheries economic research
3 Haynes JE - BSc,PhD *Officer in Charge* fisheries economic research
4 Smith PB - BAgrEc *Principal Research Officer* fisheries economic research

210 Technical Services

1 Clement RF - BSc,BA *Assistant Director*

211 Survey Design and Development Section

1 Chambers RL - BSc,MSc,PhD *Officer in Charge* survey design and development

212 Department of Science

213 Australian Government Analytical Laboratories

P.O. Box 65, Belconnen, Australian Capital Territory, 2616
Tel: (062) 524923. Telex: AA61906.

214 Research Section (NSW)
Regional Laboratory, 1 Suakin Street, Pymble, New South Wales, 2
Tel: (02) 4490111. Telex: AA73045.

1 Wells RJ - BSc,PhD,DSc *Director*
2 Diakiw V - BSc,PhD *Principal Chemist(Research)* residue method development
3 Kaslauskas R - BSc,PhD *Senior Chemist(Research)* residue method development

215 Research Section (VIC)
Regional Laboratory, 13th Floor, 11 William Street, Melbourne, Victoria, 3000
Tel: (03) 6113700. Telex: AA39266.

1 Flynn R - BSc,PhD *Senior Chemist (Research)* residue method development
2 Jenkins TC - BSc *Director*

216 Australian Institute of Marine Science
P.M.B. No. 3, Townsville, Queensland, 4810
Tel: (077) 789211. Telex: AA 47165. Cables: MARINESCI TOWNSVILLE.
FAX: 789358.

1 Baker JT - OBE *Director*
2 Andrews JC - PhD *Assistant Director*
3 Boto K - PhD coastal studies
4 Bradbury R - PhD marine systems and oceanographic studies
5 Chalker B - PhD environmental studies
6 Veron JEN - DSc reef studies

NEW SOUTH WALES GOVERNMENT

217 Department of Agriculture, New South Wales

McKell Building, Rawson Place, Sydney, 2000
Tel: (02) 217 6666. Telex: 72418. FAX (02) 2175484.

218 Central Executive

1 Knowles GH - BSc *Director General of Agriculture*
2 Francois DD - MSc,PhD *Executive Director (Fisheries)*
3 Gregory GR - BScAgr,RDOen *Deputy Director General*
4 Grimmett SG - MScAgr,PhD *Executive Director (Research and Advisory Services)*

219 Division of Animal Health, New South Wales

McKell Building, Rawson Place, Sydney, 2000
Tel: (02) 217 6666. Telex: 75809.

1 Dickinson DA - BVSc,MACVSc *Chief of Division*
2 Mylrea Dr PJ - MVSc,PhD *Deputy Chief*

220 Animal Health Services

1 Alchin C - BA,MAIAS *Pastures Protection Officer* pastures protection administration
2 Carroll SN - BVSc,MACVSc,MRCVs *Director, Animal Health Services*
3 Doust T - BVSc *Registrar, Stock Medicines* stock medicine administration
4 Gorta MR - BAgrSc,MSc *Assistant Principal Livestock Protection Officer* vertebrate pest control & noxious insect administration
5 Roberts B - WDA *Principal Livestock Protection Officer* animal disaster relief administration
6 Rolfe D - BVSc *Assistant Principal Veterinary Officer Animal Health Services* disease informations systems
7 Valentine LW - BVSc *Principal Veterinary Public Health* consumer protection

221 Disease Control

1 Doughty Dr FR - BVSc,PhD *Principal Veterinary Officer (Animal Quarantine & Exotic Diseases)* import/export animals & animal products
2 Healy BP - BVSc *Assistant Principal Veterinary Officer (Regulatory)* stock act & regulations, animal welfare
3 Locke Ms RH - BVSc *A/Assistant Principal Veterinary Officer (TB&B)*
4 Scott-Orr Ms H - BVSc,DipBact *Principal Veterinary Officer, Endemic Diseases* disease control administration
5 Vacant *Director, Disease Control*

222 Animal Health Research

1 Smeal Dr MG - MVSc,PhD *Director Animal Health Research* research planning

223 Division of Agricultural Services, New South Wales

McKell Building, Rawson Place, Sydney, 2000
Tel: 217 6666. Telex: 75809.

1 Mohr GM - BAgrEc *Research Co-ordinator* commonwealth/industry funds; administration of research
2 Stone JG - HDA *Deputy Chief, Division of Agricultural Services*
3 Witschi PA - BScAgr *Chief, Division of Agricultural Services*

224 Agricultural Engineering Centre
Roy Watts Road, Glenfield, 2167
Tel: (069) 605-1511.

1 Andrews AS - BEngAgric agricultural engineering; grain storage handling; rural energy
2 Brown GA - BEngAgric agricultural engineering; tractor/implements matching, operations and computing
3 Humphries MR - BEngAgric agricultural engineering; tractor operations and management
4 Kernebone FE - BScChemEng,M.I.C.E.U.K.C.E. alternative energy; amelioration of soils
5 Kruger IR - BEngAgric agricultural engineering; tillage research
6 Palmer AL - BEngMech agricultural engineering; tillage research
7 Quick Dr GR - DipMechEng,MS,PhD *Director of Agricultural Engineering*
8 Sanders KF - BEngAgric agricultural engineering; computer simulation

225 Biometrical Branch

1 Coote BG - BSc *Senior Biometrician* statistical consulting and computing
2 Cullis B - BSc *Biometrician* statistical consulting
3 Darnell R - BAppSc,DipBiom *Biometrician* statistical consulting
4 Dettmann EB - BScAgr *Special Biometrician* statistical consulting
5 Evans JC - MScAgr,PhD *Special Biometrician* multivariate analysis
6 Gilmour AR - BScAgr,PhD *Special Biometrician* statistical consulting, computing and animal breeding
7 Gleeson AC - BSc,MStat,PhD *Senior Research Scientist* statistical consulting and research
8 Kaldor CJ - BScAgr,MSc *Senior Biometrician* statistical consulting
9 Kirton HC - BSc,MSc *Deputy Chief Biometrician* design of experiments
10 Lau T - BA,MA,MStat *Biometrician* statistical consulting
11 Lill W - BSc *Senior Biometrician* statistical consulting
12 Murison RD - BAppSc *Senior Biometrician* statistical consulting
13 Nicholls PJ - BSc,MSc *Senior Biometrician* statistical consulting
14 Raison JM - BA *Biometrician* statistical consulting
15 Reid DD - BSc,BCom,MStat *Senior Biometrician* statistical consulting
16 Ridings HI - BSc *Senior Biometrician* statistical consulting
17 Roberts EA - BScAgr *Chief Biometrician* statistical consulting

226 Division of Plant Industries

McKell Building, Rawson Place, Sydney, 2000
Tel: (02) 2175063.

1 Claxton RA - BScAgr *Chief, Division of Plant Industries*
2 Cook LJ - BAgrSc,PhD *Director* plant production research; administration of research
3 Dawbin KW - BScAgr *Research Officer (Remote Sensing)* remote sensing research
4 Long JK - BScAgr *Deputy Chief, Division of Plant Industries*
5 McCloy KR - BSurv *Principal Officer (remote sensing)* remote sensing research
6 Seberry JA - BScAgr *Principal Officer* plant production research; administration of research

227 Division of Animal Production

McKell Building, Rawson Place, Sydney, 2000
Tel: (02) 217 5233.

1 File GC - HDA,BRurSc *Chief, Division of Animal Production*
2 Butt JA - BSc sheep and wool; goats (meat and fibre)
3 Clark AR - HDA livestock management
4 Heptonstall DB - HDA dairy industry, goats (milk)
5 Nowland WJ - WDA pigs; poultry; bees
6 Robards GE - BSc,PhD *Director, Animal Production Research*
7 Smith GJ - BScAgr,PhD *Deputy Chief, Division of Animal Production*
8 Speers P - HDA,GDE,DipSocSci beef cattle; horses; deer
9 Wilson BR - MRurSc,PhD administration of research

228 Division of Fisheries

McKell Building, Rawson Place, Sydney, 2000
Tel: (02) 217 5585.

1 Ayres PA - MIBiol,PhD,FRIPHH,MRSH *Chief of Division*
2 Malcolm Dr WB - BSc,PhD *Chief Biologist*
3 Hamer G - BSc *Principal Officer*
4 Reynolds LF - BSc,MSc *Deputy Chief of Fisheries*

229 Division of Marketing and Economic Services

McKell Building, Rawson Place, Sydney, 2000
Tel: 2175585.

1 Baker NR - BEc commodity prospects, international trade
2 Booth JF - BScAgr,BSc *Deputy Chief, Division of Marketing and Economic Services*
3 Espinas VE - BSc,MEc agricultural policy; applied economics
4 Foley GJ - BComm data collection and dissemination
5 Gellatly C - BAgrEc,MEc,PhD *Chief, Division of Marketing and Economic Services*
6 Godden DP - BAgrEc,BAgr,MEc,PhD agricultural policy; applied economics
7 Green WH - BAgrEc agricultural policy, commodity marketing
8 Griffith SJ - BAgrEc *Director of Marketing Services, Division of Marketing and Economic Services*
9 Henry LA - BEc,DipAgrEc marketing policy, market information
10 Johnston JH - BAgrEc,MEc,PhD agricultural policy, applied economics
11 Manwaring JF - BRurSc marketing research, marketing strategies
12 Marshall GR - BScAgr agricultural policy, applied economics
13 Mason G - MScAgr,DipEd agricultural policy, applied economics
14 McKenzie NH - BAgrEc agricultural policy, applied sciences
15 Pearce G - DipAppSc marketing research, marketing strategies
16 Phillips J - MScAgr agricultural policy, applied economics
17 Price DE - BAgrEc agricultural policy, applied economics
18 Razon B - BSc marketing research, marketing strategies

19 Vernon DM - BAgrEc,DipEd *Director of Economic Services, Division of Marketing and Economic Services*
20 Watkins WR - BScAgr agricultural policy, applied economics

230 At North Coast Agricultural Institute
Wollongbar, New South Wales, 2480

1 Buggie GJ - MScAgr regional and farm management economic research
2 Davis D - BAgrEc,MEc regional and farm management economic research

231 At Rangeland Management Research and Advisory Unit
Cobar, New South Wales, 2835

1 Burgess DMN - BEc regional and farm management economic research

232 At Yanco Agricultural Institute
Yanco, New South Wales, 2703

1 Jones RE - BComm,DipAgrEc regional and farm management economic research
2 Young R - BAgrEc regional and farm management economic research

233 At Regional Veterinary Laboratory
Armidale, New South Wales, 2350

1 Griffith GR - BAgrEc,MEc,PhD econometric modelling, agricultural policy

234 At Department of Agriculture, Dubbo
Dubbo, New South Wales, 2830

1 Benson RJ - BAgrEc regional and farm management economic research
2 Davies BL - BAgrEc regional and farm management economic research

235 At Department of Agriculture, Goulburn
Goulburn, New South Wales, 2580

1 Burfitt T - BAgrEc regional and farm management economic research

236 At Department of Agriculture, Gunnedah
Gunnedah, New South Wales, 2580

1 Bryant MJ - HDA,BEc regional and farm management economic research
2 Patrick I - BAgrEc regional and farm management economic research

237 Department of Agriculture, Maitland
East Maitland, New South Wales, 2323

1 Catt CBJ - BScAgr,MEc regional and farm management economic research

238 Department of Agriculture, Orange
Orange, New South Wales, 2800

1 Mullen JD - BAgrEc,MEc,PhD regional and farm management economic research
2 Vere DTR - BAgrEc,MEc regional and farm managements economic research

239 Department of Agriculture, Wagga Wagga
Wagga Wagga, New South Wales, 2650

1 Godyn DL - BScAgr,MEc regional and farm management economic research
2 Reilly TLC - BRurSc,DipAgrEc regional and farm management economic research

240 NSW Department of Agriculture Region I, North Coast

241 Agricultural Research Centre
North Coast Agricultural Institute, Wollongbar, New South Wales, 2480
Tel: (066) 240 200.

1 Allen RN - MScAgr root diseases, tropical horticultural crop pathology
2 Ayres JF - BSc,DipEd,PhD pasture utilization
3 Baigent DR - BSc,PhD ARACI soil and plant chemistry; amino acid, fatty acid and carbohydrate analysis
4 Batterham ES - MRurSc,PhD pig nutrition
5 Dettman Miss EB - BScAgr design and analysis of agricultural experiments; statistical computing
6 Fitzell RD - MScAgr tropical pathology
7 Giles LR - HDA,MRurSc pig nutrition
8 Hamilton BA - BSc,PhD dairy cow nutrition
9 Hill G - HDA *Manager*
10 Hochman Z - MScAgr pasture production
11 Leggo D - BScAgr *Regional Director of Research*
12 Murtagh GJ - MScAgr,PhD pasture physiology

13 O'Toole DK - MSc milk quality and dairy nutrition
14 Restall BJ - BSc,PhD breeding reproduction and management of goats
15 Robinson B - BAppSc agronomy, simulation studies and computer extension of research
16 Saini H - BScAgr,MScPhD plant biochemistry
17 Southwell I - MSc,PhD,ARACI organic chemistry, essential oils
18 Treverrow N - MRurSc economic entomology

242 Agricultural Research & Advisory Station
PMB Grafton, 2480
Tel: (066) 420 420.

1 Armstrong JP - HDPT,BA(SC) *Manager*
2 Barlow R - MRurSc,PhD beef cattle breeding and genetics
3 Colless JM - BScAgr maize breeding
4 Darnell R - BScAgr biometrical design and analysis
5 Desborough PJ - MScAgr crop agronomy; minimum tillage systems soyabeans
6 Hearnshaw H - BRurSc,MAgrSc,DipEd beef cattle breeding and reproduction
7 Hennessy DW - MRurSc,PhD beef cattle nutrition
8 Hughes RM - MRurSc crop agronomy, especially nutrition
9 Mears PJ - BScAgr,MAgrSc,PhD *Supervisor of Research* sub-tropical pastures and crops
10 Rowland SJ - BA,PhD freshwater aquaculture and fisheries biology
11 Virgona JL - BSc estuarine fisheries biology
12 Wilson GPM - HDA pasture species evaluation especially legumes

243 Taree Agronomy Unit
P.O. Box 253, Taree, 2430
Tel: (065) 520 880.

1 Launders TE - MScAgr pasture agronomy, beef and dairy

244 Tropical Fruit Research Station, Alstonville
P.O. Box 72, Alstonville, New South Wales, 2477
Tel: (066) 280 604.

1 Batten DJ - BScAgr,PhD tropical fruits, lychee, mango, longan etc.
2 Huett DO - MScAgr vegetable agronomy, physiology and nutrition
3 Johns G - BRurSc,PhD banana horticulture, pre-harvest physiology
4 Seccombe J - BScAg *Assistant Manager*
5 Trochoulias T - BScHort tropical fruits, macadamia, avocado, tea & coffee

245 Grafton Agricultural Research and Advisory Station
Private Mail Bag, Grafton, New South Wales, 2460
Tel: (066) 420 420.

1 McGregor PJ - BVSc herd/flock health

246 North Coast Agricultural Institute
Wollongbar, New South Wales, 2480
Tel: (066) 240 261.

1 Bryce AC - BVSc dairy cattle health

247 Regional Veterinary Laboratory
Wollongbar, New South Wales, 2480
Tel: (066) 240 261.

1 Boulton JG - BVSc diagnostic biology
2 Callinan RB - MVSc diagnostic pathology
3 Cook RW - BVSc,MS,PhD *Officer in Charge* diagnostic pathology
4 Fraser GC - BVSc diagnostic biology
5 Gill PAJ - BVSc diagnostic pathology

248 District Advisory Office Coffs Harbour
P.O. Box 530, Coffs Harbour, New South Wales, 2450
Tel: (066) 520 411.

1 Williamson GW - BVSc pesticides

249 Board of Tick Control
North Coast Agricultural Institute, Wollongbar, New South Wales, 2480
Tel: (066) 240 200.

1 Weaver RN - BVSc *Chairman*

250 Board of Tick Control Laboratory
P.O. Box 285, Lismore, New South Wales, 2480
Tel: (066) 214 384.

1 Heath AB - ARMIT cattle tick control, pesticide residues
2 Machin MV - BSc cattle tick control, pesticide residues
3 McDougal KW - DipTechSc *Officer in Charge* acaricide dips

251 NSW Department of Agriculture Region 2, New England, Hunter & Metropolitan

RESEARCH STATIONS

253 Glen Innes Agricultural Research & Advisory Station
Private Mail Bag, Glen Innes, 2370
Tel: (067) 32 1633.

1 Barwick SA - BScAgr,MScAgr *Livestock Research Officer* sheep, genetics (study leave)
2 Curll ML - WDA,MRurSc,PhD *Supervisor of Research* pasture livestock relationships
3 Dawes PD - DipAppSciAgr *Manager*
4 Dicker RW - BSc *Livestock Research Officer* energy efficiency of grazing cattle
5 Fitzgerald Dr RD - MScAgr,PhD *Senior Research Agronomist* pasture ecology
6 Fowler Dr DG - MEc,PhD *Research Scientist* sheep reproduction & management
7 Robinson Mr GG - BScAgr,MRurSci *Special Research Agronomist* ecology of native grasslands; trees on farms
8 Vacant *Livestock Research Officer* vertebrate pests

254 Gosford Horticultural Research & Advisory Station
P.O. Box 720, Gosford, 2250
Tel: (043) 28 0300.

1 Trimmer W - HDA,GDRE *Manager*
2 Bishop A - BSc,MSc *Entomologist* pest management, lucerne
3 Goodwin Dr S - MAgrSci,PhD *Senior entomologist* pest management, intensive horticulture & ornamentals
4 Lamont G - BAgSc *Research Horticulturalist* ornamental horticulture
5 Nguyen Dr QV - BEAgr,MScAgr,PhD research horticulturalist - vegetable breeding
6 Roughley Dr R - MScAgr,PhD *Supervisor of Research - Principal Research Scientist* legume bacteriology, Australian Inoculants Research and Control Service
7 Sarooshi R - BScAgr,MScAgr *Senior Research Horticulturalist* horticulture, citrus, miscellaneous crops
8 Toth JJ - BAgSc *Senior Research Agronomist* weeds in horticulture crops
9 Vacant *Research Horticulturalist* vegetable agronomy
10 Worral R - BScAgr *Research Horticulturalist* ornamental horticulture

RESEARCH UNITS

255 Armidale - Animal Genetics and Breeding Unit
University of New England, 2351
Tel: (067) 73 2055.

1 Graser Dr H - PhD *Livestock Research Officer* estimation of genetic improvement
2 Hammond Dr K - HDA,BScAgr,PhD *Unit Leader - Senior Research Scientist* estimation of genetic improvement; electronics in agriculture

256 Research Unit, Regional Veterinary Laboratory
Private Mail Bag, Armidale, 2350
Tel: (067) 72 1611.

1 Campbell Mr AG - BSc *Entomologist* eucalypt dieback research
2 Griffith Dr GR - BAgEc,PhD *Research Scientist* application of economics to animal production

257 Research Unit, Maitland East
P.O. Box 9, East Maitland, 2323
Tel: (049) 30 2444.

1 Brownlee H - BScAgr *Senior Research Agronomist* general agriculture

258 Research Unit, Scone
P.O. Box 168, Scone, 2337
Tel: (065) 45 1742.

1 Hill Dr M - BScAgr,PhD *Research Agronomist* pasture ecology

259 Hawkesbury Agricultural Unit, Richmond
P.O. Box 217, Richmond, 2753
Tel: (045) 78 2666.

1 Armstrong JP - HDPT,BA *Manager*
2 Fell Dr LR - MRurSc,PhD *Supervisor of Research - Principal Livestock Research Officer* management of stress in livestock
3 Ibrahim Dr GF - BAgrSci,PhD *Special Livestock Research Officer* microbiology

4 Ross Mr GD - BSc,MSc *Livestock Research Officer* ruminant nutrition - microbiology of feedstuffs
5 Shutt Dr D - PhD *Livestock Research Officer* cattle & sheep physiology and endocrinology
6 Thomas A - BSc *Livestock Research Officer* dairy cattle physiology
7 Wheeler Ms AM - BSc,DipEd,MSc *Special Livestock Research Officer* mineral nutrition of ruminants

260 Agricultural Station Seven Hills
P.O. Box 11, Seven Hills, 2147
Tel: (02) 622 6322.

Advisory & veterinary staff are located at this centre

1 Armstrong JP - HDPT,BA *Manager*
2 Humphris CS - ASTC *Senior Chemist* stock feeds

261 Agricultural Research Centre Tamworth
R.M.B. 944, Tamworth, 2340
Tel: (067) 67 9300.

1 Archer Dr KA - BRurSc,MRurSc,PhD *Senior Research Agronomist* pasture management
2 Barnes WC - BSCAgr *Senior Chemist* cereal chemistry
3 Crocker GE - BScAgr *Senior Research Agronomist* soil fertility
4 Crosby BJ - HDA *Manager*
5 Doyle AD - MScAgr *Senior Research Agronomist* cereal nutrition & barley breeding
6 Felton WL - HDA,BScAgr *Senior Research Agronomist* no tillage cropping systems, weed ecology
7 Ferris Dr IG - BA,MSc,PhD *Pesticide Chemist*
8 Gleeson Dr AC - BSc,MStat,PhD *Research Scientist* biometrics
9 Gunning Dr RV - BA,BSc,PhD *Entomologist* Heliothis resistance
10 Hare Dr RA - BScAgr,PhD *Senior Research Agronomist* wheat breeding
11 Herridge Dr DF - BScAgr,MScAgr,PhD *Special Bacteriologist* nitrogen fixation
12 Holford Dr ICR - BScAgr,MScAgr,PhD,DIC *Principal Research Scientist* summer crops
13 Holland JF - BScAgr *Senior Research Agronomist* summer crops
14 Hosking Dr JR - BScAgr,PhD *Entomologist* biological control of cacti
15 Lodge Dr GM - BScAgr,PhD *Research Scientist* pasture management
16 MacKay M - BA *Curator* Australian winter cereals collection
17 Marcellos Dr H - BScAgr,MScAgr,PhD *Senior Research Agronomist* grain legume agronomy
18 Martin Dr RJ - BScAgr,MScAgr,PhD *Research Agronomist* weed ecology
19 McKenzie Dr E - BSc,PhD *Chemist* cereal chemistry
20 Moore Dr KJ - BScAgr,MScAgr,PhD *Research Scientist* plant pathology
21 Morgan Dr JM - BScAgr,MSc,PhD *Research Scientist* plant physiology
22 Schwinghamer Dr MW - BScAgr,PhD *Plant Pathologist*
23 Single Dr WV - BScAgr,PhD,DIC *Consultant* physiology of frost tolerance
24 Thompson Dr JA - BScFor,MSc,PhD *Special Bacteriologist* nitrogen fixation
25 Waterhouse DB - BSc *Research Agronomist* lucerne varieties & management
26 Wetherall RS - BScAgr *Senior Research Agronomist* pasture species & varieties
27 Wilkins J - BScAgr,MSc *Livestock Research Officer* sheep fertility (study leave)
28 Wolfe Dr EC - MScAgr,PhD *Regional Director of Research*

262 Horticultural Post Harvest Laboratory Gosford
P.O. Box 335, Gosford, 2250
Tel: (043) 28 0300.

1 Morris Dr SC - BScAgr,PhD *Research Horticulturalist* postharvest treatment of vegetables
2 Rigney Dr CJ - BScAgr,MSc,PhD *Senior Research Horticulturalist* fresh fruit disinfestation
3 Wild Dr BL - BScAgr,PhD *Officer in Charge* postharvest treatment of fruit

263 At CSIRO Division of Food Research
P.O. Box 52, North Ryde, 2113
Tel: (02) 887 8333.

1 Scott KJ - BScAgr,DipEd *Senior Research Scientist* fruit & vegetable storage
2 Wade Dr NL - BScAgr,PhD *Senior Research Scientist* fruit and vegetable storage

264 Agricultural Research Station
P.M.B. Myall Vale Mail Run, Narrabri, 2390
Tel: (067) 93 1105.

1 Scott RC - WDA *Manager*
2 Allen Dr S - BSc,PhD *Plant Pathologist* plant pathology, cotton diseases
3 Constable Dr GA - MScAgr,PhD *Senior Research Scientist* cotton agronomy & physiology
4 Daniells IA - BSc *Chemist* soil physics
5 Forrester NW - BScAgr *Senior Entomologist* management of Heliothis resistance

6 Hodgson AS - MScAgr *Senior Research Agronomist* cotton agronomy
7 Rose Dr IA - BScAgr,PhD *Senior Research Agronomist* soybean breeding & agronomy
8 Spurway Dr RA - BScAgr,MRurSci,PhD,HDA *Supervisor of Research* soybean physiology & agronomy

265 Department of Agriculture
P.O. Box 546, Gunnedah, New South Wales, 2380
Tel: (067) 429200.

1 Dunn SE - BVSc,WDA *Senior Field Veterinary Officer* applied regional animal health research

266 Department of Agriculture
P.O. Box 9, East Maitland, New South Wales, 2323
Tel: (049) 302444.

1 Kennedy DJ - BVSc,MVS,MACVSc *Senior Field Veterinary Officer* applied regional animal health research

267 Department of Agriculture Region 3, Orana and Far Western

Regional Headquarters, P.O. Box 865, Dubbo, 2830
Tel: (068) 811270.

1 Southwood OR - BScAgr *Regional Director of Agriculture*
2 Benson RJ - BAgrEc applied regional and farm management research
3 Davies BL - BAgrEc applied regional and farm management
4 Harrison HT - HDA *Regional Director of Advisory Services*
5 Pryde LC - BVSc,HDA *Senior Field Veterinary Officer*
6 Wise GA - BVSc *Regional Director of Veterinary Services*

268 Trangie Agricultural Research Centre
Private Mail Bag 19, Trangie, New South Wales, 2823
Tel: (068) 8882 M87.

1 Rogan IM - BRurSc *Regional Director of Research* sheep breeding
2 Bendeich Miss T - BALibSc *Librarian*
3 Cook L - BVSc *Veterinary Officer*
4 Cooper JL - BRurSc,MSc irrigated cereal agronomy
5 Drew TP - BRurSc pasture agronomy, winter oilseeds and grain legumes
6 Esler WJ - BSc *Manager*
7 Hall DJM - BScAgr soils physics
8 Herd RM - BSc,MSc,PhD beef cattle nutrition
9 Kilgour RJ - BScAgr,PhD reproductive physiology
10 Lee GJ - BSc,MSc ruminant nutrition
11 Leighton P *Systems Analyst/Programmer*
12 McKenzie D - BNatRes,MSc soils chemistry
13 Morthorpe KJ - BRurSc lucerne seed production, pasture agronomy
14 Morthorpe Mrs L - BRurSc irrigated soybean agronomy
15 Mortimer Miss S - BSc,PhD sheep breeding
16 O'Brien P - BSc,PhD vertebrate pest control
17 Parnell PF - BRurSc beef cattle breeding
18 Perry Ms D - BSc,DipEd growth and development studies

269 Rangeland Management Research and Advisory Unit
P.O. Box 286, Cobar, New South Wales, 2835
Tel: (068) 362207.

1 Burgess D - BEcon *Economist*
2 Date W - DipAppSc *Agronomist*
3 Downing BH - BSc,PhD *Special Officer - Rangeland Management*
4 Dyce GW - BNatRes rangelands vegetation dynamics
5 Grice AC - BSc,PhD rangelands vegetation dynamics
6 Kreveld M - BVSc *Veterinary Officer*
7 Suddes H - BVSc *Special Veterinary Officer, Officer in Charge*

270 Walgett Research & Advisory Unit
P.O. Box 249, Walgett, New South Wales, 2832
Tel: (068) 281288.

1 Bellotti WD - BSc,PhD pasture establishment and evaluation
2 Watt P - BAppSc *Agronomist*

271 NSW Department of Agriculture Region 4, Central West, South East and Illawarra

Headquarters, Agricultural Research & Veterinary Centre, Forest Rd, Orange, New South Wales, 2800
Tel: (063) 636700. Telex: 63730.

1 Colman RL - MScAgr,PhD *Regional Director of Research*

272 Berry Pasture Research Unit
P.O. Box 63, Berry, New South Wales, 2535
Tel: (044) 641251.

1 Havilah EJ - BScAgr pasture agronomy
2 McLaughlin BD - BScAgr pasture agronomy

273 Bathurst Agricultural Research Station
P.O. Box 143, Bathurst, New South Wales, 2795
Tel: (063) 311988.

1 Wickson RJ - Manager
2 Koffman W - BAgrEng stone fruit varieties, weeds, post harvest
3 Menzies AR - BScAgr growth regulants, stone fruits and berries

274 Canberra Agronomy Research Unit
P.O. Box 1600, Canberra, Australian Capital Territory, 2601
Tel: (062) 465775.

1 Dann PR - BScAgr,MSc pasture and crop agronomy
2 Garden DL - BRurSc pasture agronomy

275 Condobolin Agricultural Research and Advisory Station
P.O. Box 300, Condobolin, New South Wales, 2877
Tel: (068) 952099.

1 Becke JF - HDA *Manager*
2 Banks LW - MScAgr *Supervisor of Research* crop agronomy
3 Fettell NA - MScAgr wheat cultivation
4 Kruger I - BEAgr evaluation of machinery energy requirements and efficiency for wheat production
5 Milthorpe PL - BScAgr guayule and jojoba
6 Palmer AL - BEMech evaluation of machinery energy requirements and efficiency for wheat production
7 Young RR - MRurSc medic evaluation

276 Cowra Agricultural Research Station
P.O. Box 242, Cowra, New South Wales, 2794
Tel: (063) 421790.

1 Coates TP - WDA *Manager*
2 Armstrong EL - MScAgr dryland crop agronomy
3 Fogarty NM - MSc,PhD sheep breeding; prime lambs
4 Hall DG - BRurSc,MSc sheep production
5 Holst PJ - MScAgr,PhD goat breeding
6 Mead JA - BScAgr,MAgrSc crop agronomy

277 Orange Agricultural Research and Veterinary Centre
Forest Road, Orange, New South Wales, 2800
Tel: (063) 636700.

278 Agricultural Research

1 Atkins KD - MRurSc,PhD sheep breeding
2 Auld BA - MScAgr,PhD weed ecology, mycoherbicides
3 Bower CC - BSc,PhD pome fruits
4 Campbell Mrs JF - MScHort rootstocks, pruning, training and spacing of pome fruits
5 Campbell MH - MScAgr,PhD aerial pasture establishment
6 Colman RL - MScAgr,PhD *Regional Director of Research*
7 Denney GD - BScAgr pasture utilization, nutrition for wool production
8 Dowling PM - MScAgr,PhD *crop and pasture agronomy*
9 Gilmour AR - BScAgr,PhD biometrical design and analysis
10 Greer EB - MScAgr management and infertility in pigs
11 Kemp DR - MScAgr,PhD pasture agronomy
12 Medd RW - HDA,BRurSc,PhD weed and pasture ecology
13 Michalk DL - BScAgr,PhD pasture agronomy
14 Murray Prior R - HDA *Manager*
15 Nikandrow A - MScAgr,PhD diseases of field crops, mycoherbicides
16 Penrose LJ - MScAgr diseases of deciduous fruits
17 Ridings Mrs HI - BSc,DipStat biometrical design and analysis
18 Saunders GR - BSc noxious and feral animals
19 Thwaite WG - BScAgr,MSc entomology of pome fruits
20 Williams AJ - BScAgr,PhD efficiency of wool growth

279 Veterinary Research

1 Bourke CA - HDA,BVSc neurological diseases
2 Carrigan MJ - BVSc diagnostic pathology, reproductive diseases
3 Seaman JT - BVSc,MACVS diagnostic pathology, toxicological diseases
4 Webb RF - MVSc,DipBact *Officer in Charge* parasitology

280 Orange Advisory Office
P.O. Box 53, Orange, New South Wales, 2800
Tel: (063) 638250.

1 Mullen JD - BAgrEc,MEc,PhD applied economics and market research
2 Vere DTR - BAgrEc,MEc applied economics and market research

281 NSW Department of Agriculture Region 5, Murray and Riverina

282 Dareton Agricultural Research and Advisory Station

P.O. Box 62, Dareton, New South Wales, 2717
Tel: (050) 274401.

1 Bevington Dr KB - MScAgric *Research Horticulturist* citrus nutrition and propagation
2 Kaye GD - HDA,BEc *Manager*
3 Prior Ms LD - BScAgric *Research Horticulturist* irrigation and salinity

283 Deniliquin Research and Advisory Unit

P.O. Box 29, Deniliquin, New South Wales, 2710
Tel: (058) 811362.

1 Martin RE - BRurSci *Research Agronomist (Irrigated Crops)* irrigation management of crops
2 Slavich PG - MScAg *Soil Chemist* soil salinity investigations

284 Griffith Horticultural Research and Advisory Station

P.O. Box 1087, Griffith, New South Wales, 2680
Tel: (069) 620274.

1 Freeman Dr BM - BScAgr,MSc,PhD *Senior Horticulturist (Research)* plant physiology - grapevines
2 Gillespie PD - HDA,BA *Manager*
3 Turkington CR - BScAgr,RDOe *Assistant Principal Horticulturist (Research)* vine improvement

285 Temora Agricultural Research And Advisory Station

P.O. Box 304, Temora, New South Wales, 2666
Tel: (069) 821277.

1 England JC - WDA *Manager*
2 Anderson JL - BScAgric *Agronomist (Seed Production)* pure seed production
3 Martin RH - BScAgr *Special Research Agronomist* cereal breeding
4 Roberts GL - BSc *Research Agronomist* cereal breeding

286 Wagga Wagga Agricultural Research Institute

Private Mail Bag, Wagga Wagga, New South Wales, 2650
Tel: (069) 230999. Telex: 69714.

1 Dengate HN - BSc,PhD *Director of Research*
2 Want GR - BAppSc *Manager*
3 Allender W - MAppSc *Chemist* wheat chemistry
4 Ballantyne Dr BJ - MScAgr,PhD *Special Plant Pathologist* wheat pathology
5 Brennan JP - BA *Senior Economist* applied regional and farm management research
6 Conyers MK *Chemist* soil chemistry
7 Cornish Dr PS - MScAgr,PhD *Senior Research Scientist* plant physiology
8 Cullis B - BSc *Biometrician* co-operative research and biometrical services
9 Dear BS - BRurSc *Senior Research Agronomist* pasture agronomy
10 Evans Dr J - BScAgr,PhD *Senior Bacteriologist* microbiology
11 Fisher JA - BScAgr *Senior Research Agronomist* wheat breeding
12 Glennie-Holmes M - BSc *Senior Chemist* barley chemistry
13 Helyar Dr KR - MScAgr,PhD *Senior Research Scientist* acid soils
14 Kaiser Dr AG - MScAgr,PhD *Special Livestock Research Officer* silage production
15 Khan AM - MScAgric,PhD *Senior Research Agronomist* wheat breeding
16 Knights EJ - BScAgr *Research Agronomist* chickpea breeding and plant tissue culture
17 Lemerle Ms D - BSc *Research Agronomist* weed agronomy
18 Leys Dr AR - BScAgr,PhD *Research Agronomist* weed agronomy
19 Lill WJ - BSc *Senior Biometrician* co-operative research and biometrical services
20 Mailer RJ - BAppSc *Chemist* oilseed chemistry
21 May Dr CE - MSc,PhD *Research Scientist* cytogenetics
22 Mulholland JG - MSc *Senior livestock Research Officer* fodder production
23 Murray Dr GM - BScAgr,MSc,PhD *Senior Plant Pathologist* diseases of field and fodder crops
24 Read Dr BJ - BScAgr,PhD *Senior Research Agronomist* barley breeding
25 Scott BJ - MscAgr *Senior Research Agronomist* plant nutrition and lupin selection
26 Taylor AC - BScAgr *Special Research Agronomist* cereal agronomy
27 Wratten N - BScAgr *Research Agronomist* rapeseed breeding

287 Yanco Agricultural Institute

Private Mail Bag, Yanco, New South Wales, 2703
Tel: (069) 530211. Telex: 169030.

1 Grieve Dr AM - BSc,PhD *Regional Director of Research*

2 Andrews Ms JA - BSc *Research Agronomist* agronomy of summer crops
3 Bacon PE - MScAgr *Chemist* soil chemistry
4 Batten Dr GD - MRurSci,PhD *Senior Chemist (Soil Chemistry)* nutrition and physiology of crops
5 Blakeney AB - MScAg *Chemist* cereal chemistry, rice quality
6 Cother Dr EJ - BScAgr,PhD *Research Scientist (Plant Pathology)* diseases of vegetables
7 Dawe ST - BScAg,MSc *Senior Livestock Research Officer (Sheep & Wool)* intensive lamb production from high fertility sheep
8 Ellison Ms PJ - BSc *Plant Pathologist* epidemiology and disease management
9 Heenan Dr DP - BAgSc,PhD *Research Agronomist* rice agronomy
10 Hermus Ms RC - BAgrSc *Research Horticulturist* irrigation and nutrition of vegetables
11 Hutton RJ - BScAgr,MScAgr *Research Horticulturist* horticultural crop physiology
12 James Dr DG - BSc,PhD *Entomologist* insect pests of rice and field crops
13 Jones EL - BA *Entomologist* insect pests of rice and fruit
14 Lattimore Ms ME - BRurSc *Research Agronomist* pastures evaluation
15 Lewin Dr LG - BScAgr,HDA,PhD *Research Agronomist* rice breeding
16 Logan BJ - BScAg,MSc *Research Horticulturist* potato agronomy
17 Nichols PGH - BAgSc *Research Agronomist* lucerne breeding
18 Pollock DC - BAppSci *Research Agronomist* weed biology and control in rice
19 Ryan WJ - HDA *Manager*
20 Sinclair PJ - BScAgr *Research Horticulturist* agronomy of vegetables
21 Thompson JA - MScAgr *Research Agronomist* irrigation agronomy

288 Regional Veterinary Laboratory

Wagga Wagga, New South Wales, 2650
Telex: 169030.

1 Glastonbury JRW - BVSc,MVSc,MVS,MACVs *Regional Veterinary Research Officer* Officer in Charge
2 Links IJ - BVSc,DipBact,MACVs *Veterinary Research Officer* veterinary microbiology
3 Marshall DJ - HDA,BVSc *Veterinary Research Officer* diagnostic pathology
4 Searson JE - BVSc,MVSc *Special Veterinary Research Officer* diagnostic pathology & serology
5 Whittington RJ - BVSc *Veterinary Research Officer* diagnostic pathology

289 Inland Fisheries Research Station

P.O. Box 182, Narrandera, New South Wales, 2700

1 Dirou JF - HDA *Manager*
2 O'Connor PF - BSc,MSc *Biologist*
3 Rimmer MA - BSc,MSc *Biologist*
4 Thurstan SJ - BSc *Technical Officer*

290 Biological and Chemical Research Institute, Rydalmere

Cnr. Victoria Rd & Pemberton St, PMB No 10, Rydalmere, New South Wales, 2116
Tel: (02) 630 0251.

1 Smith AM - MScAgr,PhD *Director of the Institute*

291 Chemistry Branch

Biological and Chemical Research Institute

1 Abbott TS - BScAgr,PhD soil physics
2 Ahmad N - BSc pesticide residues
3 Awad AS - MScAgr water chemistry and plant nutrition
4 Chan KY - BSc,DipNatRes,PhD soil physics
5 Cresswell GC - BSc,PhD plant nutrition
6 Critchton JR - BSc,PhD soil and land use assessment
7 Gibson TS - MScAgr,PhD plant biochemistry
8 Gilbert WS - BSc pesticide residues
9 Haddad KS - MScAgr plant nutrition
10 Higginson FR - BScAgr,PhD *Director of Chemistry*
11 Hudson AW - BSc soil fertility
12 Humphris CSM - ASTC stockfoods
13 Lawrie RA - BSc soil and land use assessment
14 Livanos G - BApplSc pesticide formulations
15 Maguire MJ - BSc physical chemistry, soil mineralogy
16 Major EJ - BSc analytical chemistry
17 McCleary BV - BScAgr,PhD cereals, carbohydrate biochemistry
18 Milham PJ - BSc analytical chemistry
19 Plues LA - Assoc Anal Chem,Grad Dip,MSc analytical chemistry
20 Riddler AMH - BSc soil and land use assessment
21 Short CC - BSc,ASTC analytical chemistry
22 Singh G - BSc,HNC pesticide residues
23 Vimpany IA - BScAgr soil fertility
24 Weir RG - BScAgr plant nutrition
25 Willats HG - MSc analytical chemistry

292 Chemistry Branch Regional Staff

1 Allender WJ - BA,MAppSc cereals, Wagga
2 Bacon PE - MScAgr rice nutrition, Yanco
3 Baigent DR - BSc,PhD analytical chemistry, Wollongbar
4 Barnes WC - BScAgr cereals, Tamworth
5 Batten GD - MRurSc,PhD soils, Yanco
6 Blakeney AB - MScAgr rice quality, Yanco
7 Conyers MK - BNatRes,MScAgr soil fertility, Wagga
8 Daniells IG - BSc soils, Narrabri
9 Ferris IG - BA,MSc,PhD pesticides, Tamworth
10 Glennie-Holmes MR - BSc cereals, Wagga
11 Hall DJM - BScAgr soils, Trangie
12 Holford ICR - MScAgr,PhD soil fertility and plant nutrition, Tamworth
13 Mailer RJ - BAppSci oilseeds, Wagga
14 McKenzie DC - BNatRes,MRurSc soils, Trangie
15 McKenzie EA - BSc,PhD cereals, Tamworth
16 Saini HS - BScAgr,MSc,PhD analytical chemistry and biochemistry, Wollongbar
17 Slavich PS - BScAgr soils, Deniliquin
18 Southwell IA - BSc,PhD essential oils, Wollongbar

293 Entomology Branch

Biological and Chemical Research Institute

1 Attia FI - BScAgr,MSc,PhD pests of stored products
2 Baker GL - BSc locusts and grasshoppers
3 Beattie GAC - BScAgr,PhD citrus pests
4 Casimir M - BScAgr *Director of Entomology*
5 Clift AD - BScAgr,PhD mushroom pests
6 Edge VE - BSc,PhD control of mites, cotton pests
7 Fletcher MJ - BSc,PhD systematic entomology, Homoptera
8 Gellatley JG - BScAgr fruit flies, pesticide recommendations and registration
9 Goodyer GJ - BScAgr insects affecting pasture and field crops
10 Greening HG - BAgrSc pests of stored products
11 Hamilton JT - BScAgr pests of vegetables
12 Holtkamp RH - BSc pests of lucerne, control of locusts
13 Hughes PB - BA,PhD sheep blowfly, insecticide resistance
14 Levot GW - BSc,PhD sheep ectoparasites, Diptera
15 Schicha E - BSc,Dr rer nat systematic entomology, mites
16 Wallbank BE - BSc,PhD pests of stored products

294 Plant Pathology Branch

Biological and Chemical Research Institute

1 Barkley MB - BS,PhD virology
2 Barkley Mrs P - MScAgr citrus and nursery diseases
3 Bertus AL - MSc plant quarantine diseases
4 Fahy PC - BScAgr,MSc,PhD bacterial plant diseases
5 Green CD - ARCS,BScSpecial Entomology,PhD *Director (Plant Pathology)* nematology
6 Kable PF - MScAgr,PhD crop disease management
7 Letham DB - MScAgr vegetable diseases
8 McLeod RW - MScAgr nematology
9 Nair NG - MSc,PhD mushroom culture and diseases, disease epidemiology
10 Pares RD - BScAgr,DIC,PhD virology
11 Priest MJ - BSc systematic mycology; herbarium
12 Roberts WP - BSc,DipCompSc,PhD soil microbiology
13 Stovold GE - BScAgr field crop and pasture diseases
14 Taylor JB - BAgrScHort,BAgrSc,PhD ornamental plant diseases
15 Walker J - BScAgr systematic mycology; herbarium
16 Wellings CR - MScAgr wheat stripe rust (on study leave)
17 Wong PTW - BScAgr,PhD cereal diseases

295 Central Veterinary Laboratory, Glenfield

Roy Watts Road, Glenfield, New South Wales, 2167
Tel: (069) 605 1511.

1 Shannon AD - BVSC,PhD *Director* immunology
2 Arzey Mrs KE - BVSc,BSc,DIPEd poultry pathology
3 Bell IG - BVSc poultry pathology
4 Boray JC - DVM,PhD parasitology
5 Bowler JK - HDA *Manager*
6 Burton RW - BVSC,MVS epidemiology
7 Chin JC - BSC,PhD immunology, microbiology
8 Cross GM - MVSC,PhD poultry health
9 Eamens GJ - BVSC bacteriology, serology
10 Griffin SJB - HNDAgr *Assistant Manager*
11 Healy PD - BVSC,PhD biochemistry, inherited diseases
12 Hornitzky MAZ - BSc,MSc bacteriology, bee diseases
13 Jones DT - BVSC diagnostic pathology
14 Kirkland PD - BVSC,PhD virology, arbovirus infections
15 Littlejohns IR - BVSC virology, pestivirus infections

16 Ross AD - BVSC,MSC,PhD *Officer in Charge* diagnostic pathology
17 Ryan Miss DP - BVSC,PhD mastitis, immunogenetics
18 Sheldrake RS - MSCAGR,PhD mucosal immunology, mycoplasmology
19 Walker KH - BVSC diagnostic pathology

296 Nutrition and Feeds Evaluation Unit

Roy Watts Road, Glenfield, New South Wales, 2167
Tel: (069) 605 1511.

1 Low Mrs SG - BRurSc,MScAgr ruminant nutrition
2 Neutze SA - BScAgr,MScAgr *Officer in Charge* ruminant nutrition
3 Oddy VH - BSc,MSc sheep nutrition

297 Brackish Water Fish Culture Research Station

Salamander Bay, New South Wales, 2301
Tel: (049) 82 1232.

1 Allan G - BSc *Biologist* prawn farming
2 Capps E - BScAgr *Biologist* purification and water classification
3 Maguire Dr GB - PhD *Research Scientist (Prawn Farming)*
4 Martin R - HDA *Manager*
5 Nell Dr J - PhD *Research Scientist; Senior Biologist* oyster nutrition; aqua culture
6 Smith I - BSc *Biologist* oyster diseases

298 Fisheries Research Institute

202 Nicholson Parade, Cronulla, New South Wales, 2230
Tel: (02) 523 6222. Telex: 76125. FAX: (02) 527 3770.

1 Mason DW - WDA *Manager*

299 Marine Exploration Section

1 Gorman TB - MSc *Senior Biologist*
2 Graham K - BSc *Biologist*

300 Marine Resources Section

1 Lau T - BA,MA,MStats *Biometrician*
2 McDonall V - BSc *Biologist* crustaceans
3 Montgomery S - MSc *Biologist* crustacean stocks
4 Reid D - BSc,BComm,MStats *Biometrician*
5 Rowling K - BSc *Biologist* trawl fish

301 Marine Recreational Fishery Section

1 Diplock J - BSc *Biologist* kingfish
2 Henry G - MSc *Biologist* snapper
3 Matthews J - BSc *Research Assistant* artificial reefs
4 Pepperell J - BSc,PhD *Senior Biologist*

302 Inland Fisheries Section

1 Battaglene S - BSc *Biologist* Burrinjuck Dam
2 Burchmore Ms J - MSc *Biologist* stream surveys
3 Faragher R - MSc *Biologist* trout assessment
4 Harris J - BVSc,PhD *Biologist* Hawkesbury river
5 Johnson HT - MSc *Biologist* environmental
6 Richardson Ms B - BA Biol Sci,MSc *Senior Biologist*

303 Environmental Studies Section

1 Bell J - MSc *Biologist* fish habitat
2 Dunstan D - BSc *Biologist* estuarine conservations
3 Gibbs - BSc *Biologist* benthic studies
4 Pollard D - BSc,PhD *Senior Research Scientist* aquatic reserves
5 Scribner E - BSc,MESc *Biologist* water quality
6 West R - MSc *Biologist* fish habitat
7 Williams R - BA,MAppSc *Senior Biologist*

304 Water Resources Commission of New South Wales

305 Research Branch

P.O. Box 952, North Sydney, New South Wales, 2060
Tel: (02) 922-0121. Telex: WATCOM 21188. Cables: Aquacomm.

1 Garman DEJ - MSc,PhDd *Acting Principal Research Officer* salinity and reclamation

COUNTRY OFFICES

306 Leeton
P.O. Box 156, Leeton, New South Wales, 2705
Tel: (069) 533677. Telex: WATLET 69013.

1 Kelly ID - BScAgr *Senior Research Officer* irrigation land evaluation; water and salt movement; water quality and salinity

307 Griffith
P.O. Box 492, Griffith, New South Wales, 2860
Tel: (069) 621133. Telex: WATLET 69013.

1 van der Lelij A - IrAgr *Research Officer* tile drainage; tube-well drainage; irrigation hydrology

308 Deniliquin
P.O. Box 205, Deniliquin, New South Wales, 2710
Tel: (058) 812122. Telex: WARCD 55435.

1 vacant

309 Forestry Commission of New South Wales

95-99 York Street, Sydney
[PO Box 2667, Sydney, New South Wales, 2001]
Tel: (02) 234 1567. Cables: Newforests, Sydney.

1 Gentle SW - BscFor,DipFor,PhD *Commissioner*
2 Brack CL - BScFor *Resources Inventory Officer*
3 Casimir PA - BScFor,DipFor *Systems Design Officer*
4 Coomber BB - BAgr *Marketing Economist*
5 Cooper-Southam F - BScFor *Marketing Systems Officer*
6 Dear PM - BScFor *Systems Design Officer*
7 Donovan RM - BScFor *Resources Inventory Officer*
8 Drielsma JH - BScFor,MF,PhD *Management Investigations Officer*
9 Dryburgh A - BScFor *Systems Design Officer*
10 Free R - BScFor,DipFor *Assistant Commissioner*
11 Hoschke FG - BScFor *Forest Protection Officer*
12 Lind P - BSc,MSc *Forest Biometrician*
13 Skidmore AK - BScFor *Systems Design Officer*
14 Squire RH - BScFor,DipFor *Officer in Charge/Lands Classification Branch*
15 Stewart JP - BScFor,DipFor *Assistant Commissioner*
16 Wilson RV - BScFor,DipFor,MF *Officer in Charge/Economics and Data Services Branch*

310 Wood Technology and Forest Research Division

Oratava Avenue, West Pennant Hills, New South Wales, 2120
Tel: (02) 871 3222. Cables: Newforests, Sydney.

1 Ades PK - BScFor *Research Officer Tree Improvement/Forest Genetics*
2 Alexiou PN - BScFor,AIWSc *Research Officer Wood Drying*
3 Bevege DI - BScFor,DipFor,PhD *Chief of Division*
4 Chapman PR - BScTech,HND *Group Leader, Timber Engineering*
5 Cornish PM - BSc,MSc,PhD *Forest Hydrologist*
6 Dowden HGM - BScFor,PhD *Group Leader, Forestry Ecology*
7 Eldridge RH - BA *Research Officer* entomology and wood destroying insects
8 Fairey KD - MSc,AStC *Research Officer* entomology
9 Gardner WD - DipTechSc *Research Officer* wood preservation
10 Gerrettson-Cornell L - DASc,MScAgr *Research Officer* pathology
11 Ghali M - BScAgrEng,GMIE *Research Officer* mechanical and physical properties of timber
12 Grant D - BScEng,GMIE *Research Officer* mechanical stress grading
13 Hartley JN - BE *Research Officer* wood drying
14 Horne RR - BScFor,DipFor *Group Leader* silviculture
15 Johnson IG - BScFor *Research Officer* tree breeding
16 Johnstone RS - BSc,ASTC,AIWSc *Research Officer* wood preservation
17 Keirle Miss RM - MScAgr,AIWSc *Research Officer* pathology
18 King D - BSc *Systems Officer*
19 Koraca Mrs J - ALAA *Librarian*
20 Krilov A - DipEng,MEngSc,PhD *Principal Research Scientist, Group Leader* forest conservation engineering
21 Lambert Mrs MJ - BSc,MSc,ASTC,ARACI *Senior Research Scientist, Group Leader* soils and nutrition
22 Lowery JM - BScFor,DipFor *Deputy Chief of Division*
23 Mackay SM - BScFor,MNatRes *Research Officer* forest hydrology
24 Nassif NM - BScAgrEng,MSc *Research Officer* wood drying
25 Robinson G - BScFor *Forester* stand dynamics
26 Ryan PJ - BNatRes,MNatRes,PhD *Research Officer* soil chemistry
27 Simpson JA - MAgrSc *Group Leader* biology
28 Stone Miss C - BAgrSc,MSc,DIC *Research Officer* forest entomologist
29 Turner J - BScFor,PhD *Senior Research Officer* nutrient cycling and tree nutrition
30 Wilkes JF - BScFor,DipSc,PhD *Research Officer* wood structure
31 Wilkins JF - BSc *Research Officer* wood structure

RESEARCH FORESTERS, COUNTRY DISTRICTS

311 Forest Office
Bathurst, New South Wales, 2795
Tel: (063) 31 2044.

1 Hall MJ - BSc *Research Officer* pine silviculture; weed control

312 Forest Office
Coffs Harbour Jetty, New South Wales, 2451
Tel: (066) 52 1822.

1 Mackowski CM - BScFor *Forester* eucalypt silviculture , ecology

313 Forest Office
Eden, New South Wales, 2551
Tel: (064) 96 1547.

1 Bridges RG - BScFor,MSc *Forester* eucalypt silviculture, ecology

314 Forest Office
Wauchope, New South Wales, 2446
Tel: (065) 851908.

1 Binns DL - BScFor *Forester* forest ecology

315 Soil Conservation Service of N.S.W. Headquarters

Export House, 22 Pitt Street, Sydney, New South Wales, 2000
[Box R201, Royal Exchange Post Office, Sydney, New South Wales, 2000]

1 Atkinson G - MSc *Soil Conservationist*
2 Charman PEV - BSc *Deputy Director (Research)*
3 Cunningham GM - BScAgr *Chief of Services*
4 Day SJ *Commissioner*
5 Edwards K - BScAgr *Research Officer*
6 Emery KA - BScAgr *Research Officer*
7 Graham OP - BSc *Soil Conservationist*
8 Morse RJ - BA,HDA *Soil Conservationist*
9 Roberts RW - BA,QDH *Director of Env. Research and Studies*

316 Cowra Research Centre
P.O. Box 445, Cowra, 2794
Tel: (063) 421811 or (063) 421202.

1 Hamilton GJ - MScAgr *Assistant Director (Research)*
2 Murphy BW - BScAgr *Research Officer*
3 Packer IJ - BScAgr *Research Officer*
4 Ryan KT - BScAgr *Research Officer*
5 Watt LA - BScAgr,BEc,MSc *Research Officer*

317 Gunnedah Research Centre
P.O. Box 462, Gunnedah, 2380
Tel: (067) 420300.

1 Crouch RJ - MScAgr *Research Officer*
2 Elliott GL - BScAgr *Assistant Director Research*
3 Lang RD - BScAgr *Research Officer*
4 Marschke GW - BSc *Soil Conservationist*
5 Rosewell CJ - BScAgr,MNatRes *Research Officer*

318 Western Region
P.O. Box 77, Condobolin, New South Wales, 2877
Tel: (068) 952033.

1 Leys JF - BSc *Soil Conservationist*
2 Walker PJ - BScAgr *Deputy Regional Director*

319 The Australian Museum

6-8 College Street, Sydney, New South Wales, 2000
[PO Box A285, Sydney South, New South Wales]
Tel: (02) 339 8111.

1 Cogger HG - MSc,PhD *Deputy Director* systematics, ecology and biogeography of Australian and Indo-Pacific reptiles; ecology of arid zone reptiles
2 Griffin DJG - MSc,PhD *Director* systematics of Decapod Crustacea

320 Division of Anthropology

1 Lampert RJ - PhD *Head of Division* prehistory of Australia
2 Meehan B - MA,PhD *Scientific Officer* material culture and subsistence of Australian Aborigines in the Northern Territory
3 Specht JR - MA,PhD *Scientific Officer* prehistory of Western New Britain

321 Division of Earth Sciences

1 Ritchie RJ - PhD *Head of Division* paleozoic fossil fishes

2 Sutherland FL - MA,PhD *Scientific Officer* mineral and gemstones, tektites, igneous minerals and rocks

322 Division of Invertebrate Zoology

1 Hutchings P - PhD *Head of Division* polychaete faunas of coral reefs, mangroves and estuaries
2 Gray MRV - MSc *Scientific Officer* systematics of Mygalomorph spiders
3 Jones AR - MSc,PhD *Scientific Officer* ecology of benthic invertebrate communities
4 Lowry JK - MSc,PhD *Scientific Officer* systematics, ecology and zoogeography of amphipods
5 McAlpine DK - MSc,PhD *Scientific Officer* systematics of Acalpytrate Diptera
6 Ponder WF - MSc,PhD *Scientific Officer* molluscan comparative anatomy and taxonomy
7 Rowe FWE - PhD *Scientific Officer* echinoderm systematics and ecology
8 Rudman WB - MSc,PhD *Scientific Officer* molluscan comparative anatomy and taxonomy

323 Division of Vertebrate Zoology

1 Hoese DF - PhD *Head of Division* systematics and ecology of fishes
2 Flannery T - PhD *Scientific Officer* systematics of recent and fossil marsupials of Australia and New Guinea
3 Greer AE - PhD *Scientific Officer* systematics, ecology and evolution of reptiles
4 Leis J - PhD *Scientific Officer* systematics and ecology of larval fishes
5 Paxton JR - MSc,PhD *Scientific Officer* systematics and biology of mid water fishes
6 Pyke G - PhD *Scientific Officer* animal ecology
7 Recher HF - PhD *Scientific Officer* animal ecology

324 Royal Botanic Gardens and National Herbarium of N.S.W.

Mrs Macquaries Road, Sydney, New South Wales, 2000
Tel: 231 8111.

1 Briggs BG - BSc,PhD *Acting Director*

325 Scientific Division - National Herbarium of N.S.W.

1 Bedford DJ - BSc *Acting Assistant Director (Scientific)* plant taxonomy *Xanthorroea*

326 Plant Taxonomy

1 Everett Ms J - BSc Poaceae, Asteraceae, Myoporaceae
2 Harden Mrs GJ - MSc rainforests, flora writing
3 Hill KD - MSc eucalypts, gymnosperms, Asdepiadaceae
4 Jacobs SWL - BScAgr,PhD Poaceae, water plants, Centrospermae
5 Powell JM - BSc,MSc,PhD pollen morphology and Epacridaceae
6 Weston PH - BSc,PhD Proteaceae, Rutaceae, Fabaceae, Orchidaceae
7 Wilson KL - BScAgr Cyperaceae, Polygonaceae, Juncaceae, Casuarinacaee
8 Wilson PG - BSc,PhD Myrtaceae, Lythraceae, *Indigofera* (Fabaceae)

327 Ecology

1 Benson D - BSc vegetation mapping, rare species conservation, fire ecology
2 Fox M - BSc,PhD vegetation mapping; disturbance ecology, human impacts

328 Living Collections Division - Horticultural Botany

1 Blaxell DF - BSc,DipAgr *Assistant Director Living Collections* plant taxonomy and horticulture; eucalypts, orchidaceae
2 Wallace B - BSc,PhD plant taxonomy and horticulture; orchidaceae; vascular epiphytes

329 Zoological Parks Board of New South Wales

Taronga Zoo, P.O. Box 20, Mosman, 2008
Tel: (02) 969 2777. Telex: AA25545. Cables: Zoology Mosman.

1 Throp JL *Director*
2 Finnie E - MVSc,MACVSc *Deputy Director*
3 Hartley WJ - FRVSc,FRCPath,MVSc,BSc *Co-Ordinator Veterinary Pathology Registry*
4 Woodside D - PhD *Curator of Mammals/Research Co-Ordinator* bat echo-location

330 State Pollution Control Commission

G.P.O. Box 4036, Sydney, New South Wales, 2001
Tel: (02) 265 8888. Telex: AA72234.

1 Standen P - ASTC *Director*

331 Chemicals and Noise Division

1 Forrest WG - BScFor,PhD *Chief*

332 Environmental Biology Branch

1 Bowen SE - BSc,PhD pollution in terrestrial ecosystems
2 Scheltema JH - BScAgr,MScAgr air pollution and horticulture
3 Thompson GB - BA,MSc,PhD *Principal Biologist*

NORTHERN TERRITORY GOVERNMENT

333 Department of Primary Production

Head Office, G.P.O. Box 4160, Darwin, Northern Territory, 5794
Tel: (089) 895511. Telex: 85127.

334 Executive

1 Hooper ADL - MSc *Director - Land Use Planning*
2 Johnston BR - DipAgr,DipExtension *Special Projects Officer*
3 Plummer PJ - BA *Deputy Secretary*
4 Saville SP - OBE,BScAgr,DipEd,DTA,MSc *Secretary*

335 Plant Industry Division

1 Sturtz JD - BAgrSc

336 Animal Industry Division

1 Fallon GR - MVSc *Chief Veterinary Officer*
2 Neumann G - BVSc,MVSc,MACVSc *Senior Veterinary Officer, BTEC Unit*
3 Newton-Tabrett DA - BVSc *Assistant Secretary*

337 Technical Services Division

1 Thistlewaite RJ - PhD,BScFor,DDDipFor *Assistant Secretary*
2 Tobin JG - BAgrSc *Chief Economist*

338 Agricultural Quarantine Service

1 Fenner TL - BAgrSc *Quarantine Advisor (Plants) - Entomology*
2 Kilduff ID - DipAgr *Executive Officer*
3 Taylor IF - DipAgrSc *Senior Agricultural Quarantine Officer*

339 Northern Region Office

Head Office, GPO Box 4160, Darwin, Northern Territory, 5794
Tel: (089) 221211. Telex: BERLAB 85346.

340 Executive

1 Allwood AJ - BAgrSc *Regional Officer*
2 Bainbridge MH - BSc *Regional Officer (Katherine)*

341 Plant Industry Division

342 Plant Production Branch

1 Baker IW - BAppSc,MAgrSc *Perrenial Crops Horticulturist* tropical fruit
2 Dasari NR - BSc,MSc,PhD *Senior Crops Officer* rice
3 Miller IL - BScAgr,DipTropAgr *Senior Agronomist* weeds

343 Plant Health Branch

1 Hamdorf IJ - BSc *Senior Chemist, Pesticide Registrar* biochemistry
2 Hansen DH - CertSeedTestingProficiency *Senior Seed analyst*
3 Pitkethley RN - BScAgr *Senior Plant Pathologist*
4 Strickland GR - BScZool *Senior Entomologist*

344 Animal Industry Division

1 Ford BD - BRurSc,MSc *Senior Animal Production Officer* buffaloes, cattle and pastures
2 Gard GP - BVSc,PhD *Senior Veterinary Virologist*
3 Kirby GWM - BRurSc,MRurSc *Principal Animal Production Officer* genetics, economics
4 Knight I - BVSc *Regional Veterinary Officer*
5 McCool CJ - BVSc,MSc *Senior Veterinary Officer (Field Research)*
6 Melville LF - BVSc,BSc,DipVetPath animal pathology

345 Economics And Marketing

1 Franettovich MJ - BAgrSc,GradDipComputing economics

346 Library

1 Chilvers MA - ALAA *Senior Librarian*
2 Strickland K - ALAA *Senior Librarian*

347 Southern Region Office

GPO Box 2134, Alice Springs, Northern Territory, 5750
Tel: (089) 522344. Telex: 81222.

348 Executive

1 Petty DR - BAgrSc *Executive Officer Administration*
2 Phillpot SC - BAgEco *Regional Director*

349 Laboratory

1 Gilfedder J - BSc virology
2 McEwan DR - BVSc *Officer in Charge, Laboratory*

350 Plant Industry Division

1 Basin GN - BSc rangeland management
2 Dance RA - BRurSc *Senior Agronomist/Animal Production Officer*
3 McEllister FV - DipAgr *Senior Technical Officer*
4 Shaw KA - BAgrSc rangeland management

351 Animal Industry Division

1 Bertram JD - BAppSc-Rt animal production
2 Locke KB - BVSc,DipAgr *District Veterinary Officer*
3 Williams OJ - MVSc *Senior Veterinary Officer (Laboratory)*

352 Conservation Commission of The Northern Territory

P.O. Box 1046, Alice Springs, Northern Territory, 5750
Tel: (089) 50 8211. Telex: AA81191.

1 Thomas WA - BVetSc,MBA *Director*

353 Resources Utilisation Unit

354 Northern Region
P.O. Box 38496, Winnellie, Northern Territory, 5789
Tel: (089) 22 0211. Telex: AA85336.

1 Haines MW *Research Officer* plantation silviculture and mensuration
2 White AE - BScFor *Executive Officer*

355 Southern Region
P.O. Box 1046, Alice Springs, Northern Territory, 5750
Tel: (089) 50 8211. Telex: AA81191.

1 Sandell P - BScFor *Project Officer* arid zone plants

QUEENSLAND GOVERNMENT

356 Bureau of Sugar Experiment Stations (BSES)

50 Meiers Road, Indooroopilly, Brisbane, Queensland
[PO Box 86, Indooroopilly, Queensland, 4068]
Tel: (07) 371 6100. Telex: 42227. Cables: Sugarbureau.

1 Ryan CC - BAgrSc,PhD *Head, Pathology Division* administration, sugarcane diseases
2 Sturgess OW - MAgrSc *Director*
3 Birch RG - MSc,PhD *Research Officer* sugarcane diseases, biotechnology
4 Chardon CWA - QDA,BEc,DipCompSc *Research Officer* EDP
5 Egan BT - BSc *Research Coordinator* administration
6 Leonard GJ - MSc,PhD *Senior Research Officer* polysaccharide chemistry
7 Leverington KC - BSc *Deputy Director*
8 Stickley BDA - MSc *Senior Research Officer* pesticide residue chemistry
9 Taylor PWJ - MScAgr *Research Officer* sugarcane diseases
10 Wegener MK - MAgrSc *Senior Information Officer* agricultural communication

357 Bundaberg

P.O. Box 651, Bundaberg, Queensland, 4670
Tel: (071) 79 3228. Telex: 49661.

1 Atherton PG - BScApp,DipSugTech *Head, Mill Technology Division* administration, chemical engineering
2 Bull RM - MSc *Research Officer* insect pests of sugarcane
3 Dick RG - BE *Research Officer* agricultural engineering
4 Hogarth DM - MScAgr,PhD *Senior Research Officer* cane breeding, varietal selection, statistical genetics, biometrics
5 Kingston G - MAgrSc *Senior Research Officer* sugarcane agronomy
6 Kirby LK - DipSugChem *Senior Research Officer* sugar chemistry
7 Noble AG - BScApp,DipSugTech *Research Officer* chemical engineering
8 Ridge DR - MAgrSc *Head, Soils & Agronomy Division* administration, soil physics, sugarcane agronomy

358 Mackay

P.O. Box 5503, Mackay Mail Centre, Mackay, Queensland, 4741
Tel: (079) 54 5100. Telex: 48505.

1 Chapman LS - BAgrSc *Senior Research Officer* sugarcane agronomy
2 Haysom MBC - DipIndChem *Research Officer* analytical chemistry

359 Ayr

P.O. Box 117, Ayr, Queensland, 4807
Tel: (077) 82 5455. Telex: 47634.

1 Ham GJ - BAgrSc *Research Officer* sugarcane agronomy

360 Tully

P.O. Box 566, Tully, Queensland, 4854
Tel: (070) 68 1488. Telex: 48912.

1 Croft BJ - BAgrSc *Research Officer* sugarcane diseases
2 Hurney AP - BAgrSc *Senior Research Officer*
3 Magarey RC - BAgrSc *Research Officer* sugarcane diseases
4 Reghenzani JR - BAgrSc,QDA *Research Officer* sugarcane agronomy

361 Meringa

P.O. Box 122, Gordonvale, Meringa, Queensland, 4865
Tel: (070) 68 1488. Telex: 48912.

1 Berding N - MAgrSc,PhD *Senior Research Officer* cane breeding, varietal selection
2 Chandler KJ - MAgrSc *Research Officer* insect pests of sugarcane
3 Skinner JC - BScAgr,PhD *Principal Research Fellow* cane breeding, varietal selection

362 Department of Forestry, Queensland

41 George Street, GPO Box 944, Brisbane, Queensland, 4001
Tel: 2242111. Cables: Forestry, Brisbane.

1 Cranny P - BScFor,DipFor *Assistant Conservator*
2 Kelly J - BScFor,DipFor *Conservator*
3 Ryan T - BScFor,DipFor *Deputy Conservator*

363 Division of Technical Services

Roma Street, P.O. Box 5, Brisbane, Queensland, 4000
Tel: (07) 2296500. Cables: Forestry, Brisbane.

1 Schaumberg JB - BScFor,DipFor *Director*

364 Forest Research Branch

1 Bacon G - BScFor,PhD *Manager*
2 Henry NB - BScFor,DipFor forest mensuration and biometrics
3 Nikles DG - BScFor,DipFor,PhD tree breeding

365 Biology Laboratory

80 Meiers Road, Indooroopilly, Queensland, 4068
Tel: (07) 3713533.

1 Bolland L - BScFor,PhD
2 Brown B - BAgrSc,PhD forest pathology
3 Wylie R - BScEnt forest entomology

366 Zoology Laboratory

80 Meiers Road, Indooroopilly, Queensland, 4068
Tel: (07) 3713533.

1 Kehl JC - BSc forest zoology

367 Atherton Research Station

P.O. Box 210, Atherton, Queensland, 4883
Tel: (07) 911844.

1 Applegate GB - BScFor,MNatRes agroforestry
2 Nicholson DI - BScFor,DipFor *Forest Research Officer* tropical rainforest silviculture

368 Dalby Research Station

P.O. Box 590, Dalby, Queensland, 4405
Tel: (074) 622022.

1 West D - BScFor cypress pine and and inland hardwoods silviculture

369 Gympie Research Centre

M.S. 483, Gympie, Queensland, 4570
Tel: (071) 822244.

1 Collings AD - BSc hydrology research
2 Ryan P - BScFor hardwoods research
3 Simpson J - BScFor soils research
4 Yule RA - BScEnt,DipFor *Principal Research Officer* silviculture

370 Timber Utilisation Branch

Roma Street, P.O. Box 5, Brisbane, Queensland, 4000
Tel: (07) 2296500. Cables: Forestry, Brisbane.

1 Bardsley J - BScFor,DipFor *Manager*
2 Bragg C - BScFor processing, seasoning and mechanics
3 Gough DK - BScFor,DipFor *Principal Utilisation Officer*
4 Greve D - BScFor timber use and information
5 Harding K - BSc wood structure and utilisation
6 Leightley LE - BSc,PhD wood chemistry and preservation

371 Division of Technical Services (L.U.& I.)

GPO Box 944, Brisbane, Queensland, 4001
Tel: (07) 2248373. Cables: Forestry, Brisbane.

1 Hinson NS - BScFor,DipFor,MS *Director*

372 Division of Forest Management (Operations)

GPO Box 944, Brisbane, Queensland, 4001
Tel: (07) 2248373. Cables: Forestry, Brisbane.

1 Pegg RE - BScFor,DipForMS *Director*

373 Division of Forest Management (Development)

GPO Box 944, Brisbane, Queensland, 4001
Tel: (07) 2248373. Cables: Forestry, Brisbane.

1 Kanowski P - BScFor,DipFor *Director*

374 Department of Lands, Queensland

375 Biological Branch

P.O. Box 36, Sherwood, Queensland, 4075
Tel: (07) 379 6611.

Research into rangeland and aquatic weeds and their control

1 Diatloff G - MAgrSc *Director* formulation and biological activity of herbicides

376 Alan Fletcher Research Station
P.O. Box 36, Sherwood, Queensland, 4075
Tel: (07) 379 6611.

1 Armstrong TR - BagrSc,BEcon *Agronomist* chemical control of weeds
2 Donnelly GP - BSc *Entomologist* biological control of weeds using insects
3 Harvey GJ - MAgrSc,BSc *Supervising Agronomist* chemical control of woody weeds
4 McFadyen PJ - MSc,PhD *Supervising Entomologist* biological control of weeds using insects
5 McFadyen RE - BA,PhD *Senior Entomologist* biological control of weeds using insects
6 Tomley AJ - BSc,QDA *Plant Pathologist* biological control of weeds using plant pathogens
7 Wild CH - BSc,PhD *Senior Entomologist* biological control of weeds using insects
8 Willson BW - BSc *Senior Entomologist* biological control of weeds using insects

377 Tropical Weeds Research Centre
P.O. Box 187, Charters Towers, Queensland, 4820
Tel: (077) 87 3300.

1 Ablin MP - BSc *Entomologist* biological control of tropical weeds using insects
2 Bolton MP - MSc *Agronomist* ecology of tropical weeds

378 Queensland Department of Primary Industries

379 Division of Administration

GPO Box 46, Brisbane, 4001
Tel: (07) 227 4111. Telex: AA41620. Cables: Agriculture, Brisbane. FAX 221 4302.

1 Alexander GI - AO,PhD,MSc,BVSc,FACVSc *Director General*
2 Oxenham BL - BAgrSc *Assistant Director-General (Research)*

380 Biometry Branch
GPO Box 46, Brisbane, 4001
Tel: (07) 224 0414.

1 Pepper Miss PM - PhD,MSc,DipAutCom *Director* modelling and simulation
2 Giles Mrs JE - BSc,BEcon design and analysis
3 O'Rourke PK - BA,BSc design and analysis
4 Swain AJ - PhD,BSc,DipAutCom *Assistant Director* analysis and program development

381 At Rockhampton
P.O. Box 689, Rockhampton, 4700

1 Blight GW - BSc,MScStud design and analysis

382 At Toowoomba
P.O. Box 102, Toowoomba, 4350

1 Mayer RJ - MSc,BSc design and analysis

383 Extension Services Branch
GPO Box 46, Brisbane, 4001
Tel: (07) 227 4111.

1 Littmann MD - MSc,MPubAd *Acting Director* extension administration and development
2 Wilson TD - BAgrSc,BEcon,MScExtEd extension evaluation
3 Wissemann AF - BSc,MAgrStud extension evaluation

384 Information and Extension Training Branch
GPO Box 46, Brisbane, 4001
Tel: (07) 224 0414.

1 Durand MRE - MRCVS *Director* information services
2 Mason WD - BA,ALAA *Librarian in Charge*

385 Research Stations Branch
GPO Box 46, Brisbane, 4001
Tel: (07) 227 4111.

1 Stonard P - BScAgr *Director* administration of research stations
2 Hay VT - DipMechEng *Assistant Director* administration of research stations
3 Wollin AS - MEngSc,BEngAgr,GradDipCompEng *Executive Engineer* computerisation of agricultural engineering

386 At Ayr
P.O. Box 591, Ayr, 4807

1 Garside AL - BAgrSc irrigation project development

387 Division of Animal Industry

GPO Box 46, Brisbane, 4001
Tel: (07) 227 4111.

1 Woolcock BA - BVSc,MACVS *Director*
2 Laws L - MVSc,BVSc,MACVS *Deputy Director*

388 Beef Cattle Husbandry Branch
GPO Box 46, Brisbane, 4001
Tel: (07) 224 0414.

1 Hopkins PS - PhD,MVSc *Director*
2 Rudder TH - BEcon,QDAH *Assistant Director*
3 Tudor GD - MAgrSc beef cattle nutrition
4 Wythes Miss JR - BRurSc beef marketing

389 Brigalow Research Station
M.S. 586, Theodore, 4719

1 Tierney TJ - MVSc,BVSc beef cattle reproduction

390 At Rockhampton
P.O. Box 689, Rockhampton, 4700

1 Holroyd RG - PhD,MSc,BVSc beef cattle reproduction

391 At Toowoomba
P.O. Box 102, Toowoomba, 4350

1 Plasto AW - BAgrSc beef cattle nutrition and intensive management

392 At Townsville
P.O. Box 931, Townsville, 4810

1 Hodge PB - MVSc,BSc beef cattle husbandry and reproduction

393 Pig and Poultry Branch
GPO Box 46, Brisbane, 4001
Tel: (07) 227 4111.

1 Byrnes RV - MSc,BSc,QDA,GradDipBusAdmin *Director*
2 Connor JK - MAgrSc,BAgrSc *Assistant Director* poultry nutrition
3 McPhee CP - PhD,MAgrSc,BAgrSc pig genetics
4 Williams KCK - PhD,BVSc,QDAH pig nutrition

394 Veterinary Services Branch
GPO Box 46, Brisbane, 4001
Tel: (07) 224 0414.

1 Wells ID - BVSc *Director*
2 Fraser IR - BVSc,QDAH diseases of livestock
3 Langford GFD - BVSc *Assistant Director* diseases of livestock
4 McCubbin IK - BVSc brucellosis and tuberculosis eradication
5 Walthall JC - BVSc,BEcon *Assistant Director* disease eradication and exotic diseases
6 Webster WR - MVSc,BVSc diseases of pigs

395 At Cairns
P.O. Box 652, Cairns, 4870

1 Clague DC - BVSc,QDAH diseases of livestock

396 At Maryborough
P.O. Box 432, Maryborough, 4650

1 Hass CR - MVSc,BVSc diseases of livestock

397 At Mount Isa
P.O. Box 1333, Mount Isa, 4825

1 Streeton TA - BVSc diseases of livestock

398 At Rockhampton
P.O. Box 689, Rockhampton, 4700

1 Hale KB - BVSc,BEcon diseases of livestock

399 At Roma
P.O. Box 308, Roma, 4455

1 Dadswell LP - MVSc,BVSc diseases of livestock

400 At Toowoomba
P.O. Box 102, Toowoomba, 4350

1 Stevens MS - BVSc diseases of livestock

401 Queensland Animal Research Institute
Fairfield Road, Yeerongpilly, 4105
Tel: (07) 892 9400.

1 Garter RJW - BSc,FRACI *Director of Animal Laboratories*

402 Biochemistry Branch

1 McEwan T - PhD,MSc,BSc,FRACI *Director*
2 Blaney BJ - BAppSc,DipIndChem mycotoxins
3 Franke FP - PhD insect attractants
4 Hurwood IS - BSc,GradDipBusAdmin,ARACI *Assistant Director* toxicology
5 Moir KW - BSc nutrition of grazing animals
6 Moore CJ - PhD,BSc fruit fly attractants
7 Moy DC - PhD,BSc insecticide residue chemistry
8 Murphy GM - MSc,BSc trace element deficiencies
9 Neil AR - PhD,BSc cattle nutrition
10 Oelrichs PB - PhD,MSc,BSc,ARACI chemistry of toxic plants

403 Pathology Branch

1 Callow LL - PhD,DVSc,BVSc *Director*
2 Baldock RC - BVSc parasite epidemiology
3 Blackall PJ - BSc bacteriology, vaccine production for pigs and poultry
4 Chung YS - PhD,DVM virus diseases of cattle
5 Connole MD - BSc animal mycology
6 Dalgliesh RJ - MVSc,BVSc *Assistant Director*
7 Dimmock Mrs CK - BSc haematology, immunology of cattle
8 Elder Miss JK - PhD,BSc,DipInfoProc epidemiology
9 Hill MWM - MSc,BVSc histopathology
10 Ketterer PJ - BVSc,DipBact diseases of pigs
11 McKenzie RA - BVSc poisonous plants
12 Rogers RJ - MVSc,BVSc,GradDipBusAdmin *Assistant Director*
13 Smeltzer TI - BSc,DipCompSc serology of bacterial diseases, computer systems
14 Storie GJ - BVSc brucellosis and tuberculosis
15 Trueman KF - MVSc,GradDipClincPath diagnostic pathology
16 Young PL - PhD,BVSc,BSc virus diseases of cattle and poultry

404 Oonoonba Animal Health Station
P.O. Box 1085, Townsville, 4811
Tel: (077) 892 9400.

1 Hoffman D - PhD,BVSc,DipTropVetMed,QDAH *Officer in Charge*
2 Norton JH - MSc,BVSc diseases of cattle
3 Parker RJ - DipMLT animal diseases
4 Thomas Miss AD - MSc,BSc *bacteriology*

405 At Tick Fever Research Centre
Grindle Road, Wacol, 4076

1 de Vos AJ - MVSc,BVSc *Officer in Charge* tick fever vaccine
2 Barry DN - PhD,BSc tick fever vaccine

406 Sheep and Wool Branch

1 Winks L - BAgrSc,GradDipBusAdmin *Director*
2 O'Sullivan BM - PhD,BVSc,GradDipBusAdmin *Assistant Director*
3 Rose Miss ME - MAgrSc,BAgrSc genetics

407 Division of Dairying and Fisheries

GPO Box 46, Brisbane, 4001
Tel: 224 0414.

1 Miller JG - BAgrSc,QDDM,GradDipBusAdmin *Director of Dairying and Fisheries*
2 Gillies Miss AJ - MAppSc,BAppSc,GradDipMan *Deputy Director (Dairy Research)*
3 Stephenson RG - BVSc sheep nutrition

408 Dairy Cattle Husbandry Branch
G.P.O. Box 46, Brisbane, 4001
Tel: (07) 224 0414.

1 Thurbon PN - BAgrSc *Director*
2 Collard RV - BAgrSc,DipAgrExt,GradDipBusAdmin dairy herd improvement
3 Quinton FW - MVSc,BVSc *Assistant Director* artificial breeding

409 At Kairi Research Station
P.O. Box 27, Kairi, 4872

1 Davison TM - MAgrSc,BAgrSc dairy cow nutrition, tropical pastures

410 At Mutdapilly Research Station
M.S. 825, Ipswich, 4305

1 Cowan RT - PhD,BAgrSc dairy cow nutrition and pasture management

411 At Rockhampton
P.O. Box 689, Rockhampton, 4700

1 Chopping GD - MAgrSc,BAgrSc pasture management

412 At Dairy Herd Improvement Laboratory
Grindle Road, Wacol, 4076

1 Andrews RJ - BAgrSc technology of milk production

413 Food Research Branch
Hamilton Laboratory Complex, 19 Hercules Street, Hamilton, 4007
Tel: (07) 268 2421.

1 Juffs HS - PhD,BAgrSc,QDDM,GradDipBusAdmin *Director of Research*
2 Aston JW - MAppSc,BAppSc flavour in cheese and other foods
3 Bowden RP - BSc processing of tropical fruit
4 Deeth HC - PhD,BSc,BEcon lipolysis, fish spoilage
5 Dommett TW - MSc,DipInfoProc,GradDipBusAdmin quality control of dairy products, computerisation
6 Dulley JR - BSc,GradDipBusAdmin *Assistant Director*
7 Fitz-Gerald Mrs CH - BSc,BA lipolysis, goats milk
8 Marschke RJ - BSc milk quality control, mastitis diagnosis
9 Mitchell GE - MAppSc,BAppSc dairy product analysis, irradiation, meat quality

414 Division of Land Utilisation

GPO Box 46, Brisbane, 4001
Tel: (07) 227 4111.

1 Crack BJ - MSc,BSc,ARACI *Director*
2 Trudgian KG - BSc,GradDipBusAdmin *Deputy Director*

415 Land Resources Branch
Indooroopilly Laboratories, Meiers Road, Indooroopilly, 4068
Tel: (07) 337 9311.

1 Briggs HS - MAgrSc,BAgrSc *Director*
2 Hughes KK - BSc *land degradation assessment*
3 McDonald RC - BAgrSc soil survey
4 Powell B - BAgrSc,QDA soil survey
5 Reid RE - MAgrSc,BRurSc land resource assessment
6 Turner EJ - BAgrSc land resource assessment
7 Vandersee BE - BAgrSc,QDA *Assistant Director*

416 At Ayr
P.O. Box 591, Ayr, 4807

1 Day KJ - MAgrSc,BAgrSc land resource assessment

417 Soil Conservation Research Branch
Indooroopilly Laboratories, Meiers Road, Indooroopilly, 4068
Tel: (07) 377 9311.

1 Coughlan KJ - PhD,MAgrSc,BAgrSc,GradDipBusAdmin *Director*
2 Swaw RJ - BAgrSc soil salinity
3 Troung PN - PhD,BAgrSc soil conservation
4 Webb AA - MAgrSc,BAgrSc,GradDipBusAdmin *Assistant Director*

418 At Biloela Research Station
P.O. Box 201, Biloela, 4715

1 Thomas GA - MAgrSc,BAgrSc soil conservation

419 At Rockhampton
P.O. Box 689, Rockhampton, 4700

1 Yule DF - PhD,MSc,BAgrSc soil conservation

420 At Queensland Wheat Research Institute
P.O. Box 5282, Toowoomba, 4350

1 Loch RJ - BA,BAgrSc soil conservation
2 Radford BJ - MAgrSc,BAgrSc surface management

421 At Toowoomba
P.O. Box 102, Toowoomba, 4350

1 Smith GD - PhD,BAgrSc soil management

422 Agricultural Chemistry Branch
Indooroopilly Laboratories, Meiers Road, Indooroopilly, 4068
Tel: (07) 337 9311.

1 Bruce RC - BSc,ARACI *Director* soil chemistry
2 Barry GA - BSc soil analysis
3 Hamilton DJ - MSc,BSc *Assistant Director* pesticide chemistry
4 Hargreaves PA - BSc,ARACI *pesticide residues*
5 Haydon GF - BSc plant analysis
6 Moody PW - BAgrSc soil fertility
7 Noble RM - MSc,BAppSc plant analysis
8 Rayment GE - DipIndChem,QDA,ARACI soil science, soil chemistry
9 Simpson BW - BSc *pesticides and plant chemistry*
10 Whitehouse MJ - BSc soil fertility
11 Williams HR - BSc plant analysis

423 At Biloela Research Station
P.O. Box 201, Biloela, 4715

1 Standley J - PhD,BSc,DipTropAgron soil fertility

424 Division of Marketing

GPO Box 46, Brisbane, 4001
Tel: (07) 224 0414.

1 Densley DRJ - BAgrSc,BEcon *Director of Marketing*
2 Hamilton CP - MAggrSc,QDAH *Deputy Director of Marketing*

425 Economic Services Branch
GPO Box 46, Brisbane, 4001
Tel: (07) 224 0414.

1 Robinson IB - BEcon *Director of Economic Services*
2 Moorhouse W - BEcon,BEcom,GradDipBusAdmin,DipInfoProc,AAUQ,AASA *Assistant Director*
3 Bartholomew RB - BEcon project analysis
4 Donnet RJ - BAgrEcon,DipFinMang,AASA financial analysis
5 Mill SJ - MAgrSc,BAgrSc,QDA marketing

426 At Atherton
P.O. Box 654, Atherton, 4883

1 Sing NC - BAgrSc,BEcon tropical fruit

427 At Bundaberg
P.O. Box 1143, Bundaberg, 4670

1 Hardman JR - BAgrSc,BEcon citrus, beef

428 At Townsville
P.O. Box 931, Townsville, 4810

1 Wicksteed LT - BEcomm,DipAgr beef cattle systems modelling

429 Standards Branch
Indooroopilly Laboratories, Meiers Road, Indooroopilly, 4068
Tel: 337 9311.

1 Mungomery WV - BScAgr *Director of Agricultural Standards*
2 Adams RH - BSc seed research

430 Marketing Services Branch
G.P.O. Box 46, Brisbane, 4001
Tel: (07) 224 0414.

1 White BJ - PhD,BEcon *Director of Marketing Services*
2 Baker RJ - BSc,BEcon,DipEd marketing research
3 Connor AF - BEcon *Assistant Director*
4 Smith KL - BEcon,MBA,AASA marketing research

431 Division of Plant Industry

G.P.O. Box 46, Brisbane, 4001
Tel: (07) 227 4111.

1 Purss GS - MScAgr,BScAgr *Director*

2 Leslie JK - PhD,BAgrSc *Deputy Director*
3 Syme JR - PhD,BSc,GradDipInfoProc *Executive Officer (Crop Protection)*
4 Ward DK - BAgrSc *Senior Technical Administrative Officer*

432 At Queensland Wheat Research Institute
P.O. Box 5282, Toowoomba, 4350

1 Clarke AL - PhD,MAgrSc *Director*

433 Agriculture Branch
G.P.O. Box 46, Brisbane, 4001
Tel: (07) 224 0414.

1 McNee DAK - BAgrSc,DipAgrExt *Director of Agriculture*

434 Agronomy Section

1 Mungomery VE - PhD,MAgrSc *Assistant Director (Agronomy)*
2 Williamson AJP - BScAgr cereals

435 At Biloela Research Station
P.O. Box 201, Biloela, 4715

1 Hunter MN - PhD,MAgrSc,BAgrSc agronomy of oilseeds
2 Walker SR - MAgrSc,BagrSc weed control in cereals and oilseeds

436 At Emerald
P.O. Box 81, Emerald, 4720

1 Hibberd DE - MAgrSc cereals, crop nutrition
2 Keefer GD - BSc cotton, irrigated crops

437 At Gatton Research Station
P.O. Box 241, Gatton, 4343

1 Greasley AS - BSc *Officer in Charge* vegetables, irrigation

438 Hermitage Research Station
via Warwick, 4370

1 Brinsmead RB - MAgrSc,BAgrSc *Officer in Charge* oilseeds and grain legumes
2 Goyne PJ - PhD,BAgrSc cereals, crop physiology
3 Henzell RG - PhD,BAgrSc sorghum breeding
4 Johnston RP - PhD,BAgrSc barley breeding
5 Marley JMT - BAgrSc cereal husbandry, weed control
6 Redden RJ - PhD,MAgrSc,BAgrSc oilseed breeding
7 Rose JL - PhD,MAgrSc,BAgrSc soybean and grain legume breeding

439 Kairi Research Station
P.O. Box 27, Kairi, 4872

1 Martin IF - BSc *Officer in Charge* maize

440 At Kingaroy
P.O. Box 23, Kingaroy, 4610

1 Dickson T - MAgrSc,BSc crop nutrition, oilseeds
2 Shorter R - PhD,MAgrSc,BAgrSc peanut breeding

441 At Mareeba
P.O. Box 1054, Mareeba, 4880

1 Hawton D - PhD,MSc,BSc cereals and oilseeds, weed control

442 At Queensland Wheat Research Institute
P.O. Box 5282, Toowoomba, 4350

1 Brennan PS - PhD,MAgrSc,BAgrSc wheat breeding
2 Strong WM - PhD,BAgrSc wheat nutrition
3 Wilson BJ - PhD,MSc,BAgrSc weed control, cereals and oilseeds
4 Woodruff DR - BSc,MAgrSc cereal physiology and modelling

443 Agrostology Section

1 Cameron DG - BScAgr pasture improvement
2 Walker B - PhD,MAgrSc,DipTropAgr,BScAgr *Assistant Director (Agrostology)*

444 At Ayr
P.O. Box 591, Ayr, 4807

1 Pressland AJ - PhD,MSc,BAgrSc pasture production

445 At Bundaberg
P.O. Box 1143, Bundaberg, 4670

1 Lee GR - BScAgr pasture improvement

446 At Charleville Pastoral Laboratory
P.O. Box 282, Charleville, 4470

1 Beale IF - PhD,MAgrSc,BAgrSc *Officer in Charge* semi arid rangelands

447 Brian Pastures Pasture Research Station
via Gayndah, 4625

1 Gramshaw D - PhD,MAgrSc,BScAgr *Officer in Charge* pasture production
2 Orr DM - MAgrSc pasture utilization

448 At Gympie
P.O. Box 395, Gympie, 4570

1 Cook BG - BAgrSc pasture improvement and nutrition
2 Loch DS - PhD,BCom,BAgrSc physiology of pastures

449 At Ipswich
P.O. Box 96, Ipswich, 4035

1 Lowe KF - MAgrSc pasture production

450 At Mackay
P.O. Box 668, Mackay, 4740

1 Bishop HG - BAgrSc,QDA pasture species evaluation

451 At Mareeba
P.O. Box 1054, Mareeba, 4880

1 Gilbert MA - BAppSc pasture species nutrition and evaluation
2 Hall TJ - BAppSc pasture species evaluation
3 Hopkinson JM - PhD,BSc physiology of pasture plants
4 Miller CP - MAgrSc,QDA pasture development

452 At Rockhampton
P.O. Box 689, Rockhampton, 4700

1 Anderson ER - MAgrSc,BAgrSc pasture production, species evaluation
2 Burrows WH - PhD,MAgrSc pasture production, ecology
3 Middleton CH - BAgrSc,QDA species evaluation
4 Scanlan JC - BAgrSc woody weed control

453 South Johnstone Research Station
P.O. Box 20, South Johnstone, 4859

1 Teitzel JK - BAgrSc,QDA pasture nutrition and management

454 At Toowoomba
P.O. Box 102, Toowoomba, 4350

1 Scattini WJ - PhD,MAgrSc pasture management and utilisation
2 Silcock RG - PhD,MSc,BAgrSc pasture production
3 Weston EJ - BAgrSc,QDA resource assessment, pasture improvement

455 At Warwick
P.O. Box 231, Warwick, 4370

1 Clarkson NM - BAgrSc pasture species evaluation

456 Botany Branch
Indooroopilly Laboratories, Meiers Road, Indooroopilly, 4068
Tel: 377 9311.

1 Johnson RW - PhD,MSc *Director* administration of Queensland Herbarium, economic botany, computers
2 Dowling RM - BSc mangrove ecology, economic botany
3 Henderson RJ - MSc,BSc taxonomy
4 McDonald TJ - MSc,BSc resource assessment
5 McDonald WJF - MAgrSc,BAgrSc vegetation mapping
6 Pedley L - BSc *Assistant Director* taxonomy of plants of northern Australia
7 Reynolds Miss ST - MSc,BSc taxonomy
8 Ross Mrs EM - BAppSc taxonomy
9 Simon BK - MSc,BSc taxonomy
10 Stanley TD - BSc taxonomy

457 Entomology Branch
Indooroopilly Laboratories, Meiers Road, Indooroopilly, 4068
Tel: (07) 377 9311.

1 Passlow T - MScAgr,BScAgr *Director*
2 Bengston M - PhD,MSc *Assistant Director* stored product pests
3 Cantrell BK - BSc taxonomy
4 Drew RAI - PhD,BSc fruit fly
5 Galloway ID - PhD,BSc taxonomy
6 Heather NW - PhD,MSc,BSc,QDA fresh fruit disinfestation
7 Houston KJ - MSc,BSc taxonomy
8 Rhodes JW - BSc apiculture
9 Sabine BNE - BSc technical liaison
10 Teakle RE - MSc,BSc insect pathology

458 At Ayr
P.O. Box 591, Ayr, 4807

1 Kay IR - BAgrSc tropical horticulture pests; rice pests

459 At Biloela Research Station
P.O. Box 201, Biloela, 4715

1 Page FD - MAgrSc,BAgrSc sorghum and cotton pests, host plant resistance

460 At Emerald
P.O. Box 81, Emerald, 4720

1 Allsopp PG - MAgrSc,BAgrSc pests of pasture, grain, oilseeds and cotton

461 At Kingaroy
P.O. Box 23, Kingaroy, 4610

1 Rogers DJ - MAgrSc pests of oilseeds and legumes

462 At Mareeba
P.O. Box 1054, Mareeba, 4880

1 Cunningham IC - PhD,MAgrSc,QDA pest management of rice, tobacco and tropical fruits

463 At Nambour and Maroochy Office and Laboratory Complex
P.O. Box 5803, Nambour, 4560

1 Smith D - BSc citrus pests
2 Waite GK - MAgrSc,BAgrSc horticultural crops and pest management

464 At Redlands Horticultural Research Station
P.O. Box 327, Cleveland, 4163

1 Hargreaves JR - MSc,BSc *pest management in vegetables and ornamentals*

465 At Rockhampton
P.O. Box 689, Rockhampton, 4700

1 Elder RJ - MAgrSc,BAgrSc,QDA pasture and grain pests; locusts

466 At Toowoomba
P.O. Box 102, Toowoomba, 4350

1 Franzmann BA - MSc,BSc pest management in vegetables and grain
2 Murray DA - MSc,BSc pest management in oilseeds and cotton
3 Twine PH - PhD,MAgrSc,BAgrSc cotton pests, *Heliothis*

467 Horticulture Branch
G.P.O. Box 46, Brisbane, 4001
Tel: (07) 224 0414.

1 Behncken GM - PhD,BAgrSc *Director of Horticulture*
2 Barke RE - MSc,BSc *Assistant Director (Research)*
3 Carroll ET - BA,BAgrSc nutrition and physiology

468 At Hamilton Laboratory Complex
19 Hercules Street, Hamilton, 4007
Tel: (07) 268 2421.

1 Brown BI - BSc post harvest investigations of fruit and vegetables
2 Peacock BC - BSc storage and transport of fruit and vegetables

469 At Granite Belt Horticultural Research Station
P.O. Box 10, Applethorpe, 4378

1 Dodd BC - BAgrSc deciduous fruits, vegetables

470 Kamerunga Horticultural Research Station
Harley Street, Redlynch, via Cairns, 4872
Tel: (070) 268 2421.

1 Watson BJ - BAgrHort bananas and tropical fruits

471 At Maryborough
P.O. Box 432, Maryborough, 4650
Tel: (071) 268 2421.

1 Chapman JC - BAppSc citrus, sub tropical fruits and vegetables

472 At Nambour and Maroochy Office and Laboratory Complex
P.O. Box 5083, Nambour, 4560

1 Chapman KR - MAgrSc physiology of sub tropical crops
2 Winks CW - BSc plant breeding, sub tropical fruits

473 At Redlands Horticultural Research Station
P.O. Box 327, Cleveland, 4163

1 Gage JF – PhD,MAgrSc,BSc vegetable agronomy
2 Hibberd AM – PhD,BAgrSc plant breeding - vegetable crops
3 Schoorl D – DAgrSc,BAgrSc,QDA container design, transport and storage

474 Plant Pathology Branch

Indooroopilly Laboratories, Meiers Road, Indooroopilly, 4068
Tel: (07) 377 9311.

1 Colbran RC – PhD,MAgrSc *Director*
2 Alcorn JL – PhD,MSc,BSc mycology
3 Dale JL – PhD,BAgrSc virology, virus diseases of sub tropical fruits
4 Diatloff A – MSc rhizobium, pastures and grain legumes
5 Greber RS – MSc virology, horticultural crops and grains
6 Hughes IK – MSc diseases of ornamentals
7 Moffett Miss ML – PhD,MSc bacterial diseases
8 Muirhead IF – PhD,MAgrSc,BAgrSc *Assistant Director* tropical fruit
9 O'Brien PC – PhD,MAgrSc,BAgrSc nematology, grain legumes
10 O'Brien RG – MAgrSc,BAgrSc diseases of vegetables
11 Pegg KG – MSc,BSc diseases of tropical fruits
12 Stirling GR – PhD,MAgrSc nematology, grain and fruits

475 At Granite Belt Horticultural Research Station
P.O. Box 10, Applethorpe, 4378

1 Heaton JB – BSc vegetables, deciduous fruits

476 At Mareeba
P.O. Box 1054, Mareeba, 4880

1 Peterson RA – BagrSc vegetables, tropical fruits

477 At Queensland Wheat Research Institute
P.O. Box 5282, Toowoomba, 4350

1 Rees RG – PhD,MAgrSc,MSc diseases of winter grain crops
2 Thompson JP – PhD,BAgrSc soil microbiology, mycorrhiza, grain crops
3 Wildermuth GB – PhD,MSc,BSc diseases of winter grain crops

478 At Toowoomba
P.O. Box 102, Toowoomba, 4350

1 Dodman RL – PhD,BAgrSc diseases of summer cereals
2 Kochman JK – PhD,BSc diseases of oilseeds

479 Queensland Museum

Gregory Terrace, Fortitude Valley, Brisbane, Queensland, 4006
Tel: (07) 52 2716.

1 Bartholomai A – MSc,PhD *Director* fossil marsupials and reptiles
2 Campbell BM – MSc *Deputy Director* taxonomy, crustacean ecology
3 Jell PA – BSc,PhD *Assistant Director* fossil arthropods and echinoderms

480 Zoology

1 Cannon LRG – BSc,MSc,PhD *Senior Curator, Lower Invertebrates* parasitology
2 Covaacevich JA – BA,MSc *Senior Curator, Reptiles* taxonomy reptilia
3 Dahms EC – BAgrSc,BSc *Senior Curator, Higher Entomology* taxonomy, biology hymenoptera
4 Davie PJ – BSc *Curator, Crustacea* taxonomy brachyra
5 Ingram GJ – BSc *Curator, Amphibians* taxonomy, ecology, amphibians, birds
6 Mather P – PhD,DSc *Senior Curator, Higher Invertebrates* ascidiacea
7 McKay RJ – BA *Curator, Ichthyology* fishes of the Great Barrier Reef, freshwater fishes of Australia
8 Molnar RE – MA,PhD *Curator, mammals* fossil and recent mammals
9 Monteith GB – BSc,PhD *Senior Curator* lower entomology: systematics and biogeography Hemiptera; ecology of rainforest insects
10 Raven RJ – BSc,PhD *Curator Arachnology* taxonomy and ecology of spiders
11 Stanisic J – BSc,MSc *Curator Molluscs* taxonomy and biogeography of terrestrial mollusca

481 Geology

1 Wade MJ – BSc,PhD *Senior Curator, Geology* precambrian fossils; tertiary foraminifera; early palaeozoic nautiloids

482 Anthropology and Archaeology

1 Quinnell MC – MA *Senior Curator, Archaeology and Anthropology* archaeological research

483 History and Technology

1 Agnew NH – MSc,PhD *Conservator* conservation of museum collections

2 Robinson DJ – BSc,PhD *Senior Curator, History and Technology* conservation and history of technology
3 Sanker IG – BSc,DipEd *Senior Curator, Industrial Technology* development of industrial technology

484 Sugar Research Institute

Nebo Road, Mackay, Queensland, 4740
[PO Box 5611, Mackay Mail Centre, Queensland, 4741]
Tel: (079) 521511. Telex: 46312. Cables: Sugasearch.

1 Allen JR – AM,BE,PhD,FTS,FIEAust,MIMechE *Director*
2 Cullen RN – BEMech,MEngSc,FIEAust *Deputy Director*

485 Chemistry and Chemical Engineering

1 Ashbolt NJ – BAgrSc,PhD,MASM chemical engineering
2 Broadfoot R – BEChem,PhD chemical engineering
3 Crees OL – MSc,ARACI chemical engineering
4 Dixon TF – BEMech,PhD,MIEAust,MAIE,MCI chemical engineering
5 Hale DJ – BScApp,BEChem chemical engineering
6 Inkerman PA – BSc,PhD,AAIFST,ARACI chemical engineering
7 Ivin PC – MSc,ASTC,ARACI chemical engineering
8 Watson LJ – BSc,MSc chemical engineering
9 Whayman E – BSc,PhD,ARACI *Chief Chemist*
10 Wright PG – MScApp,PhD,DipSugTech,ARACI *Chief Chemical Engineer*

486 Engineering

1 Edwards BP – BSc,PhD engineering
2 Everitt PG – BEngElec engineering
3 Garson CA – BEMech engineering
4 Loughran JG – BEngMech,MEngSc,MIEAust engineering
5 Mason V – BScEng,PhD,MAAS,MIEAust *Chief Research Engineering*
6 McGinn JA – DipMEE,MIEAust engineering
7 Murry CR – BEMech,PhD,MIEAust,MACS *Chief Planning Engineer*
8 Reichard SR – HND,MIEE,CEng,MIEAust engineering

487 Systems Engineering

1 James RA – DipMEE,MAIEA,FPWI
2 McWhinney W – BScApp,BEChem,MEngSc,PhD,MIEAust,ARACI *Chief Systems Engineer*
3 Pinkney AJ – BSc,BCom,DipInfProc

SOUTH AUSTRALIA GOVERNMENT

488 South Australia Department of Agriculture

25 Grenfell Street, Adelaide, South Australia
[1671, Adelaide, South Australia, 5001]
Tel: (08) 227 3000. Telex: 88422. Cables: SAGRIC.

Research laboratories, Fosters Road, Northfield, South Australia 5085.

1 Harvey PR - BVSc,DVSM,MRCVS *Deputy Director-General*
2 Radcliffe JC - BAgrSc,PhD *Director-General of Agriculture*

489 Policy and Planning Unit

1 Potter JC - BAgrSc *Director Policy and Planning*
2 Crabb DF - BAgrEc *Policy Economist*
3 O'Brien MV - BAgrSc *Senior Policy Adviser (Planning)*
4 Wickes RB - MAgrSc *Principal Officer, Research Management*

490 Division of Animal Services

1 Feagan JT - MScAgr,DipPA *Director Animal Services* administration of division

491 Animal Research Branch

1 Pell AS - BSc,MSc,MAgSc,PhD *Chief Animal Research Branch* administration of research

492 Sheep Research

1 Fleet MR - BAgrSc wool quality
2 James PJ - BAgrSc sheep blowfly
3 Ponzoni RW - BAgrSc,PhD animal breeding
4 Stafford JE - HDA,BRurSc meat sheep breeding
5 Walker SK - MAgrSc reproduction in sheep
6 Walkley JRW - MAgrSc sheep breeding

493 Pig Research

1 Davies RL - MAgrSc pig nutrition
2 Pointon AM - BVSc,MSc pig health
3 Stone BA - BAgrSc,PhD pig reproduction

494 Parafield Poultry Research Centre

230 Salisbury Highway, Parafield Gardens, South Australia, 5107
Tel: (08) 258 1244.

1 Polkinghorne RW - BAgrSc,PhD *Officer-in-charge* poultry breeding
2 Glatz PC - BAgrSc,PhD poultry physiology, poultry behaviour
3 Hughes RJ - BSc egg shell quality

495 Northfield Research Centre

Fosters Road, Northfield, South Australia, 5085
Tel: (08) 266 8333.

1 Bartsch BD - BAgrSc dairy nutrition
2 Graham ERB - MSc,PhD milk proteins, genetic variants
3 McLean DM - BRurSc,MSc milk proteins, genetic variants
4 Munro GL - BSc,DipCompSc herd testing, genetics
5 Round MH - BAgrSc sheep nutrition
6 Valentine SC - BAgrSc pasture utilization, dairy nutrition
7 Zviedrans P - BSc milk proteins, syneresis, genetic variants

496 Animal Health and Regulatory Branch

25 Grenfell Street, Adelaide, South Australia
Tel: (08) 266 8333. Telex: 88422. Cables: SAGRIC.

1 Holmden JH - BVSc *Chief Veterinary Officer* administration of division
2 Bray JH - RDA,BVSc,MACVSc poultry diseases
3 Collard JA - BVSc agriculture and veterinary chemicals
4 Dobson KJ - BVSc,MACVSc *Principal Veterinary Officer* administration of research
5 Newlands RW - BVSc sheep and cattle diseases
6 Whyte PBD - MVSc,MBA pig diseases

497 Central Veterinary Laboratories Branch

I.M.V.S., Frome Road, Adelaide, South Australia, 5000
Tel: (08) 228 7580. Telex: 82647. Cables: Medvet.

1 Davidson PAS - BSc,BVSc,MRCVS *Chief Central Veterinary Laboratories Branch* administration of branch

498 Animal Resources Section

1 Holmes MF - BVSc,MRCVS laboratory animal science
2 Kuchel TR - BVSc,MVS,DVA,MRCVS *Head of Section* animal pain and stress, laboratory animal science

499 Biochemistry Section

1 Judson GJ - BRurSc,Phd *Head of Section* biochemistry
2 Babidge P - BSc,PhD biochemistry
3 Benson TH - MSc,ARACI biochemistry
4 Koh TS - BSc biochemistry

500 Microbiology Section

1 Cargill CF - BVSc,MS,PhD *Head of Section* immunology
2 Davenport Mrs PK - BSc,MASM mycology
3 Surman PG - MSc virology

501 Parasitology Section

1 Beveridge I - BVSc,PhD *Head of Section* parasitology
2 Ford GE - MVSc,PhD,MRCVS parasitology

502 Pathology Section

1 Gardner DE - BVSc,DipMicro,PhD,MACVSc *Head of Section* mammalian pathology
2 Copland MD - BVSc,MACVSc,MRCVS clinical pathology
3 de Saram WG - MVSc,PhD mammalian pathology
4 Giesecke Mrs PR - MVSc,MRCVS mammalian pathology
5 Phillips PH - BVSc,PhD mammalian pathology
6 Tham VL - BVSc,DTVM,MRCVS avian pathology

503 Division of Plant Services

Grenfell Centre, 25 Grenfell Street, Adelaide, South Australia, 5001
Tel: (08) 227 9911. Telex: 82647. Cables: Medvet.

1 Tideman AF - BAgrSc *Director Plant Services* administration of division
2 Boyce KG - MAgrSc,PhD herbage seed physiology, production and certification
3 Till MR - BAgrSc *Acting Chief Plant Resources Branch*
4 van Velsen RJ - MAgrSc,PhD *Chief Agricultural Resources Branch*

504 Plant Resources Branch

Tel: (08) 227 9911.

505 Northfield Research Laboratories

Fosters Road, Northfield, South Australia, 5085
Tel: (08) 266 8333.

1 Alberts E - BSc virology
2 Ali SM - BAgrSc,PhD breeding for resistance to disease in peas
3 Allen PG - MAgrSc assessment of insect damage in pastures
4 Auricht GC - BAgrSc pasture genetic research
5 Bailey PT - MSc,PhD biological control of insect pests
6 Baker GJ - BAgrSc integrated pest control in vines
7 Barr AR - BAgrSc oat breeding; agronomy of oats, triticale and rye
8 Crawford EJ - RDA herbage plant introduction
9 Dahlenburg AP - BAgrSc post harvest fruit research
10 Fischle AR - HDA herbicide evaluation
11 Heap J - BAgrSc weed control in pastures and crops
12 Higgs ED - BAgrSc administration and pasture research
13 Hopkins DC - MAgrSc biological control of sitona weevil
14 Kaehne ID - MAgrSc breeding lucerne
15 Kloot PM - MAgrSc weed control in pastures and crops
16 Madge PE - BAgrSc,PhD pest control of community pests and insects
17 Maier NA - BAgrSc nutrition of horticultural crops
18 Robinson JB - BAgrSc,PhD nutrition of horticultural crops
19 Stephenson DW - BAgrSc weed control in pastures and crops
20 Swincer DE - MAgrSc,PhD horticulture plant improvement, virus indexing
21 Tugwell BL - BAgrSc post harvest physiology, fruit storage and handling
22 Wicks TJ - BAgrSc plant pathology in fruit and vegetables

506 Plant Pathology Unit

Waite Road, Urrbrae, South Australia, 5064
Tel: (08) 79 9416.

1 Dube AJ - BAgrSc,PhD diseases of field crops and pastures
2 Hirsch M - BAgrSc,PhD evaluation of seed treatments
3 McKay A - BAgrSc biology and control of ryegrass toxicity
4 Riley IT - BAgrSc,PhD annual ryegrass toxicity
5 Wallwork H - BSc,MPhil,PhD wheat leaf diseases
6 Whittle PJL - BAgrSc resistance in wheat to root rots

507 Agricultural Resources Branch

1 French RJ - RDA,MAgrSc *Chief Soil Agronomist*

508 Irrigation
25 Grenfell Street, Adelaide, South Australia, 5001
Tel: (08) 227 9911. Telex: 88422. Cables: SAGRIC.

1 Desmier RE - BSc irrigation management
2 Dubois B - RDA water management
3 Hartley RER - BAgrSc,MEngSc land use and hydrology
4 McQuade CV - BSc soil erosion
5 Meissner AP - BAgrSc environment modelling
6 Moore SD - RDA runoff and hydrology
7 Schrale G - IrAgr agro-meteorology

509 Conservation, Survey, Arid Zone

1 Davies G - BRurSc soil conservation
2 Lay BG - MSc arid zone ecology
3 Maschmedt DJ - BAgrSc soil and water conservation
4 Matheson W - MAgrSc soil conservation

510 Cropping

1 Georg D - BAGRSC Grain legume agronomy
2 Gibson PR - MAgrSc quantitative assessment of biological N_2 fixation
3 Heard TG - BAgrSc agronomy of wheat and barley
4 Higgs ED - BAgrSc cropping systema research
5 Wheeler RD - BAgrSc barley agronomy

511 Vertebrate Pest Control

1 Cooke BD - MSc,PhD rabbit control
2 Heinrich KJ - BAgrSc *Chief,Animal and plant control*
3 Henzell RP - BSc,PhD feral goat ecology and biology in arid areas
4 Mutze GJ - BSc mice
5 Sinclair RG - MAgrSc,PhD ecological effects of pest control

512 Northfield Research Laboratories
Fosters Road, Northfield, South Australia, 5085
Tel: (08) 266 8333.

1 Elliot DE - BAgrSc nitrogen and phosphorus nutrition of crops and pastures
2 Fawcett RG - RDA,BAgrSc,PhD tillage practices
3 Hannam RJ - BAgrSc trace element nutrition
4 Heanes DL - BSc methods of plant and soil analysis
5 King PM - MAgrSc water repellence, crop rotations
6 Reuter DJ - MAgrSc *Principal Research Officer* soils, administration of soils research
7 Schultz JE - BAgrSc tillage, crop rotations, nitrogen nutrition of wheat

513 Economics Division

1 Ryland GJ - QDAH,BAgrSc,MEc,PhD *Chief Agricultural Economist* administration of division

514 Economics Services

1 Oborne G - BAgrEc,PhD *Principal Agricultural Economist (Economic Services)* dryland farming systems
2 Hansen BR - BAgrEc dryland farming systems
3 Krause MA - BAgEc,MEc dryland farming systems
4 Mayers BA - BRuSci,MEc computerised farm planning and transportation studies
5 McLean GD - BAgEc horticultural policy
6 Rees RO - BAgrEc commodity forecasting

515 Farm Mangement
1

1 Trengove GC - BAgrSc,DipAgrExt,DipAgrSc *Principal Agricultural Economist (farm management)* administration of farm management
2 Inglis IR - BAgrEc farm management
3 King JL - BAgEc irrigation farm management
4 Saranyawatin P - BScAgr farm mechanisation

516 Mathematics and Computing Services Unit

1 Ellis JV - BAgrSc,BSc *Senior Biometrician* group management, systems support
2 Frenscham AB - BSc experimental design and analysis
3 Gehling WJ - BAppSc,DisCompSc data base design
4 Kenyon RV - BSc,DipCompSc experimental design and analysis
5 McCloud PI - BA,DipCompSc categorical analysis

517 Library
25 Grenfell Street, Adelaide, South Australia, 5001
Tel: (08) 227 3000. Telex: 88422. Cables: SAGRIC.

1 Deane CA - BEc,ALAA *Librarian*

518 Central Region

519 Head Office

1 Webber GD - RDA,HRDE *Chief Regional Officer* administration of region
2 Plowman DJ - MScAg,PhD *Regional Officer Research* administration of research
3 Rogers IS - MScAg vegetables
4 Ronan GS - BRurSci,MEc economics
5 Selinger MT - BScAg,RDA,DipAgExt *Regional Officer Extension* administration of extension

520 Barossa Viticultural Research Centre

1 Brown A *Senior District Officer* administration of research centre
2 Cirami RM - MSc vine improvement varieties, rootstock, grape quality
3 McCarthy MG - MScAg vine improvement, irrigation, vineyard management, grape quality

521 Kadina District Office

1 Crosby J - BScAg farming systems
2 Edwards R - BAgEc economics

522 Lenswood Research Centre

1 Whitehead ME - RDAT *Manager* administration of research centre, irrigation
2 Granger A - BScAg pome fruit
3 James P - RDAT crop improvement
4 Williams CMJ - BScAg,PhD potatoes

523 Turretfield Research Centre

1 Ninnes BA - RDA *manager* administration of research centre
2 Beale PE - MAgrSc,PhD breeding subterranean clover, pasture production
3 Cutten IN - MAgrSc sheep reproduction
4 Gifford DR - BAgrSc beef cattle management and cross breeding
5 Hughes AD - FRSC,ResDipRIC,ARACI ruminant nutrition
6 Kleeman DO - BAgrSc sheep nutrition
7 Little DL - BScAg pasture animal studies
8 Pullman AL - BAgrSc beef cattle
9 Wegener P - BScAg farming systems

524 South East Region
Box 618, Naracoorte, South Australia, 5271
Tel: (087) 647419.

1 Thompson N - DipAgr,DipV&FM *Acting Chief Regional Officer* administration of region
2 Ellis RW - BAgrSc *Principal Research Officer* research administration, beef cattle breeding

525 South East Regional Headquarters
Naracoorte, South Australia, 5271
Tel: (087) 647419.

1 Hawthorne WA - BAgrSc agronomy of protein and cereal crops
2 Hodge TJV - BAgrSc soil acidity, soil fertility
3 Lewis DC - MAgrSc soil fertility
4 MacFarlane JD - BSc trace element nutrition of crops and pastures
5 Martyn RS - BAgrSc agronomy of pastures and horticltural seed crops
6 Potter TD - BAgrSc agronomy of oilseeds, cereals and special crops

526 Struan Research Centre
Box 618, Naracoorte, South Australia, 5271
Tel: (087) 647419.

1 Cooper JD - RDA *Manager* administration of Research Centre
2 Deland MPB - MRurSc beef cattle, cross breeding
3 Earl CR - BAgrSc breeding and nutrition of sheep and beef cattle

527 Kybybolite Research Centre
Kybybolite, South Australia, 5262
Tel: (087) 642002.

1 Brown TH - BAgrSc pasture utilization and sheep management
2 Craig AD - MAgrSc pasture species evaluation for sheep grazing
3 Dunstan EA - BAgrSc administration of Research Centre, sheep fertility and nutrition
4 Pedler AP - BAgrSc lucerne cultivar evaluation, lucerne agronomy, *Rhizobium* biology

528 South East Regional Veterinary Laboratories
Box 618, Naracoorte, South Australia, 5271
Tel: (087) 64 7419.

1 Hindmarsh M - HDA,BVSc,MVS *Head of Laboratory* mammalian pathology

2 Belfield A - BAgrSc bacteriology

529 Murray Lands Region

530 Regional Headquarters
PO Box 411, Loxton, South Australia, 5333
Tel: (085) 847241.

1 Thomas GN - BAgrSc *Chief Regional Officer* administration of region
2 Cook AV - BAgEc,AssocDipVals,MSc economics

531 Loxton Research Centre
PO Box 411, Loxton, South Australia, 5333
Tel: (085) 84 7315.

1 Cole PJ - BAgrSc *Principal Research Officer* irrigation, salinity, research management

532 Crop Production

1 Gallasch PT - RDA,BAgrSc citrus
2 Gathercole FJ - BAgrSc stonefruit, almonds
3 Hobman FR - BSc new horticultural crops
4 Nicholas PR - BAgrSc viticulture
5 Warriner SA - BAgrSc vegetables

533 Irrigation, Salinity and Soils

1 Douglas T - BAgrSc soils
2 Stevens RM - BSc salinity
3 Wood BSC - MSc water management, irrigation

534 Plant Protection

1 Furness GO - MAgrSc entomology
2 Magarey PA - BAgrSc plant pathology
3 Walker GE - BSc,PhD soil pathology

535 Irrigated Crop Management Service

1 Watson KA - BAgrSc irrigation

536 Murray Bridge District Office
Murray Bridge, South Australia, 5253
Tel: (085) 32 2266.

1 McCord AM - BAgrSc *Senior Research Officer* farming systems
2 Cock GJ - BAgrSc water management, irrigation

537 Wanbi Agricultural Centre
Wanbi, South Australia, 5310
Tel: (085) 87 4151.

1 Cawthorne C administration of centre

538 Eyre Region
Regional Headquarters, PO Box 11, Port Lincoln, South Australia, 5606
Tel: (086) 82 0280.

1 Robinson GK - RDA,MScAgrExt *Chief Regional Officer* administration of region
2 Egan JP - BAgrSc *Senior Research Officer (Agronomy)* land use survey, plant nutrition and fertilizer usage

539 Cleve District Office
PO Box 156, Cleve, South Australia, 5640
Tel: (086) 28 2091.

1 Adams A - RDAT monitoring of the cropping system
2 Wetherby KG - MAgrSc land use survey and water repellent sands

540 Minnipa Research Centre
Minnipa, South Australia, 5654
Tel: (086) 80 5104.

1 Holloway RE - RDA *Manager* administration of Research Centre, soil factors affecting crop growth
2 Evans M - BAgrSc,PhD weeds

541 Botanic Gardens and State Herbarium
North Terrace, Adelaide, South Australia, 5000
Tel: (08) 228 2311.

1 Morley BD - BSc,PhD *Director and Secretary to the Board of the Botanic Gardens*

2 McAllister EJ - BSc *Assitant Director*

542 State Herbarium

1 Jessop JP - MSc,PhD *Chief Botanist* flora writing
2 Barker WR - PhD *Botanist* scrophulatiaceae
3 Bell GH - BSc *Botanist* pottiaceae
4 Chinnock RJ - MSc,PhD *Botanist* myoporaceae, ferns
5 Munir AA - BSc,MSc *Botanist* verbenaceae
6 Toelken HR - BSc,PhD *Botanist* crassula, kunzea
7 Weber JZ - DipNatSc,MSc *Botanist* orchidaceae

543 Horticultural Taxonomy and Plant Breeding

1 Christensen TJ - BAppSc *Plant Record Officer* bromeliceae
2 Haegi LAR - BSc,PhD *Horticultural Botanist* solanaceae

544 Technical and Advisory

1 Reichstein Miss TAM - CertHort *Senior Technical Officer* begoniaceae, piperaceae

545 Tree Advisory Officer

1 Whitehill JAE - CertKew,AInstPA garden history

546 Plant Pathology

1 Lee TC - BAgSc *Senior Research Officer* ornamental horticulture; *Phytophthora*

547 Library

1 Denny Miss EG - BA,ALAA

548 The South Australian Museum
North Terrace, Adelaide, South Australia, 5000
Tel: (08) 223 8911.

1 Russell LD - BSc,DiptMEd *Director*
2 Ling JK - MSc,PhD *Head Division of Natural Science*
3 Watts CHS - BSc,DPhil *Chief Scientist, Division of Natural Science*

549 Arachnology & Helminths

1 Lee DC - MSc *Senior Curator* systematics and biology of mites

550 Entomology

1 Gross GF - DSc *Principal Curator* systematics of Australian hemiptera
2 Matthews EG - BA,PhD *Senior Curator* systematics and biology of coleoptera

551 Herpetology

1 Schwaner TD - BSc,MS,PhD *Senior Curator* systematics and ecology of reptiles and amphibians

552 Ichthyology

1 Glover CJM - MSc *Senior Curator* systematics and ecology of freshwater fishes

553 Invertebrates

1 Zeidler W - MSc *Senior Curator* systematics and biology of crustaceans

554 Mammalogy

1 Baverstock PR - BSc,PhD *Principal Scientist* systematics of mammals, evolutionary biology of vertebrates differentiation of important medical and veterinary parasites by biochemical means
2 Kemper CM - BSc,PhD *Curator* systematics and ecology of mammals
3 Smith MJ - BSc,PhD *Scientist* reproduction biology of mammals
4 Watts CHS - BSc,DPhil *Chief Scientist* systematics and ecology of mammals

555 Ornithology

1 Parker SA - BSc *Curator* systematics and ecology of birds

556 Protozoa

1 Adams MA - BSc *Scientist* differentiation of important medical and veterinary parasites by biochemical means

557 Library

1 Anthony MM - MLAA *Librarian* large reference library of zoological, anthropological and mineralogical titles

558 South Australia Woods and Forest Department

PO Box 1604 G.P.O., Adelaide, South Australia, 5001
Tel: (08) 2167211. Telex: AA82231. Cables: Woodforest, Adelaide. FAX
2167246.

1 South PM - BSc,DipFor *Director and Head of Department*

559 Research and Development Division

1 Archer GR - BScFor,MScFor *Research Officer* silviculture
2 Boardman R - BScFor,MScFor *Principal Research Officer* silviculture
3 Boomsma DB - BScFor,PhD *Senior Research Officer* silviculture
4 Cowan RM - BSc,DipFor,DipFor *Assistant Director*
5 Keeves A - BSc,DipFor,DipFor *Principal Forest Management Officer*
6 Leech JW - BScFor,DipFor,MScFor,PhD *Forestry Systems Officer*
  mensuration and management systems
7 Lyons RG - BScFor *Project Officer* utilization
8 Millard IB - BScFor *Forest Resources Officer* mensuration and inventory
9 Zed PG - BScFor *Senior Operations Research Officer* operations research

560 Forest Operations Division

1 Ridley PER - DipAgr,BScAgr *Chief*
2 Bungey RS - BScFor,MScAgr *Forest Management Officer* entomology
3 Gepp BC - BScZoo,MSc *Scientific Officer* wildlife and forest recreation
4 Vacant *Director*

TASMANIA GOVERNMENT

561 Tasmania Department of Agriculture

PO Box 192B, Hobart, Tasmania, 7001
Tel: (002) 30 8022. Telex: TASAG 58333. FAX 34 1335.

1 Smith AN - BSc,MSc,PhD,CChem,FRSC *Director*

562 Animal Services

PO Box 192B, Hobart, Tasmania, 7001
Tel: (002) 30 8022. Telex: TASAG 58333.

1 Midgley CA - BAgrSc,DipAgrExt *Senior* Animal Services research programmes
2 Ryan FB - BVSc *Assistant Director*
3 Witt DJ - BVSc *Senior* Animal Services research programme

563 Extensive Animal Industries Branch

Dept. of Ag., PO Box 180, Launceston South, Tasmania, 7249
Tel: (003) 32 2101.

1 Atkins DC - BAgrSc beef husbandry
2 Banks RG - BAgrSc quantative genetics and population structure of livestock species
3 Griffiths HD - BRurSc beef husbandry
4 Hopkins DL - BAgrSc beef and sheep husbandry, carcase specification
5 King CF - BAgSc *Senior* reproduction and perinatal mortality and behaviour in sheep
6 Reid RND - BAgrSc sheep husbandry
7 Ritar AJ - BScAg,PhD prime lamb production, goat production, artificial breeding
8 Woodburn Miss J - BScAg grazing management

564 Intensive Animal Industries Section

Dept. of Ag., PO Box 180, Launceston South, Tasmania, 7249
Tel: (003) 32 2101.

1 Banks PA - BScAgr,GradDipMgt management and environment, pigs and poultry
2 Bruce JC - CDA,Compound Provender poultry
3 Cowan MR - BEc,CRAC pig nutrition and environment
4 Douglas JB - NDP,NCPH,CDP *Senior*

565 Veterinary Field Branch

Dept. of Ag., PO Box 180, Launceston South, Tasmania, 7249
Tel: (003) 32 2101.

1 Sharman ME - BVSc,QDAH *Chief*
2 Butler AR - BVSc preventive medicine and herd health
3 Elliott J - BVSc *epidemiology*
4 Gregory GG - BVSc *Senior* vertebrate pests
5 McManus TJ - BVSc preventive medicine and arboriviruses
6 Middleton M - BVSc preventive medicine and caseous lymphadenitis
7 Pitt DJ - BVSc preventive medicine and herd health
8 Statham M - BAgrSc,PhD vertebrate pests

566 Dairy Branch

Dept. of Ag., Reece House, Burnie, Tasmania, 7320
Tel: (004) 30 2202.

1 Smith JT - BAgEc,MAgSc *Chief*
2 Fulkerson WJ - BAgrSc,PhD *Senior* dairy husbandry

567 Veterinary Laboratory Branch

Mt. Pleasant Laboratories, PO Box 46, Launceston South, Tasmania, 7249
Tel: (003) 32 2101.

1 Munday BL - BVSc,MVSc,MVASc *Chief* comparative pathology, protozoology
2 Carson J - BSc,PhD microbiology of fish diseases
3 Handlinger Mrs JH - BVSc,PhD diagnostic pathology, fish diseases, reproductive pathology
4 Mason RW - BVSc,MVSc *Senior* diagnostic pathology, trace elements, avian diseases, plant and general poisoning
5 McKenna RP - AAIMLS,AIMLS immuno/serology
6 Morris TM - BScAgr,NIRD bacteriology, dairy products, environmental microbiology
7 Obendorf DL - BVSc,BAnSc,PhD diagnostic pathology and parasitology, wildlife diseases and parasites
8 Peel BF - BAgrSc diagnostic bacteriology
9 Pike BA - BAgrSc *Senior* bacteriology, dairy products, environmental microbiology
10 Veldhuis GAJ - AAIMLS histology

568 Plant Services

PO Box 192B, Hobart, Tasmania, 7001
Tel: (002) 30 8022. Telex: TASAG 58333. FAX 34 1335.

1 Martin EJ - MScAgr *Director*
2 Geard ID - MAgrSc *Assistant Director*
3 Richardson BD - BAgrSc *Senior* Plant Services research programs

569 Fruit and Ornamentals Branch

Dept. of Ag., St. Johns Ave., New Town, Tasmania, 7008
Tel: (002) 28 4851.

1 Hardy RJ - BAgrSc,HDA *Chief*
2 Boucher WD - BAgrSc,PhD stone fruit management systems
3 Cooke Mrs SL - BAgrSc floriculture
4 Jones KM - BSc spray thinning and plant growth regulators
5 Jotic P - BHortSc,DipAg fruit root stocks, pruning, plant density, new varieties
6 Peacock FS - BAgrSc vineyard management, appellation
7 Rapley PEL - BAgrSc berry fruits, nuts
8 Wilson SJ - BAgrSc *Senior* berry fruit, pome fruit, climate
9 Young Miss CE - BAgrSc varieties and management of berry fruit

570 Vegetables and Allied Crops Branch

PO Box 303, Devonport, Tasmania, 7310
Tel: (004) 24 0201.

1 Frappell BD - BAgrSc,MSc *Chief*
2 Beattie BM - BAgrSc cultivar evaluation, plant improvement
3 Chapman KSR - BAgrSc,PhD post harvest physiology
4 Chung B - BAgrSc,MPhil crop water requirements
5 Forbes JJ - HDA,BAppScBiol cultivar evaluation
6 Laurence RCN - BSc,PhD *Senior* management research programs
7 Walker MR - DipAg *Senior* management research programs

571 Chemistry and Soils Section

Mt. Pleasant Laboratories, PO Box 46, Launceston South, Tasmania, 7249
Tel: (003) 32 2101.

1 Bastick CH - BAgrSc,DipAgrEng soil drainage, management
2 Pinkard GJ - BAgrSc soil conservation
3 Temple-Smith MG - BAgrSc,PhD *Senior* soil fertility, plant nutrition
4 Wright DN - BAgrSc soil structure, management

572 Pastures and Field Crops Branch

Mt. Pleasant Laboratories, PO Box 46, Launceston South, Tasmania, 7249
Tel: (003) 32 2101.

1 Stephens JG - BSc *Chief*
2 Carpenter JA - MScAgr,PhD *Senior* pasture species
3 Hart MA - MAgrSc essential oils
4 Johnson DE - BAgrSc pasture agronomy
5 McCutcheon RH - HDA,GDE *Senior* pasture and crop agronomy
6 Mitchell PJ - BScAgr management and nutritive values, pastures
7 Orr RJ - BAgrSc pasture and field crop agronomy
8 Reid R - BA pasture agronomy
9 Rowe BA - BAgrSc pasture nutrition and establishment
10 Russell J - BSc field crop agronomy
11 Secomb DHR - RDA pasture and field crop agronomy
12 Smith RS - MScAg,DipEd small seed production and lucerne agronomy
13 Talay PR - BAgrSc pasture and field crop agronomy
14 Vertigan WA - BSc cereal breeding and grain legumes

573 Weeds Section

Mt. Pleasant Laboratories, PO Box 46, Launceston South, Tasmania, 7249
Tel: (003) 32 3101.

1 Hyde-Wyatt BH - BSc,ARCS *Senior* weeds
2 Friend DA - BAgrSc,PhD biology and ecology of weeds
3 Harradine AR - BScAgr,PhD weeds of pasture forage and perennial horticulture crops
4 Holmes JE - BScAg weeds of vegetable crops
5 Kidd Ms SJ - BAgrSc weed control

574 Entomology Section

Dept. of Ag., St. Johns Ave., New Town, Tasmania, 7008
Tel: (002) 28 4851.

1 Hill L - BSc pests of vegetables and field crops, taxonomy of Hemiptera
2 Ireson JE - BAgrSc biological control of ragwort, ecology of pasture and taxonomy of Collembola
3 Leighton Miss SM - BSc insect damage thresholds in pasture, pests of field crops
4 McQuillan PB - BAgrSc ecology of pasture insects, taxonomy Lepidoptera and Scarabaeidae
5 Rath AC - BSc,MASM insect pathology
6 Terauds A - BSc *Senior* orchard pests, integrated control, fruit disinfestation

7 Williams Mrs MA - MSc orchard mites, taxonomy of Phytosciidae and Eriophiidae

575 Plant Pathology Section
Dept. of Ag., St. Johns Ave., New Town, Tasmania, 7008
Tel: (002) 28 4851.

1 Grice Ms MS - BAgSc diseases of field crops and potatoes
2 Guy PL - BSc,PhD plant virus diseases
3 Johnstone GR - BAgSc,PhD virus diseases of vegetables, field crops and pasture plants
4 Munro D - BAgSc serology, electron microscopy of viruses
5 Sampson PJ - BAgSc *Senior* diseases of potatoes and pome fruits
6 Wiberg Miss LM - MSc vegetable diseases
7 Wong JA-L - BSc,PhD *Sclerotinia* diseases, onion white root rot, apple core rot

576 Resources
G.P.O. Box 192B, Hobart, Tasmania, 7001
Tel: (002) 30 8022. Telex: TASAG 58333.

1 Vacant *Assistant Secretary*

577 Agricultural Resources
G.P.O. Box 192B, Hobart, Tasmania, 7001
Tel: (002) 30 8022. Telex: TASAG 58333.

1 Williams PF - BAgrSc,MAgrSc,MSocScAdmin *Senior* management of research stations and external research funds

578 Computing and Biometrics
G.P.O. Box 192B, Hobart, Tasmania, 7001
Tel: (002) 30 8022. Telex: TASAG 58333.

1 Brook PWJ - BSc,DipCS *Senior* computing and related research
2 Koen TB - BSc,AgrSc biometrics, research project proposal scheme
3 Wallace SP - BSc biometrics, research project proposal scheme

579 Planning and Co-ordination
G.P.O. Box 192B, Hobart, Tasmania, 7001
Tel: (002) 30 8022. Telex: TASAG 58333.

1 Edwards GC - BAgEc,MEc,DDA *Assistant Secretary*
2 Mucha Ms CDS - BAgrSc,PhD *Research Co-ordinator* research project proposal schemes

580 Tasmanian Forestry Commission
PO Box 207B, Hobart, Tasmania, 7001
Tel: (002) 308033. Telex: TASFOR AA57204.

1 Vacant *Chief Commissioner*
2 James JO - BLLB *Commissioner Commercial*
3 Skuja AG - BSc,DipFor *Commissioner Management*
4 Watts DR - BSc,DipFor *Commissioner Private Forests and Operations*

581 Division of Engineering and Operations

1 Rushton J - DipStrucEng,MIAAust *Chief of Division*

582 Fire Management Branch

1 Mount AB - BFor,DipFor,MSc *Branch Head*

583 Division of Forest Management

1 Higgs ML - BSc,PhD *Chief of Division*
2 Candy SG - BSc,DipBiom forest biometrics
3 Vacant *Principal Research Officer*

584 Division of Marketing

1 Gardner RG - BFor,DipFor *Chief of Division*

585 Division of Private Forestry

1 DeClancy-Walsh JP - BA *Chief of Division*

586 Commercial Development Unit

1 Faircloth PL - BSc,DipEarthSc,MNatRes *Manager of Commercial Development*
2 Davies RJ - DipAg,DipAMBA,MEnvSt wood energy
3 Inglis DJ - BEc,AASA Forest Economics
4 Osborn TE - BSc,DipOR linear programming, harvest scheduling

587 Silvicultural Research and Development Branch

1 Felton KC - BSc,MF *Branch Head*
2 Elliott HJ - BSc,DipAgrEnt,PhD *Forest Entomologist*
3 Goodwin AN - BScFor eucalypt silviculture (on study leave)
4 Hickey JE - BFofSc rainforest silviculture
5 Lockett EJ - BSc eucalypt silviculture
6 Neilsen WA - BSc,MFS plantation silviculture
7 Palzer C - BFor,DipFor,PhD co-ordination of intensive eucalypt management research
8 Wardlaw TJ - BSc *Forest Pathologist*
9 Wilkinson GR - BSc plantation silviculture

588 Library

1 Anstie C - BSc,GradDipLib *Librarian*

589 Queen Victoria Museum and Art Gallery
Wellington Street, Launceston, Tasmania, 7250
Tel: (003) 316777.

1 Green RH *Curator of Zoology* biological surveys of Tasmania and off-shore islands. Extended study of effects on native fauna of clear felling forest management
2 Tassell CB - MSc *Director and Editor* Records of the Queen Victoria Museum and Art Gallery

VICTORIA GOVERNMENT

590 Department of Agriculture and Rural Affairs, Victoria

166 Wellington Parade, PO Box 500, East Melbourne, Victoria, 3002
Tel: (03) 651 7011. Telex: AA34261. FAX 03 4175440.

591 Head Office

592 Executive

1 Wright JJ - BSc,MEngSc *Chief General Manager*
2 Hore DE - BVSc,PhD,MACVSc *General Manager Operations Division*
3 Taylor MG - BAgrSc,DipAgrEcon *General Manager Rural Policy and Marketing Division*
4 Vorghese J - BA *General Manager Corporate Resources Division*

593 Rural Policy and Marketing Branches

594 Policy Development

1 Patton CT - BAgrSc,DipAgrExt *Manager Policy Development*
2 Alston J - BAgrSc,MAgrSc,PhD *Chief Policy Analyst, Economics*
3 Bailey PJ - MAgrSc *Chief Policy Analyst, Industries*
4 Elliott BR - BAgrSc,MAgrSc *Principal Analyst, Extensive Crop Products*
5 Shea PF - BAgrSc,MSc,PhD *Principal Analyst, Intensive Crop Products*
6 Sullivan J - BAgrSc,MBA,PhD *Principal Analyst, Dairying Products*
7 Truscott T - MAgricSc,PhD *Principal Analyst, Extensive Livestock Products*

595 Office of Rural Affairs

1 McClelland V - BAgrSc,MAgrSc *Manager, Office of Rural Affairs*

596 Marketing Branch

1 Naughtin J - BAgrSc,MAgrSc *Manager Marketing*
2 Berghofer DF - BAgrSc,DipAgrEcon *Manager Marketing Services*
3 Cadman R *Manager Marketing Projects*
4 Ruello N *Manager Fish Marketing*

597 Operations Branch

176 Wellington Parade, PO Box 500, East Melbourne, Victoria, 3002

1 Craven J - BVSc,MVSc,PhD *Regional Manager Inner Melbourne Region*
2 Crosbie SF - BSc,PhD *Manager Biometric Services*
3 Goddard ME - BVSc,PhD *Director Livestock Improvement (Acting)* genetics
4 Harford B - BAgrSc,MAdmin *Regional Manager North West Region*
5 Jones LP - BSc,PhD *Senior Livestock Improvement Statistician* animal breeding
6 Lawson RA - BAgrSc,MAgrSc,PhD *Regional Manager Gippsland Region*
7 McArthur J - BAgrSc,MAgrSc *Regional Manager North Central Region*
8 Muir B - BAgrSc *Regional Manager South West Region*
9 Reeves T - BSc,MAgrSc *Regional Manager North East Region*
10 Youl BS - DipAgrSc,BAppSc *Livestock Improvement Officer* dairy farm management

598 Victoria Animal Research Institute

Princes Highway, Werribee, Victoria, 3030
Tel: (03) 741 1300. Telex: AA38507.

1 Sharkey MJ - BAgrSc,PhD *Director*

599 Animal Physiology

1 Davis IF - BAgrSc reproduction/nutrition
2 Dolling M - BAgrSc,PhD hormone control and measurement of wool growth
3 Parr RA - MAgrSc reproduction-embryo and lamb survival
4 Tilbrook AJ - BAgrSc,PhD product development for chemical castration;animal welfare
5 Williams A - BAgrSc out-of-season breeding and wool growth

600 Animal Behaviour/Welfare

1 Winfield CG - MSc *Head of Section* animal behaviour, management, welfare
2 Barnett JL - BSc,PhD stress physiology and welfare
3 Cronin GM - MAgrSc,PhD pig behaviour and welfare studies
4 Eldridge GA - BAgrSc cattle handling, marketing
5 Hemsworth PH - BAgrSc,PhD pig management, behaviour

601 Ruminant Nutrition

1 Hodge RW - BAgrSc,PhD *Head of Section* nutrition and growth

2 Birrell H - MAgrSc pasture utilisation, sheep
3 McGregor BA - BAgrSc nutrition of goats
4 Watson MJ - BAgrSc nutrition intensively managed livestock

602 Intensive Industries

1 Taverner MR - BAgrSc,PhD *Head of Section* pig nutrition
2 Abu-Serewa S - BAgrSc,MSc nutrition of laying hens
3 Campbell RG - BAgrSc,PhD pig growth and development
4 Johnson RJ - BAgrSc,PhD nutrition and physiology of poultry
5 Karunajeewa H - BAgrSc,PhD nutrition of laying hens
6 King R - BAgrSc,PhD sow nutrition and reproduction

603 Agricultural Systems

1 White DH - MAgrSc,PhD,DipElecComp *Head of Section* simulation, analyses and management
2 Bowman PJ - MAgrSc simulation pastoral systems
3 McLeod C - MAgrSc simulation sheep systems
4 Weber K - BAgrSc,PhD,DipEd simulation cattle systems

604 Gilbert Chandler Institute of Dairy Technology

Princes Highway, Werribee, Victoria, 3030
Tel: (03) 7411300. Telex: AA 38507.

1 Mein GA - BAgrSc,MAgrSc,PhD *Director*
2 Broome MC - BAgrSc,MAgrSc,DipEd starters, cheese, cultured products, whey utilisation
3 Brown MR - DipAgSc,BAgrSc milking machines, milking management and mastitis
4 Craven HM - BSc dairy factory interlaboratory tests
5 Dixon BD - DipAgr,BAgrSc food processing
6 Furphy JR - BAgrSc milking machines, milking management and mastitis
7 Girguis N - BAgrSc,MAgrSc starters, cheese, cultured products, whey utilisation
8 Hickey MW - MAppSc,PhD starters, cheese, cultured products, whey utilisation
9 Krause D - BSc starters, cheese cultured products, whey utilisation
10 Kyle WSA - BSc,PhD dairy nd vegetable blends, protein, genetic engineering
11 Roginski H - MAppSc,PhD food science - starters, cheese, cultured products, whey utilisation
12 Versteeg C - MSc,PhD food processing
13 Yu RS-T - BSc,MSc,PhD dairy and vegetable blends, protein, genetic engineering

605 Animal and Irrigated Pastures Research Institute

Cooma Road, RMB 3010, Kyabram, Victoria, 3260
Tel: (058) 522 4881. Telex: AA38503.

1 Mason WK - BAgrSc,PhD *Director(Acting)*

606 Animal Production

1 Stockdale LR - BAgrSc,MAgrSc *Head of Section (Acting)* dairy cattle nutrition
2 Etheridge MO - BAgrSc digestive physiology and silage utilization
3 Lemerle C - BSc,MSc digestive physiology in ruminants
4 Moran JB - BRurSc,MRurSc,PhD dairy cattle nutrition, veal production

607 Plant Production

1 Mason WK - BScAgr,PhD *Head of Section*
2 Blaikie SJ - BAgrSc agronomy and physiology of irrigated perennial pastures
3 Kelly KB - BAgrSc agronomy and physiology of irrigated annual pastures
4 Mundy GN - DipAgrSc,BAgrSc,MAgrSc N cycling irrigated pastures
5 Pritchard KE - DipAgr fodder crop agronomy
6 Slarke RH - DipAgrSc fodder crop agronomy
7 Small DR - BAgrSc,MEnvSc soil management; nutrient cycling

608 Victoria Dairy Research Institute

Hazeldean Road, Ellinbank, Victoria
[RMB 2460, Warragul, Victoria, 3820]
Tel: (056) 261 209. Telex: AA37258.

1 Rogers GL - MAgrSc,PhD *Director (Acting)* ruminant nutrition
2 Grainger C - MAgrSc,PhD *Head of Section* dairy production computer modelling, dairy cow nutritional requirements
3 McGowan AA - MAgrSc,PhD *Head of Section* pasture production
4 Clarke T - DipAppChem analytical chemistry, pasture production and utilisation
5 Gourley CJ - BAgrSc soil fertility and soil physics
6 Moate PJ - MAgrSc dairy cow nutrition, mineral nutrition
7 O'Brien G - BAgrSc productivity of pasture legumes
8 Stewart J - MAgrSc grazing management
9 Thomas GW - BAgrSc pasture and conserved fodder management systems

609 Victoria Pastoral Research Institute

Mt. Napier Road, Hamilton, Victoria
[PO Box 180, Hamilton, Victoria, 3300]
Tel: (055) 730 900. Telex: AA55879.

1 Obst JM - RDA,MRurSc,PhD *Director* trees, crops, pasture, sheep and cattle
2 Bird PR - MDA,BSc(Agr),PhD *Head of Section* agroforestry, cattle nutrition and growth
3 Cummins LJ - BSc,MVSc,PhD *Head of Section* sheep and cattle breeding and reproduction
4 Foot JZ - MAgrSc,PhD *Head of Section* sheep nutrition and growth
5 Reed KFM - MAgrSc,PhD *Head of Section* plant production and use
6 Brockhus MA - BAgrSc animal nutrition
7 Cayley JW - BAgrSc pasture management
8 Chin JF - DipAgrSc pasture plant breeding
9 Clarke SG - BAgrSc pasture plant breeding
10 Cumming KM - DipAgrSc agroforestry
11 Cunningham PJ - BAgrSc pasture plant breeding
12 Flinn PC - ARNIT,AppChem chemist
13 Graham JF - DipAgrSc livestock breeding
14 Morgan JH - BSc,MRuralSc livestock breeding
15 Quigley PE - BAgrSc pasture management
16 Saul GR - DipAgrSc animal nutrition
17 Spiker SA - DipAgrSc livestock breeding
18 Thompson RL - DipAgrSc animal nutrition

610 Victoria Veterinary Research Institute

Park Drive, Parkville, Victoria, 3052
Tel: (03) 347 2322. Telex: AA38483.

611 Pathobiology

1 Barr DA - BVSc,MSc,PhD epidemiology, avian virology
2 Finnie JW - BVSc,MSc cns pathology, histopathology
3 Condron RJ - BVSc,DipBact ultrastructural pathology
4 Davies D - BSc,PhD molecular biology and bacterial genetics
5 Forsyth WM - BVSc bacteriology, histopathology
6 Milner AR - BSc,MSc,PhD parasite immunity and immunochemistry
7 Scott PC - BVSc,BSc molecular biology, avian and mammalian pathology

612 Attwood Veterinary Research Laboratory

Mickleham Road, Westmeadows, Victoria, 3047
Tel: (03) 333 1200. Telex: AA38484.

1 Stephens L - DipAgrSc,BVSc,MSc,PhD *Director*
2 Allen J - BVSc,PhD biochemistry
3 Azuolas J - BAppSc biochemistry
4 Campbell J - MAgrSc arboviruses
5 Campbell NJ - BVSc,PhD,MACVS parasitology
6 Chander D - MAgrSc,PhD microbiology
7 Chappel RJ - BAgrSc,PhD immunochemistry
8 Davis G - BAgrSc arboviruses
9 Davis K - BVSc virology
10 Denholm LJ - DipagrSc,BVSc,Hons,Phd mineralised tissues and metabolic bone disease, ovine peridontal disease
11 Hennessy DP - PhD,MAgrSc,BAgrSc reproductive pathology
12 Krautil F - BSc bacteriology
13 Presidente P - BSc,PhD parasitology
14 Rogerson BA - NatCertChem serology
15 Smith M - MSc antibiotic residues

613 Regional Veterinary Laboratory, Bairnsdale

Calvert Street, Bairnsdale, Victoria, 3875
Tel: (051) 522 751. Telex: AA55103.

1 Button C - BVSc *Acting Director* veterinary diagnostic pathology
2 Barton NJ - BAgrSc,MAgrSc,DDA protozoal abortion, anthelmintic resistance, control of nematode parasites
3 Browning JW - DipAppScMedTech veterinary microbiology, mastitis, antibiotic resistance
4 Colvin N - BSc biochemistry
5 Malecki JC - BSc immunocastration
6 Slee KJ - BAppScMedTech veterinary microbiology, yersiniosis in cattle
7 Stilbeck N - DipAppSc Johnes disease
8 Thomas K - BSc,PhD facial eczema, thiamine deficiency

614 Regional Veterinary Laboratory, Benalla

Hume Highway, Benalla, Victoria
[PO Box 388, Benalla, Victoria, 3672]
Tel: (057) 62 2933. Telex: AA56174.

1 Nicholls TJ - BVSc,MSc,PhD *Director*
2 Barton MG - DipMedLabTech microbiology
3 Dawkins H - BSc,PhD immunochemistry
4 Ferrier G - BSc,MSc biochemistry, immunochemistry
5 Gudkovs N - BSc aquatic parasitology and microbiology
6 Humphrey JD - BVSc,MSc,PhD aquatic pathobiology

7 Johnson B - BSc,PhD immunochemistry
8 Lancaster M - BVSc veterinary diagnostic pathology
9 Langdon JS - BVSc,PhD aquatic pathobiology
10 Mackie J - BVSc,PhD veterinary diagnosis
11 Norris M - BSc serology
12 Overend D - BAgrSc,MSc,PhD veterinary parasitology
13 Patten B - BVSc,PhD immunochemistry
14 Paynter D - BSc,PhD biochemistry
15 Rahaley RS - BVSc,MSc,PhD,NACVSc,DIPACUP veterinary pathology
16 Smits B - BVSc veterinary diagnosis
17 Spencer TL - BSc,PhD immunochemistry
18 Whiting T - BAppSc immunochemistry

615 Regional Veterinary Laboratory, Bendigo

Midland Highway, Bendigo, Victoria
[PO Box 125, Bendigo, Victoria, 3550]
Tel: (054) 484 505. Telex: AA38502.

1 Jones RJ - BVSc,MSc *Director*
2 Fahy VA - BVSc,PhD *Head, Porcine Research Unit*
3 Callinan APL - MAgrSc,PhD parasitology, epidemiology
4 Connaughton I - MAppSc neonatal mortality
5 Driesen SJ - DipAgrSc neonatal mortality, reproductive physiology
6 Hale CJ - FIMLT porcine virology
7 McCormick BM - DipBiochem,MAppSc porcine virology
8 McKenchie K - BSc microbiology
9 Millar BD - BSc,MSc bovine and porcine leptospirosis
10 Monckton RP - BSc,PhD gene probe research
11 Sims LD - BVSc porcine pathology
12 Spicer EM - MAppSc neonatal mortality, reproductive physiology
13 Williamson PL - BAppScBiol neonatal mortality

616 Regional Veterinary Laboratory, Hamilton

Ballarat Road, Hamilton, Victoria
[PO Box 406, Hamilton, Victoria, 3300]
Tel: (055) 730 700. Telex: 55878.

1 Riffkin GG - BVSc,MRCVS,PhD *Director* immunoparasitology
2 Beers PT - BVSc,PhD *Veterinary Pathologist* campylobacters
3 Edwards LD - DipMedTech bacteriology
4 McGillivery DJ - BSc,MAgrSc molecular epidemiology, footrot
5 Walsh JR - DipAppBSc,MSc peptides
6 Webber JJ - BVSc,MSc,PhD *Senior Veterinary Pathologist* bovine infertility, bacteriology
7 Yong WK - BSc,PhD *Senior Scientific Officer* immunoparasitology, footrot

617 Victoria Crops Research Institute

Natimuk Road, Horsham, Victoria, 3002
[Private Bag 260, Horsham, Victoria]
Tel: (053) 823141. Telex: AA55880.

1 Barber DA - BSC,PhD *Director*

618 Agronomy

1 Amor RL - MAgrSc,PhD *Head of Section* weeds
2 Cawood RJ - BAgrSc,MAgrSc soil/plant water relations
3 Flood R - LDA,BSc,MAgrSc,PhD crop phenology
4 Ford GW - BAgrSc,PhD soil agronomy
5 Fulton M - MAgrSc field crop agronomy
6 Gardner WK - BAgrSc,PhD field crop agronomy
7 Mahoney JE - BA,PhD grain legume agronomy
8 O'Leary G - BAgrSc,MAgrSc field crop agronomy

619 Breeding

1 Brouwer JB - MAgrSc,PhD grain legume breeding
2 Ellis SE - BAgrSc barley breeding
3 Joanides C - BAgrSc oat and triticale breeding
4 Martin PJ - BAgrSc wheat breeding
5 Salisbury PA - BAgrSc oilseed breeding
6 Young RM - BAgrSc wheat breeding

620 Cereal Chemistry

1 Archer MJ - BAgrSc,MAgrSc *Senior Cereal Chemist*
2 Panozzo J - BSc cereal chemistry

621 Plant Pathology

1 Kollmorgan JF - MAgrSc,PhD,DipEd *Senior Plant Pathologist*
2 Ballinger DJ - DipAgrSc plant pathology
3 Bretag TW - BAgrSc,MAgrSc plant pathology
4 Brown JS - BAgrSc,MAgrSc plant pathology
5 Cotterill P - BAgrSc root diseases
6 Eastwood R - BAgrSc nematology

622 Victoria Horticultural Research Institute, Knoxfield

PO Box 174, Ferntree Gully, Victoria, 3156
Tel: (03) 2209222. Telex: AA 38482.

1 Frith GJT - MAgrSc,PhD *Director*

623 Fruit Crops Section

1 Thompson WK - DipHort,MAgrSc *Head of Section* stone and pome fruit
2 Clayton-Greene K - MSc,PhD tree training, blueberries
3 El Zeftawi BM - MAgrSc raspberries, other fruit
4 Jager L - DipAgr pome fruit varieties
5 Miller P - BAgrSc growth regulators
6 Morrison B - DipAgr stone fruit, strawberries
7 Raff JW - DipAgrSc,PhD pollination, biotechnology
8 Thompson G - BAgrSc propagation, nut crops

624 Postharvest Section

1 Peggie ID - BAgrSc *Head of Section* packaging, transport
2 Brohier RL - DipAgrSc fruit maturity
3 Cumming BA - DipAgrSc cooling, packaging, transport
4 Dooley LB - BSc,MAgrSc,PhD postharvest physiology
5 Goubran FH - BAgrSc new fruits, cool storage
6 Little CR - DipAgrSc,DipAgrEng cool storage
7 Tomkins RB - BSc vegetable storage

625 Ornamentals Section

1 Richards D - MAgrSc *Head of Section* hormones, roots
2 Beardsell DV - MAgrSc reproductive biology, selection
3 Faragher JD - BAgrSc,MSc,PhD cut flowers, postharvest physiology
4 Hanger B - MAgrSc,PhD nutrition, hydroponics
5 Hutchinson JF - DipAgr tissue culture
6 James EA - BAgrSc nutrition, media
7 Wilkinson RI - DipAgr growth regulators

626 Institute for Irrigation and Salinity Research

Ferguson Road (Private Bag), Tatura, Victoria, 3616
Tel: (058) 249222. Telex: 38504.

1 Jerie RH - BAgrSc,PhD *Director (Acting)*

627 Irrigated Field Crops

1 Smith C - BSc,PhD *Head of Section*
2 Gyles OA - BSc agronomy
3 Olsson KA - PhD soil/water
4 Schultz M - BSc agronomy
5 Taylor AJ - DipHort lucerne
6 Tisdall JM - BAgrSc,PhD soil biology
7 Ugalde D - BAgrSc,PhD wheat quality
8 Whitfield D - BAgrSc,MAgrSc,PhD agronomy
9 Wilson IB - DipAgrSc soil/water

628 Horticultural Crops

1 Jerie PH - BAgrSc,PhD *Head of Section*
2 Ashcroft W - BSc tomato agronomy
3 Barrass I - BSc entomology
4 Dann IR - BAgrSc physiology
5 Flett SP - BAgrSc pathology
6 Issell LG - DipHort plant propagation
7 Mitchell PD - DipHort irrigation
8 Van Den Ende B - DipHort fruit crop horticulture

629 Salinity Research

Ferguson Road (Private Bag), Tatura, Victoria, 3616
Tel: (058) 241344. Telex: 38504.

1 West DW - BAgrSc,PhD *Head of Section*
2 Daamen C - BSc soil-water modelling
3 Heuperman AF - MScAgrEng irrigation and drainage
4 Jordan P - BAppSc *Chemist*
5 Lyle CW - BEngAgr,MEngSc soil hydrology
6 Mehanni AH - MScAgr,BSc soil chemistry
7 Myers B - BAgrSc,PhD agronomy
8 Noble CL - BAgrSc,PhD plant physiology
9 Prendergast JB - BAgrEng soil hydrology
10 Rengasamy P - MSc,PhD soil chemistry
11 Repsys AP - DipAgrSc irrigation and drainage

630 Victoria Plant Research Institute

Burnley Gardens, Swan Street, Burnley, Victoria, 3121
Tel: (03) 8101511. Telex: AA38481.

1 Smith PR - MAgrSc,PhD *Director (Acting)*

631 Entomology

1 MacFarlane JR - BAgrSc,PhD *Head of Section* horticultural entomology
2 Ridland PM - MAgrSc *Head of Section* field crops entomology
3 Williams P - BSc,DIC,PhD *Head of Section* pasture and stored products entomology
4 Berg GN - MAgrSc pasture entomology
5 Comery J - BSc diagnostics, field crops entomology
6 Goodman R - DipHort apiary
7 Horne P - BSc,PhD pesticides
8 McDonald G - BSc,MAgrSc field crops entomology
9 Mekhamer M - BAgrSc quarantine pests
10 Oldroyd B - BScAgr,PhD apiculture
11 Osmelak J - BSc horticultural entomology
12 Reinganum C - MSc invertebrate virology
13 Smith AM - BSc field crops entomology
14 Stahle PP - BSc,PhD pasture entomology
15 Williams D - BAppSc integrated pest management

632 Plant Pathology

1 Merriman PR - DIC,PhD,MIBiol *Head of Section* mycology
2 Sward RJ - BSc,PhD *Head of Section* virology
3 Beetham PR - BSc viruses of sweet potato
4 Clarke RG - DipAgr diseases of field crops and pastures; diagnostics
5 de Boer RF - BSc cereal diseases
6 Garrett RG - MSc,PhD viruses of legumes
7 Greenhalgh FC - BAgrSc root rots of pasture legumes
8 Guy GL - DAppBiol,BAgrSc quarantine plant pathology
9 Holland SA - BAgrSc viruses of stone fruits and grapevines
10 Mason AJ - BAgrSc viruses of potatoes
11 Maughan J - MAgrSc diseases of vegetables - diagnostics
12 Moran JR - BSc viruses of ornamentals
13 Osborn RK - LDA pathogen-tested potatoes and ornamentals
14 Pascoe IG - DipAgrSc,BSc fungal taxonomy
15 Porter IJ - BAgrSc soil solarization
16 Revill P - BSc fruit variety assessment
17 Richardson CV - DipAgrSc diseases of ornamentals - diagnostics
18 Washington WS - BAgrSc,MAgrSc diseases of horticultural crops - diagnostics
19 Whan JH - DDA,MAgrSc,PhD application of fungicides
20 Wimalajeewa DLSC - BSc,PhD bacteriology
21 Winoto-Suatmadji R - DrinNematology ornamentals and diagnostics

633 Seed Testing Station

Burnley Gardens, Swan Street, Burnley, Victoria, 3121
Tel: (03) 8101511. Telex: AA38481.

1 Williams A - BAgrSc *Head of Section*
2 Mebalds M - BSc seed pathology

634 Rutherglen Research Institute

Chiltern Valley Road, Rutherglen, Victoria, 3685
Tel: (060) 329208. Telex: AA56175.

1 Coventry DR - MAgrSc,PhD *Director*

635 Plant Production

1 Brooke HP - MAgrSc,PhD *Head of Section* plant production
2 Alger DJ - DDA field crops, breeding
3 Code GR - DipAgrSc weeds
4 Ellington A - MSc soil fertility
5 Haines PJ - BAgrSc field crop studies
6 Hirth JR - MAgrSc pasture research, sub-clover
7 Milne J - BAppAgr weeds
8 Roberts AM - BAgrSc pasture
9 Slatley JF - BAppSc pasture
10 Slattery WJ - MSc *Chemist*
11 Steed GR - PhD,BAgrSc soil water relationships, soil physics

636 Livestock Production Section

1 Thatcher LP - MAgrSc *Head of Section* livestock production
2 Currie JR - BRurSc,PhD animal behaviour
3 Hamilton D - MAgrSc beef cattle research
4 Kenney PA - MAgrSc grazing behaviour, nutrition
5 Reeve JL - BAgrSc sheep physiology
6 Robert GR - DipAgrSc production systems

637 Sunraysia Horticultural Centre, Irymple

Koorlong Avenue, Irymple, Victoria
[PO Box 460, Irymple, Victoria, 3498]
Tel: (050) 245603. Telex: AA55558.

1 Campbell MM - MAgrSc,PhD *Manager*
2 Taylor BK - BAgrSc,PhD *Manager Research and Development*

638 Plant Protection Section

1 Buchanan GA - MSc *Head of Section* entomology

2 Edwards ME - BAgrSc nematology, post harvest physiology
3 Emmett RW - BAgrSc plant pathology
4 Predebon S - BAgrSc post harvest
5 Stirrat S - BSc pest management programs

639 Viticulture Section

1 Hayes R - BSc dried fruit
2 Johns R - DipAgrSc dried fruit
3 Leamon K - DipAgr *Head of Section*
4 Schache M - BSc dried fruit
5 Whiting J - DipAgrSc,BSc wine grapes

640 Citrus, Tree Crops and Vegetables Section

1 Blennerhassett R - DipAgrSc citrus
2 Dimsey R - BAgrSc vegetables
3 Madge D - BAgrSc citrus
4 Thornton I - DipHort *Head of Section*

641 Irrigation/Salinity

1 Dale MO - MPhil *Head of Section* irrigation management
2 Brown M - BSc,DipEd salinity
3 Cummins T - BAppSc irrigation
4 Nagarajah S - PhD irrigation technology

642 Victoria Turf Research and Advisory Institute
Ballarto Road, Frankston, Victoria
[PO Box 381, Frankston, Victoria, 3199]
Tel: (03) 786 3311.

1 Woodcock T - DipAgrSc,BAgrSc *Officer-in-Charge* plant pathology
2 Lush WM - BAgrSc,MSc,DPhil plant physiology, population biology
3 Neylan JJ - BAgrSc soils, water use

643 Mallee Research Station
Walpeup, Victoria, 3507
Tel: (050) 941203. Telex: AA55559.

1 Griffiths JB - BSc *Officer-in-Charge*
2 Castleman GH - DipAgrSc oilseeds, triticale
3 Incerti M - BAgrSc tillage, crop rotations
4 Mock IT - DipAgrSc lupins, barley
5 Parker JKD - DipAgrSc livestock
6 Walsgott DN - DipAgrSc crop rotations, wheat

644 Potato Research Station, Toolangi
Private Mail Bag, Healesville, Victoria, 3777
Tel: (059) 629218.

1 Kellock A - BAgrSc,PhD *Officer-in-Charge* bud multiplication
2 Kirkham RP - MAgrSc,PhD potato genetics
3 McGregor GR - BAgrSc,MSc berry and nut culture
4 Strange PC - BAgrSc potato agronomy

645 Victoria Tobacco Research Station
Myrtleford, Victoria, 3737
Tel: (057) 521311.

1 Morgan MJ - BAgrSc *Officer-in-Charge (Acting)* agroforestry, tobacco agronomy
2 Baxter GG - BAgrSc tobacco crop protection
3 Bienvenu FE - DipHort,BAppSc agroforestry, mint agronomy
4 Hayes GA - BAgrSc tobacco breeding
5 Nieuwenhuis JC - DipAgrSc berries and nut crop agronomy
6 Ralph BJ - BAppSc,DipSc tobacco chemistry
7 Terpstra HJ - DipTropAgr tobacco pest control
8 Webb M - DipAgr tobacco growth regulation, peanut agronomy

646 Vegetable Research Station, Frankston
PO Box 381, Frankston, Victoria, 3199
Tel: (03) 7863311.

1 Morgans A - BAgrSc *Officer-in-Charge* herbicides
2 Kaddous FGA - BSc recycled water
3 McGeary DJ - BAgrSc agronomy
4 Morgan WC - BAgrSc,PhD seed physiologist
5 Waters CT - BAppSc crop establishment

647 Agricultural Engineering Centre
Princes Highway, Werribee, Victoria, 3030
Tel: (03) 7411300. Telex: AA38507.

1 Brown WT - BAgrE *Manager Agricultural Engineering*
2 Buckmaster D - BEng spray application
3 Foster MP - BEAgr buildings
4 Gould IV - MAgrSc,DDA fruit mechanisation
5 Huzzey JE - BEAgr tillage
6 Makin AW - BEAgr,DDA soil compaction
7 Reid PJ - BEAgr *Senior Engineering Services Officer*

8 Wills BJ - BEAgr,DipMechE spraying

648 Animal Health and Welfare Standards Branch
176 Wellington Parade, East Melbourne, Victoria
[PO Box 500, East Melbourne, Victoria, 3002]

1 White WE - MVSc,PhD *Manager Animal Health and Welfare Standards*
2 Harrison MA - BVSc,MACVSc *Principal Veterinary Officer (Extensive Livestock)*
3 Jones TE - MVSc,MACVSc,MASM *Manager Bureau of Animal Welfare*
4 Salisbury JR - MVSc,MRCVSc *Principal Veterinary Officer (Training Programs)*
5 Tweddle NE - BVSc *Principal Veterinary Officer (Brucellosis/Tuberculosis)*

649 Product Standards Branch
176 Wellington Parade, P.O. Box 500, East Melbourne, Victoria, 3002

1 Turner AJ - MVSc,PhD,MACVSc *Manager Product Standards*

650 Animal Quarantine

1 Millar HWC - BVSc *Principal Veterinary Officer*

651 Chemical Standards

1 Matthews DC - BAgrSc *Agricultural Standards Officer* agricultural chemicals, fertilizers, stockfoods

652 Dairy Quality Assurance Branch

1 Welton RL - BSc,PhD,DipEd *Director*

653 Stock Medicines Board

1 Herbert LA - BVSc *Veterinary Officer*

654 Meat Inspection Branch

1 Rees JB - BVSc,MRSVS,MACVSc *Principal Veterinary Officer*

655 Plant Standards Branch
Telex: AA 38507.

1 Kinsella MN - MAgrSc,MSc *Manager Plant Standards*
2 Blackstock JM - BAgrSc *Assistant Manager, Plant Improvement*
3 East R *Assistant Manager, Plant Quarantine*
4 Sharkey P *Assistant Manager, Plant Standards Control*

656 Victoria State Chemistry Laboratory
5 Macarthur Street, Melbourne, Victoria, 3002
Tel: (03) 651 2278.

1 Swartz PL *Director*
2 Peverill KI - MSc,PhD,ARACI *Assistant Director Primary Industry* soil plant relations, soil science
3 Stanhope RC - MSc,FRACI *Assistant Director* food technology

657 Scientific Services Section

1 Briner GP - MSc,Phd,ARACI *Senior Research Chemist* x-ray methods, clay mineralogy

658 Research Section

1 Hilliard EP - C Chem,MAgrSc,MRSC *Manager* soil and plant chemistry, biotechnology, environmental sciences
2 Badawy NS - BAgrSc land use
3 Johnstone PK - BSc,ARACI herbicide - soil relations
4 Maher JM - BAgrSc soil - land use, soil physics
5 Martin JJ - BAgrSc soil - land use
6 Minchinton IR - DipChem,ARACI infra-red spectrometry, pesticide-soil interaction
7 Minett W - BSc grain protection
8 Palmer MV - BSc,PhD biotechnology, plant physiology
9 Rothnie NE - BAppSc,PhD organic chemistry, histochemistry
10 Sang JP - DipAppChem,ARACI oilseed chemistry
11 Weston BA - BAgrSc,MAppSc environmental science, pollution studies

659 Analytical Chemistry Section

1 Fung KH - BSc,MAgrSc,MBA *Manager* analytical methodology
2 Baker TG - BFofSc,PhD forest soils, soil-plant relations
3 Curtis PR - MAppSc,ARACI instrumental analytical chemistry, environmental chemistry
4 Greenhill NB - BSc,MAgrSc soil chemistry
5 Handson PD - DipAppChem,BAppBoil plant chemistry, environmental chemistry

660 Soil and Plant Advisory Section

1 Brown AJ - DipAgrSc *Senior Soils and Plant Advisory Officer* extension services

661 Organic Chemistry Section

1 Luke BK - DipAppChem,ARACI *Manager* pesticide chemistry
2 Burke DG - BSc,PhD mass spectrometry
3 Roberts GS - DipAppChem,ARACI pesticide chemistry

662 Health Sciences Section

1 Greene NC - DipAppChem,ARACI *Manager* food technology
2 Fox M - BSc feedstuffs
3 Menz NF - DipAppChem,ARACI food technology
4 Rayner CJ - MAppChem feedstuffs, nutritional chemistry

663 Victoria Department of Conservation Forests and Lands

Head Office, 240 Victoria Parade, East Melbourne, Victoria, 3002
Tel: (03) 651 4011.

1 Eddison PA - Dip,TRP *Director-General*

664 Fisheries & Wildlife Service

250 Victoria Parade, East Melbourne, Victoria, 3002
Tel: (03) 651 4011.

1 Newman GG - BSc,MSc,MBA,PhD *Director*

665 Wildlife Management Branch

123 Brown Street, Heidelberg, Victoria, 3084
Tel: 03 450 8600.

1 Chamley WA - BSc,DPhil *Assistant Director*
2 Warneke RM - BAgrSc,MSc *Manager Wildlife Research*
3 Emison WB - BA,MSc *Principal Scientist* bird ecology
4 Loyn RH - BA,MA *Senior Scientist* forest ecology
5 Norman FI - BSc,DPhil *Principal Scientist* waterfowl research
6 Seebeck JH - BSc *Senior Scientist* small mammal ecology

666 Freshwater Fish Management Branch

123 Brown Street, Heidelberg, Victoria, 3084
Tel: (03) 450 8600.

1 Craig FG - DipFor,BSc,PhD *Assistant Director*
2 Pribble HJ - BSc,MSc *Manager Freshwater Fish Research*
3 Anderson JA - BSc,PhD *Senior Scientist* habitat requirements of fish
4 Bacher GJ - BSc *Scientist* aquatic toxicology
5 Jackson PD - BSc,PhD *Senior Scientist* habitat requirements of fish
6 Tunbridge BR - BSc *Principal Scientist* minimum streamflow requirements of fish

667 Marine Resources Management Branch

Marine Science Laboratories, Weeroona Parade, Queenscliff, Victoria, 3225
Tel: (052) 520111.

1 Byrne JL - BEc,MA *Assistant Director*
2 Bulthuis DA - BA,MSc,PhD *Manager Research Section* seagrass ecology
3 Arnott GJ - BSc,PhD zooplankton ecology
4 Axelrad DM - BSc,MSc,PhD nutrient cycling, eutrophication, phytoplankton and seagrass ecology
5 Brand GW - BSc,PhD toxicology; ecotoxicology
6 Coleman N - BSc,PhD marine biology; benthic ecology; mollusc ecology and physiology
7 Cowdell RA - BAppSci marine chemistry, micronutrients
8 Evans DD - BSc,PhD *Editor* publications
9 Fabris GJ - BAppSci,MSc *Manager Chemistry Services* trace metal chemistry
10 Garnham JS - BSc trace metal chemistry
11 Gibbs CF - BSc,PhD marine chemistry; chemical oceanography; oil pollution
12 Gwyther D - BSc,PhD fisheries science - scallops; prawns, tropical freshwater and marine fisheries
13 Hickman NJ - BSc fisheries science; mariculture; oysters; mussels
14 Hobday DK - BSc fisheries science; population studies
15 Longmore AR - BSc marine chemistry
16 McShane PE - BSc marine biology; phytoplankton
17 Mickelson MJ - BSc,PhD biological oceanography; phytoplankton; remote sensing
18 Murray AP - BSc marine organic chemistry; marine pollution
19 Palmer DH - BSc toxicology; oil dispersant testing
20 Walker TI - BSc fisheries population dynamics; shark biology
21 Wankowski JWJ - BSc,PhD fisheries science; population studies; population modelling

668 Ecology Branch

250 Victoria Parade, East Melbourne, Victoria, 3002
Tel: (03) 651 4011.

1 Rudman P - BSc,PhD *Manager*
2 Alexander DP - DipCE,MIE Aust,EWS ecological systems
3 Helman ZM - BSc,MEnvSc ecological policy
4 Lloyd HL - BEd ecological survey

669 State Forests and Lands Service

1 Smith RP - BSc,MBA,PhD *Director*

670 Von Mueller Institute

1 Dexter BD - DipFor,BScFor,MScFor *Director*

671 Research Section

1 Flinn DW - DipFor,BScFor,PhD *Manager*

672 Pathology and Entomology Section

1 Harris JA - DipFor,BScFor,MSc *Scientist* forest entomology
2 Kassaby FY - BSc,MSc *Scientist* forest pathology
3 Marks GC - BSc,MSc,PhD *Senior Scientist* forest pathology
4 Neumann FG - DipFor,BSc,MSc *Senior Scientist* forest entomology
5 Smith IW - BSc *Scientist* forest pathology and silviculture

673 Silviculture Section

1 Squire RO - DipFor,BScFor,MScFor,PhD *Principal Scientist* silviculture
2 Fagg PC - DipFor,BScFor,MScFor *Scientist* silviculture
3 Geary PW - DipFor,BScFor *Scientist* silviculture
4 Kellas JD - DipFor,BScFor,MScFor *Scientist*

674 Nutrition and Physiology Section

1 Morris JD - DipFor,BScFor,PhD *Principal Scientist* tree physiology
2 Hopmans P - BSc,PhD *Scientist* soil science and tree nutrition

675 Genetics Section

1 Pederick LA - DipFor,BScFor,PhD *Senior Scientist* tree breeding
2 Cook IO - DipFor,BScFor *Scientist* forest genetics

676 Ecology Section

1 Tolhurst KG - DipFor,BScFor *Scientist* fire ecology
2 Tomkins IB - BSc,PhD *Scientist* fire ecology and silviculture

677 Royal Botanic Gardens and National Herbarium of Victoria

Birdwood Avenue, South Yarra, Victoria, 3141
Tel: (03) 63 9424.

1 Churchill DM - BSc,MA,PhD *Director and Government Botanist* taxonomy, palaeobotany, ecology
2 Aston Miss HI - BSc *Project Scientist* taxonomy of *Nymphoides* (Menyanthaceae), Australian aquatic angiosperms
3 Cohn HM - BA,GradDipLib *Librarian* history of Australian botany
4 Conn BJ - BSc,MSc,PhD taxonomy of Labiatae, New Guinea flora
5 Filson RB - MSc taxonomy of lichens
6 Foreman DB - MSc,PhD taxonomy of Australian Proteaceae, rainforest flora (including New Guinea) taxonomy and morphology of primitive flowering plants
7 Lumley PF - MG ornamental horticulture
8 Ross JR - MSc,PhD *Principal Scientist* taxonomy of the tribe Bossiaeeae (Papilionaceae)
9 Short PS - BSc,PhD taxonomy of Australian Gnaphaliinae (Compositae: Inuleae)
10 Smith RV - BSc,DipFor taxonomy of *Grevillea* (Proteaceae), naturalised aliens

678 Victoria Land Protection Service

378 Cotham Road, Kew, Victoria, 3101
Tel: (03) 817 1381.

1 Campbell RW - BVSc *Director*

679 Research Inventory and Assessment Branch

Keith Turnbull Research Institute, Ballarto Road, Frankston, Victoria, 3199
Tel: (03) 785 0111.

1 Amos TG - BSc,PhD *Manager*

680 Research Section

681 Pest Plants Research
Keith Turnbull Research Institute, Ballarto Road, Frankston, Victoria, 3199
Tel: (03) 785 0111.

1 Combellack JH - MPhil,BTech *Team Leader* pest plant control and management, safe use and application of herbicides
2 Pritchard GH - BAgrSc biology of pest plants, control and management of pest plants
3 Richardson RG - MAgrSc plant physiology, herbicide usage, ecology and management of pest plants

682 Biological Control of Weeds Research

1 Field RP - MAgeSc,PhD *Team leader* biological control of weeds and insect pests including European and forest pests
2 Adair RJ - BSc biological control of weeds, Australian plant ecology, plant taxonomy
3 Bruzzese E - BAgrSc insects and pathogens for control of blackberry and thistles
4 Mclaren D - BSc biological control of weeds, entomology, insect/plant interactions
5 Shepherd RCH - MSc,MAgrSc biological control of spiny emex and skeleton weed

683 Pest Animals Research

1 Brown AM - BSc,PhD *Team leader* microbial control of pest animals
2 Coman BJ - MAgrSc,PhD fox, wild dog and rabbit ecology and control
3 Nolan IF - BA rabbit ecology, control, and management strategies

684 Soil, Water and Vegetation Research

1 Cummings DJ - BAgrSc soil physical characteristics, erosion processes, soil structure and soil stability
2 Lane ML - BAgrSc tree establishment
3 Papst WA - BAgrSc,MS farm management and soil conservation research, alpine ecology
4 Papworth MP - MSc,PhD physical proceses of water movement through soil, application to land management
5 van de Graaff RHM - MAgrSc,PhD soil degradation and improvement, soil and land survey
6 van Rees H - BSc,PhD rangeland research and monitoring, land degradation and soil structure, erosion processes
7 Wu AYK - MSc,PhD catchment hydrology and instrumentation
8 Zallar S - BAgrSc conservation agronomy, management system for specified land uses and reclamation, stabilization or eroded areas

685 Salinity and Hydrogeology Research

1 Jenkin JJ - MSc,PhD *Team leader* geomorphology and hydrogeology applied to land deterioration, morphotectonics, coastal studies, soil-landform relationships
2 Day CA - BSc geomorphology and hydrogeology applied to dryland salinization, catchment characterization
3 Dyson PR - DipAppSci hydrogeology, geoch

686 Inventory and Assessment Section

1 Rowan JN - BSc *Senior Project Officer*
2 Rowe RK - ADipFor,MSc *Chief Inventory and Assessment Officer*

687 Land Inventory and Assessment

1 Howe DF - BAgrSc *Team Leader* land resource inventory and assessment, extension
2 Hook RA - BSc land resource inventory and interpretation
3 Leslie TI - BSc *Senior Project Officer* methods development
4 Lorimer MS - MAgrSc land resource inventory and interpretation, landscape process research
5 Schoknecht N - BSciFor land resource inventory and interpretation
6 White LA - BAgrSc land resource inventory and interpretation

688 Water Conservation and Protection

1 Slater SJE - BSc,PhD *Team Leader* policy formulation, investigation scheduling
2 Clinnick PF - BAgrSc sediment production, run-off and water quality from agricultural and forest areas, development of softwood establishment prescriptions, investigation and determinations of land use
3 Master I - BAgrSc,PhD catchment investigations and determination of land use
4 McKinnon DM - DipMechEng,BAgrSc catchment investigation and determination of land use

689 Trees and Pest Plants

1 Lane DWA - MAgrSc *Team Leader* plant ecology, pest plant distribution and impact
2 Yugovic J - BSc taxonomy, plant ecology and systems

690 Pest Animals Inventory

1 Backholer JR - BAgrSc distribution and impact of pest animals
2 Murphy GD - BSc *Team Leader* inventory systems, pest animal abundance

691 Melbourne and Metropolitan Board of Works
PO Box 4342, Melbourne, Victoria, 3001
Tel: (03) 62 0221.

1 Marginson RD - Chairman
2 Ingersoll RJ *General Manager*

692 Systems Operations Division

693 Treatment Plants Branch
Werribee Farm, Private Bag 10, Werribee PO, Werribee, Victoria, 3030
Tel: (03) 741 9209.

1 Scott TM - BEAgr,MIEAust *Engineer, Processes and Operations*
2 Holland BJM - BAgrSc,MRurSc *Farming Services Manager* agriculture, animal health
3 Williams D - DDA,BComm *Senior Technical Officer* agriculture, environmental

694 Metropolitan Operations
Upper Yarra Region, PO Box 6, Woori Yallock, Victoria, 3139
Tel: (059) 64 7100.

1 O'Shaughnessy PJ - BScFor,DipFor *Forester - Research and Projects* forest catchment hydrology

695 Museum of Victoria

696 Division of Natural History

285-321 Russell Street, Melbourne, Victoria, 3000
Tel: (03) 669 9888.

Concerned essentially with geology and zoology

1 Vacant *Director*
2 Darragh TA - MSc,PhD,DipEd *Deputy Director* fossil mollusc taxonomy

697 Zoology

1 Boyd Ms S - BSc *Assistant Curator, Department of Invertebrate Zoology* marine bivalve molluscs
2 Dixon Miss JM - BSc *Curator, Department of Mammalogy* mammalogy
3 Gomon MF - BSc,MSc,PhD *Curator, Department of Ichthyology* fish taxonomy
4 Lu CC - BSc,MSc,PhD *Curator, Department of Invertebrate Zoology* cephalopod mollusc taxonomy
5 Neboiss A - MSc,PhD,FRES *Curator, Department of Entomology* aquatic insect taxonomy
6 Poore GCB - BSc,PhD *Curator, Department of Crustacea* crustacean taxonomy
7 Yen AL - BSc,PhD *Assistant Survey Officer* terrestial ecology

698 Rural Water Commission of Victoria

590 Orrong Road, Armadale, Victoria, 3143
Tel: (03) 508 2222. Telex: WATERCO AA30739. FAX (03)5082686.

1 Constable DJ - BCE,MIE,EWS *General Manager*

699 Technical Services Division

700 Water, Materials and Environmental Science Branch
590 Orrong Road, Armadale, Victoria, 3143
Tel: (03) 508 2222. Telex: WATERCO AA30739.

1 Drew WM - PhD,MEngSc,FRMIT,DipChemE,DipChem *Chief of Branch*
2 Bannister RW - BSc chemistry
3 Bennison GL - BSc limnology
4 Bowles Ms BA - PhD,BSc water quality assessment
5 Graham WA - PhD,BAgrSc water quality assessment
6 Kairaitis A - DipAppSc,MEng materials science
7 Martin Ms JL - BSc microbiology
8 McNeill Ms AR - BSc microbiology
9 Richards TJ - BEMet,MSc materials science
10 Richards TJ - BE;MSc Materials science
11 Tozer RR - DipAppSc water quality assessment
12 Welsh DR - BAppSc water quality assessment

701 Planning and Marketing Division

702 Investigations Branch

1 Evans RS - PhD,BSc *Principal Geologist*
2 Ife D - BSc,MSc groundwater
3 Trewhella NW - BCE groundwater and drainage

703 Operations Division

704 Irrigation Services Branch

1 Bill SM - MAppSc,BSc,DipFor scientific editing
2 Bridley SF - BAgrSc groundwater and drainage
3 Campbell Mrs BJ - BEAgr irrigation practices
4 Dickinson PJ - BAgrSc drainage
5 Fowler Miss CA - BSc scientific editing
6 Jones LD - BAgrSc irrigation and drainage practices
7 Lavis AR - BAgrSc catchment water and salt balances
8 Martin JF - BEAgr irrigation drainage practices
9 Patto PM - BSc irrigation and drainage practices
10 Poulton DC - BSc irrigation scheduling, drainage
11 Rendell RJ - BEAgr land layout
12 Robinson EP - MAgrSc,MEngSc *Chief Irrigation Officer*
13 Sayer PH - BEc,DipAgrSc,DipExpMkg economics
14 Twyford-Jones PL - BEc,DipAgr economics
15 Webster A - BSc seepage
16 Wilding JL - BAgrSc aquatic weed management
17 Zwar MJ - BEAgr irrigation and drainage practices

WESTERN AUSTRALIA GOVERNMENT

705 Western Australia Department of Agriculture

Headquarters, Baron-Hay Court, South Perth, Western Australia, 6151
Tel: (09) 368 3333.

1 Halse NJ - MScAgr *Director of Agriculture*
2 Carroll MD - PhD,BA,BScAgr *Deputy Director of Agriculture*
3 Gabbedy BJ - MVsc,MScAgr *Assistant Director of Agriculture*
4 Parkin RJ - BScAgr *Assistant Director of Agriculture*
5 Toms WJ - MScAgr *Assistant Director of Agriculture*
6 Quinlivan BJ - DScAgr,BA,FAIAS *Chief Research Liaison Officer* seed physiology

706 Animal Health Division

1 Armstrong JM - BVSc,MACVSc *Chief of Division*
2 Purcell DA - MRCVS,MVB,MA,PhD *Chief Veterinary Pathologist*
3 Allen JG - BVSc,BScc lupinosis, sheep myopathics
4 Berry PH - BVSc annual ryegrass toxicity
5 Besier RB - BVSc epidemiology and control of parasites
6 de Chaneet GC - BVSc,MVS epidemiology and control of parasites
7 Dickson J - MRCVS,DVSM pathology of plant toxicities; infectious and nutritional diseases
8 Edwards JR - BVSc,BScAgr trace elements; weaner illthrift
9 Griffiths GL - PhD poultry pathology
10 LeCras DV - BAppScMedTech serology (elisa) of bovine leptospirosis
11 Lloyd JM - BSc mycoplasma microbial pathogens
12 Main DC - BVSc flouride toxicity in pigs
13 Masters HG - ARACI trace element interaction
14 McKenzie DP - BVSc sheep diseases and production
15 Mercy AR - BVSc,MVS pig diseases and production
16 Mitchell RK - BVSc,MVS cattle diseases and production
17 Peet RL - BVSc,MS reproductive diseases and conditions
18 Petterson D - BSc,MSc lupinosis, annual ryegrass toxicity; toxic plants
19 Richards RB - BVSc,MVSc ovine, bovine footrot; polioencephalomalacia in ruminants
20 Robertson GM - BVSc,MSc,PhD *Veterinary Microbiologist*
21 Steel FVP - BSc polioencephalomalacia; vitamin E-selenium; production diseases
22 Sutherland SS - BSc bovine brucellosis; dermatophilosis of sheep
23 Vogel P - BSc,PhD annual ryegrass toxicity; toxic plants
24 Wilkinson FC - BVSc,MRCVS epidemiology and economics of diseases in cattle, sheep and pigs

707 Animal Production Division

1 Lightfoot RJ - BScAgr,PhD *Chief of Division*
2 Barker DJ - BVSc,MRCVS *Acting Principal Officer, Beef Production*
3 Althorpe D - BAppScMechEng dehydration and energy research
4 Beetson BR - BScAgr,MScAgr sheep breeding genetics
5 Carrick MJ - MSc,PhD breeding and genetics in cattle and pigs
6 Clarke P - BE,MSc materials handling, air drying
7 Croker KP - MScAgr,PhD sheep reproduction
8 Dunlop AC - BAgSc sheep nutrition
9 Fels HE - BScAgr sheep nutrition and management
10 Gherardi SG - BScAgr blowfly control management
11 Godfrey NW - MAgrSc beef production
12 Greathead KD - BScAgr beef production
13 Hodgson DJ - BSc food microbiology
14 Howe RR - BScAgr *Officer-in-Charge, Animal Breeding and Research Institute*
15 Hudson JE - BScAgr meat research, oil extraction
16 Johnson T - BScAgr meat research
17 Kelly AT - BSc dehydration and oil extraction
18 Kelly RW - BScAgr,PhD reproduction physiology
19 Lewer RP - BScAgr,PhD genetics
20 May PJ - BScAgr beef production
21 McDonald CL - BScAgr,MScAgr live sheep export
22 O'Malley PJ - BScAgr broiler research
23 Olney GR - BScAgr,DipDairyHusb dairy cattle and milking machines
24 Ralph IG - BScAgr wool research
25 Roberts DW - BScAgr *Senior Research Officer* meat and energy research
26 Rowe JB - PhD,BRurSc sheep nutrition
27 Ryan WJ - BScAgr *beef cattle production*
28 Sawyer GT - BScAgr,PhD dairy cattle physiological aspects
29 Skevington SG - BAppSc refridgeration and meat research
30 Wroth RH - MScAgr,BVSc sheep meat industry

708 Horticultural Division

1 Stynes BA - MScAgr,PhD *Chief of Division*

2 Burt JR - BScAgr vegetables and tropical fruit
3 Burton NW - BScAgr fruit
4 Devitt AC - BScAgr,DipOen *Senior Viticulturist*
5 Floyd RM - BScAgr vegetables
6 Gherardi PB - BScAgr viticulture
7 Godley GL - BScAgr,DipHort fruit
8 Gratte HVB - BScAgr vegetables
9 Hawson MG - BScAgr,DipTropAgr tropical horticulture and vegetables
10 Hill TR - BScAgr vegetables
11 Hosking DC - BSc vegetables
12 Paulin R - BScHort fruit
13 Paust G - BScAgr vegetables
14 Phillips DR - BScAgr vegetables
15 Shorter NH - BScAgr fruit
16 Ward G - BSc Fruit
17 Webb MG - BScAgr vegetables

709 Plant Production Division

1 Francis CM - BScAgr,PhD *Chief of Division*
2 Barclay IR - BScAgr,PhD plant breeding
3 Collins WJ - BScAgr,PhD genetic resources
4 Cowling WA - BScAgr,PhD plant breeding
5 Crosbie GB - AITChem grain products
6 Fisher HM - BScAgr cereal variety testing
7 Gladstones JS - BScAgr,PhD plant breeding
8 McLean RJ - BScAgr plant breeding
9 Portman PA - BScAgr plant breeding
10 Roy NN - BScAgr,PhD plant breeding
11 Tarr AW - BAppSc grain products
12 Whan BR - MA,BScAgr plant breeding

710 Plant Research Division

1 Chatel DL - BScAgr,PhD *Chief of Division*
2 Bolland MDA - BScAgr,PhD phosphorus nutrition
3 Bowden JW - BScAgr,PhD phosphate nutrition
4 Brennan RT - BScAgr trace element nutrition
5 Cripps JL - BA,BScHort,MScAgr horticulture nutrition
6 Gartrell JW - BScAgr trace elements nutrition
7 Gillespie DG - BScAgr lupin alkaloids
8 Goldspink BH - AITChem lupin alkaloids
9 Jarvis RJ - BScAgr tillage systems
10 Mason MG - BScAgr nitrogen nutrition
11 Nicholas DA - BScAgr pasture agronomy
12 Oakley AE - APTCChem radioisotopes, chemical analysis
13 Perry MW - BScAgr crop physiology
14 Poole ML - BScAgr crop agronomy
15 Porter WM - BScAgr acid soil problems
16 Rowland IG - BScAgr ley farming rotations
17 Tennant D - BScAgr,PhD soil moisture
18 Thorne C - BScAgr crop pasture system
19 Walton GM - BScAgr lupin agronomy
20 Weir RN - APTCChem wheat physiology

711 Biometrics

1 D'Antuono M - BSc biometrics
2 Spiejers Mrs EJ - BSc biometrics

712 Plant Pathology

1 Brown AGP - BScAgr,PhD *Principal Plant Pathologist*
2 Barbetti MJJ - BScAgr field crop and pasture diseases
3 Carter E - MScAgr fruit diseases
4 Khan TN - MScAgr,PhD barley foliage diseases
5 MacNish GC - BA,BScAgr,PhD cereal root rots
6 Sweetingham MW - BScAgr lupin diseases
7 Wood PMCR - BScAgr lupin diseases

713 Weed Agronomy

1 Gilbey DJ - BScAgr doublegee *Emex australis* ecology
2 Holmes J - BScAgr annual ryegrass control
3 Madin RW - BScAgr crop weeds
4 Peirce JR - BScAgr noxious weeds
5 Rutherford PA - BScAgr pesticides

714 Resource Management Division

1 Robertson GA - DPhil,BScAgr *Chief of Division*

715 Soil Conservation

1 Bligh KJ - BE surface hydrology
2 Carter DJ - BScAgr wind erosion
3 Riches JRH - BScAgr,DipAgrExt rehabilitation of disturbed land
4 Stoneman TC - MScAgr *Principal Research Officer*

716 Soil Research and Survey

1 Clark AJ - BScAgr salt land agronomy; soil surveys
2 Engel R - BScAgr soil salinity
3 George PR - BScAgr water balance; soil salinity
4 Laing IAF - BScAgr water conservation and catchment improvement
5 Malcolm CV - MScAgr salt land agronomy; environment
6 Nulsen RA - PhD salinity; soil water relations
7 Pepper RG - MScAgr clay chemistry

717 Irrigation and Drainage

1 Cole KS - BScAgr general irrigation
2 Swan B - BScAgr *Principal Irrigation Officer*

718 Rangeland Management

1 Wilcox DG - BScAgr *Principal, Rangeland Management*
2 Burnside D - BScAgr range research
3 Gardiner HG - BScAgr range research
4 Hacker RB - DrPhD range research, remote sensing
5 Holm A - MScAgr tropical dryland legume agronomy
6 Kok B - BScAgr range cattle research
7 Mitchell AA - BScAgr range research and range condition survey
8 Morrissey JG - BScAgr,DipAgrExt range research, range condition monitoring
9 Payne AL - BScAgr range condition survey
10 Pratchett D - BScAgr animal husbandry

719 Herbarium

1 Green JW - BSc,MSc,PhD *Curator* systematics of Western Australian flora, Myrtaceae
2 Aplin TEH - BSc toxic plants systematics of Western Australian flora
3 Hnatiuk RJ - MSc,PhD heathlands of Western Australian flora
4 Kenneally KF - BSc biological surveys; systematics of Western Australian flora
5 Lander NS - BSc systematics of Western Australian flora, Asteraceae, Lamiaceae
6 Marchant NG - BSc,PhD cytotaxonomy, systematics of Western Australian Flora, Droseraceae, Xyridaceae, Myrtaceae
7 Maslin BR - MSc systematics of Western Australian flora, Mimosaceae
8 Perry G - MSc systematics of Western Australian flora, Loganiaceae
9 Wilson DG - MSc systematics of Western Australian flora, Chenopodiaceae

720 Entomology Branch

1 Davis PR - BSc biological control
2 Fisher K - BSc biological control
3 Grimm M - BSc crop and pasture pests
4 Learmonth SE - BScAgr tropical entomology
5 Michael PJ - BSc crop and pasture pests
6 Monzu N - PhD ecology of sheep blowflies
7 Richards KT - MScAgr taxonomy
8 Rimes GD - BSc *Principal Entomologist* stored products pests and pest molluscs
9 Sandow JD - MSc biological control
10 Sproul AN - BScAgr horticultural entomology
11 Strickland GR - BSc tropical entomology
12 Woods WM - BScAgr crop and pasture pests

721 Marketing and Economics Branch

1 Morrison DA - PhD,BScAgr,BEc *Acting Officer in Charge*
2 Chopping MH - BScAgr,MEcon farm management
3 Harrington QN - BScAgr,MSc farm management
4 Humphry MG - BScAgr farm management
5 Kingwell RS - BScAgr policy (sector analysis)
6 Malcolm JW - BScAgr,MSc market analysis
7 O'Loughlin EJ - BEc,DipAgrEcons meat marketing
8 Roberts EJ - BScAgr,BEc machinery work study, farm management
9 Rutter KR - BScAgr,BEc marketing

722 Agricultural Protection Board

1 Hogstrom AW - MScAgr *Chief Executive Officer*

723 Vertebrate Pests Research

1 King DR - MSc,PhD myxomatosis
2 Oliver AJ - BSc,ARCS vertebrate pest control
3 Thompson PC - BSc dingoes
4 Wheeler SH - BSc,PhD rabbits

724 Western Australia Department of Conservation and Land Management

Hackett Drive, Crawley, Western Australia, 6009
Tel: (09) 3868811.

State Headquarters, Hayman Road, Como WA, 6152 Tel. (09)3676333

1 Shea SR - BScFor,PhD *Executive Director*

725 Production and Protection Research

1 Christensen P - BSc,PhD *Principal Research Officer*
2 Abbott IJ - BSc,PhD *Research Officer* entomology
3 Bartle J - BScAgric *Research Officer* inspector
4 Brown PH - BScFor *Assistant District Forest Officer* salination
5 Burrows ND - BScFor *District Forest Officer* fire
6 Butcher TB - BScFor *Senior District Forest Officer* softwood
7 Davey AS - BScFor *District Forest Officer* rehabilitation
8 De Braganza LF - BScFor *Assistant District Forest Officer* 2nd rotation trials - pine
9 Inions G *Assistant Forest Officer* karri silviculture
10 Mazanec R *Assistant District Forest Officer* genetics
11 McCaw L - BScFor *Assistant District Forest Officer* fire
12 McGrath JF - BSc,Agric,PhD *Research Officer* pine nutrition
13 Moore RW - BScFor *District Forest Officer* agroforestry
14 Shearer BL - BSc,PhD *Senior District Forest Officer* dieback
15 Siemon G - BScFor,PhD *Senior District Forest Officer* inspector
16 Stoneman GL - BScFor *Assistant District Forest Officer* fauna
17 Strelein GJ - BScFor *District Forest Officer* jarrah silviculture
18 Stukely MJ - BScAgric,PhD *Research Officer* dieback
19 Tippett J - AGSCTS *Research Officer* dieback
20 Ward DJ - BAppSc *Research Officer* computing
21 Wardell-Johnson G - BScFor *Assistant District Forest Officer* fauna

726 Western Australian Wildlife Research Centre

PO Box 51, Wanneroo, Western Australia, 6065
Tel: (09) 4051555.

1 Burbidge AA - BSc,PhD *Principal Research Officer* nature reserve selection and management, conservation of rare vertebrates, turtles
2 Burbidge AH - BSc biological survey
3 Friend G - BSc,PhD fire ecology (animals) esp. reptiles and invertebrates
4 Friend JA - BSc,PhD management of rare animals, especially numbat, macro and meso fauna, soil ecology, ecological energetics
5 Halse SA - BSc,PhD effect of salinity on ducks and aquatic invertebrates, biology of Cape Barren Geese
6 Hopkins AJM - BSc fire ecology (plants), rehabilitation of disturbed areas
7 Hopper SD - BSc,PhD conservation of plants, management of wildflower industry, vertebrate pollinators
8 Keighery GJ - BSc biological surveys (plants), alien plants, taxonomy and ecology native plants
9 Kinnear JE - BSc,MSc,PhD marsupial ecology and conservation, macropod nutrition, ecology energetics
10 Lane JAK - BSc *Senior Research Officer* estuarine birds, waterfowl, conservation of wetlands
11 McKenzie NL - BSc,MSc biological survey, selection of conservation reserves, ecology and taxonomy of bats
12 Prince RIT - BSc,Agric,PhD macropod marsupials, dugong
13 Taylor A - MSc co-ordinator of Banksia atlas

727 At Karratha

Welcome Road, Karratha, Western Australia, 6714
Tel: (091) 868290.

1 Morris KD - BSc,MSc ecology and management of islands, rodent ecology

728 At Kalgoorlie

Hannan Street, Kalgoorlie, Western Australia, 6430
Tel: (090) 212095.

1 Pearson D - BSc ecology and management of desert ecosystems, desert mammals

729 Western Australia Government Chemical Laboratories

125 Hay Street, Perth, Western Australia, 6000
Tel: (09) 325 5544.

1 Hughes HC - BSc,FRACI,MAIAS *Acting Director*
2 McLinden VJ - APTC FRACI FAORC *Acting Deputy Director*

730 Agricultural Chemistry Laboratory

1 Baseden SC - BSc,MAIAS *Acting Chief of Laboratory*
2 Allen DG - BSc phosphorus soil chemistry and plant nutrition
3 Codling BJ - BSc,DipComp,ARACI computer applications; plant and fertilizer chemistry
4 Harris DJ - PhD,ARACI lupin alkaloids
5 Jeffery RC - BSc,MAIAS soil chemistry
6 Spadek ZE - BSc amino acids, animal nutrition
7 Wilson IR - DipChem soil acidity, trace elements
8 Wilson NL - BSc trace elements, fatty acids
9 Wilson PE - Dip,ARACI trace elements, automated chemistry

731 Kings Park and Botanic Garden

West Perth, Western Australia, 6005
Tel: (09) 321 4801, (09) 321 5065.

1 Wycherley PR - OBE,BSc,PhD *Director*
2 Bennett EM - MSc,DipEd systematic botany, plant ecology
3 Bunn E - BSc micropropagation
4 Dixon IR - CHort cultivation of native plants
5 Dixon KW - BSc,PhD biology of native plants
6 Fryer RA - BScHort propagation of native plants
7 Kullman WH - DipHort *Superintendent*

UNIVERSITIES AND COLLEGES

732 Australian National University

PO Box 4, Canberra, A.C.T., 2600
Tel: (062) 195111. Cables: Natuniv Canberra.

733 Research School of Biological Sciences

1 Slatyer RO - AO,DScAgr,FAA,FRS *Professor and Director* eucalypt forest ecosystems

734 Department of Behavioural Biology

1 Mark RF - MMedSc,MB,ChB,CES,FAA *Professor and Head of Department* neurophysiology
2 Hill KG - BSc,PhD *Fellow* neurophysiology
3 Morgan IG - BSc,PhD *Fellow* neurochemistry

735 Department of Developmental Biology

1 Gunning BES - MSc,PhD,FAA,FRS *Professor and Head of Department* ultrastructure of plant cells
2 John PCL - PhD *Fellow* genetical and biological study of plant cell cycle
3 Letham DS - MSc,PhD *Senior Fellow* plant growth substances, especially cytokinins

736 Department of Environmental Biology

1 Cowan IR - MSc,PhD,FIP *Professional Fellow and Head of Department* environmental physiology of higher plants
2 Osmond CB - MSc,PhD,FAA *Professor* photosynthesis, physiological ecology
3 Badger MR - BScAgr,PhD *Fellow* biochemistry of photosynthesis and respiration
4 Farquhar GD - BSc,PhD *Senior Fellow* integration of carbon and nitrogen metabolism with plant water relations
5 Noble IR - BSc,PhD *Fellow* dynamic aspects of plant ecology

737 Department of Genetics

1 Pateman JA - BSc,PhD,MA,FRSEd,FRS *Professor and Head of Department* regulation of gene expression
2 Clarke-Walker GD - MSc,DPhil *Senior Fellow* molecular genetics of yeast
3 Creaser EH - MA,PhD *Senior Fellow* structure and evolution of enzymes
4 Doy CH - BSc,PhD,FRACI,FRSC *Senior Fellow* molecular genetics of *Aspergillus*
5 Rolfe BG - BAgrSc,PhD *Senior Fellow* bacterial regulation and nitrogen fixation
6 Shine J - BSc,PhD *Fellow* structure and function of mammalian genes

738 Department of Neurobiology

1 Horridge GA - MA,PhD,ScD,FAA,FRS *Professor and Head of Department* arthropod receptors and integration
2 Ball EE - AB,PhD *Fellow* insect sensory system
3 Blest AD - BSc,DPhil *Senior Fellow* photoreceptor membrane turnover, functional anatomy of spider eyes
4 Srinivasan MV - BE,PhD *Fellow* visual systems

739 Department of Population Biology

1 John B - MSc,PhD,DSc,BIBiol *Professor and Head of Department* population cytogenetics of insects
2 Gibson JB - BSc,PhD,MA *Professional Fellow* population genetics
3 Miklos GLG - BSc,PhD *Fellow* DNA cloning ang sequencing in eukaryotes
4 Shaw DD - BSc,MSc,PhD *Fellow* differentiation and speciation in grasshoppers

740 Molecular Biology Unit

1 Naora H - BSc,DSc *Professional Fellow and Head of Unit* molecular biology of eukaryotes

741 Taxonomy Unit

1 Watson L - MSc,FLS *Senior Fellow and Head of Unit* methodology and application

742 Virus Ecology Research Group

1 Gibbs AJ - BSc,PhD,ARCS *Senior Fellow and Head of Group* viruses of plants and lower animals

743 John Curtin School of Medical Research

Australian National University, GPO Box 334, Canberra City, A.C.T., 2601
Tel: (062) 492550. Telex: Aust 62033. Cables: Curtinschool.

1 Porter R - BMedSc,DSc,MA,BCh,DM,FAA *Professor and Director*

744 Department of Immunology

1 Morris B - BVSc,DPhil,FAA *Professor and Head of Department* developmental immunology in large animals; reproductive immunology
2 Brenan M - BSc,PhD *Research Fellow* immunology of pregnancy
3 McCullagh PJ - MD,DS,DPhil,MRCP *Senior Fellow* immunological tolerance; foetal immunology
4 Shelton JN - BVSc,PhD,FACVSc *Senior Fellow* reproductive biology
5 Simpson-Morgan MW - BVSc,PhD *Senior Research Fellow* foetal immunology and physiology; computer monitoring in experimental systems involving animals
6 Smeaton TC - BAgSc,PhD *Research Fellow* cellular immunology

745 Department of Physiology

1 Levick WR - MB,BS,MSc,FAA,FRS *Professor* visual neurophysiology
2 Close RI - MSc,PhD *Senior Fellow* biophysics of muscle
3 Henry GH - MAppSc,DSc *Senior Fellow* neurophysiology of cerebral cortex

746 Research School of Pacific Studies

Australian National University, GPO Box 4, Canberra City, 2601

747 Department of Biogeography and Geomorphology

1 Wace NM - PhD *Senior Fellow* historical ecology of naturalized plants in Australia; car-borne and person-dispersed flora in Canberra; flora of the Tristan da Cunha Islands; forestry and introduced plants on Tristan

748 Department of Economics

1 Barlow C - MBE,BSc,MS,PhD *Senior Fellow* economic aspects of tropical agriculture
2 Booth A - BA,PhD *Senior Research Fellow* economics of agricultural development in Asia (particularly Indonesia)
3 Jayasuriya S - BA,MAgrDevEc,PhD *Research Fellow* economics of agriculture in developing countries particularly rice, rubber and rice based farming systems
4 Ravallion M - BSc,MScEcon,PhD *Research Fellow* welfare and development economics in Bangladesh, India and Indonesia
5 Shand RT - MScAgr,PhD *Senior Fellow* economics of agricultural and rural development in Asia

749 Department of Human Geography

1 Brookfield HC - BA,PhD *Professor, Head of Department* land-use and farming systems analysis - social causes of land degradation, Malaysia Pacific
2 Allen BJ - MA,PhD *Research Fellow* land-use systems and rural change; adaptions to frost in high altitude tropical areas; integration of cash crops; Papua New Guinea
3 Overton JD - MA,PhD *Research Fellow* Socio-economic differentiation in rural areas; impact of cash cropping on rural societies; Fiji, Kenya

750 Faculty of Science

Australian National University, GPO Box 4, Canberra City, A.C.T., 2601
Tel: (062) 493887. Telex: AA 61670. Cables: NATUNIV Canberra.

751 Department of Biochemistry

1 Howells AJ - BAgrSc,PhD *Senior Lecturer* molecular biology of insects

752 Department of Botany

Australian National University, GPO Box 4, Canberra City, 2601

1 Warren Wilson J - MA,DPhil,DSc,FIBiol *Professor* morphogenesis of tissues; growth analysis
2 Aston MJ - MScAgr,PhD *Senior Lecturer* plant water relations; physiology of grasses for grazing
3 Brittain EG - BSc,PhD *Visiting Fellow* computer modelling of plant physiological processes
4 Carnahan JA - MSc,PhD *Senior Lecturer* vegetation survey and mapping
5 Carroll BJ - BSc,PhD *Postdoctoral Fellow* soybean genetics, nodulation, nitrate metabolism/microbiology
6 Chilvers GA - BScAgr,PhD *Senior Lecturer* mycorrhizas, fungal ecology
7 Connell SW - BSc,PhD *Lecturer* wetland ecology, plant population ecology
8 Critchley C - DiplBiol,DrRerNat *Postdoctoral Fellow* photosynthesis and photoinhibition; salt tolerance
9 Day DA - BSc,PhD,DipTchg *Lecturer* plant cell biology and nitrogen fixation
10 Delves AC - BSc,DipEd,PhD *Lecturer* plant breeding, developmental plant genetics
11 Gresshoff PM - BSc,PhD *Senior Lecturer* developmental plant genetics/nitrogen fixation/legume nodulation
12 Hayes W - BA,MB,BCh,ScD,HonLLD,HonDSc,FRCPI,FAA,FRS,FRSEd *Emeritus Professor, Visiting Fellow* bacterial genetics
13 House SM - BSc,MSc,PhD *Lecturer* breeding systems; forest population ecology
14 Paton DM - BSc,PhD *Reader* eucalyptus growth; frost resistance
15 Price GD - BSc,MSc,PhD *Postdoctoral Fellow* inorganic carbon uptake in aquatic plants

16 Turner VB - BSc,PhD *Senior Tutor* plant reproductive ecology
17 Whitecross MI - MSc,PhD *Senior Lecturer* plant cell biology

753 Department of Zoology

1 Bryant C - MSc,PhD,FIBiol *Professor, Head of Department* parasite metabolism; comparative invertebrate biochemistry
2 Barwick RE - MSc,PhD *Senior Lecturer* reptile and amphibian ecology and physiology; functional analysis of the anatomy of lower vertebrates
3 Behm CA - BSc,PhD *Senior Technical Officer* parasite metabolism; comparative invertebrate biochemistry
4 Bennet E-M - BSc,PhD *Visiting Research Fellow* parasite biochemistry
5 Butcher G - BSc,PhD *Visiting Research Fellow* immunology and pathology of malaria babesiosis
6 Clark IA - BVSc,PhD *Visiting Research Fellow* immunology and pathology of malaria and babesiosis
7 Cockburn A - BSc,PhD *Lecturer* evolutionary theory
8 Fischer M - MSc,BSc,PhD *Postdoctoral Fellow* parasite molecular biology and immunology
9 Gullan PJ - BSc,PhD *Lecturer* entomology and arachnology; coccoid systematics and anatomy
10 Happold DCD - MA,PhD *Senior Lecturer* ecology and behaviour of mammals
11 Harris VAP - BSc,PhD *Senior Lecturer* marine littoral ecology; invertebrate behaviour
12 Howell MJ - BSc,MSc,PhD *Reader* parasite molecular biology and immunology
13 Janssens PA - BSc,PhD *Senior Lecturer* comparative vertebrate endocrinology and biochemistry; developmental biology
14 Marples TG - MSc,PhD *Senior Lecturer* ecological energetics and productivity; ecology of bats
15 Nicholas WL - BSc,PhD *Reader* nematology; parasitology

754 Department of Forestry

1 Griffin DM - MA,PhD,ScD,FRSA *Professor, Head of Department* forest pathology and microbial ecology; forestry in lesser developed countries
2 Bachelard EP - BScF,MF,PhD,FIFA *Professor* tree physiology; growth and development with particular emphasis on growth substances
3 Banks JCG - BScFor,MSc,PhD *Lecturer* dendrology; forest stand dynamics and genecology
4 Byron RN - BScFor,MA,PhD *Senior Lecturer* forestry for developing countries; long term forestry sector planning; private forestry investment; trade in forest products
5 Davey SM - BSc,BScFor *Senior Tutor* forest ecology; zoology; environmental management; silviculture
6 Evans PD - BSc,PhD *Lecturer* wood science, forest products utilization
7 Field JB - BSc,PhD *Lecturer* forest soils; geomorphology; pedology; hydrogeochemistry; remote sensing; hydrology; landscape development; geology
8 Florence RG - MScFor,PhD *Reader* indigenous forest ecology; silviculture; productivity of plantation ecosystems
9 Groves KW - BSc,MSc *Senior Lecturer* harvesting, conversion and forest products
10 Hillis WE - DSc,AGInstTech,FIAWs,FIAWSc,FRACI,FTS *Visiting Fellow* wood science; interaction of forestry practices on wood quality and bark utilization
11 Pryor LD - DSc,DipFor *Emeritus Professor, Visiting Fellow* genus Eucalyptus, tropical plantation forestry
12 Shepherd KR - BScFor,PhD,FIFA *Reader* plantation silviculture; tree genetics, physiology and conservation
13 Slee MU - MA,MSc,PhD *Senior Lecturer* tree breeding; silviculture of tropical and sub-tropical plantation species
14 Stodart DM - BE,MS *Senior Lecturer* forest engineering; hydrology; ground and aerial surveying; road transport systems
15 Tanton MT - BSc,PhD,DIC,ARCS *Senior Lecturer* forest zoology; wildlife and range management; soil fauna and litter breakdown; pesticides
16 Trevett ACF - BSc,PhD *Lecturer* environmental physics; fire science and management
17 Turner BJ - BScFor,MF,DFor *Reader* application of modern analytical techniques to solution of decision-making and planning problems in the management of forest resources; remote sensing
18 Wood GB - BScFor,DipFor,PhD *Senior Lecturer* mensuration; inventory; aerial surveying

755 The Flinders University of South Australia

Bedford Park, South Australia, 5042
Tel: (08) 2753911.

1 Hancock KJ - BA,PhD,FASSA *Vice Chancellor*

756 School of Biological Sciences
Bedford Park, South Australia, 5042
Tel: (08) 2753911.

1 Acton GJ - BSc,PhD *Senior Lecturer* plant hormones
2 Baudinette RV - BSc,PhD *Associate Professor* animal physiology
3 Bull CM - BSc,PhD *Senior Lecturer* ecology
4 Elliott DC - MSc,PhD *Senior Demonstrator* plant hormones

5 Kirby GC - BSc,PhD *Senior Lecturer* population biology
6 Wells RT - BSc,PhD *Senior Lecturer* arid zone ecology
7 Whalen MA - BSc,PhD *Lecturer* plant ecology

757 Griffith University

Kessels Road, Nathan, Queensland, 4111
Tel: (07) 275 7111. Telex: AA40362. Cables: Unigriff Brisbane.

1 Sewell A - ISO,AASA,ACIS,FIMA,FIDA *Chancellor*
2 Webb LR - BCom,PhD *Vice-Chancellor*

758 School of Australian Environmental Studies

1 Bodero J - BSc *Lecturer* statistical evaluation and prediction of milk yields; animal genetics
2 Boughton WE - ME,PhD *Associate Professor* hydrology, the effects of land use on catchment hydrology
3 Christie EK - MAgrSc,PhD *Associate Professor* resource management, agriculture in developing countries
4 McDonald GT - MA,PhD *Reader* land use analysis and planning
5 McTainsh GH - MA,PhD *Lecturer* soil erosion
6 Parlange JY - PhD *Professor* modelling environmetal problems; transport processes for water solutes particularly in the soil-plant-atmosphere continuum
7 Rickson RE - MA,PhD *Senior Lecturer* farmer response to soil conservation problems; strategic planning in rural communities
8 Rose CW - BSc,BE,PhD,FIP *Professor* agricultural and environmental physics
9 Saffigna PG - MAgrSc,PhD *Senior Lecturer* soil fertility

759 School of Science

1 Abrahams G - BSc,PhD *Senior Lecturer* virology; RNA transcription by bunyaviruses in cattle
2 Clegg DE - BSc,DipEd,PhD *Senior Lecturer* analytical chemistry and toxicology; levels of free fatty acids in butter

760 James Cook University of North Queensland

Post Office, James Cook University, Queensland, 4811
Tel: (077) 814111. Telex: 47009.

761 Department of Geography

1 Bandaranaike SD - BA,PhD *Lecturer* aquaculture management and fisheries economics
2 Catt PC - BSc *Lecturer* remote sensing applications in forestry and agriculture
3 Courtenay PP - BA,PhD *Associate Professor* agriculture in economic development
4 Jackson RT - BA,MA,PhD *Professor* land use and agriculture in Papua New Guinea
5 Valentine PS - BA *Lecturer* resource and forest management

762 Department of Marine Biology

1 Choat JH - MSc,PhD *Professor and Head of Department* reef fish ecology
2 Alexander CG - BSc,MSc,PhD *Senior Lecturer* invertebrate sense organs
3 Collins JD - BSc,PhD *Senior Tutor* coral biology
4 Denton GRW - MSc,PhD *Research Fellow* marine pollution and metal toxicity
5 Hartwick RF - BA,PhD *Senior Lecturer* behaviour of marine animals, biology of cubomedusae
6 Morrissey JI - BSc,PhD *Research Fellow* coral biology
7 Pichon MM - BSc,PhD,DSc *Associate Professor* benthic ecology of coral reefs

763 Graduate School of Tropical Veterinary Science

A postgraduate and research school with studies on animal health and production in northern Australia

1 Campbell RSF - PhD,MRCVS,FRCPath,FACVSc *Professor and Head of Department*

764 Division of Animal Health

1 Burgess GW - BVSc,PhD,MACVSc *Lecturer* infectious diseases of livestock
2 Copeman DB - BVSc,MSc,PhD,FACVSc *Associate Professor* parasitology of livestock, domestic and feral animals
3 Hirst RG - BVSc,DipBact,PhD,MRCVS *Senior Lecturer* infectious diseases of livestock
4 Hutchinson GW - BVSc,MSc,PhD *Senior Lecturer* parasitology of livestock, domestic and feral animals

5 Ladds PW - MVSc,PhD,MACVSc *Associate Professor* pathology and immunopathology of animals

765 Division of Animal Production

1 Entwistle KW - BVSc,PhD,QDAH,MACVSc *Associate Professor* reproductive physiology and breeding of livestock
2 Murray RM - MVSc,PhD,QDAH *Senior Lecturer* ruminant nutrition
3 Taylor JF - BSc,PhD *Lecturer* biometrics of livestock production
4 Teleni E - BAgrSci,MSc,PhD *Lecturer* ruminant nutrition

766 Macquarie University

767 School of Biological Sciences
North Ryde, New South Wales, 2113
Tel: (02) 888 8000. Telex: MACQUNI AA22377.

1 Cooper DW - BSc,PhD *Professor of Biology, Head of School* genetics; immunology
2 Adamson DA - BScAgr,PhD *Senior Lecturer* quaternary studies
3 Adamson HY - BSc,Phd,Phd *Senior Lecturer* plant physiology
4 Barlow EWR - MRur,PhD *Senior Lecturer* plant physiology
5 Bassett JR - MSc,PhD *Senior Lecturer* animal physiology
6 Briscoe DA - BSc,PhD *Senior Lecturer* genetics and zoology
7 Burrows FJ - BSc,PhD *Senior Lecturer* ecology
8 Cairncross KD - MA,PhD *Associate Professor in Biology* animal physiology
9 Frankham R - BScAgr,PhD *Associate Professor in Biology* animal genetics
10 Gebicki JM - BSc,PhD *Associate Professor in Biology* physical biochemistry
11 Hales DF - BSc,PhD *Senior Lecturer* entomology, invertebrate zoology
12 Hiller RG - BA,PhD *Senior Lecturer* biochemistry
13 Hinde RW - BSc,PhD *Senior Lecturer* biochemistry
14 Ivantsoff V - BSc,PhD *Senior Lecturer* fish systematics
15 Johnston PG - MScAgr,PhD *Senior Lecturer* cytogenetics, biochemical genetics
16 Joss JMP - MSc,PhD *Senior Lecturer* animal physiology
17 McKay GM - BSc,PhD *Senior Lecturer* mammalian evolution
18 Oldfield RJ - BSc *Senior Lecturer* microscopy
19 Pollard I - BSc,PhD *Senior Lecturer* animal physiology
20 Reed ML - BScAgr *Senior Lecturer* plant physiology
21 Robinson ES - MSc,PhD *Associate Professor in Biology* cytogenetics; developmental biology
22 Selkirk P - BSc,PhD *Lecturer* bryology
23 Smith E - BSc,PhD *Lecturer* developmental biology
24 Stokes HW - BSc,PhD *Lecturer* genetics; microbiology
25 Tait NN - BSc,MSc,PhD *Senior Lecturer* invertebrate zoology
26 Westoby M - BSc,PhD *Senior Lecturer* ecology
27 Whalley JM - BSc,PhD *Senior Lecturer* virology
28 Williams KL - BAgrSc,PhD *Professor of Biology* biotechnology; developmental biology

768 Department of Zoology
James Cook University, Townsville, Queensland, 4811
Tel: 814530. Telex: AA47009.

1 Jones RE - BSc,PhD *Professor and Head of Department* insect population ecology and behaviour
2 Heinsohn GE - MA,PhD *Senior Lecturer* marine and terrestrial native mammals
3 Kenny RP - BSc,PhD *Associate Professor* intertidal ecology
4 Lucas JS - BSc,PhD *Associate Professor* bivalve mariculture, starfish ecophysiology
5 Marsh HE - BSc,PhD *Research Fellow* mammalian ecophysiology and behaviour
6 Milward NE - MSc,PhD *Senior Lecturer* fisheries biology
7 Pearson RG - BSc,PhD *Senior Tutor* freshwater ecology

769 La Trobe University

770 School of Agriculture
Bundoora, Victoria, 3083
Tel: 478 3122. Telex: (AA)33143.

1 Carnegie PR - BSc,PhD *Professor* animal science
2 Van Steveninck RFM - Ir Ingineur,PhD *Professor* plant and soil sciences
3 Bell CJ - BSc,PhD *Lecturer* plant and crop physiology
4 Cranwell PD - BAgrSc,MAgrSc *Senior Lecturer* physiology of digestion in young pig
5 Dumsday RG - BAgrSc,PhD *Senior Lecturer* agricultural economics
6 Edwards GW - BAgrSc *Senior Lecturer* agricultural economics
7 Foster WNM - MA,DPhil,BVMS,MRCVS *Senior Lecturer* parasitism in domestic animals
8 Gow CB - BScAgr,PhD *Lecturer* animal physiology
9 Lamp CA - MAgrSc,PhD *Senior Lecturer* plant nutrition
10 Luke RKJ - BAgrSc,PhD *Senior Lecturer* biosynthesis in micro-organisms
11 Price TV - BSc,PhD *Senior Lecturer* plant pathology

12 Quilkey JJ - BEc,PhD *Senior Lecturer* agricultural economics - policy and marketing
13 Silvapulle MJ - BSc,PhD *Lecturer* statistics and computing
14 Uren NC - BAgrSc,PhD,DipEd *Senior Lecturer* chemistry and availability of plant nutrients
15 Willatt ST - BSc,MSc *Senior Lecturer* soil physical environment and plant growth

771 Monash University
Clayton, Victoria, 3168

772 Department of Botany
Clayton, Victoria, 3168

1 Holland AA - MSc,PhD *Associate Professor* fungal taxonomy and mycorrhiza
2 Gaff DF - MSc,PhD *Senior Lecturer* desiccation tolerant grasses
3 O'Brien TP - BAgrSc,MSc,PhD,DSc *Reader* plant cell biology; plant-insect interactions

773 Murdoch University
South Street, Murdoch, W. Australia, 6150

774 School of Environmental and Life Sciences
South Street, Murdoch, W. Australia, 6150

1 Potter IC - BA,PhD *Dean and Professor* animal biology
2 Borowitzka MA - BSc,PhD *Lecturer* marine phycology
3 Bradley JS - BSc,PhD *Lecturer* animal genetics
4 Cake MH - BSc,PhD *Senior Lecturer* cell biology
5 Dell B - BSc,PhD *Lecturer* plant biology
6 Dilworth MJ - BScAgr,PhD *Professor* microbiology and biochemistry
7 Glenn AR - BSc,PhD *Associate Professor* microbiology and biochemistry
8 Lethbridge RC - BSc,PhD,DIC,ARCS *Senior Lecturer* parasite and invertebrate biology
9 Loneragan JF - BSc,PhD *Professor* plant biology
10 Macey DJ - BSc,PhD *Lecturer* animal physiology
11 McComb Ms JA - BSc,PhD *Senior Lecturer* plant cytogenetics
12 Mead RJ - BSc,PhD *Lecturer* biochemistry
13 Wooller RD - BA,PhD,CertEd *Senior Lecturer* animal ecology and ethology

775 School of Veterinary Studies
South Street, Murdoch, W. Australia, 6150

1 Howell J MCC - BVS,PhD,DVSc,FRCPath,MRCVS,MACVSc *Dean and Professor* animal pathology

776 Division of Applied Veterinary Medicine
South Street, Murdoch, W. Australia, 6150

1 Swan RA - BVSc,PhD,MPVM,MRVS *Head of Division and Professor* veterinary clinical studies
2 Bolton JR - BVSc,PhD,MACVSc *Senior Lecturer* large animal medicine
3 Buddle JR - BSc,BVSc,DVPH,MACVSc *Senior Lecturer* pig medicine and production
4 Chapman HM - BVSc,PhD *Senior Lecturer* sheep medicine and production
5 Clark WT - BVMS,PhD,FRCVS *Professor* small animal medicine and surgery
6 Cullen LK - BVSc,MA,MVSC,DVA,FACVSc,MRCVS *Senior Lecturer* anaesthesiology
7 Eger CE - BVSc,MSc,DiplSmallAnimalSurg *Senior Lecturer* small animal medicine and surgery
8 Fraser DMCK - BVSc,QDAH *Senior Lecturer* dairy cattle medicine, production and surgery
9 Hawkins CD - BVSc,PhD *Lecturer* epidemiology and state medicine
10 Mills JN - BVSc,MSc,DCP *Senior Lecturer* clinical pathology
11 Read RA - BVSc *Lecturer* small animal surgery
12 Seneviratna P - BVSc,PhD,FRCVS,MACVSc *Senior Lecturer* veterinary public health
13 Shaw SE - BVSc *Lecturer* small animal medicine
14 Sutherland RJ - BVSc,MACVSc,DACVP *Associate Professor* clinical pathology
15 Taylor EG - BVSc,QDAH,MACVSc *Senior Lecturer* beef cattle medicine and production
16 Williamson PE - BVSc,PhD *Senior Lecturer* reproduction
17 Wyburn RS - BVMS,PhD,DVR,FACVSc,MRCVS *Associate Professor* veterinary medicine and surgery - radiology

777 Division of Veterinary Biology

1 Wales RG - BVSc,PhD,DVSc,MACVSc *Head of Division and Professor* physiology
2 Cook RD - BSc,PhD *Associate Professor* histology
3 Costa ND - BAgrSc,PhD *Lecturer* biochemistry
4 Creed KE - BA,BSc,PhD *Senior Lecturer* physiology
5 Dorling PR - BSc,PhD *Senior Lecturer* toxicology and chemical pathology
6 Dunsmore JD - BVSc,PhD *Associate Professor* parasitology

7 Gawthorne JM - BSc,PhD *Associate Professor* biochemistry
8 Grandage J - BVetMed,DVR,MRCVSc *Associate Professor* anatomy
9 Huxtable CRR - BVSc,PhD,MACVSc *Associate Professor* pathology
10 Johnson KG - MRurSc,PhD *Senior Lecturer* physiology
11 Pass DA - BVSc,MSc,PhD,ACVP *Senior Lecturer* pathology
12 Penhale WJ - BVSc,PhD,DipBact,MRCVS *Associate Professor* microbiology and immunology
13 Pethick DW - BAgSc,PhD *Lecturer* biochemistry
14 Reynoldson JA - BSc,PhD *Senior Lecturer* pharmacology and chemotherapy
15 Richardson KC - BSc,BVSc,ACVM *Senior Lecturer* anatomy
16 Thompson RCA - BSc,PhD,DIC *Senior Lecturer* parasitology
17 White SS - BVMS,PhD,DA,MRCVS *Senior Lecturer* anatomy
18 Wilcox GE - BVSc,PhD *Associate Professor* virology

778 The University of Adelaide

Box 498 G.P.O., Adelaide, South Australia, 5001
Tel: (08) 228 5280. Telex: UNIVAD AA 89141.

779 Department of Botany

1 Ganf GG - PhD *Senior Lecturer and Chairman* limnology, wetlands
2 Christophel DC - PhD *Senior Lecturer and Deputy Chairman* palaeobotany; systematics; rainforest floristics
3 Lange RT - PhD *Reader* arid zone and plant ecology
4 Martin PG - PhD *Emeritus Professor* phylogeny through amino acid sequences of proteins
5 Sinclair R - PhD *Lecturer* environmental physiology
6 Smith FA - PhD *Reader* ion transport, plant nutrition and photosynthesis
7 Wiskich JT - PhD *Reader* plant biochemistry; respiration; photosynthesis and organelle physiology
8 Wollaston EM - PhD *Reader* marine plant biology; plant morphology and anatomy
9 Womersley HBS - PhD,DSc *Professor* marine plant biology
10 Wood A - PhD *Senior Tutor* plant nutrition; hydroponics

780 Department of Zoology

1 Barker S - PhD *Senior Lecturer and Acting Chairman* taxonomy of Buprestidae: ecophysiology of Australian marsupials
2 Butler AJ - PhD *Senior Lecturer* marine littoral zoology
3 Davies M - MSc *Principal Tutor* evolution of Australian frogs
4 Duckhouse DA - PhD *Senior Lecturer* taxonomy: distribution and ecology of Psychodidae (Diptera, Nematocera), especially in southern hemisphere
5 Geddes MC - PhD *Senior Lecturer* taxonomy and distribution of Australian anostraca; economics of osmotic regulation; limnology of Lake Alexandrina
6 Manwell C - PhD *Professor* application of biochemical techniques, especially zone electrophoresis, to biological problems
7 Seymour RS - PhD *Senior Lecturer* dinosaur physiology and ecology; reptile cardiovascular and respiratory physiology; insect thermoregulation; egg respiration
8 Sommerville RI - PhD *Reader* parasitism; the nature and mechanism of infection; hormonal control of growth of parasitic nematodes; biology of reproduction in invertebrates
9 Tyler MJ - MSc *Reader* evolution of Australian frogs
10 Walker KF - PhD *Senior Lecturer* ecology of River Murray with special reference to freshwater mussels and fish
11 Wells A - BSc *Principal Tutor* taxonomy of Australian Hydroptilidae (Trichoptera)
12 Williams WD - DSc *Professor* taxonomy of Australian freshwater amphipods and atyids; implications of man's changes to the inland aquatic environment; ecology of inland salt lakes

781 Waite Agricultural Research Institute, The University of Adelaide

1 Quirk Professor JP - FAIAS,FRACI,FSSA,FASA,FANZAAS,FTS,FAA *Director* soil chemistry; soil physics
2 Andrewartha HG - MAgrSc,DSc,FAA *Emeritus Professor of Zoology, Research Associate* population ecology; diapause in insects
3 Catcheside DG - DSc,MA,FRS,FAA *Emeritus Professor of Genetics, Research Associate* genetic exchange in eukaryotes, taxonomic study of mosses
4 Morgan FD - MSc,PhD *Research Associate* attractants, aggregation pheromones and ecology of insects that attack trees
5 Murray RS - BSc,PhD soil swelling and micro-structure
6 Rogers WP - MSc,PhD,DSc,FAA *Emeritus Professor of Parasitology, Research Associate* physiology of infectivity of nematodes

782 Department of Agricultural Biochemistry

1 Nicholas DJD - MA,PhD,AKC,FRIC,FRACI *Waite Professor of Agricultural Biochemistry and Chairman of Department* metabolism of inorganic nitrogen and sulphur compounds in microorganisms and plants, function of trace elements
2 Jackson JF - MSc,PhD pollen biochemistry, self-incompatibility in angiosperms
3 Jennings AC - MSc,PhD biochemistry of compounds of low molecular weight in stressed plants

4 Langridge P - BSc,PhD molecular genetics of plants and bacteria
5 Liljegren DR - BSc,PhD biochemistry of plant-insect interactions
6 Tate ME - MSc,PhD structural studies on biologically active molecules, esp. nucleotide bacteriocins
7 Wallace W - BSc,PhD biochemistry and physiology of nitrogen metabolism in plants

783 Department of Agronomy

1 Driscoll CJ - MScAgr,PhD,DSc *Waite Professor of Agronomy and Chairman of Department* cytogenetical investigations of wheat and related species
2 Carter ED - MAgrSc pasture ecology; production, management and utilization
3 Graham RD - BSc,MS,PhD trace element nutrition of cereals; physiology of copper in plants; role of plant nutrition in plant-water relations and resistance to disease
4 Knight R - BAgrSc,DipAgrSc,DipTropAgr,PhD plant breeding methodology genotype-environment interactions, breeding of faba beans
5 Lance RCM - BAgrSc barley breeding; malting quality of barley; barley agronomy
6 Powles SB - DipAppSc,MScAgric,PhD weed science, herbicide resistance, photosynthesis
7 Rathjen AJ - BAgrSc,PhD wheat breeding
8 Shepherd KW - BAgrSc,PhD wheat cytogenetics and wheat protein inheritance
9 Silsbury JH - MAgrSc,PhD photosynthesis, respiration and nitrogen fixation of legumes
10 Sparrow DHB - BSc,PhD barley breeding; malting quality of barley

784 Department of Animal Sciences

1 Setchell BP - BVSc,PhD,ScD *Waite Professor of Animal Sciences and Chairman of Department* physiology, biochemistry and endocrinology of the testis
2 Brooker JD - BSc,PhD control of gene activity in cultured animal embryos, studied using recombinant DNA technology, monoclonal antibodies; production of transgenic animals using isolated embryos; genetic manipulation of rumen microorganisms
3 Brooks DE - BAgSc,PhD development of fertilizing capacity of spermatozoa passing through the epididymis, secretory processes in the epididymis and their control, protein and lipid changes of the sperm plasma membrane during epididymal transit
4 Geytenbeek PE - BAgSc sheep production systems with particular emphasis on increased efficiency in reproduction, increased lambing frequency, environmental factors and hormonal mechanisms in the initiation of oestrus in the post-partum period
5 Hynd PI - BRurSc,PhD manipulation of the rumen ecosystem to enhance fibre degradation, microbial synthesis, and postruminal nutrient supply; genetically transformed rumen microorganisms; cell division and migration in wool follicle bulbs
6 Read LC - BAgSc,PhD role of growth factors in regulation of growth of animals; effects of growth factors in milk on gastric intestinal function
7 Sabine JR - MAgrSc,PhD endocrinological and metabolic control of fat deposition and mobilization; influence of lipids on membranes, particularly during development of cancer; goat husbandry
8 Snoswell AM - BSc,PhD control of lipid metabolism in animals particularly the roles of carnitine and choline; relationship between volatile fatty acid production and dietary fibre

785 Department of Entomology

1 Pinnock DE - BSc,PhD,DIC,ARCS *Waite Professor of Entomology and Chairman of Department* insect pathology and microbial control of insect pests
2 Austin AD - BSc,PhD systematics and biology of insects and spiders
3 Laughlin R - MA,PhD dispersal and migration of insects; ecology of thrips, stored products, beetles, mosquitoes
4 Maelzer DA - BSc,PhD integrated control of horticultural and pasture pests, population dynamics and computer modelling
5 Miles PW - MSc,PhD physiology of host resistance and saliva-host interactions in Hemiptera

786 Department of Plant Pathology

1 Wallace HR - PhD,DSc,FAA *Waite Professor of Plant Pathology and Chairman of Department* ecology, physiology and nematology
2 Carter MV - BAgrSc,MAgrSc,PhD epidemiology and control of air-borne fungi
3 Clare BG - BSc,PhD physiology of fungi and physiology of parasitism
4 Fisher JM - BScAgr,PhD physiology and resistance of nematodes
5 Francki RIB - BSc,MSc,PhD structure, multiplication and transmission of viruses
6 Kerr A - BSc,PhD,FAA *Professor* bacterial diseases and disease resistance
7 Randles JW - MAgrSc,PhD virus diseases
8 Warcup JH - MSc,PhD,ScD ecology of soil fungi

787 Department of Plant Physiology

1 Aspinall D - BSc,PhD *Chairman of Department* environmental control of plant growth and development

2 Coombe BG - BAgrSc,MAgSc,PhD,FAIAS physiology of fruit set and fruit growth
3 Jenner CF - BSc,MSc,DPhil physiological and environmental aspects of cereal grain growth and development
4 Nicholls PB - BSc,PhD cereal grain development and hormonal sensitivity of the aleurone layer
5 Paleg LG - BA,MS,PhD,DSc *Professor of Plant Physiology* hormones, membranes, sterols, stress metabolism, membrane structure and function
6 Sedgley M - BSc,PhD reproductive biology of horticultural crops

788 Department of Soil Science

1 Oades JM - BSc,PhD *Chairman of Department* iron and aluminium oxides, organic matter and soil structural stability, soil biochemistry
2 Alston AM - BSc,PhD soil fertility and nitrogen and phosphorus; root growth and nutrient uptake
3 Chittleborough DJ - MAgSc,PhD pedogenesis, soil classification, processes of textural differentiation
4 Dexter AR - MSc,PhD soil structure and strength, and their effects on root growth
5 Lewis DG - MSc,PhD soil colloid chemistry and mineralogy; phosphate fixation and mobility

789 Biometry Section

1 Mayo O - BSc,PhD,DipBusMan *Officer in Charge* quantitative and population genetics; plant breeding
2 Eckert SR - BSc,PhD,DipCompSc statistics, computing
3 Hancock TW - MAgrSc quantitative genetics; experimental design
4 Morris MM - BSc statistics
5 Street DJS - BSc,PhD experimental design

790 University of Sydney

University of Sydney, Sydney, New South Wales, 2006
Tel: (02) 692 2222. Telex: AA 26169 UNISYD. FAX (02) 692 4203.

791 Department of Agricultural Chemistry

1 Turner JF - MSc,Phd,ScD,FRSC,FRACI *Professor and Head of Department* carbohydrate metabolism in plants; enzymology
2 Caldwell RA - MSc,PhD *Lecturer* physico-chemical methods in biochemistry; metal-protein interactions
3 Copeland L - BSc,PhD *Senior Lecturer* carbohydrate metabolism in legume nodules; enzyme kinetics
4 Kennedy IR - BScAgr,PhD *Senior Lecturer* enzymology of nitrogen fixation; pesticide detoxification
5 Lees EM - BSc,PhD *Senior Lecturer* nitrogen metabolism in plants
6 Matheson NK - MSc,PhD *Associate Professor* chemistry and metabolism of carbohydrates

792 Department of Agricultural Economics

1 Fisher BS - BScAgr,PhD *Professor, Head of Department* agricultural policy; market and price analysis
2 Batterham RL - BAgrEc,MS,PhD *Senior Lecturer* agricultural finance; production economics and farm management
3 Davidson BR - MAgrSc,PhD *Senior Lecturer* production economics and land use; project appraisal, agricultural history
4 Tanner C - BScAgr *Senior Lecturer* agricultural policy; international trade

793 Department of Agricultural Genetics and Biometry

794 Genetics and Plant Breeding

1 Latter BDH - BScAgr,PhD *Professor of Plant Breeding and Head of Department* quantitative genetics
2 Darvey NL - BSc,PhD *Senior Lecturer* cytogenetics
3 McWhirter KS - BScAgr,PhD *Senior Lecturer* genetics of maize and sorghum

795 Biometry

1 O'Neill ME - BSc,PhD *Senior Lecturer* distribution theory; contingency tables

796 Plant Breeding Institute

797 at Sydney

Faculty of Agriculture, University of Sydney, Sydney, New South Wales, 2006
Tel: (02) 692 2937.

1 Latter BDH - BScAgr,PhD *Professor and Director* wheat breeding methodology
2 Darvey NL - BSc,PhD *Senior Lecturer* triticale breeding
3 McWhirter KS - BScAgr,PhD *Senior Lecturer* soybean breeding; maize breeding

798 at Narrabri

I A Watson Wheat Research Centre, PO Box 219, Narrabri, New South Wales, 2390
Tel: (067) 921588.

1 Marshall DR - BScAgr,PhD *Director of I A watson Wheat Research Centre* wheat breeding; population genetics
2 Bhatt GM - BSc,MS,PhD *Senior Plant Breeder* wheat breeding; quantitative inheritance in wheat
3 Ellison FW - MScAgr,PhD *Senior Plant Breeder* wheat breeding; breeding methodology
4 Mares DJ - BSc,PhD *Senior Cereal Biochemist* sprouting damage in wheat; cereal quality

799 at Castle Hill

PO Box 180, Castle Hill, New South Wales, 2154
Tel: (02) 634 2552.

1 McIntosh RA - MScAgr,PhD *Director of Rust Research* cytogenetics of wheat; breeding for rust resistance in wheat
2 Oates JD - BScAgr *Officer-in-Charge* seed storage; oat rusts
3 The TT - MScAgr,PhD *Senior Research Fellow* cytogenetics of wheat; breeding for rust resistance in wheat

800 Department of Agronomy and Horticultural Science

1 Mullins MG - BSc,DipAgr,PhD *Professor of Horticulture, Head of Department, Dean, Faculty of Agriculture* genetic improvement of woody plants
2 Campbell LC - BSc,PhD *Senior Lecturer* crop nutrition
3 Clemens J - BSc,PhD *Senior Lecturer* horticulture
4 De Kantzow D - BScAgr *Lecturer* tillage and remote sensing
5 Goodwin PB - MScAgr,PhD *Senior Lecturer* cell to cell communication in plants, development of native flora
6 Greenhalgh WJ - MScAgr,PhD *Senior Lecturer* fruit and vegetable production, education
7 Michael PW - BScAgr,PhD *Senior Lecturer* weed taxonomy, ecology and control
8 Pearson CJ - BScAgric,MSc,PhD *Professor of Agronomy* cereal and pasture agronomy
9 Searle PGE - MAgrSc,PhD *Senior Lecturer* agronomy
10 Sutton BG - BAgrSc,PhD *Senior Lecturer* crop physiology, irrigation agronomy

801 Department of Animal Husbandry (Camden)

University of Sydney, Camden, New South Wales, 2570
Tel: (046) 552300.

1 Annison EF - PhD,DSc,FRSC *Professor and Head of Department* ruminant nutrition and biochemistry
2 Miller BG - BVSc,PhD *Garland Senior Lecturer* endocrinology of reproduction
3 Moore NW - MAgSc,PhD *Associate Professor* sheep husbandry, physiology of reproduction

802 Dairy Husbandry Research Unit

1 McDowell GH - BAgSc,PhD *Associate Professor* immunology, physiology of lactation
2 Gooden JM - BAgSc,PhD *Senior Lecturer* biochemistry and physiology of lactation

803 University of Sydney M C Franklin Laboratory

1 Leibholz Miss JM - DAgSc,MS,PhD *Reader* pig nutrition, ruminant nutrition
2 Kellaway RC - BScHort,DipTropAg,PhD *Associate Professor and William McIlrath Fellow* ruminant nutrition

804 Poultry Foundation Research Unit

1 Balnave D - PhD,DSc,FRSC *Associate Professor* poultry nutrition and metabolism
2 Bryden WL - MRurSc,DipEd,PhD *Lecturer* monogastric nutrition, mycotoxins

805 School of Biological Sciences

University of Sydney, Sydney, New South Wales, 2006
Tel: (02) 692 2222. Telex: UNISYD 26169. (02) 692 4203.

1 Anderson DT - PhD,DSc,FRS,AO *Professor of Biology and Head of School* embryology of invertebrates
2 Allaway WG - MA,PhD *Senior Lecturer* plant physiology
3 Brown KD - MSc,PhD *Senior Lecturer* molecular genetics
4 Carolin RC - BSc,PhD *Associate Professor* taxonomy of Australian plants
5 Catanzaro D - BSc,PhD *Lecturer* eukaryote genetics
6 Crossley AC - BSc,PhD *Associate Professor* insect physiology
7 Gillies CB - MAgrSc,PhD *Senior Lecturer* ultrastructure of chromosome
8 Grigg GC - BSc,PhD *Associate Professor* physiology of amphibians and reptiles
9 Hinde RT - BSc,PhD *Senior Lecturer* marine biology

10 Lanzing WJR - DrMathsNatSc *Senior Lecturer* fish physiology and histology
11 Larkum AWD - BSc,DPhil,ARCS *Associate Professor* plant physiology
12 Martin ARH - MA,PhD *Senior Lecturer* palynology
13 Meats AW - BSc,PhD *McCaughey Senior Lecturer in Entomology* water relations in insects
14 Morris VB - BSc,PhD *Reader* developmental biology and ultrastructure of the vertebrate retina
15 Myerscough PJ - BA,Dphil *Senior Lecturer* population dynamics of plants
16 Overall RL - BSc,PhD *Lecturer* plant development
17 Sale PF - BSc,MA.PhD *Associate Professor* behaviour and ecology of reef fishes
18 Shine R - BSc,PhD *Senior Lecturer* vertebrate ecology and evolution
19 Sved JA - BSc,PhD *Reader* population genetics
20 Thomson JA - MSc,MAgrSc,PhD *Professor of Biology* genetics
21 Underwood AJ - PhD,DSc *Reader* intertidal zonation
22 Valder PG - BScAgr,PhD *Senior Lecturer* mycology
23 Walker NA - BSc,PhD *Professor* cellular biology

806 Department of Microbiology

1 Reeves PR - BSc,Phd *Professor and Head of Department* genetics and function of bacterial surface components
2 Cho KY - BSc,PhD,ARACI *Senior Lecturer* chemistry of bacterial membranes, morphogenesis of edible mushrooms and their cultivation
3 Duxbury T - BSc,PhD *Senior Lecturer* microbial ecology
4 Ferenci T - BSc,PhD *Senior Lecturer* transport mechanisms in microbes
5 New PB - BSc,PhD *Senior Lecturer* microbial ecology, nitrogen fixation

807 Laboratory of the Dairy Corporation of New South Wales

1 Davey G - BSc,MSc *Officer-in-Charge* ecology and physiology of microorganisms of public health significance in milk and dairy products
2 Murrell WG - BSc,PhD,DSc *Research Affiliate* bacterial spores, food microbiology clostridial toxin

808 Department of Plant Pathology and Agricultural Entomology

809 Plant Pathology

1 Deverall BJ - BSc,DIC,PhD *Professor and Head of Department* physiology and biochemistry of disease resistance in plants
2 Bowyer JW - BAgrSc,PhD *Senior Lecturer* viruses as causes of disease in plants
3 Burgess LW - BScAgr,DipEd,PhD *Senior Lecturer* the ecology of root infecting fungi

810 Agricultural Entomology

1 McDonald FJD - MSc,PhD *Associate Professor* insect taxonomy and ecology
2 Rose HA - MAgrSc,PhD *Senior Lecturer* toxicological studies on pesticides

811 Department of Soil Science
Tel: (02) 692 2946.

1 Collis-George N - MSc,PhD.F.R.S.Chem *Professor* infiltration; air encapsulation; seedling emergence
2 Davey BG - BScAgr,PhD *Senior Lecturer* chemistry of aluminium, manganese, phosphorus and trace elements, pollution, soil acidity
3 Geering HR - MS *Senior Lecturer* dispersion; infiltration; soil colloids
4 Gerth J - DIngAgr,PhD *Post Doctoral Fellow* trace metals in iron oxides
5 Gibson NA - MSc,PhD,MRIC,ARACI,Inorg.Chem *Research Affiliate* metal extraction; environmental analysis
6 Koppi AJ - BSc,PhD *Lecturer* microscopy; weathering and soil formation; soil survey and land evaluation
7 Yates D - BSc *Research Assistant* characterisation of soil water

812 Department of Veterinary Anatomy

1 Butterfield Professor RM - DVSc,PhD,MVSc,FACVSc *Professor, Head of Department* growth and development of meat animals
2 Farrow BRH - PhD,BVSc,FACVSc *Associate Professor* neurology - lysosomal storage diseases
3 Halnan Dr CRE - PhD,MA,VetMB,FRCVS,MACVSc *Senior Lecturer* chromosomes in domesticated animals
4 Hopwood Dr PR - PhD,BVSc *Senior Lecturer* growth and development of kangaroos
5 McCarthy Dr PH - PhD,MVSc,Drmedvet *Senior Lecturer* surface anatomy of domestic animals

813 Department of Veterinary Clinical Studies

1 Edwards MJ - DVSc,PhD,MRCVS,MACVSc *Professor of Veterinary Clinical Studies* teratology, obstetrics, infertility
2 Bellenger CR - BVSc,PhD,FRCVS *Associate Professor* orthopaedics, infertility, intestinal surgery

3 Church DC - BVSc,PhD,MACVSc *Lecturer* endocrinology, cardiology
4 Davis PE - MVSc,MRCVS *Senior Lecturer* haemotology, cardiology
5 Dixon RJ - BScVet,BVSc,PhD *Lecturer* oncology, exotic diseases
6 Egerton JR - BVSc,DipBact,MACVSc,MASM *Head and Professor of Veterinary Clinical Studies* foot rot, antibiotic therapy
7 English AW - BVSc,PhD,MACVSc *Senior Lecturer* parasitology, diseases of deer
8 Hutchins DR - BVSc,MACVSc *Superintendent and Associate Professor* surgery and medicine of horses
9 Hyne RHJ - BVSc,MACVSc *Senior Clinical Instructor* wildlife studies
10 Ilkiw JE - BVSc,PhD,DipVetAn,MACVSc *Senior Lecturer* anaesthesia, tick paralysis
11 Johnson KA - MVSc,PhD,FACVSc,FACBSc *Lecturer* orthopaedics
12 Johnston KG - BVSc,DipBact,MACVSc,MASM *Senior Clinical Veterinary Pathologist* pathology of infectious diseases, mastitis
13 Keep JM - BVSc,MRCVS,MACVSc *Superintendent and Associate Professor* wildlife studies
14 Love RJ - MVSc,PhD,MACVSc *Senior Lecturer* infertility and enteric diseases of pigs
15 McClure TJ - BVSc,PhD,MACVSc *Associate Professor* bovine nutritional infertility
16 Porges WL - BVSc,PhD,HHDA *Senior Lecturer* cardiology
17 Rawlinson RJ - BVSc,DipVetRad,FACVSc *Senior Lecturer* horse medicine, ultrasonic techniques
18 Rose RJ - BVSc,PhD,DipVetAn *Associate Professor* exercise physiology, diseases of horses
19 Watson ADJ - BVSc,PhD,FRCVS *Associate Professor* antibiotic therapy, haemopoiesis
20 Wood AKW - MVSc,PhD,DipVetRad,MACVSc *Senior Lecturer* radiology

814 Department of Veterinary Pathology

1 Canfield PJ - BVSc,PhD *Senior Lecturer* clinical pathology and wildlife pathology
2 Collins GH - BVSc,PhD,MRCVS *Senior Lecturer* veterinary parasitology and protozoology
3 Gallagher CH - PhD,DVSc,FRCPath,FACVSc *Professor, Head of Department* biochemical pathology of plant poisonings, trace metal deficiencies and chemical and UV carcinogenesis
4 Herbert L - BVSc *Lecturer* microbiology
5 Love DN - BVSc,PhD,MASM *Associate Professor* anaerobic bacterial infections and microbiology generally
6 Moore J - BVSc *Tutor* pathology
7 Oxenford C - BVSc *Tutor* clinical microbiology
8 Reeve VE - BSc,PhD *Senior Research Assistant* biochemical pathology
9 Rothwell TLW - BVSc,PhD,MACVSc *Associate Professor* pathology and parasite immunology
10 Sabine JM - MSc,PhD,MASM *Associate Professor* viral diseases
11 Sangster N - BScVet,BVSc,PhD *Research Fellow* parasitology

815 Department of Veterinary Physiology

1 Stone GM - BScAgr,PhD *Associate Professor, Head of Department* reproductive physiology, endocrinology
2 Gidley-Baird AA - BSc,PhD *Senior Lecturer* reproductive physiology, endocrinology
3 Heywood LH - BVSc,PhD *Lecturer* control of digestive function
4 Martin ICA - BVSc,PhD *Reader* artificial insemination, physiology of spermatozoa
5 Titchen DA - MA,BVSc,PhD,DSc *Professor* physiology of the digestive tract
6 White IG - PhD,DSc *Reader* biochemistry of spermatozoa

816 The University of Newcastle

New South Wales, 2308
Tel: 680 401.

817 Department of Biological Sciences

1 Jones RC - BSc,PhD *Associate Professor* reproductive physiology
2 Murdoch RN - BSc,PhD *Senior Lecturer* reproductive physiology
3 Offler CE - BSc,PhD *Lecturer* plant physiology, ecology
4 Patrick JW - BScAgr,PhD *Senior Lecturer* plant physiology, ecology
5 Roberts TK - BSc,PhD *Associate Professor* reproductive immunology
6 Rodgers JC - MSc,PhD *Lecturer* reproductive physiology
7 Rose RJ - BScAgr,PhD *Senior Lecturer* plant physiology

818 Department of Economics

1 McShane RW - BA,MA *Senior Lecturer* economics of the meat industry
2 Tisdell CA - BCom,PhD *Professor of Economics* economics of wildlife

819 Department of Geography

1 McIntyre GN - BA,MA *Lecturer* agricultural climatology
2 Turner JC - BScAgr,MS,PhD *Senior Lecturer* rain forest ecology, biogeography

820 Department of Agricultural Economics and Business Management

1 Anderson JR - MAgrSc,PhD,DEc *Professor, Head of Department* resource allocation under risk, research planning, agricultural policy, farming systems research
2 Dillon JL - BScAgr,PhD,FAASA *Professor* farm management, production economics, decision theory, farming systems research, international agriculture research, agriculture in LDCs
3 Fleming EM - BCom,MEc *Lecturer* economics of agricultural development, project appraisal and management, agricultural market development
4 Hardaker JB - MSc,PhD,NDA *Associate Professor* agricultural development in LDCs, decision analysis, quantitative methods in agricultural management
5 James PG - MA,PhD,DipAgSci *Associate Professor* agricultural policy, agricultural marketing
6 MacAulay TG - MAgSci,PhD *Lecturer* agricultural policy and agricultural price analysis and marketing; commodity forecasting
7 Makeham JP - BAgSc *Senior Lecturer* farm management, economic and social aspects of adjustment, economics of pasture establishment
8 Musgrave WF - MScAgr,PhD *Professor* agricultural policy, water resource economics, benefit-cost analysis
9 Pearse RA - MSc,PhD *Senior Lecturer* poly-period linear programming, taxation effects on farm development, economics of stocktaking rates, farm management gaming, farm record keeping and analysis.
10 Piggott RR - MAgEc,PhD *Senior Lecturer* agricultural price analysis, agricultural marketing
11 Powell RA - BCom,DipEd,PhD *Senior Lecturer* agricultural policy, regional analysis, agricultural productivity, international trade in agricultural products
12 Sinden JA - BSc,PhD *Associate Professor* environmental economics, economics of recreation and tourism, land use economics
13 Wright VE - MCom,PhD *Senior Tutor* agribusiness marketing, planning and control

821 Department of Agronomy and Soil Science

822 The University of New England

Armidale, New South Wales, 2351
Tel: (067) 73 2829. Telex: AA 66050. FAX:067-73 3122.

1 McGarity JW - MScAgr,PhD *Associate Professor, Head of Division of Soils* soil fertility, land use, pedology
2 Andrews AC - BScAgr,PhD *Lecturer* pasture adaptation
3 Blair GJ - BScAgr,PhD *Senior Lecturer* plant nutrition and soil fertility
4 Cass A - MScAgr,PhD *Lecturer* soil physics
5 Gregg PC - MRurSc,PhD *Lecturer* applied entomology, pest management, crop protection, insect ecology
6 Hoult EH - MSc,PhD *Lecturer* soil fertility
7 Jessop RS - BSc,PhD *Senior Lecturer, Head of Department* crop agronomy - cereals, oilseeds and grain legumes
8 Johnson IR - BSc,MSc,PhD *Lecturer* pasture and crop modelling
9 MacLeod DA - BSc,MA,PhD *Senior Lecturer* soil pedology, soil structure
10 Sale PWG - BRurSci,PhD *Lecturer* plant nutrition
11 Smith NG - BSc,MSc,PhD *Senior Lecturer* horticulture and plant hormone analyses

823 Animal Genetics and Breeding Unit

(A joint Unit of the NSW Department of Agriculture and the University of New England)

1 Hammond K - HDA,BScAgr,PhD *Research Leader* design of breeding programs; electronic animal identification
2 Graser H-U - DipIngagr,Drscagr *Livestock Research Officer* mixed model estimation and prediction; design of breeding programs
3 Price R - BA *Research Associate* information systems design
4 Scarth RD - MS,PhD *Senior Research Associate* mixed model estimation of genetic parameters
5 Smith SP *Research Officer* linear and non-linear theory
6 Tier B - Research Officer *mixed model computing strategies*

824 Department of Animal Science

1 Barker JSF - BAgrSc,PhD,FTS *Professor, Head of Department* population and quantitative genetics
2 Brodbeck Ms PS - MRurSc,Diped *Tutor* reproduction, growth and body composition, wool composition
3 Claxton JH - BSc,PhD *Senior Lecturer* genetics of pattern formation
4 Edey TN - MAgrSc,PhD *Associate Professor* animal reproduction
5 Gray GD - BSc,PhD *research Fellow* parasitology
6 Hill MK - MAgrSc,PhD *Senior Lecturer* wool biology and growth
7 Hinch GN - BRurSc,PhD *Research Fellow* animal behaviour and reproductive physiology
8 Kinghorn BP - BSc,PhD,DrAgr *Senior Lecturer* quantitative genetics and animal breeding
9 Knibb WR - BSc,PhD *Research Fellow* population genetics
10 Presson Ms B - BA,MS,PhD *Research Fellow* parasitology
11 Thomas RH - BA,MS,PhD *Research Fellow* population and ecological genetics

12 Thompson JM - BRurSc,MSc,PhD *Lecturer* growth, development and body composition
13 Thwaites CJ - MrurSc,PhD *Senior Lecturer* animal reproduction, lamb growth, meat studies

825 Department of Biochemistry, Microbiology and Nutrition

826 Nutrition

1 Bird SH - BScAgr,DipAg,PhD *Research Fellow* rumen protozoa and defaunation
2 Cumming RB - BVSc,PhD,MRCVS *Associate Professor* in charge of poultry studies
3 Farrell DJ - MRAC,MSA,PhD,DRurSc *Associate Professor* poultry and pig nutrition
4 Gregg K - BSc,PhD *Research Fellow* in charge of genetic manipulation of rumen microbes by recombinant DNA techniques
5 Hume ID - BScAgr,PhD,DSc *Associate Professor* marsupial metabolism
6 Leng RA - BSc,PhD,DRurSc *Professor of Nutritional Biochemistry* in charge of ruminant metabolism studies
7 Nolan JV - BRurSc,PhD *Senior Research Fellow* ruminant digestion and metabolism

827 Microbiology

1 Bauchop T - BSc,PhD *Professorial Fellow* in charge of rumen microbiology
2 Chubb RC - MSc,PhD,MASM *Senior Lecturer* poultry immunology and virology
3 Hudman F - BSc,PhD *Research Fellow* rumen bacteria
4 Stephenson SK - MSc,PhD *Associate Professor, Honorary Fellow* genotype and energy metabolism interactions in mice
5 Watson K - BSc,PhD,MASM *Associate Professor* ethanol from renewable energy resources (e.g. sugar-cane cassava)

828 Biochemistry

1 Entsch B - BSc,PhD,ARACI *Lecturer* metabolism of aromatic compounds
2 Falconer IR - BSc,PhD,DSc,CChem,FRIC,FIBiol *Professor of Biochemistry* thyroid function and lactation
3 Gerdes RG - BSc,PhD,ASTC *Senior Lecturer* biochemical toxicology
4 Runnegar MTC - BA,BSc,PhD *Research Fellow* cyanobacterial toxins
5 Sutherland TM - BSc,PhD *Senior Lecturer* ruminant metabolism

829 Department of Botany

1 Brock MA - BSc,PhD,DipEd *Lecturer* ecology of saline and freshwater wetlands
2 Brown JF - BScAgr,PhD *Associate Professor* diseases of crop plants
3 Charley JL - BScAgr,PhD *Senior Lecturer* nitrogen turnover in semi-arid plant communities and nutrient forests
4 Milburn JA - BSc,PhD,FIBiol *Professor, head of Department* water and nutrient transport within useful plants
5 Parbery IH - MA,LittB *Lecturer* microbiology of leaf surfaces
6 Prakash N - MSc,PhD *Senior Lecturer* embryology and morphology of flowering plants
7 Whalley RDB - BScAgr,PhD *Senior Lecturer* physiology and ecology of native Australian grasses; seedling vigour
8 Williams JB - BSc *Senior Lecturer* ecology, physiology and anatomy of Australian rainforest and sclerophyll plants

830 University of New South Wales

PO Box 1, Kensington, New South Wales, 2033
Tel: 697 2222. Cables: Unitech Sydney.

831 School of Microbiology

1 Barnet Mrs YM - BScAgr,PhD *Senior Lecturer* nodulation of Australian legumes
2 Blainey Mrs B - BSc,MSc *Professional Officer* microbial adhesion
3 Couperwhite I - BSc,PhD *Lecturer* polysaccharides of Azotobacter
4 Marshall KC - BScAgr,MS,PhD *Professor of Microbiology* microbial adhesion

832 Department of Wool Science

1 Kennedy JP - MSc,BSc,FAIAS *Associate Professor, Head of Department* reproductive biology, management of sheep in arid areas
2 Davies HL - BSc,PhD,FAIAS *Professor* pasture utilization, mineral nutrition of cattle
3 Filan SJ - BAgrEc,MSc *Senior Lecturer* systems analysis of management, mathematical models for forecasting agricultural output
4 James JW - BA,DSc *Associate Professor* quantitative genetics
5 King GW - BSc,PhD *Lecturer* agronomy, range management
6 McManus WR - BScAgr,PhD *Associate Professor* ruminant physiology and nutrition
7 Murray DM - BAgrSc,MRurSc,PhD *Senior Lecturer* growth and development
8 Roberts EM - MAgrSc,PhD *Associate Professor* animal breeding

9 Sinclair AN - MVSc,PhD,FRCVS,FACNS,MACVS *Senior Lecturer* dermatohistopathology, parasitology
10 Teasdale DC - BSc,MBA *Lecturer* wool science, wool metrology

833 School of Zoology

834 Animal Research

1 Dawson TJ - BRurSc,PhD *Professor, Head of School* physiology of arid zone mammals, birds
2 Augee ML - BSc,PhD *Senior Lecturer* metabolic and endocrinology physiology of mammals
3 Beal AM - DipAH,BSc,PhD *Senior Lecturer* physiology of arid zone mammals
4 Croft DB - BSc,PhD *Senior Lecturer* ecology and behaviour
5 Fox BJ - BSc,DipEd,MSc,PhD *Senior Lecturer* ecology

835 Entomology

1 Woods A - MA,FRES,MIBiol *Senior Lecturer* pest control

836 The University College, University of New South Wales

Australian Defence Force Academy, Northcott Drive, Campbell, New South Wales, 2600

837 Bushfire Research Group

838 Mathematics Department

1 Catchpole EA - BSc,MSc,PhD bushfire modelling
2 Catchpole Mrs WR - BSc,MSc bushfire modelling
3 de Mestre NJ - BSc,MSc,PhD,DipEd bushfire modelling

839 Computer Science Department

1 Anderson DH - BSc,PhD bushfire modelling

840 University of Queensland

841 Department of Agriculture

St Lucia, Queensland, 4067
Tel: (07) 377 1111. Telex: UNIVQLD AA 40315. Cables: BRISBANE UNIVERSITY.

1 Humphreys LR - MScAgr,PhD,FAIAS *Reader and Head of Department* pasture agronomy
2 Asher CJ - BScAgric,PhD *Professor* plant nutrition
3 Bell LC - BAgrSc,PhD,FAIAS *Reader* soil science
4 Blamey FPC - MScAgric,PhD *Senior Lecturer* crop agronomy
5 Byth DE - MAgrSc,PhD *Reader* plant breeding
6 Chamala S - BScAg,MSc,PhD *Senior Lecturer* agricultural extension
7 Crouch BRB - BAgrSc,MSc,PhD *Senior Lecturer* agricultural extension
8 Drynan RG - BAgrSc,PhD *Lecturer* agricultural economics
9 Edwards DG - BAgSc,PhD *Reader* plant nutrition
10 Elliott R - BSc,PhD *Lecturer* animal biochemistry and nutrition
11 Evenson JP - MSc *Principal Research Fellow* crop agronomy
12 Fischer KS - BAgrSc,PhD *Senior Lecturer* crop physiology
13 Fukai S - BAgrSc,PhD *Senior Lecturer* crop agronomy
14 Gordon GH - BSc,PhD *Lecturer* cytogenetics and genetic engineering
15 Grundon NT - MAgrSc,PhD *Research Fellow* plant nutrition
16 Gutteridge RC - MAgrSc,PhD *Research Fellow* pasture agronomy
17 Horton IF - BSc,MSc,DipAutComp *Senior Lecturer* biometrics
18 Hughes JDH - BSc,MAgrSc,PhD,ARACI *Principal Research Fellow* Coordinator, Thai Australian Prince of Songkla University Project
19 Izac AMN - MA,MSc,PhD *Lecturer* resource economics
20 Longworth JW - BScAgr,PhD *Reader* agricultural economics
21 Milton JTB - BAgrSc,PhD *Research Fellow* animal nutrition
22 Norton BW - BAgrSc,PhD *Reader* animal biochemistry and nutrition
23 Paton TR - MSc *Professional Research Fellow* soil science
24 Shelton HM - BAgrSc,PhD *Senior Lecturer/research Fellow* pasture agronomy
25 So HB - Ir,MSc,PhD *Reader* soil physics
26 Troedson RJ - BAgrSc *Senior Research Officer* crop physiology
27 Wallis ES - BRurSc *Senior Research Officer* crop agronomy
28 Waring SA - BScAgr,PhD *Senior Lecturer* soil science

842 Department of Animal Sciences & Production

1 Ternouth JH - MVSc,MSc,PhD,LDA,MRCVS *Associate Professor and Head of Department* physiology of digestion and nutrition
2 Beattie AW - BScAgr,MSc,DipAutComp *Senior Lecturer* biometrical genetics
3 Blackshaw JK - BSc,MAEd,PhD *Lecturer* applied animal behaviour
4 Carrick FN - BSc,PhD *Senior Lecturer* reproductive biology and ecology of marsupials and monotremes
5 Charles DD - MVSc,PhD,MACVSc *Associate Professor* meat science and carcase composition
6 Cummins JM - MSc,PhD *Senior Lecturer* reproductive biology, especially *in vitro* fertilisation

7 Dowsett KF - BVSc,QDAH,PhD *Senior Lecturer* equine reproduction and nutrition
8 Glover TD - MA,DSc,PhD,FRCVS,FIBiol *Professor* reproductive biology
9 Heath TJ - BVSc,MAEd,PhD,FACVSc *Professor* structure and function of the lymphatic system
10 Johnstone IP - BSc,PhD,BVSc *Lecturer* reproductive biology
11 McMeniman NP - MVSc,PhD *Senior Lecturer* nutrition of horses and ruminants
12 Pattie WA - BSc,PhD *Reader* animal breeding and livestock systems
13 Pym RAE - BRurSc,MSc,PhD *Senior Lecturer* poultry genetics
14 Woodall PF - BSc,MSc,PhD *Lecturer* ornithology and reproductive biology

843 Department of Botany

Tel: 3772731.

1 Rogers RW - BSc,PhD *Senior Lecturer and Head of Department* ecological impact of man on plant communities
2 Boughton VH - MA,PhD *Senior Tutor* plant anatomy
3 Chandler GE - BSc,PhD *Lecturer* physiology and nutrition of rain forests, mycorrhizas
4 Clifford HT - PhD,DSc,FLS *Professor* experimental and numerical taxonomy of flowering plants especially grasses and palms
5 Cribb AB - MSc,PhD *Senior Lecturer* Queensland algae
6 Doley D - MSc,Dphil,DipFor *Senior Lecturer* tree physiology, weed biology, physiological effects of atmospheric pollutants
7 Irwin JAG - MAgrSc,PhD *Senior Lecturer* plant pathology, breeding for disease resistance, mycology
8 Lamb D - MSc,PhD *Senior Lecturer* forest ecology
9 Ng BH - BSc,PhD *Senior Tutor* plant physiology
10 Slater WG - MSc,DPhil *Senior Lecturer* growth physiology of plants
11 Specht RL - MSc,PhD,DSc *Professor* structure and community physiology of plant communities
12 Yates DJ - BAgrSc,PhD *Lecturer* biometeorology, the environment of plants

844 Department of Economics

1 Kenwood AG - BA,PhD *Senior Lecturer and Head of Department*
2 Jensen RC - BEcon,MAgrEcon,AEd,PhD *Reader* regional economics
3 Lougheed AL - BA,BCom,BEcon,PhD *Reader* international trade in agricultural products
4 Macdonald S - MA,PhD *Reader* technological change and innovation in agricultural history
5 Strong SM - BAgec,MS,PhD *Lecturer* agricultural economics
6 Vadlamudi YR - BA,MA,DipAgrStat,PhD *Reader* demand and supply studies of selected products

845 Department of Entomology

1 Paterson HEH - BSc,PhD *Professor and Head of Department* biology, geology and genetics of Culicidae, Drosophilidae, Tephritidae, evolutionary theory
2 Exley EM - MSc,PhD,DIC *Senior Lecturer* biology and systematics of Apoidea; pollination
3 Reye EJ - MB,BS *Research Officer* biology, systematics and control of Ceratopogonidae
4 Rice MJ - BSc,PhD *Senior Lecturer* behaviour and physiology
5 Walter GH - BSc,PhD *Lecturer* biology of Aphelinidae; ecology of pests
6 Zalucki MP - BSc,PhD *Lecturer* ecology, modelling movement behaviour, pest management

846 Department of Microbiology

Cables: Brisbane University.

1 Atherton JG - BScAppMedSc,PhD,MASM *Reader in Virology and Head of Department* animal and bacterial virology
2 Halliday WJ - BSc,PhD,DSc *Professor of Immunology and Deputy Head of Department* immunology
3 Davis GHG - BSc,PhD *Reader* medical bacteriology
4 Doelle HW - DipBiol,Drrernat,PhD,DSc,MASM *Reader* microbial chemistry, biotechnology
5 Fuerst JA - BSc,PhD,MASM *Senior Tutor* bacterial systematics, bacterial ultrastructure
6 Hayward AC - BSc,PhD,MASM *Reader* plant bacteriology
7 MacRae IC - BSc,PhD,AAIFST,MASM *Reader* soil and applied microbiology
8 Pemberton JM - BAgrSc,PhD,MASM,GradDipEdAdmin *Senior Lecturer* bacterial genetics
9 Sly LI - BSc,PhD,MASM *Chief Professional Officer* culture collections, bacterial systematics
10 Teakle DS - MAgrSc,PhD *Senior Lecturer* plant virology

847 Department of Parasitology

1 Dobson C - BSc,PhD,DSc,FIBiol *Professor, Head of Department* immunology
2 Adams JH - BA,MS,PhD *Postdoctoral Fellow* genetics of parasites
3 Hoyte HMD - BSc,PhD *Reader* protozoa, piroplasms and other haematozoa
4 Humphery-Smith I - BSc,PhD *Postdoctoral Fellow* transmission of disease by ticks

5 Lester RJG - BSc,PhD *Lecturer* marine parasitology
6 Moorhouse DE - BA,MA,DPhil *Associate Professor* ticks and parasitic insects
7 Opdebeeck JP - MVB,PhD,MRCVS *Lecturer* molecular parasitology
8 Pearson JC - BA,MA,PhD,DSc *Reader* digenetic trematodes
9 Prociv PP - MBBS,FRACP *Lecturer* medical parasitology
10 Sprent JFA - BSc,PhD,DSc,FRCVS,FAA *Professor Emeritus* ascaridoid nematodes
11 Waddell AH - BVSc,PhD,MACVSc *Associate Professor* veterinary parasitology
12 Washington EA - BSc,PhD *Postdoctoral Fellow* genetics of resistance to infection

848 Department of Physiology and Pharmacology
St. Lucia, Queensland, 4067
Tel: 377 1111. Telex: UNIVQLD AA 40315. Cables: Brisbane University.

1 Blackshaw AW - DVSc,MACVSc *Reader and Head of Department* animal reproduction
2 Bell TK - BVSc,PhD,MACVSc *Reader* genetics of blood and milk protein
3 Bourke JR - MBBS *Senior Lecturer in Pharmacology* thyroid
4 Brown L - BPharm,PhD *Lecturer in Pharmacology* pharmacology of cardiotonic drugs
5 Bryan LJ - BPharm,MPharmst,PhD *Lecturer in Pharmacology* vasoactive amine metabolism in lung
6 Cross RB - PhD,BSc,BS,MD,FRACP *reader* renal function
7 Hamilton D - MB,ChB *lecturer* computing and examination techniques
8 Huxham GJ - MBBS *Senior Lecturer* membrane and muscle physiology
9 Kaye PL - BSc,PhD *Lecturer* reproductive biology
10 Kratzing CC - MSc,PhD *Reader* fat transport, ascorbic acid in tissue
11 Ladd RJ - BVSc,BSc,PhD,MRCVS,NIBiol *Reader in Pharmacology* anti-inflammatory drugs
12 Lipton A - BSc,PhD *Senior Lecturer* uterine smooth muscle, prolactin
13 Manley SW - BSc,MBBS,PhD *Senior Lecturer* thyroid-stimulating antibodies
14 Martin L - BSc,PhD *Senior Lecturer* reproductive biology
15 Morris LR - BSc,PhD *Lecturer* endocrinology, reproduction
16 O'Donnell SR - BSc,PhD,MPS,DSc *Reader in Pharmacology* bronchodilators and adrenoreceptors
17 Oldfield B - BSc,PhD *Lecturer* sensory neurobiology
18 Pass MA - BVSc,MSc,PhD *Reader* gastrointestinal
19 Pettigrew JD - BScMed,MSc,MBBS *Professor* neuroscience
20 Sernia C - BSc,PhD *Senior Lecturer* angiotensinogen in the brain
21 Waters MJ - BSc,PhD *Senior Professional Officer* polypeptide hormone receptors
22 Wheeler AG - BSc,PhD *Senior Tutor* reproduction, endocrinology
23 Whyte JM - BSc,PhD *Senior Lecturer in Pharmacology* venoms and naturally occurring toxic products
24 Yesberg NE - MSC,PhD *Senior Lecturer* vasopressin

849 School of Veterinary Science

1 Rex MAE - MA,VetMB,MVSc,PhD,MRCCVS,FACVS,FFARACS *Dean and Professor of Veterinary Surgery* cardio-respiratory reflexes during anaesthesia
2 Johnson ER - BVSc,BSc,PhD carcass evaluation and classification
3 Wilson BE - BVSc cancer eye in cattle

850 Department of Veterinary Medicine
Large Animal Centre, 96 Pinjarra Road, Pinjarra Hills, Brisbane, Queensland, 4069
[PO Box 125, Kenmore, Queensland, 4069]
Tel: (07) 378 2893. Telex: UNIVQLD AA 40315.

1 Moodie EW - BSc,PhD,DVM&S,MRCVS,MACVSc *Head of Department*
2 Atwell RB - BVSc,PhD,FACVSc *Senior Lecturer* canine medicine; cardiovascular and respiratory disorders
3 Baker AA - BVSc,PhD,FACVSc *Senior Lecturer, Director Pastoral Veterinary Centre* cattle reproduction
4 Cameron RDA - MVSc,PhD,MRCVS *Senior Lecturer* swine medicine and reproduction
5 Chenoweth PJ - BVSc,PhD,MACT,MRCVS *Senior Lecturer (Deputy Dean Clinical)* beef cattle reproduction
6 Copland RS - QDAH,MSc,BVSc *Lecturer* sheep medicine and production
7 Daniel RCW - BVSc,MSc,PhD,FACVSc *Reader* dairy cattle medicine (mastitis and metabolic diseases) and obstetrics
8 English PB - MVSc,PhD,MACVSc *Professor* renal and electrolyte disorders: canine medicine; antimicrobials
9 Filippich LJ - BVSc,BSc,PhD *Senior Tutor* canine medicine; fauna medicine
10 McLennan MW - BVSc,MSc,MACVSc *Lecturer* dairy cattle medicine and herd health
11 Pollitt CC - BVSc,PhD *Lecturer* equine medicine
12 Rival MD - BVSc *Temporary Lecturer* sheep reproduction
13 Smith ID - DVSc,MSc,MVSc,PhD,FACVSc,FACBSc *Senior Lecturer* small ruminant medicine and production; chronobiology and therapeutics
14 Thornton JR - BVSc,MSc,PhD *Reader* laboratory medicine; exercise physiology of equine
15 Wilkinson GT - MVSc,MRCVS,FACVSc *Reader* feline medicine; dermatology

851 Department of Veterinary Pathology and Public Health

1 Frost AJ - BVSc,PhD,MACVSc,MRCPath *Reader in Veterinary Microbiology, Head of Department* bovine mastitis and *Salmonella* infection
2 Bensink JC - BSc *Senior Lecturer* abattoir hygiene, antibiotic resistance
3 Kelly WR - BVSc,MVSc,PhD *Senior Lecturer* general and toxicological pathology
4 Seawright AA - DVSc,BSc,PhD,FRCVS,MACVS,MTCVS,FRCPath,FACVT *Reader in Veterinary Pathology* many aspects of toxicology
5 Spradbrow PB - BVSc,PhD,MACVS *Reader in Veterinary Microbiology* virology and immunotherapy of carcinomas
6 Sutton RH - BVSc,PhD,MRCVS *Lecturer* pathology, especially haematopathology
7 Watt DA - MVSc,PhD *Senior Lecturer* pathology, especially neurology
8 Winter H - MedVet,DVSc,MRCVS,MACVSc,DipAmCollVetPath *Reader in Veterinary Pathology* pathology, cattle genetic hybrids
9 Woolcock JB - BVSc,PhD,MACVSc *Senior Lecturer* equine and koala infectious diseases

852 Department of Zoology

1 Kikkawa J - BSc,Dsc *Professor and Head of Department* tropical ecology and ornithology
2 Cameron AM - BSc,PhD *Senior Lecturer* marine toxicology
3 Dwyer PD - MSc,PhD *Reader* ecology of mammals
4 Endean R - MSc,PhD *Reader* toxinology
5 Fielder DR - BSc,PhD *reader* physiology of arthropods
6 Greenwood JG - MSc,PhD *Reader* ecology of plankton, crustaceans
7 Hailstone TS - MSc,PhD *reader* benthic ecology
8 Jamieson BGM - BSc,PhD,DSc *reader* systematics and ecology of earthworms
9 Mather WB - MSc,PhD *Reader* evolution of Drosophila
10 McCallum HI - BSc,DIC,PhD *Lecturer* population dynamics, mathematical modelling
11 Thorne MJ - BSc,PhD *Reader* behaviour of arthropods
12 Thulborn RA - BSc,PhD *Senior Lecturer* vertebrate palaeozoology
13 Warburton K - BSc,PhD *lecturer* population dynamics of marine organisms

853 University of Tasmania
PO Box 252C, Hobart, Tasmania, 7001
Tel: (002) 20 2101. Telex: AA 58150. FAX: 20 2186.

854 Department of Agricultural Science

1 Beattie JA - BScAgr,PhD *Senior Lecturer, Soil Science* pedology, micromorphology of soils
2 Bhat BK - BAgrSc,MSc,PhD *Senior Research Fellow* pyrethrum breeding
3 Bray AC - HDA,BScAgr,PhD *Senior Lecturer, Animal Production* animal nutrition
4 Clark RJ - BAgrSc,PhD *Research Fellow* essential oils
5 Cruickshank RH - BSc *Professional Officer* electrophoretic studies of enzymes related to taxonomy
6 Franzmann PD - BSc,PhD *Junior Research Fellow* microbial taxonomy
7 Garland CD - BSc,PhD *Research Fellow* microbiology of aquaculture
8 Hicks LN - BAgrSc,MAgrSc *Junior Research Fellow* minimal cultivation systems
9 Kerslake MF - BAgrSc,PhD *Research Fellow* essential oils
10 Line MA - MSc,PhD *Senior Lecturer, Agricultural Microbiology* soil microbiology
11 Lovett JV - BScAgr,PhD,FIBiol *Professor of Agricultural Science* weed biology; allelopathy
12 Madden JL - BAgrSc,MS,PhD *Reader, Entomology* insect behaviour; insect ecology
13 McMeekin TA - BAgr,PhD,FIFST,FAIFST *Reader, Agricultural Microbiology* food microbiology
14 Menary RC - BSc,MS,PhD *Reader, Horticultural Science* plant nutrition; essential oil and medicine crops
15 Mendham NJ - MScAgr,PhD *Senior Lecturer, Agronomy* crop physiology and plant breeding
16 Potts WC - BSc,PhD *Junior Research Fellow* forest allelochemicals
17 Yates JJ - BScAgr,PhD *Senior Lecturer, Agronomy* pasture ecology

855 Biochemistry Department

1 Clark MG - PhD *Professor of Biochemistry and Head of Department* biochemical endocrinology
2 Colquhoun E - MBBS,BSc *Lecturer* biochemical endocrinology
3 Neville Miss MES - BSc,PhD *Senior Lecturer* biochemical endocrinology
4 Sallis JD - BSc,PhD *Reader* biochemical endocrinology

856 Botany Department

1 Crowden RK - BSc,PhD *Reader* plant biochemistry, phytochemistry, angiosperm systematics
2 Hill RS - BSc,PhD *Lecturer* rainforest ecology
3 Jackson WD - BSc,PhD *Professor of Botany* forest ecology, ecological genetics and radiation genetics
4 Mills AK - BSc *Senior Tutor* mycology

5 Murfet IC - BSc,PhD *Reader in Genetics* physiological genetics
6 Reid JB - BSc,PhD *Senior tutor* physiological genetics, ecological genetics
7 Tyler PA - BSc,PhD *Senior Lecturer* limnology

857 Physics Department

1 Newman IA - BSc,PhD *Senior Lecturer* plant biophysics
2 Scott BIH - BSc,PhD *reader* plant biophysics

858 Physiology Department

1 Maskrey M - BSc,PhD *Senior Lecturer* environmental physiology
2 Nicol SC - BSc,PhD *Senior Lecturer* environmental physiology
3 Speden RN - MSc,DPhil *Reader* cardiovascular physiology and pharmacology

859 Zoology Department

1 Hickman JL - BSc,PhD *Senior Lecturer* forest litter fauna
2 Richardson AMM - BSc,PhD *Senior Lecturer* forest litter fauna, aquatic fauna
3 Swain R - BSc,PhD *Senior Lecturer* aquatic fauna

860 University of Melbourne

Parkville, Victoria, 3052
Tel: (03) 344 4000. Telex: AA35185. Cables: UNIMELB Parkville.

861 Department of Biochemistry

1 Finch LR - BSc,PhD *Reader, Head of Department* metabolism and genetics of *Mycoplasma mycoides*
2 Davidson BE - BAgrSc,BSc,PhD *Senior Lecturer* regulation of biosynthesis of aromatic amino acids in bacteria
3 Davidson BE - BSc,PhD *Senior Lecturer* molecular biology of cheese starter strains and associated bacteriophage
4 Gayler KR - BAgrSc,PhD *Senior Lecturer* expression of genes for seed proteins in lupins and related legumes
5 Grant BR - BAgrSc,MSc,PhD *Reader* biochemistry of fungi of the genus *Phytophthora*
6 Hillier AJ - BSc,PhD *Research Fellow* molecular biology of lactic acid bacteria and their phages

862 School of Botany

1 Attiwill PM - DipFor,BScFor,PhD *Chairman* nutrient cycling and productivity of forests
2 Adams MA - BSc,PhD *Research Fellow* mineralization of nitrogen and phosphorus in forest soils
3 Ashton DH - BSc,PhD *Reader* forest ecology and description of vegetation
4 Bacic A - BSc,PhD *Research Fellow* plant cell surface polysaccharides and glycoconjugates - structural characterization and biology function
5 Calder DM - MSc,PhD *Reader* reproductive biology of native plants
6 Clarke AE - BSc,PhD *Professor* molecular mechanisms of cell recognition, particularly in fungal pathogenesis and fertilization biology
7 Guest DI - BScAgr,PhD *Lecturer* plant pathology
8 Harris PJ - BA,MA,PhD *Senior Research Fellow* study of plant wall cells; carbohydrate chemistry and biochemistry
9 Knox RB - BSc,PhD,DSc *Professor* plant reproductive and cell biology; pollen biotechnology
10 Ladiges PY - DipEd,MSc,PhD *Senior Lecturer* phylogenetic relationships and classification of eucalypts; patterns of species variation
11 Neales TF - MA,PhD,DSc *Reader* water relations of Mallee eucalypts
12 Singh MB - MSc,PhD *Research Fellow* plant reproductive and cell biology; pollen biotechnology
13 Swart HJ - DSc *Senior Lecturer* taxonomy of Australian leaf parasitising fungi with emphasis on parasites of *Eucalyptus*
14 Williams EG - BSc,PhD *Principal Research Fellow* plant tissue culture and plant breeding

863 Department of Electrical & Electronic Engineering

1 McCutchan JC - BEE,MEngSc *Chairman and Senior lecturer* electric fence systems and animal response

864 School of Agriculture and Forestry

1 Beilharz RG - MScAgrSyd,PhDIowa,MAgrSc *Chairman of School and Reader* animal breeding, animal behaviour
2 Ferguson IS - DFYale,BScF *Deputy Chairman of School and Professor* forest economics, management
3 Bren LJ - AssocDipForCres,BScFor,PhD *Lecturer* forest hydrology & forest harvesting
4 Cary JW - BAgrSc,MAgrSc *Senior Lecturer* farm management extension

5 Chalk PM - MAgrSc,PhD *Senior Lecturer* soil fertilizers
6 Connor DJ - BAgrSc,phD *Professor* plant sciences, agronomy
7 Dalling MJ - MAgrSc,PhD *Senior Lecturer, Reader* inorganic nitrogen nutrition of cereals; biochemistry of nitrogen re-use in cereals
8 Dougal EF - BSc,Mss,MS,PhD *Lecturer* wood science
9 Douglas LA - BAgrSc,AMusA,PhD *Senior Lecturer* soils, fertilizers
10 Dunkin AC - BSc,AgrSc,MScAgroc,MAgrSc *Senior Lecturer* nutrition of non-ruminants
11 Egan AR - MSCAgric,PhDw.Aust *Professor* animal production, nutrition
12 Halloran GM - BSc,DAgrSc *Senior Lecturer* plant breeding and genetics
13 Hawkins HS - BAgrSc,MA,PhD *Senior Lecturer* extension principles
14 Holmes JHG - BVSc,MVSc,PhD *Senior Lecturer* nutrition and management of tropical ruminants
15 Hutson GD - BSc,PhD *Senior Lecturer* animal behaviour
16 Lloyd AG - BEcSyd,MAgrSc *Professor* agricultural policy and marketing
17 Malcolm LR - BAgrEc *Lecturer* agricultural economics
18 Parbery DG - BAgrSc,MAgrSc,PhD *Reader* plant pathology
19 Pearce GR - BScAgric,MAgrSc,PhD *Senior Lecturer* ruminant nutrition
20 Routley MF - BComm,ALAA *Librarian*
21 Rudra Ahi Bhusan - BSc,Aifc,PhD,MISTF *Senior Lecturer* forest management; systems analysis; mensuration
22 Sands R - BScFor,PhD,A.N.U. *Reader* tree physiology
23 Simpon RJ - BAgrSc,PhD *Lecturer* agronomy of pasture species
24 Spencer RD - MSc,RUPa,DipFor,BScFor *Senior Lecturer* forest inventory, resource use and conservation
25 Sturgess NH - MAgrEc,BAgrSc *Senior Lecturer* production economics
26 Tulloh NMC - DAgrSc,PhD,FTS,FASAP,FAIAS *animal growth and body composition*
27 Turvey ND - BSc,MSc,PhD *Lecturer* forest biology
28 Watson AS - MAgrEc,PhD,BAgrSc *Senior Lecturer* agricultural policy and marketing

865 Faculty of Veterinary Science
Werribee, Victoria, 3030
Tel: (03) 741 3500.

1 Jubb KVF - BVSc,MSc,PhD,MVSc,FACVSc,DVSc,DSc,FTS *Professor of Veterinary Pathology, Dean of Faculty*

866 Department of Veterinary Clinical Sciences

1 Jeffcott LB - BVetMed,PhD,FRCVS *Professor of Veterinary Clinical Sciences, Chairman of Department*
2 Blood DC - OBE,BVSc,MVSc,FACVSc,FAIAS,LLD,AssocRCVS *Professor Emeritus*
3 Brightling P - BVSc,BAnSc,MVSc *Lecturer* dairy cattle medicine
4 Caple IW - BVSc,PhD,MACVSc *Professor of Veterinary Medicine*
5 Galloway DB - BVSc,VMD,MVSc,FRCVS,MACVSc *Senior Lecturer* animal reproduction
6 Hyland J - BVSc,PhD *Senior Lecturer* veterinary obstetrics and animal production
7 Lavelle RB - MA,VetMb,MRCVS,DVR,FACVSc *Lecturer* veterinary radiology
8 Mason TA - BVSc,MVSc,MRCVS,MACVSc *Senior Lecturer* veterinary surgery
9 Mitten RW - BVSc,DVR *Lecturer* veterinary medicine
10 Morley FWH - BVSc,PhD,DAgrSc,HDA,FACVSc,FASAP,FTS,FAIAS *Senior Research Fellow* epidemiology and preventive medicine
11 Speirs VC - MVSc,FACVSc,DrMedVet,DipACVS *Senior Lecturer* surgery
12 Studdert VP - BSc,DVM *Senior Lecturer* veterinary medicine
13 Wright PJ - BVSc,MVSc,PhD *Senior Lecturer* animal reproduction

867 Department of Veterinary Paraclinical Sciences

1 Hughes KL - MVSc,DipBact,PhD,MACVSc *Senior Lecturer, Chairman of Department* veterinary microbiology, public health
2 Arundel JH - BVSc,MVSc,AMTC *Reader* veterinary parasitology
3 Harrigan KE - BVSc *Lecturer* veterinary pathology
4 Jabara AG - MSc,PhD *Senior Lecturer* veterinary pathology
5 Jubb KVF - BVSc,MSc,PhD,MVSc,FACVSc,DVSc,DSc,FTS *Professor of Veterinary Pathology*
6 Lightowlers MW - BSc,PhD *Senior Research Officer* immunoparasitology
7 Parry BW - BVSc,PhD *Lecturer* veterinary pathology
8 Rickard MD - BVSc,PhD,DVSc *Reader* veterinary parasitology
9 Studdert MJ - BVSc,MVSc,PhD,MACVSc *Reader* veterinary microbiology
10 Sullivan ND - MVSc,PhD *Senior Lecturer* veterinary pathology
11 Whithear KG - BVSc,PhD,HDA *Lecturer* veterinary microbiology

868 Department of Veterinary Preclinical Sciences

1 Cahill RNP - MB,BS *Professor of Veterinary Biology, Chairman of Department*
2 Brandon MR - BachScAgr,PhD *Senior Lecturer* veterinary immunology
3 Lee CS - BVSc,MVSc,PhD *Senior Lecturer* veterinary anatomy
4 McLean JG - BVSc,PhD,HDA *Senior Lecturer* veterinary biochemistry
5 O'Shea JD - BVetMed,PhD,MVSc *Reader* veterinary anatomy
6 Stewart GA - BVSc,MVSc *Senior Lecturer* veterinary pharmacology

869 The University of Western Australia

Nedlands, Western Australia, 6009
Tel: (09) 380 3838. Telex: AA92992. Cables: Uniwest Perth.

870 Department of Botany

1 Atkins CA - MScAgr,PhD,DSc *Associate Professor and Head of Department* plant biochemistry and physiology
2 Bell DT - BA,PhD *Senior Lecturer* forest and terrestrial plant ecology
3 Hilton RN - MA *Senior Lecturer* systematics and ecology of larger fungi
4 James SH - MSc,PhD *Senior Lecturer* cytogenetics of native plants
5 Lawrence ME - BSc,PhD *Senior Tutor* systematics and plant cell biology
6 Loneragan WA - BSc,PhD *Lecturer* forest and terrestrial plant ecology
7 McComb AJ - BSc,PhD *Associate Professor* lake, estuaries and near-shore nutrients and ecology
8 Pate JS - PhD,DSc,FAA,FRS *Professor* plant physiology, plant nutrition and nitrogen metabolism

871 School of Agriculture

1 Lindsay DR - BScAgr,PhD *Professor and Head of Animal Science Group; Dean and Head, School of Agriculture* reproductive physiology and behaviour

872 Agricultural Economics

1 Fraser RW - BEc,MPhil,DPhil *Lecturer* economics of uncertainty, agricultural and natural resource policy
2 Lindner RK - BEC,BAgSc,PhD *Professor* economics of uncertainty and technological change
3 Reutens AJ - BA,MSc,PhD *Lecturer* marketing
4 Smith J - BA,MA,PhD *Lecturer* production risk, production economics, development

873 Animal Science

1 Mackintosh JB - MAgrSc,PhD *Lecturer* ruminant nutrition
2 Moir RJ - BScAgr,HonDScAgric,AO,FTS,FAIAS,FASAP *Emeritus Professor and Honorary Research Fellow* rumen physiology
3 Williams IH - MScAgr,PhD *Lecturer* non-ruminant nutrition

874 Agronomy Group (UWA)

1 Stern WR - MScAgr,PhD *Professor Agronomy and Head of Group* cereal ear development, competition studies and systems studies in agronomy
2 Blacklow WM - BScAgr,MSc,PhD *Senior Lecturer* weed ecology, simulation models, crop physiology
3 Boyd WJR - BScAgr,MSc,PhD,DTA *Senior Lecturer* cereal breeding, host-parasite relationships
4 Greenway H - IrAgr,PhD *Associate Professor* physiology and metabolism, particularly the effects of oxygen deficiency
5 Sedgley RH - MAgrSc,PhD *Senior Lecturer* crop biophysics and ecology, wheat and grain legumes
6 Thurling N - BSc,PhD *Senior Lecturer* oil seeds breeding, physiological genetics of yield, breeding methodology

875 Biometrics

1 De'Ath G - BSc,MSc,MEd linear and generalised linear models, experimental design, data analysis
2 Goodchild NA - BScAgr,PhD *Senior Lecturer* experimental design, multivariate analysis, plant competition

876 Soil Science and Plant Nutrition

1 Robson AD - BAgrSc,PhD *Professor, Head of Department* plant nutrition and soil fertility
2 Abbott LK - BSc,PhD *Lecturer* soil microbiology, symbiotic associations between plants and microorganisms
3 Aylemore LAG - BSc,PhD *Associate Professor* physical properties of soils and clay minerals, soil-plant-water interaction, sorption and flow in porous materials
4 Gilkes RJ - BSc,PhD *Associate Professor* mineralogy and geochemistry of soils and sediments, properties of fertilisers
5 Parker CA - BAgrSc,PhD,FAIS *Emeritus Professor* soil microbiology, microbial ecology, symbiosis and parasitism in plant roots
6 Ritchie GSP - BSc,phD *Lecturer* soil chemistry and fertility
7 Sivasithamparam K - MSc,PhD *Senior Lecturer* plant pathology, mycology, microbial ecology

877 Department of Zoology

1 Finch ME - MSc,PhD *Senior Lecturer and Head of Department* comparative vertebrate morphology and palaeontology
2 Bailey WJ - BSc,PhD *Senior Lecturer* insect acoustics
3 Black WR - BA,MSc,PhD *Senior Lecturer* experimental ecology and marine biology

4 Bradley AJ - BSc,PhD *Lecturer* marsupial endocrinology
5 Bradshaw SD - BSc,PhD *Professor* vertebrate physiological ecology and endocrinology
6 Dickman CR - BSc,PhD *Lecturer* mammal ecology
7 Edward DHD - BSc,PhD *Lecturer* freshwater insect ecology
8 Johnson MS - BA,MSc,PhD *Senior Lecturer* genetic variations in natural populations
9 Knott B - BSc,PhD *Senior Tutor* biology of freshwater crustacea
10 Packer WC - BA,MA,PhD *Senior Lecturer* animal behaviour and embryology
11 Roberts JD - BSc,PhD *Senior Tutor* vertebrate speciation

878 Australian Maritime College
Australian Maritime College, P.O. Box 986, Launceston, Tasmania, 7250

879 School of Fisheries

1 Cartwright I - BSc *Lecturer* fishing technology
2 Carver WEA *Head of School*
3 Chopin F - BSc *Lecturer* fishing technology
4 King Dr MG - BSc,MSc,PhD *Senior Lecturer* fisheries biology resource assessment and management
5 Ruello Mrs JH - BSc,MSc *Lecturer* seafood handling and processing
6 Ruello NV - BSc,MSc *Lecturer* fisheries management economics and marketing
7 Wawrowski R - MEng *Lecturer* fishing technology

880 Canberra College of Advanced Education
PO Box 1, Belconnen, New South Wales, 2616
Tel: (062) 52 2111. Cables: Colladved.

1 Parfitt RT - BPharm,PhD,ARACI,FPS,CChem,FRSChem *Principal*
2 Eadie GMCL - BA,DipEd *Secretary and Registrar*

881 School of Applied Science

882 Water Research Centre

1 Cullen P - MAgrSc,DipEd *Deputy Head of School* phosphorus exports from rural lands
2 Greenham P - BSc,PhD *Senior Lecturer* human biology
3 Norris RH - BSc,PhD,GradDipEd *Lecturer* ecology; zoology

883 Applied Science Research Centre

1 Button BJ - BA,PhD *Senior Lecturer* water use, irrigation management
2 Cartledge O - BSc,PhD *Senior Lecturer* soils
3 Jones NO - BSc,PhD *Senior Lecturer* hydrology
4 Taylor G - MSc,PhD *Senior Lecturer* soils

884 Chisholm Institute of Technology
PO Box 197, Caulfield East, Victoria, 3145
Tel: (03) 573 2222.

1 Maynard GB - BApSc,DipPubAdmin,MACS *Acting Director*

885 Division of Digital Technology

1 Davis J - DipEEng,BSc,PhD *Lecturer* non-destructive evaluation of timber products using x-ray tomography
2 Wells P - BSc,PhD,DipEd *Senior Lecturer* non-destructive evaluation of timber products using x-ray tomography

886 Department of Marketing

1 Vollmers AC - BSc,MSc,PhD *Elders IXL Fellow in Agribusiness*

887 Darling Downs Institute of Advanced Education
PO Darling Downs, Toowoomba, Queensland, 4350
Tel: (076) 312100. Telex: AA 400100. 076 301182.

888 School of Applied Science

1 Roberts BR - MScAgr,PhD *Head of School* arid zone ecology, pasture grasses, fire, soil conservation
2 Akers JW - BAgrSc,MSc,DipEd *Lecturer* microbiology, microbial breakdown of coal
3 Atkinson RK - BSc,PhD *Head of Programme* spider toxins, animal physiology
4 Lindquist JE - MChem,DipIndChem,CChem *Lecturer* water quality, groundwater monitoring, soil fertility
5 McKilligan NG - MSc,DipEd *Lecturer* pest bird control, invertebrate zoology
6 Reece IH - MSc,PhD,CChem *Principal Lecturer* vegetable fuel oil, analytical techniques
7 Russell MJ - MAgrSc,NDA,DTA *Lecturer* pasture ecology, industrial revegetation, meteorology

889 School of Arts

1 Allen ACB - BSc,PhD *Lecturer* rural geography, arid zone development
2 Elvidge NC - BA,MA *Senior Lecturer* geography of irrigation development

890 School of Business Studies

1 Earle TR - MAgrSc,DipEd,DipInfProc *Senior Lecturer* land use planning, rural organization
2 Jocumsen GL - BAgrSc,BEcon,MA *Lecturer* rural sociology and economics
3 Langdon IA - MBA,BCom,DipEd *Dean* cost benefit analysis of cheese production plant

891 School of Engineering

1 Dobney PT - BSc,PhD *Principal Lecturer* cloud physics, hail studies
2 Hancock NH - BSc,PhD *Lecturer* instrumentation, micrometeorology, erosion
3 Harris HD - BE,MEngSc,PhD *Lecturer* tillage, agricultural machinery and instrumentation, traction
4 Hilton DJ - BSc,PhD,DipMs *Lecturer* machinery, gantry mechanisation, windpower, pumps
5 O'Shea JA - BEng,MEdAdmin *Head of Programme* groundwater hydrology, runoff, groundwater recharge
6 Parsons DJ - BSc,BE *Lecturer* instrumentation, erosion
7 Rixon AJ - BScAgr,PhD *Senior Lecturer* soil science, salinity public health, soil and water management, agricultural science
8 Ruffini JL - BEngAg,MScAgEng *Research Engineer* strip cropping, erosion modelling, hydraulic and hydrologic modelling
9 Smith RJ - DipCE,BEAgr,MEngSc *Lecturer* irrigation, erosion modelling, agricultural hydrology

892 Hawkesbury Agricultural College
Richmond, New South Wales, 2753
Tel: (045) 701333.

1 Lamond RA - FCA,AASA,CPA *Chairman of the College Council*
2 Swain FG - BScAgr,PhD,FAIAS *Principal*
3 Carter OG - BScAgr,MS,PhD *Assistant Principal*

893 Faculty of Food & Environmental Sciences
Richmond, NSW, 2753
Tel: (045) 701333.

1 Baumgartner PA - BScAgr,DipTertEd,FAIFST *processed meat*
2 Bavor HJ - BSc,MSc,PhD *Lecturer* water & wastewater microbiology
3 Bushell Ms R - BAppSc,HECert *Lecturer* food & water microbiology
4 Greig RIW - DipAppSc,BAppSc,MSc *Lecturer* seafood technology
5 Nguyen MH - BE,MSc,GradDip *Lecturer* food engineering
6 Powis B - MSc *Lecturer* environmental health
7 Sarafis V - BSc *Senior Lecturer* optical physics; biology
8 Skurray GR - BSc,PhD *Senior Lecturer* nutrition
9 Wyllie SG - BSc,MSc,PhD *Senior Lecturer* flavour chemistry

894 Faculty of Agriculture & Horticulture
Richmond, NSW, 2753
Tel: (045) 701333.

1 Biggs AG - BSc,MSc,MI *Senior Lecturer* horticulture
2 Braithwaite BM - BSc *Acting Dean of Faculty*
3 Ison RL - BScAgr,PhD *Lecturer* plant production
4 Packham RG - BSc,MSc,PhD *Senior Lecturer* poultry nutrition
5 Valentine I - BSc,PhD *Senior Lecturer* soil science

895 Faculty of Business

1 Bird JG - BAgrSc,MAg,DipEdAdmin *Acting Dean of Faculty*
2 Newell G - BSc,MStats *Senior Lecturer* statistics

896 Marcus Oldham Farm Management College
Private Bag 116, Mail Centre, Geelong, Victoria, 3221
Tel: (052) 433533.

1 McConnell GR - MAgrSc *Principal*

897 Department of Farm Management

1 Blackburn AG - MA,BAgrSc,DipAgEc *Senior Lecturer in Farm Management*
2 Hacking CJ - BRurSc *Lecturer in Plant Production*
3 Inglis SW - DipFM,GDE *Senior Lecturer in Animal Production*
4 Lambert JMCK - ARMITMech,BEngAgric,MIE,AgricCert *Senior Lecturer in Agricultural Engineering*
5 Morgan SG - BAgrSc *Lecturer in Farm Management*

898 Department of Horse Management

1 Stewart JO - BVSc,MRCVS *Director of Horse Management Course*

899 Orange Agricultural College
PO Box 883, Orange, New South Wales, 2880
Tel: (063) 62 4699.

1 Napier RJ - BAgrSc,MEc,DipAgStud *Principal*

900 Animal Production Department

1 Hunt ER - BVSc,MVSc,PhD,MRCVS *Senior Lecturer* reproductive biology
2 van Gelder RA - BRurSc *Lecturer* sheep production

901 Ministry of Agriculture, Trade and Industry

902 Department of Agriculture

PO Box N-3028, Nassau
Tel: (809) 32-21277-9, 32-32511.

1 Deane CC - BA,MA,CertEd *Permanent Secretary*
2 Smith C - BSc,MSc,MBA *Senior Agricultural Officer* administrative services
3 Russell S - BSc,MSc *Director/Agriculture*
4 Bonamy P - BSc,MSc,PhD *Deputy Director/Agriculture* extension, family island development

903 Gladstone Road Agricultural Centre (GRAC)

PO Box N-3028, Nassau
Tel: (809) 32-83614, 32-83985.

904 Agricultural Extension

1 Dorsett A - BSc,MSc *Assistant Director/Agriculture* research
2 Greaves J - BSc,MBA *Senior Agricultural Economist* marketing
3 Outten Ms V - BSc,MSc *Assistant Director/Agriculture, Resident Extension Officer, Grand Bahama* integrated pest management

905 Food Technology

1 Rahming G - BSc *Chemist* physicochemical analysis of imported indigenous food and feed
2 Curry Ms P - BSc *Assistant Chemist* physicochemical analysis of imported indigenous food and feed

906 Livestock Production

1 Bowe M - BSc *Assistant Agricultural Officer, Livestock Science* cattle, sheep
2 Smith K - BSc *Agricultural Officer, Livestock Science* small ruminants

907 Crop Production

1 Price Ms C - BSc,MSc *Agriculture Officer, Resident Extension Officer, Exuma* horticulture research/onions, peanuts, soy-beans
2 Smith S - BSc *Agricultural Officer, Agronomy* tropical fruit and vegetables, legumes, cereals
3 Wilson Mrs T - BSc *Agricultural Officer* pest management

908 Bahamas Agricultural Research Centre (BARC)

1 Lightbourne A - BSc *Assistant Director/Agriculture* animal nutrition, tropical forages
2 Richardson K *Agricultural Officer, Agronomist* root and tuber crops; cereals
3 Springer F - BSc,MSc *Agricultural Officer* legume research

909 Pest Management

1 Wallace E - BSc *Agricultural Officer* management and control of plant pests and diseases, integrated pest management

910 Botanic Gardens

1 Miller B - BSc *Agricultural Officer* tropical ornamentals and hardwoods

911 Marketing

1 Ebong Mrs H - BSc,MSc *Senior Agricultural Economist*
2 Swann G - BSc *Senior Agricultural Officer*

912 Agricultural Information

1 Smith Ms B - BSc *Assistant Agricultural Officer, Agricultural Information Officer* print and electronic media

913 Animal Health

1 Campbell K - BSc,DVM *Senior Veterinary Officer* supervises department's veterinary services
2 Isaacs M - BSc,DVM *Veterinary Officer* veterinary service at G.R.A.C. and veterinary extension in New Providence
3 Lyn J - BVMS,MRCVS *Veterinary Officer* animal import permits, abattoir coordination, canine control
4 Sands B - BSc,DVM *Veterinary Officer* veterinary services at G.R.A.C.

5 Springer G - BSc,DVM *Veterinary Officer* veterinary services at B.A.R.C. and extension services for North Andros

914 Forestry Service

PO Box N-592, Nassau
Tel: (809) 32-22328.

1 Hook J - BSc *Forest Officer*

915 Department of Fisheries

PO Box N-3028, Nassau
Tel: (809) 32-31014/5.

1 Thompson R - BSc,MSc *Director/Fisheries*
2 Higgs C - BSc *Deputy Director/Fisheries*
3 Braynen M - BSc *Senior Fisheries Officer* manager fish landing complex
4 Deleveaux Jr E - BSc *Senior Fisheries Officer* fisheries law enforcement

916 Department of Cooperative Development

PO Box N-3028, Nassau
Tel: (809) 32-31014.

1 Adderley N - BSc,MBs *Director*
2 Tinker K *Cooperative Officer* development and training

917 College of The Bahamas

PO Box N-4912, Nassau
Tel: (809) 32-28550.

1 Bethell K - PhD *Principal of College*

918 Bangladesh Agricultural Research Council (BARC)

PO Box 3041, Airport Road, Farmgate, Dhaka-15
Tel: 325061-5. Telex: 642401 HID BJ.

1 Ahsan ME - MAg,PhD *Chairman*
2 Islam MZ - BSc,MA *Principal Scientific Officer (Library and Documentation)*

919 Division

1 Ahmad KU - MAg,MS *Member-Director (crops)*
2 Ahsan E - MAg,PhD *Member-Director (Agricultural Economics and Social Science)*
3 Akhanda MNA - BScAg,MENgg *Principal Scientific Officer* irrigation and water management
4 Alam MN - MScAg,PhD *Principal Scientific Officer* entomology
5 Ali MO - MSc,PhD *Member-Director (Forestry)*
6 Amin SMR - MAg,PhD *Principal Scientific Officer* rubber plantation production and management
7 Choudhury ATMSU - MScAg *Principal Scientific Officer* vegetables
8 Chowdhury MSU - MAg,PhD *Member-Director(Administration and Finance)*
9 Haq KA - MScAg,PhD *Member-Director (Agricultural Engineering)*
10 Hossain AKMA - MAg,PhD *Principal Scientific Officer* agricultural marketing
11 Husain A - BAg,MS *Director (Training)*
12 Khan MMH - MScAg,PhD *Principal Scientific Officer* entomology
13 Moniruzzaman AKM - MScAg,PhD *Director (Appropriate Agricultural Technology Cell)*
14 Paul SK - MSc,PhD *Principal Scientific Officer* fisheries
15 Quddus MA - BAg,MS *Member-Director (Planning and Evaluation)*
16 Rahim A - MSc,PhD *Director (Nutrition)*
17 Saif SMH - MEng *Principal Scientif Officer* agricultural process

920 Bangladesh Agricultural Research Institute (BARI)

PO Box 2235, Joydebpur, Gazipur
Tel: 391479. Cables: RESAGRI.

1 Rahman MM - MAg,PhD *Director-General*
2 Rahman ABMF - MA *Librarian*

921 Division

1 Islam MA - MAg,PhD *Director (Training and Communication)*
2 Mazed MA - MAgEng,PhD *Director (Support Service)*
3 Mondal MH - MAg,PhD *Director (Research)*
4 Rashid MM - MAg,PhD *Director (Vegetables)*
5 Abedin MZ - MSc,PhD *Chief Scientific Officer* agronomy and farming systems
6 Ahmed H - MAg,PhD *Chief Scientific Officer* diseases of field crops
7 Ahmed SM - MSc,PhD *Project Director (Plant Breeding)*
8 Elias SM - MScAg,PhD *Chief Scientific Officer* farm management and production economics
9 Hamid MA - MAg,PhD *Chief Scientific Officer* genetics and plant breeding
10 Hossain KMA - MS,PhD *Chief Scientific Officer* fruits and vegetables
11 Hossain SHM - MSc *Chief Scientific Officer* soil fertility and plant nutrition
12 Islam MS - MScAg,PhD *Chief Scientific Officer* chemistry, fertility and plant nutrition
13 Karim MA - MS,PhD *Chief Scientific Officer* entomology and vertebrate pest management
14 Khaleque MA - MScAg *Project Director (Agronomy and Oil Crops)*
15 Maniruzzaman AFM - MS,PhD *Chief Scientific Officer* crop physiology
16 Mazed MA - MAgEng,PhD *Chief Scientific Officer* farm machinery and power

922 Bangladesh Institute of Nuclear Agriculture (BINA)

PO Box 4, Mymensingh
Tel: 4401. Cables: INA, Mymensingh.

1 Mia MM - MSc,PhD *Director*
2 Hanif MMA - MA *Librarian*

923 Division

1 Shaikh MAQ - MScAg,PhD *Chief Scientific Officer* genetics and plant breeding; germplasm maintenance and evaluation
2 Habibullah AKM - MSc,PhD *Chief Scientific Officer* soil fertility
3 Miah AJ - MSc,DSc *Chief Scientific Officer* plant breeding and genetics
4 Rahman L - MSc,MScAg *Chief Scientific Officer* plant physiology and nutrition

924 Bangladesh Jute Research Institute (BJRI)

Manik Mia Avenue, Dhaka-7
Tel: 310953. Cables: BAJRI.

1 Rahman MA - MScAg,PhD *Director-General*
2 Mian MAR - MA *Librarian*

925 Division

1 Ali MM - MSc,PhD *Director* plant pathology, microbiology
2 Bhuiyan MAM - MSc,PhD *Director* polysaccharide degradation by mould enzymes
3 Alam MS - MSc *Principal Scientific Officer* bacteriology
4 Ali MA - MSc *Principal Scientific Officer* soil fertility
5 Arangzeb S - MSc *Principal Scientific Officer* cytology
6 Choudhury R - MScAg *Chief Agronomist* seed technology
7 Haque KS - MSc *Principal Scientific Officer* plant genetics; plant anatomy
8 Hussain M - MAg,PhD *Chief Scientific Officer*
9 Jalil AFMA - MSc *Principal Scientific Officer* entomology
10 Karim KMA - MSc *Principal Scientific Officer* fibre physics
11 Mahtabuddin M - MSc,DTA *Principal Scientific Officer* quantitative genetics; plant breeding
12 Majlish MAK - MSc,PhD *Chief Scientific Officer* soil chemistry
13 Salam MA - MSc,MSc *Chief Scientific Officer* textile dyeing and printing
14 Serajuddin ASM - MSc *Principal Scientific Officer* organic chemistry
15 Waseque M - MSc *Principal Scientific Officer* plant physiology

926 Bangladesh Rice Research Institute (BRRI)

PO BRRI, Joydebpur, Gazipur

1 Mannan MA - MSc,PhD *Director General*
2 Talukder IA - MA,MLSc *Senior Librarian*

927 Research Division

1 Islam AJMA - MS,PhD *Director (Research)*

928 Administration Division

1 Alam S - MSc,PhD *Director (Administration)*

929 Agricultural Engineering Division

1 Haq KA - MS,PhD *Chief Scientific Officer*

930 Entomology Division

1 Karim ANMR - MS,PhD *Chief Scientific Officer* entomology

931 Plant Breeding Division

1 Miah MAA - MS,PhD *Chief Scientific Officer* genetics, plant breeding and tissue culture
2 Miah NM - MScAg,PhD *Chief Scientific Officer* genetics and plant breeding
3 Nasiruddin M - MS,PhD *Chief Scientific Officer* genetics and plant breeding

932 Plant Pathology Division

1 Bakar MA - Ms *Chief Scientific Officer* bacteriology
2 Miah SA - MSc,PhD *Chief Scientific Officer* plant virology

933 Plant Physiology Division

1 Haque MZ - MScAg,PhD *Chief Scientific Officer* plant physiology

934 Farm Management Division

1 Khan AH - MAg *Chief Scientific Officer*

935 Bangladesh Tea Research Institute (BTRI)

Srimangal, Moulavibazar
Tel: 225. Cables: TRESEARCH.

1 Rashid SA - MSc,PhD *Director*
2 Sana DL - MAg,MS *Principal Scientific Officer* entomology

936 Forest Research Institute (FRI)

PO Box 273, Chittagong

1 Ali MO - MSc,PhD *Director*
2 Aktheruzzaman AFM - MSc,PhD *Divisional Officer* pulp and paper
3 Das DK - MSc,MS *Divisional Officer* wood anatomy and plant taxonomy
4 Das S - MSc *Chief Research Officer* forest management
5 Hassan MM - MS,DSc *Divisional Officer* soil genesis
6 Hoque MN - BEd,MA *Librarian*
7 Karim AHMM - BSc *Divisional Forest Officer*
8 Khan SA - DSc *Divisional Forest Officer* silviculture
9 Latif MA - MSc,PhD *Divisional Officer* wood chemistry
10 Rahman MA - MSc,PhD *Divisional Officer* forest pathology
11 Razzaque MA - MSc,PhD *Divisional Officer* forest chemistry
12 Sattar MA - MSc *Divisional Officer* wood seasoning

937 Bangladesh Agricultural University

Mymensingh
Tel: 4191-3. Cables: AGRIVARSITY.

1 Haque AKMA - MSc,PhD *Vice-Chancellor*
2 Nur A - MA,LLB *Librarian*

938 Department of Agricultural Chemistry

1 Ali MS - MSc,PhD *Associate Professor* plant nutrition
2 Ishaque M - MSc,PhD *Associate Professor* plant nutrition and pesticides chemistry

939 Department of Agricultural Finance

1 Rahman ML - MA,MS,PhD *Associate Professor* agricultural finance

940 Department of Agricultural Statistics

1 Ali MA - MA,PhD *Professor* statistics
2 Khan MZA - MSc *Associate Professor* statistics

941 Department of Agronomy

1 Eunus M - MAg,PhD *Associate Professor* agronomy
2 Gaffer MA - PhD *Associate Professor* agronomy
3 Hossain SMA - MScAg,PhD *Associate Professor* agronomy

942 Department of Anatomy and Histology
Mymensingh
Tel: 4191-3. Cables: AGRIVARSITY.

1 Hossain MI - MScVetSc,PhD *Associate Professor* reproductive biology
2 Mia MA - MSVet *Professor* veterinary anatomy

943 Department of Animal Breeding and Genetics

1 Hasnath MA - MAg,MS,PhD *Professor* physiology and reproduction

944 Department of Animal Nutrition

1 Tareque AMN - BScVetSc,PhD *Associate Professor* animal nutrition

945 Department of Aquaculture and Management

1 Islam MA - PhD *Associate Professor* fish culture and management

946 Department of Bio-Chemistry

1 Gheyasuddin S - MSc,PhD *Associate Professor* plant protein and nutrition
2 Hanif MA - MSc,PhD *Associate Professor* plant biochemistry
3 Hussain M - MSc,PhD *Professor* food and nutrition

947 Department of Botany

1 Ali MA - MSc,PhD *Professor* taxonomy

948 Department of Co-operation and Marketing

1 Hussain AMM - MA,PhD *Professor* co-operatives

949 Department of Crop Botany

1 Chowdhury SH - MSc,PhD *Professor* plant physiology
2 Haque MA - MAg,PhD *Professor* plant anatomy
3 Muhsi AAA - MSc,PhD *Professor* plant nutrition and plant metabolism

950 Department of Dairy Science

1 Mannan AKMA - BScVetScAH,MS *Associate Professor* dairy microbiology
2 Rahman MS - MS,PhD *Professor* dairy production

951 Department of Entomology

1 Ahmed M - MSc,MS,PhD *Professor* insect taxonomy

952 Department of Farm Power and Machinery

1 Farouk SM - MSAgEng,PhD *Associate Professor* processing agricultural products
2 Hussain AAM - MSAgEng,PhD *Professor* farm implement design

953 Department of Fisheries Biology and Limnology

1 Islam MA - MSc,PhD *Professor* fisheries biology

954 Department of Fisheries Technology

1 Rahman MA - MSc,MS,PhD *Associate Professor* fisheries technology

955 Department of Food Technology and Rural Industries

1 Begum JA - MSc *Associate Professor* food technology

956 Department of General Animal Science

1 Haque M - BSc,MS *Associate Professor* livestock production and meat science

957 Department of Genetics and Plant Breeding

1 Chowdhury AR - MSc,PhD *Professor* genetics and cytogenetics
2 Rahman L - MScAg,PhD *Associate Professor* plant breeding

958 Department of Horticulture

1 Faruque AHM - MAg,PhD *Associate Professor* horticulture
2 Hoque MA - MAg,PhD *Associate Professor* fruit and vegetable production
3 Hussain A - MAg,PhD *Professor* fruit and vegetable production

959 Department of Medicine and Surgery

1 Rahman A - MS *Professor* veterinary medicine

960 Department of Microbiology and Hygiene

1 Amin MM - MSC,VetSc,PhD *Associate Professor* poultry diseases
2 Sarker AJ - BScAH,MS *Associate Professor* veterinary microbiology

961 Department of Parasitology

1 Huq MM - BScAH,MS *Associate Professor* parasitology
2 Huq S - MScVetSc,MSc *Associate Professor* animal production
3 Islam AWM - MScVetSc,PhD *Associate Professor* veterinary parasitology
4 Qadir ANMA - BScAH,MSVet *Associate Professor* parasitology
5 Shaikh H - MS,PhD *Professor* parasitology

962 Department of Pathology

1 Dewan ML - MS,PhD *Professor* veterinary pathology

963 Department of Plant Pathology

1 Ahmad MU - MScAg,MSc *Associate Professor* nematology
2 Ashrafuzzaman H - BSc,PhD *Professor* rice diseases
3 Khan MAA - MSc,PhD *Professor* diseases of vegetable and fruit crops
4 Momen A - MScAg,PhD *Associate Professor* plant virology

964 Department of Physiology and Pharmacology

1 Hasan Q - MScAH,PhD *Associate Professor* pharmacology
2 Sobahan MA - MSC,VetSc *Associate Professor* pharmacology

965 Department of Poultry Science

1 Ahmed S - MScAH,PhD *Associate Professor* poultry management
2 Bulbul SM - MS,PhD *Associate Professor* poultry nutrition
3 Hamid MA - MScAH,PhD *Associate Professor* poultry production
4 Latif MA - MS,PhD *Professor* poultry science
5 Wahid MA - BScAH,MS *Associate Professor* poultry production and management

966 Department of Soil Science

1 Bhuiya MZH - MSc,PhD *Professor* soil fertility, soil microbiology
2 Eaqub M - MScAg,PhD *Associate Professor* soil chemistry, soil fertility
3 Hoque MS - MScAg,PhD *Associate Professor* soil fertility
4 Idris M - MS,PhD *Professor* soil physics and water management

967 Department of Zoology

1 Hussain L - MSc,PhD *Associate Professor* chemical cytology

968 Graduate Training Institute

1 Rahman MM - MScAH *Associate Professor* sheep breeding

969 Department of Agriculture and Fisheries

PO Box HM 834, Hamilton HM CX
Tel: 809-236-4201. Telex: 3246 BERMUDA.

1 Manuel EA - BSA,DIC *Director*
2 Burnett-Herkes JN - BScAgr,MSc,PhD *Assistant Director, Fisheries*
3 Hill Mrs P *Librarian*

970 Agriculture

1 Dow Ms RL - BSc,MSc,PhD *Senior Plant Protection Officer* plant pest and disease control
2 Burnie N - BVSc,MRCVS *Veterinary Officer* animal husbandry
3 Hilburn D - BSc,MSc,PhD *Plant Protection Officer* plant pest and disease control
4 Rowland GA - BSc *Agricultural Officer, Agronomy* variety trials and cultural practices
5 Wingate DB - BA,PhD *Conservation Officer* native and endemic flora, fauna and habitats

971 Aquarium, Museum and Zoo

1 Winchell RJ - BSc,MSc *Principal Curator* seagrass and coral culture
2 Sterrer W - PhD *Curator (Bermuda Natural History Museum)* flora and fauna of Bermuda (esp. marine invertebrates)

972 Fisheries

1 Luckhurst BE - BSc,MSc,PhD *Senior Fisheries Officer* fish stock assessment
2 Frick B - BCom,MBA *Fisheries Officer, Economist*
3 Trott CA - BSc *Fisheries Officer, Food Scientist*
4 Ward JA - BSc *Fisheries Officer, Biologist*

973 Ministry of Agriculture

Private Bag 003, Gaborone
Tel: 51177. Cables: MINAGRIC.

974 Department of Agricultural Research

Private Bag 0033, Gaborone
Tel: 52381.

1 Gollifer DE - BSc,DTA,PhD *Director*

975 Animal Production Research Unit

1 Addy BL - BSc *Coordinator* animal production, range research
2 Boitumelo W - BSc dairy
3 Farrington T - BSc,MSc *Agricultural Economist*
4 Kiflewahid B - BSc,MSc,Phd *Dairy Specialist*
5 Mahabile W - BSc animal science; farming systems
6 Masilo BS - BSc dairy
7 Mayer L - BSc,MSc,PhD *Animal Nutritionist*
8 Mlambo M - DA animal husbandry
9 Montshiwa MK - BSc,MSc,PhD *Animal Geneticist*
10 Mosimanyana BM - BSc,MSc animal nutrition
11 Mphinyane W - BSc range research
12 Muke B - DA animal husbandry
13 Senyatso E - BSc *Small Stock Officer*
14 Setshwaelo LL - BSc,MSc *Animal Geneticist*
15 Sweet J - BSc,DipTropAgric *Range Management Specialist*
16 Tacheba G - BSc range research

976 Arable Crops Division

1 Stewart-Jones W - BSc,PhD *Chief Arable Research Officer*
2 Baker D - BSc,MSc agricultural economics
3 Bernhardt C - BSc,MSc,PhD seed technology
4 Carter D - BSc,MSc,PhD agronomy
5 deMooy J - BSc,PhD agronomy
6 Gakale L - BSc,MSc plant nutrition
7 Grey R - BSc,MSc,PhD animal science
8 Heinrich G - BSc,MSc,PhD agronomy
9 Horspool DG - CDA,NDA,CTechEng,MIAgrEng *Agricultural Engineer*
10 Jones RB - BSc,MSc agronomy
11 Loos H - Diplingagr,PhD agronomy
12 Madisa ME - BSc horticulture
13 Malepa DB - BSc,MSc plant nutrition
14 Manthe CS - BSc,MSc entomology
15 Manthe M - BSc agronomy, legumes
16 Maphanyane GS - BSc,MSc agronomy, legumes
17 Mayeux A - EngAgricolTropicAgron groundnuts
18 Mazhani L - BSc,MSc *Plant Geneticist*
19 Mmolawa OB - BSc *Seed Technologist*
20 Modiakgotla E - BSc,MSc agronomy, farming systems
21 Monageng K - DA mechanization
22 Montshiwa P - BSc crop husbandry, legumes
23 Motalaote B - BSc,MSc plant pathology
24 Patrick C - BSc agronomy
25 Rachem K - MSc agricultural economics
26 Riches CR - BSc,MSc weed research
27 Roostee R - Dipling water management
28 Sebolai B - BSc biometrics
29 Wiles G - BA natural sciences
30 Worman F - BSc,MSc,PhD agricultural economics

977 Support Services

1 Marumo K - BSc agricultural engineering
2 Molapong K - BSc soil chemistry
3 Samunzala BM - DA farm manager

978 Department of Agriculture, Ministry of Development

Bandar Seri Begawan
Tel: 02-43144/8. Telex: BU 2456 PERT. Cables: Agriculture Brunei.

1 Hanafiah DBH - BSc,MSc *Director*
2 Othman M - BSc,PhD *Deputy Director*

979 Research Division

Headquarters, Kilanas Agriculture Research Centre
Tel: 02-61352/4. Cables: Agriculture Brunei.

1 Mohiddin Dr MYBH - BSc,PhD *Assistant Director and Head of Research Division* acid sulphate soils; trace elements in plants and soils
2 Alim JB - BSc *Head of Fruits Unit* tropical fruits, clonal propagation (including tissue culture), breeding, genetics and conservation
3 Lim JB - BSc,MSc *Head of Animal Husbandry Unit* animal breeding and nutrition; pasture
4 Momin Mrs PM - BSc,MSc *Head of Agronomy Unit* plant breeding - rice
5 Safar Mrs SB - BSc *Head of Vegetable Unit* vegetable technology
6 Salleh PIBPM - BSc *Head of Soil Science Units* peat soil and iron toxicity in plants
7 Saman Mrs AB - BSc,MSc,MBiol *Head of Crop Pest Unit* insect pests of crops
8 Yunton IBB - BSc,MSc *Head of Plant Pathology Units* nematology, plant pathology

CANADA

980 Agriculture Canada

981 Headquarters
Sir John Carling Building, 930 Carling Avenue, Ottawa, Ontario, K1A OC5
Tel: 613-995-8963.

982 Research Branch

983 Headquarters
Sir John Carling Building, 930 Carling Avenue, Ottawa, Ontario, K1A OC5
Tel: 613-995-7084.

984 Executive
In addition to those listed here, the executive includes the Directors General of Atlantic, Quebec, Ontario, Prairie and Pacific regions and the A/Directors General of Central Experimental Farm and the Program Coordinator Directorate. Details are given under the appropriate institutions

1 LeRoux EJ - BA,MSc,PhD,FAIC *Assistant Deputy Minister, Research*
2 Lessard JR - BA,BSc,MS,PhD *Executive Assistant*
3 Radburn LR - ACBA,RIA *Director, Administration Division*

985 Program Coordination Directorate

1 Baier W - Diplomlandwirt,DrAgr,MSc *A/ Director General, Program Coordination Directorate*
2 Bouchard R - BA,BSA,MSc,PhD *Research Coordinator* animal productivity
3 Kirkland DF *Research Coordinator* international agricultural research and development
4 Larmond E - BSc *Research Coordinator* food
5 Neish GA - BSc,PhD *Research Coordinator* protection
6 Nowland J - BA,MSc *Research Coordinator* natural resources
7 Saidak WJ - BSA,MS,PhD,FWSSA *Research Coordinator* crop productivity
8 Willis CB - BScAgr,PhD,FAIC *Research Coordinator* production (horticulture & special crops)

986 Research Program Service
K.W. Neatby Building, Central Experimental Farm, Ottawa, Ontario, K1A OC6
Tel: 613-995-5301.

1 Bélanger Y - BSc *Director*
2 Perrin JA - BSc *Head of Section* scientific editing
3 Severn A *Head of Section* administration and text processing
4 Taky J - BA,BSc,MPA *Head of Section* scientific information
5 Fettes W branch liaison
6 Halchuk C *Head of Section* art and design
7 Wilson W *Head of Section* photography

RESEARCH CENTRES, STATIONS AND FARMS

987 Central Experimental Farm
K.W. Neatby Building, Central Experimental Farm, Ottawa, Ontario, K1A OC6
Tel: 613-995-6108.

1 Cantier JJ - PhD *Director General, Central Experimental Farm*
2 Larmond E - BSc *Program Specialist*

988 Animal Research Centre
Building 60, Central Experimental Farm, Ottawa, K1A OC6
Tel: (613) 993-6002.

1 Elliot JI - BSA,MSc,PhD *Deputy Director*
2 Leger DA - BSc *Assistant to the Director*
3 Gavora JS - Ing,CSc *Deputy Director*
4 Hartin KE - DVM *Veterinarian*

989 Animal Feed Safety and Nutrition Program

1 Trenholm HL - BSc,PhD *Program Chairperson* mycotoxins, toxicology
2 Akhtar MH - BSc,MSc,PhD pesticide metabolism and residues
3 Coté LM - BSc,MSc,PhD biochemical toxicology, mycotoxins
4 Farnworth ER - BSc,MSc,PhD lipid nutrition and metabolism
5 Kramer JKG - BSc,MSc,PhD lipid chemistry and biochemistry
6 Prelusky DB - BScPharm,PhD mycotoxin metabolism, toxicology

990 Animal Waste Utilization Program

1 Patni NK - BChE,MS,PhD *Program Chairman* livestock waste utilisation, farm pollution abatement

991 Biotechnology

1 Gavora JS - Ing,CSc *Program Chairman* disease resistance genetics
2 Bernon DE - BScAgr,MSc,PhD cellular, molecular and quantitative genetics
3 Hackett AJ - DVM,MSc,PhD embryo transfer, female reproductive physiology
4 Marcus GJ - BA,PhD maternal-embryonic physiological interactions
5 Nagai J - BS,DAgr embryo manipulation, quantitative genetics
6 Sabour MP - BSc,PhD molecular and cellular genetics
7 Teather RM - BSc,PhD genetic manipulation of rumen bacteria

992 Dairy and Beef Cattle Nutrition Program

1 Ivan M - Ing,MSc,PhD *Program Chairman* nutrition and metabolism of trace minerals
2 Atwal AS - BSc,MSc,PhD forage evaluation and nutrition
3 Erfle JD - BSA,PhD rumen metabolism and nutrition
4 Hidiroglou M - DVM,DipNutr vitamin and trace mineral nutrition
5 Jenkins KJ - BSA,MSc,PhD calf nutrition
6 Mahadevan S - BSc,MSc,PhD rumen metabolism and nutrition
7 Sauer FD - DVM,MS,PhD rumen metabolism and nutrition
8 Veira DM - BSc,MSc,PhD ruminant nutrition

993 Dairy Cattle Breeding and Production Program

1 Lee AJ - BScAgr,PhD *Program Chairman* dairy cattle breeding and production: statistical methodology and breeding strategies
2 Batra TR - BVSc,MVSc,MSc,PhD dairy cattle breeding: field data analyses, disease resistance
3 Lin CY - BS,MS,PhD dairy cattle breeding: multi-trait mixed model sire and cow evaluation methods; computer modelling
4 McAllister AJ - BS,MS,PhD dairy cattle breeding and production: selection for lifetime performance and use of crossbreeding

994 Poultry Breeding Program

1 Grunder AA - BSA,MSc,PhD *Program Chairman* eggshell quality genetics, broiler breeding, goose breeding and management
2 Chambers JR - BSc,MSc,PhD broiler breeding and management
3 Fairfull RW - BSc,MSc,PhD egg stock breeding and management
4 Tsang CPW - BSc,MSc,PhD physiology: eggshell quality and egg production

995 Poultry Nutrition Program

1 Sibbald IR - BScAgr,MSc,PhD,DSc *Program Chairman* avian energetics and feedstuff evaluation
2 Cave NAG - BSc,MSc,PhD amino acids and proteins, broiler breeder nutrition and management
3 Hamilton RMG - BScAgr,MSc,PhD nutrition and physiology: eggshell quality, mycotoxins

996 Sheep Production Program

1 Ainsworth L - BSc,MSc,PhD *Program Chairman* female reproductive physiology
2 Fiser PS - BSc,MSc,PhD cryopreservation, male reproductive physiology
3 Heaney DP - BS,MS,PhD nutrition and intensive management
4 Langford GA - BSc,MSc,PhD male reproductive physiology
5 Shrestha JNB - BVScAH,MS,PhD breeding and intensive production, applied quantitative genetics

997 Swine Production Program

1 Friend DW - BSc,MS,PhD *Program Chairman* sow and piglet production; mycotoxins
2 Fortin A - BScAgr,PhD carcass evaluation: swine, sheep, poultry, cattle
3 Fraser DG - BA,PhD animal behaviour: swine, sheep, cattle

998 Biosystematics Research Centre
Central Experimental Farm, Ottawa, K1A OC6
Tel: 613-996-1665.

1 Trottier R - BSc,MSc,PhD *Director*
2 Smith IM - BSc,PhD *Assistant Director*
3 Allyson Miss S - BSc,MSc *Manager, National Identification Service (Zoology)*
4 Anderson RV - BA,MS,PhD *Head of Section* unit curator of nematoda haplolaimidae (spiral nematodes), tylenchorhynchidae (stylet nematodes), aphelenchoidea (foliar nematodes)
5 Baillargeon G - BSA,MSc cultivated crops, Brassica
6 Barr DJS - BSc,MSc,PhD zoosporic parasites of vegetable crops
7 Barron JR - BSc,MSc,PhD ichneumonidae
8 Baum BR - MSc,PhD,FRSC *Head of Section* cultivated crops, *Hordeum*
9 Behan-Pelletier VM - BSc,MSc,PhD oribatei (oribatid mites); unit curator of arachnida
10 Bissett JD - BSc,PhD conidial parasites of forage crops

11 Borkent A - BSc,MSc,PhD *Program Leader* cecidomiidae (gall midges), ceratopogonidae (luting midges)
12 Bousquet Y - BSc,MSc,PhD colydiidae, elateridae (click beetles and wire worms), and other stored-product beetles
13 Bright DE - BS,MS,PhD scolytidae (bark beetle), curculionidae (weevils)
14 Campbell JM - BS,MS,PhD *Head of Section* staphylinidae (rove beetles)
15 Catling PM - BSc,PhD sedges, aquatic plants
16 Cayouette J - MTh,MSc,PhD grass flora of Canada
17 Cody WJ - BA curator of agriculture Canada vascular plant herbarium; Canadian flora, ferns
18 Corlett MP - BA,MA,PhD ascocarpic parasites of fruit crops
19 Crompton CW - MSc weed biology - taxonomy and palynology
20 Cumming JM - BSc,PhD empididae (dance flies); dolichopodidae (long-legged flies)
21 Dalpé Y - BSc,MSc,DSc mycorrhizae
22 Dang PT - BS,MS,PhD *Seconded from Canadian Forestry Service* microlepidoptera (spruce bud-worms)
23 Ebsary BA - BSc,MSc,PhD criconematidae (ring nematodes)
24 Foottit RG - BSA,MSc,PhD unit curator of hemiptera and miscellaneous insect orders; aphids, scales and thrips
25 Gibson GAP - BSc,MSc chalcidoidae (chalcid wasps)
26 Ginns JH - BSc,MSc,PhD basidiocarpic tree wood rots
27 Goulet H - BA,BSc,MSc,PhD symphyta (sawflies)
28 Hamilton KGA - BSA,MSc,PhD cicadellidae (leaf hoppers), cercopidae (spittle bugs)
29 Hudson BNA - BSc,PhD *Seconded to Research Branch Coordination Group*
30 Lafontaine JD - BA,MS,PhD *Program Leader, lepidoptera* noctuidae (cut worm moths); unit curator of lepidoptera-trichoptera
31 Landry JF - MSc microlepidoptera of agricultural importance
32 LeSage L - BSc,PhD,MSc chrysomelidae (leaf beetles)
33 Lindquist EE - BSc,MSc,PhD acari (mites and ticks)
34 Masner L - BSc,MSc,PhD proctotrupoidea (proctotrupid wasps), sphecoidea (digger wasps), eranioidea (ensign wasps); unit curator of hymenoptera
35 Mason WRM - BSc,PhD braconidae (braconid wasps)
36 Needham SA - BA,BSc curator of national culture collection of fungi and non-medical bacteria
37 Oliver DR - BA,MA,PhD chironomidae (chironomid midges)
38 Parmelee JA - BSc,MA,PhD obligate parasites of plants (rusts, smuts, mildews); curator of national mycological herbarium
39 Redhead SA - BSc,MSc,PhD mushrooms
40 Schmid F - LicésScNat,DeScNat trichoptera (caddisflies)
41 Sharkey MJ - BSc,MSc,PhD *Program Leader, Hymenoptera* braconidae (braconid wasps)
42 Shoemaker RA - BSA,MSA,PhD *Head of Section* ascocarpic parasites of cereals
43 Small E - BA,BSc,MSc cultivated crops, medicago
44 Smetana A - MUDR,CandScBiol aquatic beetles, staphylinidae (rove beetles); unit curator of coleoptera
45 Stahevitch AE - BSc,MSc,PhD weeds, cytotaxonomy
46 Teskey HJ - BSc,MSA,PhD unit curator of diptera; tabanidae (horse flies and deer flies)
47 Vockeroth JR - BA,MA,DPhil syrphidae (flower flies), scalophagidae (dung flies), culicidae (mosquitoes)
48 Warwick SI - BSc,PhD weeds, genecology
49 Yoshimoto CM - BA,MSc,PhD chalcidoidae (chalcid wasps), cynipoidea (gall wasps)

999 Engineering and Statistical Research Centre
Building 94, Central Experimental Farm, Ottawa, Ontario, K1A OC6
Tel: 613-995-9671.

1 Voisey PW - FIMechE *Director*

1000 Structures and Mechanization Section

1 Feldman M - BE,MSc *Head of Section* manager, ERDAF contract programme
2 Heslop LC - BSc mechanization
3 Jackson HA - BScEng,MSc Canada plan service design engineer - greenhouses
4 Massé DI - BSc,MSc Canada plan service design engineer - structures
5 Munroe JA - BSA,MSc,PhD structures, crop storage
6 Phillips PA - BScAgr,MSc,PhD animal environment
7 Reid WS - BScAgr,MSc mechanization
8 Turnbull JE - BSA,MSA *Director Canada Plan Service* farm structures

1001 Energy Section

1 Van Die P - BScEng,MSc *Head of Section*
2 Levesque M - BSc,MSc crop drying
3 McLaughlin N - BSc,MSc,PhD mechanization

1002 Instrumentation and Automation Section

1 Brach EJ - DEE,DipMilElectronics *Head of Section*
2 Buckley DJ - BE,MSc electronics
3 Pon C - BTech,MEng computer systems

1003 Food Process Engineering Section

1 Timbers GE - BSA,MSA,PhD *Head of Section* food process engineering
2 LeBlanc DI - BAScEng food process engineering
3 McGinnis DS - BAScEng,MSc food process engineering
4 Sidaway-Wolf DM - BAScAgrEng,MSc food process engineering

1004 Statistical Research Section

1 Binns MR - MA,DipMathStat statistics
2 Butler G - BMath,MMath statistics
3 Hall I - CertInfoProc systems and programming
4 Hobbs JD - BSc systems and programming
5 Jui PY - BSc,MSc,PhD statistics
6 Lefkovitch LP - BSc statistics
7 Lin CS - BSc,MSc,PhD statistics
8 Nault C - BSc systems and programming
9 Thompson BK - BSc,MMath,PhD statistics
10 Watt W - BA *Head of Section* systems and programming
11 Wolynetz MS - BMath,MMath,PhD statistics

1005 Land Resource Research Centre
Central Experimental Farm, Ottawa, K1A OC6
Tel: 613-995-5011.

1 Clark JS - BSA,MSc,PhD *Director*
2 van Schaik JC - MSc,PhD *Deputy Director*

1006 Soil Classification Section

1 McKeague JA - BA,BSA,MSc,PhD genesis
2 Fox CA - BA,MSc,PhD micromorphology
3 Lévesque PEM - BSA,MSA,PhD organic soils
4 Mathur SP - BSc,Assoc,IARI,PhD organic soils
5 Reynolds WD - PhD soil physics
6 Topp GC - BSA,MS,PhD *Acting Head of Section*
7 Wang C - BSA,MS,PhD chemistry
8 Wilson G - BSc,MSc,DIC engineering
9 Wires KC - BA physical structure

1007 Land Use and Evaluation Section

1 Dumanski J - BSA,MSc,PhD *Head of Section*
2 Coote DR - MS,PhD degradation
3 Culley J - PhD water quality
4 De Jong R - BSc,MSc,PhD water use
5 Huffman E - BSc,MA land use
6 MacDonald B - BSA,MSc,PhD evaluation
7 Switzer-Howse K - BSc information

1008 Agrometeorology Section

1 van Schaik J - MSc,PhD *Head of Section*
2 Boisvert J - BSc,MSc farm weather service research
3 Bootsma A - BSc,MSc agroclimatological resources
4 Desjardins RL - BSc,MA,PhD micrometeorology
5 Dwyer LM - BSc,MSc,PhD environmental meteorology
6 Edey SM - BSc applications
7 Hayhoe HN - BSc,MS,PhD biomathematics
8 Mack AR - BSA,MSc,PhD remote sensing
9 Stewart DW - BSA,MSc,PhD crop weather modelling

1009 Soil Mineralogy Section

1 Kodama H - BSc,MSc,DSc,FCSSS *Head of Section*
2 Ihnat M - BSc,PhD,FAOAC trace element chemistry
3 Miles NM - BSc soil mineralogy
4 Ross GJ - BSA,MSc,PhD soil mineralogy
5 Singh SS - BSc,MSc,PhD soil chemistry

1010 Soil Chemistry Section

1 Schnitzer M - BSc,MSc,PhD,FCSS,FASA,FSSSA *Head of Section*
2 Ivarson KC - BSc,MSc,PhD soil microbiology

1011 Environmental Chemistry Section

1 Khan SU - BSc,MSc,PhD,CChem,FCIC,FRSC *Head of Section*
2 Behki RM - BSc,MSc,PhD pesticides molecular biology
3 Gamble DS - BSc,MSc,PhD,FCIC pesticide - soil interaction
4 Nelson S - BSc,MSc,PhD plant uptake of pesticides

1012 Soil Resource Inventory and Mapping Section

1 Valentine KWG - BSA,MSc,PhD *Head of Section*
2 Shields JA - BSA,MSc,PhD soil correlation - Great Plains
3 Tarnocai C - BSF,MS soil correlation - British Columbia, Atlantic region and North

1013 Soil Survey Units
1014 Nova Scotia Soil Survey (Truro)
1 Webb KT - BSc,MSc *Head of Unit*
2 Holmstrom DA - BSA *Party Leader*
3 Patterson GT - BSc,MSc *Party Leader*

1015 New Brunswick Soil Survey (Fredericton)
1 Rees HW - BSc *Head of Unit*
2 Fahmy SH - BSc,MSc *Party Leader*

1016 Prince Edward Island Soil Survey (Charlottetown)
1 Veer C *Senior Soil Survey Technologist*

1017 Newfoundland Soil Survey (St. John's)
1 Woodrow EF - BSc *Head of Unit*
2 Hender F - BSc *Party Leader*

1018 Quebec Soil Survey (Sainte-Foy)
1 Cossette JM - BSc *Head of Unit*
2 Grenon L - BSA *Party Leader*
3 Lamontagne L - BSc *Party Leader*
4 Nolin MC - BS,MSc *Party Leader*

1019 Ontario Soil Survey (Guelph)
1 Acton CJ - BSA,MSc,PhD *Head of Unit*
2 Jones RK - BSc,MSc *Party Leader*
3 Presant EW - BSA,MSc *Party Leader*
4 Wall GJ - BSA,PhD *Party Leader*

1020 Manitoba Soil Survey (Winnipeg)
1 Smith RE - BSA,MSc *Head of Section*
2 Eilers R - BSA,MSc *Party Leader*
3 Fraser WR - BSc,MSc *Party Leader*
4 Michalyna W - BSA,MSc,PhD *Party Leader*
5 Veldhuis H - Ing,MSc *Party Leader*

1021 Saskatchewan Soil Survey (Saskatoon)
1 Acton DF - BSA,MSc,PhD *Head of Unit*
2 Anderson AJ - BSc *Party Leader*
3 Eilers WD - BSA,MSc *Party Leader*
4 Kozak LM - BSA,MSc,PhD *Party Leader*
5 Padbury G - BSA,MSc *Party Leader*
6 Rostad HPW - BSA,MSc,PhD *Party Leader*
7 Stonehouse HB - BSA,MSc *Party Leader*

1022 Alberta Soil Survey (Edmonton)
1 Pettapiece WW - BSc,MSc,PhD *Head of Unit*
2 Brierley JA - BSc *Party Leader*
3 Coen GM - BSC,MSc,PhD *Party Leader*
4 Hiley JC - BA,MA *Party Leader*
5 Tajek J - Eng *Party Leader*
6 Walker BD - BSc,MSc *Party Leader*

1023 British Columbia Soil Survey (Vancouver)
1 Moon DE - BSA,MSc,PhD *Head of Unit*
2 Green AJ - BSA,MSc *Party Leader*
3 Selby CJ - BSc,MSc *Party Leader*
4 van Vleit LJP - BSA,MSc *Party Leader*

1024 Plant Research Centre
Agriculture Canada, Central Experimental Farm, Ottawa, Ontario, K1A 0C6
Tel: 613-995-5287.
1 de la Roche AI - BSc,MS,PhD *Director*
2 Deschênes JM - BSc,MSc,PhD *Assistant Director, Research*

1025 Cereal Crops Section
1 Burrows VD - BSA,MSc,PhD *Section Head* oat breeding
2 Frégeau J - BSc,MSc,PhD crop physiology
3 Fulcher RG - BA,MSc,PhD morphogenetics and grain quality
4 Ho KM - BSc,MSc,PhD barley breeding
5 Sampson DR - BSc,AM,PhD wheat breeding
6 Seaman WL - BSc,PhD wheat pathology

1026 Forage Crops Section
1 Hamilton R - BSc,MSA,PhD *Section Head* corn breeding
2 Basu PK - BSc,MS,PhD plant disease
3 Bolton A - BSc,MSc,PhD corn pathology
4 Faris M - BSc,MSc,PhD legume breeding
5 McElroy A - BSc,MSc,PhD grass breeding
6 Voldeng H - BSA,MSc,PhD soybean breeding

1027 Cytogenetics Section
1 Fedak G - BSA,MSc,PhD *Section Head* cereal cytology
2 Armstrong KC - BSA,MSc,PhD brome cytology and chromosome banding
3 Craig I - BA *Hordeum* cytology
4 Simmonds J - BSc,MSc,PhD cereal tissue culture

1028 Genetic Engineering Section
1 Keller WA - BSA,PhD *Section Head* cell genetics
2 Brown D - BSc,MSc,PhD developmental physiology
3 Gleddie S - BSc,MSc,PhD cell genetics
4 Miki BLA - BSc,PhD molecular genetics
5 Molnar S - BSc,PhD cell genetics

1029 Entomology Section
1 Harcourt DG - BSA,PhD,FESC *Section Head* population dynamics and integrated pest management
2 Guppy JC - BSA,MS population dynamics
3 Loan C - BS,MS,PhD parasite-host relationships
4 Meloche F - BSc,MSc integrated pest management

1030 Plant Gene Resources
1 Deschenes JM - BSc,MSc,PhD *A/Head of Section*
2 Fraleigh B - DTC plant gene resources

1031 Stress Physiology Section
1 Andrews CJ - BSc,PhD *Section Head* winter survival of cereals
2 Hope HJ - BSc,MSc,PhD mechanisms of frost resistance
3 Picman A - MSc,PhD phytochemistry
4 Pomeroy MK - BSc,PhD,MSc winter survival of cereals
5 Schneider E - BSc,BSc,PhD snow mold damage
6 Singh J - BSc,PhD freezing injury - membranes

1032 Plant Pathology Section
1 Sinha RC - BSc,MSc,PhD,DSc,FRSC *Section Head* virus and mycoplasmas and vector relationships
2 Chiykowski L - BSA,MSc,PhD leafhopper - transmitted mycoplasmas
3 Paliwal YC - BSA,MSc,PhD aphid and mite transmitted viruses

1033 Nitrogen Fixation Physiology Section
1 Barran L - BSc,PhD,MSc *Section Head Rhizobium* genetics
2 Benzing-Purdie LM - BSc,MSc,PhD carbohydrate chemistry
3 Bromfield E - BSc,PhD rhizobium ecology
4 Chan YK - BSc,PhD bacterial physiology
5 Macdowall FDH - FSc,MSc,PhD legume physiology
6 Miller RW - BSc,PhD *Rhizobium* and legume physiology
7 Watson RJ - BSc,MSc,PhD *Rhizobium* & genetics
8 Wheatcroft RGL - MA,DPhil *Rhizobium* & genetics

1034 Mycotoxins Section
1 Greenhalgh R - BSc,PhD *Section Head* analytical and natural products chemistry
2 Blackwell BA - BSc,MSc,PhD spectroscopy
3 Miller JD - BSc,MSc,PhD fungal physiology
4 Young JC - BSc,MSc,PhD analytical and natural products chemistry

1035 Electron Microscope Unit
Agriculture Canada, Central Experimental Farm, Ottawa, Ontario, K1A 0C6
Tel: 613-995-5287.
1 Haggis G - BSc,PhD electron microscopy
2 Sinha RC - BSc,MSc,PhD,DSc,FRCS *Head of Section*

1036 Atlantic Region
1888 Brunswick Street, Halifax, Nova Scotia, B3J 3J8
Tel: (902) 426-7113.
1 Martel YA - BA,BScAgr,PhD *Director General*
2 Collins WB - PhD *Program Specialist*

NEW BRUNSWICK

1037 Research Station
P.O. Box 20280, Fredericton, E3B 4Z7
Tel: 506-452-3260.
1 Vacant *Director*
2 Misener GC - BSc,MSc,PhD *Acting Director*
3 Anderson RM - MLS *Librarian*

1038 Potato Breeding
1 Tarn TR - BSc,PhD *Section Head* breeding and cytogenetics
2 Coffin RH - BSc,MSc,PhD breeding and evaluation

3 DeJong H - BA,MSc,PhD diploid breeding and genetics
4 Murphy AM - BSc,MSc disease screening
5 Seabrook JEA - BSc,MSc,PhD propagation methods
6 Tai GCC - BSc,MSc,PhD quantitative genetics

1039 Potato Pest Management

1 Singh RP - BSc,MSc,PhD *Section Head* virus diseases, viroids
2 Bagnall RH - BSc,MSc,PhD virus epidemiology and resistance
3 Boiteau G - BSc,MSc,PhD insect ecology
4 Clark MC - BSc,PhD biochemistry of disease resistance
5 Coleman WK - BSc,PhD physiology
6 King RR - BSc,MSc,PhD analytical organic chemistry
7 McKenzie AR - BSc,MSc,PhD tuber-borne pathogens
8 Pelletier Y - BSc,MSc insect/plant relationships

1040 Livestock and Crops

1 Nicholson JWG - BSc,MSc,PhD *Section Head* meat animal nutrition
2 Bélanger G - BSc,MSc forage crops
3 Burgess PL - BSc,MSc,PhD dairy cattle nutrition
4 Bush RS - BSA,MSc,PhD calf nutrition
5 McQueen RE - BSc,MSc,PhD rumen microbiology

1041 Engineering, Horticulture, Soils

P.O. Box 20280, Fredericton, E3B 4Z7
Tel: 506-452-3260.

1 McLeod CD - BASc,MASc *Section Head* agriculture mechanization engineering
2 Chow TL - BSc,PhD soil hydrology
3 Estabrooks EN - BSc,MSc tree fruits and berry crops
4 Milburn PH - BSc,MSc soils engineering
5 Misener GC - BSc,MSc,PhD harvesting and storage engineering
6 Richards JE - BSc,MSc,PhD soil fertility

1042 Hervé J. Michaud Experimental Farm

P.O. Box 667, Bouctouche, EOA 1GO
Tel: 506-743-2464.

1 Rousselle GL - BA,BScAgr,MScAgr,PhD *Superintendent* cereal and forage crops
2 LeBlanc PV - BSc,BSA vegetables
3 Luffman M - BScAgr fruits

1043 Research Station

P.O. Box 7098, St. John's, Newfoundland, A1E 3Y3
Tel: 709-772-4619.

1 Bishop GA - BEng,MASC agricultural engineering
2 Hampson MC - BSc,MSc,PhD plant diseases
3 Penney BG - BSc,MSc vegetable crops
4 Proudfoot KG - BAgr,MAgr *Acting Head* potato breeding

1044 Research Station

P.O. Box 1210, Charlottetown, C1A 7M8
Tel: 902-892-5461.

1 MacLeod LB - BScAgr,MSc,PhD,FCSSS *Director*

1045 Forage/Livestock Section

1 Kunelius HT - BSc,MSc,PhD *Head of Section* forage management
2 Andrew RG - BScAgr agricultural economics
3 Narasimhalu PR - BSc,BVSc,MS,PhD forage conservation and utilization
4 Suzuki M - BSc,PhD forage and cereal biochemistry and physiology
5 Thompson LS - BScAgr,PhD *Assistant Director* forage and potato insects
6 Winter KA - BScAgr,MSc,PhD animal nutrition

1046 Cereals Section

1 MacLeod JA - BScAgr,MSc,PhD *Head of Section* soil fertility and cereal nutrition
2 Campbell AJ - BEng,MPhil agricultural engineering
3 Carter MR - BScAgr,MSc,PhD soil tillage
4 Choo TM - BSc,PhD barley breeding
5 Edwards LM - BSc,MSc,PhD soil management
6 Gupta UC - BScAgr,MScAgr,PhD,FCSSS,FASA,FSSSA soil and plant micronutrients
7 Johnston HW - BScAgr,MSc,PhD cereal diseases
8 Martin RA - BSc,BScAgr,MSc,PhD cereal diseases
9 Nass HG - BSA,MSc,PhD cereal physiology and wheat breeding

1047 Horticulture Section

1 Cutcliffe JA - BScAgr,MSc *Head of Section* vegetable nutrition and management
2 Arsenault WJ - BScAgr tobacco management and nutrition
3 Ivany JA - BScAgr,MS,PhD field crop weed control
4 Kimpinski J - BSA,MSc,PhD nematology
5 Platt HW - BSc,PhD potato diseases

6 Read DC - BScAgr,MSc,PhD pesticide degradation; vegetable insects
7 Sanderson JB - BScAgr,MSc potato management and fertility
8 White RP - BSEd,MS,PhD soil chemistry; corn and potato nutrition and management

1048 Research Station

Kentville, Nova Scotia, B4N 1J5
Tel: 902-678-2171.

1049 Administration

1 Weaver GM - BSc,PhD *Director*
2 McRae KB - BEd,MS,PhD *Regional Statistician*
3 Meyer H - BSc *Computer Systems Programmer*
4 Miner JR - BA,MLS *Librarian*
5 Regan WD - BMath *Computer Systems Manager*

1050 Crop Production

1 Blatt CR - MSc,PhD *Head of Section* crop physiology and nutrition
2 Crowe AD - BScAgr,MSc,PhD tree fruit breeding and physiology
3 Embree CG - BScAgr,MSc *Leader* tree fruits program physiology and management
4 Hall IV - BSc,MSc,PhD *Leader* berry crops program ecology and management
5 Hicklenton PR - BSc,MSc,PhD *Leader* ornamentals program physiology and management
6 Jamieson AR - BSc,MSc,PhD berry crops breeding
7 Jones RW - BSc,MSc,PhD *Leader* agronomy program physiology and management
8 Ricketson CL - BScAgr,MS,PhD *Leader* vegetable program physiology and management
9 Webster DH - MSc,MSc,PhD soils management and nutrition

1051 Crop Protection

1 Jensen KIN - BSc,MSc,PhD *Head of Section* weed physiology
2 Anderson MG - BSc,PhD vegetable storage and cereal diseases
3 Gaul SO - BSc,MSc toxicology
4 Hardman JM - BSc,MSc,DIC,PhD insect ecology, orchard pests
5 Hildebrand PD - BSc,PhD berry crops and vegetable diseases
6 Kimball ER - BSc residue chemistry
7 Lockhart CL - BScAgr,MSc berry crops and vegetable diseases
8 Neilson WTA - BSc,MSc blueberry insects
9 Nickerson NL - BSc,MSc,PhD berry crop disease
10 Ragab MTH - BSc,MSc,PhD residue chemistry
11 Specht HB - BScAgr,MSc,PhD field crops and vegetable insects

1052 Food Processing

1 Stark R - BSc,PhD *Head of Section and Program Leader* food science
2 Jackson ED - BTech,MSc,PhD food microbiology
3 Lawrence RA - BSc,MSc,ChemEng food engineering

1053 Fruit and Vegetable Storage

1 Lidster PD - BSc,MSc,PhD *Head of Section and Program Leader* fruit storage physiology

1054 Poultry

1 Hulan HW - BScAgr,MSc,PhD *Head of Section and Program Leader* nutrition and physiology
2 Proudfoot FG - BScAgr,MSc physiology and management

1055 Experimental Farm

Nappan, Nova Scotia, BOL 1CO
Tel: (902)667-3826.

1 Calder FW - BScAgr,MSc *Superintendent* forage crops
2 Laflamme LF - BScAgr,MSc,PhD ruminant nutrition
3 Troelstra R - BScAgr beef management and nutrition
4 Van Lunen TA - BScAgr,MSc swine management and nutrition

1056 Région de Québec

Complex Guy Favreau, 200 West Dorchester Blvd., East Tower - Room 1002 - R, Montreal, Quebec, H2Z 1Y3
Tel: (514) 285-8888.

1 Jasmin JJ - BScAgr,MSc *Directeur Général*
2 Darisse F - BA,BScAgr *Specialiste en programmes, intérimaine*

1057 Centre de Recherches Alimentaires (Food Research Station)

Saint-Hyacinthe, Quebec, J2S 8E3
Tel: (514) 773-5771.

1 Riel RR - BSc,MSc,PhD *Directeur*

1058 Section industrie laitiere

1 Rolland JR - BScA,MSc,PhD *Chef de Section* ingrédients
2 Roy D - BSc,MSc,PhD bioréacteurs
3 St-Gelais D - BSc,MSc produits laitiers

1059 Genie Alimentaire Section

1 Passey C - BE,ME,DSc,MBA *Chef de Section* genie alimentaire

1060 Industrie Carnée Section

1 Delaquis P - BSc,MSc microbiologie
2 Piette JP - MSc,Ing produits carnés
3 Zardakas CG - BSc,MSc,PhD biochimie musculaire

1061 Biotechnologie Section

1 Cormier F - BSc,MSc,PhD culture de cellules végétales
2 Gauthier S - BSc,MSc,PhD protéines
3 Lee B - BSc,MSc,PhD *Chef de Section* genie génétique
4 Morin A - BSc,MSc,PhD enzymologie

1062 Industrie Végétale Section

1 Bégin A - BSc,MSc, Vivres rhéologie
2 Doyon G - BSc,MSc,Vivres *Chef de Section* conservation et emballage
3 Lapsley K - BSc,MSc produits végétaux
4 Toupin C - BSc,PhD,Vivres,MSc transfert de masse (systémes biologiques)
5 Willemot C - BSc,MSc,PhD physiologie

1063 Extraction et Arômes

1 Paré JRJ - BSc,PhD,MSc *Chef de Section* spectrometrie de mass
2 Bélanger JMR - BSc,MSc,PhD extraction

1064 Station de Recherches
P.O. Box 90, Lennoxville, P.Q., J1M 1Z3
Tel: (819) 565-9171.

1 St-Pierre JC - BScAgr,MSc,PhD *Directeur*

1065 Production de Viande

1 Fahmy MH - BScAgr,MSc,PhD *Chef de Section* génétique - porcs
2 de Passillé AMB - BSc,MSc ethologie - porcs
3 Farmer C - BScAgr,MSc,PhD régie - porcs
4 Flipot PM - BScAgr,MSc,PhD nutrition - bovins de boucherie
5 Matte JJ - BScAgr,MSc,PhD nutrition - porcs
6 Pommier SA - BSc,MSc,PhD qualité des viandes - bovins, porcs, moutons

1066 Production Laitiére

1 Roy GL - BSA,MSc,PhD *Chef de Section* génétique - bovins
2 Girard CL - BScAgr,MSc,PhD nutrition jeunes ruminants
3 Guilbault LA - BSc,MSc,DEANutr,PhD physiologie de la reproduction bovins
4 Lachance B - BScAgr,MSc nutrition jeunes ruminants
5 Lapierre H - BA,MSc physiologie et nutrition
6 Pelletier G - BScAgr,MSc,PhD physiologie de la digestion - bovins
7 Petitclerc D - BSc,BScAgr,MSc,PhD physiologie de la lactation - bovins
8 St-Pierre NR - MSc,PhD,BScAgr analyse de systémes
9 Vinet C - BScAgr,MSc nutrition bovins laitiers

1067 Production Fourragére et Sols

1 Dionne JL - BA,BScAgr,PhD *Chef de Section* fertilite des sols
2 Barnett GM - BScAgr,MSc regie et environment
3 Mason WN - BScAgr,MSc,PhD régie et qualité des plantes fourragéres
4 Pesant AR - BScAgr,MSc physique des sols

1068 Station de Recherches
P.O. Box 457, St-Jean-sur-Richelieu, Quebec, J3B 6Z8
Tel: (514) 346-4494.

1 Aubé CB - BA,BScAgr,PhD *Directeur*
2 Crête R - LSA,MSc *Directeur adjoint*
3 Otis T - BScAgr *Agente d'information*

1069 Fruits Section

1 Lareau M - BScAgr,MSc gestion des cultures
2 Bostanian NJ - BSc,MSc,PhD acarologie - entomologie
3 Bourgeois G - BScAgr,MSc modélisation
4 Granger RL - BA,LSA,MSc,PhD physiologie
5 Pelletier JR - BSc,MSc phytopathologie
6 Vincent C - BScAgr,MSc,PhD entomologie

1070 Section Légumes

1 Belcourt J - BSc,MSc,DSc entomologie
2 Benoît DL - BSc,MSc,PhD malherbologie
3 Bélair G - BSc,MSc nematologie
4 Bérard LS - BSc,MSc,PhD physiologie de la sénescence
5 Boivin G - BSc,MSc,PhD entomologie
6 Chiang MS - BScAgr,MSc,PhD biotechnologie
7 Côté JC - BSc,MSc biotechnologie
8 Hudon M - LSA,MSc entomology, grain corn
9 Landry B - BSc,MSc biotechnologie
10 Martel P - BA,BSc,PhD *Chef de Section* toxicologie
11 Senécal M - BScAgr,MSc gestion des cultures
12 Vigier B - BScAgr,MSc gestion des cultures; plantes oléagineuses

1071 Ferme Expérimentale
P.O. Box 1070, L'Assomption, Quebec, JOK 1GO
Tel: (514) 589-4775.

1072 Tabacs and Ornamentals

1 Darisse F - BA,BSc *Surintendant*
2 Arnold N - BScAgr,MSc,PhD physiologie
3 Cloutier D - BSc,MSc,PhD malherbologie
4 Lamarre M - BScAgr,MSc phytotechnie
5 Ogilvie IS - BScAgr,MSc,PhD génétique
6 Richer-Leclerc C - BScAgr,MSc gestion des plantes ornamentales

1073 Station de Recherches
2560, boul. Hochelaga, Sainte-Foy, Quebec, G1V 2J3
Tel: (418) 648-4814.

1 Bourget SJ - BScAgr,MS,PhD *Directeur*
2 Gagnon C - BA,BScAgr,MSc,PhD *Directeur adjoint*
3 Germain M - BScAgr,MSc *Agent d'information*
4 Venne P - BBibl,MLS *Bibliothéque*

1074 Amelioration et Gestion des Sols et des Plantes

1 Michaud R - BScAgr,MSc,PhD *Chef de Section* génétique des légumineuses
2 Angers D - BScAgr,MSc physique et conservation des sols
3 Bernier-Cardou M - BSc,MSc statistiques
4 Bullen MR - BScAgr,MSc,PhD génétiques des plantes
5 Dubuc JP - BScAgr,PhD génétique des céréales
6 Isfan D - BSc,PhD chimie et fertilité des sols
7 Légère A - BSc,MSc,PhD malherbologie
8 Ouellet D - BScAgr régie des céréales et des plantes horticules
9 Surprenant J - BSc,MSc,PhD génétique des graminées

1075 Phytoprotection et physiologie

1 Gagnon C - BS,BScAgr,MSc,PhD *Chef de Section* pathologie des legumineuses
2 Bissonnette N - BScAgr,MSc microbiologie
3 Bolduc R - BA,BScAgr,PhD resistance au froid - physiologie
4 Bordeleau L - BScAgr,MSc,PhD microbiologie
5 Castonguay Y - BScAgr,MSc agrométéorologie
6 Cloutier Y - BSc,MSc,PhD resistance au froid - physiologie
7 Comeau A - BSc,PhD entomologie
8 Couture L - BA,BScAgr,MSc,PhD pathologie des céréales
9 Furlan V - BSc,DSc endomycorhizes
10 Lafrenière C - BScAgr,MSc régie des plantes fourragéres, agronistéorologie
11 Lalande R - BScAgr,MSc,PhD microbiologie
12 Nadeau P - BSc,DSc,PhD resistance au froid - biochimie
13 Prévost D - BScAgr,MSc microbiologie
14 Richard C - BSc,MSc,DSc,PhD pathologie des légumineuses
15 Savoie P - BScAgr,MSc,PhD génie des plantes fourragéres

1076 Ferme Expérimentale
P.O. Box 400, La Pocatière, P. Q., GOR 1ZO
Tel: (418) 856-3141.

Under the responsibility of the Research Station of Sainte-Foy

1 Comeau JE - BScAgr,MS *Surintendant*
2 Belzile L - BScAgr,MSc plantes fourragéres
3 Frève A - BScAgr,MSc pommes de terres
4 Rioux R - BA,BScAgr,MSc herbicides et malherbologie
5 Zizka J - BA,BScAgr,MSc fertilité des sols

1077 Ferme Expérimentale
1472, rue St-Cyrille, Normandin, P. Q., GOW 2EO
Tel: (418) 274-3378.

Under the responsibilty of the Research Station at Sainte-Foy

1 Wauthy JM - BScAgr *Surintendant* bovins laitiers, fertilité des sols
2 Drapeau R - BScAgr,MSc plantes fourragéres et horticules

1078 Ontario Region

K.W. Neatby Building, C.E.F., Ottawa, Ontario, K1A OC6
Tel: (613) 995-3104.

1 Cartier JJ - BA,BSc,MSc,PhD *Director General*
2 Marriage PB - BSc,PhD *Program Specialist*

1079 Food Research Centre

Central Experimental Farm, Ottawa, Ontario, K1A OC6
Tel: 613-995-3722.

1 Tape NW - BSA,PhD *Director*
2 Sahasrabudhe MR - BSc,MSc,PhD *Assistant Director*
3 Nazarowec-White M - BSc,MSc *Food Science Services*

1080 Processing Technology

1 Collins WF - BSc,PhD phenolics chemistry
2 Jones JD - BSc,MSc,PhD,FRSC,FIFST,CChem oilseeds
3 Ma CY - BSc,MSc,PhD protein functionality
4 Oomah BD - MSc,PhD cereal processing
5 Paton D - BSc,PhD *Program Leader* processing technology
6 Weisz JN - BSc carbohydrate analysis
7 Wood PJ - BSc,PhD carbohydrate chemistry

1081 Dairy Technology

1 Emmons DB - BSA,MS,PhD *Program Leader* dairy products
2 Harwalker V - BSc,MSc,PhD protein chemistry
3 McKellar RC - BSc,MSc,PhD dairy products microbiology
4 Modler HW - BSA,MS,PhD dairy products
5 Sinha RP - BSc,MSc,PhD dairy microbiology

1082 Structure and Sensory Evaluation

1 Siddiqui IR - BSc,MSc,PhD,DSc,CChem,FRSC,FCIC *Program Leader* carbohydrates
2 Froehlich D - BSc,MSc,PhD sensory evaluation
3 Kalab M - MSc,PhD electron microscopy
4 Paquet AM - BSc,PhD amino acid and peptide chemistry
5 Poste LM - BASc,DipTech sensory evaluation
6 Yiu SH - BSc,MSc,PhD microscopy

1083 Food Safety And Nutrition

1 Lloyd LE - BSc,MSc,PhD *Program Leader* food safety and nutrition
2 Holley RA - BSc,PhD meat microbiology
3 Mullin WJ - LRIC,PhD analytical methodology
4 Nadeau L - BSc,MSc nutrient analysis
5 Robichon-Hunt LF - BSc,MSc nutrition services
6 Russell LF - BSc,MSc *(on educational leave)* nutrient analysis

1084 Research Centre

University Sub Post Office, London, Ontario, N6A 5B7
Tel: 519-679-4452.

1 Morley HV - BSc,PhD *Director*

1085 Mode of Action of Selected and Potential Insect Control Agents

1 Chefurka W - BSc,MSc,PhD *Section Head* biochemistry
2 Bolter Ms CJ - BSc,MSc fumigation - toxicology
3 Bond EJ - BSA,MSc,PhD,DIC fumigation - toxicology
4 Dumas T - DCE,MSc analytical chemistry
5 Krupka RM - BA,MA,PhD biochemistry
6 Lee TT - BSc,PhD plant biochemistry and tissue culture
7 Nagai T - ME,MSc,DSc neurophysiology
8 Roslycky EB - BSc,MSc,PhD microbiology
9 Starratt AN - BSc,PhD,FCIC chemistry - attractants and repellants
10 Steele RW - BSc,PhD neurochemistry
11 Vardanis A - BSc,MSc,PhD biochemistry

1086 Mode of Action of Selected and Potential Plant-Pathogen Control Agents

1 Ward EWB - BSc,MSc,PhD *Section Head* plant pathology: mode of action of selected and potential plant-pathogen control agents
2 Cuppels Ms DA - BA,MS,PhD phytobacteriology - microbial physiology
3 Lazarovits G - BSc,MSc,PhD plant pathology - fungicides
4 Madhosingh C - BSA,MA,PhD microbial biochemistry and pathology
5 Miller DM - BSc,MSc,PhD biophysics - fungicide selectivity
6 Stoessl A - BSc,PhD,FCIC chemistry - phytoalexins and mycotoxins
7 White GA - BA,MSc,PhD biochemistry - fungicides

1087 Soil Pesticides Section

1 Bowman BT - BSA,MSc,PhD soil physical chemistry
2 Chapman RA - BSc,MSc,PhD analytical organic chemistry
3 Harris CR - BA,MA,PhD *Section Head* insect toxicology

4 McLeod DGR - BSA,MS,PhD physiology
5 Tolman JH - BSc,PhD applied entomology
6 Tomlin AD - BA,MSc,PhD pesticide ecology
7 Tu CM - BSc,MSc,PhD microbiology
8 Turnbull SA - BSc,MSc toxicology

1088 Research Station

P.O. Box 186, Delhi, Ontario, N4B 2W9
Tel: (519) 582-1950.

1 Johnson PW - BSA,MSc,PhD *Director*
2 Cheng HH - BSc,MSc,PhD entomology
3 Court WA - BSc,MSc,PhD chemistry
4 Gayed SK - BSc,MSc,PhD plant pathology
5 Pandeya RS - BSc,MSc,PhD genetics and plant breeding
6 Rosa N - BSc,MSc,PhD plant physiology
7 Roy RC - BScAgr,MSc agronomy, new crops
8 Zilkey BF - BSA,MSc,PhD plant physiology

1089 Research Station

Harrow, Ontario, NOR 1GO
Tel: 519 738 2251.

1 Marks CF - BScAgr,MSA,PhD *Director*

1090 Crop Science Section

1 Buzzell RI - BS,PhD *Head of Section* soybean breeding
2 Buttery BR - BSc,PhD soybean physiology
3 Park SJ - BSA,MS,PhD field bean breeding
4 Teich AH - BA,BSA,MSc,PhD winter wheat breeding
5 Welacky TW - BSc,BScAgr field crop agronomy

1091 Entomology Section

1 Jaques RP - BSA,MSA,PhD *Head of Section* insect pathology

1092 Horticultural Science Section

1 Layne REC - BSc,MS,PhD *Head of Section* tree fruit breeding
2 Garton RW - BScAgr vegetable cultivar evaluation and management
3 Hunter DM - BSc,MSc orchard management
4 Kappel F - BScAgr,MSc,PhD orchard management
5 Liptay A - BSA,MSc,PhD vegetable management
6 Michelutti R - MSc viticulture
7 Papadopoulos AP - MScAgr,MScHort,PhD greenhouse management
8 Poysa VW - BScAgr,PhD vegetable breeding

1093 Plant Pathology Section

1 Jarvis WR - BSc,PhD,DIC *Head of Section* vegetable diseases
2 Anderson TR - BSA,MSc,PhD soybean and corn diseases
3 Bonn WG - BSc,MS,PhD bacterial diseases of fruit
4 Dhanvantari BN - BSc,MSc,PhD bacterial diseases of vegetables
5 Traquair JA - BSc,PhD tree fruit diseases
6 Tu JC - BSc,MSc,PhD white bean diseases

1094 Soil Science Section

1 Findlay WI - BSc,MSc,PhD *Head of Section* soil fertility
2 Jewett TJ - BScAgEng greenhouse energy engineering
3 Stone JA - BS,MS,PhD soil physics
4 Tan CS - BSc,MSc,PhD soil moisture and agrometeorology

1095 Weed Science and Chemistry Section

1 Hamill AS - BSc,MSc,PhD *Head of Section* weed science
2 Gaynor JD - BSc,MSc,PhD environmental chemistry
3 Phillips DR - BSc,MSc,PhD weed physiology
4 Weaver SE - BA,PhD weed ecology

1096 Research Station

Vineland Station, Ontario, LOR 2EO
Tel: (416) 562-4113.

1 Menzies DR - BSA,MSc,PhD *Director*

1097 Entomology

1 Pree DJ - BSA,MSc,PhD *Head of Section* toxicology
2 Broadbent AB - BSc,MSc,PhD ornamental entomology
3 Hagley EAC - BScAgr,MSc,PhD,DICTA fruit pest management
4 Stevenson AB - BScAgr,PhD vegetable pest management
5 Thistlewood HMA - BSc,MPM,PhD acarology
6 Trimble RJM - BSc,MSc,PhD bioclimatology

1098 Nematology, Chemistry and Computer Science

1 Potter JW - BSA,MSA,PhD *Head of Section* nematode ecology and chemical control
2 Chiba M - BSc,DSc,CChem,FCIC residue chemistry

3 McGarvey BD - BSc,MSc chemistry
4 Olthof THA - Ing,BScAgr,PhD host parasite relations
5 Townshend JL - BSc,MSc,DIC nematode ecology
6 Yee J - BSc,MSc,PhD mathematics and computing

1099 Plant Pathology

1 Allen WR - BA,PhD *Head of Section* fruit and soil-borne viruses
2 Biggs AR - BSc,MSc,PhD tree fruit diseases
3 Cerkauskas RF - BSc,MSc,PhD vegetable diseases
4 Matteoni JA - BSc,MSc,PhD ornamental pathology
5 Northover J - BSc,PhD,DIC fruit mycology
6 Reyes AA - BSA,MSA,PhD vegetable mycology
7 Stobbs LW - BSc,MSc,PhD grapevine viruses

1100 Smithfield Experimental Farm
Box 340, Trenton, Ontario, K8V 5R5
Tel: (613) 392-3527.

1 Miller SR - BSc,MSc,PhD *Superintendent* pomology
2 Metcalf JG tomato breeding
3 Mohr WP - BSA,MSA,PhD food processing
4 Smits NJ - BSc vegetable management
5 Warner J - BSc,MSc pomology

1101 Prairie Region

2100 Bond Street, Box 8075, Regina, Sask, S4P 4E6
Tel: (306) 780-6104.

1 Pelton WL - BSA,MSA,PhD *Director General*
2 Howarth RE - BSA,MSc,PhD *Program Specialist*
3 Knipfel JE - BSA,MSc,PhD *Program Specialist*

1102 Research Station
P.O. Box 610, Brandon, R7A 5Z7
Tel: 204-728-7234.

1 Swierstra EE - BSA,MSA,PhD *Director*

1103 Animal Science

1 Rahnefeld GW - BSc,MSc,PhD *Head of Section* beef cattle breeding
2 Castell AG - BSc,MSc,PhD swine nutrition
3 Cliplef RL - BSc,MSc,PhD meats physiology
4 Dyck GW - BSc,MSc,PhD swine reproductive physiology
5 Grandhi RR - BVSc,MSc,PhD swine nutrition
6 Grinwich DL - BSc,MSc,PhD swine reproductive physiology
7 McKay RM - BSc,BSA,MSc,PhD swine genetics

1104 Plant Science

1 Bailey LD - BSA,MSc,PhD *Head of Section* soil-plant relationships
2 Chow PNP - BSc,MSc,PhD herbicides and weed control
3 Gehl DT - BSA,MSc agronomy (ERDA)
4 Grant CA - BSA,MSc,PhD agronomy
5 Sadler JM - BSc,MSc,PhD soil fertility and plant nutrition
6 Simons RG - BSc,MSc,PhD forage agronomy
7 Therrien MC - BSc,PhD barley breeding and genetics

1105 Research Station
P.O. Box 3001, Morden, ROG 1JO
Tel: 204-822-4471.

1 McBeath DK - BSA,MSc,PhD *Director*

1106 Field Crops Section

1 Friesen GH - BSA,MSc,PhD *Head of Section* weed science
2 Ali-Khan ST - BSA,MSc,PhD breeding of field peas
3 Campbell CG - BSA,MSc,PhD breeding of buckwheat
4 Gubbels GH - BSA,MSc,PhD crop management; physiology
5 Kiehn FA - BSA,MSc new crops
6 Zimmer RC - BSc,PhD diseases of field peas

1107 Horticultural Crops Section

1 Chubey BB - BSA,MSc,PhD *Head of Section* vegetable crops
2 Collicutt LM - BSA,MSc breeding of ornamentals
3 Davidson CG - BSA,MSc breeding of woody ornamentals
4 Mazza G - BSA,MSc,PhD food science and technology
5 Rex BL - BSA,MSc potato management
6 Wall DA - BSA,MSc weed control

1108 Oilseed Crops Section

1 Kenaschuk EO - BSA,MSc,PhD *Head of Section* breeding of flax
2 Dedio W - BSA,MSc,PhD breeding of sunflower
3 Hoes JA - BSA,MSA,PhD pathology of flax and sunflowers

1109 Research Station
195 Dafoe Road, Winnipeg, R3T 2M9
Tel: 204-269-2100.

1 Atkinson TG - BSA,MSc,PhD *Director*
2 Malyk M - BSc,MSc,MLS *Librarian*

1110 Cereal Breeding Section

1 Leisle D - BSA,MSc,PhD *Head of Section* durum wheat breeding
2 Brown PD - BSA,MSc,PhD oat breeding
3 Campbell AB - BSA,MSc common wheat breeding
4 Dyck PL - BSA,MSc,PhD wheat genetics
5 Kerber ER - BSA,MSc,PhD wheat cytogenetics
6 Kovacs MIP - BSc,MSc,PhD cereal chemistry
7 Lukow OM - BSc,MSc,PhD cereal chemistry
8 McKenzie RIH - BSA,MSc,PhD triple-medium wheat breeding
9 Metcalfe DR - BSA,MSc,PhD barley breeding and genetics
10 Noll JS - BSc,PhD cereal chemistry and physiology

1111 Cereal Diseases Section

1 Rohringer R - DrScAgr *Head of Section* molecular biology of cereal rusts
2 Chong J - BSc,MSc,PhD oat crown rust
3 Haber SM - BSc,MSc,PhD cereal viruses and biotechnology
4 Harder DE - BSc,MSc,PhD oat stem rust
5 Howes NK - BSc,PhD molecular biology and biotechnology
6 Kim WK - BSc,PhD molecular biology of cereal rusts
7 Martens JW - BSc,PhD wheat stem rust
8 Nielsen JJ - DrScAgr smuts
9 Tekauz AM - BSc,MSc,PhD leaf diseases
10 Thomas PL - BSA,MSc,PhD microbial genetics, smuts

1112 Stored Products Section

1 Mills JT - BSc,PhD,DIC *Head of Section* development and control of storage molds
2 Abramson D - BSc,MSc,PhD mycotoxicology
3 Barker PS - IA,MSc,PhD biology and control of stored grain pests
4 Loschiavo SR - BSc,MSc,PhD stored grain insect biology
5 Sinha RN - BSc,PhD ecology of granary insects, mites and fungi
6 White NDG - BSc,MSc,PhD biology and control of stored product insects

1113 Integrated Pest Control Section

1 Turnock WJ - BSA,MSc,PhD *Head of Section* ecology and population dynamics
2 Ayre GL - BSA,MSA insect ecology
3 Bodnaryk RP - BA,MSc,PhD biochemistry and toxicology
4 Bracken GK - BSc,MSc,PhD physiology and behaviour
5 Gerber GH - BSA,PhD histology, physiology and behaviour
6 Lamb RJ - BSc,MSc,PhD systems biology
7 Morris ON - BSc,MSc,PhD insect pathology
8 Wylie GH - BA,PhD host-parasite relations

1114 Research Station
P.O. Box 1240, Melfort, SOE 1AO
Tel: 306-752-2776.

1 Beacom SE - BSc,MSc,PhD *Director*

1115 Animal Science and Forage Utilization

1 Beacom SE - BSc,MSc,PhD crop utilization by growing-finishing beef cattle
2 McCartney DH - BSc,MSc beef cow-calf and pasture management
3 Mir Z - BSc,MSc,PhD beef cow-calf nutrition, forage utilization

1116 Beef-forage Production Systems

1 Bittman S - BSc,MSc,PhD forage crop evaluation and production
2 Jan EZ - BSc,MSc,PhD harvesting systems
3 Malik N - BSc,MSc,PhD weed control

1117 Cereal, Oilseed and Special Crop Production

1 Bowren KE - BSA *Program Leader* crop production, weed control, tillage, crop rotations
2 Nuttall WF - BSc,MSc,PhD soil fertility, crop and plant nutrition
3 Townley-Smith L - BSc,MSc cereal, oilseed, pulse crop agronomy
4 Wright AT - BSA,MSc crop evaluation and agronomy

1118 Research Station
Box 440, Regina, S4P 3A2
Tel: 306-585-0255.

1 Townley-Smith TF - BSA,MSc,PhD *Director*

1119 Biological Control Section

1 Harris P - BSF,DIC,PhD *Head of Section* insects

2 Maw MG - BSc,MSc insects
3 Mortensen K - BSc,PhD plant pathology
4 Peschken DP - BSA,MSc,DrSciA insects

1120 Herbicide Behaviour in the Environment

1 Grover R - BSc,MSc,PhD *Head of Section* availability, mobility
2 Cessna AJ - BA,PhD residues, plants
3 Smith AE - BSc,PhD,DSc,FCIC residues, metabolism, soil

1121 Weed Physiology and Ecology Section

1 McIntyre GI - BSc,PhD *Head of Section* physiology, water relations, dormancy
2 Bowes GG - BSA,MSc,PhD range weeds
3 Hsiao AI - BSc,MSc,PhD weed seed dormancy, wildcat herbicides
4 Hume L - BSc,PhD crop losses, ecology
5 Hunter JH - BSA,PhD control - annual crops
6 Thomas AG - BSc,MSc,PhD surveys, ecology

1122 Experimental Farm
Box 760, Indian Head, SOG 2KO
Tel: 306-695-2274.

1 Boughton GR - BSA,MSc *Officer-in-Charge* seed increase
2 Derkson DA - BScAgr agronomy, special crops
3 LaFond G - BA,BSc,MSc,PhD agronomy, cereal crops

1123 Research Station
107 Science Crescent, Saskatoon, S7N OX2
Tel: 306-343-8214.

1 Hay JR - BSA,MS,PhD,FWSSA *Director*
2 Downey RK - OC,BSA,MSc,PhD,FAIC,FRSC *Assistant Director*

1124 Scientific Support

1 Howarth RE - BSA,MSc,PhD *Head of Section* legume bloat-biochemistry
2 Bentham MJ - BSc,MSc *Computer Systems Manager*
3 Kichuk DM - BFA,BA,MLS *Librarian*
4 Lynn CE - BSA *Information Officer*
5 Spurr DT - BSc,MSc,PhD *Statistician*

1125 Oilseeds

1 Downey RK - OC,BSA,MSc,PhD,FAIC,FRSC *Head of Section* oilseed breeding, canola
2 Arthur AP - BSc,MSc,PhD oilseed crop insects
3 Burgess L - BSc,MSc,PhD oilseed crop insects
4 Falk KL - BSc,MSc oilseed breeding; canola
5 Hutchenson DS - BA,BSA,PhD oilseed breeding, canola
6 McGregor DI - BSc,MSc,PhD physiology of *Brassica* spp.
7 Petrie GA - BA,MA,PhD oilseed crop diseases
8 Rakow GFW - BSA,PhD oilseed breeding, canola
9 Seguin-Swartz G - BSc,PhD cytogenetics of *Brassica*
10 Ukrainetz H - BSA soil fertility
11 Verma PR - BSc,MSc,PhD oilseed crop diseases

1126 Cereals

1 Duczek LJ - BSA,MSc,PhD cereal root diseases
2 Bailey KL - BSc,MSc cereal root diseases
3 Crowle WL - BSA,MSc cereals and crop management
4 Tinline RD - BA,MSc,PhD,FCSPP cereal root diseases

1127 Forage Crops

1 Goplen BP - BSA,MSc,PhD *Head of Section* legume breeding
2 Craig CH - BA forage crop insects
3 Gossen BD - BSc,PhD forage and low temperature diseases
4 Gruber MY - BSc,MSc plant biotechnology
5 Howarth RE - BSA,MSc,PhD legume bloat, biochemistry
6 Knowles RP - BSA,MSc,PhD,FAIC grass breeding
7 Lees GL - BSc,PhD legume bloat, plant physiology
8 Wright SBM - BSA,MSc grass breeding

1128 Insect Pest Management

1 Doane JF - BSA,MS,PhD *Head of Section* insects, ecology and behaviour
2 Elliot RH - BA,PhD insecticides
3 Erlandson MA - BSc,MSc,PhD insect virology
4 Ewen AB - BA,MA,PhD,FRES insects, ecology and pathology
5 Hinks CF - BSc,ARCS,DIC,PhD insects toxicology
6 Mason PG - BSc,MSc,PhD black flies
7 Mukerji MK - BSc,MSc,PhD insects, management
8 Olfert OO - BSc,BSA,MSc,PhD insects ecology
9 Westcott ND - BSc,PhD pesticide, chemistry

1129 Scott Experimental Farm
Box 10, Scott, SOK 4AO
Tel: 306-247-2011.

1 Kirkland KJ - BSA,MSc *Superintendent* weeds and crop management
2 Brandt SA - BSA,MSc agronomy

1130 Research Station
P.O. Box 1030, Swift Current, Saskatchewan, S9H 3X2
Tel: (306)773-4621.

1 Sonntag BH - BSA,MSc,PhD *Director*

1131 Analytical Services Section

1 Winkleman GE - BSc *Head of Section* chemist

1132 Cereal Production and Utilization Section

1 Clarke JM - BScAgr,MSc,PhD harvest physiology
2 De Pauw RM - BA,MSc,PhD wheat breeding
3 Green DG - BSA,MS,PhD cereal physiology
4 McCaig TN - BSc,MSc,PhD *A/Head of Section* quality physiology
5 Salmon RE - BSA,MSA,PhD turkey nutrition
6 Stevens VI - BScAgr,MSc,PhD turkey nutrition

1133 Engineering Section

1 Dyck FB - BE,MSc *Head of Section* equipment design
2 Jame YW - BSc,MSc,PhD irrigation
3 Steppuhn H - BSc,PhD subsurface hydrology
4 Stumborg MA - BE,MScEng energy
5 Tessier S - BScEng,MSc tillage

1134 Forage Production and Utilization Section

1 Lawrence T - BSA,MSc,PhD *Head of Section* grass breeding
2 Holt NW - BSA,PhD pasture management
3 Jefferson PG - BScAgr,MSc legume breeding and physiology
4 Waddington J - BSc,MSc,PhD pasture establishment and physiology

1135 Soils and Environment Section

1 Campbell CA - BSA,MSA,PhD *Head of Section* soil chemistry and fertility
2 Biederbeck VO - BSA,MSc,PhD soil microbiology
3 Cutforth HW - BSc,MSc,PhD agrometeorology, soil physics
4 Leyshon AJ - BSc,MSc forage fertility
5 Selles F - BSAChile,BSA,PhD soil fertility
6 Zentner RP - BSA,MSc,PhD economics

1136 Research Station
P.O. Box 29, Beaverlodge, TOH OCO
Tel: 403-354-2212.

1 McElgunn JD - BSc,MSc,PhD *Director*

1137 Cereal and Oilseed Section

1 Wolfe RI - BSA,BSc,PhD *Head of Section* cereal breeding
2 Woods DL - BPharm,MPhil,PhD rapeseed breeding

1138 Environment and Soils Section

1 Rice WA - BSA,MSc,PhD *Head of Section* microbiology
2 Arshad MA - BSc,MSc,PhD soil physics
3 Darwent AL - BSA,MSc,PhD weed control
4 Davidson JGN - BSF,MSc,PhD plant pathology
5 McKenzie JS - BSA,MSc,PhD *Acting Head* plant survival
6 Mills PF - BA micrometeorology
7 Olsen PE - BSc,MSc microbiology
8 Soon YK - BSc,MSc,PhD soil chemistry

1139 Forage Crops and Apiculture

1 Nelson DL - BSA,MSc,PhD *Acting Head of Section* apiculture
2 Fairey DT - BSc,MSc,PhD legume seed physiology
3 Fairey NA - BSc,MSc,PhD forage production
4 Liu TP - BSc,MSc,PhD apiculture pathology
5 Szabo TI - BAE,MSc,PhD apiculture

1140 Experimental Farm
Fort Vermilion, TOH 1NO
Tel: 403 927 3253.

1 Clayton GW - BSA,MSc crop production
2 Siemens B - BSA,MSc *Superintendent* forage crops

1141 Research Station
Lacombe, TOC 1SO
Tel: 403-782-3316.

1 Waldern DE - PhD *Director*

1142 Scientific Support

1 Fobert LR *Systems and Programming Manager*
2 Lamkin DP *Computer Programming*
3 McIsaac JE - BA,MLS *Librarian/Information Officer*

1143 Red Meats and Beef Production Section
1144 Red Meats Program

1 Morgan Jones SD - PhD *Head of Section* meat science
2 Doornenbal H - PhD physiology
3 Greer GG - PhD microbiology
4 Jeremiah LE - PhD food technology
5 Murray AC - PhD biochemistry
6 Sather AP - PhD swine production and management

1145 Beef Production Program

1 Newman JA - PhD beef cattle production
2 Schaefer AL - PhD physiology of meat animals, growth and behaviour
3 Tong AKW - PhD beef cattle evaluation and statistics

1146 Plant Science Section
1147 Cereal Breeding, Pathology and Forage Program

1 Piening LJ - PhD *Head of Section* cereal pathology
2 Baron VS - PhD forage physiology
3 Berkenkamp WB - PhD forage crop production and pathology
4 Kibite S - PhD barley, oat, wheat breeding

1148 Crop Management and Soils Program

1 Harker KN - BSc,MSc weed physiology
2 Malhi SS - PhD soil chemistry
3 Taylor JS - PhD crop physiology

1149 Soils and Crops Substation, Vegreville

1 McAndrew DW - PhD *Officer-in-Charge* soil fertility and management
2 Pearen JR - MSc,BSc,BEd crop physiology and management

1150 Research Station
P.O. Box 3000, Main, Lethbridge, T1J 4B1
Tel: 403 327 4561.

1 Dorrell DG - BSA,MSc,PhD *Director*
2 Bolton Ms P - BA,MLS *Librarian*
3 Chung HS - BSc *Programmer*
4 Cutler CM - BA,MLS *Area Coordinator, Library*
5 Entz T - BSc,MM *Statistician*
6 Freeze BS *Bioeconomist*
7 Kozub GC - BSc,MSc *Statistician*
8 L'Arrivee JCM - BSA,MSc,PhD *Technical Information Officer*
9 Pang DW - DipCS *Computer System Manager*
10 Schaalje GB - BS,MS *Statistician*
11 Sears LJL - BA *Scientific Officer*

1151 Animal Parasitology Section

1 Shemanchuk JA - CD,FESC,BSc,MSc *Head of Section* biting-fly ecology
2 Baron RW - BSc,PhD immunology
3 Robertson RH - BA,MSc serology
4 Shipp JL - BSc,MSc,PhD black-fly ecology
5 Taylor WG - BSP,PhD pesticide chemistry
6 Yeung JMB - BSc,PhD pesticide metabolism

1152 Animal Science Section

1 Lawson JE - BSA,MSA *Head of Section* beef cattle breeding
2 Bailey CBM - BSA,MSA,PhD animal physiology
3 Bailey DRC - BSc,PhD beef cattle breeding
4 Cheng K-J - BSc,MSc,PhD rumen microbiology
5 Coulter GH - BScAgr,PhD reproductive physiology
6 Hironaka R - BSc,MSc,PhD animal nutrition
7 McClelland LA - BSc,MSc sheep breeding
8 Mears GJ - BSc,MSc,PhD animal physiology
9 Phillippe RC - BSc rumen bacteriology
10 Rode LM - BSA,MS,PhD animal nutrition
11 Veseley JAP - BSA,MSA,PhD sheep and dairy cattle breeding

1153 Crop Entomology

1 Struble DL - BA,MA,PhD,FESC *Head of Section* insect attractants
2 Byers JR - BSA,MSc,PhD cutworm ecology
3 Charnetski WA - BSc,MSc,PhD insecticides

4 Hill BD - BSc,MSc,PhD residue chemistry
5 Johnson DL - BSc,MSc,PhD rangeland insects
6 Richards KW - BSc,MSc,PhD insect pollinators
7 Schaber BD - BSc,MSc,PhD forage crop pests

1154 Plant Pathology

1 Conner RL - BSA,PhD cereal diseases
2 Gaudet DA - BSc,MSc,PhD snow moulds
3 Huang HC - BSc,MSc,PhD *Head of Section* forage and special crops diseases
4 Nelson GA - BSc,MSc,PhD potato and bacterial diseases
5 Roberts DWA - BA,PhD cold-hardiness physiology

1155 Plant Science

1 Smoliak S - FSRM,BSc,MS *Acting Head of Section* dryland pastures
2 Allan JR - BSc,MA,PhD aquatic plant physiology
3 Bergen P - BSA,MSc,PhD weed control
4 Blackshaw RE - BSc,MSc weed control
5 Hanna MR - BSA,MSA,PhD forage legume breeding
6 Kaldy MS - BSc,MS,PhD food science
7 Kemp GA - BSc,PhD vegetable breeding
8 Lynch DR - BSc,MSc,PhD potato breeding
9 MacDonald MD - BSc,PhD corn breeding
10 Major DJ - BSc,MSc,PhD crop physiology
11 May KW - BSc,MSc,PhD barley breeding
12 Moyer JR - BSc,MSc,PhD weed control
13 Mundel HH - BSA,MS,PhD new crops; hard red spring wheat breeding
14 Sadasivaiah RS - BSc,MSc,PhD soft white spring wheat breeding
15 Thomas JB - BSc,MSc,PhD winter wheat breeding
16 Whelan EDP - BSA,MSA,PhD wheat cytogenetics
17 Willms WD - BSF,MSc,PhD rangeland ecology

1156 Soil Science

1 Lindwall CW - BSc,MSc,PhD *Head of Section* agricultural engineering
2 Beke BJ - BSA,MSc,PhD hydrology
3 Carefoot JM - BSA,MSA irrigation agronomy
4 Chang C - BSc,MSc,PhD soil physics
5 Dormaar JF - BSA,MSA,PhD organic chemistry
6 Foroud N - BSc,MSc,PhD irrigation engineering
7 Grace BW - BSc,MSc,PhD agrometeorology
8 Hynes MF - BSc,MSc,PhD microbial genetics
9 Janzen HH - BSA,PhD dryland agronomy
10 Kucey RMN - BSc,PhD soil fertility
11 MacKay DC - FCSSS,BSc,MS,PhD plant nutrition
12 Sommerfeldt TG - BSc,MSc,PhD drainage engineering

1157 Pacific Region

425-750 Cambie Street, Vancouver, B.C., V6B 4V5
Tel: 604 666 8164.

1 Thompson SC - BSc,MA,PhD,NDC *Director General*
2 Bole JB - BSA,MSc,PhD *Program Specialist*
3 River RZ - BA,MA *Research Economist*

1158 Research Station
P.O. Box 1000, Agassiz, B.C., V0M 1A0
Tel: 604 796 2221.

1 Molnar JM - BSA,MSc,PhD *Director*
2 Lockyer J - BA,MLS *Librarian*

1159 Animal Science

1 Fisher LJ - BSA,MSc,PhD *Head of Section* dairy cattle nutrition
2 Buckley WT - BS,PhD ruminant mineral biochemistry
3 Gardiner EE - BS,MS,PhD poultry nutrition
4 Hunt JR - BSA,PhD poultry physiology

1160 Crop Science

1 Perrin PW - BSc,PhD *Head of Section* postharvest physiology
2 Ames-Gottfred NP - BSA,MSA forage management
3 Freeman JA - BSA,MSA,PhD small fruits management; weed control
4 Freyman S - BScAgr,MSA,PhD weed control
5 Fushtey SG - BSc,MSc,PhD turf management
6 Keng JCW - BSc,MSc,PhD soil, water and crop management
7 Kowalenko CG - BSA,MSc,PhD soil biochemistry and fertility
8 Maurer AR - BSA,MSc physiology of vegetable crops

1161 Research Station
3015 Ord Road, Kamloops, B.C., V2B 8A9
Tel: 604 376 5565.

1 Robertson JA - BSA,MSc,PhD *Director*
2 Majak W - BSc,MSc,PhD biochemistry
3 Quinton DA - BS,PhD range management
4 Stout DG - BSA,MSc,PhD plant physiology

5 van Ryswyk AL - BSA,MSA,PhD soil science

1162 Prince George Experimental Farm
R.R. £8, RMD £6, Prince George, B.C., V2N 4M6
Tel: 604 963 9632.

1 Broersma K - BSA,MSc *A/Superintendent* forage agronomy, soils

1163 Research and Plant Quarantine Station (Saanichton)
8801 East Saanich Road, Sidney, B.C., V8L 1H3
Tel: 604 656 1173.

1 French CJ - BSc,PhD *A/Director*
2 Watson MA - BA,MLib *Librarian*

1164 Ornamentals

1 French CJ - BSc,PhD *Head of Section* woody ornamentals
2 Chiko AW - BSc,MS,PhD viruses
3 Lin WC - BSc,MSc,PhD floriculture
4 Monette JPL - BSc,PhD tissue culture

1165 Vegetables

1 van Zinderen Bakker EM - BSc,MSc *Head of Section* physiology
2 Gillespie DR - BSc,MSc,PhD entomology
3 Kempler C - BSc,MSc vegetable nutrition, kiwi fruit

1166 Plant Quarantine

1 Johnson RC - BSA *Head of Section* field services
2 Lanterman W - BSc,MPSAgr,PhD *Head of Section* laboratory services
3 Thompson D - BSc tree fruit viruses

1167 Research Station (Summerland)
P.O. Box 880, Penticton, B.C., V2A 7GL
Tel: 604 494 7711.

1 Bowden DM - BSA,MSA,PhD *Director*
2 Watson P - BA,MLS *Librarian*

1168 Entomology-Plant Pathology

1 Hansen AJ - DipAgr,MSc,PhD *Head of Section* fruit tree virus diseases
2 Angerilli NPD - BSc,PhD orchard mite control, San Jose scale
3 Banham FL - BSA stone fruit insects
4 Cossentine JE - BSc,MS,PhD integrated control
5 Dyck VA - BA,MSc,PhD management of codling moth
6 McMullen RD - BSc,MS,PhD pear pest management
7 Sholberg PL - BSc,MSc,PhD tree fruit fungus diseases
8 Utkhede RS - BScAgr,MScAgr,PhD tree fruit fungus diseases

1169 Food Processing

1 Cumming DB - BSA,MSc,PhD *Head of Section* food technology
2 Beveridge HJT - BSA,MSc,PhD food chemistry and analytical methods
3 Paulson AT - BScAgr,MSc,PhD food technology
4 Strachan GE - BSA,MSc enology and food technology

1170 Pomology and Viticulture

1 Houge EJ - BSA,MS,PhD *Head of Section* herbicides, vegetation management - nutrition
2 Lane WD - BSc,MSc,PhD apple, cherry breeding
3 Looney NE - BS,MS,PhD pomology, plant physiology, growth regulators
4 Meheriuk M - BSc,BEd,MSc,PhD fruit storage and biochemistry
5 Quamme HA - BSA,MSc,PhD orchard management, hardiness
6 Reynolds AG - BSc,MSc,PhD grape management

1171 Soil Science and Agriculture Engineering

1 Stevenson DS - BSA,MS,PhD *Head of Section* soil moisture
2 Gaunce AP - BSc,MSc,PhD pesticide and environmental chemistry
3 Hoyt PB - BSc,MS,PhD soil chemistry and management
4 Moyls AL - BASc,MS,PhD agricultural equipment - development and assessment
5 Neilsen GH - BSc,MSc,PhD soil fertility and plant nutrition
6 Parchomchuk P - BASc,MS agricultural equipment development and assessment

1172 Research Station
6660 N.W. Marine Drive, Vancouver, B.C., V6T 1X2
Tel: 604 224 4355.

1 Weintraub M - BA,PhD,FNYAS *Director*
2 Hall JW - MSc,PhD *Regional Statistician*
3 Matsumoto T - BSA,MSc,MLS *Librarian*

1173 Entomology

1 Forbes AR - BA,MS,PhD,FESC *Head of Section* virus vectors
2 Frazer BD - BSc,PhD aphid ecology
3 Raine J - BSA,MS berry insects, leafhopper vectors
4 Raworth DA - BSc,PhD biological control
5 Szeto SYS - BSc,MSc pesticide chemistry
6 Theilmann DA - BSc,MSc insect virology
7 Vernon RS - BSc,MPM,PhD vegetable insects

1174 Plant Pathology

1 Stace-Smith R - BSA,PhD,FAPS *Head of Section* raspberry viruses, virus characterization
2 Daubeny HA - BSA,MSA,PhD plant breeding, small fruits
3 De Boer SH - BSc,MSc,PhD bacterial diseases
4 Ellis PJ - BSc,MPM potato viruses
5 Martin RR - BS,PhD strawberry viruses
6 Pepin HS - BSA,MA,PhD root rots, red stele of strawberry
7 Vrain TC - DEA,PhD nematology

1175 Virus Chemistry and Physiology

1 Ragetli HWJ - Ir,PhD *Head of Section* chemistry and ultra structural cytopathology of viruses
2 Hamilton RI - BSc,MSc,PhD,FAPS virology, virus interaction, and seed transmission
3 Rochon D - BA,MSc,PhD plant viruses, molecular biology
4 Tremaine JH - BSc,MSc,PhD biophysical virology

1176 International Programs Branch
Sir John Carling Building, 930 Carling Avenue, Ottawa, Ontario, K1A OC5
Tel: 613 995 5880. Telex: 053 3283 - MDD. Cables: CANAGRIC OTT.

1 Jacques Y - MBA,LICAgr *Assistant Deputy Minister*

1177 International Trade Directorate

1 de Gorter H - PhD *Senior Analyst, Trade*
2 Dobson GJ - BSA,MSc,PhD *Director*
3 McClatchy D - PhD *Chief, Trade Analysis*

1178 Policy Branch
Sir John Carling Building, 930 Carling Avenue, Ottawa, Ontario, K1A OC5
Tel: 613 995 5880. Telex: 053 3283 - MDD. Cables: CANAGRIC OTT.

1 Claydon F *Assistant Deputy Minister*

1179 Planning, Coordination and Analysis Directorate

1 Hedley DD *Director-General*

1180 Market Outlook and Analysis Division
Sir John Carling Building, 930 Carling Avenue, Ottawa, Ontario, K1A OC5
Tel: 613 995 5880. Telex: 053 3283 - MDD. Cables: CANAGRIC OTT.

1 Gellner J *Assistant Director, Policy*
2 Hassan Z *Director*

1181 Forecasting Unit

1 Cluff M

1182 Food Markets Analysis Division

1 Davey *Associate Director*
2 Johannson E *Demand and Price Analysis*
3 Krystynak R *Regulation Analysis*
4 Migie H *Director*
5 West DA - BSA,MSc,PhD *Processing and Retailing*

1183 Commodity Coordination Directorate

1 L'Ecuyer C *Director*
2 Gellner J - BSA,MSc *Assistant Director of Policy*

1184 Horticulture

1 Anderson RW - PhD *Fruits and Vegetables*
2 Brown J *Floriculture and Ornamentals*

1185 Cattle, Hogs and Sheep

1 Boswell AM - MS *Cattle, Hogs and Sheep*
2 Chin S *Chief, Agricultural Modelling*

1186 Poultry and Eggs

1 Murray DB - BSA *Poultry and Eggs*

1187 Dairy

1 Tudor-Price R - BSA *Dairy*

1188 Special Crops

1 Longmuir N *Special Crops*

1189 Development Policy Directorate

1 Perkins B *Director-General*
2 Colwell M *Crops*
3 Eyvindson R *Director, Farm Finance and Taxation Division*
4 Jones W *Farm Income and Finance*
5 McKenzie J *Director, Production Analysis Division*
6 Rizvi S *Livestock*
7 Wise T *Farm Inputs*

1190 Food Production and Inspection Branch

Sir John Carling Building, Ottawa, Ontario, K1A OC5
Tel: (613) 995 5880.

1 Vacant - Assistant Deputy Minister

1191 Health of Animals Directorate

4th Floor Sir John Carling Building, Ottawa, Ontario, K1A OC5
Tel: (613) 992 2114.

1 Vacant - Director General

1192 Animal Pathology Division

Room 417, Halldon House, 2255 Carling Avenue, Ottawa, Ontario, K1A OY9
Tel: (613) 995 5433.

1 Willis NG - DVM,DipBAct,PhD *Director*
2 Baker MF - DVM *Associate Director, Laboratory Services*
3 Gauthier J - DVM,DipVetPrevMed
4 McNab B - DVM *Special Projects Officer*
5 Sellers RF - BA,MA,BSc,MRCVS,PhD,ScD *Special Scientific Advisor*

1193 Animal Diseases Research Institute

801 Fallowfield Road, Box 11300, Station "H", Nepean, Ontario, K2H 8P9
Tel: (613) 998 9320.

1 Stemshorn BW - BSc,BMV,PhD *Director*

1194 Laboratory Operations

1 Carriere JAJ - BA,DMV *Head*
2 Fournier J - DMV,MSc *Diagnostic Coordinator*

1195 Immunology

1 Duncan JR - BSA,DVM,MSc,PhD,ACVP *Head*
2 Cherwonogrodzky J - BSc,MSc,PhD
3 McGuire RL - BSc,MSc,PhD *Safety Officer*
4 Neilsen K - BSc,MSc,PhD
5 Ruckerbauer GM - BA,MA,DVM
6 Sugden EA - BSc,PhD
7 Wright PF - BSc,MSc,PhD

1196 Microbiology Research

1 Spencer JL - DVM,MS,PhD *Head*
2 Brooks BW - BScAgr,MSc,PhD
3 Chen S - DVM,MS avian diseases
4 Fraser ADE - BSc,MSc,PhD *NSERC Fellow*
5 Garcia MM - BSA,MSc,PhD
6 Rigby CE - BSc,MSc,PhD

1197 Microbiology Service

1 Stevens JB - BSc,DVM,MSc,PhD *Head*
2 Malkin KL - BVSc,DiplVetPM
3 Stewart RB - BSc
4 Turcotte C - DMV,MSc

1198 Pathology

1 Charlton KM - DVM,PhD,ACVP *Head*
2 Bundza A - DVM,DiplClinPath

3 Casey GA - BSc
4 Corner AH - DVM,MSc
5 Dukes TW - DVM,MS
6 Gilka F - MedVet,MVDr,CSc,PhD,ACVP
7 Tolson ND - PhD *OMNR Fellow*
8 Webster WA - BSc

1199 Reproduction

1 Hare WCD - MAhc,BSc,PhD,DVM&S,FRCVS
2 Bielanski A - PhD,DVM,DrHabilit,VM
3 Eaglesome MD - BVMS,MRCVS
4 Randall GCB - BVSc,PhD,MRCVS
5 Singh EL - BSc,MSc

1200 Serology

1 Colling D - BSc,MSc
2 Kraay GJ - BSc,MSc,PhD
3 Lacroix RP - BSc,MSc
4 Robertson FJ - BScAgric,DTA,BVMS,MRCVS
5 Samagh BS - BVSc&AH,MSc,PhD,DVM *Head*
6 Shapiro JL - DVM,IntEqSurgMed,DipPath
7 Shettigara PT - BVSc,MSc,PhD,DVM,DTVM,DECH,MRCVS

1201 Virology

1 Ide PR - DVM,MSc,PhD *Head*
2 Afshar A - DVM,PhD
3 Bouffard A - DMV,MSc,PhD
4 Bouillant AMP - DMV,MS,PhD
5 Dubuc C - DVM,MSc
6 Dulac GC - DMV,MSc,PhD
7 Girard A - DMV
8 Myers DJ - DVM,MSc
9 Sellers RF - BA,MRCVS,DVM,PhD,ScD
10 Thomas FC - DVM,PhD

1202 Animal Pathology Laboratory

10 Salem Street, P.O. Box 1410, Sackville, New Brunswick, EOA 3CO
Tel: (506) 536 0135.

1 Stevenson RG - DVM,DVSM,PhD *Director*
2 Finley GG - DVM,DDP pathology
3 Simard C - DMV,MSc virology
4 Smith HJ - BScAgr,DVM,MVSc parasitology

1203 Animal Pathology Laboratory

3000 Sicotte Street, St Hyacinthe, Quebec, J2S 2L8
Tel: (514) 773 7730.

1 Gagnon AN - DMV,MSc *Director*
2 Carrier SP - DMV,MSc *Serology Section Head*
3 Messier S - DMV,MSc,PhD bacteriology research
4 Robinson Y - DMV,DiplLA,MSc,IPSAV *Meat Safety Section Head*

1204 Animal Pathology Laboratory

620 Gordon Street, Guelph, Ontario, N1G 1Y4
Tel: (519) 822 3300.

1 Bhatia TRS - BVSc&AH,MSc,MRCVS *Director*

1205 Diagnostic Services

1 Cairns PJ - DVM infectious diseases
2 Lammerding AM - BScAgr,MSc meat safety

1206 Research

1 Clarke RC - BSc,DVM,PhD bacteriology
2 Poppe C - DVM,MSc,MSA bacteriology
3 Truscott RB - BSA,DVM,MSA,PhD bacteriology

1207 Animal Pathology Laboratory

Room 408, Federal Building, Winnipeg, Manitoba, R3C 1B2
Tel: (204) 949 2205.

1 Finlay RC - DVM *Director* serology
2 Mann ED - DVM microbiology

1208 Animal Pathology Laboratory

116 Veterinary Road, Saskatoon, Saskatchewan, S7N 2R3
Tel: (306) 975 4071.

1 Yates WDG - DVM,PhD *Director*

1209 Infectious Diseases Section

1 Forbes LB - DVM,MSc laboratory services
2 Hutchings DL - DVM,DiplVPM immunology
3 Tessaro SV - BSc,MSc,DVM wildlife diseases

1210 Chemical Residue Analysis Section

1 MacNeil JD - BSc,MSc,PhD *Section Head*
2 Fesser AC - BSc,MSc *laboratory services*
3 Korsrud GO - BSc,MSc,PhD *methods research & development*
4 Martz V - BSc *Safety Officer quality assurance*
5 Patterson JR - BSc,BEd,MSc *Senior Chemist*
6 Salisbury C - BSc *laboratory services*

1211 Animal Diseases Research Institute
P.O. Box 640, Lethbridge, Alberta, T1J 3Z4
Tel: (403) 381 8182.

1 Bohac JJ - VMD,PhD *virology*
2 Cho HJ - DMM,MSc,PhD *immunology*
3 Jericho KWF - DVM,PhD *Section Head pathology*
4 Niilo L - DVM,MSc *Section Head immunology*
5 Stockdale PHG - BVet,Med,MSc,PhD *Director*

1212 Animal Pathology Laboratory
13-3071 No 5 Road, Richmond, British Columbia, V6X 2T4
Tel: (604) 273 2713.

1 Dorward WJ - BSc,DVM,VS *Director bacteriology and immunology*
2 MacNeil AC - DVM,VS *Head Diagnostic Section*

1213 Plant Health and Plant Products Directorate
Sir John Carling Building, Ottawa, Ontario, K1A OC5
Tel: (613) 992 2114.

1 Lister EE - BScAgr,MSc,PhD *Director General*

1214 Plant Health Division

1 Hopper BE - BSc,MSc *Associate Director, Biological Services*
2 Campbell WP - BSc,PhD *Director, Special Projects*
3 Chan JYH - BSc,MSc,PhD *Chief Plant Pathology Unit*
4 Garland J - BSc,MSc,PhD *Program Entomologist, Agriculture*
5 Martin P - BSc,MSc,PhD *Plant Pathologist*
6 McDonald J - BSc,MSc,PhD *Plant Pathologist*
7 Parker D - BSc,MSc *Identification Biologist, Entomology*
8 Schmidt AC - BSc,MSc *Chief, Entomology Unit*
9 Walter D - BA,MSc,PhD *Program Entomologist, Forestry*
10 White G - BSc,MSc *Clinical Pathologist*

1215 Seed Division

1 Watson GL - BScA *Associate Director, Variety Evaluation and Licensing*
2 Borrel B - BScAgr,MScV *Associate Director, Technical Programs*
3 Duke LH - BSc *Variety Evaluation Officer*
4 Singh Y - BSc,MSc,PhD *Associate Director, Seed Standards*
5 Sisson V - BScA *Variety Rights Examiner*

1216 Feed and Fertilizer Division

1 Crober CD - BScA *Associate Director, Fertilizers*
2 Daupin F - BSc,MSc *Evaluation Officer*
3 Goldrosen A - BSc,MSc *Technical Services Officer*
4 Gordon D - BSc *Evaluation Officer*
5 Ho SK - BSc,PhD *Associate Director, Feed Evaluation*
6 Kelly D - BSc *Registration and Compliance Officer*
7 McCuaig LW - BSc,PhD *Registration and Compliance Officer*
8 Montgrain C - BScA *Registration and Compliance Officer*
9 Morrison DN - BScA *Associate Director, Feed Registration and Compliance*
10 Morrison LL - BSc,MSc *Evaluation Officer*
11 O'Brien G - BVSc *Evaluation Officer*
12 Phillips D - BSc *Registration and Compliance Officer*
13 Stewart S - BScAgr *Assistant Evaluation Officer*

1217 Food Inspection Directorate
Sir John Carling Building, Ottawa, Ontario, K1A OC5
Tel: (613) 992 2114.

1 Morrissey JB - DVM *Director General*

1218 Laboratory Services Division
Laboratory Services Building No. 22, C.E.F., Ottawa, Ontario, K1A OC6
Tel: (613) 995 4907.

1 Ashton D - BSc *seed germination and physiology*
2 Barrette JP - BSc,PhD *mass spectrometry - microcontaminants, methodology*
3 Cochrane WP - ARIC,PhD *Director*
4 Cohen H - LSc,MA,PhD *mycotoxins analytical methodology*
5 Dhaliwal MS - BSA,BSc *seed purity and taxonomy*

6 Dhesi NS - BSc,MSc,PhD *seed cultivar verification*
7 Ednie AB - BSc,MSc *seed biology*
8 Maybury RB - BSc,NDD *pesticides analytical methodology*
9 Miles W - BSC,MSc *microcontaminants, methodology*
10 Moore EJ - BSc *feed and fertilizer analytical methodology*
11 Nath EJ - BSc,MSc *food microbiology analytical methodology*
12 Neidert E - BSc,MSc *food chemistry analytical methodology*
13 Randall CJ - BSc,MSc,PhD *food chemistry analytical methodology*
14 Sahota AS - BSc,MSc *seed cultivar verification*
15 Sheppard J - BSc,MSc *seed pathology*
16 Singh J - BSc,MSc,PhD *microcontaminants, methodology*
17 Wigfield YY - BSc,MA,PhD *pesticides analytical methodology*

1219 Western Laboratory - Laboratory Services Division
102 - 11th Avenue S.E., Calgary, Alberta, T2G OX5
Tel: (403) 292 5741.

1 Onciul RT - BSc *Director*
2 Deutschmann PM - BSc *microscopy*
3 Kuervers TM - BSc *feed and fertilizer*
4 Peake AA - BSc *pesticide and food chemistry*

1220 Canadian Grain Commission

1221 Grain Research Laboratory Division
1404 - 303 Main Street, Winnipeg, Manitoba, R3C 3G8

1 Tipples KH - BSc,PhD *Director*
2 Babb JC - BSc,MSc *biometrics*
3 Clear RM - BSc *mycology*
4 Cooke L - BSA *grain biology*
5 Daun JK - BSc,MSc,PhD *oilseed research*
6 DeClercq DR - BSc *oilseeds analysis*
7 Dexter JE - BSc,MSc,PhD *milling research*
8 Gaba DG - BSc *residue analysis*
9 Hatcher DW - BSc,MSc *basic wheat research*
10 Kilborn RH *breadmaking research*
11 Kruger JE - MSc,PhD *basic wheat research*
12 LaBerge DE - MSc,PhD *malting barley research*
13 MacGregor AW - BSc,PhD *malting barley research*
14 March PR - BSc *information services*
15 Marchylo BA - BSc,MSc,PhD *malting barley research*
16 Matsuo RR - BSc,PhD *durum wheat research*
17 Mellish VJ - BSc *basic wheat research*
18 Morgan BC - BSc *applied wheat research*
19 Nowicki TW - BSc *residue analysis*
20 Panford Ms JA - BSA,MSc *analytical research*
21 Preston KR - BSc,MSc,PhD *applied wheat research*
22 Tkachuk R - BA,MSc,PhD *basic wheat research*
23 Williams PC - BSc,PhD *analytical research*

1222 Canadian Forestry Service
Headquarters, Ottawa, K1A 1G5
Tel: (819) 997-1107.

1 Mantha Ms L *Director General forestry development and communications directorate*
2 Winget Dr CH *Director General research and technical services directorate*
3 Herring R *Assistant Deputy Minister*
4 Hunt Mr P *Director General Policy, Planning and International Forestry Directorate*

1223 Petawawa National Forestry Institute
Chalk River, Ontario, K0J 1J0
Tel: (613) 589-2880.

1 Pollett FC - BSc,BA,MSc,PhD *Executive Director*
2 Walker JD - BScHons,MF *Director forest management systems*
3 Brand DG - R.P.F.BScF,PhD *silviculture, growth and yield*
4 Chatarpaul L - MS,PhD *nitrogen fixation*
5 Cheliak WM - BScF,MSc,PhD *micropropagation, molecular genetics*
6 Harrington JB - BA,MSc,PhD *forest meteorology*
7 Kourtz P - BScF,MSc,PhD *fire management systems*
8 Murray G - RPF,BScF,MSc,PhD *forest genetics*
9 Newnham RM - BSc,MF,PhD *forest management modelling*
10 Power M - BScF *forest insects and disease*
11 Van Wagner CE - BEng,BScF *fire behaviour and effects*
12 Wang BSF - BScF,MF *seed physiology*
13 Yeatman TW - RPF,DipFor,BScF,MF,PhD *forest genetics*

1224 Forest Pest Management Institute
1219 Queen St. E., P.O. Box 490, Sault Ste. Marie, Ontario, P6A 5M7
Tel: (705) 949-9461. Telex: 067-77180. Cables: FAX (705) 759-5700.

1 Green GW - BA,MSc,PhD *Director*

1225 Biorational Control Agents Program

1 Ennis TJ - BSc,MSc,PhD *Program Director* insect genetics

1226 Microbial Control Agents

1 Arif BM - BSc,MSc,PhD viral biochemistry
2 Cunningham JC - BSc,DPhil viral applications
3 Fast PG - BSc,MSS,PhD bacterial biochemistry
4 Jamieson P viral characterization
5 Kaupp WJ - BSc,DPhil viral epizootiology
6 Tyrrell D - BSc,DIC,DPhil fungal physiology
7 van Frankenhuyzen K - BSc,MSc,PhD bacterial applications
8 Wilson GG - BSA,MSc,PhD protozoology (microsporida)

1227 Biological Interactions

1 Fleming RA - BSc,PhD biological systems analysis
2 Pang A - BSc,MSc,PhD immunochemistry
3 Sohi SS - BScA,MSc,PhD insect tissue culture

1228 Physiological & Genetic Mechanisms

1 Grant GG - BSc,MSc,PhD pheromone physiology
2 Retnakaran A - BSc,MSc,PhD insect growth regulators

1229 Chemical Control Agents Program

1 Caldwell ETN - BSc,MSc *Program Director* integrated pest management

1230 Pesticide Toxicology & Efficacy

1 Cadogan BL - BSc,MSc insecticide field efficacy
2 deGroot P - BScF high value stand protection
3 Helson BV - BSc,MSc,PhD insecticide toxicology
4 Prasad R - BSc,MSc,PhD herbicide toxicology
5 Reynolds PE - BA,MScF,PhD herbicide applications
6 Turgeon J - BSc,PhD insect biology & ecology

1231 Application Technology

1 Payne N - BSc,PhD spray cloud behaviour
2 Sundaram A - BSc,MSc,PhD pesticide formulations

1232 Environmental Concerns

1 Barber KN - BSc,MSc invertebrate ecology
2 Holmes S - BSc aquatic ecosystems
3 Kingsbury PD - BA,MSc environmental impact pesticides
4 Millikin RL - BSc vertebrate ecology
5 Sundaram KMS - BSc,MSc,PhD pesticide accountability

1233 Newfoundland and Labrador Region

Bldg. 304, Pleasantville, P.O. Box 6028, St. John's, Newfoundland, A1C 5X5
Tel: (709) 772 4683.

1234 Newfoundland Forestry Centre

1 Munro JA - BScFor,MScFor,PhD *Director General*

1235 Forest Protection Research

1 Hudak J - BScFor,MScFor,PhD *Program Director*
2 Bowers W - BSc,PhD progress in forest entomology
3 Dobesberger EJ - BScF,MScF,RPF forest entomology
4 Lim KP - BSc,MSc,PhD entomology
5 Raske AG - BScFor,MSc,PhD entomology
6 Warren GR - BSc,BScF,PhD progress in forest pathology
7 West RJ - BA,BSc,MSc,PhD entomology

1236 Forest Resources Research

1 Case AB - BScFor,MScFor *Program Director*
2 Hall JP - BScFor,PhD silviculture
3 Lavigne M - BScFor,PhD silviculture
4 Meades WJ - BSc,MSc,PhD ecology
5 Newton P - BScF,MScF silviculture, mensuration
6 Roberts BA - BSc,MSc,PhD ecology
7 Robertson AW - MSc biogeography, biometeorology
8 Thompson ID - BSc,MSc,PhD ecology
9 Titus BD - BSc,PhD reforestation silviculture
10 van Nostrand RS - BScFor,MScFor silviculture
11 Wells ED - BScF,MSc peatland forestry

1237 Canadian Forestry Service-Maritimes

P.O. Box 4000, Fredericton, NB, E3B 5P7
Tel: 506 452 3571.

1 Embree DG - MSc,PhD *Program Director: Technical Services*
2 Hanusiak RE - BScF,MScF *Director General*

3 Sterner TE - BScF,MSc,PhD *Program Director: Research*

1238 Forest Biology

1 Cox RM - BSc,PhD forest reproduction
2 Mahendrappa MK - BSc,MSc,PhD forest nutrients
3 Salonius PO - BSA,MSc forest soil microbiology
4 Strunz GM - BA,PhD *Chemist*

1239 Forest Ecology

1 Eveleigh ES - BSc,MSc,PhD population dynamics
2 Huot M - BScA,MScF forest succession and regeneration

1240 Forest Economics

1 Julien FA - BScF,MScF economics research
2 Runyon KL - BScF,MScFEc,PhDFEc senior socio-economic analyst

1241 Forest Entomology

1 Eidt DC - BSA,MSA,PhD environmental studies
2 Kettela EG - BScF,MScF insect control
3 Nigam PC - BSc,MSc,PhD insecticide activity
4 Royama T - BAgr,MSc,PhD insect population studies
5 Thomas AW - BSc,MSc,PhD insect dynamics

1242 Forest Genetics

1 Bonga JM - BSc,MSc,PhD clonal propagation
2 Fowler DP - BScF,MScF,PhD genetic improvement
3 Park YS - BSF,MSc,PhD genetic improvement
4 Simpson JD - BScF,MF tree improvement

1243 Tree Physiology

1 Cameron SI - BSc,MScF tree physiology
2 Little CHA - BScF,MF,PhD wood formation
3 MacLean DA - BSc,PhD tree physiology
4 Piene H - BScB,BScF,MScB,PhD tree growth

1244 Silviculture

1 Johnston JH - BScF utilization studies
2 Ker MF - BScF,MFS forest mensuration
3 Lees JC - BScF,MF,PhD hardwood silviculture
4 Smith RF - BScF,MSc seed orchard studies
5 van Groenewoud H - BSc,MSc,ScD site classification
6 Van Raalte GD - BScF,MScF spacing and fertilization
7 West RC - BScF plantation silviculture

1245 Laurentian Forestry Centre

1055 du P.E.P.S. st., Ste-Foy, Quebec, G1V 4C7
Tel: (418) 648 7347.

1 Hardy Y - BScFor,MSc,PhD *Acting Director*

1246 Forest Protection Research Program

1247 Prediction

1 Regnie/acute/re J - BSc,MSc,PhD spruce budworm demographic behaviour, parasitism
2 Volney J - BSc host-pathogen relationship

1248 Prognostic

1 Archambault L - BSc spruce budworm impact
2 Benoit P - BScAgr,MSc,PhD entomology
3 Laflamme G - BScAppl,MSc,PhD pathology
4 Laplante JP insect taxonomy

1249 Detection

1 Jobin L - BScAgr,MSc,DSc use of pheromones in detection and control of insects in natural forest
2 Lavallée R - BScAppl,MSc insect detection in plantations

1250 Control

1 Ouellette GB - BA,PhD forest pathology
2 Perry D - BSc,PhD entomopathology
3 Quednau FW - DScN parasitism
4 Smerlis E - BScFor,MA scleroderris canker
5 Valero J - BSc residual populations of insects after using Bt

1251 Intervention

1 Laflamme G - BScAppl,MSc,PhD scleroderris canker
2 Lavallée A - BA,BScAppl,MSc,PhD protection of natural forests

1252 National Program: Forest Insects and Diseases Survey (FIDS)

1 Lachance D - BScAppl,PhD *Head* entomological diagnosis
2 Cauchon R entomological diagnosis
3 Monnier C - Tf special surveys

1253 Forest Resources Research Program

1 Boudoux M - BScAppl,MSc,PhD *Program Manager*

1254 Forest Growth

1 Belair JL - BScAgr,MSc drainage
2 Ouellet D - BScAppl,MSc,PhD stand structure
3 Trencia J - BScAppl treatments
4 Ung CH - BSc,MSc,PhD regeneration
5 Zarnovican R - BScAppl,MSc,PhD productivity

1255 Silviculture

1 Beaubien J - BScAppl,MSc remote sensing
2 Popovich S - FE plantation of hybrid poplars and hardwoods
3 Roberge M - BScAppl,MSc,PhD silviculture

1256 Genetics

1 Beaulieu J - BScApp biometry, provenance
2 Corriveau AJ - BScAppl,PhD tree improvement, genecology (white spruce white pine)
3 Daoust G - BScAppl tree improvement (Norway Spruce) material

1257 Regeneration

1 Daoust AL - BSc,MSc,PhD physiology, hardening of seedlings
2 Delisle C - BScAppl,MSc plantations
3 Girouard RM - BScAgr,MSc,PhD physiology, root system
4 Gonzales A - BA,MSc,PhD soil chemistry, hardening of seedlings

1258 National Program: Long Range Transport of Airborne Pollutants (LRTAP)

1 Boutin R - BScAppl waterlogging of nutrients
2 Phu TD - BScAgr,MSc,PhD effect of acidic deposit on organic matter
3 Robitaille G - BScAppl,MSc,PhD effect of acidic deposit on the cycle of bioelements

1259 Great Lakes Forestry Centre
P.O. Box 490, Sault Ste. Marie, Ontario, P6A 5M7
Tel: 705 949 9461. Telex: 067 77180.

1 Cayford JH - BScF,MF *Director General*
2 Haig RA - BScF *Deputy Director General*
3 Huntley GD - BScF,MScF *Chief Planning, Evaluation and Research Services*
4 Morrison IK - BScF,MScF,PhD *Research Manager, Environmental Forestry and Fire*
5 Plexman CA - BA,BLS,DipEd *Chief, Communications Services*
6 Riley LF - BScF *Research Manager, Forest Resources*
7 Sullivan CR - BSc,MSA,PhD *Program Director, Research and Technical Services*

1260 Ecology and Physiology

1 Hogan GD - BSc,MSc,PhD *Senior Scientist* tree physiology
2 Jeglum JK - BSc,MSc,PhD black spruce autecology; forest ecosystem classification
3 Sims RA - BSc,MSc,PhD forest site classification, ecology and remote sensing

1261 Entomology

1 Harvey GT - BSc,MSc,PhD *Senior Scientist* insect physiology and biological control
2 Lyons DB - BSc,MSc biometeorology
3 Lysyk TJ - BSc,MS,PhD population dynamics
4 Nealis VG - BSc,MSc,PhD natural mortality factor
5 Sanders CJ - BA,MSc,PhD behaviour and ecology
6 Wallace DR - BSc,MSc,PhD behaviour and taxonomy

1262 Fire

1 Lynham TJ - BScF wildfire ecology
2 McRae DJ - BScF,MSc prescribed burning
3 Stocks BJ - BScF,MScF *Unit Head, Forest Fire Research* forest fire behaviour

1263 Forest Economics

1 Andersen S - BScF forest sector profiles
2 Johnson JD - BScF,MBA economic studies

3 Smyth JH - BScF,MScF *Chief* socio-economic analysis and planning

1264 Forest Pathology

1 Basham JT - BSc,MA,PhD decays and deterioration
2 Dumas JT - BSc,MScF,PhD stem and root decays
3 Takai S - BSc,DAgr biochemistry of resistance mechanisms
4 Whitney RD - BScF,MF,PhD *Senior Scientist* root and butt rots

1265 Insect and Disease Survey

1 Gross HL - BSc,MSc,PhD disease survey sampling and assessment methodology
2 Howse GM - BScF,PhD *Unit Head, Forest Insect and Disease Survey*
3 Meating JH - BSc,MSc evaluation of insect control operations
4 Myren DT - BS,MS,PhD forest pathology, mycology
5 Nystrom KL - BSc insect identification
6 Roden DB - BScF,MSc insect damage appraisal
7 Syme PD - BA,MSc,PhD insect taxonomy

1266 Mensuration

1 Payandeh B - LF,MF,PhD growth and yield modelling

1267 Silviculture

1 Fleming RL - BA,MScF upland black spruce seeding
2 Groot A - BScF,MScF lowland black spruce seeding
3 Haavisto VF - BScF,MSc black spruce seed production; technology transfer
4 Scarratt JB - BSc,PhD containerized planting stock production
5 Sutton RF - BSc,MSc,PhD *Senior Scientist* regeneration silviculture
6 von Althen FW - BScF,MSc hardwood and conifer reforestation
7 Wood JE - BScF vegetation control, outplanted container stock performance

1268 Mechanization of Silviculture

1 Leblanc JD - BScF,MScF biological performance of mechanized silviculture equipment
2 Smith CR - BScF *Unit Head, Mechanization of Silviculture* silviculture equipment evaluation
3 Sutherland BJ - BScF silviculture equipment evaluation

1269 Soils

1 Foster NW - BScF,MSc,PhD soil process, impacts of atmospheric pollutants
2 Nicholson JA - BScF,MS,PhD hydrologic processes, impacts of atmospheric pollutants

1270 Library

1 Burt SB - BA *Librarian*

1271 Northern Forestry Centre
5320 - 122 Street, Edmonton, Alberta, T6H 3S5
Tel: 403-435-7210. Telex: 037-2117.

1 Kiil AD - BScFor,MScFor *Regional Director General*

1272 Forest Resources Research

1 Waldron RM - BScFor,MScFor *Program Director*

1273 Nursery Management and Tree Improvement

1 Edwards IK - BSc,MSc,PhD *Project Leader* nursery soils
2 Apps MJ - BSc,MSc,PhD radiochemistry; ecological modelling
3 Dymock IJ - BSc,MSc,PhD tree physiology
4 Harvey EM - BScA,MScFor nursery management
5 Klein JI - BSc,MSc,PhD tree improvement

1274 Regeneration and Plantation Management

1 Brace LG - BScFor,MScFor *Project Leader* silviculture; vegetation management
2 Corns IGW - BSc,MSc,PhD site productivity
3 Holland WD - BScA soils and land classification
4 Ives WGH - BScA,MSc plantation assessment, mortality; entomology

1275 Stand Productivity and Forest Inventory

1 Bella IE - BScFor,MFor,PhD *Project Leader* growth and yield
2 Baker J - BSA,MSc,PhD soil chemistry
3 Grewal HS - BScFor,MFor forest mensuration
4 Hall RJ - BScFor,MScFor remote sensing
5 Moore WC - BSc,MSc remote sensing
6 Ondro WJ - ForEng,MScFor forest yield economics
7 Yang RC - BScF,MSc,PhD growth and yield

1276 Forest Protection Research

1 Malhotra SS - BScA,MSA,PhD *Program Director*

1277 Environmental Impact of Toxic Substances and Vegetation Management

1 Sidhu SS - BSc,MSc,PhD *Project Leader*
2 Feng JCY - BSc,MSc herbicide residue chemistry
3 Kalra YP - BScA,MSc,MSc analytical chemistry
4 Maynard DG - BScA,MSc,PhD soil chemistry

1278 Forest Insect and Disease Surveys and Management Systems

1 Cerezke HF - BSc,MSc,PhD *Project Leader* entomology, forest pest surveys
2 Hiratsuka Y - BSc,MSc,PhD forest pathology
3 Moody BH - BScFor,MScFor,PhD pest damage appraisal
4 Volney WJA - BScFor,MScFor,PhD pest damage appraisal
5 Wong HR - BSA,MSc,PhD insect taxonomy

1279 Wetlands, Climate and ENFOR

1 Zoltai SC - BScFor *Project Leader* peatland ecology; land classification
2 Hillman GR - BScFor,MSc,PhD wetlands drainage
3 Johnson JD - BSc,MSc plant ecology
4 Singh T - BA,MSc,PhD biometrics; biomass; climate

1280 Forest Extension

1 Powell JM - BSc,MSc,PhD *Program Director*

1281 Computing Services

1 Chow W - BSc *Manager Computing Centre* computing systems
2 Irwin AW - BSc scientific programmer, analyst
3 Paradis AJ - BA,BSc scientific programmer, analyst

1282 Fire Research Management Systems

1 Barney RJ - BSc,MSc,PhD *Project Leader*
2 Alexander ME - BSc,MSc fire behaviour
3 Chrosciewicz A - BScFor fire behaviour effects and regeneration
4 Delisle GP - BScFor,MScFor fire ecology
5 Lee BS - BScFor fire systems research
6 McAlpine RS - BScFor prescribed burning

1283 Forest Hydrology and Microclimate Research

1 Swanson RH - BSc,MSc,PhD *Project Leader* watershed management
2 Bernier PY - BScFor,MSc,PhD watershed management

1284 Technology Transfer and Information

1 Newstead RG - BScFor,MSc *Project Leader* technology transfer
2 Boughton BJ - BA scientific editor
3 Robinson DJS - BSc,MLS librarian
4 Samoil JK - BA,BEd scientific editor & publisher

1285 Regional Development

1 Heit MJ - BA,MA,PhD *Program Director*

1286 Manitoba District Office

1 McQueen JA - BScFor *District Manager*
2 Ball WJ - BScFor *Extension Research Coordinator* silviculture
3 Daoust-Savoie ME - BSc *Insect and Disease Specialist*
4 Hirsch KG - BScFor *Fire Specialist*
5 Steele TW - BScFor,MSc *Forest Economist*

1287 Saskatchewan District Office

1 Farrell JJ - BScFor *District Manager*
2 DeGroot WJ - BScFor *Fire Specialist*
3 Gardner AC - BScFor,MScFor *Research and Extension Coordinator* silviculture
4 Newman M - BScFor *Renewal and Intensive Management Coordinator*

1288 Forestry Development

1 Price HS - BScFor *Chief, Development Coordination*
2 Amirault PA - BScFor,MScFor *Insect and Disease Specialist, Alberta Agreement*
3 Côté RM - BSc forest development

1289 Resource Economics and Statistics

1 Boylen DM - BA,MA *Project Leader* forest economics
2 DeFranceschi JP - BA forest economics

3 Kuhnke DH - BScFor forest statistics
4 Mauch AM - BS,MSc forest economics
5 Williamson TB - BScFor,MSc forest economics

1290 Pacific Forestry Centre
506 W. Burnside Road, Victoria, B.C., V8Z 1M5
Tel: 604-388-0600.

1 Macdonald DR - BScFor *Director General*

1291 Forest Protection Research

1 Miller GE - BSc,MSc,MPM,PhD *Program Director* forest protection research

1292 Entomology

1 Harris JWE - BSF,MSc,PhD forest entomology: biological control
2 Hulme MA - BSA,MSc,PhD forest entomology: biological control
3 Moeck HA - BSc,MSc,PhD forest entomology: bark beetles
4 Safranyik L - BScFor,MF,PhD forest entomology: bark beetles
5 Sahota TS - BSc,MSc,PhD forest entomology: endocrinology
6 Shepherd RF - MSF,MSc,PhD forest entomology: defoliators
7 Shore TL - BSc,PhD forest entomology: pest populations
8 Thomson AJ - BSc,PhD forest entomology: biological systems analysis

1293 Forest Pathology

1 Bloomberg WJ - BScFor,PhD forest pathology: root diseases
2 Funk A - BA,PhD forest pathology: cultural studies of fungi
3 Hunt RS - BSc,PhD forest pathology: blister rusts
4 Morrison DJ - BSF,MSc,PhD forest pathology: root rots
5 Sutherland JR - BSc,MSc,PhD forest pathology: nursery diseases
6 Whitney HS - BSA,MSc,PhD,PDF forest pathology: microorganisations

1294 Support Services

1 Peet FG - BA,MA,PhD physics: digital image processing
2 White EE - BSc,MSc,PhD biochemical genetics: blister rusts

1295 Forest Resources Research

1 Dobbs RC - BSc,MFor,PhD *Program Director* forest resources research

1296 Land Classification

1 Oswald ET - BA,MA,PhD plant ecology: forest land classification
2 Senyk JP - BScFor vegetation analysis

1297 Silviculture

1 Arnott JT - BScFor,MScFor regeneration silviculture
2 Eis S - DipForEng,PhD forest ecology and silviculture
3 Meagher MD - BSF,MScF,PhD forest genetics silviculture

1298 Soils

1 Marshall VG - BSc,MSc,PhD entomology, soil fauna

1299 Support Services

1 Manville JF - BSc,PhD chemistry: chemicals in ecosystem
2 Pannekoek WJ - BSc,PhD computer programming: software support
3 Partridge J - BSc computer operations: systems development
4 Simmons CS - BSc,MSc,PhD applied statistics: statistical analysis

1300 Survey and Mensuration

1 Barclay H - BSc,MSc,PhD mensurational analysis: modelling
2 Lee YJ - BScFor,MF,PhD mensuration: remote sensing

1301 Tree Biology

1 Brix H - MF,PhD tree physiology
2 Edwards DGW - BScFor,MF,PhD tree and seed physiology
3 McMinn RG - BA,MS,PhD forest ecology

1302 Forest Environment Research

1 Winston DA - BSc *Program Director* forest environment research

1303 Environmental Forestry

1 Hetherington E - BASc,MA,PhD forest hydrology
2 Smith RB - BSc,MF,PhD impact of forest practices
3 Stanek W - MScFor,DipF,PhD environmental forestry

1304 Fire Research

1 Hawkes BC - BScF,MSc fire ecology: prescribed fire

2 Lawson BD - BSF,MScFor forest fire research
3 Silversides RH - BASc,MA,PhD fire research: meteorology

1305 Forest Insect and Disease Survey

1 Van Sickle GA - BScFor,MSc,PhD *Head of Unit*
2 Hopkins JC - BSc,PhD *Head of Herbarium* pathology
3 Humble L - BSc,MSc,PhD *Head of Insectary* entomology

1306 Pest Damage Appraisal

1 Alfaro RI - BScF,MPM,PhD entomology: quantification of pest damage

1307 Forestry Development and Relations

1 Edwards JA *Program Director* forestry development and relations

1308 Economics

1 Manning GH - BSc,MS,PhD forest economics: economic policy
2 Nicholson W - BSc,PhD forest economics
3 White WA - BSc,MA,PhD economics and policy research

1309 Technology Transfer

1 de Lestard JPG - BSF *Liaison Officer* technology transfer
2 Edwards JC - BScF *Technology Transfer Officer*

1310 Planning & Program Implementation

1 Skinner JG - BS,MA program planning
2 Ulrich VG - BSc *Implementation Officer*

1311 Library Services

1 Solyma AO - BA,BLS *Librarian*

1312 Public Information Program

1 Teske EL - BAJour media and public relations

1313 Scientific and Technical Editing

1 Glover SG - BSc *Scientific Editor*

1314 Environment Canada

1315 Canadian Wildlife Service

Headquarters, Place Vincent Massey, 351 St. John Blvd, Ottawa, K1A OE7
Tel: (819) 997-1301.

1 Clarke HA - BSc,MASc *Director General*
2 Keith JA - BSc,MA,MSc *Director (Wildlife Research and Interpretation Branch)*
3 Patterson JH - BSc,MSc,PhD *Director (Migratory Birds Branch)*
4 Cooch FG - BA,MS,PhD *Senior Research Scientist*
5 Dauphiné TC - BSc,MSc *Co-ordinator (Threatened Species and Transboundary Wildlife)*
6 Filion FL - BA,MA *Co-ordinator (Social Studies Division)*
7 Gaston AJ - BSc,MSc,PhD *Co-ordinator (Seabird Research)*
8 Marshall K - BSc,MSc *Head (Toxic Chemicals Program Section)*
9 Norstrom RJ - BSc,PhD *Head (Environmental Chemistry Section)*
10 Peakall DB - PhD *Head (Wildlife Toxicology Section)*
11 Price I - BHSc,MSc *Co-ordinator (Latin-American Program)*
12 Wendt S - BSc,MSc,PhD *Chief (Populations and Surveys Division)*

1316 Atlantic Region

P.O. Box 1590, Sackville, New Brunswick, EOA 3CO
Tel: (506) 536-3025.

1 Finney GH - BA,PhD *Director*
2 Bateman Miss MC - BSc,MSc *Biologist* ornithology - waterfowl
3 Erskine AJ - BSc,MA,PhD *Division Chief* migratory bird management
4 Hicklin PW - BA,BSc,MSc *Biologist* ornithology - seaducks
5 Johnson BC - BSc,MSc *Biologist* ornithology - environmental assessment
6 Parker GR - BSc,MSc *Research Scientist* ornithology - waterfowl
7 Prescott WH - BSc,MSc *Division Chief* wildlife and habitat conservation
8 Sealy Mrs J - BA *Librarian*
9 Smith AD - BSc,MSc *Biologist* ornithology - habitat assessment

1317 Local Office

c/o Bedford Institute of Oceanography, P.O. Box 1006, Dartmouth, Nova Scotia, B2Y 4A2
Tel: (902) 426-3274.

1 Brown RGB - BA,DPhil *Research Scientist* ornithology - seabirds
2 Hiscock EHJ - BScFor,MScFor *Manager* special projects

3 Lock AR - BA,BEd,MSc,PhD *Biologist* ornithology - seabirds
4 Nettleship DN - BSc,MSc,PhD,PdF *Research Scientist* ornithology - seabirds

1318 Local Office

c/o Department of Biology, Dalhousie University, Halifax, Nova Scotia, B3H 4J1
Tel: (902) 426-6356.

1 Kerekes JJ - BSc,MSc,PhD *Research Scientist* acid rain

1319 Local Office

Pleasantville, Building 303, P.O. Box 9158, St. John's, Newfoundland, A1A 2X9
Tel: (709) 772-5585.

1 Elliot RD - BSc,MSc,PhD *Biologist* ornithology - seabirds
2 Goudie RI - BSc *Biologist* ornithology - waterfowl

1320 Local Office

At New Brunswick, P.O. Box 400, Fredericton, New Brunswick, E3B 4Z9
Tel: (506)452-3086.

1 Pearce PA - BScFor,MScFor *Biologist* toxic substances

1321 Ontario Region

1725 Woodward Drive, Ottawa, K1A OE7
Tel: (613)998-4693.

1 Curtis SG - BA,MA *Regional Director*

1322 Migratory Birds

1 Blokpoel H - BSc,MSc,PhD *Biologist* ornithology
2 Carreiro JF - BA *Head Habitat and Ecological Assessment*
3 Dennis DG - BSc,MSc *Biologist* bird population
4 Lévesque H - BSc,MSc *Biologist* habitat and ecological assessment
5 McCullough GB - BSc *Biologist* ecological assessment
6 McNicol D - BSc,MSc *Biologist* acid rain
7 Morrison RIG - BSc,PhD *Research Scientist* ornithology
8 Planck J - BSc *Biologist* ecological assessment
9 Ross RK - BSc,MSc *Biologist* bird populations
10 Welsh DA - BSc,MSc,PhD *Biologist* bird populations
11 Weseloh DV - BSc,MSc,PhD *Biologist* toxic chemicals - herring gulls

1323 Quebec Region

Canadian Wildlife Service, 1141 Route de l'Église, C.P. 10 100 Ste-Foy, Quebec, G1V 4H5
Tel: (418) 648-3914.

1 Cinq-Mars J - BSc,MSc *Regional Director*
2 Vincent J - BSc,MSc *Principal Adviser*

1324 Division for Management of Migratory Bird Populations

1 Drolet CA - BSc,MSc *Head*
2 Bordage D - BSc *Biologist* bird surveys
3 Bourget A - MSc,MSc *Head* bird population surveys
4 Chapdelaine G - BSc *Biologist* sea birds and bird colonies
5 Gauthier J - BSc,MSc *Biologist* terrestrial bird populations
6 Reed A - BSc,MSc,PhD *Principal Scientific Adviser*

1325 Division for Wildlife and Habitat Conservation

1 Lemieux S - BSc *Head*
2 DesGranges JL - BSc,MSc,PhD habitat research
3 Laporte P - BSc,MSc *Biologist* poisons and threatened species
4 Lehoux D - BSc *Biologist* environmental and impact studies
5 Mercier Y - BSc,MBA *Biologist and Administrator* management of national wildlife reserves
6 Ringuet I - BSc,MSc *Biologist* protection and management of habitats

1326 Western and Northern Region

1327 Regional Office

Twin Atria Building, 2nd Floor, 4999 - 98 Avenue, Edmonton, Alberta, T6B 2X3
Tel: (403) 420-2525. Telex: 037-3699.

1 Kerr GR - BSc,MSc *Regional Director*

1328 Migratory Birds and Threatened Species Conservation Division

1 Scotter GW - BSc,MSc,PhD *Chief of Division*
2 Carbyn LN - BSc,MSc,PhD *Research Scientist* mammalogy

3 Fyfe RW - BA *Research Scientist* ornithology
4 Holroyd GL - BA,MSc,PhD *Head Threatened Species*
5 Kuyt E - BA,MA *Biologist* ornithology
6 Miller F - BSc,MSc *Research Scientist* mammalogy
7 Reynolds H - BSc,MSc *Biologist* mammalogy
8 Stirling IG - BSc,MSc,PhD *Research Scientist* mammalogy
9 Telfer ES - BEd,BSc,MSc *Research Scientist* mammalogy
10 Thomas D - BA,PhD *Research Scientist* mammalogy

1329 Impact Assessment Section

1 Barry TW - BA,LLB,MSc,PhD *Research Scientist* ornithology
2 Dickson DL - BSc *Biologist* ornithology
3 Dickson HL - BSc *Biologist* ornithology
4 Edwards RGW - BSc,PhD *Head, Assessment*
5 Prach RW - BSc *Biologist* ecology

1330 Habitat Section

1 McKeating G - BA,MA *Head, Habitat* ornithology
2 Poston B - BSc,MSc *Biologist* ornithology
3 Shandruk LJ - BSc,MSc *Biologist* range management
4 Trottier G - BSc,MSc *Biologist* range management

1331 Migratory Birds Management Division

1 Goodman AS - BSA,MSc *Chief of Division*

1332 Population Management

1 Turner BC - BSc,MSc *Biologist* ornithology

1333 Yellowknife Office (Ballanca Building)
P.O. Box 637, Yellowknife, N.W.T., X1A 2N5
Tel: (403) 920-8531. Telex: 037-3699.

1334 Migratory Birds and Threatened Species Conservation Division

1 McCormick K - BSc,MSc *Biologist* ornithology

1335 Migratory Birds Management Division

1 Cole RC - BSc,MSc *Biologist* ornithology

1336 Prairie Migratory Birds Research Centre
115 Perimeter Road, Saskatoon, Saskatchewan, S7N OX4
Tel: (306) 975-4087.

1337 Migratory Birds and Threatened Species Conservation Division

1 Adams GD - BSc,MSc,PhD *Research Scientist* ornithology
2 Dzubin AX - BA,MA *Research Scientist* ornithology
3 Kerbes RH - BSc,MSc *Biologist* ornithology
4 Millar JB - BSc,MSc,PhD *Biologist* ornithology
5 Taylor PS - BSc,MSc *Biologist* ornithology

1338 Migratory Birds Management Division

1 Brace RK - BSc,MSc *Head, Population Management*
2 Driver ED - BSc,MSc *Research Scientist* ornithology
3 Gollop JB - BA,MSc,PhD *Research Scientist* ornithology
4 Johns BW - BSc *Biologist* ornithology
5 Nieman D - BSc,MSc *Biologist* ornithology
6 Sugden L - BA,MSc,PhD *Biologist* ornithology
7 Townsend GH - BSc,MSc *Biologist* ornithology
8 Woodsworth E - BSc,MSc *Biologist* ornithology

1339 Winnipeg Office (Freshwater Institute)
501 University Crescent, Winnipeg, Manitoba, R3T 2N6
Tel: (204) 949-5259.

1340 Migratory Birds and Threatened Species Conservation Division

1 Boothroyd P - BSc *Biologist* ornithology
2 Rakowski P - BSc,MSc *Biologist* ornithology

1341 Migratory Birds Management Division

1 Caswell D - BSc,MSc *Biologist* ornithology
2 Hochbaum G - BSc,MSc,PhD *Biologist* ornithology

1342 Pacific and Yukon Region
P.O. Box 340, Delta, B.C., V4K 3Y3
Tel: (604) 946-8546.

1 Martell AM - BSc,MSc,PhD *Regional Director*

1343 Migratory Birds Management

1 Butler RW - BSc,MSc *Biologist* non-game birds
2 Kaiser GW - BSc,MSc *Biologist* marine birds
3 McKelvey RW - BSc,MSc *Biologist* waterfowl populations
4 Savard JP - BSc,MSc,PhD *Biologist* waterfowl ecology
5 Smith GEJ - BSc,MA,PhD *Research Scientist* biometrics
6 Vermeer K - BSc,MSc,PhD *Research Scientist* avian ecology

1344 Wildlife Conservation

1 Boyd WS - BA,BSc,MSc *Biologist* habitat ecology
2 Dennington MD - BSc *Biologist* habitat conservation
3 Retfalvi LI - BSF,MF *Biologist* habitat management
4 Russell DE - BSF,MF *Biologist* northern ecology
5 Trethewey DEC - BAEd,MSc *Biologist* environmental assessment
6 Wetmore SP - BSc,MSc *Biologist* forest birds
7 Whitehead PE - BSc,MSc *Biologist* toxicology

1345 Statistics Canada

Tunney's Pasture, Ottawa, K1A OT6
Tel: (613) 990-8722.

1 Fellegi IP - MA,PhD *Chief Statistician of Canada*

1346 Agriculture Division

1 Alain B - BSc *Statistician* atlantic provinces
2 Andrusiak GW - BSA,MSc *Assistant Director*
3 Beelen G - BA,MSc *Economist* farm expenditures and related estimates
4 Beyouti MA - BAEcon *Economist* census of agriculture
5 Bollman RD - BSA,MSc,PhD *Research Economist* micro-data analysis
6 Breese Ms G-A - BScAgr *Economist* Census of Agriculture
7 Burroughs D - BScAgr,BEd *Economist* dairy estimates
8 Burroughs R - BSA *Project Manager* census of agriculture
9 Code GO - BSc *Senior Statistician* field crops
10 Davidson G - BScMath *Project Officer* Census of Agriculture
11 Dion Mrs M - BS,BScSoc,MSoSocAgr *Project Officer* census of agriculture
12 Dobbins RN - BA *Remote Sensing Image Analyst* crops
13 Dupuis AC - BA,MA *Economist* grain markets
14 Elward MJ - BSc *Statistician* field crops
15 Fitzgerald-McDonald Mrs H - BA,MA *Economist/Cheif* livestock and animal products
16 Frederick Ms J - BAEcon *Project Officer* census of agriculture
17 Freeman RW - BScAgr *Statistician* farm register
18 Freeman WG - BSA *Project Manager* census of agriculture
19 Gauthier Mr L - BScAgrEcon *Economist* farm income
20 Gray Mrs K - BASc *Economist* crops
21 Hamilton E - BScAgrEcon *Statistician/Economist* farm income and related areas
22 Heimbecker J - BAPolSc,MBA *Economist* census of agriculture
23 Hickson AB - BSc,MSc,PhD *Senior Economist* grain markets
24 Holmes Ms M - BA,BCOMM *Economist* farm income
25 Johnson C - BScAgr,MScAgrEcon *Economist* cattle and sheep estimates
26 Jones ME - BA *Economist/Manager* census of agriculture
27 Kemp LD - BSc *Statistician* horticulture crops
28 Khaja S - BED,BScAgr,MScAgr,MAEcon *Economist* census of agriculture
29 Koroluk R - BComm,MScAgr *Economist* eggs, poultry and furs
30 Korporal K - MAGeo,BAGeo *Economist/Analyst* remote sensing
31 Kshatriya Ms S - BScAgrEcon,MAEcon *Economist* farm income
32 Leblanc-Cooke J - BSc *Economist* farm cash receipts and income
33 Leckie Ms E - BScAgrEcon *Economist* farm income and related areas
34 Lennox GA - BSA *Statistician* field crops
35 Lys P - BAEcon,MAEcon *Economist/Chief* farm income and related areas
36 Macartney LK - BA,MSc *Economist* horticulture
37 Mathieson AS - BSc *Senior Statistician* horticulture crops
38 McLaughlin B - BScAgr *Senior Statistician* Atlantic Provinces
39 Meszaros P - BSCMATH *Economist* livestock and animal products
40 Ménard M - BSc *Economist* taxation data research and analysis
41 Miller AK - BScAgr *Statistician/Chief* systems co-ordination and production
42 Murphy L - BScAgrEcon *Economist* farm income
43 Murray P - BSc *Economist* farm prices and indexes
44 Plourde JLMR - BSc *Remote Sensing Image Analyst* crops and remote sensing analysis
45 Rosien BE - BScAgr *Project Manager* livestock unit
46 Seyadoussane S - BAInterEcon *Economist* census of agriculture
47 Shumsky MT - BScAgr *Senior Statistician* Prairie Provinces
48 Snider BC - BSc *Statistician* field crops
49 Stevens PL - BScAgr *Statistician* national farm survey
50 Trant MJ - BA,MSc *Economist/Chief of Crops Section* crops economics
51 Walker P - BScAgr *Economist* pig estimates

PROVINCES OF CANADA

1347 British Columbia Ministry of Agriculture and Fisheries

Head Office, Parliament Buildings, Victoria, B.C., V8W 2Z7
Tel: (604) 387-5121.

1348 Field Operations

1 Miller RJ *A/Assistant Deputy Minister*
2 Arrand JC *Director, Crop Protection Branch*
3 Bertrand RA *Director, Soils Branch*
4 Freed DE *Director, Rural Organisations and Services Branch*
5 Hewitt Dr MP *Chief Veterinarian, Veterinary Services*
6 Wiebe WW *Director, Field and Special Crops Branch*
7 Windt TA *Director, Engineering Branch*
8 Macdonald Dr KR - *Supervisor,Veterinary Diagnostic Lab*
9 Van Lierop W *Supervisor, Soil, Feed and Tissue Testing Lab*

1349 Policy Development and Economics

1 Matviw DM *Executive Director*

1350 Economics Branch

1 Geldart G *Regional Farm Economist*
2 Gubbels P *Agricultural Economist*
3 Hurd L *Regional Farm Economist*
4 Keay R *Regional Farm Economist*
5 Peterson T *Manager, Farm Business Management*

1351 Farmland Resources Branch

1 Dragushan G *Land Policy Analyst*
2 Sasaki H *Land Resource Analyst*

1352 Policy Analysis and Co-ordination

1 Bomford LE *Senior Policy Analyst*
2 Schildroth J *Senior Policy Analyst*

1353 Statistics and Quantitative Analysis

1 Aiyadurai MW *Quantitative Analyst*
2 McAra PK *Research Officer*

1354 British Columbia Ministry of Forests and Lands

1355 Research Branch

1450 Government Street, Victoria, B.C., V8W 3E7
Tel: (604) 387-6721.

1 Baker TE - BScAgr,PhD *Acting Director*
2 Armleder H - MF integrated resource management (Williams Lake, B.C.)
3 Benskin HJ - BSc,MF forest renewal
4 Binder W - MSc,PhD tree physiology
5 Brockley R - MF forest renewal (Vernon, B.C.)
6 Cameron I - MF biometrics
7 Carlson M - MSc,PhD tree improvement (Vernon, B.C.)
8 Chatwin J - BSc,MSc integrated resource management
9 Cheng J - MSc,PhD ecology and earth sciences (Kamloops, B.C.)
10 Comeau P - BSc,PhD forest renewal
11 Draper D - BSc,PhD silviculture (Prince George, B.C.)
12 Eastham A - MSc tree physiology
13 Ebell LF - BSA,PhD tree physiology
14 Errico D - MSc decision aids
15 Goudie J - MSc biometrics
16 Hamilton E - MSc integrated resource management
17 Heaman JC - MA,MSc tree improvement
18 Herring LJ - BSF forest renewal (Prince George, B.C.)
19 Homoky SGJ - BSF integrated resource management
20 Jaquish B - MF tree improvement (Vernon, B.C.)
21 Johnstone WD - BSF,MF,PhD forest renewal (Vernon, B.C.)
22 Kiss GK - BSK,MF tree improvement (Vernon, B.C.)
23 Klinka K - ForEng,PhD integrated resource management (Vancouver, B.C.)
24 Kovats M - BSF,MSF forest productivity and decision aids
25 Krumlik G - MSc,PhD forest renewal
26 Leadem C - BSc,PhD tree physiology
27 MacKinnon A - BSF,MSF integrated resource management
28 McLeod A - BSF,MScAgr integrated resource management (Prince George, B.C.)
29 McNay S - MSc integrated resource management
30 Meidinger D - MSc integrated resource management
31 Mitchell KJ - BSF,MF,PhD forest productivity and decision aids

32 Mitchell WR - BScAgr integrated resource management (Kamloops, B.C.)
33 Newman R - BSc integrated resource management (Kamloops, B.C.)
34 Nicolson A - BSc,MSc integrated resource management
35 Nuszdorfer F - BSc integrated resource management (Vancouver, B.C.)
36 Nyberg B - BSc,MSc integrated resource management (Vancouver, B.C.)
37 Omule S - MSc,PhD forest productivity
38 Page R - MSc integrated resource management
39 Pendl FT - BSF forest renewal (Vancouver, B.C.)
40 Pojar JJ - BSC,MSc,PhD integrated resource management (Smithers, B.C.)
41 Pollack J - BSF,MF forest renewal (Smithers, B.C.)
42 Ross S - MSc,PhD tree physiology
43 Russell J - MF tree improvement (Mesachie Lake, B.C.)
44 Schwab J - BSF integrated resource management (Smithers, B.C.)
45 Simpson DG - BSc tree physiology (Vernon, B.C.)
46 Spittlehouse D - MSc,PhD integrated resource management
47 Still G - BSF,MSc integrated resource management
48 Sutherland C - MS forest renewal (Williams Lake, B.C.)
49 Thompson CF - BSF,MF forest renewal (Nelson, B.C.)
50 Toews D - BSc,MF ecology and earth sciences (Nelson, B.C.)
51 Trowbridge R - BScAgr integrated resource management (Smithers, B.C.)
52 van den Driessche R - MF,PhD tree physiology
53 Vyse A - BSc,MF forest renewal (Kamloops, B.C.)
54 Watt WJ - BScAgr,MSc integrated resource management (Williams Lake, B.C.)
55 Webber JE - BSF,PhD tree physiology
56 Wikeem B - BSc,MSc,PhD integrated resource management (Kamloops, B.C.)
57 Wilford D - BSF,MF integrated resource management (Smithers, B.C.)
58 Woods J - MF tree improvement (Mesachie Lake, B.C.)
59 Woon C - MF tree improvement
60 Yeh FCH - BSc,PhD tree improvement
61 Ying CC - BSc,MF,PhD tree improvement

1356 Alberta Agriculture

J.G. O'Donoghue Building, 7000-113 Street, Edmonton, Alberta, T6H 5T6
Tel: (403)427-2727.

1 McEwen HB - MSc *Deputy Minister*
2 Dent WJ - MSc *Assistant Deputy Minister field services*
3 Hanna HJ - DVM *Assistant Deputy Minister production*
4 Mehr BD - BA *Assistant Deputy Minister marketing*
5 Radke CD - MA *Assistant Deputy Minister planning and development*

1357 Field Services Sector

1 Berkan RF - BSA *Regional Director: Region IV - Vermilion*
2 Forrest RS - BEng *Head: Engineering Services Branch*
3 Hall WA - BSA *Regional Director: Region III - Red Deer*
4 Knapp JA *Regional Director: Region VI - Fairview*
5 Kontz R - BSc *Head: Agriculture and Community Services Branch*
6 Leavitt F - BSA *Head: Agricultural Services Section*
7 Leavitt IM - BSc *Director: Rural Services Division*
8 Myers SE - MSc *Head: Home Economics Branch*
9 Tackaberry JB - BSA *Regional Director: Region V - Barrhead*
10 Werner GW - BSA *Regional Director: Region II - Airdrie*
11 Wismer WW - BSA *Head: Agriculture Education Branch*
12 Youck RT - BSA *Head: 4-H Branch*
13 Young DR *Regional Director: Region I - Lethbridge*

1358 Marketing Sector

1 Glover DL - BSA *Director: Marketing Services Division*
2 Wulff C - MBA *Executive Director: Market Development*
3 Hanna SG - MBA *Head: Business Analysis Branch*
4 Kadis VW - PhD *Director: Food Laboratory Services Branch*
5 Rosario JA - PhD *Executive Director: Trade Policy Secretariat*
6 Tiwari NP - PhD *Special Project Supervisor: Food Laboratory Services Branch - O.S. Longman Building*

1359 Planning and Development Sector

1 Cameron MA - BSA *Director: Economic Services Division*
2 Loree JW - MSc *Head: Farm Business Management Branch*
3 Ross CB - PhD *Head: Production and Resource Economics Branch*
4 Walker D - MSc *Head: Market Analysis Branch*

1360 Production Sector

1 Basarab J - PhD *Beef Management Specialist: Beef Cattle & Sheep Branch*
2 Bolingbroke DO - DVM *Head: Regional Veterinary Laboratory (Airdrie)*
3 Bristow ED *Head: Dairy Production Branch*
4 Church JL *Acting Director: Animal Health Division*
5 Goettel AW - MSc *Head: Soils Branch*
6 Grainger GD - BSA *Director: Alberta Tree Nursery & Horticultural Centre*
7 Harries WN - BVSc *Head: Regional Veterinary Laboratory (Lethbridge)*

8 Herbert CW *Head: Regulatory Services*
9 Macyk DA *Director: Plant Industry Division*
10 Maurice D - MSc *Supervisor: Weed Research & Development*
11 Milligan JD - BSA *Head: Horse Industry & Feeder Assoc. Branch*
12 Nagge WT - DVM *Head: Regional Veterinary Laboratory*
13 Nation W *Head: Veterinary Pathology Branch*
14 Penny DC - MSc *supervisor: soil fertility*
15 Price MK - BSA *Head: Crop Protection Branch*
16 Schuld F - MSc *Head: Pork Industry Branch*
17 Schute RH *Head: Animal Health Management Branch*
18 Summers GW - DVM *Head: Meat Hygiene Branch*
19 Sydness TE *Head: Poultry Branch*
20 Walden K *Head: Dairy Processing Branch*
21 Weisenburger R - MSc *Head: Beef Cattle & Sheep Branch*

1361 Research and Resource Development Sector

1 Barlott P - MSc *Head: Conservation and Development Branch*
2 Birch A - PhD *Head: Land Use Branch*
3 Christian RG *Executive Director: Research Division*
4 Colgan BL - BSA *Director: Irrigation and Resource Management Division*
5 Hartman GD - BSA *Manager: Irrigation Secretariat*
6 Paterson B - MSc *Head: Drainage Branch*
7 Pungor AE - BSA *Head: Irrigation Branch*
8 Teklemariam Y *Manager: Research Division*

1362 Farming for the Future Council
H202 J.G. O'Donoghue Building, 7000-113 Street, Edmonton, Alberta, T6H 5T6

1 McEwen HB - MSc *Chairman*
2 Teklemariam Y *Secretary*

1363 Alberta Horticultural Research Center
Bag Service 200, Brooks, Alberta, T0J OJO

1 Krahn T - MSc *Director*
2 Esau R - MSc *Weed Control Specialist*
3 Gaudiel R - PhD *Head: Special Crops & Agronomy*
4 Howard R - PhD *Head: Crop Protection & Utilization Section*
5 Ragan P - PhD *Vegetable Specialist*
6 Schaupmeyer CA - MSc *Head: Horticulture Production Section*
7 Soehngen U - PhD *Research Horticulturist*

1364 Food Processing Development Centre
6309-45 Street, Leduc, Alberta, T9E 2Y7
[PO BOX 1217, Leduc, Alberta]

1 Schroder DJ - PhD *Head, Food Processing Development Centre*
2 Davies PH - MSc *Manager: Agri-Food Processing Development Centre*
3 Fierheller M - MSc *Food Research Scientist*
4 Lutz S - BSA *Food Research Scientist*
5 Phillipchuck GE - PhD *Supervisor, Agri-Food Processing Development Centre*

1365 Alberta Agricultural Research Institute

1 Christian RG *Executive Director*
2 McEwen HB *Vice-Chairman*

1366 Crop Research Program
Alberta Agriculture, Agriculture Building Bag Service H47 5718-56 Avenue, Lacombe, Alberta, TOC 1SO

1 Cooper D - BSA *Head: Field Crops Branch*
2 Helm J - PhD *Head: Research Section*
3 Salmon DF - PhD *Plant Breeder: Field Crops Branch*

1367 Saskatchewan Department of Agriculture

Walter Scott Building, Regina, Saskatchewan, S4S OB1
Tel: (306)787-5170. Telex: 071-2474.

1 Drew JL - BSA,MSc *Deputy Minister*
2 Bowd LES - BA,MBA *Assistant Deputy Minister*
3 Kramer SL - BA *Assistant Deputy Minister*

1368 Economics Branch
Walter Scott Building, Regina, Saskatchewan, S4S OB1
Tel: (306)787-5961.

1 Maley DL - BA,MSc *Director*
2 Christophesson D livestock marketing
3 Cushon HN - BSA,MSc stabilization, market analyses
4 Demyen MJ - BA,MA policy
5 Deyong ER - BSA,MSc transportation and income forecasting
6 Evans KW - BA,MSc trade
7 Magnusson Ms JO - BSc,MSc policy, diversification

8 Ross M - BSA livestock tax credits
9 Scott TG - BA,MA policy
10 Sigurdson DG - BSA,MSc grain marketing
11 Zepp JE - BSA,MSc statistics

1369 Information Services Branch
Walter Scott Building, Regina, Saskatchewan, S4S OB1
Tel: (306)787-5151.

1 Stewart Mrs MAH - BA,BLS librarian

1370 Saskatchewan Research Council
15 Innovation Blvd, Saskatoon, Saskatchewan, S7N 2X8
Tel: (306)933-5400. Telex: SARECO 074-2484.

1 Hutch J - BSc,PEng *President*

1371 Research and Development Branch

1 Husband WHW - BSc,MSc,PhD *Vice President* research management

1372 Resources Division

1 Arnold RA - BSc,MSc,PhD *Director* chemical analysis

1373 Biomass Resources Program

1 Coxworth E - BA,MA,PhD *Manager* agricultural products and energy
2 Sosulski K - MSc,PhD agricultural products and energy

1374 Buildings and Energy Technology Program

1 Eyre D - BE,MSc manager

1375 Environment Division

1 Maybank J - BSc,MSc,PhD *Director* atmospheric physics

1376 Aquatic Biology Program

1 Swanson S *Manager*

1377 Terrestrial Ecology Program

1 Abouguendia Z - PAg,BSc,MSc *Manager* range ecology

1378 Atmospheric Processes Program

1 Shewchuk S - BSc,MSc,PhD *Manager* acid rain studies

1379 Administration and Finance

1 MacKay G - BA,CRM *Vice President*

1380 Analytical Services

1 Smithson G - PEng,BSc,MSc,PhD *Manager* analytical research

1381 Prairie Agricultural Machinery Institute
P.O. Box 1900, Humboldt, Saskatchewan, SOK 2AO
Tel: (306) 682-2555.

1 Broad B - BSc,MSc *Acting Director*

1382 Humboldt Station
P.O. Box 1150, Humboldt, Saskatchewan, SOK 2AO
Tel: (306) 682-5033.

1 Wassermann J - PEng *Station Manager*

1383 Portage la Prairie Station
P.O. Box 1060, Portage La Prairie, Manitoba, R1N 3C5
Tel: (204) 239-5445.

1 Omichinski M - BSc *Station Manager*

1384 Veterinary Infectious Disease Organization
124 Veterinary Road, Saskatoon, Saskatchewan, S7N OWO
Tel: (306)966-7465.

1 Acres SD - DVM,MPVM,PhD *Director*
2 Babiuk LA - BSA,MSc,PhD *Associate Director (Research)* rotavirus and coronvirus
3 Hodgman PG - BScAgr *Executive Officer*
4 Lawman MJP - BSc,PhD *Research Scientist* immunology
5 Parker MD - BSc,MSc,PhD *Research Scientist* recombinant DNA technology
6 Potter AA - BSc,PhD *Research Scientist* bacterial genetics
7 Sabara MIJ - BSc,MSc,PhD *Research Scientist* molecular biology

8 van den Hurk J - BSc,MSc *Research Scientist* hemorrhagic enteritis of turkeys
9 van drunen Littel - van den Hurk S - BSc,MSc,PhD *Research Scientist* molecular virology
10 Watts TC - DVM *Co-Ordinator, Animal Support Services*
11 Willson PJ - BA,MS,DVM *Research Veterinarian - Clinical Medicine* epidemiology
12 Wilson SH - PVM,DVM,MPVM *Research Veterinarian* epidemiology

1385 Manitoba Natural Resources

1386 Forestry Branch
300-530 Kenaston Boulevard, Winnipeg, Manitoba, R3N 1Z4
Tel: (204) 945-7989.

1 Rannard CD - BScF,MScFM *Director of Forestry*

1387 Forest Management Section
1 Lamont RH - BScF,MF *Chief of Section*
2 Becker GD - BScF *forest inventory*
3 Keobouasone K - BScF *forestry geographic information service*
4 Middlebró W - BScF *Timber Management and Development*
5 Peterson G - BScF *forest growth and yield*
6 Tuinhof NB - BScF *Forestry Computer Analyst*

1388 Forest Protection Section
1 Munro GP - BSc *Chief of Section*
2 Beaubien YB - BSc *pests and 2nd generation forests*
3 Knowles KR - BSc *Survey Head*
4 Pines IL - BSc *dutch elm disease*

1389 Forestry Administration
1 Gascoyne V *Head, Timber Sales Office*

1390 Silviculture Section
1 Yarn LG - BScF *Chief of Section*
2 Ardron G - BScF *Silviculture Forester*
3 Cameron RG - BScF *Manager, Pineland Forest Nursery*
4 Segaran S - BScF,MSc *Forester Tree Improvement and Forest Genetics*
5 Tornblom SR - BScF *Manager, Clearwater Forest Nursery*

1391 Regional Forestry Programs
300-530 Kenaston Boulevard, Winnipeg, Manitoba, R3N 1Z4
Tel: (204) 945-7989.

1 Kaczanowski S - BScF *Regional Forester, Eastern*
2 Lamb D - BScF *Regional Forester, Northern*
3 McColm G - BScF *Regional Forester, Western*
4 Vogel K - BScF *Regional Forester, Southern*

1392 Ontario Ministry of Agriculture and Food

Legislative Buildings, Queen's Park, Toronto, Ontario, M7A 1A3
Tel: (416) 965-1044.

1 Switzer CM - BSAAgr,MSc,PhD *Deputy Minister*
2 Burak R - BAPolSci *Assistant Deputy Minister* finance and administration
3 Collin GH - BScAgr,MSc,PhD *Assistant Deputy Minister* marketing
4 Rennie JC - BSAAgr,MSc,PhD *Assistant Deputy Minister* technology and field services

1393 Economics Branch
Legislative Buildings, Toronto, Ontario, M7A 1B3
Tel: (416) 965-1064.

1 Seguin R - BA,MSc *Director*
2 Lentz G - BSc,PhD *Manager, Policy Development*
3 Wind F - BScAgr,MS *Manager, Economic Information Systems*
4 Boyd W - BA,MA *Senior Policy Advisor*
5 Jaeger MJ - BSc,MSc *Senior Policy Advisor*
6 Poyntz P - BSc,MBA,MPA *Senior Policy Advisor*
7 Buth D - DipBA,MSc *Policy Advisor*
8 Caine R - BScAgr,MScAgr *Policy Advisor*
9 Campbell B - BSc *Economist*
10 Contini A - BSc,BSA *Policy Advisor*
11 Crozier T - BScAgr,MSc *Policy Advisor*
12 Fearon G - BSc,MSc *Economist*
13 Fernandes C - BS *Statistician*
14 Fisher GA - BSA *Financial Analyst*
15 King J - BS,MSc *Statistician*
16 McGee B - BSc,MSc *Statistician*
17 McKibbon ED - BSc,CGA *Economist*

18 Ong J - BSc,MS,PhD *Economist*
19 Roller N - BSc,MSc *Statistician*
20 Thomson D - BSc,MSc *Policy Advisor*
21 Tmej L - BS *Economist*

1394 Education and Research Division
Guelph Agriculture Centre, Box 1030, Guelph, Ontario, N1H 6N1
Tel: (519) 823-5700.

1 McLaughlin RJ - BScAgr,MSc,PhD *Executive Director*
2 Boyd KG - BScEng,MScEng *Manager, Education and Research Programs*

1395 Rural Organizations and Services Branch
Guelph Agriculture Centre, Box 1030, Guelph, Ontario, N1H 6N1
Tel: (519) 823-5700.

1 Hagarty JJ - BScAgr,MSc *Director*
2 Fleming PF - BScAgr *Manager, Rural Organisations*
3 Holding Miss LM - BScHEc *Manager, Special Programs*
4 McMahon RT - BScAgr *Manager, Agriculture Employment and Safety*
5 Underwood Mrs AL - BScAgr *Manager, Program Development*

1396 Ontario Agricultural Museum
Box 38, Milton, Ontario, L9T 2Y3
Tel: (416) 878-8151.

1 Wiley JA - BSA *General Manager*
2 Bennett Miss SL - BA *Research and Reference Librarian*

1397 Horticultural Research Institute of Ontario
Vineland Station, Ontario, LOR 2EO
Tel: (416) 562-4141.

1 Eady FC - PhD *Director*
2 Lauro EM - PEng *Agricultural Engineer*
3 Wanner JA - MLS *Librarian*

1398 Horticultural Experiment Station
Vineland Station, Ontario, LOR 2EO
Tel: (416) 562-4141.

1 Ingratta FJ - PhD *Chief Research Scientist*
2 Blom TJ - PhD *greenhouse floriculture*
3 Chong C - PhD *ornamental plants*
4 Cline RA - PhD *plant nutrition*
5 Fisher KH - MSc *viticulture*
6 Miles NW - PhD *peaches*
7 Rinker DL - PhD *mushrooms*

NEWFOUNDLAND

1 Tehrane G - PhD *stone fruit and pears*

PRINCE EDWARD ISLAND

1399 Horticultural Products Laboratory
Vineland Station, Ontario, LOR 2EO
Tel: (416) 562-4141.

1 Chudyk RV - PhD *Chief Research Scientist*
2 Chu CL - PhD *post harvest physiology*
3 Edinger WD - PhD *oenology*

NOVA SCOTIA

1 Fuleki T - PhD *biochemistry*
2 Gray VP - BA *food technology and methodology*
3 Smith RB - PhD *post harvest physiology and storage*
4 Wang SL - PhD *food science, microbiology*

1400 Muck Research Station
Kettleby, Ontario, LOG 1JO
Tel: (416) 775-3783.

1 Valk M - BSA *Director* vegetable production on organic soils

1401 Horticultural Experimental Station
Box 587, Simcoe, Ontario, LOR 2EO
Tel: (519) 426-7120.

1 Loughton A - MSc *Director* asparagus, cole crops, new crops
2 Brammall RA - PhD *vegetable breeding and pathology*
3 Dale A - PhD *berry fruits*
4 Elfving DC - PhD *apples*
5 McKeown AW - PhD *processing tomatoes, potatoes*
6 O'Sullivan J - PhD *vegetable herbicides, peppers, pickling cucumbers*

1402 Veterinary Laboratory Services Branch
Guelph Agriculture Centre, Box 1030, Guelph, Ontario, N1H 6N1
Tel: (519) 823-5700.

1 Henry JNR - DVM,MSc *Director*

1403 Brighton Laboratory
Brighton, Ontario, KOK 1HO
Tel: (613) 475-1632.

1 Galt DE - DVM,DipPath *Lab. Head*

1404 Guelph Laboratory
Guelph, Ontario, N1H 6R8
Tel: (519) 823-8800; ext. 4500.

1 Thomson GW - DVM,MSc *Lab. Head*

1405 Huron Park Laboratory
Huron Park, Ontario, NOM 1YO
Tel: (519) 228-6691.

1 Josephson GK - DVM,DipPATH *Lab. Head*

1406 Kemptville Laboratory
Kemptville, Ontario, KOG 1JO
Tel: (613) 258-3411.

1 Lusis PI - DVM,MSc *Lab. Head*

1407 New Liskeard Laboratory
New Liskeard, Ontario, POJ 1PO
Tel: (705) 647-6701.

1 Jolette J - DVM,MSc *Lab. Head*

1408 Ridgetown Laboratory
Ridgetown, Ontario, NOP 2CO
Tel: (519) 674-5456.

1 Harden FJ - DVM,DUPH *Lab. Head*

1409 Agricultural Laboratory Services (Pesticide Laboratory)
c/o University of Guelph, Guelph, Ontario, N1G 2W1
Tel: (519) 823-8800; ext. 4825.

1 Frank R - BSc,MSc,PhD,DTA *Director*
2 Braun HE - BSc,MSc *Chief Analytical Chemist*
3 Clegg SB - BSc *Analytical Chemist* organic analysis
4 Rasper Mrs J - BSc,MSc *Analytical Chemist*
5 Ripley BD - BSc *Analytical Chemist* organic analysis
6 Stonefield KI - BA,BSc,MSc *Analytical Chemist* inorganic analysis

1410 Centralia College of Agricultural Technology
Huron Park, Ontario, NOM 1YO
Tel: (519) 228-6691.

1 Jamieson JD - BSA,MSc *Principal*

1411 Agronomy Section
Huron Park, Ontario, NOM 1YO
Tel: (519) 228-6691.

1 O'Toole JJ - BScAgr,MSc *Head of Agronomy Research*
2 Forrest RE - BScAgr,MSc *Agronomy Research*

1412 Business Management Section
Huron Park, Ontario, NOM 1YO
Tel: (519) 228-6691.

1 Stephens JR - BSc,MSc *Head of Business Research*

1413 Agricultural Engineering Services
Huron Park, Ontario, NOM 1YO
Tel: (519) 228-6691.

1 Weeden JK - BScAgr,MSc *Head of Agricultural Engineering Research*

1414 Kemptville College of Agricultural Technology
Kemptville, Ontario, KOG 1JO
Tel: 613/258-3411.

1 Curtis JD - BSA,MSc,PAg *Principal*
2 Clarke SP - BScEng,PEng engineering, manure gas
3 Curnoe WE - BScAgr,MSc *Head of Plant Science Section* agronomy, horticulture, soils
4 Cuthbertson HE - BScAgEng,PEng engineering, drainage

5 Drevjany LA - PhDNutr calf nutrition and management
6 Gillespie JR - BE,MSc,PEng *Head of Agricultural Engineering Services Section* engineering
7 Hoekstra GJ - BScBio,MSc agronomy, corn and cereals
8 Hooper GS - BScAgr,MSc *Head of Animal Science Section* beef cattle nutrition and management
9 Johnson PL - BScHEc,MSc food sensory evaluation
10 Langenberg WJ - BScAgr,MSc vegetable crops, maple syrup
11 MacDonald DW - BScAgr agronomy, weed control of field crops
12 Madill JE - BScAgr,MSc agronomy, forages, soybeans, canola
13 McEwen PL - BScAgr,MSc dairy cattle breeding; swine
14 Morrison RL - BScAgr,MSc farm management
15 Morton JG - BScAgr fruit crops, entomology
16 Mullen KW - BScEng,PEng engineering
17 Pickard MG - BA,MEd *Head of Business Management Section* farm management
18 Plue PS - BScEng,PEng engineering, grain drying
19 Reasbeck LM - BScHEc,MEd,RPDt *Head, Food Service and Technology Section* food sensory evaluation, recipe development
20 Sanderson JL - BAAHEc,RPDt food sensory evaluation
21 Stone RP - BScEng,PEng engineering
22 Toombs MR - BEng,MSc,PEng engineering, computers
23 Wheeler EE - BScAgr,MSc dairy cattle nutrition, management, breeding

1415 New Liskeard College of Agricultural Technology
Box G, New Liskeard, Ontario, POJ 1PO

1 Allen WR - BScAgr,MSc,PAg *Principal*

1416 Agricultural Engineering

1 Garland GA - BScEng,MScEng,PEng *Head of Section*
2 Kirik M - BScAgr,BAppSc,PEng

1417 Animal Science

1 Callaghan BD - BSc,MSc,PhD *Head of Section* reproductive physiology
2 Gumprich PS - BScAgr dairy cattle studies
3 MacLaren LA - BScAgr,MSc sheep breeding and management, pasture arrangement

1418 Agronomy

1 Skepasts AV - BScAgr,MSc,PAg *Head of Section* cultivar testing and management studies
2 Hughes BR - BScAgr,MSc horticultural crops, small fruit studies
3 Rowsell JG - BScAgr,MSc soil management

1419 Farm Business Management

1 Hope DG - BScAgr,MSc *Head of Section* marketing strategies and farm management

1420 Ridgetown College of Agricultural Technology
Ridgetown, Ontario, NOP 2CO
Tel: (519) 6740 5456.

1 Taylor DW - BScAgr,MSc *Principal*
2 Ablett GR - BSc,MSc crops
3 Baldwin CS - BSA,MSA,PhD soils
4 Beattie D - BSc,MS economics
5 Brimner JH - BSA,MSc,PhD chemistry
6 Brown RH - BSA,MSc biology
7 Doidge BR - BA,MA economics
8 Johnston RW - BSA,MSc soils
9 McLaren AD - BSA,MSA crops
10 Meuhmer JK - FE,MSc horticulture
11 Morris JR - BSA,MSc livestock
12 Pitblado RE - BScAgr,MSc,PhD biology
13 Ross RW - BScAgr,MBA economics
14 Shaw JE - BScAgr,MSc biology
15 Smid AE - BScAgr,MSc,PhD crops
16 Sojak M - BSA,BASc engineering
17 Stevenson CK - BSA,MSc soils
18 Underwood JA - BSA,BASc,MSc engineering
19 Vander Wielen ALM - BScAgr,MSc livestock

1421 Collége de Technologie Agricole et Alimentaire d'Alfred
Box 580, Alfred, Ontario, KOB 1AO
Tel: 613-679-2218. Cables: 613-679-2250.

1 Paulhus MJ - BScAgr *Principal*

1422 Animal Science

1 Goubau C - BScAgr *Head of Section* animal nutrition

1423 Business Management

1 Demers JA - BScAgr *Head of Section* rural socio-economics

1424 Agricultural Engineering and Plant Science

1 Weil C - MScAgrEng *Head of Section* soil and water
2 Choiniére Y - BScAgrEng ventilation livestock buildings
3 Daoust G - BScAgr field crops
4 DeMaisonneuve C - MScAgr horticultural crops

1425 Library and Communications

1 Frappier R - MEd *Head of Section* information services

1426 Ontario Ministry of Natural Resources

Queen's Park, Toronto, Ontario, M7A 1W3
Tel: (416) 965-2704.

1 Tough G *Deputy Minister*

1427 Forest Resources Group

1 Goodman JF - RPF

1428 Ontario Tree Improvement and Forest Biomass Institute

Headquarters, Maple, Ontario, LOJ 1EO
Tel: (416) 832-2761.

1429 Tree Improvement and Biomass Section

1 Buchert G - BScFor,MScFor,PhD *Principal Scientist* physiological genetics

1430 Forest Biomass Unit

1 Anderson HW - BScFor,MScFor *Leader* hardwood silviculture, biomass (clonal selection)
2 Jaciw P - BScFor silviculture upland hardwoods
3 Papadopol C - BSc,MSc,PhD forest ecophysiology

1431 Tree Improvement Unit

1 Buchert G *Leader*
2 Ho R - BSc,MSc,PhD tissue culture, flower induction, micro propagation
3 Raj Y - BSc,MSc,PhD tissue culture
4 Skeates DA - BScFor,MScFor tree seeds

1432 Forest Ecology and Silviculture Section

1433 Silviculture and Tree Biology Unit

1 Glerum C - BScFor,MScFor,PhD *Leader* tree physiology (frost hardiness, dormancy)
2 Colombo S - BScFor,MScFor tree physiology (planting stock quality)
3 Koppenaal RS - BAMScFor planting stock quality
4 Menes PA - BScFor,MScFor planting stock quality
5 Odlum KD - BSc planting stock quality
6 Stroempl G - DiplForstwirt hardwood silviculture

1434 Ontario Tree Improvement and Forest Biomass Institute Field Units

1435 At Thunder Bay

Box 2960, Thunder Bay, Ontario, P7B 5G5
Tel: (807) 767-1607.

1 McClain KM - BScFor,MScFor,PhD *Leader* tree physiology
2 Forslund R - BM statistics
3 MacDonald GB - BScFor,MScFor boreal mixed wood silviculture
4 Morris DM - BScFor,MScFor boreal mixed wood silviculture
5 Weingartner DH - BScFor,MScFor aspen silviculture

1436 At Sault Ste. Marie

Box 490, Sault Ste. Marie, Ontario, P6A 5M7
Tel: (705) 949-9461.

1 Balakrishnan N - BScFor,MScFor,PhD statistician
2 Gordon AG - BScFor,PhD *Leader* spruce genecology, nutrient cycling productivity

1437 At Midhurst

Midhurst, Ontario, LOL IXO
Tel: (705)737-5510.

1 Hutchison RE *Station Manager*

1438 Nova Scotia Research Foundation Corporation

1439 Biology Division

P.O. Box 790, Dartmouth, Nova Scotia, B2Y 3Z7
Tel: (902)424-8670. Telex: 019-22719.

1 Hellenbrand K - PhD *Director* pesticides
2 Boyle CD - MSc mycorrhizal nitrogen fixation
3 Reade AE - PhD fermentation, conversion of wood wastes to animal feed

UNIVERSITIES

1440 The University of British Columbia

1441 Faculty of Forestry

270-2357 Main Mall, Vancouver, B.C., V6T 1W5
Tel: 604-228-2727.

1 Kennedy RW - BS,MF,PhD,FIAWS,FIWSc *Professor and Dean of Faculty* wood properties and uses
2 Kozak A - BSF,MF,PhD *Professor and Associate Dean of Faculty* biometrics
3 Munro DD - BSF,MF,PhD *Professor and Director of University Research Forests* growth and yield

1442 Department of Forest Resources Management

1 Smith JHG - BSF,MF,PhD,RPF *Professor and Head* forest management
2 Ballard TM - BSF,MF,PhD *Professor* forest soils
3 Chambers AD - BSF,MF,PhD,RPF *Associate Professor* forest land management
4 Dooling PJ - BA,BPE,MA,PhD *Associate Professor* recreation and forest landscape analysis
5 Gaudreau R - BLA,MScA *Assistant Professor* forest landscape management
6 Golding DL - BSc,MS,PhD,RPF *Associate Professor* watershed management
7 Haley D - BSc,MF,PhD,RPF *Associate Professor* forest economics
8 Marshall PL - BScF,MScF,PhD *Assistant Professor* mensuration
9 Murtha PA - BScF,MS,PhD,MCASI *Professor* photogrammetry and remote sensing
10 Pearse PH - BSF,MA,PhD,RPF *Professor* economics
11 Reed FLC - BA,MA *NSERC/Indutrial Professor of Forest Policy*
12 Thirgood JV - BSc,MF,PhD,RPF *Professor* forest policy and land reclamation
13 Woodham RJ - BA,MS,PhD *Associate Professor* photogrammetry, remote sensing and computer science

1443 Department of Forest Sciences

1 Lavender DP - BSc,MSc,PhD *Professor and Head* regeneration and silviculture
2 Bunnell FL - BSF,PhD *Professor* wildlife population dynamics
3 Feller MC - BSc,MSc,PhD *Assistant Professor* fire ecology
4 Kimmins JP - BSc,MS,MPhil,PhD *Professor* forest ecology and nutrient cycling
5 Klinka K - ForEng,PhD,RPF *B.C. Forest Service Adjunct Professor of Forest Ecology (full time)*
6 Lester DT - BSc,MF,PhD *Poldi Bentley/NSERC Industrial Professor of Forest Genetics and Tree Improvement*
7 McLean JA - MSc,PhD *Professor* forest entomology
8 Northcote TG - MA,PhD *Professor* stream ecology
9 Sziklai O - DiplForEng,MF,PhD,RPF *Professor* forest genetics and tree seed
10 van der Kamp BJ - BSF,PhD *Associate Professor* forest pathology
11 Weetman GF - BScF,MF,PhD,RPF *Professor* silviculture
12 Worrall JG - BSc,BSF,MF,MPhil,PhD *Associate Professor* tree physiology

1444 Department of Harvesting and Wood Science

1 Barrett JD - BASc,PhD,FIAWS *Professor and Head* wood engineering
2 Avramidis S - BSc,MSc,PhD *Assistant Professor* wood/moisture relationships
3 Howard AF - BS,MFS,MS,PhD *Assistant Professor* operations research
4 Paszner L - BSF,MF,PhD *Professor* wood chemistry and pulping
5 Ruddick JNR *NSERC Industrial Professor of Wood Preservation* wood preservation
6 Tait DEN - BSc,MSc,PhD *Assistant Professor* simulation and modelling
7 Wilson JW - MS,PhD *Professor and Director of Graduate Studies* forest products utilization
8 Young GG - BASc,MASc,PEng *Associate Professor* forest harvesting systems

1445 Faculty of Agricultural Sciences

Room No. 248, 2357 Main Mall, Vancouver, B.C., V6T 2A2
Tel: 604-228-2620.

1 Richards JF - BSA,MSc,PhD,PAgr *Dean of the Faculty*
2 Lowe LE - BA,MA,MSc,PhD *Associate Dean*

1446 Department of Agricultural Economics

1 Barichello RR - BSA,MA,PhD *Associate Professor* rural development
2 Graham JD - BSc,MSc,PhD *Associate Professor and Head* production economics
3 Hazeldine T - BA,MA,PhD *Associate Professor* marketing and regional economics
4 Kennedy G - BA,MSc,PhD,PAgr *Associate Professor* agricultural policy
5 Short CC - BScAgr,MSc,PhD *Assistant Professor* farm management, natural resource economics

1447 Department of Bio-Resource Engineering

1 Lo VK - BS,MSc,PhD *Professor* food process engineering
2 Chieng ST - BS,MSc,PhD *Associate Professor* soil and water engineering
3 Staley LM - BASc,MSc,PEng *Professor* structures and environment
4 Zahradnik JW - BS,MS,PhD,PEng *Professor* aquaculture

1448 Department of Animal Science

1 Blair R - BSc,PhD,DSc,PAg *Professor and Head* animal nutrition
2 Beames R MCD - BAgrSc,MAgrSc,PhD,PAgr *Associate Professor* nutrition
3 Cheng KM - BS,MS,PhD *Associate Professor* genetics; behaviour; reproduction
4 Fitzsimmons RC - BS,MS,PhD *Associate Professor* embryology
5 Hart LE nutritional biochemistry, poultry
6 Krishnamurti CP - BVSc,MVSc,PhD,PAgr *Professor* physiology
7 March BE - BA,MSA,FAIC,FRSC,FPSA,PAgr *Professor (Part time)* physiology; nutrition
8 Owen BD - BScAgr,MSc,PhD,PAgr *Professor* nutrition and physiology
9 Peterson RG - BS,MS,PhD *Associate Professor* breeding and genetics
10 Shackleton DM - BSc,MSc,PhD *Associate Professor* behaviour and ecology
11 Shelford JA - BSA,MSc,PhD,PAgr *Associate Professor* nutrition and physiology
12 Tait RM - BSc,PhD,PAgr *Associate Professor* nutrition and production

1449 Department of Food Science

1 Powrie WD - BSc,MA,PhD,PAgr *Professor and Head* food chemistry
2 Kitts DD - BScAgr,MSc,PhD,NSERC *University Research Fellow* toxicology, animal metabolism
3 Nakai S - BSc,PhD *Professor* food chemistry
4 Richards JF - BSA,MSc,PhD,PAgr *Professor and Dean of the Faculty* food chemistry
5 Skura BJ - BSc,MSc,PhD *Associate Professor* food microbiology
6 Townsley PM - RSA,MS,PhD *Professor* food fermentation
7 Tung MA - BSA,MSA,PhD,PAgr *Professor* food process science
8 Vanderstoep J - BSA,MSA,PhD,PAgr *Associate Professor* nutritional chemistry and food toxicology

1450 Department of Plant Science

1 Runeckles VC - BSc,PhD,PAgr,DiplImpColl,FRSA,MBCSLA *Professor and Head* stress physiology, air pollution
2 Copeman RJ - BSc,PhD *Associate Professor* plant pathology and bacteriology
3 Eaton GW - BSA,PhD,PAgr,FRHS *Professor* pomology, biometrics
4 Haggerstone B landscape architecture
5 Holl FB - BSc,MSc,PhD *Associate Professor* agronomy, genetics
6 Isman MB - BSc,MSc,PhD *Assistant Professor* entomology
7 Jolliffe PA - BSc,PhD *Associate Professor* environmental physiology
8 Mooney PF - BMUS,MLA *Assistant Professor* landscape architecture
9 Myers JH - BSc,MS,PhD *Associate Professor* entomology, biological control
10 Norton C - BSc,MSc,PhD *Assistant Professor* plant and seed physiology, micropropagation, tissue culture
11 Paterson DD - BSc,MLA *Assistant Professor and Director of BLA Program* landscape architecture
12 Pitt MD - BS,MS,PhD *Associate Professor* range ecology and management
13 Quayle M - BLA,MLA *Assistant Professor* landscape architecture
14 Shaw M - BSc,MSc,PhD,DSc,FAPS,FLS,FRSC,PAgr *University Professor* host - parasite relationships
15 Upadhyaya M - BSc,MSc,MA,PhD *Assistant Professor* weed science

1451 Department of Soil Science

1 Lavkulich LM - BSc,MSc,PhD,PAgr *Professor and Head* land use; resource management; classification and genesis
2 Ballard TM - BSF,MF,PhD *Professor* forest soils
3 Berch SM - BSc,MSc,PhD,NSERC *University Research Fellow* soil biology, mycorrhizae

4 Black TA - BSA,MSc,PhD *Professor* biometeorology; soil physics and hydrology
5 Bomke AA - BSc,MS,PhD,PAgr *Assistant Professor* soil fertility
6 deVries J - BSc,MSA,PhD *Associate Professor* soil physics and hydrology
7 Lowe LE - BA,MA,MSc,PhD *Professor and Associate Dean* soil chemistry and biochemistry
8 Murtha PA - BScFor,MS,PhD *Professor* photogrammetry and remote sensing
9 Novak M - BEng,MSc,PhD *Assistant Professor* soil and water conservation, soil physics, microclimatology
10 Schreier HP - BA,MSc,PhD *Associate Professor* land classification; resource evaluation

1452 Simon Fraser University

1453 Department of Biological Sciences

Burnaby, British Columbia, V5A 1S6
Tel: 604 291-4475. Telex: 043-54614. FAX: 291-4581.

1 Srivastava LM - BSc,MSc,PhD *Professor, Chairman of Department* structure and physiology of higher plants; hormonal regulation of plant growth and development; electron microscopy
2 Beirne BP - BSc,MA,MSc,PhD *Professor Emeritus* principles of pest management; ecology of crop insects; biological and integrated controls
3 Belton P - BSc,PhD,ARCS *Associate Professor* sensory physiology of pests; attraction and repulsion of mosquitoes and other biting and stinging insects; the use of sound by pests and in pest management
4 Borden JH - BSc,MSc,PhD *Professor* entomology; chemical ecology and pest management of forest and agricultural insects
5 Brooke RC - BSF,MF,PhD *Associate Professor* plant ecology, microenvironmental relationships; physiological ecology
6 Finlayson T - BA *Professor Emerita* insect systematics especially of parasitoid larvae; biological control
7 Fisher FJF - BSc,MSc,PhD *Professor* plant ecophysiology; biosystematics; evolutionary ecology
8 Lister GR - BSc,PhD *Assistant Professor* plant physiology; physiological ecology of forest trees
9 Mackauer JPM - DrPhilNat *Professor, Director Centre for Pest Management* biological control; insect parasitism, especially of aphids; genetics and evolution of host-parasite systems; biosystematics of aphid parasites
10 Mathewes RW - BSc,PhD *Associate Professor* palynology and paleoecology, pollen morphology in relation to plant taxonomy
11 McClaren M - BEd,PhD *Associate Member* genetics of fungi, particularly of higher Basidiomycetes; fungal ecology
12 Rahe JE - BSA,PhD *Professor* plant pathology; disease control, biochemistry and physiology of host-parasite interaction
13 Roitberg BD - BSc,MSc,PhD *Assistant Professor* behavioural ecology and population dynamics; insect dispersal; host-parasite interactions; fruit parasite systems
14 Vidaver WE - BA,PhD *Professor* photosynthetic energy conversion; tissue culture micropropagation, forest tree physiology
15 Webster JM - BSc,PhD,DSc,ARCS,DIC,FIBiol *Professor* plant and insect nematology; parasitology
16 Winston ML - BA,MSc,PhD *Associate Professor* entomology, social insects; apiculture; bee morphology and systematics

1454 University of Alberta

Edmonton, Alberta, T6G 2P5
Tel: 403 432-4931.

1455 Faculty of Agriculture and Forestry

1 Hawkins MH - PhD *Associate Dean* external relations
2 Murphy PJ - PhD *Associate Dean* forestry
3 Pawiuk S - PhD *Associate Dean* research
4 Tyrchniewicz E - PhD *Dean*

1456 Department of Agricultural Engineering

1 Leonard J - PhD *Professor and Chairman* instrumentation, microprocessors
2 Anderson AW - MSc *Associate Professor* operations research
3 Bach L - PhD *Assistant Professor* wood science and forest products
4 Chanasyk DS - PhD *Associate Professor* soils physics
5 Domier KW - PhD *Professor* power and machinery
6 Feddes JJR - MSc *Assistant Professor* structures and environment
7 Harrison HP - PhD *Professor* farm machinery
8 Kosicki K - PhD *Assistant Professor* forest engineering
9 LeMaguer M - PhD *Professor* food engineering
10 Lindwall CW - PhD *Associate Academic Staff* farm machinery
11 McQuitty JB - MSc *Professor* farm buildings
12 Micko M - PhD *Professor* wood science
13 Rapp E - MSc *Professor* soil and water

1457 Department of Animal Science

1 Price MA - PhD *Professor and Chairman* livestock growth, meat production

2 Aherne FX - PhD *Professor* swine production
3 Baracos VE - PhD *Assistant Professor (NSERC Fellow)* protein metabolism
4 Berg RT - PhD *Professor* animal genetics
5 Christopherson RJ - PhD *Professor* animal physiology
6 Church TL - DVM *Honorary Adjunct Professor* animal preventive medicine
7 Fredeen HT - PhD *Honorary Adjunct Professor* animal genetics
8 Hardin RT - PhD *Professor* biometrics and poultry genetics
9 Hudson RJ - PhD *Professor* wildlife management and productivity
10 Kennelly JJ - PhD *Professor* dairy production
11 Makerechian MH - PhD *Professor* animal genetics
12 Mathison GW - PhD *Professor* animal nutrition, beef cattle
13 Robinson FE - PhD *Assistant Professor* poultry production
14 Salmon RK - PhD *Assistant Professor* animal biotechnology
15 Sauer WC - PhD *Associate Professor* swine nutrition
16 Sim J - PhD *Associate Professor* poultry technology
17 Thompson JR - PhD *Professor* animal biochemistry
18 Young BA - PhD *Professor* animal physiology

1458 Department of Entomology

1 Mitchell BK - PhD *Professor and Chairman* physiology, behaviour
2 Ball GE - PhD *Professor* systematics, morphology, evolution, zoogeography
3 Craig DAC - PhD *Professor* morphology, histology
4 Evans WG - PhD *Professor* ecology, behaviour
5 Gooding RH - ScD *Professor* insect biochemistry, medical entomology
6 Heming BS - PhD *Professor* morphology, development
7 Spence JR - PhD *Assistant Professor* ecology systematics

1459 Department of Food Science

1 Wolfe FH - PhD *Professor and Chairman* food chemistry
2 Hadziyev D - PhD *Professor* food chemistry
3 Jackson H - PhD *Professor* food and industrial microbiology
4 Jelen P - PhD *Professor* food processing
5 LeMaguer M - PhD *Professor* food engineering
6 Ooraikul B - PhD *Professor* food processing
7 Ozimek L - PhD *Associate Professor* Dairy Technology Chair
8 Palcic M - PhD *Assistant Professor (AHFMR Scholar)* enzymology
9 Phillipchuk GE - MSc food processing/engineering
10 Schroder DJ - PhD food processing, food microbiology
11 Sporns P - PhD *Associate Professor* food chemistry
12 Stiles ME - PhD *Professor* food microbiology, joint appointment with faculty of home economics

1460 Department of Forest Science

1 Beck JA - PhD *Professor and Chairman* forest resources management and economics
2 Achuff P - PhD *Adjunct Academic Staff* vegetation and conservation ecology
3 Bach L - PhD *Adjunct Academic Staff* wood science and forest products
4 Barney RJ - PhD *Adjunct Academic Staff* vegetation and conservation ecology
5 Butler JR - PhD *Professor* wildland recreation, management and planning, environmental education
6 Cheliak W - PhD forest genetics
7 Clark JD - BScF *Professor* industrial forest management
8 Dancik BP - PhD *Professor* dendrology and forest tree improvement
9 Drew J - PhD *Adjunct Academic Staff* reforestation, silviculture
10 Heidt JD - PhD *Associate Professor* outdoor recreation and resource economics
11 Hellum AK - PhD *Professor* silviculture
12 Higginbotham KO - PhD *Associate Professor* forest ecology and tree physiology
13 Hiratsuka Y - PhD forest pathology
14 Holroyd G - PhD *Adjunct Academic Staff* endangered species
15 Kosicki K - PhD *Assistant Professor* forest engineering
16 Lieffers VJ - PhD *Assistant Professor (NSERC Fellow)* forest ecology
17 Murphy PJ - PhD *Professor* forest recreation, forest policy and administration, forest fire control and use
18 Navratil S - PhD *Adjunct Academic Staff* forest soils, forest pathology and reforestation
19 Pharis R - PhD *Adjunct Academic Staff* forest genetics, physiology
20 Pickford SG - PhD *Adjunct Academic Staff* forest fire science
21 Rothwell RL - PhD *Professor* forest hydrology
22 Scotter GW - PhD *Adjunct Academic Staff* forest wildlife
23 Swanson RH *Adjunct Academic Staff* forest hydrology
24 Thorpe TA - PhD *Adjunct Academic Staff* forest genetics, biotechnology
25 Titus SJ - PhD *Associate Professor* resource measurements and mensuration
26 Wein RW - PhD *Professor and Director, Boreal Institute of Northern Studies*
27 Woodard PM - PhD *Associate Professor* forest fire management
28 Yeh FC - PhD *Associate Professor* forest genetics

1461 Department of Plant Science

1 Briggs KG - PhD *Professor and Chairman* plant breeding

2 Bailey AW - PhD *Professor* range ecology and management
3 Blenis PV - PhD *Assistant Professor* plant diseases
4 Flanagan A - PhD *Assistant Professor* oilseed physiology and biotechnology
5 Hiruki C - PhD *Professor* virology
6 Horak A - PhD *Assistant Professor* plant physiology
7 King JR - PhD *Assistant Professor* forages
8 Knowles NR - PhD *Assistant Professor* vegetable crops
9 Langridge WHR - PhD *Professor* plant genetics
10 Spencer MS - PhD *University Professor* plant biochemistry
11 Szalay AA - PhD *Professor* plant biotechnology, molecular genetics
12 Tewari JP - PhD *Professor* plant pathology
13 Vanden Born WH - PhD *Professor* herbicides and weed physiology
14 Walton PD - PhD *Professor* plant breeding

1462 Department of Rural Economy

1 Leroli ML - PhD *Professor* farm management, agricultural policy
2 Phillips WE - PhD *Professor and Chairman* econometrics, natural resources, economics
3 Adamowicz W - MSc *Assistant Professor* econometrics
4 Anderson AW - MSc *Associate Professor* operations research
5 Apedaile LP - PhD *Professor* econometrics, rural development
6 Bauer L - PhD *Professor* farm management and production econometrics
7 Chase-Wilde L - MSc,PhD *Assistant Professor* agricultural and resource policy
8 Constantino L - PhD *Assistant Profesor* forest economics
9 Gill DS - PhD *Professor* rural sociology, agricultural extension
10 Hawkins MH - PhD *Professor* marketing, business management
11 Murney GA - PhD *Honorary Adjunct Professor* finance economics
12 Murri DG - MS *Associate Professor* rural sociology
13 Novak FS - MSc *Assistant Professor* production economics
14 Percy HB - PhD *Honorary Adjunct Professor* economic development, forest economics
15 Veeman MM - PhD *Professor* marketing
16 Veeman TS - PhD *Professor* international development, natural resources economics

1463 Department of Soil Science

1 McGill WB - PhD *Professor and Chairman* soil biochemistry
2 Chanasyk DS - PhD *Associate Professor* soil physics
3 Coen GM - PhD *Associate Academic Staff* soil genesis and classification
4 Crown PH - PhD *Associate Professor* remote sensing, soil/land evaluation
5 Dudas MJ - PhD *Professor* soil mineralogy and chemistry
6 Juma NG - PhD *Assistant Professor* soil microbiology
7 Nyborg MPK - PhD *Professor* soil fertility
8 Pawiuk S - PhD *Professor* soil genesis and classification
9 Pettapiece WW - PhD *Associate Academic Staff* soil genesis and mineralogy
10 Pluth DJ - PhD *Professor* forest soils
11 Robertson JA - PhD *Professor* general agriculture soil fertility and management

1464 University of Saskatchewan

1465 College of Agriculture
Saskatoon, Saskatchewan, S7N OWO
Tel: (306)966-4050.

1 Rennie DA - BSA,PhD *Dean of Agriculture*

1466 Department of Agricultural Economics

1 Furtan WH - BSA,MSc,PhD *Professor and Head of the Department*
2 Brown JA - BSA,MS *Professor*
3 Brown WJ - BSA,MS *Associate Professor*
4 Devine DG - BSA,MSc,PhD *Associate Professor*
5 Fulton ME - BSA,MS,BA,MA,PhD *Assistant Professor*
6 Kulshrestha SN - BSc,MSc,PhD *Professor*
7 Lee GE - BSA,MSc,PhD *Professor*
8 Nicholson RC - BSA,MSc,PhD *Professor*
9 Rosaasen KA - BSA,MSc *Associate Professor*
10 Schmitz A - BSA,MSc,MA,PhD *Professor*
11 Schoney RA - PhD *Research Scientist*
12 Spriggs JD - BAgr,MSc,PhD *Associate Professor*
13 Storey GG - BSA,MSc,MA,PhD *Professor*
14 Van Kooten GC - BSc,MA,PhD *Associate Professor*

1467 Department of Agricultural Engineering

1 Norum DI - BE,MSc,PhD *Professor and Head of Department*
2 Barber EM - BSc,MSc,PhD *Associate Professor*
3 Berg JB - BE *Assistant Professor*
4 Campbell RJ - BSA *Assistant Professor*
5 Dodds RA - DiplAgr *Assistant Professor*
6 Fritzler BA - BSA *Assistant Professor*
7 Gray DM - BSA,MSA,PhD *Professor*
8 Kent GL - BE *Assistant Professor*
9 Milne WG - BSA *Assistant Professor*

10 Moysey EB - BE,MS *Professor*
11 Sebulsky ME - BSA *Assistant Professor*
12 Sohansanj S - BS,MS,PhD *Associate Professor*
13 Strayer RC - BE *Professor*
14 Zoerb GC - BE,MSc,PhD *Professor*

1468 Department of Animal and Poultry Science

1 Christensen DA - BSA,MSc,PhD *Professor and Head of the Department*
2 Bell JM - BSc,MSc,PhD *Professor*
3 Christison GI - BSc,MS,PhD *Professor*
4 Classen HL - BSA,MSc,PhD *Associate Professor*
5 Cohen RDH - BSc,PhD *Associate Professor*
6 Crawford RD - BSA,MS,PhD *Professor*
7 Laarveld B - Ing,MSc,PhD *Associate Professor*
8 Nicholson HH - BSA,MSA,PhD *Professor*
9 Schmutz SM - DVM *Assistant Professor*
10 Thacker PA - BSc,MSc,PhD *Associate Professor*
11 Williams CM - BSA,MSA,PhD *Professor*

1469 Department of Applied Microbiology and Food Science

1 Jones GA - BSc,MSc,PhD *Professor and Head of the Department*
2 Caldwell DE - BS,PhD *Associate Professor*
3 Humbert ES - BSA,MSc,PhD *Professor*
4 Ingledew WM - WM *BSc,PhD Professor*
5 Khachatourians GG - BA,MA,PhD *Professor*
6 Low NH - BSc,PhD *Assistant Professor*
7 McCurdy AR - BS,MS,PhD *Associate Professor*

1470 Department of Crop Science and Plant Ecology

1 Harvey BL - BSA,MSc,PhD *Professor and Head of the Department*
2 Austenson HM - BSA,MSc,PhD *Professor*
3 Baker RJ - BSA,MSc,PhD *Senior Research Scientist*
4 Bhatty RS - BSc,MSc,PhD *Senior Research Scientist*
5 Devine ND *Assistant Professor*
6 Drew BN - BSA,MSc *Professional Research Associate*
7 Foster RK - BSA,MSc *Research Scientist*
8 Fowler DB - BSA,MSc,PhD *Senior Research Scientist*
9 Gusta LV - BSA,MSc,PhD *Senior Research Scientist*
10 Holm FA - BSA,MSc *Associate Professor*
11 Hughes GR - BSc,MS,PhD *Associate Professor*
12 Jana S - BAgr,MSc,PhD *Professor*
13 Knott DR - BSA,MS,PhD *Professor*
14 McHughen AG - BSc,PhD *Professional Research Associate*
15 Redmann RE - BA,MS,PhD *Professor*
16 Ripley EA - BSC,MS *Professor*
17 Romo JT - BSc,MSc,PhD *Assistant Professor*
18 Rossnagel BG - BSA,MSA,MSc,PhD *Research Scientist*
19 Rowland GG - BSA,PhD *Research Scientist*
20 Scoles GJ - BSc,MSc,PhD *Associate Professor*
21 Simpson GM - BASc,MASc,PhD *Professor*
22 Slinkard AE - BS,MS,PhD *Senior Research Scientist*
23 Sosulski FW - BSA,MS,PhD *Professor*

1471 Department of Horticulture Science

1 Stushnoff C - BSA,MSc,PhD *Professor and Head of the Department*
2 Maginnes EA - BSc,MS,PhD *Professor*
3 Waterer DR - BSc,MSc,PhD *Assistant Professor*

1472 Department of Soil Science

(Saskatchewan Institute of Pedology)

1 Stewart JWB - BSc,BAgr,PhD *Professor and Head of the Department*
2 Acton DF - BSA,MSc,PhD *Associate Professor*
3 Anderson DW - BSA,MSc,PhD *Senior Research Scientist*
4 Bettany JR - BSc,PhD *Professor*
5 de Jong E - L1,PhD *Professor*
6 Eilers WD - MSc *Research Scientist*
7 Germida JJ - BS,MS,PhD *Associate Professor*
8 Henry JL - BSA,MSc *Professor*
9 Huang PM - BSc,MSc,PhD *Professor*
10 Kozak LM - BSA,MSc,PhD *Research Scientist*
11 McKercher RB - BA,BSA,MSc,PhD *Professor*
12 Mermut AR - PhD *Research Scientist*
13 Padbury GA - BSA,MSc *Research Scientist*
14 Rennie DA - BSA,PhD *Professor*
15 Rostad HPW - BSA,MSc,PhD *Research Scientist*
16 St Arnaud RJ - BSA,MSc,PhD *Professor*
17 Stonehouse HB - BSA,MSc *Research Scientist*
18 Tiessen H - PhD *Research Scientist*

1473 Members From Other Faculties

1 Barber EM - BSc,MSc,PhD *Associate Professor* agricultural engineering
2 Gerrard CD - BSc,MPhil,PhD *Associate Professor* economics
3 Goldie H - MSc *Professor* microbiology
4 Grant DR - BSA,PhD *Professor* chemistry

5 Gupta IC - BSc,MSc *Associate Professor* physics
6 Hass GW - BSA,BEd,MCEd *Associate Professor* extension
7 Hobin BA - BSc *Extension Specialist*
8 Manns JG - BSA,MSc,PhD *Professor* veterinary physiology sciences
9 Morrall RAA - BA,MA *Professor* biology
10 Norum DI - BE,MSc,PhD *Professor* agricultural engineering
11 Parkinson DJ - BA,MA,PhD *Assistant Professor* english
12 Randell RL - BSA,MSc,PhD *Associate Professor* biology
13 Renaut RW - BSc,PhD *Associate Professor* geological sciences
14 Rhodes CS - BSc,DVM,MSc *Professor* herd medicine and theriogenology
15 Senior JB - BSc,PhD *Associate Professor* chemistry

1474 Western College of Veterinary Medicine

Saskatoon, Saskatchewan, S7N OWO
Tel: (306)966-7448.

1 Hamilton GF - DVM,PhD,DD *Professor and Dean* orthopedic surgery
2 Butler RS - DVM,MCEd *Professor* continuing veterinary education and extension
3 Greenough PR - MRCVS,FRCVS *Professor* veterinary instructional communications, morphology of lameness
4 Latshaw WK - BSc,DVM,MS,PhD *Acting Head and Associate Dean* embryology
5 Manns JG - BSA,MSc,PhD *Professor and Dean, Graduate Studies and Research* endocrinology, ruminant metabolism and oestrus synchronization

1475 Department of Veterinary Anatomy

1 Flood PF - BVSc,MSc,PhD,MRCVS *Professor and Head of Department* reproduction in ungulates
2 Clayton HM - BVMS,PhD *Associate Professor* equine locomotion
3 Hrudka F - MVD,MedVetExot,DrSc *Professor* spermatology
4 Leach DH - BSc,MSc,PhD *Associate Professor* equine locomotion
5 Oliphant LW - BS,MS,PhD *Professor* cellular ultrastructure, ecology of raptors

1476 Department of Veterinary Anaesthesiology, Radiology & Surgery

1 Cribb PH - BSc,MRCVS *Professor and Head of Department* general anaesthesiology
2 Bailey JV - BVSc,MVSc *Associate Professor* large animal surgery
3 Barber SM - DVM *Associate Professor* equine surgery
4 Bauer MS - BSc,DVM *Assistant Professor* small animal surgery
5 Farrow CS - BS,DVM *Professor* diagnostic radiology, thoracic disease
6 Ferguson JG - BSc,DVM,MSc *Associate Professor* food animal surgery
7 Fowler JD - BS,DVM,MVSc *Associate Professor* small animal surgery
8 Fretz PB - VMD *Professor* equine surgery
9 McMurphy RM - BS,DVM *Assistant Professor* general anaesthesiology
10 Pharr JW - BS,DVM,MSc *Professor* diagnostic radiology and ultrasound
11 Presnell KR - DVM,MSc *Professor* orthopedic surgery (small animals)

1477 Department of Veterinary Internal Medicine

1 Radostits OM - DVM,MSc *Professor and Head of Department* bovine medicine and general large animal medicine
2 Meric SM - DVM *Assistant Professor* small animal medicine
3 Naylor JM - BSc,BVSc *Professor* general bovine and equine medicine, particularly metabolic nutrition support, fluid therapy and respiratory disease
4 Panciera P *Associate Professor* small animal medicine
5 Petrie L - BVMS,PhD *Associate Professor* bovine respiratory and calf diseases
6 Post K *Associate Professor* reproduction; dermatology; general small animal medicine
7 Rubin SI - BSc,DVM,MS *Assistant Professor* renal and urinary tract disease; gastroenterology; fungal diseases
8 Smart ME - DVM,PhD *Professor* nutritional problems in large animals
9 Townsend HGG - BSc,DVM,MSc *Associate Professor* large animal internal medicine, particularly equine cardiology and neurology

1478 Department of Herd Medicine and Theriogenology

1 Rhodes CS - BSc,DVM,MSc *Professor and Head of Department* swine herd health
2 Armstrong KR - DVM *Professor* dairy herd health problems and mastitis
3 Barth AD - DVM,MVSc *Professor* diagnostic spermatology, including frozen semen and evaluation
4 Bell RJ - DVM,MVSc *Assistant Professor* equine medicine and equine theriogenology
5 Bristol FM - BVSc,MSc *Professor* theriogenology, particularly equine
6 Cates EF - BSc,DVM,PhD *Professor* bovine reproduction (male)
7 Gudmundson J - DVM,MVSc *Associate Professor* food animal practice
8 Haigh J - BVMS,MSc *Professor* exotic animals, particularly reproduction in wild ungulates
9 Janzen ED - BA,DVM,MVS *Professor* field investigation of beef cattle disease
10 Mapletoft RJ - DVM,MS,PhD *Professor* general bovine reproduction, particularly repeater cows, embryo transfer and genetics defects

11 Mechor G - BSc,DVM *Assistant Professor* general large animal medicine
12 Ribble CS *Assistant Professor* field investigation of beef cattle diseases
13 Smart JN - BSA,DVM *Associate Professor* dairy herd health

1479 Department of Veterinary Microbiology

1 Acres SD - DVM,MPVM,PhD *Professor* epidemiology
2 Allen JR - BSc,MA,VetMB,MVSc,PhD *Professor* parasitology
3 Babiuk LA - BSA,MSc,PhD *Professor* immunology and biotechnology
4 Chirino-Treso JM *Assistant Professor* bacteriology
5 Durham PJK - BVSc,MVSc,PhD *Associate Professor* clinical virology
6 Haines DM - DVM,MPh *Associate Professor* immunology
7 Iversen JO - BSc,DVM,MPH,PhD *Professor* epidemiology
8 Misra V - BSc,PhD *Associate Professor* virology
9 Polley LR - BVSc,PhD,MRCVS *Associate Professor* parasitology
10 Saunders JR - DVM,DVPH,PhD *Professor and Head of Department* clinical microbiology
11 Tabel H - DVM,MSc,PhD *Associate Professor* immunology

1480 Department of Veterinary Pathology

1 Doige CE - DVM,PhD *Professor, Head of Department and Associate Dean (Research)* anatomic pathology; skeletal diseases and hepatobiliary disease
2 Jackson ML - DVM,MVetSc *Associate Professor* clinical pathology
3 Leighton FA - DVM,PhD *Associate Professor* anatomic pathology: pathology of diseases caused by toxins; wildlife diseases
4 Mills JHL - BSA,DVM,MS,PhD *Professor* anatomic pathology; skin and neoplasia
5 Olfert ED - BA,DVM,MSc *Associate Professor* laboratory animal medicine
6 Riddell C - BSc,DVM,MSc,PhD *Professor* avian pathology
7 Rosseaux CG - BVSc,PhD *Assistant Professor* anatomic pathology, toxicopathology
8 Schiefer HB - DVM,PhD *Professor and Director, Toxicology Research Centre* anatomic pathology and toxicology
9 Searcy GP - DVM,MSc,PhD *Professor* clinical pathology particularly haematology and hemopoietic pathology
10 Wobeser GA - BSA,MSc,DVM,PhD *Professor* anatomic pathology: wildlife and fish diseases

1481 Department of Veterinary Physiological Sciences

1 Sisodia CS - BVSc,MS,PhD,ABVT *Professor and Head of Department* toxicology and pharmacology
2 Blakley BR - BSc,DVM,MSc,PhD *Professor* toxicology
3 Chaplin RK - BS,MS,PhD *Associate Professor* ruminant gastrointestinal physiology
4 Crichlow EC - BSc,MSc,PhD *Associate Professor* neurophysiology
5 Forsyth GW - BSA,MSc,PhD *Professor* biochemistry of E. *coli* enterotoxins
6 Gupta VS - BSc,MSc,PhD *Professor* pharmacology, anti-viral chemotherapy
7 Hamilton DL - BSc,DVM,PhD *Professor* gastrointestinal physiology
8 Papich M - DVM,MS *Assistant Professor* clinical pharmacology
9 Rawlings NC - BSc,MS,PhD *Associate Professor* reproduction and endocrinology
10 Roe WE - BSc,DVM,MS,PhD *Professor* gastrointestinal physiology

1482 University of Manitoba

1483 Faculty of Agriculture

300 Agriculture Bldg., Winnipeg, Manitoba, R3T 2N2
Tel: (204) 474-9380.

1 Flaten D *Director, School of Agriculture*
2 McGinnis RC *Dean of Faculty*
3 Meadows DJ *Communications and Continuing Education Officer*
4 Naimark A *President of the University*

1484 Department of Agricultural Economics and Farm Management

1 MacMillan Prof. JA - PhD *Professor and Head* resource economics, agribusiness
2 Arthur Prof. LM - PhD *Professor* resource economics and environmental economics
3 Beaton NJ - PhD *Associate Professor* marketing, price analysis
4 Bjarnason HF - PhD *Adjunct Professor (with Department of External Affairs (Ottawa))* marketing
5 Boyd M *Assistant Professor* marketing, price analysis
6 Carter Prof. CA - PhD *Professor Adjunct* trade, grain marketing
7 Coyle B *Assistant Professor* international trade, econometrics, production economics
8 Framingham Prof. CF - PhD *Professor* farm management, production economics, agricultural law
9 Gilson Prof. JC - PhD *Professor* policy, agricultural finance
10 Heads Prof. J - PhD *Professor* transportation, regional development
11 Jeffrey SR *Assistant Professor* farm management, production economics
12 Josephson RM - MSc *Associate Professor* farm management

13 Kraft Prof. DF - PhD *Professor and Associate Head* production economics, resource economics
14 Loyns Prof. RMA - PhD *Professor* marketing, policy, consumer economics
15 Prentice BE - PhD *Assistant Professor* agribusiness, marketing, transportation
16 Sabourin SR - MSc *Research Associate* production economics
17 Sinclair Prof. S - PhD *Professor Emeritus* policy
18 Tangri Prof. OP - PhD *Professor* marketing, economic development, international trade
19 Tyrchniewicz Prof. EW - PhD *Professor* transportation, policy, economic development
20 Wilson AG - PhD *Professor* transportation, grain marketing
21 Yeh Prof. MH - PhD *Professor* econometrics, production economics

1485 Department of Agricultural Engineering

1 Bulley Prof. NR *Professor and Head* waste management, processing and lighting design
2 Abramson D - PhD *Adjunct Professor* grain storage
3 Britton MG - PhD *Associate Professor* structures and environment
4 Buchanan Prof. LC - BSAE *Professor* power and machinery, materials handling
5 Jayas DS - MSc *Assistant Professor* instrumentation, crop processing
6 Johnson DJ - BSA *Lecturer* power and machinery
7 Kawamoto H - PhD *Research Associate* grain storage
8 Kitson CI - BSc,AE *Research Associate* grain drying
9 Laliberte Prof. GE - PhD *Professor* soil and water
10 Muir Prof. WE - PhD *Professor* grain storage
11 Penkava FF - PhD *Associate Professor* soil and water
12 Philp JD - BSA *Lecturer* power and machinery
13 Putnam JG - DiplAg *Lecturer* power and machinery, farm structures
14 Sinha RN - PhD *Honorary Professor* grain storage
15 Tennenhouse F - BSA *Associate Professor* power and machinery
16 Townsend JS - PhD *Professor* power and machinery
17 Turra D - MSc *Research Associate* grain storage
18 Wold CE - DiplAg *Lecturer* power and machinery

1486 Department of Animal Science

1 Parker Prof. RJ - PhD *Professor and Associate Dean* genetics
2 Boila RJ - PhD *Assistant Professor* nutrition
3 Campbell Prof. LD - PhD *Professor* nutrition (poultry)
4 Crow GH - PhD *Assistant Professor* genetics
5 Devlin Prof. TJ - PhD *Professor* nutrition (ruminant)
6 Dyck GW - PhD *Adjunct Professor* reproductive physiology, (Agriculture Canada)
7 Grinwich DL - PhD *Adjunct Professor* reproduction physiology (Agriculture Canada)
8 Guenter W - PhD *Associate Professor* poultry
9 Ingalls Prof. JR - PhD *Professor* nutrition (ruminant)
10 Kennedy AD - PhD *Assistant Professor* physiology
11 Marquardt Prof. RR - PhD *Professor* biochemistry - toxicology
12 Olthoff JC - PhD *Research Associate* animal genetics
13 Palmer Prof. WM - PhD *Professor and Associate Head* physiology (reproductive)
14 Rahnefeld GW - PhD *Adjunct Professor* animal genetics, (Agriculture Canada)
15 Rotter B - PhD *Research Associate* poultry
16 Shariff MA - PhD *Research Associate* trace mineral metabolism
17 Slominski B - PhD *Research Associate* poultry nutrition
18 Stanger Prof. NE - DVM *Professor* pathology and anatomy
19 Stothers Prof. SC - PhD *Professor and Head of Department* swine
20 Wittenberg KM - PhD *Assistant Professor* nutrition (forage utilization)

1487 Department of Entomology

1 Bodnaryk RP - PhD *Adjunct Professor* insect physiology and biology (Agriculture Canada)
2 Brust Prof. RA - PhD *Professor and Head of Department* biting fly ecology, medical entomology
3 Dosdall L - PhD *Research Associate* biting fly ecology, Canada Biting Fly Centre
4 Flannagan JF - MSc *Adjunct Professor* aquatic insect ecology (Canada Department of Fisheries and Oceans)
5 Galloway MM - PhD *Assistant Professor, Director, Canada Biting Fly Centre* biting fly ecology and behaviour
6 Galloway TD - PhD *Associate Professor* veterinary entomology, aquatic insects
7 Gerber GH - PhD *Adjunct Professor* histology, physiology and behaviour (Agriculture Canada)
8 Holliday NJ - PhD *Associate Professor* pest management crop insects, ecology
9 Jay Prof. SC - PhD *Professor* bee biology
10 Lamb RJ - PhD *Adjunct Professor* systems biology (Agriculture Canada)
11 Loschiavo SR - PhD *Honorary Research Professor* stored grain insect biology and behaviour (Agriculture Canada)
12 MacKay PA - PhD *Associate Professor* insect physiology, aphid ecology and behaviour
13 Morris ON - PhD *Adjunct Professor* biological control of insects (Agriculture Canada)

14 Neill GB - PhD *Ajunct Professor* insects and diseases of nursery trees, PFRA (Agriculture Canada)

15 Robinson AG - PhD *Emeritus Professor* aphid taxonomy

16 Rosenberg DM - PhD *Adjunct Professor* aquatic insect ecology (Canada Department of Fisheries and Oceans)

17 Roughley RE - PhD *Assistant Professor* water beetle systematics

18 Sinha RN - PhD *Honorary Research Professor* ecology of granary insects, mites and fungi (Agriculture Canada)

19 Volney WJ - PhD *Adjunct Professor* forestry

20 Watters FL - PhD *Honorary Research Professor* stored product insect control (Ret.) (Agriculture Canada)

21 Wylie HG - PhD *Adjunct Professor* host parasite relations (Agriculture Canada)

1488 Department of Food Science

1 Murray Prof. ED - PhD *Professor and Head of Department* food proteins and industrial microbiology

2 Biliaderis CG - PhD *Assistant Professor* food chemistry

3 Blank G - PhD *Assistant Professor* food microbiology and fermentation

4 Blicq DM - BSc *Research Associate* industrial microbiology

5 Bushuk W - PhD *Professor* cereal chemistry

6 Chubey BB - PhD *Adjunct Professor* food chemistry and processing (Agriculture Canada)

7 Gallop Prof. RA - PhD *Professor* process systems

8 Henderson Prof. HM - PhD *Professor* food biochemistry

9 Hydamaka AW - MSc *Lecturer* water and waste management

10 Ismond MAH - PhD *Assistant Professor* food chemistry

11 Mazza G - PhD *Adjunct Professor* food chemistry and processing (Agriculture Canada)

12 McEwen TJ - PhD *Adjunct Professor* food chemistry and processing (Canadian Food Products Development Centre)

13 Muller C - BSc *Research Associate* food microbiology

14 Nesbitt JM - PhD *Adjunct Professor* dairy processing (University of Manitoba)

15 Neuman MR - MSc *Research Associate* cereal chemistry

16 Ng PKW *Research Associate* cereal chemistry

17 Penner TO *Research Associate* cereal chemistry

18 Pereira Prof. RR - PhD *Professor* dairy science and food microbiology

19 Sapirstein H - PhD *Professor* cereal chemistry

20 Scanlon M - PhD *Postdoctoral Fellow* cereal chemistry

21 Schal TS - MSc *Instructor* dairy science

22 Timbers GE - PhD *Adjunct Professor* food engineering (Agriculture Canada)

1489 Department of Plant Science

1 Evans Prof. LE - PhD *Professor and Head of Department* plant breeding

2 Ballance GM - PhD *Associate Professor* cereal chemistry

3 Bernier Prof. CC - PhD *Professor* plant pathology

4 Brown PD - PhD *Adjunct Professor* plant breeding (Agriculture Canada)

5 Brûlé-Babel A - PhD *Assistant Professor* plant breeding

6 Chu CC *Visiting Scientist* plant physiology

7 Clark Prof. KW - PhD *Professor* forage crops

8 Dahiya J - PhD *Research Associate* plant pathology

9 Daun J *Adjunct Professor* grain research lab, Can. Grain. Comm.

10 Dronzek Prof. BL - PhD *Professor and Associate Head* cereal chemistry

11 Friesen L - MSc *Research Associate* weed research

12 Hatton T - PhD *Research Associate* biochemistry

13 Helgason SB - PhD *Professor Emeritus* plant breeding

14 Hill Prof. RD - PhD *Professor* biochemistry

15 Huang H - PhD *Adjunct Professor* plant pathology (Agriculture Canada)

16 Kruger JE - PhD *Adjunct Professor* cereal chemistry (Grain Research Laboratory, Canadian Grain Commission)

17 LaCroix Prof. LJ - PhD *Professor* vegetable crops

18 Lamari L - PhD *Research Associate* plant pathology

19 Larter Prof. EN - PhD *Professor* plant breeding and genetics

20 Lenz LM - MSc *Associate Professor* ornamental horticulture

21 MacGregor AW - PhD *Adjunct Professor* cereal chemistry (Grain Research Laboratory, Canadian Grain Commission)

22 Martens JW - PhD *Adjunct Professor* plant pathology (Agriculture Canada)

23 McVetty PBE - PhD *Associate Professor* oil seed crops

24 Mohapatra S - PhD *Assistant Professor (joint appointment with immunology)* immunology

25 Morrison IN - PhD *Professor* weed research

26 Palmer Prof. CE - PhD *Professor* plant physiology

27 Pinnisch R - MSc *Research Associate* oil seed research

28 Pritchard MK - PhD *Associate Professor* horticulture

29 Remphrey WR - PhD *Assistant Professor* horticulture

30 Rimmer SR - PhD *Associate Professor* plant pathology

31 Scarth R - PhD *Assistant Professor* oil seed crops

32 Silvanovich A *Research Associate* plant physiology

33 Stefansson Prof. BR - PhD *Professor Emeritus* oil seed crops

34 Stobbe Prof. EH - PhD *Professor* agronomy

35 Storgaard Prof. AK - PhD *Professor* forage crops

36 Tai Prof. W - PhD *Professor* cytogenetics

37 Tekauz AM - PhD *Adjunct Professor* plant pathology (Agriculture Canada)

38 Thomas PL - PhD *Adjunct Professor* plant breeding (Agriculture Canada)

39 Tkachuk R - PhD *Adjunct Professor* cereal chemistry (Grain Research Laboratory, Canadian Grain Commission)

40 Walichnowski A - MSc *Research Associate* pulse crops, nitrogen fixation

41 Williams PC - PhD *Adjunct Professor* cereal chemistry (Grain Research Laboratory, Canadian Grain Commission)

42 Woodbury W - PhD *Associate Professor* crop physiology

43 Zuzens D - MSc *Research Assistant* wheat breeding

1490 Department of Soil Science

1 Racz Prof. GJ - PhD *Professor and Head of Department* soil chemistry

2 Burnett D *Research Associate* agrometeorology

3 Cho Prof. CM - PhD *Professor* soil physical chemistry

4 Choudhry GG - PhD *Assistant Professor* environmental chemistry and soil organic matter chemistry

5 Goh TB - PhD *Assistant Professor* chemistry and mineralogy

6 Grenier M - BSA *Research Associate* soil fertility

7 Hargrave A - BSA *Research Associate* soil conservation

8 Ridley AO - MSc *Professor* soil fertility and management

9 Sawatsky N - MSc *Research Associate* soil fertility

10 Shaykewich Prof. CF - PhD *Professor* soil physics and agrometeorology

11 Soper Prof. RJ - PhD *Professor* soil chemistry and fertility

12 Toews EJ - BSA *Research Associate* soil fertility

13 Webster Prof. GRB - PhD *Professor* pesticide chemistry

14 Yeomans JC - PhD *Assistant Professor* soil microbiologist

15 Zwarich MA - PhD *Associate Professor* soil genesis and classification

1491 Associated Staff - Agriculture Canada

1 Eilers RG - MSc *Pedologist* soil classification

2 Fraser W - MSc *Pedologist* soil genesis

3 Michalyna W - PhD *Pedologist* soil genesis

4 Smith RE - MSc *Pedologist, Adjunct Professor, Head of Soil Survey Unit* soil genesis

5 Veldius H - MSc *Pedologist* forest soils

1492 Associated Staff - Manitoba Department of Agriculture

1 Haluschak P - MSc *Pedologist* soil chemistry

2 Hopkins LA - MSc *Pedologist* soil classification

3 Langman MN - MSc *Pedologist* soil classification

4 McGill KS - MSc *Director of Soil Testing Laboratory* soil testing

5 Mills GF - MSc *Pedologist, Adjunct Professor* soil classification

6 Podolsky GPW - BSc *Pedologist* soil classification

1493 Research and Other Staff

1 Green BJ - BSA *Research Associate* soil fertility and management

2 Kenyon BE - BSA *Research Associate* soil management

3 Muir DCG - PhD *Adjunct Professor* aquatic environmental chemistry, Fresh Water Institute, Environment Canada

4 Onofrei C - MSc *Research Associate* land evaluation

5 Toews E - BSA *Research Associate* soil fertility-organic soils

6 Walley FL - MSc *Research Associate* soil fertility

1494 Brock University

St Catharines, Ontario, L2S 3A1
Tel: (416) 688-555 0.

1495 Department of Biological Sciences

St. Catharines, Ontario, L2S 3A1
Tel: (416) 688-5550.

1 Manocha Prof. MS - BSc,MSc,PhD *Professor* host-parasite interaction at cellular and molecular levels

2 Bown Prof. AW - BSc,PhD *Professor* mechanisms of plant growth

3 Bruce DH - BSc,PhD *Assistant Professor* distribution of excitation energy in photosynthesis

4 Dickman MD - BA,MSc,PhD *Professor* limnology and paleolimnology

5 Ursino DJ - BA,MSc,PhD *Associate Professor* soyabean photosynthesis, translocation and N2-fixation

1496 Wilfred Laurier University

Waterloo, Ontario, N2L 3C5
Tel: (519)884-1970.

1497 Department of Biology

Waterloo, Ontario, N2L 3C5
Tel: (519)884-1970.

1 Watson Prof. WY - BA,PhD *Professor and Chairman of Department* bryophytes, systematics

2 Fieldes MA - BSc,PhD *Assistant Professor* biochemical genetics

3 Hayashida K - BA,MSc,PhD *Associate Professor* developmental biology, histology

4 Kott E - BA,PhD *Professor* taxonomy of lampreys and freshwater fishes, population biology of freshwater fishes

5 McCauley Prof. RW - BA,PhD *Professor* thermal studies of freshwater fishes

6 Peirson DR - BSc,MSc,PhD *Associate Professor* plant physiology

1498 McMaster University

1280 Main Street West, Hamilton, L8S 4K1
Tel: 416-525-9140.

1499 Department of Biology

1280 Main Street West, Hamilton, L8S 4K1
Tel: 416-525-9140.

1 Threlkeld Prof. SFH - MSc,PhD *Professor and Chairman* drosophila behavioural genetics
2 Bayley Prof. ST - BSc,PhD *Professor* molecular biology of human adenoviruses
3 Chen TT - BSc,MA,PhD *Associate Professor* molecular and developmental genetics; cell biology
4 Davidson Prof. D - BSc,DPhil *Professor* cell proliferation; analysis of cell populations; cytology and cell development
5 Davies DM - BA,PhD *Professor Emeritus*
6 Dingle AD - BSc,MSc,PhD *Associate Professor* cell differentiation; organelle development
7 Finan TM - BSc,MSc,PhD *Assistant Professor* nitrogen fixing root nodule formation, plant bacterial interactions
8 Graham Prof. FL - BSc,PhD *Professor* viral gene action
9 Jensen DEN - MA,PhD *Associate Professor* host-parasite relationship
10 Kershaw Prof. KA - BSc,PhD,DSc *Professor* ecology of sub-arctic lichen heath
11 Lott Prof. JNA - BSc,MS,PhD *Professor* seed ultrastructure and development
12 Mak Prof. S - MSc,PhD *Professor* molecular virology, cell biology
13 McDonald DJ - BSc,MSc,PhD *Associate Professor* comparative animal physiology
14 Miller JJ - BA,PhD *Professor Emeritus*
15 Morton Prof. RA - MS,PhD *Professor* structure and function of cytochromes
16 Nurse CA - BESc,PhD *Assistant Professor* neuron-target interactions; electrophysiology
17 O'Donnell MJ - BSc,PhD *Assistant Professor* electrophysiology; water and ion transport
18 Oaks Prof. A - BA,MA,PhD *Professor* amino acid biosynthesis in corn seedlings
19 Prevec Prof. LA - MA,PhD *Professor* RNA viruses - lytic and oncogenic
20 Rainbow Prof. AJ - BSc,MSc,PhD *Professor* radiation biology and DNA repair mechanisms in mammalian cells and viruses
21 Rollo CD - BSc,MSc,PhD *Assistant Professor* behavioural ecology of terrestrial molluscs
22 Singh Prof. RS - BSc,MSc,PhD *Professor* population genetics, evolutionary biology
23 Sorger Prof. GJ - BSc,MSc,PhD *Professor* biochemical genetics of inorganic nitrogen assimilation
24 Takahashi Prof. I - BA,MSA,PhD *Professor* bacterial sporulation and genetic transduction
25 Westermann Prof. JEM - BSc,MA,PhD *Professor* histology; blood cells in lower vertebrates
26 Wood Prof. CM - BSc,MSc,PhD *Professor* animal physiology

1500 University of Guelph

1501 Ontario Agricultural College

Guelph, Ontario, N1G 2W1
Tel: (519)824-4120.

1 McEwen FL - BSc,MSc,PhD *Professor and Dean* applied entomology
2 Blackburn Prof. DJ - BSA,MS,PhD *Professor and Director of Diploma Course* rural development, technology transfer
3 Jenkinson GM - BSc,MSc *Associate Professor and Assistant to the Dean* animal science
4 Stone JB - BSc,MSc,PhD *Professor and Associate Dean* production systems, dairy
5 Young WS - BSc,MSc,PhD *Professor and Co-ordinator of Agricultural Extension and Continuing Education* field crop production

1502 Faculty of Agriculture

Guelph, Ontario, N1G 2W1
Tel: (519)824-4120.

1503 Department of Animal and Poultry Science

Guelph, Ontario, N1G 2W1
Tel: (519)824-4120.

1 Allen OB - BSc,MSc,PhD *Associate Professor* statistician
2 Atkinson J - BSc,MSc,PhD *Assistant Professor* digestive physiology, poultry
3 Ball RO - BSc,MSc,PhD *Assistant Professor* meat science, growth & metabolism
4 Barbut S - BS,MS,PhD *Assistant Professor* meat science
5 Bowman GH - BSA,MSc,PhD *Professor* production systems, swine
6 Buchanan-Smith JG - BS,MS,PhD *Professor* nutrition, ruminant
7 Burnside EB - BSA,MSA,PhD *Professor* genetics, dairy
8 Burton JH - BSA,MS,PhD *Associate Professor* nutrition, ruminant, equine
9 Etches RJ - BSc,MSc,PhD *Professor and Chairman* endocrinology, poultry

10 Gibbins A - BSc,MSc,PhD *Associate Professor* molecular genetics
11 Grieve DG - BSA,MSA,PhD *Professor* nutrition, dairy
12 Hacker RR - BS,MS,PhD *Professor* environmental physiology, swine
13 Hurnik JF - BA,MA,AgrPhD *Professor* ethol. beef, dairy, swine
14 Jenkinson GM - BSA,MSc *Associate Professor* assistant to Dean OAC
15 Kennedy BW - BSc,MSc,PhD *Professor* animal breeding, swine & dairy
16 King GJ - DVM,MSc,PhD *Professor* mammalian reproductive physiology
17 Leeson S - MPhil,PhD *Professor* nutrition, monogastric
18 McBride BW - BSc,MSc,PhD *Assistant Professor* nutrition, ruminant
19 McMillan I - BSc,MSc,PhD *Associate Professor* animal breeding
20 Milligan LP - BSc,MSc,PhD *Professor and Dean, Research* animal biochemistry
21 Moccia RD - BSc,MSc *Assistant Professor* pathology of farm fishes
22 Morrison WD - BSA,MSc,PhD *Professor* environmental physiology, poultry & swine
23 Mowat DN - BSc,MSc,PhD *Professor* production systems, beef nutrition
24 Schaeffer LR - BS,MS,PhD *Professor* genetics, dairy
25 Smith C - BSc(Agr),MSc,PhD,DSc *Professor* animal breeding strategies
26 Squires EJ - BSc,MSc,PhD *Assistant Professor* biochemistry and molecular biology
27 Stone JB - BSA,MS,PhD *Professor and Associate Dean OAC* dairy production & nutrition
28 Swatland HJ - BSc,MSc,PhD *Professor* meat science, growth & metabolism
29 Usborne WR - BS,MS,PhD *Professor* meat science
30 Walton JS - BSc,PhD *Associate Professor* mammalian reproductive physiology
31 Wilton JW - BSA,MSc,PhD *Professor* genetics, beef
32 Young LG - BSA,MSA,PhD *Professor* nutrition, swine

1504 Department of Horticultural Science

Guelph, Ontario, N1G 2W1
Tel: (519)824-4120.

1 Proctor JTA - BSc,MS,PhD *Professor* pomology
2 Dixon M - BSc,MSc,PhD *Assistant Professor* plant water relations
3 Eggens JL - BSc,BSA,MSc,PhD *Professor* turf physiology
4 Grodzinski B - BSc,MSc,PhD *Associate Professor* herbaceous perennials physiology
5 Lougheed EC - BSA,MSA,PhD *Professor* post-harvest physiology
6 Lumis GP - BS,MS,PhD *Associate Professor* ornamentals
7 Murr DP - BA,MA,PhD *Associate Professor* growth regulators
8 Ormrod DP - BSA,PhD *Professor and Dean, Graduate Studies* environmental physiology and air pollution studies
9 Riekels JW - BS,MS,PhD *Associate Professor* nutrition
10 Shattuck VI - BSc,MSc,PhD *Assistant Professor* plant breeding
11 Shelp BJ - BA,MA,PhD *Assistant Professor* plant nutrition
12 Souza-Machado V - BSA,MS,PhD *Associate Professor* weed control
13 Sullivan JA - MSc,PhD *Assistant Professor* strawberry & raspberry breeding
14 Thompson JE - BSA,PhD *Professor and Chairman* senescence and post harvest physiology and biochemistry
15 Tsujita MJ - BSc,MSc,PhD *Professor* greenhouse floriculture

1505 School of Engineering

Guelph, Ontario, N1G 2W1
Tel: (519)824-4120.

1 Ogilvie JR - BSc,MSA,PhD *Professor and Director* agricultural structures and systems
2 Bilanski WK - BSA,MS *Profesor* material science and farm machinery
3 Chisholm PS - BASc,MASc *Associate Professor* hydrology and hydraulics
4 Dickinson WT - BSA,BASc,MSA,PhD *Professor* hydrology and soil erosion
5 Farazdaghi H - MSc,PhD *Associate Professor* mechanization systems & tillage
6 Hayward G - BASc,MASc,PhD *Assistant Professor* process control, biotechnology
7 Iwaniw M - BSc,MSc,PhD *Assistant Professor* thermal environmental engineering
8 Jofriet JC - DipCivilEng,MASc,PhD *Professor* structural engineering
9 Meiering AH - BS,PhD *Associate Professor* food engineering
10 Mittal G - BSc,MSc,PhD *Associate Professor* food engineering
11 Negi SC - BE,MSc,PhD *Assistant Professor* agricultural structures
12 Otten L - BASc,MASc,PhD *Professor* grain drying; solar energy
13 Pall-Rudra R - BSc,MS,PhD *Assistant Professor* drainage & irrigation, water management
14 Southwell PH - AMEMechE,AFRAES *Associate Professor* energy and farm power
15 Stammers WN - BSA,MSA,PhD *Associate Professor* systems analysis and hydrology
16 Whiteley HR - DIC,BSc,MSc,PhD *Associate Professor* hydrology
17 Zelin S - BASc,MA,PhD *Associate Professor* biological engineering; electronics

1506 Department of Environmental Biology

Guelph, Ontario, N1G 2W1
Tel: (519)824-4120.

1 McIver SB - BA,MSc,PhD *Professor and Chairperson* entomology
2 Alex JF - BSA,MSc,PhD *Professor* plant taxonomy
3 Barron GL - BSc,MSA,PhD *Professor* mycology

4 Boland GJ - BSc,MSc,PhD *Assistant Professor* plant pathology
5 Burpee LL - BA,MSc,PhD *Associate Professor* plant pathology
6 Collins-Thompson DL - BSc,MSc,PhD *Professor* food microbiology
7 Cunningham JD - BSA,MSA *Professor* industrial and dairy microbiology
8 Edgington LV - BS,PhD *Professor* plant pathology
9 Ellis CR - BSc,MSc,PhD *Professor* entomology
10 Fletcher RA - BSc,MSc,PhD *Professor* plant physiologist
11 Gordon AM - BScF,PhD *Assistant Professor* agroforestry
12 Hall JC - BSc,MSc,PhD *Assistant Professor* weed science
13 Hall R - BAgrSc,PhD *Professor* plant pathology
14 Hofstra G - BSA,MSA,PhD *Professor* plant ecology
15 Kaushik NK - BSc,MSc,PhD *Associate Professor* stream ecology
16 Kevan PG - BSc,PhD *Associate Professor* apiculture
17 Laing JE - BSc,MSc,PhD *Professor* entomology
18 Lee H - BSc,PhD *Assistant Professor* microbiology
19 Marshall SA - BScAgr,MSc,PhD *Associate Professor* entomology
20 Otis GW - BSc,PhD *Associate Professor* apiculture
21 Scott-Dupree CD - BSc,MPM,PhD *Assistant Professor* apiculture
22 Sears MK - BS,MS,PhD *Associate Professor* entomology
23 Solomon KR - BSc,MS,PhD *Associate Professor* environmental toxicology
24 Stephenson GR - BS,MS,PhD *Professor* weed science
25 Surgeoner GA - BSc,MSc,PhD *Professor* livestock pests
26 Sutton JC - BSc,PhD *Professor* plant pathology
27 Trevors JT - BSc,MSc,PhD *Associate Professor* soil & environmental microbiology

1507 Department of Crop Science
Guelph, Ontario, N1G 2W1
Tel: (519)824-4120.

1 Tanner JW - BSA,MSA,PhD *Professor* crop physiology
2 Anderson GW - BSA,MSA *Associate Professor* weed control
3 Beversdorf WD - BS,MS,PhD *Professor and Chairman* oilseed breeding
4 Bowley SR - BSc,MSc,PhD *Assistant Professor* forage breeder
5 Christie BR - BSA,MSA,PhD *Professor* forage crop breeding
6 Clark EA - BSc,MSc,PhD *Assistant Professor* forage physiology
7 Falk DE - BSc,MSc,PhD *Assistant Professor* barley, oats and winter barley breeding and genetics
8 Gamble EE - BSA,MSA,PhD *Professor* cereal production and seed science
9 Hume DJ - BSA,MSA,PhD *Professor* crop physiology
10 Hunt LA - BSc,MSc,PhD *Professor* wheat breeding and physiology
11 Kannenberg LW - BS,MS,PhD *Professor* corn breeding
12 Kasha KJ - BSc,MSc,PhD *Professor* cytogenetics
13 McKersie BD - BSc,PhD *Associate Professor* plant physiology
14 Michaels TE - BA,MS,PhD *Assistant Professor* plant breeding
15 Pauls PK - BA,MA,PhD *Assistant Professor* biotechnology
16 Stockskopf NC - BSc,MSc,PhD *Professor* winter wheat breeding
17 Swanton CS - BSc,MSc *Assistant Professor* weed control
18 Tollenaar M - LdbIng,PhD *Assistant Professor* crop physiology
19 Upfold RA - BSc,MSc *Associate Professor* extension agronomist
20 Vyn TJ - BSc,MSc *Assistant Professor* crop production

1508 Department of Food Science
Guelph, Ontario, N1G 2W1
Tel: (519)824-4120.

1 Arnott DR - BSA,BS,PhD *Professor* food processing
2 Barbut S - BS,MS,PhD *Assistant Professor* poultry products technology; meat microbiology; food irradiation
3 Bullock Prof. DH - BSA,MSA,PhD *Professor* food processing
4 deMan Prof. JM - ChemEng,PhD *Professor* food chemistry
5 Duitschaever CL - BSc,MSc,PhD *Associate Professor* food microbiology
6 Ferrier LK - BSA,MSc,PhD *Associate Professor* food and dairy chemistry; food preservation and processing
7 Goff HD - BSc,AS,PhD *Assistant Professor* dairy science
8 Hill AR - BSc,MSc,PhD *Assistant Professor* milk proteins; cheese technology; dairy chemistry and processing
9 Kakuda YY - BSc,MSc,PhD *Assistant Professor* vitamin analysis; chemistry of nitrosamines
10 Rasper VF - BS,PhD *Professor* food chemistry
11 Stanley DW - BS,MS,PhD *Professor* food chemistry
12 Swatland HJ - BSc,MSc,PhD *Professor* meat quality; muscle growth; optoelectronic systems to measure food quality; intracellular mapping of enzyme activity
13 Usborne WR - BSc,MSc,PhD *Professor and Acting Chairman* meat science; sensory evaluation techniques
14 Yada RY - BSc,MSc,PhD *Assistant Professor* protein structure-function relationships; food related enzymes; postharvest physiology of fruits and vegetables

1509 Department of Land Resource Science
Guelph, Ontario, N1G 2W1
Tel: (519)824-4120.

1 Bates TE - BSA,MS,PhD *Professor* soil management
2 Beauchamp EG - BSc,MSc,PhD *Professor* soil plant relations
3 Brookfield ME - BSc,PhD *Associate Professor* geology
4 Brown DM - BSA,MSA,PhD *Professor* agrometeorology
5 Chesworth W - BSc,MSc,PhD *Professor* geochemistry

6 Elrick DE - BSA,MS,PhD *Professor* soil physics
7 Evans LJ - BSc,PhD *Associate Professor* soil genesis & clay mineralogy
8 Gillespie TJ - BSc,MA,PhD *Professor* agrometeorology
9 Groenevelt PH - BSc,MSc,PhD *Professor* soil conservation, soil structure
10 Hilts SG - BA,MA,PhD *Associate Professor* research management, environmental planning
11 Kachanoski RG - BSc,MSc,PhD *Assistant Professor* extension, soil & water conservation
12 Kay BD - BSA,MSc,PhD *Professor and Chairman* soil physical chemistry
13 King KM - BSA,MS,PhD *Professor* agrometeorology
14 Martini IP - PhD *Professor* sedimentology
15 McBride R - BSc,MSc,PhD *Assistant Professor* soil science & agricultural land use planning
16 Miller MH - BSA,MS,PhD *Professor* soil-plant relations
17 Protz R - BSA,MSc,PhD *Professor* soil genesis and classification
18 Sheard RW - BSA,MSA,PhD *Professor* plant nutrition
19 Thomas RL - BSc,MSc,PhD *Professor* soil biochemistry
20 Thurtell GW - BSA,MSA,PhD *Professor* agrometeorology
21 Voroney RP - BSc,MSc,PhD *Assistant Professor* soil biology, soil management

1510 School of Landscape Architecture
Guelph, Ontario, N1G 2W1
Tel: (519)824-4120.

1 Kehm WH - BSLA,MLA *Professor and Director*
2 Brown RD - BSc,MLA,PhD *Assistant Professor* research; application of biophysical sciences; computer applications
3 Chanasyk V - BSc,MLA *Professor* design; planting design, successional landscapes; housing for the elderly
4 Man CRJ - BArch,MLA *Professor* architecture, site planning, housing, land development, project and urban design
5 Milliken JD - BLA,MLA *Professor* techniques for land planning, operational aspects of city planning; landscape design theory
6 Nelischer M - BSc,MLA *Associate Professor* design, construction and land development
7 Stoltz RR - BS,MLA *Associate Professor* site planning; new town planning and design, landscape construction
8 Taylor J - BSLA,MLA *Associate Professor* park & recreation planning, urban open space policy planning, land planning
9 Wester LM - BSLA,MLA *Associate Professor* design teaching methodology; nightime landscape design; graphic communication; perception in design

1511 Department of Agricultural Economics and Business
Guelph, Ontario, N1G 2W1
Tel: (519)824-4120.

1 Martin LJ - BSc,MSc,PhD *Professor and Chairman* agricultural marketing
2 Bates MR - MBA *Assistant Professor* finance
3 Braithwaite WM - BA,MBA *Professor* management accounting
4 Brinkman GL - BS,MSc,PhD *Professor* rural development economics
5 De Roo A - BSc,MBA *Associate Professor* operations management
6 Fox GC - BSc,MSc,PhD *Assistant Professor* production economics
7 Funk TF - BSc,MSc,PhD *Professor* marketing research
8 Goddard EW - BSc,MSc,PhD *Assistant Professor* agricultural marketing
9 Harling KF - BSc,MSc,PhD *Associate Professor* business policy
10 Howard WH - BA,MS,PhD *Assistant Professor* farm management
11 Hutchison JG - BSA,MSc *Associate Professor* farm taxation
12 Larue B - DEC,BSc,PhD *Assistant Professor* trade policy
13 Meilke KD - BSc,PhD *Professor* price analysis
14 Pfeiffer WC - BS,PhD *Associate Professor* bio-economics
15 Phillips TP - BSc,MSc,PhD *Professor* production economics
16 Rodd RS - BA,MSc *Professor* regional economics
17 Stonehouse DP - BSc,MSc,PhD *Associate Professor* production economics
18 van Vuuren W - BA,MA,PhD *Associate Professor* land economics
19 Waldron MW - BSc,MS,PhD *Professor* continuing education administration
20 Warley TK - BSc,MSc *Professor* trade policy

1512 Department of Rural Extension Studies
Guelph, Ontario, N1G 2W1
Tel: (519)824-4120.

1 Moore GAB - BA,BD,MA,PhD *Professor and Chairman* rural development & educational technology
2 Blackurn DJ - BSA,MS,PhD *Professor and Director, Diploma Course* rural development, technology transfer
3 McCreary EK - BA,MA,EdD *Assistant Professor* rural development
4 Pfeiffer WC - BS,PhD *Associate Professor* rural development, technology transfer
5 Pletsch DH - BSA,MSc,PhD *Professor* rural development, technology transfer
6 Shute JCM - BA,MA,PhD *Professor* international and community development
7 Waldron Prof. MW - BSc,MS,PhD *Professor* continuing education administration

1513 Ontario Veterinary College

Guelph, Ontario, N1G 2W1
Tel: (519)824-4120.

1514 Department of Biomedical Sciences
Guelph, Ontario, N1G 2W1
Tel: (519)824-4120.

1 Liptrap Prof. RM - DVM,MVS,PhD *Professor* reproductive physiology
2 Porter DG - BSc,BVetMed,MSc,PhD,DSc,MRCVS *Professor and Chairman* reproductive physiology
3 Atwal Prof. OS - BVSc,MS,PhD *Professor* reticuloendothelial physiology
4 Basrur Prof. PK - BSc,MSc,PhD *Professor* cell biology, cytogenetics
5 Betteridge Prof. KJ - BVSc,MVSc,PhD *Professor* embryo biology and transfer
6 Bhatnagar Prof. MK - BSc,BVSc&AH,MSc,PhD *Professor* histochemistry, histology
7 Black Prof. WD - DVM,MSc,PhD *Professor* pharmacology, drug metabolism
8 Boermans HJ - DVM,MSc,PhD *Assistant Professor* toxicology
9 Chapman HW - BVetSc,MS,PhD *Associate Professor* digestive physiology
10 Coulon PD - BSc,MSc,DVM,PhD *Assistant Professor* pharmacology
11 Croy PD - DVM,PhD *Associate Professor* anatomy, immunology of pregnancy
12 Downie Prof. HG - DVM,MVSc,MS,PhD *Professor* cardiovascular physiology
13 Fisher KRS - BSc,MSc,PhD *Assistant Professor* development cytology, tissue function
14 Geissinger Prof. HD - DVM,MSc,PhD *Professor* pathophysiology
15 Gentry Prof. PW - BSc,PhD *Professor* blood coagulation
16 Goldberg MT - BSc,PhD *Assistant Professor* toxicology
17 Grovum WL - BSA,PhD *Associate Professor* ruminant physiology
18 Harris WH - DVM,MSc,PhD *Associate Professor* neonatal physiology
19 Johnstone IB - DVM,MSc,PhD *Associate Professor* blood and cardiovascular physiology
20 King WA - BSc,MSc,PhD *Associate Professor* embryo biotechnology
21 Partlow GD - BSc,MSc,PhD *Assistant Professor* anatomy
22 Raeside Prof. JI - BSc,MS,PhD *Professor* reproductive physiology
23 Rieger D - BSc,PhD *Assistant Professor* energy metabolism of embryos
24 Robinson Prof. GA - BSc,MSc,PhD *Professor* radiobiology
25 Summerlee AJS - BSc,BVSc,PhD *Associate Professor* neuroendocrinology
26 Yamashiro S - MSc,PhD *Associate Professor* histology, histochemistry

1515 Department of Clinical Studies
Guelph, Ontario, N1G 2W1
Tel: (519)824-4120.

1 Pascoe PJ - BVSc,DipACVA *Associate Professor* anesthesiology
2 Allen DG - DVM,MSc,DipACVIM *Associate Professor* small animal medicine, cardiology
3 Atilola MAO - DVM,MSc,PhD *Assistant Professor* radiology
4 Baird JD - BVSc,PhD *Associate Professor* mycotoxicoses in large animals
5 Binnington AG - DVM,DipVetSurg,MSc *Associate Professor* experimental surgery
6 Butler Prof. DG - DVM,MSc,PhD *Professor and Chairman* gastroenterology
7 Cockshutt JR - DVM,MSc *Assistant Professor* small animal surgery
8 Downey RS - DVM,MSc *Professor and Assistant Dean* radiology
9 Dyson D - DVM,DVSc,DipACVA *Assistant Professor* anesthesiology
10 Holmberg DL - BS,DVM,MVSc,DipACVS *Associate Professor* small animal surgery
11 Horney Prof. FD - DVM,DipACVS *Professor* experimental surgery
12 Hurtig MB - DVM,MVSc *Assistant Professor* large animal surgery
13 Kruth SA - BA,DVM,DipACVIM *Associate Professor* internal medicine
14 Livesey MA - BVM,MSc *Assistant Professor* large animal surgery
15 McDonell Prof. WN - DVM,MSc,MRCVS,PhD,DipVetAnes *Professor, Director, Veterinary Teaching Hospital* anaesthesiology
16 Miller CW - DVM,MVSc,DipACVS *Assistant Professor* small animal surgery
17 O'Grady MR - DVM,MSc,DipACVIM *Assistant Professor* cardiology
18 Parent JM - BSc,DVM,MVSc,DipACVIM *Associate Professor* neurology
19 Parker WM - DVM,DipVetMed *Associate Professor* dermatology
20 Pennock Prof. PW - BS,DVM,MS,DipACVR *Professor* radiology
21 Pook HA - BVSc,MVSc *Assistant Professor* small animal medicine
22 Shappell KK - BS,DVM,MS *Assistant Professor* large animal surgery
23 Smith-Maxie LL - DVM,MSc *Associate Professor* neurology
24 Staempfli H - DVM,DRVetMed *Assistant Professor* large animal medicine
25 Summer-Smith Prof. G - BVSc,MSc,FRCVS *Professor* musculoskeletal diseases
26 Tremblay RRM - BSc,DVM,DVSc *Assistant Professor* large animal medicine
27 Trout R - BS,DVM,PhD *Assistant Professor* large animal surgery
28 Viel L - DVM,MSc,PhD *Assistant Professor* large animal medicine
29 Williams MM - DVM,DipACVO *Assistant Professor* ophthalmology
30 Willoughby Prof. RA - DVM,PhD,DipACVIM *Professor* respiratory function

1516 Department of Pathology
Guelph, Ontario, N1G 2W1
Tel: (519)824-4120.

1 Miller Prof. RB - BSc,DVM,PhD *Professor and Chairman of the Department* reproductive pathology
2 Barker IK - DVM,MSc,PhD *Associate Professor* wildlife diseases, enteric pathology
3 Ferguson HW - BVM&S,PhD *Associate Professor* fish pathology
4 Fernando Prof. MA - MB,BS,PhD *Professor* parasitology
5 Geraci Prof. JR - BSc,VMD,PhD *Professor* fish and marine mammal pathology
6 Hayes MA - BVSc,PhD *Associate Professor* anatomic pathology
7 Hulland Prof. TJ - DVM,PhD *Professor* musculoskeletal pathology
8 Hunter DB - DVM,MSc *Associate Professor (Joint Appointment with Clinical Studies)* avian, poultry
9 Jacobs RM - BSc,DVM,PhD *Associate Professor* clinical pathology
10 Julian RJ - DVM,DipVetMed *Associate Professor* poultry pathology
11 Little Prof. PB - DVM,MS,PhD *Professor* neuropathology
12 Lumsden Prof. JH - DVM,MSc,DipVetMed *Professor* clinical pathology, cytopathology
13 McCraw Prof. BM - BA,MA,PhD *Professor* parasitology
14 Nielsen NO - DVM,PhD *Dean* pathology
15 O'Brien PJ - BSc,DVM,MS,PhD *Assistant Professor* clinical pathology
16 Percy Prof. DH - DVM,MSc,PhD *Professor* laboratory animal medicine & pathology, toxicology
17 Slocombe Prof. JOD - DVM,PhD *Professor* parasitology
18 Stirtzinger T - DVM,DipVetMed,PhD *Assistant Professor* clinical pathology
19 Valli Prof. VEO - DVM,MSc,PhD *Professor* hemopoietic pathology
20 Wilcock BP - BA,DVM,MS,PhD *Associate Professor* enteric pathology
21 Yager JA - BVSc,PhD *Associate Professor* respiratory pathology

1517 Department of Population Medicine
Guelph, Ontario, N1G 2W1
Tel: (519)824-4120.

1 Bateman KG - DVM,MSc *Assistant Professor* health management of beef cattle
2 Bonnett BN - DVM *Assistant Professor* epidemiology
3 Bucknell BC - DVM,MSc *Assistant Professor* theriogenology
4 Etherington SW - DVM,MSc,DipLASurg *Assistant Professor* theriogenology
5 Friendship RM - DVM,MSc *Assistant Professor* swine
6 Johnson WH - DVM,MSc *Associate Professor* theriogenology
7 Leslie KE - DVM,MSc *Associate Professor* large animal medicine
8 Luescher VA - DVM,PhD *Assistant Professor* ethology
9 Martin Prof. SW - DVM,MSc,MPVM,PhD *Professor and Chairman* epidemiology
10 McEwen SA - DVM,DVSC *Assistant Professor* food hygiene
11 McKeown DB - DVM *Associate Professor* ethology
12 Miniats Prof. OP - DVM,MSc *Professor* gnotobiology
13 Physick-Sheard PW - BVSc,MSc,DipVetSurg,MRCVS *Associate Professor* large animal medicine, cardiology
14 Sandals WCD - DVM *Assistant Professor* large animal medicine
15 Willan AR - BED,MSc,PhD *Assistant Professor* biostatistics
16 Wilson Prof. MR - BVSc,PhD,MRCVS *Professor* swine health management

1518 Department of Veterinary Microbiology and Immunology
Guelph, Ontario, N1G 2W1
Tel: (519)824-4120.

1 Meek AH - DVM,MSc,PhD *Associate Professor and Acting Associate Dean* epidemiology
2 Derbyshire Prof. JB - BSc,PhD,MRCVS *Professor* virology
3 Gyles Prof. CL - DVM,MSc,PhD *Professor* bacteriology
4 Johnson RP - BVSc,MVSc,PhD *Assistant Professor* immunology
5 Lang GH - DrVet,CertBact,MVSc *Associate Professor* virology
6 Prescott JF - MA,VetMB,PhD,MRCVS,FRCVS *Associate Professor* bacteriology
7 Rosendale Prof. S - DVM,DVSc *Professor* bacteriology
8 Shewen PE - DVM,BSc,MSc,PhD *Assistant Professor* immunology
9 Thorsen Prof. J - DVM,DipBact,PhD,MRCVS *Professor* virology
10 Wilkie Prof. BN - DVM,PhD *Professor* immunology

1519 The University of Western Ontario

London, Ontario
Tel: (519)679-2111.

1 Pederson KG - BA,MA,PhD *President and Vice-Chancellor*
2 Weldon DB - BA *Chancellor*

1520 Department of Plant Sciences
London, Ontario
Tel: (519)679-2111.

1 Bancroft Prof. JB - BA,PhD *Professor and Dean of the Faculty of Science* plant virology
2 Brown LM - BS,MSc,PLD *Assistant Professor* osmoregulation

3 Cavers Prof. PB - BSA,DIC,PhD *Professor* plant ecology
4 Cook Prof. FS - BA,PhD *Professor* plant physiology
5 Cummins JE - BSc,PhD *Associate Professor* fungal genetics
6 Day Prof. AW - BSc,PhD *Professor and Chairman* fungal genetics
7 Fahselt D - BA,PhD *Associate Professor* biochemical systematics
8 Greyson Prof. RI - BA,MSc,PhD *Professor* plant growth, development
9 Hayden DB - BSc,PhD *Associate Professor* plant physiology
10 Hopkins WG - BA,PhD *Associate Professor* plant physiology
11 Huner NPA - BSc,MSc,PLD *Associate Professor* photosynthesis, low temperature acclimation
12 Jancey RC - BSc,MSc,PhD *Associate Professor* taxonomy
13 Lachance MA - BA,BSc,MSc,PhD *Associate Professor* plant genetics, speciation
14 Maun MA - MA,MSc,PhD *Associate Professor* plant ecology
15 McKeen Prof. WE - BSc,MSc,PhD *Professor* plant pathology
16 McLarty DA - BA,PhD *Professor Emeritus* algal eutrophication
17 Orloci Prof. L - BSF,MSc,PhD *Professor* statistical ecology
18 Phipps Prof. JB - BSc,PhD *Professor* numerical taxonomy
19 van Huystee Prof. RB - BA,MA,PhD *Professor* plant physiology
20 Walden Prof. DB - BA,MSc,PhD *Professor* plant genetics
21 Wilson Prof. DG - BA,MA,PhD *Professor* plant physiology

1521 Department of Zoology
London, Ontario
Tel: (519)679-2111.

1 Locke Prof. M - MA,PhD,ScD,FRSC *Professor and Chairman of Department* ultrastructural studies on developing insect cells
2 Ankney CD - BSc,PhD *Associate Professor* synecology; egg weight variation in snow geese
3 Atkinson BG - BA,PhD *Associate Professor* endocrinological regulation of macromolecular events in developmental biology and genetics
4 Battle HI - BA,MA,PhD,LLD,DSc *Professor Emeritus* hereditary limb deformities and their embryological significance
5 Bourns Prof. TKR - BA,MA,PhD *Professor* host-parasite relationships of trematodes
6 Caveney Prof. S - BSc,DPhil *Professor* intercellular communication in a positional field
7 George JA - BSA,MSA,PhD *Professor Emeritus* non-pesticide insect control
8 Green Prof. RH - BS,PhD *Professor* statistical ecology applied to the structure of aquatic communities
9 Handford PT - BA,DPhil *Associate Professor* inter- and infra-specific differentiation in foraging ecology and behaviour in piscivorous birds
10 Judd WW - BA,MA,PhD *Professor Emeritus* gall producing insects; bog populations
11 Keenleyside Prof. MHA - BA,MA,PhD *Professor* fish behaviour
12 Kidder GM - BA,PhD *Associate Professor* gene transcription in relation to embryonic determination
13 McMillan Prof. DB - BSc,MSc,PhD *Professor* studies in comparative vertebrate histology and ultrastructure
14 Millar Prof. JS - BSc,MSc,PhD *Professor* mammalian reproductive strategies
15 Ogilvie DM - BSc,MSc,PhD *Associate Professor* environmental physiology
16 Owen MD - BSc,PhD *Associate Professor* physiology of insect venoms
17 Podesta RB - BSc,PhD *Associate Professor* membrane biology - tropical medicine
18 Purko J - BA,MSc,PhD *Associate Professor* developmental biology
19 Roth RR - MSc,PhD *Assistant Professor* endocrinology; nutrition and environmental biology
20 Scott DM - BSc,MSc,PhD *Professor Emeritus* population dynamics and dispersal of song-birds
21 Shivers RR - AB,MA,PhD *Associate Professor* cell biology of nervous systems
22 Singh SM - BSc,MSc,PhD *Associate Professor* population genetics
23 Steele Prof. JE - BSc,MSc,PhD *Professor* endocrine control of metabolism in insects
24 Tam WH - BSc,PhD *Associate Professor* role of placenta and uterus in regulation of ovarian function
25 Wiebe JP - BSc,PhD *Associate Professor* Leydig and Sertoli cells of the testis; enzymes and hormone studies related to steroidogenesis

1522 University of Toronto

1523 Faculty of Forestry
Toronto, Ontario, M5S 1A1
Tel: (416)978-3548.

1 Carrow Prof. JR - BScF,MSc,PhD,RPF *Dean and Professor* forest entomology
2 Nautiyal Prof. JC - BSc,AIFC,MF,PhD *Associate Dean and Professor* forest economics
3 Aird Prof. PL - BScAgr,MS,PhD,RPF,FE *Professor* forest policy, conservation of renewable resources, parks and recreation
4 Andresen Prof. JW - BS,PhD *Professor* urban forestry
5 Balatinecz Prof. JJ - BScF,MF,PhD *Professor* wood properties, processing and forest products
6 Bendell Prof. JF - BA,PhD *Professor* population dynamics in forest habitats
7 Blake TJ - BScFor,DipFor,MF,STB,PhD *Associate Professor* silviculture, eco-physiology

8 Buckingham Prof. FM - BScF,MF,DF *Professor* forest watershed management, dendrohydrology, biometeorology
9 Carleton TJ - BSc,MSc,PhD *Assistant Professor* forest ecology
10 Farrar JL - BScF,MF,PhD,RPF *Emeritus Professor* tree physiology
11 Hubbes Prof. M - DipIngAgr,DrDipIngAgr *Professor* host-parasite relationships in forest diseases, biochemistry, resistance
12 Keenan FJ - BASc,MASc,PhD,PEng *Associate Professor* timber engineering and wood construction
13 Love Prof. DV - BScF,MF,RPF *Professor (part-time)* forest management, land use planning
14 Martell DL - BASc,MASc,PhD *Associate Professor* forest fire management
15 Nordin VJ - BA,BScF,PhD *Emeritus Professor* international and tropical forestry
16 Roy Prof. DN - BSc,MSc,PhD,FRSC *Professor* chemistry of host-parasite relationships, plant biochemistry, wood chemistry
17 Smith Prof. VG - BScF,MScF,PhD,RPF *Professor* forest biometry, prediction equations in land use planning and forest product measurement
18 Timmer V - BScF,MScF,PhD *Associate Professor* forest soils
19 Vlcek Prof. J - BScF,MASc,DSc,PEng *Professor* analytical photogrammetry, digital terrain mapping, remote sensing
20 Wayman M - BA,MA,PhD,FCIC,PEng *Emeritus Professor* pulp and paper industry, wood chemistry
21 Zsuffa L - BScF,PhD,RPF *Associate Professor* tree breeding and genetics

1524 Lakehead University
Thunder Bay, Ontario, P7B 5E1
Tel: (807) 343-8110.

1525 School of Forestry
Thunder Bay, Ontario, P7B 5E1
Tel: (807) 343-8110.

1 Naysmith JK - PhD,RPF,BScF,MFS *Director* resource management policy and management
2 Benson Prof. CA - BScFor,MScFor,RPF *Professor* forest management
3 Brown Prof. KM - BScFor,MScFor,PhD *Professor* dendrology, growth and yield
4 Carmean W - BS,MS,PhD *Associate Professor* soils
5 Clarke R - BScFor,MScFor,RPF *Associate Professor* silviculture
6 Cumming HG - BA,MSc,PhD *Associate Professor* fish and wildlife
7 David EJ - DepForEng,RPF *Associate Professor* forest mechanization
8 Day Prof. RJ - BA,MA,MScFor,RPF *Professor* silviculture
9 Farmer R - BSc,MS,PhD *Professor* biometrics, forest economics
10 Kayll AJ - BSF,MF,PhD,RPF *Professor* ecological effects of fire
11 Knowles P - BS,MS,PhD *Assistant Professor* forest biology, genetics
12 Murchison HG - BScFor,MScFor,RPF *Associate Professor* mensuration, forest economics and evaluation
13 Parker Prof. W - BA,MSc,PhD *Professor* genetics, dendrology
14 Prévost YH - BSc,MSc,PhD *Assistant Professor* forest entomology, computer applications
15 Pulkki R - BScF,MScF,LScF,Doctor of Forestry,RPF *Associate Professor* harvesting and transportation
16 Richardson DJ - ForTechDip,BScFor,MScFor *Associate Professor* mensuration, road location and construction
17 Setliff E - BSc,MF,PhD *Associate Professor* pathology
18 Tanz JS - BSc,MF,RPF *Lecturer* forest management, photogrammetry, GIS
19 Westbroek H - BScFor *Associate Professor* photogrammetry
20 Yang KC - BScFor,MScFor,MSc *Associate Professor* wood technology
21 Zingel S - DrRerNat *Associate Professor* soils

1526 Carleton University
Colonel By Drive, Ottawa, Ontario, K1S 5B6
Tel: 613 564 3871.

1527 Department of Biology
Colonel By Drive, Ottawa, Ontario, K1S 5B6
Tel: 613 564 3871.

1528 Ecology and Systematics
Colonel By Drive, Ottawa, Ontario, K1S 5B6
Tel: 613 564 3871.

1 Barlow CA - MA,PhD entomology
2 Damman H - BSc,PhD plant/insect interactions
3 Howden HF - BS,MS,PhD entomology
4 Lambert JDH - BSc,MSc,PhD tropical plant ecology
5 Merriam HG - BSA,PhD systems ecology
6 Nesbitt HHJ - BA,MA,PhD,DSc,FLS,FRES,FZS entomology
7 Peck SB - BSc,MSc,PhD entomology
8 Weatherhead PJ - BSc,MSc,PhD animal ecology and behaviour

1529 Molecular and Cellular Biology
Colonel By Drive, Ottawa, Ontario, K1S 5B6
Tel: 613 564 3871.

1 Chaly N - BSc,MSc,PhD cell ultrastructure
2 Iyer VN - BSc,MSc,PhD molecular genetics

3 Lee PE - MSc,PhD insect transmitted viruses
4 McCully ME - BSA,MSc,PhD plant development
5 Wyndham RC - BSc,MSc,PhD environmental microbial waste clean-up

1530 Physiology and Biochemistry
Colonel By Drive, Ottawa, Ontario, K1S 5B6
Tel: 613 564 3871.

1 Gardner DR - BSc,PhD neurobiology and pesticides
2 Joy KW - BSc,PhD plant nitrogen metabolism
3 Sinclair J - BSc,PhD photosynthesis and pesticides
4 Webb JA - BSc,PhD carbohydrate translocation
5 Wightman F - BSc,PhD plant hormone physiology

1531 University of Ottawa

Ottawa, Ontario, K1N 6N5
Tel: (613)564-2336.

1532 Department of Biology
Ottawa, Ontario, K1N 6N5
Tel: (613)564-2336.

1 Armstrong Prof. JB - BSc,PhD *Professor* developmental biology
2 Arnason JT - BSc,PhD *Associate Professor* plant biology; insect plant interactions
3 Bonen L - BSc,PhD *Assistant Professor* molecular biology
4 Brown Prof. DL - BSc,MSc,PhD *Professor* cell biology
5 Currie DJ - BSc,PhD *Assistant Professor* freshwater ecology
6 Fenwick Prof. JC - MSc,PhD *Professor* comparative endocrinology
7 Hickey DA - BSc,PhD *Associate Professor* molecular genetics
8 Houseman J - BSc,MSc,PhD *Assistant Professor* insect physiology
9 Johnson DA - BSc,PhD *Assistant Professor* molecular biology, nitrogen fixation
10 Keddy PA - BSc,PhD *Associate Professor* plant ecology and wetlands vegetation
11 Kushner Prof. DJ - BSc,MSc,PhD *Professor* microbiology, physiology of halophilic and salt-tolerant bacteria
12 McBurney MW - BSc,MSc,PhD *Associate Professor* molecular, cellular and cancer biology
13 Moon Prof. TW - BSc,MA,PhD *Professor and Chairman* comparative physiology and biochemistry
14 Morris CE - BSc,PhD *Associate Professor* excitable cell physiology
15 Nozzolillo Prof. C - MA,PhD *Professor* phytochemistry and plant physiology
16 Perry SF - BSc,PhD *Assistant Professor* comparative respiratory and circulatory physiology
17 Philogène Prof. BJR - BSc,MSc,PhD,FESC *Professor and Dean, Faculty of Science* ecophysiology, entomology
18 Pick FR - BSc,PhD *Assistant Professor* physiological ecology of algae, energy flow in aquatic systems
19 Picman J - MSc,PhD *Associate Professor* behavioural ecology
20 Qadri Prof. SU - MSc,PhD *Professor* ichthyology
21 Weinberger Prof. P - BSc,MSc,PhD *Professor* plant physiology and ecotoxicology

1533 Université de Montréal

1534 Faculté de Médecine Vétérinaire
3 200, rue Sicotte, Case postale 5000, Saint-Hyacinthe, Quebec, J2S 7C6
Tel: (514) 773-8521.

1 Roy RS - MV,PhD doyen
2 Bisaillon A - MV,MSc secrétaire de la Faculté
3 Dallaire A - MV,MSc vice-doyen aux études
4 Lamothe P - MV,MSc vice-doyen aux affaires cliniques
5 Phaneuf L - MV,PhD responsable de l'education permanente

1535 Département de médecine
3 200, rue Sicotte, Case postale 5000, Saint-Hyacinthe, Quebec, J2S 7C6
Tel: (514) 773-8521.

1 Blais D - DMV *Professeur adjoint* anesthésiologie
2 Bonneau Prof. N - DMV,MScV *Professeur titulaire* chirurgie des petits animaux
3 Bouchard E - DMV,IPSAV *Professeur adjoint* médecine bovine, en congé
4 Breton L - DMV,MScV *Professeur agrégé* radiologie
5 Cécyre Prof. A - DMV,MS *Professeur titulaire* nutrition
6 Chalifoux A *Directeur*
7 Chalifoux Prof. A - DMV,MScV *Professeur titulaire* pathologie et microbiologie vétérinaires
8 Couture Y - DMV *Professeur agrégé*
9 D'Allaire S - DMV,MSc,PhD *Professeur adjoint* anatomie et physiologie vétérinaires
10 Di Fruscia R - DMV,DipACVIM *Professeur adjoint* médecine interne
11 Guay P - DMV,MS *Professeur titulaire* obstétrique et reproduction
12 Harvey D - DMV,MSc *Professeur adjoint* sciences cliniques vétérinaires
13 Lamothe P - DMV,MScV *Professeur titulaire* médecine
14 Larouche Y - DMV,GradDipl *Professeur agrégé* médecine des grands animaux

15 Lavoie J-P - DMV,IPSAV *Chargé d'enseignement* médecine équine, en congé
16 Marcoux M - DMV,MScV *Professeur titulaire* médecine
17 Martineau G - DMV,DESS *Professeur agrégé* production porcine
18 Olivieri M - DMV *Professeur agrégé* chirurgie
19 Paradis M - DMV,IPSAV,MScV *Professeur adjoint* médecine des petits animaux, reproduction des petits animaux
20 St-Pierre H - DMV,MScV *Professeur titulaire* anatomie et physiologie vétérinaires
21 Vaillancourt D - DMV,IPSAV,DipACT,MSc *Professeur adjoint* médecine et chirurgie; thériogénologie
22 Vrins A - DMV,IPSAV *Professeur adjoint* médecine équine; médecine interne

1536 Département d'anatomie et physiologie animales
3 200, rue Sicotte, Case postale 5000, Saint-Hyacinthe, Quebec, J2S 7C6
Tel: (514) 773-8521.

1 Blouin A *Directeur*
2 Barrette D - DMV,MSc *Professeur agrégé* biochimie
3 Bhérer J - LSA,MSc *Professeur adjoint* biologie
4 Bisaillon A - DMV,MScV *Professeur titulaire* anatomie et physiologie vétérinaires
5 Blouin A - DMV,PhD *Professeur agrégé* pharmacologie
6 Dallaire A - DMV,MScV *Professeur titulaire* éthologie
7 Delorme B - BSc,MSc,DSc *Professeur agrégé* biologie expérimentale; biochimie
8 DeRoth L - DMV,PhD *Professeur titulaire* biomédecine
9 Garon O - DMV,PhD *Professeur titulaire* biologie
10 Larivière N - DMV,MSc *Professeur agrégé* anatomie et physiologie vétérinaires
11 Phaneuf LP - DMV,PhD *Professeur titulaire* physiologie
12 Piérard J - DMV,MSc *Professeur titulaire* anatomie
13 Tremblay A - DMV,DSc *Professeur titulaire* biochimie

1537 Département de pathologie et microbiologie
3 200, rue Sicotte, Case postale 5000, Saint-Hyacinthe, Quebec, J2S 7C6
Tel: (514) 773-8521.

1 Fontaine M *Directeur*
2 Beauregard Prof. M - DMV,MSc *Professeur titulaire* pathologie
3 Bigras-Poulin M - DMV,MSc,PhD *Professeur adjoint* épidémiologie
4 Cousineau G - DMV,DV *Professeur titulaire* pathologie
5 ElAzhary Prof. Y - DMV,MSc,PhD *Professeur titulaire* pathologie et microbiologie vétérinaires; virologie
6 Fairbrother JM - BVSc,PhD *Chercheur adjoint* laboratoire de diagnostic
7 Fontaine Prof. M - DMV,MScV,PhD *Professeur titulaire* pathologie et microbiologie; pathologie
8 Fréchette Prof. J - DMV,MSc *Professeur titulaire* parasitologie
9 Hard J - MSc,PhD génie génétique
10 Higgins Prof. R - DMV,DSc *Professeur titulaire* microbiologie
11 Jacques M - BSc,MSc,PhD *Chercheur adjoint* microbiologie et immunologie
12 Khittoo G - BSc,MSc,PhD *Chercheur Adjoint* virologie
13 Lagacé Prof. A - DMV,PhD *Professeur titulaire* pathologie
14 Lallier Prof. R - BSc,PhD *Professeur titulaire* zoologie, sc. bio-cellulaires
15 Lariviére Prof. S - DMV,PhD *Professeur titulaire* microbiologie
16 Mittal KR - BVSc,PhD *Chercheur adjoint* microbiologie et immunologie
17 Morin Prof. M - DMV,PhD,DipACVP *Professeur titulaire* pathologie
18 Roy Prof. RS - DMV,MScV,PhD *Professeur titulaire* pathologie et microbiologie; immunologie
19 Silim Prof. A - DMV,PhD *Professeur agrégé* virologie
20 Simard Prof. B - DMV,MS *Professeur titulaire* biologie de la faune

1538 Centre de recherche en reproduction animale
3 200, rue Sicotte, Case postale 5000, Saint-Hyacinthe, Quebec, J2S 7C6
Tel: (514) 773-8521.

1 Chartrain I *Chercheur Adjointe* reproduction animale
2 Goff AK - BSc,PhD *Chercheur adjoint* physiologie
3 Hunter RHF - BSc,PhD,NDA,DipAgric *Chairman* reproductive physiology

1539 Université du Quebec a Montréal

Succ. "A", C.P. 8888, Montréal, Quebec, H3C 3P8
Tel: (514) 282-3000.

1540 Departement des Sciences Biologiques
Succ. "A", C.P. 8888, Montréal, Quebec, H3C 3P8
Tel: (514) 282-3000.

1 Gingras J - Doctorat 3iéme cycle *Directeur du département* ethologie des insectes d'importance agricole et médicale
2 Alary JG - PhD etudes de substituts au fumier comme amendements et d'extraits d'alcaloïdes á certains herbicides
3 Boileau S - MSc etudes de substituts au fumier comme amendements et d'extraits d'alcaloïdes á certains herbicides
4 Chavalier D - Doctorat en Sciences evaluation toxicologique de pesticides utilisés en agriculture
5 Coderre D - PhD ecologie des invertébrés terrestres; etude comparée des aphidiphages généralistes et spécialistes en culture céréalière et de légumineuses

6 De Oliveira D - PhD ecologie des insectes; pollinisation en agriculture et relations insecte-plante

7 Fournier M - PhD imports des pesticides utilisés en agriculture sur la résistance immunitaire

8 Krzystyniak K - PhD impacts des pesticides utilisés en agriculture sur la resistance immunitaire

9 Maulette Y - PhD impacts de la defoliation chez les plantes. Effets de polluants atmospheriques sue les commautés d'insectes (dépérissement des établiéres)

10 Sarhan F - PhD physiologie et biochimie végétales - adaptation et endurcissement au froid du blé

11 Smoragiewicz W biodégradation des déchets organiques. Fermentation propionique. Analyse des toxines secrétées par des moisissures pathogénes dans les fourrages moisis

12 Tabaeizadeh Z - PhD Génie génétique (biotechnologie) et amélioration des plantes

13 Tourneur JC - Doctorat d'Université equilibres hôtes-entomophages en agriculture; etude de la diapause des coccinellidae aphidiphages du maïs

14 Trottier B - PhD Evaluation toxicologique de pesticides utilisés en agriculture

1541 McGill University

1542 Faculty of Agriculture
Macdonald College, 21111 Lakeshore Road, Ste. Anne de Bellevue, H9X 1CO

1 Buckland RB - PhD *Dean*
2 Couture M - MSc *Associate Dean*
3 Gerols J - PhD *Associate Dean*
4 Stewart RK - PhD *Associate Dean*

1543 Department of Food Science and Agricultural Chemistry
Macdonald College, 21111 Lakeshore Road, Ste. Anne de Bellevue, H9X 1CO

1 Marshall WD - PhD *Associate Professor* chemistry of pesticides, environmental pollutants and food toxicants
2 Alli I - PhD *Assistant Professor* plant proteins and forage preservation
3 Amer V - PhD *Auxiliary Professor* dairy science and technology
4 Anastassiadis Prof. PA - PhD *Professor Post-retirement* glycosaminoglycans, mucopolysaccharide-peptide complexes of the domestic fowl
5 Baker Prof. BE - DSc *Professor Post-retirement* milk and plant proteins, dairy chemistry, dietary fiber, silage fermentation
6 Ecobichon Prof. DJ - PhD *Professor* toxicology
7 Kermasha S - DSc *Assistant Professor* food biochemistry;enzymology
8 Ramaswamy H - PhD *Assistant Professor* food processing and food rheology
9 Rowles W - PhD *Professor Emeritus*
10 Smith JP - PhD *Assistant Professor* food preservation under modified atmospheres, food microbiology
11 van de Voort FR - PhD *Associate Professor and Department Chairman* food analysis, extrusion
12 Yaglagan Y - PhD *Assistant Professor* chemistry of flavour development

1544 Department of Agricultural Economics
Macdonald College, 21111 Lakeshore Road, Ste. Anne de Bellevue, H9X 1CO

1 Coffin HG - PhD *Associate Professor and Chairman of the Department* production and marketing
2 Al-Zand OA - PhD *Associate Professor* agricultural marketing and policy
3 Baker LBB - MSc *Faculty Lecturer* farm management and agri-business
4 Couture MJ - MSc *Faculty Lecturer* agricultural economics, farm management
5 Green C - PhD *Auxiliary Professor* industrial organization
6 Gunjal KR - PhD *Assistant Professor* finance, production economics and economic development
7 Hassan ZA - PhD *Auxiliary Professor* marketing and demand analysis
8 Henning JC - PhD *Assistant Professor* trade and marketing
9 Thomassin PJ - PhD *Assistant Professor* natural resource economics, microrheology and policy analysis

1545 Department of Agricultural Engineering
Macdonald College, 21111 Lakeshore Road, Ste. Anne de Bellevue, H9X 1CO

1 Barrington S - PhD *Assistant Professor* agricultural structures, animal waste management and storage
2 Bolduc G - MSc *University Lecturer* drainage systems, land improvement; buildings
3 Broughton Prof. RS - PhD *Professor* hydrology, irrigation and drainage
4 Kok R - PhD *Associate Professor* food engineering, fermentation, computer applications in food and agriculture
5 Madramootoo C - PhD *Assistant Professor* hydrology, drainage, drip irrigation

6 McKyes E - PhD *Associate Professor* soil mechanics, off-road vehicles, traction, compaction and soil physics; organic waste utilisation
7 Norris ER - PhD *Associate Professor and Department Chairman* agricultural machinery; physical properties of biological materials
8 Prasher SO - PhD *Assistant Professor* hydrology, soil physical properties, drainage and irrigation
9 Raghavan GSV - PhD *Professor* soil mechanics, compaction, storage of vegetables and fruits, grain drying
10 Timbers G - PhD *Auxiliary Professor* food engineering, heat transfer, quality of fruits and vegetables, physical and chemical properties of foods

1546 Department of Animal Science
Macdonald College, 21111 Lakeshore Road, Ste. Anne de Bellevue, H9X 1CO

1 Downey BR - DVM,PhD *Associate Professor and Chairman of the Department* physiology
2 Block E - PhD *Assistant Professor* nutrition, physiology
3 Chavez E - PhD *Associate Professor* nutrition
4 Cue RI - PhD *Assistant Professor* animal genetics
5 Donefer Prof. E - PhD *Professor* nutrition
6 Fischer PWF - PhD *Auxiliary Professor* nutrition
7 Garino EJ - MSc *Faculty Lecturer* nutrition
8 Hayes JF - PhD *Associate Professor* animal genetics
9 Kuhnlein U - PhD *Associate Professor* biochemical genetics
10 Laguë PC - PhD *Associate Professor* physiology
11 Lloyd LE - DSc *Professor Emeritus* nutrition
12 Monardes H - PhD *Assistant Professor* dairy cattle genetics
13 Moxley Prof. JE - PhD *Professor Emeritus* animal genetics
14 Ng Kwai Hang KF - PhD *Associate Professor* biochemistry
15 Phillip LE - PhD *Assistant Professor* nutrition
16 Randall GCB - PhD *Auxiliary Professor* physiology
17 Robitaille G - PhD *Assistant Professor* biochemistry
18 Sanford LM - PhD *Associate Professor* physiology
19 Touchburn Prof. SP - PhD *Professor* nutrition
20 Turner J - PhD *Assistant Professor* physiological genetics

1547 Department of Entomology
Macdonald College, 21111 Lakeshore Road, Ste. Anne de Bellevue, H9X 1CO

1 McFarlane Prof. JE - PhD *Professor and Chairman of the Department* insect biochemistry, insect physiology
2 Boivin G - PhD *Auxiliary Professor* vegetable pest management
3 Dunphy GB - PhD *Assistant Professor* insect pathology and physiology
4 Hill SB - PhD *Associate Professor* soil ecology, acarology
5 Kevan DKMCE - PhD *Emeritus Professor* soil zoology; taxonomy and general biology of Orthoptera
6 Lewis DJ - PhD *Associate Professor* biting flies; aquatic entomology
7 Sanborne PM - PhD *Assistant Professor* insect systematics, Hymenoptera, Coleoptera
8 Stewart Prof. RK - PhD *Professor* insect ecology, pest management
9 Vickery VR - PhD *Emeritus Curator, Lyman Museum* taxonomy and ecology of Orthoptera, apiculture
10 Vincent C - PhD *Auxiliary Professor* fruit pest management
11 Yule Prof. WN - PhD *Professor* insect pathology and toxicology, economic entomology

1548 Department of Microbiology
Macdonald College, 21111 Lakeshore Road, Ste. Anne de Bellevue, H9X 1CO

1 Archibald F - PhD *Assistent Professor*
2 Blackwood AC - DSc *Professor Emeritus* microbial physiology, fermentations
3 Idziak Prof. ES - DSc *Professor* food microbiology, public health bacteriology
4 Ingram Prof. JM - PhD *Professor and Department Chairman* microbial biochemistry, bacterial cell wall and enzyme biochemistry
5 Knowles Prof. R - PhD *Professor* soil microbiology, nitrogen fixation and other nitrogen transformations
6 MacLeod RA - PhD *Professor Emeritus* microbial biochemistry, marine microbiology
7 Martindale DW - PhD *University Researcher* molecular genetics of unicellular eukaryotes
8 Niven DF - PhD *Associate Professor* bioenergetics, Haemophilus disease of swine
9 Tibelius KH - PhD *Assistant Professor* hydrogenase of *Azospirillum* and cloning of hydrogenase genes

1549 Department of Plant Science
Macdonald College, 21111 Lakeshore Road, Ste. Anne de Bellevue, H9X 1CO

1 Klinck Prof. HR - PhD *Professor and Chairman of the Department* cereal crop breeding and physiology
2 Bain JF - PhD *Assistant Professor and Director of McGill Herbarium* taxonomy of vascular plants
3 Bullen MR - PhD *Auxiliary Professor* cytogenetics, chromosome banding in *Phleum*

4 Buszard DJI - PhD *Associate Professor* fruit crop production and physiology
5 Coulman BE - PhD *Associate Professor* forage crop breeding and management
6 Donnelly D - PhD *Assistant Professor* physiology of horticultural crops, tissue culture
7 Estey RH - DSc *Professor Emeritus* nematodes, mycorrhizae and disease of roots
8 Fanous MA - PhD *Associate Professor* agricultural statistics
9 Grant Prof. WF - PhD *Professor* cytogenetics, chemical mutagenesis
10 Ivany JA - PhD *Auxiliary Professor* weed biology, herbicide activity
11 Kushalappa AC - PhD *Assistant Professor* epidemiology, crop disease modelling
12 Lawson NC - PhD *Associate Professor* forage crop management
13 Mather D - PhD *Assistant Professor*
14 Munos-Rivas A - PhD *Assistant Professor* fungal genetics; recombinant DNA technology
15 Peterson JF - PhD *Associate Professor* plant virus diseases, characterization of viruses
16 Reeleder RD - PhD *Assistant Professor* vegetable diseases, biological control
17 Sackston Prof. WE - PhD *Professor Post-retirement* biology and epidemiology of plant diseases
18 Smith DL - PhD *Assistant Professor* crop physiology, nitrogen transfer
19 Sparace SA - PhD *Assistant Professor* plant physiology, biochemistry, host-pathogen interactions
20 Steppler HA - DSc *Professor Emeritus* forage crop management
21 Stewart DW - PhD *Auxiliary Professor* soil-plant-atmosphere modelling
22 Stewart KA - PhD *Associate Professor* vegetable crop production and physiology
23 Warwick SI - PhD *Auxiliary Professor* systematic/genecological studies of weeds
24 Waterway MJ - MSs *Curator of McGill Herbarium* plant taxonomy and evolutionary biology
25 Watson Prof. AK - PhD *Professor* weed biology, biological and chemical weed control

1550 Department of Renewable Resources

Macdonald College, 21111 Lakeshore Road, Ste. Anne de Bellevue, H9X 1C0

1 Titman RD - PhD *Associate Professor* wildlife biology, water fowl ecology
2 Austin JL - PhD *Associate Professor*
3 Barthakur N - PhD *Associate Professor* agricultural meteorology, biometeorology and agricultural physics
4 Bider Prof. JR - PhD *Professor* wildlife biology, vertebrate ecology
5 Bird DM - PhD *Assistant Professor* raptor biology
6 Clark RG - PhD *Auxiliary Professor*
7 Dion HG - DSc *Professor Emeritus* soil chemistry
8 Gerols J - PhD *Associate Professor* french
9 Hendershot WH - PhD *Assistant Professor* soil geomorphology, soil genesis
10 Jones ARC - MSc *Associate Professor* woodlot management
11 MacKenzie Prof. AF - PhD *Professor* soil chemistry, soil fertility
12 Mehuys G - PhD *Associate Professor and Department Chairman* soil physics
13 Murkin HR - PhD *Auxiliary Professor* wetland ecology
14 O'Halloran I - PhD *Assistant Professor* soil fertility, nutrient cycling, soil plant relationships
15 Schuepp PH - PhD *Assistant Professor* transfer processes of heat and mass in the environment
16 Seymour NR - PhD *Auxiliary Professor* waterfowl biology
17 Smith TG - PhD *Auxiliary Professor* arctic marine mammal ecology
18 Whoriskey FG - PhD *Assistant Professor* fisheries biology

1551 Universite Laval

1552 Faculté des Sciences de l'Agriculture et de l'Alimentation

Pavillon Paul-Comtois, Université Laval, Québec, G1K 7P4
Tel: 418 656-3145.

1 Trudel Prof. MJ - BScAgr,PhD *Professeur titulaire et Doyen de la Faculté*
2 Carel Prof. M - DipIngAgr,CertEconAgr,DScEcon *Professeur titulaire et Vice-doyen à l'enseignement*
3 Désilets D - BScApp,MSc *Professeur agrégé et Vice-doyen à la recherche*
4 St-Laurent Prof. GJ - BScAgr,MSc,PhD *Professeur titulaire et Vice-doyen à l'extension*

1553 Departement d'Economie Rurale

Pavillon Paul-Comtois, Université Laval, Québec, G1K 7P4
Tel: 418 656-5912.

1 Ghersi Prof. G - DipIng,MSc,DScEcon *Professeur agrégé et directeur du département* économie agro-alimentaire
2 Calkins Prof. PH - BSc,MSc,PhD *Professeur titulaire* programmation mathématique, finances et risques, économie de la production
3 Carel Prof. M - DipIngAgr,CertEconAgr,DScEcon *Professeur titulaire* développement agricole et rural

4 Debailleul G - DipIngAgr,DEtPolit *Professeur assistant* agriculture et ressources naturelles, économie politique de l'agriculture
5 Dufour Prof. JC - BScAgr,MSc,PhD *Professeur titulaire* marketing agro-alimentaire
6 Dumais Prof. B - BScAgr,MScComDipCompt *Professeur titulaire* comptabilité
7 Gouin Prof. DM - BScAgr,MSc,PhD *Professeur adjoint* politiques agricoles; économic laitière
8 Lent Prof. RJ - BScEcon,MEcon,PhD *Professeur adjoint* commercialisation et fonctionnement des marchés des produits agricoles
9 Levallois Prof. R - DipIng,MSc *Professeur agrégé* gestion d'entreprises agricoles et informatique en gestion
10 Morisset Prof. M - BScSoc,MScSoc,DEtat *Professeur agrégé* politiques économiques
11 Romain Prof. R - BScApp,MSc,PhD *Professeur agrégé* financement et économie de la production
12 Saint-Louis Prof. R - LScCom,DipEtSup,PhD *Professeur titulaire* analyse des politiques et du commerce international agro-alimentaire
13 Wampach Prof. JP - DipIngAgr,LScEcon,PhD *Professeur titulaire* économie du développement agricole

1554 Departement de Genie Rural

Pavillon Paul-Comtois, Université Laval, Québec, G1K 7P4
Tel: 418 656-3173.

1 Boily Prof. R - BScAgr,MSc,PhD *Professeur agrégé et directeur du département* systèmes hydrauliques et tracteurs, énergie, contrôle et instrumentation
2 Adam Prof. R - BScAgr,MSc *Professeur suppléant* machinism agricole
3 Désilets Prof. D - BScApp,MSc *Professeur agrégé* traitement et manutention des grains, génie alimentaire
4 Lagacé Prof. R - BScAgr,MSc,PhD *Professeur agrégé* drainage, irrigation et conservation des sols
5 Marquis Prof. A - BScAgr,MSc,PhD *Professeur agrégé* bâtiments agricoles et environnement des animaux
6 Thériault Prof. R - BSpec,DIng *Professeur agrégé* mécanisation des récoltes et automatisation des opérations en serres
7 Thibald Prof. G - BScAgr,MSc *Professeur suppléant* structures de bâtiments de ferme et entreposage

1555 Departement de Nutrition Humaine et de Consommation

Pavillon Paul-Comtois, Université Laval, Québec, G1K 7P4
Tel: 418 656-2578.

1 Bouchard Prof. A - BScAgr,MSc,PhD *Professeur titulaire et directeur du département* vulgarisation agricole
2 Delisle Prof. P - BScDom,BPéd,MSc *Professeur agrégé* consommation
3 Desrosiers Prof. T - BSc,MSc,PhD *Professeur adjoint* chimie des aliments
4 Dufour-Shakrani Prof. F - BScDom,MSc,PhD *Professeur agrégé* économie de la consommation
5 Durand Prof. G - BEd,BArts,MSc,PhD *Professeur agrégé* méthodes de recherches et textiles
6 Jacques Prof. A - BSc,MSc,PhD *Professeur adjoint* biochimie des aliments
7 Locong Prof. A - BPharm,MArts,DSc *Professeur agrégé* nutrition thérapeutique
8 Normand Prof. MS - BScDom,BScDroit *Professeur assistant* législation et consommation
9 Raymond Prof. M - BScDom,MSc *Professeur agrégé* sciences des aliments
10 Roberge Prof. AG - BScDom,DSc *Professeur titulaire* neurochimie et nutrition
11 Rousseau Prof. R - BScDom,MSc *Professeur agrégé et Secrétaire de la Faculté* gestion des services d'alimentation
12 Savoie Prof. L - BScBioc,MSc,PhD *Professeur titulaire* nutrition humaine
13 Turgeon-O'Brien Prof. H - BSc,MSc,PhD *Professeur adjoint* nutrition humaine
14 Zee Prof. J - MSc,DSc *Professeur titulaire* sciences des aliments

1556 Departement de Phytologie

Pavillon Paul-Comtois, Université Laval, Québec, G1K 7P4
Tel: 418 656-2165.

1 Gendron Prof. G - BScAgr,MSc,PhD *Professeur titulaire et directeur du département* régie des plantes industrielles
2 Asselin Prof. A - BPh,BScAgr,MSc,PhD *Professeur titulaire* phytopathologie et virologie
3 Beeraj Prof. RD - BSc,PhD *Professeur agrégé* anatomie, systématique et biologie florale
4 Bédard Prof. R - BScAgr,MSc,PhD *Professeur titulaire* horticulture
5 Charest Prof. PM - BSc,PhD *Professeur adjoint* biologie cellulaire végétale
6 Dansereau Prof. B - BScHort,MSc,PhD *Professeur agrégé* horticulture ornementale
7 Dion Prof. P - BScAgr,MSc,PhD *Professeur agrégé* microbiologie
8 Dostaler Prof. D - BSc,MSc,PhD *Professeur adjoint* épidémiologie et maladies des plantes
9 Dubé Prof. PA - BScAgr,MSc,PhD *Professeur titulaire* bioclimatologie
10 Gauthier Prof. R - BScBio,MSc,PhD *Professeur agrégé* taxonomie des végétaux

11 Girard Prof. JM - BPh,BScAgr,MScAgr *Professeur titulaire* statistiques appliquées et simulation des cultures fourragères
12 Gosselin Prof. A - BSc,PhD *Professeur adjoint* horticulture
13 Lachance Prof. RA - BScAgr,MSc,PhD *Professeur titulaire* microbiologie
14 Lavoie Prof. V - BScAgr,MSc,DScNat *Professeur titulaire* écologie végétale
15 Leroux Prof. G - BSc,MSc,PhD *Professeur agrégé* malherbologie
16 Ola'h Prof. GM - DipScAgr,DSc *Professeur titulaire* mycologie et microscopie électronique
17 Pauzé Prof. FJ - LScAgr,PhD *Professeur titulaire* anatomie-morphologie végétale
18 Payette Prof. S - BScAgr,LLGéog,MSc,DSc *Professeur agrégé* écologie végétale
19 Rioux Prof. JA - BScAppl,DIng,MSc,DIng *Professeur agrégé* sciences des plantes et horticulture ornementale
20 St-Pierre Prof. CA - BScAgr,MSc,PhD *Professeur titulaire* amélioration génétique des plantes et production des céréales
21 Therrien Prof. HP - BScAgr,MSc,PhD *Professeur titulaire* physiologie végétale
22 Trudel Prof. MJ - BScAgr,PhD *Professeur titulaire* physiologie végétale et horticulture

1557 Departement de Sciences et Technologie des Aliments
Pavillon Paul-Comtois, Université Laval, Québec, G1K 7P4
Tel: 418 656-3854.

1 Martin Prof. GB - BSc,MSc,DSc *Professor titulaire et directeur du département; technologie des produits marins*
2 Amiot Prof. J - BScAgr,MSc,DipNutHum,PhD *Professeur agrégé* technologie alimentaire et nutrition
3 Arul Prof. J - BSc,MSc,PhD *Professeur adjoint* fruits et légumes
4 Boudreau Prof. A - LScAgr,PhD *Professeur titulaire* corps gras alimentaire et produits céréaliers
5 Castaigne Prof. F - DipIngAgr,DSc *Professeur titulaire* procédés industriels alimentaires
6 Goulet Prof. J - BScAgr,PhD *Professeur titulaire* microbiologie alimentaire, fermentations et analyse des aliments
7 Lacroix Prof. C - DIng,PhD *Professeur adjoint* sciences et technologie du lait, génie alimentaire
8 Paquin Prof. P - BSc,MSc,PhD *Professeur adjoint* chimie des aliments
9 Picard Prof. G - BScBio,MSc,PhD *Professeur titulaire* chimie physique, analyse et transformation des aliments
10 Simard Prof. RE - BScAgr,MSc,PhD *Professeur titulaire* microbiologie alimentaire

1558 Departement des Sols
Pavillon Paul-Comtois, Université Laval, Québec, G1K 7P4
Tel: 418-656-2842.

1 Laverdiére Prof. MR - BScAgr,MSc,PhD *Professeur agrégé* conservation et utilisation des sols
2 Antoun Prof. H - BSc,MSc,PhD *Professeur agrégé* sols
3 Bentz Prof. A - IngAgrRT,IngAgrGR,DSc *Professeur titulaire* physique des sols
4 Cailler Prof. M - Mgeo,DEtApp,DipTrly *Professeur agrégé et Directeur du Département* genèse et classification des sols
5 Cescas Prof. MP - BScAgr,MSc,PhD *Professeur titulaire* chimie des sols
6 Parent Prof. LE - BScApp,MSc,PhD *Professeur agrégé* fertilité des sols
7 Visser Prof. SA - DipIng,PhD,DSc *Professeur titulaire* microbiologie et biochimie des sols

1559 Departement de Zootechnie
Pavillon Paul-Comtois, Université Laval, Québec, G1K 7P4
Tel: 418-656-3514.

1 Bernier Prof. JF - SCsAg,MSc,PhD *Professor adjoint* nutrition animale
2 Belzile Prof. R - BScAgr,MSc,PhD *Professeur titulaire* alimentation animale
3 Dufour Prof. JJ - BScAgr,MSc,PhD *Professeur titulaire* production porcine, physiologie de la reproduction bovine et porcine
4 Laforest Prof. JP - BScAgr,MSc *Professeur assistant* physiologie animale
5 Lemay Prof. JP - LiScAgr,MSc,DSc *Professeur titulaire* physiologie de la reproduction, production caprine et ovine, hippologie
6 Martin Prof. JJ - LScAgr,MSc *Professeur agrégé* aviculture
7 Minvielle Prof. F - DipIngAgr,MSc,PhD *Professeur agrégé et Directeur de Département* génétique animale
8 Paré Prof. JP - LiScAgr,MSc,PhD *Professeur titulaire* génétique
9 Seoane Prof. JR - BScAgr,MSc,PhD *Professeur titulaire* nutrition et physiologie animale
10 St-Laurent Prof. GJ - BScAgr,MSc,PhD *Professeur titulaire* nutrition animale

1560 Dalhousie University
Halifax, Nova Scotia
Tel: (902)424-3515.

1561 Department of Biology
Halifax, Nova Scotia
Tel: (902)424-3515.

1 Kimmins Prof. WC - PhD *Professor and Chair* plant virology (host resistance mechanisms); cell wall synthesis
2 Angelopoulos EW - MS,PhD *Associate Professor* ultrastructure and microtubules of trypanosomatidae; ultrastructure of *Trichomonas tenax*
3 Boutilier RG - MSc,PhD *Assistant Professor* respiratory and cardiovascular physiology in lower vertebrates
4 Brown Prof. RG - MSc,PhD *Professor* microbial metabolism of carbohydrates and their chemical derivatives
5 Cameron Prof. ML - MSc,PhD *Professor* invertebrates fine structure (especially insects) and virus research
6 Chapman Prof. ARO - PhD *Professor* experimental autecology and taxonomy of seaweeds particularly the larger Phaeophyta
7 Doyle Prof. RW - MSc,PhD *Professor* intra-population variability in shallow and deep-living marine organisms: growth and nutrient-uptake kinetics of competing cell populations
8 Farley Prof. J - MSc,PhD *Professor* parasitology; history of biology
9 Fentress Prof. JC - BA,PhD *Professor* major appointment in psychology; behavioural biology
10 Freedman Prof. B - MSc,PhD *Professor* forest ecology
11 Garside Prof. ET - MA,PhD *Professor* thermal, osmoregulative physiology and structural analysis in fishes
12 Haley Prof. LE - MSA,PhD *Professor* embryonic development of birds; genetics of oysters
13 Hall Prof. BK - PhD,DSc *Professor* developmental biology, cell differentiation, organ regulation
14 Harvey MJ - PhD *Associate Professor* plant biosystematics and biogeography; plant ecology
15 Hicks GS - MSc,PhD *Associate Professor* plant development, morphogenesis, histology, organ culture, regulatory mechanisms
16 Kamra Prof. OP - MS,PhD *Professor* radiation genetics and chemical mutagenesis, cytogenetic effects of food additives and insecticides
17 Lane Prof. PA - MSc,PhD *Professor* community ecology; structure and stability of plankton communities; competition-predation phenomena; subtropication
18 Lee RW - MA,PhD *Associate Professor* genetics; characterization and regulation of chloroplast DNA in *Chlamydomonas*
19 MacRae TH - MSc,PhD *Assistant Professor* microtubule synthesis
20 McBride RP - MSc,PhD *Associate Professor* microbial ecology disturbed habitats
21 McLaren Prof. IA - MSc,PhD *Professor* population, evolutionary ecology of zooplankton, birds and sea mammals
22 Newkirk GF - PhD *Associate Professor (Research)* aquaculture genetics (especially of cultured molluscs); aquaculture nutrition, reproduction, stock management
23 Novitsky JA - BSc,PhD *Associate Professor* microbial ecology
24 O'Dor Prof. RK - PhD *Professor* endocrinology and reproductive physiology; particularly of cephalopods and other molluscs; evolution of regulatory systems
25 Ogden Prof. JGIII - MA,PhD *Professor* environmental distribution of Sr90, radiocarbon dating; pollen stratigraphy
26 Patriquin DG - MSc,PhD *Associate Professor* plant physiology
27 Rose MR - MSc,PhD *Associate Professor* life-history evolution; genome evolution; theoretical population biology
28 Scheibling RE - BSc,PhD *Assistant Professor* community ecology of rocky subtidal zone; disease and mass mortality of sea urchins
29 Sirard Prof. MA - DMV,PhD *Professeur adjoint* biotechnologies animales
30 Vining Prof. LC - MSc,PhD *Professor* microbial biochemistry; biosynthesis of antibiotics; metabolic controls
31 von Maltzahn Prof. KE - MS,PhD *Professor* regeneration and reproduction in lower plants
32 Whitehead H - MA,PhD *Assistant Professor (Research)* population ecology of whales
33 Willison JHM - PhD *Associate Professor* cell biology
34 Wright JJ - BSc,PhD *Assistant Professor* control of gene expression
35 Zouros Prof. E - MSc,PhD *Professor* population and ecological genetics; the genetical basis of adaptation and evolution

1562 Nova Scotia Agricultural College
Truro, NS, B2N 5E3
Tel: 902-895-1571.

1 MacRae HF - BScAgr,MSc,PhD *Principal*

1563 Department of Agricultural Engineering
Truro, NS, B2N 5E3
Tel: 902-895-1571.

1 Adams J - BSc,MSc *Associate Professor and Head of Department* farm structures
2 Adsett JF - BScAgrEng,MScE *Assistant Professor* farm machinery
3 Allen DA - BSc,MSc *Associate Professor* ventilation and storage
4 Cunningham JD - BSA,BE,MASc *Associate Professor* materials handling
5 Desir FL - BScAgrEng,MSc *Assistant Professor* cropping machinery

6 Havard PL - BScAgrEng,MSc *Associate Professor* microprocessors and controls
7 Madani SA - BSc,MSc,PhD *Assistant Professor* soil and water
8 Rifai MN - MSc,PhD *Assistant Professor* power and machinery

1564 Department of Animal Science
Truro, NS, B2N 5E3
Tel: 902-895-1571.

1 Cock Prof. LM - BSc,MSc,PhD *Professor and Head of Department and Chairman Research Committee*
2 Anderson DM - BSA,MSc,PhD *Professor* swine management and nonruminant nutrition
3 Connor ML - BSc,MSc,PhD *Associate Professor* reproductive physiology
4 Crober Prof. DC - BSc,MSc,PhD *Professor* poultry management
5 Firth NL - BSc,MSc,PhD *Assistant Professor* animal products
6 Fredeen AH - BSA,MSc,PhD *Assistant Professor* ruminant nutrition
7 Mathewson WG - BScAgr,DTA,MSc *Associate Professor* beef and sheep management
8 Patterson DL - BSc,MSc,PhD *Assistant Professor* animal genetics
9 Tennessen T - BA,BSc,MSc,PhD *Assistant Professor* animal behaviour

1565 Department of Biology
Truro, NS, B2N 5E3
Tel: 902-895-1571.

1 McFadden Prof. LA - BScAgr,MSc,PhD *Professor and Head of Department* plant pathology
2 Eaton LJ - BSc,MSc *Associate Professor* blueberry nutrition and management
3 Gray AB - BSc,MSc *Assistant Professor* plant pathology
4 Le Blanc JR - BA,BSc,PhD *Associate Professor* entomology - integrated pest management
5 Olson AR - BA,MSc,PhD *Associate Professor* structural botany
6 Sampson MG - BSc,BScAgr,MSc *Assistant Professor* weed biology
7 Stratton GW - BScAgr,MSc,PhD *Associate Professor* environmental microbiology

1566 Department of Chemistry & Soils
Truro, NS, B2N 5E3
Tel: 902-895-1571.

1 MacConnell HM - BScAgr,MSc *Associate Professor and Head of Department*
2 Brewster GR - BA,MSc,PhD *Assistant Professor* pedology
3 MacLean KS - BSc,MSc *Associate Professor* trace elements
4 Miller JC - BScAgr,MSc *Assistant Professor* soil physics
5 Robinson Prof. AR - BScAgr,MSc,PhD *Professor* fertility in dairy cattle
6 Warman PR - BSc,MSc,PhD *Professor* soil biochemistry

1567 Department of Economics and Business Management
Truro, NS, B2N 5E3
Tel: 902-895-1571.

1 Tait JC - BScAgr,MSc *Associate Professor and Head of Department*
2 Brennan JJ - BScAgr,MSc *Assistant Professor* economics of production
3 Grant KG - BA,MA,PhD *Assistant Professor* market analyses
4 Stackhouse SJB - BScAgr,MSc *Associate Professor* computers in farm management

1568 Department of Plant Science
Truro, NS, B2N 5E3
Tel: 902-895-1571.

1 Caldwell CD - BSc,MSc,PhD *Associate Professor and Head of Department* cereal production
2 Bubar Prof. JS - BScAgr,MS,PhD *Professor* crop evaluation
3 Daniels RW - BSc,MSc *Associate Professor* turf production
4 Fraser J - BSc,MSc,PhD *Assistant Professor* cold hardiness in legumes
5 Haliburton TH - BScAgr,MS *Associate Professor* vegetable production
6 Ju H - BSc,MSc,PhD *Associate Professor* nutrition in horticultural crops
7 Nowak J - MSc,PhD,PhDHabil *Associate Professor* plant biotechnology

1569 Technical University of Nova Scotia

P.O. Box 1000, Halifax, Nova Scotia, B3J 2X4
Tel: (902) 429-8300. Telex: (TUNS) 019-21566.

1 Callaghan JC *President*

1570 Department of Agricultural Engineering
P.O. Box 1000, Halifax, Nova Scotia, B3J 2X4
Tel: (902) 429-8300. Telex: (TUNS) 019-21566.

1 Watts Prof. KC - BSAAgrEng,MSc,PhDMechEng *Professor, Head* farm power and machinery
2 Ben-Abdallah N - BScAgrEng,MASc,PhD *Associate Professor* building environmental studies

3 Burney JR - BScAgrEng,MScEng,PhD *Associate Professor* soil and water conservation
4 Ghaly AE - BScEngAgr,MScEng,PhD *Associate Professor* agricultural waste management
5 Wilkie KI - BEngAgr,MScEng,PhD *Associate Professor* instrumentation

1571 Canadian Institute of Fisheries Technology
P.O. Box 1000, Halifax, Nova Scotia, B3J 2X4
Tel: (902) 429-8300. Telex: (TUNS) 019-21566.

1 Ablett RF - MSc,PhD *Assistant Professor* food science and aquaculture technology; blueberry processing
2 Ackman Prof. RG - MSc,DIC,PhD *Professor* edible fats and oils
3 Bligh Prof. EG - MSc,PhD *Professor, Director* food process technology
4 Gill TA - MSc,PhD *Associate Professor* food chemistry; computerized thermal processing of foods; animal waste digestion
5 Merritt Prof. JH - BEng,MSc *Professor* food refrigeration and frozen storage; food smoking and drying

1572 Mount Allison University

Sackville, New Brunswick
Tel: (506) 364-2500.

1573 Biology Department
Sackville, New Brunswick
Tel: (506) 364-2500.

1 Fensom DS - BaSc,CChem,FRSC *Professor Emeritus* translocation and applied electrophysiology of frost hardiness and water stress
2 Harries Prof. H - BSc,MSc,PhD *Professor* land use and forest ecologist
3 Ireland RJ - BSc,PhD *Associate Professor* plant nitrogen metabolism
4 Thompson Prof. RG - BSc,MSc,PhD *Professor* translocation and use of ^{11}C and ^{13}N in plant physiology

1574 University of New Brunswick

1575 Faculty of Forestry
Bag Service No. 44555, Fredericton, New Brunswick, E3B 6C2
Tel: (506) 453-4501.

1 Baskerville GL - BScF,MF,PhD *Dean*
2 VanSlyke AL - BScF,MF *Assistant Dean*
3 Dickson A - BSc,MS,PhD *Co-ordinator of Continuing Education*

1576 Department of Forest Engineering
Bag Service No. 44555, Fredericton, New Brunswick, E3B 6C2
Tel: (506) 453-4501.

1 Rickards Prof. EJ - DiplEng,DiplMan *Professor and Chairman* industrial engineering
2 Bjerkelund Prof. TC - BScF *Professor* tree harvesting systems and equipment
3 Douglas RA - BASc,PhD *Assistant Professor* soil mechanics
4 Meng Prof. CH - BSc,MScF,PhD *Professor* forest operations research
5 Needham TD - BSc,MSc,PhD *Assistant Professor* applied silviculture
6 Robak EW - BScFE,MBA *Associate Professor* forest operations administration
7 Sebastian Prof. LP - DiplForEng,MS,PhD *Professor* wood science; anatomical, physical and mechanical properties of wood
8 Short Prof. CA - BScE,MScE *Professor* mechanical aspects of timber harvesting; woodchip harvesting and combustion
9 Smith I - BScE,MSc,PhD *Assistant Professor* wood structures

1577 Department of Forest Resources
Bag Service No. 44555, Fredericton, New Brunswick, E3B 6C2
Tel: (506) 453-4501.

1 Easley Prof. AT - BScF,MScF,PhD *Professor and Chairman* outdoor recreation
2 Arp PA - BSc,PhD *Associate Professor* forest soils and environmental chemistry
3 Baskerville Prof. GL - BScF,MF,PhD *Professor* forest ecology
4 Clements SE - BSc,MSc,PhD *Assistant Professor* forest economics
5 Daugharty DA - BScF.MScF *Senior Instructor* forest hydrology
6 Dickison Prof. RBB - BSc,MA *Professor* climatology; forest meteorology and forest hydrology
7 Dickson Prof. A - BSc,MS,PhD *Professor* extension forestry and communications
8 Jamnick MS - BScF,MScF,PhD *Assistant Professor* forest management science
9 Jordan GA - BScF,MScF *Associate Professor* computer applications in forestry
10 Keppie Prof. DM - BS,MS,PhD *Professor* wildlife ecology
11 Krause Prof. HH - DiplFor,PhD *Professor* forest soils and tree nutrition
12 Methren IR - BScF,PhD *Professor* silviculture and fire
13 Morgenstern Prof. EK - BScF,MScF,DrrerNat *Professor* forest tree genetics & breeding
14 Oliver FS - BSc,MScF *Associate Professor* photointerpretation and remote sensing

15 Powell Prof. GR - BSc,MSc,PhD *Professor* tree development and reproduction
16 Quiring DTW - BSc,PhD *Assistant Professor* forest entomology
17 Roberts MR - BS,MS,PhD *Assistant Professor* forest ecology and silviculture
18 Savidge RA - BScF.MScF,PhD *NSERC Univ. Research Fellow* tree physiology and biotechnology
19 Schneider Prof. MH - BS,MS,PhD *Professor* wood science; products and wood-polymer interactions; wood chip combustion
20 Sebastian Prof. LP - DiplForEng,MS,PhD *Professor* wood science; anatomical, physical and mechanical properties of wood
21 VanSlyke Prof. AL - BScF,MF *Professor* forest measurements and statistical methods
22 Whitney NJ - BSc,MSc,PhD,BD *Professor* mycology and phytopathology

1578 Faculty of Science
Bag Service No. 45111, Fredericton, New Brunswick, E3B 6E1
Tel: (506) 453-4584. Telex: 014-46-202.

1579 Department of Biology
Bag Service No. 45111, Fredericton, New Brunswick, E3B 6E1
Tel: (506) 453-4584. Telex: 014-46-202.

1 Riding Prof. RT - BS,MS,PhD *Chairman and Professor* plant morphogenesis
2 Burt Prof. MDB - BSc,PhD *Professor* biology of platyhelminths and parasite ecology
3 Cashion Prof. PJ - BSc,PhD *Professor* biochemistry; nucleic acid
4 Coombs DH - BA,PhD *Associate Professor* virology and molecular biology
5 Cowan Prof. FBM - BA,MA,PhD *Professor* ultrastructure and osmoregulation; transport epithelia
6 Cumming Prof. BG - BSc,PhD *Professor* germination, growth and development in plants
7 Dilworth Prof. TG - BSA,MSc *Professor* wildlife ecology
8 Hagen Prof. DW - BS,MA,PhD *Professor* population genetics and biology, fish
9 Keppie DM - BS,MS,PhD *Associate Professor (Forestry and Biology)* wildlife ecology, populations
10 Krause Prof. MO - BSc,MS,PhD *Professor* control of gene expression in normal and virus-transformed cells; role of small RNAs in cell differentiation
11 Lynch WH - BSc,PhD *Associate Professor* physiology of aquatic psychrotrophic bacteria
12 MacKinnon BM - BSc,PhD,URF reproductive morphology and physiology of parasitic helminths
13 McKenzie Prof. JA - BA,MA,PhD *Professor* ethology and biochemical systematics of fishes
14 Paim Prof. U - BA,PhD *Professor* comparative and environmental physiology; aquaculture
15 Seabrook Prof. WD - BSc,MSc,PhD *Professor* insect neurophysiology and behaviour
16 Shore JS - BSc,PhD *Assistant Professor* evolutionary genetics and systematics of plants
17 Sivasubramanian Prof. P - BSc,MSc,PhD *Professor* endocrinology and developmental neurobiology of insects
18 Smith BP - BSc,MSc,PhD,URF parasite ecology
19 Taylor Prof. ARA - BA,PhD *Professor* marine phycology; biology of seagrasses
20 Wein Prof. RW - BSA,MSc,PhD *Professor* northern plant ecology; fire ecology
21 Whitney Prof. NJ - BSc,MSc,PhD,BD *Professor* mycology; phytopathology
22 Wiggs Prof. AJ - BSc,MSc,PhD *Professor* physiology of poikilothermic vertebrates
23 Yoo Prof. BY - BSc,MS,PhD *Professor* cell biology of plant development

CYPRUS

1580 Ministry of Agriculture and Natural Resources

Head Office, Nicosia
Tel: 02-30-2247. Telex: 4660 MINAGRI CY. Cables: MINAGRI. 445156.

1581 Department of Agriculture

Nicosia
Tel: 02-30-2250. Telex: 4660 MINAGRI. Cables: MINAGRI.

1 Phocas C - BScAgr,MScAgr *Director*
2 Serghis L - MScAnSc *Head Animal Husbandry*

1582 Agronomy

1 Constantinou D - BScAgr,DiplComprRegDev *Head of Section* agronomy
2 Joannides G - BScAgr,DiplAgron agronomy
3 Xenophontos E - BScAgr,MScPlBreed seed production

1583 Land Use

1 Grivas G - IngAgr *Head of Section*
2 Chattalos C - DepAgrIng,DrAgrSoilSc plant nutrition
3 Koudounas C - BScAgr,MScSoils soils
4 Michaelides P - BScAgrEng,MSc,Soil&WaterUse soil conservation

1584 Horticulture

1 Kyriacou I - BScHort,MAgr *Head of Section*
2 Charis C - BSc,CertOlProd olive culture
3 Papandreou T - BScAgr,MScHort horticulture
4 Patsalos K - BScAgr,MScCrop,PhD vegetables
5 Stavrides S - BScAgr horticulture
6 Theofilou Mrs V - BScAgr,CertFlor floriculture

1585 Water Use

1 Kalimeras P - BScAgr *Head of Section*
2 Zemenides S - HNDMechEng water use

1586 Extension

1 Neocleous G - BScAgr,DipAgrExt *Head of Section*

1587 Analytical Laboratories

1 Melifronides J - BScAgr,DIC,PhDApplEnt *Head of Section*

1588 Plant Protection

1 Zyngas J - BScAgr,MScHort,PhDPlPath *Head of Section*
2 Aziz-Demova Mrs M - BScAgr plant pathology
3 Christofi S - MScAgr,MScInsTax insect taxonomy
4 Gavriel J - BScAgr,DrAgr virology
5 Krambias A - BScAgr,MScEnt,PhDEnt entomology
6 Markoullis G - BScAgr,MScAgr weed control

1589 Agricultural Economics

1 Archimandritou Mrs M - BScAgr,PGDiplPrAnal agricultural economics
2 Camelaris G - MSc,AgrEco *Head of Section*
3 Alevra Rea Mrs - BScEcon agricultural economics
4 Herodotou C - BScAgr,PlProt,MScAgr agricultural economics
5 HjiMichael S - BScAgr,DiplPrPl agricultural economics
6 Loizides C - MScAgrEcon agricultural economics
7 Stavrou P - BScAgr,MScAgrEc agricultural economics

1590 Animal Improvement & Nutrition

1 Lysandrides P - BScAgr *Ag. Head of Section*
2 Antoniou T - BScAgr,PhD animal nutrition
3 Constantinou A - DiplIngAgr,DiplAnBr,DrAgrAnBr animal improvement

1591 Cattle & Dairy

1 Liasides A - BScAgr *Officer-in-charge*

1592 Swine

1 Pratsos K - BScAgr,PhDProd *Officer-in-charge*

1593 Poultry

1 Charalambous K - BScAgr,MScPoulProd *Officer-in-charge*

1594 Sheep & Goats

1 Morphakis K - BScAgr,MScAnNutr *Officer-in-charge*

1595 Viticulture & Oenology

1 Roumbas N - BScAgr,DipVit *Officer-in-charge*
2 De Martinez Mrs P - BScChem oenology
3 Frangos A - BScNatSc,DipOen oenology

1596 Agricultural Research Institute

Nicosia
Tel: 02-40-3431. Telex: 4660 MINAGRi. Cables: ARI.

1 Serghiou CS - BSc,PhDEntom *Director*

1597 Agricultural Economics

1 Papachristodoulou S - BScAgr,MSc,PhD *Officer-in-Charge* agricultural economics
2 Panayiotou G - BScAgr,MSc agricultural economics
3 Papayiannis C - BScAgr,MSc agricultural economics

1598 Agronomy

1 Hadjichristodoulou A - BScAgr,MSc,PhD *Officer-in-Charge* cereals improvement
2 Della Mrs A - BScAgr,MSc cereals improvement
3 Droushiotis DN - BScAgr,MSc fodders and pastures
4 Josephides C - BScAgr wheat improvement
5 Papastylianou J - BSc,PhD agronomy

1599 Animal Production

1 Economides S - BScAgr,MSc,PhD *Officer-in-Charge* animal nutrition
2 Hadjipanayiotou M - BScAgr,PhD animal nutrition
3 Mavrogenis AP - BScAgr,MSc,PhD animal breeding
4 Papachristoforou C - BScAgr,MSc animal breeding

1600 Chemistry

1 Hadjidemetriou D - MScChem *Officer-in-Charge* analytical chemistry and pesticide residue analysis

1601 Horticulture

1 Vakis NJ - BScAgr,MSc,PhD *Officer-in-Charge* vegetables and post-harvest horticulture
2 Aziz IH - BSc,PhD viticulture
3 Kokkalos TI - BScHort,MSc pomology and post-harvest horticulture
4 Myrianthousis TS - RDOen,BScOen viticulture and oenology

1602 Plant Protection

1 Americanos PG - BScHort,MPhil weed control
2 Ioannou N - BScAgr,MSc,PhD plant pathology
3 Iordanou N - BScAgr,MSc economic entomology
4 Kyriacou Miss A - BA,PhD virology
5 Orphanides GM - BScEnt,PhD biological control
6 Philis J - BScAgr,DIC,MPhil plant nematology

1603 Soils and Water Use

1 Orphanos PI - BScAgr,PhD *Officer-in-Charge* soil fertility and plant nutrition
2 Eliades G - BScAgr,MSc water use
3 Metochis C - BScAgr,MSc water use
4 Papadopoulos I - MScHort,PhD plant nutrition and reuse of waste water
5 Stylianou Y - BScAgr,MSc water use

1604 Department of Forests

Nicosia
Tel: 02-30-2261. Cables: FOREST CYPRUS.

1 Leontiades LI - BScFor,PostGradFor,MEOU *Director of the Department of Forests*
2 Iacovides MC - DegFor,DipFor *Senior Conservator of Forests*
3 Pantelas VS - DegFor,DipEcon,MScResMang *Senior Conservator of Forests*
4 Peonides LE - DegFor,DipPhoIntFor *Senior Conservator of Forests*
5 Theophanous S - DegFor,PostGradFor *Senior Conservator of Forests*

1605 Department of Veterinary Services

Head Office, P.O. Box 2006, Nicosia
Tel: 02-40-2135. Telex: 4660 MINAGRI. Cables: CHIVET CYPRUS.

1 Polydorou K - DVM *Director* Echinococciasis research
2 Andreou K - DVM,MScTox *Vet. Officer A'* biological products research
3 HjiSavvas T - DVM *Senior Veterinary Officer* IBR blue tongue research
4 Loucaides P - DVM *Vet. Officer A'* investigation of mortality in neonatal lambs and kids
5 Neophytou G - DVM,MSc,PhD *Vet. Officer A'* oestrus synchronisation in ewes and goats research
6 Orphanides A - DVM *Vet. Officer Class I* survey of blood samples from slaughtered pigs for Brucellosis
7 Papaprodromou M - DVM leptospirosis, toxoplasmosis, Q fever and epizootic abortions investigation
8 Toumazos P - MVSc,DVM,MPh abortions investigation; scrapie investigation

1606 Department of Fisheries

Tagmatarchou Pouliou 5-7, Nicosia
Tel: 02-30-3279. Telex: 4660 MINAGRI. Cables: MINAGRI CY.

1 Demetropoulos A - MSc *Head of Department*

1607 Research Division

1 Livadas R - MSc *Head of Division* stock assessment
2 Athanassiadou Mrs L *Head of Section* oceanography; heavy metals
3 Hadjichristophorou Miss M *Head of Section* benthos; ecology; conservation
4 Loizides L - MSc *Head of Section* marine pollution
5 Hadjistephanou N - MSc stock assessment; crustacea

1608 Pelagic Fish Division

1 Economou E - MSc *Head of Division* pelagic fish
2 Condeatis D pelagic fish

1609 Fish Culture and Inland Water Management Division

1 Stephanou Mrs D *Head of Division* fish culture
2 Anastasiades G *Head of Section* fresh water fish culture
3 Georgiou G *Head of Section* marine aquaculture

FIJI

1610 Ministry of Primary Industries

PO Box 358, Suva

1 Yarrow R - BVetSc *Permanent Secretary*
2 Patel N - BAgricSc *Deputy Secretary for Primary Industries*

1611 Training and Communication Section

1 Umar M - DTA,BAgriSc,CertRuralExt,MSc *Head of Section*

1612 Economic Planning and Statistics Section

1 Basha MI - BAgriSc *Head of Section*

1613 Extension Division

1 Ratuvulu L - DTA,BAgricSc *Director*

1614 Animal Health and Production Division

1 Singh DR - BVSc,MVSc *Director*
2 Tabunakawai N - BVSc,MVSc *Principal Veterinary Officer*

1615 Research Division

Koronivia Research Station, PO Box 77, Nausori

1 Sivan P - BAgricSc,MSc,PhD *Director* tropical root crops breeding and agronomy

1616 Koronivia Research Station

1617 Crop Production

1 Reddy N - BSc *Head of Agronomy* rice breeding and agronomy
2 Duve RN - BSc,MSc,PhD *Head of Chemistry* food chemistry
3 Kumar J - BAgricSc,MSc plant pathology
4 Kumar K - BSc,MAgricSc entomology

1618 Animal Production

1 Ranacou E - BAgricSc *Principal Agricultural Officer, Pastures* tropical pastures
2 Krishna M - BSc,MSc tropical pastures

1619 Sigatoka Research Station

1620 Crop Production

1 Iqbal M - BAgricSc,MSc *Head of Horticulture* vegetables and fruits agronomy

1621 Animal Production

PO Box 77, Nausori

1 Chand S - BAgric tropical pastures

1622 Fisheries Division

Tel: 361122.

1 Cavuilati ST - BSc,MSc *Director*
2 Sewak S - BSc,DipFisheriesMgt *Principal Fisheries Officer, Technical Service*

1623 Drainage and Irrigation Division

Tel: 312355.

1 Nath V - DTA,BSc,MSc *Director*

1624 Forestry Department

PO Box 2218, Suva
Tel: 23833. Cables: Forestry.

1 Yabaki KT - BScFor *Conservator of Forests*
2 Chang A - BScFor,DipFor *Deputy Conservator*
3 Swarup R - BScFor *Deputy Conservator*

1625 Research Division

Cables: Forestry.

1626 Timber Research

Cables: Forestry.

1 Alston AS - BSc *Principal Utilisation Officer*

1627 Fiji College of Agriculture

Tel: 47044.

1 Rigamoto MM - BScAgr,DipHomeEcon *Principal*

1628 Fiji Sugar Corporation Ltd.

PO Box 283, Dominion House, Suva
Tel: 313455. Telex: 2119.

1 Ali AR - AASA,CA,ANZIM,FBIM *Head - Managing Director*

1629 The Sugar Cane Research Centre

Fiji Sugar Corporation Ltd., PO Box 63, Lautoka
Tel: 60800. Telex: FJ5175.

1 Murti K - BSc,MA *Head - Research Manager*
2 Prasad CS - MA,PhD,CertAppSc *Research Officer*

GAMBIA

1630 Ministry of Agriculture

Central Bank Building, Banjul

1631 Department of Agriculture

1 Alphu Marong *Agronomist* rice
2 Boughton D *Economist* management economics
3 Cole MA *Agronomist* general agronomy
4 Cox A crop improvement
5 Mbenga M *Agronomist* upland cereals
6 Owens S *Agronomist* maize agronomy
7 Senghore T *Agronomist* cropping systems
8 Sompo Ceesay NMS *Assistant Director* research management

1632 Crop Protection Services

1 Jagne DCA *Director* post harvest/storage entomology
2 Bruce-Oliver S *Scientific Officer* entomology of growing crops
3 Mbenga EO *Scientific Officer* food storage
4 Sagnia SB *Senior Scientific Officer* entomology of growing crops
5 Trawally BB *Scientific Officer* phytopathology

1633 Department of Forestry

1 Taal BM - BSc,MSc *Director* forestry
2 Bojang F - BSc,MSc *Assistant Director* forestry
3 Danso AA - BSc *Forestry Officer* forestry

1634 Department of Fisheries

1 Cham M aqua culture, oyster culture research
2 Jones A fishing gear technology
3 Lloyd-Evans Miss A fisheries statistics/stock assessment
4 N'jie AE fisheries post harvest technology
5 Ndene JB fisheries stock assessment and post harvest technology

1635 Council for Scientific and Industrial Research

PO Box M.32, Accra
Tel: 777651. Cables: Sciences, Accra.

1 Butler RGJ - BSc,MSc,PhD,FInstP,FIEE *Director-General*
2 Koli SE - BSc,PhD *Coordinator of Agricultural Research*
3 Adu A - BSc,MIBiol *Secretary*
4 Villars JA - FLA,MSc *Deputy Librarian* information science

1636 Animal Research Institute

PO Box 20, Achimota, Accra
Tel: 664301 Ext.63 777661 Ext.47. Cables: Animres.

1 Sapong D - BSc,MSc *Acting Director* studies of the abundance and distribution of helminth parasites of farm animals in Ghana; use of the traditional medicine in the control of helminth parasites of farm animals
2 Aggrey EK - BSc,MPhil evaluation of agricultural residues and agro-industrial by-products as feed for ruminants (sheep and goats)
3 Bempong IA - BSc,MSc studies on growth rate of calves on whole milk, milk replacer and dry calf-starter
4 Boa-Amponsem K - BSc,DipASci,MPhil studies on heat tolerance in chicken
5 Bruce T - BSc,MS protozoon parasites in livestock with special reference to (i) coccidia (ii) babesiosis, theileriosis anaplasmosis (iii) the control of cattle ticks by use of acaricidal properties
6 Kwansa W - BSc,MSc causes of preweaning mortality in piglets on commercial farms
7 Monney BG - BSc,PhD a study of livestock production systems
8 Okantah SA - BSc,PhD cattle breeding research
9 Okine J - BSc mass breeding of tsetse for an eventual use in an integrated eradication/control program for north of Ghana
10 Osei-Somuah A - BSc,DVM effects of local herbs and chemicals on mange in pigs
11 Rhule SWA - BSc,PGDip,PhD digestibility studies on sheanuts cake with pigs

1637 Crops Research Institute

PO Box 3785, Kumasi
Tel: 6221-2. Cables: Cropsearch.

1 Addison EA - BSc,MSc *Acting Director* phytosanitary, general plant pathology

1638 Food storage Section

1 Ofosu A - BSc,MSc *Head of Section* food storage, preservation of maize on the cob in farmers' cribs and improvement of rural storage methods

1639 Plant Pathology Section

1 Heming OB - BSc,PhD,DIC *Head of Section* plant nematology
2 Twumasi JK - BSc,MSc,PhD general plant pathology

1640 Horticulture Section

1 Abutiate WSY - BSc,MSc *Head of Section* fruit crops - citrus, mango, pear, pineapples
2 Agble F - BSc,MSc,PhD breeding and agronomy of tomatoes and onions

1641 Plant Breeding Section

1 Twumasi-Afriyie S - BSc,MSc *Head of Section* maize breeding
2 Ampong JE - BSc,MSc maize breeding
3 Asafo-Adjei B - BSc,MSc cowpea breeding
4 Atuahene-Amankwa G - BSc,MSc cowpea breeding
5 Badu-Araku B - BSc,MSc maize breeding
6 Dekuku RMC - BSc,MSc rice breeding
7 Marfo KO - BSc cowpea, groundnut breeding
8 Mercer-Quashie H - BSc,MSc sorghum, soyabeans, millet breeding
9 Sallah PYK - BSc,MSc maize breeding

1642 Agronomy Section

1 Kissiedu AFK - BSc,MSc *Head of Section* maize,cassava, cowpea, sweet potato and yam agronomy
2 Aflakpui HKS - BSc maize, cassava and cowpea
3 Amankwatia YO - BSc,MSc bast fibres
4 Ampong-Nyarko K - BSc,MSc,PhD weed science
5 Asafu-Adjei JN - BSc,MSc maize, cassava and cowpea
6 Nyamekye A - BSc,MSc soil fertility
7 Quao EO - BSc,MSc tobacco

1643 Entomology Section

1 Duodu YA - BSc,MSc,PhD *Head of Section* biological control of insect pests and weeds
2 Afun VJ - BSc insect pests of cowpea
3 Braimah H - BSc biological control of insect pests and weeds
4 Domber T - BSc insect pests of cowpea

1644 Plant Genetic Resources Section

1 Bennet-Lartey SO - BSc,MSc *Head of Section* plant introduction and exploration

1645 Agricultural Statistics

1 Marfo KA - BSc *Head of Section* experimental statistics
2 Dapaah SO - BSc experimental statistics

1646 Publications and Library Section

1 Glime TK - BA *Head of Section* scientific information

1647 Agricultural Economics Section

1 Ibrahim ARS - BSc,MSc *Head of Section* farm surveys

1648 Microbiology Section

1 Dakora FB - BSc,MSc *Head of Section* nitrogen fixation
2 Kaleem Mrs F - BSc,MSc nitrogen fixation

1649 Food Research Institute

PO Box M.20, Accra
Tel: 777330. Cables: FoodSEARCH.

1 Eyeson KK - BSc,DipNut,DipFd,MRSH,AIFST *Director*

1650 Analysis Division

1 Halm Miss M - BSc,MSc *Acting Head of Division* food storage
2 Asibey-Berko E - BSc,PhD nutrition and biochemistry
3 Erskine-Ampah Mrs C - BSc,MSc food microbiology
4 Hayford Miss A - BSc food chemistry and microbiology
5 Johnson PNT - BSc food chemistry and food storage
6 Kafui Mrs K - BSc nutrition and biochemistry
7 Lokko Mrs P - BSc,MSc nutrition and biochemistry
8 Plahar WA - BSc,MSc,PhD food chemistry
9 Tete-Marmon J - AIMLT food technology and microbiology
10 Vomotor KA - BSc food storage

1651 Processing Division

1 Abbey LD - BSc fish and fish products
2 Amoah-Awua WKA - BSc,MAppSc fruits and vegetables biochemistry
3 Andah Mrs A - BSc,MPhil cereal and cereal products
4 Ata JABA - BSc,MSc,PhD oil and oil seeds
5 Dziedzoave N - BSc rootcrops processing
6 Nerquaye-Tetteh Mrs G - BSc fish and fish products
7 Tetteh E - BSc meat and meat products

1652 Engineering Division

1 Lartey BL - BSc,DipFdTech,MARSH,AIFST,MGhIE *Head of Division* food engineering
2 Mensah BA - BSc,MSc food engineering

1653 Economics and Consumption Division

1 Dovlo Mrs FE - BSc,PhD *Head of Division* food consumption and planning
2 Osei-Yaw Mrs A - BSc,MSc food consumption and planning
3 Quartey CK - BSc agricultural economics

1654 Information Division

1 Ankrah EK - BSc,MSc *Acting Head of Division* food chemistry; food storage
2 Streetor Miss M - ALA *Chief Library Assistant*
3 Ameh K - DipJourn public relations

1655 Pilot Plant Division

1 Dei-Tutu J - BSc,DipFdSc,MSc,PhD *Head of Division* fruits and vegetables processing

1656 Administration

1 Musey JE - BSc science administration; economics

1657 Institute of Aquatic Biology
PO Box 38, Achimota, Accra
Tel: 775511. Cables: Aquabi.

1 Odei MA - BSc,DIC,PhD,MIBiol *Director* schistosomiasis

1658 Limnology Division

1 Antwi LAK - Bsc,Msc *Head of Division* analytical chemistry
2 Biney CA - BSc,MSc,DipMP pollution; chemistry, coastal and inland waters

1659 Aquatic Plants Division

1 DeGraft-Johnson KAA - BSc,MSc,DipEnvM *Head of Division* aquatic weeds control

1660 Fisheries Division

1 Ofori JK - BSc,MSc,DipAquaTech *Head of Division* aquaculture (tin fish)
2 Abban EK - BSc,MSc fish ecology and genetics
3 Dankwa HR - BSc fish ecology in rivers

1661 Parasitology/Microbiology Division

1 Ofori Miss JC - BSc,MSc *Head of Division* microbiology, biological control agents
2 Opoku AA - BSc,DipEnvM entomology (simuliidae)

1662 Hydrobiology

1 Samman J - BSc,MSc,PhD *Head of Division* invertebrate fauna in rivers (Microcrustacea)
2 Amakye JS - BSc,MSc aquatic entomology (Tricoptera)

1663 Oil Palm Research Centre
PO Box 74, Kade
Tel: 2, Kusi.

1 Wonkyi-Appiah JB - BSc,PhD *Principal Research Officer and Acting Head of Centre* plant breeding
2 Ahiekpor EK - BSc,MSc *Research Officer and Head of Section* plant breeding
3 Nantwi A - BSc,MSc *Senior Research Officer and Head of Section* plant physiology; agronomy
4 Boateng O - BSc *Assistant Research Officer* crop science
5 Dery S - BSc *Acting Head of Entomology* plant protection
6 Ofori-Asamoah - BSc,PhD *Research Officer* soils; plant nutrition

1664 Soil Research Institute
Private Post Bag, Academy PO, Kwadaso-Kumasi
Tel: 2353/4. Cables: Chiefsoil.

1 Halm AT - BSc,MSc *Director* soil chemistry; soil microbiology

1665 Soil Genesis, Survey and Classification

1 Asiama RD - BSc,MSc *Head of Division* soil genesis, survey and classification
2 Agyili P - BA,MSc,DipAFI soil genesis

1666 Soil Chemistry and Mineralogy Division

1 Kanabo I - BSc,MSc *Head of Division* soil chemistry

1667 Soil Fertility Division

1 Dennis EA - BSc,MSc *Head of Division* soil chemistry and fertility

1668 Soil Conservation and Erosion Control Division

1 Mensah-Bonsu - BSc,MSc *Head of Division* soil conservation and erosion control

1669 Soil Microbiology

1 Halm AT - BSc,MSc *Head of Division* soil microbiology

GOVERNMENT MINISTRIES

1670 Ministry of Agriculture
PO Box M.37, Accra
Tel: 665421. Cables: Minag.

1 Atta-Konadu VA - BSc,MSc *Chief Director* agricultural economics; planning, monitoring and evaluation

1671 Training and Manpower Department

1 Coleman KA - BSc *Director and Head of Department*
2 Hayford KA - BSc *Principal Agricultural Officer* agricultural education

1672 Crop Services Department

1 Quartey-Papafio HK - BSc *Director and Head of Department*
2 Biney TB - BSc *Deputy Director*

1673 Irrigation Development Authority

1 Frimpong K - BSc,MSc,MGhIE *Chief Executive*
2 Wiafe K - BSc,DipWSAS,MGhIE *Deputy Chief Executive*

1674 Engineering Division

1 Opoku-Mensah A - BSc,MSc,MGhIE *Director and Head of Division*
2 Apio S - BSc *Agricultural Engineer*
3 Attipoe BC - BSc,MSc *Agricultural Engineer*
4 Obuobi ALS - BSc,MGhIE *Director Engineering*

1675 Agronomy Division

1 Affram MAK - BSc *Acting Director* agronomy

1676 Agricultural Economics and Marketing Department
PO Box M.37, Accra
Tel: 665421. Cables: AGRICOSTATS.

1 Obeng-Boampong Mrs F - BSc *Deputy Director and Acting Head of Department* agricultural economy
2 Asante A - BSc,MSc,PhD *Deputy Director* agricultural economy and marketing
3 Lutterodt Mrs F - BSc,MSc *Deputy Director* agricultural economy and marketing
4 Mensah Miss P - BSc,MSc *Deputy Director* agricultural economy and marketing

1677 Animal Husbandry Department

1 Akwensivie SA - BSc *Acting Director* animal husbandry
2 Etse DA - BSc,MSc *Assistant Director* animal husbandry
3 Hesse HH - BSc *Assistant Director* animal husbandry
4 Idrisu A - BSc *Assistant Director* animal husbandry
5 Kwatia SM - BSc,MSc *Assistant Director* animal husbandry

1678 Fisheries Department

1 Dowuona VN - BSc,MSc *Director of Fisheries*
2 Armah M - BSc,MSc *Deputy Director of Fisheries*
3 Asafo CK - BSc *Deputy Director of Fisheries*
4 Denyoh FMK - BSc,MSc *Deputy Director of Fisheries*
5 Mensah MA - BSc *Deputy Director of Fisheries*

1679 Veterinary Services Department
PO Box M.161, Accra

1 Gyening KO - DVSc,MRCVS,DTVM,CertRabiesDiag,CertPoulProd *Director* tropical veterinary medicine and poultry production
2 Kankoh AG - DrMedVet,DTVM *Deputy Director* tropical veterinary medicine
3 Nyarko D - BVMS,MRCVS,MSc,CertAbattoirMgmt *Deputy Director* veterinary parasitology
4 Obinim JK - DVM,DVSM *Deputy Director* veterinary pathology
5 Adjogble HR - BVSc,MVSc *Regional Veterinary Officer* avian medicine
6 Aduamah FK - BVM,DTVSc *Senior Veterinary Investigation Officer* veterinary pathology
7 Agyen-Frempong N - BVM,DTVM,MSc *Regional Veterinary Officer* tropical veterinary medicine
8 Aklaku IK - DVM,MSc *Head, Laboratory Services* veterinary microbiology
9 Amanfu WK - DVM,MSc *Senior Veterinary Investigation Officer* veterinary microbiology
10 Ansah JBB - DVM,DTAPH *Principal, Veterinary College* tropical veterinary medicine
11 Appiah SN - DVM,MSc *Senior Veterinary Investigation Officer* veterinary microbiology

12 Atuora DOC - BS,DVM,DTVM *Regional Veterinary Officer* tropical veterinary medicine

13 Doku CK - BSc,DVM,Cert Tsetse and Trypanosomiasis Cont *Head, Tsetse and Trypanosomiasis Unit* tsetse and trypanosomiasis control

14 Enchill J - MVSc,DTVM,CertTecEd *Regional Veterinary Officer* tropical veterinary medicine

15 Gandaa HB - BScAgr,BVSc,DTVM,MSc *Regional Veterinary Officer* tropical veterinary medicine

16 Koney EBM - BSc,DVM,MSc *Senior Veterinary Officer* tropical veterinary medicine

17 Kpodo D - BVMS,MSc *Senior Veterinary Officer* tropical veterinary medicine

18 Laryea AM - BVSc,DTVM *Regional Director of Agriculture* tropical veterinary medicine

19 Ofosu SA - DVM,MSc *Regional Veterinary Officer* tropical veterinary medicine

20 Opoku-Pare GA - DVM,MSc *Veterinary Investigation Officer* avian medicine

21 Quayson K - BVM,MSc *Senior Veterinary Officer* tropical veterinary medicine

22 Taylor JK - BVSc,MRCVS,DVSM,CertAbattoirMgmt veterinary public health

23 Yelewere GE - BVSc,DTVM *Regional Veterinary Officer* tropical veterinary medicine

STATE CORPORATIONS AND ORGANISATIONS

1680 Cocoa Research Institute
Head Quarters, PO Box 8, Tafo-Akim
Tel: 51. Cables: Direscao.

1 Martinson Mrs VA - BSc,PhD *Executive Director* cytogenetics
2 Manu M - BSc,MSc *Research Secretary*
3 Tetteh EK - ALA *Librarian*

1681 Agronomy Division

1 Ampofo ST - BSc,MSc,PhD *Head of Division* agronomy
2 Brew KM - MSc,DipStats statistics
3 Osei-Bonsu K - MSc,PhD agronomy

1682 Entomology Division

1 Owusu-Manu E - MSc,PhD *Head of Division* entomology
2 Asante SK - MSc entomology
3 Brew AH - MSc entomology
4 Mensah RK - BSc entomology

1683 Physiology/Biochemistry Division

1 Adomako D - BSc,PhD *Head of Division* biochemistry
2 Frempong EB - BSc,PhD plant physiology
3 Okraku-Ayeh F - BSc biochemistry

1684 Plant Breeding Division

1 Amponsah JD - BSc,MSc *Head of Division* plant breeding
2 Adu-Ampomah Y - BSc,PhD plant breeding
3 Boamah-Adomako - BSc,MSc plant breeding
4 Enuson JK - BSc,DipStats biometrics

1685 Plant Pathology

1 Dakwa JT - MSc,PhD *Head of Division* plant pathology
2 Danquah DA - MSc,PhD mycology
3 Ollennu LAA - MSc,PhD virology
4 Owusu GK - BSc,MAgrSc,PhD virology

1686 Soil Science Division

1 Ahenkorah Y - BSc,MSc,PhD *Head of Division* soil chemistry
2 Appiah MR - MSc,PhD soil chemistry
3 Halm BJ - BSc,MSc,PhD soil chemistry

1687 Forest Products Research Institute
University PO Box 63, Kumasi
Tel: 53 Ext.400 Kumasi 5873. Cables: Forsearch.

1 Addo-Ashong FW - BSc,MSc *Director* wood identification, properties of plantation grown woods

1688 Protection Division

1 Atuahene SKB - BSc,MSc,PhD *Head of Division* ecology and biology of potentially dangerous forest pests in Ghana; bee biology
2 Ofosu-Asiedu A - BSc,MSc,PhD mycorrhiza; durability of wood in service; mushrooms

1689 Silviculture/Management Division

1 Britwum SPK - BSc,MSc *Head of Division* regeneration of the natural forest; afforestation of degraded lands
2 Arthur Mrs B - BSc,MSc degradation of agricultural waste; *Terminalis ivorensis* decline
3 Brockman-Amissah J - BSc,MSc agroforestry
4 Cobbinah J - BSC,MSc,PhD plant derived insecticides; snail cultivation

1690 Utilization Division

1 Jetuah FK - BSc,MSc,PhD *Head of Division* essential oils and tannins; formaldehyde formulation
2 Ampong FFK - BSc,MSc diffusion treatment; railway sleepers; marine timbers; transmission poles
3 Dzam B - BSc,MSc briquettes from wood waste
4 Gyimah Mrs A - BSc,MSc seed technology; tree improvement
5 Ofori J - BSc,MSc,DIC,PhD,AIWSc drying and seasoning characteristics of fencing posts; durability of fencing post woods
6 Sekyere D - BSc,MSc,PhD pulping characteristics of exotic indigenous and agricultural waste

UNIVERSITIES

1691 University of Ghana
PO Box 25, Legon, Accra
Tel: 775381-6. Cables: UNIVERSITY.

1 Akilagpa S - LLB,LLM,LLMJSD *Vice-Chancellor*
2 Benneh G - BA,PhD *Pro-Vice-Chancellor*

1692 Faculty of Agriculture
PO Box 68, Legon, Accra
Tel: 775381 Ext.8229.

1 Doku EV - BSc,PhD *Professor and Dean of Faculty* plant breeding

1693 Agricultural Economy and Farm Management Department

1 Dadson JA - BSc,MSc,MA,PhD *Acting Head of Department* agricultural planning and development; farm management; agricultural finance
2 Alhassan R - BSc,MSc,PhD *Lecturer* soil and water management
3 Baah SO - BA,MA agricultural economy
4 Baidu J - BSc,MSc *Lecturer in Agriculture*
5 Fosu J - BSc,MSc agricultural economy
6 Fosu KY - BSc,MPhil agricultural economy
7 Obben J - BSc,MA agricultural economy

1694 Agricultural Engineering Department

1 Al-hassan S - BSc,MSc *Acting Head of Department* soil and water management
2 Bani RJ - BSc,MSc *Co-ordinator*

1695 Agricultural Extension Department

1 Geker JK - BSc,MSc,PhD *Acting Head of Department* communication; visual aids, adult education
2 Fiadjoe FYM - BSc,MSc social economics; factors affecting farmers
3 Ocloo Mrs V - BSc,MSc agricultural extension

1696 Animal Science Department

1 Assoku RK - BVMS,MRCVS,PhD *Professor and Head of Department* animal health; immunology; microbiology
2 Aboagye Mrs - BSc,MSc animal breeding
3 Ahunu BK - BSc,MSc,PhD tropical agriculture
4 Anna Mrs RB - BSc,MSc ruminant nutrition
5 Asamoah L - BSc,MSc animal husbandry and nutrition
6 Awotwi B - MSc,DrMedVet animal health; physiology; veterinary dermatology
7 Baafi-Yeboah S - MVS,DrMedVet *Senior Lecturer* veterinary surgery
8 Fleischer JE - BSc,MSc,PhD animal husbandry
9 Okine EK - BSc,MSc animal science; animal nutrition

1697 Crop Science Department

1 Blay Mrs E - BSc,MSc,PhD *Head of Department* horticulture
2 Doku EV - BSc,PhD *Professor and Dean* plant breeding
3 Adjetey JA - BSc,MSc plant pathology
4 Akroffi D - BSc,MSc,PhD crops breeding
5 Sorwli FK - BSc,PhD crops breeding

1698 Home Science Department

1 Nsarkoh Mrs J - BSc,MSc *Acting Head of Department* food and nutrition
2 Anokwa Mrs C - BSc,MSc extension methods
3 Francois Mrs EM - BSc,MSc nutrition

4 Hevi-Yiboe Mrs L - BSc,MSc food and nutrition
5 Osei-Opare Mrs A - BSc,MSc home science extension
6 Williams Mrs N - BSc,MSc human development

1699 Soil Science Department

1 Acquaye DK - BSc,PhD *Professor and Head of Department*
2 Laryea KB - BSc,MSc,PhD soil physics
3 Thompson EJ - BSc,MS,PhD soil chemistry
4 Yentume SD - BSc,MSc,PhD soil physics

1700 Agricultural Research Station (Kpong)
PO Box 9, Kpong
Tel: 3. Cables: UNIVERSITY, Kpong.

1 Oteng JW - MSc,PhD *Officer-in-Charge* soil chemistry
2 Amaning-Kwarteng K - BSc,MSc agronomy
3 Arthur PF - BSc,MS plant breeding
4 Darkwa EO - BSc,MSc nutrition of crop plants
5 Nyalemegbe KS - BSc,MSc plant breeding
6 Oti-Boateng C - BSc,MSc agronomy

1701 Legon Research Station
PO Box 25, Legon

1 Williams GES - BSc,MSc,PhD *Research Officer-in-Charge* poultry nutrition
2 Adjei MB - BSc,MSc,PhD animal husbandry and nutrition
3 Ayebo DA - BSc,MSc,PhD animal husbandry; dairy microbiology
4 Canacoo EA - BSc,DVM animal health; immunology; microbiology
5 Dzakuma JM - BSc,MSc animal husbandry and nutrition
6 Nazie A - BSc,MSc animal breeding and ruminant nutrition
7 Pessey DE - BSc,PhD pasture production and conservation

1702 Kade Research Station
PO Box 43, Kade

1 Afreh-Nuamah K - BSc,MSc *Research Officer-in-Charge* mineral nutrition of crop plants
2 Asamoah-Appiah S - BSc,MSc agronomy of citrus, avocado, mango and other food crops
3 Osei JK - BSc,MSc cover crops and rubber agronomy

1703 Faculty of Science

1704 Biochemistry Department
PO Box 54, Legon
Tel: 775381 Ext.8825.

1 Oduro KK - BSc,MSc,PhD *Acting Head of Department* protein, peptide structure and function
2 Acquah RA - BSc,MSc,PhD carbohydrates and energy metabolism
3 Addy Mrs ME - BSc,PhD enzymes and metabolism
4 Ashong JO - BSc,MSc protein foods
5 Gbewonyo WSK - BSc,MSc enzymes, metabolism
6 Gyang FN - BSc,PhD enzymes defects
7 Larway PF - BSc,PhD carbohydrates and energy metabolism
8 Osei Mrs YD - BSc,MSc mechanism of drug excretion
9 Rodrigues FK - BSc,PhD enzymes and cell metabolism

1705 Botany Department
PO Box 55, Legon
Tel: 775381 Ext.825.

1 Laing E - BSc,PhD *Head of Department* genetics and biochemistry
2 Acheampong Miss E - BSc,MSc,PhD taxonomy and ecology
3 Amoako-Nuama Mrs CE - BSc,MSc,PhD anatomy and ecology
4 Markwei Miss CM - BSc,MSc *phycology and algae*
5 Mensah Mrs A - BSc,MSc,PhD anatomy and ecology
6 Odamtten GT - BSc,MSc quantitative ecology

1706 Nutrition and Food Science
PO Box 134, Legon
Tel: 775381 Ext.8820.

1 Orrace-Tetteh R - BSc,PhD *Associate Professor and Head of Department* infant weaning foods; proteins and protein requirements; energy needs and nutrition
2 Bediako-Amoah Mrs B - BSc,PhD nutrient composition of Ghanaian foods
3 Sefa-Dedeh S - BSc,MSc,PhD food composition, dietary standards, determination of nutrient needs, vitamins

1707 Department of Zoology

1 Vanderpuye CJ - BSc,MSc,PhD *Associate Professor and Head of Department* bioecology of *Heteroptera*
2 Adjei EL - BSc taxonomy of Orthopteroids
3 Attah P - BSc,MSc ecology of animal parasites
4 Attuquayefio DK - BSc,MSc hormones in metamorphosis of **grasshoppers**

5 Coker WZ - BSc,MSc,PhD insecticide resistance
6 Hughes B - BSc bioecology of *Heteroptera*
7 Nyorku G - BSc,MSc ecology of fish parasites

1708 Volta Basin Research Project

1 Bannerman RLK - DipBiol,PhD *Research Fellow-in-Charge* agricultural extension
2 Amoah Mrs C - BSc,MSc extension services and education
3 Cobbina J - BSc,MSc,PhD communication and adult education

1709 University of Science and Technology

University Post Office, Kumasi
Tel: 5351/5360. Cables: Kumasitech.

1 Kwami FO - DipIMG,DrING,TU *Vice-Chancellor*

1710 Department of Agricultural Economics & Farm Management

1 Asante-Kwatia DC - BScAgr,DipAgrEcon,MScAgrEcon *Dean & Head of Department* resources economics, agricultural finance and marketing, investment analysis
2 Anim RK - BSc,MSc,PhD extension, orientation, organization and functions, programming
3 Baffour-Senkyire JK - BSc,MSc,PhD extension, methodology; groups dynamics
4 Famiyeh JA - BSc,AgricNgtCert,FEDI agricultural project planning and monitoring
5 Nani-Nutakor JM - BSc,MSc agricultural development; production economics
6 Osei-Agyemang K - BSc,LIDM computer programming

1711 Department of Agricultural Engineering

1 Djokoto IK - BSc,MSc,PhD,ASAE *Head of Department* power and machinery
2 Agodzo SK - BSc,MSc soil and water engineering
3 Aklaku ED - DipIng farm structures and environmental control
4 Cenneper PJ - MSc,IngAgr project management
5 Darko JO - BSc,MSc crop storage and primary processing
6 Twum A - BSc,MSc,CEng,MGhIE,MI power and machinery

1712 Department of Animal Science

1 Kese AG - MSc,PhD,DTA *Head of Department* poultry nutrition
2 Alhassan WS - BSc,MSc,PhD animal nutrition and physiology
3 Atuahene Mrs CC - MSc poultry nutrition
4 Buadu MK - BVMS,MRCVS,MSc animal breeding
5 Donkor A - BSc non-conventional feed
6 Gyawu P - BSc,MSc,PhD reproductive physiology
7 Kabuga JD - BSc,MSc poultry nutrition and breeding
8 Okai DB - BSc,MSc,PhD animal nutrition; swine production
9 Olympio S - MSc,PhD poultry nutrition and breeding
10 Osafo ELK - BSc,MSc ingredients
11 Tuah AK - MSc,PhD animal nutrition

1713 Crops Science Department

1 Osafo DM - BSc,MSc,PhD *Head of Department* crop physiology
2 Ablorh FS - BA,MSc soil genesis and classification
3 Adjepong-Yamoah A - BSc grassland husbandry
4 Ankomah AB - BSc,PhD soil chemistry
5 Asare E - BSc,MSc crop physiology
6 Baffoe-Bonnie E - BSc,MSc,PhD soil physics
7 Boakye DB - BSc,MSc,PhD agricultural entomology
8 Hemeng Mrs BMS - BSc,MSc,DIC nematology
9 Nsowah GF - BSc,DipAgrSc,PhD plant breeding
10 Ofori FA - BSc,MSc plant pathology
11 Quansah C - MSc,PhD soil management
12 Quashie-Sam SJ - BSc,MSc,PhD crop physiology
13 Safo EY - BSc,MSc,PhD soil fertility
14 Safo-Kantankah O - BSc,MSc,PhD crop physiology and plant breeding

1714 Department of Horticulture

1 Atubra OK - BSc,MSc *Head of Department* fruit crops physiology
2 Awuah RT - BSc,MSc,PhD ornamental horticulture
3 Blankson EJ - MSc,LicInstPR landscape design
4 Boateng PY - BSc vegetable crops
5 Mensah THS - BSc ornamental horticulture
6 Olympio Mrs NS - MSc,PhD vegetable crops physiology
7 Sekyi PE - DipHort,MSc landscape design
8 Timbo Mrs GM - BSc,MSc vegetable crops physiology
9 Worman JC - BSc,MSc,DTA,MIBiol,FLS,DrAgr vegetable crops physiology; pineapple agronomy; ornamental horticulture

1715 University of Cape Coast

Cape Coast
Tel: Cape Coast 2440/9. Cables: University.

1 Dickson KB - BA,PhD *Vice-Chancellor*
2 Dadson BA - BSc,MSc,PhD *Pro-Vice-Chancellor*

1716 School of Agriculture

1 Asamoa GK - BSc,MSc,PhD *Associate Professor and Director* soil genesis, classification, land evaluation and land use planning

1717 Animal Science Division

1 Ibrahim A - BSc,MSc *Head of Division* animal physiology, nutrition, husbandry and environmental physiology of ruminants
2 Adumua-Bosman - BSc,MSc,PhD fisheries
3 Asamoah KA - BSc,MSc biochemistry, animal nutrition and breeding, improvement of nutrient levels in food crops
4 Enchil J - NVSc,CertVSc,DTVN,CertEd veterinary surgery
5 Odei FNA - BSc,MAgrSc animal nutrition, ruminant husbandry and pasture

1718 Crop Science Division

1 Haizel KA - BSc,DTA,PhD *Professor and Head of Department* crop ecology, weed biology and control
2 Acquah G - BSc,MSc plant breeding, grain legume improvement and varietal response to fertility
3 Addo-Quaye AA - BSc,MSc,PhD crop science
4 Carson AG - BSc,MSc,PhD agronomy, weed science
5 Nakaar NW - BSc forestry
6 Opoku-Asiamah Y - BSc,MSc plant protection, studies on pests and diseases of vegetable crops, evaluation of pesticides
7 Tetteh JP - BSc,MSc,PhD plant breeding and agronomy

1719 Soil Science Division

1 Asmah AE - BSc,MSc *Assistant Head* soil chemistry, fertility and plant nutrition

1720 Agricultural Engineering Division

1 Lamptey DL - BSc,MSc *Acting Head of Department* effect of controlled environment on the growth and development of farm animals; performance of various materials used as storage structures, soil and water conservation
2 Bani RJ - BSc,MSc agricultural engineering
3 Karikari TJ - BSc,MSc hydrology, irrigation and soil and water conservation

1721 Agricultural Economics Division

1 Micah JA - BSc,MSc,PGDipPA *Head of Division* agricultural economics
2 Gerrar F - BSc,MSc agricultural economics
3 Honny LA - BSc,MSc,PGDipPA project analysis

1722 Agricultural Extension Division

1 Saah NK - DipAgr,BScAgr,MA,PhD *Head of Division* agricultural education and extension

1723 Botany Department

1 Buxton Mrs GNT - BSc,MSc,PhD *Acting Head of Department* whole plant physiology
2 Amewowor DHAK - BSc,PhD microbial ecology; mycology
3 Asare-Boamah NK - BSc,MSc,PhD plant physiology
4 Donkor Mrs VA - BSc,MSc plant physiology
5 Oyaye EC - BSc,MSc,PhD ecology

1724 Department of Zoology

1 Eyeson KN - BSc,PhD *Associate Professor and Head of Department* biology of lagoon *Tilapia*
2 Ameyaw-Akumfi C - BSc,MSc,PhD biological science
3 Monney KA - BSc,MSc biological science
4 Quartey SO - BSc,MSc,PhD biology of freshwater shrimps
5 Yankson K - BSc,MSc,PhD biology of lagoon oyster
6 Yeboah S - BSc,PhD biological science

1725 Department of Geography

1 Nkunu AW - BA,MA *Acting Head of Department* food production and food culture; marketing of agricultural produce

1726 Centre for Development Studies

1 Amonoo E - PhD *Head of Department* crop marketing

2 Afful KN - BSc,MSc,PhD cost-benefit analysis of rural fisheries
3 Boakye JKA - BA,MSc,PhD evaluation of agricultural extension policies
4 Honny LA - BSc,MSc,PGDipPA agricultural project analysis
5 Micah JA - BSc,MSc,PGDipPA agricultural economics

GUYANA

1727 Ministry of Agriculture

Regent & Vlissengen Roads, PO Box 1001, Georgetown
Tel: 02 68568. Cables: MINFLAM.

1 McAndrew NC - MSc *Deputy Chief Agricultural Officer* crops and livestock

1728 Crop Services Division

Ministry of Agriculture, Mon Repos, East Coast Demerara

1 Ross Mrs P - MSc *Head of Division* seed technology
2 Kalimootoo Mrs C - BSc *Agricultural Officer Crop production* vegetable crops
3 Lewis Mrs J - BSc *Agricultural Officer Crop Protection* quarantine
4 Ramkobair D - BSc *Agricultural Officer Crop Protection* entomology
5 Saywack H - BSc *Agricultural Officer Crop Production* orchard crops
6 Summer Miss B - BSc *Agricultural Officer Crop Production* vegetable crops

1729 Animal Services Division

PO Box 1001, Georgetown
Tel: 02 68714. Cables: MINFLAM.

1 McPherson VOM - PhD *Head of Division* animal production

1730 Livestock Services

1 Francis Mrs CA - MSc *Senior Livestock Officer, Livestock Production* swine
2 Nurse GA - MSc *Technical Manager, Livestock Production* nutrition
3 Pierre MA - DVM *Veterinary Officer* dairy
4 Pilgrim Miss KA - BSc,DVM *Veterinary Officer* wildlife

1731 Veterinary Diagnostic Services

Mon Repos, East Coast Demerara
Tel: 020 2176/8.

1 Applewhaite LM - BVSc,MRCVS,DTVM,MSc *Director* parasitology
2 Craig-Clarke Mrs IE - DVM,MSc *Veterinary Officer* clinical pathology
3 Joseph Mrs BM - BSc microbiology
4 Reid HAC - DVM,MSc *Senior Veterinary Officer* pathology

1732 Extension Services

Agriculture Inservice Training and Communication Centre, Mon Repos, East Coast Demerara
Tel: 020 2804.

1 Dayaram E - BSc,MPhil *Head of Division*
2 George Miss B - BSc *Agricultural Officer* coconuts
3 Masilamany P - MSc *Senior Agricultural Officer* crop production technology
4 McDonald A - BSc *Agricultural Officer* crop production technology
5 Singh T - BSc *Agricultural Officer* extension training
6 Trotz R - BSc *Agricultural Officer* technology transfer
7 Wilson C - BSc *Agricultural Officer* communication methodology

1733 National Agricultural Research Institute
Central Agricultural Station, Mon Repos, East Coast Demerara
Tel: 020 2881/3 or 47 2842.

1734 Headquarters

1 Granger MA - PhD *Director*
2 Forde Mrs B - MSc *Head-of-Unit, Coastal Plains* root & tuber crops
3 Simpson LA - PhD *Head-of-Unit, Intermediate Savannahs* soils physics

1735 Research

1 Cumberbatch N - MSc pasture agronomy and conservation
2 Gibbons R - PhD genetic resources
3 Jeyandran Mrs G - MSc plant breeding and crop introduction
4 Jones G - MSc plant pathology
5 Livan M - MSc soil management & soil productivity
6 Maheswaran Mrs K - MSc rice agronomy
7 Moses O - BSc biometrics
8 Munroe L - BSc entomology (biological pest control)
9 Nelson E - MSc agricultural economics/farming systems
10 Rahat D - BSc farm mechanization
11 Trotman Mrs A - MSc soil microbiology
12 Tulsieram L - MSc plant breeding (rice)

1736 Guyana Sugar Corporation Limited
22 Church Street, Georgetown
Tel: 66171. Cables: SUGARCANE.

1737 Head Office

1 Cadawallader JA - DipAgrEng field machinery
2 Davis Jnr HB - PhD soil physics; biochemistry
3 Friday N - HTD,HTN,DipAgrEng experimental field equipment
4 Garrett-James B - MSc plant nutrition
5 Naraine DS - BSc drainage and irrigation

1738 Guyana Sugar Experimental Station

1 Cummings M - MSc varietal breeding and selection
2 Dookun R - MSc varietal breeding and selection
3 Fung B - BSc varietal breeding and selection

1739 Plant Protection Unit

1 Dasrat B - BSc plant protection
2 Huntley S - BSc entomology

1740 Other Crops

1 Forte F aquaculture
2 Piggott J - BSc diversified crop management

1741 Field Experimentation and Extension

1 Behari B - BSc
2 Bishundial B - BSc
3 Cummings S - BSc
4 Johnson W - BSc
5 Poonai A - BSc,BA
6 Singh B - BSc

1742 National Padi and Rice Grading Centre
Cowan Street, Kingston, Georgetown
Tel: 02 58618.

1 Croal K - BScBiol *General Manager*
2 Cumberbatch HE - MAAgrSc management practices in rice
3 Hazel R - BSc education and standards

117

1743 Agriculture & Fisheries Department

393 Canton Road, 12th - 14th Floors, Kowloon
Tel: 3-7332211. Cables: AGFISH HONG KONG.

1 Lee HY - JP,BSc,MSc,PhD,MIBiol,MESA *Director*

1744 Agriculture

1 Yip KS - BVSc,MRCVS *Assistant Director (Agriculture)*

1745 Crop Division

1 So Ms LS - BSc,MSc,DIC *Senior Agricultural Officer* crop development, control of pesticides and plant importation
2 Fong CW - BSc,MSc,FRES,MIBiol *Agricultural Officer* pest control
3 Ho KL - BSc,MSc *Agricultural Officer* disease control
4 Ho KY - BSc,MSc,MIBiol *Agricultural Officer* horticulture
5 Kwong Miss ML - BSc,PhD *Agricultural Officer* plant nutrition, soil and fertilizers
6 Lau CSK - MSc,MIBiol,CBiol,AIMLS *Agricultural Officer* pesticides

1746 Agricultural Development Division

1 Yip SM - BSc,DipHort *Senior Agricultural Officer* agricultural extension and co-operative societies
2 Tse KL - BSc,DipAgEcon *Agricultural Officer* agricultural economics and development

1747 Livestock Division

1 Cheng HN - BVSc *Senior Veterinary Officer* animal health, husbandry and regulatory work
2 Bousfield RB - BVSc *Veterinary Officer* rabies control and animal enforcement work
3 Burrows DJG - BSc,BVMed *Veterinary Officer* animal import and quarantine
4 Cheung SM - BSc,CertEd,BVSc *Veterinary Officer* animal health
5 Lam Miss MP - BSc,DipAgr *Agricultural Officer* poultry husbandry
6 Liu KK - BSc,BVMS *Veterinary Officer* rabies control, animal import and quarantine
7 Lo KS - BSc,BVSc *Veterinary Officer* veterinary laboratory work
8 Lo Mrs YY - BSc,NDP *Agricultural Officer* pig husbandry

1748 Conservation and Country Parks

1 Iu KC - MSc *Assistant Director (Conservation & Country Parks)*

1749 Conservation Division

1 Cheung MK - BSc *Senior Forestry Officer* multiple land use and conservation
2 Chan PK - BSc *Forestry Officer* fauna conservation
3 Lay CC - BSc,MSc *Forestry Officer* conservation education, countryside management
4 Leung HM - BSc *Ecologist* ecology and plant taxonomy
5 Leung KL - BSc,MSc *Forestry Officer* plant propagation, countryside management

1750 Country Parks Division

1 Lau SP - BSc,MSc *Senior Forestry Officer* overall development and management of country parks
2 Chan JK - BSc *Forestry Officer* countryside recreation management and amenity woodland silviculture with special reference to country parks
3 Chan TO - BSc,BForSc *Forestry Officer* countryside recreation management and amenity woodland silviculture with special reference to country parks
4 Lai CW - BSc,MSc *Forestry Officer* countryside management and forest protection
5 Yeung KK - RanCert *Forestry Officer* countryside recreation management and amenity woodland silviculture with special reference to country parks

1751 Countryside Development Division

1 Lam WK - BForSc *Forestry Officer* countryside recreation planning and landscape management with special reference to country parks
2 Leung CH - BA,BSc *Forestry Officer* conservation education
3 Tang Miss YK - BA,MSc *Forestry Officer* countryside recreation planning and landscape management with special reference to country parks
4 Wong FY - BA,MSc *Senior Forestry Officer* countryside recreation planning and landscape management with special reference to country parks

1752 Fisheries

1 Lee KC - BSc,MSc,PhD *Assistant Director(Fisheries)*

1753 Fisheries Management and Marketing Division

1 Cheung SH - BA,TTC(GTC) *Senior Fisheries Officer* fisheries extension
2 Au SN - BA,DipCoopDev,DipMS *Fisheries Officer (Ag.)* fisheries co-operative and credit
3 Cheng KW - BA,DipDevEcon *Fisheries Officer* fisheries economics and statistics
4 Chung T - BSc,MSc *Fisheries Officer* fish marketing
5 Yim TY - BSc,DipFisheriesManagement *Fisheries Officer* fisheries liaison and extension

1754 Capture Fisheries Division

1 Wu SS - BSc,MPhil,PhD *Senior Fisheries Officer (Ag.)* fisheries ecology and marine pollution
2 Leung WY - BA *Fisheries Officer* fish stock assessment
3 Mak MS - BSc,MSc,PhD *Fisheries Officer* marine fouling; fish physiology
4 Sham CH - BSc,DipFishingGearTechnology *Fisheries Officer* fishing gear technology

1755 Aquaculture Fisheries Division

1 Chan KW - BSc,CertFisheriesOfficers'Course *Senior Fisheries Officer (Ag.)* fisheries extension & management
2 Mok TK - BSc *Fisheries Officer* fish culture development
3 Sin WC - BSc,CertFisheries *Fisheries Officer* marine fish culture, regulating and licensing
4 Wong C - MSc,DipAgric&FishMngmt,DipTropSubTropAgric,MINPEng *Fisheries Officer* marine fish culture, regulating and licensing
5 Wong Ms PS - BSc,MPhil,PhD *Fisheries Officer* fisheries environment and red tides

1756 Department of Agricultural Research and Education

1757 Indian Council of Agricultural Research

Krishi Bhavan, New Delhi, 110001
Tel: 388991-7. Telex: 031-62249 ICAR IN. Cables: AGRISEC.

1 Randhawa NS - MScAgr,PhD *Director-General and Secretary to the Government of India* soil science
2 Abrol IP - MSc,PhD *Deputy Director General* soil salinity
3 Acharya RM - MSc,PhD *Deputy Director General* animal breeding
4 Paroda RS - MSc,PhD,FWA *Deputy Director General* plant breeding
5 Prasad C - MSc,PhD *Deputy Director General* agricultural extension
6 Rao MV - MSc,PhD *Special Director General* plant breeding
7 Singh M - MSc,PhD *Deputy Director General* agricultural education
8 Adlakha SC - MVSc,PhD *Assistant Director General* animal husbandry
9 Alam A - MSc,PhD *Assistant Director General* agricultural engineering
10 Alvi SAA - MTech,PhD *Senior Scientist* farm machinery and power
11 Arora CL - MVSc,PhD *Assistant Director General* animal breeding and production
12 Arvindan M - MVSc *Senior Scientist* animal physiology
13 Balasubramaniam A - MScAgr *Senior Scientist* plant breeding
14 Bhatia PC - MScAgr,PhD *Senior Scientist* soil and water management
15 Bhatt VS - MSc,PhD *Director* publication and information
16 Bodade VN - BScAgr,AssocIARI *Senior Scientist* agronomy
17 Bose PC - BSc *Information Systems Officer* information management
18 Chadha KL - MSc,PhD *Deputy Director General* horticulture
19 Chopra SC - MVSc,PhD *Senior Scientist* sheep breeding
20 Choudhary BN - MSc,PhD *Senior Scientist* agricultural economics
21 Dehadrai PV - MSc,PhD *Deputy Director General* fisheries
22 Dutta CP - MScAgr *Senior Scientist* horticulture
23 Ganguli NC - MSc,DSc *Senior Scientist* biochemistry
24 Jain SC - MVSc,PhD *Assistant Director General* dairying and animal products technology
25 Jain TC - MScAgr,PhD *Assistant Director General* agronomy
26 Kamal MY - MSc,PhD *Senior Scientist* fisheries science
27 Kaul GL - MScAgr,PhD *Assistant Director General* horticulture
28 Kempanna C - BScAgr,AssocIARI,PhD,FASc *Assistant Director General* commercial crops production
29 Khan E - MSc,PhD *Senior Scientist* palaeontology
30 Maji CC - MSc,PhD *Assistant Director General* agricultural economics
31 Makhija OP - BScAgr,AssocIARI,PhD *Senior Scientist* plant breeding
32 Malhotra MN - MVSc *Senior Scientist* veterinary parasitology
33 Manjhi S - MSc,PhD *Senior Scientist* agronomy
34 Mathur PB - MVSc,PhD *Assistant Director General* animal husbandry
35 Moitra S - BSc,BLibSc *Librarian* library science
36 Murti PSRC - MVSc *Senior Scientist* animal health
37 Natarajan C - MVSc,PhD *Assistant Director General* animal health
38 Nayar BB - MA *Senior Scientist* agricultural statistics
39 Patil BD - MSc,PhD *Assistant Director General* plant breeding
40 Ram Phal - MSc *Senior Scientist* horticulture
41 Rastogi BK - MSc *Senior Scientist* agricultural economics
42 Sadaphal MN - MSc,PhD *Assistant Director General* agronomy
43 Sahni VM - MScAgr,PhD *Senior Scientist* plant breeding
44 Sankar L - MSc,PhD *Assistant Director General* plant breeding
45 Seth SL - MSc,PhD *Senior Scientist* agronomy
46 Sharma DC - MSc *Senior Scientist* agricultural extension
47 Singh A - MSc,PhD *Senior Scientist* plant breeding
48 Singh A - MSc,PhD *Senior Scientist* agronomy
49 Singh K - MVSc,PhD *Senior Scientist* animal nutrition
50 Singh RN - MScAgr,PhD *Senior Scientist* agronomy
51 Sinha AK - MScAgr,PhD *Senior Scientist* soil science
52 Sriram TA - BScAgr *Senior Technical Officer* horticulture
53 Upadhyay UC - MScAgr,PhD *Assistant Director General* agronomy
54 Velayutham M - MSc,PhD *Assistant Director General* soil science

1758 Indian Agricultural Research Institute

New Delhi, 110012
Tel: 581493. Cables: KRISHIPUSA.

1 Michael AM - BScAgrEngg,MTech,PhD *Director* soil and water conservation engineering
2 Goswami NN - PhD *Joint Director and Dean (Education)* soil chemistry and soil fertility
3 Singh SN - PhD *Joint Director (Extension)* extension
4 Venkataraman GS - PhD *Joint Director (Research)* microbiology

1759 Division of Agronomy

1 Prasad R - MScAg,PhD *Head*
2 Arora PN - BScAg,AssocIARI,PhD *Senior Agronomist* vegetable crops
3 Bhardwaj RBL - BScAg,AssocIARIM,PhD *Professor of Agronomy*
4 Mahendra Pal - MScAg,PhD *Senior Agronomist* cropping systems
5 Saraf CS - MScAg,PhD *Senior Agronomist* pulse crops

6 Seth J - MScAg,PhD *Senior Agronomist* soil fertility
7 Singh G - BVSc,MSc *Senior Scientist* animal husbandry
8 Singh KN - MScAg,PhD *Senior Agronomist* rice
9 Sinha MN - MScAg,PhD *Senior Agronomist* soil fertility, radio tracer technique

1760 Division of Agricultural Chemicals

1 Roy NK - MSc,PhD *Head* design discovery and development of pesticides
2 Agnihotri NP - MSc,PhD pesticide residues
3 Devkumar C - MSc plant origin pesticides
4 Dikshit AK - MSc,PhD pesticide residues
5 Dureja Miss P - MSc,PhD photometabolism of pesticides
6 Handa SK - MSc,PhD *Project Coordinator* pesticide residues
7 Jain HK - MSc,PhD pesticide residues
8 Khazanchi R - MSc,PhD design, discovery and development of pesticides
9 Kulshrestha Mrs G - MSc,PhD pesticide residues
10 Madhuban G - MSc,PhD pesticide residues
11 Parmar BS - MSc,PhD pesticide formulation
12 Saxena DB - MSc,PhD plant origin pesticides
13 Singh RP - MSc,PhD plant origin pesticides
14 Tomar SS - MSc,PhD plant origin pesticides
15 Walia S - MSc,PhD photometabolism of pesticides

1761 Division of Agricultural Economics

1 Sirohi AS - MA,MS,PhD *Head* agricultural economics
2 Atteri BR - MScAgr agricultural economics
3 Bahl DK - MA agricultural statistics and computer programming
4 Dass SR - MA agricultural economics
5 Goel SK - MA agricultural statistics and computer programming
6 Guglani PL - MA,PhD agricultural economics
7 Haque T - MA,PhD agricultural economics
8 Kumar P - MSc,PhD agricultural economics
9 Mathur VC - MA agricultural economics
10 Mruthyunjaya - MSc,PhD agricultural economics
11 Muralidharan MA - MA agricultural economics
12 Puran Chand - MA agricultural economics
13 Sharma BM - MSc,PhD agricultural economics
14 Singh C - MA,PhD agricultural economics
15 Singh I - MSc,PhD agricultural economics
16 Singh K - MSc,PhD agricultural economics
17 Singh RP - MSc,PhD agricultural economics
18 Srivastava Mrs D - MA,MS,PhD agricultural economics

1762 Division of Agricultural Engineering

1 Ilyas SM - MSc,PhD *Head* post harvest engineering of cereals, pulses, oilseeds and spices; utilisation of solar energy in crop processing
2 Adlakha SK - MTech farm power and energy planning
3 Chandra P - MSc,PhD plant environment control; process simulation
4 De D - MTech tillage and traction
5 Kalra MS - MSc biogas technology, rural energy systems
6 Kamble HG - MSc crop threshing
7 Kaul KK - BTech field plot machinery
8 Khanna SK - MSc crop harvesting; dryland implements
9 Kulkarni SD - MTech food legume and rice processing and agricultural structures
10 Nag D - MTech oil seed and rice processing
11 Panwar JS - MSc,PhD *Professor* systems analysis; tillage; energy management
12 Sahu SD - MTech design of hand tools; energy requirement in agricultural production
13 Samuel DVK - MTech oilseed processing and dairy engineering
14 Saxena JP - MTech weeding tools; threshing machines; dry land implements
15 Sharma AK - MSc,PhD tuber planting equipment; anaerobic digestion
16 Sharma HS - MTech grain drying and storage; UHT processing
17 Singh A - MTech food grain cleaning and grading; storage
18 Singhal OP - PhD energy in agriculture and rural sector
19 Sirohi NPS - MTech seed drills; tillage
20 Srivastava AP - MTech,PhD paddy sowing equipment
21 Tandon SK - MSc,PhD minimum tillage, sowing and planting equipment, plant protection equipment
22 Wadhwa AK - MTech tillage and traction; ergonomics
23 Wanjari OD - MTech solar grain drying; grain storage

1763 Division of Agricultural Physics

1 Das DK - MSc,PhD *Head* agricultural physics
2 Bhavanarayana M - MSc,PhD soil physics
3 Chakravarty NVK - MSc,PhD agro-meteorology
4 Chaudhari KG - PhD soil physics
5 Chopra Mrs UK - MSc soil physics
6 Gupta RP - MSc,PhD soil physics; soil structure co-ordinator
7 Joshi RC - MSc,PhD soil physics
8 Khawas B - MSc biophysics

9 Panda BC - MSc,PhD biophysics
10 Rao TV - MSc,PhD soil-mineralogy
11 Rao YN - PhD soil physics
12 Sarma KSS - MSc,PhD soil physics
13 Sastry CVS - MSc,PhD agroclimatology
14 Sastry PSN - MSc agrometeorology
15 Singh G - PhD soil physics
16 Subbarao YV - BE,ME electronics
17 Tomar KP - MSc,PhD soil-mineralogy

1764 Division of Biochemistry

1 Mehta SL - MSc,PhD *Head* plant molecular biology
2 Barat GK - DSc *Professor* biochemistry
3 Dongre AB - PhD biological nitrogen fixation and protein quality
4 Duggal SK - PhD nutrition and biochemistry
5 Gupta YP - MSc,PhD nutrition
6 Johari RP - PhD molecular biology and plant biochemistry
7 Kapoor HC - PhD molecular biology
8 Koundal KR - PhD molecular biology
9 Lodha ML - MSc,PhD protein metabolism, biological nitrogen fixation
10 Madaan TR - PhD molecular biology and biochemistry
11 Misra BK - PhD nutrition
12 Santha Mrs IM - PhD molecular biology
13 Sharma ND - PhD plant molecular biology
14 Sikka KC - BSc,AssocIARI nutrition
15 Singh P - MSc,PhD inorganic nitrogen metabolism and photorespiration
16 Singh SP - PhD nutrition

1765 Division of Entomology

1 Bhatia SK - MSc,PhD,AssocIARA *Head* insect resistance to insecticide
2 Anand Mrs CM - MSc,PhD insect physiology - nutrition
3 Anand RK - MSc,PhD insect taxonomy - coleoptera
4 Atma Ram - MSc,PhD biological control of pests
5 Bhanotar RK - MSc,PhD locust
6 Bhatnagar RK - MSc,PhD ornithology
7 Bisht DS - MSc,PhD honey bees
8 Chandra Dinesh - MSc,PhD insect ecology
9 Chaudhari S - MSc,PhD insect pathology
10 Dhingra Mrs S - MSc,PhD insect toxicology
11 Doharey KL - MSc,PhD floriculture entomology
12 Farooqui SI - MSc,PhD insect taxonomy - hymenoptera
13 Gautam RD - MSc,PhD biological control of pests
14 Ghai Miss S - MSc,PhD insect taxonomy - acarina
15 Gujar GT - MSc,PhD insect physiology - insecticide metabolism
16 Gupta GP - MSc,PhD cotton pests
17 Gupta Mrs M - MSc,PhD insect physiology - nutrition
18 Gupta SL - MSc,PhD insect taxonomy - lepidoptera
19 Katiyar KN - MSc,PhD cotton pests
20 Katiyar RN - MSc,PhD extension entomology
21 Kishore P - MSc,PhD millet pests
22 Kumar S - MSc insect ecology
23 Kundu GG - MSc,PhD soyabean pests
24 Lal OP - MSc,PhD vegetable pests
25 Mahto Y - MSc,PhD oilseed pests
26 Marwaha KK - MSc,PhD maize pests
27 Mehrotra KN - MSc,PhD *Professor of Eminence* insect physiology - insecticides metabolism
28 Naim M - MSc,PhD honey bees
29 Pant JC - MSc,PhD insect physiology - nutrition
30 Panwar VPS - MSc,PhD maize pests
31 Parshad B - MSc,Phil,PhD,AssocIARI biological control of pests - predators
32 Paul AVN - MSc,PhD biological control of pests
33 Phadke KG - MSc,PhD oilseed pests
34 Phokela Miss A - MSc,PhD insect physiology - insecticide metabolism
35 Prasad SK - MSc,PhD oilseed pests
36 Rai S - MSc,PhD vegetable pests
37 Ramakrishnan N - MSc,PhD insect pathology
38 Ramakrishnan Mrs U - MSc,PhD insect taxonomy - rhynchota
39 Ramamurthy VV - MSc,PhD insect taxonomy - curculionidae
40 Rao PJ - MSc,PhD insect physiology - insecticide metabolism
41 Sarup P - MSc,PhD,AssocIARI maize pests
42 Saxena JD - MSc,PhD wheat pests
43 Saxena PN - MSc insect toxicology
44 Saxena VS - MSc,PhD insect toxicology
45 Sharma VK - MSc,PhD oilseed pests biocontrol
46 Siddiqui KH - MSc,MPhil,PhD maize pests
47 Singh DS - MSc,PhD insect toxicology
48 Singh RP - MSc,PhD insect toxicology
49 Singh Mrs S - MSc,PhD storage pests
50 Singh SP - MSc,PhD pulses pests
51 Singh VS - MSc,PhD wheat pests
52 Singh VS - MSc,PhD barley pests
53 Singh Y - MSc,PhD pulses pests
54 Sircar P - MSc,PhD insect toxicology
55 Srivastava BGD - MSc,PhD insect physiology - nutrition
56 Srivastava KL - MSc,PhD insect pathology
57 Srivastava KM - MSc,PhD vegetable pests
58 Srivastava KP - MSc,PhD sorghum pests
59 Srivastava ML - MSc,PhD sorghum pests
60 Srivastava VS - MSc,PhD insect toxicology
61 Srivastava YNS - MSc,PhD locust
62 Subramaniam B - MSc,PhD insect physiology - insecticide metabolism
63 Sukhani TR - MSc,PhD sorghum pests
64 Tiwari LD - MSc,PhD insect physiology - insecticide metabolism
65 Verma KK - MSc,PhD sorghum pests
66 Verma Mrs S - MSc,PhD insect toxicology - residues
67 Yadav TD - MSc,PhD storage pests

1766 Division of Nematology

1 Dasgupta DR - MSc,PhD *Head*
2 Gill JS - MSc,PhD
3 Khan EU - MSc,PhD
4 Sethi CL - MSc,PhD
5 Swarup G - MSc,AssocIARI,MS

1767 Division of Mycology and Plant Pathology

1 Grewal JS - MSc,PhD *Head* diseases of pulse crops

1768 Mycology

1 Agarwal DK - MSc,PhD,FPSI mycology, taxonomy of fungi (herbarium)
2 Bahl Mrs N - MSc,FPSI research on edible mushrooms
3 Choudhry PN - MScAg,PhD taxonomy of fungi
4 Kapoor JN - MSc,AssocIARI,PhD taxonomy of fungi and mushroom research
5 Lal Miss SP - MSc,PhD Indian *Penicillium* species and storage of fungi
6 Sarbhoy AK - MSc,DPhil,FLS,FPSI mycology and taxonomy of fungi

1769 Fungal Pathology

1 Bhowmik TP - MSc,PhD plant pathology (diseases of oil seed crops)
2 Chatterjee SC - MSc,PhD wheat diseases
3 Dhanraj KS - BScAg,AssocIARI,PhD virus diseases on crop plants
4 Kandhari Mrs J - MSc diseases of millets
5 Kapoor IJ - MScAgr,DPhil fungal diseases of tomato and carnation
6 Khan QZ - MSc pathology
7 Kulshreshtha DD - MSc,PhD diseases of pulse crops
8 Kumar D - MSc,PhD diseases of trees and mycorrhizae
9 Mathur SB - MSc,PhD diseases of millets
10 Murty Mrs VK - MSc diseases of guava and other fruits
11 Pal M - MSc,PhD diseases of pulse crops
12 Payak MM - MSc,PhD maize pathology
13 Phatak KD - MSc extension pathology
14 Pillai Mrs PK - MSc,PhD diseases of wheat and barley
15 Sahni ML - AssocIARI,PhD fungicides: plant pathology
16 Sen Mrs B - MSc,PhD fungal diseases of vegetable crops
17 Sharma RC - MSc,PhD corn diseases
18 Sharma SK - MSc,PhD rusts of cereals and groundnut
19 Singh - MSc diseases of oilseed crops
20 Singh DV - MSc,PhD plant pathology
21 Singh GR - MSc,PhD fungal diseases of rice
22 Singh RH - MScAg,PhD,FPSI plant pathology
23 Srivastava KD - MSc,PhD plant pathology
24 Vir D - MSc,PhD pathology of post harvest diseases

1770 Virology

1 Ahlawat YS - MSc,PhD citrus virus diseases, mycoplasma diseases
2 Anand GPS - MSc,PhD groundnut virus diseases
3 Basu AN - MSc,DPhil aphidology, virus entomology
4 Chenulu VV - AssocIARI,PhD plant virology
5 Dhingra KL - MSc,MPhil,PhD plant viruses including temperate fruit trees, pulse crop and soyabean
6 Giri BK - MScAg,PhD characterization of viruses infesting bittergourd and tomato including mosaic syndrome
7 Gupta Miss MD - MSc,PhD virus diseases of ornamental plants, tissue culture and detection of seed borne viruses
8 Jain RK - MSc rice virus diseases
9 Mishra MD - MSc,PhD,AssocIARI virus diseases of rice; tissue culture
10 Niazi FR - MSc,PhD virus and mycoplasma diseases of rice
11 Pandey PK - MSc,PhD viral diseases of citrus, vegetable and groundnut
12 Ramachandran Mrs P - MSc,PhD seed borne viruses and virus diseases
13 Rangaraju R - MSc,PhD virus diseases of pulse crops
14 Varma A - MSc,PhD virus characterization; mycoplasmal diseases
15 Verma VS - MSc,PhD tissue culture of plant viruses
16 Vishwanath SM - MSc,PhD preparation of antisera against crop plants, chilli and some vegetable viruses and citrus virus diseases

1771 Bacteriology

1 Durgapal JC - MSc investigations on bacterial diseases of rice
2 Jindal JK - MSc,MPhil,PhD bacterial diseases of pulse crops
3 Pandey KR - MSc,PhD bacterial diseases of rice
4 Singh RP - MScAg,PhD plant bacteriology
5 Trivedi BM - MSc bacterial diseases of vegetable crops
6 Verma JP - PhD,DSc plant bacteriology

1772 Division of Fruits and Horticultural Technology

1 Bose PC - PhD *Horticulturist* propagation and orchard management of mango and guava
2 Goswami AM - PhD propagation, rootstock and orchard management of citrus
3 Jindal PC - PhD grape breeding
4 Khanna RC - MSc salt tolerance in mango; orchard management in citrus
5 Khurdiya DS - PhD post harvest technology of fruits and vegetables
6 Kishore DK - PhD fruit physiology and micropropagation of fruit crops
7 Maini SB - PhD post harvest technology of fruits and vegetables
8 Majumder PK - PhD breeding and propagation of mango and guava
9 Nigam KB - MSc nursery management, multiplication and distribution of fruit crops
10 Pandey SN - PhD physiology and breeding of grapes; registrar, international registration authority for mango
11 Roy SK - PhD *Project Coordinator* post harvest technology of fruit and vegetables
12 Saxena SK - PhD rootstock and orchard management of citrus
13 Sethi Mrs V - PhD post harvest technology of fruits and vegetables
14 Sharma BB - PhD water management, orchard management and breeding of citrus
15 Sharma HC - PhD fruit physiology and micropagation of fruit crops
16 Sharma YK - PhD propagation, rootstock and breeding of guava

1773 Division of Vegetable Crops

1 Ahluwalia KS - PhD cole crops breeder
2 Gill HS - PhD cole crops breeding
3 Kalda TS - PhD brinjal (aubergine) and pumpkin breeder
4 Mishra JP - PhD melon breeder
5 More TA - PhD cucumber and melon breeder
6 Pal N - PhD onion and carrot breeder
7 Sesadhri VS - PhD melon breeding
8 Sharma JC - PhD melon breeding
9 Sharma RK - PhD beans breeder
10 Sharma RR - PhD pea breeder
11 Singh N - PhD onion breeding and physiological studies
12 Singh R - PhD cole crops breeder
13 Sirohi PS - PhD gourds and pumpkin breeder
14 Tewari RN - PhD tomato breeding
15 Verma VK - PhD pea and luffa breeder and physiological studies

1774 Division of Floriculture and Landscaping

1 Dohare SR - MSc,PhD *Head* floriculture
2 Kaicker US - MSc,PhD floriculture
3 Singh B - MSc,PhD floriculture

1775 Division of Microbiology

1 Gaur AC - PhD,DSc *Head* soil microbiology/biological nitrogen fixation and organic recycling
2 Apte RG - PhD mass multiplication of bacteria on *Azotobacter*
3 Balasundaram VR - PhD microbiological aspects of soybean cultivation; studies on rhizobium root nodulation of tropical legumes
4 Gaur YD - PhD studies on rhizobium and root nodulation of tropical legumes
5 Goyal SK - PhD morphology and physiology of algae with special reference to nitrogen fixation
6 Gulati SL - PhD applied microbiology and fermentation
7 Jauhri KS - PhD studies on carriers for culture; techniques for inoculation of seed and distribution of cultures
8 Kaushik BD - PhD biology of blue green algae in relation to rice cultivation
9 Kavimandan SK - PhD studies on association between microorganisms and crop plants in the rhizosphere and phyllosphere in relation to crop growth and production
10 Magu SP - PhD organic matter recycling and interaction of pesticides with microorganisms
11 Mathur RS - PhD microbiological aspects of organic matter decomposition and recycling of farm and city wastes
12 Rewari RB - PhD biological nitrogen fixation
13 Roychoudhry Mrs P - PhD biology of blue green algae in relation to rice
14 Sadasivam KV - PhD studies on *Rhizobium*; microbiological decomposition and recycling of agricultural and city wastes
15 Sen AN - PhD biological and symbiotic nitrogen fixation; serological studies
16 Shende ST - PhD agricultural microbiology
17 Singh CS - PhD legumes root nodulation interaction with other microorganisms
18 Tilak KVBR - PhD biological nitrogen fixation; *Rhizobium* and cereal nitrogen fixation
19 Vinod kumar - PhD microbiological aspects of pulse cultivation

1776 Division of Seed Science and Technology

1 Agarwal PK - MSc,PhD *Head* seed technology
2 Dadlani Mrs M - MSc,PhD *Plant Physiologist* seed physiology
3 Gaur A - MSc,PhD *Plant Pathologist* seed pathology
4 Sharma GC - MSc,PhD seed entomology

5 Verma MM - MSc,PhD,DipSeedTech seed technology

1777 Water Technology Centre

1 Arya YC - BSc,MSc water and soil engineering
2 Bhattacharya AK - MSc,PhD agricultural engineering
3 Kaul RK - MSc,PhD water and soil engineering
4 Paliwali KV - MSc,PhD *Professor* soil science
5 Rajput TBS - MTech water and soil engineering
6 Sarma PBS - ME,PhD water and soil engineering
7 Sinha SK - MSc,PhD *Professor of Eminence* plant physiology

1778 Biotechnology Centre

1 Chopra VL - MSc,PhD *Professor Eminence* investigations on somaclonal variation from basic and applied angle; molecular mechanism of diseases resistance
2 Sushil Kumar - MSc,PhD *Professor* biotechnology; molecular biology of plant-microbe interaction and of other adaptive mechanisms in bacteria
3 Gupta N - MSc,PhD cellular mutagenesis and selection of variants of agricultural value in cell cultures
4 Joshi BC - MSc,PhD *National Fellow* induction of homologous recombination between the chromosomes of wheat and rye through chromosome 5b manipulations
5 Kashyap LR - MSc,PhD molecular genetic studies in free living nitrogen-fixing bacteria azotobacters
6 Majid R - MSc,PhD protoplast culture and somatic hybridization
7 Pai RA - MSc,PhD lukaryotic genome organization and expression
8 Purnima Mrs - MSc,PhD tissue culture propagation of genotypes of agricultural relevance
9 Raina SK - MSc,PhD gametoclonal and somaclonal variation in cereals especially rice
10 Sharma RP - MSc,PhD lukaryotic genome organization and expression
11 Srinivasan - MSc,PhD molecular biology

1779 Agro Energy Centre
New Delhi, 110012
Tel: 581493. Cables: KRISHIPUSA.

1 Banerjee Mrs M - MSc,PhD *Microbiologist* microbiology
2 Khamankar YG - MSc,PhD *Geneticist* genetics
3 Wattal Mrs D - MSc,PhD *Plant Physiologist* plant physiology

1780 All India Coordinated Research Project on Long Term Fertilizer Experiments

1 Nambiar KKM - MScAgr,PhD *Project Coordinator* planning, implementation and coordination of the long term fertilizer experiments

1781 All India Coordinated Pulses Improvement Project (South)

1 Salimath PM - MScAgr,PhD *Scientist In-Charge* pulse breeding

1782 All India Coordinated Maize Improvement Project

1 Singh J - MSc,PhD *Project Coordinator* maize breeding
2 Agarwal KM - MSc,PhD agricultural statistics
3 Daya N - MSc,PhD maize agronomy
4 Gupta NP - MSc,PhD maize breeding
5 Mathur LM - MSc,PhD maize entomology
6 Singh I - MSc,PhD maize breeding
7 Singh NN - MSc,PhD maize breeding
8 Singh SB - MSc,PhD maize breeding

1783 All India Coordinated Floriculture Improvement Project

1 Malik RS - MSc,PhD *Project Coordinator* floriculture
2 Dadlani NK - MSc,PhD

1784 Farm Operation Service Unit

1 Kaul KK - BScAgrEngg&Tech *Post Graduate Head* farm mechanisation

1785 Division of Plant Physiology

1 Sirohi GS - MSc,PhD *Head* plant physiology
2 Bagga AK - MSc,PhD plant physiology
3 Baldev B - MSc,PhD plant physiology
4 Bhardwaj SN - MSc,PhD plant physiology
5 Bhargava SC - MSc,PhD *Professor* plant physiology
6 Deshmukh PS - MSc,PhD plant physiology
7 Ghildiyal MC - MSc,PhD plant physiology
8 Nanda Mrs R - MSc,PhD plant physiology
9 Ojha RJ - MScAgr,PhD plant physiology
10 Pandey M - MSc,PhD plant physiology
11 Raghuveer P - MSc,PhD plant physiology
12 Sabra Miss A - MSc,PhD plant physiology
13 Saini AD - MSc,PhD plant physiology

14 Sengupta UK - MSc,PhD plant physiology
15 Shukla DC - MSc,PhD plant physiology
16 Srivastava GC - MScAgr,PhD plant physiology
17 Tomar DPS - MSc,PhD plant physiology
18 Tomar OPS - MSc plant physiology
19 Uprety DC - MSc,PhD plant physiology
20 Ved Prakash - MSc plant physiology
21 Wasnik KG - MSc,PhD plant physiology
22 Wattal PN - MSc,PhD plant physiology

1786 Nuclear Research Laboratory

1 Tiwari PN - BSc,MSc,PhD *Senior Scientist* physics
2 Abrol YP - BSc,MSc,PhD plant physiology and biochemistry
3 Ananda Kumar P - MSc nitrogen metabolism and photorespiration in higher plants
4 Arora RP - BSc,MSc,PhD soil physics fertilizer and water use studies
5 Chahal SS - BSc,MSc nuclear magnetic resonance spectroscopy
6 Chandrasekharan H - BSc,MSc isotope hydrology
7 Chatrath MS - BSc,MSc,PhD fungal diseases and fungicides
8 Chatterjee SR - BSc,MSc,PhD protein, amino acid, nitrogen and sulphur metabolism
9 Deb DL - BScAgr,MSc,PhD soil fertility
10 Gambhir PN - BSc,MSc,PhD application of NMR and allied techniques in agricultural research
11 Garg AK - BSc,MSc,MPhil,PhD radiation entomology
12 Ghosh Mrs G - BScAgr,MSc,PhD soil fertility
13 Ghosh SK - BScAgr,MSc,PhD clay mineralogy and soil fertility
14 Gupta JP - MSc,MPhil,PhD plant pathology fungicides
15 Mathur JMS - BSc,MSc,PhD biochemistry-lipid biochemistry
16 Mitra DK - BSc,MSc,PhD plant pathology-viral and mycoplasmal diseases of plants
17 Moharir AV - BSc,MSc,PhD fibre science biophysics (cotton structure and properties)
18 Nair TVR - BA,MSc,PhD plant physiology (nitrogen metabolism)
19 Pande PC - BSc,MSc,PhD biochemistry
20 Prakash N - MSc bio-physics (biophysical studies of plants under water and temp stress)
21 Prasad HH - MSc,PhD use of radioactive isotope and radiation for entomological investigations
22 Raman TS - MS,MSc,PhD comparative biochemistry of isoprenoids and sterols
23 Roy MK - BSc,MSc,PhD food irradiation
24 Sachdev MS - BSc,MSc soil fertility and fertilizer use
25 Sachdev Mrs P - BSc,MSc soil fertility and fertilizer use
26 Sastry LVS - MSc,PhD intermediary metabolism and biochemistry
27 Sethi GR - BSc,MSc metabolism of insecticides-use of nuclear tools in insect pest management
28 Srivastava KN - BSc,MSc,PhD improvement of protein quality and quality in pulses

1787 Research, Planning and Coordination
New Delhi, 110012
Tel: 581493. Cables: KRISHIPUSA.

1 Sharma AK - AssocIARI,PhD taxonomy, bionomics and ecology of hymenopterous parasitoids of insect pests of crops, research coordination, monitoring and evaluation

1788 IARI Regional Station
Indore, Madhya Pradesh
Tel: 5021. Cables: KRISHIPUSA.

1789 Division of Agronomy

1 Singh AK - MSc *Head* wheat agronomy

1790 Division of Genetics

1 Matai BH - MSc,PhD,AssocIARI *Zonal Wheat Coordinator* planning, coordination and implementation of wheat improvement in central zone
2 Ijardar JS - MScAgr,PhD spring and winter wheat improvement
3 Varma PK - MSc,PhD germplasm maintenance and evaluation of *Triticum durum* and *Triticum aestivum*

1791 Division of Physiology

1 Ruwali KN - MSc,PhD stress physiology in wheat

1792 IARI Regional Station
Flowerdale, Shimla-2
Tel: 2662. Cables: KRISHIPUSA.

1 Nagarajan S - PhD *Head* wheat pathology
2 Bahadur P - PhD wheat pathology
3 Nayar SK - PhD wheat pathology

1793 IARI Regional Station
Pune-5
Tel: 57801. Cables: KRISHIPUSA.

1 Summanwar AS - MScAgr,PhD *Head* investigation on virus diseases of crops
2 Chavan VM - MScAgr investigations on insect vectors of plant virus diseases
3 Ram RD - MScAgr investigation on virus diseases of crops

1794 IARI Regional Station
Katrain, Kulu-175129, Himachal Pradesh
Cables: KRISHIPUSA.

1 Verma TS - MScAgr,PhD *Head and Senior Scientist* vegetable breeding
2 Chaudhary ML - MScAgr,PhD improvement of bulbous ornamentals
3 Joshi S - MSc,PhD vegetable breeding
4 Kapoor KS - MScAgr,PhD vegetable pathology
5 Thakur PC - MScAgr,PhD vegetable breeding

1795 IARI Regional Station
Rajendranagar, Hyderabad-500030
Tel: 48349. Cables: KRISHIPUSA.

1 Rana BS - MScAgr,PhD *Head/Senior Sorghum Breeder* research management;breeding resistant varieties, hybrids and male steriles
2 Achuta Rao K - MSc,PhD efficient utilization of nutrients
3 Anantharaman PV - AssocIARI plant nutrient partitioning
4 Arunachalam V - MA,PhD biometrical genetics of groundnut
5 Bala Ravi S - MScAgr,PhD pigeon-pea breeding
6 Bandyopadhyay A - MScAgr,PhD biometrical genetics of groundnut
7 Indira Mrs S - MSc host plant resistance to diseases
8 Kaul Mrs S - MSc sorghum breeding
9 Pankaja Reddy Miss R - MScAgr,PhD groundnut breeding
10 Potty VP - MScAgr,PhD nitrogen fixation
11 Prasad MVR - MScAgr,PhD groundnut breeding
12 Rao MH - MScAgr,PhD pigeon-pea cropping systems
13 Reddy MK - MScAgr,PhD disease resistance in groundnut
14 Singh BU - MSc host plant resistance to insects

1796 IARI Regional Station
Amartara Cottage, Cart Road, Simla, Himachal Pradesh, 171004
Tel: 2561. Cables: KRISHIPUSA.

1 Randhawa SS - MSc,PhD *Head*

1797 IARI Regional Station
Kalimpong, West Bengal, 734301
Cables: KRISHIPUSA.

1 Chakraborty NK plant pathology

1798 IARI Regional Station
Wellington, Tamil Nadu, 643231
Tel: 6463. Cables: KRISHIPUSA.

1 Menon MK - MSc *Head*
2 Brahma RN - MSc *Plant Pathologist*

1799 IARI Regional Station
Pusa, Bihar
Cables: KRISHIPUSA.

1 Jha MP - PhD plant breeding
2 Farooqui OR - MSc plant breeding
3 Kumar A - MSc agronomy
4 Manshra Ram - PhD horticulture
5 Narula PN - PhD plant breeding
6 Singh BP - MSc plant breeding
7 Singh KMP - PhD cytogenetics
8 Singh RK - PhD cytogenetics
9 Singh TB - PhD plant pathology
10 Srivastava OP - PhD cytogenetics

1800 Indian Agricultural Statistics Research Institute
Library Avenue, New Delhi, 11012
Tel: 587121. Cables: AGRIRESTA.

1 Prem Narain - PhD,DSc,FSS,FNA *Director* statistical genetics
2 Agrawal Miss R - PhD crop forecasting and sample survey
3 Bahuguna GN - MSc,MA agricultural statistics
4 Banerjee AK - PhD sample survey and design of experiments
5 Bapat SR - MSc crop forecasting techniques
6 Bassy GS - MSc sample survey methodology
7 Bhatia VK - MSc,DipAgrStat,PhD genetical statistics
8 Choudhary HB - MA agricultural statistics
9 Dixit UN - PhD agricultural economics
10 Doshi SP - MSc agricultural statistics computer programming
11 Garg LK - MSc agricultural statistics
12 Gopalan R - MA agricultural statistics and computer application

13 Gupta KC - MSc agricultural statistics and computer programming
14 Gupta SS - MSc sample survey techniques
15 Jain JP - MA,PGDip,MS,PhD statistical genetics
16 Jain RC - MSc,DipPG,PhD crop yield forecasting methodology
17 Kaistha AC - MSc agricultural statistics and electronic data processing
18 Malhotra JC - MA,StatDip genetical statistics
19 Mathur SN - MA computer sciences and agricultural statistics
20 Mehrotra PC - MSc,PhD sample surveys
21 Nigam AK - MSc,PSCC,PhD design of experiments and survey sampling
22 Ohri NK - PhD sample surveys and design of expts
23 Pandey RK - MSc,PhD agricultural economics
24 Pillai SS - PhD *Joint Director* systems analysis
25 Prajneshu - PhD *Sr Professor* statistical ecology
26 Rai SC - MSc,PGDipStat agricultural statistics, non-parametric inference
27 Rustogi VS - MSc,StatDip sample surveys
28 Sahni ML - MA agricultural statistics and computer programming
29 Saxsena BC - MA,PhD sample surveys
30 Shanti Sarup - MA,DipStat agricultural statistics
31 Sharma BS - PhD genetical statistics
32 Sharma VK - MA,MSc,PhD econometrics and design of experiments
33 Singh B *MA,PhD* management and production economics
34 Singh J - MA,PhD agricultural statistics
35 Singh R - PhD *Sr Professor* sample surveys
36 Tyagi KK - MSc,PGDipAgrilStat,PhD sample surveys
37 Wahi SD - MSc agricultural statistics

1801 Central Rice Research Institute

Cuttack, Orissa, 753 006
Tel: 20020. Telex: 676-220. Cables: RICE CUTTACK.

1 Patnaik S - MSc,PhD *Director*
2 Seetharaman R - BSc,MSc,PhD *Officer on Special Duty*

1802 Division of Agronomy

1 Manna GB - BSc,MScAgr,PhD *Head of Division* weed control
2 Asthana DC - BScAgr,MScAgr,DipITT,PhD water management; fertility management
3 Das KC - BSc,MScAgr,PhD *Project Coordinator* ORP on water management, cropping systems
4 Jha KP - MScAgr,PhD water management; cropping systems on farm research
5 Meena NL - BScAgr,MScAgr,PhD *Chief Scientist* water management
6 Murthy BTS - BScAgr,MScAgr weed control
7 Padalia CR - BScAgr,MScAgr fertility management

1803 Agricultural Economics Section

1 Ram S - BScAgr,MScAgr,PhD economics of rice cultivation

1804 Division of Agricultural Engineering

1 Mahapatra B - BSc,MSc *Head of Division* electronics and instrumentation
2 Behera AK - BScAgrEng,MTech agricultural structures and process engineering
3 Choudhury AK - BScAgrEng,MTech soil and water conservation engineering
4 Kurup GT - BTech,MTech,PhD agricultural structures and process engineering
5 Mishra P - BScAgrEng,MTech agricultural structures and process engineering
6 Mohapatra PC - BScEng,MTech soil and water conservation engineering
7 Pangasa RK - BSc,IETE electronics and instrumentation
8 Parida BC - BScAgrEng,MTech farm machinery and power
9 Paul DK - BTech soil and water conservation engineering

1805 Agricultural Statistics

1 Rawlo S - BA,MA agricultural statistics
2 Rao AVS - BSc,MSc agricultural statistics

1806 Division of Biochemistry and Rice Technology

1 Lodh SB - BSc,MSc,DPhil *Head of Division* nutrition and enzymology
2 Sahay MN - BSc,MSc,PhD biochemistry

1807 Division of Entomology

1 Kulshrestha JP - BSc,MSc *Head of Division* integrated pest management (chemical control)
2 Dani RC - BSc,MSc insect physiology
3 Mathur KC - BSc,MSc,PhD biological control
4 Mishra BC - BSc,MSc,PhD host plant resistance
5 Padhi Mrs G - BSc,MSc,PhD insect ecology
6 Prakash A - BSc,MSc stored grain pests
7 Prasad JS - BSc,MSc,PhD nematology
8 Rajamani S - BSc,MSc,PhD insect ecology
9 Rao PSP - BSc,MSc,PhD insect ecology
10 Rao VN - BSc,MSc biological control
11 Rao YRVJ - BSc,MSc,PhD biological control
12 Rao YS - BSc,MSc,PhD nematology

13 Samal Mrs P - BSc,MSc,PhD host plant resistance
14 Sasmals S - BSc,MSc chemical control

1808 Division of Extension, Communication and Training

1 Rao MV - BScAgr,MScAgr *Head of Division* agronomy; extension; training
2 Apparao G - BScAgr,MScAgr,PhD extension; training; interpersonal communication

1809 Division of Genetics Resources

1 Sharma SD - BSc,MSc,PhD *Head of Division* cytogenetics and plant genetic conservation
2 Dhua SR - BScAgr,MScAgr plant breeding
3 Ghorai DP - BSc,MSc,PhD plant breeding
4 Murthy AK - BSc,MSc plant genetic conservation
5 Panda A - BScAgr,MScAgr plant breeding

1810 Division of Plant Breeding and Genetics

1 Roy JK - BSc,MSc,PhD *Head of Division* plant breeding and genetics
2 Chaudhury D - BSc,MSc,PhD plant breeding and genetics
3 De RN - BScAgr,MScAgr,PhD plant breeding and genetics
4 Gangadharan C - BSc,MSc,PhD plant breeding and genetics
5 Jachuck PJ - BSc,MSc plant breeding and genetics
6 Misra RN - BSc,MSc,PhD cytogenetics
7 Patnaik NKC - BSc,MSc plant breeding and genetics
8 Prasad K - BSc,MSc plant breeding and genetics
9 Rao GJN - BSc,MSc,PhD cytogenetics and genetics
10 Rao MJBK - BSc,MSc plant breeding and genetics
11 Ratho SN - BScAgr,MScAgr,PhD plant breeding and genetics
12 Sarma NP - BSc,MSc,PhD cytogenetics; genetics; mutation breeding
13 Sreedharan PN - BSc,MSc plant breeding and genetics
14 Srinivasalu K - BSc,MSc plant breeding
15 Srivatasava DP - BSc,MSc,PhD plant breeding and genetics

1811 Division of Plant Pathology

1 Gangopadhyay S - BScAgr,MScAgr,PhD *Head of Division* fungal diseases
2 Anjaneyulu A - BScAgr,MScAgr,PhD viral diseases
3 Baktavatsalam G - BScAgr,MScAgr,PhD viral diseases
4 Devadattu S - BSc,MSc,PhD bacterial diseases
5 Dua Mrs U - BScAgr,MScAgr fungal diseases
6 Kauraw LP - BScAgr,MScAgr,PhD fungal diseases
7 Mohanty SK - BSc,MSc,PhD viral diseases
8 Nayak P - BScAgr,MScAgr bacterial diseases
9 Padhi B - BScAgr,MScAgr,PhD fungal diseases
10 Premalata Datha Mrs A - BSc,MSc,PhD bacterial diseases; fungal diseases
11 Reddy PR - BSc,MSc,PhD bacterial diseases
12 Row KVSRK - BSc,MSc,PhD fungal diseases, bacterial diseases
13 Singh UD - BScAgr,MScAgr fungal diseases
14 Sridhar R - BScAgr,MScAgr,PhD physiological studies
15 Tiwari SN - BScAgr,MScAgr fungal diseases
16 Veeraraghavan J - BSc,MSc,PhD fungal diseases

1812 Division of Plant Physiology

1 Bhattacharjee DP - BSc,MSc,DPhil *Head of Division* excess water stress
2 Murthy KS - BScAgr,MScAgr,PhD *National Fellow* photosynthesis and productivity
3 Nayak SK - BSc,MSc low light stress
4 Ramakrishnayya G - BSc,MSc moisture stress
5 Rao CN - BSc,MSc,PhD solar energy utilization
6 Sahu G - BSc,MSc,PhD climate and rice

1813 Division of Soil Science and Microbiology

1 Sethunathan N - BSc,MSc,PhD *Head of Division* microbiology of soils; pesticide microbiology
2 Adhya TK - BSc,MSc pesticide microbiology; biological nitrogen fixation
3 Bhadrachalam A - BSc,MS agricultural chemistry
4 Chakravarti SP - BSc,MScAgr,PhD soil fertility; plant nutrition
5 Das RN - BSc,MSc,PhD soil science; agricultural chemistry
6 Misra AK - BSc,PhD soil fertility, plant nutrition; heavy metal pollution
7 Mohanty SK - BScAgr,MScAgr,PhD soil chemistry; soil fertility; chemistry of submerged soils
8 Nayak PK - BSc,MScAgr,PhD soil science; agricultural chemistry
9 Panda D - BScAgr,MScAgr soil science; agricultural chemistry
10 Patnaik S - BSc,MSc,MS,PhD soil science; agricultural chemistry
11 Rao VR - BSc,MSc,PhD biological nitrogen fixation
12 Samanthay RN - BSc,MSc soil science; agricultural chemistry
13 Sarkunan V - BScAgr,MScAgr,PhD soil science; agricultural chemistry
14 Singh PK - BSc,MSc,PhD microbiology

1814 Central Rainfed Lowland Rice Research Station

1 Shukla SN - BSc,MSc,PhD *Officer In Charge*
2 Pande KL - BSc,MSc,PhD plant breeding and genetics

1815 Central Rainfed Rice Research Station
Hazaribag, Bihar, 825301

1 Chauham VS - MSc,PhD *Officer-in-Charge* plant breeding and genetics
2 Chauhan JS - BSc,MSc,PhD plant breeding and genetics
3 Mishra GN - BSc,MScAgr cropping systems
4 Shukla VD - BScAgr,MScAgr,PhD viral diseases

1816 Indian Institute of Sugarcane Research
Lucknow, Uttar Pradesh, 226 002
Tel: 52572, 51673. Telex: 535 - 309. Cables: GANNANUSANDHAN.

1 Singh Kishan - MScAgr,PhD *Director*
2 Srivastava SC - MSc,PhD *Project Coordinator (Sugarcane)*
3 Srivastava SN - MScAgr,PhD *Librarian*

1817 Agronomy
1 Alam M - BScAgr,AssocIARI *Head of Division* sugarcane agronomy
2 Bhattacharyya BB - MScAgr sugarcane agronomy
3 Chauhan RS - MScAgr,PhD sugarcane agronomy
4 Motiwale MP - MScAgr,PhD sugarcane agronomy
5 Prasad S - MScAgr,PhD sugarcane agronomy
6 Varma RS - MScAgr sugarcane agronomy
7 Yadav RL - MScAgr,PhD sugarcane agronomy

1818 Soil Science and Agricultural Chemistry
1 Yadav DV - MScAgr,PhD *Head of Division*

1819 Botany and Breeding
1 Tripathi BK - MScAgr,PhD *Head of Division*

1820 Plant Physiology and Bio-Chemistry
1 Madan VK - MSc,PhD *Head of Division*

1821 Plant Pathology
1 Misra SR - MScAgr,PhD *Head of Division* sugarcane pathology
2 Agnihotri VP - MSc,PhD sugarcane pathology
3 Lal R - MScAgr sugarcane pathology
4 Lal S - MScAgr sugarcane pathology
5 Pandey S - MScAgr,PhD sugarcane pathology
6 Ram RS - MScAgr,PhD sugarcane pathology
7 Rashid A - MSc,PhD sugarcane pathology
8 Shukla US - MScAgr,PhD sugarcane pathology
9 Singh N - MScAgr,PhD sugarcane pathology
10 Singh RP - MSc,PhD sugarcane pathology
11 Sinha OK - MScAgr,PhD sugarcane pathology
12 Srivastava SN - MScAgr,PhD sugarcane pathology

1822 Entomology
1 Varma A - MSc *Head of Division* sugarcane insect pests
2 Chandra J - MSc sugarcane insect pests
3 Mehrotra AK - MSc sugarcane insect pests
4 Tewari RK - MScAgr,PhD sugarcane insect pests
5 Tiwari NK - MScAgr,PhD sugarcane insect pests
6 Yadav RA - MScAgr,PhD sugarcane insect pests

1823 Agricultural Engineering
1 Das FC - MTech *Head of Division*

1824 Agricultural Economics and Statistics
1 Lal J - MScAgr,PhD

1825 Sugarcane Breeding Institute
Coimbatore, Tamil Nadu, 641 007
Tel: 37641-37645. Telex: 855-383. Cables: GANNAPRAJANAN.

1 Mohan Naidu K - MScAgr,PhD *Director* plant physiology

1826 Division of Agronomy
1 Srinivasan TR - MScAgr,PhD *Head of Division*
2 Rakaiyappan P - MScAgr soil science
3 Sundra B - MScAgr agronomy

1827 Division of Agricultural Chemistry
1 Chiranjivi Rao KC - MSc,PhD *Head of Division* agricultural chemistry

1828 Division of Agricultural Entomology
1 David H - MA,PhD *Head of Division*

1829 Division of Genetics and Cytogenetics
1 Srinivasan TV - MSc,PhD *Head of Division*
2 Jalaja Mrs NC - MSc,PhD

1830 Division of Plant Breeding
1 Natarajan BV - MSc,PhD sugarcane breeding
2 Bhagyalakshmi KV - MSc,PhD sugarcane breeding
3 Nagarajan R - MScAgr,PhD sugarcane breeding
4 Palamichamy K - MScAgr,PhD sugarcane breeding
5 Ramana Rao TC - MSc,PhD sugarcane breeding
6 Saki BK - MSc,PhD sugarcane breeding
7 Sanakaranarayanan P - MSc,PhD sugarcane breeding
8 Subramanayam KN - MSc sugarcane breeding

1831 Division of Plant Pathology
1 Alexander KC - MScAgr,PhD *Head of Division* sugarcane pathology
2 Madhusudhana Rao M - MSc sugarcane pathology
3 Mohan Raj D - MScAgr,PhD sugarcane pathology
4 Padmanabhan P - MScAgr,PhD sugarcane pathology

1832 Division of Plant Physiology
1 Singh S - MScAgr,PhD *Head of Division* sugarcane physiology
2 Rao P - MScAg,PhD sugarcane physiology
3 Reddy S - MSc,PhD sugarcane physiology
4 Venkatramana - MSc,PhD sugarcane physiology

1833 Section of Nematology
1 Mehta Miss UK - MSc,PhD *Head of Section* sugarcane nematology

1834 Section of Agricultural Statistics
Coimbatore, Tamil Nadu, 641 007
Tel: 37641-37645. Cables: GANNAPRAJANAN.

1 Shanmughasundram - MScAgr *Head of Section* statistics

1835 SBI Regional Centre
240-A Model Town, Karnal, Haryana, 132001

1 Balasunderam N - MSc,PhD *Head of the Centre* genetics and cytogenetics
2 Misra A - MScAgr,PhD agronomy
3 Prakasam N - MScAgr plant pathology
4 Surinder Kumar - MSc,PhD plant breeding

1836 SBI Regional Centre
Civil Station Post, Cannanore, Kerala, 670 002

1 Vijayan Nair N - MSc *In Charge of Research Centre* genetics and cytogenetics

1837 Central Institute for Cotton Research
Post Bag No. 125, GPO, Nagpur - 1, Maharashtra
Tel: 26323. Telex: 0715-267. Cables: KAPASSANTHAN NAGPUR.

1 Mannikar ND - MSc,PhD *Director*

1838 Soil Science and Agricultural Chemistry
1 Mannikar ND - MSc,PhD *Head*

1839 Agronomy
1 Bonde WC - MSc *Head* cotton agronomy
2 Mudholkar NJ - MSc cotton agronomy

1840 Plant Pathology
1 Sheo Raj - MSc,PhD *Head* cotton pathology
2 Meshram MK - MSc cotton pathology
3 Mukewar PM - MSc,PhD cotton pathology
4 Teneja NK - MSc cotton pathology

1841 Crop Improvement
1 Narayanan SS - MSc,PhD *Head*

1842 Plant Breeding
1 Bhat MG - MSc,PhD cotton breeding
2 Singh P - MSc,PhD cotton breeding
3 Singh VV - MSc cotton breeding

1843 Cytogenetics

1 Kothandaraman R - MSc cotton cytogenetics

1844 Entomology

1 Jayaswal AP - MSc,PhD *Head* cotton entomology
2 Bhamburkar MW - MSc,PhD cotton entomology
3 Singh OP - MSc,PhD cotton entomology

1845 Plant Physiology and Biochemistry

1 Sahay RK - MSc,PhD *Head* cotton physiology
2 Rao MRK - MSc,PhD cotton physiology
3 Singh D - MSc,PhD cotton physiology

1846 CICR Regional Station

Coimbatore - 3, Tamil Nadu
Tel: 31240. Telex: 0715-267. Cables: KAPASSANTHAN.

1 Basu AK - MSc,PhD *Project Coordinator and Head* cotton breeding
2 Bhat JG - MSc,PhD plant physiology
3 Henry S - MSc plant breeding
4 Kannan A - MSc,PhD plant pathology
5 Krishnamurthy R - MSc plant breeding
6 Krishnashwamy R - MSc cytogenetics
7 Meenakshisundram PC - MSc,PhD soil science
8 Natarajan K - MSc,PhD entomology
9 Premsekar S - MSc,PhD plant breeding
10 Seshadri V - MSc agronomy
11 Seshadrinthan AR - MSc,PhD plant physiology
12 Shanmugham K - MSc agronomy
13 Srinivasan K - MSc plant breeding
14 Sundramurthy VT - MSc,PhD entomology
15 Surulivelu T - MSc,PhD entomology

1847 CICR Regional Station

Sirsa, Haryana
Tel: 20428. Telex: 0715-267. Cables: KAPASSANTHAN.

1 Gaul BL - MSc,PhD *Head* agronomy
2 Banerjee SK - MSc,PhD entomology
3 Duhoon SS - MSc,PhD plant breeding
4 Narula AM - MSc,PhD plant pathology
5 Singh J - MSc plant breeding

1848 Cotton Technological Research Laboratory

Post Bag 16640, Adenwala Road, Matunga Bombay, Mahrashtra, 400019
Tel: 4127273,7274,7275,7276. Cables: TECHSEARCH.

1 Sundram V - MSc,PhD,FTI *Director*

1849 Quality Evaluation Division

1 Munshi VG - MSc,PhD,FTI fibre science
2 Iyer Mrs JK - MSc agricultural statistics
3 Oka PG - MSc fibre science and technology
4 Petkar BM - MSc fibre science and technology
5 Ukidve AV - MSc cotton technology

1850 Physics Division

1 Patil NB - MSc,PhD fibre science and technology
2 Aravindanath S - MSc,PhD fibre science and technology
3 Chidambareswaran PK - MSc,PhD fibre science and technology
4 Hussain GFS - MSc fibre science and technology
5 Iyer Mrs BP - MSc,PhD fibre science and technology
6 Krishna Iyer KR - MSc,PhD fibre science and technology
7 Nachane RP - MSc fibre science and technology
8 Paralikar KM - MSc,PhD,FRMS fibre science and technology
9 Ramesh Babu N - BE,MTech microprocessor technology
10 Sreenivasan S - MSc,PhD fibre science and technology

1851 Mechanical Processing and Ginning Division

1 Parthasarathy MS - MText,MScTech spinning and ginning technology
2 Rao Ramamurthy TN - BSc,BScTech spinning and wool science
3 Srinathan B - BScText,MScText spinning, ginning and knitting

1852 Chemistry Division

1 Bhatt Miss IG - MSc,PhD cellulose chemistry radiation chemistry, cottonseed oil
2 Iyer Mrs V - MSc,MPhil cellulose chemistry
3 Raje Miss CR - MSc finishing of textile and cellulose chemistry

1853 Microbiology and Biochemistry Division

1 Khandeparakar VG - MSc,PhD biosynthesis, agricultural microbiology and bio-technology
2 Balasubramanya RH - MSc,PhD agricultural microbiology and biotechnology
3 Bhatwadekar Mrs SP - MSc biochemistry of cellulose

1854 Jute Agricultural Research Institute

Nilgunj, Barrackpore, West Bengal, 743101
Tel: 174,773,726. Cables: JUTESEARCH.

1 Mukherju N - MSc,PhD *Director* plant pathology

1855 Research Division

Nilgunj, Barrackpore, West Bengal, 743101
Tel: 174,773,726. Cables: JUTESEARCH.

1856 Agronomy

1 Mitra PC - MSc,PhD multiple cropping
2 Biswas DK - MScAgr agronomy
3 Biswas GC - MScAgr agronomy
4 Pal MK - MSc,PhD extension
5 Pathak S - MSc,PhD extension
6 Ray AR - MSc general agronomy

1857 Plant Breeding and Genetics

1 Basak SL - MScAgr,PhD breeding techniques
2 Chatterjee SK - MSc bio-physics
3 Gupta D - MScAgr plant breeding
4 Kumar D - MScAgr,PhD plant breeding
5 Mitra GC - MSc,PhD plant breeding
6 Mondal Miss N - MScAgr,PhD plant breeding
7 Paria P - MScAgr,PhD genetics
8 Saha A - MSc,PhD plant breeding
9 Shome A - MScAgr,PhD tissue culture
10 Sinha MK - MScAgr,PhD plant breeding

1858 Soils and Microbiology

1 Roy PK - MSc,PhD soil test and crop response
2 Mondal BC - MScAgr soil science
3 Pal HP - BSc,AIC soil science
4 Saha NN - MScAgr,PhD soil science

1859 Agricultural Entomology

1 Singh B - MSc integrated pest control
2 Das LK - MSc,PhD agricultural entomology
3 Mullick RN - MScAgr agricultural entomology
4 Pandit NC - MSc,PhD agricultural entomology
5 Pradhan SK - MScAgr agricultural entomology

1860 Mycology and Plant Pathology

1 Mishra CBP - MScAgr,PhD control of disease of fibre
2 Gosh SK - MScAgr,PhD plant pathology
3 Mazumidar Miss M - MSc,PhD plant pathology
4 Mondal RK - MScAgr,PhD plant pathology
5 Shome A - MScAgr plant pathology

1861 Statistics Section

1 Sasmal BC - MSc,PhD experimental design and analysis
2 Katyal V - MA design and analysis of field experiments
3 Samanta BK - MSc design and analysis of field experiments

1862 Plant Physiology Section

1 Bisaria AK - MSc,PhD chemical regulation of growth
2 Bhattacharyya AC - MSc physics
3 Palit P - MSc,PhD plant physiology

1863 Agricultural Engineering Section

1 Mondal TC - MSc,PhD farm machinery and power

1864 Agricultural Meteorology Section

1 Rao PV - MScTech,PhD agricultural meteorology

1865 Operational Research Project

1 Mishra RN - MScAgr soil chemistry
2 Rahman A - MScAgr agronomy

1866 All India Coordinated Research Project on Jute and Allied Fibres

1 Mukherjee N - MSc,PhD plant pathology
2 Bhattacharya AK - MScAgr agronomy
3 Dutta AN - MSc plant breeding
4 Dutta P - MSc agricultural stations

5 Maita SN - MSc,PhD plant breeding

1867 Research Stations

1868 Sunhemp Research Station
Pratapgarh, Uttar Pradesh

1 Prakash G - MSc plant breeding
2 Gupta BN - MSc agronomy

1869 Sisal Research Station
Bamra, Sambalpur, Orissa

1 Singh K - MSc agronomy

1870 CNJSM Farm
Bud Bud, Burdwan, West Bengal

1 Singh AP - MSc plant breeding

1871 Jute Technological Research Laboratories
12, Regent Park, Calcutta, West Bengal, 700 040
Tel: 72-2538. Telex: 2966. Cables: JUTLAB.

1 Pandey SN - MSc,PhD *Director*

1872 Textile Technology

1 Mazumder A - MTech textile manufacturing
2 Ghosh SK - BScTech,LTT,ATA textile manufacturing
3 Sengupta P - BSc,BScTech textile manufacturing

1873 Chemical Technology

1 Day A - MSc,PhD organic chemistry

1874 Non-Woven Technology

1 Debnath CR - MTech,PhD textile manufacturing
2 Samajpati Mrs S - MSc,PhD organic chemistry

1875 Chemistry

1 Sen SK - MSc,PhD organic chemistry
2 Bhadury SK - MSc,PhD organic chemistry

1876 Physics

1 Roy PK - MSc,PhD physics
2 Bag SC - MSc,PhD physics
3 Das BK - MTech,PhD physics
4 Mukherjee AK - MSc,PhD physics

1877 Microbiology

1 Bhattacharyya SK - MSc,PhD microbiology
2 Basak MK - MSc,PhD microbiology

1878 Design, Development and Maintenance

1 Paul D - ME engineering

1879 Central Tobacco Research Institute
Rajahmundry - 5, Andra Pradesh
Tel: 3271-3274. Telex: 0492-205. Cables: TABSEARCH.

1 Charie MS - MSc *Director* entomology

1880 Research Division

1881 Agronomy
Rajahmundry - 5, Andra Pradesh
Tel: 3271-3274. Telex: 0492-205. Cables: TABSEARCH.

1 Singh KD - MScAgr agronomy
2 Harishukumar P - MScAgr,PhD agronomy
3 Raghavaiah CV - MScAgr agronomy
4 Rao MS - MScAgr agronomy

1882 Agricultural Entomology

1 Prasad GR - MSc entomology
2 Krishnananda N - MSc,PhD entomology
3 Rao RSN - MSc entomology
4 Sitaramaiah S - MScAgr entomology

1883 Agricultural Statistics

1 Rao DVS - MSc,PhD statistics
2 Murthy NS - MA,PSCC statistics

1884 Agricultural Chemistry and Soil Science

1 Murthy VK - MScAgr,PhD soil chemistry

1885 Biochemistry and Plant Physiology

1 Rao BVK - MSc,PhD chemistry
2 Bangarayya M - MScAgr physiology
3 Gopalakrishna CVSSV - MSc,PhD chemistry
4 Murthy PSN - MSc,PhD chemistry
5 Ushasree KV - MSc,MPhil bio-chemistry

1886 Genetics and Breeding

1 Rao K - MSc cytogenetics
2 Narayanan AI - MSc cytogenetics
3 Ramavarma KT - MScAgr cytogenetics
4 Rao CV - MSc,PhD plant breeding
5 Rao RVS - MSc,PhD plant breeding
6 Subhashini Mrs U - MSc cytogenetics
7 Venkateswarlu T - MSc cytogenetics

1887 Plant Pathology

1 Nagarajan K - MScAgr,PhD plant pathology
2 Raju CA - MScAgr,PhD plant pathology
3 Reddy TSN - BSc,AssocIARI plant pathology

1888 Quality Cell

1 Murthy SR - MSc,PhD agricultural chemistry

1889 Library

1 Rao R - BSc,AssocDRTC library science

1890 Research Stations

1891 CTRI Research Station
Dinhata, Cooch Behar District, West Bengal, 736 135

1 Tripata SN - MScAgr,PhD agronomy

1892 CTRI Research Station
Guntur, Andhra Pradesh, 522 004

1 Sitaramachari T - MScAgr plant breeding
2 Rao MUM - MScAgr agronomy

1893 CTRI Research Station
Hunsur, Karnataka, 571 105

1 Ramakrishnayya BV - MSc,PhD soil chemistry
2 Devappa K - MSc agronomy
3 Hussaini SS - MSc,PhD nematology
4 Moses JSL - MScAgr plant breeding
5 Rao SV - MSc,PhD plant breeding
6 Shenoi MM - MSc,PhD plant pathology

1894 CTRI Research Station
Kandukur, Andhra Pradesh, 523 105

1 Satyanarayana KV - MSc,MS,PhD cytogenetics

1895 CTRI Research Station
Pusa, Samasthipur District, Bihar, 848 125

1 Sinha R - MSc,PhD agronomy
2 Dwivedi SSL - MScAgr agronomy
3 Pandey RG - MScAgr,PhD plant breeding

1896 CTRI Research Station
Vedasandur, Anna District, Tamil Nadu, 624 710

1 Krishnamurthy SK - MSc agronomy
2 Lakshminarayana R - MSc,PhD plant breeding

1897 Central Plantation Crops Research Institute
Kasaragod, Kerala, 670 124
Tel: 94. Telex: 0842 248 CASCIN. Cables: RESEARCH KASARAGOD.

1 Nair MK *Director*
2 Bhaskara Rao EVV - PhD *Project Coordinator* cashew
3 Edison S - PhD *Project Coordinator* spices
4 Jayasanakar NP - PhD *Joint Director*
5 Mandal RC - PhD *Senior Scientist*
6 Rethinam P - PhD *Project Coordinator* coconut and arecanut
7 Shama Bhat K - PhD *Joint Director*
8 Venugopal K - PhD *Joint Director*

1898 Agronomy

  1 Mohammed Yusuf - PhD *Head of Division*

1899 Agricultural Economics

  1 Das PK - PhD *Head of Division*

1900 Animal Sciences

  1 Bhattacharyya AR - MVSc,FRVCS *Head of Division*

1901 Agricultural Statistics

  1 Jacob M - MPhil *Head of Division*

1902 Biochemistry

  1 Nagaraja KV - PhD *Head of Division*

1903 Entomology

  1 Pillai GB - MSc *Head of Division*

1904 Genetics and Plant Breeding

  1 Iyer RD - PhD *Head of Division*

1905 Horticulture

  1 Singh R - PhD *Head of Division*
  2 Swamy KKN - PhD

1906 Microbiology

  1 Ghai SK - PhD *Head of Division*

1907 Nematology

  1 Koshy PK - PhD *Head of Division*

1908 Plant Pathology

  1 Nambiar KKN - PhD *Head of Division*

1909 Plant Physiology

  1 Patil KD - PhD *Head of Division*

1910 Rodent Ecology

  1 Advani RR - PhD

1911 Soil Science

  1 Biddappa CC - PhD *Head of Division*

1912 Technology

  1 Madhavan K - BSc,IETE

1913 Indian Institute of Horticultural Research

Post Box No. 8025, Bangalore, Karnataka, 560080
Tel: 362486,367909,360483. Telex: BG-534. Cables: HORTSEARCH.

  1 Pandey RM - MSc,PhD *Director* fruit physiology and micropropagation of fruit crops

1914 Division of Fruit Crops

  1 Singh R - MScAgr,PhD *Head of Division* grape breeding
  2 Agarwal PK - MSc,PhD breeding of citrus and banana
  3 Iyer CPA - MSc,PhD breeding of mango, guava and papaya
  4 Jalikop SH - MScAgr breeding of semiarid fruits
  5 Kohli RR - MScAgr,PhD agrotechniques for mango, guava and papaya
  6 Prakash GS - MScAgr training, pruning and rootstocks in grapes
  7 Reddy BMC - MScAgr agrotechniques for citrus, banana and pineapple
  8 Shikhamany SD - MScAgr,PhD grape nutrition
  9 Subramanyam MD - MSc,PhD breeding of mango, guava and papaya

1915 Division of Vegetable Crops

  1 Dutta OP - MScAgr,PhD *Head of Division* cucurbits, okra, eggplant breeding
  2 Anand N - MScAgr,PhD tomato, sweet pepper breeding
  3 Deshpande AA - MSc chillies breeding
  4 Naik LB - MScAgr vegetable seed production
  5 Pandey SC - MScAgr,MPhil,PhD cauliflower, cabbage breeding
  6 Pathak CS - MSc,PhD onion, carrot breeding
  7 Prabhakar BS - MScAgr,PhD vegetable cropping systems
  8 Shukla V - MScAgr fertilizer management in vegetable crops
  9 Tikoo SK - MSc,PhD tomato breeding
  10 Vani A - MSc beans breeding

1916 Division of Ornamental Crops

  1 Singh F - MSc,PhD *Head of Division* orchid breeding and tissue culture
  2 Bhat RN - MSc,PhD breeding ornamental plants, rose
  3 Karihaloo JL - MSc,PhD breeding of ornamental plants, *Hippeastrum gloriosa*
  4 Mukhopadhyay A - MSc,PhD agrotechnique of ornamental plants, rose, gladiolus
  5 Negi SS - MSc,PhD breeding of ornamental plants, gladiolus, chrysanthemum
  6 Raghava SPS - MSc,PhD breeding of ornamental plants, chrysanthemum, marigold

1917 Division of Medicinal and Aromatic Crops

  1 Krishnan R - MSc,MTech,PhD *Head of Division* breeding of *Catharanthus roseus* and steroid-bearing *Solanum* species
  2 Srivastava HC - MSc,PhD breeding of *Jasminum grandiflorum*
  3 Vasanthakumar T - MSc breeding of patchouli and scented geranium

1918 Division of Post Harvest Technology

  1 Amba Dan - MScHort,PhD *Head of Division* fruit and vegetable processing
  2 Ethiraj S - MAgrSc,PhD enology and food in microbiology
  3 Gopalakrishna Rao KP - MAgrSc fresh vegetable handling and storage
  4 Krishnamurthy S - MSc,PhD fresh fruit handling and storage
  5 Suresh ER - MSc enology and food microbiology

1919 Division of Plant Pathology

  1 Singh SJ - MScAgr,PhD *Head of Division* vegetable virology and electron microscopy
  2 Chacko CI - MScAgr,PhD ornamental pathology
  3 Kishun R - MScAgr,PhD bacteriology-fruit and vegetables
  4 Maholay MN - MSc,PhD seed pathology
  5 Raghavendra Rao NN - MScAgr,PhD betelvine pathology
  6 Rewal RD - MScAgr,PhD fruit pathology
  7 Sharma SR - MScAgr,PhD fruit virology
  8 Sridhar TS - MSc vegetable pathology
  9 Tewari RP - MSc,PhD mycology
  10 Ullasa BA - MScAgr,PhD post harvest pathology

1920 Division of Entomology and Nematology

  1 Narayanan K - MScAgr,PhD *Head of Division* insect pathology
  2 Jayanth KP - MSc weed biological control
  3 Krishnamoorthy A - MScAgr biological control
  4 Krishnamoorthy PN - MScAgr vegetable crops
  5 Mani M - MScAgr biological control
  6 Mohan KS - MSc insect pathology
  7 Parvatha Reddy PP - MScAgr,PhD nematology
  8 Singh SP - MScAgr,PhD biological control
  9 Srinivasan K - MScAgr,PhD vegetable crop entomology
  10 Sujaya U - MSc medicinal and ornamental crop entomology
  11 Tandon PL - MScAgr,PhD fruit crop entomology
  12 Tewari GC - MScAgr,PhD vegetable crop entomology
  13 Verghese A - MScAgr fruit crop entomology

1921 Division of Plant Physiology and Biochemistry

  1 Dore Swarmy R - MSc,PhD *Head of Division* plant biotechnology, tissue culture of horticultural crops
  2 Anand L - MSc pesticide residue analysis in plants and environment
  3 Ananthanarayanan TV - MSc biochemistry of differentiation and genetic transformation of plant cells in culture; plant biotechnology
  4 Awasthi MD - MScAgr,PhD pesticide residue analysis in plants, soil and environment
  5 Challa P - MSc,PhD chemical weed control, herbicide effects on nutrient uptake in fruit crops
  6 Chandravadana MV - MSc,PhD secondary plant metabolites
  7 Leela D - MSc,PhD chemical weed control, herbicide residue analysis in vegetable crops
  8 Mathai CK - MScAgr,PhD environmental physiology and canopy architecture of fruit trees
  9 Murti GSR - MScAgr,PhD plant hormones in growth and development
  10 Pal RN - MSc,PhD physiology of tree fruit crops
  11 Selvaraj Y - MScAgr,PhD aroma biogenesis and ripening enzymes in fruits
  12 Srinivasa Rao NK - MScAgr,PhD physiology of water stress, source-sink relationships

1922 Division of Soil Science

  1 Bhargava BS - MScAgr,PhD *Head of Division* leaf analysis
  2 Edward Raja M - MScAgr,PhD micronutrient studies
  3 Hariprakasha Rao M - MScAgr,PhD potassium nutrition
  4 Hegde DM - MScAgr,PhD water management

5 Iyengar BRV - MSc,PhD root activity & fertilizer use efficiency
6 Onkarayya H - MScAgr,PhD mycorrhiza
7 Palaniappan R - MScAgr,PhD soil salinity studies
8 Sukhada M - MSc,PhD nitrogen fixation

1923 Division of Plant Genetic Resources

1 Alexander MP - MAgrSc,PhD *Head of Division* pollen storage
2 Doijode SD - MAgrSc,PhD seed storage
3 Ganeshan S - MSc pollen storage

1924 Division of Economics and Statistics

1 Ramachander PR - MA,DipStat *Head of Division* biometrical genetics; experimental designs
2 Biswas SR - MSc computer modelling designs, software development
3 Rao GSP - MSc,DipStat insect ecology; fruit tree experimentation
4 Srinivasan VR - MSc epidemiology, software development
5 Subramanyam KV - MScAgr,PhD economic aspects of production and marketing of horticultural crops

1925 Division of Extension

1 Pal AB - MScAgr,PhD *Head of Division* agricultural extension
2 Krishna Rao D - MSc fish and fisheries
3 Rao TR - MSc agricultural extension
4 Reddy YN - MScAgr,PhD propagation
5 Seshadri K - MScAgr,PhD plant pathology

1926 Agricultural Engineering Section

1 Mandhar SC - BSc,MTech farm machinery and power

1927 Central Horticultural Experiment Station
Chethalli, Kodagu District, 571248
Tel: 35. Cables: HORTUS.

1928 Horticulture

1 Ravishankar H - MScAgr fruit culture

1929 Agricultural Entomology

1 Bhumannavar BS - MScAgr fruit entomology

1930 Plant Pathology

1 Narayanappa M - MScAgr fruit pathology

1931 Central Citrus Experiment Station
Gonikoppal, Kodagu District, 571213
Tel: 74. Cables: CITRUSEARCH.

1932 Horticulture

1 Nanaya KA - MScAgr,Hort,HHS,AHS fruit culture

1933 Plant Pathology

1 Manjunath KL - MSc fruit virology

1934 Central Horticultural Experiment Station
Shukla Colony, Hinoo, Ranchi, Bihar, 834002
Tel: 8231. Telex: 0625-210. Cables: HORTSEARCH.

1935 Horticulture

1 Singh DP - MScAgr,PhD vegetable breeder
2 Dubey KC - MScAgr,PhD fruit breeder
3 Ranvir Singh MSCAGR *vegetable breeder* vegetables, breeding
4 Singh HP - MScAgr fruit breeder

1936 Soil Science

1 Kotur SC - MScAgr soil science

1937 Central Horticultural Experiment Station
Ambavadi, Civil Lines, Godhra, 389001
Tel: 2611. Cables: HORTSEARCH.

1938 Horticulture

1 Raturi GB - MSc,PhD fruit culture
2 Purohit AG - MSc,PhD fruit breeder
3 Vijay OP - MScAgr vegetable breeder

1939 Plant Pathology

1 Jagadish Chandra - MScAgr,PhD fruit pathology

1940 Agricultural Entomology

1 Bagle BG - MScAgr fruit entomology
2 Krishna Kumar NK - MScAgr vegetable entomology

1941 Citrus Research Station
c/o National Bureau of Soil Survey and Land Use Planning, Seminary Hills, Nagpur, Maharashtra, 440006
Tel: 32386. Telex: 0715-262. Cables: SOILANDBRU.

1942 Horticulture

1 Dhander DG - MSc,PhD fruit culture

1943 Central Institute of Horticulture for Northern Plains
B-217, Indra Nagar, Lucknow, Uttar Pradesh, 226016

1 Srivastava RP - MScAgr,PhD *Director* horticulture
2 Abbas SR - MSc,PhD entomology
3 Joshi OP - MScAg,PhD soil science
4 Kalra SK - MSc,PhD post harvest technology
5 Mishra AK - MSc,PhD plant pathology
6 Om Prakash - MSc,PhD plant pathology
7 Rajput MS - MScAgr soil science
8 Sharma S - MSc entomology
9 Singh BP - MScAgr,PhD post harvest technology
10 Sinha GC - MScAg,PhD horticulture
11 Srivastava RP - MSc,PhD entomology
12 Tandon DK - MSc,PhD biochemistry
13 Yadav IS - MScAg,PhD horticulture

1944 Central Potato Research Institute
Simla, Himachal Pradesh, 171001
Tel: 3118. Telex: 0391-240 MOOL-IN. Cables: POTATOSEARCH.

1 Nayar NM - AssocIARI,PhD *Director* plant breeding
2 Sharma KP - MScAgr,AssocIARI,PhD *Project Coordinator* plant breeding

1945 Division of Genetics

1 Misra PC - MScAgr *Head* genetics and cytogenetics
2 Chandra R - MSc,PhD economic botany
3 Garg KC - MSc,MPhil genetics and cytogenetics
4 Gaur PC - MSc,MPhil,PhD plant breeding
5 Pandey SK - MSc economic botany

1946 Division of Biochemistry and Physiology

1 Sukumaran NP - MSc *Head* plant physiology
2 Bannerjee VN - MSc plant physiology
3 Bansal KC - MSc plant physiology

1947 Division of Crop and Soil Science

1 Grewal JS - MScAgr,PhD *Head* soil fertility
2 Dahiya PS - MSc agricultural economics
3 Malhotra VP - MSc agricultural statistics
4 Sharma HC - MSc agricultural statistics
5 Sharma RC - MScAgr,PhD soil science
6 Sood MC - MScAgr soil physics
7 Trehan SP - MSc soil fertility

1948 Division of Plant Pathology

1 Shekhawat GS - MSc,PhD *Head* plant pathology
2 Bhattacharyya SK - MScAgr,DPhil plant pathology
3 Dhingra MK - MScAgr plant pathology
4 Garg ID - MSc,PhD plant pathology
5 Khurana SMP - MSc,PhD plant pathology
6 Singh BP - MScAgr,MPhil plant pathology
7 Singh MN - MScAgr,MPhil,PhD plant pathology
8 Singh RA - MScAgr,PhD plant pathology
9 Singh RP - MScAgr,PhD plant pathology
10 Singh S - MScAgr,PhD plant pathology

1949 Division of Entomology and Nematology

1 Prasad KSK - MScAgr,PhD *Head* agricultural nematology
2 Chandla VK - MScAgr agricultural entomology
3 Kishore R - MScAgr agricultural entomology
4 Misra SS - MScAgr agricultural entomology

1950 Division of Seed Production

1 Chauhan HS - MSc plant breeding

1951 Central Potato Research Station
Kufri
Tel: 3118. Telex: 0391-240. Cables: POTATOSEARCH.

1 Singh S - MSc,PhD plant pathology

1952 Central Potato Research Station
Jalandhar
Tel: 3118. Telex: 0391-240. Cables: POTATOSEARCH.

1 Rana MS - MScAgr,PhD genetics and cytogenetics
2 Kang GS - MSc,PhD plant breeding
3 Saini SS - MScAgr agronomy
4 Verma AK - MScAgr,PhD plant pathology

1953 Central Potato Research Station
Modipuram
Tel: 3118. Telex: 0391-240. Cables: POTATOSEARCH.

1 Singh JP - MSc seed technology
2 Anand SK - MSc plant breeding
3 Dwivedi R - MSc,PhD plant pathology
4 Gupta A - MScAgr agronomy
5 Jeswani MD - MScAgr,PhD plant pathology
6 Lal SS - MScAgr agronomy
7 Misra JB - MSc biochemistry
8 Raj BT - MScAgr agricultural entomology
9 Sharma VC - MScAgr,PhD plant pathology
10 Singh SV - MScAgr,PhD economic botany
11 Verma KD - MSc,PhD agricultural entomology

1954 Central Potato Research Station
Patna
Tel: 3118. Telex: 0391-240. Cables: POTATOSEARCH.

1 Saxena AP - MScAgr agricultural entomology
2 Janardanjee - MSc agricultural entomology
3 Khushu CL - MSc genetics and cytogenetics
4 Kushwah VS - MScAgr,PhD agronomy
5 Prasad B - MSc,PhD plant pathology
6 Rai RP - MSc,PhD plant pathology
7 Ram G - MScAgr agricultural entomology
8 Singh L - MScAgr plant physiology
9 Singh SN - MScAgr,PhD soil fertility

1955 Central Potato Research Station
Gwalior
Tel: 3118. Telex: 0391-240. Cables: POTATOSEARCH.

1 Singh C - MSc,PhD agronomy
2 Khanna RN - MScAgr,PhD plant pathology
3 Singh DS - MScAgr,PhD plant pathology

1956 Central Potato Research Station
Rajgurunagar
Tel: 3118. Telex: 0391-240. Cables: POTATOSEARCH.

1 Akhade MN - AssocIARI agronomy
2 Kaley DM - MScAgr plant physiology

1957 Central Potato Research Station
Mukteswar
Tel: 3118. Telex: 0391-240. Cables: POTATOSEARCH.

1 Upreti GC - MSc plant pathology

1958 Central Potato Research Station
Shillong
Tel: 3118. Telex: 0391-240. Cables: POTATOSEARCH.

1 Lal L - MScAgr,PhD agricultural entomology

1959 Central Potato Research Station
Darjeeling
Tel: 3118. Telex: 0391-240. Cables: POTATOSEARCH.

1 Gopal J - MScAgr genetics and cytogenetics

1960 Central Potato Research Station
Ootacamund
Tel: 3118. Telex: 0391-240. Cables: POTATOSEARCH.

1 Gajaraja CP - MScAgr plant breeding
2 Amalraj SFA - MSc plant breeding
3 Arora RK - MSc,PhD plant pathology
4 Singh DB - MSc,PhD agricultural nematology

1961 Central Soil and Water Conservation Research and Training Institute
Dehradun, Uttar Pradesh, 248195
Tel: 24213. Telex: 0595-237. Cables: SOILCONS.

1 Dhruva Nayana VV - MSc,PhD *Director*
2 Agarwal MC - MScAgr statistics
3 Arora YK - MSc,PhD horticulture
4 Bhardwaj SP - MSc,PhD soil physics and soil and water conservation
5 Dadhwal KS - MScAgr soil fertility
6 Gupta RK - AssocIARI forestry
7 Kalyan Singh - MSc,PhD soil science
8 Mohan SC - BScEngg soil fertility
9 Patnaik US - MTech soil and water conservation engineering
10 Pratap Nayain - MSc,PhD soil physics and soil and water conservation
11 Puri DN - BScAgr,DDR,DipWH plant studies
12 Ram Babu - MSc,PhD statistics
13 Sastry G - MTech soil and water conservation engineering
14 Sewa Ram - MSc,PhD agronomy
15 Sharda VN - MTech soil and water conservation engineering
16 Singh G - MSc,PhD *Chief Scientist* agronomy
17 Singh PN - MSc,PhD soil physics and soil water conservation
18 Tomar VPS - MScAgr forestry
19 Tyagi PC - PhD plant breeding

1962 Research Centre, Agra
Agra

1 Bhushan LS - MSc,PhD soil physics and soil and water conservation
2 Bankey Lal MSCAGR horticulture
3 Nitant HC - MScAgr,DSc soil science
4 Om Prakash PHD agronomy

1963 Research Centre, Bellary
Bellary

1 Rama Mohan Rao MS - PhD soil physics and soil and water conservation
2 Jayaram NS - MScAgr soil physics and soil and water conservation
3 Silvarajan S - MA agricultural economics

1964 Research Centre, Chandigarh
Chandigarh

1 Mittal SP - MSc agronomy
2 Agnihotri Y - MScMath,MScStat agricultural statistics
3 Grewal SS - MScAgr,PhD soil physics and soil and water conservation
4 Varma B - MSc soil physics and soil and water conservation
5 Arjun Prasad - MSc agronomy
6 Chandra Prakash - MTech soil and water conservation engineering
7 Prasad SN - MScAgr agronomy
8 Singh R - MScAgr,PhD soil science

1965 Research Centre, Ootachmund
Ootacamund

1 Samraj P - MSc forestry
2 Jaya Kumar M - BScAgr soil and water conservation engineering
3 Seshachalam N - MScAgr agronomy

1966 Research Centre, Vasad
Vasad

1 Chinnamani S - BSc,MS,DDR soil physics and soil and water conservation
2 Singh H - PhD agronomy
3 Subhash Chandra - MScAgr,PhD soil physics and soil and water conservation

1967 Research Centre, Ganiyar, Rewari
Ganiyar, Rewari

1 Sharma AK - PhD agronomy

1968 Central Soil Salinity Research Institute
P.O. Karnal, Haryana, 132001
Tel: 3874, 2856. Cables: SALINITY, Karnal.

1 Mondal RC - MSc,PhD *Director*

1969 Soils and Agronomy Division

1 Khosla BK - MSc,PhD *Head of the Division* soil and water conservation
2 Anand Swarup - MScAgr,PhD soil chemistry
3 Ashok Kumar - MScAgr,PhD agronomy
4 Batra Mrs L - MSc microbiology
5 Bhargav GP - MSc,PhD pedology
6 Chhabra R - MSc,PhD soil chemistry
7 Gupta RK - MScAgr,PhD soil chemistry
8 Mehta KK - MSc,PhD soil chemistry

9 Rao DLN - MSc,PhD soil chemistry
10 Samra JS - MSc,PhD soil chemistry
11 Sharma DP - MScAgr,PhD soil physics
12 Sharma DR - MSc,PhD soil physics
13 Sharma RC - MScAgr pedology
14 Singh KN - MSc,PhD agronomy
15 Singh M - MScAgr,PhD soil chemistry
16 Singh SB - MScAgr,PhD soil chemistry
17 Tomar OS - MSc,PhD agronomy

1970 Genetics and Plant Physiology

1 Rana RS - MSc,PhD *Professor of Eminence* genetics/cytogenetics
2 Bhattacharya RK - MSc,PhD *Head of Division* plant breeding
3 Dua RP - MScAgr,PhD plant breeding
4 Gill KS - MScAgr,PhD plant physiology
5 Qadar Ali - MSc,PhD plant physiology
6 Sharma SK - MScAgr,PhD plant physiology
7 Singh KN - MScAgr,PhD plant breeding
8 Sinha TS - MScAgr,PhD plant breeding

1971 Project Coordinating Cell Division

1 Rajput RK - MScAgr,PhD *Project Coordinator* agronomy; water management
2 Aggarwal A - MTech soil and water conservation engineering
3 Rao GGSN - MSc agricultural meteorology
4 Sharma BM - MScAgr,PhD soil chemistry
5 Singh SN - MScAgr agronomy

1972 Engineering Division

1 Gupta SK - MScAgrEngg,PhD soil and water conservation engineering
2 Luthra SK - BScEngg electronics and instrumentation
3 Rao KVGK - MTech,PhD soil and water conservation engineering
4 Singh OP - MTech,PhD soil and water conservation engineering
5 Tyagi NK - MTech,PhD *Head of the Division* soil and water conservation engineering

1973 Agricultural Economics Division

1 Joshi PK - MScAgr,PhD *Head of Division* agricultural economics

1974 CSSRI Research Station

P.O. Canning Town, District 24 Paraganas, West Bengal, 743329

1 Bandyopadhyay AK - MSc,PhD *Officer-in-Charge* soil physics and soil and water conservation
2 Bal AR - MSc,PhD plant physiology
3 Bandyopadhya BK - MSc,PhD soil chemistry
4 Biswas CR - MScAgr agronomy
5 Dutt SK - MSc plant physiology
6 Sen HS - MSc,PhD soil physics and soil and water conservation

1975 Central Arid Zone Research Institute

Jodhpur, Rajasthan
Tel: 21984. Cables: SHUSHAKSHETRA.

1 Shankaranaryan KA - MSc,PhD *Director* remote sensing techniques

1976 Division of Basic Resources Studies

1 Dhir RP - AssocIARI,PhD
2 Chatterji PC - MSc geology
3 Choudhari JS - MScAgr,PhD pedology
4 Kumar Suresh - MSc economic botany
5 Ram Balak - MA,LLB,DipPhotoInt geography
6 Saxena SK - MSc economic botany
7 Sen AK - MA,DipPhotoInt geography
8 Shankar V - MSc,PhD botany
9 Singh S - MA remote sensing techniques
10 Vangani NS - AMIE(Civil) water and soil engineering
11 Vats PC - MA,PhD geography

1977 Division of Plant Studies

1 Dass HC - BScAgr,MSc,PhD *Post Doctoral Fellow, Head of Division*
2 Amal Raj VAM - MSc economic botany
3 Bankar GJ - MSc horticulture
4 Bhansali RR - MSc,PhD plant pathology
5 Ghanim Abu - MSc,PhD organic chemistry
6 Gupta A - MSc biochemistry
7 Harsh LN - MSc,PhD silviculture
8 Jain BL - MScAgr,PhD soil physics and soil water conservation
9 Jindal SK - MScAgr,PhD plant breeding
10 Kackar NL - MScAgr genetics
11 Maurya S - MSc economic botany
12 Mertia RS - MSc,PhD economic botany
13 Parihar DR - MSc,PhD agricultural entomology
14 Satya Vir - MSc,PhD agricultural entomology
15 Sharma SK - MSc economic botany

16 Singh KC - MScAgr agronomy
17 Solanki KR - MScAgr,PhD plant breeding
18 Vashishtha BB - MScAgr horticulture

1978 Division of Animal Studies

1 Mittal JP - MVSc,PhD *Head of Division* animal physiology
2 Ghosh PK - MSc,DPhil animal physiology
3 Khan MS - MSc,PhD animal biochemistry
4 Rana BD - MSc,PhD animal ecology

1979 Division of Wind Power and Solar Energy Utilisation Studies

1 Gupta JP - MSc,PhD *Head of Division* physics
2 Joshi NL - MScAgr,PhD agronomy
3 Pande PC - MSc,PhD physics
4 Ramakrishna - MScTech,PhD,DipAppStat climatology
5 Thanvi KP - MSc agricultural statistics

1980 Division of Soil-Water-Relations Studies

1 Lahiri AN - MSc,PhD *Head of Division*
2 Aggarwal RK - MSc,PhD soil chemistry/fertility
3 Garg BK - MSc,PhD plant physiology
4 Gupta JP - MSc,PhD soil physics and soil water conservation
5 Kathju S - MSc,PhD plant physiology
6 Rao AV - MScAgr,PhD microbiology
7 Singh SD - MSc,PhD agronomy
8 Vyas SP - MSc,PhD plant physiology

1981 Division of Agricultural Economics and Statistics

1 Kalla JC - MScAgr,PhD *Head of Division* agricultural economics
2 Diwakar GD - MScAgr,PhD agricultural economics
3 Gupta BS - MSc,PSCC agricultural economics
4 Krishna GVSR - MSc,PhD agricultural statistics
5 Purohit ML - MScAgr,PhD agricultural economics

1982 All India Coordinated Research Project for Dryland Agriculture

1 Daulay HS - MScAgr agronomy
2 Bhati TK - MScAgr agronomy
3 Kumar D - MSc,PhD plant breeding/genetics
4 Singh HP - MScAgr,PhD soil physics and soil water conservation

1983 All India Coordinated Project for Millet Improvement

1 Saxena MBL - MSc,PhD plant breeding
2 Yadav RPS - MScAgr agronomy

1984 All India Coordinated Research Programme on Rodent Control

1 Jain AP - MSc,PhD rodent ecology

1985 Professor of Eminence Scheme

1 Prakash I - MSc,PhD *Prof. of Eminence*
2 Joshi PP - MSc organic chemistry

1986 Central Research Institute for Dryland Agriculture

Hyderabad, Andhra Pradesh, 500 659
Tel: 239177. Telex: 155-6252. Cables: DRYLAND.

1 Singh RP - MScAgr,AssocIARI,PhD *Director*

1987 Resource Management

1 Venkateswaralu J - MScAgr,PhD *Head of Division* soil fertility
2 Atwal JS - BTech farm machinery development
3 Das SK - MScAgr,PhD soil fertility
4 Gupta PD - MTech,PhD tillage studies
5 Korwar GR - MScAgr alternate land use system
6 Sriram C - MTech,DSc tillage studies
7 Thayagaraj CR - MTech energy studies in farm sector
8 Victor US - MSc,PhD agro-climatology of dryfarming areas of India
9 Vijaylakshmi K - MSc,PhD efficient land use
10 Vittal KPR - MScAgr,PhD soil-plant-water relationship

1988 Crop Science

1 Balasubramanian V - MScAgr,PhD *Head of Division* water stress physiology
2 Das ND - MSc control of root wilt in castor

IN India

3 Gangadhar Rao D - MSc agro-forestry system
4 Hanumantha Rao C - MSc,PhD breeding of sorghum for *Striga* resistance
5 Subba Reddy G - MScAgr,PhD intercropping systems

1989 Design and Analysis

1 Ramanatha Chetty CK - MSc *Head of Division* experimental design for agro-forestry and intercropping
2 Narayana Reddy M - MSc experimental design for cropping systems
3 Purnachandra Rao K - MScAgr,PhD risk, finance and marketing in dryfarming
4 Reddy YVR - MScAgr,PhD impact of dryland farming on farmers economy

1990 Transfer of Technology

1 Singh M - MScAgr *Head of Division* effectiveness of communication methods
2 Bhaskara Rao UM - MSc crop weather relationship
3 Maruthi Shankar GR - MSc statistical modelling of soil test data
4 Prabhanjan Rao SB - MA on-farm evaluation of dryland technology
5 Prasad MS - MScAgr
6 Ramana Rao BV - MSc *Project Coordinator* agro-meteorology
7 Reddy KCK - MScAgr,PhD *In Charge Coordinator* soil test and fertilizer response
8 Sharma S - MTech watershed hydrology
9 Singh SP - MScAgr,PhD *Project Coordinator* intercropping system
10 Vishnumurthy T - MScAgr,PhD operational research constraint analysis

1991 Central Institute of Agricultural Engineering
Nabi Bagh, Berasia Road, Bhopal, Madhya Pradesh, 462018
Tel: 63618, 66756. Telex: 0705-273 CIAE IN. Cables: KRIYANTRA.

1 Ojha TP - MTech,PhD *Director* agricultural engineering, research management

1992 Crop Production Engineering Division

1 Rajput DS - MTech,PhD *Head of Division* farm machinery and power
2 Mishra AK - PhD soil science
3 Singh VV - MTech,PhD farm machinery and power

1993 Post Harvest Engineering Division

1 Kachru RP - MTech,PhD *Head of Division* agricultural structure and processing engineering
2 Bisht BS - MTech,MBA agricultural structure and processing engineering

1994 Agricultural Energy and Power Division

1 Shyam M - MTech *Head of Division* farm machinery and power

1995 Agro-Industrial Extension Division

1 Singh CR - MTech,PhD *Head of Division* farm machinery and power

1996 CIAE-IRRI Industrial Extension Project on Small Farm Machinery for Rice Cultivation

1 Datt P - MTech,PhD *Head of Project* farm machinery and power

1997 All India Coordinated Scheme on Farm Implements and Machinery Coordinating Cell - CIAE

1 Bisen HS - MTech farm machinery and power
2 Devani RS - MTech,PhD *Project Coordinator* farm machinery and power
3 Prasad J - MTech farm machinery and power

1998 All India Coordinated Scheme on Post Harvest Technology Coordinating Cell - CIAE

1 Chouksey RG - MTech agricultural structure and processing engineering
2 Dakshinamurthy A - MSc plant pathology
3 Nimje PM - MSc,PhD agronomy
4 Shukla BD - PhD *Project Coordinator* agricultural structure and processing engineering

1999 Indo-US Project on Soybean Processing and Utilization Coordinating Cell - CIAE

1 Ali N - MTech,PhD *In-Charge Project Director* agricultural structure and processing engineering
2 Gandhi AP - MSc,PhD biochemistry

2000 All India Coordinated Scheme on Intensive Testing of Power Tillers and Development of New Machinery - CIAE

2001 Coordinated and Research Project on Renewable Sources of Energy for Agriculture and Agro Based Industries, Coordinating Cell - CIAE

1 Maheshwari RC - MTech,PhD *Project Coordinator* farm machinery and power
2 Singh HP - MTech agricultural structure and processing engineering

2002 All India Coordinated Scheme on Draft Animal Power Coordinating Cell - CIAE

1 Yadav JC - MSc farm machinery and power
2 Srivastava NSL - MTech *Project Coordinator* farm machinery and power

2003 Indian Lac Research Institute
Namkum, Ranchi, Bihar, 834010
Tel: 22715. Telex: 0625-266. Cables: LACCUM.

1 Kapil RP - MSc,PhD *Director*

2004 Agronomy and Plant Genetics

1 Kumar P - MSc,PhD *Head of Division*
2 Singh BP - MSc *Agronomist*

2005 Entomology

1 Das BB - MSc *Head of Division*
2 Chauhan NS - MSc
3 Naqvi AH - MSc
4 Ramani R - MSc,MPhil

2006 Technology

1 Gupta PC - MSc,PhD *Head of Division*

2007 ICAR Research Complex for North Eastern Hills Region
Cedar Lodge, Shillong, Meghalaya, 793 003
Tel: 26878. Telex: 237-214.

1 Abraham MJ - MSc,PhD plant breeding
2 Awasthi RP - MScAgr agronomy
3 Azad Thakur NS - MSc entomology
4 Barwal RN - MSc,PhD entomology
5 Bhrigubanshi S - MSc,PhD soil science
6 Bishwa Nath - MSc,PhD *Joint Director* entomology
7 Bujarbaruah KM - MSc,PhD animal production
8 Chaudhury SK - MSc,PhD plant breeding
9 Choudhury RG - MSc plant pathology
10 Dhiman KR - MSc,PhD plant breeding
11 Dwivedi RN - MSc agronomy
12 Gangawar SK - MSc,PhD entomology
13 Ghai JC - MSc,PhD animal production
14 Ghei JC - MSc,PhD animal genetics and breeding
15 Ghosh SS - MSc,PhD animal production
16 Gupta DK - MSc,PhD plant pathology
17 Gupta HS - MSc,PhD plant breeding
18 Gupta PN - MSc horticulture
19 Gupta SC - MSc,PhD horticulture
20 Kochhar S - MSc,PhD plant breeding
21 Laskar S - MSc,PhD *Joint Director* soil science
22 Maiti CR - MSc,PhD animal health
23 Mandal BK - MSc,PhD fisheries
24 Medhi RP - MSc horticulture
25 Mishra KC - MSc,PhD animal health
26 Munda GC - MScAgr,PhD agronomy
27 Nanda SK - MSc,PhD poultry science
28 Pandey MB - MSc,PhD Krishi Vigyan Kendra
29 Parthasarathi VA - MSc,PhD horticulture
30 Patel CS - MScAgr agronomy
31 Prasad RN - MSc,PhD *I/C Director*
32 Rai M - MSc,PhD horticulture
33 Rai RN - MSc,PhD *Joint Director* agronomy
34 Rajkhowa C - MSc animal parasitology
35 Ram M - MSc soil science
36 Rao KVP - MSc,PhD *Joint Director* soil science
37 Rao NS - MSc entomology
38 Roy S - MSc,PhD entomology
39 Sardana S - MSc plant breeding
40 Satpathy KK - MSc operational research project on lift irrigation
41 Saxena SC - MSc,PhD animal production
42 Sharma AC - MSc plant breeding
43 Sharma BK - MSc plant breeding
44 Sharma KD - MSc water management

131

45 Sharma YP - MSc,PhD plant pathology
46 Shukla RP - MSc,PhD entomology
47 Singh A - MSc agricultural engineering
48 Singh BN - MScAgr,PhD water management
49 Singh MD - MSc agricultural engineering
50 Singh RD - MSc agronomy
51 Srivastava LS - MSc plant pathology
52 Upadhyaya RC - MSc *Joint Director* horticulture
53 Varma A - MSc,PhD animal nutrition
54 Varma ND - MSc,PhD disease diagnostic laboratory
55 Varma RN - MSc,PhD plant pathology
56 Vyas NL - MSc,PhD horticulture
57 Yadav DS - MSc horticulture

2008 Vivekananda Parvatiya Krishi Anusandhan Shala
Almora, Uttar Pradesh, 263 601
Tel: 2208. Cables: VILAB.

1 Koranne KD - MSc,PhD *Director*

2009 Crop Improvement

1 Joshi HC - MSc wheat, minor millets and vegetable breeding
2 Dube SD - MSc,PhD tissue culture, nutritional stress
3 Jag Shoran - MSc,PhD wheat breeding
4 Mani VP - MSc maize breeding
5 Sharma AK - MSc pulse and oilseed breeding

2010 Crop Production

1 Bhatnagar VK - MSc,PhD water requirement of hill crops; soil management
2 Sharma ER - MSc,PhD soil water and temperature management; fodder and grasslands
3 Srivastava RC - MTech run off recycling

2011 Crop Protection

1 Sharma AK - MSc,PhD seed pathology; powdery mildew

2012 Social Sciences

1 Bisht KKS - MSc computer programming

2013 Central Agricultural Research Institute for A & N Islands
Port Blair, (Andaman and Nicobar Islands)
Tel: 2576. Cables: KRISHIKENDRA.

1 Singh NT - MScAgr,PhD *Director* soil science
2 Ahlawat SPS - MSc,PhD poultry breeding
3 Dorairaj K - MSc marine capture and culture fisheries
4 Gangwar B - MScAgr,PhD agronomy
5 Kumar V - MScAgr,PhD soil science
6 Michael Raj S - MScAgr,PhD soil fertility management
7 Mongia AD - MScAgr soil science
8 Pal RN - MVSc,PhD animal nutrition
9 Ramesh CR - MSc,PhD plant pathology
10 Sharma TVRS - MScAgr plant breeding
11 Singh BP - MScAgr horticulture
12 Singh S - MScAgr,PhD horticulture
13 Soundararajan R - MSc marine capture and culture fisheries
14 Tripathi KP - MTech soil-water conservation engineering

2014 Indian Grassland and Fodder Research Institute
Jhansi
Tel: 833, 908. Telex: 326 GFR 241 JI. Cables: GHASANUSANDHAN.

1 Singh Panjab - MScAgr,PhD *Director*

2015 Plant Improvement

1 Yadava RBR - MScAgr,PhD *Head of Division* physiology
2 Ahmad ST - MSc,PhD pathology
3 Bhagmal - MScAgr,PhD plant breeding
4 Choubey RN - MScAgr,PhD plant breeding
5 Gupta MG - MSc,PhD genetics
6 Gupta SK - MScAgr,PhD plant breeding
7 Gupta SR - MSc,PhD economic botany
8 Jain RK - MSc,PhD nematology
9 Pandey KC - MSc entomology
10 Shukla GP - MScAgr,PhD plant breeding
11 Tripathi SN - MSc,PhD genetics and cytogenetics
12 Zadoo SN - MSc,PhD genetics and cytogenetics

2016 Agronomy and System Synthesis

1 Gill AS - MScAgr *Head of Division* agronomy
2 Menhilal - MScAgr,PhD agronomy

2017 Soil Science

1 Hazra CR - MScAgr,PhD *Head of Division* soil science
2 Pahwa MR - MSc,PhD microbiology
3 Rawat CR - MScAgr agronomy
4 Singh Dashrath - MScAgr,PhD soil science
5 Tyagi RK - MA,PhD geography

2018 Plant Animal Relationship

1 Gupta RK - MSc,PhD *Head of Division* organic chemistry
2 Kumar P - MVSc,PhD animal nutrition
3 Majoomdar AB - MSc organic chemistry
4 Pachauri VC - MVSc,PhD animal nutrition
5 Rekib A - MVSc,PhD animal nutrition
6 Singh AP - MScAgr,PhD agricultural chemistry
7 Updahyay RS - MSc livestock production and management
8 Varshney JP - MVSc,PhD veterinary medicine and surgery

2019 Economics and Biometrics

1 Handa DP - MA agricultural statistics
2 Sreenath PR - MA agricultural statistics
3 Yadav IPS - MScAgr agricultural economics

2020 Seed Technology

1 Singh RP - MScAgr,PhD *Head of Division* agronomy
2 Faruqui SA - MSc,PhD entomology
3 Singh OP - MScAgr,PhD agronomy
4 Singh SN - MSc,PhD pathology
5 Sinha NC - MSc,PhD physiology
6 Tomar PS - MScAgr,PhD agronomy

2021 Weed Ecology and Control

1 Pandey RK - MScAgr,PhD agronomy
2 Dutta TR - MSc,PhD physiology

2022 Agricultural Engineering and Post Harvest Technology

1 Singh J - MTech,PhD *Head of Division* agricultural engineering
2 Varshney RB - BTech agricultural engineering

2023 Grassland Management

1 Kanodia KC - MSc,PhD *Head of Post* economic botany
2 Jayan PK - MSc,PhD economic botany
3 Rai P - MScAgr,PhD agronomy
4 Trivedi BK - MSc,PhD economic botany

2024 Agro-Silvipasture

1 Dabroy R - MSc,PhD *Head of Division* economic botany
2 Chauhan DS - MSc,PhD agronomy
3 Gupta VK - MSc,PhD plant breeding
4 Pathak PS - MSc,PhD economic botany
5 Singh RC - MSc agroforestry

2025 Extension, Communication and Training

1 Lokhande MR - MScAgr,PhD *Head of Division* extension

2026 Sub-Station

1 Misri BK - MSc,PhD *Centre In Charge* economic botany

2027 National Dairy Research Institute
Karnal, Haryana, 132001
Telex: 0396-204 NDRI. Cables: DAIRYSEARCH.

1 Nagarcenkar R - MSc,PhD *Director*
2 Patel RK - MSc,PhD *Joint Director*

2028 Division of Dairy Cattle Genetics and Breeding

1 Gurnani M - MSc *Head*

2029 Animal Genetics and Breeding

1 Balakrishnan CR - MSc,PhD
2 Chawla DS - MSc,PhD
3 Goswami SL - MSc,PhD

4 Joshi BK - MSc,PhD
5 Tripathi VN - MSc,PhD

2030 Livestock Production and Management

1 Sarma PA - MSc,PhD
2 Tomar OS - MSc,PhD
3 Nagpaul PK - MSc,PhD
4 Pachlag SV - MSc,PhD
5 Verma GS - MSc,PhD

2031 Division of Dairy Cattle Nutrition

1 Sharma DD - MSc,PhD *Head of Division*

2032 Animal Nutrition

1 Gupta BN - MSc,PhD
2 Chhabra Mrs A - MSc
3 Chopra RC - MSc,PhD
4 Kurar CK - MSc,PhD
5 Prasad T - MSc,PhD
6 Rai SN - MSc,PhD
7 Singh GP - MSc,PhD
8 Singhal KK - MSc,PhD
9 Srivastava A - MSc,PhD
10 Walli TK - MSc,PhD

2033 Agronomy

1 Singh A - MSc,PhD

2034 Division of Dairy Cattle Physiology

1 Madan ML - MSc,PhD *Head of Division*

2035 Animal Physiology

1 Ludri RS - MSc,PhD
2 Prakash BS - MSc,PhD
3 Upadhaya RC - MSc,PhD

2036 Animal Reproduction

1 Jain GC - MSc,PhD

2037 Division of Dairy Chemistry

1 Jain MK - MSc,PhD *Head of Division*
2 Bector BS - MSc,PhD
3 Bhatia KL - MSc,PhD
4 Bindal MP - MSc,PhD
5 Joshi VK - MSc,PhD
6 Malik RC - MSc,PhD
7 Mathur MP - MSc,PhD
8 Mathur ON - MSc,PhD
9 Singh A - MSc,PhD
10 Singhal OP - MSc,PhD
11 Tandon KC - MSc,PhD

2038 Division of Biochemistry

1 Anand SR - MSc,PhD *Head of Division*
2 Deodhar AD - MSc,PhD
3 Dutta SM - MSc,PhD
4 Gandhi KK - MSc,PhD
5 Kansal VK - MSc,PhD
6 Pandey RS - MSc,PhD

2039 Division of Dairy Bacteriology
Karnal, Haryana, 132001
Telex: 0396-204 NDRI. Cables: DAIRYSEARCH.

1 Prasad DN - MSc,PhD *Head of Division*
2 Batish VK - MSc,PhD
3 Chander H - MSc,PhD
4 Ghodekar DR - MSc,PhD
5 Malik RK - MSc,PhD
6 Mathur DK - MSc,PhD
7 Neelakanthan S - MSc,PhD
8 Rattan Chand - MSc,PhD
9 Singh K - MSc,PhD
10 Singh RS - MSc,PhD
11 Sinha RN - MSc,PhD

2040 Division of Dairy Technology
Karnal, Haryana, 132001
Telex: 0396-204 NDRI. Cables: DAIRYSEARCH.

1 Gupta SK - MSc,PhD *Head of Division*

2041 Dairy Technology
Karnal, Haryana, 132001
Telex: 0396-204 NDRI. Cables: DAIRYSEARCH.

1 Balachandran R - MSc,PhD
2 Arora KL - MSc,PhD
3 Ladkani LG - MSc,PhD
4 Mathur LN - MSc,PhD
5 Patel AA - MSc,PhD
6 Patel GR - MSc,PhD
7 Patel RS - MSc,PhD
8 Rajor RB - MSc,PhD
9 Rajorhia GS - MSc,PhD
10 Singh J - MSc,PhD
11 Singh S - MSc,PhD
12 Tewari BD - MSc,PhD

2042 Chemical Engineering
Karnal, Haryana, 132001
Telex: 0396-204 NDRI. Cables: DAIRYSEARCH.

1 Zaidi AH - MSc,PhD

2043 Division of Dairy Economics and Statistics
Karnal, Haryana, 132001
Telex: 0396-204 NDRI. Cables: DAIRYSEARCH.

1 Rawat BS - MSc,PhD *Head of Division*

2044 Agricultural Economics
Karnal, Haryana, 132001
Telex: 0396-204 NDRI. Cables: DAIRYSEARCH.

1 Singh R - MSc,PhD
2 Gupta JN - MSc
3 Singh K - MSc,PhD
4 Solanki RS - MSc,PhD

2045 Agricultural Statistics
Karnal, Haryana, 132001
Telex: 0396-204 NDRI. Cables: DAIRYSEARCH.

1 Singh B - MSc,PhD
2 Aggarwal SB - MSc,PhD

2046 Division of Agricultural Extension
Karnal, Haryana, 132001
Telex: 0396-204 NDRI. Cables: DAIRYSEARCH.

1 Kherde RL - MSc,PhD *Head of Division*
2 Anand Mrs U - MSc,PhD
3 Dubey VK - MSc,PhD
4 Ram Kumar - MSc,PhD
5 Tyagi KC - MSc,PhD
6 Verma OS - MSc,PhD

2047 Division of Dairy Engineering
Karnal, Haryana, 132001
Telex: 0396-204 NDRI. Cables: DAIRYSEARCH.

1 Agrawala SP - MSc,PhD *Head of Division*
2 Sarma SC - MSc,PhD

2048 Section of Operational Research Project
Karnal, Haryana, 132001
Telex: 0396-204 NDRI. Cables: DAIRYSEARCH.

1 Singh CB - MSc,PhD
2 Mahi Pal - MSc,PhD

2049 Section of Farm Mechanisation
Karnal, Haryana, 132001
Telex: 0396-204 NDRI. Cables: DAIRYSEARCH.

1 Arora SP - MSc,PhD *Professor of Eminence and Coordinator*
2 Punj ML - MSc,PhD *Project Coordinator*

2050 Southern Regional Station
Adugodi, Bangalore, Karnataka
Tel: 579661,2,3,4. Telex: 0845-8397. Cables: DAIRYSEARCH.

1 Sampath SR - MSc,PhD *Head*
2 Ananda Kumar M - MSc,PhD agricultural economics
3 Balakrishnan M - MSc,PhD livestock production and management
4 Bhanumurthy JL - MSc dairy technology
5 Bhavdasan MK - MSc,PhD dairy chemistry
6 Chinnaiya GP - MSc,PhD animal reproduction
7 Gawali SR - MSc livestock production and management
8 Kulkarni S - MSc dairy technology
9 Mulay CA - MSc,PhD dairy technology
10 Narayanan KM - MSc,PhD dairy chemistry

11 Natarajan AM - MSc,PhD dairy bacteriology
12 Prasad SAD - MSc electronics and instruments
13 Ramamurthy MK - MSc,PhD dairy chemistry
14 Ranganathan K - MSc agricultural extension
15 Rao AS - MSc animal nutrition
16 Rao SVN - MSc,PhD agricultural extension
17 Sampath KT - MSc animal nutrition
18 Sarma PV - MSc,PhD animal physiology
19 Sinha MN - MSc,PhD agricultural extension
20 Sitaramaswamy J - MSc agricultural statistics
21 Vijaylakshmi Miss S - MSc agricultural economics

2051 Eastern Regional Station
P.O. Kalyani District Nadia, West Bengal
Tel: 264. Cables: DAIRYSEARCH.

1 Basu SB - MSc,PhD *Head*
2 Ganguly TK - MSc,PhD soil science

2052 Indian Veterinary Research Institute
Izatnagar, Uttar Pradesh, 243 122
Tel: 74069. Telex: 0577-205. Cables: VETEX/SERAIZATNAGAR.

1 Bhat PN - MVSc,PhD *Director*
2 Kunzru ON - MS,PhD *Joint Director*
3 Mallick BB - BSc,CVSc,DSc *Joint Director*
4 Mohanty GC - MVSc,PhD *Joint Director*
5 Rao BU - MVSc,PhD *Joint Director*

2053 Animal Genetics

1 Bhat Mrs PP - MSc,PhD
2 Koul GL - MVSc,PhD
3 Mazumdar NK - MVSc,PhD
4 Mishra RR - MVSc,PhD
5 Nautyal LP - MVSc,PhD
6 Raina BL - MVSc,PhD
7 Rao MK - MSc,PhD
8 Rawat Mrs S - MSc,PhD
9 Satish Kumar - MVSc,PhD
10 Sidhu NS - MSc,PhD
11 Taneja VK - MSc,PhD

2054 Animal Nutrition

1 Agarwal ON - MSc,PhD
2 Joshi DC - MVSc,PhD
3 Khan MY - MVSc,PhD
4 Nath K - MSc,PhD
5 Singh UB - MVSc,PhD,DSc

2055 Animal Physiology

1 Joshi BC - BVSc,PhD,DSc
2 Mahapatro BB - MS,PhD
3 Pandey JK - MVSc,PhD
4 Sanwal PC - MVSc,PhD

2056 Animal Reproduction

1 Benjamin BR - MSc,PhD
2 Gupta SK - MVSc,PhD
3 Sahni KL - MVSc,PhD

2057 Avian Diseases

1 Sah RL - MVSc,PhD
2 Verma KC - MVSc,PhD

2058 Biochemistry

1 Singh LN - MVSc,PhD
2 Srivastava AK - MSc,PhD

2059 Biophysics

1 Yashroy RC - MSc,PhD

2060 Biostatistics

1 Garg RC - MSc

2061 Epidemiology

1 Khera SS - MVSc,PhD

2062 Livestock Products and Technology

1 Sharma N - MVSc,PhD,DSc
2 Yadav PL - MSc,PhD

2063 Agricultural Economics

1 Singh P - MSc,PhD

2064 Poultry Science

1 Pani PK - PhD

2065 Veterinary Bacteriology and Virology

1 Bansal MP - MVSc,PhD
2 Bansal RP - MVSc,PhD
3 Gupta BK - PhD
4 Gupta BR - PhD
5 Khanna PN - DVM,CSc
6 Kulshreshtha SB - MVSc,PhD
7 Lal SM - MVSc,PhD
8 Mehrotra ML - MVSc,PhD
9 Mishra HN - MVSc,PhD
10 Nanda YP - MVSc,PhD
11 Pandey AK - MSc,PhD
12 Rai A - MVSc,PhD
13 Sen AK - MVSc,PhD
14 Sikdar A - MVSc,PhD
15 Yadav MP - MVSc,PhD
16 Yadav VK - MVSc,FRVAC

2066 Veterinary Pathology

1 Paliwal OP - FRVCS
2 Parihar NS - MVSc,PhD
3 Pradhan HK - MVSc,PhD
4 Prasad MC - MVSc,PhD
5 Sah RL - MVSc,PhD

2067 Pharmacology

1 Gupta PK - MVSc,PhD
2 Lal J - MSc,PhD
3 Sashtry MS - MSc,PhD

2068 Veterinary Public Health

1 Sen GP - MVSc,PhD

2069 Veterinary Surgery

1 Bhargava AK - PhD

2070 Veterinary Parasitology

1 Chandra Ravi - MVSc,PhD
2 Dhar DN - MVSc,PhD
3 Malviya HC - MSc,PhD
4 Sharma NN - MSc,PhD
5 Sharma Deorani VP - MVSc,PhD
6 Shrivastava RVN - MVSc,PhD
7 Shrivastava VK - MVSc,PhD
8 Subramanian G - MSc,PhD

2071 Central Institute for Research on Buffaloes
Sirsa Road, Hissar, Haryana, 125500
Tel: 4513. Telex: 0345-204 CIRBIN. Cables: BUBALINEX.

1 Mudgal VD - MSc,PhD *Director* energy and protein metabolism in buffaloes

2072 Animal Nutrition

1 Dahiya SS - MSc,PhD roughage utilization

2073 Buffalo Management

1 Singh M - MSc,PhD buffalo management
2 Singh R - BVSc,MS buffalo breeding

2074 Reproduction Physiology

1 Jindal SK - MSc reproductive physiology

2075 Central Sheep and Wool Research Institute
Sirsa Road, Avikanagar, Rajasthan, 304501
Tel: 63 AVIKANAGAR. Telex: 036-315 CSWRIN. Cables: SHEEPINST MALPURA.

1 Chaudhary AL - MS,PhD *Director*

2076 Division of Animal Genetics and Breeding

1 Mahajan JM - MVSc,PhD *Head of Division*
2 Arora AL - MSc,PhD

3 Bapna DL - MSc agricultural statistics
4 Satya P - MVSc livestock production and management

2077 Division of Nutrition

1 Bhatia DR - MVSc,PhD *In Charge*
2 Dhiman PC - MSc agronomy
3 Karim SA - MVSc animal nutrition
4 Ravindra Kumar - MSc,PhD biochemistry
5 Sehgal JP - MSc animal nutrition
6 Singh NP - MSc,PhD animal nutrition

2078 Division of Physiology

1 Kalra DB - MS,PhD *Head*
2 Kaushish SK - MSc,PhD animal reproduction

2079 Division of Wool Science

1 Chopra SK - MSc fibre physics
2 Gupta NP - Btext textile manufacture
3 Parthasarthy S - MSc,PhD fibre physics
4 Patni PC - Mtext textile manufacture
5 Pokharna AK - Mtext textile manufacture
6 Singh US - MSctext textile chemistry

2080 Division of Extension

1 Sharma SC - MVSc,PhD *Head*
2 Ram Ratan - MSc livestock production and managements

2081 Meat Science Section

1 Prasad VSS - MVSc,PhD *In Charge* livestock products technology

2082 Animal Health Section

1 Dubey SC - MVSc,PhD veterinary microbiology
2 Dash PK - MVSc veterinary medicine
3 Srivastava CP - MVSc veterinary pathology

2083 Statistics

1 Rawat PS - MSc *In Charge*

2084 Technical and Monitoring

1 Bohra SDJ - MSc,PhD *In Charge* animal genetics and breeding

2085 CSWRI Arid Region Campus
Bikaner, Rajasthan
Tel: 4731 Bikaner. Cables: KARAKULDIV Bikaner.

1 Singh M - MVSc,PhD *Head* animal nutrition
2 Bhagwan PSK - MVSc veterinary pathology
3 Mudgal RD - MSc agronomy
4 Sahahi MS - MVSc animal genetics and breeding
5 Singh VK - MVSc animal genetics and breeding

2086 Division of Fur Animal Breeding
Garsa, Kulu, Himachal Pradesh
Cables: FURDIV BHUNTAR.

1 Singh RN - MVSc,PhD *Head* animal genetics and breeding
2 Malhi RS - MVSc textile manufacture
3 Swain MN - MVSc,PhD livestock production and management

2087 Southern Regional Research Centre
Mannavanur, Tamil Nadu
Cables: SHEEPSTN KODAIKANAL.

1 Rai AK - MVSc,PhD *Officer In Charge* animal physiology

2088 Central Institute for Research on Goats
Makhdoom, P.O. Farah, Mathura, Uttar Pradesh, 281 122
Tel: 4809. Cables: VETEX-FARAH.

1 Bhattacharyya NK - BSc,MVSc,PhD *Director* embryo technology

2089 Goat Genetics and Breeding

1 Khan BU - MVSc,PhD performance studies and improvement in productivity and reproductivity in Indian goats
2 Roy R - MVSc,PhD study on cross bred dairy cattle and performance studies in goats

2090 Goat Reproduction and Physiology

1 Agrawal KP - MVSc,PhD deep freezing of semen

2091 Goat Nutrition and Feed Resources Development

1 Ogra JL - MSc,PhD goat rumen microbes, identification and evaluation of goat feed and fodder
2 Sharma K - MSc fodder production and pasture improvement
3 Singh N - MSc,PhD goat rumen microbes trace elements

2092 Environment Physiology and Shelter Management

1 Singh K - MSc,PhD goat housing in semi-arid climate and physiology of adaptation

2093 Goat Production Transfer Technology Division

1 Kaul PN - MVSc,PhD transfer of technology from laboratory to land

2094 Goat Health Section

1 Singh N - MVSc,PhD epidemiology of goat microbial diseases
2 Vihan VS - MVSc,PhD enteritis and pneumonia in goat and sheep neonatal mortality

2095 Central Avian Research Laboratories
Izatnagar, Uttar Pradesh, 243 122
Tel: 74332, 74230. Telex: 0577-206 CARI IN. Cables: Poultry Izatnagar.

1 Panda B - BSc,MS,PhD *Director*

2096 Centre of Advance Studies

1 Panda B - BScVet,MS,PhD *Director* poultry products technology
2 Shrivastava AK - BSc,BVSc&AH,MVSc quail nutrition
3 Singh RP - BVSc&AH,MVSc quail products technology
4 Thomas PC - BVSc&AH,MVSc quail breeding

2097 Planning, Monitoring and Coordination

1 Mohapatra SC - BVSc&AH,MS,PhD *Head of Division* poultry breeding for egg and meat
2 Ayyagari V - BSc,BVSc&AH,MVSc poultry breeding
3 Khare SP - BSc,MA,MSc economics of poultry production
4 Rao GV - GMVC,AssocIVRI

2098 Department of Physiology and Reproduction

1 Panda JN - BVSc&AH,MS,PhD *Head of Division* physiology of avian reproduction
2 Mudgal RP - BSc,MSc,PhD

2099 Department of Avian Genetic Resources

1 Singh H - BVSc&AH,MVSc,PhD *Head of Division* blood groups in chickens

2100 Department of Genetics and Breeding

1 Johari DC - BVSc&AH,MVSc,PhD *Head of Division* poultry breeding for eggs
2 DevRoy AK - BVSc&AH,NDPH,MVSc poultry breeding for meat
3 Panda BK - BVSc&AH,MVSc avian health control
4 Sharma RP - BVSc&AH,MVSc,PhD poultry breeding for meat
5 Singh BP - BVSc&AH,MVSc poultry breeding for meat

2101 Department of Nutrition and Feed Technology

1 Sadagopan VR - BVSc&AH,MVSc,PhD *Head of Division* poultry nutrition
2 Johri TS - BVSc&AH,MVSc,PhD aflatoxin
3 Shrivastava HP - BVSc&AH,MVSc nutrition of unconventional poultry feeds
4 Shyamsunder G - BVSc&AH,MVSc poultry nutrition
5 Verma SVS - BVSc&AH,MVSc,PhD nutrition of agro industrial byproducts

2102 Department of Poultry Housing and Management

1 Ahuja SD - BVSc&AH,AssocIVRI *Head of Division* quail breeding and management
2 Agarwal SK - BVSc&AH,MVSc quail breeding and management
3 Sachdev AK - BVSc&AH,MVSc incubation and hatching

2103 Department of Post Harvest Technology

1 Verma SS - BVSc&AH,MVSc *Head of Division* poultry products technology
2 Mahapatra CM - BVSc&AH,MVSc poultry meat products technology
3 Pandey NK - BVSc&AH,MVSc egg products technology

2104 Department of Technology Transfer

1 Sinha SP - BVSc&AH,MVSc *Head of Division*

135

2105 Central Inland Capture Fisheries Research Institute

Barrackpore-743101, Parganas, West Bengal, 24
Tel: Bkp-161, 322, 388 (Direct). Cables: FISHSEARCH.

1 Jhingran AG *Director* ecology of inland waters, riverine fisheries, reservoir fisheries, estuarine fisheries and inland aquaculture

2106 Technical Cell

1 Das P - MSc,PhD *Head of Cell* freshwater aquaculture, inland fisheries extension and training, fish genetics and physiology
2 Kuldip Kumar - MSc fish culture, ecology and control of insects; cryogenic preservation of fish spermatozoa

2107 Extension, Training, Information and Documentation Division

1 Bhaumik U - MSc fisheries extension, pesticide pollution in paddy fields
2 Halder DD - MSc *Head of Division* breeding and culture of brackish fishes and shellfishes, fisheries extension and training, information and documentation in inland fisheries
3 Pandit PK - MSc fisheries extension, pesticide pollution in paddy fields

2108 Brackishwater Fish Culture Section

1 Ghosh A - MSc estuarine fisheries, brackishwater fish culture, sewage-fed fisheries and paddy-cum-fish culture
2 Chattopadhyay GN - MSc,PhD soil chemistry in fishery waters
3 Das KM - MSc,PhD digestive physiology of fishes, sex reversal studies in fishes
4 Ghosh A - MSc,PhD digestive physiology of fishes, physiology of fish reproduction

2109 Biochemistry Section

1 Mukhopadhyay PK - MSc,PhD reproductive physiology of carps, fish nutrition, pesticide pollution

2110 Economics Section

1 Paul S - MA,LLB fisheries economics

2111 Engineering Cell

1 Mukherjee AB - AMIE fish farm engineering
2 Saha C - BCE fish farm engineering

2112 Central Institute of Freshwater Aquaculture

Dhauli, P.O. Kausalyagang, District Puri, Bhubaneswar, Orissa, 751002

1 Sinha VRP - MSc,PhD *Head of Division* freshwater aquaculture research and management
2 Ayyappan S - MSc freshwater ecology and aquatic microbiology and cage culture of carps
3 Bhanot KK - MSc estuarine fisheries
4 Bhanot Mrs KK - MSc paddy-cum-fish culture and fish nutrition
5 Chatterjee DK - AIC soil and water chemistry of fish ponds
6 Das CR - MSc,PhD freshwater fish culture
7 Das RC - MSc fish breeding
8 Dey RK - MSc fish diseases
9 Dilip Kumar - MSc carp culture and breeding; fish pathology
10 Gupta SD - MSc carp culture and breeding
11 Jaitly PN - MSc carp seed production and carp culture
12 Jana RK - MSc carp breeding, hybridization and gynogenesis
13 Jena S - MSc plankton and productivity in fish ponds
14 Khan HA - MSc freshwater fish culture
15 Kowtal GV - MSc brackishwater fish culture and hatchery management
16 Mishra BK - MSc fish diseases
17 Mohanty SN - MSc,PhD carp culture and carp nutrition
18 Patnaik S - MSc biology and control of aquatic weed
19 Ram KJ - MSc,PhD freshwater prawn culture
20 Ranadhir M - MA,MSc fisheries economics
21 Rao GRM - MSc,PhD carp culture and carp seed production
22 Rao NGS - MSc freshwater aquaculture
23 Reddy PVGK - MSc fish hybridization and gynogenesis
24 Rout M - MSc fisheries statistics
25 Sahoo AK - MSc carp culture
26 Singh BN - MSc,PhD fish culture and fish nutrition
27 Sukumaran KK - MSc freshwater aquaculture
28 Tripathi SD - MSc freshwater aquaculture

2113 Riverine & Lacustrine Fisheries Division

24, Pannalal Road, Allahabad, Uttar Pradesh, 211002

1 Sehgal KL - MSc,PhD *Head of Division* capture and culture fisheries of Indian uplands, rivers and lakes
2 Gupta RA - MSc riverine fish catch statistics and population dynamics in reservoirs
3 Khan MA - MSc,PhD riverine fisheries, pen/cage culture in reservoirs
4 Mohamed MP - MSc,PhD biochemistry and fish physiology
5 Mukherjee GN - MSc,PhD riverine fisheries and culture of catfishes
6 Panwar RS - MSc,PhD pollution investigations in riverine and reservoir environments
7 Ravish Chandra - MSc riverine fisheries
8 Saxena RK - MSc riverine fisheries and pen and cage culture
9 Seth RN - MSc culture of catfishes
10 Singh B - MSc riverine fisheries survey
11 Singh SP - MSc culture of catfishes and riverine fisheries survey
12 Sree Prakash - MSc,PhD breeding and culture of freshwater prawns
13 Srivastava GN - MSc culture of catfishes; breeding and culture of prawns
14 Srivastava KP - MSc riverine spawn collection, survey on riverine fisheries, pen/cage culture in reservoirs

2114 Estuarine Fisheries Division

Barrackpore, West Bengal, 743101

1 Saigal BN - MSc,PhD capture and estuarine fisheries, information and documentation in inland fisheries
2 Bagchi MM - AIC soil and water chemistry, pollution investigation in rivers
3 Das MK - MSc livestock-cum-fish culture, fish diseases and control
4 Das RK - MSc water chemistry and microbiology in fishery waters
5 Ghosh AK - MSc fish diseases and control
6 Ghosh BB - AIC industrial, agricultural and domestic pollution related to fisheries
7 Joshi HC - MSc pesticide pollution in aquatic environment
8 Mitra PM - DipStat riverine fish catch statistics
9 Mukhopadhyay MK heavy metal and pesticide pollution in rivers
10 Pal RN - MSc fish pathology and diseases
11 Saha SB - MSc plankton and productivity in fishery waters
12 Sugunan VV - MSc reservoirs ecology and fisheries, carp culture in jute-retting ponds, culture of inland molluscs
13 Unnithan VK - MSc,PhD fish culture in jute-retting waters, endocrinology of carp breeding, culture of inland molluscs
14 Vinci Mrs GK - MSc reservoir ecology and fisheries, carp culture in jute-retting ponds, culture of inland molluscs

2115 Bangalore Research Centre of CICFRI

No. 51, 8th Cross, 7th Main Road, Malleswaran, Bangalore, Karnataka, 560003

1 Govind BV - MSc reservoir and tank fisheries and cage culture of carps
2 Ayyar SP - MSc,PhD air breathing fish culture and cage culture
3 Kumaraiah P - MSc culture of air breathing fishes
4 Ramaprabhu T - MSc,PhD biology and control of aquatic weeds in fish culture

2116 Bhagalpur Research Centre of CICFRI

Khanjarpur, Beatson Road, Bhagalpur, Bihar, 812001

1 Laal AK - MSc,PhD ecology and fisheries of rivers

2117 Bhimtal Research Centre of CICFRI

District Nainital, Uttar Pradesh, 263136

1 Joshi CB - MSc,PhD capture fisheries, fisheries of upland regions

2118 Bilaspur Research Centre of CICFRI

Roara Section, Bilaspur, Himachal Pradesh, 174001

1 Rao YR - MSc,PhD ecology and fisheries of freshwater reservoirs
2 Kaushal DK - MSc ecology and fisheries of freshwater reservoirs

2119 Calcutta Research Centre of CICFRI

39, Rabindra Sarani, (3rd Floor), Calcutta, West Bengal, 700073

1 Saha GN - AIC ecology and fishery management in bheries, fertilization in fish ponds
2 Thakurta SC - AIC soil and water chemistry in bheries

2120 Eluru Research Centre of CICFRI

Sastri Bhavan, DMC Home Street, Kandukuruvari Thota, Patehbad, Eluru-2, West Godavari District, Andhra Pradesh

1 Gopalakrishnayya C - MSc ecology and fishery management in freshwater reservoirs
2 Ramakrishniah M - MSc,PhD ecology and fishery management in freshwater reservoirs

2121 Gauhati Research Centre of CICFRI

Natun Sarania, Gauhati, Assam, 781003

1 Singh DN - MSc,PhD riverine and reservoir fisheries
2 Yadava YS - MSc ecology and fisheries of beels, livestock-cum-fish culture

2122 Kakdwip Research Centre of CICFRI
District 24-Paraganas, Kakdwip, West Bengal, 743347

1 Bhowmik ML - MSc,PhD brackishwater fish farming
2 Chakraborti RK - AIC soil and water chemistry of brackishwater fish ponds
3 Das SR - MSc breeding of brackishwater fishes and prawns
4 Nath D - MSc soil and water chemistry of estuaries and brackishwaters

2123 Kakinada Research Centre of CICFRI
16-23-1 Sambamurthi Nagar, Kakinada, Andhra Pradesh, 533001

1 Rao KJ - MSc,PhD carp and carp seed production culture
2 Rao RM - MSc carp culture and carp seed production
3 Subrahamanyam M - MSc,PhD breeding culture of freshwater prawns

2124 Kalyani Research Centre of CICFRI
B/11/226, Central Avenue East, Kalyani, District Nadia, West Bengal

1 Mondal AK - MSc,PhD frog farming

2125 Karnal Research Centre of CICFRI
P.O. Saidapura (CSSRI), Karnal, Haryana, 132001

1 Mishra DN - MSc carp culture and carp seed production
2 Shah KL - MSc carp culture and carp seed production
3 Tyagi BC - MSc carp culture and carp seed production

2126 Central Institute of Brackishwater Aquaculture
1, Karaneswarar Koil Street, (Near All India Radio), Mylapore, Madras, Tamil Nadu, 600004

1 Gopinathan K - MSc ecology and fisheries of freshwater reservoirs
2 Gupta BP - MSc,PhD soil and water chemistry of freshwater reservoirs
3 Prasadam RD - MSc,PhD pen and cage culture of brackishwater fishes and prawns
4 Radhakrishnan S - MSc,PhD algae culture, productivity studies in estuaries
5 Ramakrishna KV - MSc,PhD breeding of brackishwater fishes, pen and cage culture of brackishwater fishes and prawns
6 Raman K - MSc culture of freshwater and brackishwater fishes, biology and culture of marine prawns, estuarine
7 Rao AVP - MSc breeding and seed production of penaeid prawn
8 Rao LH - MSc,PhD breeding and seed production of freshwater and penaeid prawns
9 Srinivasagam S - MSc breeding and culture of estuarine crabs

2127 Muzaffarpur Research Centre of CICFRI
House No. 113, Ward No. 27, Damuchak, Muzaffarpur, Bihar, 842001

1 Rai SP - MSc,PhD soil and water chemistry in mans, pen culture of carps

2128 Operational Research Project
Kalyani, West Bengal

1 Chakraborty DP - AIC chemistry of fish culture ponds
2 Das NK - MSc breeding and culture of mullets and prawns
3 Sinha M - MSc,PhD composite fish culture and carp seed production, biology of carps

2129 Pollachi Research Centre of CICFRI
10, Chakrapani Iyer Street, Venkatesa Coloney, Pollachi, Tamil Nadu, 642001

1 Abraham M - MSc,PhD ecology and fisheries of freshwater reservoirs
2 Aravindakshan PK - MSc carp culture and carp seed production
3 Kaliamurthy M - MSc,PhD carp culture and carp seed production
4 Murugesan VK - MSc breeding and culture of air-breathing fishes
5 Selvaraj C - MSc training and demonstration in fish culture and ecology and fisheries of freshwater reservoirs

2130 Patna Research Centre of CICFRI
2-C(A), Peoples Cooperative, (2nd Floor), Kankarbagh, Patna, Bihar, 800020

1 Kumar D - MSc breeding and culture of air-breathing fishes

2131 Pune Research Centre of CICFRI
c/o Assistant Director of Fisheries, Sadasiv Sadan, 873, Bhandarkar Institute Road, Deccan Gymkhana, Patna, Maharashtra, 411004

1 Pisolkar MD - MSc,PhD ecology and fisheries of freshwater reservoirs
2 Rao PLN - MSc carp culture and carp seed production

2132 Puri Research Centre of CICFRI
No. 12, MIG Quarters, Water Works Road, Puri-2, Orissa

1 Pillai SM - MSc,PhD breeding, hatchery management and culture of brackishwater fishes and prawns

2 Rajyalakshmi Mrs T - MSc,DSc breeding and hatchery management of brackishwater prawns and fishes; riverine and estuarine fisheries
3 Ravichandran P - MSc breeding, hatchery management and culture of brackishwater fishes and prawns

2133 Rahara Research Centre of CICFRI
8, Station Road, Khardah, District 24-Paraganas, West Bengal, 743186

1 Banerjee RK - AIC,PhD soil and water chemistry of fishery waters
2 Chakraborti PK - MSc brackish water paddy-cum-fish culture, tilapia culture, mangrove ecology
3 Dutta AK - MSc carp breeding and seed production, sewage-fed-fish culture and paddy-cum-fish culture
4 Mukhopadhyay SK - MSc,PhD carp breeding and seed production, sewage-fed-fish culture and paddy cum-fish culture
5 Naskar KR - MSc,PhD utilization of aquatic weeds in fish culture mangrove ecology
6 Roy AK - MSc statistical analysis of fish culture operation
7 Saha SK - AIC soil and water chemistry of fishery waters
8 Sen PR - MSc,PhD freshwater fish culture and breeding
9 Sengupta A - BE brackishwater farm engineering, aeration in fish ponds

2134 Rihand Research Centre of CICFRI
c/o Assistant Director of Fisheries, Rihand, P.O. Turra, District Mirzapur, Uttar Pradesh, 231221

1 Desai VR - MSc,PhD fish biology, reservoir fisheries survey and management

2135 Srinagar Research Centre of CICFRI
Harwan, Srinagar, Jammu and Kashmir, 191123

1 Moza Usha - MSc,PhD nutrition of coldwater fishes, productivity of coldwater lakes
2 Raina HS - MSc,PhD breeding and culture of coldwater fishes and ecology of coldwater lakes
3 Shyamasundar - MSc,PhD breeding and culture of coldwater fishes and ecology of coldwater lakes
4 Vass KK - MSc,PhD breeding and culture of coldwater fishes and ecology of coldwater lakes

2136 Central Marine Fisheries Research Institute
Cochin, Kerala, 682031
Cables: CADALMIN, ERNAKULAM.

1 Alagarswami K - MA *Head of Division* molluscan research and culture
2 James PSBR - MSc,PhD *Director* fishery biology, fish taxonomy and fin fish culture - marine mammalogy
3 Nair PVR - MSc,PhD *Joint Director and Head of Division* phytoplankton productivity
4 Algaraja K - MSc,PhD fisheries statistics
5 Balan V - MSc sardine fishery biology
6 Bande VN - MSc demersal fisheries
7 Chandrika V - MSc,PhD marine microbiology
8 Chennubhotla VSK - MSc,PhD seaweed resources and culture
9 Diwan AD - MSc,PhD fish and shellfish physiology
10 Eapen KJ - MSc,PhD animal genetics
11 Easterson DCV - MSc nutrition and bioenergetics
12 George J - MSc,PhD fish and shellfish genetics
13 George KC - MSc,PhD fish egg and larvae
14 George MK - MSc,PhD breeding and genetics of finfish
15 Jacob T - MSc fisheries
16 Krishnan L - MSc,PhD breeding of finfish
17 Kulasekhara Pandian S - MSc,PhD culture of live feed
18 Kunjukrishna Pillai V - MSc marine pollution
19 Lakshminarayana A - MSc,PhD prawn breeding
20 Mary Manissery K - MSc prawn culture and resources
21 Mathew KJ - MSc,PhD euphasid resources
22 Murty AVSS - MSc fisheries ecology
23 Muthu MS - MSc prawn culture
24 Nandakumar G - MSc crustaceans
25 Narayana Kurup K - MSc,PhD fishery statistics
26 Narayana Pillai V - MSc,PhD fishery oceanography
27 Noble A - MSc mackerel fishery biology
28 Panikkar KKP - MSc fishery economics
29 Parameswaran Pillai P - MSc,PhD tuna fisheries
30 Paul Raj R - MSc,PhD fish and prawn nutrition
31 Pillai NN - MSc prawn culture
32 Ponniah AG - MSc,PhD fish genetics
33 Rajagopalan MS - MSc mangrove ecology
34 Rajan KN - MSc crustacean resources
35 Ramamirtham CP - MSc chemical oceanography
36 Rengarajan K - MSc zooplankton studies
37 Sadananda Rao D - MSc chemical oceanography
38 Sathiadas R - MSc fishery economics
39 Scariah KS - MSc,PhD fishery statistics
40 Sehara DBS - MSc fishery economics
41 Selvaraj GSD - MSc mangrove ecology
42 Sivakami S - MSc,PhD fishery biology
43 Somasekaran Nair KV - MSc,PhD fishery biology
44 Subbaraju G - MScTech remote sensing
45 Suseelan C - MSc crustacean fisheries

46 Thomas MM - MSc,PhD fisheries extension
47 Vedavyasa Rao MSc,PhD - crustacean fisheries; physiology nutrition
48 Venkatanarayana Rao K - MSc pelagic fisheries - resources

2137 Regional Centre of CMFRI
Marine Fisheries P.O., Mandapam Camp, Tamil Nadu, 623520
Cables: CADALMIN, MARINE FISHERIES.

1 Bensam P - MSc,PhD fish eggs and larvae
2 Kaliaperumal N - MSc,PhD seaweed culture
3 Lal Mohan RS - MSc,PhD finfish culture
4 Mahadevan S - MSc molluscan fisheries
5 Ramadoss K - MSc molluscan fisheries
6 Thiagarajan R - MSc pelagic fisheries

2138 Madras Research Centre of CMFRI
9, Commander-in-chief Road P.O., Madras, Tamil Nadu, 600 105
Cables: CADALMIN, MADRAS.

1 Devados P - MSc pelagic fishery resources
2 Devarajan K - MSc crustacean culture
3 Girijavallabhan KG - MSc zooplankton
4 Gnanamuttu JC - MSc demersal resources
5 James DB - MSc,PhD crustacean resources
6 Kathirvel M - MSc prawn breeding and culture
7 Kuthalingam MDK - MSc,PhD pelagic fisheries
8 Mohan Raj G - MSc finfish culture
9 Nammalwar P - MSc,PhD finfish culture
10 Radhakrishnan EV - MSc lobster culture
11 Rajagopalan M - MSc seaturtle conservation
12 Ramamurthy S - MSc,PhD crustacean resources
13 Rengarajan K - MSc mussel culture
14 Sarvesan R - MSc cephalopod resources
15 Silas EG - MSc,PhD,DSc *Officer on special duty* Central Institute for Aquaculture - oceanic fisheries resources - development
16 Sundaram PTM - MSc demersal fisheries
17 Vijayakumar M - MSc crustacean nutrition and pathology
18 Vivekanandan E - MSc,PhD demersal resources

2139 Karwar Research Centre of CMFRI
P.O. Box No. 5, Karwar, North Kanara
Cables: CADALMIN, KARWAR.

1 Annigiri GG - MSc pelagic fisheries
2 Dhulkhed MH - MSc pelagic fisheries

2140 Tuticorin Research Centre of CMFRI
90, North Beach Road, Tuticorin-1, Tamil Nadu
Cables: CADALMIN, TUTCORIN.

1 Ameer Hamsa KMS - MSc finfish culture
2 Dharmaraj S - MSc,PhD pearl oyster culture
3 Gopinathan CP - MSc,PhD phytoplankton culture
4 Marichamy R - MSc finfish culture
5 Mohamed Kasim H - MSc,PhD crustacean resources
6 Muthiah P - MSc oyster culture
7 Nagappan Nayar K - MSc molluscan resources and culture
8 Rajamani M - MSc,PhD prawn culture
9 Rajapandian ME - Msc oyster culture
10 Sam Bennet P - MSc demersal fisheries
11 Satyanarayana Rao K - MSc,PhD oyster biology and culture
12 Shanmugam S - MSc fish and prawn seed
13 Siraimeetan PON - MSc tuna fisheries

2141 Vizhinjam Research Centre of CMFRI
Vizhinjam, Trivandrum, Kerala, 695521
Cables: CADALMIN, VIZHINJAM.

1 Appukuttan KK - MSc molluscan fisheries
2 Gopalakrishna Pillai N - MSc pelagic fisheries
3 Gopinadha Pillai CS - MSc,PhD corals and coral reefs
4 Mukundan C - MSc demersal fisheries
5 Nair PNR - MSc,PhD demersal fisheries
6 Prabhakaran Nair K - MSc cephalopod resources
7 Rani Ms MJ - MSc zooplankton
8 Thomas PA - MSc,PhD spongest goryonids

2142 Calicut Research Centre of CMFRI
West Hill, Calicut, Kerala, 673 005
Cables: CADALMIN, CALICUT.

1 Gopinatha Menon N - MSc,PhD demersal fisheries
2 Kumaran M - MSc fish eggs and larvae
3 Kuriakose PS - MSc,PhD mussel culture
4 Lazarus S - MSc,PhD finfish culture
5 Yohannan TM - MSc pelagic fisheries

2143 Mangalore Research Centre of CMFRI
P.O. Box No. 244, Bolar, Mangalore, 575001
Cables: CADALMIN, MANGALORE.

1 Pai MV - MSc,PhD pelagic fisheries
2 Sukumaran KK - MSc crustacean resources
3 Syda Rao G - MSc,PhD molluscan fisheries

2144 Waltair Research Centre of CMFRI
Waltair, Visakhapatnam, Andhra Pradesh, 530 003
Cables: CADALMIN, WALTAIR.

1 Appa Rao T - MSc,PhD demersal fisheries
2 Luther G - MSc,PhD pelagic resources
3 Radhakrishna K - MSc,PhD primary productivity
4 Sudhakara Rao G - MSc crustacean fisheries

2145 Kakinada Research Centre of CMFRI
21/44-11-9, Kolipalli Road, Kakinada, Andhra Pradesh, 533 002
Cables: CADALMIN, KAKINADA.

1 Lalitha Mrs D - MSc,PhD crustacean fisheries
2 Narasimham KA - MSc clam culture
3 Sriramachandra Murty V - MSc,PhD demersal fisheries

2146 Bombay Research Centre of CMFRI
148, Army and Navy Building, 2nd Floor, M.G. Road, Bombay, Maharashtra, 500 023
Cables: CADALMIN, BOMBAY.

1 Deshmukh VM - MSc pelagic fisheries
2 Kagwade PV - MSc,PhD crustacean fisheries
3 Kuber Vidyasagar - MSc cephalopod resources
4 Kurian A - MSc pelagic fisheries

2147 Veravel Research Centre of CMFRI
opp. Octroi Naka, Talala Road, Veraval, Gujarat, 362265
Cables: CADALMIN, VERAVEL.

1 Lipton AP - MSc fish and fishery science

2148 Central Institute of Fisheries Technology
Willingdon Island, Matsyapuri P.O., Cochin, Kerala, 682029
Tel: 6845. Telex: 0885-440. Cables: FISH TECHNOLOGY.

1 Nair MR - MSc *Director*

2149 Biochemistry and Nutrition Division
1 Devadas K - MSc,PhD *Scientist-in-Charge*
2 Mukundan MK - MSc,PhD
3 Stephen J - MSc,PhD

2150 Craft Division
1 Nair NUK - MSc,PhD
2 Ravindran K - MSc,DSc *Scientist-in-Charge*

2151 Engineering Division
1 Namboodiri KS - BScEngg
2 Pillai SA - MTech *Scientist-in-Charge*
3 Sivadas TK - MSc,PhD

2152 Extension, Information and Statistics Division
1 Kandoran MK - MSc *Scientist-in-Charge*

2153 Gear Division
1 Panicker PA - MSc *Scientist-in-Charge*

2154 Microbiology Division
1 Iyer KM - MSc *Scientist-in-Charge*
2 Surendran PK - MSc,PhD

2155 Processing and Packaging Division
1 Chinnamma G - MSc,PhD
2 Gopakumar K - MSc,PhD *Scientist-in-Charge*
3 Govindan TK - MSc,PhD
4 Lakshmanan PT - MSc,PhD
5 Nair KGR - MSc,PhD

2156 Research Centre of CIFT
Bunder Road, Veraval, Gujarat, 362265
Tel: 20297. Telex: 0163-202-CIFT. Cables: MATSYAOUDYOGIKI.

1 Viswanathan Nair PG - MSc,PhD

2157 Research Centre of CIFT
D-No-54-2-9, Yesuvari St, Jagannaickpur, Kakinada, Andhra Pradesh, 533002
Tel: 4436. Telex: 0461-229-CIFT. Cables: MATSYAOUDYOGIKI.

1 Rao CCP - MVSc,PhD Scientist-in-Charge

2158 Research Centre of CIFT
Burla P O 768017, Sambalpur District, Orissa
Tel: 19. Cables: MATSYAOUDYOGIKI.

1 Khan AA - MSc Scientist-in-Charge

2159 Research Centre of CIFT
162 Sassoon Dock, Colaba, Bombay, Maharashtra, 400005
Tel: 213892. Cables: FISHPROCES.

1 James A - MSc,PhD Scientist-in-Charge

2160 Research Centre of CIFT
2nd Floor, 'Shanta', 18th June Road, St. Inez, Panaji, Goa, 403001
Tel: 5905. Cables: MATSYAOUDYOGIKI.

1 Mathai TJ - MSc Scientist-in-Charge

2161 Research Centre of CIFT
Beach Road, West Hill, Calicut, Kerala, 673005
Tel: 76607. Cables: CARE CADALMIN.

1 Nair TSU - MSc Scientist-in-Charge

2162 Central Institute of Fisheries Education
P.O. Box No. 7392, Bombay, 400 058
Tel: 571446/7/8-573404. Telex: 11-71476.

1 Sreekrishna Y - MSc,PhD Director fisheries biology; hatchery and seed rearing systems; fisheries management

2163 Ocean Management and Marine Fisheries Division

1 Devraj N - MSc,PhD Head of Division marine stock assessment
2 Ravindranathan V - MSc oceanography; limnology

2164 Inland Fisheries Division
P.O. Box No. 7392, Bombay, 400 058
Tel: 571446/7/8-573404. Telex: 11-71476.

1 Kohli MPS - MSc,PhD Head of Division inland fisheries; air breathing fish culture

2165 Aquaculture Division

1 Sinha KPSR - MSc fish breeding; hatchery

2166 Fishery Technology

1 Sreekrishna Y - MSc Head of Division fishing boats and gear
2 Patil MR - MSc chemical oceanography; vessel management

2167 Fish Processing Technology

1 Ramananda RD - MSc,PhD fish biochemistry; processing product development

2168 Project Planning and Evaluation

1 Ghosh KK - MSc Head of Division fisheries project planning; quantitative management; statistical systems softwear
2 Biradhar RS - MSc fisheries statistics; stock assessment

2169 Agricultural Economics

1 Rao PS - MA,MCom Head of Division fisheries economics
2 Subha Rao PV - MA co-operation and marketing

2170 Agricultural Extension

1 Rao M - MSc Head of Division transfer of technology

2171 Central Fisheries Extension Training Centre
5/60 Kaushlya Nagar, Saroor Nagar, Hyderabad, Andhra Pradesh, 500035

1 Reddi DV - MSc Principal aquaculture extension

2172 Inland Fisheries Training Centre
Barrackpore, West Bengal

1 Chondar SL - MSc,PhD aquaculture
2 Maheshwari UK - MSc,PhD inland fisheries
3 Singh K - MSc,DFSc Principal inland fisheries management

2173 Regional Training Centre
Chinhat, Lucknow, Uttar Pradesh

1 Chowdhary DK - MSc Principal operational research project management
2 Mitra SD - MSc fish breeding and culture
3 Rastogi ML - MSc freshwater fish culture

2174 Brackish Water Fish Farm
Kakinada

1 Gopal Krishna G - MSc brackish water fish farm management

2175 Fresh Water Fish Farm
Pawerkheda

1 Somalingam J - MSc fresh water fish farm management

2176 Central Tuber Crops Research Institute
Sreekariyam, Trivandrum, Kerala, 695 017
Tel: 8551. Telex: 0884-247 ROOT IN. Cables: TUBERSEARCH.

1 Balagopalan C - MSc,PhD Head of Division post harvest technology
2 Ghosh SP - MSc,PhD Director
3 Nayar GG - MSc,PhD Head of Division genetics and plant breeding
4 Pal TK - MA,PhD Head of Division economics, statistics & extension
5 Pillai NG - MSc,PhD Head of Division agronomy and soils
6 Ramanujam T - MSc,PhD Head of Division physiology and biochemistry
7 Sadasivan Pillai K - MSc,PhD Head of Division agricultural entomology
8 Thankappan M - MSc,PhD Head of Division plant pathology
9 Bala Nambisan Mrs - MSc,PhD biochemistry
10 Gangadharan Nair N - MSc,PhD plant pathology
11 Gopalakrishnan Nair P - MSc,PhD agricultural chemistry
12 Gopalakrishnan Nair S - MSc plant breeding
13 Indiramma Mrs P - MSc plant physiology
14 Kabeerathumma Mrs S - MSc,PhD soil science
15 Malathi Miss V - MSc,PhD plant pathology
16 Mohan Kumar B - MSc,PhD soil fertility
17 Mohankumar CR - MSc agronomy
18 Muraleedharan Nair G - MSc agronomy
19 Narayana Moorthy S - MSc,PhD organic chemistry
20 Prabhakar M - MSc agronomy
21 Rajamma Mrs P - MSc,PhD agricultural entomology
22 Rajendran PG - MSc,PhD genetics
23 Ramakrishnan Nayar TV - MSc agronomy
24 Ravindran CS - MSc agronomy
25 Samuel Jos - MSc,PhD genetics and cytogenetics
26 Vasudevan K - MSc,PhD plant breeding
27 Vijaya Bai Miss K - MSc,PhD genetics and cytogenetics

2177 Regional Centre of CTCRI
316-A, Kharvel Nagar, Bhubaneswar, Orissa, 751 001
Tel: 50337. Cables: TUBERSEARCH.

1 Naskar SK - MSc,PhD plant breeding
2 Singh DP - MSc,PhD plant breeding
3 Varma SP - MSc,PhD Project Coordinator and Scientist-in-Charge

2178 National Bureau of Plant Genetic Resources (IARI Campus)
New Delhi, 110012
Tel: 586908, 589214. Cables: Germplasm.

1 Arora RK - MSc,PhD Director

2179 Conservation Division

1 Chaudhary Miss NK - MSc,DipAgrStat design of field experiments
2 Khanna PP - MSc In Charge plant genetic resources conservation
3 Verma Mrs K - MSc seed technology and seed identification

2180 Germplasm Exchange Division

1 Singh BP - MSc,PhD Head of Division introduction of plant genetic resources and genetics and plant breeding
2 Kumar B - MSc introduction of germplasm with special reference to cereals and millets
3 Singh R - MSc,PhD introduction of plant genetic resources and breeding of pulse crops

2181 Evaluation Division

1 Thomas TA - MSc Head of Division germplasm collection and evaluation of pulses, vegetables and miscellaneous economic plants

2 Dabas BS - MSc,PhD plant breeding genetics; germplasm evaluation of grain legumes; vegetables and bulb crops
3 Mathur PN - MSc,DIIT,PhD plant breeding and genetics, forage crops
4 Mithal SK - MSc,PhD plant breeding and genetics
5 Pareek SK - MSc agronomy
6 Singh B - MSc,PhD plant breeding and genetics of cereals and pulses
7 Singh R - MSc evaluation of oil seed crops

2182 Medicinal and Aromatic Plants Section

1 Maheshwari ML - MSc,PhD *Head of Section* phytochemistry of medicinal and aromatic plants
2 Gupta R - MSc,PhD *Project Coordinator* economic botany and chemotaxonomy
3 Kidwai MA - MSc,PhD chemical and analytical aspects of medicinal and aromatic plants
4 Mandal S - MSc,PhD biochemistry of proteins
5 Sethi KL - MSc,PhD plant breeding and genetics of medicinal and aromatic plants
6 Singh BM - MSc evaluation of medicinal and aromatic plants
7 Srivastava VK - MSc,PhD chemical and analytical aspects of medicinal and aromatic plants

2183 Exploration Division

1 Chandel KPS - MSc,PhD plant exploration, breeding, conservation, evaluation and utilization
2 Kazim M - MSc plant exploration
3 Nayar Miss ER - MSc biosystematics, ecology and numerical taxonomy

2184 Quarantine Division

2185 Plant Pathology Section

1 Ram Nath - MSc *Head of Division, Plant Quarantine and Sectional Head* seed pathology and quarantine
2 Agarwal PC - MSc,PhD seed pathology and quarantine
3 Lambat AK - MSc,PhD seed pathology and quarantine
4 Majumdar A - MSc,PhD seed pathology and quarantine
5 Varshney JL - MSc,PhD seed pathology and quarantine

2186 Entomology Section

1 Verma BR - MSc,PhD *Head of Section* insect toxicology, fumigation storage entomology; quarantine

2187 Nematology Section

1 Sanwal KC - MSc,PhD *Head of Section* nematode systematics and morphology, quarantine
2 Mathur VK - MSc,PhD nematode taxonomy, biology; quarantine

2188 Regional Stations

2189 Akola

1 Patel DP - MSc,PhD *Head of Station* germplasm collection, evaluation and maintenance of crops suited to Deccan Plateau

2190 Almora

1 Pant KC - MSc *Head of Station* germplasm collection and evaluation of crops suited to mid hills of Himalayas

2191 Amravati

1 Kawalkar TG - MSc *Head of Station* germplasm collection, evaluation and maintenance of crops suited to Deccan Plateau

2192 Hyderabad

1 Varaprasad KS - MSc,PhD *Head of Station* nematode biology and control; quarantine

2193 Jodhpur

1 Chopra DP - MSc *Head of Station* germplasm collection, evaluation and maintenance of agri-horticultural crops suited to arid and semiarid regions
2 Bhandari DC - MSc,PhD germplasm collection of crops in arid and semiarid regions

2194 Shillong

1 Sharma BD - MSc,PhD *Head of Station* germplasm collection, evaluation and maintenance of crops suited to north-eastern hill region
2 Hore DK - MSc,PhD germplasm collection of crops from north-eastern hills and plains

2195 Shimla

1 Joshi BD - MSc,PhD *Head of Station* germplasm collection, evaluation and maintenance of agricultural crops suited to temperate regions
2 Rathore DS - MSc,PhD germplasm evaluation and maintenance of horticultural crops suited to temperate regions

2196 Trichur

1 Muralidharan VK - MSc *Head of Station* germplasm collection, evaluation and maintenance of crops suited to hot humid tropics
2 Valayudhan KC - MSc germplasm collection, evaluation and maintenance of crops suited to hot humid tropics

2197 National Bureau of Soil Survey and Land Use Planning
Amravati Road, Nagar, Maharashtra
Tel: 32386. Telex: 0715-262 NBSL IN. Cables: SOILANDBRU.

1 Sehgal JL - MSc,MS,DSc *Director*

2198 Pedology Division

1 Kalbande AR - AssocIARI,PhD *Head of Division* soil micromorphology
2 Deshpande SB - AssocIARI,PhD soil genesis
3 Pal DK - MSc,PhD clay mineralogy
4 Venugopal KR - MSc soil micromorphology

2199 Soil Correlation and Classification Division

1 Pofali RM - MA *Head of Division* soilscape relationships
2 Batta RK - MTech soil and water engineering
3 Hirekerur LR - MSc,PhD soil correlation
4 Sohan Lal - MSc soil taxonomy
5 Yadav SC - MSc agronomy

2200 Remote Sensing Division

1 Saxena RK - MSc *Head of Division* remote sensing

2201 Land Use Planning Division

1 Bhaskar KS - MSc agronomy

2202 Cartography Division

1 Singh AMP - MA *Head of Division* cartography

2203 Documentation Centre

1 Patil YM - BSc,BLibSc,ADR *Documentation Officer* library and information services

2204 Regional Centre
Nagpur, Seminary Hills, Nagpur
Tel: 34615, 34622. Cables: SOILCORE.

1 Gaikawad ST - MSc,PhD *Head of Centre* soil survey and mapping
2 Challa O - MSc,PhD soil survey and mapping
3 Chaturvedi A - MA cartography
4 Gajbhiye KS - MSc,PhD soil analysis
5 Sharma JP - MSc,PhD soil survey and mapping

2205 Regional Centre
Bangalore, Post Box No. 2487
Tel: 362242. Cables: SOILCORE.

1 Naga Bhushana SR - MSc *Head of Centre* soil survey and mapping
2 Prabhakara - MA *cartography*
3 Rajendran - MSc,PhD soil science
4 Sarma VAK - MSc,PhD soil genesis and soil analysis
5 Shankaranarayana HS - MSc soil survey training

2206 Regional Centre
IARI Campus, New Delhi
Tel: 586837. Cables: SOILCORE.

1 Rana KPC - MSc *Head of Centre* soil survey and mapping
2 Naidu LGK - MSc agronomy
3 Sachdeva CB - MSc,PhD soil survey and mapping
4 Sukhpal Singh - MSc,PhD soil survey and mapping
5 Tarsem Lal - MA cartography

2207 Regional Centre
233 NSC Bose Road, Calcutta
Tel: 725602. Cables: SOILCORE.

1 Thampi CJ - MSc,PhD *Head of Centre* soil analysis
2 Maji AK - MSc,PhD soil survey and mapping

3 Sarkar D - MSc,PhD soil survey and mapping
4 Singh DS - MA,BEd cartography

2208 Regional Centre
Assam Agricultural University Campus, Jorhat, Assam
Tel: 35. Cables: SOILCORE.

1 Chamuah GS - MSc,PhD *Head of Centre* soil survey and mapping

2209 Regional Centre
35 Relief Society, Panigate, Baroda
Tel: 557463. Cables: SOILCORE.

1 Shyampura RL - MSc,PhD *Head of Centre* soil survey and mapping

2210 National Bureau of Animal Genetic Resources and National Institute of Animal Genetics
NDRI Campus, Karnal, Haryana, 132001
Tel: 3918. Telex: 0396-204 NDRI. Cables: GENETICS.

1 Nair PG - BVSc,PhD *Director*

2211 Cytogenetics
1 Sahai R - MSc,PhD *Head*
2 Bhatia Miss S - MSc,PhD

2212 Immunogenetics and Bio-Chemical Genetics
1 Prasad SK - MSc,PhD *Head*
2 Vijay Shankar - MSc,PhD

2213 National Bureau of Animal Genetic Research
1 George M - MSc,PhD

2214 National Research Centre on Equines
Sirsa Road, Hissar, Haryana, 125 001
Tel: 4170. Telex: 0345 - 204.

1 Uppal PK - BVSc,MVSc,PhD *Director*

2215 National Bureau of Fish Genetic Resources
B-209, Mehdauri Colony, Teliyer ganj Allahabad, Uttar Pradesh, 211004
Tel: 56627. Cables: FISHBUREAU.

1 Jhingran AG - MSc,DPhil,FNASC *Project Director* riverine and lacustrine fisheries

2216 Fish Biology
1 Kapoor D - MSc fish culture
2 Mahanta PC - MSc fish breeding and culture

2217 Biochemistry
1 Hajra A - MSc biochemistry

2218 National Research Centre for Groundnut
Timbawadi, P.O. Junagadh, Gujarat, 362015
Cables: GNUTSEARCH.

1 Reddy PS - MScAgr,PhD *Director*

2219 Genetic Resources Section
1 Bhagat NR - MScAgr,PhD *Head of Section* management of genetic resources of groundnut

2220 Plant Breeding Section
1 Basu MS - MScAgr *Head of Section* breeding for earliness and resistance to biotic and abiotic stresses

2221 Cytogenetics Section
1 Tiwari SP - MScAgr,PhD *Head of Science* utilization of wild species of the section *Arachis* for groundnut improvement
2 Sen P - MScAgr,PhD utilization of wild species other than section *Arachis* for groundnut improvement

2222 Plant Pathology Section
1 Ghewande MP - MSc,PhD *Head of Section* epidemiology of foliar diseases and aflatoxin studies in groundnut

2223 Plant Physiology Section
1 Dwivedi RS - MScAgr,PhD *Head of Section* mineral nutrition and post harvest physiology
2 Joshi YC - MSc water stress physiology

2224 Microbiology Section
1 Kulkarni JH - MScAgr,PhD nitrogen fixation in relation to *Rhizobium* host and environment

2225 Biochemistry Section
1 Nagaraj G - MSc,PhD oil biogenesis; oil content and quality evaluation in groundnut

2226 National Centre for Mushroom Research and Training
Chambaghat, Solar, Himachal Pradesh, 173213
Tel: 851.

1 Rai RD - MSc,PhD post harvest technology and morphogenesis of mushrooms
2 Sohi HS - MSc,PhD,FNA *Officer on Special Duty*

2227 National Research Centre on Camel
Jorbeer, P.O. Shivabari, Bikaner, Rajasthan, 334 001

1 Khanna ND - BVSc&AH,AIVRI *Project Director*

2228 National Academy of Agricultural Research Management
Rajendranagar, Hyderabad, Andhra Pradesh, 500 030
Tel: 48394, 48395. Telex: 0155-665.

1 Anwar MM - MScAgr,PhD *Associate Professor*
2 Balaguru T - MScAgr,PhD *Associate Professor*
3 Devesh Kishore - MScAgr,PhD *Professor*
4 Gaddagimath RB - MA,BLibSc *Librarian*
5 Raghavendra Rao V - MScAgr,PhD *Professor*
6 Rajagopalan M - MAStat,PhD *Professor*
7 Raman KV - MSc,PhD *Director*
8 Reddy S - MScAgr *Associate Professor*
9 Sabarathnam VE - MScAgr *Associate Professor*
10 Saha SN - MScAgr,PhD *Professor*
11 Samanta RK - MScAgr,PhD *Associate Professor*
12 Subba Rao KV - MSc,PhD *Professor*
13 Venkateswarlu K - MScAgr,PhD *Associate Professor*

2229 All India Coordinated Wheat Improvement Project
Wheat Project Directorate, IARI Campus, New Delhi

1 Agrawal RK - MSc,PhD plant breeding
2 Bisht SS - MSc,PhD plant breeding
3 Chaudhary HB - MSc,PhD plant breeding
4 Goel LB - MSc,PhD plant pathology
5 Gupta JP - MSc,PhD agronomy
6 Hans VK - MSc,PhD plant breeding
7 Malik BS - MSc,PhD plant breeding
8 Mathur HC - MSc,PhD plant pathology
9 Menon TCM - MSc,PhD agricultural statistics
10 Naqvi SMA - MSc,PhD plant breeding
11 Pandey HN - MSc,PhD plant breeding
12 Sethi AP - MSc,PhD plant breeding
13 Singh RP - MSc,PhD agronomy
14 Singh SD - MSc,PhD plant pathology
15 Sinha VC - MSc,PhD plant pathology
16 Tandon JP - MSc,PhD *Project Director* plant breeding

2230 Directorate of Rice Research
Rajendranagar, Hyderabad, Andhra Pradesh, 500030
Tel: 48036. Telex: 155 6739. Cables: RICEIN.

1 Bentur JS - MSc,PhD
2 Gosh A - MSc,PhD
3 Hakim KL - MSc,PhD
4 Kalode MB - MSc,PhD entomology
5 Krishnaiah K - MSc,PhD entomology
6 Krishnaiah NV - MSc,PhD
7 Mahajan RK - MSc,PhD
8 Murty VVS - MSc,PhD plant breeding
9 Narayana Reddy - MSc,PhD
10 Padmaja Rao Mrs S - MSc,PhD
11 Pasalu IC - MSc,PhD
12 Prasada Rao U - MSc,PhD
13 Rama Rao K - MTech agricultural engineering
14 Ramchand - MSc,PhD
15 Rao AV - MSc statistics

16 Reddy APK - MSc,PhD plant pathology
17 Reddy PC - MSc,PhD
18 Sadasiva Reddy C - MSc,PhD
19 Sharma SK - MSc,PhD agronomy
20 Shastry MVS - MSc,PhD
21 Shinde JE - MSc,PhD *Project Director*
22 Singh SP - MSc,PhD
23 Subbaiah SV - MSc,PhD
24 Venkataraman S - MSc,PhD
25 Venkateshwarlu B - MSc,PhD plant physiology

2231 Directorate of Pulses Research
Kanpur, Uttar Pradesh, 208024
Tel: 241264, 245146. Cables: PULSESRES.

1 Chandra S - MSc,PhD *Project Director* genetics and plant breeding

2232 Division of Genetics and Plant Breeding
Kanpur, Uttar Pradesh, 208024
Tel: 241264, 245146. Cables: PULSESRES.

1 Asthana AN - MSc,PhD pulse crops breeding
2 Dube SD - MSc pigeonpea breeding
3 Lal J - MSc,PhD cytogenetics and tissue culture
4 Mandal BB - MSc cytogenetics
5 Mishra SP - MSc,PhD mungbean breeding
6 Singh N - MSc,PhD chickpea breeding
7 Tripathi DP - MSc,PhD breeding of fieldbean and urdbean

2233 Division of Agronomy

1 Ali M - MSc,PhD soil water management
2 Kushwaha BL - MSc,PhD cropping systems
3 Singh NB - MSc,PhD crop management
4 Singh SN - MSc,PhD soil water management

2234 Division of Entomology

1 Ahmed R - MSc,PhD biological pest control
2 Dias CAR - MSc stored grain pests
3 Lal SS - MSc,PhD host plant resistance
4 Sachan JN - MSc,PhD economic entomology
5 Yadava CP - MSc insect ecology

2235 Division of Plant Pathology

1 Amin KS - MSc,PhD epidemiology and host plant resistance
2 Singh RA - MSc,PhD host plant resistance and quarantine

2236 Division of Physiology and Bio-Chemistry

1 Singh DN - MSc,PhD production physiology

2237 Division of Statistics and Economics

1 Rao KVS - MSc,PhD experiment designs
2 Tyagi VP - MSc,PhD agricultural finance

2238 Division of Engineering

1 Lall RR - MTech structures and designs

2239 DPR Regional Station
Regional Centre, P.B. No. 4, Morar, Gwalior - 6, Madhya Pradesh

1 Varshney JG - MSc,PhD cropping systems

2240 Directorate of Oilseeds Research
Rajendranagar, Hyderabad, Andhra Pradesh, 500 030
Cables: SODHTELMAN.

1 Ankineedu G - MScAgr *Project coordinator* plant breeding
2 Chowdhery R - MSc,PhD entomology
3 Kumar PR - MSc,PhD *Project Coordinator* plant breeding
4 Maiti S - MSc,PhD plant pathology
5 Rai M - MSc,PhD *Project Coordinator* plant breeding
6 Ramachandran M - MSc,PhD plant breeding
7 Rao VR - MSc,PhD *Project Coordinator* plant breeding
8 Sastry JA - MSc statistics
9 Sastry KS - MSc,PhD plant pathology
10 Sharma SM - MSc,PhD *Project Coordinator* plant breeding
11 Sindagi SS - MSc *Project Coordinator* plant breeding
12 Singh M - MSc,PhD agricultural chemistry
13 Singh V - MSc,PhD entomology
14 Yadava TP - MSc,PhD *Project Director* plant breeding

2241 Andhra Pradesh Agricultural University
Rajendranagar, Hyderabad, Andhra Pradesh, 500 030
Cables: AGRIVERSITY.

1 Appa Rao A - PhD *Vice Chancellor*

2242 Department of Agronomy

1 Ankaiah S - MScAgr *Agronomist* tobacco
2 Binjam B - MScAgr,PhD *Agronomist* fodder production
3 Burhanuddin M - MScAgr,PhD *Agronomist* weed agronomy
4 Chari AS - MScAgr,PhD *Agronomist* sugarcane
5 Hussain MM - MScAgr,PhD *Senior Scientist* fodder production and cropping system
6 Kondep SM - MScAgr,PhD *Associate Professor* weed science
7 Latchanna A - MScAgr,PhD *Associate Professor* weed science
8 Mirja WA - MScAgr *Agronomist* farm management
9 Mumtabujuddin K - MScAgr,PhD *Agronomist* fertilizer agronomy
10 Murthy VS - MScAgr,PhD *Senior Scientist* rice agronomy
11 Raghavulu P - MScAgr,PhD *Associate Director of Research* dry farming
12 Rajan MSS - MScAgr *Senior Scientist* water management
13 Raju MS - MScAgr *Agronomist* cropping system
14 Ramanaiah NV - MScAgr,PhD *Associate Professor* proto agronomy
15 Ramaseshaiah K - MScAgr,PhD *Agronomist* water management
16 Rangamaninoir KTV - MScAgr,PhD *Agronomist* water management
17 Rao AY - MScAgr,PhD *Meteorologist* meteorology
18 Rao IVN - MScAgr,PhD *Agronomist* meteorology
19 Rao KL - MScAgr,PhD *Agronomist* sugarcane
20 Rao KN - MScAgr *Weed Control Officer* weed science
21 Rao MN - MScAgr *Associate Professor* fertilizer management
22 Rao MR - MScAgr,Phd *Senior Scientist* sugarcane
23 Rao PG - MScAgr *Associate Professor* intercropping
24 Rao PVR - MScAgr *Agronomist* water management (sugarcane)
25 Rao SK - MScAgr,PhD *Agronomist* water management
26 Reddy AS - MScAgr,PhD *Agronomist* millet agronomy
27 Reddy BB - MScAgr,PhD *Agronomist* water management
28 Reddy BB - MScAgr,PhD *Associate Professor* water management
29 Reddy CS - MScAgr,PhD *Agronomist* groundnut agronomy
30 Reddy GB - MScAgr,PhD *Associate Professor* dry farming
31 Reddy GV - MScAgr,PhD *Professor* water management
32 Reddy KA - MScAgr,PhD *Associate Director of Research* water management
33 Reddy MGRK - MScAgr,PhD *Associate Professor* crop management
34 Reddy MN - MScAgr,PhD *Senior Scientist* water management
35 Reddy PR - MScAgr *Associate Professor* dryland farming
36 Reddy SC - MScAgr,PhD *Professor* crop management
37 Reddy SN - MScAgr,PhD *Associate Professor* rice and cotton agronomy
38 Reddy SR - MScAgr,PhD *Professor* water management and groundnut economy
39 Reddy SR - MScAgr,PhD *Agronomist* millet agronomy
40 Reddy TMM - MScAgr *Associate Professor* fertilizer management
41 Reedy NV - MScAgr,PhD *Agronomist* rice
42 Sagar GC - MScAgr,PhD intercropping
43 Sarma DA - MScAgr,PhD *Agronomist* water management
44 Satyanarayana V - MSvAgr,PhD *Professor* soil fertilizer management
45 Siva Sankara V - MScAgr,PhD *Agronomist* fertilizer management
46 Venkateswarlu MS - MScAgr,PhD *Agronomist* crop production

2243 Department of Plant Breeding

1 Anantasayanam K - MScAgr *Associate Professor* groundnut;sunflower
2 Devid V - MScAgr *Plant Breeder* cotton
3 Gopalan B - MScAgr,PhD *Plant Breeder* sesamum
4 Jagdish CA - MScAgr,PhD *Professor* millet and rice
5 Jayaraj D - MScAgr *Deputy Director of Research* rice
6 Kulkarni N - MScAgr,PhD *Plant Breeder* millets and rice
7 Murthy PSN - MScAgr *Plant Breeder* rice
8 Narayan K - MScAgr *Plant Breeder* millets and maize
9 Narayana D - MScAgr,PhD *Senior Scientist* sorghum
10 Peraiah A - MScAgr *Specialist* rice
11 Ramaseshaiah B - MScAgr,PhD *Plant Breeder* cotton and millets
12 Ranganadhacharyulu N - MScAgr,PhD *Cotton Breeder* cotton
13 Rao CB - MScAgr *Rice Breeder* rice
14 Rao CR - MScAgr,PhD *Associate Professor* rice
15 Rao GB - MScAgr *Specialist* rice
16 Rao GM - MScAgr,PhD *Professor* rice
17 Rao GPP - MScAgr *Plant Breeder* vegetables
18 Rao JVR - MScAgr,PhD *Plant Breeder* sugarcane
19 Rao KKP - MScAgr,PhD *Associate Director of Research* sugarcane
20 Rao KP - MScAgr,PhD *Associate Professor* alfalfa cucumber
21 Rao MMKD - MScAgr,PhD *Senior Scientist* millets and pulses
22 Rao MV - MScAgr *Plant Breeder* mesta
23 Rao MVK - MScAgr,PhD *Associate Professor* groundnut
24 Rao PJ - MScAgr *Plant Breeder* millets
25 Rao PV - MScAgr *Plant Breeder* millets
26 Rao RVA - MScAgr *Plant Breeder* fibre crops
27 Ratnakar B - MScAgr *Plant Breeder* castor
28 Reddy BM - MScAgr,PhD *Plant Breeder* cotton
29 Reddy CR - MScAgr,PhD *Senior Scientist* groundnut
30 Reddy CS - MScAgr,PhD *Associate Professor* sorghum

31 Reddy GV - MScAgr *Plant Breeder* rice
32 Reddy MR - MScAgr *Plant Breeder* millets
33 Reddy MV - MScAgr,PhD *Professor* groundnut and rice
34 Reddy NS - MScAgr,PhD *Senior Scientist* rice
35 Reddy PN - MScAgr,PhD *Associate Professor* groundnut
36 Reddy PR - MScAgr,PhD *Plant Breeder* mesta
37 Sahab H - MScAgr *Plant Breeder* millets
38 Sarma MY - MScAgr *Research Officer* maize
39 Satyanarayan A - MScAgr,PhD *Senior Scientist* pulses
40 Sriramulu C - MScAgr,PhD *Associate Director of Research* millet and pulses
41 Subbarayudu VC - MScAgr *Plant Breeder* millets
42 Subrahmanyam D - MScAgr *Associate Professor* rice
43 Venkateswarlu P - MScAgr *Plant Breeder* millets and pulses
44 Venkateswarlu S - MScAgr,PhD *Plant Breeder* pulses

2244 Department of Horticulture

1 Babu RS - MScAgr,PhD *Associate Professor* fruits, vegetables and cashew
2 Khan MAR - MScAgr *Floriculturist* fruits, grapes
3 Kulkarni V - MScAgr,PhD *Horticulturist* fruits, mango
4 Murthy NS - MScAgr,PhD *Vegetable Breeder* vegetables
5 Murthy PR - MScAgr,PhD *Horticulturist* cashew, spices and fruits
6 Padmanabhan V - MScAgr *Associate Professor* vegetables,banana and grapes
7 Prasad PK - MScAgr,PhD *Associate Professor* vegetables
8 Rao DVS - MScAgr,PhD arid fruit crops
9 Rao LJ - MScAgr,PhD *Associate Professor* vegetables
10 Rao MR - MScAgr,PhD *Principal and Professor* processing of fruits and vegetables
11 Rao MV - MScAgr,PhD *Horticulturist* arecanut,coconut and mango
12 Rao NS - MScAgr *Associate Professor* cashew;fruits and vegetable processing
13 Rao PV - MScAgr,PhD *Horticulturist* vegetables and cashew
14 Rao PVSN - MScAgr *Associate Professor* fruits and mango
15 Rao RR - MScAgr,PhD *Senior Scientist* vegetables and cashew
16 Rao TS - MScAgr *Horticulturist* spices;coconut
17 Reddy KS - MScAgr *Horticulturist* mango and cashew
18 Reddy KVS - MScAgr,PhD *Horticulturist* vegetables
19 Reddy VP - MScAgr *Horticulturist* fruits, nursery practices
20 Sarma KN - MScAgr *Horticulturist* fruits, coconut
21 Satyanarayana G - MScAgr,PhD *Senior Scientist* fruits
22 Suryanarayana V - MScAgr,PhD *Associate Professor* mango;coconut

2245 Department of Plant Pathology

1 Balasubramanian KA - MSc,PhD,DSc *Principal and Professor* sorghum pathology
2 Mohiuddin S - MScAgr,PhD *Associate Professor* fungicides
3 Moses GJ - MScAgr,PhD *Plant Pathologist* bacteriology of rice
4 Narasimham V - MScAgr *Plant Pathologist* rice pathology
5 Pandu SR - MScAgr,PhD *Plant Pathologist* plant virology
6 Raj RB - MScAgr *Plant Pathologist* plant pathology
7 Raju AS - MScAgr *Plant Pathologist* chillies diseases
8 Raju DG - MScAgr,PhD *Associate Director* virology
9 Raju KS - MScAgr *Plant Pathologist* microbiology
10 Rao GK - MScAgr *Plant Pathologist* sorghum pathology
11 Rao GV - MScAgr,PhD *Plant Pathologist* plant bacteriology
12 Rao HSN - MScAgr,PhD microbiology
13 Rao KC - MScAgr *Associate Professor* sorghum pathology
14 Rao NG - MScAgr,PhD *Associate Professor* virology
15 Rao PA - MScAgr,PhD *Plant Pathologist* betelvine pathology
16 Ratnan CV - MScAgr,PhD *Plant Pathologist* rice fungal diseases
17 Reddy AGR - MScAgr,PhD *Associate Professor* microbiology
18 Reddy KR - MScAgr,PhD *Professor* seed pathology
19 Reddy KS - MScAgr,PhD *Associate Professor* virology
20 Reddy MS - MScAgr,PhD *Principal and Professor* mycology
21 Reddy TB - MScAgr,PhD *Plant Pathologist* fruit crop pathology
22 Sankarlingam S - MScAgr *Plant Pathologist* maize pathology
23 Satyanarayana Y - MScAgr *Plant Pathologist* sugarcane pathology
24 Sitaramaiah K - MScAgr,PhD *Associate Professor* plant nematodes

2246 Department of Entomology

1 Ahmed A - MScAgr,PhD *Professor* rice nematology and economic entomology
2 Ayyanna T - MScAgr,PhD *Entomologist* entomology
3 Azam KM - MScAgr,PhD *Professor* ecological and pest management
4 Aziz SA - MScAgr,PhD *Entomologist* biological control
5 Krishna JG - MScAgr,PhD *Professor* insect toxicology
6 Kumar T - MScAgr,PhD *Entomolgist* insect pest management and maize
7 Naidu KPS - MScAgr *Associate Professor* economic entomology
8 Paul MD - MScAgr *Entomologist* sorghum
9 Raitra KC - MSc,PhD *Associate Professor* fisheries
10 Raju NBT - MScAgr,PhD *Associate Professor* nematology
11 Rao BN - MScAgr,PhD *Entomologist* residue analysis
12 Rao BS - MScAgr *Associate Professor* nematology and economic entomology
13 Rao CS - MScAgr,PhD *Associate Professor* insect toxicology and grain entomology
14 Rao DM - MScAgr chillies and cotton

15 Rao DVS - MScAgr *Entomologist* groundnut
16 Rao KT - MScAgr,PhD *Entomologist* nematology and pulses
17 Rao NH - MScAgr,PhD *Entomologist* entomology
18 Rao PK - MScAgr,PhD *Professor* insect taxonomy
19 Rao PR - MScAgr,PhD *Entomologist* rice
20 Rao PS - MScAgr,PhD *Entomologist* toxicology
21 Rao SV - MScAgr *Entomologist* chillies and economic entomology
22 Rao VLVP - MScAgr,PhD *Entomologist* rice and sugarcane
23 Reddy AS - MScAgr,PhD *Senior Scientist* cotton
24 Reddy DD - MScAgr,PhD *Associate Professor* nematology and oilseeds
25 Reddy GPV - MSc,PhD *Associate Professor* insect physiology and toxicology
26 Subbaratnam GV - MScAgr,PhD *Associate Professor* economic entomology
27 Zaheeruddin SM - MScAgr,PhD *Entomolgist* sugarcane and rice

2247 Department of Plant Physiology

1 Murthy PS - MScAgr,PhD *Plant Physiologist* botany
2 Murthy SR - MScAgr *Associate Professor* horsegram
3 Narayana V - MScAgr,PhD *Plant Physiologist* oilseeds plant physiology
4 Narayanan A - MScAgr,PhD *Professor* pigeonpea physiology
5 Raj AS - MScAgr,PhD *Professor* millet physiology
6 Ramaiah B - MScAgr,PhD *Plant Physiologist* sugarcane
7 Rao BR - MScAgr *Specialist* sugarcane
8 Rao DVM - MScAgr,PhD *Plant Physiologist* cotton plant physiology
9 Rao GVH - MScAgr,PhD *Associate Professor* millet plant physiology
10 Rao IVS - MScAgr,PhD *Associate Professor* groundnut plant physiology
11 Rao JS - MScAgr,PhD *Associate Professor* wheat plant physiology
12 Rao LM - MScAgr *Associate Professor* agril botany groundnut
13 Rao PH *Associate IARI Specialist* sugarcane
14 Reddy PR - MScAgr,PhD *Associate Director of Research* groundnut physiology
15 Sarma PS - MScAgr,PhD *Associate Professor* soyabean plant physiology
16 Singh BG - MScAgr *Associate Professor* mungbean plant physiology

2248 Department of Soil Science and Agricultural Chemistry

1 Mohankumar KC - MScAgr *Microbiologist* microbiology
2 Padmanabham M - MScAgr,PhD *Soil Scientist* micronutrients
3 Pillai RN - MScAgr,PhD *Professor* soil microbiology
4 Prasad AR - MScAgr,PhD *Soil Physicist* soil fertility
5 Raju AS - MScAgr,PhD *Associate Professor* soil fertility
6 Raju KV - MScAgr,PhD *Chief Scientist* micronutrients
7 Raman Mrs S - MSc *Analytical Chemist* physical chemistry
8 Ramesham M - MScAgr *Soil Scientist* soil fertility
9 Rao AP - MScAgr,PhD *Soil Scientist* soil chemistry
10 Rao IVS - MScAgr,PhD *Associate Director* micronutrients
11 Rao MS - MScAgr,PhD *Soil Physicist* soil physics
12 Rao MVP - MScAgr,PhD *Senior Scientist* radioisotopes
13 Reddy DS - MScAgr,PhD *Associate Professor* soil physics
14 Reddy KC - MScAgr,PhD *Soil Scientist* soil fertility
15 Reddy KS - MScAgr *Associate Professor* soil fertility
16 Santaram MV - MScAgr,PhD *Professor* soil microbiology
17 Sathe A - MScAgr,PhD *Soil Chemist* radioisotopes
18 Sriramulu A - MScAgr *Soil Scientist* soil physics
19 Subbaiah VV - MScAgr *Soil Scientist* micronutrients

2249 Department of Agricultural Engineering

1 Rao PVN - DMLT,MS,PhD *Senior Scientist* farmpower machinery
2 SivaRao KSVV - BTech *Agricultural Engineer* soilwater engineering
3 Sukumaran CR - DIIT,PhD *Associate Professor* agricultural processing

2250 Department of Agricultural Economics

1 Chowdhary KR - MScAgr,PhD *Agricultural Economist* resource economics
2 Nagabhushanam TDG - MScAgr,PhD *Professor* resource economics
3 Naidu MR - MScAgr,PhD *Agricultural Economist* agricultural finance
4 Parthasarathy PB - MA,PhD *Professor* resource economics
5 Raju VT - MScAgr,PhD *Professor* resource economics
6 Rao AN - MScAgr,PhD *Associate Professor* agricultural marketing
7 Rao AP - MScAgr *Agricultural Economist* agricultural finance
8 Reddy BS - MScAgr,PhD *Agricultural Economist* agricultural marketing
9 Reddy TG - MScAgr,PhD *Associate Professor* resource economics
10 Singh S - MScAgr,PhD *Agricultural Economist* agricultural finance
11 Sitharaman S - MScAgr,PhD *Associate Professor* agricultural finance
12 Swamy GM - MScAgr,PhD *Agricultural Economist* agricultural finance
13 Tej Bahadur - MA,PhD *Associate Professor* regional economics

2251 Department of Extension

1 Reddy KJ - MScAgr,PhD *Professor* rural sociology
2 Rao GN - PhD *Professor* experiments designs
3 Reddy NY - MScAgr,MS,PhD *Professor* agricultural extension (audio-visual aids)
4 Reddy SV - MScAgr,PhD *Professor* agricultural extension and communication

5 Satyanarayana C - MScAgr,PhD *Professor* agricultural extension (communication)

2252 Department of Statistics and Mathematics

1 Kulkarni BS - PhD *Associate Professor* applied statistics

2253 Faculty of Home Science

2254 Department of Foods and Nutrition

1 Geervani P - MSc,PhD *Professor* nutrition
2 Khader Mrs V - MSc,PhD *Associate Professor* nutrition
3 Rao KC - MSc,PhD *Professor* nutrition
4 Reddy Mrs MU - MSc,PhD *Associate Professor* nutrition
5 Sarojini Mrs G - MSc *Associate Professor* nutrition
6 Shard Mrs D - MSc,PhD *Associate Professor* nutrition
7 Vimala Mrs V - MSc,PhD *Associate Professor* nutrition

2255 Department of Textiles and Clothing

1 Jacob Mrs M - MSc *Associate Professor* textiles
2 Vatsala Mrs R - MSc,PhD *Professor* textiles

2256 Department of Physiology and Microbiology

1 Mani MR - MSc zoology

2257 Faculty of Veterinary Science

Rajendranagar, Hyderabad, Andhra Pradesh, 500 030
Cables: AGRIVERSITY.

2258 Department of Anatomy

1 Hafeezuddin N - MVSc *Professor* anatomy
2 Singh UP - MVsc,PhD,FRVCS *Professor* anatomy

2259 Department of Physiology

1 Murthy ASN - MVSc,PhD *Research Officer* animal physiology
2 Prasad A - MVSc,PhD *Associate Professor* animal physiology
3 Reddy JN - MVSc,PhD *Associate Professor* animal physiology
4 Reddy KK - MVSc,PhD *Associate Professor* animal physiology
5 Reddy RR - MSc,PhD *Professor* animal physiology
6 Yadgiri B - MVSc *Professor* animal physiology

2260 Department of Pharmacology

1 Janardhan A - MVSc,PhD *Associate Professor* pharmacology
2 Murthy MK - MVSc *Associate Professor* pharmacology
3 Sivaramakrishnaiah K - MVSc *Associate Professor* pharmacology

2261 Department of Veterinary Pathology

1 Mohender M - BVSc,MS *Associate Professor* veterinary pathology
2 Mohiuddin KM - MVSc,PhD *Associate Professor* verterinary pathology
3 Murthy ASN - MVSc,PhD *Associate Professor* veterinary pathology
4 Naidu NRG - MSc,PhD *Associate Professor* veterinary pathology
5 Rao PR - MVSc,PhD,FRVCS *Professor* veterinary pathology
6 Reddy MV - MVSc *Associate Professor* veterinary pathology
7 Yadgirkar G - MVSc,FRVCS *Associate Professor* veterinary pathology

2262 Department of Parasitology

1 Muralidharan SRG - MVSc,PhD *Associate Professor* parasitology
2 Rao BV - MSc,PhD *Professor* parasitology
3 Reddy KR - MVSc,PhD *Associate Professor* parasitology
4 Satyanaraya Charyulu N - MVSc,PhD *Associate Professor* parasitology

2263 Department of Microbiology

1 Babu TS - MSc,PhD *Associate Professor* veterinary microbiology
2 Rao TS - MVSc,PhD *Associate Professor* veterinary microbiology
3 Sarma BJ - MSc,PhD *Associate Professor* veterinary microbiology

2264 Department of Medicine

1 Chaudhary PC - MSc,PhD *Professor* veterinary medicine
2 Gaffar A - MSc,PhD *Professor* veterinary medicine
3 Karunanidhi PS - MVSc,PhD *Associate Professor* veterinary medicine
4 Raghavan RS - MS *Professor* veterinary medicine
5 Suryanarayana C - MVSc,PhD *Associate Professor* veterinary medicine

2265 Department of Surgery

1 Ramakrishna O - MVSc,PhD,FRVAE *Professor* surgery
2 Rao RLN - MVSc,PhD *Professor* surgery

2266 Department of Gynaecology and Animal Reproduction

1 Rao AR - MS,PhD *Dean of Post Graduate Studies* animal reproduction
2 Rao PN - MVSc,PhD *Professor* reproductive physiology
3 Reddy KK - MVSc,PhD *Associate Professor* gynaecology
4 Reddy VS - MVSc *Associate Professor* gynaecology

2267 Department of Genetics and Animal Breeding

1 Jayaramakrishna V - MVSc,PhD *Professor* genetics and animal breeding
2 Krishnamacharyulu E - MVSc,PhD *Associate Professor* genetics and animal breeding
3 Murthy AS - MVSc,PhD *Associate Professor* genetics and animal breeding
4 Rao H - MVSc,PhD *Associate Professor* genetics and animal breeding
5 Rao VD - MVSc,PhD *Associate Professor* genetics and animal breeding
6 Reddy CE - MSc,PhD *Associate Professor* genetics and animal
7 Reddy KK - MVSc,DipStat,PhD *Associate Professor* genetics and animal breeding
8 Reddy VRC - MVSc,PhD *Associate Professor* genetics and animal breeding
9 Sarma PLN - MVSc,PhD *Associate Professor* genetics and animal breeding

2268 Department of Meat Science and Technology

1 Reddy GR - MSc,PhD *Associate Professor* meat science
2 Reddy MS - MSc,PhD *Associate Professor* meat science
3 Varadarajulu P - MSc,PhD *Professor* meat science

2269 Department of Poultry Science

1 Rao PV - MVSc,PhD *Professor* poultry science
2 Reddy CV - MS,PhD *Associate Professor* poultry science
3 Reddy DN - MS,PhD *Associate Professor* poultry science
4 Reddy VR - MSc,PhD *Associate Professor* poultry science
5 Reddy VR - MSc,DipStat,PhD *Associate Professor* poultry science
6 Siddiqui SM - MS *Professor* poultry science
7 Subbarayudu D - MSc *Professor* poultry science

2270 Department of Dairy Science

1 Jairam BT - MSc,PhD *Associate Professor* animal breeding
2 Naidu KN - MVSc,PhD *Associate Professor* animal breeding
3 Rao GN - MVSc,PhD *Associate Professor* dairy science
4 Rao MR - MVSc,PhD *Professor* physiology
5 Sastry PM - MVSc *Professor* dairy science

2271 Department of Animal Nutrition

1 Krishna N - MVSc,PhD *Associate Professor* animal nutrition
2 Mudaliar ASR - MS *Professor* genetics and wool grading
3 Naidu MM - MVSc,PhD *Associate Professor* animal nutrition
4 Prasad DA - MVSc,PhD *Professor* animal nutrition
5 Raghavan GV - MVSc,PhD *Professor* animal breeding
6 Rao AM - MVSc,PhD *Associate Professor* biochemistry
7 Rao VGK - MVSc,PhD *Associate Professor* animal nutrition
8 Rao ZP - MSc,DASc *Associate Professor* animal nutrition
9 Sivayya K - MVSc,PhD *Associate Professor* animal nutrition

2272 Department of Food and Fodder Technology

1 Reddy BS - MVSc,PhD *Associate Professor* animal nutrition
2 Reddy MJ - MVSc,PhD *Professor* animal nutrition
3 Reddy RR - MVSc,PhD *Associate Professor* animal nutrition

2273 Department of Fishery Science

1 Prasad GS - MSc,PhD *Associate Professor* fishery science
2 Rao KG - MSc,PhD *Associate Professor* fishery science
3 Rao MB - MSc,PhD *Professor* fishery science
4 Reddy KK - MSc,PhD *Associate Professor* entomology
5 Reddy OR - MSc,PhD,DFSc *Associate Professor* fishery science

2274 Birsa Agricultural University

Ranchi, Bihar, 834 006
Cables: AGRIVERSITY.

1 Kundra JC *Vice Chancellor*

2275 Faculty of Veterinary Science and Animal Husbandry

1 Mishra HR - MS,PhD *Dean*

2276 Teaching and Research Department

2277 Anatomy
1 Sinha RD - MS,PhD *Head*
2 Prasad G - MScAH,PhD,FRVAC *Associate Professor*
3 Prasad J - PhD *Associate Professor*

2278 Animal Breeding and Genetics
1 Mishra HR - MS,PhD *Head*
2 Pandey RS - MVSc,PhD *Associate Professor*
3 Singh RL - MScAH,PhD *Associate Professor*

2279 Animal Nutrition
1 Gupta BS - MVSc,PhD
2 Sinha RRP - MScAH,PhD *Associate Professor*
3 Srivastava JP - MVSc,PhD *Senior Scientist*

2280 Animal Physiology
1 Pandey JN - MVSc,PhD
2 Prasad SP - MVSc,PhD *Associate Professor-cum-Senior Scientist*

2281 Animal Production
1 Singh R - MScAH,PhD *Associate Professor*

2282 Apiculture and Pisciculture
1 Shamsuddin M - MSc,PhD *Head*

2283 Gynaecology
1 Singh BK - MVSc,DrMed *Vet*
2 Sinha AK - MVSc,PhD *Associate Professor*

2284 Medicine
1 Verma BB - MPVM,PhD
2 Sinha RP - MVSc,DSc *Associate Professor*
3 Thakur DK - MVSc,PhD *Associate Professor*

2285 Microbiology
1 Sahay BN - MVSc,PhD
2 Soman JP - MVSc,PhD
3 Tiwary BK - MSc,PhD *Associate Professor*

2286 Parasitology
1 Ansari MZ - MVSc,PhD *Associate Professor*
2 Basu A - MVSc,PhD *Associate Professor*

2287 Pathology
1 Chauhan MVS - MS,PhD
2 Jha GJ - MVSc,PhD *Associate Professor*

2288 Pharmacology
1 Banerjee NC - MS,PhD *Head*
2 Singh RCP - MVSc,PhD *Associate Professor*
3 Yadav KP - MVSc,PhD *Associate Professor*

2289 Veterinary Public Health and Epidemiology
1 Narayan KG - CSc

2290 Veterinary Surgery
1 Khan AA - MVSc,PhD,FRVA
2 Sahay PN - MVSc,PhD *Associate Professor*

2291 Gujarat Agricultural University

Sardar Krushinagar, District Banaskantha, Gujarat, 385506

1 Ahlawat RPS - MSc,PhD *Deputy Director of Research* agronomy
2 Janakiraman K - MS,PhD *Director of Research and Dean PG Studies* reproductive biology
3 Pandya DN - MSc,PhD *I/c Director* extension education
4 Parthasarthy R - MA,IAS *Vice Chancellor*
5 Patel CB - MSc,PhD *I/c Deputy Director of Extension Education* entomology
6 Patel JC - MSc,PhD *Deputy Director of Research* agronomy
7 Patel RJ - MSc *Registrar* agricultural economics
8 Patel RS - MSc,PhD *I/c Deputy Director of Extension Education* plant pathology
9 Pateni SV - MSc *I/c Deputy Director of Extension Education* extension education
10 Sheth RD - MSc,PhD *Deputy Director of Research* agricultural microbiology
11 Vyas AP - MSc,PhD *Deputy Director of Research* animal breeding and genetics
12 Vyas HN - MSc,PhD *Deputy Director of Research* entomology
13 Walnunjkar WG - PhD *Deputy Director of Research* agronomy

2292 Faculty of Agriculture
1 Canzaria MV - MSc,PhD *Principal* soil science
2 Mehta HC - MSc,PhD *Principal* soil science
3 Pandya DN - MSc,PhD *Principal* extension education
4 Patel RC - MSc,PhD *I/c Principal* entomology

2293 Department of Agronomy
1 Parmar MT - MSc,PhD *Professor of Agronomy* agronomy
2 Patel BA - MSc *Professor of Agronomy* agronomy
3 Patel RB - MSc,PhD *Professor of Agronomy* agronomy

2294 Department of Plant Breeding
1 Amin VC - MSc,PhD *Professor of Botany* plant breeding
2 Patel SA - MSc,PhD *Professor of Botany* plant breeding
3 Shukla PT - MSc,PhD *Professor of Botany* plant breeding

2295 Agriculture Chemistry
1 Patel CA - MSc,PhD *Professor of Agricultural Chemistry* agricultural chemistry
2 Patel KA - MSc,PhD *Professor of Agricultural Chemistry* agricultural chemistry
3 Patel MS - MSc,PhD *Professor of Agricultural Chemistry* agricultural chemistry
4 Patel NK - MSc,PhD *Professor of Agricultural Chemistry* agricultural chemistry

2296 Agricultural Engineering
1 Mallik M - PhD *Professor of Soil and Water Management* agricultural engineering
2 Panda RN - MTech,PhD *Professor of Agricultural Engineering* agricultural engineering (soil science)
3 Tripathi SK - PhD *Professor of Rural Engineering* agricultural engineering

2297 Entomology
1 Nittal VP - MSc,PhD *Professor* entomology
2 Patel RC - MSc,PhD *Professor* entomology
3 Shah AH - MSc,PhD *Professor* entomology

2298 Extension Education
1 Halyal KG - MSc,PhD *Professor* extension education
2 Patel BT - MSc,PhD *Professor* extension education
3 Waghmare SK - MSc,PhD *Professor* extension education

2299 Horticulture
1 Bhatt JP - MSc,PhD *Professor* horticulture
2 Chundawat BS - MSc,PhD *Professor* horticulture
3 Katrodia JS - MSc,PhD *Professor* horticulture

2300 Plant Pathology
1 Dange SRS - MSc,PhD *Professor* plant pathology
2 Patel DJ - MSc,PhD *Professor* plant pathology
3 Vaisnay MU - MSc,PhD *Professor* plant pathology
4 Vala DG - MSc,PhD *Professor* plant pathology

2301 Statistics
1 Vaishnav MR - MSc,PhD *Professor* agricultural statistics

2302 Faculty of Veterinary Science
1 Jhala VM - MVSc,PhD *Principal* bacteriology
2 Patel MR - MSc,PhD,FRVCS *Principal* surgery

2303 Veterinary Anatomy
1 Panchmukhi BG - MScVet,PhD *Professor of Anatomy* veterinary anatomy
2 Vyas KN - MScVet,PhD *Professor of Anatomy* veterinary anatomy

2304 Animal Genetics and Breeding
1 Dhandha OP - MSc,PhD *Professor of Animal Genetics and Breeding* animal production

2 Patel MN - MVSc,PhD *Professor of Animal Genetics and Breeding* animal genetics and breeding
3 Shukla RK - MVSc,PhD *Professor of Animal Genetics and Breeding* animal genetics and breeding

2305 Animal Nutrition

1 Panda MB - MSc,PhD *Professor of Animal Nutrition* bio-chemistry/animal nutrition

2306 Animal Husbandry

1 Dave AD - MVSc,PhD *Professor of Animal Husbandry* livestock production and management

2307 Animal Bacteriology

1 Dholakia PN - MVSc,PhD,FRVCS *Professor of Bacteriology* veterinary bacteriology

2308 Clinic

1 Oja SC - MVSc,PhD *Professor of Clinic* veterinary surgery

2309 Gynaecology

1 Dugvakar YG - MVSc,PhD *Professor of Gynaecology* reproduction biology
2 Kodagali SB - MVSc,PhD,FRVCS *Professor of Gynaecology* veterinary gynaecology

2310 Medicine

1 Dave MR - MVSc,PhD *Professor of Veterinary Medicine* veterinary medicine

2311 Parasitology

1 Avsthi BL - MVSc,PhD *Professor of Parasitology* veterinary pathology
2 Pathakar DK - MVSc,PhD *Professor of Parasitology* veterinary parasitology

2312 Veterinary Pathology

1 Hiranjal DD - MVSc,PhD *Professor of Veterinary Pathology* veterinary pathology
2 Kaul PL - MVSc,PhD *Professor of Veterinary Pathology* veterinary pathology

2313 Veterinary Physiology

1 Prabhu GA - MVSc,PhD *Professor of Veterinary Physiology* veterinary bio-chemistry

2314 Livestock Production and Management

1 Shukla KP - MVSc,PhD *Professor of Livestock Production and Management* reproductive biology

2315 Veterinary Surgery

1 Tadkod DM - MVSc,PhD *Professor of Surgery* surgery

2316 Faculty of Dairy Science

1 Verma IS - PhD *Principal* dairy technology

2317 Dairy Economics

1 Patel RD - MA,PhD *Professor* economics

2318 Dairy Chemistry

1 Sharma RS - MSc,PhD *Professor* dairy chemistry

2319 Process Engineering

1 Siripurapu SCB - DT,PhD *Professor of Process Engineering* food-production engineering

2320 Dairy Engineering

1 Shah US - ME,PhD *Professor of Dairy Engineering* mechanical engineering

2321 Dairy Micro Biology

1 Dave JM - MSc,PhD *Professor of Dairy Microbiology* dairy bacteriology

2322 Dairy Technology

1 Upadhyay KG - MSc,PhD *Professor of Dairy Technology* dairy technology

2323 Faculty of Home Science

1 Shah Miss JG - ME,PhD *Professor of Home Science* home science education

2324 Extension Education

1 Patel HL - MSc,PhD *Professor of Extension* sociology
2 Patel HN - MS *Principal* extension education
3 Patel NR - MSc,PhD *Training Organiser* agronomy
4 Patel ZG - MSc,PhD *Training Organiser* agronomy

2325 Research Scientists (Crops)

1 Chakraborty MK - MSc,PhD *Scientist* biochemistry
2 Dalal KC - MSc,PhD *Scientist* cytogenetics;plant breeding
3 Dangarwala RT - MSc,PhD *Scientist* micronutrient;agricultural chemistry
4 Dave MR - MSc,PhD *Scientist* millet breeding
5 Desai KB - MSc,PhD *Scientist* sorghum;agricultural botany
6 Desai ND - MSc,PhD *Scientist* paddy breeding
7 Goswami KP - MSc,PhD *National Fellow* soil science;agronomy
8 Jadon BS - MSc,PhD *Scientist* wheat breeding
9 Jaisani BG - MSc,PhD *Scientist* tobacco breeding
10 Maheta VM - MVSc,PhD *Scientist* reproductive biology
11 Maliwal GL - MSc,PhD *Scientist* agricultural chemistry
12 Mehta KG - MSc,PhD *Scientist* plant physiology
13 Mehta NP - MSc,PhD *Scientist* cotton breeding
14 Mehta PM - MSc,PhD *Scientist* soil science;agricultural chemistry
15 Patel BG - MSc *Scientist* oilseed breeding
16 Patel BN - MSc,PhD *Scientist* tobacco;plant pathology
17 Patel CL - MSc,PhD *Scientist* soil science;agricultural chemistry
18 Patel JR - MSc,PhD *Scientist* forage breeding
19 Patel MS - MSc,PhD *Scientist* plant breeding
20 Radadia NS - MSc,PhD *Scientist* livestock production and management
21 Raman S - MSc,PhD *Scientist* soil and water management;agricultural chemistry
22 Shah RM - MSc,PhD *Scientist* pulses breeding
23 Shukla PC - MVSc,PhD *Scientist* animal nutrition
24 Thakkar LG - MSc *Scientist* agricultural engineering

2326 Harayana Agricultural University

Hissar, Haryana
Tel: 2640. Cables: AGRIVERSITY.

1 Kataria LD *Vice Chancellor*

2327 Directorate of Research

1 Chamola SD - PhD *Scientist* economics
2 Chaudhary BD - PhD *Scientist* plant breeding
3 Chaudhary FS - PhD *Scientist* statistics
4 Hooda RS - PhD *Scientist* crops
5 Pandey RN - PhD *Agricultural Economist* agril economics
6 Singh M - PhD *Additional Director of Research*
7 Tyagi RPS - PhD *Director of Research*

2328 College of Agriculture

Hissar, Haryana
Tel: 2640. Cables: AGRIVERSITY.

1 Chaudhary MS - PhD *Associate Dean*
2 Grover RK - PhD *Dean*

2329 College of Basic Science and Humanities

1 Chaudhary JB - PhD *Dean* genetics

2330 Department of Agronomy

1 Bhan VM - PhD *Professor* weed science
2 Faroda AS - PhD *Professor* agronomy
3 Malik BS - PhD *Professor*

2331 Department of Soils

1 Agarwal MC - PhD *Professor* agricultural engineering
2 Aggarwal RP - PhD *Professor*
3 Chaudhary ML - PhD *Professor*
4 Karwasra SPS - PhD *Professor*
5 Khanna SS - PhD *Professor*
6 Manchanda HR - PhD *Professor*
7 Nath J - PhD *Professor*
8 Poonia SR - PhD *Professor*
9 Rajpal - PhD *Professor*

2332 Department of Microbiology

1 Dadarwal KR - PhD *Professor*
2 Lakshaminarayana - PhD *Professor*
3 Mishra MM - PhD *Professor*
4 Tauro P - PhD *Professor*

2333 Department of Chemistry and Biochemistry

1 Singh R - PhD *Professor*
2 Dhindsa KS - PhD *Professor*
3 Kathpal TS - PhD *Pesticide chemist*
4 Sheoran IS - PhD *Plant Physiologist*
5 Wagle DS - PhD *Professor*

2334 Department of Horticulture

1 Chauhan KS - PhD *Professor*
2 Daulta BS - PhD *Horticulturist*
3 Godara NR - PhD *Horticulturist*
4 Gupta OP - PhD *Associate Professor* post harvest technology
5 Gupta PC - PhD *Plant Pathologist*
6 Singh D - PhD *Extension Specialist*
7 Singh R - PhD *Horticulturist*
8 Yamdagni R - PhD *Professor*

2335 Department of Plant Breeding

1 Mor BR - PhD *Professor* cotton breeding
2 Arora SK - PhD *Senior Bio-Chemist* plant breeding
3 Arya AS - PhD *Associate Professor* statistics
4 Beniwal CR - PhD *Associate Professor* plant breeding
5 Bhola AL - PhD *Associate Professor* agronomy
6 Bishnoi LK - PhD *Associate Professor* agronomy
7 Chaudhary RK - PhD *Associate Professor* plant breeding
8 Chauhan MS - PhD *Associate Professor* plant pathology
9 Chauhan Mrs R - PhD *Associate Professor* entomology
10 Dahiya BS - PhD *Associate Professor* plant breeding
11 Dang JK - PhD *Associate Professor* plant pathology
12 Dass B - PhD *Associate Professor* chemistry
13 Gupta KR - PhD *Associate Professor* plant breeding
14 Gupta SK - PhD *Associate Professor* chemistry-biochemistry
15 Jain DK - PhD *Associate Professor* plant breeding
16 Jalali BL - PhD *Senior Plant Pathologist* plant breeding
17 Kairon MS - PhD *Chief Scientist* cotton breeding
18 Kapur RL - PhD *Senior Breeder* bajra breeding
19 Kaushik CD - PhD *Associate Professor* plant pathology
20 Khurana AD - PhD *Associate Professor* entomology
21 Khurana AL - PhD *Associate Professor* microbiology
22 Khurb SS - PhD *Associate Professor* entomology
23 Kumar P - PhD *Associate Professor* plant breeding
24 Lather BPS - PhD *Associate Professor* plant breeding
25 Lodhi GP - PhD *Senior Breeder* sorghum breeding
26 Luthra YP - PhD *Associate Professor* chemistry
27 Narwal SS - PhD *Associate Professor* agronomy
28 Panwar DVS - PhD *Associate Professor* plant breeding
29 Saharan GS - PhD *Associate Professor* plant pathology
30 Saini ML - PhD *Associate Professor* plant breeding
31 Sharma GD - PhD *Associate Professor* plant breeding
32 Sharma PD - PhD *Associate Professor* entomology
33 Singh D - PhD *Associate Professor* plant breeding
34 Singh D - PhD *Associate Professor* plant breeding
35 Singh DP - PhD *Associate Professor* plant breeding
36 Singh H - PhD *Associate Professor* plant breeding
37 Singh I - PhD *Associate Professor* plant breeding
38 Singh IP - PhD *Associate Professor* plant breeding
39 Singh M - PhD *Associate Professor* agronomy
40 Singh S - PhD *Associate Professor* plant breeding
41 Singh SV - PhD *Senior Plant Pathologist* plant breeding
42 Singh VP - PhD *Senior Scientist* pulses breeding
43 Taneja AD - PhD *Associate Professor* bio-chemistry
44 Thakur DP - PhD *Professor* plant pathology
45 Tomar RPS - PhD *Associate Professor* plant breeding
46 Tomar YS - PhD *Associate Professor* plant breeding
47 Tyagi CS - PhD *Associate Professor* plant breeding
48 Verma OPS - PhD *Associate Professor* agronomy
49 Yunus M - PhD *Associate Professor* plant breeding

2336 Department of Genetics

1 Chowdhury JB - MSc,PhD *Professor*
2 Kumar PR - PhD *Coordinator (Rapeseed and Mustard)* genetics and plant breeding
3 Luthra OP - MSc,PhD *Associate Professor* genetics
4 Maherchandani N - MSc,Assoc.IARI,PhD *Professor* cytogenetics
5 Rana OPS - MSc,PhD *Associate Professor* genetics
6 Sareen PK - MSc,PhD *Associate Professor* genetics
7 Sharma DR - MSc,PhD *Associate Professor* genetics
8 Singh KP - MSc,PhD *Associate Professor* genetics
9 Tripathi ID - MSc,PhD *Associate Professor* genetics
10 Vashishat RK - MSc,PhD *Associate Professor* genetics
11 Yadava JS - PhD *Geneticist* rapeseed and mustard

2337 Department of Plant Pathology

1 Thakur DP - PhD *Professor*
2 Ahuja SC - PhD *Plant Pathologist*
3 Chand JN - PhD *Director of Publication*
4 Chauhan MS - PhD *Plant Pathologist*
5 Dang JK - PhD *Plant Pathologist*
6 Grover RK - PhD *Dean College of Agriculture*
7 Gupta PC - PhD *Plant Pathologist*
8 Jalali BL - PhD *Senior Plant Pathologist*
9 Kaushik CD - PhD *Plant Pathologist*
10 Parashar RD - PhD *Senior Plant Pathologist*
11 Raiu B - PhD *Plant Pathologist*
12 Rishi N - PhD *Plant Pathologist*
13 Saharan GS - PhD *Plant Pathologist*
14 Satyavir - PhD *Senior Plant Pathologist*
15 Srivastava MP - PhD *Extension Specialist*
16 Suhag LS - PhD *Plant Pathologist*
17 Tripathi NN - PhD *Associate Professor*

2338 Department of Entomology

1 Chaudhary JP - PhD *Professor*
2 Chhillar BS - PhD *Entomologist*
3 Chopra NP - PhD *Professor*
4 Dass S - PhD *Extension Specialist*
5 Kushwaha KS - PhD *Entomologist (Rice)*
6 Naresh JS - PhD *Entomologist*
7 Verma AN - PhD *Senior Entomologist*
8 Yadav PR - PhD *Insect Ecologist*

2339 Department of Nematology

1 Bhatti DS - PhD *Professor*
2 Dalal MR - PhD *Associate Professor*
3 Gupta PC - PhD *Associate Professor*

2340 Department of Mathematics and Statistics

1 Srivastava OP - PhD *Professor*
2 Kaushik LS - PhD *Associate Professor* statistics
3 Puri PD - PhD *Associate Professor* statistics
4 Singh U - PhD *Associate Professor* statistics

2341 Department of Agricultural Meteorology

1 Bishnoi OP - MSc,PhD *Associate Professor*

2342 Department of Extension Education

1 Sharma RK - PhD *Professor*
2 Bhatti DS - PhD *Director* nematology
3 Kumar K - PhD *Professor*
4 Nand H - PhD *Associate Professor*
5 Nirwal RS - PhD *Extension Scientist* communication

2343 Department of Agricultural Economics

1 Gangwar AC - PhD *Associate Professor*
2 Gupta DD - PhD *Associate Professor*
3 Nandal DS - PhD *Associate Professor*
4 Panghal BS - PhD *Associate Professor*
5 Singh H - PhD *Associate Professor*
6 Singh IJ - PhD *Professor*

2344 Department of Sociology

1 Sharma ML - PhD *Professor*
2 Dak TM - PhD *Associate Professor* research
3 Punia RK - PhD *Associate Professor* rural sociology

2345 Department of Agricultural Engineering

1 Gupta RSR - MTech *Senior Agricultural Engineer*
2 Jain ML - BTech,MS *Professor*
3 Sharma DN - MSc *Research Engineer*
4 Singh P - M Tech,PhD *Associate Professor*

2346 Department of Foods and Nutrition

1 Chauhan BM - MSc,PhD *Scientist* biochemistry
2 Kapoor AC - MSc,PhD *Associate Professor* nutritional biochemistry
3 Mehtra Mrs U - MSc,PhD *Associate Professor*

2347 Project-cum-Plan Formulation

1 Pandey UK - PhD *Scientist*
2 Singh Z - PhD *Director*

2348 Department of Zoology

1 Mathur RB - MSc,PhD *Associate Professor*
2 Mathur Mrs S - MSc,PhD *Associate Professor*

2349 Krishi Gyan Kendra

1 Bagga RK - PhD *agronomy*
2 Bhaskar VV - PhD *dairy extension*
3 Chaudhary MK - PhD *agricultural economics*
4 Gill PS - PhD *Associate Professor agronomy*
5 Kadian OP - PhD *Coordinator plant pathology*
6 Singh BP - PhD *Coordinator*
7 Singh JP - PhD *Associate Professor*
8 Singh K - PhD *horticulture*
9 Singh KP - PhD *Coordinator*

2350 College of Animal Sciences

1 Yadava IS - PhD *Dean*

2351 Department of Animal Nutrition

1 Bhatia SK - PhD *Scientist ruminant nutrition*
2 Gopal Krishna - PhD *Animal Nutritionist animal nutrition*
3 Gupta PC - PhD *Scientist animal nutrition*
4 Mandokhot VM - PhD *Scientist animal nutrition*
5 Pradhan K - PhD *Professor animal nutrition*
6 Saxena VP - PhD *Associate Professor non-ruminant nutrition*
7 Sharma VP - PhD *Scientist poultry*
8 Singh K - PhD *Extension specialist animal science*
9 Vidyasagar - PhD *Professor animal nutrition*

2352 Department of Livestock Production and Management

1 Singh RA - PhD *Professor*
2 Aggarwal CK - PhD *Associate Professor*
3 Chaudhaury SR - PhD *Associate Professor*
4 Procer RD - PhD *Associate Professor*
5 Sastry NSR - PhD *Associate Professor*
6 Sengupta BP - PhD *Chief Scientist*
7 Singh N - PhD *National Fellow*

2353 Department of Veterinary Pathology

1 Paul Gupta RK - MVSc,PhD *Professor*
2 Bhatia KC - MVSc,PhD *Associate Professor*
3 Kharole MU - MVSc,PhD *Associate Professor*
4 Sadama JR - MVSc,PhD *Animal Pathologist*
5 Verma PC - MVSc,PhD *Nutritional Pathologist*

2354 Department of Veterinary Public Health and Epidemiology

1 Kaushik RK - PhD *Professor*
2 Balami DK - PhD *Associate Professor*
3 Garg DN - PhD *National Fellow*
4 Mahajan SK - PhD *Associate Professor*
5 Mahipal SK - PhD *Associate Professor*
6 Mandokho Mrs U - PhD *Associate Professor*
7 Purohit VD - PhD *Associate Professor*
8 Satija KC - PhD *Associate Professor*
9 Singh J - PhD *Associate Professor*

2355 Department of Veterinary Microbiology

1 Jain NC - PhD *Professor*
2 Ahuja KL - PhD *Epidemiologist*
3 Chaturvedi GC - MVSc,PhD *Associate Professor*
4 Goel MC - PhD *Professor immunology*
5 Kalra SK - PhD *Associate Professor*
6 Monga DP - PhD *Associate Professor*
7 Prasad S - PhD *Senior Virologist*
8 Srivastava RN - PhD *Scientist virology*
9 Tewari SG - PhD *Scientist*

2356 Department of Animal Breeding

1 Balaine DS - PhD *Professor*
2 Chopra SC - PhD *Professor*
3 Goswami RP - PhD *Associate Professor*
4 Kanaujia AS - PhD *Associate Professor*
5 Khanna RNS - PhD *Associate Professor*
6 Kumar J - PhD *Professor*
7 Rathi SS - PhD *Associate Professor*
8 Sharma JS - PhD *Professor*
9 Sharma RK - PhD *Associate Professor*
10 Singh B - PhD *Associate Professor*

2357 Department of Surgery and Radiology

1 Nigam JM - PhD *Professor*
2 Krishnamurthy D - PhD *Scientist*

2358 Department of Veterinary Medicine

1 Sharma RD - MSc,PhD *Chief Scientist*
2 Bansal SR - MVSc,PhD *Extension Specialist*
3 Bhardwaj RM - MVSc,PhD *Associate Professor*
4 Gupta SL - MVSc,PhD *Scientist*
5 Kapur MP - MVSc,PhD *Research Officer*
6 Shanti Sarup - MVSc,PhD *Associate Professor*
7 Singh RP - MVSc,PhD *Professor*

2359 Department of Anatomy and Histology

1 Dhingra LD - PhD *Professor*
2 Nagpal SK - PhD *Associate Professor*
3 Sharma DN - PhD *Associate Professor*
4 Singh Y - PhD *Professor*

2360 Department of Veterinary Parasitology

1 Banerjee DP - MVSc,PhD *Professor*
2 Bhatnagar PK - MVSc,PhD *Associate Professor*
3 Chhabra MB - MVSc,PhD *Associate Professor*
4 Gupta RP - MVSc,PhD *Senior Research Officer*
5 Ruprah NS - MVSC,PhD *Professor*

2361 Department of Veterinary Pharmacology

1 Garg BD - MVSc,PhD *Professor*
2 Rana RD - MVSc,PhD *Associate Professor*
3 Uppel RP - MVSc,PhD *Associate Professor*

2362 Department of Animal Products Technology

1 Panda PC - MVSc,PhD *Professor*
2 Kapoor CM - MSc,PhD *Associate Professor*

2363 Department of Gynaecology and Obstetrics

1 Khar SK - DSc *Research Officer*
2 Suresh Chandra - PhD *Associate Professor*
3 Verma SK - PhD *Associate Professor*

2364 Department of Veterinary Public Health

1 Chandiamani NK - PhD *Professor*

2365 Himachal Pradesh Krishi Vishwa Vidyalaya

Palampur, Himachal Pradesh
Tel: 176062. Cables: HIMKRISHI.

2366 Department of Agronomy and Agrometeorology

1 Singh CM - PhD *Professor*
2 Sharma HL - PhD *Associate Professor*
3 Singh B - PhD *Associate Professor*
4 Thakur RC - PhD *Associate Professor*

2367 Department of Soil Science and Water Management

1 Minhas RS - PhD *Professor*
2 Acharya CL - PhD *Associate Professor*
3 Bishnoi SK - PhD *Soil Chemist*
4 Chokor IS - PhD *Agronomist*
5 Dohroo RK - MSc *Agriculture engineer*
6 Mahajan KK - PhD *Extension Specialist*
7 Sankhyan SD - PhD *Associate Professor*
8 Sharma PK - PhD *Associate Professor*
9 Verma SD - PhD *Chief Scientist*
10 Verma TS - PhD *Scientist*

2368 Department of Plant Breeding

1 Sethi GS - PhD *Professor*
2 Gupta VP - PhD *Associate Professor*
3 Katoch DC - PhD *Plant Breeder*

2369 Department of Agriculture Economics

1 Moorti TV - PhD *Professor*
2 Mehta P - PhD *Associate Professor*
3 Vashist GD - PhD *Associate Professor*

2370 Department of Home Science

1 Kalia M - PhD *Professor*

2371 Department of Animal Science

1 Narang MP - PhD *Professor*
2 Bhowmik KBD - DSc *Professor*
3 Katoch BS - PhD *Professor*
4 Manuja NK - PhD *Associate Professor*

2372 Department of Mycology and Plant Pathology

1 Singh BM - PhD *Professor*

2373 Department of Entomology and Apiculture

1 Kashya NP - PhD *Associate Professor*

2374 Department of Vegetable Crops and Floriculture

1 Sharma JK - PhD *Associate Professor*

2375 Department of Pomology and Fruit Technology

1 Mishra SK - PhD *Associate Professor*

2376 Department of Basic Science

1 Sharma OP - PhD *Professor*
2 Bansal GL - PhD *Associate Professor* plant physiology
3 Chachra SP - PhD *Associate Professor* physics
4 Kaistha BL - PhD *Associate Professor* statistics and mathematics
5 Mahajan NC - PhD *Associate Professor* chemistry
6 Sharma SK - PhD *Associate Professor* genetics

2377 Regional Research Station, HPKVV

1 Dawa T - PhD *Senior Plant Breeder* plant breeding
2 Desh Raj - PhD *Scientist* entomology
3 Kalia NR - PhD *Plant Breeder* plant breeding
4 Kanwar BB - PhD *Scientist* soil science
5 Sharia JJ - PhD *Scientist* agronomy
6 Sharma PD - PhD *Scientist* soil science
7 Sharma RK - PhD *Scientist* agricultural economics
8 Sharma SC - PhD *Wheat Breeder* wheat breeding
9 Sood AK - PhD *Scientist* plant pathology
10 Thakur JR - PhD *Scientist* plant breeding

2378 Regional Research Station, HPKVV

1 Bhalla SK - PhD *Chief Scientist*

2379 Regional Research Station
PO Box 262, Shalimar, Dhaulakaun, Sirmaur, 190001

1 Gupta RD - PhD *Extension Specialist* soil science
2 Kalia V - PhD *Plant Breeder* plant breeding
3 Sharma PK - PhD *Scientist* soil science
4 Singh A - PhD *Scientist* agricultural economics
5 Singh LN - PhD *Chief Scientist*
6 Singh M - PhD *Entomologist* entomology
7 Verma BR - PhD *Plant Pathologist* plant pathology

2380 Rice Research Sub-Station
Malan, District Kangra, Himachal Pradesh, 190001

1 Bassi K - PhD *Associate Professor* agronomy
2 Katoch PC - PhD *Rice Breeder* plant breeding
3 Kaul BK - PhD *Entomologist* entomology
4 Sharma OP - PhD *Scientist* plant pathology

2381 Research Sub-Station
Akrot, District Una, Himachal Pradesh, 190001

1 Verma SP - PhD *Agronomist* agronomy

2382 Oil Research Sub-Station
Akrot, Kangra, Himachal Pradesh, 190001

1 Rana ND - PhD *Plant Breeder* plant breeding

2383 Crop Research Station
Berthin, District Bilaspur, Himachal Pradesh, 190001

1 Sood BC - PhD *Plant Breeder* plant breeding

2384 Crop Research Sub-Station
Sundernagar, District Mandi, Himachal Pradesh, 190001

1 Sood BR - PhD *Scientist* agronomy

2385 Seed Production Unit, HPKVV
Palampur, District Mandi, Himachal Pradesh, 190001

1 Singh HB - PhD *Seed Production Scientist* plant breeding

2386 Directorate of Research, HPKVV
Palampur, District Mandi, Himachal Pradesh, 190001

1 Sharma CR - PhD *Microbiologist* microbiology
2 Thakur DS - PhD *Deputy Director* agricultural economics

2387 Sher-E-Kashmir University of Agricultural Sciences and Technology
PO Box 262, Shalimar, Shrinagar, Jammu and Kashmir, 190001

1 Ahmad A - BVSc,MS,PhD *Vice Chancellor*
2 Dar KM - MScAgr,PhD *Deputy Director and Senior Scientist* extension education
3 Dar MA - BSc Agr,MScAgr *Director* extension education
4 Gafoor MA - MVSc,PhD *Dean Faculty of Veterinary Sciences and Animal Husbandry*
5 Gupta MP - MScAgr,PhD *Chief Training Organiser and Chief Scientist* extension education
6 Parimoo MM - MSc *Deputy Director* information
7 Sharma HC - MScAgr,PhD *Director* research
8 Tahir HM - MScAgr *Project Planning and Monitoring Officer and Dean Faculty of Agriculture*

2388 Plant Breeding and Genetics Division

1 Allahrang - PhD *Associate Professor* plant breeding
2 Khajuria MR - PhD *Associate Professor* plant breeding
3 Sandhu JS - PhD *Professor*
4 Zargar GH - PhD *Associate Professor* plant breeding

2389 Plant Pathology Division

1 Ghulam-ud-din S - PhD *Associate Professor* plant pathology
2 Mir NA - PhD *Associate Professor* plant pathology
3 Mohammed N - PhD *Associate Professor* plant pathology
4 Putoo BL - PhD *Associate Professor* plant pathology

2390 Entomology Division
Jammu and Kashmir, Shalimar, Post Box No. 262, Srinagar, Jammu and Kashmir, 190001

1 Bhat AH - PhD *Associate Professor* entomology
2 Masoodi MA - PhD *Associate Professor* entomology
3 Revo CL - PhD *Associate Professor* entomology

2391 Agronomy, Soils and Statistics

1 Mir AH - PhD *Associate Professor*
2 Shah MH - PhD *Associate Professor* agrostology
3 Wani MY - PhD *Associate Professor* soil science

2392 Pomology Division

1 Nagoo GA - PhD *Associate Professor* horticulture
2 Qureshi AS - MSc *Associate Professor* horticulture
3 Sofi AA - PhD *Associate Professor* horticulture

2393 Olericulture Division

1 Tanki MI - PhD *Associate Professor* vegetables

2394 Post-harvest Technology/Agricultural Engineering

1 Syed M - BE *Associate Professor* agricultural engineering

2395 Agro-forestry Division

1 Sagwal SS - MSc *Associate Professor* forestry

2396 Regional Research Station
Khudwani, Kashmir, 190001

1 Ganai BA - PhD *Senior Scientist* agronomy
2 Kaw RN - PhD *Senior Scientist* plant breeding

2397 College of Agriculture/Regional Research Station

1 Banday AH - MScAgr *Associate Professor* plant breeding
2 Dhar MK - PhD *Associate Professor* agricultural economics
3 Kaul AK - PhD *Associate Professor* plant pathology
4 Khan GM - PhD *Associate Professor In Charge Associate Dean*
5 Mir S - PhD *Associate Professor* soils
6 Mukoo KN - PhD *Associate Professor* extension education
7 Sheikh GA - PhD *Associate Professor* entomology

2398 Regional Agricultural Research Station

1 Chowdhry MR - PhD *Chief Scientist* water management
2 Gupta BR - MScAgr *Senior Scientist* plant pathology
3 Gupta RD - PhD *Deputy Director* extension education
4 Gupta RS - MScAgr *Senior Extension Specialist* plant pathology
5 Kotwal DR - PhD *Senior Scientist* entomology
6 Prabhakar - PhD *Senior Scientist* soils
7 Sharma YR - PhD *Senior Scientist* plant pathology
8 Singh H - PhD *Chief Scientist* agronomy
9 Singh J - PhD *Senior Scientist* plant breeding

2399 Regional Horticultural Research Station

1 Singh JS - PhD *Chief Scientist* horticulture

2400 Division of Sericulture

Mirgund, Kashmir, 190001

1 Das BC - PhD *Professor* sericulture
2 Trag AR - PhD *Associate Professor* sericulture

2401 Faculty of Veterinary Sciences and Animal Husbandry

Alusteng, Srinagar, 190001

1 Pampori NA - PhD *Associate Professor* biochemistry
2 Rashid A - PhD *Associate Professor* microbiology
3 Teli AA - MVSc *Associate Professor* animal production
4 Wani GM - PhD *Associate Professor* reproduction

2402 University of Agricultural Sciences

Gandhi Krishi Vigyana Kendra Campus, Bangalore, 560065
Tel: 366753. Telex: 0845-8393-UASK-IN. Cables: UNIVAGRIS.

1 Patil SV - PhD *Vice Chancellor*

2403 Department of Agronomy

1 Biradar BM - MScAgr *Associate Professor* agronomy
2 Bommegowda A - PhD *Senior Seed Research Officer* seed production
3 Chandrashekar B - PhD *Agronomist* weed management and crop production
4 Chandrashekaraiah AN - MScAgr *Associate Professor* agro-forestry
5 Gidnavar VS - MScAgr *Agronomist* soil management
6 Goudreddy BS - PhD *Associate Professor* crop production
7 Gumasta SK - PhD *Agrostologist*
8 Hosamani MM - PhD *Senior Agronomist* crop production
9 Kachapur MD - MScAgr *Agronomist* soil management
10 Krishnamurthy K - PhD *Director of Research*
11 Krishnegowda KT - PhD *Agronomist* crop production
12 Kulkarni KR - PhD *Associate Director of Research* cropping systems and water management
13 Mumegobvola MK - MScAgr *Agronomist* crop production
14 Patil BN - MScAgr *Scientist* chemical weed control
15 Patil VS - PhD *Regional Associate Director* water management
16 Prabhakara STK - PhD *Agronomist* weed science
17 Radder GD - PhD *Chief Scientist* dry farming
18 Ramachandra Prasad TV - MScAgr *Agronomist* growth analysis and weed control
19 Ramanagowda P - PhD *Agronomist* soil and water conservation
20 Setty RA - PhD *Rice Agronomist* crop production
21 Shanthamallaiah NR - MScAgr *Senior Scientist* irrigation and water management
22 Singlachar NA - PhD *Regional Associate Director* soil fertility
23 Yadahalli - MScAgr *Scientist* crop production and fertility

2404 Department of Chemistry and Soils

1 Anantha Narayana R - PhD agricultural chemistry
2 Badanur VP - PhD *Soil Physicist* soil physics
3 Balakrishna Rao K - PhD *Regional Associate Director* plant nutrition and soil fertility
4 Doddamani VS - PhD *Soil Scientist*
5 Gajanan GN - MScAgr *Soil Chemist* soil fertility environmental pollution
6 Hadimani AS - PhD *Regional Associate Director* soil survey and genetics and classification
7 Kavalappa BN - MScAgr *Soil Scientist* soil fertility
8 Krishnappa AM - PhD *Associate Professor* soil fertility

9 Manjunathaiah HM - PhD *Associate Professor* soil chemistry
10 Pandurangaiah L - MScAgr *Soil Scientist* watershed management
11 Parashivamurthy AS - PhD *Chief Soil Scientist* soil fertility chemistry
12 Parvathappa HC - PhD *Soil Physicist* soil physics
13 Patil C - PhD *Soil Physicist* soil science
14 Reddy S - PhD *Soil Scientist* soil chemistry
15 Sathyanarayana T - PhD *Chief Scientist* soil physics
16 Sivamurthy SC - PhD *Soil Chemist* soil chemistry
17 Suseela Devi L - PhD *Soil Chemist* soil science
18 Viswanath DP - PhD *Soil Scientist* soil science

2405 Department of Agricultural Botany

1 Ahmed RS - PhD *Plant Scientist* plant breeding and genetics
2 Avadhani KK - PhD *Plant Scientist* plant breeding
3 Bhat BN - PhD *Plant Scientist* genetics and plant breeding
4 Chandrappa HM - PhD *Associate Professor* plant breeding
5 Channakrishnaiah KM - MScAgr *Geneticist* plant breeding and genetics
6 Chikkadevaiah - PhD *Sugarcane Breeder* plant breeding and genetics
7 Giriraj K - PhD *Senior Sunflower Breeder* genetics and plant breeding
8 Gopalakrishna Rao M - PhD *Plant Scientist* genetics and plant breeding
9 Habib AF - PhD *Oilseed Breeder* plant breeding
10 Jayaramaiah - MScAgr *Ragi Breeder* genetics and plant breeding
11 Joshi MS - BScAgr *Sorghum Breeder* plant breeding
12 Kadappa SN - PhD *Senior Scientist* genetics and plant breeding
13 Manjunath A - PhD plant breeding and genetics
14 Nagaraja Shetty MV - PhD *Chief Scientist* genetics and plant breeding
15 Naidu BS - PhD *Regional Associate Director* genetics and plant breeding
16 Parameshwar NS - MSc *Professor* plant breeding
17 Patil SA - PhD *Breeder* plant breeding
18 Patil SJ - PhD *Maize Breeder* genetics and plant breeding
19 Polica Patil CS - PhD *Breeder* genetics and plant breeding
20 Rajashekaraiah S - BSc *Maize Breeder*
21 Shambulingappa - PhD *Senior Scientist* cytogenetics and plant breeding
22 Talwar TS - PhD *Cotton Breeder* genetics and plant breeding
23 Thirumalachar DK - MSc *Senior Scientist* genetics and plant breeding
24 Veerappa KB - PhD *Plant Breeder* genetics and plant breeding
25 Vidyachandra B - MSc *Rice Breeder* plant breeding
26 Virupakashappa K - PhD *Geneticist* genetics and plant breeding
27 Viswanath SR - MScAgr *Soybean Breeder* plant breeding and genetics

2406 Department of Agricultural Entomology

1 Gavigowda - MScAgr *Associate Professor* economic entomology
2 Gowda P - PhD *Chief Scientific Officer* entomology
3 Gubbaiah - MScAgr *Entomologist* economic entomology
4 Hiremath - PhD *Associate Professor* agricultural entomology
5 Krishnaprasad - MScAgr *Associate Professor* agricultural entomology
6 Kulkarni KA - PhD *Entomologist* entomology
7 Patil BV - PhD *Entomologist* pest management
8 Rajagopal D - PhD *Associate Professor* soil biology and ecology
9 Siddappaji - MScAgr *Associate Professor* sericulture
10 Thimmaiah G - BScAgr *Entomologist* integrated pest management
11 Vishakantaiah M - MScAgr *Scientist* economic entomology

2407 Department of Seed Technology

1 Aswathiah B - PhD *Seed Research Officer* seed technology

2408 Department of Plant Pathology

1 Anahosur KH - PhD *Pathologist* mycology
2 Anilkumar TB - PhD *Pathologist* plant bacteriology
3 Bidari VB - PhD *Scientist* fungal and viral diseases
4 Hiremath PC - PhD *Plant Pathologist*
5 Hiremath RV - PhD *Plant Pathologist* mycology
6 Nanjegowda D - PhD *Nematologist* nematology
7 Sannegowda - PhD *Plant Pathologist* plant bacteriology
8 Seshadri VS - PhD *Pathologist* mycology
9 Siddaramaiah AL - MScAgr *Pathologist* plant virology
10 Sudarsan Rao AN - PhD *Associate Professor* fungal pathology
11 Viswanath S - MSc *Virologist* virology
12 Yaraguntaiah RC - PhD *Chief Scientific Officer* plant virology

2409 Department of Horticulture

1 Bankapur VM - PhD *Horticulturist* horticulture
2 Khan MM - PhD *Professor* horticulture
3 Narayana Reddy MA - PhD *Horticulturist* pomology
4 Pattanna Shetty HV - PhD *Regional Associate Director* horticulture
5 Rao MM - PhD *Horticulturist* horticulture
6 Uthaiah BC - PhD *Associate Professor* plantation crops
7 Vijayakumar N - PhD *Horticulturist* pomology

2410 Department of Crop Physiology

1 Janardhan KV - PhD *Physiologist* crop physiology
2 Rao PSR - PhD *Senior Physiologist*
3 Vajranabhaiah SN - PhD *Plant Physiologist* crop physiology

2411 Department of Agricultural Microbiology

1 Gopala Gowda HS - MScAgr *Microbiologist* biological nitrogen fixation
2 Hegde SV - PhD *Scientist* biological nitrogen fixation
3 Raj J - PhD *Microbiologist* biological nitrogen fixation
4 Siddarame Gowda TK - PhD *Microbiologist* biological nitrogen fixation

2412 Department Agricultural Engineering

1 Belgaumi MI - MScAgr *Associate Professor* soil and water conservation engineering
2 Chennappa TC - ME *Agricultural Engineer* soil and water engineering
3 Krishna Murthy KS - MScAgr *Research Engineer* agricultural process engineering
4 Subhash SK - MS *Drainage Engineer* soil and water conservation engineering

2413 Department of Agricultural Economics

1 Chandrashekar GS - MScAgr *Associate Professor* agricultural economics
2 Gopalakrishna Hebbar B - MScAgr *Associate Professor* agricultural economics
3 Lokamanya DS - MScAgr *Associate Professor* agricultural economics
4 Surya Prakash S - MScAgr *Associate Professor* agricultural economics

2414 Department of Aquaculture

1 Devaraj KV - PhD *Chief Scientific Officer* inland fisheries
2 Sathyanarayana Rao GP - PhD *Fisheries Officer* aquaculture

2415 Department of Fishery Biology

1 Rajagopal KV - MSc *Associate Professor* fishery biology and fishery culture

2416 Department of Animal Genetics and Breeding

1 Basavaiah P - PhD *Geneticist* genetics and breeding
2 Lokanath GR - PhD *Poultry Geneticist* genetics and breeding
3 Paramshivaiah BM - MVSc *Associate Professor* animal genetics and breeding
4 Pratap Kumar KC - PhD *Poultry Geneticist* genetics and breeding
5 Ramappa BS - PhD *Geneticist* genetics and breeding
6 Uthappa IM - PhD *Chief Scientific Officer* animal production

2417 Department of Animal Nutrition

1 Krishnappa P - PhD *Professor* dairy husbandry

2418 Department of Home Science

1 Gowramma TS - PhD *Scientist* food and nutrition

2419 Department of Statistics

1 Jagannath MK - MSc *Statistician* biometrics and agricultural statistics
2 Vijayamma R - MSc *Associate Professor* statistics

2420 Department of Zoology

1 Srihari K - PhD *Senior Zoologist* rodent biology

2421 Advanced Centre for Research on Black Cotton Soils

1 Murthy ASP - MScAgr,PhD *Professor* clay mineralogy and soil chemistry
2 Ananthanarayana R - MScAgr,PhD *Associate Professor* agricultural chemistry
3 Doddamani VS - MScAgr,PhD *Associate Professor* soil physics
4 Gidnavar VS - MScAgr *Associate Professor* crop production; soil fertility
5 Kumathe SS - MTech *Engineering Associate Professor* drainage engineering
6 Manjunathaiah HM - MScAgr,PhD *Associate Professor* micronutrients
7 Reddy S - MScAgr,PhD *Associate Professor* soil science

2422 Jawaharlal Nehru Krishi Vishwa Vidyalaya

Krishinagar, Jabalpur, 482004
Tel: 23771,23772. Cables: KRISIVISWA.

1 Arya SV - MScAgrEngg *Vice-Chancellor*
2 Dubey SN - MScAgr,PhD *Director of Instruction*
3 Sharma DK - MA,PhD *Director of Extension Services*
4 Shukla VP - MA,PhD *Dean Faculty of Agriculture*
5 Verma GP - MSrAgr,PhD *Director of Research Services*

2423 Directorate of Research Services

1 Chourasia RR - MScAgr,PhD *Scientist* agricultural economics
2 Khot VN - MScAgr *Deputy Director Research*
3 Mehta RS - MTech *Senior Scientist* agricultural engineering
4 Mishra BL - MScAgr,PhD *Professor* agricultural economics
5 Tiwari AK - MVsc *Deputy Director Research*

2424 Zonal Agricultural Research Station/College of Agriculture, Jabalpur

Krishinagar, Jabalpur, 482004
Tel: 23771,23772. Cables: KRISIVISWA.

1 Choubey SD - MScAgr,PhD *Professor* agronomy
2 Khare MN - MScAgr,PhD *Professor* plant pathology
3 Mishra RS - MScAgr,PhD *Professor* agricultural economics
4 Nair PKR - MScAgr,PhD *Professor* pomology and fruit preservation
5 Sharma YK - MScAgr,PhD *Professor* food science
6 Singh CB - MScAgr,PhD *Senior Scientist* plant breeding
7 Singh K - MScAgr,PhD *Professor* vegetable crops
8 Sinha SB - MScAgr,PhD *Professor* soil science
9 Tiwari DK - MSc,PhD *Professor* agricultural botany
10 Vaishampayan SM - MScAgr,PhD *Senior Scientist* entomology
11 Verma AK - MScAgr *Professor* agricultural extension
12 Agrawal GD - MScAgr,PhD *Scientist* food science
13 Agrawal RK - MScAgr,PhD *Professor* entomology
14 Bansal KN - MScAgr,PhD *Associate Professor* soil science
15 Beohar ABL - MScAgr,PhD *Scientist* plant breeding
16 Dave GS - MScAgr,PhD *Associate Professor* plant pathology
17 Dubey JN - MScAgr,PhD *Associate Professor* soil science
18 Dubey SK - MScAgr,PhD *Scientist* agronomy
19 Gajendragadkar GR - MScAgr,PhD *Scientist* soil science
20 Gargav VP - MScAgr,PhD *Professor* entomology
21 Gogulwar NM - MScAgr,PhD *Associate Professor* agronomy
22 Gour SS - MScAgr,PhD *Professor* agricultural economics
23 Jain HC - MA,PhD *Associate Professor* agricultural economics
24 Janoria MP - MScAgr,PhD *Associate Professor* plant breeding
25 Kashiv RC - MScAgr,PhD *Scientist* agricultural economics
26 Katiyar OP - MScAgr,PhD *Scientist* entomology
27 Khatri AK - MScAgr,PhD *Scientist* entomology
28 Khurchania SP - MScAgr,PhD *Scientist* agronomy
29 Mehta SK - MScAgr,PhD *Scientist* agricultural botany
30 Mishra BL - MSc,PhD *Scientist* statistics
31 Narsinghani VG - MScAgr,PhD *Scientist* plant breeding
32 Nayak ML - MScAgr,PhD *Scientist* plant pathology
33 Nema MK - MScAgr,PhD *Scientist* horticulture
34 Odak SC - MScAgr,PhD *Associate Professor* entomology
35 Pandey RP - MScAgr,PhD *Scientist* horticulture
36 Rathore GS - MScAgr,PhD *Scientist* soil science
37 Seshagiri A - MSc,PhD *Scientist* statistics
38 Sharma KK - MScAgr,PhD *Scientist* plant breeding
39 Sharma RK - MTech *Scientist* agricultural engineering
40 Singh SP - MScAgr,PhD *Dean*
41 Tawar ML - MScAgr,PhD *Associate Professor* plant breeding
42 Tembhare BR - MScAgr,PhD *Scientist* soil science
43 Tiwari RK - MScAgr *Scientist* agricultural botany
44 Tomar GS - MScAgr,PhD *Scientist* agro-forestry
45 Upendra Shanker - MTech *Scientist* agro-meteorology
46 Verma LN - MScAgr,PhD *Associate Professor* soil science
47 Vyas SC - MScAgr,PhD *Associate Professor* plant pathology

2425 College of Agricultural Engineering, Jabalpur

1 Lal Chand - MSc,PhD *Professor* mathematics and statistics
2 Chandak ML - MSc,PhD *Professor* mathematics
3 Gupta VK - MSc,PhD *Professor* physics
4 Katakwar CR - ME *Associate Professor* civil engineering
5 Naryan HR - MScAgrEngg *Associate Professor* mechanical engineering
6 Raje SR - MTech *Professor* farm machinery and power
7 Tyagi HR - PhD *Dean* agricultural engineering

2426 College of Veterinary Science and Animal Husbandry, Jabalpur

1 Agrawal MC - MVSc,PhD *Associate Professor* parasitology
2 Awadhiya RP - MVSc *Associate Professor* pathology
3 Bajpai LD - MVSc,PhD *Associate Professor* animal nutrition
4 Datta IC - MVSc,PhD *Professor* biochemistry
5 Dave BK - MVSc,PhD *Professor* animal production and management
6 Dhawedkar RG - MVSc,PhD *Associate Professor* microbiology
7 Dhingra MM - MVSc,PhD *Scientist* piggery
8 Katpatal BG - MVSc,PhD *Professor* animal breeding and genetics
9 Khan AG - MVSc,PhD *Senior Scientist* poultry
10 Kharche KG - MVSc *Associate Professor* obstetrics and gynaecology
11 Kushwaha HS - MVSc,PhD *Associate Professor* biochemistry
12 Malik MR - MVSc,PhD *Associate Professor* anatomy
13 Mandloi AK - MSc,PhD *Associate Professor* fisheries
14 Moitra AK - MVSc,PhD *Associate Professor* microbiology
15 Netke SP - MVSc,PhD *Dean*
16 Pandey SK - MVSc,PhD *Associate Professor* surgery
17 Parashar GC - MVSc,PhD *Professor* pharmacology
18 Parekh HKB - MVSc,PhD *Senior Scientist* cattle

19 Patle BR - MVSc,PhD *Professor* animal nutrition
20 Rao KNP - MVSc *Associate Professor* medicine
21 Sahastrabudhe VK - MVSc,PhD *Associate Professor* parasitology
22 Saxena SK - MVSc,PhD *Professor* physiology
23 Shah HL - MVSc,PhD *Professor* parasitology
24 Shrivastava AM - MVSc,PhD *Professor* anatomy
25 Shrivastava HOP - MVSc,PhD *Associate Professor* parasitology
26 Shrivastava PN - MVSc *Associate Professor* pharmacology
27 Singh VP - MVSc,PhD *Scientist* animal production and management
28 Sone JL - MVSc *Associate Professor* medicine
29 Veged JL - MVSc,PhD *Professor* pathology

2427 College of Agriculture, Indore

1 Bichoo SL - MScAgr,PhD *Scientist* entomology
2 Choudhari AK - MScAgr,PhD *Scientist* agricultural economics
3 Dabholkar AR - MScAgr,PhD *Scientist* plant breeding
4 Garg KP - MScAgr,PhD *Senior Scientist* agronomy
5 Gupta RK - MScAgr,PhD *Senior Scientist* soil science
6 Kapoor KN - MScAgr,PhD *Associate Professor* entomology
7 Karanjkar SV - MScAgr,PhD *Associate Professor* agricultural economics
8 Kushwaha JS - MScAgr,PhD *Associate Professor* plant breeding
9 Nigam PK - MScAgr,PhD *Scientist* agricultural botany
10 Raghwanshi RKS - MScAgr,PhD *Scientist* agronomy
11 Rajput GS - MTech *Scientist* agricultural engineering
12 Sawant AR - MScAgr,PhD *Scientist* plant breeding
13 Saxena HA - MVSc *Scientist* veterinary science
14 Sharma RA - MScAgr,PhD *Scientist* soil science
15 Sharma RA - MScAgr,PhD *Associate Professor* agronomy
16 Shinde CB - MScAgr *Scientist* entomology
17 Shrivastava NC - MTech *Senior Scientist* agricultural engineering
18 Shroff VN - MScAgr,PhD *Associate Director Research*
19 Sisodia NS - MScAgr,PhD *Associate Professor* plant breeding
20 Tulsi MJS - MScAgr,PhD *Scientist* agronomy

2428 College of Veterinary Science & Animal Husbandry, Mhow

1 Adwal SC - MVSc,PhD *Associate Professor* physiology
2 Bhaid MU - MVSc *Associate Professor* animal nutrition
3 Chouhan RAS - MVSc *Associate Professor* surgery
4 Johar KS - MVSc,PhD *Dean*
5 Khan FH - MVSc,PhD *Associate Professor* animal breeding and genetics
6 Nakra SK - MVSc *Associate Professor* surgery
7 Pandit RK - MVSc,PhD *Associate Professor* surgery
8 Pathak PN - MVSc,PhD *Professor* microbiology
9 Richariay VS - MVSc *Associate Professor* pathology
10 Saxena HK - MVSc *Associate Professor* animal production and management
11 Singh BN - MVSc *Associate Professor* animal production and management
12 Sisodia RS - MVSc,PhD *Associate Professor* medicine
13 Supekar PG - MVSc *Associate Professor* medicine
14 Tanwani SK - MVSc,PhD *Associate Professor* microbiology
15 Tiwari GP - MVSc,PhD *Associate Professor* anatomy

2429 College of Agriculture, Rewa

1 Gupta SK - MTech *Scientist* agricultural engineering
2 Gupta US - MScAgr,PhD *Scientist* soil science
3 Khare KB - MScAgr,PhD *Scientist* agronomy
4 Mishra GP - MScAgr,PhD *Scientist* plant breeding
5 Pandey RC - MScAgr,PhD *Senior Scientist* agronomy
6 Pandey RP - MScAgr,PhD *Scientist* agronomy
7 Rai MM - MScAgr,PhD *Dean*
8 Sharma CP - MScAgr *Scientist* horticulture
9 Sharma MM - MScAgr,PhD *Senior Scientist* agronomy
10 Singh P - MScAgr,PhD *Scientist* agronomy
11 Tiwari BL - MScAgr,PhD *Scientist* horticulture
12 Tiwari BS - MScAgr *Scientist* agricultural botany
13 Tiwari BS - MScAgr *Associate Professor* agricultural botany
14 Tripathi RP - MScAgr,PhD *Associate Professor* agricultural economics

2430 College of Agriculture, Sehore

1 Agrawal SC - MScAgr,PhD *Scientist* plant pathology
2 Bhandarkar DM - MTech *Scientist* agricultural engineering
3 Bhide UK - MScAgr,PhD *Scientist* soil science
4 Chourasia RK - MScAgr,PhD *Scientist* plant pathology
5 Gupta BM - MScAgr *Associate Professor* agricultural botany
6 Gwal HB - MScAgr,PhD *Scientist* agronomy
7 Jain BL - MScAgr *Scientist* economics
8 Khan RA - MScAgr,PhD *Associate Director Research*
9 Lal MS - MScAgr,PhD *Senior Scientist* plant breeding
10 Mohd. Niyamtullah - MScAgr *Scientist* agricultural botany
11 Namdeo SL - MScAgr,PhD *Scientist* soil science
12 Nema DP - MScAgr,PhD *Dean*
13 Sharma HK - MScAgr,PhD *Scientist* plant breeding
14 Sharma SR - MScAgr,PhD *Scientist* agronomy
15 Shrimal HR - MScAgr,PhD *Scientist* agronomy
16 Singh OP - MScAgr,PhD *Scientist* entomology
17 Singh PP - MScAgr,PhD *Scientist* agronomy

18 Singh Suman K - MScAgr,PhD *Scientist* food science
19 Thakur RC - MScAgr,PhD *Scientist* entomology
20 Tripathi JN - MScAgr *Associate Professor* agricultural economics

2431 College of Agriculture, Gwalior

1 Beohar DC - MScAgr *Associate Professor* agricultural botany
2 Bhadoria SS - MScAgr,PhD *Scientist* plant breeding
3 Chouhan DVS - MScAgr,PhD *Associate Professor* agronomy
4 Chouhan GS - MScAgr,PhD *Scientist* plant breeding
5 Dhamdhare SV - MScAgr,PhD *Scientist* entomology
6 Gupta KP - MScAgr,PhD *Scientist* agronomy
7 Mishra MK - MScAgr,PhD *Associate Director Research*
8 Mishra US - MScAgr,PhD *Professor* entomology
9 Nema MG - MScAgr,PhD *Scientist* agricultural economics
10 Sharma SG - MScAgr,PhD *Scientist* soil science
11 Shrivastava MP - MScAgr,PhD *Scientist* plant breeding
12 Singh D - MScAgr,PhD *Dean*
13 Singh GP - MScAgr,PhD *Associate Professor* plant breeding
14 Singh TP - MScAgr *Associate Professor* horticulture
15 Tembhare BR - MScAgr,PhD *Scientist* soil science
16 Thakur SS - MScAgr,PhD *Professor* extension
17 Tiwari AS - MScAgr,PhD *Senior Scientist* plant breeding
18 Tripathi SN - MScAgr *Associate Professor* horticulture

2432 College of Agriculture, Raipur

1 Agrawal BB - MScAgr,PhD *Scientist* plant breeding
2 Agrawal KC - MScAgr,PhD *Scientist* plant pathology
3 Agrawal SK - MScAgr,PhD *Scientist* plant breeding
4 Ali MA - MA,PhD *Scientist* statistics
5 Baghel SS - MScAgr,PhD *Dean*
6 Bisen DC - MScAgr,PhD *Senior Scientist* soil science
7 Chandrawanshi BR - MScAgr,PhD *Senior Scientist* agronomy
8 Dave AB - MScAgr,PhD *Associate Professor* soil science
9 Gouraha RS - MScAgr,PhD *Scientist* extension
10 Jaggi IK - MScAgr,PhD *Scientist* soil science
11 Khan MI - MScAgr,PhD *Scientist* agricultural botany
12 Kittoor SU - MScAgr,PhD *Scientist* entomology
13 Marothia DK - MScAgr,PhD *Scientist* agricultural economics
14 Pal AR - MScAgr,PhD *Associate Professor* soil science
15 Rathod KL - MScAgr,PhD *Associate Director Research*
16 Shastri ASRAS - MSc,PhD *Scientist* agro-meteorology
17 Shrivastava MN - MScAgr,PhD *Scientist* plant breeding
18 Shrivastava OS - MScAgr,PhD *Associate Professor* entomology
19 Shrivastava PS - MScAgr,PhD *Senior Scientist* plant breeding
20 Shrivastava SK - MScAgr,PhD *Scientist* plant pathology
21 Shukla TC - MScAgr *Associate Professor* agricultural botany
22 Singh BP - MScAgr,PhD *Scientist* plant pathology
23 Singh MP - MScAgr *Associate Professor* horticulture
24 Singh Ram - MScAgr,PhD *Scientist* agronomy
25 Tiwari MD - MScAgr *Associate Professor* horticulture
26 Verma VP - MTech *Scientist* agricultural engineering

2433 Zonal Agricultural Research Station, Powarkheda

1 Deshkar SM - MScAgr,PhD *Senior Scientist* agronomy
2 Kailasia DS - MTech *Scientist* agricultural engineering
3 Khare PD - MScAgr,PhD *Scientist* agronomy
4 Raghu JS - MScAgr,PhD *Senior Scientist* agronomy
5 Sharma HC - MScAgr,PhD *Scientist* plant pathology
6 Sheopuria RR - MScAgr,PhD *Scientist* plant breeding
7 Tiwari YD - MScAgr,PhD *Scientist* soil science
8 Yadav BP - MTech *Scientist* agricultural engineering

2434 Regional Research Station, Ambikapur

1 Bajpai RP - MScAgr,PhD *Scientist* agronomy
2 Gaurung BS - MVSc *Scientist* animal nutrition
3 Mishra KL - MTech *Scientist* agricultural engineering
4 Pandey RL - MScAgr,PhD *Scientist* plant breeding
5 Sharma RB - MScAgr,PhD *Senior Scientist* agronomy
6 Shukla BN - MScAgr *Scientist* plant pathology
7 Singh PN - MScAgr,PhD *Scientist* horticulture
8 Singh PN - MScAgr,PhD *Scientist* agricultural economics
9 Thakur BS - MScAgr,PhD *Scientist* entomology
10 Tomar AS - MScAgr,PhD *Scientist* soil science

2435 Regional Research Station, Jagadalpur

1 Bhandarkar JN - MScAgr,PhD *Scientist* agronomy
2 Jain PK - MVSc *Scientist* animal husbandry
3 Joshi GC - MScAgr *Scientist* horticulture
4 Sahu RK - MScAgr *Scientist* soil science
5 Solanki SS - MScAgr,PhD *Senior Scientist* agronomy
6 Tiwari RD - MScAgr *Scientist* plant breeding

2436 Regional Research Station, Tikamgarh

1 Gupta OP - MScAgr,PhD *Scientist* soil science
2 Jakhmola SS - MScAgr,PhD *Scientist* entomology
3 Saran RN - MScAgr,PhD *Senior Scientist* agronomy

4 Sharma HC - MTech *Scientist* agricultural engineering
5 Verma ML - MScAgr,PhD *Scientist* plant pathology
6 Yadav LN - MScAgr,PhD *Scientist* plant breeding

2437 Regional Research Station, Chhindwara

1 Dubey KC - MScAgr,PhD *Scientist* horticulture
2 Goydani BM - MScAgr,PhD *Scientist* plant breeding

2438 Regional Research Station, Khandwa

1 Deshmukh SC - MScAgr,PhD *Scientist* agronomy
2 Deshpande RR - MScAgr,PhD *Scientist* entomology
3 Jain RK - MScAgr *Scientist* agronomy
4 Mandloi KC - MScAgr,PhD *Senior Scientist* plant breeding
5 Mandloi KK - MScAgr,PhD *Scientist* plant breeding
6 Shrivastava RC - MScAgr *Scientist* horticulture

2439 Special Agricultural Research Station, Dindori

1 Dhagat AK - MScAgr,PhD *Scientist* agronomy
2 Dubey BP - MScAgr,PhD *Scientist* agronomy
3 Sharma ND - MScAgr,PhD *Scientist* plant pathology
4 Shrivas SR - MScAgr,PhD *Scientist* plant breeding

2440 Special Agricultural Research Station, Sagar

1 Deole JY - MScAgr *Scientist* entomology
2 Jain SK - MScAgr *Scientist* plant breeding
3 Purohit JP - MScAgr *Scientist* agronomy
4 Tomar VS - MScAgr,PhD *Senior Scientist* soil science

2441 Special Agricultural Research Station, Morena

1 Kaushal Kishore - MTech *Scientist* agricultural engineering
2 Kushwaha PS - MScAgr *Scientist* agronomy
3 Verma OP - MScAgr,Phd *Senior Scientist* soil science

2442 Special Agricultural Research Station, Mandsaur

1 Nigam KB - MScAgr *Scientist* plant breeding

2443 Special Agricultural Research Station, Kharggone

1 Bhalla PL - MScAgr,PhD *Senior Scientist* horticulture
2 Jagtap JG - MScAgr,PhD *Scientist* plant breeding

2444 Agricultural Research Station, Bilaspur

1 Agrawal RP - MScAgr,PhD *Senior Scientist* agronomy
2 Chandrakar BS - MScAgr *Scientist* agronomy
3 Joshi BS - MScAgr,PhD *Scientist* agronomy
4 Kashyap RL - MScAgr,PhD *Scientist* horticulture
5 Katre RK - MScAgr,PhD *Senior Scientist* soil science
6 Pandya KS - MScAgr,PhD *Scientist* agronomy
7 Sharma PN - MScAgr,PhD *Scientist* extension
8 Shrivastava SS - MScAgr,PhD *Scientist* horticulture

2445 Agricultural Research Station, Waraseoni

1 Bisen MS - MScAgr,PhD *Scientist* plant breeding
2 Tripathi RS - MScAgr,PhD *Senior Scientist* agronomy

2446 Agricultural Research Station, Jaora

1 Vaidya MS - MScAgr *Scientist* agronomy

2447 Agricultural Research Station, Jhabua

1 Deshpande WR - MScAgr,PhD *Senior Scientist* extension

2448 Agricultural Research Station, Chhatarpur

1 Deshpande AL - MScAgr,PhD *Scientist* plant pathology
2 Khare KB - MScAgr *Scientist*

2449 Agricultural Research Station, Guna

1 Deokant - MTech *Scientist* agricultural engineering

2450 Agricultural Research Station, Harda

1 Upadhyaya SP - MScAgr,PhD *Scientist* agronomy

2451 Agricultural Research Station, Khaperkheda

1 Dhimole OP - MScAgr *Scientist* agronomy

2452 Agricultural Research Station, Bohani

1 Sharma RK - MScAgr,PhD *Scientist* agronomy

2453 Agricultural Research Station, Bari

1 Kool YM - MTech *Scientist* agricultural engineering

2454 Agricultural Research Station, Sidhi

1 Baghel AS - MScAgr *Scientist* agricultural economics
2 Patel RK - MTech *Scientist* agricultural engineering
3 Singh RP - MScAgr *Scientist* plant breeding

2455 Agricultural Research Station, Shahdol

1 Namdeo KN - MScAgr,PhD *Scientist* agronomy

2456 Agricultural Research Station, Dhar

1 Shrivastava UK - MScAgr *Scientist* agronomy

2457 Agricultural Research Station, Vidisha

1 Jain PC - MScAgr *Scientist* agronomy
2 Singh GR - MScAgr,PhD *Scientist* plant breeding

2458 Marathwada Agricultural University

Parbhani, Maharashtra, 431402
Tel: 301 PBX. Cables: UNIVERSITY.

1 Rao NBP - Vice Chancellor

2459 Agronomy

1 Raikhelkar SV *Head, Department of Agronomy*
2 Borulkar DN *Senior Agronomist* agronomic research
3 Dhoble MV *Agronomist* dryland agriculture
4 Giri AN *Seed Production Specialist*
5 Jadhav GS *Associate Professor* agronomy
6 Joshi PK *Agricultural Research Officer* adaptive research
7 Katare RA *Agronomist* pulses
8 Khan QA *Agricultural Research Officer* adaptive research
9 Kuberkar SY *Agricultural Research Officer* adaptive research
10 Kurlekar VG *Director of Seeds* seed technology
11 Lomte MH *Agronomist* sorghum
12 Londge VD *Agronomist* water management
13 Nankar JT *Associate Director*
14 Nirval BG *Deputy Director*
15 Pawar KR *Director of Research*
16 Ramakrishna Rao G *Associate Professor* meteorology
17 Rodge RP *Chief Scientist* water management
18 Shaikh AR *Deputy Director* research
19 Shelka VB *Professor of Agronomy* agronomy
20 Shelke DK *Agronomist* weed management
21 Yadav MV *Agronomist* weed management

2460 Animal Husbandry and Dairying

1 Biabani SZ *Principal*
2 Kamble VJ *Deputy Director* research
3 Rotte SG *Professor* animal husbandry and dairying

2461 Agricultural Botany

1 Borikar ST *Seed Research Officer*
2 Kulkarni LP *Sorghum Pathologist*
3 Nandanwankar KG *Sorghum Breeder*
4 Nerkar YS *Head Department of Botany*
5 Patil RA *Junior Sorghum Breeder*
6 Quadar MA *Plant Breeder*
7 Reddy VG *Soybean Breeder*
8 Salunke MR *Professor of Botany*
9 Thombre PG *Cotton Specialist* cotton research

2462 Agricultural Chemistry and Soil Science

1 Ghonsikar CP *Associate Dean and Principal*
2 Lande MG *Soil Physicist*
3 Malewar GU *Head Department of Chemistry*

2463 Agricultural Economics

1 Ashturkar BW *Professor of Economics*
2 Borle JN *Deputy Director*
3 Rajmane KD *Head Department of Economics*

2464 Extension

1 Dongaonkar NW *Professor of Extension*
2 Nandapurkar GG *Head of Department of Extension*
3 Sangle GK *Director of Extension Education*

2465 Entomology

1 Chandurwar RD *Entomologist*
2 Deshpande AD *Associate Professor* entomology
3 Deshpande SV *Operational Research Officer*
4 Dhanorkar BK *Associate Professor* entomology
5 Mahajan SV *Entomologist*
6 Pawar VM *Senior Insect Toxicologist*
7 Puri SN *Head Department of Entomology*
8 Raodeo AK *Associate Dean* instruction
9 Shivpuje PR *Associate Professor* entomology

2466 Plant Pathology

1 Bindu KJ *Principal*
2 Garud TB *Pathologist*
3 Godbole GM *Microbiologist*
4 Mali VR *Head Department of Plant Pathology*
5 Mayee CD *Professor of Plant Pathology*
6 Pawar NB *Pathologist*
7 Zote KK *Pathologist*

2467 Horticulture

1 Ansarwadekar KW *Associate Professor*
2 Ballal AL *Officer In Charge*
3 Budrukkar ND *Horticulturist*
4 Deshmukh PA *Head Department of Horticulture*
5 Khedkar DM *Associate Professor*
6 Patil VK - Director of Instruction and Dean

2468 Agricultural Engineering

1 Quadri SA *Agricultural Engineer* water management
2 Rapte SL *Professor of Agricultural Engineering*

2469 Faculty of Veterinary and Animal Sciences, Parbhani

1 Anantwar LG *Associate Professor* medicine
2 Bhokre AP *Associate Professor* surgery
3 Deshpande BB *Professor* pathology
4 Deshpande KS *Professor* genetics and breeding
5 Gaffar MA *Associate Dean and Principal*
6 Gaffar MA *Professor* animal nutrition
7 Kulkarni MN *Professor* microbiology
8 Lembhe AF *Professor* dairy science
9 Mahajan BS *Associate Professor* extension
10 Panchabhai VS *Associate Professor* surgery
11 Paraonkar DR *Professor* gynaecology
12 Sakhar PG *Professor* animal management
13 Samad A *Associate Professor* medicine
14 Sawant MK *Professor* physiology
15 Shastri UV *Professor* parasitology
16 Shingatgiri RK *Professor* anatomy
17 Vadlamudi VP *Associate Professor* pharmacology
18 Wagh KR *Professor* pharmacology

2470 Faculty of Agricultural Technology, Parbhani

1 Gunjal BB *Associate Professor* food engineering
2 Ingle UM *Associate Dean and Principal*
3 Kulkarni DN *Associate Professor* food science
4 Wankhede DB *Professor* bio-chemistry and nutrition

2471 Faculty of Home Science, Parbhani

1 Harode Miss S *Associate Dean and Principal*
2 Rohini Devi Mrs P *Associate Professor* food nutrition
3 Snehalata Reddy Mrs N *Professor* food and nutrition

2472 Department of Entomology

1 Puri SN - MScAgr,PhD
2 Bilapate GG - MScAgr,PhD *Reader* insect ecology
3 Deshpande AD - MScAgr,PhD *Professor* insect pathology
4 Deshpande SV - MSc,PhD taxonomy
5 Dhanorkar BK - MScAgr,PhD *Associate Professor* toxicology
6 Pawar VM - MSc,PhD *Senior Toxicologist*
7 Sarkate MB - MScAgr *Reader* insect parasitology
8 Shivpuje PR - MScAgr,PhD *Associate Professor* taxonomy
9 Tayade DS - MScAgr,PhD *Reader* sericulture

2473 Orissa University of Agriculture and Technology

Bhubaneswar, Orissa, 351 003
Cables: AGRITECH.

1 Ramamurthy K *Vice Chancellor*

2474 Faculty of Veterinary Science and Animal Husbandry

1 Panda NC - BVSc & AH,Assoc.IVRI,MS,PhD *Dean*

2475 Department of Pathology

1 Das B - MVSc *Virologist*
2 Nayak BC - PhD *Professor*
3 Patnaik GM - PhD *Reader*
4 Rao AG - MVSc *Chief Epidemiologist*
5 Rao AT - PhD *Reader*

2476 Department of Bacteriology

1 Dutta NK - BVSc & AH,MS,PhD,MIPHA *Reader*
2 Kar BC - MVSc *Epidemiologist*
3 Panda SN - GVSc,Assoc.IVRI,FIPHA *Professor*
4 Sahu BN - BVSc,MS,PhD *Reader*

2477 Department of Gynaecology

1 Mohanty BN - PhD *Professor*

2478 Department of Animal Breeding and Genetics

1 Mishra SC *Poultry Geneticist*
2 Patra BN - PhD *Reader*
3 Pattnaik GR - MVSc *Reader*

2479 Department of Surgery

1 Bose VSC - MVSc *Reader*
2 Mitra AK - PhD *Professor*
3 Ray AK - MVSc *Reader*

2480 Department of Medicine

1 Mishra A - PhD *Reader*
2 Ray SK - PhD *Reader*
3 Tripathy SB - PhD *Professor*

2481 Animal Nutrition

1 Panda NC - BVSc & AH,Assoc.IVIR,MS,PhD *Professor*
2 Sahu BK - MVSc *Reader*

2482 Department of Animal Nutrition

1 Mishra M - MS,PhD *Professor*
2 Nayak JB - BVSc & AH,MVSc,PhD *Reader*

2483 Department of Physiology, Pharmacology and Bio-Chemistry

1 Mishra MS - BVSc,DVM,CSc,SG *Reader*
2 Mishra SN - MSc,PhD *Reader*
3 Murthy TL - BVSc,MSc

2484 Department of Anatomy

1 Dash LN - PhD *Reader*

2485 Department of Parasitology

1 Mishra SC - PhD *Professor*
2 Panda DN - MVSc *Reader*

2486 Faculty of Agriculture

1 Mishra A - PhD *Dean*

2487 Department of Breeding and Genetics

1 Mishra PK - MScAgr *Potato Breeder*
2 Mohanty HK - MScAgr,PhD *Professor*
3 Mohapatra BK - MScAgr,PhD *Reader*
4 Samalo BN - MScAgr,PhD *Professor*
5 Satapathy D - BScAgr,Assoc.IARI *Chief Scientist* dryland agriculture

6 Sinha SK - MSc,PhD *Professor*

2488 Department of Agronomy

1 Bhol BB - MScAgr,PhD *Reader*
2 Hati N - MScAgr,PhD *Reader*
3 Lenka D - MSc,PhD *Professor*
4 Misra SN - MSc,PhD *Professor*
5 Prusty JC - MScAgr,PhD *Group Leader* demonstration programme

2489 Department of Plant Physiology

1 Das KP - MScAgr,PhD *Reader*
2 Ghosh BK - MScAgr,PhD *Reader*
3 Parichha PC - MScAgr,PhD *Professor*

2490 Department of Extension Education

1 Kar LN - Assco.IARI,MS *Reader*
2 Tripathy A - MScAgr,PhD *Professor*

2491 Department of Nematology

1 Das SN - MScAgr,PhD,FBS,FPSI *Professor*
2 Padhi NN - MScAgr,PhD,FPSI *Reader*
3 Ray S - MScAgr,PhD *Reader*

2492 Department of Economics

1 Chakraverty ML - MScAgr,PhD *Reader*
2 Dasgupta HK - MScAgr,PhD *Professor*
3 Hota BC - MSc,PhD *Professor*
4 Mallick SC - MScAgr,PhD *Reader*
5 Singh JP - MScAgr,PhD *Reader*

2493 Department of Chemistry

1 Acharya N - MSc,PhD *Associate Professor*
2 Misra C - Assoc.IARI,PhD *Professor*

2494 Department of Entomology

1 Maity BK - MScAgr *Senior Scientist*

2495 Department of Plant Pathology

1 Narain A - MScAgr,PhD *Professor*
2 Das CM - MScAgr,PhD *Reader*
3 Das SR - MScAgr *Senior Plant Pathologist*
4 Mishra D - MScAgr,PhD *Reader*
5 Mohantry AK - MSc,PhD *Professor*
6 Mohanty NN - MScAgr,PhD *Professor*
7 Rath GC - MSc,PhD *Professor*

2496 Department of Horticulture

1 Das GC - MScAgr,PhD *Professor*
2 Mishra RS - MScAgr,PhD
3 Singhsamanta PK - MScAgr *Reader*

2497 Department of Soils and Agricultural Chemistry

1 Mitra GN - PhD *Professor*
2 Acharya N - PhD *Associate Director*
3 Behera B - PhD *Associate Director*
4 Das PK - PhD *Soil Physicist*
5 Jena D - PhD *Senior Scientist*
6 Lushrauk - PhD *Reader*
7 Misra BK - PhD *Senior Scientist*
8 Misra C - PhD *Professor*
9 Misra SN - PhD *Senior Chemist*
10 Nanda SSK - PhD *Associate Dean*
11 Panda M - PhD *Operational Research Officer*
12 Panda N - PhD *Professor*
13 Pradhan NK - PhD *Research Officer*
14 Sahu D - PhD *Associate Director*
15 Sahu OK - PhD *Reader*
16 Sahu SK *Senior Chemist*
17 Senapati HK *Residue Analyst*

2498 Department of Agricultural Entomology

1 Panda N - PhD *Professor*
2 Satpathy JM - PhD *Dean*
3 Das AC - MScAgr *Entomologist*
4 Das MS - MScAgr *Reader-Cum-Entomologist*
5 Jena BC - MScAgr *Senior Entomologist*
6 Maiti BK - MScAgr *Senior Entomologist*
7 Mishra B - MScAgr *Senior Entomologist*
8 Mohapatra HK - MScAgr *Senior Entomologist*

9 Nanda UK - MScAgr *Reader*
10 Patnaik NC - PhD *Reader*
11 Ram S - MScAgr *Senior Entomologist*
12 Rout GD - PhD *Professor*
13 Samalo AP - PhD *Reader*
14 Senapati B - PhD *Associate Director*
15 Shi N - MScAgr *Senior Entomologist*

2499 Rajasthan Agricultural University

Bikaner, Rajasthan, 334 001
Cables: UNIVERSITY UDAIPUR.

1 Nag KN *Vice-Chancellor*

2500 Department of Agronomy

1 Bajpai MR *Associate Professor*
2 Bhargava SS *Associate Professor* millet improvement
3 Dhakar LL *Professor*
4 Dungarwal HS *Professor* water management
5 Garg OP - *Associate Professor* minor irrigation and water use
6 Gour BL *Associate Professor* maize agronomy
7 Gupta OP *Professor*
8 Gupta PC *Professor*
9 Jain PM *Associate Professor* opium poppy
10 Katole NS *Associate Professor* water management
11 Khan IA *Associate Professor* pulse improvement
12 Porwal BL *Associate Professor* sorghum improvement
13 Porwal NK *Associate Professor* soyabean research
14 Rathore SS *Professor* dryland agriculture
15 Rawal DR *Associate Professor* dryland research
16 Shaktawat MS *Reader*
17 Sharma BL *Professor*
18 Sidhu BS *Associate Professor* water management
19 Singh G *Associate Professor* management of salt affected soil
20 Singh RP *Associate Professor* maize
21 Verma JK *Professor* water management

2501 Department of Limnology & Fisheries

1 Rajbanshi VK - BSc,MSc,PhD,FZS,FRMS *Associate Professor & Head of Department* fish biology, aquaculture, limnology, fish nutrition, environmental biology
2 Durve VS - BSc,MSc,PhD *Associate Professor* limnology, fishery biology, fish nutrition, molluscan fishery, marine biology, aquatic toxicology

2502 Department of Plant Breeding and Genetics

1 Bhandari MMC *Associate Professor* opium poppy
2 Bhardwaj RP *Associate Professor* moth improvement
3 Bhatnagar SK *Associate Professor* maize breeding
4 Bhatnagar SM *Associate Professor* barley improvement
5 Bhatnagar VK *Associate Professor* wheat breeding
6 Bhatt BN *Associate Professor* spices and cashewnut
7 Dashora SL *Associate Professor* moth improvement
8 Dave RV *Associate Professor* breeder, seed production
9 Gandhi SM *Professor* seed technology
10 Hussain I *Associate Professor* rural energy
11 Jain KK *Professor* pulse improvement
12 Jain R *Associate Professor* wheat improvement
13 Joshi P *Professor* sesamum
14 Joshi VN *Associate Professor* maize breeding
15 Menon Miss U *Associate Professor* bajra improvement
16 Nathawat KS *Professor* cotton improvement
17 Sanghi AK *Professor*
18 Sarda RP *Associate Professor* dryland research
19 Sharma GS *Associate Professor* sorghum improvement
20 Sharma JP *Associate Professor* soybean research
21 Sharma RC *Associate Professor* bajra improvement
22 Sharma RK *spices and cashewnut*
23 Singh AK *Associate Professor* groundnut-mustard
24 Tripathi RS *Associate Professor* paddy improvement
25 Verma RS *Associate Professor* bajra improvement
26 Vyas KL *Associate Professor* millet improvement

2503 Department of Agricultural Chemistry and Soil Science

1 Deo R *Associate Professor* water management
2 Sanghi CL *Associate Professor* agricultural chemistry
3 Somani LL *Associate Professor* improvement of soil physical conditions
4 Srinivas *Associate Professor* dryland agriculture
5 Talati NR *Associate Professor* water management
6 Vyas KK *Professor* management of salt affected soil

2504 Department of Plant Pathology

1 Agnihotri JP *Professor*
2 Kothari KL *Professor* maize pathology
3 Sharma LC *Associate Professor* pulse improvement

4 Tyagi RNS *Associate Professor* pulse improvement

2505 Department of Entomology

1 Bhardwaj SC *Associate Professor*
2 Gokhle VG *Associate Professor* orintology
3 Kaul CK *Associate Professor* pulse improvement
4 Kavadia VS *Professor*
5 Khan RM *Professor* wheat improvement
6 Mathur BN *Associate Professor* bajra improvement
7 Saxena RC *Associate Professor* maize entomology
8 Sharma LS *Associate Professor* sorghum improvement
9 Sharma SK *Professor*
10 Singh V *Associate Professor* pesticide residue
11 Vaish OP *Associate Professor* cotton improvement

2506 Department of Biochemistry

1 Mathur PN *Associate Professor*
2 Namawati GC *Associate Professor*
3 Simlot MM *Professor*

2507 Department of Nematology

1 Mathur BN *Associate Professor* wheat improvement
2 Yadav BS *Professor*

2508 Department of Agricultural Statistics
8

1 Agrawal BL *Associate Professor*
2 Mathur HCK *Professor*
3 Mathur LBL *Associate Professor* pulse improvement

2509 Department of Plant Physiology

1 Chaturvedi SN *Associate Professor* seed technology
2 Kumar A - *Associate Professor*
3 Mathur JR *Professor*
4 Muralia RN *Associate Professor*
5 Singh K *Associate Professor*

2510 Department of Agricultural Economics

1 Choudhary SR *Associate Professor*
2 Kumar A *Associate Professor*
3 Pareek SN *Associate Professor*

2511 Department of Horticulture

1 Pathak SP *Associate Professor*
2 Pundrik KC *Associate Professor* floriculture
3 Verma SS *Associate Professor* fruit improvement

2512 Department of Extension Education

1 Chaupawat JS *Associate Professor* operational research project
2 Khan PM *Associate Professor* operational research project
3 Mundra SN *Associate Professor* extension education
4 Paliwal SC *Associate Professor* operational research project
5 Sharma BM *Professor*
6 Vyas DS *Associate Professor* operational research project

2513 Agricultural Engineering

1 Ali A *Associate Professor* water management
2 Bhatt YC
3 Mathur AN *Associate Professor* solar energy
4 Mishra AL *Associate Professor* water management
5 Singh J *Associate Professor* post harvest technology
6 Yadav RK *Associate Professor* management of salt affected soil

2514 Animal Science

1 Gahlot AK *Associate Professor* livestock management
2 Gupta MK *Associate Professor* veterinary science
3 Jain LS *Associate Professor* buffalo breeding

2515 Home Science

1 Choudhary Mrs M *Associate Professor* food and nutrition

2516 Tamil Nadu Agricultural University

Coimbatore, Tamil Nadu, 641 003
Tel: 35461. Telex: 855 360 TNAU IN. Cables: FARMVAR.

1 Rajagopalan V - PhD *Vice Chancellor*
2 Shanmugam N - PhD *Registrar*

2517 Agricultural College and Research Institute, Coimbatore

2518 Faculty of Agriculture

1 Guruswamy Raja VD - PhD *Dean*
2 Ramaswami KR - PhD *Dean Postgraduate Studies*
3 Gopalakrishnan V - MSc *Associate Professor* agricultural statistics
4 Murugesan M - PhD *Professor* agricultural statistics

2519 Department of Biochemistry

1 Sadasivam S - PhD *Professor* biochemistry
2 Thayumanavan B - PhD *Associate Professor* biochemistry

2520 Directorate of Research, Agriculture

1 Balasubramanian M - PhD *Director*
2 Guruswamy M - PhD *Associate Professor* soil science and agricultural chemistry
3 Radha Krishna T - MSc *Associate Professor* statistics
4 Rajendran K - MScAgr *Associate Professor* agricultural economics
5 Ramaswamy C - PhD *Professor* agronomy

2521 Water Technology Centre, Coimbatore

1 Augustine Selvaseelan D - PhD *Associate Professor* soil science and agricultural chemistry
2 Kulandaivelu R - PhD *Professor* agronomy
3 Padma Kumari O - BScEngg *Associate Professor* soil and water conservation
4 Radhakrishnan R - PhD *Associate Professor* crop physiology
5 Ramanathan G - PhD *Professor* soil science and agricultural chemistry
6 Sivanappan RK - MTech *Director*
7 Swaminathan LP - MScAgr *Associate Professor* agricultural economics

2522 Department of Agricultural Processing

1 Devadas CT - MEAgr *Associate Professor* post harvest technology
2 Gohandapani L - MScEngg *Associate Professor* processing
3 Natarajan CT - PhD *Associate Professor* agronomy

2523 Department of Agro Industries

1 Balasubramanian M - PhD *Associate Professor* farm implements
2 Sreenarayanan VV - PhD *Associate Professor* farm implements

2524 Department of Soil and Water Conservation

1 Chandrasekaran D - PhD *Professor*

2525 Department of Agricultural Structures

1 Karai Gowder KR - MEAgr *Professor*

2526 College of Agricultural Engineering

1 Subramaniyan V - PhD *Dean*

2527 Department of Bio-energy

1 Swaminathan KR - MTech *Professor*
2 Kunhi MKT - ME *Associate Professor* bio-energy
3 Murugesan V - MEAgr *Associate Professor* energy requirement
4 Rajasekaran P - PhD *Associate Professor* agricultural microbiology
5 Ramanathan M - MEAgr *Associate Professor* wind energy
6 Ramaswamy K - PhD *Associate Professor* agricultural microbiology
7 Sampatharajan A - MEAgr *Associate Professor* bioenergy
8 Vijayaraghavan NC - MEAgr *Associate Professor* solar energy

2528 Department of Farm Machinery

1 Karunanithi R - MEAgr *Professor*
2 Job TV - MEAgr *Associate Professor* power tillers
3 Rangaswamy K - MEAgr *Associate Professor* farm implements

2529 Zonal Research Centre

1 Manian R - ME *Professor* farm implements

2530 School of Genetics

1 Antony Raj S - PhD *Associate Professor* pulses
2 Ayyasamy MK - MSc *Associate Professor* statistics
3 Balasubramaniam A - MScAgr *Associate Professor* millet agronomy
4 Balasubramanian M - PhD *Professor* breeding pulses
5 Dorairaj SM - PhD *Professor* genetics
6 Doraiswamy S - PhD *Associate Professor* pulses pathology
7 Henry Louis I - PhD *Associate Professor* cytogenetics

8 Jawad Hussain HS - MScAgr *Associate Professor* breeding pulses
9 Jayaprakasam M - MScAgr *Associate Professor* pulses biochemistry
10 Jayaraman N - MScAgr *Associate Professor* breeding millets
11 Kadambavanasundaram M - PhD *Professor* breeding cotton
12 Manamohanlal S - MScAgr *Associate Professor* breeding rice
13 Manimekalai G - PhD *Associate Professor* tissue culture
14 MD Alikhan W - MScAgr *Associate Professor* breeding cotton
15 Mohamed Sheriff N - PhD *Associate Professor* oil seeds
16 Mohanasundram K - MScAgr *Associate Professor* breeding sorghum
17 Muralidharan V - MScAgr *Associate Professor* breeding pulses
18 Murugesan S - PhD *Associate Professor* cotton entomology
19 Mylswamy V - MSCAgr *Associate Professor* breeding cotton
20 Natarajan S - PhD *Associate Professor* rice pathology
21 Palaniswamy GA - PhD *Associate Professor* breeding rice
22 Palaniswamy S - PhD *Professor* breeding millets
23 Palaniswamy S - PhD *Professor* breeding rice
24 Pallikonda PRK - MSCAgr *Associate Professor* breeding pulses
25 Pothiraj P - PhD *Associate Professor* cotton agronomy
26 Rajagopal CK - MScAgr,PGDip *Professor* breeding pulses
27 Rajan AV - MScAgr *Associate Professor* pulses
28 Rangaswamy M - PhD *Associate Professor* breeding rice
29 Sadasivam R - MScAgr *Associate Professor* pulse physiology
30 Selvaraj V - PhD *Associate Professor* breeding pulses
31 Sree RSR - PhD *Director* plant breeding and genetics
32 Sridharan CS - MScAgr *Associate Professor* breeding oilseeds
33 Suresh S - MScAgr *Associate Professor* breeding millets
34 Thangamuthu GS - PhD *Associate Professor*
35 Velusamy R - PhD *Associate Professor*

2531 Department of Forage Crops

1 Chandrasekaran NR - MScAgr *Professor and Head*
2 Seshadri P - MScAgr *Associate Professor* agronomy and forage
3 Subramaniam S - PhD *Associate Professor* breeding fodder crops
4 Surendran C - PhD *Associate Professor*
5 Vijendra Das LD - PhD *Associate Professor*

2532 Directorate of Soil and Crop Management Studies

1 Subramanian S - PhD *Director*

2533 Soil Science Department

1 Kothandaraman GV - PhD *Professor*
2 Francis HJ - PhD *Associate Professor* soils
3 Helkiah J - PhD *Associate Professor* fertilizers
4 Kumaraswamy K - PhD *Associate Professor* isotopes
5 Manickam TS - PhD *Professor* soil science
6 Mathan KK - PhD *Associate Professor* soil physics
7 Nagarajan R - MScAgr *Associate Professor* organic chemistry
8 Perumal R - PhD *Associate Professor* soil test
9 Ravikumar V - PhD *Professor* soil physics
10 Savithri Mrs P *Associate Professor* fertilizers
11 Selvakumari G - PhD *Associate Professor* soils

2534 Department of Agronomy

1 Palaniappan SP - PhD *Professor*
2 Balasubramanian N - MScAgr *Associate Professor* crop cultivation
3 Chinnasamy KN - PhD *Associate Professor*
4 Gopalasamy N - PhD *Associate Professor*
5 Kempu CN - MScAgr *Associate Professor* weed control
6 Sennaian P - MScAgr *Professor* agronomy
7 Shanmugasundaram VS - PhD *Associate Professor*

2535 Horticulture Faculty

1 Irulappan I - PhD *Associate Professor* vegetables

2536 Department of Spices

1 Md Abdul Kader JBM - PhD *Professor*
2 Mohideen KM - PhD *Associate Professor*

2537 Department of Floriculture

1 Nanjan K - PhD *Associate Professor* floriculture
2 Rangaswami P - MSc *Associate Professor* vegetables

2538 Department of Olericulture

1 Thamburaj S - MScAgr *Associate Professor* vegetables

2539 Department of Pomology

1 Sundararajan S - MScAgr *Professor*
2 Kulasekaran M - MSc *Professor* horticulture
3 Veeranah L - PhD *Associate Professor* crop physiology
4 Vijayan KP - MScAgr *Associate Professor* crop physiology

2540 Department of Seed Technology

1 Karivaratharaju TV - PhD *Professor*
2 Selvaraj JA - PhD *Associate Professor* seed technology

2541 Department of Agricultural Microbiology

1 Oblisamy G - PhD *Professor*
2 Kandasamy D - PhD *Associate Professor* agricultural microbiology
3 Kannaiyan S - PhD *Associate Professor* Azolla
4 Krishnan SP - MScAgr *Associate Professor* biological nitrogen
5 Purushothaman D - PhD *Associate Professor*
6 Rangarajan R - PhD *Professor* microbiology

2542 Department of Crop Physiology

1 Nataraja Rathinain RN - PhD *Professor and Head*
2 Moosa Sheriff M - PhD *Associate Professor*

2543 Directorate of CARDS

1 Balakrishnan V - MScAgr economic structure
2 Kandaswamy A - PhD *Director*
3 Subramanian SR - PhD *Professor* rural economics
4 Sudarsanam R - MA,MEd,MSc,PhD *Associate Professor*

2544 Department of Agricultural Extension and Rural Sociology

1 Perumal G - PhD *Professor*
2 Nanjaiyan K - PhD *Associate Professor*
3 Ranganathan G - MScAgr *Associate Professor*
4 Sripal KB - PhD *Associate Professor*

2545 Department of Agricultural Economics

1 Aiyasamy PK - PhD *Professor*
2 Krishnamoorthy S - PhD *Associate Professor*
3 Palanisamy K - PhD *Associate Professor*
4 Ramasamy C - PhD *Associate Professor* growth linkage
5 Srinivasan N - PhD *Associate Professor* marketing management
6 Varadarajan S - PhD *Professor* marketing management

2546 Directorate of Plant Protection Studies

1 Jeyaraj S - PhD *Director*
2 Radha Mrs NV - PhD *Associate Professor* sericulture

2547 Department of Plant Pathology

1 Jeyarajan R - PhD *Professor*
2 Bhaskaran R - PhD *Associate Professor* pathology
3 Naryanaswamy P - PhD *Professor* rice virus
4 Sivaprakasam K - PhD *Associate Professor* mushrooms
5 Venkata Rao A - PhD *Associate Professor* biological control

2548 Department of Agricultural Entomology

1 Mohanasundaram M - PhD *Professor* entomology
2 Ayyavoo R - PhD *Associate Professor* plant protection
3 Paramasivam KS - MScAgr *Associate Professor* breeding rice
4 Rabindra RJ - PhD *Associate Professor* plant protection
5 Raghupathy A - MScAgr *Associate Professor* pesticides
6 Rajukannu K - PhD *Associate Professor* pesticides
7 Rangarajan AV - MA *Professor* plant protection
8 Subramanian A - MScAgr *Associate Professor* entomology
9 Sundarababu PC - PhD *Professor* rice entomology
10 Vijayaraghavan S - MA *Associate Professor* insects

2549 Department of Nematology

1 Vadivelu Mrs S - PhD *Professor and Head*
2 Gunasekaran CR - MScAgr *Associate Professor* nematology
3 Lakshmanan PL - PhD *Associate Professor* nematology

2550 Agricultural College and Research Institute
Madurai, Tamil Nadu, 625 104
Tel: 42451.

1 Kumaraswami T - PhD *Dean*

2551 Department of Agronomy

1 Purushothaman S - PhD *Professor* water management studies
2 Ramiah S - PhD *Professor*
3 Subramanian CP - MScAgr *Associates Professor* plant pathology
4 Subramanian P - MScAgr *Associate Professor*

2552 Agricultural Botany

1 Soundarapandian G - PhD *Professor*
2 Devarathinam AA - PhD *Associate Professor* agricultural botany
3 Nagarajan M - PhD *Associate Professor* crop physiology
4 Rajasekaran S - PhD *Associate Professor*
5 Ramalingam C - MScAgr *Associate Professor* seed technology
6 Vaithialingam R - PhD *Associate Professor* crop physiology

2553 Soil Science and Agricultural Chemistry

1 Gopalswami A - PhD *Professor*
2 Govindasamy M - PhD *Associate Professor*

2554 Department of Agricultural Entomology

1 Gopalan M - PhD *Professor*
2 Gunathilagaraj K - PhD *Associate Professor* insect pathology
3 Janarthanan R - PhD *Associate Professor* entomology
4 Vasudevamenon PP - BSc *Associate Professor* nematology
5 Venugopal M - PhD *Associate Professor* rice entomology

2555 Department of Plant Pathology

1 Mariappan V - PhD *Professor*
2 Narasimhan V - PhD *Associate Professor* rice virus
3 Subramaniam KS - MScAgr *Associate Professor* pathology

2556 Department of Horticulture

1 Doraipandian A - PhD *Associate Professor*
2 Subbiah R - PhD *Associate Professor*

2557 Department of Agricultural Extension and Rural Sociology

1 Muthiah M - PhD *Professor*
2 Annamalai R - PhD *Associate Professor* extension

2558 Department of Agricultural Economics

1 Heenakshisundram V - PhD *Professor and Head*
2 Sivananthan M - PhD *Associate Professor*
3 Srinivasan R - PhD *Associate Professor*

2559 Department of Agricultural Microbiology

1 Balasubramanian A - PhD *Professor*
2 Muthukrishnan P - MScAgr *Associate Professor* microbiology
3 Padmanabhan C - PhD *Associate Professor* environmental biology

2560 Department of Physical Science

1 Subbiah S - PhD *Associate Professor* soil science and agricultural chemistry
2 Rangaswamy R - MScStat *Associate Professor* statistics

2561 Department of Animal Husbandry

1 Lakshmipathi V - MVSc *Associate Professor* animal nutrition

2562 Department of Food Science and Nutrition

1 Neelakandan S - PhD *Professor* food technology
2 Palanisamy A - PhD *Associate Professor* plant pathology

2563 Agricultural Research Station, Bhavanisagar

1 Chamy A - PhD *Professor*
2 Gopalan A - PhDAgr *Associate Professor* agricultural botany
3 Iruthayaraj MR - PhD *Professor* agronomy;water management studies
4 Raja Mrs IM - MScAgr *Associate Professor* soil science and agricultural chemistry
5 Rangiah PK - PhD *Associate Professor* agronomy
6 Selvaraj KV - PhD *Associate Professor* agronomy
7 Sukumaran N - PhD *Associate Professor* fisheries
8 Sundaram N - MScAgr *Professor* agricultural botany
9 Thangamani Narayanaswamy Mrs C - MScAgr *Associate Professor* plant pathology
10 Thiagarajan NM - MScAgr *Associate Professor* soil science and agricultural chemistry
11 Vadivelu KK - PhD *Associate Professor* seed technology

2564 Regional Research Station, Paiyur

1 Narayanan A - PhD *Professor*
2 Alagesan V - MScAgr *Associate Professor* agricultural extension
3 Azhakiamanavalan RS - PhD *Associate Professor* horticulture
4 Balasubramanian G - PhD *Associate Professor* agricultural entomology

5 Navakodi K - BScAgr *Associate Professor* breeding
6 Palaniappan TA - BScAgr *Professor* agricultural botany
7 Rajendran P - MScAgr *Associate Professor* agronomy
8 Ramachandran K - PhD *Associate Professor* crop physiology
9 Ramasamy N - PhD *Associate Professor* horticulture
10 Ramasamy R - MScAgr *Associate Professor* agronomy
11 Ramaswamy R - MSc *Associate Professor* plant pathology
12 Vaidhyanathan P - PhD *Associate Professor* plant breeding
13 Viswanathan G - MScAgr *Associate Professor* soil science and agricultural chemistry

2565 Horticulture Research Station, Kodaikanal

1 Balakrishnan R - PhD *Professor*
2 Chockalingam P - MSc *Associate Professor* horticulture
3 Natarajan K - PhD *Associate Professor* entomology
4 Ramanathan S - MScAgr *Associate Professor* soil science and agricultural chemistry
5 Sampath V - BSc,DIH *Associate Professor* horticulture

2566 Tamil Nadu Agricultural University Research Centre, Salem

1 Sundaresan R - PhD *Associate Professor*
2 Santhanam G - MScAgr *Associate Professor* entomology
3 Vasudevan P - MScAgr *Associate Professor* soil science and agricultural chemistry

2567 National Pulses Research Centre, Pudukkottai

1 Rathinasamy R - PhD *Professor* breeding pulses
2 Arjunan G - PhD *Associate Professor* plant pathology
3 Damodaran A - MScAgr *Associate Professor* agronomy
4 Jahangir KS - PhD *Professor* breeding pulses
5 Nadanam N - PhD *Associate Professor* agronomy
6 Natarajan T - MScAgr *Associate Professor* microbiology
7 Shanmugam M - MScAgr *Associate Professor* soil science and agricultural chemistry
8 Subba Rao PV - MScAgr *Associate Professor* entomology

2568 Horticultural Research Station, Periyakulam

1 Sambandamurthi S - PhD *Professor* horticulture
2 Dharmalingam C - PhD *Associate Professor* seed technology
3 Durairaj P - PhD *Associate Professor* plant pathology
4 Suthanthirapandian IR - MScAgr *Associate Professor* horticulture
5 Thangaraj T - MSc *Associate Professor* food science

2569 Agricultural Research Station, Kovilpatti

1 Iyyemperumal S - PhD *Professor*
2 Balasubramanian TN - MScAgr *Associate Professor* agronomy
3 Muthuvel P - PhD *Associate Professor* soil science and agricultural chemistry
4 Ranganathan AK - MScAgr *Associate Professor* agronomy
5 Sreesamulu VS - PhD *Professor* soil science and agricultural chemistry
6 Subbaraman N - PhD *Associate Professor* plant breeding
7 Sundaresan N - PhD *Associate Professor* breeding

2570 Tamil Nadu Rice Research Institute, Aduthurai

1 Chelliah S - PhD *Director*
2 Abdul Kareem A - PhD *Professor* agricultural entomology
3 Arokiaraj A - MScAgr *Associate Professor* agronomy
4 Giridharan S - MScAgr *Associate Professor* agricultural botany
5 Lewin Devasahagam H - MScAgr *Associate Professor* plant pathology
6 Ramanathan KM - PhD *Professor* soil chemistry
7 Saivaraj K - PhD *Associate Professor* entomology
8 Sankaran S - PhD *Professor* agronomy
9 Sheik Dawood A - MScAgr *Associate Professor* agronomy
10 Sivasubramaniam V - PhD *Professor* rice
11 Subramanian M - PhD *Associate Professor* plant breeding
12 Thangaraj M - PhD *Associate Professor* crop physiology
13 Thiagarajan CP - PhD *Associate Professor* seed technology
14 Thiagarajan P - MScAgr *Associate Professor* soil science and agricultural chemistry
15 Uthamasamy S - PhD *Associate Professor* entomology
16 Vairavan S - MSc *Associate Professor* agricultural botany

2571 Coconut Research Station, Veppankulam

1 Ramanathan P - PhD *Professor*
2 Ramadoss N - MScAgr *Associate Professor* plant pathology
3 Ravindran TS - PhD *Associate Professor* genetics

2572 Regional Research Station, Virdhachalam

1 Bhat V - MScAgr *Professor*
2 Kandasamy G - MScAgr *Associate Professor* agricultural botany

3 Loganathan S - PhD *Associate Professor* soil science and agricultural chemistry
4 Mahadevan NR - PhD *Associate Professor* entomology
5 Muthusamy M - PhD *Associate Professor* plant pathology
6 Palanichamy KP - PhD *Associate Professor* horticulture
7 Ramalingam SR - PhD *Professor* cytogenetics
8 Ramdoss G - MScAgr *Associate Professor* agronomy
9 Srinivasan PS - PhD *Associate Professor* crop physiology

2573 Kumaraperumal Farm Science Centre, Trichy

1 Kadirvel AK - BScAgr *Principal*
2 Baskaran TL - MScAgr *Associate Professor* plant pathology
3 Letchoumanane S - MScAgr *Associate Professor* entomology
4 Ramakrishnan S - PhD *Associate Professor* agronomy
5 Sathiamurthy S - MScAgr *Associate Professor* horticulture

2574 Regional Research Station, Aruppukottai

1 Subbarayalu M - PhD *Professor*
2 Anchanam Alagiapillai O - MScAgr *Associate Professor* horticulture
3 Marimuthu N - PhD *Associate Professor* veterinary
4 Natrajan B - PhD *Associate Professor* agricultural economics
5 Rengasamy P - PhD *Associate Professor* breeding
6 Sadayappan S - PhD *Associate Professor* agronomy
7 Seethogarnan K - MScAgr *Associate Professor* plant pathology
8 Subramanain TL - PhD *Associate Professor* soil science
9 Thandapani V - PhD *Associate Professor* plant pathology
10 Vijayakumar R - MVSc *Associate Professor* animal husbandry

2575 Soil Salinity Research Centre, Trichy

1 Rajamannar A - PhD *Professor*
2 Ayyamperumal A - PhD *Associate Professor* botany
3 Krishnamurthy VS - PhD *Associate Professor* soil science and agricultural chemistry
4 Ramakrishnan MS - PhD *Associate Professor* agronomy
5 Thiagarajan TM - MScAgr *Associate Professor* soil science and agricultural chemistry

2576 Crop Pattern Scheme, Tirunelveli

Madurai, Tamil Nadu, 625104
Tel: 42451.

1 Narayanan TR - MScAgr *Associate Professor*
2 Ramakrishnan G - PhD *Associate Professor* plant pathology

2577 Soil and Water Management Research Institute, Thanjavur

1 Kandaswamy P - PhD *Professor*
2 Robinson L - MScAgr *Associate Professor* agronomy

2578 All India Coordinated Research Project on Betelvine Diseases, Velur

1 Marimuthu T - PhD *Associate Professor*

2579 University Research Centre, Vellore

1 Subbaiah KK - PhD *Professor*
2 Balakrishna VK - MScAgr *Associate Professor* forage
3 Fazlullah Khan AK - PhD *Associate Professor* breeding
4 Jagannathan NT - MScAgr *Associate Professor* agronomy
5 Ramaiah M - PhD *Associate Professor* plant pathology
6 Ranganathan K - PhD *Associate Professor* plant pathology

2580 Cotton Research Station, Srivilliputhur

1 Prasad MN - PhD *Professor*
2 Dason AA - MScAgr *Associate Professor* agronomy
3 Mohamed Hanifa A - MScAgr *Associate Professor* entomology
4 Paulas D - MScAgr *Associate Professor* horticulture
5 Subramanian S - MScAgr *Associate Professor* crop physiology

2581 Horticultural Research Station, Thadiyankudisai

1 Arumugam R - PhD *Professor*
2 Anbu S - MScAgr *Associate Professor* horticulture
3 Devarajan L - PhD *Associate Professor* soil science and agricultural chemistry
4 Parameswaran S - PhD *Associate Professor* entomology
5 Parthasarathy R - BE,MSc *Associate Professor* agricultural engineering
6 Rajamanickam G - MA,BEd *Associate Professor* agricultural extension
7 Ramasamy KP - PhD *Associate Professor* agronomy
8 Subbaraja KT - PhD *Associate Professor* plant virology

2582 Forestry Research Station, Mettupalayam

1 Sivaraman MR - PhD *Professor*

2 Dasthakir MG - PhD *Associate Professor* agro-forestry
3 Jambulingam R - MSc *Associate Professor* new tree species
4 Vinai Rai RS - PhD *Associate Professor* agro-forestry

2583 Horticultural Research Station, Yercaud

1 Seemanthini Ramadass MB - PhD *Professor*
2 Azeez Basha A - MScAgr *Associate Professor* entomology
3 Narasimhan CRL - PhD *Associate Professor* soil science and agricultural chemistry
4 Sayed S - PhD *Associate Professor* horticulture

2584 Agricultural Research Station, Aliyaranagar

1 Rajasekaran R - MScAgr *Associate Professor*

2585 Agricultural College, (Killikulam) Tirunelveli

1 Kandaswamy TK - PhD *Dean*
2 Chandrasekaran A - MScAgr *Associate Professor* soil science and agricultural chemistry
3 Dhanakodi CV - MScAgr *Associate Professor* breeding
4 Ebanesar Raja E - PhD *Associate Professor* animal health
5 Krishnamohan G - PhD *Associate Professor* plant pathology
6 Krishnarajan J - PhD *Associate Professor* agronomy
7 Krishnaswamy R - PhD *Associate Professor* soil science and agricultural chemistry
8 Mohamed Ali A - PhD *Professor* agronomy
9 Rajakrishnamoorthy V - MEAgr *Associate Professor* agricultural engineering
10 Sivasamy N - MScAgr *Associate Professor* breeding
11 Swamiappan M - PhD *Associate Professor* entomology
12 Vedamuthu PGB - MScAgr *Associate Professor* horticulture

2586 Fisheries College, Tuticorin

1 Jegatheesan G - PhD *Dean*
2 Selvaraj P - MScAgr *Associate Professor* fishery economics
3 Sundarajan V - PhD *Associate Professor*
4 Venkata Ramanujam K - PhD *Associate Professor* fish culture

2587 Paddy Research Station, Ambasamudram

1 Subramanian M - PhD *Professor*
2 Andi K - MScAgr *Associate Professor* soil science and agricultural chemistry
3 Balasubramanian S - MScAgr *Associate Professor* horticulture
4 Chockalingam S - MScAgr *Associate Professor* sugarcane research
5 Dharmalingam V - MScAgr *Associate Professor* botany
6 Gopal Rao TK - MScAgr *Associate Professor* water management
7 Mahalingam B - BScAgr *Associate Professor* botany
8 Mahalingam B - BScAgr *Associate Professor* botany
9 Mannel WW - BScAgr *Associate Professor* breeding rice
10 Mathar SA - MScAgr *Associate Professor* plant pathology
11 Padmanaban MD - MScAgr *Associate Professor* entomology
12 Padmanabhan D - MScAgr *Associate Professor* plant pathology
13 Pappiah CM - PhD *Associate Professor* multiple cropping experiments
14 Parameswaran P - MScAgr *Associate Professor* agronomy
15 Ramachandaran TK - PhD *Professor* operational research
16 Ramamurthi TG - BScAgr *Associate Professor* experiment on cultivators field
17 Ranganathan TB - PhD *Professor* poultry research
18 Robinson JG - MScAgr *Associate Professor* agronomy
19 Sakunthala Miss VA - MA,MSc *Associate Professor* sugarcane breeding
20 Sivakumar CV - PhD *Associate Professor* nematology
21 Swaminathan K - MScAgr *Associate Professor* experiment on cultivators field
22 Swaminathan R - BScAgr *Associate Professor* botany
23 Velayutham B - MScAgr *Associate Professor* nematology
24 Veluswamy P - PhD *Professor* sugarcane research
25 Venugopalan S - BScAgr *Associate Professor* agronomy
26 Veradarajan G - BScAgr *Associate Professor* nematology

2588 Madras Veterinary College
Madras, Tamil Nadu, 600 007

1 Kothandaraman P - PhD *Dean*

2589 Department of Anatomy

1 Sheshadri VK - BVSc,MSc *Professor* anatomy
2 Gopal MS - PhD *Associate Professor* anatomy

2590 Department of Nutrition

1 Shanmugasundram S - MVSc *Professor* animal nutrition
2 Jaganathan V - PhD *Associate Professor* animal nutrition

2591 Department of Animal Genetics
1 Ulaganathan V - PhD *Professor* animal genetics
2 Ramachandran S - PhD *Associate Professor* animal genetics
3 Sethu Madhavan V - PhD *Associate Professor* animal genetics

2592 Department of Animal Disease Investigation and Control
1 Mahalingam P - PhD *Professor* bacteriology
2 Koteeswaran A - MVSc *Associate Professor* microbiology
3 Mallihaswaran K - MVSc *Associate Professor* dairy science
4 Rahamathulla Khan GA - MVSc *Associate Professor* pathology

2593 Department of Clinics
1 Balu A - PhD *Professor*
2 Neduseralathan B - MVSc *Associate Professor* obstetrics and gynaecology
3 Parthasarathy KR - MVSc *Associate Professor* pathology
4 Rajan TSS - PhD *Associate Professor* therapeutics
5 Richard NG - MVSc *Associate Professor* surgery

2594 Department of Dairy Science
1 Abedullahkhan MM - PhD *Associate Professor* dairy science
2 Earnest J - PhD *Associate Professor* dairy science
3 Sadasivam P - PhD *Associate Professor* dairy science

2595 Department of Extension
1 Jayachandran T - PhD *Professor* dairy science
2 Selvaraj KM - MVSc *Associate Professor* meats

2596 Department of Livestock Production and Management
1 Shanmugasundaram S - PhD *Professor* poultry science
2 Thiyagarajan M - MVSc *Professor* hygiene

2597 Department of Microbiology and Biochemistry
1 Masilamani RR - PhD *Professor*
2 Janakiraman D - PhD *Professor*
3 Moses JS - PhD *Professor* microbiology
4 Padmanabhan VD - PhD *Associate Professor*
5 Venkatesan RA - PhD *Professor*
6 Venugopal AT - PhD *Associate Professor*

2598 Department of Meats
1 Arumugam MP - MVSc *Professor* animal nutrition
2 Govindarajan CV - MVSc *Associate Professor* meats
3 Kosalaraman VR - MVSc *Associate Professor* meats
4 Shanmugam AM - MVSc *Associate Professor* meats

2599 Department of Lab Animal Medicine
1 Venkatesan ES - MVSc *Associate Professor* medicine

2600 Department of Obstetrics and Gynaecology
1 Quayam SA - PhD *Professor* obstetrics and gynaecology
2 John D - PhD *Associate Professor* obstetrics and gynaecology
3 Pattabiraman SR - PhD *Associate Professor* obstetrics and gynaecology

2601 Department of Parasitology
1 Lalitha CM - MSc *Professor* parasitology
2 Anandhan R - PhD *Associate Professor* parasitology
3 Balasundaran S - PhD *Associate Professor* parasitology
4 Joseph SA - PhD *Associate Professor* parasitology
5 Kosi TJ - PhD *Associate Professor* parasitology
6 Rajavelu G - PhD *Associate Professor* parasitology

2602 Department of Pathology
1 Sundararaj A - PhD *Professor* pathology

2603 Department of Pharmacology
1 Natarajan R - MSc *Professor* pharmacology
2 Jayasunder S - PhD *Associate Professor* pharmacology

2604 Department of Preventive Medicine
1 Raghavan N - MVSc,DVSM,BPH *Professor*
2 Rathinam S - PhD *Associate Professor* public health
3 Thanappapillai M - MVSc *Associate Professor* preventive medicine

2605 Department of Physiology
1 Govindaraj R - PhD *Professor* physiology
2 Ahmed M - PhD *Associate Professor* biochemistry

2606 Department of Poultry Science
1 Jayaprasad AI - PhD *Professor* poultry science
2 Kumararaj R - MSc *Associate Professor* poultry science
3 Vedanayakam K - PhD *Associate Professor* animal nutrition
4 Viswanathan K - MVSc *Associate Professor* poultry science

2607 Department of Statistics
1 Venkatesan R - PhD *Professor* statistics
2 Radhakrishnan T - MScStat *Associate Professor* statistics

2608 Department of Surgery
1 Godfrey D - MVSc *Professor* surgery
2 Diwan MMMS - D *Associate Professor*

2609 Department of Therapeutics
1 Gnanaprakasam V - PhD *Professor*
2 Mohanalingam U - MVSc *Associate Professor* dairy science
3 Prabhakaran R - MSc *Associate Professor* dairy economics
4 Venkataraman R - MVSc *Associate Professor* therapeutics

2610 Poultry Research and Development Centre, Madurai
1 Renga Reddy P - MSc *Associate Professor* veterinary

2611 Kumaraperumal Farm Science Centre, Trichy
1 Ganesh D - PhD *Associate Professor* animal genetics

2612 Livestock Research and Development Centre, Tirunelveli
1 Balachandran S - MVSc *Associate Professor* animal genetics

2613 Livestock Research and Development Centre, Erode
1 Subramaniam AC - MVSc *Associate Professor* therapeutics
2 Rajasekaran J - MVSc *Associate Professor* obstetrics and gynaecology

2614 Livestock Research and Development Centre, Dharmpuri
1 Ramasamy PS - MVSc *Associate Professor* animal nutrition

2615 Livestock Research and Development Centre, Nagercoil
1 Albert A - MVSc *Associate Professor* microbiology

2616 Livestock Research and Development Centre, Tirupur
1 Ramachandran PV - PhD *Associate Professor* parasitology

2617 Poultry Research and Development Centre, Vellore
1 Sundaresan K - PhD *Associate Professor* poultry science

2618 Department of Animal Husbandry, Coimbatore
1 Manickam R - PhD *Associate Professor* preventive medicine

2619 Mechery Sheep Research Station, Potteneri
1 Venkatakrishnan R - PhD *Professor* animal nutrition

2620 Poultry Research and Development Centre, Namakkal
1 Narahari D - PhD *Associate Professor* poultry science

2621 Poultry Research and Development Centre, Pudukkottai
1 Nagarajan R - MVSc *Associate Professor* dairy science

2 Thanagaraj TM - MVSc *Director of Research* obstetrics and gynaecology
3 Venkataiyan S - MVSc *Director of Research* physiology
4 Viswanathan V - MVSc *Associate Professor* animal nutrition

2622 Poultry Research Station, Nandanam

1 Sundararasu V - PhD *Professor* poultry science

2623 Sheep Breeding Research Station, Sandyanallah

1 Natarajan N - PhD *Professor* animal genetics
2 Viswanathan MR - MVSc *Associate Professor* animal nutrition

2624 Livestock Research Station, Kattupakkam

1 Jothiraj S - MVSc *Professor* animal nutrition
2 Kathaperumal V - MVSc *Associate Professor*

2625 Livestock Research and Development Centre, Thanjavur

1 Thirumalai S - MVSc *Associate Professor* animal nutrition

2626 Veterinary College and Research Institute, Namakkal

1 Ramamurthi R - PhD *Dean*
2 Kandasamy N - MVSc *Associate Professor* genetics
3 Rajagopalan G - MVSc *Associate Professor* animal nutrition
4 Venkatakrishnan A - MVSc *Associate Professor* anatomy
5 Vijayakumar M - MScAgr *Associate Professor* horticulture

2627 Livestock Research Station, Kattupakkam

1 Muthu B - MVSc *Professor* animal nutrition
2 Bagavandoss - MScAgr *Associate Professor* botany
3 Balaraj CTA - MVSc *Associate Professor* parasitology
4 Balasubramanian R - MVSc *Associate Professor* therapeutics
5 Chandrasekaran K - MVSc *Associate Professor* dairy science
6 Dharmaraj FK - MVSc *Associate Professor* animal nutrition
7 Ernest A - MVSc *Associate Professor* therapeutics
8 Eswaran BM - PhD *Associate Professor* animal genetics
9 Jayaraman VS - MVSc *Associate Professor* animal nutrition
10 Jeyarajan S - MVScAgr *Associate Professor* animal genetics
11 Krishnan AR - PhD *Associate Professor* animal genetics
12 Palanisamy V - MScAgr *Associate Professor* horticulture
13 Ramiah KS - MVSc *Associate Professor* animal genetics

2628 Poultry Research and Development Centre, Trichy
Madras, Tamil Nadu, 600 007

1 Gajemdram K - MVSc *Associate Professor* poultry science

2629 Chandra Shekhar Azad University of Agriculture and Technology

Kanpur, Uttar Pradesh
Tel: 24089. Cables: AGRIVARSITY KANPUR.

1 Bharadwaj MBL - MSc,PhD *Vice-Chancellor*

2630 Faculty of Agriculture

1 Bhan S - MScAgr,PhD *Professor* soil conservation and water management
2 Goyal RD - MScAgr,PhD *Professor* seed technology
3 Kushwaha NS - MScAgr,PhD *Professor* animal husbandry
4 Mathur YK - MScAgr,PhD *Professor* entomology
5 Pathak AN - MScAgr,PhD *Professor* soils and agricultural chemistry
6 Rathi KS - MScAgr,PhD *Professor* agronomy
7 Saxena HK - MSc,PhD *Professor* crop physiology
8 Singh HG - MScAgr,PhD *Dean*
9 Singh HN - MScAgr,PhD *Professor* genetics and plant breeding
10 Singh RI - MScAgr,PhD *Professor* economics and statistics
11 Srivastava GP - MSc,PhD *Professor* bio-chemistry
12 Srivastava RK - MSc,PhD *Professor* agricultural engineering
13 Swaroop J - MSc,PhD *Professor* plant pathology
14 Willington M - MA,PhD *Professor* home science
15 Yadav JP - MScAgr,PhD *Professor* agricultural extension
16 Katiyar RP - MScAgr,PhD *Senior Scientist* pulses
17 Mehrotra DN - MScAgr,PhD *Director* research
18 Singh V - MScAgr,PhD *Professor* fertilizers
19 Srivastava AN - MScAgr,PhD *Senior Scientist* oilseeds
20 Tewari KN - MScAgr,PhD *Senior Scientist* soil salinity

2631 Faculty of Veterinary College, Mathura
Kanpur, Uttar Pradesh
Tel: 24089. Cables: AGRIVARSITY KANPUR.

1 Ahluwalia SS - MVSc,MS,PhD *Professor* parasitology
2 Barsaul GS - MVSc,PhD *Professor* veterinary animal nutrition
3 Chandra G - MVSc,PhD *Professor* veterinary anatomy
4 Dwevedi JN - MVSc,FRCGS,PhD *Professor* pathology
5 Pandey MD - MVSc,MRCVS,PhD *Professor* physiology
6 Pant HC - MVSc,PhD *Professor* gynaecology and reproduction
7 Pathak RC - MS,MVSc,PhD *Professor* bacteriology and virology
8 Rai P - MVSc,PhD *Professor* medicine
9 Singh BP - MS,PhD *Dean*
10 Srivastava RK - MVSc,PhD *Professor* biochemistry

2632 Aligarh Muslim University

Aligarh, Uttar Pradesh, 202 001
Tel: 4984. Cables: AGRIVARSITY KANPUR.

2633 Zoology Department
Aligarh, Uttar Pradesh, 202 001
Tel: 4984. Cables: AGRIVARSITY KANPUR.

2634 Entomology

1 Khan MA - MSc,PhD *Professor*
2 Agarwal MM - MSc,PhD *Professor* taxonomy of insects
3 Aziz SA - MSc,PhD *Professor* ecology of insects
4 Khan NH - MSc,PhD *Professor* physiology of insects

2635 Botany Department
2636 Nematology

1 Saxena SK - MSc,PhD *Professor* ecology of nematodes
2 Alam MM - MSc,PhD *Reader* ecology of nematodes
3 Hussain SI - MSc,PhD *Reader* taxonomy interaction
4 Jairajpuri MS - MSc,PhD *Professor* taxonomy of nematodes
5 Khan AM - MSc,PhD *Professor* ecology of nematodes
6 Khan MW - MSc,PhD *Reader*
7 Khan SH - MSc,PhD *Reader* taxonomy of nematodes
8 Siddiqui ZA - MSc,PhD *Reader* ecology of nematodes

2637 Plant Pathology

1 Mahmood K - MSc,PhD *Professor* virology
2 Prakash D - MSc,PhD *Reader* seed pathology

2638 University of Delhi

Delhi, 110007
Tel: 2918983.

2639 Department of Botany

1 Mukerji KG - PhD *Professor* mycology;plant pathology;microbiology
2 Babu CR - PhD *Professor* ecology and population genetics
3 Bhatnagar SP - PhD *Professor* gymnosperm and forest sciences
4 Bhojwani SS - PhD *Professor* tissue culture of crop plants;biotechnology
5 Dakshini KMM - PhD *Professor* ecology and population genetics
6 Ganapathy PS - PhD *Professor* tissue culture of crop plants;biotechnology
7 Gupta SC - PhD *Professor* tissue culture of crop plants;biotechnology
8 Maheshwari SC - PhD *Professor* tissue culture of crop plants;biotechnology
9 Raghavan MRV - PhD *Professor* phycology and nitrogen fixation
10 Ram MHY - PhD *Professor* tissue culture of crop plants;biotechnology
11 Rangaswami NS - PhD *Professor* tissue culture of crop plants;biotechnology
12 Rashid A - PhD *Professor* tissue culture of crop plants;biotechnology
13 Sachar RC - PhD *Professor* biochemistry
14 Shivanna KR - PhD *Professor* tissue culture of crop plants;biotechnology

2640 University of Hyderabad

Hyderabad, Andhra Pradesh, 500134
Tel: 31902.

2641 School of Life Sciences

1 Subba Rao K - PhD *Professor and Dean of the School* neurochemistry
2 Radhakrishanan AN - DIISc,PhD *Professor* enzymology
3 Radhakrishnamurthy CH - MSc,PhD *Reader* animal physiology
4 Raghavendra AS - MSc,PhD *Reader* plant physiology
5 Ramamurty PS - MSc,PhD,DiplnGerman *Professor* entomology reproductive physiology of insects
6 Reddy AR - MSc,PhD *Reader* biochemical and developmental genetics
7 Reddy PRK - MSc,PhD *Professor* reproductive endocrinology
8 Sharma RP - MSc,PhD *Reader* photobiology
9 Singh HN - MSc,PhD *Professor* nitrogen fixation

10 Subrahmanyam NC - MScAgr,PhD *Reader* genetics and plant molecular biology
11 Suryanarayana T - MSc,PhD *Reader* molecular biology
12 Thampan RV - MSc,PhD *Reader* reproductive endocrinology
13 Vijayan EV - MSc,PhD *Reader* neuroendocrinology

2642 University of Kalyani

Kalyani, West Bengal, 741235
Tel: 220,286,477,478.

2643 Faculty of Science

2644 Department of Zoology

1 Chakravorty S - MSc,PhD *Reader* entomology

2645 Punjabi University

Patiala, Punjab
Tel: 73261-63.

1 Johal SS - MA,PhD *Vice-Chancellor*

2646 Department of Botany

1 Bir SS - MSc,PhD *Professor* cytogenetics
2 Dargan JS - MSc,PhD *Reader* plant pathology
3 Gill BS - MSc *Professor* plant breeding
4 Kumar P - MSc,PhD *Reader* plant physiology
5 Saini SS - MSc *Professor* plant pathology
6 Sharma M - MSc,PhD *Reader* angiosperm taxonomy
7 Sharma TA - MSc,PhD *Reader* microbiology
8 Sidhu M - MSc,PhD *Reader* cytogenetics
9 Trivedi ML - MSc,PhD *Reader* morphology
10 Vasudeva SM - MSc,PhD *Reader* pteridology

2647 University of Manipur

Manipur, Imphal, 759003
Tel: 859 and 756. Cables: MANVERSITY.

1 Mahale KJ - MA,LLM,DLit *Vice-Chancellor*
2 Singh HT - MSc,PhD,FLS,FZSI *Dean of School of Sciences*

2648 Department of Life Science

Manipur, Imphal, 759003
Tel: 859 and 756. Cables: MANVERSITY.

1 Singh HT - MSc,PhD,FLS,FZSI *Professor* fishery and fish biology
2 Kumar S - MSc,PhD *Professor* genetics and plant physiology
3 Prasad B - BSc,PhD,FESI *Associate Professor* entomology
4 Singh TK - MSc,PhD,FESI *Associate Professor* entomology

2649 Manipur Agricultural College

Irdisemba, Imphal

1 Singh SB - MScAgr,Phd *Associate Professor and Principal* horticulture
2 Singh AI - MScAgr,PhD *Associate Professor* agronomy

2650 University of Meerut

Meerut, Uttar Pradesh
Tel: PBX 75021-24.

1 Mohan R *Vice-Chancellor*

2651 Department of Zoology

1 Pandey KC - MSc,PhD,DSc *Professor* fisheries

2652 Bose Institute

93/1, Acharya Prafulla Chandra Road, Calcutta, 700009
Tel: 35-7073,35-2402,35-2403,35-6619,35-6702,35-6790. Telex: 021-2646. Cables: BOSTUTE.

1 Biswas BB - MSc,PhD,FNA,FASc *Director* plant and protozoal gene organization;expression and cloning;myoinositol phosphates metabolism in plants

2653 Department of Chemistry

1 Barua AK - MSc,PhD *Professor* chemistry of plant products
2 Bhattachayya P - MSc,PhD *Reader* natural product chemistry with special reference to chemical taxonomy
3 Chakrabarti P - MSc,PhD *Professor* structure and functions of biomembranes of diverse origins
4 Chakraborty DP - MSc,PhD,DSc *Professor* chemistry of natural products and related compounds
5 Dutta J - MSc *Professor* lipid chemistry

6 Ghosh A - MSc,PhD *Reader* lipid chemistry
7 Ghosh D - MSc,PhD *Reader* prostaglandin biosynthesis
8 Sinha NK - MSc,PhD *Professor* physical, chemical and structural aspects of bioactive proteins

2654 Department of Biochemistry

1 Bhattacharyya B - MSc,PhD *Reader* protein chemistry
2 Biswas S - MSc,PhD *Reader* plant motility
3 Ghosh S - MSc,PhD *Professor* recombinant DNA; nitrogen fixation and plant-microbe interaction
4 Mandal NC - MSc,PhD *Reader* bacteriophage lambda genetics
5 Mandal RK - MSc,PhD *Professor* plant and animal gene organization

2655 Department of Botany

1 Bose S - MSc *Fil dr Professor* mutation and breeding
2 Chanda S - MSc,Dr rer nat,FPns,FPnS,FPbS,FCAI,FIAS *Professor* palynology; environmental biology in relation to allergy and plant pathology; pollen immunochemistry
3 Das A - MSc,PhD *Reader* crop genetics
4 Ganguly SN - PhD,DSc,FIC,CChem,FRSC *Reader* chemistry and physiology of plant growth substances
5 Gupta S - MSc,PhD *Reader* development biology
6 Majumdar B - MSc,PhD *Reader* genetical improvement of rice by hybridization
7 Mukherjee BB - MSc,PhD *Reader* environmental control of plant growth and development
8 Roychowdhury AK - MSc,PhD *Reader* genetic differentiation in Indian populations

2656 Department of Microbiology

1 Chakrabarty SL - MSc,PhD *Professor* biodegradation; antibiotics and microbial enzymes
2 Chandra AL - MSc,PhD *Reader* antibiotics waste leather utilization
3 Das A - MSc,PhD *Reader* strain development for industrial use for the production of organic acids
4 Mishra AK - MSc,PhD *Professor* biohydrometallurgy and microbial technology
5 Nanda G - MSc,PhD *Reader* amino acids production from soil isolated bacteria
6 Samamta TB - MSc,PhD *Reader* structure, function, sequence homology study of Cytochrome P450 in fungal microsomes metabolizing xenobiotics

2657 Department of Biophysics

1 Banerjee A - MSc,PhD *Reader* dynamics of macromolecular structure and model building; x-ray crystallography

2658 Department of Physiology Laboratory

1 Medda AK - MSc,DPhil,FNASc *Professor* comparative thyroid and sex hormone physiology
2 Ray AK - MSc,PhD *Reader* mechanism of thyroid hormone action; thyroid hormone biogenic amines interaction in animals

2659 Tissue Culture Laboratory

1 Sen SK - MScAgr,PhD,DSc *Professor* plant genetics; genetic engineering of crop plants

2660 Indian Grain Storage Institute

PO Box 10, Hapur, Uttar Pradesh
Tel: 2496. Cables: KHADYAKHOJ.

1 Lal S - MScAgr,PhD *Director*

2661 Engineering Research Division

PO Box 10, Hapur, Uttar Pradesh
Tel: 2496. Cables: KHADYAKHOJ.

1 Birewar BR - BE Agr *Senior Agriculture Engineer*

2662 Pest Control and Storage Research Division

1 Srivastava JL - MScAgr,PhD *Deputy Director*

2663 Microbiology Research Division

1 Srivastava SK - MSc *Deputy Director*

2664 Maharashtra Association for the Cultivation of Science

Law College Road, Pune, Maharashtra, 411004
Tel: 56357,53683. Cables: MACSCIENCE.

1 Godbole SH - MSc,PhD *Director*

2665 Department of Genetics and Plant Breeding

1 Patil VP - MSc,PhD *cytology, genetics and breeding of crop plants*
2 Rao VSP - MSc,PhD wheat cytogenetics breeding, plant breeding methodology

2666 Department of Mycology and Plant Pathology

1 Rao VG - MScAgr,PhD systematics, physiology and cytology of fungi

2667 Nimbkar Agricultural Research Institute
PO Box 23, Phaltan, Maharashtra, 415 523
Tel: 336,396,577. Cables: NIMSEED.

1 Nimbkar BV - MS *President*

2668 Agricultural Division

1 Nimbkar N - PhD *Director*
2 Basarkar CD - PhD *Entomologist* entomology
3 Deshmukh AK - PhD *Plant Breeder* safflower and sunflower
4 Ghanekar AR - MSc *Plant Physiologist* plant physiology
5 Josheph RD - MSc *Microbiologist* microbiology
6 Zende NJ - MSc *Botanist* botany
7 Ziroee G - MScAgr *Plant Breeder* cotton

2669 Fredrick Institute of Plant Protection and Toxicology
Padappai, Tamil Nadu, 601301
Tel: 487173. Telex: MS 041-7547 and 7432. Cables: COROCOMBI.

1 David BV - MScAgr,PhD *Director*

2670 Department of Toxicology

1 Balakrishnamurthy PB - MSc,PhD genetic toxicology
2 Chandran VR - BVSc *Deputy Director* toxicology
3 Giri DK - MVSc neurotoxicity in chicken
4 Murthy MV - MSc pharmacology; pesticide inhalation toxicity evaluations
5 Samanna HC - MSc,MVSc clinical biochemistry
6 Susheela M - MSc,PhD reproduction and teratogenicity studies in rats
7 Viraraghavan K - BVSc veterinary pathological evaluations

2671 Department of Weed Science
Padappai, Tamil Nadu, 601301
Tel: 487173. Telex: MS 041-7547 and 7432. Cables: COROCOMBI.

1 Theetharappan TS - MScAgr,PhD herbicides evaluation
2 Valliappan K - MScAgr *Scientist* herbicides evaluation

2672 Department of Entomology

1 Kandasamy C - MSc,PhD agricultural entomology
2 Clement P - MScAgr *Scientist* agricultural entomology

2673 Department of Plant Pathology

1 Ravichandran V - MSc,MPhil,PhD plant pathology
2 Philomena PA - MSc,PhD *Scientist* plant tissue culture

2674 Department of Pesticide Chemistry

1 Sundararajan R - MScAgr,PhD pesticide residue analysis

2675 Instrumentation Section

1 Kandasamy NR - MSc,PhD analytical and maintenance services

2676 Paddy Processing Research Centre
Tiruvarur, 610001
Tel: 2576. Cables: RESEARCH.

1 Vasan BS *Project Head* post harvest technology of rice preservation
2 Pillaiyar P - MScAgr,MISE *Deputy Manager* post harvest rice technology

2677 Regional Research Laboratory, CSIR
Canal Road, Jammu, 180001
Tel: 6368,6383,5540;PBX7641,7651. Telex: 377-231. Cables: REGLAB.

1 Chopra CL - PhD *Acting Director*
2 Dutt AK - MSc arboriculture
3 Grewal Mrs S - MSc tissue culture
4 Gupta Mrs S - MSc plant physiology
5 Jamwal PS - BSc plant survey
6 Kak SN - PhD plant breeding
7 Kapur BM - BScAgr publication and library
8 Kaul BL - PhD plant breeding

9 Khosla SN - PhD weed science
10 Raina BL - PhD food technology
11 Sarin YK - PhD plant ecology
12 Sastry KSM - PhD applied mycology
13 Saxena BP - MSc insect physiology
14 Sharma BM - PhD plant survey
15 Sobti SN - PhD plant breeding
16 Srivastava JB - MSc fisheries
17 Thakur RN - PhD applied mycology

2678 Branch Laboratory
Sanat Nagar, Srinagar, 190005
Tel: 77392,77401. Cables: REGLAB.

1 Dhar AK - PhD plant breeding
2 Kachroo JL - PhD mushroom cultivation
3 Kaul MK - PhD plant survey
4 Kaul TN - PhD mushroom cultivation

2679 Regional Research Laboratory, CSIR
Bhubaneswar, Orissa, 751013
Tel: 51638. Cables: RESEARCH.

1 Dutta PK - MScAgr,PhD *Project Coordinator* medicinal aromatic and economic plants

2680 Tea Research Association
6,Netaji Subhas Road, Calcutta, 700001
Tel: 583277,566406,569151.

1 Barooah HP - MBA *Chairman*

2681 Tocklai Experimental Station
Jorhat, Assam, 785008
Tel: Jorhat 54.

1 Banerjee B - MSc,MS,PhD,FAZ,FRES *Acting Director*

2682 Soils and Meteorology Department

1 Dey SK - BSc,AssocIARI tea plant nutrition, water management
2 Sen A - MScAgr,PhD tea plant nutrition
3 Singh B - MTech,PhD water technology

2683 Botany Department

1 Bezbaruah HP - MSc,PhD plant breeding; genetic engineering
2 Manivel L - MScAgr,PhD plant physiology
3 Singh ID - MScAgr,MSc,PhD plant breeding

2684 Agronomy Department

1 Sinha MP - MSc,PhD agronomy

2685 Plant Protection Department

1 Das SC - MSc *In Charge* pest and disease control

2686 Biochemistry Department

1 Dev Choudhury MN - MSc,PhD enzymology
2 Mahanta PK - MSc,PhD flavour technology
3 Ravindranath SD - MSc,PhD *Biochemist* biochemistry of tea
4 Ullah MR - MSc,PhD chemical withering (desiccation)

2687 Engineering Research and Development Department

1 Boruah TC - MSc tea machinery development

2688 Statistics Department

1 Biswas Ajit K - MSc methodology of experimentation; surveys
2 Biswas Asim K - BSc methodology of experimentation

2689 Agricultural Economics and Planning Units

1 Awasthi RC - MCom,LLB,PhD *Agricultural Economist* studies on tea economics
2 Sarkar AR - BSc planning and co-ordination

2690 Central Sericultural Research and Training Institute
Berhampore, West Bengal
Tel: BHP-46. Cables: SILKBOARD.

1 Subba Rao G - MSc,PhD *Director*

2691 Central Sericultural Research and Training Institute
Srirampuram, Manandavadi Road, Mysore-8
Tel: 21406,20757. Telex: 0846-203 CSRI IN. Cables: SILKBOARD.

1 Jolly MS - ASS,DNSc,PhD *Director*

2692 Sericulture

1 Baig M - MSc *Senior Research Officer* silkworm pathology
2 Basavaraj HK - MSc *Senior Research Officer* bivoltine breeding
3 Benchamin KV - MSc,PGDS *Deputy Director* silkworm physiology
4 Benchamin KV - MSc,PGDS *Deputy Director* seed technology
5 Biram Saheb NM - MSc *Senior Research Officer* grainage
6 Kumar P - MSc *Senior Research Officer* entomology
7 Nagaraj CS - MSc *Deputy Director* bivoltine breeding
8 Nagaraju J - MSc *Senior Research Officer* multivoltine breeding
9 Noamani MKR - MSc *Joint Director*
10 Vijayaraghavan K - MSc *Deputy Director* multivoltine breeding
11 Vindhya Mrs GS - MSc *Senior Research Officer* silkworm rearing technology and innovation

2693 Moriculture

1 Basavanna HM - BSc *Deputy Director* chemistry
2 Choudhury PC - MSc *Deputy Director* mulberry agronomy
3 Chowdhury SK *Deputy Director* reeling and fibre technology
4 Dandin SB - MSc,PhD *Deputy Director* irrigated mulberry breeding and genetics
5 Giridhar Mrs K - MSc *Senior Research Officer* statistics
6 Govindaiah - MSc *Senior Research Officer* mulberry agronomy
7 Sikdar AK - MSc *Deputy Director* rainfed mulberry breeding and genetics
8 Srinivasan EB - MSc *Senior Research Officer* chemistry
9 Suryanarayana N - MSc *Joint Director*

2694 Regional Sericultural Research Station
No. 22, II Stage, Koramangala Layout, Bangalore, Karnataka, 560034
Tel: 563270. Cables: SILKBOARD.

1 Rajanna L - MSc *Senior Research Officer*
2 Sarathchandra B - MSc,PhD *Joint Director*

2695 Regional Sericultural Research Station
PO Box 21, Coonoor, Tamil Nadu, 643101
Tel: 6413. Cables: SILKBOARD.

1 Satyanarayana R - MSc *Senior Research Officer*

2696 Regional Sericultural Research Station
6/682, Ramanagar, Ananthapur, Andhra Pradesh, 515004
Tel: 2864. Cables: SILEXT.

1 Kannantha V - MSc *Senior Research Officer*
2 Madhava Rao YR - MSc *Deputy Director*

2697 Regional Sericultural Research Station
425-A, 515, Kolada Beedhi, Chamarajanagar, Karnataka, 571313
Tel: 152. Cables: SILEXT.

1 Chandrasekhar DS - MSc *Senior Research Officer*
2 Raghavendra M - MSc *Senior Research Officer*
3 Thyagaraja BS - MSc,PhD *Joint Director*

2698 Regional Sericultural Research Station
No. 51, Court Road, Salem, Tamil Nadu, 636007
Tel: 67184. Cables: SILEXT.

1 Gopinath OK - MSc *Senior Research Officer*
2 Thangavelu K - MSc,PhD *Joint Director*
3 Veeraiah TM - MSc *Deputy Director*

2699 Coffee Board
PO Box 5366, No 1, Dr Ambadkar Veedhi, Bangalore, 560001

1 Warrior SK - IAS *Chairman*

2700 Central Coffee Research Institute
Coffee Research Station P.O., Chikmangalur District, Karnataka, 577117
Tel: 43.

1 Bhat PK - MScAgr *Head of the Division of Entomology/Nematology* insect pests suppression
2 Ramaiah PK - MScAgr,PhD *Director*
3 Rao WKM - MScAgr *Joint Director and Head of Division of Agronomy*
4 Sreenivasan MS - MSc *Head of the Division of Botany* plant improvement, horticulture, biometrics
5 Nirmala K - MSc *Mycologist* plant diseases and their control
6 Venkata Raman D - MSc *Plant Physiologist* crop physiology, waste utilisation

2701 Coffee Research Sub-Station
Chethalli, Kodagu, Karnataka
Tel: 24.

1 Azizuddin M - BSc *Officer-in-Charge* agronomy

2702 Regional Research Station
Kalpetta, Wynad, Kerala
Tel: 256.

1 Govindarajan TS - MSc *Deputy Director* mycology

2703 Regional Coffee Research Station
Raghavendra Nagar, Chintapalli Taluk, Andhra Pradesh
Tel: 26-Exch Narasipatnam.

1 Gopal HN - PhD *Deputy Director* research to standardise location specific technology

2704 Regional Coffee Research Station
Thandikudi, Madurai District, Tamil Nadu

Location Specific Studies

1 Ramachandran M - MSc *Scientific Officer* crop improvement

2705 Regional Coffee Research Station
Diphu, Assam

Location Specific Studies

1 Srinivasan CS - MSc,PhD

2706 Rubber Research Institute of India
Kottayam, 686009
Tel: 8311. Telex: 888 205 RUBR IN. Cables: RUBRBOARD.

1 Sethuraj MR - MScAgr,PhD *Director*
2 Jayarathnam K - MScAgr,PhD *Deputy Director* mycology and plant pathology
3 Mathew M - MScAgr *Deputy Director* agronomy and soil science
4 Nair VKG - BSc,MLibSc *Documentation Officer*
5 Panikkar AON - MSc,PhD *Deputy Director* botany
6 Radhakrishna Pillay PN - MSc *Joint Director*
7 Thomas EV - MSc,PhD *Deputy Director* chemistry and rubber technology
8 Vijayakumar KR - MSc,PhD *Deputy Director* plant physiology and exploitation

2707 Central Experiment Station
Chethackal, 689672
Tel: Edamon 43.

1 Raghunathan Nair N - BScAgr *Senior Superintendent*

2708 Research Complex for North Eastern Region
Research Complex (NE Region), RG Barua Road, Guwahati, 781003

1 Potty SN - MScAgr,PhD *Project Co-ordinator*

2709 Regional Research Station
Agartala, 799006
Tel: 5287.

1 Krishna Kumar AK - MScAgr *Deputy Director*

2710 Gujarat Fisheries Aquatic Sciences Research Institute
Okha, District Jamnagar, Gujarat, 361350

1 Chhaya ND - MSc,MSU *Deputy Commissioner* fishery biology
2 Bhaskaran M - MSc *Research Officer* marine biology
3 Dave KG - BA,Dip *Fisheries Assistant Director* inshore survey and gear technology
4 Mahavanshi TN - MSc *Assistant Research Officer* aquaculture and hatchery (prawn)
5 Parmar KG - BSc *Assistant Director* fish processing
6 Raval YB - MSc *Assistant Research Officer* pollution
7 Trivedi CR - MSc *Assistant Research Officer* fisheries algae, remote sensing

2711 Gujarat Fisheries Aquatic Sciences Research Station, Sikka

1 Patel MI - MSc *Research Officer* molluscan culture

2712 Gujarat Fisheries Aquatic Sciences Research Centre, Porbandar

1 Trivedi YA - MSc *Assistant Director* pomfret and bacteriology

2713 Gujarat Fisheries Aquatic Sciences Research Centre, Veraval
Okha, District Jamnagar, Gujarat, 361350

1 Bardai MK - BE *Supervisor* boat design and manufacture
2 Solomon PP - BSc *Supervisor* inland fisheries and gear technology

2714 International Crops Research Institute for the Semi-Arid Tropics (ICRISAT)
Patancheru, Andhra Pradesh, 502 324
Tel: Hyderabad 224016. Telex: 0152-203. Cables: CRISAT.

ICRISAT's mandate is to improve the yield, stability, and food quality of sorghum, pearl millet, chickpea, pigeonpea and groundnut crops basic to life in the semi-arid tropics and to develop farming systems that will make maximum use of the human and animal resources and the limited rainfall of the region.

1 Swindale LD - MS,PhD *Director General*
2 Ambrose SP - BSc,IAS *Principal Government Liaison Officer*
3 Goon M - MBA *Principal Administrator*
4 Gunasekera BCG - PhD *Principal Soil and Water Scientist*
5 Kanwar JS - MScAgr,PhD,FNA *Director of Research*
6 Phillips SJ - BS,EdM *Special Assistant to the DG for Education Affairs*
7 Reddy MSS - MScAgr,PhD *Scientist - II, Office of DR*

2715 Sorghum Improvement Program

1 Mughogho LK - DIC,PhD *Acting Program Leader, Principal Plant Pathologist*

2716 Breeding

1 Agarwal BL - MScAgr,PhD *Plant Breeder*
2 Mukuru SZ - MS,PhD *Principal Plant Breeder*
3 Murty DS - MSc,PhD *Plant Breeder*
4 Reddy BVS - MSc,PhD *Plant Breeder*
5 Vaidya PK - MScAgr,PhD *Plant Breeder*

2717 Entomology

1 Leuschner K - PhD *Principal Cereal Entomologist*
2 Sharma HC - MSc,PhD *Entomologist*
3 Taneja SL - MScAgr,PhD *Entomologist*

2718 Pathology

1 Bandopadhyay R - MScAgr,PhD *Plant Pathologist*
2 Pande S - MScAgr,PhD *Plant Pathologist*

2719 Physiology

1 Peacock JM - PhD *Principal Plant Physiologist*
2 Seetharama N - MSc,PhD *Plant Physiologist*
3 Soman P - MSc,PhD *Plant Physiologist*

2720 Millet Improvement Program

1 King SB - MSc,PhD *Acting Program Leader, Principal Plant Pathologist*

2721 Breeding

1 Chavan SB - MScAgr,PhD *Plant Breeder*
2 Rai KN - MSc,PhD *Plant Breeder*
3 Singh P - MScAgr,PhD *Plant Breeder*
4 Talukdar BS - MScAgr,PhD *Plant Breeder*
5 Witcombe JR - PhD *Principal Plant Breeder*

2722 Microbiology

1 Krishna KR - MScAgr,PhD *Microbiologist*
2 Lee KK - MS,PhD *Principal Cereal Microbiologist*
3 Wani SP - MScAgr,PhD *Microbiologist*

2723 Pathology

1 Singh SD - MSc,PhD *Plant Pathologist*
2 Thakur RP - MScAgr,PhD *Plant Pathologist*

2724 Physiology

1 Alagarswamy G - MScAgr,PhD *Plant Physiologist*
2 Bidinger FR - PhD *Principal Plant Physiologist*
3 Mahalakshmi V - MSc,PhD *Plant Physiologist*

2725 Pulses Improvement Program

1 Faris DG - MSA,PhD *Coordinator, Asian Grain Legume Program*
2 Nene YL - MScAgr,PhD *Program Leader, Principal Plant Pathologist*

2726 Agronomy

1 Ae N - MS,PhD *Associate Microbiologist*
2 Ariharaj J - BS *Associate Physiologist*
3 Chauhan YS - MSc,PhD *Agronomist, Physiology*
4 Johansen C - PhD *Principal Agronomist*
5 Kumar Rao JVDK - MScAgr,PhD *Agronomist, Microbiology*
6 Okada K - MS,PhD *Assistant Microbiologist*
7 Rupela OP - MSc,PhD *Agronomist, Microbiology*
8 Saxena NP - MScAgr,PhD *Agronomist, Physiology*

2727 Chickpea Breeding

1 Gowda CLL - MScAgr,PhD *Plant Breeder*
2 Kumar J - MSc,PhD *Plant Breeder*
3 Sethi SC - MSc,PhD *Plant Breeder*
4 Singh O - MScAgr,PhD *Plant Breeder*
5 Van Rheenen HA - MSc,PhD *Principal Plant Breeder*

2728 Pigeonpea Breeding

1 Gupta SC - MScAgr,PhD *Plant Breeder*
2 Jain KC - MScAgr,PhD *Plant Breeder*
3 Saxena KB - MScAgr,PhD *Plant Breeder*
4 Sharma D - MScAgr,PhD *Senior Plant Breeder*
5 Singh L - MScAgr,MSc,PhD *Principal Plant Breeder*

2729 Entomology

1 Lateef SS - MScAgr,Doc Agr *Entomologist*
2 Reed W - PhD *Principal Entomologist*
3 Sithanantham S - MScAgr,PhD *Entomologist*

2730 Pathology

1 Ghanekar AM - MScAgr,PhD *Plant Pathologist*
2 Haware MP - MScAgr,PhD *Plant Pathologist*
3 Reddy MV - MScAgr,PhD *Plant Pathologist*

2731 Groundnut Improvement Program

1 McDonald D - PhD *Acting Program Leader, Principal Plant Pathologist*

2732 Breeding

1 Dwivedi SL - MScAgr,PhD *Plant Breeder*
2 Reddy LJ - MScAgr,DIIT,PhD *Plant Breeder*
3 Vasudeva Rao MJ - MScAgr,PhD *Plant Breeder*

2733 Cytogenetics

1 Moss JP - DPhil *Principal Cytogeneticist*
2 Sastri DC - MSc,PhD *Cytogeneticist*
3 Singh AK - MSc,PhD *Cytogeneticist*

2734 Entomology

1 Amin PW - MSc,PhD *Entomologist*
2 Ranga Rao GV - MScAgr,PhD *Entomologist*
3 Wightman JA - MSc,PhD *Principal Entomologist*

2735 Pathology

1 Mehan VK - MSc,PhD *Plant Pathologist*
2 Reddy DVR - MScAgr,PhD *Principal Plant Virologist*
3 Subramanyam P - MSc,PhD *Plant Pathologist*

2736 Physiology

1 Nageswar Rao RC - MScAgr,PhD *Plant Physiologist*
2 Nambiar PTC - MSc,PhD *Microbiologist*
3 Ramraj VM - MSc,PhD *Plant Physiologist*
4 Williams JH - Mphil,Dphil *Principal Plant Physiologist*

2737 Resource Management Program

1 von Oppen M - MSc,PhD *Program Leader, Principal Economist*

2738 Agronomy

1 Burford JR - MAgSc,PhD *Principal Soil Chemist*
2 Hong CW - MSc,PhD *Principal Soil Chemist*
3 Huda AKS - MScAgr,PhD *Agroclimatologist*
4 King ABS - PhD *Cropping Systems Entomologist*
5 Laryea KB - MSc,PhD *Principal Soil Physicist*
6 Natarajan M - MSc,PhD *Agronomist*
7 Ong CK - PhD *Cropping Systems Agronomist*
8 Pawar CS - MSc,PhD *Entomologist*
9 Rao MR - MScAgr,PhD *Agronomist*
10 Reddy MS - MS,PhD *Agronomist*
11 Rego TJ - MSc,PhD *Soil Scientist*
12 Sahrawat KL - MSc,PhD *Soil Scientist*

13 Singh P - MSc,PhD *Soil Scientist*
14 Van Den Beldt RJ - PhD *Principal Agronomist*
15 Virmani SM - MScAgr,PhD *Principal Agroclimatologist*

2739 Engineering

1 Athavale RN - MSc,PhD *Senior Hydrologist*
2 Awadhwal NK - BTech,PhD *Agricultural Engineer/Soil Physicist*
3 Bansal RK - MTech *Agricultural Engineer*
4 Pathak P - MTech *Agricultural Engineer*
5 Sachan RC - MTech,PhD *Agricultural Engineer*
6 Schutt A - BiolEngg *Assistant Engineer, Soil Fertility*
7 Singh S - MScAgr,PhD *Soil Scientist*
8 Srivastava KL - MTech *Agricultural Engineer*
9 Takenaga T - DrAgrSc *Principal Agricultural Engineer*

2740 Economics

1 Ghodake RD - MScAg,PhD *Economist*
2 Jodha NS - MA,PhD *Sr. Economist*
3 Mueller RAE - Dr.Sc.Agr *Principal Economist*
4 Singh RP - MSc Ag,PhD *Economist*
5 Walker TS - MS,PhD *Principal Economist*

2741 Support Programs

2742 Genetic Resources

1 Appa Rao S - MScAgr,DIIT,PhD *Botanist*
2 Mengesha MH - MSc,PhD *Program Leader, Principal Germplasm Botanist*
3 Prasada Rao KE - MScAgr *Senior Botanist*
4 Pundir RPS - MScAgr,PhD *Botanist*
5 Ramanatha Rao V - MScAgr,PhD *Botanist*
6 Remanandan P - MSc,PhD *Botanist*

2743 Biochemistry

1 Jambunathan R - MSc,PhD *Principal Biochemist*
2 Singh U - MSc,PhD *Biochemist*
3 Sivaramakrishnan S - MSc,PhD *Biochemist/Plant Physiologist*
4 Subramaniam V - MScAgr,PhD *Biochemist*

2744 Fellowships and Training

1 Das Gupta SK - MScAgr,PhD *Senior Training Officer*
2 Diwakar B - MScAgr,PhD *Senior Training Officer*
3 Nagur T - MScAgr,PhD *Senior Training Officer*
4 Oswalt DL - MS,PhD *Principal Training Officer*
5 Singh F - MScAgr,PhD *Training Officer*

2745 Statistics

1 Gilliver B - Dip Stat *Principal Statistician*
2 Singh M - MSc,PhD *Statistician*

2746 Computer Services

1 Estes JW - MS *Computer Services Officer*

2747 Plant Quarantine

1 Varma BK - MScAgr *Chief Plant Quarantine Officer*

2748 Information Services

1 Fuccillo DA *Head Information Services*
2 Feakin Ms SD - BscAgr *Research Editor*
3 Wills JB - MA *Research Editor*

2749 Library and Documentation Services

1 Haravu LJ - MA,Assoc.Docum. *Manager*

2750 ICRISAT Sahelian Centre
B.P. 12404, Niamey, Niger
Tel: 72-25-29. Telex: 5406 NI. Cables: ICRISAT Niamey.

1 Gibbons RW - BS,Dip.Agr.Sc *Director ISC & W.A. Programs*
2 Anand Kumar K - MSc,PhD *Principal Millet Breeder*
3 Fussell LK - BScAgr *Principal Millet Agronomist*
4 Klaij MC - MSc,PhD *Principal Soil and Water Engineer*
5 Nwanze KF - MSc,PhD *Principal Entomologist*
6 Okiror SO - MS,PhD *Principal Millet Breeder*
7 Renard C - Dr.Sc.Agr *Principal Agronomist* cropping systems
8 Sivakumar MVK - MScAgr,PhD *Principal Agroclimatologist*
9 Werder J - MS,Dr.Sc.Tech *Principal Millet Pathologist*

2751 SADCC/ICRISAT
P.O. Box 776, Bulawayo, Zimbabwe
Tel: 79563. Telex: 3570 ZIMBABWE. Cables: MATGRIC, BULAWAYO.

2752 Sorghum and Millet Improvement Program
Zimbabwe

1 Bisht DS - BSc *Farm Development Specialist*
2 de Milliano WAJ - PhD *Principal Plant Pathologist*
3 Gupta SC - MScAgr,PhD *Principal Millet Breeder*
4 House LR - MS,PhD *Project Manager*
5 Obilana AB - MS,PhD *Principal Sorghum Breeder*

2753 ICRISAT
B.P. 4881, Ouagadougou (via Paris), Burkina Faso
Tel: 335995. Telex: BF. Cables: OUAGADOUGOU.

1 Lohani SN - PhD *Principal Millet Breeder*
2 Matlon PJ - MPA,PhD *Principal Production Economist*
3 Pattanayak CM - MSc,PhD *Team Leader regional sorghum coordinator*
4 Ramiah KV - MSc,PhD *Principal Cereal Breeder Striga*
5 Thomas MD - MSc,PhD *Principal Sorghum Pathologist*

2754 ICRISAT Chitedze Research Station
Private Bag 63, Lilongwe, Malawi
Tel: 76-72-26. Telex: 4466 UNDEVPRO MALAWI. Cables: UNDEVPRO, LILONGWE.

1 Bock KR - PhD *Team Leader, Principal Groundnut Pathologist*
2 Nigam SN - MScAgr,PhD *Principal Groundnut Breeder*

2755 ICRISAT Mali Program
c/o American Embassy, B.P. 34, Bomako (via Paris), Mali
Tel: 22-61-66 (Sotuba). Telex: 459 ILCA, BOMAKO.

1 Scheuring JB - MS,PhD *Team Leader, Principal Cereal Breeder*
2 Shetty SVR - MScAgr,PhD *Principal Agronomist*

2756 ICRISAT C/o OAU/STRC
J.P. 31 SAFGRAD, P.O. Box 30786, Nairobi, Kenya
Tel: 338544. Telex: 22893, NAIROBI. Cables: IBAR, NAIROBI.

1 Guiragossian V - MSc,PhD *Coordinator for Sorghum and Millets*

2757 UNDP/ICRISAT
P.O. Box 913, Khartoum, Sudan
Telex: 22214 UNDP, KHARTOUM. Cables: RESEARCH, ELOB.

1 Jain RP - MSc,M Phil,PhD *Principal Millet Breeder*

2758 ICARDA
P.O. Box 5466, Aleppo, Syria
Tel: 50465. Telex: ICARDA 31206 SY. Cables: ICARDA, SYRIA.

1 Singh KB - MScAgr,PhD *Principal Chickpea Breeder*

2759 CIMMYT
Apdo. Postal 6-641, Londres 40, Mexico 6, D.F., Mexico
Tel: 5854355. Telex: 1762060,CIMMYT. Cables: CENCIMMYT, MEXICO.

1 Paul CL - MScAgr,PhD *Principal Sorghum Agronomist*

2760 Pakistan Agricultural Research Council
National Agricultural Research Centre, P.O. Box 2125, Islamabad, Pakistan
Tel: 827468. Telex: 5604 PARC PK. Cables: AGRESCOUNCIL,ISLAMABAD.

1 Chacko MJ - MSc,Assoc IARI,PhD *Entomologist-in-Charge* investigations on natural enemies of insect pests and noxious weeds
2 Rahman MS - MScAgr,MS,PhD *Principal Chickpea Breeder/Pathologist*
3 Ramaseshiah G - BSc,BScAgr,DSc *Entomologist* investigations on natural enemies of insect pests and noxious weeds

2761 Government of Himachal Pradesh

Shimla, Himachal Pradesh
Tel: 3089. Cables: HINVETY, SHIMLA.

1 Mehta VL - BVSc,NDAH *Director* animal husbandry

2762 Government of Kerala

Vikas Bhavan Trivandrum, Kerala
Tel: 64480.

2763 Directorate of Agriculture

1 Suseelam P *Director of Agriculture* agriculture development
2 Jayasankaran CN *Additional Director of Agriculture* agricultural extension

2764 Directorate of Animal Husbandry

1 George V *Directorate of Animal Husbandry* animal health
2 Tharanath K *Director* animal health and veterinary
3 Govindankutty P *Additional Director* animal health
4 Nair Padmanashan DV *Chief Disease Investigation Officer* animal diseases

2765 Government of Madhya Pradesh

Bhopal, Madhya Pradesh
Tel: 66864. Cables: AGRI.

2766 Directorate of Agriculture

1 Mehta NN *Director of Agriculture* agricultural development

2767 Government of Manipur

Sanjenthong, Imphal, Manipur
Tel: 20166. Cables: DIRAGRIMPUR.

1 Singh NB *Director of Agriculture* planning research and implementation of agriculture programme
2 Devi KM *Agricultural Economist* agricultural economics and statistics
3 Kirankumar Singh RK *Potato Specialist* plant breeding
4 Sharma BH *Seed Analyst* seed multiplication
5 Singh AI *Deputy Director* extension
6 Singh AJ *Deputy Director of Agriculture* information
7 Singh AKU *Deputy Director* planning
8 Singh HK *Maize Breeder* plant breeding
9 Singh KI *Soil Chemist* soil testing
10 Singh KS *Special Officer* planning
11 Singh LM *Additional Director of Agriculture* extension
12 Singh MG *Deputy Director* mechanisation
13 Singh NY *Rice Breeder* plant breeding
14 Singh RKT *Project Officer* agricultural extension
15 Singh SA - PhD *Plant Protection Officer* plant protection

2768 Government of Nagaland

Kohima, Nagaland
Tel: 2546. Cables: AGRILAND.

2769 Department of Agriculture

1 Imchen T - BScAgr *Director*
2 Biswakarma G - MScAgr *Agronomist*
3 Chuba K - MScAgr *Horticulturist*
4 Sema Y - MScAgr *Joint Director*

2770 Government of Punjab

17 Bays Building, Sector 17, Chandigarh, Punjab
Tel: 31251. Cables: CHIEFFORESTSPONSAB.

2771 Conservator of Forests Office
17 Bays Building Sector 17, Chandigarh
Tel: 31251. Cables: CHIEFFORESTSPONSAB.

1 Chowdhry YP - MScAgr,AIFC,IFS *Additional Chief Conservator of Forests* forestry
2 Dogra AS - MA,MSc *Deputy Conservator of Forests* forestry; fortran programming
3 Rai MP - MSc,AIFC *Conservator of Forests* forestry
4 Sidhu - BScAgr,AIFC,IFS *Chief Conservator of Forests* forestry

2772 Government of Rajasthan

Jaipur, Rajasthan

2773 Department of Forest

1 Sahu YK *Assistant Conservator of Forests* tiger project
2 Singh T *Assistant Director* wild life

2774 Government of Tamil Nadu

P.B. No. 749, Madras, Tamil Nadu
Tel: 450199. Telex: 041 6716MEEN IIX. Cables: FISHERIES.

2775 Department of Fisheries
PB No 749, Madras
Tel: 450199. Telex: 041 6716MEEN 11X. Cables: FISHERIES.

1 Joseph KC - MSc *Joint Director of Fisheries* chemistry

2776 Research Division

1 Krishnan P - MSc *Deputy Director of Fisheries* mariculture
2 Muthuswamy R - MSc *Assistant Director Fisheries* marine biology

3 Rajappan K - BSc,DFSc,MTech *Assistant Director of Fisheries: mariculture*
4 Soundararaj R - BSc *Assistant Director of Fisheries* fresh water biology
5 Thangadurai AJ - BSc,DFPT *Assistant Director of Fisheries* hydro-biology
6 Venkatanarasinha Pillai - BSc,DFSc *Assistant Director of Fisheries* fisheries technology

2777 Government of Uttar Pradesh

Matsya Bhavan, Faizabad Road, Lucknow, Uttar Pradesh

2778 Department of Fisheries

1 Johri NK - MSc *Director* fisheries

2779 Department of Forests
Lucknow, Uttar Pradesh
Tel: 44968.

1 Chandra G *Chief Conservator Forests* forest management
2 Chaturvedi AN *Additional Chief Conservator Forests* research and development
3 Lohani DN *Principal Chief Conservator* forestry
4 Srivastava KB *Chief Conservator Forests* social forestry
5 Srivastava SS *Chief Conservator Forests (Hills)* forestry

2780 Government of West Bengal

Directorate of Forests P-16 India Exchange Place Extn, New CIT Building, 3rd Floor, Calcutta, 700 073

2781 Department of Forests

1 Basu PK - MScAgr *Divisional Forest Officer* silviculture
2 Roy Chaudhury PC - BSc,IFS *Conservator of Forests* forestry
3 Roy Chaudhury PK - BSc,IFS *Divisional Forest Officer* silviculture

2782 Government of Arunachal Pradesh

P.O. Itanagar, Arunchal Pradesh, 791 11
Tel: 361. Cables: CHIEFOR.

2783 Forest Department

1 Roy SB - BSc,IFS *Chief Conservator of Forests* forestry

2784 Silviculture Division

1 Deori ML - MSc,IFS *Divisional Forest Officer* silviculture
2 Haridasan K - MSc,PhD *Systematic Botanist*
3 Singh NB - MSc *Forest Geneticist*

2785 Orchid Research and Development Centre

1 Hegde SN - MSc,PhD *Orchidologist*
2 Rao AN - MSc,PhD *Assistant Orchidologist*

2786 Government of Union Territory Chandigarh

UT Secretariat, Sector - 9, Chandigarh
Tel: 41877,26076.

2787 Animal Husbandry Department

1 Kalha DS *Director*
2 Garg RC - BVSc & AH surgery
3 Kocher JC - BVSc & AH *Officer-in-Charge* animal husbandry
4 Kohli AVS - BVSc & AH pet animals
5 Sharma RM - BVSc & AH frozen semen technique
6 Singh VP - MA *Research Officer* animal husbandry; statistics

2788 Government of Goa, Daman & Diu

Vidhyut Bhavan, 4th Floor, Panaji-Goa, 403 001

2789 Directorate of Agriculture

1 Duarteum - MScAgr *Director of Agriculture*

2790 Union Territory of Lakshadweep

Kavaratti, 673555

2791 Directorate of Fisheries
Kavaratti, 673555

1 Verghese G - BA,Dip *Director* fisheries technology; fishery science

2792 Government of Meghalaya

Additional Secretatiate Building, Shillong, 793 004
Tel: 280 (PABX).

2793 Directorate of Agriculture

1 Syiemlich BJ - BScAgr *Director* agriculture

2794 Government of Pondicherry

Pondicherry
Tel: 5277. Cables: DIRECTORAGRICULTURE.

2795 Department of Agriculture

1 Jones SEA - BScAgr *Director of Agriculture*
2 Franchinathan SP - MScAgr,PhD *Additional Director of Agriculture*
3 Zacharia PP - BScAgr *Additional Director* biogas

2796 Govind Ballabh Pant University of Agriculture and Technology

Pantnagar, Uttar Pradesh, 263 145

1 Narain K - MSc *Vice-Chancellor*
2 Singh IP - PhD *Pro-Vice-Chancellor*
3 Ramatirth - MSc,MLibSc *Librarian*

2797 Directorate of Communication and Publications

1 Singh SS - PhD *Director*

2798 Directorate of Research

1 Modgal SC - PhD *Director* agronomy
2 Omanwar PK - PhD *Joint Director* soil science
3 Singh A - PhD *Joint Director* plant breeding
4 Singh V - PhD *Joint Director* agronomy

2799 Crop Research Centre

1 Dwivedi JL - PhD *Assistant Director* plant breeding

2800 Horticultural Research Centre

1 Lal G - PhD *Director* pomology and ornamental horticulture

2801 Livestock Research Centre

1 Chauhan SS - PhD *Assistant Director* animal nutrition
2 Singh SP - PhD *Associate Director* veterinary pathology

2802 Poultry Research Centre

1 Singh H - PhD *Assistant Director* animal breeding

2803 Fisheries Research and Training Centre

1 Singh CS - PhD *Senior Research Officer* limnology and fisheries

2804 Hill Campus, Ranichauri

1 Gautam PL - PhD *Joint Director* plant breeding
2 Lal M - MTech *Senior Research Officer* irrigation and drainage
3 Nautiyal MC - PhD *Senior Research Officer* horticulture

2805 Research Centre, Daurala

1 Shahe HN - PhD *Joint Director*
2 Singh B - MScAgr *Assistant Director* agronomy

2806 Cotton Research Centre, Bulandshahr

1 Sharma VP - BScAgr *Soil Scientist*
2 Singh UP - PhD *Officer I/C* botany

2807 Rice Research Centre, Nagina

1 Mishra OP - PhD *Assistant Agronomist* agronomy

2808 Rice Research Centre, Majhera

1 Ram B - PhD *Senior Scientist* horticulture

2809 Research and Testing Station, Pauri

1 Chauhan JS - BTech *Assistant Engineer*

2810 Directorate of Extension

1 Singh MP - PhD *Director Extension* extension

2811 College of Agriculture

1 Chaudhary RP - PhD *Dean* animal breeding

2812 Department of Agronomy

1 Sharma KC - PhD *Professor* soil fertility management and wheat agronomy
2 Chandel AS - PhD *Associate Professor* mineral nutrition and soybean
3 Gupta PC - PhD *Professor* seed technology
4 Kumar A - PhD *Associate Professor* crop production
5 Lal B - PhD *Professor* irrigation agronomy
6 Lal P - PhD *Associate Professor* agricultural climate and irrigation
7 Misra RD - PhD *Professor* irrigation, plant water relationship
8 Saini SK - PhD *Senior Research Officer* sugarcane
9 Sharma GL - PhD *Agronomist* biofertilizer
10 Singh C - PhD *Professor* extension education
11 Singh G - PhD *Associate Professor* weed control
12 Singh JN - PhD *Professor Mentha* agronomy and weed control
13 Singh M - PhD *Professor* crop physiology and productivity
14 Singh NB - PhD *Senior Research Officer* energy requirement and wheat agronomy
15 Singh NP - PhD *Senior Research Officer* crop production and nutrition
16 Singh OP - PhD *Associate Professor* weed control
17 Singh PP - PhD *Professor* soil fertility and sugarcane
18 Singh V - PhD *Associate Professor* agronomy of forage crops and grassland
19 Singh Y - PhD *Senior Scientist* soil and water management

2813 Department of Agricultural Communication

1 Doshi JK - PhD *Professor* social anthropology
2 Kumar A - ME *Associate Professor* electronics and communication
3 Ramchandani G - PhD *Associate Professor* communication research sociology and organization communication
4 Saxena MP - PhD *Professor* agricultural communication
5 Singh JN - PhD *Associate Professor* radio and TV
6 Singh K - PhD *Associate Professor* extension education
7 Trikha RN - MSc *Associate Professor* agricultural journalism
8 Tripathi JD - PhD *Associate Professor* agricultural extension and communication development

2814 Department of Agricultural Economics

1 Pandey VK - PhD *Professor* agricultural marketing and price analysis
2 Prasad B - PhD *Professor* rural banking and agricultural cooperation
3 Sharma AN - PhD *Professor* rural banking
4 Sharma JS - PhD *Professor* agricultural finance
5 Sharma VK - PhD *Professor* resource economics
6 Singh LR - PhD *Professor* farm management and production economics

2815 Department of Animal Science

1 Verma ML - PhD *Professor* ruminant nutrition biochemistry
2 Agrawal IS - PhD *Associate Professor* animal nutrition
3 Banerjee AK - PhD *Professor* animal breeding
4 Kerishna ST - PhD *Professor* animal breeding
5 Naik DG - PhD *Professor* animal nutrition
6 Prasad RB - PhD *Associate Professor* animal breeding
7 Sharma OP - PhD *Associate Professor* animal breeding
8 Singh KS - PhD *Professor* animal nutrition
9 Singh M - PhD *Associate Professor* animal nutrition
10 Singh R - PhD *National Fellow* animal breeding
11 Singh V - PhD *Associate Professor* animal breeding
12 Tiwari AD - PhD *Professor* animal nutrition
13 Verma SK - PhD *Associate Professor* animal breeding

2816 Department of Entomology

1 Khare BP - PhD *Professor* insect economics and stored production entomology
2 Bhattacharya AK - PhD *Professor* insect physiology
3 Khalsa MS - PhD *Professor* insect ecology and behaviour
4 Pathak PK - PhD *Professor* insect toxicology and host plant relationship
5 Sachan GC - PhD *Senior Research Officer* insect physiology and insect toxicology
6 Sehgal VK - PhD *Senior Research Officer* host plant relationship, pulse entomology, economics and taxonomy
7 Sharma VK - PhD *Professor* economic entomology, insect toxicology, insect resistance
8 Singh G - PhD *Associate Professor* insect symbiology; insect pests of fruits and vegetables; biological control

2817 Department of Food Science and Technology

1 Verma NS - PhD *Professor* dairy technology and food chemistry

2 Chauhan GS - PhD *Senior Research Officer* technology of cereals, legumes and oilseeds
3 Mital BK - PhD *Professor* food microbiology
4 Nath N - PhD *Associate Professor* food processing

2818 Department of Horticulture

1 Ram S - PhD *Professor* pomology and ornamental horticulture
2 Lal G - PhD *Senior Research Officer* olericulture, breeding
3 Singh R - PhD *Professor* pomology
4 Singh RP - PhD *Professor* olericulture
5 Srivastava BK - PhD *Associate Professor* olericulture
6 Tewari JP - PhD *Associate Professor* pomology

2819 Department of Plant Breeding

1 Pandya BP - PhD *Professor* genetics and plant breeding and pulses breeding
2 Agrawal BD - PhD *Professor* maize breeding
3 Agrawal RL - PhD *Professor* seed technology
4 Khan AQ - PhD *Senior Research Officer* sugarcane breeding
5 Mani SC - PhD *Senior Research Officer* genetics and breeding cereal crops
6 Mishra SN - PhD *Professor* plant breeding and plant genetics
7 Pandey MP - PhD *Senior Research Officer* plant breeding and genetics
8 Ram H - PhD *Senior Research Officer* plant genetics, soybean breeding
9 Singh B - PhD *Rape Seed Breeder* oilseeds breeding
10 Singh DP - PhD *Associate Professor* disease research and mutation breeding
11 Singh IS - PhD *Senior Research Officer* triticale breeding and genetics
12 Singh R - PhD *Professor* plant breeding and plant genetics
13 Singh TB - PhD *Associate Professor* wheat breeding and genetics
14 Tyagi DVS - PhD *Associate Professor* cytogenetics and plant breeding

2820 Department of Plant Pathology

1 Mukhopadhyay AN - PhD *Professor* biological control, sugarbeet diseases
2 Agarwal VK - PhD *Professor* seed pathology
3 Dwivedi TS - PhD *Associate Professor* physiological production; plant pathology; pulse pathology
4 Gupta MC - PhD *Associate Professor* sugarbeet pathology
5 Kolte SJ - PhD *Professor* oilseed crop diseases
6 Lal S - PhD *Professor* maize bacterial disease
7 Singh A - PhD *Professor* cereal diseases, physiology and genetics
8 Singh KP - PhD *Senior Research Officer* sugarcane
9 Singh RA - PhD *Professor* fungal and bacterial diseases of rice, mycology
10 Singh RP - PhD *Mycologist* mycology
11 Sitaramaiah K - PhD *Associate Professor* plant nematology
12 Thapliyal PN - PhD *Professor* soybean, fungicides

2821 Department of Soil Science

1 Ghildyal BP - PhD *Professor* soil physics
2 Singh TA - PhD *Professor* soil fertility/fertilizer use
3 Gangwar MS - PhD *Professor* soil chemistry/micronutrients
4 Ghosh D - PhD *Associate Professor* soil chemistry
5 Mishra B - PhD *Associate Professor* soil fertility and fertilizer use
6 Pareek RP - PhD *Senior Research Officer* soil microbiology
7 Prakash O - PhD *Senior Research Officer* soil water management and agro-forestry
8 Rathore TR - PhD *Senior Research Officer* soil survey/soil physics
9 Sachan RS - PhD *Associate Professor* soil chemistry and fertility
10 Sharma AK - PhD *Associate Professor* soil pedology
11 Singhania RA - PhD *Associate Professor* soil chemistry/fertility
12 Tomar VS - PhD *Soil Physicist* soil physics/soil and water management
13 Tripathi RP - PhD *Soil Physicist* soil water management and conservation

2822 College of Veterinary Science

1 Singh IP - PhD *Dean* veterinary microbiology

2823 Department of Microbiology and Public Health

1 Singh VB - Professor *veterinary bacteriology*
2 Garg SK - PhD *Associate Professor* veterinary virology
3 Misra DS - PhD *Associate Professor* veterinary microbiology and veterinary public health
4 Sharma VD - PhD *National Fellow* veterinary bacteriology and mycology

2824 Department of Pathology

1 Singh NP - PhD *Professor* veterinary pathology
2 Sharma UK - PhD *Associate Professor* veterinary pathology
3 Singh SP - PhD *Senior Research Officer* veterinary pathology

2825 Department of Physiology

1 Sud SC - PhD *Professor* physiology

2 Lakhchaura BD - PhD *Associate Professor* microbiology/endocrinology
3 Mohsin M - PhD *Associate Professor* poultry physiology;reproduction physiology and neurophysiology
4 Singh SP - PhD *Associate Professor* physiology of mineral elements climatic physiology

2826 Department of Pharmacology and Toxicology

1 Bahga HS - PhD *Professor* pharmacology
2 Sharma LD - PhD *Associate Professor* pharmacology

2827 Department of Anatomy and Histology

1 Prakash P - PhD *Professor* veterinary anatomy and histology
2 Tewari AN - PhD *Associate Professor* veterinary anatomy

2828 Department of Gynaecology and Obstetrics

1 Maurya SN - PhD *Associate Professor* gynaecology and obstetrics
2 Tripathi SS - PhD *Senior Research Officer* animal reproduction and artificial insemination
3 Vadhnere SV - MVSc *Associate Professor* animal gynaecology

2829 Department of Parasitology

1 Bhatia BB - PhD *Professor* veterinary parasitology
2 Gaur SNS - PhD *Associate Professor* veterinary parasitology

2830 Department of Medicine

1 Joshi HC - PhD *Professor* veterinary medicine
2 Pachauri SP - PhD,DSc *Associate Professor* veterinary medicine

2831 Department of Surgery and Radiology

1 Kumar A - PhD *Professor* veterinary surgery; veterinary anaesthesia
2 Sharma VK - MVSc *Associate Professor* veterinary surgery and radiology

2832 Department of Veterinary Clinics

1 Saxena OP - PhD *Associate Professor* veterinary clinical surgery

2833 College of Technology

1 Dhol AK - BME,MS *applied mechanic*

2834 Department of Farm Machinery and Power Engineering

1 Singh B - PhD *Professor* farm machinery and power operation
2 Singh KN - PhD *Professor* farm machinery and power operation
3 Thakur TC - PhD *Associate Professor* farm machinery and power and applied soil mechanics

2835 Department of Post Harvest Process and Food Engineering

1 Singh BPN - MS *Professor* process and food engineering
2 Gupta DK - PhD *Professor* process engineering
3 Narain M - PhD *Professor* process and food engineering
4 Saxena RP - MTech *Professor* agricultural process and food engineering and farm structures

2836 Department of Irrigation and Drainage Engineering

1 Chauhan HS - PhD *Professor* irrigation and drainage engineering
2 Jaiswal CS - PhD *Chief Scientist* irrigation and drainage engineering
3 Shukla KN - PhD *Professor* irrigation and drainage engineering

2837 Department of Soil and Water Conservation Engineering

1 Rastogi RA - PhD *Professor* soil and water conservation engineering
2 Das G - PhD *Professor* soil and water engineering

2838 College of Home Science

1 Shinde SJ - MSc *Dean* clothing and textile

2839 Department of Foods and Nutrition

1 Usha Miss MS - MSc *Professor* food and nutrition
2 Bhattacharya Mrs L - MSc *Associate Professor* food and nutrition
3 Kumar Mrs A - MSc *Associate Professor* food and nutrition

2840 College of Basic Sciences and Humanities

1 Sharma D - DPhil *Dean* physical chemistry

2841 Department of Biological Sciences

1 Chaturvedi C - DPhil *Professor* mycology
2 Pant RC - PhD *Associate Professor* plant physiology
3 Rathore VS - PhD *Associate Professor* plant physiology
4 Singh KP - PhD *Associate Professor* helminthology, plant nematology

2842 Department of Fisheries

1 Singh CS - PhD *Senior Research Officer* limnology and fisheries

2843 Department of Biochemistry

1 Garg GK - PhD *Associate Professor* bacterial sporulation; molecular biology and genetics engineering
2 Misra DP - PhD *Senior Research Officer* biochemistry
3 Tandon SM - PhD *Senior Research Officer* microbiology

2844 Department of Microbiology

1 Johri BN - PhD *Professor* microbial physiology
2 Rana RS - PhD *Associate Professor* microbial physiology
3 Verma J - MSc *Associate Professor* bacterial metabolism

2845 All India Coordinated Projects Unit

1 Bhatnagar PS - PhD *Project Coordinator* soybean and sugarbeet breeding
2 Agnihotri AK - PhD soil science
3 Agrawal BD - PhD plant breeding
4 Bhattacharya - PhD entomology
5 Chaturvedi C - PhD biological science
6 Chaubey HS - PhD plant pathology
7 Chaudhary RP - PhD animal science
8 Chauhan HS - PhD agricultural engineering
9 Das G - PhD agricultural engineering
10 Garg GK - PhD biochemistry
11 Garg GK - PhD biochemistry
12 Garg SK - PhD microbiology and public health
13 Gaur SNS - PhD veterinary parasitology
14 Gupta DK - PhD agricultural engineering
15 Gupta DK - PhD agricultural engineering
16 Gupta MN - PhD agricultural engineering
17 Gupta MN - PhD agricultural engineering
18 Johri BN - PhD microbiology
19 Khare BP - PhD entomology
20 Kumar A - PhD surgery and radiology
21 Mishra B - PhD soil science
22 Mishra RD - PhD agronomy
23 Misra DS - PhD microbiology and public health
24 Misra RD - PhD agronomy
25 Pandya BP - PhD pulse breeding
26 Pareekh RP - PhD soil science
27 Pareekh RP - PhD soil science
28 Rahal MS - PhD animal science
29 Rathi YPS - PhD plant pathology
30 Sharma GL - PhD agronomy
31 Sharma VD - PhD microbiology and public health
32 Sharma VD - PhD microbiology and public health
33 Sharma VK - PhD agricultural economics
34 Singh A - PhD plant breeding
35 Singh A - PhD plant pathology
36 Singh A - PhD plant pathology
37 Singh A - PhD plant breeding
38 Singh A - PhD plant breeding
39 Singh B - PhD plant breeding
40 Singh B - PhD agricultural engineering
41 Singh B - PhD agricultural engineering
42 Singh B - PhD plant breeding
43 Singh BPN - PhD agricultural engineering
44 Singh BPN - PhD agricultural engineering
45 Singh G - PhD agronomy
46 Singh G - PhD agronomy
47 Singh M - PhD agronomy
48 Singh M - PhD agronomy
49 Singh R - PhD plant breeding
50 Singh R - PhD animal science
51 Singh RA - PhD plant pathology
52 Singh RP - PhD horticulture
53 Singh RP - PhD horticulture
54 Singh RP - PhD horticulture
55 Singh RS - PhD plant pathology
56 Singh RS - PhD plant pathology
57 Singh Y - PhD agronomy
58 Singh Y - PhD agronomy
59 Sitaramaiah K - PhD plant pathology
60 Thapliwal PN - PhD plant pathology
61 Tripathi RP - PhD soil science
62 Tripathi RS - PhD physics
63 Verma ML - PhD animal science
64 Verma SK - PhD animal science

2846 Mahatma Phule Agricultural University

Rahuri, Distt. Ahmednagar, Maharashtra, 413 722
Tel: 16. Cables: KRISHIVID.

1 Ajri DS - MScAgr *Head* agricultural entomology
2 Belhe ND - MScAgr,PhD animal science
3 Chaudhari KG - MScAgr,PhD horticulture
4 Daftardar SY - MScAgr,PhD *Head* agricultural chemistry
5 Dhongade MP - MScAgr,PhD agricultural economics
6 Jadhav DR - MScAgr agriculture extension
7 Kadam SS - MScAgr,PhD *Head* biochemistry
8 Kanawade LR - MTech,AgrEngg agricultural process engineering
9 Patil ND - MScAgr,PhD *Director of Institute and Dean, Faculty of Agriculture*
10 Shinde PA - MScAgr,PhD *Head* plant pathology
11 Sinnarkar VV - MTech,AgrEngg agricultural engineering
12 Thorat AK - MScAgr food science and technology
13 Bapat DR - MScAgr,PhD *Senior Sorghum Breeder*
14 Bhat NR - MScAgr,PhD *Onion Breeder*
15 Boramanikar PK - MScAgr *Cotton Pathologist* cotton improvement project
16 Darekar KS - MScAgr,PhD *Nematologist*
17 Desai SN - MScAgr,PhD *Project Director* adaptive research
18 Desale JS - MScAgr *Forage Breeder*
19 Deshmukh RB - MScAgr,PhD *Pulses Breeder*
20 Dhonde PW - MScAgr *Cotton Agronomist*
21 Hande YK - MScAgr,PhD *Professor* plant pathology
22 Kadam BK - MScAgr *Sorghum Agronomist*
23 Kale KM - MScAgr,PhD *Senior Scientist* cattle project
24 Kale PN - MScAgr,PhD *Professor* horticulture
25 Khade KK - MScAgr *Chief Scientist (W.M.P.)*
26 Khuspe TS - MScAgr,PhD *Director of Extension Education*
27 Kokate AS - MScAgr,PhD *Professor* horticulture
28 Lawande KE - MScAgr *Senior Vegetable Breeder*
29 Lukade GM - MScAgr *Sorghum Pathologist*
30 Magar SS - MScAgr,PhD *Soil Physicist*
31 Mali AR - MScAgr *Cotton Entomologist*
32 Mehetre SS - MScAgr,PhD *Cotton Physiologist*
33 Mote UN - MScAgr,PhD *Sorghum Entomologist*
34 Patil RB - MScAgr *Seed Research Officer*
35 Patil VK - MVSc,PhD *Geneticist* goats project
36 Pawar BB - MScAgr *Professor* agricultural botany
37 Pawar MD - MScAgr *Director of Communication*
38 Purkar JK - MScAgr,PhD *Chief Seeds Officer*
39 Somawanshi RB - MScAgr,PhD *Associate Professor* agricultural chemistry
40 Sonar KR - MScAgr,PhD *Soil Chemist*
41 Sonawane RB - MScAgr *Cotton Breeder*
42 Surve DN - MScAgr,PhD *Professor* agronomy
43 Thete RY - MScAgr,PhD *Associate Professor* agricultural botany
44 Ugale SD - MScAgr,PhD *Bajra Breeder*
45 Umarani NK - MScAgr,PhD *Head Department of Agronomy*

2847 College of Agriculture
Pune

1 Bhite HS - MScAgr,PhD *Associate Professor* agricultural extension
2 Bhore DP - MScAgr *Professor* horticulture
3 Bhujbal BG - MScAgr,PhD *Floriculturist*
4 Desnmukh SS - MScAgr *Principal*
5 Jadhav RV - MTech,AgrEngg *Professor* agricultural engineering
6 Kadam Mrs VC - MScAgr *Officer* cotton improvement project
7 Kale AM - MScAgr,PhD *Professor* animal science
8 Katwate MT - MScAgr *Principal*
9 Kharse YT - MScAgr *Associate Professor* agricultural economics
10 Mane VS - MScAgr *Associate Professor* agronomy
11 More WD - MScAgr,PhD *Professor* plant pathology
12 Murty TK - MScAgr *Entomologist* millet project
13 Parkale DG - MScAgr,PhD *Professor* agricultural economics
14 Patil BC - MScAgr,PhD *Associate Dean*
15 Patil MM - MScAgr,PhD *Professor* agricultural botany
16 Patil PL - MScAgr,PhD *Agricultural Bacteriologist*
17 Patil SSD - MScAgr *Geneticist*
18 Pokharkar RN - MScAgr *Professor* agricultural entomology
19 Thorat SS - MScAgr,PhD *Professor* agricultural extension

2848 College of Agriculture
Dhule

1 Mogal BH - MScAgr,PhD *Associate Dean*
2 Dalaya VP - MScAgr *Professor* agricultural entomology
3 Ekbote MV - MScAgr *Professor* plant pathology
4 Hinge BJ - MScAgr,PhD *Professor* agricultural economics
5 Joshi AC - MScAgr *Professor* agronomy
6 Kale SP - MScAgr,PhD *Professor* agricultural chemistry
7 Kibey MB - MScAgr *Professor* agricultural extension

8 Manke BS - MScAgr,PhD *Professor* agricultural botany
9 Naphade AS - MScAgr *Professor* horticulture
10 Patil BM - MTech,AgrEngg *Professor* agricultural engineering
11 Purandare SL - MScAgr *Professor* animal science and dairy science

2849 College of Agriculture
Kolhapur

1 More BB - MScAgr,PhD *Associate Dean*
2 Dhumal PM - MScAgr,PhD *Professor* plant pathology
3 Gajare BP - MScAgr *Professor* agricultural entomology
4 Jadhav RB - MScAgr *Professor* horticulture
5 Mohite BV - MScAgr,PhD *Professor* agronomy
6 Narawade VS - MScAgr *Professor* animal science and dairy science
7 Patil BR - MScAgr,PhD *Associate Director*
8 Patil MD - MScAgr,PhD *Professor* agricultural chemistry
9 Patil RG - MScAgr,PhD *Principal*
10 Rade VM - MScAgr *Professor* agricultural extension
11 Shah BS - MScAgr *Professor* agricultural botany
12 Shingte MB - MScAgr *Professor* agricultural engineering

2850 Dryland Agricultural Research Station
Near D.A.V. College, Solapur, Maharashtra

1 Koregave BA - MScAgr,PhD *Chief Scientist*
2 Kadam SK - MScAgr *Soils Specialist and Associate Director*

2851 Sugarcane Research Station
Padegaon, Post Nira, Distt. Satara

1 Jadhav LH - MScAgr *Sugarcane Breeder*
2 Jadhav SB - MScAgr,PhD sugarcane research station

2852 Pulses and Oilseeds Research Scheme
Pandharpur, Distt. Solapur

1 Patil SSD - MScAgr *Agronomist*

2853 Wheat Research Station
Niphad, Distt. Nasik

1 Bangal DB - MScAgr,PhD *Wheat Physiologist*
2 More DC - MScAgr *Wheat Specialist*

2854 Wheat Research Station
Mahabaleshwar, Distt. Satara

1 Mutkekar ML - MScAgr *Wheat Rust Mycologist*

2855 Oilseeds Research Station
Jalgaon

1 Deokar AB - MScAgr *Oilseeds Specialist*
2 Tendulkar AV - MScAgr *Senior Scientist*

2856 Agricultural Research Station
Wadgaon (Maval), Distt. Pune

1 Kalke SD - MScAgr *Officer I/C*

2857 All India Co-Ordinated Research Project on Agro-Meterology
Solapur

1 Patil CB - MScAgr,PhD *Agro Meteorologist*

2858 Scheme for Operational Research Project (IDA assistance for dryland Agriculture)
Solapur

1 Biradar SC - MScAgr *Agronomist*

2859 All India Co-Ordinated Research Project on Dryland Agriculture
Solapur

1 Dangat SB - MScAgr,PhD *Associate Professor* agricultural economics
2 Kulkarni NR - MScAgr *Agronomist*
3 Patil AJ - MScAgr *Soil Physicist*
4 Patil MW - MScAgr *Plant Breeder*
5 Pol PS - MScAgr,PhD *Geneticist*
6 Wayase US - MScAgrEngg *Agricultural Engineer*

2860 National Agricultural Research Project for Scarcity Zone
Solapur

1 Ghorpade DS - MScAgr *Associate Professor* crop physiology

2 Jadhav AS - MScAgr *Associate Professor* agronomy
3 Kamat MS - MScAgr *Associate Professor* soil microbiology

2861 All India Co-Ordinated Project on Safflower
Solapur

1 Patil PS - MScAgr,PhD *Geneticist*
2 Patil RC - MScAgr,PhD *Geneticist*

2862 Sugarcane Research Station
Padegaon, Post Nira, Distt. Satara

1 Kadam BK - MScAgr *Agronomist*
2 Karawande PD - MScAgr *Agronomist*
3 Patil VA - MScAgr *Soil Scientist*
4 Vaidya BR - MScAgr *Bio-Chemist*
5 Wali MK - MScAgr *Sugarcane Entomologist*

2863 Agricultural Research Station
Jalgaon, Post Nira, Distt. Satara

1 Bhole GR - MScAgr *Groundnut Breeder*
2 Deore SK - MScAgr *Agricultural Chemist*
3 Dhumal VS - MScAgr *Assistant Entomologist*
4 Garse TL - MScAgr *Agronomist*
5 Jadhav BR - MScAgr *Associate Professor* microbiology
6 Narkhede BN - MScAgr,PhD *Safflower Breeder*
7 Patel DM - MScAgr *Sesamum Breeder*
8 Patil BA - MScAgr *Horticulturist*
9 Patil MB - MScAgr *Plant Pathologist*
10 Patil RS - MScAgr *Superintendent*

2864 Agricultural Research Station
Karad, Distt. Satara

1 Nimbalkar VS - MScAgr *Associate Professor* plant breeding

2865 Agricultural Research Station
Mohal, Distt. Solapur

1 Bowa BD - MScAgr *Associate Professor* plant breeding

2866 Agricultural Research Station
Niphad, Distt. Nasik

1 Desale SC - MScAgr *Hybrid Wheat Research Officer*

2867 Adaptive Research in Irrigated Areas under MWUP Khadakwasala Project
Patas, Distt. Pune

1 Chavan CD - MScAgr *Agricultural Research Officer*

2868 Agricultural Research Station
Digraj, Distt. Sangli

1 Pawar AM - MScAgr *Associate Professor* plant pathology

2869 Agricultural Research Station
Lonawale, Distt. Pune

1 Newase AG - MScAgr *Rice Pathologist*

2870 Sugarcane Seed Foundation Programme
Pravaranagar, Distt. Admednagar

1 Medhane NS - MScAgr *Seed Production Officer*

2871 Grape Research Station
Pimpalgaon, Baswant, Distt. Nasik

1 Kaulgud SN - MScAgr *Horticulturist*

2872 Ministry of Agriculture

P.O. Box 480, Kingston 6
Tel: 927-1731-50.

2873 Data Bank and Evaluation Division

P.O. Box 480, Kingston
Tel: 927-9831.

1 Allman A - BSc rural development
2 Bernard H - BSc rural development
3 Campbell J - BSc,MSc *Director*
4 Innerarity F - BSc *Head of Evaluation and Rural Development*
5 McLean A - BSc rural development
6 Ramdon R - BSc agricultural statistics

2874 Agricultural Planning and Policy Review Unit

1 Binnie A - BSc *Head Macro Economics*
2 Boyne L - DipAgri,BSc *Director*
3 Lindo N - DipAgri,BSc *Head Farm Management*
4 Munroe J - DipAgri,BSc,MSc *Head Data Analysis*

2875 Production and Extension Department

1 Frazer H - DipAgri,BSc *Deputy Director/Vegetable Agronomist*
2 Irving DB - DipAgri,BSc *Director*

2876 Department of Science, Technology and Research

P.O. Box 480, Kingston 6
Tel: 927-9831.

1 Lowe HIC - BSc,MSc,PhD *Executive Director*
2 Johnson J - BSc,MSc,MBa *Research Coordinator/Management*
3 McClenan V - BSc,DipEd environment

2877 Research and Development Division

P.O. Box 480, Kingston 6
Tel: 927-9831.

1 van Whervin LW - BSc,MSc *Director*
2 Baker RJ - BSc,MSc,PhD *Agronomist/Soil Chemist*
3 Collins BE - BSc *Principal Research Officer - Plant Protection*

2878 Bodles Agricultural Research Station (Southern Region)

Old Harbour P.O., St Catherine
Tel: 983-2267.

2879 Crop Research

1 Dester A - BSc *Acting Director*
2 Blake RA - BSc *Agronomist* root crops
3 McGlashan D - BSc *Agronomist* vegetables
4 Young F - BSc,PhD *Head Post Entry Quarantine Station*

2880 Farming Systems

1 Dehaney J - DipAgri *Junior Agronomist*
2 Martin-Lawrence Z - DipAgri *Junior Agronomist*
3 McBean J - BSc livestock development
4 Reid C - BSc *Chief Agronomist*

2881 Livestock Research

1 Jennings PG - BSc,PhD *Principal Research Officer* livestock/animal nutrition
2 Brown A - BSc cattle breeding development
3 Douglas F - BSc dairy cattle breeding
4 Holness Mrs JA - BSc,MSc animal breeding and genetics
5 Lawrence BH - BSc dairy cattle breeding and husbandry
6 McDonald DC - BSc small stock production (goat, sheep, rabbits)
7 Miller D - BSc sheep and goat husbandry
8 Muschette J - BSc,MSc swine nutrition

2882 Plant Protection

1 Ellis DW - BSc,MSc *Director* entomology
2 Creighton Miss C - BSc entomology
3 Edman FL - BSc plant pathology
4 Harvey E - BSc nematology
5 Mais Mrs M - BSc,MSc plant pathology

6 Murray R - BSc entomology
7 Rhodes L - BSc,MSc entomology

2883 Grove Place Agricultural Research Station

Mandeville PO, Manchester
Tel: 962-2502.

1 Hemmings O - BSc beef cattle and sheep breeding
2 Logan JL - BSc pasture agronomy and utilization
3 Maxwell JD - BSc *Director* beef cattle breeding and husbandry

2884 Montpelier Agricultural Research Station

Montpelier PO, St. James

1 Lee Hew EG - BSc,MSc *Director/Agronomist*

2885 Natural Resources Conservation Department

P.O. Box 305, Kingston 10
Tel: 923-6965, 923-5166.

1 Anderson M - BSc,PhD *Director* environmental management
2 Cunningham C - DipAgri watershed management
3 Foster E - BSc resource management
4 Gayle D - BSc,MSc pesticides and pollution management
5 Strong Y - BSc wildlife management

2886 Fisheries Division

Marcus Garvey Drive, Kingston
Tel: 923-8811.

1 Alexander L - BSc,MSc *Acting Director, Inland Fisheries* oysterculture
2 Hanson C - BSc oysterculture
3 Haughton M - BSc *Deputy Director* fish biology
4 Kong G - BSc development offshore fisheries
5 Moo-Young R - BSc,MSc *Director* aquaculture
6 Rodney V - BSc fisheries biology and pollution
7 Thompson D - BSc oysterculture

2887 Forest Department

4 Hillman Road, Kingston 8
Tel: 924-1997.

1 Jones RS - BSc *Director*
2 Porter K - BSc fuelwood development

2888 Veterinary Division

P.O. Box 480, Kingston
Tel: 927-6475.

1 Bryan LH - DVM *Director*
2 Boothe GA - DVM *Deputy Director*
3 Grant G - DVM diagnostic serology

2889 University of the West Indies

Mona, Kingston 7
Cables: UNIVERS.

2890 Department of Botany

P.O.Box 8, Mona, Kingston
Tel: 927-0753.

1 Sidrak GH - BSc,MSc,PhD *Professor and Head*
2 Coates-Beckford P - BSc,DIC,MSc,PhD plant pathology and nematology
3 Devi Prasad PV - BSc,MSc,PhD algal physiology
4 Donelly P - BSc,MSc soil microbiology
5 Iremonger S - BSc,PhD plant ecology
6 Potluri - BSc,MSc,PhD soil microbiology
7 Vaidya K - PhD cytogenetics

2891 Department of Zoology

P.O. Box 12, Kingston
Tel: 927-1661.

1 Goodbody I - MA,PhD *Professor, Director Marine Science Unit*
2 Aiken K - MSc fisheries
3 Bacon P - BSc,PhD marine ecology
4 Bogel P - BSc,PhD reptiles
5 Freeman B - PhD reader in animal ecology
6 Garraway E - PhD entomology
7 Greenfield M - BSc,PhD physiology
8 Mansingh A - BSc,PhD entomology
9 Robinson RD - PhD parasitology
10 Steele RD - PhD *Head of Department* fisheries

2892 Caribbean Agricultural Research and Development Institute

P.O. Box 113, Mona, Kingston 7
Tel: 927-6531.

1 Grant JE - BSc agricultural engineering
2 Reid Mrs JC - BSc,MSc *Head of Unit* entomology

2893 Banana Board

P.O. Box 602, Kingston
Tel: 922-5347.

1 Gonzalves RA - BSc *General Manager/Technical Director*
2 Conie J - BSc entomology
3 Dixon J - BSc,MSc *Director Research* botany
4 Heath S - BSc,MSc *Nematologist* entomology

2894 Coconut Industry Board

P.O. Box 204, Kingston 10
Tel: 926-1770. Cables: CONUT BOARD.

1 Barrant CI - BSc,MSc *Director of Research*
2 Nelson L - BSc *Plant Breeder*
3 Wilson T - BSc *Agronomist*

2895 Scientific Research Council of Jamaica

P.O. Box 350, Kingston 6
Tel: 927-1771-4. Cables: SCIENTIST.

1 Taylor G - BSc,MSc,PhD *Executive Director*
2 Brown M - BSc food technology
3 Burrowes J - BSc,MSc food technology
4 Butterfield D - BSc,MSc chemical process engineering
5 Coke LB - BSc,PhD *Director Research & Development* crop physiology
6 Crawford E - BSc analytical chemistry
7 Earle P - BSc chemical engineering
8 Fairclough A - BSc chemical engineering
9 Gordon L - BSc microbiology
10 Jefferson D - BSc algal culture
11 Kerr H - BSc food technology
12 Lynch A - MSc food technology
13 McLaughlin W - BSc,PhD microbiology
14 Mitchell S - BSc tissue culture
15 Mortley C - BSc biogas
16 Shackleford D - BSc chemical engineering
17 Thomas A - BSc analytical chemistry
18 Thompson C - MSc mineral anaylsis
19 Williams L - BSc entomology

2896 Sugar Research Institute

Kendal Road, Mandeville P.O.
Tel: 962-2441 or 962-1287. Cables: S.I.R.I..

1 Shaw MEA - BSc *Director*
2 Brown A - BSc agronomy
3 Easy M - BSc *Extension Agronomist*
4 Falloon T - BSc,MSc *Deputy Director* entomology
5 Fearon C - BSc soil fertility
6 Stanford D - BSc agronomy
7 Summers D - BSc *Agricultural Engineer*
8 Thompson H - DICTA,BSc *Analytical Chemist*

2897 Ministry of Agriculture and Livestock Development

P.O. Box 30028, Nairobi
Tel: 720030. Cables: MINAG.

2898 Scientific Research Division

2899 Head Office

1 Chabeda AEO - BSc,MSc *Chief Research Officer*
2 Ondieki JJ - BScAgr,MSc *Chief Research Officer*
3 Chema S - BVM,MSc,PhD *Deputy Director (Livestock Department)*
4 Cheruiyat HK - BSc,MSc,PhD *Assistant Director (Animal Health)*
5 Gatougi PM - BVM,MSc *Assistant Director (Animal Health)*
6 Hiaga G - BSc,MSc *Assistant Director (Soil/Water Management)*
7 Kamau CR - BSc,MSc *Assistant Director (Regional Research Centres)*
8 Majisu BN - BSc,MSc,PhD *Director*
9 Matata JBW - BSc,PGD,DIP(Biometry) *Deputy Director (Planning, Manpower Development and Training)*
10 Miyogo M - BA,CPA *Deputy Director (Planning, Finance and Administration Department)*
11 Rutto JK - BSc,MSc *Assistant Director (Crops)*
12 Wachira D - BSc,MSc,PhD *Assistant Director (Animal Production)*
13 Wangia Mrs S *Programme Officer*
14 Wapakala WW - BSc,MSc *Deputy Director (Crops, Soil/Water Management Department)*

2900 Beef Research Station

Lanet, P.O. Box 1275, Nakuru
Tel: 85324.

1 Kevelenge JEE - BScAgr,MScAn,Prod *Officer-in-Charge, Senior Animal Nutritionist* agro byproduct utilization

2901 Animal Nutrition

1 Onyange TA - BScAgric,MScAnProdn *Animal Nutritionist* beef cattle nutrition in intensive feeding

2902 Pasture and Fodder Agronomy

1 Kieng'atti RK - BScBot *Fodder Agronomist*
2 Kinuthia HK - BSc,MScPastAgro *Head of Pasture Agronomy* sweet potato tops

2903 Animal Breeding

1 Indetie D - BScAgric *Animal Breeder* beef cattle crosses

2904 Cotton Research Station Kibos

P.O. Box 1490, Kisumu
Tel: 035-2938.

Improvement of varieties and agronomic methods in cotton and rainfed rice, research on pest control in cotton and rice

1 Opondo RM - BScAgr,MSc *Senior Research Officer-in-Charge*

2905 Cotton Research

1 Ashiono GB - BScAgr cotton agronomy research
2 Kouko WO - BSc soil fertility, cotton agronomy
3 Mambiri AM - BSc,MSc control of cotton pests
4 Ndong'a MFO - BSc,MSc integrated pest control in cotton
5 Ombakho GA - BSc,MSc cotton breeding
6 Pamba WN - BSc cotton agronomy

2906 Rainfed Rice Research

1 Gerroh CO - BSc soil chemistry; soil fertility; rice agronomy
2 Malinga YWK - BSc,MSc control of pests of rice

2907 Mariakani Animal Husbandry Research Station

P.O. Box 30, Mariakani
Tel: 035-2938.

1 Bero Ogora R - MSc
2 Njuguna PRM - BSc
3 Njunie MN - BScAgric,MSc

2908 Coast Agricultural Research Station

P.O. Box 16, Kikambala
Tel: 485526/4853131 MOMBASA.

2909 Tree Crops and Horticulture Section

1 Abubaker AS - MScAg *Director, Head of Section* coconut and cashewnut agronomy
2 Saha HM - BScAgr horticulture (vegetables)
3 Wataaru RK - BSc coconut breeding
4 Wekesa Miss E - BScAgr horticulture (fruit vegetables)

2910 Crop Protection

1 Seif AA - MSc *Head of Section* plant pathology
2 Gitonga W - BSc plant entomology
3 Kiarie MK - BSc plant entomology
4 Mulli RZ - BScMSc *Incharge of Matuga Research Station* plant entomology
5 Mwangi SFM - BSc plant pathology
6 Mwangi VK - BSc plant entomology
7 Ouko JO - BSc plant pathology

2911 Cereals and Legumes

1 Kamau GM - BSc *Head of Section* maize agronomy
2 Chivatsi WS - BScAgr maize breeding
3 Gacheru Miss EN - BScAgr *maize agronomy*
4 Kiuru PD - BScAgr rice agronomy
5 Muli MB - BScAgr legumes agronomy
6 Odhiambo GD - BScAgr maize agronomy

2912 Root and Tuber Section

1 Tipape JL - BScAgr *Head of Section* agronomy

2913 Sugarcane Section

1 Jagathesan D - PhD *Head of Section* sugarcane breeding
2 Jamoza JE - BScAgr sugarcane breeding

2914 Livestock and Pastures

1 Mureithi JG - BSc *Head of Section* animal production
2 Abwao SM - BSc,MSc veterinary
3 Ocholla LW - BSc pastures

2915 Farming Systems Section

1 Odupa Miss AZ - BScAgr *Farm Systems Economist*

2916 Biometrics Section

1 Islam AS - BSc,MSc statistics
2 Ngae GN - BSc statistics

2917 Garissa Agricultural Research Station

P.O. Box 230, Garissa
Tel: 100.

2918 Forage/Animal Production

1 Bauni SM - BSc,MSc forage livestock management

2919 Crop Production

1 Kaguma AM - BScAgr crop agronomy

2920 Crop Protection

1 Muasya WN - BEdSc crop breeding

2921 Msabaha Agricultural Research Station

P.O. Box 344, Malindi
Tel: 100.

1 Githunguri CM - BAgricSc *Agronomist and Officer-In-Charge* root and tuber crops
2 Gitonga W - BAgricSc *Entomologist* cotton and other fibres

2922 Nyandarua Agricultural Research Station

Private Bag, Ol Joro, Orok
Tel: 225.

2923 Animal Research

1 Munyi DN - AScAgric,BScAgric

2924 Crop Research

1 Gathama SK - BScAgric
2 Kanampiu NN - BScAgric

2925 Nyanza Agricultural Research Station
P.O. Box 523, Kisii
Tel: KISII 20527.

General research on cereals, legumes, vegetables, fruits, oil seed crops, pasture/fodder and animal production

1 Nzabi AW - MScAgr *Officer-In-Charge*

2926 Maize Agronomy

1 Macharia CN - BScAgr *Head of Section, Maize Agronomist*

2927 Animal Production

1 Sunyai SO - MSc *Head of Section* animal production - pasture and zero grazing studies

2928 Plant Pathology

1 Gitau PN - BSc *Head of Section* pathologist

2929 Plant Entomology

1 Otieno JO - BSc *Head of Section, Entomologist*

2930 Plant Breeding

1 Wasike VW - BScAgr *Head of Section, Groundnut Breeder*

2931 Biometry

1 Kidula NL - BSc *Head of Section, Biometrician*

2932 Perkerra Agricultural Research Station
P.O. MARIGAT, VIA NAKURU

2933 Dryland Farming Section

1 Agufana AU - BSc
2 Barno JS - BSc
3 Michieka DO - MScAgric

2934 National Agricultural Research Centre, Muguga

Muguga

2935 Research Divisions
P.O. Box 30148, Nairobi
Tel: 0154-32880/1-6.

1 Ngundo BW - BScBio,PhDPlNem *Director*
2 Seme JO - MSc germplasm conservation

2936 Chemistry Division
Tel: 0154-32880/1-6.

1 Okalebo JR - BSc,MScSChem *Division Head* cereal and legume phosphate nutrition

2937 Soil Fertility and Plant Nutrition Section

1 Esilaba AO - BScAgr,MScSChem cereal and legume mineral nutrition
2 Mangale N - BSc,MScSAgrChem *Section Head* cowpea mineral nutrition

2938 Soil Microbiology Section

1 Karanja Mrs NK - BScAgr,MScSMicro *Section Head* biological nitrogen fixation

2939 Entomology and Biological Control Division

1 Mailu AM - BScAgr,MScEnt,PhDEnt *Division Head* pigeon pea insect pests; white sugarcane scale

2940 Armyworm Forecasting Section
Headquarters, P.O. Box 30148, Nairobi
Tel: 0154-32880/1-6.

No senior research workers

2941 Commonwealth Institute of Biological Control Section

1 Robertson IAD - BScAgr *Section Head* cassava green mite

2 Kairo MTK - BSc coffee *Icerya pattersoni*
3 Markham RH - BA,PhDEnt cassava green mite; cereal stemborer; lygus parasite
4 Murphy Mrs RJ - BSc cereal stemborer
5 Murphy ST - BScAgr,PhDEnt,DIC,MIBiol coffee green scale; coffee berry borer
6 Nang'aya FLO - BSc cereal stemborer
7 Oduor GI - BSc coffee green scale

2942 Library Division

1 Kaiyare DN - CLS Cert Doc *Division Head* librarian

2943 Physics Division

1 Kahumbura JM - DipChem,MScSPhy *Division Head* conservation tillage

2944 Agricultural Economics Section

2945 Agricultural Engineering Section

1 Mutua JM - DipAgrEng,BScAgrEng *Section Head* farm tool mechanization

2946 Agricultural Meteorology Section

1 Lenga Miss FK - BScAgr,MScAgrMet,PhDAgrMet water production functions for maize, beans and their intercrop
2 Mugah JO - BSc,MScAgrMet *Section Head* sunflower and soil moisture; water and nitrogen effects on sunflower yield

2947 Soil Physics Section
Nairobi

1 Kilewe AM - Bed,MScSPhy *Section Head* rainfall erosivity factors; soil erodibility characteristics; crop cover and residue management on runoff and soil loss; erosion and soil productivity relationship; fertilizers and manure in restoring productivity of eroded land; soil loss tolerance relationship

2948 Statistics Section

2949 Plant Pathology and Nematology Division

1 Theuri JM - BEd,MPhil PlV *Ag Division Head* cowpea yellow mottle virus; groundnut rossette virus; sweet potato virus

2950 Maize Breeding Section

1 Mutinda CJM - BScAgr maize breeding
2 Mutisya FM - BScAgr,MSc,PhD maize breeding
3 Odhiambo MO - BScAgr,MSc,PhD maize breeding

2951 Nematology Section

no senior research worker

2952 Plant Pathology Section

1 Njuguna JGM - BSc,MScPlPath *Section Head* sweet potato virus

2953 Tissue Culture Section

1 Wambugu Mrs FM - BSc,MScTC *Section Head* tissue culture of maize and millets; in vitro potato tuberlet production

2954 Plant Quarantine Division

1 Okioga DM - BScAgr,MScPlPath,PhDPlPath,DIC *Division Head* plant quarantine; inventory of plant introduction; seed pathogen detection test

2955 Genetics Section

1 Kamau HN - BSc,MScPlG *Section Head* citrus organ culture and conservation; bean germplasm regeneration; evaluation and conservation

2956 Virology Section

1 Mathu Mrs RW - BSc *Section Head* plant virus indexing; identification of virus in *Alstromeria* flowers

2957 National Agricultural Research Laboratories

P.O. Box 14733, Nairobi
Tel: 48211.

1 Muchena FN - BSc *Director*

2958 Soil Science and Plant Nutrition Section

1 Qureshi JN - BSc,MScAg *Senior Soil Chemist* soil fertility and plant nutrition
2 Ayaga GO plant analysis
3 Gachoka JF *Soil Chemist* sulphur in soils
4 Gatei JN - BSc,MSc *Soil Mineralogist* clay mineralogy
5 Gikonyo Mrs EW - BSc soil analysis
6 Irambu EM - BSc plant analysis
7 Irungu Mrs GN - BSc soil analysis
8 Kanyanjua SM - BSc *Agronomist* fertilizer trials
9 Kathuli P - BSc fertilizers, manures and water analysis
10 Kibunja Mrs CN - BSc soil microbiology
11 Kungu RM - BSc feed and foodstuff analysis
12 Maina Miss GN - BSc,MSc *Biochemist* food science
13 Mburu DN - BSc,MSc *Agricultural Chemist* weed science
14 Muriuki Mrs AW - BSc *Agronomist*
15 Njoroge JN - BSc soil analysis
16 Nyotu HG - BSc *Agricultural Chemist*
17 Oduor POS - BSc,MSc *Soil Chemist* potassium in soils
18 Wakiiru GM - BSc soil analysis
19 Wanjau FM - BSc,MSc *Soil Chemist* analytical chemistry

2959 Irrigation and Drainage Research

1 Njihia CM - BSc,MSc *Senior Research Officer, Head of Programme*
2 Kamau PA - BSc,MAgrSc *Soil Physicist* crop water use
3 Mochiemo GO - BSc *Drainage Scientist*
4 Mugwanja PN - BSc *Agronomist*
5 Muteru SW - BScAgrEng land and water management
6 Mwangi JN - BScAgrEng land and water management
7 Ndegwa GM - BSc irrigation water management
8 Radiro MPO - BSc water management/soil physics
9 Sijali IV - BSc water management/soil physics

2960 Soil Survey Section

1 Aore WW - BSc,DipSoilSurv *Soil Surveyor*
2 Ekirapa AE - BSc,CertRemoteSensing,DipIrri *Soil Management Officer*
3 Gachene CKK - BSc,MSc *Soil Surveyor*
4 Gatahi MM - BSc,MSc *Land Evaluation Officer*
5 Gicheru PT - BSc,DipSoilSurv *Soil Surveyor*
6 Kamoni PT - BSc,MSc *Land Evaluation Officer*
7 Kariuki JN - BSc *Soil Chemist*
8 Kilambya DW - BA,DipIntegSurv *Agro-Economist*
9 Kiome RLM - BSc,MSc *Soil Surveyor*
10 Kithome M - BSc *Soil Chemist*
11 Macharia PN - BSc,DipRuralSurv *Vegetation Surveyor*
12 Mwichabe S - BSc,DipRuralSurv *Vegetation Surveyor*
13 Ndaraiya FM - BSc *Agro-Climatologist*
14 Ochieng' NP - MSc *Soil Chemist*
15 Okoth PFZ - BSc,DipSoilSurv *Soil Surveyor*
16 van Engelen VWP - MSc *Soil Surveyor*
17 Wamicha WN - BSc,MSc *Soil Surveyor*
18 Wanjogu NS - BSc,DipSoilSurv *Soil Surveyor*
19 Wanjohi JK - BSc *Soil Surveyor*
20 Waruru BK - BSc *Soil Surveyor*
21 Weeda A - MSc *Teamleader, KSSP* soil management and land evaluation
22 Wokabi SM - BSc,DipSoilSurv,MSc *Soil Surveyor*

2961 Plant Pathology Section

1 Duggal Miss A - BScAgr *Plant Pathologist (Advisory Services)*
2 Gatumbi Mrs R - BSc *Plant Pathologist, Mycological Herbarium*
3 Kimani Mrs VW - BSc *Plant Pathologist*
4 Makini Mrs F - BScBot *Plant Pathologist, Ornamentals*
5 Michieka AO - MSc *Potato Bacteriologist*
6 MKalama Miss T - BScAgr *Plant Pathologist* rice diseases
7 Munene Mrs MW - BScAgr *Bacteriologist*
8 Njoroge WJG - BSc,MSc *Plant Pathologist, Herbarium*
9 Njuguna Mrs L - BSc *Research Officer* nematology
10 Odhiambo B - BSc,MSc *Virologist*
11 Odhiambo GW - BSc *Plant Pathologist*
12 Ogallo LJ - BSc,MSc *Nematologist*
13 Osano Mrs A - BSc,MSc *Plant Pathologist* fungicide resistance

2962 Entomology Section

1 Kibata GN - BSc,Dip.P1Prot,MSc *Senior Entomologist*
2 Chepkwony AK - BSc *Entomologist* crop storage
3 Gichuhi MA - BSc *Entomologist* taxonomy
4 Kabira PN - BSc,MSc *Entomologist, Vegetables*
5 Kambo CMG - BSc *Plant Inspector*
6 Kimoli Mrs MK - BSc *Entomologist* crop storage
7 Koech S - BSc *Entomologist* crop storage
8 Macharia CM - BSc *Entomologist, Field Crops*
9 Muinamia CK - BSc,MSc *Entomologist* pesticide residues
10 Mutai Mrs LA - BSc,MSc *Chemist* pesticide residues
11 Mutambuki SK - BSc *Entomologist* crop storage
12 Nderitu JH - BEd,DipP1Prot,MSc *Entomologist, Legumes*
13 Ngaah Miss CO - BSc *Entomologist* crop storage
14 Ngatia J - BSc,MSc *Chemist* pesticide residues
15 Ngugi Mrs JN - BSc *Entomologist* crop storage
16 Njeri Miss R - BSc *Chemist* pesticide residues
17 Njihia SN - BSc *Plant Inspector*
18 Olubayo Mrs FM - BScAgr,MSc *Entomologist* crop storage
19 Rabuor JO - BSc *Chemist* pesticide residues
20 Sikuku RO - BSc,MSc *Chemist* pesticide residues
21 Syarra GRO - BSc *Plant Inspector*
22 Wekesa PW - BSc *Entomologist* crop storage

2963 Biometric Section

1 Lukibisi FB - BSc,MSc *Biometrician* design and analysis of field experiments
2 Milikau RL - BSc,DipStatistics *Biometrician* design and analysis of field experiments
3 Mwangi JN - BSc,MSc *Biometrician* design and analysis of field experiments
4 Njoga MK - BSc,MSc *Biometrician* design and analysis of field experiments

2964 National Agricultural Research Station

P.O. Box 450, Kitale

2965 Maize Breeding

1 Chumo JJ - BEd maize breeding
2 Kamidi Mrs - BScAgric maize breeding
3 Laboso AK - BSc maize breeding
4 Mulaa Mrs MA - BSc entomology
5 Muthoka DK - BScAgricMSc *Director* maize breeding
6 Ngesa H - BSc maize breeding
7 O.Ochor T - BScMSc maize breeding
8 Ochieng JAW - BScMSc maize breeding
9 Ondongo JO - BScAgric maize breeding
10 Walulu RS - BSc maize breeding
11 Wanjekeche EW - BScAgric maize breeding

2966 Maize Agronomy

1 Akwayah MW - BSc maize agronomy
2 Kiiya WW - BSc maize agronomy
3 Lukano P - BScAgric maize agronomy
4 Mwania NM - BScAgric,MSc maize agronomy
5 Ngugi SM - BSc maize agronomy
6 Ochieng JO - BScAgric maize agronomy
7 Olang Dr AD - PhD maize agronomy
8 Ondieki BO - BSc maize agronomy
9 Onyango Miss RMA - BScAgric maize agronomy
10 Wekesa Mrs F - BSc maize agronomy

2967 Food Technology

1 Gikonyo GM - BScDipMSc food technology
2 Mwangi Miss TK - BSc food technology
3 Nkonge C - BScAgri,Msc food technology

2968 Pasture

1 Gwayumba WE - BScAgric,MSc pasture
2 Mbugua EK - BScAgric pasture
3 Mukisira EA - BScAgric,MSc pasture
4 Mwendia CW - BScAgricMSc pasture
5 Namasake JF - BSc pasture
6 Okemo Miss R - BScAgric,MSc pasture
7 Wandera JL - BScAgric,MSc pasture

2969 Biometrics

1 Kamidi RE - BScDip biometrics
2 Nasila JN - BSc biometrics
3 Ngeny JMA - BSc biometrics

2970 Farm System Economics

1 Shiluli MC - BSc farm system economics

2971 National Animal Husbandry Station

P.O. Box 25, Naivasha

2972 Animal Breeding

1 Ahuya CO - BSc animal breeding
2 Muhuyi WB - MSc animal breeding
3 Mwai AO - BSc animal breeding
4 Ole Sinkeet SN - MSc animal breeding
5 Wanambacha JW - BSc,MSc,PG,Dip animal breeding

2973 Animal Nutrition

1 Chemitei VSE - BSc animal nutrition
2 Irungu KRG - MSc animal nutrition
3 Kiprop JR - BSc animal nutrition
4 Kiragu Mrs JW - BSc,MSc animal nutrition
5 Munyi DN - BSc animal nutrition
6 Waithake MM - BSc animal nutrition

2974 Animal Production

1 Kayongo SB - BSc,MSc animal production
2 Muia JK - BSc animal production
3 Talam S - BSc,MSc animal production

2975 Socio-Economics

2976 National Dryland Farming Research Station Katumani (NDFRS)

P.O. Box 340, Machakos
Tel: 21506/7/8. Cables: KATUMANI.

The NDFRS is responsible for research on rainfed semi-arid agriculture in areas receiving 500-800mm rainfall annually.

1 Kusewa PK - BScAgr,MScAgr,DipTAgr Director of Station pasture agronomy

2977 Agronomy

1 Ariithi KC - BScAgr Research Officer cropping systems agronomy
2 Ikombo BM - MScAgr,DipSSc,PhD Research Officer soil fertility
3 Itabari JK - BSc,MScAgr Head of Section, Research Officer crop physiology; weed science
4 Songa Mrs JM - BScAgr Research Officer entomology
5 Songa WA - BScAgr Research Officer plant pathology
6 Wafula BM - DipAgr,BScAgr Research Officer agroclimatology and crop physiology

2978 Animal Production

1 Elungatta EE - BScAnSc Head of Section, Research Officer animal nutrition
2 Karachi M - BScAgr,MScAgr,PhD Senior Research Officer pasture agronomy
3 Menin LK - BSc Research Officer pasture agronomy
4 Wandera FP - BSc,MScAgr Research Officer pasture agronomy

2979 Crop Improvement

1 Ragwa LR - BScEd,MScAgr Head of Section, Research Officer millet breeding
2 Kamau JW - BSc,DipSeed Tech Research Officer genebank and seed technology
3 Kanyenji B - BSc Research Officer sorghum breeding
4 Kiarie Mrs AW - BScEd,MScAgr Research Officer crops improvement
5 Mugo SN - BScAgr Research Officer maize breeding
6 Muthoka SM - BScAgr,MScAgr Research Officer minor pulses improvement
7 Muyonga CK - DipAgr,BScAgr Research Officer seed production
8 Ngugi ECK - BSc,MScAgr Research Officer cowpea breeding
9 Njoroge K - BSc,MScAgr,PhD Research Officer maize breeding
10 Omanga PA - BScAgr,MScAgr Research Officer pigeonpea breeding
11 Rono WK - BScAgr,MScAgr Research Officer dry beans improvement

2980 Farming Systems

1 Muhammed L - BScAgr,MScAgr Econ,ICSA Head of Section, Research Officer farm management economics
2 Audi PO - DipAgr,BScAgr Research Officer farm management economics
3 Gatheru M - BSc Statistical Officer statistical design & analysis
4 Kimotho L - DipAgr,BScAgr Research Officer farming systems agronomy
5 Muinde K - BSc Statistical Officer statistical design & analysis
6 Nguluu SN - DipAgr,BScAgr Research Officer farming systems agronomy

2981 Soil and Water Management

1 Kinama JM - BScAgr,DipS&WCons Head of Section, Research Officer in Soil Water Conservation
2 Kitheka SK - BSc Research Officer soil and water relations
3 Okwach GE - BScAgr,MScAgr Research Officer soil and water relations

2982 ACIAR Dryland Project

1 Jones RK - BAgrSc,PhD Joint Project Leader soil fertility; plant nutrition
2 Beattie WM - MAIAS Experimentalist
3 Keating BA - BAgrSc,PhD Crop and Pasture Agronomist
4 Ockwell AP - BAgrSc,PhD Agricultural Economist

2983 FAO Dryland Project

1 Thairu DM - BSc,MScAgr Team Leader, Agronomist
2 Shakoor A - BSc,MSc,PhD Plant Breeder
3 Tessema S - BSc,MSc,PhD Animal Nutritionist

2984 National Potato Research Station

P.O. Box 338, Limuru
Tel: KARURI 40998, 40664, 40665.

1 Njoroge IN - NScAgr,MScAgr Director pathology

2985 Agronomy Section

1 King'ori Miss B - BScAgr,MScAgr agronomy
2 Wambugu Mrs L - BScAgr agronomy

2986 Breeding Section

1 Kamau CK - BScAgr potato breeding

2987 Farming Systems Economics Section

1 Gitunge FN - BScAgr extension
2 Munene Mrs S - BScAgr,MScEcon economics

2988 Food Technology Section

1 Kabira JN - BScFoodTech,MScFoodTech food technology

2989 Pathology Section

1 M'Murungi JM - BScAgr,MScAgr pathology
2 Njenga DN - BScAgr pathology

2990 Seed Production Section

2991 National Pyrethrum and Horticultural Research Station

P.O. Box 100, Molo
Tel: 41 & 177.

1 Ottaro GWM - BSc,MSc Director of the Station

2992 Pyrethrum Breeding Section

1 Ikahu JK - BSc,MSc Pyrethrum Varietal Breeder
2 Kipserem LK - BSc Pyrethrum Clonal Breeder

2993 Pyrethrum Agronomy Section

1 Ngugi CW - BScAgr Pyrethrum Agronomist
2 Ouma Mrs JP - BScAgr Pyrethrum Agronomist

2994 Horticulture Section

1 Obat JO - BSc Vegetables and Temperate Fruits Breeder
2 Ouma G - BSc Vegetable and Temperate Fruits Agronomist

2995 Plant Protection Section

1 Anyango JJ - BSc Entomologist

2996 Soil Science Section

1 Mwan gi JA - BSc variation of soil chemistry with measurable physical quantities

2997 National Range Research Station, Kiboko

P.O. Box 12, Makindu
Tel: Makindu 14.

1 Mirandattu BAJ - BSc,MSc,PhD Director

2998 Range Animal Nutrition and Management

1 Kamau PN - BSc,MSc nutrition of small stock
2 Kibet PF - BSc,MSc nutrition of range cattle
3 Mbui MK - BSc,MSc grazing management
4 Mnene WJ - BSc,MSc nutrition of range cattle
5 Musimba NKR - BSc,MSc,PhD range livestock nutrition
6 Shaabani SB - BSc,MSc livestock feed analysis

2999 Range Ecology and Management

1 Ali AR - BSc,MSc range plant eco-physiology
2 Gitunu AMM - BSc range management
3 Maranga EK - BSc,MSc range plant ecology
4 Mbakaya DS - BSc,MSc grazing management
5 Musembi DK - BSc,MSc agrometeorology
6 Omwono FO - BSc range botany
7 Too DK - BSc,MSc range plant ecology

3000 National Seed Quality Control Service

P.O. Box 1679, Nakuru
Tel: 037-42001, 85-314/5.

3001 Administration

1 Kiplagat JKB - BScAgr,MSc *Ag. Director*
2 Kamugira F - BScAgr
3 Wabwoto BAN - BScAgr,MSc

3002 Inspectorate

1 Amatike WO - BScAgr seed certification: agricultural crops
2 Ngoroi EH - BScAgr seed certification: agricultural crops
3 Obiero HM - BScAgr seed certification: agricultural crops
4 Ongudi JO - BScAgr seed certification: agricultural crops
5 Opiyo AMMO - BScAgr seed certification: horticulture
6 Otukho SRO - BScAgr seed certification: horticulture

3003 Seed Analytical and Health Testing

1 Kimani Miss EW - BScAgr investigation on pathological problems on seed
2 Mwaura LK - BScAgr *Head of Seed Analysis Laboratory*
3 Ndirangu Mrs LW - BScAgr investigation on pathological problems on seed
4 Ochuodho JO - BScAgr investigation on pathological problems on seed
5 Oloo WO - BScAgr investigation on pathological problems on seed

3004 Variety Description

1 Kimotho Miss VM - BScAgr distinctness uniformity stability
2 Kingori Miss SW - BScAgr distinctness uniformity stability of beans
3 Kute Mrs CA - BScAgr distinctness uniformity stability of maize
4 Mukiri G - BEdSc,MSc distinctness uniformity stability
5 Shuma JM - BScAgr distinctness uniformity stability of wheat
6 Thagana WM - BScAgr distinctness uniformity stability of sorghum

3005 National Sugar Research Station, Kibos

P.O. Box 1221, Kisumu
Tel: 2293.

1 Osoro MO - MSc,PhD *Director*

3006 Breeding and Variety Selection Section

3007 Agronomy Section

1 Abuom R - BScAgric soil fertility investigations
2 Amolo AJ - BScAgric *soil fertility and plant nutrition*
3 Chebosi PK - BSc,DIP sugarcane variety spacing
4 Ochola PN - BSc soil and plant tissue analysis
5 Oduor J - BSc chemistry

3008 Plant Pathology Section

1 Kariaga MG - BEdSce effect of virus diseases on sugarcane yield
2 Ongoma GIB - BScAgric,MScPath effect of hot water treatment on sugarcane smut pathogen

3009 Plant Physiology Section

3010 Extension, Economics and Biometrics

1 Olumbe JWK - DipAgric,BScAgric,MScAgron *Head, Extension Specialist*
2 Elivaha PR - BScStat biometrical design and analysis
3 Wawire NW - BAEcon *Agricultural Economist*

3011 Sugar Technology Section

1 Majani KA - BScMET,ASSCT,AIST sugarcane and sugar quality analysis
2 Omanga EA - BSc sugarcane and sugar quality analysis

3012 Apicultural Research Branch

National Beekeeping Station, P.O. Box 30028, Nairobi
Tel: 564324, 56301, 564302.

1 Kigatiira IK - MSc,PhD *Head Beekeeping Branch* honeybee behaviour; migrations; swarming, absconding; queen rearing
2 Abayo GK - BSc,DipAgric,CertApic honeybee breeding
3 Anam O - BSc hive product chemistry; properties of honey and beeswax
4 Asiko Miss G - BSc bee behaviour; colony dynamics
5 Gikungu Mrs G - BVM honeybee health; honeybee diseases
6 Kagio ASM - BSc honeybee behaviour; Queen/worker relationship in a colony
7 Kahenya - BSc honeybee botany; taxonomy of melliferous flora in Kenya
8 Kasolia DS - BSc hive products processing and quality control
9 Katunge Miss GN - BSc honeybee botany, nectar production
10 Mbaya KS - MSc,DipEd honeybee breeding; taxonomy; queen rearing
11 Mbithi Mrs JR - BSc hive products - chemistry, quality control
12 Moinde J - BED honeybee breeding; honeybee genetics
13 Muriithi W - BSc honeybee health; honeybee pests and predators
14 Muya BI - BScDipAgric,CertApic extension methods & training aids
15 Mwangi JF - BSc bee botany, pollination of crops by bees and pesticide effects on pollinators
16 Ndiwa JN - BED beekeeping equipment; equipment designs; improvement and manufacture
17 Onyango JA - BSc,DipAgric extension methods and training aids

3013 Department of Veterinary Services

3014 Head Office

National Veterinary Research Centre, Kabete, P.O. KABETE, KENYA, Kabete
Tel: 592231.

1 Injairu RM - BSc,MSc *Director*
2 Koske JK - BVM,DTVM,MSc *Chief Veterinary Investigation Officer*
3 Waghela S - BVSc,MSc *Assistant Director of Veterinary Services (Research)*

3015 National Veterinary Research Centre, Mugaga

P.O. Box 32, Karuri
Tel: 32106/7.

1 Kariuki DP - BVS,MSc *Director*
2 Vacant - Deputy Director

3016 Bacteriology Division

Contagious bovine and caprine pleuropneumonia, brucellosis and leptospirosis research and vaccine production

1 Mboya K - BSc,MSc *Head of Division* brucellosis
2 Litamoi JKK - BVM,MSc caprine mycoplasma
3 Mulira G - BVM,DipBact caprine mycoplasma
4 Muriu DN - BVM,MSc caprine mycoplasma
5 Muthomi EK - BVM,MSc bovine and caprine mycoplasmas
6 Simam PK - BVM,MSc bovine mycoplasma
7 Wakhusama SW - BVM,MSc leptospirosis

3017 Helminthology Division

Nematode, cestodes and fascioliasis research

1 Onyango-Abuje Mrs JA - BSC,MSc,PhD *Head of Division* cysticercosis
2 Wachira TM - BVM,MSC hydatidosis
3 Wamae LW - BVM,MSc fascioliasis

3018 Pathology Division

Pathology and immunology

1 Rumberia RM - BVM,MSc pathology and immunology of malignant catarrhal fever
2 Wesonga HO - BVM,MSc pathology and immunology of contagious bovine pleuropneumonia

3019 Protozoology Division

Epidemiology, immunology and research on control of theileriosis and other protozoa parasites research

1 Mutugi JJ - BVM,MSc,PhD *Head of Division* epidemiology
2 Awich JR - BSc,MSc immunology
3 Groocock C - BVetMB,PhD immunology
4 Lesan Mrs AK - BVM,MSc immunology
5 Linyonyi A - BVM,MSc chemotherapy

6 Manyasi D - BSc,MSc immunology
7 Mbogo SK - BVM,MSc immunology
8 Ndungu SG - BVM,MSc epidemiology
9 Ngumi Miss PN - BVM,MSc immunology
10 Nyanyi Mrs MBE - BSc immunology
11 Young AS - BSc,PhD epidemiology

3020 Virology Division

Rinderpest, malignant catarrhal fever and other viral diseases and vaccine production.

1 Mirangi PK - BVM,MSc *Head of Division* vaccine production and immunology
2 Ireri RG - BVM,MSc immunology
3 Mbugua N - BVM,MSc immunology
4 Rossiter PB - BVetMB,PhD epidemiology
5 Wafula JS - BVM,MSc vaccine production, immunology and epidemiology
6 Wamwayi HM - BVM,MSc immunology

COMMODITY RESEARCH FOUNDATIONS

3021 Coffee Research Foundation
P.O. Box 4, Ruiru
Tel: 21092/3,22652/3 THIKA.

1 Opile' WR - BScAgric,MSc,PhD *Director of Research*

3022 Research Sections

3023 Experimental Agronomy Section

1 Njoroge JM - BScAgric,MSc *Head of Section, Senior Research Officer (Agronomy), Agronomist*
2 Waweru MG - CRG SCT I field experimentation

3024 Breeding Section

1 Owuor JBO - BSc,MSc *Head of Section* resistance breeding, multiplication

3025 Chemistry Section

1 Michori PK - BSc,MSc *Deputy Director, Soil Chemist* soil fertility & plant nutrition
2 Maroko JB - BSc,MSc *Pesticide Chemist*
3 Mburu JN - BSc,MSc *Soil Chemist*

3026 Economics Section

1 Nyoro JK - BSc,MSc *Agricultural Economist*

3027 Entomology Section

1 Wanjala ME - BScAgric,MSc *Head of Section, Senior Research Scientist (Entomology)* coffee entomology
2 Kinuthia Miss MW - BSc *Entomologist*

3028 Pathology Section

1 Masaba Mrs DM - BSc,MSc *Head of Section, Senior Research Officer (Pathology)* plant pathology
2 Kairu GM - BSc,MSc *Bacteriologist (Coffee)*
3 Muthangya PM - BSc,MSc plant pathology

3029 Physiology Section

1 Kiara JM - BScAgric,MSc *Head of Section, Senior Research Officer (Crop Physiologist)* weed control; water use
2 Gathaara MPH - BSc,MSc *Environmental Physiologist*

3030 Tea Research Foundation of Kenya
P.O. Box 820, Kericho

1 Othiero CO - BSc,MSc,PhD *Director*

3031 Research Department

3032 Agronomy

1 Mwakha E - BSc,MSc

3033 Botany

1 Njuguna CK - BSA,MSA

3034 Chemistry

1 Owuor PO - BSc,MSc,PhD
2 Wanyoko - BSc,MSc

3035 Crop Environment

1 Obaga SO - BSc

3036 Plant Protection

1 Onsando JM - BSc,MSc
2 Sudoi V - BSc,MSc

3037 Ministry of Tourism and Wildlife

3038 Kenya Marine and Fisheries Research Institute
P.O. Box 81651, Mombasa
Telex: 21115.

1 Allela SO - BSC,DipAqua,DipResM *Director*

3039 Kiaumu Laboratory (Freshwater)
P.O. Box 1881, Kisumu
Tel: 40126,40129.

1 Ochieng' JI - BEdSc,MSc *Head*

3040 Limnology/Environmental Science

1 Kibaara DN - BEd,DipLimn phytoplankton studies
2 Ochieng' E - BSc,MSc trace elements and pesticide levels in natural waters
3 Ochumba PBO - BEd,MSc physical limnology
4 Onyari J - BSc,MSc heavy metal studies in fish, invertebrate and sediments

3041 Fish Technology Unit

1 Nyakundi R - BSc nutritive value of fish and utilization of Nile perch
2 Ogunja JC - MSc post harvest losses and quality assessment in fish

3042 Fish Biology Unit

1 Asila A - BEd,Dip *Statistics* stock assessment of fish
2 Awino A - MSc biology and ecology of Tilapia spp.
3 Maithya J - BSc fish parasitology
4 Ogari J - BSc,MSc abundance and distribution of Nile perch eggs and larvae

3043 Aquaculture Unit

1 Getabu A - BEd,MSc improved fish culture techniques
2 Kuria FK - BSc,MSc artificial propagation of riverine fishes
3 Ochieng' JI - BEd,MSc fry production studies

3044 Gear Technology Unit

1 Okollo Capt. HO *Master (inland waters)* experimental design of trawl and trammel nets

3045 National Museums of Kenya

3046 Herbarium
P.O. Box 45166, Nairobi
Tel: 742161.

3047 Taxonomic Botany

1 Kabuye Miss CHS - BSc *Botanist in Charge* plant identification esp. Gramineae, Cyperaceae and Malvaceae
2 Beentje HJ - BSc,MSc,PhD *Senior Botanist* plant identification Kenya trees and shrubs
3 Ndiangui N - BSc,MSc *Botanist* plant identification esp. Leguminosae

3048 Division of Natural Sciences
P.O. Box 40658, Nairobi
Tel: 742161.

3049 Entomology Section

1 Ritchie M - BSc,D/CPhD *Head of Section* insect identification esp. Orthoptera and Acridroidea
2 Bagine RKN - BEd,MSc *Entomologist* termite taxonomy and identification
3 Clifton MP *Entomologist* Lepidoptera identification; Noctuid moth taxonomy

3050 Library

1 Otike J - MLS,DipLib *Librarian*

3051 National Irrigation Board Operational Research and Training Project

P.O. Box 30372, Nairobi
Tel: 722590.

1 Ogombe IJO - BScAgr,MSc *Senior Research Officer*
2 Roelofsen B - BSclrr,MSclrr *Agronomist, Team Leader*

3052 Ahero Irrigation Research Station
P.O. Box 1961, Ahero

3053 Agronomy

1 Wabuke Mrs GAW - DipAgr *Assistant Research Officer*

3054 Entomology

1 Okhoba MM - BSc *Senior Research Officer*

3055 Irrigation and Drainage

1 Lamay PK - BScAgrEng *Research Officer*

3056 Mwea Irrigation Research Station
P.O. Box 80, Wanguru
Tel: 3 & 4.

1 Kiragu AW - DipAgr *Assistant Research Officer*

3057 Hola Irrigation Research Station
P.O. Box 5, Hola
Tel: 62.

3058 Agronomy

1 Barasa EO - BScAgrMSc *Senior Research Officer*

3059 Entomology

1 Agot MO - BSc *Research Officer*

3060 Irrigation and Drainage

1 Mulwa RK - BScAgr,Diplrr *Research Officer*

UNIVERSITIES

3061 University of Nairobi

PO Box 30197, Nairobi
Tel: 334244.

3062 College of Agriculture and Veterinary Sciences

1 Maloiy GMO - MBS,PhD,DSc,FIBiol *Principal, Professor of Animal Physiology*

3063 Department of Clinical Studies
3064 Medicine Section

1 Buoro IBJ - BVM,MVSc *Lecturer* small animal medicine
2 Karioki DI - BVM,MSc *Assistant Lecturer* porcine medicine
3 Kiptoon JC - BVSc,MSc,PhD *Associate Professor* ruminant medicine
4 Maribei JM - BVSc,MSc,PhD *Senior Lecturer* ruminant medicine
5 Mulei CM - BVM,MVSc *Lecturer* ruminant medicine
6 Nyamwange S - BVM,MSc *Assistant Lecturer* ruminant medicine
7 Price JE - BVM,MS,PhD *Senior Lecturer* small animal medicine
8 Wamukoya JPO - BVSc,MSc,DipAH *Senior Lecturer* ruminant medicine

3065 Surgery Section

1 Byagagaire SD - BVM,MSc *Lecturer* equine surgery
2 Mbiuki SM - BVM,MS,PhD *Senior Lecturer* ruminant surgery
3 Mbugua SW - BVM,MS *Senior Lecturer* small animal surgery
4 Mwangi JAN - BVM,MSc *Assistant Lecturer* ruminant surgery

3066 Reproduction and Obstetrics Section

1 Agumbah GJO - BVM,MSc,PhD *Senior Lecturer* ruminant reproduction

2 Munyua SJM - BSc,BVM,MPhil *Lecturer* ruminant reproduction
3 Mutiga ER - BVM,MS,PhD *Senior Lecturer* farm animal reproduction
4 Ogaa JS - BVSc,DVM *Senior Lecturer* goat and equine reproduction

3067 Department of Range Management

1 Karue CN - BScAgri,London,MS,PhD *Professor* animal nutrition
2 Ndung'u JN *Chief Technician* science laboratory technology, chemistry and biochemistry
3 Ng'ethe Dr JC - BSc,MS,PhD range ecology
4 Njoka TJ - BScAgri,MS,PhD range management

3068 Department of Food Technology and Nutrition

1 Alemu EK - BSc,MSc,PhD *Lecturer* antinutrients from cowpeas
2 Imungi JK - BSc,MSc,PhD *Lecturer* leaf proteins extraction
3 Mbugua SK - BSc,MSc,PhD *Lecturer* coffee pulp utilisation
4 Siegenthaler EJ - Dipl,Ing,Agr,PhD *Professor* bacterial load on restaurant counter cloths

3069 Department of Veterinary Anatomy

1 Oduor-Okelo D - BVSc,MSc,Dr.Vet.Med,PhD *Senior Lecturer and Chairman* reproductive anatomy
2 Bhattacharjee - BSc,MSc,PhD development and cytoarchitecture of retina
3 Maina JN - BVM,PhD *Lecturer* respiratory anatomy of the lungs and accessory respiratory organs
4 Mwangi DK - BVM,Dr.Vet.Med cardiovascular anatomy
5 Warui C - BVM,PhD *Tutorial Fellow* renal anatomy

3070 Department of Agricultural Engineering

1 Thomas DB - BA,DipAgr,MScAgrEng *Senior Lecturer and Chairman of Department* soil conservation
2 Lenselink KJ - MSc *Lecturer* irrigation
3 Muchiri G - BSc,MSc *Senior Lecturer* dryland tillage and energy
4 Mwaura EN - DipAgrEng,BScAgrEng,MScAgrEng,PhD *Lecturer* processing and energy
5 Oudman L - BSc,MSc *Lecturer* animal draft power
6 Some DKA - BSc,MSc,PhD *Lecturer* processing and storage

3071 Department of Animal Production

1 Said AN - DipAnimHusb,BVM,PhD *Associate Professor and Chairman of Department* forage & arable farm by-products utilization
2 Abate A - BSc,MSc,PhD *Lecturer* ruminant nutrition and feed evaluation
3 Carles AB - Vet.MB,MA,PhD *Senior Lecturer* performance of ruminant production systems
4 Collins EL - BSc,MSc,PhD *Lecturer* applied animal breeding in cattle and poultry
5 Itulya S - BSc,MSc,PhD quantitative genetics
6 Kipngeno WAK - BSc,MSc,PhD *Lecturer* statistics; biometrical design and analysis of experiments
7 Mbugua PN - BSc,MSc,PhD *Senior Lecturer* animal nutrition and production, poultry
8 Mitaru BN - BSc,MSc,PhD *Lecturer* nutritional evaluation of feedstuffs
9 Rege JEO - BSc,MSc,PhD *Lecturer* sire evaluation prediction and estimation methods
10 Schwartz HJ - MSc,PhD pastoral livestock production systems; feeding behaviour
11 Wanyoike MM - BSc,PhD *Senior Lecturer* cattle production systems

3072 Department of Soil Science

1 Anyango B - BSc,MSc *Tutorial Fellow* microbiology; biological nitrogen fixation
2 D'Costa VP - BSc,PhD *Senior Lecturer* pedology, soil survey and land evaluation; soil fertility
3 Keter JK - BSc,MSc *Senior Lecturer* soil testing, plant analysis and characterization of soils
4 Keya SO - BSc,PhD *Professor* biological nitrogen fixation
5 Kinyali SM - BSc,MSc soil physics; irrigation & drainage
6 Mbuvi J - MSc,PhD *Lecturer* soil survey & classification; land evaluation
7 Mochoge B - BSc,MSc,DipEd *Senior Lecturer* nitrogen behaviour in soil ecosystems and fertility
8 Muriuki S - BSc,MSc *Assistant Lecturer* soil chemistry and soil fertility
9 Ssali HB - BSc,MSc,PhD *Senior Lecturer* soil fertility and plant nutrition; biological nitrogen fixation

3073 Department of Animal Physiology

1 Kamau JMZ - BVM,MSc,PhD *Senior Lecturer and Chairman of Department* energy metabolism and environmental physiology
2 Gombe S - BSc,BVM,PhD *Associate Professor* physiology of reproduction
3 Kanui TI - BVM,PhD *Lecturer* electrophysiology
4 Maloiy GMO - MBS,PhD,DSc,FIBiol *Professor* renal and environmental physiology
5 Wathuta EM - BVM,PhD *Lecturer* digestive physiology

3074 Department of Public Health, Pharmacology & Toxicology

1 Gathuma JM - BVSc,MSc,PhD *Associate Professor* epidemiology of animal diseases, parasitic zoonoses
2 Kagiko MM - BVM,MSc,PhD *Lecturer* food-borne diseases, immunoparasitology
3 Kahunyo JKM - BVM,MSc *Lecturer* pesticide toxicology
4 Kamanga-Sollo EIP - BVM,MSc,PhD *Postdoctoral Fellow* immunoparasitology
5 Kang'ethe EK - BVM,MSc *Lecturer* speciation of meat and meat products
6 Kayihura MM - BVSc,MSc,Dipl.Microbiol,& FoodHyg *Visiting Lecturer* Meat and food hygiene
7 Lkken P - LicOdon't,PhD *Visiting Associate Professor* pharmacology of anti-inflammatory drugs
8 Lindqvist KJ - DVM,MSc,PhD,FRCPath *Visiting Professor* microbiology; immunology; immunoparasitology
9 Maitho TE - BVM,MSc,PhD *Lecturer* pesticide toxicology, pharmacology of anti-inflammatory drugs
10 Maritim AC - BVM,MSc,PhD *Postdoctoral Fellow* pharmacokinetics of drugs
11 Mitema SEO - BVM,MS,PhD *Lecturer* toxicology
12 Muchiri DJ - BVM,MSc *Lecturer* plant toxicology
13 Nderu FMK - BVM,MPVM *Lecturer* epidemiology of animal diseases
14 Njeru FM - BVM,MSc *Lecturer* immunoparasitology

3075 Department of Veterinary Pathology and Microbiology

1 Mugera GM - DipVetSci,MSc,PhD *Professor* pathology and chemotherapy with medicinal plants
2 Bebora LC - BVM,MSc *Lecturer* avian salmonellosis
3 Gathumbi PK - BVM *Assistant Lecturer* toxicity of medicine plants
4 Gitao CG - BVM,MSc *Lecturer* characterisation of animal pox viruses
5 Kagunya DKJ - BVM,MSc *Lecturer* animal pathogenesis of bacterial infections
6 Kamau JA - BVM,MSc *Lecturer* pathology of toxic plants
7 Kimeto BA - BSc,DVM *Senior Lecturer* fine structure in bovine theileriosis
8 Kimoro CO - BVM *Tutorial Fellow* caprine fascioliasis immune responses
9 Kuria JKN - BVM,MSc *Tutorial Fellow* pathogenesis Corynebacterium pseudotuberculosis
10 Maingi NE - BVM *Assistant Lecturer* caprine trypanosomiasis pathogenesis
11 Munyua WK - Dip.Vet.Sci,MSc,PhD *Senior Lecturer* bovine theileriosis and trypanosomiasis immunology
12 Mwangi DM - BVM *Assistant Lecturer* caprine trypanosomiasis immunology
13 Ngatia TA - BVM,MSc *Lecturer* bovine mastitis
14 Njiro SM - BVM,MSc,PhD *Lecturer* pathology and immunology with medicinal toxic plants
15 Nyaga JM - BVM,MSc *Lecturer* characterisation of newcastle disease virus strains
16 Nyaga PN - BVM,MPVM,PhD *Associate Professor & Chairman* pathogenesis of viral infections, chemotherapy with medicinal plants
17 Wandera JW - Dip.Vet.Sci,MSc,PhD *Associate Professor* domestic animal pneumonias

3076 Jomo Kenyatta College of Agriculture & Technology

P.O. Box 62000, Nairobi
Tel: 0151-22646/8.

1 Githaiga JM - BEd,MSc *Principal*
2 Orie GA - BVSc,MPA,Cert.Univ.Adm,PhD *Deputy Principal*

3077 Department of Agriculture Engineering

1 Weru SS - DipAgrEng,BScAgrEng,MScAgrEng *Principal Lecturer, Head of Department* farm power and farm mechanization
2 Akeng'a RB - DipAgrEng,BScAgr *Assistant Lecturer, Deputy Head of Department* farm power
3 Matsui H - MScAgrEng *Japanese Volunteer* farm mechanization
4 Mwangi GG - BScFarmMach,BScAgrMech,Cert.Ped *Lecturer* agricultural processing, biogas generation and utilization
5 Torisu R - BScAgrEng,MScAgrEng,DAgrEng *Japanese Expert* farm mechanization

3078 Department of Food Processing

1 Kenji GM - BSc *Assistant Lecturer, Head of Department* food analysis
2 Akimoto T - BSc *Japanese Volunteer* food microbiology
3 Aranishi M - BSc *Japanese Volunteer* food chemistry
4 Gichuru SG - BVetMed *Assistant Lecturer* food microbiology
5 Kiiyukia C - BSc *Lecturer* food quality control
6 Moturi SA - MScFoodTech *Assistant Lecturer* food engineering
7 Mwanjumwa Miss L - BSc,MSc *Senior Lecturer* vegetable and food chemistry
8 Sugiyama T - BSc,MScAgr *Japanese Expert and Team Leader* food processing

9 Tsukamoto S - BSc,MSc *Japanese Expert* food chemistry and baking quality of various flours
10 Wandati Mrs TW - BSc *Lecturer* baking quality of various flours

3079 Department of Horticulture

1 Kahangi Mrs EM - BSc,Hort,MSc,Agron *Principal Lecturer, Dean Agriculture/Head of Department* indigenous vegetables, tissue culture techniques for rapid multiplication of plants
2 Fujime Y - BScMScDAgr *Japanese Expert* tissue culture techniques for rapid multiplication of plants. Indigenous vegetables
3 Gichuki EM - BSc *Lecturer, Assistant Head of Department* grain storage
4 Moriya H - BSc *Japanese Expert* differentiation of papaya sexes
5 Ngumi Miss VW - BEd *Assistant Lecturer* indigenous vegetables
6 Tanaka U - BScAgr *Japanese Volunteer* soil fertility in semi-arid areas
7 Watako AO - BSc *Assistant Lecturer* factors influencing flower growth (Chrysanthemum sp)

INTERNATIONAL RESEARCH INSTITUTIONS

3080 International Council for Research in Agroforestry (ICRAF)

P.O. Box 30677, Nairobi
Tel: 29867. Telex: 22048. Cables: ICRAF.

ICRAF is an autonomous, non-profit, international council whose objects are to increase the social, economic and nutritional well-being of peoples of developing countries through the promotion of agroforestry systems designed to achieve better land use without detriment to the environment, to encourage and support research and training relevant to agroforestry systems, to facilitate the collection and dissemination of information relevant to such systems and to assist in the international co-ordination of agroforestry.

1 Lundgren BO - MSc,PhD *Director*
2 Arnold HR - BSc,MSc *Forestry*
3 Baumer M - PhD *range management and marginal lands*
4 Darnhofer T - PhD *Bioclimatologist/Agrometeorlogist*
5 Depommier D - BSc,MSa *Forestry Engineer*
6 Djimde M - BSc,MSc,PhD *livestock production*
7 Gatama D - BSc *climatology*
8 Hoekstra DA - BSc,MSc *agricultural economics*
9 Huxley PA - BSc,PhD,FIB,FRSA *Horticulturist/Agronomist*
10 Kamau IN - BSc,MSc *Ecologist/Senior Research Assistant*
11 Kanani SS - BSc,MSc,MA *Geographer/Director's Assistant*
12 Kwesiga F - BSc,PhD *Forester*
13 Labelle R - BSc,MSc *Information Officer*
14 Minae S - BSc,MSc,PhD *Agricultural Economist*
15 Muraya P - BSc *computer science*
16 Nair PKR - BSc,MSc,PhD,DrSc *Agronomist/Soil Scientist*
17 Ngugi DN - BSc,PhD *Agronomist*
18 Ntiru RC - BA *Publications Officer*
19 Odra J - BSc,MSc,PhD *Crop Physiologist*
20 Oduol P - BSc,MSc *Forester*
21 Raintree JB - BA,MA,PhD *Ecological Anthropologist*
22 Rocheleau D - PhD *Geographer/Systems Ecologist*
23 Scherr S - MSc,PhD *agricultural economics*
24 Sibanda H - BSc,PhD *Environmentalist*
25 Torres F - BSc,PhD *range management; livestock production*
26 von Carlowitz P - *Forester*
27 Wahome JN - BSc,MSc *Training Assistant* animal science
28 Waiguru PN - BEd,DipCompSci
29 Wambugu D - BSc,MSc *Agronomist*
30 Wood PJ - BA,MA *Forester*
31 Young A - PhD,DSc *land evaluation; soil science*
32 Zulberti E - MSc,PhD *Training Officer*

3081 International Laboratory for Research on Animal Diseases (ILRAD)

P.O. Box 30709, Nairobi
Tel: 592311 Nairobi. Telex: 22040 'ILRAD. Cables: ILRAD, Nairobi,Kenya.

3082 Administration

1 Gray AR - DSc,PhD *Director General*
2 Doyle JJ - PhD *Director of Research*
3 Rowe PR - PhD *Director of Administration*
4 Lloyd BC - FCA *Financial Controller*

3083 Parasitology/Trypanosomiasis Laboratory

1 Gardiner PR - PhD cell biology, *T.vivax*
2 Knowles G - PhD protozoan genetics
3 Nantulya VM - PhD *immunology, T. vivax*

3084 Biochemistry/Molecular Biology Laboratory

1 Grab DJ - PhD cell biology
2 Lonsdale-Eccles J - PhD enzymology
3 Ole Moi Yoi - MD clinical biochemistry
4 Young JR - PhD molecular biology

3085 Cell Biology Laboratory

1 Fish WR - PhD trypanosome biochemistry
2 Hirumi H - DM,SC trypanosome culture

3086 Immunology Laboratory

1 Barry W - PhD cellular immunology
2 Black SJ - PhD immunology, trypanosomes
3 Shapiro SZ - PhD immunochemistry

3087 Parasitology/Theileriosis Laboratory

1 Dolan TT - PhD epidemiology, *Theileria*
2 Morzaria S - PhD protozoology, *Theileria*
3 Musoke AJ - PhD immunology, *Theileria*

3088 Pathology Laboratory

1 Morrison WI - PhD immunopathology
2 Paling RW - DVM epidemiology, trypanosomiasis
3 Teale AJ - PhD immunogenetics

3089 Electron Microscopy

1 Ito S - PhD electron microscopy

3090 Tick Unit

1 Voigt WP - DVM,DMEDVET parasitology, ticks

3091 Tsetse Vector Unit

1 Moloo SK - PhD parasitology, tsetse

3092 Central Core

1 Hinson C - MSc culture media, radiobiology

3093 Experimental Animal Units/Diagnostic Laboratory

1 Jordt T veterinary medicine

3094 Wildlife Diseases Project

1 Grootenhuis J - PhD parasitology, wildlife diseases

3095 Training and Information

1 Lenahan JK - PhD *Training and Outreach Officer*
2 Umbima WE - MA,ALA *Librarian*
3 Westley SB - BA *Writer/Editor*

3096 Support Services

1 Holt MW - CertEngMIMechMIMar *Chief Engineer*
2 Lobo MA - MSc *Electronics Engineer*

MALAWI

3097 Ministry of Agriculture

PO Box 30134, Lilongwe, 3
Tel: 733 300.

1 Mwandemere HK - PhD *Chief Agricultural Research Officer*
2 Manda DRB - BSc *Deputy Chief Agricultural Research Officer*
3 Makato CJA - Cert *Research Liaison Officer*
4 Ndisale Mrs BR - MSc *Research Economist*
5 Ntokotha EM - PhD *Senior Agricultural Research Officer*
6 Pindani CW - BA *Administration Officer*

3098 Department of Agricultural Research

3099 Chitedze Agricultural Research Station
PO Box 158, Lilongwe
Tel: 767222/225.

1 Mkamanga GY - BSc,MSc,PhD *National Research Coordinator (Cereal Crops) and Officer-in-Charge*

3100 Agricultural Economics, Statistics and Data Processing

1 Chirembo AB - BSc,MSc *Commodity Team Leader (Biometrics)*
2 Kisyombe FW - BSc *Senior Biometrician*
3 Lupiya GW - BSc *Computer Programmer*
4 Mwenda ARE - BSc *Agricultural Economist*

3101 Adaptive Research Team

1 Nyirenda FM - DipAgr,BSc *Agronomist, National Research Coordinator (Adaptive)*
2 Bulla GM - BSc *Socio-Economist*
3 Chanika CSM - BSc *Agronomist* on course
4 Chisambiro L - BSc *Agronomist* on course
5 Kalumbi DMA - BSc *Socio-Economist*
6 Kaonga WT - BSc *Agronomist*
7 Moyo CC - BSc *Agronomist*
8 Mughogho WCC - DipAgr *Agronomist*
9 Mwafulirwa ES - BSc *Agronomist*
10 Mwango ES - BSc,MSc *Socio-Economist*
11 Ndengu JD - BSc *Socio-Economist*
12 Nyondo FKKM - BSc,MSc *Socio-Economist*
13 Phiri BSC - BSc *Socio-Economist*

3102 Cereal Crops

1 Mkamanga GY - BSc,MSc,PhD *National Research Coordinator (Cereal Crops)*
2 Chibambo M - DipAgr *Senior Technical Officer (Roots and Tubers)*
3 Mnyenyembe PH - BSc,MSc *Wheat Breeder*
4 Munthali Mrs FC - BSc,MSc *Wheat Agronomist*
5 Ngwira LDM - BSc,MSc *Maize Commodity Team Leader*
6 Ngwira Mrs P *Plant Pathologist* on course
7 Nhlane WG - BSc *Maize Breeder*
8 Nyirenda N - BSc *Agronomist (Wheat and Barley)*
9 Sibale Mrs EM - BSc,MSc *Maize Breeder*
10 Zambezi BT - BSc,MSc,PhD *Maize Breeder*

3103 Crop Storage

1 Chimbe CMK - BSc,MSc *Crop Storage Entomologist*

3104 Entomology

1 Chimaliro ES - CertLabTech *Technical Officer (Entomology)*

3105 Farm Mechanisation Unit

1 Kumwenda WF - BSc *Farm Machinery Officer*
2 Singa DD - BSc *Commodity Team Leader (FMU)*

3106 Livestock/Pastures

1 Dzowela BH - BSc,MSc,PhD *National Research Coordinator (Livestock/Pastures)*
2 Chintsanya NCC - BSc *Animal Production Research Officer*
3 Kumwenda MLK - BSc *Dairy Scientist* on course
4 Mhango Miss CR - DipAgr *Technical Officer (Pasture)*
5 Mtukuso AP - BSc *Small Ruminants Scientist*
6 Munthali JTK - BSc,MSc,PhD *Senior Animal Nutritionist*
7 Zimba AWC - BSc,MSc *Animal Breeder*

3107 Oilseeds, Legumes and Fibres

1 Sibale PK - BSc,MSc,PhD *National Research Coordinator (Oilseeds)*
2 Chiyembekeza AJ - BSc,MSc *Groundnut Breeder*
3 Kisyombe CT - BSc,MSc *Senior Groundnut Pathologist*
4 Maliro CE - BSc,MSc *Groundnut Agronomist*

3108 Soil Microbiology

1 Chibwana PAD - BSc *Microbiologist*
2 Khonje DJ - BSc,MSc *Soil Microbiologist* on course

3109 Soil Fertility & Plant Nutrition

1 Saka AR - BSc,MSc,PhD *Commodity Team Leader*
2 Banda P - BSc,MSc *Soil Surveyor*
3 Kadyampakeni WA - CertLabTech *Chief Laboratory Technician*
4 Matabwa CJ - BSc,MSc *Soil Scientist*

3110 Bvumbwe Agricultural Research Station
PO Box 5748, Limbe
Tel: 662206. Cables: Agrisearch, Limbe.

1 Maida JHA - BAgrSc,PhD,FCSM *Deputy Chief Agricultural Officer and Officer-in-Charge* soil fertility and chemistry
2 Chapola GM - BSc,MSc,PhD *Assistant Chief Agricultural Research Officer* plant pathology and quarantine
3 Panje PP - BSc,MSc *Assistant Chief Agricultural Research Officer* soil physics
4 Chikonda MM - BSc *Senior Agricultural Research Officer* entomology
5 Chilembwe EHC - BScAgr,MScHort decidious fruits and plant propagation
6 Chilima ZW *Chief Technical Officer* soil fertility and plant nutrition
7 Chisenga L - DipLabTech *Senior Technical Officer* soil physics
8 Chithambo GSG - BSc soil fertility and chemistry - on course
9 Chizala CT - DipAgr,BScAgr tropical fruits and spices
10 Daudi AT - DipAgr,BScAgr,MSc nematology
11 Gondwe WT - BScAgr,MScHort vegetable crops
12 Kambale GJS - DipAgr *Technical Officer* plant quarantine
13 Mainjeni Mrs CE - DipAgr *Technical Officer* plant pathology
14 Mainjeni DD - DipAgr *Technical Officer* coffee agronomy
15 Makina DW - DipLabTech *Technical Officer* nematology
16 Muwalo ES - DipAgr,CertTropCrop *Senior Technical Officer* crop storage
17 Nkhoma Miss DD - BSc soil fertility and chemistry - on course
18 Nsanjama MN - DipAgr,BScAgr,MSc coffee agronomy
19 Ntuana LR - DipAgr *Senior Technical Officer* cereals agronomy
20 Phiri IMG - BScAgr,MScHort tree nut crops
21 Phiri Mrs VC - DipAgr *Technical Officer* seed technology
22 Simwaka FHA - CertTropCrop,CertPlQuar *Senior Technical Officer* phytosanitary inspection
23 Soko HN - BAgrSc *Senior Agricultural Research Officer* grain legumes
24 Thindwa Miss HP - BSc,MSc macadamia entomology - on course

3111 Makoka Agricultural Research Station
P. Bag 3, Thondwe
Tel: Thondwe 251.

1 Nyirenda GKC - BSc,MSc,DIC,PhD *Cotton Commodity Team Leader: Chief Cotton Entomologist*
2 Banda MHP - BScAgr,MSc *Cotton Breeder*
3 Chigwe CFB - DipAgr,BScAgr,MSc,CertIICTesting,PhD *Senior Cotton Breeder*
4 Kapeya E - BScAgr,MSc *Cotton Entomologist*
5 Kaunda AH - Chief Technical Officer *biometrics*
6 Luhanga WW - BSc,CertPestMan *Armyworm Entomologist*
7 Matabwa ES - UCAT *Technical Officer* cassava and sweet potato
8 Mchowa JW - BSc,MSc *Groundnut Entomologist*
9 Sauti RFN - DipAgr,BScAgr,CertRootCropProd *Rootcrops Commodity Team Leader, Agronomist/Breeder*
10 Sauti-Phiri LL - BSc,MSc *Senior Cotton Agronomist*

3112 Ngabu Agricultural Research Station
PO Box 48, Ngabu
Tel: Ngabu 33 426244.

1 Kumwenda Mrs ME - DipAgr *Officer-in-Charge* sorghum millet breeding
2 Chirwa RM - BScAgr,MSc
3 Kumwenda JDT - BSc,MSc *Sorghum Agronomist*

3113 Kasinthula Agricultural Research Station
PO Box 28, Chiikwawa

1 Mzembe CP - DipAgr,BSc,PhD *Irrigation Coordinator*
2 Chintu EM - DipAgr,BSc *Sorghum Millet Breeder*

3114 Baka Agricultural Experimental Station
PO Box 97, Karonga
Tel: Karonga 58, 35.

3115 Baka Research Station

1 Lwesya AY - DipAgr,CertSorghum/PerealMilletBreed *Officer-in-Charge* agronomy
2 Chunga BE - DipAgr,CertRiceProd agronomy
3 Likoswe AA - DipAgr seed technology
4 Maguru KR - DipAgr adaptive agronomic research

3116 Meru Sub-Research Station
PO Box 12, Chitipa
Tel: Chipita 26.

1 Luhana SC - DipAgr adaptive agronomic research
2 Zgambo EM - CertAgrTech

3117 Lunyangwa Research Station
PO Box 59, Mzuzu
Tel: 332633.

1 Gadabu AD - BSc,MSc,PhD *Officer-in-Charge* entomology
2 Chilimba ADC - DipAgr roots and tuber crops and wheat agronomy
3 Kumwenda RLN - DipAgr rice agronomy
4 La Croix EAS - BSc,MIBiol crop protection
5 Mughogho HM - DipAgr maize agronomy

3118 Chitala Research Station
PO Box 315, Salima
Tel: Salima 0 - 1211 or 0 - 1221.

1 Muwowo AED - DipAgr *Chief Technical Officer; Officer-in-Charge Chitala Research Station* agronomy

3119 Soil Survey Unit
PO Box 92, Lilongwe
Tel: 720344.

1 Lowole MW - BSc,DipAgrE,MSc *Head of Unit ans Senior Soil Surveyor* soil survey and land evaluation
2 Banda P - BSc,MSc *Soil Surveyor* soil survey

3120 Lifuwu Rice Research Station
PO Box 102, Salima

3121 Rice Crop Research Unit

1 Chaseta AL - DipAgr *Senior Technical Officer* seed technology
2 Kumwenda AS - BScAgr rice breeding - on course

3122 Agricultural Development Division

3123 Blantyre Agricultural Development Division (A.D.D.)
PO Box 30227, Chichiri, Blantyre, 3
Tel: 672 022.

3124 Adaptive Research Team

1 Kawonga WT - BScAgr *Team Leader* crop agronomy
2 Ndengu JD - BScAgr socio economics

3125 Liwonde Agricultural Development Division
P.Bag 3, Liwonde
Tel: Liwonde 532410.

3126 Adaptive Research Section - Management Unit

1 Werner J - DipIngAgr *Research Officer Head of Section*
2 Longwe AWF - BSc *Research Officer*

3127 Department of Veterinary Services and Animal Industry
Head Office, PO Box 30372, Lilongwe, 3

1 Thyangathyanga Dr GAA - BVM,DPVM *Chief Veterinary Officer*
2 Mkandawire Dr RCJ - BVM,MSc *Assistant Chief Veterinary Officer (Research Coordinator)*

3128 Central Veterinary Laboratory

1 Mateyu LR - BVM,MSc *Laboratory Leader*
2 Chimera B - BVM *Veterinary Investigation Officer (Bacteriology)*
3 Haresnape JM - BSc,MSc,PhD *Virologist*
4 Kamwoch SP - BVM *Veterinary Investigation Officer (Protozoology)*
5 Ngulube BN - BVM *Principal Veterinary Officer (Protozoology)*
6 Saeluzika JG - BSc,PhD *Biochemist*

3129 Blantyre Regional Laboratory

1 Mwalwanda ST - BVSc,DipTropVetMed *Senior Veterinary Officer (Laboratory)*

3130 Animal Industry Section

1 Kasowanjete MBB - BSc,MSc,DipAnimSc *Animal Breeder*
2 Sill O - PhD *Animal Breeder*

3131 Ministry of Forestry and Natural Resources
Private Bag 350, Lilongwe 3

3132 Fisheries Research

3133 Fisheries Research Unit
PO Box 27, Monkey Bay
Tel: 6, Monkey Bay.

1 Lewis DSC - BSc,PhD *Senior Research Officer* biology of fishes
2 Mapila S - BSc *Senior Fisheries Research Officer* limnology and fish biology
3 Alimoso S - BSc *Fisheries Research Officer* fish biology
4 Kachinjika OV - BSc,MSc *Fisheries Research Officer* fish biology
5 Magasa J - BSc *Fisheries Research Officer* fish biology
6 Tweddle D - BSc *Fisheries Research Officer* fish biology

3134 Fish Farming Research Centre
PO Box 44, Domasi
Tel: 531215.

1 Msiska OV - BSc,MSc *Senior Research Officer* aquaculture
2 Rashid BB - BSc *Fisheries Research Officer* aquaculture

3135 Forestry Research Institute of Malawi
PO Box 270, Zomba
Tel: 522866.

1 Barnett DW - BSc,PhD zoology, entomology
2 Chipompha NWS - BSc,MSc pathology
3 Kanaji B - DipAgr,BScFor
4 Kayambazinthu D - BSc,MScForEV
5 Masamba CR - DipAgr,BScFor
6 Ngulube M - BSc,MScFor
7 Sitaubi LA - BSc,MAgrSc
8 Venkatesh CS - BSc,MSc,PhD genetics

3136 Geological Survey Research
Headquarters, PO Box 27, Zomba
Tel: 522166,522670,522368.

1 Chatupa JC - BSc,MSc *Chief Geologist (National Head)* coal petrology
2 Mndala AT - BSc,DIP *Assistant Chief Geologist* glass sands, mineral exploration
3 Phiri FR - BSc,MSc *Principal Geologist* industrial minerals
4 Chikusa C - BSc geological mapping and mineral exploration
5 Chisale RTK - BSc geological mapping and mineral exploration
6 Gondwe PC - BSc geological mapping and mineral exploration
7 Kaphwiyo CE - BSc geochemical exploration
8 Madhlopa A - BSc ceramic raw material
9 Masamba WR - BSc geochemistry
10 Mshali RS - BSc coal analysis
11 Nkhoma JES - BSc geological mapping and mineral exploration
12 Piper DP - BSc non-metallic mineral evaluation in the basement
13 Senzani F - BSc geological mapping and mineral exploration

3137 Malawi Tobacco Research Authority

3138 Head Office
PO Box 418, Lilongwe
Tel: 722266.

1 Tsonga EW - BSc,MSc *General Manager*

3139 Research Division
Headquarters, Kandiya Research Station, PO Box 418, Lilongwe
Tel: 722266.

1 Chipala E - BSc,MSc,PhD *Research Co-ordinator*
2 Bernard MP - BSc,DipAgricTech,MIAgricEng *Senior Tobacco Engineer*
3 Chaongola GH - BSc *Research Liaison Officer*
4 Chilemba EMJ - DipAgric *Research Liaison Officer*
5 Chimoyo HM - BSc *Senior Plant Breeder*
6 Chisambiro RRZ - DipAgric *Acting Burley Tobacco Agronomist*
7 Manondo HS - DipLabTech *Assistant Research Chemist*
8 Mwandira CK - BSc,PGD,MME *Tobacco Engineer*
9 Ng'ambi TBS - BSc *Plant Breeder*

3140 Mwimba Flue-Cured Tobacco Research Station
PO Box 224, Kasungu
Tel: 200318/200322 (253418/253442).

1 Gondwe WK - BSc,MSc *Senior Agronomist and Officer-in-Charge*
2 Manduwa JDD - BSc *Agronomist*

3141 Tea Research Foundation of Central Africa
PO Box 51, Mulanje
Tel: Thornwood 224, 218, 255.

1 Grice WJ - MA,DipAgr *Director*
2 Clowes MSTJ - BSc,MSc,PhD *Assistant Director Research*
3 Fernie AT - AMBIM *Assistant Director Administration*
4 Eyyüboğlü HY - MSc *Productivity Officer*
5 Johnson AL - MA,MIMechE,CEng *Process Engineer*
6 Kamanga KDE - BSc *Tea Seed Oil Research Officer*
7 Kayange CW - DipAgr *Senior Horticulturist*
8 Malenga NEA - BSc *Senior Agronomist*
9 Mkwaila B - BSc,MSc *Senior Advisory Officer, Entomologist*
10 Nyirenda HE - BSc,MSc *Senior Plant Breeder*
11 Phiri GSN - BSc *Agronomist*
12 Rattan PS - BSc,MSc,PhD *Senior Plant Pathologist and Entomologist*
13 Velthuis Mrs AG - MSc *Librarian*
14 Whitehead DL - MSc,PhD *Biochemist*

UNIVERSITIES AND COLLEGES

3142 University of Malawi

3143 Bunda College of Agriculture
PO Box 219, Lilongwe
Tel: 721-455 and 721-231. Cables: Bundagric, Lilongwe.

1 Chimphamba Professor BB - BSc,MSc,PhD,FLS *Principal*
2 Kasomekera ZM - BSAE,MSEngr,PhD *Vice-Principal, Lecturer, Head Agricultural Engineering Dept.*
3 Namponya CR - MPhil,FLA *College Librarian*
4 Sichinga CA - BScAgric,MSc *Farm Manager*

3144 Faculty of Agriculture

1 Ngwira TN - BSc,UCE,MSc,PhD *Dean*

3145 Agricultural Engineering

1 See Vice-Principal *Lecturer and Head of Department*
2 Asiedu JJ - DipIng,PhD *Lecturer*
3 Kamanga MH - DipEngr,BSAE *Assistant Lecturer*
4 Makoko MS - BScAgric,MSc *Lecturer*
5 Mwendera EJ - BSc,PGDip,MSc *Lecturer*
6 Mwinjilo ML - DipEngr,BS,MSc *Lecturer*
7 Ngwira SC - DipEngr,BSMechEngr,MSc *Lecturer*
8 Simango DG - DipEngr,BSAgricMech,MSAgricEngr *Lecturer*

3146 Crop Production

1 MacColl D - BSc,PhD *Senior Lecturer and Head of Department*
2 Mughogho LK - BSc,DIC,PhD *Professor*
3 Bokosi J - BScAgric,MSc *Lecturer*
4 Chiyenda SS - BSc,MSStats,PhD *Lecturer*
5 Kanyama-Phiri GY - DipAgric,BS,MS *Lecturer*
6 Kwapata MB - DipAgric,BS,MS *Lecturer*
7 Lungu NF - BSc,PhD *Reader*
8 Lupwayi NZ - BScAgric,DipSoilSc,MSc *Lecturer*
9 Materechera SA - BScAgric,MSc *Lecturer*
10 Mkandawire ABC - BScAgric,MS *Lecturer*
11 Mloza Banda HR - BScAgric,MSc *Lecturer*
12 Msuku WAB - DipPHI,BScAgricBot,MSPlantPath,PhD *Lecturer & Director of Bean/Cowpea Research*
13 Saka VW - BScPlantPath,MSc,PhDPlantNem *Senior Lecturer*

3147 Livestock Production

1 Makhambera TPE - BS,MS,PhD *Reader and Head of Department*
2 Kaminjolo JS - BVSc,DrMedVet *Professor*
3 Banda JW - BScAgric,MSc *Lecturer*
4 Kamwanja LA - DipAgric,BS,MS,PhD *Lecturer*
5 Mfitilodze MW - BVM,PhD *Lecturer*
6 Mtimuni JP - BS,MSc,PhD *Senior Lecturer*
7 Mumba PPEM - BSC,BSc,MScR'dg *Lecturer*
8 Ngwira TN - BSc,UCE,MSc,PhD *Senior Lecturer*
9 Nyirenda HGC - DipPoultHusb,BS,MS *Lecturer*
10 Phoya RKD - DipAgric,BS,MS,PhD *Lecturer*

3148 Rural Development

1 Mkandawire RM - BSocSc,MA,MS,PhD *Lecturer & Head of Department*
2 Khaila SW - DipAgric,BSAgric,MS *Lecturer*
3 Mataya CS - BScAgric,MSc *Lecturer*

4 Nankumba JS - BScAgric,MSAgricEcon *Lecturer*
5 Ng'ong'ola DH - BScAgric,MScAgricEcon *Lecturer*
6 Nothale DW - BScAgric,DipAgricSc,MSc *Lecturer*
7 Phiri CD - BScAgric,DipAgricEcon,MSc *Lecturer*

3149 Home Economics/ Human Nutrition

1 Mtimuni Mrs BM - DipAgric,BSc,PhDNutrition *Lecturer & Head of Department*
2 Lonje Miss V - BScAgric *Staff Associate*
3 Mafuleka Miss MM - DipAgric,BS,MS *Lecturer*
4 Mkandawire Mrs PMM - BSHomeEc,MARurSocDev *Lecturer*

3150 Chancellor College
PO Box 280, Zomba
Tel: 522 222. Cables: Chancoll, Zomba.

1 Jedegwa VBJ - BAPubAdmin
2 Kadzamira ZD - AB,PhD *Principal*
3 Mvalo WES - BA,MSc *College Registrar*
4 Uta JJ - MLS,ALA *College Librarian*

3151 Faculty of Science

1 Msonthi JD - BSc,PhD,MRCS,CChem *Dean*

3152 Biology

1 Sambo EY - BSC,MSc,PhD *Senior Lecturer & Head of Department*
2 Berrie GK - BSc,PhD *Professor*
3 Chiotha SS - BEd,BSc,MSc *Lecturer*
4 Khonga EB - BSc,MS *Lecturer*
5 Meredith Miss HM - BSc,MSc,CertEd *Reader*
6 Munthali DC - BSc,DIC,MSc,PhD *Senior Lecturer*
7 Noble RP - BSc,MSc,PhD *Lecturer*
8 Patterson JW - BA,MSc,PhD *Senior Lecturer*
9 Telford SR - BSc,PhD *Lecturer*

3153 Herbarium
PO Box 280, Zomba
Tel: 522 222. Cables: Chancoll, Zomba.

1 Seyani JH - BScAgric,MSc,DPhil,MIBiol *Senior Lecturer/Curator*

3154 Mathematics

3155 Statistics

1 Caple CI - BScEcon,MSc *Lecturer*
2 Emerson MJ - BSc,MSc,FIS *Lecturer*
3 Madise Miss NJ - BSc,BSc,MSc *Lecturer*

3156 Computer Science

1 Heller Miss JL - BA *Assistant Lecturer*
2 Kalua PP - BSc,MSc *Lecturer*
3 Kumwenda Miss CVM - BSc *Staff Associate*

3157 Faculty of Social Science

1 Kandoole BF - BSocSc,MA,PhD *Dean*

3158 Economics

1 Chipeta C - BScEcon,MA,PhD,LAAI *Professor & Head of Department*
2 Chipande GHR - BSocSc,PhD *Lecturer*
3 Harawa RD - BSocSc,BAEcon,MScEcon *Lecturer*
4 Kaluwa BM - BSocSc,DipEcon,MA *Lecturer*
5 Kandoole BF - BSocSc,MA,PhD *Lecturer*
6 McGraw RL - BCom,LicentiateAcc,CA,MBA *Lecturer*
7 Mkwezalamba MME - BSocSc,BSocScEcon,DipEcon,MA *Lecturer*
8 Ngalande-Banda EE - BSocSc,MSc *Lecturer*
9 Thorne M - BA,BAD,MA *Senior Lecturer*

MALAYSIA

GOVERNMENT DEPARTMENTS

3159 Department of Agriculture, Sarawak

State Complex, Kuching, Sarawak
Tel: 54111. Cables: Pertanian, Kuching.

1 Kong TCJ - BAgrSc *Director*
2 Tan CCA - BAgrSc *Deputy Director*

3160 Research Division

1 Sim ES - BSc,MSc *Assistant Director*
2 Ngui STT - BAgrSc *Senior Research Officer*

3161 Agronomy

1 Chai CC - BAgrSc (upland) hill rice
2 Chia JLM - BAgrSc,DipAgrSc field crops
3 Jaman O - BAgrSc sago & on-farm research
4 Kueh HS - BAgrSc crops in peat soil
5 Lai KF - BAgrSc on-farm research
6 Lau CY - BHortSc coffee, tea, cashew nut, clove & nutmeg
7 Ng TT - BAgrSc cropping systems & research information documentation
8 Ringot RR - BAgrSc on-farm research
9 Tang THD - BAgrSc coconut and cocoa
10 Voon BH - BAgrSc fruits and vegetables
11 Wong C - BAgrSc wet (flooded) rice
12 Wong TH - BAgrSc pepper

3162 Biometrics

1 Fatimah O - BAgrSc biometry

3163 Botany

1 Amin DP - BAgrSc pepper breeding
2 Kadin E - BSc pepper physiology
3 Lim LL - BAgrSc weed science
4 Murni S - BAgrSc rice breeding
5 Rita M - BSc,PhD peanuts
6 Rosmah J - BSc pepper breeding
7 Sim SL - BSc,MSc,PhD pepper breeding

3164 Chemistry

1 Sim CYP - BSc,DipEd *Senior Chemist*
2 Chin SP - BSc soil chemistry & plant nutrition
3 Lim ET - BSc *Soil Chemist*
4 Tan SL - BAgrSc sago chemistry

3165 Chemical Engineering

1 Lau YTG - BSc sago processing

3166 Plant Protection

1 Eng L - BAgrSc *Plant Pathologist* virology & microbiology
2 Kueh TK - BAgrSc,MSc *Senior Plant Pathologist* pepper diseases
3 Leong CTS - BAgrSc,MAgrSc *Entomologist* applied entomology
4 Megir G - BSc *Entomologist* pest control
5 Ron Ah Goh - BAgrSc *Entomologist* pest control
6 Teo CH - BAgrSc,MSc *Plant Pathologist* crop diseases

3167 Soil Survey

1 Federick HT - BAgrSc soil survey
2 Lau JU - BSc soil survey
3 Lim CP - BSc,MSc *Senior Soil Surveyor*
4 Loi KS - BSc,MSc soil survey
5 Teng CS - BAgrSc,MSoilSc soil survey
6 Tie YL - BSc soil survey

3168 Fisheries

1 Ang MH - BSc,MSc lake & riverine fisheries
2 Litis BA - BAppSc hydroelectric - reservoir fisheries studies
3 Pang SY - BSc fresh water giant prawns

3169 Veterinary

1 Chia PC - DVM investigation and observation trial on animal diseases
2 Hsiung KY - BVSc,MRCVS *Senior Veterinary Officer* feeding trials on rabbit and poultry
3 Sia CHF - DVM investigation and observation trial on animal diseases
4 Wei HKJ - BAgrSc conducting research on animal production

3170 Department of Veterinary Services and Animal Industries, Sabah

7th Floor, Menera Wisma Khidmat, Jalan Belia, Kota Kinabalu, Sabah, 88000
Tel: 088-55259. Telex: MA80235. Cables: DIRVET.

1 Mokhtar IB - BAppSc,MSc *Director*

3171 Animal Production

1 Bacon AJ - MAVet,MB,DVSM,MRCVS *Senior Veterinary Officer* cattle breeding and buffalo multiplication
2 Daniel CTF - BAppSc,MSc *Animal Husbandry Officer* swine and poultry - performance evaluation and production
3 Salleh ASA - DAPH,BAppSc *Senior Animal Husbandry Officer* dairy cattle breeding; smallholder production and processing

3172 Diagnostic and Research Division

P.O. Box 59, Tg. Aru, Kota Kinabalu
Tel: 088-51235.

3173 Research Centre
Kepayan, Kota Kinabalu, Sabah
Tel: 088-51233.

1 Phin CS - BVM,DTVM,MVM *Senior Veterinary Officer* disease research

3174 Pasture Research

Feed Mill, Kepayan, Kota Kinabalu, Sabah
Tel: 088-51233.

1 Tai CC - BScAgri,MRSc *Pasture Agronomy* feed and fodder production

3175 Malaysian Agricultural Research and Development Institute

G.P.O. Box 12301, Kuala Lumpur
Tel: 50774 Kuala Lumpur. Telex: MA 37115.

1 Mohd. Yusof bin Hashim *Director General*
2 Ahmad Zaharuddin b. Idrus - PhD *Chief Research Officer* food chemistry
3 Thirunakarasu N - PhD *Chief Research Officer* genetics
4 Ahmad Zamzam b. Mohamed - Ph.D. *Director, Central Analysis Laboratory*
5 Abd. Razak b. Shaari - Ph.D. physiology of durians and mango
6 Chan Ying Kwok - M.Sc. breeding & selection of papaya, pineapple and anona
7 Ding Ting Huung - M.Agric.Sc. commercial vegetable production, landscape architecture, crop physiology
8 Khelikuzaman Merra b. Hussain - M.Agric.Sc. breeding & selection of root crops and other ornamental plants
9 Lim Heng Jong - M.Sc. component technology testing, transfer of package technology
10 Mohd. Shamsudin b. Osman - M.Sc. insect pests of mango, durian and duku/langsat
11 Nik Mohd. Masdek b. Nik Hassan - M.Sc. *Fruit Virologist*
12 Nuraini bt. Ibrahim - M.Sc. breeding & selection of orchids, roses, carnations and chrysanthemum
13 Rukayah bt. Aman - M.Sc. flowering physiology of mango, duku/langsat, mangosteen and collection of rare fruits
14 Saharan b. Hj. Anang - Ph.D seed production - foundation seeds
15 Siti Hawa bt. Jamaluddin - M.Phil. breeding & selection of banana and bread fruits/cempedak
16 Tan Swee Lian - B.Agric.Sc.;M.Agric.Sc. breeding & selection of root crops; cultural practices, crop nutrition
17 Vijaysegaran a/l Shanmugam - M.Sc. the study of fruit flies
18 Wong Lock Jam - M.Agric.Sc. breeding and selection of potential crops
19 Yaacob b. Doon - Ph.D. *Fruit Virologist*
20 Zainal Abidin b. Mohamad - Ph.D. breeding & selection of durians, mango and guava

3176 at Ibu Pejabat, MARDI

1 Adinan b. Husin - M.Sc. *Director, Food Technology Division, MARDI*
2 Tay Tian Hock - Ph.D. *Director, Technical Services*
3 Ahmad Tajuddin b. Zainuddin - M.Sc. *Director, Livestock Division*
4 Ahmad Tasir b. Lope Pihie - M.Sc. *Director, Promotion Technology Division*
5 Mohd. Hashim b. Noor - Ph.D. *Director, The Social and Techno-Economic Division, MARDI*
6 A. Wahid b. Sulaiman - M.Agric.Sc. selection and breeding of goats and sheep
7 Ab. Rahim b. Muda - M.Sc. biology, ecology and control of insect pests during storage of cereals
8 Abd. Rahman b. Mohd. Yasin - M.Sc. selection and breeding of goats and sheep
9 Abd. Rashid b. Ahmad - M.Sc. soil chemistry
10 Abd. Shukor b. Abd. Rahman - M.Sc. storage and packing house operation and quality control of vegetable crops
11 Abdul Aziz b. Bidin - Ph.D. soil enrichment based on crop modelling
12 Abdul Karim b. Sidam - Ph.D. study of root-knot nematode attack on plants
13 Abdul Malik b. Johan - M.Sc. integrated farms of paddy and cocoa
14 Abdul Munir Jaafar - M.Sc. research method and statistical analysis for agricultural research
15 Abdullah b. Hassan - M.Sc. post harvest physiology, storage and handling of tropical fruits
16 Abu Bakar Hj. Hussin - Ph.D. food quality control
17 Abu Othman b. Abd. Rahman - M.Sc. advisory services and co-operative food manufacturers
18 Ahmad Robin Wahab b. Ahmad Nordin - B.Agric.Sc. study leave
19 Ahmad Shokri b. Othman - M.Agric.Sc. research method and statistical analysis for agricultural research
20 Ahmad Zaharuddin bin Idrus - Ph.D. secondment to Kementerian Sains Teknologi dan Alam Sekitar
21 Ailin Ton bte Isahak - M.Sc. editing, planning, evaluation research, promotion and sales of publications, public relations
22 Aminuddin b. Yusoff - Ph.D. pedology of paddy soils
23 Annie Chin Nee Sim Hooi Guat - M.Sc. quality of Malaysian cocoa beans
24 Arasu NT - Ph.D. genetic - induced breeding for multiplication of plants
25 Ayob b. Sukra - M.Sc. modification of agricultural machinery
26 Berahim b. Hj. Ithnin - M.Sc. collection and analysis of data from agricultural industries
27 Cheah Uan Boh - M.Sc. studies of environmental effect of pesticides on microflora and the environment
28 Chen Chin Peng - M.Agric.Sc. feeding of goats, leucaena, setaria, and MARDI digitaria
29 Chew Boon Hock - Ph.D. resistance of pest and plant diseases
30 Chia Joo Suan - M.Sc. heavy metals in foods
31 Dhiauddin Mohd. Nour - M.Sc. dehumidified storage on rice
32 Faridah Abdul Aziz - B.Agric.Sc. food development of fruit products
33 Fazlil Ilahi b. Abd. Wahab - M.Eng. mechanical mobility of agricultural machinery
34 Ghuah Eng Chong - MSc food development and quality of fish and meat products
35 Ghulam b. Mohammed Hashim - Ph.D. soil improvement
36 Haji Ahmad b. Hj. Ab. Wahab - M.Sc. chemical analysis and management
37 Hamdzah b. A. Rahman - M.Sc. small food-industries development and entrepreneurship
38 Hassan b. Mat Daud - M.Sc. studies of genetics population
39 Heng Ching Kok - Ph.D. plant chemical extracts for use as pesticide
40 Heong Kong Luen - M.Sc. the utilization of system analysis in agriculture
41 Hussain b. Hj. Zakaria - M.Sc. small food industries development
42 Hussan b. Abdul b. Kadir - M.Sc. microbiological control of pests in vegetables
43 Hussein Abdul Rahman - M.Sc. food quality control of Integrated Food Marketing Programme
44 Jainudin b. Abdullah - M.Sc. upgrading food quality of local food manufacturers
45 Jalaludin b. Jipelos - Ph.D. intra and inter-agencies communication and development of MARDI
46 Jamal b. Hj. Said - M.Sc. collection and analysis of data for paddy, cocoa and tobacco crops
47 Jumali b. Suratman - M.Sc. plant chemical exhaust as pesticides
48 Kamal Adzham b. Kamarudin - Ph.D. ecology of pathogens of vertebrates
49 Lam Peng Fatt - M.Ac. postharvest physiology, storage and handling of tropical fruits
50 Lee Boon Siew - Ph.D. taxonomy, ecology and biological control of fungi
51 Lee Chong Soon - M.Sc. research method and statistical analysis for agricultural research
52 Liang Juan Boo - M.Sc. energy and protein requirement during growth in beef cows and buffaloes
53 Liew Kai Leon - M.Phil. plant nutrition
54 Lim Guan Soon - Ph.D. integrated control and management of plant pest control
55 Loke Wai Hong - Ph.D. physiology and control of plant diseases by semi-chemical means
56 Lum Keng Yeang - Ph.D. molecular and microbiological plant disease control

57 Mat Isa b. Awang - Ph.D. aflatoxin in local food and agricultural produce
58 Md. Eusof b. Abd. Jamak - M.Sc. breeding and selection of dairy friesian cross
59 Md. Jusoh b. Mamat - Ph.D. application of chemical pesticides
60 Mohamad b. Halib - Ph.D. dry-land paddy growing areas and rural development
61 Mohamad Hanif b. Mohamad Jamil - Ph.D. recombinant DNA for multiplication of plant materials
62 Mohd. Aziz b. Dollah - Ph.D. physiological needs (water retention and perspiration) of KK and cross cows
63 Mohd. Ismail b. Abdullah - M.Sc. handling, primary processing and packaging of fish products; freezing and frozen storage of fruits
64 Mohd. Jaafar b. Daud - M.Sc. animal feed byproducts of palm oil, paddy and maize
65 Mohd. Najib b. Mohd. Amin - M.Sc. pasture - agronomy and soil enrichment
66 Mohd. Nordin b. Mohd. Som - Ph.D. processing of local fruits and vegetables
67 Mohd. Yunus Jaafar - B.Sc.(Maths) research method and statistical analysis for agricultural research
68 Mohd. Yusof b. Abdullah - M.Sc. eco-physiological research
69 Mohd. Zahari b. Abu Bakar - M.Sc. soil chemistry of acid sulphate soil
70 Mohd. Zaki b. Hj. Ghazali - Ph.D. agricultural climatology
71 Nathan XJ - M.Agric.Sc. intra and inter-agencies communication and development of MARDI
72 Nik Fuad b. Nik Mohd. Kamal - Ph.D. economic studies of agricultural and livestock industries for development and improvement
73 Noraini Hj. Mohd. Khalid - M.Sc. study leave
74 Ong Ching Ang - M.Agric.Sc. virology taxonomy at molecular level and viral diseases
75 Ong Hwee Keng - B.Agric.Sc. management practice of piglets
76 Ong Seng Hock - M.Sc. side-effects of pesticides applications, and environmental disturbances
77 Ooi Hoi Seng - M.Eng. modification of agricultural machinery and processing
78 Othman b. Hassan - M.Sc. advisory & extension services to small scale food industries
79 Ramadasan Krishnan - M.Sc. eco-physiology and administration of cocoa research programme
80 Ramli b. Man - M.Sc. advisory services and training for local food manufactures
81 Rohani Md. Yon - M.Sc. handling, storage and grading of rice
82 Salma bt. Idris - M.Sc. eco-physiology and maintenance of plant herbarium
83 Seet Chin Puan - M.Sc. management of duck layers, quail and chicken quality
84 Senawi b. Dato' Mohamed Tamin M - Ph.D. in vitro multiplication of breeding materials for cocoa, coconut and fruits
85 Shaaban b. Shahar - Ph.D. economics and marketing of Malaysian cocoa crop production
86 Siti Petimah bte Hj. Hassan - M.Sc. editing and publication
87 Sivarajasingam a/l Sitampalam - Ph.D. selection of dairy breeds
88 Subari b. Shibani - M.Sc. study leave
89 Suhaimi Masduki - M.Ac. scaling-up production of traditional fermented foods
90 Wan Johari b. Wan Daud - M.Sc. training and advisory services for entrepreneurs of food business and upgrading of quality of traditional foods
91 Wan Rahimah Wan Ismail - M.Sc. advisory and extension services to small scale food industries
92 Wong Chee Yoong - M.Sc. statistical data for paddy, acceptance of modern varieties and costing of paddy input versus output
93 Wong Choi Chee - M.Sc. ecophysiology of pastures for feed
94 Wong Nan Chong - M.Sc. pedological studies of Malaysian soils
95 Yeoh Quee Lan - B.Sc. scaling-up production of traditional fermented foods
96 Yeong Shue Woh - M.Sc. poultry and Peking ducks fed from agricultural byproducts; egg quality and amino acids content of feed
97 Zahara Merican - M.Sc. quality characteristics of pepper and scaling-up production of traditional fermented foods
98 Zahari b. Radi - B.Econs. advisory services for economics and marketing in fruit crops
99 Zaidah Idris - M.Sc. food quality of traditional food
100 Zainol b. Abdul Aziz - M.Sc. eco-physiology and technology of seed production
101 Zanariah bt. Jiman - M.Sc. food quality controls

3177 at Jalan Kebun, Selangor

1 Ho Ban Lee - Ph.D. bacterial and virus disease management of lowland vegetables
2 Lee Soo Ann - Ph.D. weed control in fruit orchards
3 Lim Weng Hee - Ph.D. crop protection on ornamental plants
4 Mah Shock Ying - M.Sc. disease management of lowland vegetables
5 Melor bte. Rejab - M.Sc. breeding and selection of lowland non legume vegetables
6 Ramlah bt. Mohamad - M.Sc. fertilizer of banana and pineapple
7 Ramli b. Mohd. Noor - Ph.D. breeding and selection of lowland vegetables - legumes
8 Vimala Purushotaman - B.Agric.Sc;M.Agric.Sc. cultural practices; crop nutrition

9 Yeoh Kiat Choon - M.Sc. crop mechanization in vegetable production, maize, groundnut, tea and cassava

3178 at Seberang Perai, Pulau Pinang

1 Che Embi b. Yusof - M.Agric.Sc. *Director, Paddy Research Division*
2 Ajimilah Nyak - M.Sc. study on quality and grading of rice
3 Chan Seak Khean - M.Agric.Sc. cultural practices and crop nutrition; breeding and selection of lowland vegetable
4 Chen Yoke Hwa Chin Chong - M.Sc. breeding and selection of paddy varieties for paddy irrigated areas
5 Cheong Ah Wah - M.Phil. mechanised crop cultural practices
6 Habibuddin b. Hashim - M.Sc. insect-host relationship
7 Lam Yuet Ming - M.Sc. bionomics and control practices of rice field rats
8 Lee Choo Kiang - Ph.D. breeding and selection of cereal crops and legumes crops
9 Leong Chee Onn - M.Phil. commercial production; evaluation of annual crops - paddy based
10 Mohamad b. Osman - Ph.D. development of non-conventional, explorative breeding methods
11 Mohd. Arif b. Junus - Ph.D. soil amelioration techniques and practices
12 Mohd. Shahrin b. Yob - M.Sc. terminal irrigation facilities development
13 Mooi Kok Chee - M.Sc. component technology testing; cost of production of asparagus
14 Othman b. Omar - Ph.D. breeding and selection of selected paddy varieties
15 Supaad b. Mohd. Amin - Ph.D. host-pathogen-environment interactions
16 Tay Chan Yong - M.Sc. qualification of ecophysiological and physical response to constraints

3179 at Cameron Highlands, Pahang

1 Ko Weng Wah - M.Sc. temperate fruits; pathology of lemon
2 Mohd. Shukor b. Nordin - M.Sc. breeding & selection of highland vegetables
3 Rahman b. Syed Hj. Abd. Rashid SA - Ph.D. insect pest management of highlands vegetables

3180 at Keluang, Johor

1 Abd. Rahman Azmil b. Datuk Idris - M.Sc. crop nutrition and planting systems
2 Basery b. Mohamad - Ph.D. natural grass feeding for cows and metabolism
3 Chong Woon Sum - M.Sc. weed management in vegetables; cultural practices - planting systems
4 Eng Pei Kong - M.Agric.Sc. seeds of plants of animal pastures and fodders
5 Loh Chow Fong - M.Sc. pest and disease management
6 Mohd. Arif b. Omar - Ph.D. breeding and improvement of ruminant breeds for meat
7 Mohd. Ghawas b. Maarof - M.Sc. breeding & selection of spices and beverages
8 Mohd. Sukri b. Hj. Idris - M.Sc. the performance of palm oil and cocoa waste as animal feed on Kedah, Kelantan and other cattle breeds

3181 at Pontian, Johor

1 Haron b. Hj. Ashari - M.Sc. component technology testing, transfer of package technology
2 Kamarudin b. Ambak - M.Sc. studies on the physico-chemical structure of peat soils
3 Leong Ah Chye - M.Sc. cultural practices and cropping systems; cost of production of asparagus

3182 at Gajah Mati, Kedah

1 Baharudin b. Baharom - M.Sc. cultural practices in noble cane
2 Omar b. Hj. Md. Isa - M.Sc. evaluation of annual crops - paddy based
3 Siti Zainab bt. Ramawas - M.Sc. breeding & selection of sugarcane and potential crops

3183 at Kuala Trengganu, Trengganu

1 Ahmad Anwar b. Ismail - Ph.D. weed management
2 Hasbullah b. Mamat - M.Sc. breeding & selection of silkworm; pest and disease management
3 Noor Rawi bin Abu Bakar - M.Sc. evaluation of annual crops - bris based; breeding & selection of mulberry

3184 at Sg. Baging, Pahang

1 Chai Tiang Bian - M.Hort.Sc. vegetative propagation of cashew nuts and duku/langsat
2 Salleh b. Hassan - M.Sc. breeding & selection of cashew, rambutan and duku/langsat

3185 at Unit Bentong, Pahang

1 Lim Soong Pek - M.Agric.Sc. appropriate agricultural technology

3186 at Kemaman, Trengganu

1 Zainuddin b. Jaafar - B.Agric.Sc. *Fruit Virologist*
2 Abdullah b Othman - M.Sc. horticulture of cocoa and coconut

3187 at Jeram Pasu, Kelantan

1 Norlia bt. Yunus - M.Agric.Sc. fertilizer & agronomy of musk melon and mango

3188 at Kubang Keranji, Kelantan

1 Chang Poon Min - M.Sc. identification of rice insects, insect pathogens and their importance
2 Syed Mohd. b. Syed Ibrahim - M.Agric.Sc. plant physiology and agronomy of musk melon; appropriate technology

3189 at Hilir Perak, Perak

1 Denamany s/o Ganggadharan - M.Sc. fertilization of cocoa plants
2 Ibrahim b. Abdullah - Ph.D. secondment to PORIM
3 Jamadon b. Bahari - M.Sc. breeding and rehabilitation of cocoa and coconut
4 Jamaludin b. Salim - M.Sc. cocoa pest management
5 Mamot b. Said - Ph.D. quality of cocoa, fruits and beans
6 Mansor b. Puteh - M.Sc. study leave
7 Mohamad Musa b. Md. Jamil - Ph.D. plant disease and disease management of coconut in the nursery and field
8 Shaharuddin b. Saamin - M.Sc. study leave
9 Shahrir b. Shamsudin - M.Sc. fermentation studies of Malaysian cocoa beans
10 Sivapragasam a/l Annamalai - B.Agric.Sc. quality and storage of cocoa and fruits
11 Syed Kamaruddin b. Syed Wazir - Ph.D. physiology of cocoa-plants for growth multiplication and increase of yield
12 Tey Chin Chong - M.Phil. plant pathology of cocoa/coconut

3190 at Bukit Ridan, Pahang

1 Goh Hock Swee - M.Sc. horticulture and management of cocoa and coconut crops
2 Nordin b. Yusof - M.Sc. production of sperms from buffaloes fed with enriched diets for breeding purposes

3191 at Pasir Puteh, Kelantan

1 Ghani b. Senik - M.Sc. development and improvement of beef and poultry product
2 Salma bte Omar - M.Sc. advisory and extension services to small scale food industries

3192 at Tampoi, Johor

1 Norhidayat bte Abdul Rahman Alkaf S - M.Sc. product development from coconut

3193 at Telong, Kelantan

1 Zulkifly b. Mohd. Zain - B.Agric.Sc. *Director, Tobacco Research Division*
2 Abdullah b. Che Tengah - M.Sc. soil fertility, rehabilitation and soil management of tobacco
3 Khairuddin b. Yaakob - M.Sc. agronomic practices in tobacco production
4 Khairuddin b. Yaakob - MSc. agronomic practices in tobacco production
5 Mat Daham b. Mohd. Daud - M.Sc. handling, storage and mechanization of tobacco industries
6 Yuan Pak Mun - M.Sc. control of nematodes by chemical and crop rotation

3194 at Jerangau, Terengganu

1 Kamarudin b. Harun - MSc irrigation: sprinkler system for tobacco
2 Wahab b. Ismail - M.Sc. agronomic practices for cocoa and coconut crops in West Pahang State

3195 at Tanjung Karang Selangor

1 Abdul Rahman b. Abu Bakar - M.Sc. to quantitatively describe and characterise the agroclimatic environment of the rice growing areas
2 Ismail b. Abu Bakar - M.Sc. pedology
3 Mohamed b. Mohd. Salleh - Ph.D. insect - host relationship
4 Tham Kah Cheng - M.Agric.Sc. soil fertility

3196 at Parit, Perak

1 Joy Varughese - Ph.D. production of breeder and foundation seeds
2 Sani b. Kimi - M.Sc. resource potential identification and exploitation

3197 at Krian, Perak

1 A. Wahab b. Abd. Hamid - Ph.D. breeding and selection of paddy varieties for paddy irrigated areas

3198 at Alor Star, Kedah

1 Minhad b. Salleh - M.Sc. productive cropping patterns

3199 at Pejabat KPMN, Kedah

1 Alias b. Ismail - M.Sc. extension and agricultural development in Kedah State
2 Normiyah bt. Rejab - M.Sc. transfer, promotion and development of technology for Kedah

3200 at Pejabat KPMN, Perak

1 Mahamud b. Shahid - Ph.D. transfer, promotion and development of agriculture for the state of Perak
2 Mohd. Jelani b. Bahaudin - B.Agric.Sc. transfer, promotion and development of agriculture for the state of Perak

3201 at Pejabat KPMN, Selangor

1 Joseph Samy - Ph.D. transfer, promotion and development of agriculture for the state of Selangor
2 Nasrun Baginda Abd. Hamid - M.Sc. transfer, promotion and development of agriculture for the state of Selangor

3202 at Pejabat KPMN, Negeri Sembilan

1 Muhamad b. Harun - B.Agric.Sc. transfer, promotion and development of agriculture for the state of Negri Sembilan
2 Muhamed b. Yaacob - M.Sc. transfer, promotion and development of agriculture for the state of Negri Sembilan

3203 at Pejabat KPMN, Melaka

1 Siti Doyah bt. Othman - B.Sc. transfer, promotion and development of agriculture for the state of Melaka

3204 at Pejabat KPMN, Terengganu

1 Abdul Rahman b. Daud - M.Sc. transfer, promotion and development of agriculture for the state of Terengganu

3205 at Pejabat KPMN, Johor

1 Abdul Rahman b. Harun - M.Agric.Sc. transfer, promotion and development of agriculture for the state of Johor

3206 at Pejabat KPMN, Kelantan

1 Mohd. Azmi b. Ibrahim - M.Sc. transfer, promotion and development of agriculture for the state of Kelantan

3207 at Kuala Linggi, Melaka

1 Ting Chaong Chaang - M.Sc. soil chemistry of acid sulphate soil

3208 at Kundang, Selangor

1 Othman b. Abu Bakar - MSc pedology of biro soil and tintailings

3209 The Forest Research Institute of Malaysia (FRIM)

Kepong, Selangor, 52109 KL
Tel: 03-6262633, 03-6262152. Telex: FRIM-MA-27007. Cables: UTAN.

1 Salleh MN - BScFor,DipFor,DipPhoto-Interp,MSc,PhD *Director General*
2 Ng SP - BSc,PhD *Dep. Director General*

3210 Forestry Division

1 Ng SP - BSc,PhD *Director* botanist
2 Kochummen KM - BSc,DipPlantTaxon *Botanist*
3 Manokaran N - BSc,MSc *Ecologist*
4 Tho YP - BSc,PhD *Entomologist*
5 Wan RWM - DipAgr,BScFor,MSc *Mensuration Officer*
6 Zakaria I - DipFor,BScFor,MSc *Tree Improvement Officer*

3211 Forest Products Division

1 Cheah LC - BScFor *Research Evaluation Officer*
2 Choo KT - BSc *Wood Seasoning Officer*
3 Daljeet KS - BSc *Wood Protection Officer*
4 Ho KS - BE,MAgrDev *Primary Machining Officer*
5 Hong LT - BSc,MSc *Mycologist*
6 Kong Mrs SE - BSc,DipLib *Librarian*
7 Norhara Ms H - BSc *Public Relation Officer*
8 Rahmah Ms WR - BSc *Publication Officer*
9 Shukari Midon M - BSc *Timber Engineering Officer*
10 Wong WC - BSc,MAgrDev *Director, Wood Chemist*

3212 Rubber Research Institute of Malaysia

PO Box 10150, Kuala Lumper 01-02
Tel: 467033. Telex: 30369.

Research staff details not available at time of going to press.

UNIVERSITIES & COLLEGES

3213 Universiti Kebangsaan Malaysia

Bangi, Selangor, 43600 UKM
Tel: 8250001. Telex: UNIKEB MA 31495. Cables: "UNIKEB".

1 Datuk Abdul Hamid AR - AM,MBBS,DA,FFARCS,FFARACS *Vice Chancellor*
2 Jalani BS - BAgSc,PhD *Deputy Vice-Chancellor (Academic and Research)*

3214 Faculty of Life Sciences

1 Mohd. Sanusi J - BAgSc,PhD *Dean*

3215 Department of Biochemistry

1 Rohani Abdul K - BSc,MSc,PhD *Lecturer, Head of Department* nutritional evaluation of oil palm and legume protein
2 Mohd. Noor E - BSc,MSc,PhD *Associate Professor Deputy Dean* control of rubber biosynthesis; mechanism of pathogenicity in Melioidosis (Pseudomonas pseudomallei)
3 Tan Soon C - BSc,MSc,PhD *Associate Professor* post harvest physiology of tropical fruits (bananas)
4 Halimah AS - BSc,MSc *Lecturer* post harvest physiology of tropical fruits (guava)
5 Hasidah Mohd. S - BSc,MSc *Lecturer* protein studies of legumes
6 Othman O - BSc,MSc,PhD *Lecturer* lignocellulose degradation
7 Rahmah M - BSc,PhD *Lecturer* tissue culture
8 Sahidan S - BSc,MMedSc *Lecturer* medically important plant extracts - effect on erythrocytes
9 Salmijah S - BSc,PhD *Lecturer* nitrogenous compounds in cocoa
10 Zainon MA - BSc,PhD *Lecturer* post harvest physiology of tropical fruits (mango)

3216 Department of Botany

1 Abdullah Karim AG - BAgSc,MSc,PhD physiology of plant productivity; tissue culture
2 Hamid L - BAgSc,MSc,PhD plant physiology; post harvest physiology
3 Ismail S - BSc,MSc weed biology; pesticide residues

3217 Department of Food Science & Nutrition

1 Abdullah A - BSc,MAg,PhD *Lecturer, Head of Department* food quality - sensory evaluation
2 Abdul Rahman S - BSc,MSc,PhD *Lecturer* clinical nutrition
3 Babji AS - BSc,MSc,PhD *Associate Professor* meat science
4 Embong MS - BAgSc,MSc,PhD *Associate Professor* food chemistry and biochemistry
5 Hassan O - SmSn,MSc,PhD *Lecturer* dietary fiber
6 Ishak S - SmSn,MSc,PhD *Lecturer* food chemistry and biochemistry
7 Karim AN - BSc,MSc *Lecturer* nutrition
8 Mohd. Noor MS - DVM,MSc,PhD *Lecturer* nutrition/energy
9 Nik Daud NI - BAgSc,MSc,PhD *Lecturer* folic acid
10 Nitisewojo P - Ms,PhD *Lecturer* food chemistry and biochemistry

3218 Department of Genetics

1 Jalani S - BAgSc,PhD *Professor* cytogenetics and plant breeding
2 Mahani Mansor C - BSc,PhD *Associate Professor, Head of Department* cytogenetics and genetics of Drosophila
3 McCreary KG - BSc,PhD *Professor* mushroom breeding
4 Saidi M - BSc,MS,PhD *Associate Professor* animal breeding and statistics
5 Zakri AH - BSc,MSc,PhD *Associate Professor* quantitative genetics and plant cell genetics
6 Abdul MM - BAgSc,MAgSc,PhD *Lecturer* animal breeding
7 Ahmed MMM - BSc,MS,PhD *Lecturer* genetics of plant disease resistance
8 Farida HS - BSc,MSc,PhD *Lecturer* molecular biology
9 Fauzi D - SmSn,PhD *Lecturer* fungal genetics
10 Mohd. AS - BSc,MSc,PhD *Lecturer* microbial and molecular genetics
11 Roznah H - BSc,MSc,DPhill *Lecturer* immunogenetics

3219 Department of Microbiology

1 Abdul Jalil AK - DVM,PhD *Lecturer* solute waste disposal - composting
2 Ismail A - BAgSc,MSc,PhD *Lecturer* plant viral disease
3 Mohamed Anuar O - DVM,MSc *Lecturer* animal disease
4 Mohammed O - BSc,MAgSc,PhD *Lecturer* plant pathology
5 Mohd SJ - BAgSc,PhD *Lecturer* microbial insecticide

6 Nor Muhammad M - BSc,PhD *Lecturer* microbial insecticide
7 Wan Mohtar Wan Y - BSc,PhD *Lecturer* solid substrates fermentation, celluloses

3220 Department of Zoology

1 Zainal Abidin Abu H - BSc,PhD *Head of Department* parasitology
2 Mohd Nordin H - BAgSc,PhD *Professor* animal behaviour
3 Abdul Wahab K - BSc,PhD *Lecturer* respiratory physiology
4 Alias K - BSc,MSc,PhD *Lecturer* endocrinology
5 Arbain K - BSc,MSc,PhD *Associate Professor* parasitology
6 Davison GWH - BSc,PhD *Lecturer* bird ecology and behaviour
7 Maimon A - BSc,MSc,PhD *Lecturer* nutritional physiology
8 Maryati M - BSc,PhD *Lecturer* entomology
9 Mohd Salleh MS - BSc,MS *Lecturer* entomology
10 Nik Mat D - BSc,MS *Lecturer* ichthyology
11 Noran M - BSc,MS *Lecturer* developmental physiology
12 Othman R - BSc,MSc *Lecturer* aquatic biology
13 Rohani A - BSc,MSc,PhD *Lecturer* parasitology
14 Sharifah FA - BSc,MSc,PhD *Lecturer* entomology
15 Zubaid AA - BSc,MSc *Lecturer* bat taxonomy

3221 Faculty of Science & Natural Resources

Locked Bag No 62, Sabah
Tel: 88996 Kota Kinabula.

1 Murtedza M - BSc,PhD *Associate Professor, Dean* water quality; toxic organics in environment

3222 Department of Biology

1 Ridzwan H - BSc,MS,PhD *Lecturer, Head Department* animal physiology and nutrition
2 Ghazally I - BSc,MSc,PhD *Associate Professor* studies on *Pseudomonas pseudomallei*; water quality
3 Haleem R - BSc,PhD *Lecturer, Deputy Dean* studies on campylobacters; water quality
4 Jumaat A - BSc,MSc *Lecturer* Dipterocarpus and *Rafflesia*

3223 Department of Chemistry

1 Rohani P - BSc,PhD *Lecturer, Head of Department* natural products - humic acid; pesticide studies
2 Siraj O - BSc,PhD *Lecturer, Deputy Dean* natural products from lichens and seaweeds

3224 Department of Marine Science

1 Zainuddin U - BSc,MSc *Lecturer, Head of Department* marine toxicology
2 Norhadi I - BSc,MSc *Lecturer* primary production and aquaculture (giant clams)

3225 University of Malaya

Lembah Pantai, Kuala Lumpur
Tel: 575466 (12 saraf); 567819 (saraf terus). Telex: MA 37453. Cables: UNIVSEL.

3226 Department of Botany

1 Ahmad Nawawi b. Hj. Ayob - BSc,BAgr,PhD *Professor* mycology; plant pathology
2 Shen TC - BSc,MSc,PhD *Professor & Head of Department* nitrogen metabolism
3 Abdul Aziz bin Ahmad Shah - BSc,MSc,PhD *Lecturer* plant ecology
4 Amru Nasrulhaq-Boyce - BSc,PhD *Lecturer* plant biochemistry
5 Chin SC - BSc,MS,MPhil,MSc,PhD *Lecturer* plant taxonomy; ethnobotany
6 Halijah Ibrahim - BSc,PhD *Lecturer* experimental taxonomy
7 Jones DT - BA,MA *Lecturer* plant taxonomy
8 Kuthubutheen AJ - BSc,PhD *Associate Professor* mycology
9 Lim AL - BSc,MSc,PhD *Lecturer* plant anatomy & embryology
10 Mohamed Abdul Hj. Mohamed - BSc,PhD *Lecturer* bryology & pteridology
11 Muhamad bin Zakaria - BSc,PhD *Lecturer* phytochemistry
12 Nair H - BSc,PhD *Associate Professor* postharvest physiology
13 Ong KC - BSc,PhD *Lecturer* wastewater treatment; fermentation
14 Palaniappan VM - BSc,MSc,PhD *Associate Professor* plant ecology
15 Ratnasabapathy M - BSc,MSc,MIBiol *Associate Professor* phycology; limnology
16 Tung HF - BSc,PhD *Associate Professor* plant growth and development; weed control

3227 Department of Genetics & Cellular Biology

1 Mak Chai - BSc,PhD *Senior Lecturer* plant breeding
2 Shaiful AAA - BSc,PhD *Lecturer* rumen microbiology; nitrogen fixation
3 Teoh SB - BSc,PhD *Senior Lecturer* plant cytogenetics
4 Wan Khadijah - BSc,MSc *Lecturer* animal breeding

3228 Department of Zoology

1 Siew YC - MSc,DIC,PhD *Associate Professor, Head of Department* insect physiology
2 Abdul Aziz Hj. Kechil - BSc,PhD animal nutrition and physiology
3 Chua TH - MSc,PhD *Associate Professor* agricultural entomology
4 Fong FW - MSc forest ecology
5 Green PT - BSc,PhD avian ecology and behaviour
6 Habsah Abdul Kadir - BS,MS agricultural entomology
7 Khoo SG - BSc,PhD *Associate Professor* agricultural and forest entomology
8 Kiew BH - BSc,PhD *Associate Professor* animal ecology & conservation
9 Kumar AS - MSc,PhD mangrove ecology
10 Lee SS - BSc,PhD *Associate Professor* agricultural entomology; insect physiology
11 Mah YL - BSc,PhD vertebrate behaviour and ecology
12 Mohd. Sofian Azirun - BSc,PhD agricultural entomology
13 Ng SM - BSc,PhD agricultural and forest entomology
14 Ramli Abdullah - MSc,PhD animal nutrition and reproduction
15 Wells DR - BSc,PhD *Associate Professor* avian biology

3229 Universiti Sains Malaysia

Minden, Gelugar, Penang
Tel: 883822. Telex: 40254.

Has a strong school of Biological Sciences but details of research staff not available at time of going to press

3230 Universiti Pertanian Malaysia

43400 Serdang, Selangor
[Unipertama, Sungai Besi]
Tel: (03) 9486101 -10(10 lines). Telex: Serdang Main Campus: UniperMA 37454.

1 Dato' Nayan bin Ariffin Professor - DipAgric,BS,MS,PhD *Vice Chancellor*

3231 Department of Agricultural Economics

1 Gibbons Prof ET - BSc,MSc,PhD *Professor* agricultural marketing
2 Ahmad Shuib - BSc,MSc *Lecturer* recreational economics
3 Chew Tek A - BAgricSc,MPhil,PhD *Associate Professor* agricultural economics
4 Eddie Chiew Fook Chong - BAgricSc,MAgric *Lecturer* production economics
5 Fatimah bte Mohd Arshad - BEcons,MSc,PhD *Lecturer* agricultural marketing
6 Ismail bin Abd. Latif - BSc,MSc *Lecturer* agricultural marketing
7 Mahfoor bin Haji Harron - BSc,MSc *Lecturer* farm management
8 Mohd Ghazali bin Mohayidin - BSc,MSc,PhD *Lecturer, Deputy Dean of Faculty* production economics
9 Mohd Zainal Abidin bin Tambi - BAgricEcons,MEcons *Lecturer* livestock marketing

3232 Department of Agronomy & Horticulture

1 Mohd Khalid bin Mohd Nor - DipAgric,BSc,MSc,PhD *Associate Professor, Deputy Vice-Chancellor (Development)* biometry
2 Khalip bin Abdul Raffar - BAgricSc,DipEdu,MAgricSc *Lecturer, Deputy Dean of Faculty* seed technology; cropping systems
3 Mohammad bin Mohd Lassim - DipAgric,BSc,MSc,PhD *Associate Professor, Head of Department* seed technology
4 Abdul Halim bin Haji Hashim - DipAgric,BS,MS,ProDipPlantTechM'gt *Lecturer* crop production
5 Adam bin Puteh - DipAgric,BS,MS *Lecturer* seed technology
6 Ali Hassan bin Haji Md. Ali - BAgricSc,MAgricSc *Lecturer* crop production
7 Asiah binti Abdul Malek - DipAgric,BS,MS *Lecturer* ornamentals
8 Cheah Chooi Hwa - BSc,MSc,PhD *Associate Professor* genetics; cytogenetics
9 Chin Hoong Fong - FIBiol,MAgricSc,PhD *Professor* seed technology
10 Hassan bin Mad - BAgric,BS,MS *Lecturer* pomology
11 Hassan bin Mohd Rashid - DipAgric,BS,MS *Lecturer* ornamentals
12 Hor Yue Luan - BAgricSc,MAgricSc,PhD *Lecturer* seed technology
13 Ibrahim bin Mohd Yusof - DipAgric,BS,MS,PhD *Associate Professor* pomology
14 Lim Eng Siong - DipAgric,BS,MS,PhD *Associate Professor* plant breeding
15 Maheran binti Abdul Aziz - BAgricSc,MSc *Lecturer* tissue culture
16 Mihdzar bin Abdul Kadir - DipAgric,BS,MS *Lecturer* plant breeding
17 Mohammad bin Md Ali - DipAgric,BSc,MS,PhD *Associate Professor* plant physiology
18 Mohd Idris bin Zainal Abidin - DipAgric,BS,MS,PhD *Lecturer* plant breeding
19 Mohd Ridzwan bin Abdul Halim - BAgricSc,MAgricSc *Lecturer* pasture agronomy
20 Mohd Said bin Saad - BAgricSc,MS *Lecturer* plant breeding
21 Mohd Zain bin Hj. Abdul Karim - BAgricSc,MAgricSc,PhD *Professor* crop physiology
22 Mustafa Kamal bin Md Shariff - DipAgric,BS,MS *Lecturer* landscape design
23 Raja a/l Amarthalingam - BAgricSc,MS *Lecturer* weed control

24 Raja Muhammad bin Raja Harun - DipAgric,BS,MS,PhD *Associate Professor* plant physiology
25 Rosli bin Mohamad - DipAgric,BS,MS,PhD *Lecturer* pesticide toxicology; weed science
26 Saleh bin Kadzimin - DipAgric,BS,MS,PhD *Lecturer* ornamentals
27 Singh S - DipAgric,MS *Lecturer* rice agronomy
28 Siti Aishah binti Hassan - DipAgric,BS,MS *Lecturer* olericulture
29 Siti Hajar binti Ahmad - DipAgric,BS,MS,PhD *Lecturer* vegetable physiology
30 Syed Mohd Zain bin Syed Hassan - BSc,MSc *Lecturer* weed biology
31 Thohirah Lee binti Abdullah - DipAgric,BS,MS *Lecturer* ornamentals
32 Wan Chee Keong - DipAgric,BS,MS,PhD *Associate Professor* pomology
33 Wan Mohamad bin Wan Othman - DipAgric,BAgricSc,MAgricSc,PhD *Lecturer* pasture crops
34 Wong Kai Choo - BAgricSc,PhD *Associate Professor* crop physiology
35 Yap Thoo Chai - BSc,MS,PhD *Professor* plant breeding; biometry

3233 Department of Animal Science

1 Syed Jalaludin bin Syed Salim - BVSc & AH,MPhil,PhD *Professor, Deputy Vice-Chancellor (Academic)* poultry nutrition
2 Abdul Salam bin Abdullah - DVM,MSc,PhD *Associate Professor, Deputy Dean of Faculty* pharmacology
3 Kassim bin Hamid - BAgricSc,PhD *Associate Professor, Head of Department* environmental physiology
4 Abdul Razak Alimon - BSc,DipScAgr,MSAgric *Lecturer* ruminant nutrition
5 Adnan bin Sulong - DVM,MS,PhD *Lecturer* lactation physiology
6 Ahmad Basir Pravee Vijchulata - BS,MS,PhD *Lecturer* dairy production
7 Alex Tuen - BSc,MSc *Lecturer* ruminant production
8 Haji Mustapha Mamat - BAgrocSc,DipTropAgroPasture,MAgricSc *Lecturer* agrostology
9 Hew Peng Yew - DVM *Lecturer* animal mineral nutrition
10 Kassim Hamid - BAgricSc,PhD *Associate Professor* environmental physiology
11 Kuan King Kai - DVM,MS *Lecturer* swine science
12 Mak Tian Kwan - BSc,MS *Lecturer* animal genetics
13 Mohamed Hilmi Haji Abdullah - DVM,MVSc,PhD *Lecturer* animal histology
14 Mohd Ali bin Rajion - BSc,MSc *Lecturer* foetal physiology
15 Mohd Mahyuddin bin Mohd Dahan - DVM,MSc,PhD *Associate Professor: Dean Faculty of Food Science and Technology* meat science
16 Oh Beng Tatt - DipAgr,BS,MS *Associate Professor* poultry science
17 Osman Awang - DipAgric,BS,MS *Lecturer* dairy science
18 Ramlah Abdul Hamid - DipAgric,BS,MS *Lecturer* poultry science
19 Samuel C - DVSc,DipAgr,MAgrSc,PhD *Associate Professor* ruminant production
20 Tan Hock Seng - BSc,BVetMed,PhD,MRCVS *Associate Professor* animal reproductive physiology
21 Tengku Azmi Tengku Ibrahim - DVN,MVSc *Lecturer* animal histology
22 Tuan Ariffeen Bongso - BVSc,MSc,PhD *Associate Professor* animal genetics
23 Vidyadaran MK - DVM,MVS *Lecturer* veterinary anatomy
24 Wan Nordin Wan Mahmood - DVM,MVSc *Lecturer* animal histology
25 Zainal Aznam b Mohd Jelan - DVM,MSc *Lecturer* ruminology

3234 Department of Food Science

1 Zaiton bte Hassan - DipAgric,BS,MS *Head of Department* traditional fermented food; microbial spoilage of food
2 Mohamad Salleh b Ismail - BSc,PhD *Associate Professor* fermentation technology
3 Augustin MA - BSc,PhD *Lecturer* fats and oils; analysis of food; enzymes; post harvest pesticides
4 Azizah bte Osman - BSc,PhD *Lecturer* post harvest physiology and handling of fruits and vegetables
5 Baharuddin b Abd Ghani - BS,MS *Lecturer* industrial microbiology
6 Gulam Rusul Rahmat Ali - DVM,MSc *Lecturer* food pathogens
7 Hasanah bte Mohd Ghazali - BSc,MSc *Lecturer* immobilized enzyme; analysis of food
8 Jinap bte Selamat - DipSc&Ed,BSc,MS *Lecturer* food flavours
9 Mohamad Nordin b Abdul Karim - BSc,MSc *Lecturer* nutrition: protein evaluation
10 Mohd Nasir bin Azudin - BSc,PhD *Lecturer* cereal technology
11 Mohd Salleh bin Ismail - BSc,PhD *Lecturer* fermentation technology
12 Noranizam bte Mat Long - BSc,PhD *Lecturer* microbiology of food spoilage
13 Suhaila binti Mohamed - BSc,PhD *Lecturer* biophysical chemistry of food, protein chemistry, rheology. polysaccharides (gums & gels)

3235 Department of Food Technology

1 Mohd Yusof bin Abu - DipAgric,BS,MS,PhD *Head of Department* thermal processing, food irradiation; dairy technology
2 Asiah bte Mohd Zain - BSc,MSc *Associate Professor* infant food technology
3 Abdullah b Abu Bakar - DipAnHlth&Prod,BSA,MS,PhD *Lecturer* meat and fish technology
4 Asbib b Ali - BSc,MSc *Lecturer* food engineering - dehydration and dehydration kinetics
5 Asiah bte Mohd Zain - BSc,MSc *Lecturer* infant food technology
6 Dzulkifly b Mat Hashim - BSc,MSc *Lecturer* food engineering - food rheology and solar energy

7 Jamilah bte Bakar - BSc,MS *Lecturer* food packaging technology
8 Mohd Ali b Hassan - BSc,MSc *Lecturer* food engineering - low temperature processing and energy from agric. waste
9 Russly b Abdul Rahman - BSc,MSc *Lecturer* food engineering - canning technology
10 Salmah bte Yusof - DipAgric,BS,MS *Lecturer* beverage technology and food quality assessment
11 Yaakob b Che Man - DipFoodTech,BSc,NSc *Lecturer* convenient and high protein low cost foods
12 Yu Swee Yean - BSc,PhD *Lecturer* meat and fish technology

3236 Department of Forest Management

1 Abdul Manap bin Ahmad - BSc,BSc,DipFor,PhD *Professor; Director Inst. of Consultancy* forest pathology
2 Yusof bin Hadi - BSc,MF,PhD *Head of Department* operational research
3 Ashari bin Muktar - DipFor,BS,MS *Associate Professor* wood science; forest management
4 Ahmad Said bin Sajap - DipAgric,BS,MS *Lecturer* forest entomology
5 Awang Noor bin Abdul Ghani - BS,MS *Lecturer* forest economics
6 Faizah bte Abood Haris - BSc,MSc *Lecturer* forest products protection
7 Lai Food See - BA,DipEd,MSc *Lecturer* watershed management
8 Lee Su See - BSc,MS *Lecturer* forest pathology
9 Rusli bin Mohd - BS,MPhil *Lecturer* administration forest policy
10 Shukri bin Mohamed - BS,MS *Lecturer* industrial forestry
11 Wan Sabri bin Wan Mansor - DipFor,BS,MS *Lecturer* forest recreation
12 Yusuf bin Hadi - BS,MFS,PhD *Lecturer* forest management planning

3237 Department of Forest Production

1 Khamis bin Awang - BSc,PhD *Lecturer: Dean of Faculty* ecophysiology; agroforestry
2 Mohd Basri bin Hamzah - BSc,MSc *Associate Professor, Deputy Dean of Faculty* indigenous silviculture
3 Mohd Zin bin Jusoh - DipAgric,BS,MS *Associate Professor* wood science forest management
4 Abdul Rahman bin Md Derus - DipFor,BS,MS *Lecturer* forest operation, engineering
5 Doraisingam Manikan - BS,MS *Lecturer* forest harvesting
6 Ismail bin Hashim - BS,MS *Lecturer* wood processing and operations
7 Jalaluddin bin Harun - BS,MS *Lecturer* wood chemistry
8 Jamaluddin bin Basharuddin SMD - MSc *Lecturer* forest ecology
9 Lim Meng Tsai - BSc,MSc,PhD *Lecturer* forest ecology
10 Mohd Hamami bin Sahri - BSc,MS *Lecturer* sawmilling and wood processing
11 Razali bin Abd Kader - DipFor,BS,MS *Lecturer* composite wood
12 Sheikh Ali bin Sheikh Abod - BSc,MSc *Lecturer* plantation silviculture
13 Sulong bin Ibrahim - DipAgric,BS,MS,DipForSur *Lecturer* forest inventory

3238 Department of Human Development Studies

1 Hasnah binti Udin - CertEdu,BS,MEd *Associate Professor, Head of Faculty* early childhood education/child development
2 Chee Heng Leng - BA,MS *Lecturer* public health and nutrition
3 Halimah binti Haji Ahmad - CertEdu,CertHomeEcon,BS,MS *Lecturer* consumer & family economics
4 Huang Soo Lee - DipAgric,BS,MPH *Lecturer* population & family planning
5 Jariah binti Haji Masud - BEcon,MS *Lecturer* family economics
6 Khor Geok Lim - BScAgric,MSc *Lecturer* nutrition
7 Laily binti Haji Paim - BS,MS *Lecturer* family economics & home management
8 Maznah binti Ismail - BSc,MS *Lecturer* nutritional biochemistry
9 Mohd Nasir bin Mohd Taib - BS,MS *Lecturer* human physiology
10 Nawalyah binti Abdul Ghani - BS,GradDip,MS *Lecturer* clinical nutrition
11 Normah binti Hashim - DipAgric,BS,MS *Lecturer* food science & nutrition
12 Rokiah binti Mohd. Yusof - DipAgric,BS,MS *Lecturer* food & nutrition
13 Zailina bt Hashim - BA,MS *Lecturer* environmental health
14 Zaitun binti Yassin - DipAgric,BS,MS *Lecturer* food & nutrition
15 Zamaliah binti Mohd Marjan - DipAgric,BS,MS *Lecturer* food & nutrition

3239 Department of Plant Protection

1 Varghese G - BSc,MS,PhD,FLS *Professor* soil-borne pathogens
2 Abdul Rahman bin Abdul Razak - BAgric,BAgricSc,MPhil,PhD *Associate Professor* plant nematology
3 Abdul Ghani bin Ibrahim - DipAgric,BAgricSc,MPhil,PhD *Associate Professor* applied entomology, biological control
4 Dzolkifli bin Omar - DipAgric,BS,MS *Lecturer* insect toxicology
5 Gurchan Singh a/l Harnam Singh - BSc,MSc *Associate Professor* pests of mango
6 Hiryati binti Abdullah - DipAgric,BS,MS *Lecturer* phytobacteriology
7 Inon binti Sulaiman - DipAgric,BS,MS *Lecturer* plant virology
8 Joseph Bong Choong Fah - DipAgric,BS,MS *Lecturer* entomology
9 Kamaruzaman bin Sijam - DipAgric,BS,MS,PhD *Lecturer* phytobacteriology
10 Khoo Khay Chong - BAgricSc,PhD *Associate Professor* insect ecology
11 Lim Tong Kwee - BAgricSc,MAgricSc,PhD *Associate Professor* epidemiology & disease control
12 Makhdzir bin Mardan - DipAgric,BS,MS *Lecturer* apiculture

13 Mohamad Zakaria bin Hussin - DipAgric,BS,MS *Lecturer* insects of field crops
14 Mohamed bin Hj. Muid - BSc,MSc,PhD *Lecturer* host-plant parasite relationships
15 Mohammed Matthieu bin Abdullah MV - DrAgricSc *Lecturer* insect pathology
16 Mohd Nor bin Shamsuddin - BS,MS *Lecturer* pest management
17 Mohd Yusoff bin Hussein - DipAgric,BS,MS,PhD *Associate Professor* integrated pest management and biological control of insects
18 Noorma binti Osman - DipAgric,BS,MS *Lecturer* pests of stored crops
19 Rita Muhammad Awang - Drs,MSc *Lecturer* insect physiology
20 Rohani binti Ibrahim - DipAgric,BS,MS,PhD *Lecturer* taxonomy of fruit flies
21 Sariah binti Meon - BAgric,BS,MS,PhD *Lecturer* disease physiology
22 Siti Hajijah binti Hj. AS - BAgric,MSc *Lecturer* technology crop protection
23 Sulaiman bin Hanapi - Ph.D *Lecturer* entomology
24 Tan Tee How - BSc,MSc *Lecturer* agricultural microbiology
25 Wan Zainum binti Hj. Wan Nik - BAgricSc,MAgricSc *Associate Professor* seed pathology
26 Yusof bin Ibrahim - DipAgric,BS,MS *Lecturer* entomology, vegetable pests
27 Zainal Abidin bin Mior Ahmad - BAgrSc,MAgrSc *Lecturer* plant pathology
28 Zakaria bin Sidek - BS,MS *Lecturer* plant virology

3240 Department of Soil Science

1 Wan Sulaiman bin Wan Harun - BSc,MSc,DrAgrSc *Associate Professor Dean of Faculty* soil physics & soil conservation
2 Zulkifli bin Haji Shamsuddin - DipAgric,BS,MS *Associate Professor, Head of Department* soil microbiology
3 Othman bin Yaacob - DipAgric,BAgrSc,MAgrSc,PhD *Professor* soil fertility and management
4 Abdul Jabar bin Mohd Kamal - DipAgr,BS,MS *Associate Professor* soil fertility and management
5 Ahmad Husni bin Mohd Haniff - DipAgric,BS,MS *Lecturer* soil fertility & management
6 Aminuddin bin Hussin - DipAgric,BS,MS *Lecturer* soil chemistry
7 Annuar bin Abdul Rahim - DipAgric,BS,MS *Lecturer* soil fertility & management
8 Azizah bte Hashim - BSc,MSc *Lecturer* soil microbiology
9 Hamdan bin Jol - BAgrSc,MSc *Lecturer* soil genesis, classification & clay mineralogy
10 Jamal bin Talib - DipAgric,BS,PhD *Lecturer* soil physics & soil management
11 Mohd Hanafi bin Musa - BScEd,MSc *Lecturer* soil & plant analysis
12 Mohd Khanif bin Yusop - DipAgric,BS,MS,DrAgrSc *Lecturer* soil chemistry
13 Mohd Mokhtaruddin bin Abdul Manan - BAgricSc,MSc,DrAgrSc *Lecturer* soil physics & soil conservation
14 Mok Chak Kim - DipAgric,PhD *Associate Professor* soil physics & soil conservation
15 Nik Mokhtar bin Nik Wan - BSc,MSc *Lecturer* soil genesis, classification & soil clay mineralogy
16 Paramananthan a/l Selliah - BSc,DipTropAgron,DSc *Associate Professor* soil genesis, classification & clay mineralogy
17 Radziah bte Othman - BS,MS *Lecturer* soil microbiology
18 Rosenani binti Abu Bakar - BAgrSc,MSc *Lecturer* soil chemistry
19 Shahbuddin bin Mohd Fiah - DipAgric,BS,MS *Lecturer* soil fertility & management
20 Shamsuddin bin Jusop - BSc,MSc,DSc *Lecturer* soil genesis, classification & clay mineralogy
21 Sharifuddin bin Haji Abdul Hamid - DipAgric,CertE,BS,MS,DrAgrSc *Associate Professor* soil & plant analysis
22 Siti Dauyah @ Zauyah binti Darus - BS,MSc *Lecturer* soil genesis, classification & clay mineralogy
23 Zaharah binti Abdul Rahman - DipAgric,DipAdvSoilSc,BS,MS,DrAgrSc soil chemistry

3241 Department of Veterinary Clinical Studies

1 Mohd Razeen Jainudeen - BVSc,MSc,PhD *Head of Department, Professor Animal Reproduction*
2 Abdul Aziz Saharee - BVSc&AH,BVSc,MSc,MRCVS *Lecturer* ruminant medicine
3 Aini Ideris - DVM,MVS *Lecturer* avian medicine
4 Bashir Ahmad Fateh Mohamed - DVM,MVS *Lecturer* equine surgery & medicine
5 Che Teh Fatimah Nachiar Iskandar - DVM,DipLargeAnMed *Lecturer* ruminant medicine
6 Nadzariah Cheng Abdullah - DVM,DipMed&Surg *Lecturer* small animal medicine
7 Nadzri Salim - DVM,MVS *Lecturer* large animal medicine and epidemiology
8 Rashid Ibrahim - DVM,DipSmallAnSurg *Lecturer* small animal surgery
9 Sharifuddin Abdul Wahab - DVM,MS *Lecturer* animal reproduction
10 Too Hing Lee Henry - BVSc&AH,BVSc,MVSc *Lecturer* swine medicine

3242 Department of Veterinary Pathology & Microbiology

1 Abdul Latif bin Ibrahim - DVM,MSc,PhD *Professor, Dean of Faculty* virology
2 Mohd Fadzil bin Haji Yahya - DAP&E,BVSc,MSc,MRCVS *Associate Professor, Head of Department*
3 Sheikh Omar Abdul Rahman - BVSc,MVSc,MRCVS *Associate Professor* veterinary pathology
4 Abdul Rahim Mutalib - DVM,MS *Lecturer* laboratory animal medicine
5 Abdul Rahman Aziz - DVM,MVSc *Lecturer* veterinary pathology
6 Abdul Rani bin Bahaman - DVM,BSc,MPhil *Lecturer* veterinary mycology
7 Chooi Kum Fai - DVM,DipVetPath *Lecturer* veterinary pathology
8 Mohd Shah bin Haji Abdul Majid - BVSc,MVS,MSc *Lecturer* bacteriology
9 Rasedee bin Abdullah - BSc,MSc *Lecturer* clinical biochemistry
10 Rehana Abdullah Sani - DVM,PhD *Lecturer* parasitology
11 Saleha Abdul Aziz - DVM,DipVetPrevMed *Lecturer* veterinary public health
12 Shaik Mohd Amin Babjee - BVSc,MSc,PhD *Lecturer* protozoology
13 Tham Kok Mun - DVM,PhD *Lecturer* non-ruminant virology
14 Ungku Chulan Ungku Mohsin - BVSc,MVS *Lecturer* avian pathology

3243 Ministry of Agriculture, Fisheries and Natural Resources

Reduit
Tel: 54-1091.

1 Lutchmeenaraidoo K - BScAgr *Chief Agricultural Officer*
2 Brunet LR - BScAgr *Principal Agricultural Officer - Crop*
3 Monty MJL - BSc *Principal Agricultural Officer - Development*
4 Mosaheb MF - BVSc,BScAH,MSc *Principal Agricultural Officer Livestock*

3244 Agronomy Division

1 Naidu SN - BScAgr,MSc *Head of Division*
2 Boodoo MY - BScAgr tea
3 Cangy CL - BScAgr mushrooms
4 Hanoomanjee P - BScAgr tea
5 Heerasing JM - BScAgr root crops
6 Jhundoo CR - BScAgr fodder
7 Li Kwan Pun LCK - BScAgr cereals
8 Ramahotar K - BScAgr tobacco
9 Ramnauth K - BSc,MSc biometry
10 Rojoa H - BScAgr,PhD pulse crops

3245 Horticulture Division

1 Rajkomar I - BScAgr,MSc *Head of Division*
2 Bachraz DY - BScAgr temperate fruit crops
3 Gopaul N - BScAgr,MSc tissue culture
4 Lan Chow Wing KF - BScAgr vegetables
5 Reetoo D - BScAgr spices and ornamentals
6 Seeruttun S - BScAgr onion and garlic
7 Veerappa V - BScAgr,MSc vegetables

3246 Plant Pathology Division

1 Bucha J - BSc,MScBot *Head of Division*
2 Beni Madhu SP - BSc bacteriology
3 Chinappen M - BSc nematology
4 Dossa MI - BSc,MScMicrobio virology
5 Pillay M - BSc mycology

3247 Entomology Division

1 Joomaye A - BSc,MScEnt *Head of Division*
2 Abeeluck D - BSc integrated control of stable flies
3 Permalloo S - BSc fruit fly control
4 Ramsamy M - BSc,MPhil,MIB,PhD stored pests, citrus pests
5 Seewooruttun SI - BSc,MSc vegetable pests

3248 Agricultural Chemistry Division

1 Lam Thuon Mine LC - BSc *Head of Division*
2 Chintaram D - BSc pesticide residues
3 Coonjan N - BSc soils
4 Lalljee B - BSc,MSc plants, fruits
5 Mohungoo AH - BSc feedstuffs
6 Ramanjooloo NN - BSc plants
7 Sheik Fareed V - BSc,PhD pesticide residues

3249 Veterinary Division

1 Abdoolah AMS - DVM
2 Bhuwanee VB - BVSc&AH
3 Bissessur H - MSc
4 Boodhoo R - MSc
5 Deenoo R - BVSc&AH
6 Dooky C - BVSc&AH
7 Groodoyal VB - BVScAH mycology and tick borne diseases
8 Jaumally MR - BVScAH sero-diagnostic procedures
9 Jhoree H - MSc
10 Juggessur VS - BVSc&AH
11 Juhangeer A - BVSc&AH
12 Jusrut D - BVSc
13 Mandarbaccus J - BVSc&AH
14 Meenowa D - BVSc&AH
15 Mohadeb T - BVSc
16 Neerunjun B - BVSc&AH
17 Prayag IL - BVM
18 Sibartie D - BVM,MVM *Head of Division*

3250 Animal Production Division

1 Hulman B - BSc,MPhil *Head of Division* beef cattle nutrition

2 Boodoo A - BScAnimSc animal physiology
3 Gaya H - MSc,BSc dairy cattle
4 MaPoon LK - BScAgr small ruminants
5 Naidoo G - BVScAH dairy cattle
6 Naseeven R - BScAgr,MSc beef cattle
7 Rajkomar B - BScAgr poultry
8 Teeluck P - BScAgr sheep
9 Topsy JK - MSc small ruminants

3251 Dairy Chemistry Division

1 Mardamootoo P - MSc,PhD *Head of Division* mastitis vaccine
2 Moraby MR - BScAg
3 Ng Man Sun JFS - BSc,MSc dairy technology
4 Rangasamy PN - MSc mastitis studies

3252 Agricultural Engineering Division

1 Satar AR - BSc,MIAE *Head of Division*
2 Punchoo BE
3 Purmanund S - BSc

3253 Land Use and Projects Division

1 Ramtohul M - BSc,MSc *Head of Division*
2 Domun MVD - BSc

3254 Extension Services Division

1 Mauree SP - BSc,MAgrSt *Head of Division*
2 Ragobur D - BSc

3255 Forestry Services

Curepipe
Tel: 86-4966.

1 Owadally AW - BSc *Conservator of Forests*
2 Appiah S - BSc,MSc
3 Dulloo ME - BSc
4 Mungroo Y - BSc wildlife
5 Paupiah SA - BSc

3256 Fisheries Division

Port Louis
Tel: 01-2091.

1 Munbodh M - BSc,MSc *Head of Division*
2 Chiniah V - BSc
3 Goorah D - BSc,MSc
4 Jehangeer MI - BSc,MSc
5 Samboo CR - MSc
6 Venkatasami A - MSc

3257 Mauritius Sugar Industry Research Institute

Head Office, Reduit
Tel: 54-1061. Telex: 4477 SUGTRAC IW.

1 Ricaud C - BSc,PhD *Director*
2 Julien R - BSc,PhD *Assistant Director*

3258 Agricultural Chemistry

1 Deville J - BSc,MSc,DipAgr,CChem,MRSC
2 Ng Kee Kwong KF - BSc,PhD

3259 Biometry

1 Lim Shin Chong LCY - BSc,MSc

3260 Botany

1 Soopramanien GC - BSc,MSc,PhD

3261 Cane Breeding

1 Domaingue R - BSc
2 Peerun Z - BSc
3 Ramdoyal K - BSc,DEA,DocteurIngenieur

3262 Cultural Operations and Weed Agronomy

1 Chaillet G - BSc,DipAgr
2 Hardy M - DipAgr

3 McIntyre G - BSc,DipAgr,MIBiol

3263 Entomology

1 Govendasamy M - BS,DipAgr

3264 Food Crop Agronomy

1 Govinden N - BSc,MSc

3265 Plant Pathology

1 Autrey JC - BSc,MSc,PhD,MIBiol

3266 Sugar Technology

1 Maudarbaccus R - BSc,MScEng,MIChem
2 Ng Kwing King J - BS,MAIChE
3 Randabel M - DipAgr
4 Yin Lun Wong Sak Hoi - BSc,MSc

3267 Economics

1 Berthelot B - BA

3268 University of Mauritius

3269 School of Agriculture
University of Mauritius, Reduit
Tel: 54-1041. Telex: 4621 UNIM.

1 Manrakhan J - BSc,MSc *Vice Chancellor*
2 Peerally AP - BSc,MSc,PhD *Professor, Head of School*

3270 Division of Agricultural Botany

1 Peerally AP - BSc,MSc,PhD *Professor* plant pathology, mushroom science
2 Lutchmeah RS - PhD *Lecturer* mycology
3 Vencatasamy DR - BSc,MSc,PhD *Senior Lecturer* rhizobiology

3271 Division of Agricultural Chemistry

1 Bhoojedhur S - BSA,MSc,PhD *Professor* soil chemistry
2 Kasenally AS - BSc,PhD *Professor* inorganic chemistry, biotechnology
3 Choolun R - BSc,PGCE,MSc,PhD *Senior Lecturer* biochemistry
4 Gurib A - PhD *Lecturer* biochemistry, marine chemistry
5 Ramjuthun R *Assistant Lecturer* physical chemistry, marine chemistry

3272 Division of Agricultural Management
University of Mauritius, Reduit
Tel: 54-1041. Telex: 4621 UNIM.

1 Manrakhan J - BSc,MSc *Professor* agricultural economics
2 Mundil HK - BSc,MSc *Lecturer* agricultural management
3 Unmole H - BSc,MSc *Senior Lecturer* agricultural engineering

3273 Division of Agricultural Zoology

1 Facknarh S - BSc,MSc *Lecturer* entomology, marine science
2 Fagoonee I - BSc,MSc,PhD *Associate Professor* entomology, pestology marine resources

3274 Division of Animal Production

1 Jotee D - PhD *Lecturer* animal nutrition
2 Ramchurn R - BVSc,MSc *Senior Lecturer* veterinary science
3 Ramkisson J - BSc,MSc *Senior Lecturer* animal nutrition

3275 Division of Crop Production

1 Goburdhun S - BSc,MSc,PhD *Senior Lecturer* potato tuber conservation
2 Osman AM - BSc,DTA *Associate Professor* leucaena research
3 Ruhee KC - BSc,MSc *Senior Lecturer* horticulture

3276 Ministry of Agriculture, Trade, Lands & Housing

Plymouth

1 Michael F - BScAgric *Director of Agriculture*
2 Swanston G - BSc,BVM *Chief Veterinary Officer*
3 Gerald C - BScAgric,MScAgric *Agricultural Officer, Livestock*

3277 Caribbean Agricultural Research & Development Institute (Montserrat Country Unit)

C/O Ministry of Agriculture, PO Box 272, Plymouth
Tel: 809-491-5694. Cables: "CARDINST".

1 Fletcher RE - BSc,MSc *Country Team Leader & Agronomist* farming systems research & development; food crop, vegetable and pasture agronomy
2 Weekes S - BScAgric *Agronomist*

3278 Ministry of Agriculture and Fisheries

PO Box 2526, Private Bag, Wellington
Tel: 720 367. Telex: MAF WN NZ31532.

3279 Primedia

1 Thompson PD - BSc,BBS *Manager*
2 Stoddart Mrs HM - BS,NZLAC,ANZLA *Chief Librarian*

3280 Advisory Services Division

PO Box 2526, Gillingham House, 101-103 The Terrace, Wellington
Tel: (04) 720-367. Telex: NZ 31532.

1 Hercus JM *Director*
2 Riddell RJ - BAgrSc *Deputy Director*
3 Watts BB *Superintendent, Pesticides Board*
4 Crump DK - MAgrSc *CAO (Agronomy)*
5 Findlay RM - *CAO (Horticulture)*
6 Foley AJ *Pesticides Board Scientist*
7 Harrington GJ *CAO (Agricultural Engineering)*
8 Hedley Dr J - DipHortSc,PhD *Chief Plant Health Officer*
9 Ivess RJ - BHortSc,DipHortSc *CAO (Plant Exports)*
10 Lunn DW - MHortSc *Pesticides Board Scientist*
11 Lysaght PA - BAgrSc *CAO (An Hus)*
12 Walker AB *CAO (Economics)*
13 Walton GM - MSc *CAO (Extension)*

3281 Lynfield Plant Protection Centre

131 Boundary Road, Blockhouse Bay
[PO Box 41, Auckland]
Tel: (09) 676 024; 676 026.

1 Baker RT - PhD *entomology*
2 Barber CJ *nematology*
3 Boesewinkel HJ *pathology*
4 Cowley JM - PhD *entomology*
5 Dance HM *pathology*
6 Hampton MD *botany*
7 Hill CF - PhD *pathology*
8 Kleinpaste RH *entomology*
9 Manson DCM *entomology*
10 Rainbow AF - PhD *Officer in Charge*
11 Young BR *virology*

3282 Levin Plant Protection Centre

Kimberley Road, Private Bag, Levin
Tel: (069) 87059.

1 Grbavac N *pathology*
2 Somerfield KG - PhD *Officer in Charge (Entomology)*
3 Soteros JJ *pathology*
4 Wong SK - PhD *entomology*

3283 Lincoln Plant Protection Centre

Lincoln College, P.O. Box 24, Lincoln
Tel: (03) 252-811.

1 Harvey IC - PhD *Officer in Charge* pathology
2 Harper GM *nematology*
3 Mohammed NA - PhD *pathology*
4 Stephenson BP - PhD *entomology*

3284 Seed Industry Section

3285 Seed Testing Station - Palmerston North

301 Church Street, P.O. Box 609
Tel: (063) 68-079.

1 Scott DJ *Officer in Charge*
2 Finnerty AM *supervising seed analyst*
3 Hampton JG *pathology*
4 Johnson DW *bacteriology*
5 Liew RSS *pathology*
6 Young KA *plant physiology*

3286 Agricultural Research Division

Private Bag, Wellington
Tel: 720-367. Cables: MAF WN NZ31532.

1 Hutton JB - MAgrSc,PhD,FNZIAS *Director*
2 Bilbrough GW - BAgrSc *Scientific Liaison Officer*
3 Crosbie PJ - BSc *Systems Modeller*
4 Darkey CKG - BSc,MSc,PhD *Systems Modeller*

5 Keenan BR - BAgrSc,BD *Assistant Director (Divisional Services)*
6 Kettle PR - BSc,PhD *Scientific Technical Advisor*
7 Scott RS - BAgrSc,PhD *Assistant Director (Soil and Plants)*
8 Whiteford CM - BSc,NZCS *Scientific Services Officer*
9 Wright DE - MSc,PhD,DSc,FNZIC *Assistant Director (Animals)*
10 Yates PA - BSc *Systems Modeller*

3287 Northern South Island Region

Research Group Headquarters, Canterbury Agriculture and Science Centre,
P.O. Box 24, Lincoln
Tel: 252-021.

1 Joblin ADH - BAgrSc,DipTropAgr *Director*

3288 Agronomy

1 Gallagher JN - BSc,PhD *Leader* crop physiology
2 Martin RJ - BSc,DTA,PhD crop agronomy
3 Stephen RC - BAgrSc crop agronomy, soil fertility

3289 Biometrics

1 Dyson CB - BSc *Leader*
2 Baird DB - BSc
3 Saville DJ - MA,MSc

3290 District Scientists

1 McLeod CCS - BAgrSc *Officer in Charge Adair Agricultural Research Station* pastures
2 Morton JD - MAgrSc pastures
3 Trought MCT - NDA,BSc,PhD *Officer in Charge Marlborough Primary Production Research Centre* horticulture

3291 Horticulture

1 Stoker R - BSc,PhD *Leader* crop irrigation; export horticulture crops
2 Haynes RJ - BHortSc,PhD soil fertility

3292 Liaison

1 Ryde DH - BAgrSc

3293 Management

1 Hayman JM - Leader *irrigation agronomy*
2 Bray AR - BAgrSc,PhD sheep carcass composition and nutrition
3 Geenty KG - MAgrSc,DipWool,PhD sheep nutrition and body composition, wool production
4 Munro JM - BAgrSc,PhD sheep nutrition, body composition and reproduction
5 Taylor AR - BAgrSc,MSc,PhD irrigation and tree growth; irrigation efficiency

3294 Pastures

1 Vartha EW - MAgrSc,PhD *Leader* pasture management
2 Francis SM - BAgrSc pasture cultivars, quality, production, persistence and composition
3 Radcliffe JE - MSc,PhD pasture; shelter; fodder trees

3295 Plant Protection

1 Williams JM - MSc,PhD *Leader* vertebrate pest ecology and management
2 Bell J - BSc vertebrate pest ecology and management
3 Bourdot GW - BAgrSc,DipAgrSc,PhD weed biology; weed science
4 Butler JHB - BAgrSc,PhD weed science
5 Goldson SL - MSc,PhD insect pest ecology
6 Jackson TA - BHortSc,MSc,PhD microbial control; Scarabaeidae

3296 Soil Fertility

1 Rickard DS - MSc,MNZIC *Officer in Charge: Winchmore Irrigation Research Station, Leader* soil fertility and irrigation
2 Fitzgerald PD - BSc agricultural meteorology, computer applications
3 Morrison RS - BSc,PhD incubation soil testing; chemical analyses; fertilisers
4 Nguyen ML - MAgrSc soil fertility; soil chemistry
5 Williams PH - BAgrSc

3297 Southern South Island Region

Invermay Agricultural Centre, Private Bag, Mosgiel
Tel: 3809, Mosgiel.

1 Allison AJ - MAgrSc,PhD *Regional Director*

3298 Agronomy

1 Cossens GG - BSc,DipAOSM *Leader* tussock grasslands; irrigation; cropping
2 Brash DW - BAgrSc horticulture; orchard nutrition and irrigation
3 Hall IR - BSc,PhD pasture agronomy; mycology
4 Jacyna T - MSc,PhD horticulture; pip and stonefruit production
5 Keoghan JM - MAgrSc,PhD pasture species evaluation
6 Lowther WL - MAgrSc,PhD legume establishment and microbiology
7 Smallfield BM - BAgrSc cropping systems

3299 Entomology

1 Barratt BIP - BSc,PhD biology and control of insect pests
2 Stewart KM - BSc,PhD biology and control of insect pests; bee biology

3300 Soil Fertility

1 Sinclair AG - BSc,PhD *Leader* soil sulphur; computer modelling
2 Boswell CC - MAgrSc,MPhil,PhD evaluation of sulphur fertilisers
3 Dolby R - BSc,PhD soil testing services; analytical methods
4 Floate MJS - BSc,PhD nutrient cycling and fertility of upland soils
5 Greenwood PB - BAgrSc irrigation; cropping systems; soil physics
6 Metherell AK - BAgrSc soil fertility; trace element studies

3301 Energy and Environment

1 Stewart DJ - MSc,PhD biogas production research and development

3302 Animal Production

1 Davis GH - MAgrSc *Leader* management and breeding of prolific sheep
2 Hawker H - BSc,PhD wool metrology; sheep nutrition and wool production
3 Montgomery GW - BAgrSc,PhD endocrinology of sheep reproduction
4 Owens JL - MRurSc management and evaluation of prolific sheep

3303 Animal Nutrition

1 Drew KR - MAgrSc,PhD *Leader* deer management; venison production
2 Fennessy PF - MAgrSc,PhD nutrition and growth of sheep and deer
3 Haines SR - BSc,PhD immunoendocrinology; chemistry
4 Lord EA - BSc lean lamb production
5 Mackintosh CG - BVSc,PhD health and disease control in deer
6 Moore GH - BSc,PhD production and management of deer
7 Suttie JM - BSc,PhD deer physiology; biotechnology

3304 Animal Health Laboratory

1 Mason PC - BSc,PhD parasitology of domestic animals
2 Orr MB - BVM&S,MRCVS,PhD diseases in sheep, goats and deer

3305 Biometrics

1 Johnstone PD - BSc,DipPhChem *Leader* biometrics; EDP
2 Dodd KG - BSc,PhD biometrics; EDP
3 Littlejohn RP - MSc,PhD biometrics; EDP

3306 Regional Information Centre

Invermay Agricultural Centre, Private Bag, Mosgiel
Tel: 3809.

1 Buxton DAL - BAgrSc *Senior Information Officer* research/extension liaison; communication development
2 Winter JR - MA,DipNZLS *Librarian* regional library services

3307 Tara Hills High Country Research Station

Private Bag, Omarama
Tel: 830, Omarama.

1 Douglas MH - BAgrSc *Officer in Charge* agronomy; pasture species evaluation; irrigation
2 Allan BE - MAgrSc pasture agronomy; grazing management

3308 Woodlands Research Station

Woodlands, RD1, Invercargill
Tel: 393-012, Invercargill.

1 Risk WH - MAgrSc *Officer in Charge* pasture and crop agronomy
2 McDonald RC - BAgrSc,PhD pasture management; forage cropping; strip mine restoration
3 McEwan JC - BSc nutrition and management of sheep
4 Struthers GA - BVSc,MACVSc diagnosis of tuberculosis in deer

3309 Levin Horticultural Research Centre

Private Bag, Levin
Tel: (069) 87-059.

1 Powell CL - BSc,PhD *Director*
2 Swain DJ - MScAgr,DipHortSc *Scientific Liaison Officer*
3 Mayclair RM - BA,DipNZLS *Librarian*

3310 Vegetable Production

1 Bussell WT - MSc,PhD *Leader* vegetable agronomy
2 Robb AR - BHortSc asparagus physiology

3311 Fruit Production

1 Burge GK - BHortSc,MSc,DipPlSc *Leader* pomology
2 Irving DE - MSc,PhD pomology
3 Kingston CM - BHortSc berryfruit, greenhouse grapes
4 Porter LA - MAgrScHort strawberries

3312 Ornamental Crops

1 Butcher SM - BEd,MSc *Leader* plant physiology
2 Bicknell RA - BSc environmental physiology
3 Dennis DJ - BAgScHort,PhD crop agronomy

3313 Soil Science and Plant Nutrition

1 Prasad M - BSc,MA,PhD *Leader* soil chemistry, plant nutrition
2 Swain DJ - MScAgr,DipHortSc soil physics

3314 Post-harvest Physiology

1 Lill RE - BAgrSc,PhD *Leader* fruit and vegetable storage
2 Carpenter A - MSc,PhD post-harvest entomology
3 Downs CG - MSc cut flower transport
4 Given NK - MSc fruit ripening biochemistry
5 King GA - BSc,PhD fruit and vegetable storage
6 O'Donoghue EM - BSc biochemistry

3315 Crop Protection

1 Cox TI - BScAgr *Leader* weed control
2 Cheah LH - MSc,PhD plant pathology

3316 Ruakura Animal Research Station

Private Bag, Hamilton
Tel: (071) 62-839. Telex: NZ31532.

1 Jury KE - MSc,BA *Director*

3317 Analytical Biochemistry

1 Peterson AJ - BSc,PhD *Leader* reproductive physiology
2 McIntosh JT - BSc,DPhil,ANZIMLT salivary protein

3318 Animal Management

1 Donnelly PE - BAgrSc,DPhil *Leader* protein extraction from pasture - associated animal production research
2 Watson TG - MSc,PhD *Deputy Leader* gastrointestinal nematode parasitism in domestic ruminants
3 Adam JL - MSc,PhD nutrition and grazing management for farmed deer
4 Asher GW - MSc,PhD reproductive physiology and management of farmed deer
5 Reardon TF - MSc,Agr nutritional requirements for grazing beef cattle and dairy goats; calf rearing
6 Smith JF - MSc,Agr,PhD reproductive physiology; immunology and nutritional control of reproduction in sheep

3319 Biophysics

1 Woolford MW - MSc,DPhil *Leader* milking techniques, laser light scattering
2 Phillips DSM - MSc milking techniques

3320 Dairy Science Group

1 Bryant AM - MAgr,Sc,PhD *Leader* dairy cow nutrition and management, feed production systems
2 Carruthers VR - BAgrSc bloat; passage of digesta
3 L'Hullier PJ - BAgrSc sward structure, tiller dynamics, plant-dairy cow interface
4 McDonald RM - BSc,BE,MIPENZ protein extraction equipment and economics
5 Newth RP - NZCS,MNZIC chemical analysis of biological material
6 Thom ER - MAgrSc,DPhil sward competition, pasture management, cultivar evaluation
7 Vaughan SR - BE(CHEM)PhD biomass refining industrial cellulose fermentation

3321 Engineering Development Group

1 Mills RA - CENG,MIMECH,MIPENZ *Leader* mechanical process systems, design and development
2 Jordan RB - BE,PhD *Deputy Leader* electronic and microcomputer systems development
3 Sherlock RA - BSc,PhD,FInstP,FNZIP,MIEEE sensor, instruction and control systems

3322 Genetics

1 Clarke JN - MAgrSc,PhD *Leader* sheep evaluation and selection
2 Johnson DL - MSc,PhD *Deputy Leader* biometrics - various components and prediction of breeding values
3 Baker RL - MAgrSc,PhD sheep and beef cattle evaluation and selection
4 Bodin L - MAnBio,DipAgronGen,DipEngAgron (NRAC Fellow): (NRAC visitor from France) genetics of sheep reproduction
5 Hohenboken W - MSc,PhD (NRAC Fellow from Oregon,USA) evaluation of breeding resources
6 Morris RL - BA,PhD beef cattle and sheep breeds and selection
7 Nicoll GB - MAgrSc,PhD (seconded to department of lands & survey): sheep, beef and goat breeding programmes
8 Parratt AC - BSc,PhD lean lamb production, beef selection

3323 Growth Physiology

1 Bass JJ - BSc,PhD *Leader* endocrinology of growth
2 Payne E - BSc,PhD *Deputy Leader* lipid metabolism
3 Davis SR - BSc,MSc,PhD lactation and growth physiology
4 Hodgkinson SC - MPhil immunobiochemistry of growth
5 Upreti GC - BSc,MSc,PhD biochemistry, facial eczema and lipids
6 Wolff JE - MAgrSc,PhD foetal and postnatal growth

3324 Meat Group

1 Kirton AH - MAgrSc,PhD,DSc,FNZIAS *Leader* carcass composition, quality and classification
2 Butler-Hogg B - BAgrSc,PhD *Deputy Leader* carcass meat quality in beef and other species

3325 Mycotoxic Diseases

1 Towers NR - BSc,PhD *Leader* facial eczema control, mineral nutrition (Mg,Na)
2 Munday R - BSc,PhD *Deputy Leader* sporidesmin biochemistry and toxicity mechanisms, facial eczema content
3 Campbell AG - MAgrSc,NDD,FNZIAS hereditable resistance to facial eczema and ryegrass staggers
4 Gallagher RT - MSc,PhD toxin chemistry, immunochemistry, vaccine development
5 Mortimer PH - MA,BVSc,MRCVS plant toxicity, actiology and control of ryegrass staggers
6 Smith BL - BAg,BVS,DipMicro,DPhil facial eczema control, sporidesmin pharmacokinetics, bracken carcinogenicity

3326 Physiology

1 Macmillan KL - MAgrSc,PhD *Leader* reproduction in dairy cattle, somatic cell counting, seasonal supply, sahiwal cattle
2 Kilgour R - MA,DPhil,FNZPSS *Deputy Leader* animal behaviour, operant studies, lamb survival, animal welfare
3 Day AM - BVSc,MRCVS induced calving, post-partum ansestrum, oestrous cycle control
4 Pearson AJ - MSc,DPhil possum nutrition, reproduction and fur production

3327 Ruakura Soil and Plant Research Station

1 Cullen NA - BAgrSc *Director*

3328 Soil Fertility

1 Quin BFC - PhD *Chief Scientist*
2 Cornforth IC - PhD *Group Leader* plant nutrition
3 O'Connor MB - MAgrSc *Group Leader* fertilizer and trace elements; in the field
4 Steele KW - MAgr,PhD *Group Leader* microbiology and nutrient cycling
5 Bonish PM - MSc microbiology
6 Brown MW - FRIC plant & analytical chemistry
7 Campkin R - MAgrSc microbiology; sulphur research
8 Clark CJ - MNZIC plant nutrition
9 Di Menna M - MSc,PhD microbiology; fungi research
10 Edmeades DC - MSc,PhD fertiliser and trace elements, laboratory
11 Haystead A - BSc,PhD nutrient cycling
12 Hill RA - PhD microbiology; fungi research
13 Ledgard SF - BAgrSc microbiology
14 Lee A - BSc microbiology; sulphur research
15 Perrott KW - MSc,PhD nutrient cycling
16 Rajan SSS - MSc,PhD fertilizer and trace elements, laboratory
17 Sarathchandra SU - BSc,PhD nutrient cycling
18 Saunders WMH - MSc,PhD microbiology; sulphur research
19 Sherrell CG - MAgrSc fertilizer and trace elements, laboratory
20 Smith GS - MSc,PhD plant nutrition
21 Sutton MM - MSc,PhD plant and analytical chemistry
22 Toxopeus MRJ - MAgrSc fertiliser and trace elements; in the field
23 Watkinson JH - MNZIC,PhD fertilizer and trace elements

3329 Entomology and Plant Protection

1 Pottinger RP - MAgr,PhD *Chief Scientist*

2 Clayton JS - BSc,PhD *Group Leader* aquatic plants
3 East R - BSc,PhD *Group Leader* insect control and organic chemistry
4 Rahman A - MSc,PhD *Group Leader* plant science
5 Baars JA - BSc,DRL pasture agronomy
6 Coffey BT - MSc,PhD aquatic plants
7 Goold GJ - MAgrSc pasture agronomy
8 Holland PT - BSc,PhD pesticide residue and organic chemistry
9 Lauren DR - BSc,PhD pesticide residue and organic chemistry
10 Poole PR - BSc,PhD pesticide residue and organic chemistry
11 Prestidge RA - MSc,PhD general entomology
12 Robertson LN - DipSc,PhD general entomology
13 Rohitha BH - MAgrSc,PhD general entomology
14 Tomkins AR - BHortSc,PhD pesticide residue and organic chemistry
15 Watson RN - MSc,PhD scarabs research

3330 Horticulture

1 Wood FH - MAgrSc,PhD *Chief Scientist*
2 Hopping ME - MHortSc,PhD *Group Leader* kiwifruit physiology
3 Brown NS - MAgrSc horticulture; manutuke
4 Buwalda JG - BAgrSc,PhD nutrition and physiology
5 Douglas JA - MAgrSc horticultural agronomy
6 Henzell RF - BSc,PhD growth regulators
7 Judd MJ - MSc water relationships and shelter
8 King PD *Group Leader* protected cropping
9 Klinac DJ - BSc,PhD management and plant health
10 McAneney KJ - BSc,PhD water relationships and shelter
11 McCormick SJ - BSc vegetable and field crops
12 Smart RE - MSc,PhD viticulture
13 Wilson GJ - BAgrSc,MSc vegetable crops

3331 Wallaceville Animal Research Centre

Wallaceville Animal Research Centre, Upper Hutt
Tel: (04) 286-089. Cables: VETLAB, Upper Hutt.

1 Parle JN - MSc,PhD *Director*
2 Lattey JM - BSc *Scientific Liaison Officer*
3 Vandervorst EOFJ - NZLA Cert,PhD *Librarian*

3332 Apiculture

1 Clinch PG - BSc *Head of Section* disease control; kiwifruit pollination; product quality

3333 Bacteriology and Central Services

1 Skerman TM - MSc,PhD *Head of Section* footrot; bacterial physiology

3334 Biometrics

1 Roberts MG - MSc,PhD *Head of Section* experimental design; mathematical modelling
2 Morrison LM - BA statistical data analysis

3335 Epidemiology

1 Kane DW - BVSc,DipMicrobiol *Head of Section* ill-thrift: periodontal diseases in sheep
2 Phillips RA - BSc,PhD ill-thrift

3336 Experimental Pathology

1 Thurley DC - BVetMed,PhD *Head of Section* periodontal diseases in sheep; ill-thrift
2 Harris A - BSc electron microscopy
3 Herceg M - MSc,DVSc goat diseases

3337 Experimental Serology

1 Ris DR - BSc,DipAgr *Head of Section* hydatids; Johne's Disease

3338 Growth

1 Allsop TF - MSc,PhD *Head of Section* metabolic diseases; growth and nutritional studies
2 Moore LG - BSc,PhD growth
3 Pfeffer AT - BSc,BVSc growth and metabolism
4 Rogers KM - BSc,MSc,PhD growth and metabolism

3339 Hydatids

1 Gemmell MA - BVSc *Officer in Charge (Taieri)* hydatids
2 Heath DD - MRurSc,PhD *Head of Section* sheep measles; hydatids
3 Lawson JR - BSc,PhD hydatids

3340 Immunology and Experimental Biology

1 Jonas WE - BVSc,DipMicrobiol,PhD *Head of Section* toxoplasmosis; sarcocystis
2 Hicks BR - MSc,PhD toxoplasmosis; sarcocystis

3341 Meat Monitoring and Pesticide Residues

1 Solly SRB - MSc *Head of Section* residue detection and species verification programme
2 Clear MH - BSc diagnostic toxicology
3 Erasmunson AF - MSc,PhD diagnostic toxicology
4 Lock JW - MSc heavy metal and residue detection

3342 Parasitology

1 Brunsdon RV - MSc,PhD *Head of Section* nematode control in ruminants
2 Bisset SA - MSc nematode control in ruminants
3 Douch PGC - MSc,PhD helminth immunity
4 Heath ACG - BSc,PhD epidemiology; ectoparasites
5 Vlassoff A - BSc nematode control in ruminants; anthelmintic resistance

3343 Reproductive Physiology

1 McNatty KP - MSc,PhD *Head of Science* endocrinological control; multiple ovulation programme
2 Henderson KM - BSc,PhD endocrinological control; multiple ovulation
3 Truman P - BSc,MSc,PhD prolificacy

3344 Trace Element Biochemistry and Radiochemistry

1 Millar KR - MSc,PhD *Head of Section* mineral nutrition and metabolism
2 Osborn PJ - BSc,PhD biochemistry; radiochemistry
3 Turner JC - MSc,PhD biochemistry; radiochemistry

3345 Virology

1 Davies DH - BVSc,PhD *Head of Section* pneumonia; pleurisy
2 Buddle BM - BVSc,DipMicrobiol,PhD goat diseases

3346 Corporate Services Division

Private Bag, Wellington
Tel: 720-367.

3347 Biometrics Section

PO Box 2526, Wellington
Tel: 720-367.

1 Robertson TG - MSc *Manager*
2 Green Miss RM - BSc,DipORS biometrics
3 Harte DS - MSc quality assurance
4 Jowett JH - MSc sample survey design
5 Kissling RC - MSc biometrics
6 Viggers Mrs EAM - BA,RSSDip biometrics

3348 Fisheries Management Division

Private Bag, Wellington
Tel: 720-367.

1 Cunningham BT - BSc *Director*
2 Walshe KAR - BSc *Assistant Director*
3 Akroyd JM - BSc *Chief Fishery Management Officer (Shellfish)*
4 Cooper RD - MSc *Senior Fishery Management Officer (FMP)*

3349 Auckland Region

Ministry of Agriculture & Fisheries, Private Bag, Auckland
Tel: 794-700.

1 Boyd RO - MSc *Regional Fishery Management Officer*

3350 Central Region

Ministry of Agriculture & Fisheries, Private Bag, Hastings
Tel: 87-125.

1 Henriques PH - PhD *Regional Fishery Management Officer*

3351 Challenger Region

Ministry of Agriculture & Fisheries, Private Bag, Nelson
Tel: 81-069.

1 Kilner A - MSc *Regional Fishery Management Officer*

3352 Southern Region

Ministry of Agriculture & Fisheries, Private Bag, Dunedin
Tel: 740-333.

1 Voller R - MSc *Regional Fishery Management Officer*

3353 Ministry of Forestry

P.O. Box 1610, Wellington
Tel: 751-569. Telex: NZ31012. Cables: Forestry.

1 Ballard R - MAgSci,PhD *Secretary*
2 Bassett C - BSc,DipFor,PhD *Assistant Secretary (Research)*
3 O'Loughlin C - PhD,MSc,DipFor *Chief Director of Research*

3354 Forest Research Institute

Private Bag, Rotorua
Tel: 475-899. Telex: NZ21080. Cables: FRESTRA.

1 Kininmonth JA - BSc,BA,PhD *Regional Director*
2 Anderson BJ - ANZLA,NZLACert *Librarian in Charge*
3 Andrew IA - BSocSc,MSc *Research Field Leader of Computing*
4 Griffith JA - BA *Chief Science Editor*
5 Hays SEA - BSc *Editor*
6 Klitscher KR - BA *Information Officer*
7 Tarlton GL - BA,MLS *Editor*
8 Whitehouse LJ - BSc,MSc,DipSc *Editor*

3355 Forest Health and Improvement Division

1 Wilcox MD - BSc,MA,PhD *Director*

3356 Forest Health

1 Bain J - BSc,MSc biological control of *Paropsis*; quarantine
2 Chou CKS - BSc,MSc,PhD diplodia dieback; cypress canker
3 Dick MA - BSc disease diagnosis: herbarium
4 Gadgil PD - PhD,BScAgr,MScAgr *Research Field Leader* Dothistroma needle blight and Cyclaneusma needle-cast
5 Hood IA - BSc,MSc armillaria root-rot; decay in indigenous trees
6 Kay MK - MSc insects affecting eucalypts

3357 Marketing and Planning Services

1 Carson SD - BSc,MS,PhD *Research Cooperative Manager*
2 Plackett PA - BA,MA,PhD *Technology Marketing Manager*
3 Theron MJ - BSc,BCom,BA,DipEd,DBL *Technology Planning and Evaluation Manager*
4 Williams FJN - BSc,BForSci *Software Manager*

3358 Genetics and Tree Improvement

1 Shelbourne CJA - BA,MSc,PhD tree breeding
2 Burdon RD - BSc,BA,PhD *Research Field Leader* quantitative genetics; gene resource
3 Ecroyd CE - BSc herbarium curator
4 Johnson GR - BS,MS,PhD tree breeding

3359 Indigenous Forest Management

1 Herbert JW - BSc,MSc,PhD *Research Field Leader* forest ecology and silviculture
2 Bergin DO - BSc,MSc silviculture
3 Burns BR - BSc,MSc,DipResManag forest inventory; forest ecology
4 Innes JG - BSc,MSc animal ecology
5 Leathwick JR - BForSc forest ecology
6 Shaw WB - BSc,MSc mapping special reserves; forest ecology

3360 Propagation and Early Growth

1 Menzies MI - BSc,BScFor,PhD *Research Field Leader* seedling quality and field performance; field evaluations of vegetative propagules
2 Aitken-Christie J - NZCS micropropagation of juvenile trees
3 Horgan KJ - BSc micropropagation of mature trees
4 Smith DR - BSc,PhD cell culture and embryogenesis

3361 Silvicultural Equipment Research Group

1 Trewin ARD *Group Leader* nursery and establishment options
2 Mason EG - BScFor young tree stability; cultivation equipment; biological assessment

3362 Plant Protection Chemistry

1 Zabkiewicz JA - BSc,PhD *Group Leader* biochemistry of herbicides and pheromones; pampas grass
2 Coker A - BSc biochemistry; nitrogen metabolism
3 Ray JW - BSc herbicide application technology

3363 Forest Management and Resources Division

1 Tustin JR - BSc,MA *Director*

3364 Economics and Social Science

1 Bourke IJ - DipBusAdmin,MAgrSc,PhD *Research Field Leader* marketing and international trade
2 Aldwell PHB - DipAg,DipTch,BEd,MA study leave
3 Horgan GP - BSc,MCom application of microeconomics to forestry issues
4 Katz A - BForSc,PhD trade studies
5 Maplesdon FM - BForSc strategic studies
6 Revington LJ - BA,MA regional impact; social aspects of development
7 Smith BNP - BSc,MPhil,PhD social impact of forestry development
8 Wilson PA - BSocSc applied economic issues

3365 Exotic Forest Management

1 James RN - BSc,BScFor,DPhil *Research Field Leader* special purpose species
2 Calderon SS - IngFor special purpose species mensuration
3 Carson MJ - BForSc,PhD silviculture and tree breeding
4 Deadman HM - BForSc special purpose species; eucalyptus spp silviculture
5 Evison DC - BA,BForSc agroforestry
6 Glass BP - BForSc study leave
7 Knowles RL - NZFSRangerCert leader Agroforestry Special Project; radiata silviculture
8 Maclaren JP - BA,BForSc agroforestry and special purpose species
9 Manley BR - BForSc,PhD radiata pine conversion studies, forest estate modelling
10 McGregor MJ - BSc,DipORS computer programming specialist; statistical analysis
11 Nicholas ID - BForSc special purpose species; black walnut; *Acacia* sp
12 Somerville AR - BForSc conversion planning; log quality assessment and resource evaluation
13 Threadgill JA - BSc forest estate modelling
14 Todoroki CL - BA computer programming; conversion studies
15 Whiteside ID - BSc,DipFor conversion planning; effect of log and tree quality characteristics on conversion and timber grades

3366 Forest Mensuration and Management Systems

1 Goulding CJ - BScFor,PhD *Research Field Leader* mensuration and management systems
2 Deadman MW - BSc,MSc forest management information system
3 Dunningham AG - BCom forest management information system
4 Garcia O - IngFor,MSc,PhD growth models; forest estate models; sawing simulation; biomathematics
5 Pilaar CH - BSc,BA forest management information system
6 Shula RG - BSFor mensuration of forest energy; growth modelling
7 Tennent RB - BForSc,PhD growth models; data collection strategy

3367 Harvest Planning

1 Murphy GE - BScFor,PhD *Research Field Leader* harvesting system analysis
2 Clement BC - BE,MA specialist computer assistance; computer-based harvest planning packages
3 Firth JG - BScFor,MS development of aerial photogrammetry as a planning and management tool; soil disturbance trials
4 Gilchrist A - IngFor,PhD development of harvesting simulation model and costing routines
5 Terlesk CJ Harvesting systems analysis; stand reorganisation trials
6 Twaddle A - BForSc value maximisation; log sorting and segregation; production studies of harvesting systems

3368 Soils and Site Amendment

1 Hunter IR - BA,MScFor,DipSocSc *Research Field Leader* economic evaluation of fertiliser response, particularly in established stands; site productivity relationships
2 Dyck WJ - BSF,MF effect of management practice on ground and stream-water quality; nutrient cycling studies
3 Gadgil RL - BSc,PhD biological nitrogen fixation; mycorrhizal suppression of litter decomposition
4 Knight PJ - BSc nutrient requirements of special purpose species; maintenance of fertility in forest nurseries
5 Skinner MF - BAgrSc,PhD silvicultural prescriptions for establishment of radiata pine

3369 Tree Physiology and Stand Productivity

1 Rook DA - BScFor,MSc,PhD *Research Field Leader* environmental factors affecting growth; CO_2 exchange
2 Beets PN - BSc,MForSc,PhD drymatter allocation of forest stands; effects of silviculture and nutrition; systems modelling
3 Grace JC - BSc,MSc,DPhil computer modelling; radiation interception
4 Whitehead D - BA,PhD water use by *P. radiata* stands; crown and canopy development; radiation interception and dry matter production by forest crops

3370 Wood Technology Division

1 Butcher JA - BSc,PhD *Director* timber, technology

3371 Wood Materials and Biotechnology

1 Mackie KL - BSc,PhD *Research Field Leader* wood pretreatment; chemistry
2 Burton RJ - BE study leave
3 Clark TA - BTech,PhD fermentation; wood pretreatment
4 Dawson BSW - BSc,PhD
5 Donaldson LA - BSc,MSc
6 Franich RA - MSc,PhD
7 MacFarlane PN - BTech,PhD effluent treatment
8 McDonald AG - BSc,DipSc,MSc
9 Singh AP - BSc,MSc,PhD electron microscopy
10 Suckling ID - BSc,PhD lignin chemistry; wood plastic components

3372 Wood Products

1 Plackett DV - BSc,PhD *Research Field Leader* surface coatings
2 Booker RE - BSc,MSc,PhD drying physics
3 Burton RJ - BE,PhD
4 Collins MJ - BE,MSc timber engineering, grading, standards
5 Lomax TD - BSc,MSc,PhD adhesives
6 Walford GB - BE,PhD timber engineering, grading, housing

3373 Pulp and Paper Research Organisation of New Zealand

1 Duffy GG - BSc,ASTC,PhD *Director* pulp suspension processing
2 Uprichard JM - BA,PhD *Research Field Leader* chemical pulping; wood chemistry
3 Allison RW - BSc,PhD chemical pulping
4 Corson SR - BE,DrIng,DipMgmt mechanical pulping
5 Fullerton TJ - BSc,MSc,PhD *Director* chemical pulping
6 Kibblewhite RP - BSc,MSc,PhD paper and fibre properties
7 Lloyd JA - BSc,MSc pulping; extractives
8 Miller PR - BE paper properties
9 Richardson JD - BE mechanical pulping

3374 Wood Processing

1 Cown DJ - BScFor,PhD *Research Field Leader* wood quality; saw milling
2 Hedley ME - BSc,PhD *Research Field Leader* preservative evaluations; fungicide
3 Archer KJ - BSc,MSc fungus/wood preservation interactions
4 Bergervoet AJ - BForSc study leave
5 Cox O - BA,BForSc wood processing economics
6 Doyle J - BSc,PhD quality control in sawmilling
7 Drysdale JA - DIC mycology; preservatives
8 Haslett AN - BForSc hardwood drying and utilisation
9 Miller WR - BE dryer design
10 van Wyk JL - BSc,BEng,BCom,MCom,DCom sawmill simulation
11 Vinden P - BSc,MSc,DIC,PhD treatment processes
12 Ward NH - BE computer simulations

3375 Forest Research Institute

P.O. Box 31-011, Christchurch
Tel: 517099. Telex: NZ4909. Cables: FRESTRA.

3376 Protection Forestry Division

1 Pearce AJ - BSc,MSc,PhD *Director*

3377 Forest Animal Ecology

1 Coleman JD - BSc,MSc,PhD *Research Field Leader* possum ecology and management
2 Batcheler CL - BSc,MA montane ecology; animal management
3 Challies CN - BSc,BScF,PhD deer management
4 Fraser KW - BSc deer census and management; recreational hunting
5 Fredric BJ - BSc,DipSc Computer analysis and modelling
6 Hickling GJ - BSc,PhD possum ecology and management
7 Morgan DR - BSc control and census methodology of possums
8 O'Reilly-Nugent G - BSc,MSc deer census; recreational hunting
9 Parkes JP - BSc,MSc ecology and management of goats and hares; hare biology
10 Spurr EB - BSc,PhD bird survey and ecology
11 Warburton B - BSc,MSc possum census and management; wallaby control

3378 Forest Land-use Impacts

1 Jackson RJ - BSc,PhD *Research Field Leader* forest land use hydrology; soil moisture
2 Fahey B - BA,MA,PhD soil erosion; forest practices
3 Kelliher FM - BSc,MS,PhD forest process modelling
4 Marden M - BSc,MSc,PhD erosion processes; terrain zoning
5 Phillips CJ - BSc,MSc study leave
6 Rowe LK - BSc,DipAgrSc stream water quality

3379 Forest and Grassland Ecology

1 Benecke U - BSCFor,MAgrSc,DrOecPubl *Research Field Leader* ecophysiology
2 Allen RB - BForSc study leave
3 Espie PR - BAgrSc grassland ecology; soil-plant relationships
4 Evans GR - BAgSc forest and grassland biomass studies
5 Hall GMJ - BSc,MCom computing, ecophysiology & vegetation analysis
6 Hollinger DY - BA,PhD forest physiological ecology
7 McCracken IJ - BSc,BA,PhD climate and tree growth physiology
8 Payton IJ - BSc,PhD autecology
9 Rose AB - BSc community ecology
10 Stewart GH - MSc,PhD stand ecology

3380 Forest Management and Productivity

1 Nordmeyer AH - BAgrSc,MAgrSc site improvement
2 Balneaves JN forest establishment and weed control
3 Davis MR - BAgrSc,MAgrSc legumes; plant nutrition; legume/tree interaction

4 Franklin DA - BSc *Research Field Leader* silviculture and management
5 Kelland CM - BAgrSc soil nutrients; soil development; root biomass
6 Ledgard NJ - BSc,MSc site stabilisation; tree productivity; silviculture

3381 Fisheries Research Division

P.O. Box 297, Wellington
Tel: 861029.

The Division's work concerns the assesment of living aquatic resources and their habitats, the provision of advice on options for management, and research on stock enhancement and aquaculture. Marine staff are not listed here.

1 Allen RL - PhD *Director*
2 Colman JA - PhD *Assistant-director (marine)*
3 McKoy JL - PhD *Assistant-director (marine)*
4 Beardsell MF - PhD *Scientific liaison officer*
5 Williams Miss JC - NZLAC *Librarian*

3382 Freshwater Fisheries Section
P.O. Box 8324, Riccarton, Christchurch
Tel: 488939.

1 McDowall RM - PhD *Assistant-director (freshwater)* native fishes; systematics

3383 Salmon Research

1 Todd PR - PhD *Leader* salmon; eels; native fish
2 Field-Dodgson MS - MSc salmon culture
3 Flain M - BSc *salmon and trout; rivers and lakes*
4 Hopkins CL - MA salmon; fish production
5 Unwin MJ - MSc salmon; fisheries resources

3384 Freshwater Resource Assessment & Research

1 Graynoth E - BSc *Leader* recreational fisheries; effects of power development schemes
2 Davis Ms SF - BSc fisheries resources; effects of water abstraction
3 Glova GJ - PhD effects of water abstraction
4 Jellyman DJ - PhD effects of water abstraction; eels; native fish
5 Jowett IP - BE environmental simulation modelling; hydrology
6 Sagar PM - MSc low flow effects; benthic fauna

3385 Freshwater Resource Assessment & Research, Wellington
P.O. Box 297, Wellington
Tel: 861029.

1 Tierney Ms LD - BSc *Leader* fisheries resources
2 Hicks BJ - MSc fisheries resources, effects of water abstraction
3 Richardson Ms J - BSc fisheries resources; effects of water abstraction

3386 Freshwater Resource Assessment & Research, Rotorua
P.O. Box 951, Rotorua
Tel: 479579.

1 McCarter NH - MSc *Leader* recreational fisheries
2 Hayes J - PhD effects of mining
3 Mitchell CP - MSc native fish; grass carp
4 Rowe DK - MSc recreational fisheries

3387 Freshwater Resource Assessment & Research, Oamaru
P.O. Box 6, Oamaru
Tel: 48248.

1 James GD - MSc *Leader* effects of power development schemes

3388 Department of Scientific and Industrial Research
PO Box 1578, Wellington
Tel: 729979. Telex: 3276 RESEARCH. Cables: RESEARCH.

3389 Biotechnology Division
DSIR, Private Bag, Palmerston North
Tel: 68019. Telex: NZ31285 Palmsir. Cables: Biochem, Palmerston North.

1 Porter NG - BSc,PhD *Officer in Charge* physiology of crop plants
2 Robertson JG - MSc,PhD *Director*
3 Shaw GJ - BSc,PhD *Programme Leader (Food chemistry)* mass spectrometry of organic compounds
4 Reay PF - BSc,PhD *Assistant Director* growth and development of plants
5 Russell GB - BSc,PhD *Assistant Director* chemistry and role of natural products
6 Ulyatt MJ - MAgrSci,PhD *Assistant Director* ruminant digestion and animal nutrition
7 Asmundson RV - BA,MSc,PhD anaerobic microbial metabolism
8 Biggs DR - BSc,PhD biochemistry of natural products

9 Birch EJ - BSc,PhD organic chemistry
10 Body DR - MSc lipid composition and chemistry
11 Broad TE - BVSc,MSc,PhD *Programme Leader (Animal gene technology)* animal cell culture and genetic studies
12 Burlingame BA - BS *Programme Leader (Food composition and nutritional information)* nutritional information
13 Dellow DW - BSc,PhD animal digestion
14 Dommisse EM - BSc plant cell manipulation
15 Forrest JW - BSc,PhD gene mapping
16 Grace ND - MAgrSc mineral nutrition
17 Grant MR - BSc plant molecular biology
18 Greenwood DR - BSc,PhD biochemistry of plant insect interactions
19 Gurnsey MP - BSc,DipSci controlled release of bioactive agents
20 Harris PM - MHSc,PhD *Programme Leader (Animal nutrition and physiology)* animal tissue metabolism
21 James KAC - MSc,PhD monogastric nutrition
22 Joblin KN - MSc,PhD *Programme Leader (Applied microbiology)* minerals and microbial metabolism
23 John A - MS,PhD ruminant digestion and animal nutrition
24 Johnson CB - MSc,PhD lipid composition and chemistry
25 Jones WT - BSc,PhD *Programme Leader (Immunology and antibody production)* protein chemistry and monoclonal antibodies
26 Kelly WJ - BSc,PhD genetics of anaerobic bacteria
27 Lane GA - BSc,PhD *Programme Leader (Biochemistry of plant and insect interactions)* chemistry and role of natural products
28 Lee J - MSc,PhD minerals in plants
29 Monro JA - BSc,PhD carbohydrate chemistry
30 Patel JJ - MSc,PhD rhizobium phages
31 Pearce PD - MSc,PhD oligonucleotide synthesis
32 Reynolds PHS - MSc,PhD *Programme Leader (Plant gene expression and manipulation)* enzymology
33 Rowan DD - BSc,PhD chemistry and role of natural products
34 Stewart Scott I - MSc,PhD animal cytogenesis
35 Tapper BA - BSc,PhD plant secondary metabolite biosynthesis
36 Visser FR - MSc,PhD food chemistry
37 Vlieg P - BSc,PhD fish composition
38 Waghorn GC - BAgrSc,MAgrSci,PhD ruminant digestion and animal nutrition
39 Webster DW - BSc,NZCS computing science
40 Welch RAS - BAgr,BSc,PhD *Programme Leader (Controlled release technology)* controlled release of bioactive agents

3390 Biochemical Processing Centre

1 Boland MJ - BSc,PhD *Officer in Charge* biochemical processing
2 Frude MJ - BSc,BE biochemical engineering
3 Kennedy LD - MSc,PhD biochemical processing
4 Otter DE - MSc,PhD biochemistry
5 Rodriguez SB - BA,PhD biochemistry of fermentation

3391 Electron Microscope Laboratory

1 Craig AS - NZCS,BSc,PhD quantitative electron microscopy

3392 Auckland Industrial Development Division
P.O. Box 2225, Auckland
Tel: 34-116. Telex: MARC 21623.

1 Beach AD - MIP application of electronic techniques to meat industry
2 Billing DP - NZCS ethylene control in storage of produce
3 Clist RS - PhD computer imaging techniques for fruit measurement
4 Finch GR - NZCS in transit fruit damage
5 McDonald B - NZCS cooling, storage and transport of produce
6 Penman DW - ME computer imaging in horticultural grading
7 Simpson JN - MIMechE mechanical systems for horticultural grading

3393 Botany Division, DSIR
Lincoln, Private Bag, Christchurch
Tel: 252-511. Telex: LINSIR NZ4703. Cables: Plantlab Christchurch.

1 Harris W - BSc,PhD *Director* grassland ecology; genecology
2 Wardle P - MSc,PhD,FRSNZ *Assistant Director, Ecology Group Leader* book on NZ vegetation
3 Thomson AD - MSc,PhD plant science history; biographies; bibliographies; plant virology

3394 Ecology Section

1 Meurk CD - BSc,PhD ecology and management of native grasslands; ecology and biogeography of subantarctic and mountain regions
2 Molloy BPJ - MSc,PhD,DipAgr quaternary and fire history of vegetation; production of field guides; taxonomy of several groups; floristic and vegetation analyses
3 Partridge TR - BSc,PhD vegetation succession; quantitative ecology; estuaries and wetlands

3395 Palynology and Morphology Section

1 McGlone MS - MSc,PhD *Biosystematics Group Leader (Palynology and Morphology plus Taxonomy Sections)* quaternary botany; paleoecology; climate history; biogeography
2 Patel RN - BSc,MSc,DipFor wood anatomy
3 Philipson Mrs MN - MSc,PhD ultrastructure of plant reproduction and secretion; plant endophyte relationship; plant embryology

3396 Taxonomy Section

1 Edgar Miss E - BA,MSc,FLS monocotyledons especially grasses
2 Fife AJ - MSc,PhD taxonomy of cryptogams, especially mosses, liverworts and lichens
3 Garnock-Jones PJ - BSc,PhD Caryophyllaceae, Asteraceae, Brassicaceae, Ranunculaceae and Hebe
4 Given DR - BSc,PhD,FLS Asteraceae, Rosaceae and Pteridophyta;endangered species; island and arctic-alpine floras; geothermal floras
5 Macmillan Miss BH - NZCS *Assistant Herbarium Keeper* taxonomy especially Acaena and bryophytes
6 Parsons MJ - MSc,PhD *Herbarium Keeper* New Zealand marine macro-algae
7 Sykes WR - BSc,NDH naturalised and cultivated dicotyledons; introduced gymnosperms; flora of Pacific Islands
8 Webb CJ - BA,BSc,PhD *Section Leader* reproductive biology naturalised dicotyledon flora; taxonomy of Asteraceae, Apiaceae, native Fabaceae

SUBSTATIONS

3397 Kaikohe

P.O. Box 194, Kaikohe
Tel: 566. Telex: MARC NZ21623. Cables: Plantlab.

1 Clunie NMU - BSc,MSc,PhD reserve surveys in Northland; botanical inventory and resource studies; ecology of scrublands and forests

3398 Auckland

DSIR, Private Bag, Auckland
Tel: 893-660. Telex: MARC NZ21623. Cables: Plantlab.

1 Esler AE - MAgrSc wild plants of urban Auckland

3399 Rotorua

C/-Forestry Research Institute, Private Bag, Rotorua
Tel: 475-899. Telex: FRESTRA NZ21080. Cables: Plantlab.

1 Clarkson BD - MSc,PhD reserve surveys in Central North Island; succession on volcanic soils; late Quaternary vegetation history of Taranaki; Egmont National Park survey

3400 Havelock North

DSIR, Private Bag, Havelock North
Tel: 778-196.

1 Walls GY - BSc ecology of islands; plant/animal relationships; vegetation survey and conservation

3401 Wellington

C/-Soil Bureau,DSIR, Private Bag, Lower Hutt
Tel: 673-119. Telex: PHYSICS NZ3814. Cables: Plantlab.

1 Atkinson IAE - MSc,PhD *North Island Ecology Section Leader* soil processes; ecological processes in Tongariro National Park; ecology of islands; plant/animal relationships
2 Kelly GC - MSc survey of scenic reserves in Nelson, North Cape and on Chatham Islands

3402 Nelson

DSIR, Private Bag, Nelson
Tel: 81082.

1 Williams PA - MSc,PhD *South Island Ecology Section Leader* vegetation succession; scrub weeds; vegetation of Marlborough, N.W. Nelson

3403 Dunedin

DSIR, Private Bag, Dunedin
Tel: 774-050.

1 Allen RB - BSc,DipSc,PhD ecology of forest vegetation of Otago and Southland; reserve surveys
2 Johnson PN - BSc,DipSc,PhD wetland plants; sand dunes; flora and vegetation of Otago and Southland
3 Lee WG - BSc,DipSc,PhD soil/plant relationships, especially on ultramafic areas; animal/plant relationships; Southland Reserve Survey

3404 Chemistry Division

Private Bag, Petone
Tel: 666-919. Cables: PHYSICS NZ3814.

1 Leary GJ - MSc,PhD,FNZIC *Director* lignin chemistry

2 Crump DR - MSc,PhD mammalian pheromone chemistry
3 Foo LY - MSc,PhD tannin chemistry
4 Furneaux RH - MSc,PhD synthesis of plant growth regulators, algal polysaccharides
5 Hemmingson JA - MSc,PhD lignin chemistry
6 Markham KR - MSc,PhD,FRSNZ,FNZIC chemotaxonomy and flavonoid chemistry
7 Morgan KR - BSc,PhD NMR studies on wood
8 Porter LJ - MSc,PhD,FNZIC tannin chemistry
9 Tyler PC - BSc,PhD synthesis of plant growth regulators

3405 Crop Research Division

DSIR, Private Bag, Christchurch
Tel: Chch 252-511. Telex: NZ4703.

1 Dunbier MW - MAgrSc,PhD *Director*
2 Wilson DR - BAgrSc,PhD physiology of crop growth
3 Wratt GS - BSc *Senior Advisory Officer*

3406 Biological Research and Support

1 Wynn-Williams RB - MAgrSc,PhD *Group Leader* lucerne, crop evaluation
2 Admiraal JV - BAgrSc cereal agronomy
3 Cross RJ - MAgrSc crop genetic resources
4 Jamieson PD - BSc,PhD agrometeorology
5 Sturrock JW - BSc,DipAgr,PhD agrometeorology

3407 Food and Biotechnology/Commercial Development

1 Sanderson FR - MAgrSc,PhD *Group Leader* epidemiology of cereal diseases
2 Ashby JW - BSc,PhD *Group Leader* science management/industry liaison
3 Bezar HJ - BAgrSc Scientific Liaison Officer
4 Christey MC - BAgrSc novel tissue culture techniques for plant breeding
5 Conner AJ - BSc,PhD novel tissue culture techniques for plant breeding
6 Dommissee EM - BSc novel tissue culture techniques for plant breeding
7 Ferguson JD - BSc,PhD novel tissue culture techniques for plant breeding
8 Grant JE - BSc,PhD transformation systems for grain legumes
9 Harvey WI - BTech(Food) food technology
10 Lammerink J - IrAgr herbs, garlic, brassicas
11 Lancaster JE - BSc flavour analysis, onions
12 Mann JD - MSc,PhD analysis of food chemicals
13 Palmer JA - BScHort new crops, tree crops
14 Pickering RA - MSc wide hybrids in cereals
15 Porter NG - BSc,PhD analysis of food chemicals
16 Quazi MH - MScAgr,PhD plant genetics

3408 Plant Breeding and Selection

1 Jermyn WA - MAgrSc,PhD *Group Leader* field pea breeding
2 Coles GD - BSc barley breeding
3 Falloon PG - BAgrSc,PhD asparagus breeding
4 Genet RA - MHortSc potato breeding
5 Griffin WB - BSc,PhD wheat breeding
6 Lammerink JP - BFoodTech food technology
7 Lewis IK - DipHortSc,MSc,PhD strawberry breeding
8 Scott RE - DipAgr,MAgrSc garden pea breeding
9 Wright DSC - PhD wheat, triticale and oat breeding

3409 Pukekohe

Cronins Road, RD1, Pukekohe
Tel: 86 119.

1 Anderson JAD - MAgrSc *Officer in Charge* potato breeding
2 Bansal RK - BSc,PhD maize and sweetcorn breeding
3 Grant DG - MHortSc curcubit and onion breeding

3410 Palmerston North

DSIR, Private Bag, Palmerston North
Tel: 68 019. Telex: 31 285.

1 McEwan JM - MSc,PhD *Officer in Charge* cereal breeding, sprouting in wheat

3411 Blenheim

Marlborough Research Centre,DSIR, Private Bag, Blenheim
Tel: 89 059.

1 Malone MT - MHortSc vegetable breeding and agronomy

3412 Riwaka

DSIR, Old Mill Road, RD3, Motueka
Tel: 89 106.

1 Alspach PA - BSc,MSc biometrics
2 Beatson RA - BAgrSc,MAgrSc,PhD breeding hops, tobacco, apricots
3 Hall HK - MSc hybrid berry and raspberry breeding
4 Smale PE - BHortSc,MHortSc *Officer in Charge* berry fruit; tree crop culture

3413 Gore

DSIR, Private Bag, Gore
Tel: 84 015.

1 Armstrong KW *Officer in Charge* cereal breeding
2 Gowers S - BSc,PhD brassica breeding

3414 Ecology Division

Private Bag, Lower Hutt
Tel: WN 694 859.

1 Crawley MC - PhD *Director*
2 Brockie RE - PhD forest ecology, ecology of possums
3 Campbell DJ - MSc forest vegetation, plant ecology
4 Cockrem JF - PhD reproductive physiology of birds
5 Cowan PE - PhD behaviour and physiology of possums
6 Daniel MJ - MSc *Section Leader* forest ecosystems, bats
7 Efford MG - PhD population dynamics, possums
8 Fitzgerald BM - PhD predatory prey dynamics of rodents and feral cats
9 Flux JEC - PhD population studies of starlings
10 Hay JR - PhD environmental comment, bird ecology
11 Pritchard AD - BSc scientific editor
12 Robertson HA - PhD bird ecology
13 Rudge MR - PhD *Deputy Director, Business Manager* agricultural ecology, land use, goats

3415 Havelock North

Ecology Division, Private Bag, Havelock North
Tel: HS 778 196.

1 Langham NPE - PhD feral cat ecology
2 McLennan JA - PhD brown kiwi ecology and conservation

3416 Nelson

Ecology Division, Private Bag, Nelson
Tel: 81081.

1 Clout MN - PhD *Section Leader* forest ecology, coadaptation of frugivorous and podocarp birds
2 Moller H - PhD forest ecology, foraging strategy of nectar feeding birds
3 Taylor RH forest ecology, population dynamics, penguins, conservation of parakeets
4 Wilson PR - PhD forest ecology, ecology and conservation of kaka

3417 Entomology Division

120 Mt Albert Road, DSIR, Private Bag, Auckland
Tel: 893660. Telex: NZ21623. Cables: Entlab. 863-330.

1 Longworth JF - MSc *Director* Auckland
2 Rhodes Ms WB *Scientist, policy and planning* Auckland

3418 Agriculture Section

1 Sutherland ORW - BSc,PhD *Head of Section* insect behaviour; plant resistance mechanisms; Auckland
2 Burgess Ms EPJ - MSc sensory physiology and behaviour of pasture pests: Auckland
3 Donovan BJ - MSc,PhD pollination of lucerne and horticultural crops, systematics of native bees, biological control of German wasp
4 Dymock Ms J - BSc plant resistance to pasture pests: Palmerston North
5 Farrell JAK - BSc,PhD cereal aphid damage and control; host plant relationships and virus transmission: Lincoln
6 Gnanasunderam Ms C - BSc chemistry of insect/host plant interaction; insect defence secretions: Auckland
7 Hill RH - BSc,PhD biological control of gorse: Lincoln
8 Macfarlane RP - MAgrSc,PhD pollination of field and horticulture crops by *Bombus* spp.; systematics of Asilinae: Lincoln
9 McGregor PG - BSc insect pests of cereals: Palmerston North
10 Pearson WD - BSA,MSc,PhD pests affecting legume seed production: Lincoln
11 Syrett Ms P - MAgrSc biological control of broom, ragwort, and St John's wort: Lincoln

3419 Horticulture Section

1 Wearing CH - BSc,PhD *Head of Section* integrated pest management implementation; postharvest disinfestation: Auckland
2 Batchelor TA - MSc postharvest disinfestation, irradiation (overseas until 1988)
3 Cameron PJ - BSc,PhD biological control, integrated control; lucerne aphids and noctuids in field crops: Auckland
4 Charles JC - MSc integrated pest control in berry fruit and kiwifruit: Auckland
5 Clearwater JR - MSc,PhD identification and field application of insect pheromones: Auckland
6 Dentener PR - Ir postharvest disinfestation; irradiation: Auckland
7 Ferguson AM - MAgrSc kiwifruit pollination; control of vegetable pests: Lincoln
8 Foster SP - MSc,PhD postharvest disinfestation; insect pheromone chemistry, and biosynthesis: Auckland

9 Hayes Ms AJ - PhD modelling predator/prey mite interactions: Auckland
10 Hill MG - BSc,PhD biological and chemical control of greedy scale on kiwifruit; phenological modelling: Auckland
11 Markwick Mrs NP - BSc,PhD insecticide resistance in predatory mites: Auckland
12 Martin NA - BSc,PhD integrated control of greenhouse crop pests; earthworms: Auckland
13 McLaren Mrs GF - MHort,PhD integrated control of stone fruit pests: Clyde
14 Singh P - BSc,AIRI,PhD insect rearing management, dietetics, nutritional physiology. Auckland
15 Steven D - BSc,PhD ecology and control of kiwifruit pests; insect dispersal: Auckland
16 Suckling DM - BSc,DipBioTech,PhD leafroller insecticide resistance and resistance management; pheromones: Lincoln
17 Thomas WP - MSc *Scientist in Charge, Lincoln* integrated control in berry and pip fruit; ecology and control of leafrollers: Lincoln
18 Wadell Ms BC - MSc postharvest disinfestation: Auckland
19 Walker JTS - MHortSc,PhD integrated control of pip and stone fruit pests; acarology: Havelock North

3420 Insect Pathology Section

1 Scotti PD - BSc,PhD *Head of Section* small RNA viruses of insects; virus diagnostic methods; cell culture: Auckland
2 Anderson DL - PhD honeybee diseases: Auckland
3 Chilcott CN - PhD *Bacillus thuringiensis*: Auckland
4 Crawford AM - BSc,PhD baculoviruses; molecular biology; cell culture: Auckland
5 Grandison GS - BSc,PhD plant parasitic nematodes of economic crops: Auckland
6 Malone Ms LA - MSc,PhD insect protozoa and microsporidia: Auckland
7 Marshall JW - MSc,PhD plant parasitic nematodes of South Island economic crops: Lincoln
8 Mercer CF - BAgrSc,MSc plant parasitic nematodes of pasture: Palmerston North
9 Wigley PJ - MAgrSc,DPhil ecology of insect pathogens; *Bacillus thuringiensis*; general diagnosis: Auckland
10 Wouts WM - Ir,PhD biological control of insect pests by nematodes; taxonomy of plant and insect parasitic nematodes: Auckland

3421 Systematics Section

1 Holloway Ms BA - MSc,PhD *Head of Section* systematics of higher Diptera; Anthribidae and Lucanidae (Coleoptera): Auckland
2 Craw RC - BSc,PhD systematics of Coleoptera (Curculionodea) and Lepidoptera (Geometridae): Auckland
3 Crosby TK - BSc,PhD *Curator* systematics of aquatic insects and Nematocera: Auckland
4 Dugdale JS - BSc systematics of Lepidoptera; Tachinidae; Cicadidae: Auckland
5 Duval CT - BSc *Editor* 'Fauna of New Zealand' series: Auckland
6 Maddison PA - BSc,PhD Pacific Island insect pests and other fauna: Auckland
7 Morales Ms CF - MSc systematics of Hemiptera: Auckland
8 Ramsay GW - MSc,PhD systematics of Acari and Orthoptera: Auckland

3422 Grasslands Division

Private Bag, Palmerston North
Tel: 68-019. Telex: PALMSIR 31285. Cables: Grasslands. 61130.

1 Lancashire JA - BSc,MAgrSc *Director*
2 Baker MJ - BSc *Scientific Editor*
3 Gaynor DL - MAgrSc *Commercial Manager*
4 Watt PC - BA,DipBusAdmin *Scientific Liaison Officer*

3423 Agronomy

1 Hay RJM - MAgrSc *Leader* pasture management, species performance
2 Charlton JFL - BSc,PhD legume establishment on hill country, legume evaluations
3 Forde Mrs MB - MSc,PhD biosystematics, taxonomy and plant introductions
4 Hume DE - BAgrSc pasture species evaluation
5 Hunt WF - MAgrSc,PhD physiology of pasture production

3424 Pasture Ecology

1 Field TRO - MSc,PhD *Leader* physiology of pasture production
2 Brock JL - MSc legume growth and nitrogen fixation in the field
3 Campbell BD - BAgrSc ecophysiology of pastures
4 Keogh RG - BSc pasture and animal ecology

3425 Plant Breeding

1 Rumball W - MSc,PhD *Leader* cocksfoot, browntop, phalaris, *Bromus*, amenity grasses, sainfoin, red clover
2 Burgess RE - MSc,PhD ryegrass, *Lotus* spp
3 Caradus JR - MSc,PhD low phosphorus tolerance in white clover, white clover root development

4 Easton HS - MAgrSc,DresSciNat lucerne, *Festuca*, ryegrasses
5 Van den Bosch J - BAgrSc white clover pests and diseases, plant introductions
6 Woodfield DR - BAgrSc dryland white clover and other legumes

3426 Molecular Genetics

1 Ronson CW - BSc,PhD *Rhizobium* genetics
2 Sanders PM - BSc plant molecular genetics
3 Scott Miss AG - BSc tissue and cell culture
4 White DWR - BSc,PhD *Leader* somatic cell genetics, tissue culture

3427 Plant Nutrition

1 Dunlop J - MSc,PhD *Leader* physiology of salt uptake by plants
2 Hart AL - BSc,PhD phosphorus uptake and transport in plants
3 Hay MJM - MAgrSc competitive relationships in plant
4 Pritchard MW - MSc analytical chemistry and nutrient speciation in plants and soil

3428 Soil Fertility

1 Crush JR - MSc,PhD *Leader* nitrogen fixation
2 Ball PR - MAgrSc,PhD nitrogen economy of grazed pastures
3 Carron RA soil science and nutrient cycling

3429 Turf Research

1 Evans PS - MSc,PhD *Leader* turf management
2 Murphy JW - MAgrSc turf management

3430 Aorangi (Intensive flat land Farm)

1 Cosgrove GP - MAgrSc pasture ecology and animal/plant interactions
2 Clark DA - MAgrSc *Leader* pasture utilisation and management

3431 Ballantrae (Hill Country Farm)

1 Lambert MG - MAgrSc *Leader* hill pasture ecology, nutrient cycling
2 Barker DJ - MAgrSc pasture production and evaluation, dry hill country
3 Betteridge K - MAgrSc hill country goat research and phosphate nutrition of dairy cows
4 Chapman DF - BAgrSc hill pasture ecology
5 McKay AD - BAgrSc,PhD soil fertility

3432 Seed Production

1 Rolston MP - MAgrSc,PhD *Leader* herbage seed production, weed ecology and control
2 Hare MD - DipVFM,MArgSc herbage seed production
3 Kevern ER - DipVFM multiplication and distribution of certified breeders seed

3433 Kaikohe

P.O. Box 194, Kaikohe

1 Rumball PJ - MAgrSc pasture and animal production
2 Andrews WGK - MAgrSc *Leader* nitrogen cycling, autumn lambing
3 Richardson MA - BAgrSc soil science and livestock interactions
4 Woods PW - MAgrSc dry hill country pasture ecology

3434 Lincoln

Private Bag, Christchurch

1 Scott D - MSc,PhD *Leader* ecology, tussock grasslands
2 Brown KR - BSc grass seed production
3 Clifford PTP clover seed production
4 Hoglund JH - MAgrSc nitrogen cycling
5 Pollock KM - BSc,PhD high country pasture evaluations

3435 Gore

Private Bag, Gore

1 Stevens DR - BAgrSc pasture ecology, dry hill country
2 Widdup KH - MSc *Officer in Charge* plant breeding

3436 Division of Horticulture and Processing

DSIR, Private Bag, Auckland

1 Bieleski RL - MSc,PhD,FRSNZ *Director*
2 Burns DJW - BSc,PhD *Assistant Director* food processing
3 Davison RM - MSc,PhD *Assistant Director* kiwifruit; plant growth substances

3437 Pomology

1 Campbell JM tropical agriculture; consultancies

2 Glucina PG - MSc stone fruits; persimmons
3 Lowe RG - BSc kiwifruit pomology
4 McArtney SJ - MHortSc apple pomology
5 McLaren GF - MHortSc,PhD stonefruit pomology, breeding
6 Patterson KJ - MSc,PhD physiology and crop management, feijoas
7 Popenoe J - MSc,PhD apple and kiwifruit pomology;growth regulators
8 Snelgar WP - MSc,PhD physiology of kiwifruit
9 Tustin DS - BHortSc,PhD apple and nashi pomology
10 Volz RK - BSc kiwifruit and apple pomology

3438 Fruit Breeding

1 Atkinson RG - MSc apple molecular genetics and tissue culture
2 Fraser LG - MSc,PhD kiwifruit tissue culture and molecular genetics
3 Hammett KRW - BSc,PhD pepino breeding; plant variety rights
4 McNeilage MA - BSc,PhD kiwifruit breeding and genetics
5 Pringle GJ - BSc tamarillo genetics and breeding
6 Seal AG - BSc,PhD *Section Head* fruit breeding and genetics
7 White AG - BHortSc pipfruit breeding

3439 Postharvest Physiology

1 Hopkirk G - MHSc,PhD fruit physiology and storage, kiwifruit and avocado
2 Lallu N - BSc,PhD fruit physiology, growth, development and storage, kiwifruit and ethylene; Asian Pear storage (overseas)
3 Lay Yee M - BHort,PhD fruit fumigation, molecular biology of fruit
4 MacRae EA - BSc,PhD fruit physiology and storage; persimmon storage; chilling injury
5 Watkins CB - MSc,PhD *Section Head* fruit physiology and storage, pip fruits

3440 Horticultural Processing

1 Heatherbell DA - BSc,PhD *Section Head* juice technology, wine research
2 Lodge N - MSc (overseas secondment, Singapore)
3 McMath KL - BHSc sensory evaluation of foods
4 Perera CO - MSc,PhD fruit flavours, fruit processing
5 Wong M - BTech chemistry of fruit processing

3441 Fish Technology

1 Wong RJ - MSc,PhD *Section Head* chemistry of fish products
2 Fletcher GC - BSc microbiology of fish products
3 Jerrett AR - BSc fish physiology
4 McDonald GA - BTech surimi, fish processing
5 Ryder JM - BSc,PhD fish processing, fish biochemistry
6 Scott DN - BTech (Food) fish quality and storage
7 Wilson NDC - BSc,DipFoodTech fish processing (located at Nelson - P.O. Box 5114, Port Nelson, Nelson)

3442 Horticultural Physiology

1 Beever DJ - MSc,PhD postharvest pathology; storage physiology
2 Ferguson AR - MSc,PhD biology of the genus *Actinidia*; history of plant domestication; plant nutrition
3 Ferguson IB - BSc,PhD calcium physiology; senescence and low temperature physiology; nutrient uptake and transport
4 Marsh KB - MPhil,PhD plant nutrition
5 Mitchell RE - BSc,PhD toxins of phytopathogenic bacteria
6 Roughan PG - MSc,PhD *Section Head* plant lipid biochemistry; chilling sensitivity in plants
7 Young H - BSc,PhD flavour and aroma chemistry; chemistry of biologically active compounds; gas chromatography and mass spectrometry
8 White AG - BHortSc *Section Head* pipfruit breeding (located at Havelock North)

3443 Riwaka Research Station

DSIR, R.D. 3, Motueka, Nelson

1 Smale PE - MHortSc *Section Head* berryfruit
2 Alspach PA - BSc,BPhil agronomy, annual food crops; biometrics
3 Beatson RA - PhD plant breeder, tobacco, hops
4 Rohrbach JF - BChemEng processing

3444 Industrial Processing Division

Private Bag, Petone
Tel: (04) 690-441. Telex: NZ 31678. Cables: CHEMDIV, Petone. (04) 690-132.

1 Beanland E - MSc *Director*

3445 Chemical Engineering

1 Smith DKW - BE,PhD *Group Leader* horticultural processing, natural product processing

3446 Biochemical Engineering

1 Davies RJ - PhD *Section Leader* fermentation, natural products processing

3447 Division of Water Sciences

Division created October 1988. Details or research staff not available at time of going to press.

3448 Division of Land and Soil Sciences

Private Bag, Lower Hutt
Tel: 673-119. Telex: PHYSICS NZ 3814. Cables: SOBUR, Wellington.

1 Milne JDG - BSc,MSc,PhD *Director* science management, loess stratigraphy; urban soil surveys; suitability of land for viticulture
2 Wells N - BScSp *Deputy Director* industrial use of soil clays

3449 Agronomy

1 Widdowson JP - MAgrSc,PhD *Section Leader* available sulphur, fertility of Pacific Island soils, aluminium toxicity, land restoration
2 Hart PBS - MAgrSc,PhD maintenance of available nitrogen; land restoration; organic-matter turnover
3 Hume LJ - BAgrSc,MSc fertility of Pacific Island soils; nutrition of citrus; nitrogen fixation in legumes; aluminium toxicity

3450 Soil Analysis

1 Lee R - BSc,PhD *Section Leader* major nutrient studies; pedological processes, aluminium in soils, exchange processes in soils
2 Searle PL - LRIC,MSc method development; instrumentation; sulphur in soils

3451 Soil Biology and Biochemistry

1 Tate KR - MSc,PhD,FNZIC *Section Leader* soil organic matter; carbon and major nutrient cycling; microbial biomass; humic substances
2 McColl Miss HP - MSc earthworms and nutrient cycling under pasture; staphylinid beetles
3 Orchard Miss VA - BSc,PhD microbiology; microbial ecology, actinomycetes; microbial activity
4 Ross DJ - MSc,PhD,DipBact,FNZIC *guest worker* microbial biomass; soil enzymes; nitrogen availability
5 Sparling GP - BSc,PhD soil microbial biomass; nitrogen and phosphorus cycling; soil organic matter; microbial activity
6 Speir TW - MSc,PhD soil enzymes and plant nutrient supply; microbial biomass; phosphorus and sulphur, waste utilisation and disposal on to soil
7 West AW - PhD microbial biomass and activity
8 Yeates GW - BSc,PhD,DSc biological and agricultural significance of nematodes

3452 Soil Chemistry and Mineralogy

1 Childs CW - BSc,PhD,FNZIC *Section Leader* mineralogy and chemistry of iron in soils; Moessbauer spectroscopy
2 Churchman GJ - BSc,PhD,FNZIC clay minerals; soil physical properties
3 Claridge GGC - MSc,PhD clay minerals; Antarctic and tropical soils; clays for ceramics; electron microscopy
4 Eden DN - BSc,MPhil,PhD sand/silt minerals; loess
5 Parfitt RL - BSc,MS,PhD plant soils interactions; soil formation; surface chemistry
6 Theng BKG - BAgrSc,PhD,FNZIC clay-organic interactions, surface chemistry

3453 Soil Engineering and Physics

1 Thomas RF *Section Leader* soil physical and engineering properties; soil classification
2 Cook FJ - BSc,DipAgrSc,MPhil flow properties of fluids through porous media
3 Jacquet D - MSc,PhD engineering properties of volcanic ash soils
4 McQueen DJ - MSc soil structure, soil compaction; topsoil removal and replacement. land rehabilitation
5 Miller NG - BSc soil strength characteristics; computer applications
6 Ross CW - BAgrSc,PhD conservation tillage; land restoration; turf culture; soil structure;soil compaction
7 Schafer GJ - MSc database entry
8 Stephenson WR - BSc dynamic properties of soils; computer applications; electronics;earthquakes
9 Watt JPC - BAgrSc,MS field studies on soil water; irrigation

3454 Scientific Services

1 Furkert RJ - JP,MSc,PhD,FNZIC *Science Manager, Central*
2 Davin Mrs JE - BSc,DipNZLS *Librarian in Charge*
3 Giltrap DJ - MSc,DPhil computer techniques in soil survey
4 Isaacs DC - BAgrSc scientific editing
5 Kinloch DI - MSc soil classification; scientific services

6 Russell PC - BHortSc scientific editing

3455 Soil Resources

1 Clayden B - BSc *Chief Pedologist* soil classification, soil survey methods
2 Bruce JG - MSc *Chief Soil Correlator* soil correlation and classification; loess
3 Leamy ML volcanic ash soils; soil classification

SUBSTATIONS

3456 Hamilton

N.Z. Soil Bureau, DSIR, C/-Ruakura Agricultural Research Centre, Private Bag, Hamilton
Tel: 63166.

1 Orbell GE - MSc *District Soil Scientist* Waikato soils, Pacific Island soils
2 McLeod M - BSc soils of Hauraki lowlands, Pacific Island soils
3 Singleton PL - MSc Waikato soils, land restoration after mining

3457 Rotorua

N.Z. Soil Bureau, DSIR, C/-Forest Research Institute, Private Bag, Rotorua
Tel: 475-899. Telex: FRESTRA NZ 21080. Cables: FRESTRA, Rotorua.

1 Rijkse WC - Ing *District Soil Scientist* soil survey, yellow brown pumice soils, forest soils

3458 New Plymouth

N.Z. Soil Bureau, DSIR, P.O. Box 4043, New Plymouth East
Tel: 803585.

1 Purdie BR - BSc *District Soil Scientist* soils of Taranaki; soils of Manukau City; soil taxonomy
2 Palmer RWP - MSc soils of Taranaki; soils of South Westland

3459 Havelock North

N.Z. Soil Bureau, DSIR, Private Bag, Havelock North
Tel: 778-196 Napier.

1 Griffiths E - BSc *District Soil Scientist* soils of Hawke's Bay; irrigation surveys
2 McLaughlin B - BA,MPhil soils of Heretaunga Plains

3460 Palmerston North

N.Z. Soil Bureau, DSIR, Private Bag, Palmerston North
Tel: 68019. Telex: PALMSIR NZ 31285. Cables: GRASSLANDS, Palmerston North.

1 Wilde RH - BSc,MSc *District Soil Scientist* Manawatu and Wanganui soils; soil morphology; soil interpretation; Pacific Island soils
2 Shepherd TG - MSc soils of Manawatu; tephra characterisation

3461 Wellington

N.Z. Soil Bureau, Private Bag, Lower Hutt
Tel: 673-119. Telex: PHYSICS NZ 3814. Cables: SOBUR, Wellington.

1 Leslie DM - MSc *Science Manager, International Development* technical aid coordinator; Pacific Island soils and land use;agrotechnology transfer
2 Mew G - MA *Soil Correlator, Wellington* forestry interpretation; gley soils and podzols; soil correlation; town and country planning
3 McDonald WS - MSc computer applications for soil science
4 Percival HJ - MSc,PhD,FNZIC *Science Manager, Marketing* soil solution/mineral equilibria; x-ray diffraction
5 Wilson AD - MSc *Science Manager (Districts)* soils of Northland, land evaluation

3462 Nelson

N.Z. Soil Bureau, DSIR, Private Bag, Nelson
Tel: 81082.

1 Campbell IB - MSc *District Soil Scientist - Nelson and Marlborough* soils of Nelson and Marlborough; steepland soils
2 Laffan MD - MAgrSc soils of Marlborough; soil erosion processes; Pacific Island soils
3 Vincent KW - BSc soils of Wairarapa and Marlborough

3463 Lincoln

N.Z. Soil Bureau, DSIR, Private Bag, Christchurch
Tel: 252-511. Telex: LINSIR NZ 4703. Cables: PLANTLAB Christchurch.

1 Webb TH - MAgrSc *District Soil Scientist* soils of Canterbury; soil characterisation
2 Trangmar BB - MSc,PhD surveys of soil and erosion on Port Hills; computer techniques in soil surveys and fertility studies

3464 Dunedin

N.Z. Soil Bureau, DSIR, Private Bag, Dunedin
Tel: 774-051.

1 Beecroft FG - BSc,BA,DipNatRes *District Soil Scientist* soil survey; soil morphology and interpretation for land use

2 Hewitt AE - DipAgrSc,BSc,PhD *National Coordinator for Soil Classification* soil classification
3 McIntosh PD - BSc,PhD soils of Southland; soil fertility of hill country; soil genesis in loess
4 Smith SM - BA,MApplSc soil survey; soil interpretation; soil-vegetation relationships; hill soils

3465 Plant Diseases Division

Mt Albert Research Centre, DSIR, Private Bag, Auckland
Tel: (09) 893 660. Telex: MARC NZ21623. Cables: PLANTLAB.

1 Brook PJ - MSc,PhD *Director* fungus diseases

3466 Bacteriology

1 Hale CN - MSc,PhD *Section Head Rhizobium* ; bacterial diseases; field crops
2 Crowhurst RN - MSc bacterial rots, agricultural crops
3 Watson DRW - BSc,DIC bacterial diseases
4 Young JM - MSc,PhD bacterial diseases, bacterial taxonomy

3467 Virology

1 Forster RLS - MHortSc,PhD *Section Head* virus diseases, legumes
2 Guilford PJ - MSc virus diseases, disease resistance
3 Morris BAM - MSc,PhD virus diseases, disease resistance

3468 Mycology

1 Beever RE - MSc,PhD *Section Head* fungus physiology and genetics
2 Buchanan PK - BSc,PhD fungus taxonomy
3 Hallett IC - BSc,PhD electron microscopy; ultrastructure of fungi
4 Johnston PR - MSc fungus taxonomy, Ascomycetes
5 McKenzie EHC - BSc,PhD fungus taxonomy, Hyphomycetes
6 Sharrock KR - MSc,PhD plant disease physiology
7 Templeton MD - BSc,PhD plant disease physiology

3469 Plant Protection

1 Fullerton RA - BAgrSc,PhD *Section Head* fungus diseases, cereals and tropical crops
2 Fowler M - MSc,PhD soilborne diseases, field crops
3 Hartill WFT - BSc,PhD epidemiology and disease control
4 Hawthorne BT - MAgrSc,PhD fungus diseases, vegetables and field crops
5 Horner IJ - MSc soilborne diseases, tree crops
6 Menzies SA - MAgrSc,PhD fungus diseases, vegetable crops
7 Pak HA - MSc epidemiology of fungus diseases
8 Pennycook SR - BSc,PhD fungus diseases, fruit crops
9 Robertson GI - MSc,PhD soilborne diseases, Phycomycetes

SUBSTATIONS

3470 Palmerston North
Palmerston North Research Centre, DSIR, Private Bag, Palmerston North

1 Latch GCM - MAgrSc,PhD *Officer in charge* fungus diseases, pastures
2 Skipp RA - BTech,PhD fungus diseases, pasture legumes

3471 Lincoln
Lincoln Research Centre, DSIR, Private Bag, Christchurch

1 Cromey MG - MSc,PhD fungus diseases, field crops
2 Beresford RM - MSc,PhD fungus diseases, field crops
3 Falloon RE - MAgrSc,PhD *Officer in Charge* fungus diseases, field crops

3472 Plant Physiology Division

Private Bag, Palmerston North
Tel: (03)68019. Telex: NZ31285 PALMSIR. Cables: PLANTPHYS.

1 Kerr JP - MAgrSc,PhD *Director* agronomy, agricultural meteorology

3473 Plant Physiology Group

1 Warrington IJ - MHortSc *Group Leader* environmental physiology
2 McPherson HG - MAgrSc,PhD crop physiology

3474 Crop Physiology

1 Slack CR - BSc,PhD *Group Leader* lipid biochemistry, cereal quality
2 Brooking IR - BSc crop growth and development
3 Davies LJ - MAgrSc,PhD agronomy, plant development
4 Eagles HA - BAgrSc,PhD plant breeding
5 Hardacre AK - MSc plant physiology
6 Haslemore RM - BSc,DipSci,PhD cereal quality

3475 Plant Cell Physiology

1 Laing WA - BSc,PhD *Programme Convenor* photosynthesis, photorespiration
2 Christeller JT - MSc,PhD cell physiology, enzymology
3 Cohen D - MAgrSc,PhD plant tissue culture

3476 Environmental Physics

1 Clothier BE - BSc,PhD *Programme convenor* soil physics, plant water use
2 Gandar PW - MAgrSc,PhD plant growth environment interactions
3 McNaughton KG - BSc,PhD micrometeorology, plant water use
4 Spriggs TW - MSc,PhD environmental physics

3477 Agricultural Systems

1 Taylor AO - BSc,PhD forage agronomy and utilisation

UNIVERSITIES

3478 University of Auckland

Private Bag, Auckland
Tel: 737-999.

3479 Department of Botany

1 Brown JMA - BSc,PhD *Associate Professor, Head of Department* physiological ecology, particularly of freshwater plants; pathways of photosynthesis and influence of limiting factors such as CO_2, nutrients and light; ecology of underwater vegetation; photosynthesis of crop plants
2 Braggins JE - MSc,PhD *Senior Lecturer* systematics of bryophytes and pteridophytes, spore morphology
3 Dromgoole FI - BSc,MSc,PhD *Senior Lecturer* physiological ecology of marine and freshwater plants; primary productivity in aquatic environments; environmental control of photosynthesis
4 Gould KS - BSc,PhD *Lecturer* positional information in plant development; *in vitro* culture of plant organs; mathematical modelling of the growth of plant shoots, leaves, flowers and fruit
5 Harris PJ - MA,PhD *Lecturer* plant cell walls: their morphology, chemical composition, biosynthesis, degradation and function; cytochemical localisation of cell-wall components; cell walls in relation to ruminant nutrition; carbohydrate chemistry and biochemistry
6 Jensen LCW - AB,MSc,PhD *Senior Lecturer* morphology and cytology of growth differentiation in plants, including ultrastructural studies of cell division
7 Lovell PH - BSc,PhD *Professor* developmental plant physiology, senescence and ethylene physiology; adventitious root formation, germination and seedling development; weeds; vegetative spread; horticultural plants
8 Mitchell ND - MA,MSc,PhD *Senior Lecturer* quantitative techniques for vegetation surveys; autecology of northern 'broadleaf' species; interpretation of species distribution patterns
9 Murray BG - BSc,PhD *Senior Lecturer* population cytogenetics, plant breeding systems and reproductive biology
10 Ogden J - MSc,PhD *Senior Lecturer* forest ecology, community dynamics and dendrochronology, population dynamics and reproductive strategies of herbaceous plant species
11 Pearson MN - BSc,PhD *Lecturer* pathology of tropical and sub-tropical crops, plant virology
12 Segedin BP - MSc,PhD *Senior Lecturer* taxonomy of fungi, in particular the Agaricales; and Hyphomycetes; fungi in forest litter; fungi in streams; fungi in thermal soils
13 Stewart A - BSc,PhD *Lecturer* mechanisms of plant resistance to fungal diseases; biological control of plant pathogens
14 Taylor GM - MSc *Senior Tutor* taxonomy of gilled fungi

3480 Department of Zoology

1 Young EC - MSc,PhD *Professor* entomology
2 Cowley DR - MSc,PhD *Senior Lecturer* entomology

3481 University of Canterbury

Private Bag, Christchurch
Tel: 482009. Telex: UNICANT NZ 4144.

1 Bowron RH - LLB *Chancellor*
2 Brownlie AD - MCOM *Vice-Chancellor*

3482 Department of Geography

1 Johnston WB *Professor, Head of Department* geography and public policy, especially rural and environment
2 Bedford RD *Senior Lecturer* rural population
3 Cant RG *Reader* rural development
4 Johnston DC *Senior Lecturer* rural transport

3483 Department of Geology

1 Cameron Miss AA - MSc *Lecturer* living and fossil foraminifera
2 Mackinnon DI - BSc,PhD *Senior Lecturer* living and fossil brachiopods; cret-tert palynology
3 Swanson KM - NZCS,BSc *Technical Officer* living and fossil ostracodes

3484 Department of Plant and Microbial Sciences

1 Butterfield BG - PhD,DSc,FIWSc,FRMS,FIAWSc *Head of Department* wood structure and properties
2 Burrows CJ - MSc,PhD,DSc *Reader* plant ecology
3 Cole ALJ - BSc,PhD,DIC,ARCS *Reader* plant diseases, fungal biochemistry
4 Greenfield LG - BSc,DPhil *Senior Lecturer* soil microbiology
5 Kelly D - BSc,PhD *Lecturer* weed ecology
6 Leung DWM - BSc,PhD *Lecturer* tissue culture
7 Lloyd DG - BSc,PhD,FRSNZ *Professor* reproductive biology
8 Mahanty HK - BSc,MSc,PhD *Senior Lecturer* recombinant DNA and cloning of colicin producing plasmids
9 Timmerman GMT - BS,MA,PhD *Lecturer in vitro* RNA synthesis and processing
10 Walker JRL - BSc,PhD,FNZIC *Reader* biology of phenolics

3485 Department of Zoology

1 Clark WC - MSc,PhD,DIC *Professor* parasitic nematodes
2 Blair D - BSc,PhD *Senior Lecturer* parasitic flatworms
3 Davison W - BSc,PhD *Senior Lecturer* fish physiology and musculature
4 Duncan KW - BSc,PhD *Senior Lecturer* systematics, ecological physiology, biometrics
5 Field LH - MA,PhD *Senior Lecturer* neurophysiology
6 Forster ME - BSc,PhD *Senior Lecturer* fish physiology
7 Jackson RR - BSc,PhD *Senior Lecturer* spider behaviour
8 Johns PM - MSc *Senior Lecturer* entomology
9 Marsden ID - BSc,PhD *Senior Lecturer* marine biology
10 Mason D - BSc,PhD *Lecturer* cell physiology
11 McLay CL - BSc,PhD *Senior Lecturer, Head of Department* marine crustacean ecology
12 McLean IG - MSc,PhD *Senior Lecturer* behavioural ecology of vertebrates
13 Sin FYT - BSc,PhD *Senior Lecturer* genetics
14 Stout VM - MSc,PhD *Reader* freshwater ecology and limnology
15 Taylor HH - BSc,PhD *Senior Lecturer* physiology, ionic regulation
16 Winterbourn MJ - MA,PhD *Reader* stream biology
17 Wisely VM - MSc *Lecturer* developmental biology

3486 Active Emeriti

1 Knox GA - BA,MSc,FRSNZ *Professor* marine ecology
2 Pilgrim RLC - MSc,PhD *Professor* flea larvae taxonomy
3 Warham J - BSc,MSc,PhD,DSC *Reader* sea-birds

3487 School of Forestry

1 Sweet GB - BSc,DipFor,PhD,DSc *Professor, Head of Department* forestry, genetics & physiology of NZ indigenous forests
2 Allen JD - BSc,PhD *Senior Lecturer* forest pathology & microbiology
3 Bilek EM - MBA,PhD *Lecturer* forest economics
4 Fieber WF *Lecturer* forest engineering and harvesting
5 Mead DJ - BSc,BForSc,PhD *Senior Lecturer* forest silviculture & nutrition
6 Norton DA - BSc,PhD *Lecturer* plant ecology and palaeoecology
7 O'Reilly RA - BScF, MForSc *Senior Lecturer* forest engineering & harvesting
8 Walker JCF - MA,PhD *Senior Lecturer* wood properties, preservation & drying, roundwood
9 Whyte AGD - BScFor,PhD,DipFor *Reader* forest management & planning

3488 Lincoln College

Canterbury
Tel: (03) 252-811.

1 Ross BJ - MAgrSc *Principal*
2 Wright Sir A - KBE *Chairman of the Council*

3489 Department of Agricultural Economics and Marketing

1 McArthur ATG - BScAgric,MAgrSc,PhD *Reader, Head of Department* economics of animal breeding
2 Cosgriff PW - BAgSc *Senior Lecturer*
3 Holland LA - BA *Senior Lecturer* social surveys
4 Lamb CG - BComAg *Lecturer* marketing
5 Lattimore RG - MAgrSc,PhD *Reader* international trade
6 Rayner AC - BCom,MSocSc *Professor* natural resource economics
7 St Hill RL - BCom *Lecturer* macroeconomics
8 Ward BD - MA,MLitt *Senior Lecturer* econometrics
9 Willis FM - BBS,MBA *Lecturer*
10 Wood JW - BScAgric,MSA *Senior Lecturer* regional economics

11 Woods LD - MAgrSc *Senior Lecturer* marketing economics
12 Zwart AC - BAgrSc,MSc,PhD *Professor* marketing

3490 Department of Agricultural Engineering

1 Ward GT - BSc,PhD *Professor, Head of Department* solar energy; protected environments
2 Cherry NJ - BSc,PhD *Senior Lecturer* meteorology; wind energy
3 Chilcott RE - BScEng,MSc,DIC *Senior Lecturer* wind energy conversion systems; agricultural traction
4 Dakers AJ - ME *Senior Lecturer* agricultural wastes; irrigation; small hydro
5 Davies TR - MSc,PhD *Reader*
6 Douglass B - BSc,MA *Senior Lecturer* soil conservation
7 Lindsay GG - BSc *Senior Lecturer*
8 MacKenzie DW - BE *Senior Lecturer*
9 Painter DJ - BE,PhD *Reader* water resources; hydrology; irrigation

3491 Department of Agricultural Microbiology

1 Mulcock AP - MSc,PhD *Professor, Head of Department* microbial ecology
2 Close RC - MSc,PhD *Reader* plant pathology
3 Gaunt RE - BSc,PhD *Senior Lecturer* plant pathology
4 Noonan MJ - BSc,PhD *Senior Lecturer* microbial ecology
5 Skellerup Mrs MV - BSc *Tutor* microbiology; plant pathology

3492 Animal Sciences Group

3493 Animal and Veterinary Sciences Group

Canterbury
Tel: (03) 252-811.

1 Irvine CHG - BVSc,BSc,DSc,FACVSc *Professor Emeritus* endocrinology; reproduction; exercise physiology
2 Sykes AR - PhD,DSc,FRSA *Professor, Head of Department* nutrition; mineral metabolism; parasitism
3 Barrell GK - BSc,DipSci,PhD *Senior Lecturer* endocrinology; reproduction; bone growth
4 Beatson PR - BAgrSc *Lecturer* animal breeding and production
5 Elvidge DG - MAgrSc *Senior Lecturer* sheep production
6 Familton AS - BVSc,MACVSc *Senior Lecturer* disease parasitology; trace elements
7 Hughes TP - MAgrSc *Senior Lecturer* pastures; intake; dairy production
8 Nicol AM - BSc,MAgrSc,PhD *Senior Lecturer* beef cattle growth; nutrition
9 Poppi DP - MAgrSc,PhD *Senior Lecturer* ruminant nutrition; intake; parasitism
10 Thompson KF - BAgrSc,CertWool *Senior Lecturer* sheep production; grazing management
11 Young MJ - BAgSc *Lecturer*

3494 Centre for Computing and Biometrics

1 Anderson DH - BSc *Lecturer* computing in agriculture
2 Baird J - BAgSc,MApplSc *Senior Computer Consultant*
3 Brown MSC - BSc *Programmer/analyst* man-machine interface
4 Frampton Ms ER - BSc,DipAgrSc *Tutor* entomology
5 Love BG - BSc,MAgrSc,DipEd *Senior Lecturer* experimental design
6 McKinnon AE - BE,PhD *Senior Programmer/Analyst* biological modelling
7 McLennan TJ - BSc *Lecturer*
8 Robson AB - BSc,MSc *Senior Lecturer* biological modelling
9 Sedcole JR - MAgSc,PhD *Senior Lecturer*

3495 Centre for Resource Management

University of Canterbury, Private Bag, Christchurch
Tel: 482 009 ext. 8660 or 252 811 ext. 598.

1 Hayward JA - MAgrSc,PhD *Director* hydrology
2 Sharp BMH - BAgrEcon,MSc,PhD,DipAgr,DipVFM *Deputy Director* resource economics
3 Abrahamson MA - MAgrSc,DipMgt *Research Officer* management soil
4 Baines JT - BE,DipTchg *Senior Research Officer* energy
5 Gough JD - BSc,MCom *Research Officer* resource economics
6 Gray MM - LTh *Senior Research Officer*
7 Hide R - MSc *Research Officer* economics irrigation
8 Kerr GN - MA *Research Officer* resource economics
9 Kerr IGC - BAgrSc *Senior Research* land management
10 Leathers KL - BS,MSc,PhD *Principal Research Officer* resource economics
11 MacIntyre AA - BSc,PhD *Lecturer*
12 O'Connor KF - BA,BAgrSc,PhD *Professor of Range Movement* land management
13 Robertson BT - DipAg,DipVFM *Senior Research Officer* information
14 Taylor CN - MSc,PhD *Research Officer*
15 Taylor CN - BA,MSc,PhD *Research Officer* social science
16 Ward JC - BScHons,MSc,PhD *Research Officer* marine biology
17 White EG - MHortSc,PhD *Principal Research Officer* insect populations
18 Williams TA - BSc,DipJourn *Research Officer*
19 Wright J - BSc,MSc *Research Officer* energy resources

3496 Department of Entomology

1 Penman DR - MAgrSc,PhD *Professor, Head of Department* pest management; biological control
2 Chapman RB - MAgrSc *Senior Lecturer* insecticides; resistance management
3 Emberson RM - BSc,PhD *Senior Lecturer* taxonomy; biological control of weeds
4 Scott RR - MSc,PhD *Senior Lecturer* ecology; conservation of invertebrates
5 Waller JB - BSc,PhD *Senior Lecturer* toxicology; stored products pests

3497 Department of Farm Management

University of Canterbury, Private Bag, Christchurch
Tel: 482 009 ext. 8660.

1 Juchau RH - BCom,BEd,MA,FASA,FNZIM,CPA,ACA *Professor* private and public financial management
2 Binnie SWG - ANZIV *Lecturer* computers in valuation
3 Boyd TP - MSc,SCV,ANZIM *Senior Lecturer*
4 Bywater AC - BSc,PhD *Professor & Head of Department* agricultural systems
5 Clark MB - MCom,ACA *Senior Lecturer* accounting and taxation in farming
6 Croft CS - BAgrCom,ANZIV *Senior Lecturer* rural land valuation
7 Fleming PH - BAgrSc *Lecturer* farm management and extension
8 Frengley GAG - MAgrSc,PhD,DipVFM *Reader* farmer attitudes and investments
9 Frew SA - BComAg *Computer Consultant* computer software development
10 Gaul PF - BAgrSc *Lecturer*
11 Gow NG - BAgrSc *Senior Lecturer* farm finance
12 Harris MR - DipAgr,DipFM *Computer Consultant* computer software development
13 Hay D - BCom *Lecturer*
14 Kerdemelidis D - MCom *Lecturer*
15 Lay JK *Lecturer* farm decision making with computers
16 McCrea PR - BComAg *Lecturer* agricultural resource management
17 McIntosh JB - LLB *Lecturer* agricultural law
18 Moorhead EG - BAgSc,ANZIV *Senior Lecturer*
19 Nahkies PB - VALPROFURB,DipVPM *Lecturer* building cost analysis
20 Newman DL - BAgrSc,BCom,ANZIV *Senior Lecturer* financial management
21 Nuthall PL - MAgrSc,PhD *Reader* computers in farm management
22 Oliver PR - BCom,DipAg,DipVFM,ACA *Research Officer* computer software for farmers
23 Pittaway SF - BAgrSc *Senior Lecturer* irrigation water pricing
24 Plank RD - BACant,DipVFM,ANZIV *Senior Lecturer* farming systems development
25 Ryde BJP - DipVFM,ANZIV *Senior Lecturer* tropical cropping systems
26 Wright AE - LLB *Lecturer* agricultural law

3498 Department of Horticulture, Landscape and Parks

1 Rowe RN - BAgrSc,MSc,PhD *Professor, Head of Department* stonefruit agronomy, post harvest; root physiology
2 Crowder RA - BSc *Senior Lecturer* vegetable agronomy; biological husbandry
3 Edwards RA - DipHortSc *Lecturer*
4 Farr DJ - BHortSc *Lecturer* cut flowers; protected cropping
5 Jackson DI - MAgrSc,PhD *Reader* grape agronomy, physiology
6 Morley-Bunker MJS - MSc *Lecturer* sub-tropical and tropical fruit agronomy
7 Stevens RB - BHortSc *Senior Lecturer*
8 Taylor JO - MA *Senior Lecturer* amenity horticulture; turf; aboriculture
9 Thiele GF - MHortSc *Reader*
10 Thomas MB - MHortSc,PhD *Senior Lecturer* nursery production; nutrition
11 Tipples RS - MA,PhD *Senior Lecturer*

3499 Landscape Architecture Section

1 Borthelmeh MR - BHortSc,DipLA,ANZILA *Studio Tutor*
2 Densem GH - BA,DipLA,ANZILA *Senior Lecturer*
3 Kirby VG - BA,DipTRP,DipLA,MRTPI ALI *Lecturer*
4 Lucas DJ - BSc,DipLA,ANZILA *Teaching Fellow*
5 Swaffield SR - MA,DipLD,DipTP *Senior Lecturer and Head of Section*

3500 Department of Parks, Recreation and Tourism

1 Cushman G - MSc,PhD,DipPE *Reader and Head of Department*
2 Devlin PJ - MA,PhD,DipTchg,FRIH *Senior Lecturer*
3 Shuttleworth JJ - MA,PhD,DLC,DipTchg *Senior Lecturer*
4 Simmons DG - BSc,MApplSc *Lecturer*
5 Simpson CS - MA,DipTchg,ATCL *Lecturer*
6 Taylor AR - MA,DipTchg,DipCRS,MIPRA *Senior Lecturer*

3501 Department of Plant Science

1 White JGH - MAgrSc,PhD *Professor and Head of Plant Science* agronomy
2 Andrews MHG - BSc,PhD *Lecturer*

3 Askin DC - BAgrSc,PhD *Lecturer* pasture agronomy
4 Daly GT - MAgrSc,PhD *Senior Lecturer* ecology
5 Fautrier AG - BSA,BSc,PhD *Senior Lecturer* genetics
6 Field RJ - BSc,PhD *Professor of Plant Science* plant physiology, weed science
7 Hill GD - RD,MSc *Senior Lecturer* crop agronomy; plant breeding
8 Jarvis P - BSc,PhD *Senior Lecturer* plant physiology, botany
9 Love BG - BSc,MAgrSc,DipEd *Senior Lecturer* biometrics
10 Lucas RJ - BAgrSc,DipTchg *Senior Lecturer* experimentation; agronomy
11 McKenzie BA - BS,DipAgrSc *Senior Lecturer* crop agronomy
12 McNeil DL - BSc,PhD *Senior Lecturer*
13 Scott WR - MAgrSc,PhD *Senior Lecturer* crop agronomy
14 Smetham ML - BSc,MAgrSc *Senior Lecturer* pasture agronomy

3502 Rural Development and Extension Centre

1 Fleming ND - MA *Senior Education Officer*
2 Tate GF - BA,MSc,DipVFM *Senior Lecturer* extension methodology

3503 Department of Soil Science

1 Swift RS - BSc,PhD,FRSC *Professor, Head of Department* soil fertility, soil organic matter, trace elements
2 Adams JA - BSc,PhD *Senior Lecturer* soil fertility, forest soils
3 Buchan GD - MSc,PhD *Senior Lecturer*
4 Cameron KC - BSc,PhD *Lecturer* soil fertility
5 Campbell AS - MSc,PhD *Reader* clay mineralogy, soil chemistry
6 Goh KM - MAgrSc,PhD *Reader* soil fertility, soil organic matter
7 Harrison R - BSc,PhD *Lecturer*
8 Kemp RA - MSc,PhD *Lecturer* pedology, micromorphology
9 Livingstone LG - MAgrSc,PhD *Senior Lecturer* soil physics, instrumental analysis
10 McLaren RG - BSc,PhD *Senior Lecturer* soil fertility, trace elements
11 Sherlock RR - MSc,PhD *Lecturer* soil fertility, soil chemistry
12 Tonkin PJ - BSc,PhD *Senior Lecturer* pedology, geomorphology

3504 Department of Wool Science

1 Wilkinson BR - MAgrSc,PhD *Reader, Head of Department* wool production and fleece biology
2 Cottle DJ - BSc,PhD *Senior Lecturer*
3 Ross DA - MAgrSc,PhD *Professor* technical wool marketing and processing
4 Sanderson RH - DipAgrSc *Lecturer*

3505 University of Otago

P.O. Box 56, Dunedin
Tel: 771-640.

1 Valentine JA - BCom,FCA *Chancellor*
2 Irvine ROH - MD,Drhc,FRCP,FRACP,FRSA *Vice-Chancellor*

3506 Department of Biochemistry

1 Peterson GB - MSc,MA,DPhil *Professor, Head of Department* DNA chemistry; molecular genetics
2 Carne A - BSc,PhD *Lecturer* protein chemistry
3 Carrington JM - BSc,PhD *Lecturer* metabolism
4 Cutfield JF - MSc,PhD *Senior Lecturer* protein chemistry
5 Emerson GW - MSc,PhD *Senior Lecturer* metabolism
6 Farnden KJF - MSc,PhD *Senior Lecturer* nitrogen fixation
7 Forrester IT - MSc,PhD *Senior Lecturer* biotechnology; sperm biochemistry
8 Grigor MR - MSc,PhD *Senior Lecturer* lipid biochemistry
9 Hill DF - BSc,PhD *Research Officer* molecular biology
10 Lamont IL - BSc,DPhil *Lecturer* molecular biology and genetics of bacteria
11 Poulter RTM - BSc,PhD *Senior Lecturer* fungal genetics
12 Reeve AW - MSc,PhD *Senior Research Officer* cancer research
13 Russell DW - MSc,PhD *Senior Lecturer* plant biochemistry
14 Smith MG - MSc,PhD *Associate Professor* viral molecular biology
15 Stockwell PA - BSc,PhD *Research Officer* applications of computer science to molecular biology
16 Sullivan PA - MSc,PhD *Associate Professor* fungal cell walls; metabolism
17 Tate WP - MSc,PhD *Associate Professor* molecular biology
18 Thompson MT - BSc,PhD *Senior Lecturer* metabolism
19 Trotman CNA - BTech,PhD developmental biochemistry
20 Wilkins RW - MSc,PhD *Director, Molecular Carcinogenesis Laboratory* cancer research

3507 Department of Botany

1 Bannister P - BSc,PhD *Professor, Chairman of Department* physiological ecology
2 Hall IR - BSc,PhD *Honorary Lecturer* mycorrhizas
3 Jameson PE - BSc,PhD,MNZIC *Lecturer* plant tissue culture
4 Mark AF - MSc,PhD,FRSNZ *Professor* plant ecology, nature conservation
5 Wilson JB - BSc,PhD *Senior Lecturer* plant community structure

3508 Department of Clinical Biochemistry

1 Sneyd JGT - BMedSc,MB,ChB,PhD,FRACP *Professor, Head of Department* insulin and diabetes
2 Loten EG - BMedSc,MB,ChB,PhD *Senior Lecturer* mechanism of insulin
3 Rennie PIC - MB,ChB,MD *Senior Lecturer* metabolism in obese mice
4 Watts C - BSc,PhD,MNZIC *Associate Professor* enzymology of glycogen metabolism

3509 Department of Microbiology

1 Jones D - BSc,PhD *Professor, Head of Department* biotechnology; physiology of bacteria
2 Gibbins BL - MSc,PhD *Senior Lecturer* fungal immunology
3 Griffin JFT - BA,PhD *Senior Lecturer* immunology of deer
4 Kalmakoff J - BSc,PhD *Associate Professor* biological control
5 Loutit MW - MSc,PhD *Professor* heavy metals, bacteria and food chains
6 Pillai JS - MSc,PhD *Associate Professor* biological control
7 Smith JMB - MSc,PhD *Associate Professor* fungal infections
8 Tagg JR - MSc,PhD *Senior Lecturer* oral streptococci
9 Tannock GW - BSc,PhD *Senior Lecturer* microbiota of gut

3510 Medical Research Council of New Zealand Virus Research Unit

1 Austin FJ - MSc,PhD influenza virus
2 Maguire T - MSc,PhD dengue fever
3 Mercer A - BSc,PhD ORF virus
4 Robinson AJ - BVSc,PhD,Dip Microbiol *Director* ORF virus

3511 Department of Physiology
P.O. Box 56, Dunedin
Tel: 771-640.

1 Hubbard JI - PhD,MD *Professor of Neurophysiology* neuroendocrinology
2 Bray JJ - PhD *Senior Lecturer* neuroimmunology
3 Day TA - PhD *Lecturer* central control of autonomic system
4 Harris AJ - PhD *Associate Professor* development of nerve and muscle
5 Leader JP - PhD *Associate Professor of Physiology* epithelia
6 Macknight ADC - MD,PhD,FRSNZ *Professor of Physiology* epithelia
7 Mills RG - PhD *Senior Lecturer* autonomic pharmacology and physiology

3512 Department of Radiology

1 Doyle TA - MBChB,MD,DDR,BA,DipObst,FRACR *Professor, Head of Department* lung and oesophageal carcinoma

3513 University of Waikato

Private Bag, Hamilton
Tel: 62-889.

1 Drayton MJ - MBE,MA,DipEd *Chancellor*
2 Malcolm WG - MA,PhD *Vice-Chancellor*

3514 Department of Biological Sciences

1 Silvester WB - MSc,PhD *Professor, Head of Department* nitrogen fixation; nutrient cycling
2 Adam KD - DiplBio,PhD plant population biology and genetics
3 Green TGA - MA,PhD *Senior Lecturer* crop photosynthesis and water relations, horticultural plants
4 McLeay LM - MAgSc,PhD *Reader* ruminant digestive physiology
5 Molan PC - BSc,PhD *Senior Lecturer* artificial insemination
6 Musgrave DR - BSc,PhD *Lecturer* genetics

3515 Victoria University of Wellington

Private Bag, Wellington
Tel: 721-000.

1 McGrath JJ - QC,LLM *Chancellor*
2 Holborow LA - MA,BPhil *Vice Chancellor*

3516 Department of Biochemistry

1 Clark AG - MSc,PhD,FNZIC *Reader* enzymic mechanisms of insecticide metabolism and resistance
2 Jordan TW - MSc,PhD *Reader, Chairman of Department* biochemistry and genetics of disease resistance in livestock

3517 Department of Agriculture
Private Bag, Wellington
Tel: 721-000.

1 Kirby JM - BA,PhD *Senior Lecturer* rural and regional planning
2 Willis RP - BA,MA *Senior Lecturer* general agriculture

3518 Research Project on Economic Planning

1 Philpott BP - MA,MCom,ACA *Macarthy Professor of Economics* general equilibrium modelling of agriculture
2 Evans LT - PhD *Senior Lecturer* microeconomics, econometrics

3519 New Zealand Meteorological Service

P.O. Box 722, Wellington

1 Hickman JS - BSc *Director*

3520 Research Division

1 Thompson D - PhD *Assistant Director, Research*
2 Wratt DS - PhD *Superintendent* boundary layer processes and applications
3 Goulter SW - MSc *Biometrician* frost forecasting
4 McGann RP - MSc *Meteorologist* agricultural applications of boundary layer meteorology
5 Revfeim KJA - PhD *Meteorologist*
6 Salinger J - PhD *Meteorologist* agricultural meteorology
7 Stainer RD - MSc *Superintendent Training* forest meteorology
8 Steiner JT - PhD *Meteorologist* weather modification
9 Taylor BF - PhD *Manager Climate Data Centre* satellite data applications to agriculture
10 Wratt DS - PhD *Superintendent* boundary layer processes and applications

3521 Dynamic Meteorology Section

1 Kidson JW - ScD *Superintendent*
2 Renwick J - BSc *Meteorologist* statistical weather forecasting

3522 Atmospheric Physics & Chemistry Section

1 Larsen HR - PhD *Meteorologist*
2 Fisher GW - BSc *Meteorologist* microphysics of fogs and sprays

3523 Scientific Information Section

1 Heine RW - PhD *Superintendent* evapotranspiration, agricultural meteorology

3524 Services Division

1 Maunder J - PhD *Assistant Director, Services* real time application of weather data
2 Neale AA - MSc *Chief Forecaster*
3 Smith RM - BSc *Chief* Wellington Weather Centre

3525 Auckland Weather Centre
Level 3, 30 Ponsonby Road, Auckland

1 Hessell JWD - BSc *Chief*

3526 Christchurch Weather Centre
P.O. Box 14004, Christchurch Airport

1 Osborne HAL - MSc,MS *Chief*

3527 New Zealand Dairy Board

3528 Livestock Improvement Division
Cnr Ruakura and Morrinsville Roads, Private Bag, Hamilton
Tel: 62-809. Telex: NZ21246 ANSERHN. Cables: "Dapmark". (071) 62-428.

1 Whittaker WG - General Manager

3529 Research and Development Section

1 Bishop S - PhD *Scientist*
2 Breier G - MAgrSc *Scientist*
3 Jackson RG - MAgrSc *Scientist*
4 Shannon P - BScAgr,DSc *Research Leader*
5 Wickham BW - PhD *Research and Development Controller*

3530 Extension and Survey

1 Bartlett KR - MAgrSc *Extension and Survey Controller*

RESEARCH ASSOCIATIONS

3531 Cawthron Institute
Private Bag, Nelson
Tel: 82319. (054) 69464.

1 Robertson G - CEng,BE(Chem),MIChemE,MIPENZ *Chief Executive Officer*

3532 Biotechnology

1 Kaspar HF - MSc,PhD *Programme Leader* microbial N transformations, marine biotechnology
2 Broderick AJ - BSc,PhD microbial fermentations
3 Cooke MD - BSc,PhD transferable bacterial antibiotic resistance
4 Grant WD - BSc,PhD bacteriolytic enzymes, marine fungi
5 Mountfort DO - BSc,PhD methanogenic bacteria, methylotrophic bacteria

3533 Marine Microbial Ecology

1 Gillespie PA - MSc,PhD *Programme Leader* microbial processes, pollution effects
2 MacKenzie AL - BSc (Hons) phytoplankton ecology

3534 New Zealand Dairy Research Institute
Private Bag, Palmerston North
Tel: 74 129. Telex: NZ31312. Cables: Daisearch.

1 Page G - MRSC,PhD *Director*
2 Fryer TF - BSc,PhD *Assistant Director*
3 Hobman PG - BTech *Assistant Director*
4 Lawrence RC - BSc,PhD,FNZIC *Assistant Director*
5 Sanderson WB - BAgrSc,PhD *Assistant Director*
6 Baldwin AJ - BE,MSc engineering services
7 Barnett JW - MSc,PhD effluent technology
8 Bergerson AK - BTech milk powders and drying
9 Bissell TG - BE,PhD milkfat
10 Bloore CG - BE,MSc,PhD milk powders and drying
11 Boston GD - BSc analytical chemistry
12 Bruynel TR - BSc microbiology
13 Caddick CG - BSc starter technology
14 Cant PAE - BSc,PhD milkfat & butter
15 Clark JN - BTech effluent technology
16 Clark JN - BTech applied biochemistry
17 Cleland HR - BTech whey products
18 Cocup RO - LCG,CertEd food ingredients
19 Conaghan EF - NZCS analytical chemistry
20 Creamer LK - MSc,PhD,FNZIC protein chemistry
21 Croft SE - MSc food ingredients
22 Crow VL - BSc,PhD applied biochemistry
23 Davey GP - MSc,PhD microbial genetics
24 Downard CJ - BSc whey products
25 Dunlop FP - BTech,DipDySc&Tech cheese technology
26 Elston PD - DipDyTech casein products
27 Ennis BM - BTech applied biochemistry
28 Fayerman AM - BTech,DipTech food ingredients
29 Flint SH - MSc general microbiology
30 Gray IK - MAgrSc analytical chemistry
31 Haggett TOR - MSc analytical chemistry
32 Haylock SJ - BSc,PhD casein products
33 Heap HA - NZCS starter technology
34 Hewitt SA - BTech,PhD whey products
35 Higgins JJ - BTech,ME whey products
36 Hoppe GK - MSc whey products
37 Hoten JC - BTech milkfat & butter
38 Huffman LM - BA,MS,PhD whey products
39 Humphrey RS - BSc,PhD whey products
40 Illingworth D - MRSC milkfat & butter
41 Janssen PWM - BE,PhD engineering services
42 Jarvis AW - MAgrSc,PhD microbial genetics
43 Keen AR - BSc,PhD flavour
44 Knighton DR - DipSc,MSc,PhD protein chemistry
45 Lamont PS - BHSc product use and evaluation
46 Limsowtin GKY - DipAg,DipDyTech,BSc microbial genetics
47 Lochore JC - DipDyTech milkfat & butter
48 Loh DW - BE casein products
49 Lorimer SD - BSc,PhD flavour
50 Lovell-Smith JER - BE engineering services
51 Maccoll AJ - MHSc product use & evaluation
52 MacGibbon AKH - BSc,PhD milkfat & butter
53 Mackereth BE - DipDySc&Tech milk powders & drying
54 Martley FG - BSc,PhD cheese technology
55 Matheson AR applied mathematics
56 McKenna AB - BTech aseptic processing
57 McQueen JM - BTech casein products
58 McRobert AB - Tech(CEI) engineering services
59 Mercer WB - BE,DipDyScTech milkfat & butter
60 Mills OE - MSc,PhD flavour
61 Morison KR - BE,PhD engineering services
62 Motion RL - BSc,PhD protein chemistry
63 Munster RE - MSc general microbiology
64 Newstead DF - MSc,PhD aseptic processing
65 Norris R - BSc,PhD milkfat & butter
66 O'Carroll DK - BTech casein products
67 O'Connell MT - DipDyTech whey products
68 Oolders TT - BE effluent technology
69 Palfreyman KR - BTech,DipDySc&Tech milkfat & butter
70 Patchett RJ - NZCE,REA engineering services
71 Pearce KN - MSc,PhD protein chemistry
72 Pearce LE - MSc,PhD microbial genetics
73 Philpott BA - BTech milkfat and butter
74 Radford JB - DipFieldTech,BAgrSc effluent technology
75 Richardson BC - MSc,PhD protein chemistry
76 Richardson RK - MSc analytical chemistry
77 Roeper J - Ing casein products
78 Salmons WN - NZCS effluent technology
79 Schollum LM - BSc,PhD whey products
80 Shaker RR - MSc cheese, technology
81 Singh H - MSc,PhD protein chemistry
82 Smart JC - BSc,PhD applied biochemistry
83 Southward CR - MSc,FNZIC casein products
84 Sowry DA engineering services
85 Talboys FW - DipDyTech *Production Superintendent*
86 Thomas TD - MSc,PhD applied biochemistry
87 Tokley RP - DipDyTech,CMicro milk powders & drying
88 Towler C - BSc product use & evaluation
89 Tremain LR - BMus,MA,DipNZLS library & information
90 Turner KW - MSc cheese technology
91 Tuttiett PT - BSc applied mathematics
92 Ward DD - BSc,PhD flavours
93 Wards SM - MHSc product use & evaluation
94 Webby PS - DipDyTech milk powders & drying
95 Wiles PG - BE,PhD engineering services
96 Wilson RD - MSc flavour

3535 New Zealand Logging Industry Research Association Inc.
P.O. Box 147, Rotorua
Tel: (073) 87-168. Telex: FRESTRA NZ 21080. C/- (073) 479-380.

1 Galbraith JE - BEMech,MSCForEng *Director*
2 Gaskin JE *Assistant Director*
3 Prebble RL
4 Raymond K - BForSc
5 Vaughan LW - BSc,BScFor

3536 New Zealand Leather and Shoe Research Association
Private Bag, Palmerston North
Tel: 82-108.

Note: The Association provides R & D and support services to three industrial member sectors - the freezing works industry, the tanning industry and the footwear manufacturing industry.

1 Passman A - BSc,DipNLC *Director*
2 Charlton JH - BSc analysis
3 Cooper S - PhD ovine leather technologies
4 Halligan G - BSc microscopy
5 Jackways GD - BTech woolskin technologies
6 Mason IG - BTech,MSc effluent treatment/prevention

3537 New Zealand Meat & Wool Boards' Economic Service
P.O. Box 5179, Wellington
Tel: 722-178.

1 Taylor NW - MAgrSc,FNZSFM,FNZIAS *Director*
2 Davison RM - BAgrCom *Chief Economist*
3 Burtt AN - BAgrEcon *Research Economist*
4 Jamieson ID - BAgrSc *Economist*
5 Speirs BS - BAgrSc *Senior Economist*

3538 Napier Office
P.O. Box 617, Napier
Tel: 354-629.

1 Warner AO - DipVFM *Senior District Officer*

3539 The Meat Industry Research Institute of New Zealand (Inc.)
P.O. Box 617, Hamilton
Tel: +64 71 56159. Telex: NZ21470. Cables: MIRINZ Hamilton. +64 71 53833.

1 Wright DE - MSc,PhD,DSc *Director*

3540 Science

1 Chrystall BB - BAgrSc,MS,PhD *Head of Science Division* meat processing, meat quality, quality assurance
2 Archibald RJ - MSc,PhD processed meats, new meat products

3 Bell RG - BSc,PhD processed meats and water microbiology
4 Devine CE - MSc,PhD stunning and slaughter, physiology and muscle structure
5 Gibson H - BTech processed meats, consumer ready products
6 Gill CO - BSc,PhD meat microbiology, microbial physiology
7 Hagyard CJ - NZCSc meat quality, sensory evaluation
8 Lowry PD - MSc microbial physiology, microbiological quality control
9 Moore VJ - MS,PhD colour and appearance of meat
10 Sephton S - BSc meat processing
11 Tan KH - MSc,PhD microbial biochemistry
12 Watson PJ - BTech processed meats, new meat products
13 West J - BSc,MSc nutritional composition of foods
14 Young OA - BSc,PhD biochemistry, muscle structure

3541 Process Technology

1 Fleming AK - BE,DrIng,MIPENZ *Head of Process Technology Division* industrial refrigeration
2 Brown GI - BE tallow processing
3 Cooper RN - BTech,MSc,PhD waste abatement, recovery and treatment
4 Donnison AM - BHSc,BSc,DPhil microbiology of effluent treatment
5 Fox GR - BTech anaerobic digestion of meat plant effluent
6 Loeffen MPF - BE computer systems engineering, process control
7 Pham QT - BE,PhD heat and mass transfer in foodstuffs, refrigeration
8 Russell JM - BSc,PhD irrigation and anaerobic treatment of meat wastes, waste-water analysis
9 Swan JE - MTech,PhD by-products technology and process engineering
10 Tarley PJ - BTech
11 van Oostrom A - MSc composting of solid wastes; effluent nutrient stripping with aquatic plants

3542 Development Engineering

1 Longdill GR - BE,PhD *Head of Development Engineering Division* technology transfer, meat processing
2 Authier JF - Ir meat processing, carcass dressing automation
3 Wickham G - BE boning technology

3543 Information and Advisory Service

1 Bentley GR - BAgrSc *Head of Information and Advisory Service* extension, information, promotion
2 Tobin Miss PAO'H - NZLACert *Head Librarian* library services

3544 Wool Research Organisation of New Zealand (Inc.)

3545 Laboratories and Pilot Plant, Lincoln

Private Bag, Christchurch
Tel: 252-421. Telex: NZ4608. Cables: WRONZ. 252-717.

1 Simpson WS - MSc,PhD,FNZIC,FTI *Director*
2 Carnaby GA - BSc,PhD,ATI textile physics
3 Early DB - BE instrumentation and control
4 Edmunds AR - MSc,PhD,FNZEI wool measurement and marketing
5 Hammersley MJ - MA wool measurement
6 Ingham PE - BSc,PhD,MNZIC wool chemistry
7 Jamieson RG - MSc,MNZIC woolscouring
8 Lappage J - MSc,PhD,FTI textile engineering
9 McKinnon AJ - MSc,PhD,FNZIC textile chemistry
10 Orwin DFG - MAgrSc,PhD,MNZIAS wool growth
11 Ross DA - MAgrSc,PhD textile education

GOVERNMENT DEPARTMENTS

3546 Federal Ministry of Science and Technology

PMB 12793, 9 Kofo Abayomi Road, Lagos
Cables: SCTECH.

1 Iyamabo DE - BScFor,BSc,MF *Coordinating Director* silviculture and tree improvement

3547 Agricultural Sciences Department

1 Adetunji SA - BSc,MSc,PhD *Director*

3548 Arable Crops Division

1 Onuorah PE - BSc,MSc,PhD *Assistant Director*
2 Damagun Mrs - BSc *Scientific Officer II*
3 Ogbuehi SN - BSc,BS,PhD *Chief Scientific Officer*
4 Sarumi MB - BSc,MSc *Principal Scientific Officer*

3549 Tree Crops Division

1 Odegbaro OA - BSc,MSc,PhD *Assistant Director*

3550 Fisheries Division

1 Adegoke AS - BSc,MSc,PhD *Chief Scientific Officer*

3551 Manpower Training Division

1 Efionayi JAB - BSc,MSc,Eds,PhD,DipAVC *Chief Scientific Officer*

3552 Veterinary & Livestock Division

1 Oloko OA - BVM,MVSc,PhD *Chief Scientific Officer*

3553 Animal Production Division

1 Adebambo VO - BSc,PhD *Chief Scientific Officer*
2 Udo-Aka Mrs MI - BSc,MSc *Chief Scientific Officer*

3554 AERLS Division

1 Chiaha SC - BScMed,PhD *Chief Scientific Officer*

3555 Others

1 Sagua VO - BSc,PGCert,MIBiol *Director (Special Duties I)* downstream fishery
2 Aliyu A - DVM,MPVM *Director (Special Duties II)*
3 Abiola AA - BSc,MSc,PhD,MIBiol *Assistant Chief Officer*
4 Aribisala Mrs AO - BSc,PhD *Chief Scientific Officer*

3556 Ibadan Office

PMB 5382, Moor Plantation, Ibadan

1 Onweluzo BSK - BSc,MSc *Assistant Director* forestry division

3557 Library, Documentation and Information Division

1 Adigun Mrs OA - BSc,MED,DIPLib librarianship, information science

RESEARCH INSTITUTES

3558 Cocoa Research Institute of Nigeria

Private Mail Bag 5244, Ibadan
Tel: 0-22-41 2430,41 2400.

Conducts research into cocoa, coffee, cola, cashew and tea cultivation and utilization

1 Olatoye ST - BSc,PhD *Director* plant physiology
2 Adegbola MOK - BSc,MS,PhD *Assistant Director* plant pathology
3 Filani GA - BSc,PhD *Assistant Director* plant pathology
4 Williams JA - BSc,MS,PhD *Assistant Director* plant breeding

3559 Agronomy Division

1 Oladokun MAO - BSc,MSc,PhD *Head of Division* agronomy
2 Adenikinju SA - BSc,MS agronomy
3 Adeyemi AA - BSc,MSc agronomy

4 Olukotun OA - BSc agronomy

3560 Entomology Division

1 Omole MM - MSc,PhD *Head of Division* entomology
2 Idowu OL - BSc,DIC,PhD entomology
3 Igboekwe DA - BSc,MSc entomology
4 Ndubuaku TCN - BSc,MSc entomology
5 Okelana FA - BSc,MPhil entomology

3561 Plant Breeding Division

1 Esan EB - BSc,MS,PhD *Head of Division* plant breeding
2 Badaru K - BSc,MPhil plant breeding
3 Ojo AA - BSc,MS,PhD plant breeding

3562 Plant Pathology Division

1 Olunloyo OA - BSc,PhD *Head of Division* plant pathology
2 Ladeinde OA - BSc,MSc plant pathology
3 Odebode AC - BSc plant pathology

3563 Soils and Chemistry Division

1 Egbe NE - BSc,AICTA *Head of Division* soil chemistry
2 Ayodele EA - MSc soil chemistry
3 Chude VO - BSc,MSc,PhD soil chemistry
4 Famuyiwa O - BSc,MA,PhD microbiology
5 Obatolu CR - BSc,MSc soil chemistry

3564 Library Division

1 Oladele BA - BLs,MLs *Librarian* librarianship; information science

3565 Federal Institute of Industrial Research Oshodi

P.M.B.21023, Ikeja Airport, Lagos
Tel: 521010. Cables: Applied Ikeja.

1 Koleoso OA - BSc,MSc,PhD *Director of Research* management of industrial & technological research and development, research on industrial utilisation of vegetable oils and fats; flavours and essences, development of appropriate processing technology, product development, technology transfer and consultancy

3566 Food Wastes Utilisation

1 Oniwinde AB - MSc,PhD *Assistant Director (Lab)* biotechnology and toxicological studies
2 Kuboye Mrs AO - BSc,MSc *Assistant Chief Research Officer* production of industrial yeast from molasses
3 Ayanleye TA - BSc,MSc *Principal Research Officer* production of industrial enzymes and glucoside syrup
4 Lufadeju CB - BSc *Senior Research Officer* the production of sparkling wine from palm wine

3567 Engineering

3568 Protein Research & Nutrition

1 Omoloye OG - BSc development of protein based miscellaneous foods
2 Ozumba Mrs AU - BSc,MSc nutritional improvement and sensory evaluation of processed foods

3569 Micro-Biological Services

1 Solomon Mrs HM - Bsc,PhD *Head of Unit* preservation of fresh fruits
2 Akinola Mrs SO - BSc,MSc development and production of microbiological media
3 Dike Mrs E - BSc,MSc microflora of Nigerian fermented foods
4 Ojo Miss G - HND,BSc table wine from fruits

3570 Analytical Services

1 Akinsoji Mrs GM - MIST chemical analytical service, analysis of water, food
2 Ekpo MU - HND,AIST analytical chemistry
3 Irele Mrs PT - BSc,MSc management of analytical services

3571 Textile Research

1 Aladeselu I - BSC,MSc,ATI,GChem,PhD *Head of Unit* polymer and industrial chemicals
2 Balogun MD - HND,ATI traditional weaving technology
3 Ikpeama A - ATI,MSc,PhD cold water starch and adhesives

4 Odutola MO - BSc,ATI dyeing and quality control

3572 Library Information and Documentation

1 Sodipe RO - ALA *Head of Unit*
2 Adenaike BO - BSc,MSc documenting, retrieving and disseminating scientific and technological information
3 Aduwo Mrs MA - BSc,MLS patent documentation and industrial abstracts
4 Glover Miss OA - BSc,MLS documenting, retrieving and disseminating scientific and technological information

3573 Fruits, Vegetables, Flavours & Essences

1 Osinowo FAO - BSc,MSc,PhD *Head of Unit* product development of flavouring condiments, chemical and biochemical evaluation of Nigerian herbs and spices
2 Meadows AB - HND,MSc development of flavour concentrate from local raw materials, ginger concentrate
3 Rowland AK - MSc rubberseed oil

3574 Grains, Roots and Tubers

1 Onyekwere OO - BS,MS *Assistant Director and Head of Unit* grains, roots and tubers
2 Jibogun Mrs AC - BSc,MSc sorghum malt/sorghum beer
3 Olatunji O - BSc,PhD research and development in legume processing and utilization

3575 Forestry Research Institute of Nigeria
P.M.B.5054, Ibadan
Tel: 414022,414441,414073. Telex: 31207,Ibadan. Cables: Forsearch.

1 Kio PRO - BSc,MSc,PhD,DipFor *Director & Professor* mensuration
2 Dada GOB - BScFor,MScCA *Head of Division* extension/research liaison and information services
3 Odeyinde MA - BSc,MSc *Assistant Director* mycology

3576 Research Divisions

3577 Ecology Division

1 Gbile ZO - BSc,MSc,PhD,DipTax *Head of Division* taxonomy
2 Okorie PE - BSc,MSc ecology
3 Ola-Adams BA - BSc,MSc ecology
4 Onyeachusim HD - BSc,MSc taxonomy
5 Soladoye MOA - BSc,PhD taxonomy

3578 Education and Training Division

1 Ero II - BScFor,MScF,PhD *Head of Division* ecology/land use planning/forestry education
2 Adegbola PO - BSc,MSc,PGDWildlife fire ecology/wildlife

3579 School of Forestry, Ibadan
P.M.B.5054, Ibadan
Cables: Forsearch.

1 Oguntala AB - BSc,PhD *Principal* ecology
2 Abu JE - BScFor,MScAgrEcon forest economics
3 Adebayo O - BScFor,MScAgrEcon forest economics
4 Famuyide OO - BScFor,MScAgrEcon forest economics
5 Ikojo HA - BScFor,MSc mycology
6 Okafor CC - BSc,MSc pathology

3580 School of Forestry, Jos
P.M.B. 2019, Jos

1 Omiyale O - BScFor,MSc,PhD,DipFor *Principal* mensuration
2 Adebisi LA - BScFor,MSc silviculture
3 Omoayena BO - BScAgr tree breeding/ mensuration
4 Oyeleye B - BScFor,MSc biometrics
5 Shiyam CA - BScFor landscaping

3581 School of Wildlife Management, New Bussa
P.M.B. 268, New Bussa

1 Ekeke BA - BScAgr,MSc *Principal* silviculture/wildlife
2 Meduna AJ - BScFor wildlife
3 Onadeko SA - BScFor wildlife extension
4 Onyeanusi AE - BSc wildlife/range management
5 Osakwe ME - BScAgr wildlife pathology

3582 School of Forest Mechanisation, Afaka, Kaduna
P.M.B.2273, Afaka, Kaduna

1 Nwokedi CI - BSc,MScAgrEng *Project Manager* forest mechanisation

3583 Forest Economics and Sociology Division
P.M.B. 5054, Ibadan
Tel: 41 4022,41 4441, 41 4073. Telex: 31207, Ibadan. Cables: Forsearch.

1 Nwonwu FOC - BSc,MSc,PhD *Head of Division* forest economics
2 Bakare BO - BScAgr agricultural economics

3584 Forest Products Research Division

1 Ademiluyi EO - BScFor,MFR *Head of Division* timber engineering
2 Adenuga BO - BScFor,MScAgrExt mechanical wood processing
3 Badejo SOO - BSc,MFR timber engineering
4 Fajimi MA - BScFor,MSc timber seasoning
5 Ifebueme SC - BScFor,MSc wood preservation
6 Ogbogu GU - BScFor,MSc wood preservation
7 Omolodun NO - BScFor,MSc wood anatomy

3585 Hardwood Improvement Project Division

1 Ladipo DO - BSc,MIBiol,PhD *Head of Division* physiological genetics
2 Awoleye FO - BSc,MSc,PhD tree breeding
3 Oni O - BSc,MSc physiological genetics

3586 Library and Documentation Division

1 Adedigba YA - ALA,MALibandInfMagt *Head of Division* information use and networks
2 Adeboye JA - BLS,MLS,GCEd cataloguing and classification
3 Ette Mrs EN - BA,MLS,PGDLib abstracting and indexing
4 Ogbo Miss FC - BA,MALangArts translation

3587 Pathology and Entomology Division

1 Akanbi MO - BSC,MSc,PhD *Head of Division* entomology
2 Alebiodu IB - BSc,MSc entomology
3 Ashiru MO - BSc,MSc entomology
4 Fatuga B - BSc,MSc pathology
5 Obisesan KD - BA,MSc pathology
6 Ogungbenro BM - BSc,MSc pathology

3588 Planning and Biometrics Division

1 Adedoyin Mrs BI - BSc,CompSc *Head of Division* computer science
2 Bosah J - BScCompStats computer science/statistics

3589 Soils and Tree Nutrition Division

1 Kadeba O - BSc,MScSoilSc *Head of Division* tree nutrition
2 Aluko AP - BSc,MPhilSoilSc tree nutrition
3 Fagbenro JA - BScAgr,MScSoilSc,MScFor soil chemistry
4 Orimoyegun SOO - BSc,MSc soil chemistry

3590 Tree Crop Production Division

1 Ekwebelam SA - BScFor,MSc,PhD *Head of Division* silviculture
2 Adeola AO - BScFor,MF,MEd,PhD,PGDE agro forestry/social forestry
3 Obiaga PC - BScAgr,MSc silviculture/weed control
4 Odigie GAE - BScFor,MPhil silviculture/plant/water/soil relation
5 Oji NO - MScFor silviculture
6 Somade Mrs AF - BSc,MSc seeds technology

3591 Tree Improvement Division

1 Okoro Miss OO - BSc,MPhil *Head of Division* physiology
2 Akoun JA - BScAgr tree breeding
3 Oduwaiye EA - IngGrad,MScFor,DipFor physiology

3592 Research Stations

3593 Savanna Forestry Research Station, Samaru
P.M.B. 1039, Samaru, Zaria
Tel: 069-32591.

1 Ogigirigi MA - BSc,MSc *Head of Station* physiology
2 Adegbehin JO - BScFor,MSc,PhD mensuration
3 Ezenwa MIS - BScAgr,MSc soil survey
4 Gbadegesin RA - BScAgr,MSc pathology
5 Igboanugo ABI - BScAgr,PhD physiology
6 Momodu AB - BScFor,MSc soil physics
7 Nweze IO - BScMechEng forest mechanisation
8 Otegbeye GA - BScAgr,MSc physiology
9 Shado MB - BScFor,MSc silviculture

3594 Shelterbelt Research Station, Kano
P.M.B. 3229, Kano, Zaria

1 Ojo GOA - BScFor *Head of Station* silviculture
2 Adekiya FY - BSc,MF,PhD silviculture
3 Onyewuotu LOZ - BSc,MSc agro climatology
4 Ujah JE - MScAgr soils

3595 Eastern Forestry Research Station, Umuahia
P.M.B. 7011, Umuahia

1 Umeokafor EAU - BScFor,MScForMgt *Head of Station* pathology
2 Okeke AI - BScAgr silviculture

3596 Swamp Forest Research Station, Port-Harcourt
P.M.B. 6202, Onne, Port-Harcourt

1 Oddo AE - BScFor,MSc *Head of Station* silviculture

3597 Moist Forest Research Station, Sapoba
PO Box 2444, Benin City

1 Adebagbo CFAO - BScFor,MSc *Head of Station* silviculture

3598 National Animal Production Research Institute, Shika
P.M.B. 1096, Shika, Zaria
Tel: 32596 PBX. Cables: NAPRI,ZARIA,NIGERIA.

1 Nuru S - BVSc,MPVM,MRCVS,PhD *Director & Professor* livestock reproductive diseases
2 Olayiwole MB - BScAgr,MSc,PhD *Deputy Director & Professor* ruminant nutrition

3599 Research Division

3600 Animal Reproduction & Artificial Insemination

1 Oyedipe EO - DVM,MSc,PhD *Head of Department* livestock reproductive physiology
2 Adenowo TK - DVM livestock reproductive physiology
3 Bale JO - BSc,MSc,PhD livestock reproductive diseases (bacteriology)
4 Bawa EK - DVM livestock reproductive microbiology
5 Dawuda PN - DVM artificial insemination
6 Eduvie LO - DVM,MSc,PhD livestock reproductive physiology
7 Ekpe G - BSc,MSc livestock reproductive physiology
8 Marire BO - DVM livestock reproductive physiology
9 Nuru S - BVSc,MPVM,MRCVS,PhD reproductive diseases
10 Osinowo AO - BSc,MSc,PhD animal reproductive physiology
11 Pathiraja N - DVM,PhD swine reproductive physiology
12 Rekwot P - DVM artificial insemination
13 Sekoni VO - DVM,MSc,PhD artificial insemination
14 Voh AA - DVM,MSc artificial insemination

3601 Livestock Economics and Rural Sociology

1 Okaiyeto PO - BSc,MSc *Acting Head of Department* sociology
2 Adesipe YM - BSc,MSc livestock market economy
3 Ahmed HU - BSc,MSc livestock extension sociology
4 Gefu JO - BSc,MSc livestock rural sociology
5 Okon S - BScAgr,MSc livestock production economy

3602 Cattle Production

1 Umoh JE - BScAgr,MPhil,PhD *Head of Department* ruminant nutrition
2 Abubakar BY - BScAgr,MSc,PhD livestock breeding
3 Adamu AM - BSc,MSc,PhD nutrition
4 Alhassan WS - BSc,MSc,PhD nutrition
5 Buvanendran V - BVSc,MSc,PhD *Planning and Monitoring Officer* genetics and breeding
6 Ehoche OW - BScAgr,MSc ruminant nutrition
7 Johnson AO - BScAgr,MSc,PhD cattle breeding
8 Lufadeju EA - BSc,MSc ruminant nutrition
9 Olayiwole MB - BScAgr,MSc,PhD ruminant nutrition
10 Oni OO - DVM livestock breeding

3603 Poultry and Swine

1 Offiong SA - BSc,MSc,PhD *Head of Department* poultry husbandry
2 Akanbi O - BSc,MSc,PhD poultry nutrition and reproductive physiology
3 Jegede JO - BSc swine nutrition
4 Ogundipe JO - BScAgr,MSc,PhD poultry husbandry
5 Omeje SA - BSc,MSc,PhD poultry breeding
6 Sekoni AA - BSc poultry nutrition
7 Tegbe TSB - BSc,MSc,PhD swine nutrition

3604 Pasture Agronomy

1 Mzamane MN - BSc,MSc,PhD *Acting Head of Department* pasture agronomy
2 Agishi EA - BScAgr,MSc,PhD pasture agronomy
3 Ariba OO - BSc irrigated pasture agronomy
4 Kera BS - BSc rangeland management
5 Okeagu MU - BSc,MSc pasture agronomy
6 Onifade OS - BScAgr,MSc pasture agronomy
7 Otchere EO - BSc,MSc,PhD pasture and forages specialist
8 Tanko RJ - BSc rangeland management

3605 Small Ruminants

1 Adu IF - BSc,PhD *Head of Department* small ruminant nutrition
2 Lakpini CAM - DVM,MSc nutrition
3 Odeniyi A - DVM small ruminant husbandry
4 Osuhor C - BSc nutrition
5 Taiwo BB - BSc,PhD small ruminants breeding

3606 Animal Health

1 Jagun AG - DVM,MSc *Head of Unit* surgery / clinician
2 Adewuyi AA - DVM surgery / clinician

3607 Library

1 Adewole SO - MLSc *Librarian*

3608 National Horticultural Research Institute, Ibadan
P.M.B. 5432, Ibadan
Tel: 022-412490. Cables: NIHORT.

1 Adeyemi SAO - BSc,MSc *Director* entomology
2 Adedoyin SF - BSc,MSc extension services
3 Adefolaju Mrs PI - BSc,MLs *Librarian*

3609 Vegetable Division

1 Denton OA - MSc,PhD *Head of Division* vegetable breeding
2 Olufolaji Mrs OA - BSc,MPhil plant physiology

3610 Fruits Division

1 Adelaja BA - BSc,MPhil *Head of Division* plant physiology

3611 Citrus Division

1 Amih CA - BSc,MSc *Head of Division* citriculture

3612 Chemistry and Utilization Division

1 Usoroh NJ - BSc,PhD *Head of Division* weed science
2 Akhigbe GO - BSc,MPhil biochemistry
3 Taylor Mrs OA - BSc,PhD biochemistry

3613 Crop Protection Division

1 Ogunlana MO - BSc,PhD *Head of Division* entomology
2 Adebanjo A - BSc,MPhil plant pathology
3 Agunloye O - BSc,MSc entomology
4 Anno-Nyakko - BSc,PhD virology

3614 Agricultural Economics and Statistics Division

1 Ogunfidodo A - BSc,MSc biometrics

3615 Agricultural Engineering Division

1 Asoegwu SN - BSc,MSc agricultural engineering

3616 National Root Crops Research Institute
Umudike, Umuahia, Imo State
[PMB 7006, Umudike]
Tel: 088-220188. Cables: Agrisearch, Umudike.

The National Root Crops Research Institute, Umudike, has the mandate to conduct research into the production and products of root and tuber crops (cassava, yam, Irish potato, cocoyam, sweet potato, ginger, etc) of economic importance. It further ensures that agricultural research findings from the various research centres are collected, interpreted and made available to farmers in the four eastern states of Nigeria, in the form that they can use them for increased and efficient agricultural production

1 Ene LSO - BScAgr,PhD *Director* cytogenetics

3617 Research Section
Headquarters, Umudike, PMB 7006, Umudike, Umuahia, Imo State
Tel: 088-220188. Cables: Agrisearch, Umudike.

1 Okoli OO - BScAgr,MScAgron,PhDCrpSc *Assistant Director (Research)*

3618 Cassava Programme

1 Okeke JE - BScAgr,PhDAgron *Coordinator* soil nutrition
2 Echendu TNC - BScAgr,PhDPlantSc entomology
3 Emehute JKU - BSc,MScZoo entomology
4 Nnodu EC - BScBot,MSc,PhD plant pathology - post harvest losses
5 Nwokedi PM - BSc,MScEng agricultural engineering
6 Oti E - BSc,MScBioChem food biochemistry
7 Ugwu BO - BSc,MPhilAgrEcon production & marketing economics

3619 Yam Programme

1 Alozie SO - BScAgrBiochem,MScBiochem,PhD *Toxin Coordinator* storage biochemistry
2 Agbo FMO - BScBot,MSc,PhDAgron genetics and plant breeding
3 Enyinnaya AM - DipAgr,BScAgr,MSc,PhDAgron production systems
4 Igwilo NH - BScAgr,PhDAgron growth and environment physiology
5 Nwakiti AO - BScBot,MSc,PhD plant pathology
6 Obi WU - BScAgrEng,MScEng crop processing and storage

3620 Irish Potato Programme
Headquarters, Vom, PMB 4, Vom-Jos, Plateau State

1 Ifenkwe OP - BSc,PhDAgron *Co-ordinator* crop science
2 Nwokocha HN - BSc,MScBot plant pathology
3 Okonkwo JC - BScPlantSoilSc,MSc,PhDAgron seed production and storage

3621 Cocoyam Programme
Umudike, Umuahia
Tel: 088-220188. Cables: Agrisearch, Umudike.

1 Arene OB - BABiol,MScPlantPath,PhDCropSc *Coordinator* plant pathology
2 Ijioma Mrs BC - BSc,MScMicrobiol food and industrial microbiology
3 Mbanaso Mrs ENA - BScBot,MPhilPlantPhy plant tissue culture
4 Orkwor GC - BScPlantSoilSc,MScAgron weed science

3622 Sweet Potato Programme

1 Nwinyi SCO - DipAgr,BScChem,MScSoilSc *Coordinator* soil fertility
2 Chikwendu CC - BScBot,MSc,PhDAgr plant breeding
3 Ogbuehi CRA - DipAgr,BSc,MSc,PhDAgron horticulture
4 Okorocha Mrs EA - BScZoo,MScBot phytonematology

3623 Farming Systems Research Programme

1 Unamma RPA - BScPlantSoilSc,DipCropProt,MSc,PhDAgron *Coordinator* weed science
2 Anuebunwa FO - BSc,MScAgrEcon marketing economics
3 Ezulike TO - BSc,MPhilAgrBiol economic entomology
4 Ikeorgu JEG - BScCropSc,MSc,PhDAgron farming systems
5 Nnoke FN - BScPlantSoilSc,MPhilSoilSc,DipAgron soil science
6 Ogbonnaya JC - BScPlantSoilSc,MScCropSc agronomy
7 Ohiri AC - BSDScChem,MSc,PhDAgron soil physics
8 Udealor A - BScBot,MScCropSc plant physiology

3624 Ginger and other Root Crops Programme

1 Njoku BO - BScChem,MScSoilSc *Co-ordinator* soil chemistry, crop nutrition
2 Ezeh NOA - BSc,MScAgrEcon industrial/managerial economics
3 Okwuowulu PO - DipAgr,CI,MIBiol assimilate biology
4 Onyia GOC - BSc,PhDBiochem storage biochemistry, processing

3625 Agricultural Extension Research Liaison Services Section
Headquarters, Umudike, P.M.B. 7006, Umudike, Umuahia, Imo State
Tel: 088-220188. Cables: Agrisearch, Umudike.

1 Odurukwe SO - BSc,MScPlantSoilSc,PhDAgron *Coordinator* soil fertility
2 Chinaka CC - BScPlantSoilSc,MScAgron crop production
3 Enyinnia T - BSc,MSc,PhDPlantPath chemical pest control
4 Ibe MU - BScAgr,DipAgr,MScEd training
5 Iloka AW - BScAgr monitoring and evaluation
6 Iwueke CC - BScAgrEcon,MScAgrEd communication
7 Ogidi CW - DipAgrEng,BEd technical education

3626 Training Section (Federal School of Agriculture)
Headquarters, Umudike, P.M.B. 7006, Umudike, Umuahia, Imo State
Tel: 088-220188. Cables: Agriscol, Umudike.

1 Igbokwe MC - DipAgr,BScPlantSoilSc,PhDAgron *Principal* soil fertility
2 Agbakoba AM - DipAgr,BScAgr,MScAnSc animal production and management
3 Agu TC - BEng,MEng soil and water conservation
4 Ano AO - BScChem,MScAnChem soil chemistry
5 Ejiofor MAN - BScBot,MSc,PhDMicrobiol food microbiology and processing
6 Eluagu LS - DipAgr,BAgr,MScAgrEcon marketing economics
7 Ibedu MA - BScSoilSc,MScAgron soil chemistry
8 Melifonwu AA - BScPlantSoilSc,MScAgron weed science
9 Nwachukwu EC - BScBot,MScCropSc plant breeding
10 Nwadikom GI - BEng,MEng farm power and machinery; processing
11 Nwauzor EC - BScBiol,PhDPlantPath nematology
12 Oji Mrs MA - DipEd,BSc,MScHomeEc food science and nutrition
13 Ujoh SCU - DipAgr,BScAnSc,DipEd *Dean of Students Affairs*
14 Ukpabi UJ - BSc,MScBiochem food and plant biochemistry

3627 Nigerian Institute For Oil Palm Research
Head Office, P.M.B. 1030, Benin City

1 Ataga DO - BScChem,PhDSci *Director*

3628 Agricultural Economics Division

1 Eneh FK - BScAgr,Econs *Assistant Chief Research Officer*
2 Inedia G - BScEcons,MScEcons *Senior Research Officer*
3 Orewa SI - BSc,MSc,PhDAgrEcons *Assistant Chief Research Officer*
4 Udom DS - BScAgr,PhDAgrEcons *Assistant Chief Research Officer*

3629 Agronomy Division

1 Onwubuya II - BScPlSci,PhD *Chief Research Officer*
2 Otulu SN - BSc,MSc,PhDWdSci *Principal Research Officer*
3 Udosen CV - BScBot,MPhilAgrBiol *Senior Research Officer*

3630 Agricultural Engineering Research Division

1 Babatunde OO - BScAgrEngr,MScAgrEngr *Research Agric Engineer I*
2 Badmus GA - BScAgrEngr,MScAgrEngr *Senior Agric Research Engineer*
3 Ilechie CO - BScAgrEngr,MScAgrEngr *Senior Agric Research Engineer*
4 Okoli JU - BScMechEngr *Research Agric Engineer I*

3631 Biochemistry Division

1 Okiy DA - BScBioch,MScFSci,MPhilFsi,PhDAppChem *Assistant Chief Research Officer*
2 Abigor RD - BScBioch,MScBioch *Research Officer I*
3 Bafor Mrs ME - BScChem,MScFSci *Principal Research Officer*
4 Enonuya DOM - BScBioch,MScBioch *Senior Research Officer*

3632 Chemistry Division

1 Omoti U - BScChem,MScSlSci,PhDSolSci *Chief Research Officer/Programme Leader Oil Palm*
2 Aghimien AE - BScSolSci,MSc,PhDSolChem *Senior Research Officer*
3 Aisueni NO - BSc,MSc,PhDMicrob *Senior Research Officer*
4 Ejedegba BO - BScAgr,MScBioch *Research Officer I*
5 Isenmila AE - BScSoilSc,MScSoilPhysics *Research Officer I*

3633 Entomology Division

1 Agwu SI - BScZool,PhDDicAppEnt *Assistant Director/Programme*
2 Aisagbonhi CI - BScFor,MPhilEntom *Research Officer*
3 Appiah FO - BScAgr,MPhilEntom *Senior Research Officer*
4 Enobakhare DA - BScZool,MSc,PhDEntom *Senior Research Officer*

3634 Extension, Research, Liaison and Services Division (ERLSD)

1 Eseigbe RA - MScExten *Chief Scientific Officer/Acting Programme Leader* extension & training

3635 Library Division

1 Egbune Mrs E - MAHis,DipLib *Principal Librarian*

3636 Physiology Division

1 Remison SU - BScCrSci,PhDPlEcol/Physiol *Chief Research Officer/Acting Programme Leader* coconut
2 Olumekun VO - BScBot,MScAgrBiol/CrPhsl *Research Officer I*

3637 Plant Breeding Division

1 Otedoh MO - BScBot,PGD,PhD *Chief Research Officer/Programme Leader* raphia
2 Akpan EEJ - BScBot,PGDAgr,MSc *Assistant Chief Research Officer* plant breeding/genetics
3 Odewale JO - BSc,MSc,PhDBot *Senior Research Officer*
4 Okolo EC - BScAgr,MSc,PhDBot *Senior Research Officer*
5 Okwuagwu Mrs CO - BSc,MSc,PhD *Principal Research Officer*

3638 Pathology Division

1 Odigie EE - BScMBiol,MScPlPath *Senior Research Officer*
2 Ojomo EE - BScBiol,MScMBiol *Senior Research Officer*
3 Omamor Mrs IB - BScBot,PhDPlant Pathology *Principal Research Officer*
4 Oritsejafor JJ - BScBot,MScAgrBiol,PhDAgrBiol *Principal Research Officer*
5 Oruade-Dimaro EA - BScBot,MScAgrBiol *Senior Research Officer*

3639 Statistics Division

1 Nnabuchi SE - HNDAgr,DipAgrStat,MScStat *Chief Research Officer*

3640 Nigerian Institute for Trypanosomiasis, Kaduna
Head Office, P.M.B. 2077, Kaduna
Tel: 210271, 210292. Cables: NITR, Kaduna.

1 Magaji Y - BSc,MSc *Director*
2 Onyiah JA - BSc,PhD *Assistant Director*

3641 Entomology Division

1 Gregory WG - BSc,PhD *Head of Division*
2 Ajala TO - AIST *Principal Laboratory Technologist*
3 Offor Mrs II - BSc,MSc *Assistant Chief Research Officer* tsetse pheromone
4 Omoogun GA - BSc,MSc *Assistant Chief Research Officer* applied entomology, tsetse ecology

3642 Pathology Division

1 Edeghere HUF - BSc,PhD,DIC *Head of Division* diagnosis and pathogenesis of African trypanosomiasis
2 Babalola ML - DVM,MVM *Senior Veterinary Research Officer* pathophysiology of parasitic infections
3 Falope OO - DVM *Senior Veterinary Research Officer*
4 Olatunde DS - AIST,AIMW *Chief Technologist*
5 Uduebho MO *Senior Laboratory Technologist*

3643 Library and Information Division

1 Thompson GA - BSc,PGDLS *Head of Division*

3644 Branch Office
P.M.B. 3, Vom, Plateau State

1 Agu WU - DVM,PhD *Officer-in-Charge*

3645 Biochemistry Division

1 Emeh JK - BSc,MSc *Head of Division*
2 Akanji MA - BSc,MSc *Senior Research Officer*

3646 Parasitology Division

1 Ekejindu GOC - BSc,PhD *Head of Division*
2 Uzoiggwe RN - BSc,MSc *Senior Research Officer*

3647 Veterinary and Livestock Studies Division

1 Agu WE - DVM,PhD *Head of Division*
2 Kalu AU - DVM,MPhil *Principal Veterinary Research Officer*

3648 Rubber Research Institute of Nigeria
P.M.B. 1049 (Iyanomo), Benin City

1 Fasina AB - BSc,PhD *Assistant Director* polymer science
2 Odetola JA - BSc,MSc,PhD *Assistant Director* plant physiology
3 Okaisabor EK - BSc,PhD *Director* plant pathology

3649 Agronomy Division

1 Agbaka AC - BSC,MSc agronomy
2 Nwankwo BA - BSc,MSc,PhD botany
3 Sagay GA - BSc,MSc agronomy

3650 Extension and Research Liaison Division

1 Aigbekaen EO - BSc,MSc agricultural extension and rural sociology
2 Idemudia PAE - BSc,MSc extension education
3 Uraih OBC - BSc,DipAgrExt,MSc,PhD

3651 Plant Breeding Division

1 Olapade EO - BSc,MSc,PhD plant breeding

3652 Plant Pathology and Microbiology Division

1 Begho ER - BSc,MPhil plant pathology
2 Igeleke CL - BSc,MSc plant pathology
3 Otoide VO - BSc,MSc microbiology

3653 Quality Control Division
P.M.B. 1049 (Iyanomo), Benin City

1 Fasina AD - BSc,PhD polymer science

3654 Soils and Plant Nutrition Division

1 Awa AA - BSc,PhDAgr biochemistry and animal nutrition (animal science)

2 Onuwaje OU - BSc,MSc soil chemistry
3 Salawu RA - BSC,MPhil soil science
4 Ugwa - BSc,MSc *Head Akwete Substation* soil science

3655 Nigerian Stored Products Research Institute, Lagos
P.M.B. 12543, Lagos
Tel: 862653 863321.

1 Agboola SD - BSc,MSc *Director*
2 Oyeniran JO - BSc,PhD *Assistant Director*

3656 Research Division
P.M.B. 12543, Lagos
Tel: 862653 863321.

1 Akinwande Miss O - BSc,MSc post harvest microbiology
2 Aror E - BSc,MSc post harvest microbiology; microbiology
3 Mejule Mrs FO - BScAgr,MSc pests and pest control; entomology
4 Onyekwelu IU - BSc,MSc,PhD biochemical deterioration, biochemistry

3657 Ibadan Branch
P.M.B. 5044, Ibadan
Tel: 413122.

1 Okoye WI - BSc *Head of Unit* analysis of oils
2 Adesida MA - BSc,PhD extension
3 Akano Mrs DA - BSc,MSc post harvest microbiology
4 Kuku Mrs FO - BSc microbiological deterioration of palm kernel
5 Meretiwon B - BSc,MSc extension of research findings
6 Onwuzulu OC - BSc,MSc studies of storage structures
7 Ubani Mrs ON - BSc,MSc biochemical studies
8 Williams JO - BSc,MSc,PhD pest and pest control

3658 Kano Branch
P.M.B. 3032, Kano
Tel: 3372.

1 Sowunmi O - BSc,PhD *Head of Unit* biochemical studies of stored food
2 Abiodun AA - BSc,MSc extension of research findings
3 Babarinsa FA - BSc,MSc studies on storage structure improvement
4 Ibrahim MH - BSc,MSc post harvest microbiology
5 Ndam ON - BSc,MSc insect pests

3659 Sapele Branch
P.M.B. 4065, Sapele
Tel: 42602.

1 Opadokun JS - BSc,MSc *Head of Unit*
2 Dabor Miss DO - BSc,MSc biochemist
3 Ogali Mrs EL - BSc,MSc microbiology
4 Olatunde GO - BSc,MSc biology of pests

3660 Port Harcourt Branch
P.M.B. 5063, Port Harcourt
Tel: 335514/335403/330009.

1 Akinnusi OA - BSc *Head of Unit* studies on storage structures improvement
2 Barbar FO - BSc,MSc post harvest microbiology
3 Roberts SI - BSc chemical analysis of post harvest crops

UNIVERSITIES & COLLEGES

3661 Ahmadu Bello University
P.M.B. 1044, Samaru, Zaria

Includes the Institute of Agricultural Research, Samaru. Details of research staff not available for this edition.

3662 University of Ibadan

3663 Department of Agriculture
Tel: 400550–400614 & 46246-7. Cables: University Ibadan.

1 Egunjobi Prof JK - BSc,PhD *Professor* agricultural biology
2 Youdeowei Prof A - BSc,PhD *Professor* agricultural biology
3 Adedipe NO - BScAgr,MSc,PhD agricultural biology
4 Adeniji MO - BSc,MSc,PhD agricultural biology
5 Adesiyan SO - BScAgr,PhD agricultural biology
6 Egunjobi Mrs OA - BSc,MSc,PhD agricultural biology
7 Ekpo EJA - BScAgr,MSc,PhD agricultural biology
8 Fawole B - BScAgr,MSc,PhD agricultural biology
9 Fawole I - BSc,MPhil,PhD agricultural biology
10 Funmilayo O - BSc,PhD agricultural biology
11 Ikotun T - BScAgr,PhD agricultural biology
12 Ivbijaro MF - BScAgr,PhD agricultural biology
13 Ladeinde TAO - BScAgr,PhD agricultural biology
14 Odebiyi JA - BScAgr,PhD agricultural biology

15 Ogunyemi Mrs S - BSc,MPhil,PhD agricultural biology
16 Osisanya EO - BSc,PhD agricultural biology
17 Tayo TO - BScAgr,PhD agricultural biology

3664 Kainji Lake Research Institute
P.M.B. 6006, New Bussa, Kwara State

1 Ayeni JSO - PhD *Director* wildlife ecology

3665 Limnology Division
P.M.B. 6006, New Bussa, Kwara State

1 Adeniji HA - MSc *Head of Division* hydrobiology
2 Mbagwu IG - BSc hydrobiology
3 Ogbondeminu FS - BSc bacteriology
4 Ovie SI - MSc hydrobiology

3666 Fisheries Division
P.M.B. 6006, New Bussa, Kwara State

1 Ita EO - MSc *Head of Division* fisheries/biology
2 Balogun JK - MSc fisheries biology
3 Bankole NO - BSc fisheries biology
4 Eyo AA - MSc fish processing
5 Okoye FC - MSc aquaculture
6 Omorinkoba WS - BSc fisheries biology
7 Otubusin SO - MSc aquaculture
8 Sado EK - MTech aquaculture

3667 Agriculture (Irrigated) Division
P.M.B. 6006, New Bussa, Kwara State

1 Erinne EC - MSc *Head of Division* irrigation engineering
2 Abolaji JO - BSc agronomy
3 Awoyemi MD - MSc crop protection

3668 Wildlife/Range Division
P.M.B. 6006, New Bussa, Kwara State

1 Okaeme AN - MSc *Head of Division* wildlife parasitology
2 Agbelusi EA - BSc wildlife ecology
3 Ayorinde KL - MSc animal production
4 Daddy F - MSc range management
5 Ibiwoye TII - MSc small ruminant parasitology
6 Obot EA - PhD aquatic macrophyte ecology
7 Oyatogun MOO - MSc agroforestry

3669 Socio-Economics/Extension Division
P.M.B. 6006, New Bussa, Kwara State

1 Ayande JO - MSc *Head of Division* socio-economics of resettlements
2 Bello S - BSc extension services

3670 Rivers State University of Science & Technology
P.M.B. 5080, Port Harcourt
Tel: 335808, 335823. Cables: RIVERSTECH.

1 Orshi His Highness JA - BL.LLD *Chancellor*
2 Banigo EOI - BSc,PhD,DipChemEngr *Professor, Vice-Chancellor*

3671 Department of Soil Science, Crop Science and Forestry

1 Isirimah NO - BSc,MSc,PhD *Acting Head of Department* soil fertility; biochemistry
2 Abell SO - BSc,MSc,PhD *Lecturer* ecology
3 Burutolu EPA - BSc,MSc *Lecturer* soil fertility
4 Douglas DC - BSc,PhD *Senior Lecturer* crop production; horticulture
5 Ebong UU - BSc,DipAgr *Senior Lecturer* plant breeding
6 Ijeh-Tarila AG - BSc,MSc *Lecturer* crop management
7 Loganathan P - BSc,MSc,PhD *Associate Professor* soil chemistry
8 Nokoe S - BSc,MF,PhD *Senior Lecturer* inventory and biometrics
9 Ojanuga AG - BScAgr,MSc,PhDPed *Professor* soil survey; land evaluation
10 Opuwaribo EE - BSc,PhD *Senior Lecturer* soil chemistry, fertility
11 Ossom EM - BSc,MPhil,PhD *Lecturer* plant physiology
12 Sutton PM - BSc,MSc,PhD *Senior Lecturer* pedology
13 Tufuor K - BSc,MSc,PhD *Senior Lecturer* silviculture and plant breeding
14 Wahua TAT - BSc,MSc,PhD *Senior Lecturer* farming systems
15 Zuofa K - DipLand-Wirt,DoktorAgrar *Senior Lecturer* crop production

3672 Department of Food, Animal Science

1 Oji UI - BSc,MSc,PhD *Senior Lecturer, Acting Head of Department* nutrition
2 Achinewhu SC - BSc,MSc,PhD *Senior Lecturer* food toxicology; food chemistry

3 Akpapunam MA - BSc,MSc,PhD *Lecturer I* food chemistry
4 Alawa JP - BSc,MPhil,PhD *Lecturer II* animal production
5 Deedua IW - BSc,MSEH,PhD *Lecturer II* biochemical, human nutrition
6 Denenu EO - BSc,PhD *Senior Lecturer* biochemical engineering
7 Giami SY - BSc,MSc *Assistant Lecturer* food microbiology
8 Ibiama Mrs E - BSc,MSc *Lecturer I* food chemistry; food analysis
9 Mepba HD - HND,MSc *Assistant Lecturer* food fermentation; fruits and vegetable processing
10 Monsi A - BSc,MSc,PhD *Senior Lecturer* avian reproductive physiology
11 Orok EE - BSc,MSc *Assistant Lecturer* animal reproductive physiology
12 Oruwari BM - BSc,MSc,PhD *Lecturer II* animal physiology

3673 Department of Agricultural Economics and Extension

1 Jaja JS - BSc,MSc,PhD *Acting Head of Department* co-operatives
2 Acquah BK - BSc,MSc,PhD *Lecturer* farm management and production economics
3 Chukuigwe EC - BSc,MSc *Lecturer* public sector analysis and agricultural finance
4 Ekine DI - BSc,MSc *Lecturer* agricultural marketing
5 Emah GN - BSc,MSc,EdS *Lecturer* agricultural education
6 Jumbo I - BSc,MSc *Lecturer* farm management and production economics
7 Nkanta DM - BSc,MSc *Lecturer* agricultural development
8 Sutton JUM - BSc,MSc *Lecturer* agricultural communication
9 Yorama MT - BSc,MSc,EdD *Lecturer* agricultural education

3674 Institute of Agricultural Research and Training

1 Achinewhu SC - BSc,MSc,PhD *Acting Director* fermentation technology; biotechnology
2 Ndegwe NA - BSc,PhD *Head of Research Department* maize, cassava, yam and egusi melon agronomy
3 Alagoa CN - HND studies on mangrove oyster and marine pollution
4 Daniel-Kalio LA - BSc,MSc,PhD *Head of Section* agronomy and pest control in okra and soya-bean
5 Ikpe F - MSc winged bean agronomy
6 Jaja ET - BSc yam and winged bean agronomy
7 Mishran HK - BSc,MSc,PhD soil potassium studies
8 Ogbonda C - BSc,MSc studies on phytoplankton
9 Pathak VN - BSc,MSc,PhD *Head of Section* soil fertility and fertilizer use studies
10 Singh SP - BSc,MScAgr *Head of Section* socio-economic and adoption survey
11 Torunana JMA - BSc,MSc soil fertility improvement studies
12 Wariboko C - BSc,MSc,PhD maize breeding

3675 University of Ife

3676 Institute of Agricultural Research and Training
Head Office, P.M.B. 5029, Moor Plantation, Ibadan
Tel: 312861. Cables: Instagric., Ibadan.

1 Olaloku EA - BSc,MSc,PhD *Director*

3677 Research Division

1 Sobulo RA - BSc,MAgrSc,PhD *Principal Research Fellow, Assistant Director* soil science
2 Taylor TA - BSc,PhD,FRES,FIBiol,FSAN entomology

3678 Farming Systems Programme

1 Omidiji MO - BSc,PhD *Head of Programme* plant breeding
2 Adelana BO - BSc,PhD agronomy
3 Adenle VO - BSc,PhD pathology
4 Akande M - BSc,MSc,PhD animal ecology
5 Akinlosotu TA - BSc,DIC,PhD entomology
6 Daramola AM - BSc,PhD *Principal Research Fellow* entomology
7 Oyedokun JB - BSc,MSc agronomy

3679 Food Legumes Improvement and Production Programme

1 Ogunfowora AO - BSc,DIC,PhD *Head of Programme* nematology
2 Afolabi NO - BSc,MSc agronómy
3 Ogunbodede BA - BSc,MPhil plant breeding
4 Omueti O - BSc,MSc,PhD biochemistry
5 Oyekan PO - BSc,MSc,PhD plant pathology
6 Raji JA - BSc,MPhil plant breeding
7 Soyinka SA - BSc,PhD virology

3680 Industrial Crops Improvement and Production Programme

1 Obajimi AO - BSC,MSc,PhD *Head of Programme* plant breeding agronomy
2 Ogunremi EA - BSc,PhD agronomy

3681 Soil, Water Management Programme

1 Ojo-Atere JO - BSc,MSc,PhD *Head of Programme* soil pedology
2 Ajuwon SO - BSc,MSc,PhD soil physics
3 Awonaike KO - BSc,PhD microbiology
4 Banjoko VA - BSc,PhD soil chemistry
5 Osiname OA - BSc,MSc,PhD soil science

3682 Livestock Improvement and Production Programme

Head Office, P.M.B. 5029, Moor Plantation, Ibadan
Tel: 312861. Cables: Instagric., Ibadan.

1 Adebowale EA - BSc,PhD *Head of Programme* animal nutrition
2 Adebambo OA - BSc,PhD animal breeding
3 Ogundola FI - BSc,PhD dairy science
4 Olaloku EA - BSc,MSc,PhD dairy production
5 Oluokun JA - BSc,MSc,PhD mineralogy

3683 Services Division

1 Osikanlu YOK - BSc,MSc,Phd *Acting Assistant Director* plant pathology

3684 Soil Survey and Testing Service

1 Olomu EI - BSc,MSc *Head of Service* soil science
2 Makanjuola JS - BSc,MSc soil science

3685 Seed Production and Testing Service

1 Osikanlu YOK - BSc,MSc,PhD *Head of Service* plant pathology
2 Ayoade SK - HigherDipAgr,CertSeedTech *Principal Agricultural Superintendent*
3 Kogbe JOS - BSc,PhD agronomy
4 Majaro OO - HigherDipAgr *Assistant Chief Agricultural Superintendent*

3686 Agricultural Extension Research Liaison Service

1 Fadare TA - BSc,MSc *Head of Service* entomology

3687 Statistics Unit

1 Onakade AD - BSc,MSc *Head of Unit* biometrics
2 Amusan JA *Assistant Chief Agricultural Superintendent*

3688 Communications Unit

1 Animasawun BO - HigherDipAgr *Principal Agricultural Superintendent*

3689 Training Division

1 Akintunde FO - BSc,MSc *Assistant Director (Training)* poultry science

3690 School of Agriculture, Ibadan

1 Ositelu GS - BSc,MSc *Principal of School* animal nutrition
2 Adegbulugbe TA - BSc,MSc agricultural engineering
3 Adetayo BO - BSc,MSc agronomy
4 Adetoro AO - BSc,MSc,PhD soil science
5 Akindele AA - BSc,MSc,MNSAE,MNSE agricultural engineering
6 Alaka RT - BSc,PGDE dairy technology
7 Dada FA - BSc,PGDEMed agronomy
8 Daramola DS - BSc,MPhil soil science
9 Lanipekun AG - BSc,MSc agricultural economics
10 Ogundipe MA - BEd,MSc home economics and extension
11 Oguntoyinbo FI - BSc,PGDE,MSc agricultural chemistry
12 Oyeneye SA - BSc,MSc agricultural economics
13 Sowonola OA - BSc,MSc nutrition

3691 School of Agriculture, Akure

1 Adewunmi AI - BSc,PhD *Acting Principal of School* agricultural extension
2 Adebayo SI - BSc,PGDE agronomy
3 Ononogbu CE - DVM animal breeding

3692 School of Animal Health & Husbandry, Ibadan

1 Crockett EC - BVMS,MRCVS,MSc *Principal of School* parasitology
2 Archebamen CPE pathology
3 Faniyan AA - BVSc surgery
4 Ikusika O - DVM microbiology
5 Jolaosho AA - DVM pharmacology
6 Nwufoh KJ - DVM skin diseases
7 Ogbonna JU - BSc,MSc animal nutrition
8 Oguntoyinbo OG - DVM,MSc avian medicine
9 Popoola O - DVM medicine

3693 University of Nigeria, Nsukka

Nsukka

3694 Faculty of Veterinary Medicine

Nsukka
Tel: 042/771911. Cables: Nigersity Nsukka.

3695 Department of Veterinary Pathology and Microbiology

1 Mohan K - BVSc,AH,PhD *Associate Professor and Head of Department* veterinary bacteriology; mycoplasma of small ruminants
2 Chineme CN - DVM,PhD *Professor* experimental and comparative pathology
3 Kamalu Mrs BP - DVM *Lecturer* haematology
4 Komolafe OO - DVM,MSc,PhD *Senior Lecturer* virology, parvo-viruses
5 Oboegbulem S - DVM,MS *Senior Lecturer* veterinary public health, salmonellosis
6 Okoye JOA - DVM,MPhil *Lecturer* gumboro disease
7 Onyekweodiri EO - DVM,FRCVS *Lecturer* general pathology and meat inspection
8 Shoyinka SVO - DVM,MSc *Lecturer* dermatophilosis
9 Uzoukwu M - BVMS,MS,PhD *Professor* comparative pathology

3696 Department of Veterinary Medicine

1 Uzoukwu M - BVMS,MS,PhD *Professor, Head of Department* experimental and comparative medicine, comparative pathology
2 Chukwu CC - DVM,MSc,PhD *Lecturer* sero-epidemiological studies of animal brucellosis in Nigeria
3 Emehelu CO - DipAH&H,DVM *Lecturer* studies on helminthiasis among indigenous dogs in Nsukka, Anambra State
4 Okolo MIO - BVetMed,MSc,MRCVS *Lecturer* epidemiology of schistosomiasis in human and animal population in Anambra State
5 Orajaka LJE - DipAH&H,DVM *Lecturer* pathogenicity of *E.coli* isolates from dead-in-shell
6 Uche EU - DVM,MSc *Lecturer* studies on canine trypanosomiasis in Anambra State
7 Ugochukwu EI - DVM,MPhil *Lecturer* pathogenicity and antibiogram of bacterial isolates from diarrhoeic goats
8 Wosu LO - BVM,MSc,PhD *Senior Lecturer* development of techniques for quick diagnosis of rinderpest and peste de petit ruminants (PPR) in Nigeria

3697 Department of Veterinary Parasitology and Entomology

1 Ikeme MM - BSc,PhD,DipBiol *Professor, Head of Department* veterinary helminthology
2 Chiejina SC - DVMS,PhD,MRCVS *Senior Lecturer* veterinary parasitology
3 Hegde KS - BSc,MS,PhD *Professor* veterinary parasitology
4 Madubunyi LC - BSc,MS,PhD *Senior Lecturer* veterinary entomology

3698 The Polytechnic, Ibadan

Head Office, Satellite Campus, Iree
Tel: 410151-4. Cables: POLYTECHNIC, IBADAN.

1 Dina SO - BSc,MSc *Head of Department* cowpea entomology
2 Adeyeye A - BSc,MSc *Lecturer I* residue analysis
3 Ogundiya MO - BSc,MSc *Lecturer I* breeding
4 Oladiran AO - BSc,MPhil,PhD *Senior Lecturer* cowpea pathology

3699 The Polytechnic of Sokoto State, Birnin Kebbi

Head Office, P.M.B. 1034, Birnin Kebbi
Tel: 97. Cables: Polysok.

1 Abdullahi AN - BSc,CPEIP,MA,FBIM,FIPC *Rector*
2 Gulma MA - BEngr,PhD,FIIE,MIEE,MSESM,MCORN *Director CST*

3700 Department of Agricultural Engineering

1 Saley SD - HND,PGD,MSc,MBA,MIAgrE *Head of Department* brown sugar processing plant
2 Boama IY - BSc,VVITA *Lecturer* low head diaphragm pump for irrigation
3 Mohammed MA - HND,MSc,MCIT *Lecturer* local materials for drip irrigation
4 Ojediran JO - BSc,MSc,MCIT *Lecturer* solar drier

3701 Department of Sciences

1 Ajikobi SO - BSc,PGD,MSTAN *Head of Department* design of solar incubator

3702 Central Research Committee

1 Gulma MA - BEng,PhD,FILE,MIEE,MSESN,MCOREN *Director* solar energy and evapotranspiration

INTERNATIONAL INSTITUTIONS

3703 International Institute of Tropical Agriculture
Oyo Road, P.M.B. 5320, Ibadan
Tel: 413224/413315/413440. Telex: TROPIB NG 31417 and TDS IBA NG 20311 (Box 015). Cables: TROPFOUND IKEJA.

3704 Administration
1 Hartmans EH - PhD *Director General*
2 Okigbo BN - PhD,DSc *Deputy Director General*
3 Akintomide MA - BS *Director for Administration*
4 McDonald LJ - BA,LLB *Computer Manager*
5 Pendleton JW - PhD *Deputy Director General research*

3705 Grain Legume Improvement Program
Oyo Road, P.M.B. 5320, Ibadan
Tel: 413224/413315/413440.

1 Singh SR - PhD *Program Director*
2 Aggarwal VD - PhD *Plant Breeder Burkina Faso*
3 Bello L - PhD *Plant Breeder*
4 Dashiell K - PhD *Plant Breeder*
5 Elsayed FA - PhD *Plant Breeder*
6 Emechebe AM - PhD *Pathologist (Visiting Scientist)*
7 Gowman M - MS *Entomologist*
8 Hohenberg J - PhD *Microbiologist*
9 Jackai LEN - PhD *Entomologist*
10 Kueneman EA - PhD *Regional Coordinator for Latin America, Brazil*
11 Muleba N - PhD *Physiologist, Agronomist, Burkina Faso*
12 Ntare BR - PhD *Regional Coordinator for the Sahel Region, Niger*
13 Odekunle GA *Senior Research Technician*
14 Ogundimu HT *Executive Secretary*
15 Oyekan P - PhD *Pathologist (Visiting Scientist)*
16 Pandey RK - PhD *Regional Coordinator for Southeast Asia, Philippines*
17 Root WR - PhD *Plant Breeder, Zaria, Nigeria*
18 Shannon DA - PhD *Agronomist*
19 Shoyinka SA - PhD *Breeder; Pathologist, Zaria, Nigeria*
20 Singh BB - PhD *Plant Breeder*
21 Suh J - PhD *Entomologist, Burkina Faso*
22 Watt EE - PhD *Plant Breeder, Brazil*
23 Wiedijk F - PhD *Entomologist*

3706 Maize Improvement Program
1 Bjarnason M - PhD *Plant Breeder and Acting Program Director*
2 Dabrowski ZT - PhD *Entomologist*
3 Dangi OP - PhD *Plant Breeder NCRE, Cameroon*
4 Diallo A - PhD *Maize Breeder, Burkina Faso*
5 Efron Y - PhD *Program Director (on sabbatical leave at CIMMYT)*
6 Everett L - PhD *Plant Breeder NCRE Cameroon*
7 Fajemisin JM - PhD *Pathologist; Breeder*
8 Khadr FH - PhD *Plant Breeder, Zaria, Nigeria*
9 Kim SK - PhD *Plant Breeder*
10 Mareck JH - PhD *Plant Breeder*
11 Ogunremi LT - PhD *Agronomist*
12 Parkinson VO - PhD *Pathologist*
13 Rodriguez M - PhD *Agronomist and Team Leader, SAFGRAD, Burkina Faso*
14 Talleyrand H - PhD *Agronomist, NCRE, Cameroon*

3707 Rice Improvement Program
1 Abifarin AO - PhD *IITA Liaison Scientist at WARDA, Liberia*
2 Akibo-Betts D - PhD *Entomologist*
3 Alam MS - PhD *Entomologist*
4 Alluri K - PhD *Agronomist; Breeder*
5 Dobson R - PhD *Pathologist*
6 Janakiram D - PhD *Plant Breeder, NCRE, Cameroon*
7 John VT - PhD *Pathologist*
8 Kaung Z - PhD *IRRI Liaison Scientist and Acting Director*
9 Masajo TM - PhD *Plant Breeder*
10 Navasero EP - MS *Rice Quality Specialist*
11 Nguu NV - PhD *Agronomist*
12 Roy AC - PhD *Agronomist, NCRE, Cameroon*

3708 Root and Tuber Improvement Program
1 Hahn SK - PhD *Program Director*
2 Almazan AM - PhD *Biochemist; Food Technologist*
3 Alvarez MN - PhD *Plant Breeder*
4 Asnani VL - PhD *Plant Breeder*
5 Brockman FE - PhD *Agronomist and Project Leader, PRONAM, Zaire*
6 Caveness FE - PhD *Nematologist*
7 Hammond W - MSc *Entomologist*
8 Haug T - Ir *Entomologist*
9 Hennessey RD - PhD *Entomologist, Zaire*
10 Herren HR - PhD *Entomologist*
11 Lema KM - PhD *Entomologist*
12 Loehr B - PhD *Entomologist, Brazil*
13 Neuenschwander P - PhD *Entomologist*

14 Ng SYC - MS *Tissue Culturist*
15 Otoo JA - PhD *Breeder; Agronomist*
16 Pandey SJ - PhD *Extension Agronomist, Zaire*
17 Pfeiffer HJ - Ir *Agronomist and Project Leader, CNRCIP, Cameroon*
18 Schulthess F - Ir *Entomologist*
19 Sicely EM - PhD *Interim Project Coordinator for Biological Control*
20 Sullivan D - PhD *Entomologist (Visiting Scientist)*
21 Theberge RL - PhD *Pathologist*
22 Varela AM - BSc *Entomologist, Brazil*
23 Whyte JA - PhD *Plant Breeder, Cameroon*
24 Yaninek JS - MS *Entomologist*

3709 Farming Systems Program
1 Juo ASR - PhD *Program Director, Soil Scientist and Coordinator of Research on Wetland Production Systems*
2 Akobundu IO - PhD *Weed Scientist (On Sabbatical Leave)*
3 Ashraf M - PhD *Agricultural Economist*
4 Atayi EA - PhD *Agricultural Economist and Chief of Party, NCRE, Cameroon*
5 Ay P - PhD *Socioeconomist*
6 Balasubramanian V - PhD *Agronomist, Rwanda*
7 Evers A - Ir *Water Management Engineer, Bida, Nigeria*
8 Ezumah HC - PhD *Agronomist*
9 Garman C - MS *Agricultural Engineer*
10 Gebremeskel T - PhD *Agricultural Economist*
11 Ghuman BS - PhD *Soil Scientist, Bendel State, Nigeria*
12 Gunneweg H - Ir *Water Management Engineer, Benin Republic*
13 Hahn ND - PhD *Socioeconomist and Head of Socioeconomics Unit*
14 Hulugalle N - PhD *Soil Scientist*
15 Kang BT - PhD *Soil Scientist; Agronomist*
16 Kikafunda-Twine J - PhD *Agronomist, NCRE, Cameroon*
17 Kiwamu CC - BSc *Research Assistant*
18 Kosaki T - PhD *Soil Scientist*
19 Lal R - PhD *Soil Physicist and Coordinator of Research on Upland Production Systems*
20 Lawson TL - PhD *Agroclimatologist and Program Director, Benin Substation, Cotonou*
21 McHugh D - MS *Extension Agronomist, NCRE, Cameroon*
22 Mughogho SK - PhD *Agronomist, IFDC*
23 Mulongoy K - PhD *Microbiologist*
24 Mutsaers HJW - PhD *Agronomist and Coordinator of On-Farm Research*
25 Navasero NC - BS *Agricultural Engineer*
26 Nweke F - PhD *Agricultural Economist, University of Nigeria, Nsukka*
27 Palada MC - PhD *Agronomist*
28 Poku JA - PhD *Weed Scientist*
29 Price M - PhD *Agronomist and Project Manager FSR/OSR, Rwanda*
30 Swennen R - PhD *Agronomist, Onne, Nigeria*
31 van der Kruijs ABM - Ir *Soil Scientist, Onne Nigeria*
32 Van Elslande AHS - Ir *Soil Scientist*
33 Vogel WO - PhD *Agricultural Economist*
34 Vuylsteke D - Ir *Plant Physiologist, Onne Nigeria*
35 Wilson GF - PhD *Agronomist*

3710 International Cooperation and Training Program
1 Terry ER - PhD *Program Director*
2 Gasser H - PhD *Principal Program Planning Officer*
3 Uriyo A - PhD *Principal Training Officer*

3711 Research Support Units
Research Support Units consist of: Farm Management, Analytical Services Lab, Genetics Resources Unit Biometrics and Virology Unit

3712 Farm Management
1 Couper DC - MS *Farm Manager*

3713 Genetic Resources Unit
1 Ng NQ - PhD *Head and Plant Geneticist*
2 Ajala MO - MS *Research Associate*
3 Davids M - Ir *Plant Scientist*

3714 Virology Unit
1 Rossel HW - Ir *Virologist*
2 Thottappilly G - PhD *Virologist*

3715 Analytical Services Laboratory
1 Pleysier JL - PhD *Head*

3716 Biometrics
1 Nguyen KN - PhD *Biometrician*

3717 Documentation, Information, and Library Program
1 Lawani SM - PhD *Assistant Director and Head*

2 Akande SB - MLS *Cataloger*
3 Azubuike AA - MLS *Bibliographer*
4 Babaleye SOT - MCA *Communications Associate*
5 Gambier GA - Lic *Head of Interpretations and Translations Unit*
6 Ibekwe GO - BA *Principal Librarian*
7 Keyser JE - BS *Head of Publications Unit*

PAPUA NEW GUINEA

3718 Department of Agriculture and Livestock

P.O. Box 417, Konedobu
Tel: 21 4699. Telex: 22143. Cables: AGRIC, KONEDOBU.

1 Toreu B *Chief Research Administrator* agricultural chemistry
2 Bailey J *Principal Chemist* agricultural chemistry
3 Dori F *Entomologist* entomology
4 Gubag K *Chemist* agricultural chemistry
5 Hadfield W *Plant Physiologist* plant physiology
6 Kerage J *Chemist* agricultural chemistry
7 Kula G *Research Director* plant pathology
8 MacFarlane M *Chemist* soil chemistry
9 Muthappa B *Principal Pathologist* plant pathology
10 Philemon E plant pathology
11 Sherington J *Principal Biometrician* biometry
12 Simin A biometry
13 Tomlinson D *Pathologist* plant pathology

3719 Land Use Office
P.O. Box 1863, Konedobu, National Capital District

1 Freyne D *Chief Land Use Officer*
2 Wayi BM *Senior Land Use Officer*
3 Andrew MI *Land Use Officer*
4 Daink F *Land Use Officer*
5 Grose CJ *Land Use Officer*
6 Howlett D *Land Use Officer*
7 Huria IS *Land Use Officer*
8 Igua P *Land Use Officer*
9 Leonard H *Land Use Officer*
10 Siri MW *Land Use Officer*
11 Varvaliu A *Land Use Officer*

3720 Bubia Agricultural Research Station
P.O Box 1639, Lae, Morobe Province
Tel: 42 4933.

1 Menggenang A *Agronomist* grain crop agronomy
2 Akus W food crop agronomy; farming systems
3 Arura M entomology; farming systems
4 Cox P plant pathology; farming systems
5 Gimbol C economist; farming research
6 Isaacson B economist; farming systems research
7 King G *Team Leader* food crop agronomy; farming systems
8 Tumana C food crop agronomy; farming systems

3721 Lowlands Agricultural Research Station
P. O. Box Keravat, Rabaul, East New Britain Province
Tel: 926251.

1 Sitapai T *Officer In Charge* cocoa
2 Kurika L *Agronomist* food crop agronomy; farming systems
3 Ling G *Agronomist* cocoa farming systems
4 Moxon J *Entomologist* entomology

3722 Highlands Agricultural Research Station
P.O. Box 384, Aiyura, Via Kainantu, Eastern Highlands Province
Tel: 771133/771009.

1 Levett M *Officer In Charge* food crop agronomy; farming systems

3723 Kuk Agricultural Research Station
P.O. Box 339, Mount Hagen, Western Highlands Province
Tel: 551377/551335/551325.

1 Wohuinangu J *Officer In Charge* agronomy
2 Gunther M agronomy; farming systems
3 Kanua M *post-harvest agronomy; farming systems*
4 Kokoa P plant pathology; farming systems
5 Masamdu R entomology; farming systems
6 Thistleton B entomology; farming systems
7 Waruba M *Agronomist* agronomy

3724 Plant Introduction and Quarantine Station
P.O. Box 339, Laloki, Central Province
Tel: 281068.

1 Kambuou R pulse crops agronomy
2 Ovia A bananas

3 Rangai S food crop agronomy; farming systems
4 Warisaiho C post harvest; food storage

3725 Oil Palm Research Association
Kimbe

1 Prior R *Entomologist* entomology

3726 Livestock Division

3727 National Veterinary Laboratory, Kila Kila
P.O Box 6372, Boroko
Tel: 21-7011. Telex: NE 23076.

1 Cridland SW - BRurSc,PhD animal nutrition; veterinary biochemistry
2 Nunn MJ - BVSc,MSc,MACVSc,MASM,MIBiol,GDipMgt *Chief Veterinary Research Officer* veterinary pathology; veterinary preventive medicine
3 Onaga I - BSc trace elements
4 Owen IL - BSc,PhD veterinary parasitology
5 Pangumen M - BSc food microbiology
6 Patten BE - BVSc veterinary microbiology
7 Puana I - BAg serology

3728 Monogastric Research Centre
P.O Box 1434, Lae
Tel: 42-1022.

1 Bakau B - BAg poultry husbandry
2 Bilong G - BAg,GradDipPoulTech poultry nutrition
3 Moat M - BAg poultry husbandry
4 Nani W - BAg,GDipSciAg poultry husbandry
5 Wenge K - BAg,Dipd'App pig husbandry
6 Yayabu R - BAg poultry breeding

3729 Pastoral Research Centre, Erap
P.O Box 1984, Lae
Tel: 44-9374.

1 Bannik A - BAg pasture agronomy
2 Galgal K - BAg ruminant nutrition
3 Gama JC - BScAnProd cattle husbandry
4 Gwaiseuk W - BAg,MAgSc ruminant nutrition, cattle breeding
5 Maika C - BAg pasture agronomy
6 Raurela M - BAg pasture agronomy
7 Tupper GJ - BAgSc,MAgSc,DipEd *Senior Animal Production Officer* pasture agronomy; ruminant production

3730 Sheep Research Centre, Menifo
P. O. Box 766, Goroka
Tel: 72-7865.

1 Kauge A - BAg ruminant nutrition
2 Manua P - BAg pasture agronomy
3 Pitala J - BAg animal genetics
4 Sivasupramanian S - BSAgric,MScAgric,MAgricSt *Senior Animal Production Officer* pasture agronomy
5 Vano H - BAg,MAgSt sheep husbandry

3731 Investment Division
DAL Headquarters, P.O. Box 417, Konedobu

1 Banguinan P - BSc *Deer Ecologist* deer ecology

3732 Department of Forests
P.O. Box 134, Bulolo, Morobe Province
Tel: 44 5248. Telex: NE 22360. Cables: FORENTOM.

1 Roberts Dr H - BSc,PhD *Principal Research Officer* forest entomology; under bark pests of Eucalyptus
2 Daur PP - BSc,MPM *Research Worker* forest entomology, defoliators of pines
3 Howcroft NS - DipFor *Research Worker* plantation silviculture; tree breeding; seed technology; agro-forestry
4 Mukiu J - BSc *Research Worker* forest pathology, heart rot
5 Nangi W - BSc,DipFor *Research Worker* mycology; mushroom farming
6 Wake Y - BSc *Research Worker* plantation silviculture, tree breeding, seed technology

3733 University of Papua New Guinea
P.O. Box 320, University, National Capital District
Tel: 253900. Telex: NE 22366.

3734 Faculty of Science
P.O. Box 320, University
Tel: 253900. Telex: NE 22366.

1 Hill L - BSc,PhD *Dean*

3735 Department of Biology
P.O. Box 320, University
Tel: 253900. Telex: NE 22366.

1 Osborne P - BSc,MSc,PhD *Head of Department* freshwater plant ecology and taxonomy; effects of nutrient enrichment of lakes
2 Burrows I - BSc,PhD *Lecturer* biological deterioration of industrial materials and stored products, upgrading waste materials and recycling
3 Hill L - BSc,PhD *Senior Lecturer* animal physiology; environmental planning and resource management; traditional knowledge and utilization of resources of PNG communities
4 Hopkins HCF - MA,PhD *Lecturer* plant systematics; pollination and dispersal of flowering plants
5 Hopkins MJG - BA,PhD *Lecturer* entomology; animal and plant interactions
6 Mowbray DL - BSc,PhD *Senior Lecturer* effects of chemical pollutants on the environment; integrated pest control; pesticide residues, toxicology and regulation
7 Pernetta Dr JC - BA,MA,DPhil *Senior Lecturer* population ecology; vertebrate biology; traditional knowledge and utilization of resources by PNG communities; taxonomy of reptiles and mammals; wildlife utilization
8 Saulei S - BSc,MSc,PhD *Lecturer* aquatic botany and ecology; plant population and ecology; quantitative ecology

3736 University of Technology
Lae, Morobe

3737 Department of Agriculture
1 Enyi BAC - BSc,MSc,PhD,FIBiol *Head of Department*
2 Benoit A - BSc,MSc,DSc *Lecturer*
3 Das DK - BSc,MSc,PhD *Senior Lecturer*
4 Gurnah AM - BSc,PhD *Senior Lecturer*
5 Kaiulo JV - BAg,MS *Lecturer*
6 Kargbo CS - BSc,MS,PhD *Lecturer*
7 Kesavan V - BSc,MSc,PhD *Associate Professor*
8 Kohun PJ - BAg,MAg *Lecturer*
9 Komolong M - BAg *Assistant Lecturer*
10 Krebs GL - DipAppSci,BSc,PhD *Lecturer*
11 Malik RC - BVSc,MSc,PhD *Senior Lecturer*
12 Manus P - BAg *Assistant Lecturer*
13 Merrett PJ - BSc,PhD *Lecturer*
14 Nath S - BSc,MTech,PhD *Lecturer*
15 Vele P - BAg *Assistant Lecturer*

3738 Department of Forestry
1 Johns RJ - MSc *Head of Department*
2 Hilton RGB - BScF,MSc *Senior Lecturer*
3 Liew TC - BSc,MSc *Lecturer*
4 Mercer CWL - BSc,MSc *Lecturer*
5 Siaguru P - BSc,MSc *Assistant Lecturer*
6 Tisseverasinghe AEK - BSc,MA *Senior Lecturer*
7 Vatasan GS *Senior Lecturer*
8 Vatasan N - DipFor *Lecturer*

3739 University of Sierra Leone

Private Mail Bag, Freetown
Tel: 26859, 27340. Cables: unisal.

3740 University Research and Development Services

University of Sierra Leone, Private Mail Bag, Freetown
Tel: 26859, 27340. Cables: UNISAL.

1 Davidson OR *Director*

3741 Food and Agriculture Division

1 George JB - PhD *Head of Division* post harvest biology
2 Amara DS - PhD soil fertility and conservation
3 Bangura AM - PhD agricultural extension - rural development
4 Barrie IS - MSc agricultural economics
5 Dahniya MT - PhD farming systems - root crop agronomy
6 Davies-Grant LBB - MSc ruminant production
7 Gbani AB - PhD biometry
8 Gboku MLS - MPhil agricultural extension
9 Jackson REA - MSc weed science
10 Jambawai MJ - MSc mechanization of tropical crops
11 Jones TAOC - DVM animal nutrition and animal diseases
12 Kaindaneh PN - MS production economics
13 Kaitibie ADS - MSc plant pathology
14 Kamara CS - PhD soil physics and soil management
15 Kamara IS - MSc agroclimatology
16 Kandeh HBS - PhD demography - population and agriculture
17 Karim-Sesay JA - MSc mechanization and crop processing - renewable energy
18 Kuyembeh NG - MSc appropriate technology/mechanization
19 Lahai AC - PhD plant pathology - field and storage diseases of rice
20 Lakoh AK - PhD agricultural extension - farming system
21 Lappia JNL - PhD farming systems, rural development and project evaluation
22 Leigh JS - PhD meat technology
23 Matturi AS - DSc animal nutrition
24 Ndoleh AS - MSc plant physiology of legumes
25 Rashid-Noah AB - PhD irrigation and hydrology
26 Rhodes ER - PhD soil fertility and cropping systems
27 Roberts AT - MPhil agricultural marketing
28 Roberts Mrs J - PhD entomology and transmission of virus
29 Sandi A - MSc rural development; farm labour supply and role of women in agriculture
30 Sesay A - PhD plant physiology - legumes
31 Sesay K - MS post harvest engineering
32 Taylor WE - PhD entomology, crop protection and management
33 Tommy JL - PhD rural-urban migration; rural financial markets; post harvest loss assessment

SOLOMON ISLANDS

3742 Ministry of Agriculture and Lands

Head Office, P.O. Box G13, Honiara
Tel: 21511. Telex: PRIMUS HQ 66311.

1 Manira A *Permanent Secretary Of Agriculture and Lands*

3743 Agriculture Division

P.O. Box G13, Honiara
Tel: 21621 Ext 34. Telex: RUSEP HQ 66417.

1 Rea GA - MSc,NDA *Under-Secretary (Agriculture)*

3744 Agricultural Quarantine Department
Tel: 21621 ext.29.

1 Eta C - DTA *Senior Field Officer*

3745 Extension Department
Tel: 21621 ext.33.

1 Abana W *Principal Field Officer* extension
2 Iamae E *Chief Field Officer*
3 Kaenili P - CTA *Senior Field Officer* education and training

3746 Land Use Development
Tel: 21621 ext.55.

1 Launder J - BA *Farm Management Advisor*
2 Qaloboe F - DTA *Principal Field Officer*

3747 Veterinary/Livestock Department
Tel: 21621 ext.67.

1 (vacancy) *Chief Veterinary Officer*

3748 Veterinary Laboratory
Tel: 23039/23062.

1 McQueen C - BVSc,MRCVSc,DTVM *Principal Veterinary Officer*

3749 Tenevatu Farm
Tel: 31130.

1 Wate D *Senior Livestock Officer*

3750 Research Department
Dodo Creek Research Station, P.O. Box G13, Honiara
Tel: 31111. Telex: RUSEP HQ 66417.

1 Abington J - BSc,DipAgrSc *Chief Research Officer*

3751 Agricultural Economics Section

1 Mackey E - BSc,MSc *Principal Research Officer*
2 Maeba F - BAg *Research Officer*

3752 Agronomy Section

1 Caiger S - BSc,MAgSc *Principal Research Officer* field crops
2 Hancock IR - BSc,DTA,PhD *Principal Research Officer* farming systems
3 Henderson CP - BSc,MSc *Research Officer* ethnobotanical survey
4 Oliouou MM - BSc *Research Officer* tree crops
5 Rua J - BAg,DTA *Research Officer* farming systems
6 Samu J - BSc *Research Officer* field crops
7 Trewren KR - BSc *Prinicpal Research Officer* tree crops

3753 Entomology Section

1 Pauku LR - BAg *Research Officer*
2 Williams CT - BSc,PhD *Research Officer*

3754 Plant Pathology Section

1 Liloqula Ms R - BSc,MSc *Principal Research Officer and Officer-in-Charge*
2 Saelea J - BAg,DTA *Research Officer*

3755 Soils and Plant Nutrition Section

1 Cheatle RJ - BA,MSc,CertEd,FRGS *Principal Research Officer*
2 Wairiu M - BAg *Research Officer*

3756 Library

1 Basile Mrs S *Librarian/Information Officer*

3757 Tenaru Field Experiment Station
P.O. Box G13, Honiara
Tel: 31167. Telex: RUSEP HQ 66417.

1 Honiseu J - CTA *Assistant Field Officer*

3758 Fote Field Experiment Station
P.O. Box 18, Auki, Malaita Province
Tel: 40247. Telex: RUSEP HQ 66417.

1 Wainioamena A - CTA *Assistant Field Officer*

3759 Ministry of Natural Resources

Head Office, P.O. Box G24, Honiara
Tel: 21521. Telex: SOLFISH HQ 66306.

1 Siapu G *Permanent Secretary of Natural Resources*

3760 Fisheries Division

1 Joachim R *Chief Fisheries Officer*
2 Diake S - BSc *Fisheries Officer* biological research
3 Kitira B - BSc *Fisheries Officer* resource research
4 Maruyama C - BSc *Assistant Fisheries Officer* computer analyst
5 Nichols P - BSc,MSc *Senior Fisheries Officer* research
6 Rawlinson N - BSc *Assistant Fisheries Officer* baitfish research
7 Smith M - BSc *Assistant Fisheries Officer* baitfish population analysis
8 Susurua R - DipFish *Principal Fisheries Officer* development
9 Wata A - BSc *Fisheries Officer* project evaluation

3761 Fisheries Laboratory
Tel: 23188.

3762 Forestry Division

1 Gaviro S - BSc *Chief Forest Officer*
2 Aupai A - CertFor *Senior Forest Officer* plantation
3 Bauro S - BSc *Assistant Forest Officer* timber
4 Bevani A - CertFor *Senior Forest Officer* management
5 Doliano E - DipFor *Senior Forest Officer* timber control and utilization
6 Kwanairara ED - DipFor *Principal Forest Officer*
7 Mauriasi R - CertFor *Senior Forest Officer* utilization
8 Sirutee B - CertFor *Senior Forest Officer* extension

3763 Herbarium
Tel: 22370.

3764 Forest Research Station
P.O. Box G79, Munda, Western Province
Telex: SOLFISH HQ 66306.

1 Chaplin GE - BSc *Senior Forest Officer* research
2 Iputu S - BSc *Senior Forest Officer* research

3765 Rural Services Project
P.O. Box 24, Honiara
Tel: 21200. Telex: RUSEP HQ 66417.

1 Walaodo E - DTA,MSc *Project Director*

3766 Forum Fisheries Agency
P.O. Box 629, Honiara
Tel: 21124. Telex: FORFISH HQ 66336.

1 Muller DAP - *Director*

3767 Livestock Development Authority
P.O. Box 525, Honiara
Tel: 21650.

1 Luilamo G - *General Manager*

3768 Levers Solomons Ltd
Head Office, Yandina, Russell Islands
Cables: Levers Yandina.

1 Friend D *Managing Director*
2 Leng AS - BSc,DipAgrSc *Research Officer*

3769 Solomon Islands Plantations Ltd
Head Office (Ngalimbiu), P.O. Box 350, Honiara
Tel: 31121/21412. Telex: SIPL HQ 66323.

1 Woodhead B - BSc,DTA,OBE *General Manager*
2 Rankine I - BSc,MSc *Research Officer*

3770 Solomon Islands College of Higher Education

3771 School of Natural Resources
Panatina Campus, P.O. Box G23, Honiara
Tel: 30242.

1 Kinne Sr R - BSc,PhD *Head of School of Natural Resources*

3772 National Agricultural Training Institute

P.O. Box 18, Auki, Malaita Province
Tel: 40247.

1 Cogger RJ - BScAg,DipEd *Principal*

3773 Department of Agriculture

Peradeniya
Tel: 08-88157. Cables: Agriculture.

1 Gunawardane SDIE - BSc,MSc,PhD *Director*

3774 Agriculture Research Division

P.O. Box 5, No 1, Sarasavi Mawatha, Peradeniya
Tel: 08-88030. Cables: Agriculture Research.

1 Herath HME - BSc,MSc,PhD *Deputy Director Research*
2 Amarasiri SL - BSc,MSc,PhD *Addl. Deputy Director Research*
3 Devasabai K - BSc,MSc agronomy
4 Jauffer MM soil science
5 Lecamwasam A - BScAgr,DipAgr agronomy
6 Pathirana Mrs DA - MScAgr plant breeding
7 Vignarajali N - BSc,MSc,PhD farming systems research

3775 Central Agricultural Research Institute

Gannoruwa, Peradeniya
Tel: 08-88011, 88012, 88013. Cables: Agriculture Research.

1 Fernando MHJP - BSc,PhD,DIC *Deputy Director (Research)*

3776 Division of Botany

1 Jayawardene SDG - BSc,MSc,PhD *Head of Division*
2 De Silva KPU - BSc,MSc root crops
3 Hindagala CB - BScAgr,MSc plant breeding
4 Joseph KDSM - BSc,MSc rice agronomy
5 Mendis HM - BSc,MSc tissue culture
6 Peiris Mrs R - BSc,MSc rice agronomy
7 Samarajeewa PK - BSc,MSc bio-techniques

3777 Division of Entomology

1 Peries IDR - BSc,MSc *Head of Division*
2 De Alwis HM - BSc,MSc rice thrips and biocontrol of vegetable pests
3 Ekanayake Mrs HMRK - BSc,PhD nematology
4 Nugaliyadda Miss MM - BSc,MSc nematology & banana pests
5 Wijesekera GAW - BSc economic threshold levels of rice pests; natural enemies of rice pests

3778 Division of Agricultural Chemistry

1 Abeykoon V - BSc soil fertility
2 Chandrasekara AI - BSc pesticide chemistry
3 Gunapala N - BSc soil fertility
4 Jinadasa DM - BSc,MSc soil fertility
5 Lathiff MA - BSc soil chemistry
6 Marikkar SS - BSc,MSc soil fertility
7 Nagarajah S - BSc,MSc,PhD *Head of Division*
8 Seneviratne R - BSc,MSc,PhD soil chemistry
9 Wettasinghe A - BSc pesticide chemistry
10 Wijesundera Mrs S - BScAgr soil science

3779 Division of Plant Pathology

1 Bandara WMSM - BSc taxonomic classification (root and tuber)
2 Bandaranayake CM rice pathology
3 Fernando WPD - BSc,MSc soil pathology
4 Haberakade Miss RR - BSc rice pathology
5 Jayanandarajah Mrs P - BSc seed pathology
6 Sivakadachchan P - BSc,PhD,MSc *Plant Pathologist*
7 Udugama SSTTA - BSc,MSc bacterial & fungal work

3780 Division of Food Science

1 Hussain Mrs SF - BSc,MSc *Food Technologist*
2 Wijeratne DBT - BSc food science technology

3781 Division of Horticulture

1 Pathmasiri Mrs IS - BSc,MSc *Officer in Charge*
2 Fernando KKS - BSc,MSc tissue culture
3 Medagoda Mrs I - BSc agronomy and breeding
4 Perera Miss A - BSc,MSc horticulture
5 Radokrishnan R - BSc agronomy

3782 Division of Biometry

1 Premajayantha V - BSc biometry
2 Tirimanne AM - BSc

3783 Division of Plant Quarantine

1 Sivanathan P - BSc,PhD *Chief Plant Quarantine Officer*
2 Abeykoon MB plant quarantine
3 de Silva RSY - BSc,MSc microbiology

3784 Central Library

1 Gunasekera Mrs SC - BLs *Chief Librarian*
2 Ratnavibushana DY - BScAgr *Head Library Services*

3785 Agricultural Research Centre

Maha-Illuppallama

1 Sikurajapathy M - MSc,PhD *Deputy Director Research* crop science
2 Amarasinghe AAL - BSc,MSc weed science
3 Ariyaratne HP - BScAgr,MSc plant breeding
4 Arulandhy V - BSc,MSc,PhD breeding
5 Bandara PT - BScAgr soil science
6 Dharmasena PB - BSc,MSc soil conservation & land use
7 Fonseka HHD - BScAgr plant breeding
8 Hettiarachchi K - BScAgr plant breeding
9 Jayawardena SM - BScAgr agronomy
10 Keerthiratne GB - BScAgr,MSc irrigation agronomy
11 Keerthisena RSK - BScAgr,MSc soil physics
12 Meththananda KA - BScAgr adaptive research
13 Muthukudaarachchi DH - BScAgr,MPhil plant breeding
14 Nayakakorala HB - BScAgr soil physics
15 Pathirana Mrs DA - MScAgr plant breeding
16 Perera BMK - MScAgr,PhD agronomy
17 Roonage MA - BScAgr water management
18 Samarasinghe Mrs M - MSc,BSc adaptive research
19 Wasantha Chitral GM - BScAgr adaptive research
20 Weerakoon WL - BSc,MSc conservation farming
21 Wijeratne Banda PM - BScAgr entomology
22 Wijethilake LC - BSc,MSc plant pathology
23 Zoysa Mrs IJ DE - BScAgr,PhD plant pathology

3786 Agricultural Research Centre

Diyatalawa Road, Bandarawela
Tel: 057-2520, 2499. Cables: Agriculture Research Bandarawela.

1 Yogaratnam V - BSc,PhD,MSc *Deputy Director Research* agronomy
2 Abeyratne KWSD - BSc plant breeding
3 Abeysinghe A - BScAgr adaptive research
4 Alahakoon PW - BSc agronomy
5 Ariyaratna HM - BSc breeding
6 De Silva KL - BScAgr agricultural economics
7 Jayasundera JMP - BSc soil science
8 Tharamarajah SK - BScAgr,MSc horticulture
9 Wahab MNJ - BScAgr,MSc agronomy
10 Wahundeniya Mrs I - BScAgr entomology
11 Wahundeniya WMBK - BScAgr cereal agronomy
12 Wijewardene JDH - MSc soil science

3787 Agricultural Research Station

Sita-Eliya, Nuwara Eliya
Tel: 052-2615.

1 Abeytunge KGW - BScAgr,MScAgr soil-plant-water relations
2 Abeytunge Mrs SL - BScAgr,MScAgr agronomy & plant breeding
3 Sirisena DN - BScAgr soil chemistry

3788 Agricultural Research Station

Rahangala, Boralanda

1 Herat LG - BSc,MSc plant breeding vegetable crops

3789 Agricultural Research Station

Uva-Kumbukkana, Monaragala

1 Bandara WMSM - BSc agronomy
2 Nilmalgoda ERUW - BScAgr,MSc horticulture

3790 Agricultural Research Centre

Angunukolapelessa

1 Balasuriya I - BScAgr,MSc *Deputy Director Research* agronomy
2 Abeyratna Miss MDRS - BSc entomology
3 Andrahennadi CP - BScAgr cotton breeding
4 Bandara PMS - BScAgr agronomy
5 Dayananda MP - BSc,MSc soil science
6 Dharmasiri LC - BScAgr agronomy and weed science
7 Jayamanne PB - BSc grain legume breeding
8 Jayasena KW - BScAgr,PhD plant pathology
9 Karunathilaka WAK - BScAgr land management
10 Kularatna WHD - MScEcon economics
11 Munasinghe Mrs G - BScAgr water management

12 Sirisena JA - BScAgr horticulture
13 Weerasena LA - BScAgr,MSc oil crops breeding
14 Wijesundara Mrs S - BScAgr soil science
15 Wijewickrama PJA - BScAgr oil seed agronomy

3791 Rice Research Station
Ambalantota

1 Abeysekera SW - BSc plant breeding and rice agronomy

3792 Agricultural Research Centre
Bombuwela, Kalutara
Tel: 034-22673.

1 Gunathilake GA - BScAgr,MScAgr *Deputy Director*
2 Abeysiriwardena DS DE Z - BScAgr rice breeding
3 Bandara WMJ - BSc soil chemistry
4 Dassanayake Mrs EM - BScAgr fruits
5 Dayananda MP - BSc soil science
6 de Silva PAP - BSc rice agronomy
7 Delpachitra Mrs ND - BScAgr entomology
8 Duragajaweera Mrs E - MSc adaptive research
9 Harischandra Mrs SN - BScAgr roots & tubers
10 Jayawickrama HD - BScAgr agronomy
11 Kahandawala JBDS - BScAgr roots & tubers
12 Mendis VW - BSc entomology
13 Mirihagalla KL - BScAgr vegetables
14 Mithrasena AHG - BScAgr adaptive research
15 Senanayake L - BScAgr vegetables
16 Senanayake N - BScAgr,MSc agronomy weed science
17 Wanigasooriya SC - BScAgr physiology
18 Weerasuriya Miss JD - BScAgr pesticide chemistry
19 Wickramasinghe DL - BScAgr,MSc plant pathology
20 Wijesundara C - BScAgr agronomy, fertilizer

3793 Rice Research Station
Bentota

1 Jayawickrama HD - BScAgr rice agronomy breeding

3794 Agricultural Research Centre
Iranamadu Junction, Kilinochchi
Tel: 327.

1 Vivekanandan AS - BSc,MSc,PhD *Deputy Director Research*
2 Dhuruvasangary V - MSc,MPhil agronomy, water management
3 Kanagasunderam Miss S - BScAgr breeding
4 Mylvaganam Miss P - BSc entomology
5 Nagendran A - BSc,MPhil economics
6 Ponnambalavanar Mrs P - BSc,MSc breeding
7 Ponnudurai S - BSc,MSc breeding, cereals
8 Poornampillai Mrs CD - BSc breeding
9 Regunathan Mrs B - BSc,MPhil agronomy, oil crops
10 Senthinathan A - BSc,MSc agronomy, cropping systems
11 Velauthapillai Mrs V agronomy

3795 Rice Research Station
Paranthan
Tel: 301 Kilinochchi.

1 Pooranampillai JR - BSc,MPhil agronomy weed science
2 Sivayogeswaran Mrs S - BSc rice breeding
3 Srilingam Mrs R - BSc,MPhil rice breeding
4 Vivekanandan AS - BSc,MSc,PhD rice agronomy

3796 Agricultural Research Station
Murunkan

1 Ravel MJ - BSc entomology
2 Yogaratnam P - BSc,MSc agronomy, rice breeding

3797 Agricultural Research Station
Thirunelvely, Jaffna

1 Kugathasan T - BScAgr,MScAgr *Officer-in-Charge* grain legumes
2 Balasubramaniam EM - BSc weed science
3 Emerson BN - BSc,MSc agronomy
4 Jayapathy S - BScAgr,MScAgr fruit crops
5 Kanaganayagam N agronomy

3798 Agricultural Research Centre
Girandurukotte

1 Upasena SH - PhD *Deputy Director Research* research management, cropping systems
2 Bandaranayake MAW - BScAgr water management, irrigation agronomy
3 Bentota A - BSc agronomy
4 Dhanapala MG - BSc entomology
5 Fernando MFSW water management
6 Peiris KHS - BSc agronomy
7 Somadasa LP - BScAgr cereals
8 Weerasinghe ASP - BSc plant pathology

9 William MDHM agronomy

3799 Agricultural Research Station
Aralaganwila

1 Samarakoon S - BScAgr,MPhilAgr *Officer in Charge* agronomy cropping systems
2 Ariyaratne Mrs I - BSc,MSc plant protection
3 De Silva SHAA - BScAgr water management
4 Ekanayake EMDSN - BScAgr agronomy
5 Jeewananda Miss GASS - BScAgr agronomy
6 Katipearachchi Y entomology
7 Ranasinghe LL - BScAgr agronomy
8 Weerasinghe P - BScAgr soil science
9 Weerasinghe Miss SWK - BScAgr agronomy

3800 Agricultural Research Centre
Makandura, Gonawila (NWP)

1 Somapala H - DipEngAgro,MSc,MAg,PhD *Deputy Director Research* soil physics
2 Jayakody DPP - BSc intercropping
3 Kuruppu Mrs NPU - BSc pathology
4 Raphael BS agronomy
5 Raphael Mrs FMG - BSc(Agr) horticulture
6 Sriskandadas Mrs M - BSc,MSc pathology
7 Subasinghe SMC - BSc,MSc,PhD entomology
8 Sulaiman Mrs SFM - BSc,MSc entomology
9 Weerakoon WMW - BScAgr rice agronomy
10 Welikala N - BSc,MSc agronomy and agro-climatology

3801 Agricultural Research Station
Kalpitiya

1 Kuruppuarachchi DSP - BSc water management

3802 Agricultural Research Station
Walpita Via Mirigama

1 Stanley WDL - BSc,MScAgr soil physics

3803 Central Rice Breeding Station
Batalagoda, Ibbagamuwa
Tel: 801. Cables: Agresearch.

1 Senadhira D - BSc,MSc,PhD *Deputy Director Research* rice breeding
2 Bandaramanike Mrs MSBWM - BScAgr agronomy, entomology
3 Dhanapala MP - BSc,MSc rice agronomy/breeding
4 Dissanayake DMN - BSc pathology
5 Fahim M - BSc rice breeding
6 Kandasamy S - BScAgr,MSc agronomy
7 Kudagamage C - BSc,MSc entomology
8 Nugaliyadda L - BSc,MSc entomology
9 Pathinayake BD - BSc rice breeding
10 Rajapaske RMT - BSc,MPhil rice breeding/agronomy
11 Sandanayake CA - BSc rice breeding
12 Sarthaj IZ - BSc cereal chemistry
13 Wickramasinghe DB - BSc soil science

3804 Division of Horticulture
Gatambe, Peradeniya
Tel: 08-88075, 88076.

1 Vaz CR DE - BScAgr,PhD *Additional Deputy Director Horticulture, Head of Division Horticulture*
2 Pinto MER - BScAgr,MSc *Additional Deputy Director Horticulture*
3 Dharmasena CD - BScAgr,MScAgr soybean co-ordinator, agronomist
4 Dissanayake LD - BSc agronomy
5 Heekenda HMS - BSc horticulture
6 Hewavitharana Mrs P - BSc,MSc *Food Technologist*
7 Mudannayake Mrs C - BSc,MPhil microbiology
8 Samaratunga H - BSc,MSc agronomy (fruits and vegetables)
9 Sarananda KH - BScAgr horticulture
10 Siriwardana TDW - BSc,DipFoodTech,MSc *Food Technologist*
11 Soyza AOC - BScAgr varietal improvement
12 Sundaralingam S - BSc horticulture
13 Thirukumaran P - BSc,MPhil soil microbiology

3805 Division of Land and Water Use

1 Somasiri S - PhD *Head of Division, Soil Scientist, Land Use Specialist*
2 Krishnarajah P - MSc *Head of Soil Conservation, Soil Conservation Specialist*
3 Gangodawila D - MSc geomorphology
4 Handawela J - BSc,MSc,PhD soil chemistry
5 Kannangara Mrs RPK - MSc agroclimatology
6 Ratnayake RMK soil science
7 Sumanaratne HD - MSc irrigation agronomy
8 Wickramasinghe A - MSc soil & water engineering

3806 Division of Soil Conservation

Botanical Gardens, Peradeniya
Tel: 08-88180. Cables: Soils.

1 Karunatilaka UP soil conservation
2 Krishnarajah P - BScAgr,MSc *Head of Division*
3 Selvarajah A soil conservation

3807 Division of Systematic Botany

National Herbarium, Royal Botanical Gardens, P.O. Box 15, Peradeniya
Tel: 08-88053.

1 Jayasuriya AMP - MSc,MPhil,PhD,FLS *Systematic Botanist and Curator of the National Herbarium*

3808 Royal Botanical Gardens
Peradeniya
Tel: 08-88088.

1 Jayasuriya AHM - PhD *Systematic Botanist*
2 Sumithraarachchi DB - MSc,FLS *Superintendent*

3809 Royal Botanical Gardens
Hakgala
Tel: 052-2182.

1 Wijesundara DSA - BScBot *Curator*

3810 Royal Botanical Gardens
Henarathgoda
Tel: 033-2316.

1 Dhanasekera DMUB - BScBot *Curator*

3811 Division of Agricultural Economics and Projects

Department of Agriculture, Peradeniya
Tel: 08-88206.

1 Ranaweera NFC - BScAgr,MA,MSc,PhD *Deputy Director and Head of Division* agricultural development policies; project analysis; farming systems
2 Agalawatta MB - BScAgr smallholder economics
3 Atapattu NK - BScAgr smallholder economics
4 Balasuriya G - BScAgr smallholder economics
5 Gunawardena JATP - BScAgr,MSc production economics
6 Hafi AAB - BScAgr resource economics
7 Hemaratne HA - BAEcon smallholder economics
8 Mahrouf ARM - BScAgr project coordinator
9 Samaratunga PA - BScAgr,MSc production economics
10 Senanayake Mrs BMAC - BScAgr,MSc production economics
11 Silva GAC - BScAgr,MSc agricultural development policies; socio economics; survey methodology
12 Suraweera DEF - BScAgr,MSc,PhD production economics; project management
13 Wickramasinghe YM - BScAgr,MSc production economics

3812 National Institute of Plantation Management

MDH Jayawardhana Mawatha, Athurugiriya
Tel: 561232.

1 Bodhidasa HDT *Chairman*
2 Wickremasinghe RL - BSc,PhD *Director* training faculty
3 Bandara NK - BSc *Deputy Director* training faculty
4 Bandara DMK - BSc *Training Specialist in Management*
5 Fonseka KBM - BSc,MBA,ACIMA *Training Specialist in Accounting and Financial Management*
6 Premachandra WM - BA,MSc *Training Specialist in Agricultural Economics*

3813 Forest Department

P.O. Box 509, Colombo 2
Tel: 32251. Cables: Forests.

1 Nanayakkara VR - BSc,MA,FIBiol *Conservator of Forests*

3814 Forestry Research Division

1 Bandara PDMGD - BSc plant protection
2 Udawatta R - BSc,MSc soil science
3 Vivekanandan K - BSc,PhD,FIBiol *Chief Research Officer*
4 Weerawardana NDR - BSc,MSc agroforestry

3815 Forestry Extension and Education Branch

1 Dayananda KT - BSc,MSc *Deputy Conservator of Forests* forestry education & extension

3816 Coconut Development Authority

P.O. Box 1628, Colombo 1
Tel: 21025-28. Telex: 21217 COCOBOD-CE. Cables: 'COCOBOARD' Colombo.

3817 Economic Research Division
Coconut Development Authority, 320, T B Jayah Mawatha, Colombo 10
Tel: 598289-90.

1 Perera UVH - BAEconStatis,DipSmallScaleIndustries *Director, Head of Section*

3818 Department of Animal Production and Health

15 Old Galaha Road, Peradeniya
Tel: 08-88195. Cables: Veterinary.

1 Siriwardene JAdeS - BVSc,DipAgrSc,PhD *Director*

3819 Veterinary Research Institute
Gannoruwa, Peradeniya
Tel: 08-88311,88312,88125.

1 De Alwis MCL - BVSc,PhD *Deputy Director*

3820 Division of Animal Breeding

1 Jalatge EFA - BVSc,DipAH *Head of Division* animal genetics
2 Jayeruban MG - BVSc

3821 Division of Animal Nutrition

1 Ganegoda GAP - BVSc,DipNutr,MSc *Head of Division* poultry and pig nutrition
2 Guneratne SP - BVSc poultry nutrition
3 Ranawana SSE - BVSc,PhD ruminant nutrition, minerals

3822 Division of Pasture and Fodder Crops

1 Jayawardane ASP - DVM,DipTropAgr *Head of Division* biological nitrogen fixation, sylvo-pastoral studies
2 Pieris H - BVSc forage utilization, animal production studies

3823 Division of Bacteriology

1 Balasuriya BRU - BVSc animal pathology
2 Wanasinghe DD - DVM,PhD mastitis, clostridial diseases
3 Wijewardana Mrs TG - BVSc haemorrhagic septicaemia

3824 Division of Parasitology

1 Weilgama DJ - BVSc,MVSc,PhD *Head of Division* tick borne diseases

3825 Division of Poultry Pathology

1 Ranasinghe GKMS - BVSc bacterial diseases of poultry
2 Wickremasinghe R - BVSc,MSc *Head of the Division* viral diseases of poultry

3826 Division of Reproductive Disorders

1 Mohamed AR - BVSc,PhD *Head of the Division* non-infectious reproductive disorders

3827 Division of Virology

1 Jayawardena GWL - BVSc other viral diseases
2 Kodituwakku Mrs S *Head of Division* foot and mouth disease

3828 Winged Bean Institute
Kundasale

1 Alles WS - BSc,MSc,PhD *Senior Research Scientist*
2 Babu AGC - BSc agronomy
3 Bandara MHSK - BSc entomology
4 De Soyza S - BSc microbiology
5 Dias D - BSc utilization

6 Gunasekera M - BSc genetics & plant breeding
7 Jayasekera SJBA - PhD *Research Officer in Charge*
8 Parthipan S - BSc,MPhil agronomy
9 Ramanayake S - BSc,MPhil genetics & plant breeding

3829 Ceylon Institute of Scientific and Industrial Research

P.O. Box 787, CISIR, Colombo
Tel: 598620. Telex: 21248 MINID CE CISIR. Cables: CISIR.

1 Jansz ER - BSc,PhD *Director*

3830 Research Division

3831 Agro Industries Section

1 Jayasinghe S - BSc,MSc forestry

3832 Division of Virology

1 Fernando WWHS - BSc,PhD *Head of Division* foot and mouth disease
2 Jayawardana GWL - BVSc
3 Kodituwakku Mrs SN - BVSc

3833 Division of Reproductive Disorders

1 Mohamed AR - BVSc,PhD non infectious infertility and physiology of reproduction
2 Pieris GS - BVSc,MVSc *Head of Division* infectious infertility

3834 Division of Poultry Diseases

1 Ranasinghe Mrs GKMC - BVSc poultry diseases and vaccines
2 Wickramasinghe R - BVSc,MSc *Head of Division* poultry diseases

3835 Library

1 Wijesekera Mrs VS - BLSc *Librarian*

3836 Department of Minor Export Crops

No. 1095, Peradeniya Road, Peradeniya
Tel: 08-88363, 08-88364.

1 Kathiragamathaiyah S - BSc,PhD *Director*

3837 Research Station
Matale
Tel: 0662-391.

1 Wickramasinghe PJ - BScAgr,MSc *Senior Assistant Director (Research)*

3838 Division of Agronomy

1 Abeykoon AMD - BScAgr agronomy
2 Mathavan R - BScAgr,MPhilAgr agronomy
3 Premaratne WHE - DipAgr,BScAgr,MScAgr,MSc mixed cropping
4 Sritharan R - BScAgr,MPhilAgr agronomy

3839 Division of Biometry and Statistics

1 Duradharshi HLA - BSc *Head of Division* research data analysis

3840 Division of Entomology

1 Perera HAS - BSc,MPhilAgr *Head of the Division* entomology

3841 Division of Genetics and Crop Improvement

1 Gurusinghe P DE A - BScAgr,MPhilAgr,PhD *Head of Division* breeding
2 Samuel Miss MRA - BSc,MScAgr,MPhilAgr,PhDAgr breeding

3842 Food Technology Section

1 Curtis RG - BSc,MSc postharvest biology and processing of fruits and vegetables
2 Genetileke KG - BSc,MSc,PhD tissue culture; postharvest biology and processing of fruits and vegetables
3 Rajapaske D - BSc cereal science and technology
4 Wilson S - BSc,PhD postharvest biology and processing of fruits and vegetables

3843 Industrial Microbiology Section

1 Dharmasiri MAN - BSc postharvest technology, physiological plant pathology, cultivation and preservation of edible mushroom

3844 Natural Products Section

1 Ranatunga J - BScAgr plant tissue culture, agronomy

3845 Natural Resources Energy & Science Authority

47/5, Maitland Plane, Colombo 7
Tel: 596771-3.

1 Jayawardene RP - MBBS,MD,FRCP,FACC *Director General*

3846 Scientific Affairs Division

1 De Silva MAT - BSc,MSc *Deputy Director General* micronutrients in plant nutrition; R & D management and evaluation
2 Ferdinandez DEF - BSc,MSc *Director-Scientific Affairs* crop ecology; research management
3 Wijesinghe LCAS - BSc,MA *Additional Director General* biomass energy consumption surveys

3847 Coconut Research Institute
Bandirippuwa Estate, Lunuwila
Tel: 030-3795.

1 Liyanage DV - BSc,PhD *Chairman*
2 Wettasinghe DT - BScAgr,PhD *Director*
3 Mahindapala R - BSc,MSc,PhD *Deputy Director*
4 Jayasekera Mrs C - BSc *Officer-in-Charge Plant Physiology Unit and Assistant Botanist*
5 Jeganathan M - BSc,MPhil *Head, Division of Soil and Plant Nutrition*
6 Kanagaratnam P - BScAgr,PhD,DIC *Head, Division of Crop Protection*
7 Karunaratne Mrs SM - BSc,MSc *Officer-in-Charge Tissue Culture Unit and Botanist*
8 Liyanage Mrs LVK - BScAgr,MSc *Head, Division of Agronomy*
9 Mathes DT - FIS,BSc,DipStat,DipBio *Officer-in-Charge Biometry Unit and Biometrician*
10 Perera MJC - ALA *Librarian*
11 Ratnayake PAN - BSc *Officer-in-Charge Division of Processing Research and Assistant Technologist*
12 Wickramaratne MRT - BSc,PhD,DIC *Head, Division of Genetics and Plant Breeding*

3848 Sugarcane Research Institute
Colombo Office: 7 Esther Ave., Off Park Road, Colombo 5
Tel: 585869. Telex: 22217.

1 Wickremasinghe RL - BSc,PhD,FIChem,CChem *Chairman*

3849 Research Station
Uda Walawe

1 Dharmawardene MWN - BSc,PhD *Executive Director* sugarcane tissue culture
2 Basnayake BWMJ - BSc,MSc *Research Officer* sugarcane breeding
3 Basnayake Mrs WVS - BSc *Research Officer* sugarcane tissue culture
4 Chandrasena G - BSc,MSc *Research Officer* sugarcane fermentation
5 Jayaratne DL - BSc,MSc *Research Officer* sugarcane pathology
6 Jayawardene GM *Superintendant, Experimental Farm*
7 Kumarasinghe NC - BSc,MSc *Research Officer* sugarcane entomology and nematology
8 Rajapakse TBJ - BSc *Research Officer* sugarcane machinery
9 Ranjith SA - BSC,MSc *Research Officer* sugarcane chemistry and nutrition
10 Ratnaweera U - BSc,MSc *Principal Research Officer* sugarcane agronomy
11 Sanmuganathan K - BSc,MSc *Research Officer* sugarcane water management and irrigation agronomy
12 Sivanathan A - BSc,MSc *Research Officer* sugarcane agronomy
13 Sunil HK - BSc,MSc *Research Officer* sugarcane agronomy and extension
14 Weerasinghe SK - BSc,DipStat *Research Officer* biometry and data processing
15 Wijesuriya A - BSc *Research Officer* sugarcane breeding

3850 Rubber Research Institute
Dartonfield, Agalawatta
Tel: 034-71426.

1 Liyanage A des - BAgrSc,PhD *Director*
2 Karunaratne SW - BSc,MSc,ARIC,ANCRT,IPRI,FIC *Deputy Director*
3 Jayasekera NEM - BScAgric,PhD *Head, Department of Genetics & Plant Breeding*
4 Samaranayake Mrs ACI - BSc,PhD *Head, Department of Plant Science*
5 Attanayake DPSTG - BScAgric *Asst. Geneticist & Plant Breeder*
6 Coomarasamy A - BSc,PhD *Head of Polymer Chemistry Section*
7 de Silva KGK - BSc,MSc,PhD *Rubber Chemist*
8 Dharmaratne Mrs SC - BSc *Asst. Geneticist & Plant Breeder*
9 Dissanayake DMAP - BScAgric *Asst. Soil Chemist*

10 Fernando WSE - BSc,MSc,PhD *Head of Rubber Technology & Development Section*
11 Gunatilleke MDRJ - BSc,MSc *Asst. Rubber Chemist*
12 Jayaratne AHR - BSc,MSc *Asst. Plant Pathologist*
13 Jayasinghe CK - BSc,MSc *Asst. Plant Pathologist*
14 Jayaweera JPN - BSc *Asst. Rubber Chemist*
15 Kalyani Miss NMV - BSc *Asst. Rubber Chemist*
16 Karunaratne SW - BSc,MSc,ARIC,ANCRT,FPRI,FIC *Head of Department* rubber chemistry; rubber technology
17 Liyanage A DES - BAgrSc,PhD *Head, Department of Plant Pathology*
18 Liyanage Mrs NIS - BScAgric,MSc,PhD *Asst. Plant Pathologist*
19 Nugawela RCWMRA - BSc,MSc *Botanist*
20 Perera EDIH - BSc *Asst. Rubber Chemist*
21 Perera MCS - BSc,MSc,PhD *Rubber Chemist*
22 Ranasinghe Miss MS - BSc *Asst. Botanist*
23 Rodrigo VHL - BScAgric *Research Asst. (Intercropping)*
24 Samarappuli Mrs MLA - BScAgric *Asst. Soil Chemist*
25 Samaraweera MKSA - BSc,MSc,PhD *Agricultural Chemist*
26 Seneviratne Mrs P - BSc *Asst. Botanist*
27 Seneviratne Mrs P - BSc *Asst. Rubber Chemist*
28 Senewiratne WMG - BSc *Asst. Rubber Chemist*
29 Silva LBK - BSc *Research Asst.*
30 Silva S DE - BSc *Asst. Agricultural Economist*
31 Talgaswatta H - BSc *Asst. Agricultural Economist*
32 Tillekeratne LMK - BSc,MSc,PhD *Head of Specifications & Analysis Section*
33 Weeraratne BP - BSc *Asst. Rubber Chemist*
34 Wickremasinghe Mrs SI - BSc *Asst. Biochemist*
35 Wickremasinghe WN - BSc,DipStat,MSc *Biometrician*
36 Yapa PAJ - BSc,PhD *Head of Biochemistry Section*
37 Yogaratnam N - BAgrSc,PhD *Head, Department of Soils & Plant Nutrition*

3851 Department of Irrigation

P.O. Box 1138, Jawatta Road, Colombo 7
Tel: 586427. Cables: 'DIRRIG'.

1 Dimantha S - BScAgr,MSc *Head, Land Use Division* soil science; land use planning; water management
2 Dassanayake AR - BScAgr *Asst. Soil Chemist* soil science
3 De Silva CGR - BSc,MSc *Asst. Soil Chemist* soil science
4 Samaranayake SAPK - BSc *Asst. Soil Chemist* soil science
5 Senerath A - BScAgr *Asst. Soil Chemist* soil science
6 Weerasekara WD - BSc,MSc *Asst. Soil Chemist* soil science

3852 Department of Census and Statistics

No. 6, Albert Crescent, P.O. Box 563, Colombo 7
Tel: 92988.

3853 Division of Agriculture
No. 3/1, Rajakeeya Mawatha, Colombo 7
Tel: 597591, 597592, 597646.

1 Yasasiri AADC - BA *Additional Director, Head of the Division*
2 Atapattu DP *Statistican* estimation of the paddy rice area
3 De Silva Mrs AP - BSc estimation of area and production of subsidiary food crops
4 Dissanayake AMU - BA *Assistant Director* census of agriculture
5 Handagama A - BSc *Statistician* cost of production surveys
6 Sangarapillai S - BSc *Deputy Director, Deputy Head of the Division*
7 Satharasinghe AJ - BSc *Statistician* estimation of (rice) paddy production

3854 Agrarian Research & Training Institute
P.O. Box 1522, 114, Wijerama Mawatha, Colombo 07
Tel: 596981, 598539, 598540, 598541. Cables: AGRATI.

1 Alwis HPGJ - BA *Director*
2 Abeysekera WAT - BScAgr,MScAgrEcon,PhD *Head of Agricultural Resource Management Division*
3 Chandrasiri LRA - BA,MA *Head of Marketing and Food Policy Division*
4 Perera ULJJ - BA,MA,PhD *Research Associate*
5 Sanmugam T - MScPostGradDiploma *Head of Statistics and Data processing Division*
6 Wanigaratne RD - BA,MA,PhD *Head of Agricultural Planning and Evaluation Division*
7 Abeyratne FADESM - BScAgr,MScAgr,Econ *Research and Training Officer*
8 Ranasinghe W - ALA *Librarian*
9 Sepala ACK - BA,BPhil,PostGradDipAgrEcon,MA *Research and Training Officer*
10 Wickramasinghe G - BPhil,MA *Head of Irrigation, Water Management and Agrarian Relations Division*
11 Wijayaratne CM - BScAgr,MAAgrEcon,PhD *Research and Training Officer*

3855 Tea Research Institute
Head Office, St. Coombs, Talawakele
Tel: Hatton 601; Nuwara Eliya 8244;8385-86.

1 Sivapalan P - BSc,PhD,FIBiol *Director*
2 Kulasegaram S - BSc,PhD,MIBiol *Deputy Director Research*
3 Munindradasa M - BA,LLB *Deputy Director (Administration)*

3856 Division of Advisory & Extension Service/Talawakelle

1 Jayakody JAAM - BSc agricultural economics
2 Navaratne DK - BSc agricultural advisory & extension
3 Rajasingham CC - DIP agricultural advisory & extension
4 Somaratne A - BSc,MSc *Acting Head*
5 Wanigasundera WADP - BSc,PhD agricultural advisory & extension
6 Wimaladharma S - DIP agricultural advisory & extension

3857 Low Country Station, Ratnapura

1 Samarakoon HH *Senior Advisory Officer/Officer-in-Charge*

3858 Research Advisory & Extension Centre, Kandy

1 Bandaranayake JIH *Senior Advisory Officer/Officer-in-Charge*
2 Samarajeewa S agricultural advisory & extension

3859 Advisory & Extension Centre, Talgampola

1 Jayasinghe HD - DIP *Senior Advisory Officer/Officer-in-Charge*

3860 Advisory & Extension Centre, Deniyaya

1 Jayasooriya SG *Senior Advisory Officer/Officer-in-Charge*

3861 Advisory & Extension Centre, Passara

1 Sundaralingam S - BSc *Senior Advisory Officer/Officer-in-Charge*

3862 Division of Agricultural Chemistry

1 Ananthacumaraswamy Mrs S soil chemistry
2 Fernando V soil chemistry
3 Jayman TCZ soil chemistry
4 Krishnapillai S - BSc,PhD *Acting Officer-in-Charge* soil chemistry
5 Saman Hettiarachchi - BSc soil chemistry
6 Wimaladasa GD - BSc soil science and analytical chemistry

3863 Division of Agronomy

1 Ananthacumaraswamy A - BSc,MPhil soil physical properties
2 Basnayake Mrs AK - MScAgr soil physical properties and water relations
3 Ekanayake PB - BScAgr soil fertility
4 Wadasinghe G - BSc,MSc *Acting Head, Ce.* growth analysis

3864 Division of Biochemistry

1 Abeysinghe ISB - BSc plant biochemistry
2 Amarakoon AMT - BSc plant biochemistry
3 Herath N - BSc plant biochemistry
4 Liyanage Mrs AC - MSc plant biochemistry

3865 Division of Entomology

1 Amarasinghe Ms DL - BSc entomological research
2 Thirugnanasuntheran K - BSc,MPhil *Acting Officer-in-Charge* insect ecology and pest management
3 Vitarana Mrs SI - BSc,MSc,DIC *Entomological/Acting Head* applied entomology

3866 Division of Nematology

1 Dahanayake RB - BSc nematode physiology
2 Gnanapragasam Ms NC - BSc,MPhil,PhD *Nematologist and Head of Division* management of plant parasitic nematodes; studies on insect nutrition

3867 Division of Plant Pathology

1 Arulpragasam PV - BSc,MPhil *Plant Pathologist and Acting Head of Division* transmittance of viral diseases of tea; tissue culture
2 Seneviratne Ms P - BSc tissue culture
3 Wickramasinghe UK - BSc microbiology of tea leaf surface; spoilage of black tea

3868 Division of Plant Physiology, Plant Propagation and Plant Breeding

1 Anandappa Ms TI - BSc,MSc,PhD *Plant Breeder and Head of Division* clonal selection; breeding
2 Kathiravetpillai A - BSc,PhD clonal selection; hormone physiology; tea nursery techniques; fuel trees

3 Krishnapillai S - BSc,PhD mineral nutrition, nitrification/denitrification
4 Sekar Ms - BSc plant breeding
5 Shanmugarajah V - BSc hormone physiology, tea nursery techniques, fuel trees
6 Solomon HR clonal selection and identification

3869 Division of Technology

1 Ediriweera DP - BScEng,MSc,CEng,MIE tea processing and engineering
2 Kandappah C - BSc tea processing
3 Mohamed MTZ - BSc tea processing (Working Overseas)
4 Samarasingham S tea processing
5 Samaraweera DSA - BScEng,MScEng,PhD,DIC *Technologist and Head of Division* tea processing and biochemical engineering
6 Thevathasan A - BSc tea processing
7 Willarachchi HK - BScEng tea processing and engineering

3870 Statistics Section

1 Thevathasan T - BSc,DipAppStat,FSS

3871 St Coombs & Lamiliere Estate, Talawakelle

1 Gunasekera D *Superintendent*

3872 St Joachim Estate, Ratnapura

1 Seevaratnam LA - BSc *Superintendent*

3873 The International Irrigation Management Institute

IMII,Digana Village Via Kandy
Tel: (08) 74274-74334. Telex: 22318 IIMIHQ CE.

3874 Liaison Office

P.O. Box 2075, 64A Jawatte Road, Colombo 7
Tel: (01) 586135.

1 Berthery D - MSc *Agricultural Engineering*
2 Groenfeldt D - PhD *Social Anthropologist*
3 Kikuchi M - PhD *Agricultural Economist*
4 Martin E - PhD *Agricultural Economist*
5 Merrey D - PhD *Anthropologist*
6 Miranda SM - PhD *Agricultural Engineer*
7 Panabokke CR - PhD *Agronomist*
8 Rao PS - PhD *Systems Scientist*

UNIVERSITIES

3875 University of Peradeniya

Peradeniya

3876 Faculty of Agriculture

1 Gunasena HPM - BScAgr,PhD *Dean, Professor of Crop Science*

3877 Department of Crop Science

1 Gunasena HPM - BScAgr,PhD,MIBiol *Senior Lecturer & Head of Department*
2 Senanayake YDA - BScAgr,MSc,PhD *Professor of Crop Science* genetics
3 Peiris BCN - BScAgr,MSc *Asst. Lecturer* seed physiology
4 Perera DGAH - BScAgr,MFR *Asst. Lecturer* agroforestry
5 Premaratne KP - BScAgr,MSc,PhD *Lecturer* plant nutrition
6 Rajapakse NC - BScAgr,MSc *Asst. Lecturer* postharvest physiology
7 Rajapakse Mrs S - BScAgr,MSc horticulture
8 Ranamukaarachchi SL - BScAgr,MSc *Asst. Lecturer* cropping systems
9 Sangakkara UR - BScAgr,PhD *Lecturer* crop ecology
10 Thattil RO - BSc,MSc,PhD *Asst. Lecturer* biometry

3878 Department of Animal Science

1 Rajaguru ASB - BScAgr,AH & DT,MSc *Professor and Head of Department* monogastric nutrition
2 Cyril HW - BScAgr,MSc *Asst. Lecturer* product technology
3 Perera Mrs ERK - BScAgr,MSc *Asst. Lecturer* environmental physiology
4 Perera N - BScAgric,MSc,PhD
5 Premaratne S - BScAgr,MSc,PhD ruminant nutrition
6 Ravindran V - BScAgr,MSc *Asst. Lecturer* monogastric nutrition

3879 Department of Agricultural Biology

1 Bandara JMRS - BScAgr,PhD *Senior Lecturer & Head of Department* plant pathology
2 Herath HMW - BScAgr,MSc,PhD *Professor of Agricultural Biology* plant physiology
3 Ahangama Mrs D - BScAgr,MSc,PhD *Lecturer* entomology

4 Bandara DC - BScAgr,MSc,PhD *Asst. Lecturer* plant physiology
5 Galappatti Mrs JS - BA,MSc,PhD *Lecturer* pathology
6 Perera ALT - BScAgr,MSc *Asst. Lecturer* plant breeding
7 Ranasinghe MASK - BSc,MSc,PhD *Lecturer* entomology
8 Sumanasinghe VAD - BScAgr,MSc,PhD *Asst. Lecturer* plant taxonomy
9 Wijegunasekera HNP - BScAgr,MPhil *Lecturer* entomology

3880 Department of Food Science and Technology

1 Ravindran Mrs G - BScAgr,MSc,PhD *Lecturer* quality control & food analysis
2 Samarajeewa U - BSc,PhD *Lecturer* food microbiology

3881 Department of Soil Science

1 Thenabadu MW - BScAgr,MSc,PhD *Professor & Head of Department* soil fertility
2 Jayakody AN - BSc,MSc,PhD *Lecturer* soil science
3 Keerthisinghe G - BSc,MSc,PhD *Lecturer* plant nutrition & soil fertility
4 Mapa RB - BScAgr,MSc,PhD *Lecturer* soil science
5 Yapa LG - BScAgr,MSc,PhD *Lecturer* soil physics

3882 Department of Agricultural Economics & Extension

1 Herath HMG - BScAgr,MSc,PhD *Lecturer & Head of Department* farm management
2 Abeygunawardena P - BScAgr,MSc,PhD
3 Bogahawatte C - BScAgr,MS,PhD *Lecturer* agricultural marketing
4 Jayatilake MWAP - BScAgr,MSc *Asst. Lecturer* rural sociology
5 Pinnaduwage S - BScAgr,MScMS,PhD *Senior Lecturer* agricultural development & extension
6 Sappideen TB - BScAgr,MSc *Asst. Lecturer* land economics
7 Sivayoganathan C - BScAgr,MAgrSt,PhD *Lecturer* extension education & communication
8 Wickramasuriya HVA - BScAgr,MSc *Asst. Lecturer* agricultural education
9 Zuhair SMM - BScAgr,MSc *Asst. Lecturer* agricultural finance

3883 Department of Agricultural Engineering

1 Basnayake BFA - BScAgrEng *Asst. Lecturer* energy technology
2 Goonasekera KGA - BScAgr,MSc *Asst. Lecturer* hydrology
3 Gunawardena ENR - BScAgr,MSc *Asst. Lecturer* soil & water management
4 Jayasekera Mrs AA - BScAgr,MSc *Asst. Lecturer* onfarm water management
5 Jayatissa DN - BScAgr *Asst. Lecturer* harvest handling machinery
6 Punidadas P - BScAgr,MSc *Asst. Lecturer* food engineering

3884 Maha-Illuppallama Sub Campus

1 Alvapillai P - BScAgr *Agri. Teaching Asst.* agricultural engineering
2 De Silva Mrs C - BScAgr,MSc *Agri. Teaching Asst.* animal science
3 Jayaweera Mrs CS - BScAgr,MSc *Asst. Lecturer* weed science
4 Keerthisinghe JP - BScAgr *Agri. Teaching Asst.* crop science
5 Malkanthi Menike Miss R - BScAgr *Agri. Teaching Asst.* crop science
6 Rambanda M - BScAgr,MScAgrEng *Asst. Lecturer* field machinery
7 Somaratne HM - BScAgr *Agri. Teaching Asst.* agricultural economics & extension
8 Tharmaraj J - BScAgr *Agri. Teaching Asst.* animal science
9 Thiruchelvam S - BScAgr,MSc *Asst. Lecturer* production economics
10 Thirumavithana SC - BScAgr *Agri. Teaching Asst.* crop science
11 Wijeyathungam K - BScAgr *Agri. Teaching Asst.* crop science

3885 Library

1 Fernando JV - ALA *Assistant Librarian*
2 Mudannayake Miss I - BA,ALA *Assistant Librarian*
3 Premachandra GS - BScAgr *Farm Manager*

3886 Faculty of Science
Peradeniya

1 Dias HW *Dean*

3887 Department of Botany

1 Dassanayake MD - BSc,MA,PhD *Emeritus Professor* flora of Sri Lanka
2 Kulasooriya SA - BSc,PhD,FIBiol,FNA *Professor of Microbiology* soil microbiology; nitrogen fixation
3 Adikaram NKB - BSc,PhD *Lecturer* post harvest fruit pathology
4 Balasubramaniam Prof S - BSc,PhD *Head of Department* phytochemistry; ecology and ecophysiology of mangroves and rainforests
5 Gunatilleke CVS - BSc,PhD *Senior Lecturer* tropical forest ecology; reproductive biology
6 Gunatilleke IAUN - BSc,PhD *Senior Lecturer* tropical forest microbiology; reproductive biology
7 Kumarasinghe RMK - BSc,PhD,DIC *Senior Lecturer* legume nodulation

3888 Department of Zoology

1 Breckenridge Prof WR - BSc,PhD *Head of Department* molluscs of agricultural and medical importance
2 Amerasinghe FP - BSc,PhD *Senior Lecturer* entomology, especially mosquito systematics & ecology
3 De Silva Mrs DN - BSc,MSc *Lecturer* fisheries and aquaculture
4 De Silva KHGM - BSc,PhD *Senior Lecturer* fisheries and aquaculture
5 De Silva Mrs PK - BSc,PhD *Senior Lecturer* fisheries and aquaculture
6 Edirisinghe Mrs JP - BSc,PhD *Senior Lecturer* entomology, especially biological control of agricultural pests
7 Wijekoon Mrs S - BSc,PhD,DIC *Senior Lecturer* plant and soil nematodes of agricultural importance

3889 Post Graduate Institute of Agriculture

Old Galaha Road, Peradeniya
Tel: 08-88219, 08-88318.

1 Senanayake YDA - BScAgr,MS,PhD *Director*

3890 University of Jaffna

PO Box 57, Thirunelvely, Jaffna
Tel: Jaffna 22481/23626.

1 Ganesalingam VK - BSc,MSc,PhD *Professor, Head of Department* entomology
2 Chitravadively K - BSc,MSc,RNDr,PhD *Lecturer* marine biology
3 Rajendram GF - BA,MA,MSc,PhD *Senior Lecturer* entomology
4 Selvarajah Mrs N - BSc,MSc *Senior Lecturer* limnology

3891 Department of Botany

1 Kandiah S - BSc,PhD *Senior Lecturer, Head of Department* plant physiology and horticulture
2 Saverimuthu AMT - BSc,PhD *Lecturer* eco-physiology
3 Saverimuthu Mrs N - BSc,MPhil *Asst. Lecturer* plant pathology
4 Sundaresan RVS - BSc,MSc *Senior Lecturer* biochemical virology
5 Theivendirarajah K - BSc,PhD *Professor* microbial genetics

3892 University of Ruhuna

3893 Faculty of Agriculture

Mapalana, Kamburupitiya
Tel: 041-2300.

1 Samarawickrama G - MBBS,MD,PhD *Vice Chancellor*
2 Pathirana KK - BVSc,MSc,PhD *Dean*

3894 Department of Agronomy

1 Weerasinghe KDN - MSc,PhD *Head of Department* irrigation & salinity
2 Alwis PLAG - BScAgr *Asst. Lecturer* agricultural engineering
3 Amarasekera DABN - BScAgr *Asst. Lecturer* crop science
4 Dayatilleke G - BScAgr *Asst. Lecturer* crop science
5 Goonewardene SFBN - BScAgr,MSc *Asst. Lecturer* soil science
6 Hemachandra DL - BScAgr *Asst. Lecturer* plant pathology
7 Kotelawela J - BScAgr,MSc *Senior Lecturer* horticulture
8 Pathirana R - MSc,PhD *Lecturer* genetics and improvement of oil seeds
9 Rajapaske R - BScAgr,MScAgr,MS,PhD *Lecturer* entomology; biological control of pests
10 Senanayake G - BScAgr *Asst. Lecturer* genetics & plant breeding
11 Senaratne R - BScAgr,MPhil,PhD *Lecturer* biological nitrogen fixation
12 Serasinghe PSJW - BScAgr *Asst. Lecturer* plantation crops
13 Wijeratne Mrs V - BScAgr,MSc *Asst. Lecturer* food technology
14 Wijesuriya SDE - BScAgr,MSc *Asst. Lecturer* agricultural soil and water engineering

3895 Department of Agricultural Economics and Extension

1 Amarasinghe O - BScAgr *Asst. Lecturer* agricultural economics and extension
2 Charles SK - BScAgr,PhD *Senior Lecturer and Head of Department* farm management
3 Dharmasena KH - BScAgr *Asst. Lecturer* microeconomics
4 Wijeratne M - BScAgr,MSc *Asst. Lecturer* agricultural economics and extension
5 Zoysa AMP DE - BScAgr *Asst. Lecturer* agricultural economics

3896 Department of Animal Science

1 Pathirana KK - BVSc,MSc,PhD *Professor and Head of Department* forages and nontraditional livestock feeds
2 Goonewardene WWDA - BScAgr *Asst. Lecturer* animal genetics
3 Serasinghe Ms RT - BScAgr *Asst. Lecturer* agrostology
4 Wickramsinghe Ms ID - BVSc *Asst. Lecturer* reproductive physiology

3897 University of Kelaniya

3898 Faculty of Science

Kelaniya

1 Balasooriya I - BSc,PhD,FIBiol *Professor of Botany and Vice Chancellor* soil microbiology
2 Fernando EFW - BSc,MSc,PhD,DIC *Senior Lecturer in Zoology* entomology
3 Fernando IVS - BSc,MSc,PhD,DIC *Senior Lecturer in Zoology* entomology
4 Gunasekera SA - BSc,PhD,DIC,FIBiol *Associate Professor of Botany and Head, Dept. of Botany* mycology and plant pathology
5 Jayawardene Mrs RMD - BSc,MSc,PhD *Lecturer in Botany* taxonomy
6 Wickremesinghe LKG - BSc,PhD *Senior Lecturer in Chemistry* insect-plant interactions
7 Widanapathirana S - BSc,MSc,PhD,DipMicro&Biotech *Professor of Microbiology* microbiology

3899 Eastern University

Vantharumoolai, Chenkaladi
Tel: 065-2473.

1 Arudpragasam KD - BSc,PhD *Director*

3900 Faculty of Agriculture

1 Sandanam S - BSc,PhD *Dean*

3901 Department of Agronomy

1 Ahamed AN - BSc,MSc *Lecturer* agricultural economics
2 Iqbal MCM - BSc,ETH-Diplom *Assistant Lecturer* plant breeding
3 Raveendranath S - BSc,PhD,DIC *Lecturer* applied entomology
4 Sabesan K - MSc,PhD *Lecturer* soil fertility
5 Sandanam S - BSc,PhD *Senior Lecturer* crop science
6 Sivasubramaniam S - BSc,PhD,DIC,FIChem *Senior Lecturer, Head of Department* soil chemistry

TANZANIA

3902 Ministry of Agriculture and Livestock Development

3903 Directorate of Research & Training Headquarters

PO Box 9071, Dar es Salaam
Tel: 27231.

1 Ilmolelian LL - BSc,BVMS,DAP&E *Director of Research and Training*
2 Kikopa R - DrMedVet,DTM *Chief Research Officer Livestock*
3 Shao FM - BSc,MSc,PhD *Chief Research Officer (Crops) Plant Pathologist*
4 Awasi MBH - BVSc,DipAnimHealth&Prod,CertVet
5 Lutkamu Mrs M - BSc,MSc soil science
6 Makundi PJ - BSc *Agricultural Economist*
7 Malima Mrs VF - BSc,MSc botany
8 Mbonika RJ - BSc,MSc soil science
9 Ruzika EN - BSc *Agricultural Economist*

3904 Dakawa Agro-Scientific Research Centre

PO Box 972, Morogoro

1 Iranga G - BSc maize agronomy
2 Kanyeka ZL - BSc,MSc rice breeding
3 Kiula B - BSc maize agronomy
4 Mtwaenzi HH - BSc rice agronomy
5 Nyambo Miss DB - BSc,MSc soil science
6 Zakayo J - BSc rice agronomy

3905 Horticulture Research and Training Institute Tengeru

PO Box 1253, Arusha
Tel: Duluti 94.

1 Butter J - DipAgr *Project Coordinator*
2 Lymo Miss V - BSc horticulture research
3 Ngaiza Mrs T - BSc,MSc soil science
4 Schippers R - DipAgr vegetable seed production
5 Swai IS - BSc,MSc *Plant Pathology National Horticulture Research Coordinator*
6 Swai RE - BSc,MSc horticultural research extension

3906 National Coconut Development Project

PO Box 6226, Dar es Salaam

1 Balingasa EN - PhD coconut breeding
2 Barley H - MSc agricultural extension
3 Bauer E - MSc *Technical Services*
4 Bryant R - MSc plantation management
5 Chipungahelo Mrs G - BSc coconut agronomy
6 Hubert K - PhD *Project Coordinator*
7 Kabonge PMD - BSc coconut breeding
8 Kaiza D - BSc *Plant Pathology*
9 Kinyawa PL - BSc *Counterpart Project Coordinator*
10 Kullaya A - MSc plant breeding
11 Lameck D - MSc agricultural extension
12 Mapunda F soil science
13 Mogaeka SM - BSc,MSc agricultural economics
14 Mombar EA - MSc planning
15 Mpunami Miss A - BSc plant pathology
16 Munthali AD - DipAgr *Project Coordinator*
17 Ngowe AF - BSc coconut agronomy
18 Ngowi J - BSc,MSc soil science
19 Ntimbwa T - BSc agricultural extension
20 Romney D - MSc coconut agronomy
21 Schuiling M - PhD plant pathology
22 Seguni Z - BSc entomology
23 Temu A - BSc agricultural extension

3907 Tanzania Agricultural Research Organization (TARO) HQ

PO Box 9761, Dar es Salaam
Tel: 44247.

1 Semuguruka GH - BSc,MSc,PhD *Director of Research*
2 Chota Miss EA - BSc farming systems research
3 Cloutier PE - BSc *Research/Extension Adviser*
4 Kaaya Mrs JE - BSc information and documentation
5 Lev L - PhD farming systems research; production economics
6 Madulu JD - BSc,MSc agronomy
7 Makere Mrs CM - BSc farming systems research
8 Mmbaga Miss ZS - BSc farming systems research
9 Polepole MN - BSc,MA research planning and evaluation

10 Sungusia D - BSc,MSc *National Co-ordinator* farming systems research

3908 TARO - Kibaha

PO Box 30031, Kibaha
Tel: 2016/7.

1 Haki JM - BSc,MSc,PhD *Director, National Sugarcane Research Coordinator* sugarcane breeding
2 Bararama FM - MSc soil science
3 Kaijage CB - BSc statistics
4 Kuwite Miss CA - BSc,MSc plant pathology
5 Massawe Mrs SC - BSc sugarcane breeding
6 Minja Miss EM - BSc,MSc entomology

3909 TARO - Lyamungu

PO Box 3004, Moshi
Tel: 4411.

1 Kessy DL - BSc,MSc *Director, National Coffee Research Coordinator* coffee agronomy
2 Akulumuka VCA - BSc farming systems research
3 Andersen B - MSc coffee breeding
4 Bardner R - MSc coffee entomology
5 Dunstan WR - BSc,PhD armyworm
6 Gondwe Miss B - BSc,MSc bean pathology
7 Koinange EMK - BSc,MSc *Coordinator, National Bean Research Programme* bean breeding
8 Kullaya IK - BSc,MSc soil fertility
9 Kundya Miss I - BSc coffee pathology
10 Matowo PR - BSc,MSc maize agronomy
11 Mbuya OS - BSc bean agronomy
12 Mcharo EY - BSc coffee entomology
13 Meda RMA - BSc chemistry
14 Mmbaga MET - BSc bean agronomy
15 Mmbaga Mrs TM - BSc cereals agronomy
16 Naylor PG - PhD coffee breeding
17 Ngulu SF - BSc,MSc coffee pathology
18 Nyange EN - BSc coffee breeding
19 Ringia DI - BSc farming systems research
20 Samkyi TMF - BSc farming systems research
21 Whittle AM - PhD coffee pathology

3910 TARO - Ilonga

PO Box Kilosa, Kilosa
Tel: Kilosa 49.

1 Mitawa GM - BSc,MSc,PhD *Director* sorghum agronomy
2 Assenga JA - BSc,MSc grain legume breeding
3 Bunyecha FNK - BSc farming systems research
4 Chambo HS - BSc millet breeding
5 Chambuya RL - BSc,MSc legume entomology
6 Chillagane EA - MSc agricultural engineering; farming systems research
7 Fivawo Miss NC - BSc,MSc sunflower breeding
8 Guutazi AS - BSc sorghum and millet entomology
9 Kabahaula YB - BSc engineering
10 Kabissa JCB - BSc,MSc cotton entomology
11 Katuli NCR - BSc land use management
12 Lamboll D - BSc,MSc *Co-ordinator Sunflower Research Project* sunflower agronomy
13 Mansuetus ASB - BSc maize pathology
14 Mbowe FFA - BSc,MSc *Grain-Legume Agronomy Coordinator, National Grain-Legume Research Programme*
15 Mduruma Mrs ZO - BSc,MSc maize breeding
16 Mhando E - BSc soils
17 Mlay EE - BSc,MSc agronomy; irrigation
18 Mligo JK - BSc,MSc grain-legume breeding
19 Monyo ES - BSc sorghum agronomy
20 Moshi AJ - BSc,MSc,PhD *Co-ordinator, National Maize Research Programme* maize breeding
21 Mshana Miss DE - BSc cotton breeding
22 Mushi CS - BSc sorghum agronomy
23 Mvungi ED - MSc rice agronomy
24 Mwanjali - BSc farming systems research
25 Nyanga Miss J - MSc maize agronomy
26 Saadan HM - BSc,MSc *Co-ordinator, National Sorghum Research Programme* sorghum and millet breeding
27 Sumari WLM - BSc farming systems research

3911 TARO - Katrin

Private Bag, Ifakara
Tel: 78 Ifakara.

1 Mgonja AP - BSc,MSc,PhD *Director* rice pathology
2 Ching'ang'a HM - BSc,MSc *Coordinator Rice Research Programme* rice breeding
3 Kigalu JM - BSc,MScAgrEng rice irrigation
4 Kuhupi Mrs A - BSc,MSc rice breeding
5 Mbapila JC - BSc rice entomology

6 Mgonja Mrs MA - BSc,MSc
7 Msomba SW - BSc rice breeding
8 Muliahela LME - BSc rice agronomy
9 Myovella GG - MSc soil fertility
10 Pillai KG - BSc,MSc,PhD *Regional Coordinator: IRRI-IRTP* rice agronomy
11 Shigazu Miss JSE - BSc rice pathology

3912 TARO - Marikitanda
PO Amani, Tanga

1 Ndamugoba D - BSc,MSc *Coordinator, Tea Research Programme* tea agronomy
2 Ndamugoba Mrs IK - BSc,MSc bean breeding

3913 TARO - Maruku
PO Box 127, Bukoba
Tel: 533.

1 Mbwana ASS - BSc,MSc *Director, Coordinator, Banana Research Programme* banana nematology
2 Marandu EF - BSc coffee breeding
3 Mkamba Mrs IJ - BSc,MSc soil chemistry

3914 TARO - Mlingano
Private Bag, Ngomeni Tanga
Tel: 8 Ngomeni Tanga.

1 Haule KL - BSc,MSc *Director, Co-ordinator, National Soil Service* soil survey and land evaluation
2 de Wijs M - MSc soil and plant analysis
3 Harrop JF - BSc,MSc soil fertility
4 Ikeera T - BSc,MSc sisal agronomy
5 Ikkera Mrs ST - BSc,MSc soil chemistry
6 Kalumuna Miss MC - BSc soil fertility
7 Kimaro D - BSc soil survey and land evaluation
8 Kiwambo BJ - BSc,PhD soil chemistry, soil analysis
9 Kiwovelle JJ - BSc,MSc soil chemistry
10 Lema Miss LF - BSc sisal pathology
11 Magogo JMP - BSc soil survey & land evaluation
12 Mgema WGH - BSc cereals agronomy and breeding
13 Mgogo SE - BSc,MSc soil survey & land evaluation
14 Mlay Miss B - BSc sisal breeding
15 Msafiri HB - BSc *Co-ordinator, Sisal Research Programme* sisal agronomy
16 Musisa NS - BSc soil fertility
17 Mwanyika EE - BSc agricultural engineering
18 Myaka FA - BSc maize agronomy
19 Niemeyer JKW - BSc,MSc soil survey
20 Nzabayanga SI - BSc soil survey
21 Shenkalwa S - MSc soil fertility
22 Shirima OS - BSc soil chemistry; resource-efficient farming methods
23 Vahaye G - BSc,PhD *Co-ordinator, Resource Efficient Farming Methods and Agro-Forestry*
24 Van Barneveld GW - MSc *Soil Resources Specialist*
25 Van Kekem AJ - MSc soil survey & land evaluation

3915 TARO - Naliendele
PO Box 509, Mtwara
Tel: 2085.

1 Shomari SH - BSc,MSc *Director, Cashew Pathology Co-ordinator, Cashew Research Programme*
2 Chambi JY - BSc,MSc *Co-ordinator, Oilseeds Research Programme* sesame breeding
3 Kafiriti EM - BSc oilseed agronomy
4 Mwenda FK - BSc,MSc groundnut breeding
5 Ndunguru DJ - BSc cashew agronomy
6 Ngatunga ELM - BSc,MSc soil fertility
7 Raya MD - BSc,MSc oilseed crop protection
8 Sijaona MER - BSc cashew breeding

3916 TARO - Selian
PO Box 6024, Arusha
Tel: 3883/3181. Telex: 42106.

1 Mosha AS - BSc,MSc *Director, Co-ordinator Wheat Research Programme* wheat breeding
2 Antapa PL - BSc,MSc agricultural engineering
3 Chatwin SC - BSc soil survey
4 Loewen-Rudgers LA - PhD wheat soil fertility
5 Lyimo SD - BSc,MSc wheat agronomy
6 Maeda Miss E - BSc wheat agronomy
7 Mkunga MJ - BSc land use/management
8 Modestus WK - BSc,MSc soil fertility
9 Moyer J - PhD weed agronomy
10 Ndondi RV - BSc,PhD wheat breeding
11 Nyakyi AS - BSc,MSc wheat agronomy
12 Sariah Mrs A - BSc wheat breeding
13 Warren FS - PhD wheat agronomy

3917 TARO - Tumbi
PO Box 306, Tabora
Tel: 2431.

1 Kalemela KT - BSc,MSc *Director, Co-ordinator, Tobacco Research Programme* tobacco agronomy
2 Elia W - BSc,MSc maize agronomy
3 Kaihura FS - BSc soil fertility
4 Kanoni MP - MSc tobacco agronomy
5 Katundu JM - BSc tobacco entomology
6 Lema MN - BSc,MSc soil science
7 Mfinanga ES - BSc tobacco agronomy
8 Mkangwa CZ - BSc soil chemistry
9 Titus WLM - BSc tobacco breeding

3918 TARO - Ukiriguru
PO Box 1433, Mwanza
Tel: 40596/7.

1 Msabaha MAM - BSc,MSc,PhD *Director, Co-ordinator, National Root & Tuber Crops Improvement Programme* root and tuber crops breeding
2 Akonnay HB - BSc maize agronomy
3 Gama B - BSc,MSc soil chemistry
4 Giles J - PhD cotton breeding
5 Kapingu MPK - BSc,MSc *Co-ordinator, Cotton Research Programme* cotton breeding
6 Kibani THM - BSc,MSc cotton pathology
7 Kipokola TP - MSc cotton agronomy
8 Mowo JG - BSc,MSc cotton agronomy
9 Mtambo K - BSc root and tuber crops pathology
10 Ndibaza ER - BSc root and tuber crops agronomy
11 Nyambo Miss BT - BSc,MSc cotton entomology
12 Parez G - MSc cotton extension, social economics

3919 TARO Research Workers attached to the Agricultural Centre
PO Box 400, Mbeya
Tel: 2116.

1 Haule SJ - BSc maize agronomy
2 Marandu WYF - BSc,MSc,PhD maize breeding

3920 Uyole Agricultural Centre Mbeya
PO Box 400, Mbeya
Tel: 22116/7.

1 Liwenga JM - BSc,PhD *Director, Plant Physiologist*
2 Akarro FM - BSc,MSc animal health
3 Haule S - BSc maize agronomy
4 Kabunga DA - BSc plant protection
5 Kamasho JA - BSc,MSc soil science
6 Kiffaro GG - BSc animal nutrition
7 Kiliaki C - MSc agro-economics
8 Kirway TN - BSc,MSc soil science
9 Kwiligwa EE - BSc,MSc agricultural engineering
10 Ley GC - BSc,MSc soil science
11 Lutende DKL - BSc,MSc agricultural engineering
12 Lyimo NG - MSc agronomy
13 Macha CA - BSc crop production
14 Madata Mrs C - BSc,MSc plant breeding
15 Madata GS - BSc,MSc,PhD *Chief Research Officer* animal nutrition
16 Maganga ET - BSc food technology
17 Maimu ZY - BSc crop production
18 Mallya GA - BSc,MSc entomology
19 Manang EZ - MSc plant breeding
20 Marandu WY - BSc,MSc,PhD maize breeding
21 Mayona GM - BSc,MSc maize agronomy
22 Mayona TJ - BSc pasture seed production
23 Mbwile KP - BSc,MSc animal nutrition
24 Mchau KW - BSc,MSc animal science
25 Meela EE - BSc crop production
26 Mghgho RMK - BSc crop production
27 Miettenen EERO - MSc plant breeding
28 Mindasi FMS - MSc chemistry; applied science of materials in chemistry
29 Minja Miss AEJ - BSc animal nutrition
30 Mkina M - BSc crop production
31 Mkuchu Mrs MM - BSc crop production
32 Morris ER - MSc wheat breeding
33 Msaki SP - MSc pedology, land evaluation
34 Mukurasi N - MSc pasture and forage seed production
35 Mussei AN - BSc agricultural economics
36 Mwalyego FM - MSc plant pathology
37 Mwanbene ROF - BSc,MSc *Principal Research Officer* plant breeding
38 Nsemwa LTH - BSc,MSc plant pathology
39 Nyamora AM - BSc,MSc horticulture
40 Ponjee CKJ - BSc,MSc agricultural economics
41 Shayo FM - BSc,MSc beans pathology
42 Shetto RM - BSc,MSc agricultural mechanics
43 Shiyumbi JKK - BSc animal husbandry
44 Temu AE - BSc,MSc agronomy

3921 Tanzania Pesticide Research Institute
PO Box 3024, Arusha
Tel: 3557/8.

1 Simon B - BSc,PhD *Director* plant pathology
2 Abdallah Mrs RSS - BSc *Seed Pathologist*
3 Bujulu J - MSc plant pathology
4 Chogo JB - BSc,MSc *Chemist*
5 Ijani ASM - BSc,MSc plant pathology
6 Kachecheba JL - MPhil *Horticulturist*
7 Kaoneka BS - BSc *Chemist*
8 Kimaro Mrs EM - BSc zoology, botany
9 Kisampfu Mrs VA - BSc zoology, botany
10 Lyatuu HA - DipEd,BSc,MSc plant protection
11 Macha Miss ES - BSc *Botanist*
12 Macha Mrs MPS - BSc *Chemist*
13 Manyanza D - BSc zoology, botany
14 Maradufu AN - MA,PhD *Chief Research Officer* chemistry
15 Matee JJ - BSc *Agronomist*
16 Matemu Mrs DP - BSc chemistry
17 Mbise JT - DipEd,BSc,MSc plant protection
18 Mbise S - MSc entomology
19 Mosha FW - MSc,PhD entomology
20 Moshi AO - BSc,MPhil agricultural chemistry
21 Muangirwa CJ - BSc,MSc,DIC entomology
22 Mununa Miss F - BSc zoology, botany
23 Mushi CSK - BSc,MSc bio-aeronautics
24 Mwamiwa A - BSc chemistry
25 Mziray WR - BSc *Botanist*
26 Ngowi Mrs AVF - BSc chemistry
27 Nguma JFM - BVM,MVSc veterinary medicine
28 Ringo DFP - BSc,MSc entomology
29 Said JA - MSc agricultural entomology
30 Tanganyika DN - BSc *Chemical Engineer*
31 Uronu Mrs AB - BSc botany; statistics
32 Uronu BEMA - DipEd,BSc agricultural entomology

3922 Tanzania Livestock Research Organisation Headquarters
PO Box 6910, Dar es Salaam
Tel: 63191.

1 Macha AM - BVMSc,PhD,MRCVS *Director General*
2 Tarimo CS - BSc,MSc,CertRadiolsotope *Director of Research*

3923 Animal Diseases Research Institute
PO Box 9254, Dar es Salaam
Tel: 63191.

1 Mella PNP - BVSc *Director* pathology
2 Jaggadi MM - BSc,MSc chemistry
3 Kaale MA - BSc,MSc acaricide chemistry
4 Kagaruki Mrs LK - BSc entomology
5 Kuundy DJ - MVM bacteriology
6 Loreta K - CanMedVet,PhD virology
7 Lweno Mrs MF - BSc helminthology
8 Mbwambo HA - MVM protozoology
9 Mhama JRL - DVM,MSc bacteriology, vaccines
10 Minja MMJ - MVM toxicology
11 Mkonyi PAA - BVM protozoology
12 Msami HMH - BVM pathology
13 Msangi AR - BSc helminthology
14 Mugoyo JW - MVM pathology
15 Ngereza ARH - MVM virology
16 Njau BC - BVM,MSc protozoology

3924 Veterinary Investigation Centre
PO Box 3081, Arusha
Tel: 3566.

1 Nyange JFC - BVM,DipVetPath *Officer-in-charge* pathology
2 Mbise AM - MVM

3925 Veterinary Investigation Centre
PO Box 290, Iringa
Tel: 2154.

1 Maiseli NNG - BVM,MSc *Officer-in-Charge*
2 Kessy Miss HJ - BVSc

3926 Veterinary Investigation Centre
Private Bag, Mpwapwa
Tel: 38.

1 Kitali IM - BVM,DipPath *Officer-in-Charge*

3927 Veterinary Investigation Centre
PO Box 129, Mwanza
Tel: 2083.

1 Msanga JF - BVM,DipTVM,MSc *Officer-in-Charge*
2 Msangi DJA - BVM

3928 Veterinary Investigation Centre
PO Box 121, Mtwara
Tel: 2327.

1 Nsengwa GRM - BVM
2 Otam MM - MVM

3929 Veterinary Investigation Centre
PO Box 73, Tabora
Tel: 2293.

1 Mtei BJ - MVSc,PhD *Officer-in-Charge*
2 Massale YM - DipAnimProd,BVSc

3930 Livestock Production Research Institute
Private Bag, Mpwapwa
Tel: 21.

1 Mpiri DB - BSc,DipAnimProd,MSc *Officer-in-Charge*
2 Das SM - BVScandHealth
3 Goodchild AV - BVSc,PhD,MRCVS
4 Kabatange Miss M - BSc,DipAnimProd,MSc
5 Kasonta JS - BScAgric,MScAgric,DipAgricExt
6 Kitalyi Mrs AJ - BScAgric,DipAnimProd,MSc
7 Kusekea ML - BScAgric,DipAgricSc,MAgricSc
8 Mero RN - DipAgric,BScAgric
9 Msanga YN - DipAgric,BScAgric
10 Msechu JKK - BScAgric,DipAgricSc,MScAgric
11 Ngaiza Miss MG - MScAnimHusb
12 Rukantabula AA - MSc
13 Rushalaza VG - BScAgric
14 Sempeho G - DipAgricIngenieur,MScAgric,PhDAgricEcon
15 Wella EB - DipEduc,MSc,PhD

3931 Livestock Research Centre
West Kilimanjaro
Tel: 3.

1 Kajuna AR - DipRangeMgnt,BSc,SRM *Officer-in-Charge*
2 Bee JKA - BScAgric,MScAgric
3 Mandari ATG - DipCropProd,BScAgric
4 Mandari GCH - CertinGoatandSheepHusb,BScAnimSc,MScAnimGenetics

3932 Livestock Research Centre
PO Box 5016, Tanga
Tel: 2009.

1 Lugenja MMS - DipAgric,BScAgric,MSc
2 Mbulwa MHS - BScAgric
3 Mkiwa Miss FEJ - BScAgric
4 Msanga YM - BScAgric,DipAnimSc
5 Mulangila RCT - BSc
6 Sendalo DSC - CertVetSc,BScAgric

3933 Livestock Research Centre
PO Box 3021, Malya
Tel: 15.

1 Mkonyi JI - DipRangeMgnt,BScAgric,DipAnimProd,MScAgric *Officer-in-Charge*
2 Kapinga PEH - BScAgric,MSc
3 Shayo CM - BScAgric

3934 Pasture Research Centre
PO Box 5, Kongwa
Tel: 21.

1 Masasi AP - BScAgric,CertRangeMgnt *Officer-in-Charge*
2 Msangi RB - BScAgric

3935 Tsetse Research Institute
PO Box 1026, Tanga
Tel: 2577.

1 Gao MK - BSc entomology
2 Kiwia NE - BSc
3 Mkishe FW - BSc
4 Msofee MVPS - BSc

3936 Ministry of Agriculture, Lands, & Food Production

St. Clair Circle, Port-of-Spain, Trinidad & Tobago
Tel: 622-1221. Telex: MINAGRI.

1 Alleyne EP - DICTA,BSc,MSc,PhD *Permanent Secretary*
2 Dookeran M - DICTA,MSc *Chief Technical Officer*

3937 Agricultural Planning Division

Trinidad & Tobago

1 Rudder W - BSc,MSc *Director*

3938 Extension Division

Trinidad & Tobago

1 Nelson H - BSc,MSc *Director*

3939 Apiaries Unit

St. Augustine Nurseries, via Curepe PO, St. Joseph, Trinidad & Tobago
Tel: 663-2687.

1 Hallim MK - BSc,MS *Entomologist, Officer-in-Charge*

3940 Research Division

Central Experimental Station, via Arima PO, Centeno, Trinidad & Tobago
Tel: 646-4334/7, 646-1646.

1 Barrow RM - DICTA,MSc,PhD *Director of Research*
2 Indalsingh TJ - BScAgr,MSc *Agronomist* seed research

3941 Crop Research Subdivision

Central Experimental Station, via Arima PO, Centeno, Trinidad & Tobago
Tel: 646-4334/7, 646-1644.

1 Bart A - BScAgr,MSc *Deputy Director of Research, Crops*

3942 Agronomy Section
Trinidad & Tobago

1 Alleyne Mrs V - BScAgr,BScHort,MSHort coffee
2 Andrews L - BScAgr,MSc fruit crops
3 Bocage Mrs C - BScAgr fruit crops
4 Bridgemohan P - BScAgr weed agronomy
5 Gonsalves Mrs C - BScAgr cocoa
6 Griffith RI - BScAgr,MScAgron vegetable crops
7 Harper Mrs M - BScAgr cocoa
8 Henry E - BScAgr fruit crops
9 Hosein F - BScAgr,DipMgt cereal crops
10 Indalsingh TJ seed research
11 Knox J - BSc *fruit crops*
12 Mohammed MES - BScAgr,MSc plant breeding
13 Mooleedhar V - BScAgr,MSc cocoa
14 Ragoonath Miss J - BSc banana agronomy
15 Ramdin D - BScAgr cereal crops
16 Rampersad I - BScAgr,MSc forage crops
17 Seecharan D - BScAgr cereal crops
18 Seesahai A - BScAgr root crops
19 Sinanan Miss P - BScAgr vegetable crops

3943 Crop Protection Section
Trinidad & Tobago

1 Baksh N - BScAgr,MSc citrus virology
2 Bala G - BScNatSc,MScAgr plant nematology
3 Dilbar A - BScAgr crop entomology
4 Esack Miss Y - BScNatSc pesticide residues in food
5 Ferguson Mrs G - BScAgr crop entomology (annual crops)
6 Fortune M - BScAgr plant pathology
7 Hosein F - BScAgr plant nematology
8 Hosein H - BScAgr plant quarantine
9 Jones M - BSc,MSc,Bfa crop entomology
10 Laurence G - BSA,MSc *apibotany, entomology*
11 Lewis S - BScAgr plant quarantine
12 McComie Miss L - BSc,MBc grasshoppers
13 Persad GC - BScAgr,MSc plant pathology
14 Rajanauth G - BScAgr,MPh plant pathology (annual crops)
15 St Hill A - BScAgr plant pathology

3944 Farm Management Section
Trinidad & Tobago

1 Burgess K - BScAgr
2 Lee Fook U - BScAgr

3945 Post Harvest Technology Section
Trinidad & Tobago

1 Mohammed M - BScAgr,MSc postharvest technology, vegetables
2 Mootoo Mrs M - BScAgr,MSc postharvest technology, fruit crops
3 Rosales Miss J - BScAgr postharvest technology, root crops and flowers

3946 Soil, Land Capability and Chemistry Section
Trinidad & Tobago

1 Bridgelal Mrs P - BScAgr soil chemistry
2 Gabriel L - BA,MAgrSc,PhD analytical chemistry
3 Maharaj H - BScAgr soil survey, land use
4 Wilson HW - BScAgr,MScSoils soil chemistry

3947 Library
Trinidad & Tobago
Tel: 646-1643.

1 Hee Houng Mrs M - ALA,MS *Librarian*

3948 Coconut Research Unit
Trinidad & Tobago

1 Griffith RSE - DICTA,MSc,DSc *Director of Coconut Research*
2 Daly Mrs P - BScNatSc pests of coconut

3949 Livestock Research Subdivision
Aripo Livestock Complex, via Arima PO, Aripo, Trinidad & Tobago
Tel: 667-3448.

1 Howard SB - DICTA,DipMgt,BSc,MSc,PhD *Deputy Director of Research, Livestock*
2 Borde G - BS,MSc,DVM reproductive physiology
3 Bute R - BS,MSc animal breeding
4 Davis FE - DECFI,BSc,DTA,PhD forage agronomy
5 Dindial P - BSc,MSc livestock management
6 Ramlal H - BSc,DECFI,MSc animal nutrition

3950 Forestry Division

Long Circular Road, Port-of-Spain, Trinidad & Tobago
Tel: 62-23217,62-24521.

1 Ramdial BS - BScFor,MFpr,PhD *Conservator of Forests*
2 Dardaine S - BScFor,DipMan *Deputy Conservator of Forests*

3951 Watershed Management
Trinidad & Tobago

1 Lackhan NP - BSc,DipWaterMan hydrology, silviculture
2 Meade K - BScFor
3 Ramsaran C - BScFor

3952 Research Section
Trinidad & Tobago

1 Ramnarine S - BScFor forest research
2 Sandy R - BScFor forest research

3953 National Parks Section
Trinidad & Tobago

1 Bickram R - BScFor,DipFor recreation
2 Fizool S - BScFor

3954 Wildlife Section
Trinidad & Tobago

1 Gyan Miss N - BSc wildlife management
2 Hislop G - BScWildlifeMan wildlife management
3 James Miss C - BSc wildlife biology
4 Kerrcher Mrs G - BScWildlifeMan wildlife management
5 Ramdial BP - BSc,DipEd wildlife ecology

3955 Fisheries Division

Ministry of Agriculture, Lands and Food Production, St Clair Circle, Port-of-Spain, Trinidad & Tobago
Tel: 62-21221.

1 La Croix M - BSc,DipMan *Director*

3956 Caribbean Fisheries, Training and Development Institute

Western Main Road, PO Box 1150, Port-of-Spain, Chaguaramas, Trinidad & Tobago
Tel: 634-4275.

1 Jordan C - BSc,MSc,DipMan *Principal* marine fisheries research
2 Chin Yuen Kee Z - BSc,MSc,MMA
3 Fabres B - BSc
4 Gooriesingh K - BA,MSc

3957 Bamboo Grove Fish Farm

Bamboo Grove, Valsayn, Trinidad & Tobago
Tel: 662-5208.

1 Jobity A - BSc,MSc aquaculture

3958 Animal Health Division

Ministry of Agriculture, Lands and Food Production, St Clair Circle, Port-of-Spain, Trinidad & Tobago
Tel: 622-2072.

1 Moe VG - DICTA,BSc,DVM *Director, Animal Health*

3959 Veterinary Diagnostic Laboratory

Havelock Street, Curepe, Trinidad & Tobago
Tel: 662-5678.

1 Cazabon H - BSc,MSc,DVM *Head of Section*

3960 National Animal Disease Centre

Caroni North Bank Road, Centeno, Trinidad & Tobago

1 Borde G - MSc,DVM sub-fertility in dairy cattle
2 Caeser ER - DICTA,DVM,DipLAMS seasonal and geographical distibution of screw worm myiasis in Trinidad and Tobago, control of babesiosis and anaplasmosis
3 Campbell MD - BScAgr,DVM,MVSc cryptosporidiosis in sheep in Tobago, survey of seasonal distribution of mastitis

3961 Agricultural Services Division

St Augustine Nurseries, via Curepe PO, St Joseph, Trinidad & Tobago
Tel: 663-4774.

1 Jones E - DICTA,PhD *Director of Agricultural Services*

3962 Horticulture Subdivision

La Reunion Propagation Station, via Arima PO, Centeno, Trinidad & Tobago
Tel: 646-2651.

1 Pegus J - BSc,MSc,PhD,DIC *Technical Officer* horticulture

3963 Chaguaramas Agricultural Development Project

Cottage 'A', Tucker Valley, Chaguaramas, Trinidad & Tobago
Tel: 634-4285.

1 Cross L - BScAgr,MSc *Deputy Director of Agricultural Services*

3964 Project Implementation Unit

Mausica Road, Centeno, via Arimo P.O., Trinidad & Tobago
Tel: 646-3240.

1 Timothy E - BScAgrEcon,MScAgr,PhD *Director*

3965 Blenheim Sheep Project

Tobago House of Assembly, Department of Agriculture, Scarborough, Trinidad & Tobago
Tel: 639-2441, 639-2428, 639-2279.

1 Keens-Dumas MJ - BSc,MSc *Project Manager*

3966 Caroni Research Station

Waterloo Road, Carapichaima, Trinidad & Tobago
Tel: 673-0027/9. Cables: Tricaroni-P.O.S..

1 Carr TWA - DICTA,BSc,PhD *Director of Research*
2 Buxo DA - DICTA agronomy of food and tree crops

3 des Vignes WG - BSc,MSc entomology; biological control
4 Donelan AF - BSc,MSc agronomy; sugar cane variety evaluation
5 Georges JEW - BSc,MSc soil physics; rice agronomy
6 Mahadeo CR - BSc,MA,PhD entomology; biological pest control
7 Manuel RL - BSc,MSc,PhD entomology; chemical sugar cane pest control
8 Mason GF - BSc,MSc,PhD agronomy of sugar cane; herbicides
9 Ramkhelawan RR - BSc agricultural engineering
10 Rampersad EM - BSc,MSc plant pathology
11 Shand CR - BSc,PhD soil fertility; leaf and soil analysis

3967 Caribbean Agricultural Research and Development Institute

University of the West Indies Campus, St Augustine, Trinidad & Tobago
Tel: 645-1205/7.

1 Parasram S - BSc,MSc,PhD *Executive Director*
2 Forde STCM - BSA,MSA,PhD *Director, Research and Development*
3 Ahmed B - BSc,MSc,PhD *Agronomist*
4 Ali A - BSc,MSc *Economist*
5 Ali A - BSc,MSc *Pesticide Chemist*
6 Collins P - BSc,PhD *Agronomist*
7 Haque SQ - BSc,MSc,PhD *Virologist*
8 Harricharan HC - DipAgr,BSc,MSA,PhD *Animal Productionist*
9 Lauckner FB - BSc,FIS,FSS *Biometrician*
10 Osuji PO - BSc,MSc,PhD,MIBiol,DipMngtStudies *Animal Nutritionist*
11 Phelps RH - BSA,MSA,PhD *Plant Pathologist, Head of Unit*
12 Sanchez J - BSc *Statistician*
13 Singh ND - DipAgr,BSc,MSc,PhD *Nematologist*
14 Thompson P - BA,MSc *Documentalist*
15 Walmsley D - BSc,MSc,PhD *Soil Chemist*

3968 Caribbean Industrial Research Institute

Tunapuna PO, Trinidad, Trinidad & Tobago
Tel: 662-7161/5 or 663-4161/2. Telex: 24438 CARIRI WG. Cables: CARIRI TRINIDAD.

1 Charles H - BSc,MSc *Director*

3969 Agro-Technology Products and Services Division

Trinidad & Tobago

1 Shah L - BSc,MS,MSc *Divisional Manager* solid substrate fermentation, drying (pneumatic)
2 Felmine J - BSc,MPhil chemical analysis of animal feeds; mycotoxins
3 Imbert M - BSc,MSc,PhD medicinal plants, herbs and spices
4 Lashley D - BSc,MSc food products and process development

3970 Sugarcane Feeds Centre

Pokhor Road, Longdenville, Trinidad & Tobago
Tel: 665-9967.

1 Neckles FA - BScAgr,MScAgr,DipMngtStudies *Project Officer*

3971 Integrated Animal Production

Trinidad & Tobago

1 Brown Miss JA - BScAgr dairy nutrition; heifer replacement rearing
2 Charles KW - BScAgr,BScBot farm management, sugar cane and crop production
3 Garcia GW - BScAgr ruminant production and nutrition; forage protein
4 Khan Miss NZ - BScAgr calf nutrition and management
5 Lallo CHO - BScAgr,MScAgr cattle, sheep and goat nutrition
6 Nancoo DA - BScAgrEng engineering aspects of production
7 Taylor DT - BScAgr,MScAgrEcon economic evaluation studies; model building

3972 Institute of Marine Affairs

PO Box 3160 Carenage Post Office, Carenage, Trinidad & Tobago
Tel: (809)-634-4292,4293,4294. Cables: MARINAF.

1 Masson NJ - MOM,PhD,DIC *Director*
2 Ramsaroop D - PhD *Deputy Director*

3973 Natural Resources Division

Trinidad & Tobago

1 Hudson D - BScAg *Manager* sedimentology
2 Bachew S - BSc geology
3 Dass Mrs S - MPhil fisheries biology
4 De Souza G - BSc geology
5 Gabbadon P - BSc aquaculture
6 Hubbard R - BSc marine ecology
7 Julien-Flüs Mrs M - BSc fisheries biology
8 Laydoo R - BSc coral reef ecology
9 Lee Lum Mrs L - BES geography
10 Lewis N - BSc geology
11 Maingot Miss J - BSc fisheries biology
12 Ramcharan E - MSc,PhD wetlands ecology
13 Sturm M - PhD fisheries biology

3974 Environmental Quality Division
Trinidad & Tobago

1 Siung-Chang Mrs A - PhD *Manager* marine ecology
2 Agard J - MSc pollution ecology
3 Gobin Miss J - BSc benthic ecology
4 Hoyte-Haynes Mrs P - BSc physical oceanography
5 Norman P - MSc pollution chemistry

3975 Marine and Environmental Analytical Laboratory Division
Trinidad & Tobago

1 Foster Miss F - MSc *Ag Manager* pesticide residues
2 Boodoosingh M - BSc hydrocarbons
3 Heileman L - BSc,MSc trace metals
4 Mohammed A - BTech nutrients
5 Rondon-Jeffrey Mrs C - BSc hydrocarbons

3976 Coastal Area Planning and Management Division
Trinidad & Tobago

1 McShine Miss H - MSc *Ag Manager* environmental impact assessment
2 Bertrand Mrs D - BSc environmental geology
3 Gerald L - BA land use, photointerpretation
4 Joseph M - MSc resource economics
5 Manwaring G - BA economics
6 Romano H - BSc geology

3977 Socio Economic and Legal Division
Trinidad & Tobago

1 Palmer N - JD *Manager* environmental law
2 Goodridge J - JD environmental law

3978 Information Services Division
Trinidad & Tobago

1 Duncan A - BSc,MEd *Ag Manager* public education and information; publications
2 Ottley Mrs T - BA,MSc library services

3979 University of The West Indies

St Augustine, Trinidad & Tobago
Tel: 663-1359.

1 Preston AZ - JP,LLB,FCA,FCCA,FCIS,FREconS *Vice Chancellor*
2 Richards GM - CMT,MSc,PhD,MRST,AMCST,FINST,PT,MAPE,CEng *Professor, Pro-Vice-Chancellor, Principal*
3 Gumbs FA - BSc,MSc,PhD *University Dean, Faculty of Agriculture*

3980 Department of Agricultural Economics and Farm Management
Trinidad & Tobago

1 Birla SC - MSc,MSc,PhD production economics; farm management
2 Pemberton CA - BScAgr,MSc,PhD production economics; small farm development; managerial processes; marketing of agricultural commodities
3 Rankine LB - BSc,MSc,PhD production economics; resource economics; marketing of agricultural commodities
4 Singh RH - BSc,MSc,PhD natural resources planning, evaluation, management; land use planning

3981 Department of Agricultural Extension
Trinidad & Tobago

1 Henderson TH - DICTA,MSc,PhD *Director of Caribbean Agricultural Extension Project (CAEP)* based in Dominica
2 Barker STCP - BSc,MSc,CAEP *Research Fellow in Agricultural Extension and Leeward Islands Programme Leader* based in Antigua
3 Campbell D - BScAgr,MSc,DEA,PhD *Outreach Lecturer and Windward Islands Programme Leader* based in St Lucia
4 Dolly DI - BSc,MSc *Lecturer in Agricultural Extension*
5 Gomes PI - MA,PhD *Lecturer in Agricultural Sociology and Head of Department*
6 Meiller L - BSc,MSc,PhD *Visiting Research Fellow in Agricultural Communications*
7 Seepersad J - BScAgr,MSc,PhD *Lecturer in Agricultural Extension*

3982 Cocoa Research Unit
Trinidad & Tobago

1 Kennedy AJK - BSc,KSc,PhD *Head of Unit* genetic resources and breeding programme
2 Gonsalves C - BSc,MSc *Visiting Research Fellow* cocoa agronomy
3 Sreenivasan TN - BSc,MSc,PhDAssocIARI cocoa pathology; screening for resistance; biological control

3983 Department of Livestock Science
Trinidad & Tobago

1 Archibald KAE - DICTA,MSc,PhD *Lecturer* forage production and utilization
2 Mbatya PBA - CVS,DAH,BScAgr,MSc,PhD *Temporary Lecturer* forage production and utilization; ruminant production and nutrition; pig and poultry production and nutrition
3 Rastogi RK - BVScAH,MVSc,PhD *Senior Lecturer* cattle, sheep and rabbit breeding and production
4 Williams HE - MRCVS,DVM,MSc,PhD *Professor, Head of Department* parasitic, viral, and tick-borne diseases of livestock; livestock industry studies
5 Youssef FG - BScAgr,PhD *Senior Lecturer and Coordinator of Animal Nutrition Laboratory* animal nutrition and disorders; establishment of nutrition laboratories

3984 Department of Plant Science and Biochemistry
Trinidad & Tobago

1 Spence JA - BSc,PhD,DipAgrSc,DTA *Professor of Botany and Head of Department*
2 Antoszewski R - MSc,PhD *Senior Lecturer* biochemistry
3 Ariyanayagam R - BScAgr,MSc,PhD *Senior Lecturer* botany
4 Barnes RF - BSc,ARCS *Senior Lecturer* botany
5 Comeau PL - BA,BSc,MSc,PhD *Lecturer* botany
6 Comeau SY - BSc *Lecturer* taxonomy
7 Dabydeen S - BSc,MS,PhD *Lecturer* botany
8 Davidson JB - BSc,PhD *Lecturer, Subject Leader* biochemistry
9 Duncan JE - BSc,PhD *Senior Lecturer* botany
10 Elango FN - BScAgr,MSc,PhD *Lecturer* plant pathology
11 McDavid CR - BSc,PhD *Senior Lecturer* botany
12 Pollard G - BSc,PhD *Lecturer* entomology
13 Sirju-Charran G - BSc,PhD *Lecturer* botany
14 Wilson MB - BSc,PhD *Lecturer* biochemistry

3985 Department of Soil Science
Trinidad & Tobago

1 Ahmad N - DICTA,AICTA,MSc,PhD *Professor, Head of Department* soil chemistry, pedology; soil management; soil fertility
2 Donawa AL - BScNatSc,MSc,PhD soil and food microbiology
3 Griffith SM - BSc,MSc,PhD soil chemistry; soil fertility; environmental and marine chemistry; interactions of organics in the soil
4 Gumbs FA - BSc,MSc,PhD *University Dean, Faculty of Agriculture* soil physics; soil conservation; soil management; irrigation; drainage
5 Lindsay JI - DipAgr,BSc,MSc soil management; tillage; moisture and soil conservation; soil physics

3986 Department of Crop Science
Trinidad & Tobago

1 Ferguson TU - BSc,PhD *Senior Lecturer, Head of Department* crop production; grain legumes; root crops; vegetables; agronomy
2 Asnani V - BSc,MSc,PhD *Research Fellow* plant breeding in tropical root crops
3 Bekele I - BSc,MSc *Lecturer* biometrics; application of quantitative methods in agriculture
4 Braithwaite RAI - BSc,PhD *Senior Lecturer* crop production, weed science
5 Fletcher LM - BSc,PhD *Assistant Lecturer* biometrics; quantitative growth analysis and mathematical modelling
6 Harvey WO - BSc,MAgEng,PhD *Lecturer* agricultural engineering; crop mechanisation - sugar cane, cocoa, root crops, vegetables; computer-aided irrigation design, solar crop drying, design of farm structures and controlled environment facilities, computer applications to agriculture, energy systems
7 Raj Kumar D - BSc,MSc,PhD *Lecturer* crop production; tropical fruit, tropical landscaping and ornamentals, citrus
8 Wickham LD - BSc,PhD *Lecturer* crop production; post harvest physiology
9 Wilson LA - BSc,MSc,PhD *Professor* crop science; pre- and post-harvest physiology, crop utilization and storage

3987 Ministry of Agriculture and Forestry

PO Box 102, Entebbe
Tel: 20981/10. Cables: "NATURE ENTEBBE".

3988 Directorate of Research

1 Mukasa-Kiggundu AA - BSc,EAMSc *Chief Agricultural Research Officer* soil science, soil chemistry

3989 Kawanda Research Station

1 Sakira WA - BScAgric,MSc *Acting Director of Research, Agronomist* research administration
2 Rubaihaye Mrs EB - BSc,MSc *Principal Research Officer* botany - maize
3 Asea PEA - BSc,MSc *Scientific Officer* chemistry - soil science
4 Ayo C - MSc,PhD *Scientific Officer* agronomy, horticulture
5 Bafokuzara DN - BSc,MSc *Research Officer* nematology - vegetables
6 Bahule Miss J - BSc *Scientific Officer* botany - coffee
7 Bakanwangiraki B - BScAgric *Scientific Officer* agronomy - beans
8 Bigirwa C - BSc *Scientific Officer* botany - maize
9 Birikunzira B - BScAgric,MSc *Scientific Officer* coffee pathology
10 Ddamulira J - BSc *Research Officer* botany - soybeans
11 Hakiza Mrs G - BScAgric,MSc *Scientific Officer* coffee pathology
12 Hazika JH - BSc,MSc *Research Officer* botany - maize
13 Jjemba PK - BScAgric *Scientific Officer* soil science
14 Kabeere Miss F - BSc,MSc *Research Officer* botany - seed pathology
15 Kabuye MS - BScAgric *Scientific Officer* agronomy - soybeans
16 Karamura Mrs D - BSc,MSc *Scientific Officer* plant taxonomy
17 Karamura EB - BSc,MSc *Scientific Officer* storage entomology
18 Kashaija IN - BSc *Scientific Officer* nematology - bananas
19 Kato G - BSc *Scientific Officer* botany - coffee
20 Kavuma JB - BScAgric *Senior Research Officer* agronomy - soybeans
21 Kawesa JS - BScAgric *Scientific Officer* horticulture
22 Kibirige-Sebunya I - BSc,MSc,PhD *Senior Research Officer* botany - coffee
23 Kisakye Miss J - BScAgric *Scientific Officer* agronomy - beans
24 Kyetere DT - BSc *Scientific Officer* botany- maize
25 Laker H - BScAgric *Scientific Officer* plant protection
26 Luweka Mukasa - BSc,MSc *Scientific Officer* entomology, horticulture
27 Malagala Mwanga RD - BScAgric *Scientific Officer* botany - root crops
28 Male-Kayiwa Mrs S - BSc,MSc *Scientific Officer* botany - beans
29 Mangheni WO - BSc *Scientific Officer* botany - maize
30 Mudega-Gibugonyi - BSc *Scientific Officer* botany - rice
31 Mugalu S - BSc *Scientific Officer* entomology - biological control
32 Mugenyi G - BSc *Scientific Officer* entomology - bananas
33 Mukulu Miss G - BSc *Scientific Officer* entomology - coffee
34 Musana Mrs MM - BScAgric,MSc *Scientific Officer* botany- beans
35 Musoli PS - BSc *Scientific Officer* botany - coffee
36 Nabasirye Miss N - BSc,MSc *Scientific Officer* biometrics
37 Namaganda Miss J - BSc *Scientific Officer* nematology - bananas
38 Nassuuna Miss NS - DipLibrarianship *Librarian*
39 Ocen Ayer JR - BSc,MSc *Scientific Officer* entomology - stem borers
40 Okoth VOA - BSc,MSc,PhD *Scientific Officer* virology - maize
41 Ongaya ML - BScAgric *Scientific Officer* agronomy - *Solanum* potatoes
42 Opio Mrs AF - BSc,MSc *Scientific Officer* plant pathology
43 Oree A - BSc *Scientific Officer* entomology - beans
44 Sekyewa C - BSc *Scientific Officer* horticulture
45 Sendiwanyo EV - MSc *Scientific Officer, Soil Surveyor* soil survey
46 Sengooba Mrs T - BSc,MSc *botany - bean pathology*
47 Silim Mohammed N - BSc,MSc *Scientific Officer* entomology - biological control
48 Tibaijuka B - MSc,PhD *Scientific Officer* beans physiology
49 Tumushabe GT - BSc *Scientific Officer* entomology - cocoa
50 Turyamureeba MG - BSc *Scientific Officer* botany - *Solanum* potatoes
51 Tushemereirwe WK - BSc *Scientific Officer* plant pathology
52 Wagoire Wamala W - BSc *Scientific Officer* botany - wheat & barley
53 Wajja-Musukwe TN - BSc,MSc *Scientific Officer* agronomy - maize
54 Wetala MPE - BSc,MSc *Scientific Officer* agronomy, weeds - coffee
55 Zziwa Miss NCN - BSc *Scientific Officer* botany - legumes

3990 Serere Research Station

1 Mwaule YW - BScAgric,MSc *Director of Research, Horticulturist*
2 Acidiria M - BSc *Scientific Officer* horticulture
3 Anyii CP - BSc,MSc *Scientific Officer* agronomy - pastures
4 Areke TEE - BSc *Scientific Officer* botany - legumes
5 Bua A - BSc *Scientific Officer* agronomy - oil crops
6 Busolo Bulafu - BSc *Scientific Officer* botany - groundnuts
7 E.Pieru G - MScAgric,PhD *Scientific Officer* entomology - legumes
8 Esere N - BSc *Scientific Officer* botany - sorghum & millet
9 Kkizza W - BSc,MSc *Scientific Officer* botany - sorghum & millet
10 Koma-Alimu FX - BSc,MSc *Research Officer* chemistry - soil science
11 Mafulira T - BA *Senior Research Officer* biometrics, soil science

12 Makumbi Zake - BSc,MSc *Scientific Officer* botany - sorghum & millet
13 Molo R - BSc *Scientific Officer* entomology - vertebrate pests, cereals
14 Mwoga VN - BSc,MSc *Scientific Officer* chemistry - bananas
15 Ochieng-Mbuya G - BSc *Scientific Officer* agronomy - farming systems
16 Ociti P. Obwoya C - BSc *Scientific Officer* agronomy - root crops
17 Odogola WR - BSc *Agricultural Engineer* engineering - solar energy
18 Odongo B - BSc *Scientific Officer* entomology - rootcrops
19 Okello R - BSc,MSc *Scientific Officer* botany - sorghum & millet
20 Okello-Ekochu E - BSc *Scientific Officer* horticulture
21 Okurut Akol FH - BSc *Scientific Officer* entomology, vertebrate pests, cereals
22 Onenanyoli AHA - MScAgric *Scientific Officer* agronomy - oil seeds
23 Oryakot J - BSc *Scientific Officer* agronomy - sorghum & millet
24 Otim-Nape GW - BScAgric *Scientific Officer* pathology - root crops
25 Serunjogi LK - BSc *Scientific Officer* botany - cotton
26 Wamajje DW - BSc,MSc *Senior Agricultural Officer* food technology
27 Wanyera NW - BScAgric,MSc *Scientific Officer* botany - root crops

3991 Namulonge Research Station

1 Kintukwonka A - BSc,MSc *Acting Director of Research, Soil Chemist*
2 Odongo JWC - BSc,MSc *Principal Research Officer* agricultural meteorology
3 Ayiseni DAR - BScAgric,MSc *Scientific Officer* botany - cotton
4 Beteisa-Hibenye D - BSc,MSc *Scientific Officer* botany - cotton
5 Busuulwa LN - BScAgric *Research Officer* botany - cotton
6 Gumisiriza G - BScAgric *Scientific Officer* botany - legumes
7 Imanywoha J - MSc *Scientific Officer* agronomy - root crops
8 Kabaija E - *Scientific Officer* animal nutrition
9 Kabirizi Miss JM - BScAgric *Scientific Officer* agronomy - pastures
10 Kangire A - BScAgric *Scientific Officer* botany - cotton
11 Lutaya AK - BSc,MSc *Scientific Officer* meteorology - cotton
12 Magunda MK - BScAgric,MSc *Scientific Officer* soil physics
13 Maiteki GW - BSc,MSc *Scientific Officer* entomology - cotton
14 Napulu CWN - BScAgric,MSc *Scientific Officer* agronomy - pastures
15 Ogwang J - BSc,MSc *Scientific Officer* entomology - cotton
16 Olanya JCO - BSc *Scientific Officer* livestock
17 Serwadda VN - BSc *Scientific Officer* livestock nutrition
18 Wansoloba KM - BSc *Scientific Officer* botany - cotton

3992 Nakawa Forestry Research Station

1 Oloya M - BSc,DICAppEntom,PostgradDipFor *Principal Research Officer, Officer-in-Charge* entomology - insect pests of introduced pines and cypress plantations; administration
2 Akode S - BScFor,PostgradDipFor *Scientific Officer (on MSc course Mekerere)* silviculture - nursery practice and seedling production: problems associated with natural regeneration and forestry practice in Bwamba
3 Carvalho J - BScFor,PostgradDipForestResSurvey *Senior Utilization Officer, Head Forest Products Section, Deputy Officer-in-Charge of the Station* wood processing including logging working properties, saw operation and maintenance; wood energy including charcoal production and utilization
4 Esegu JFO - BScFor,MSc *Scientific Officer* tree-breeding - comparative studies and selection of species suitable for agro-forestry or forestry or community forestry; breeding and improvement of indigenous and exotic broad-leaved tree species
5 Kamugisha JR - BScFor,MSc *Scientific Officer* ergonomics - use of ergonomic principles in improving working conditions, and labour productivity in sawmilling, and the forestry related industries
6 Kiwuso P - BScFor *Scientific Officer (on MSc course)* pathology - impact of root diseases on development of plantation species
7 Ktiyo PWD - BScFor,MSc *Scientific Officer* wood utilization - wood preservation by chemical treatment; natural durability of indigenous tree species
8 Muhairwe CK - BScFor *Scientific Officer (on MSc course in Forest Resource Management (UK))* forestry resource management - problems associated with marketing of forest produce
9 Musoke Mrs R - BScFor,MSc *Scientific Officer, Acting Head of the Silviculture Research Section* tree-breeding - seed collection storage, and quality control; breeding and improvement of conifers; seed orchard establishment and maintenance
10 Ndimukulaga JP - BScFor *Scientific Officer* wood utilization - harvesting and logging techniques; protection of logs from degrade before processing
11 Ochaki JR - BScFor,MSc *Scientific Officer* wood products utilization - strength properties of indigenous natural high forest tree species and plantation grown species; seasoning properties of plantation species; utilization of minor forest products including gum arabic
12 Okello A - BScFor *Scientific Officer* entomology - natural durability of timbers to borer attack; use of indigenous shrubs or trees as termite repellents in Eucalyptus plantations; the biology and ecology of some stem borers and girdlers of Eucalyptus
13 Okorio J - BScFor *Scientific Officer (on MSc course in Tanzania)* silviculture - stand development and productivity of various plantation species

14 Sewanywa SP - BScFor *Scientific Officer* wood utilization - improvement of charcoal production techniques through the use of earth kilns, portable kilns, brick kilns or retorts; promotion of charcoal utilization for both domestic and industrial use

15 Turyareba Miss P - BScFor *Scientific Officer* wood utilization - development of energy saving charcoal and woodfuel cooking stoves; survey of indigenous shrubs and trees commonly used by community as fuelwood

16 Turyatunga FP - BScFor,MSc *Scientific Officer* resource management - a review of current management of natural high forest practices in Uganda

17 Wamala J - BScFor *Scientific Officer (on insect taxonomy course in U.K.)* entomology, pathology - monitoring of potential insect and disease pests of plantations

This chapter has separate sections for "Great Britain", "England & Wales", "Scotland" and "Northern Ireland". But first come international organizations plus institutions covering all four countries.

3993 CAB International

Wallingford, Oxfordshire, OX10 8DE
Tel: 0491 32111. Telex: 847964 (COMAGG G). Telecom Gold/Dialcom: 84: CAU001. Fax: 0491 33508.

CAB International (CABI), formerly the Commonwealth Agricultural Bureaux, is a not-for-profit organization owned and administered by its member governments, currently 29 in number. Established in 1928, CABI exists to provide information, scientific and development services for agriculture and allied disciplines throughout the world.

3994 Executive Council

1 Dato' Ahmad Yunus *Chairman (Malaysia)*
2 Allsop PAD *(Guyana)*
3 Anegbeh PJE *(Nigeria)*
4 Ballah-Conteh Mrs A *(Sierra Leone)*
5 Calvin-Smith Mrs V *(Trinidad & Tobago)*
6 Danial bin Haji Hanafiah *(Brunei Darussalam)*
7 Farrer KF *(Australia)*
8 Dilmahamood BM *(Mauritius)*
9 Jackson Miss M *(Jamaica)*
10 Kachimera P *(Malawi)*
11 Kang'e CM *(Kenya)*
12 Karunasiri PG *(Sri Lanka)*
13 Mogotsi C *(Botswana)*
14 Mutondo S *(Zambia)*
15 Nyanduga Mrs SB *(Tanzania)*
16 Ocana BJ *(Uganda)*
17 Papanicolaou J *(Cyprus)*
18 Rao SN *(India)*
19 Robbins S *(Dependent Territories)*
20 Shahjahan Majumder M *(Bangladesh)*
21 Shannon Dr DWF *(United Kingdom)*
22 Shindler D *(Observer, Canada)*
23 Stubbs Mrs S *(The Bahamas)*
24 Sung Kangwai *(Fiji)*
25 To be nominated *(The Gambia)*
26 To be nominated *(New Zealand)*
27 To be nominated *(Papua New Guinea)*
28 To be nominated *(Solomon Islands)*
29 Undenge S *(Zimbabwe)*
30 Vago Dr J *(Hungary)*
31 Yaw Osei *(Ghana)*

3995 Headquarters

1 Mentz D - BAgSci,BEcon,DDA *Director General*
2 Metcalfe JR - MA,PhD,DIC,DTA *Director Information Services*
3 Greenland DJ - MA,DPhil,DAgSc,FIBiol *Director Scientific Services*
4 Martin T - FCA *Director Administrative Services*
5 Gilmore JH - BSc,MACS *Director Systems*
6 Ogbourne CP - BSc,PhD *Deputy Director Information Services*

3996 Development Services

1 Heydon DL - BSc,DipFBA,MBIM *Head, Development Services*
2 Affonso MAR - BA *Project Officer* LDC project identification and administration
3 Gooch PS - BSc,MIInfSc *Project Officer (Information)*

3997 Pesticide Databank

1 Ivens GW - MA,BSc,DPhil *British Crop Protection Council Liaison Officer*

3998 Library Services Centre

Silwood Park, Buckhurst Road, Ascot, Berkshire, SL5 7TA
Tel: 0990 872747. Telex: 9312102256(DDG). Telecom Gold/Dialcom: 84:CAU016. Fax: 0990 872901.

1 Hamilton CJ - ALA *Library Services Manager*

3999 Regional Office - Asia

P.O. Box 11872, 50760 Kuala Lumpur, Malaysia
Tel: 2552111. Telex: 28031. FAX: 2551888.

1 Song Chil Lee *Regional Representative*
2 Ho Thian Hua *Scientific Officer (Information)*

4000 Regional Office - Caribbean and Central America

Gordon Street, Curepe, Trinidad
Tel: 662 4173. Telex: 24438 CARIRI.

1 Spence JA - BSc,PhD,DipAgricSci,DTA *Regional Representative*

4001 CAB International Institute of Biological Control

4002 Headquarters

Silwood Park, Buckhurst Road, Ascot, Berkshire, SL5 7TA
Tel: 0990 872999. Telex: 9312102255 (BC G). Telecom Gold/Dialcom: 84:CAU015. Fax: 0990 872901.

1 Hassell MP - MA,DPhil,DSc,FRS,FIBiol *Consultant Director*
2 Spence JA - BSc,PhD,DipAgrSci,DTA *Consultant Director*
3 Greathead DJ - BSc,ARCS,PhD,DSc,FIBiol *Director* biological control
4 Waage JK - AB,PhD,DIC *Chief Research Officer* applied ecology

4003 Information Service

1 Fry Ms JM - BSc *Information Officer* natural enemy databank
2 Girling DJ - BSc,MSc,MIBiol *Information Officer* biological control information

4004 UK Project Staff

1 Evans HC - BSc,MSc,PhD *Pathologist* pest and weed pathogens
2 Fowler SV - BA,DPhil *Entomologist* weeds
3 Mills NJ - BSc,PhD *Scientist-in-charge* forest insects
4 Moore D - BSc,PhD *Entomologist* tropical pests
5 Prior C - BA,PhD *Pathologist* insect and plant pathogens

4005 CIBC European Station

1 Chemin des Grillons, CH-2800 Delémont, Switzerland
Tel: 066 22 12 57. Cables: Biocontrol, Delémont.

1 Carl KP - DnatSc *Scientist-in-charge* agricultural pests
2 Schroeder D - DnatSc *Entomologist* weeds

4006 CIBC Indian Base

c/o Post Bag 2484, Hebbel Agricultural Farm P.O., Bangalore, 560024, India
Tel: (0812) 331961. Cables: Biocontrol, Bangalore.

1 Chacko MJ - MSc,AssocIARI,PhD *Scientist-in-charge* insect pests

4007 CIBC Kenya Station

c/o Kenya Agricultural Research Institute, P.O. Box 30148, Nairobi, Kenya
Tel: 0154 32883/4/5 Ext 229. Telex: 22040 ILRAD (for CIBC). Cables: Agforg, Nairobi.

1 Cock MJW - BSc,ARCS,PhD,DIC *Scientist-in-charge* general biological control
2 Murphy ST - BSc,PhD,DIC *Entomologist* coffee pests

4008 PARC-CIBC Station

P.O. Box 8, Rawalpindi, Pakistan
Tel: 842347. Telex: 5948/5949 PCORP BIOCONTROL. Cables: Biocontrol, Rawalpindi.

1 Mohyuddin AI - MSc,PhD *Scientist-in-charge* general biological control

4009 CIBC Malaysian Station

P.O. Box 11872, 50760 Kuala Lumpur, Malaysia
Tel: 2552111. Telex: 28031. FAX: 2551888.

1 Ooi PAC - BSc,MSc *Entomologist* tropical pests

4010 CIBC Caribbean and Latin American Station

Gordon Street, Curepe, Trinidad, W.I.
Tel: 662 4173. Telex: 24438 CARIRI. Cables: Biocontrol, Port of Spain, Trinidad.

1 Baker P - BSc,PhD *Scientist-in-charge* tropical pests

4011 CAB International Institute of Entomology
56 Queen's Gate, London, SW7 5JR
Tel: 01 584 0067. Telex: 9312102251 IE G. Cables: ENTINFO,LONDON
SW7. Fax: 01 581 1676.

Identification Service: c/o British Museum (Natural History), Cromwell
Road, London SW7 5BD. Tel.:01-589 6323

1 Mound LA - PhD,DSc *Consultant Director*
2 Harris KM - BSc,DSc,DAS,DTA,FIBiol *Director*
3 Betts CR - BSc, PhD *Scientific Co-ordinator*
4 Booth RG - BA,PhD *Taxonomist* Coleoptera
5 Cox ML - BSc,PhD *Taxonomist* Coleoptera
6 Holloway JD - MA,PhD *Taxonomist* Lepidoptera
7 Macfarlane D - BSc *Taxonomist* Acarina
8 Madge RB - MS,PhD *Taxonomist* Coleoptera
9 Polaszek AB - BSc,PhD *Taxonomist* Hymenoptera
10 Stonedahl GM - MS,PhD *Taxonomist* Hymenoptera
11 Walker Ms AK - MIBiol *Taxonomist* Hymenoptera
12 White IM - BSc,PhD *Taxonomist* Diptera
13 Wilson MR - BSc,PhD *Taxonomist* Hemiptera

4012 CAB International Mycological Institute
Ferry Lane, Kew, Surrey, TW9 3AF
Tel: 01 940 4086. Telex: 9312102252 MI G. Fax: 01 332 1171.

1 Hawksworth Professor DL - PhD,DSc,FLS,CBiol,FIBiol *Director*
Ascomycotina
2 Bell EA - BSc,MA,PhD,FRIC *Consultant Director*

4013 Identification and Taxonomic Services Division

1 Sutton BC - PhD,DSc *Head, Taxonomic and Identification Services*
Division Deuteromycotina, esp. Coelomycetes
2 Bradbury JF - PhD, DIC *Bacteriologist* plant pathogenic bacteria
3 Brayford D - PhD *Mycologist* Cylindrocarpon, Fusarium
4 Bridge PD - PhD *Mycologist* biochemical systematics
5 Cannon PF - PhD *Mycologist* Ascomycotina, esp. Sordariales
6 David JC - MSc *Mycologist* Hyphomycetes, esp. *Cladosporium,*
Alternaria
7 Hall G - PhD *Mycologist* Oomycetes, esp. *Pythium* and *Phytophthora*
8 Kirk PM - PhD *Mycologist* Zygomycotina, dematiaceous Hyphomycetes
9 Lawrence Z - PhD *Aspergillus, Penicillium*
10 Minter DW - PhD *Mycologist* Ascomycotina, esp. Rhytismatales
11 Mordue JEM - PhD *Mycologist* Uredinales and Ustilaginales, sclerotial
fungi
12 Punithalingam E - PhD,DIC *Mycologist* Coelomycetes
13 Sivanesan A - PhD *Mycologist* Ascomycotina, esp. Dothideales
14 Waller JM - PhD *ODA Liaison Officer* tropical plant pathology
15 Williams MAJ - PhD *Mycologist* moniliaceous Hyphomycetes

4014 Library and Information Services Division

1 Dring Mrs VJ - BSc *Scientific Information Officer* Editor, Biodeterioration
Abstracts
2 Hudson KJ - BSc,MSc *Librarian*

4015 Culture Collection and Industrial Services Division

1 Allsopp D - PhD,CBiol,MIBiol,MIInfSc *Head, Culture Collection and*
Industrial Services Division industrial mycology and biodeterioration
2 Kelley J - PhD,CBiol,MIBiol *Industrial Services Officer* industrial mycology
and biodeterioration
3 Paterson RRM - MSc,PhD *Microbial biochemist*
4 Platt R - BA *Database Administrator* Culture Collection database
5 Smith D - PhD,DIC,CBiol,MIBiol *Curator, Culture Collection* preservation
of fungi
6 Yaghmaie PA - BSc *Scientific Officer* biochemical screening

4016 CAB International Institute of Parasitology
395A Hatfield Road, St. Albans, Hertfordshire, AL4 0XU
Tel: 0727 33151. Telex: 9312102254 IP G. Cables: Helmab,St.Albans.

1 Webbe Professor G - DSc *Consultant Director* helminths of medical
importance; schistosomiasis research
2 Hague NGM - PhD *Consultant Director* control of plant parasitic
nematodes
3 Muller RLM - PhD *Director* helminths of veterinary and medical
importance; filariasis and dracunculiasis

4017 Division of Helminth Taxonomy

1 Khalil LF - PhD,D.A.P.&E. *Taxonomist* parasites of veterinary importance;
parasites of game animals; parasites of fish
2 Gibbons LM - PhD *Taxonomist* nematodes of veterinary importance;
nematodes of artiodactylids; nematodes of rodents
3 Jones A - PhD *Taxonomist* helminths of veterinary importance; taeniid
cestodes of domestic and game animals; amphistomes of ruminants and
fishes

4018 Division of Plant Nematode Taxonomy

1 Siddiqi MR - DSc *Taxonomist* plant parasitic nematodes
2 Hunt DJ - PhD,FRES *Taxonomist* plant and insect nematodes
3 Nobbs J - BSc,PhD *Taxonomist* taxonomy, heteroderids

4019 Onchocerciasis Research Unit

1 Townson S - PhD chemotherapy and control of human onchocerciasis
2 Connelly Ms C - BSc in vitro drug assays; culture techniques
3 Dobinson AR screening of chemotherapeutic agents in helminthology
cryopreservation of Helminths

4020 Tropical Nematology Advisory and Research Unit

1 Bridge J - PhD tropical plant nematology, overseas surveys and
consultancies, research and advisory work, training
2 Page SLJ - PhD tropical plant nematology, research, overseas surveys,
extension and training (on secondment to Zimbabwe)
3 Plowright RA - PhD plant parasitic nematodes of rice

4021 Information Services

4022 Bureau of Agricultural Economics
Wallingford, Oxfordshire, OX10 8DE
Tel: 0491 32111. Telex: 847964 (COMAGG G). Fax: 0491 33508.

1 Bellamy Miss MA - BScEcon,DipAgEcon,MIInfSc *Managing Editor*
2 Cairns Miss J - BScAgrEcon *Scientific Information Officer*
3 Cummings Miss SJR - BA *Scientific Information Officer*
4 Espinasse Mrs C - BA *Scientific Information Officer*
5 Khadar ID - Doctorat de 3eme Cycle,BSc(Econ),CESAT *Scientific*
Information Officer
6 Kubisz Miss KM - BSc(Econ) *Scientific Information Officer*
7 Leighfield Mrs MA - ALA,AMTS,MIInfSc *Scientific Information Officer*
8 Osborn Miss J - BScAgrEcon *Scientific Information Officer*

4023 Bureau of Animal Breeding and Genetics
Wallingford, Oxfordshire, OX10 8DE
Tel: 0491 32111. Telex: 847964 (COMAGG G). Fax: 0491 33508.

1 Turton JD - BSc,MRCVS,DTVM *Managing Editor and Chief Managing*
Editor, Animal Sciences
2 Berrill Mrs AB - BSc,FIInfSc *Scientific Information Officer*
3 Carr WR - FRIC *Scientific Information Officer*
4 Fowler Mrs AA - FK *Scientific Information Officer*
5 Gaffney B - BA,MSc *Scientific Information Officer*
6 Reford R - BSc,MSc *Scientific Information Officer*
7 Rostron A - BSc *Scientific Information Officer*

4024 Bureau of Animal Health
Wallingford, Oxfordshire, OX10 8DE
Tel: 0491 32111. Telex: 847964 (COMAGG G). Fax: 0491 33508.

1 Phillips GD - BSc,PhD,BVSc,MRCVS *Managing Editor*
2 Allsopp CE - BSc *Scientific Information Officer*
3 Bamford S - BSc,PhD *Scientific Information Officer*
4 Brooks JAH - MSc,DIC,FRES *Scientific Information Officer*
5 Cooper Ms M - BSc *Scientific Information Officer*
6 Darbyshire HBA - BVSc,PhD,FRCVS *Scientific Information Officer*
7 Davis AJ - BSc,MSc,PhD,FRES *Scientific Information Officer*
8 Gascoigne Mrs E - FIL *Scientific Information Officer*
9 Gordon CW - BA,MIInfSc *Scientific Information Officer*
10 Gotts MG - MA *Scientific Information Officer*
11 Hails MR - BSc,DipAgEcon,MIInfSc,MIBiol *Scientific Information Officer*
12 Halsall Miss JL - MSc,MIInfSc *Scientific Information Officer*
13 Johnson AW - BVSc,MRCVS *Scientific Information Officer*
14 Johnson CD - CChem,MRSC *Scientific Information Officer*
15 Neenan PMP - BSc,HDip in Ed *Scientific Information Officer*
16 Pittaway AR - BSc,MSc,DIC *Scientific Information Officer*
17 Pozniak GI *Scientific Information Officer*
18 Taylor R - BSc,MIBiol *Scientific Information Officer*
19 Woods Ms S - MRCVS *Scientific Information Officer*

4025 Bureau of Dairy Science and Technology
Wallingford, Oxfordshire, OX10 8DE
Tel: 0491 32111. Telex: 847964 (COMAGG G). Fax: 0491 33508.

1 Wilson PD - BA,DPhil *Managing Editor*
2 Bird SP - BSc,PhD *Scientific Information Officer*
3 Davies JM - BSc,APMI *Scientific Information Officer*
4 Howells BW - BSc,BD *Scientific Information Officer*
5 Leeson AD - BA *Scientific Information Officer*
6 Livesley DE - BSc *Scientific Information Officer*
7 Ross Mrs JL - BSc *Scientific Information Officer*

4026 Bureau of Animal Nutrition
Wallingford, Oxfordshire, OX10 8DE
Tel: 0491 32111. Telex: 847964 (COMAGG G). Fax: 0491 33508.

1 Dodsworth EA - BTech,MSc,PhD,MIBiol *Managing Editor*
2 Allibone JE - BSc,MSc *Scientific Information Officer*

3 Church T - BSc *Scientific Information Officer*
4 Fleming DJ - BSc,MSc,MIBiol *Scientific Information Officer*
5 Hankin RG - MA,DipEd,MIInfSc *Assistant Director*
6 Manson KP - BSc,SRD *Scientific Information Officer*
7 Ochia BA - BSc,MSc,PhD,MIInfSc,MIBiol *Scientific Information Officer*
8 Stewart DWH - MA,BSc *Scientific Information Officer*
9 Walker Mrs MW - MA *Scientific Information Officer*

4027 Forestry Bureau
Wallingford, Oxfordshire, OX10 8DE
Tel: 0491 32111. Telex: 847964 (COMAGG G). Fax: 0491 33508.

1 Becker KM - BS,PhD *Managing Editor*
2 Biggs Mrs BEM - BSc *Scientific Information Officer*
3 Elbourn CAD - MSc *Scientific Information Officer*
4 Haynes REH - MA,DPhil *Scientific Information Officer*
5 Langdon KE - BSc,MIL *Scientific Information Officer*
6 Petrokofsky GS - BSc *Scientific Information Officer*
7 Ridout LM - BA,PhD *Scientific Information Officer*

4028 Bureau of Horticulture and Plantation Crops
Wallingford, Oxfordshire, OX10 8DE
Tel: 0491 32111. Telex: 847964 (COMAGG G). Fax: 0491 33508.

1 Bhat KKS - BSc,MSc,Docteur ès Sciences *Managing Editor*
2 Cousins Mrs DJ - BSc *Scientific Information Officer*
3 Cox Miss JEK - BSc *Scientific Information Officer*
4 Doroszenko AM - BSc,PhD *Scientific Information Officer*
5 Hardman Miss EE - MSc *Scientific Information Officer*
6 Jordan JW - FIL,KewCert *Scientific Information Officer*
7 Rendell-Dunn Ms AJ - BSc *Scientific Information Officer*

4029 Bureau of Pastures and Field Crops
Wallingford, Oxfordshire, OX10 8DE
Tel: 0491 32111. Telex: 847964 (COMAGG G). Fax: 0491 33508.

1 Wightman P - BSc *Managing Editor and Chief Managing Editor (Plant Sciences)*
2 Armstrong J - MA,Dip.Agric,DTA *Scientific Information Officer*
3 Boyce R - BSc,MIInfSc *Scientific Information Officer*
4 Hall AM - BA,PhD *Scientific Information Officer*
5 Hill Miss RE - BSc *Scientific Information Officer*
6 Hill Miss SE - BSc *Scientific Information Officer*
7 Kisz Miss JM - BSc *Scientific Information Officer*
8 Mahal MS - BSc,PhD *Scientific Information Officer*

4030 Bureau of Plant Breeding and Genetics
Wallingford, Oxfordshire, OX10 8DE
Tel: 0491 32111. Telex: 847964 (COMAGG G). Fax: 0491 33508.

1 Watkins R - DipAgr,BSc,PhD,FIHort *Managing Editor*
2 Chambers D - BSc,MSc *Scientific Information Officer*
3 Cooper BF - MA,PhD *Scientific Information Officer*
4 Hemming DJ - BSc,PhD *Scientific Information Officer*
5 Jellis Mrs S - MA *Scientific Information Officer*
6 Loder JE - MSc *Scientific Information Officer*
7 Powell Miss JM - BA *Scientific Information Officer*
8 Praciak AM - BSc *Scientific Information Officer*
9 Turner GJ - BSc,PhD *Scientific Information Officer*

4031 Bureau of Soils
Wallingford, Oxfordshire, OX10 8DE
Tel: 0491 32111. Telex: 847964 (COMAGG G). Fax: 0491 33508.

1 Butters B - MSc,DipAgrSc,DTA *Managing Editor*
2 Carroll DM - BSc *Scientific Information Officer*
3 Creasey Mrs R - BSc *Scientific Information Officer*
4 Eldridge DJ - BSc *Scientific Information Officer*
5 Greenslade Miss JL - BSc *Scientific Information Officer*
6 Taylor HR - BSc *Scientific Information Officer*

4032 Bureau of Crop Protection
Wallingford, Oxfordshire, OX10 8DE
Tel: 0491 32111. Telex: 847964 (COMAGG G). Fax:0491 33508.

1 Scott PR - MA,PhD *Managing Editor*
2 Basu P - BSc *Scientific Information Officer*
3 Brooks Mrs EDL - BSc,BAgr *Scientific Information Officer*
4 Brown Mrs SC - BA *Scientific Information Officer*
5 Brunt Miss JA - MSc *Scientific Information Officer*
6 Codd Miss TM - BSc,PhD *Scientific Information Officer*
7 Larkham Miss JJ - BA *Scientific Information Officer*
8 Mayall JL *Scientific Information Officer*
9 Siddiqi SF - BSc *Scientific Information Officer*
10 Viney GA - MSc *Scientific Information Officer*
11 Whittaker Mrs A - MSc *Scientific Information Officer*
12 Winton-Lezcano Mrs SA - MSc *Scientific Information Officer*

4033 International Food Information Service
Lane End House, Shinfield, Reading, Berkshire, RG2 9BB
Tel: (0734) 883895. Telex: 847204 (DSIFIS G).

The International Food Information Service (IFIS) is an independent organization sponsored jointly by CAB International (UK), the Bundesministerium für Ernahrung, Landwirtschaft und Forsten (FRG), The Institute of Food Technologists (USA), and the Centrum voor Landbouwpublikaties en Landbouwdocumentatie (Netherlands), and administered by a Management Committee representing the four sponsors.

1 Schützsack U - PhD *Joint Managing Director*
2 Brookes H - MA *Joint Managing Director/Editor*
3 Mann EJ - LLD,NDD,CDD *Senior Adviser*
4 Amor Mrs D - BSc *Scientific Information Officer*
5 Godin Mrs VJ - BSc *Scientific Information Officer/Deputy Editor*
6 Hall DI - MA *Scientific Information Officer*
7 Kalbskopf G - dipl Dok *Scientific Information Officer*
8 Lang Mrs A - BSc *Scientific Information Officer*
9 Latymer Z - PhD *Scientific Information Officer*
10 Loeffel Ms E - BSc *Scientific Information Officer*
11 May Mrs J - BSc *Scientific Information Officer*
12 Merryweather LM - PhD *Scientific Information Officer*
13 Sears Mrs CR - BTech *Scientific Information Officer*
14 Watts Mrs LJ - BSc *Scientific Information Officer*
15 Wheatcroft AJD - BSc *Scientific Information Officer*

4034 International Bee Research Association
Hill House, Chalfont St Peter, Gerrards Cross, SL9 0NR
Tel: (0753) 885011.

1 Adey M - PhD *Director*
2 Bradbear N - PhD *Information Officer for Tropical Apiculture*
3 Day R - BSc *Scientific Information Officer (part-time)*
4 Jones T *Chairman of Council Tropical Development & Research Institute London*
5 Lowe DG - BSc,MIBiol *Senior Scientific Information Officer, Editor "Apicultural Abstracts"*
6 Walker AP - MA *Scientific Information Officer (part-time)*

4035 Department of the Environment
Headquarters, 2 Marsham Street, London, SW1P 3EB
Tel: 01-212 3434.

4036 Building Research Establishment
Princes Risborough Laboratory, Princes Risborough, Aylesbury, Buckinghamshire, HP17 9PX
Tel: (084 44) 3101.

1 Baker JM - BSc,FIBiol,FIWSc *Head of Laboratory*

4037 Timber and Protection Division

1 Morgan JWW - BSc,PhD,FRIC,FIWSc *Head of Division*
2 Bravery AF - BSc,DIC,PhD *Head of Biodeterioration Section*
3 Brazier JD - BSc,DSc,FIWSc *Head of Properties of Materials Section*
4 Hodson J - BSc,PhD,CChem,MRSC *Head of Environment Section*
5 Miller ER - BSc,PhD *Head of Finishes Section*
6 Orsler RJ - ARIC,PhD *Head of Preservation Section*
7 Smithies JN - MSc,BTech *Head of Sawmilling Section*

4038 Components and Structures Division

1 Carruthers JFS - BSc,CEng,MIMechE,FIWSc *Head of Components and Structure Division*
2 Beech JC - BSc *Head of Flat Roofs and Sealants Section*
3 Covington SA *Head of Components Quality and Performance Section*
4 Fewell AR - HNC *Head of Structural Design and Performance Section*

4039 Natural Environment Research Council
Headquarters, Polaris House, North Star Avenue, Swindon, SN2 1EU
Tel: 0793-40101. Telex: 444293 ENVRE G.

1 Fish H - CBE *Chairman*
2 Bowman JC - BSc,PhD,FiBiol *Secretary*
3 Briden Professor J *Director of Earth Sciences*
4 Tinker PBH - *Director of Terrestrial and Freshwater Science*
5 Woods Professor JD *Director of Marine Science*

4040 NERC Scientific Services

Holbrook House, Station Road, Swindon, SN1 1DE
Tel: 0793-40101. Telex: 444293 ENVRE G.

1 Hinde BJ *Director*

4041 Research Vessel Services

No 1 Dock, Barry, South Glamorgan, Wales, CF6 6UZ
Tel: 0446-737451. Telex: 497101 RVBASE G.

1 Skinner LM *Head of Service*

4042 British Geological Survey

Nicker Hill, Keyworth, Nottingham, NG12 5GG
Tel: 06077-6111. Telex: 378173 BGSKEY G.

1 Lumsden GI *Director*
2 Bloomfield Dr DK *Director-Programmes D* programmes overseas and global information system
3 Evans WB *Director-Programmes B* geological surveys - England South of Ribble-Tees line and Wales
4 Haworth Dr RT *Director Geophysics*
5 Hull JH - *Director-Programmes C energy projects on the UK Continental Shelf*
6 Kelk B *Director Information and Central Services*
7 Smith EG *Director Geology*

4043 British Geological Survey Programme A

Murchison House, West Mains Road, Edinburgh, Scotland, EH9 3LA
Tel: 031-667-1000. Telex: 727343 SEISED G.

1 Galois Dr RW *Director-Programmes A* geological surveys - Scotland and England North of Ribble - Tees line

4044 British Geological Survey Geochemistry

64 Gray's Inn Road, London, WC1X 8NG
Tel: 01-242-4531. Telex: 262199 BGSCLR G.

1 Moore PJ - Director Geochemistry

4045 Institute of Hydrology

Maclean Building, Crowmarsh Gifford, Wallingford, Oxfordshire, OX10 8BB
Tel: 0491-38800. Telex: 849365 HYDROL G.

1 McCulloch JSG *Director*

4046 Institute of Marine Biochemistry

St Fitticks Road, Torry, Aberdeen, Scotland, AB1 3RA
Tel: 0224-875695. Telex: 849365 HYDROL G.

1 Cowey CB *Director*

4047 Institute for Marine Environmental Research

Prospect Place, The Hoe, Plymouth, Cornwall, PL1 3DH
Tel: 0752-21371.

1 Bayne BL *Director*

4048 Institute of Oceanographic Sciences

Brook Road, Wormley, Nr Godalming, Surrey, GU8 5UB
Tel: 042879-4141. Telex: 858833 OCEANS G.

1 Laughton AS - FRS *Director*

4049 Bidston Observatory
Institute of Oceanographic Sciences, Bidston, Birkenhead, Merseyside, L43 7RA
Tel: 051-653-8633.

1 Dyer KR - Assistant Director

4050 Institute of Terrestrial Ecology

Administrative Headquarters, Monks Wood Experimental Station, Abbots Ripton, Huntingdon, PE17 2LS
Tel: (04873) 381. Telex: 32416.

1 Jeffers JNR - FIS,MBIM,MBiol *Director*

4051 Directorate

1 Dempster JP - BScZool,DICEnt,PhD,DSc
2 Jeffers JNR - FIS,MBIM,MBiol
3 Last Professor FT - DScPlantPath,ARCSBot,FRSE

4052 Monks Wood Experimental Station - ITE
Abbots Ripton, Huntingdon, PE17 2LS
Tel: (04873) 381. Telex: 32416.

4053 Vertebrate Ecology - Monks Wood

1 Newton I - BScZoolBotChem,PhD birds of prey and pollution
2 Stebbings RE - MIBiol,PhD population ecology of bats

4054 Invertebrate Ecology - Monks Wood

1 Davis BNK - BScZool,PhD phytophagous insects
2 Dempster JD - BScZool,DICEnt,PhD,Dsc *Entomology*
3 Pollard E - BScHortSc,DipNatCons *Entomology*
4 Welch RC - BScEnt,ARCS,DICEnt,PhD coleoptera

4055 Animal Function - Monks Wood

1 Moriarty F - BScZool,ARCS,DICEnt,PhD pollutants in aquatic animals

4056 Plant Community Ecology - Monks Wood

1 Hooper MD - BScBot,PhD environmental impact assessment
2 Sheail J - BAGeog,PhD nature conservation in Britain (ITE historian)
3 Wells TCE - BScAgrBot grasslands ecology

4057 Data and Information - Monks Wood

1 Lakhani KH - BScStat,FIS,Bar at Law statistics/computing
2 Mountford MD - BAMaths,DipStat statistics/computing

4058 Banchory Research Station - ITE
Hill of Brathens, Glassel, Banchory, Kincardineshire, AB3 4BY
Tel: (03302) 3434. Telex: 32416.

4059 Vertebrate Ecology

1 Harris MP - PhD puffin research
2 Kruuk H - PhDZool animal populations
3 Mitchell B - BScZool,PhD red deer populations
4 Moss R - BScBiochem,PhD avian research
5 Staines BW - BScZool,PhD red deer populations
6 Watson A - BScZool,DSc,PhD avian research

4060 Soil Science

1 Miles J - BScBot,PhD vegetation research
2 Miller GR - BScBot,PhD vegetation studies

4061 Bangor Research Station - ITE
Penrhos Road, Bangor, Gwynedd, LL57 2LQ
Tel: (0248)36 4001. Telex: 32416.

4062 Plant Biology - Bangor

1 Perkins DF - BScBot,PhD plant physiology

4063 Plant Community Ecology - Bangor

1 Hill MO - MAMaths,DICStat,PhD forestry
2 Shaw MW - BScFor forest research

4064 Soil Science - Bangor

1 Ball DF - BAGeol,MA,PhD soil chemical variability
2 Hornung M - BScGeol,PhD effect of forest management on soil

4065 Data and Information - Bangor

1 Milner C - BScAgr,PhD statistics/computing/remote sensing
2 Radford GL - BScBot,MScEco statistics and computing
3 Wyatt BK - BScChem,PhD statistics/computing/remote sensing

4066 Culture Centre of Algae & Protozoa
36 Storey's Way, Cambridge, CB3 0DT
Tel: (0223) 61378. Telex: 32416.

1 Baker Dr JR - DScGen,PhD,MA,FIBiol
2 Morris Dr GJ - BTec,PhD
3 Page Dr FC - PhD systems of amoebae

4067 Edinburgh Laboratory - ITE
Bush Estate, Penicuik, Midlothian, EH26 0QB
Tel: 031 445 4343.

4068 Vertebrate Ecology - Bush

1 Maitland PS - BScZool,PhD freshwater ecology

4069 Plant Biology - Bush

1 Cannell MGR - BScAgrBot,PhD forest physiology
2 Fowler D - BScEnvPhs,PhD,FRMetS acid rain
3 Leakey RRB - HBaAgr,CBaAgr,BSc,PhD tropical forestry
4 Longman KA - BScFor,PhD plant physiology
5 Unsworth M - BSc,Phs,PhD dry deposition

4070 Data and Information - Bush

1 Smith IR - BScCivEng statistics/computing

4071 Furzebrook Research Station - ITE
Wareham, Dorset, BH20 5AS
Tel: (09295) 51518.

4072 Invertebrate Ecology - Furzebrook

1 Elmes GW - BScZool,PhD population dynamics of ants
2 Merrett P - BScZool,PhD spider populations
3 Morris MG - MAZool,PhD entomology
4 Thomas JA - MANatSc,PhD *Entomologist*
5 Ward LK - BScZool,PhD heathland invertebrates
6 Webb NRC - BScZool,PhD heathland ecology

4073 Vertebrate Ecology - Furzebrook

1 Goss-Custard JD - BScZool,PhD wading birds

4074 Plant Biology - Furzebrook

1 Chapman SB - BScBot,PhD heathlands
2 Gray AJ - BABiolPhil,PhD coastal management

4075 Merlewood Research Station - ITE
Grange-over-Sands, Cumbria, LA11 6JU
Tel: (04484) 2264.

4076 Vertebrate Ecology - Merlewood

1 Lowe VPW - MAZool taxonomy of squirrels and deer

4077 Plant Biology - Merlewood

1 Callaghan TV - BScBot,PhD renewable energy sources

4078 Plant Community Ecology - Merlewood

1 Brown AHF - BScFor research on woodland vegetation
2 Bunce RGH - BScBot,PhD semi-natural woodland classification
3 Horrill AD - BScBot,PhD vegetation changes in grazing
4 Sykes JM - MAFor,MScFor monitoring vegetation

4079 Soil Science - Merlewood

1 Harrison AF - BScBot,BAgrSc,PhD nutrient cycling
2 Howard PJA - BScBot,MScEcol soil microbiology

4080 Data and Information - Merlewood

1 Jeffers JNR - FIS,MBIM,MBiol statistics and research management
2 Lindley DK - BScFor,MScMaths statistics/computing

4081 Chemistry and Instrumentation - Merlewood

1 Allen SE - BScGen,HNC chemistry

4082 British Antarctic Survey
High Cross, Madingley Road, Cambridge, CB3 0ET
Tel: 0223-61188. Telex: 817725 BASCAM G.

1 Laws RM - CBE,FRS *Director British Antarctic Survey and Sea Mammal Research Unit*

4083 Sea Mammal Research Unit

1 Harwood Dr J *Officer-in-Charge*

4084 Grant Aided Associations

4085 Freshwater Biological Association
The Ferry House, Far Sawrey, Ambleside, Cumbria, LA22 0LP
Tel: 09662-2468/9.

1 Clark RT *Director*

4086 Windermere Laboratory
The Ferry House, Far Sawrey, Ambleside, Cumbria, LA22 0LP
Tel: 09662-2468/9.

1 Kinsman DJJ *Assistant Director*

4087 River Laboratory
East Stoke, Wareham, Dorset, BH20 6BB
Tel: 0929-462314.

1 Berrie AD *Officer-in-Charge*

4088 Marine Biological Association of the United Kingdom
The Laboratory, Citadel Hill, Plymouth, PL1 2PB
Tel: 0752-21761.

1 Denton Professor EJ - CBE,FRS *Director*

4089 Scottish Marine Biological Association
Dunstaffnage Marine Research Laboratory, P.O. Box 3, Oban, Argyll, PA34 4AD
Tel: 0631-62244. Telex: 776216 MARLAB G.

1 Currie Professor RI *CBE Director*

4090 Overseas Development Administration

Eland House, Stag Place, London, SW1E 5DH
Tel: 01-213 3000. Telex: 263907/8 LDNG. Cables: Ministrant.

4091 Headquarters

1 Davies JC - PhD,BSc,MIBiol *Deputy Chief Natural Resources Adviser (Research)*
2 Wilson AT - BSc,DipAgr,DIA,MIBiol *Chief Natural Resource Adviser*
3 Davies JC - PhD,BSc,MIBiol *Deputy Chief Natural Resources Adviser (Research)*
4 Goldsack JR - BSc,DipAg,DTA *Deputy Chief Natural Resources and Principal Agricultural Adviser*
5 Bell RD - BSc,FIMechE,FIAgrE *Agricultural Engineering Adviser*
6 Cooper RA - BVSc,MSc,MRCVS *Animal Health Adviser (Temporary)*
7 Freeland GG - BVMS,MSc,MRCVS *Animal Health Adviser*
8 Grimwood BE - BSc,PhD,CChem,MRIC *Assistant Agricultural Adviser*
9 Hansell JRF - BSc *Assistant Agricultural Adviser*
10 Heffer DW - FCISA,CSD,AMBIM *Cooperatives Advisers*
11 Howard WJ - BSc,BPhil *Forestry Adviser*
12 Kemp RW - MA,BA *Forestry Adviser*
13 Smith RW - BSc,MIBiol *Agricultural Adviser (Research)*
14 Stobbs AR - VRD,BSc,FRGS *Agricultural Adviser*
15 Stoneman J - BSc,FIBiol *Fisheries Adviser*
16 Tarbit J - BSc,PhD,MIBiol *Fisheries Adviser*
17 Tuley P - BSc,DAgSci,DTA,MIBiol,FLS *Agricultural Adviser*

4092 Development Divisions

4093 British Development in the Caribbean
Carlisle House, Hincks Street, PO Box 167, Bridgetown, Barbados, West Indies
Tel: 4262190/4297199. Telex: 2236.

1 Tainsh AJ - BSc,MScAgExt *Natural Resources Adviser*

4094 British Development Division in Southern Africa
PO Box 30059, Capital City, Lilongwe 3, Malawi
Tel: Lilongwe 731544. Telex: 4727.

1 Salmon DJ - NDA,MSc *Natural Resources Adviser*
2 Watson MF - BSc,DTA *Natural Resources Adviser*

4095 British Development Division in East Africa
c/o British High Commission, Bruce House, Standard Street, Nairobi, Kenya [30465]
Tel: 335944. Telex: NAIROBI 22219.

1 Trotman DA - NDA,MScAgExt *Natural Resources Adviser*
2 Weare PR - BSc,DTA,MIBiol *Senior Natural Resources Adviser*

4096 British Development Division in South East Asia
c/o British Embassy, Ploenchit Road, Bangkok, Thailand
Tel: 2527161/9.

1 Waddell RL - BSc,DipAgr,DTA *Senior Natural Resources Adviser*
2 Wilson MJ - BSc,DTA *Natural Resources Adviser*

4097 British Development Division in the Pacific (Suva)
Vanua House, Victoria Parade, Private Bag, Suva, Fiji
Tel: 311944. Telex: FJ2289.

1 Beales RW - BScZool,MIBiol *Regional Fisheries Adviser*
2 Warren JB - BSc,Agric,DTA *Natural Resource Adviser*

4098 Overseas Development Natural Resources Institute (ODNRI)

Central Avenue, Chatham Maritime, Chatham, Kent, ME4 4TB
Tel: 0634 880088. Telex: 263907/8 LDN G.

A scientific unit of the Overseas Development Administration, which carries out research and development in the renewable resources field. [At the time of going to press some Departments are at distant locations but by September 1989 all are expected to be located at Chatham, except the Termite Biology and Advisory Service and the Radar Unit at Malvern.]

1 Perfect TJ - MSc *Deputy Director*
2 Robertson CA - BSc,MSc,PhD *Deputy Director*
3 Beattie Mr GA *Director*
4 Jones T - OBE,BSc,ARCS,FIBiol *Deputy Director*
5 Nabney J - BSc,PhD *Deputy Director*

4099 Central Services

1 Disney JG - BSc,DCC *Head of Department*

4100 Library Section

1 Wright JA - ALA *Head of Library and Information Services*

4101 Animal Products and Feeds Department
4102 Fish

1 Ames GR - BSc,PhD,CChem,FRSC *fish technology*
2 Clucas I
3 Howell B
4 King D
5 O'Leary C
6 Parry R
7 Rogers J
8 Wood C

4103 Meat, Dairy Products, Hides and Skins

1 Duff A
2 Hector D
3 Kenny M
4 Leach I
5 Silverside D

4104 Animal Feeds

1 Bainton J
2 Carlaw P
3 Panigrahi S
4 Plumb V
5 Thorne P
6 Watson I
7 Wood J

4105 Mycotoxins

1 Coker R
2 Dutta A
3 Ganguli R
4 Jewers K - BSc,PhD *mycotoxins*
5 Jones B
6 Medlock V
7 Nagler M

4106 Plant Food Commodities Department

1 Cooke RD - BSc,PhD *Head of Department*

4107 Oilseeds and Edible Nuts

1 Broadbent JA *oilseeds and edible nuts*
2 Hammonds T
3 Harris RV - BSc,PhD *oilseeds and edible nuts*

4 Head S
5 Southwell K
6 Swetman T

4108 Beverage Crops & Miscellaneous

1 Baker Mr DM *beverage crops and miscellaneous*
2 McDowell I
3 Robinson J
4 Tomlins K

4109 Microbiology & Fermentation

1 Fuchs R
2 Nicolaides L
3 Phillips S
4 Reilly Mr PJA *microbiology and fermentation*
5 Stephens T
6 Twiddy D
7 Vasconcelos A
8 Westby A

4110 Fruit, Vegetables & Root Crops

1 Behn K
2 Burbage L
3 Hughes P
4 Jeger Dr MJ *fruit, vegetable and root crops*
5 Johanson A
6 Moore D
7 O'Brien G
8 Plumbley R
9 Poulter N
10 Proctor Ms FJ - BSc *fruit, vegetables and root crops*
11 Rickard J
12 Wainwright H

4111 Forestry

4112 Non Food Commodities Department
Central Avenue, Chatham Maritime, Chatham, Kent, ME4 4TB
Tel: 0634 880088.

1 Green JHS - BSc,PhD,CChem,FRSC *Head of Department*

4113 Pulp, Paper & Vegetable Fibres

1 Palmer ER *pulp, paper and vegetable fibres*

4114 Spices, Essential Oils & Other Plant Extractives

1 Barrow M
2 Coppen J
3 Dann AE - CChem,MRSC *spices, essential oils and other plant extractives*
4 Green CL - BSc,PhD *spices, essential oils and other plant extractives*
5 Milchard M
6 Wood A

4115 Insect Products

1 Adams P
2 Beevor P
3 Cork A
4 Gough A
5 Hall DR - BA,PhD *insect products*
6 Mullings J

4116 Economics, Social and Statistics
Central Avenue, Chatham Maritime, Chatham, Kent, ME4 4TB
Tel: 0634 880088.

1 Hebblethwaite MJ - BSc,MSc,MA *Head of Department*

4117 Marketing Research and Systems

1 Coote C
2 Trotter B

4118 Plant Commodities Economics

1 Anand N
2 Baulch B
3 Greenhalgh Dr P *plant commodities economics*
4 Hone A
5 Winter JD - BScAgr *food commodities economics*

4119 Animal Commodities Economics

1 Boustead PJ *food commodities economics*
2 Coulter J
3 Gordon A

4 Marter AD - BA animal commodities economics
5 Silvey D

4120 Pest and Vector Economics

1 Barrett J
2 Iles Mr MJ pest and vector economics
3 Walker J

4121 Resource Assessment and Planning

1 Barrett T
2 Errington M
3 Grimble RJ - BSc,MSc,PhD resource assessment and planning

4122 Statistics and Econometrics

1 Daplyn P
2 Green M
3 Green Miss SM - MA,DipStat statistics

4123 Applied Biology Department ODNRI

Central Avenue, Chatham Maritime, Chatham, Kent, ME4 4TB
Tel: 0634 880088.

1 Sands WA - BSc,MSc,DSc,ARCS,DIC,FIBiol *Head of Department*

4124 Termite Biology and Taxonomy

Dept. of Entomology, British Museum (Natural History), Cromwell Road, London, SW7
Tel: 01-589-6323.

1 Bacchus Dr S termite biology and taxonomy

4125 Larvicides

1 Brown N
2 Fenwick A - PhD larvicides
3 Goll Dr P larvicides

4126 Grasshopper Taxonomy and Ecology

1 Jago ND - BSc,PhD,ARCS grasshopper, taxonomy and ecology
2 Ritchie M
3 Rowley J

4127 Moth Pheromones

1 Bettany B
2 Campion DG - PhD,FRES moth pheromones
3 Chamberlain D
4 Critchley B
5 Downham M
6 McVeigh E
7 McVeigh L
8 Russell D

4128 Applied Ecology Department ODNRI

Central Avenue, Chatham Maritime, Chatham, Kent, ME4 4TB
Tel: 0634 880088.

1 Wood TG - BSc,PhD *Head of Department*
2 Reynolds D
3 Riley J - DPhil,CEng radar entomology and flight
4 Smith A

4129 Biogeography and Meteorology

Chatham

1 Crummay F
2 Haggis M
3 Pedgley DE - BSc biogeography and meteorology
4 Pender J
5 Tucker M

4130 Heliothis Ecology and Control

1 Holt J
2 King ABS - BSc,PhD heliothis ecology and control

4131 Termite Ecology and Control

1 Black H
2 Cowie Dr RH termite ecology and control
3 Logan J
4 Pearce M

4132 Physiology and Behaviour

1 Armes N

2 Cooter RJ - PhD physiology and behaviour

4133 Pest and Vector Management Department ODNRI

Central Avenue, Chatham Maritime, Chatham, Kent, ME4 4TB
Tel: 0634 880088.

1 Magor Dr JI *Head of Department (Acting)*

4134 Pest Bionomics

1 Abisgold J
2 Dewhurst C
3 Fishpool L
4 Page B
5 Rosenberg Dr LJ pest bionomics

4135 Crop Pest Management

1 Bigger M - MSc,FRES crop pest management
2 Ingram R
3 Maslen Dr NR crop pest management
4 Ward Ms A crop pest management
5 Whitwell A

4136 Vector Ecology and Control

1 Allsopp R
2 Cheke B
3 Millest A
4 Torr S

4137 Pesticide Impact Assessment Group

1 Crick H
2 Douthwaite B
3 Grant Dr IF pesticide impact assessment

4138 Insect Crop Relations

1 Cook A
2 Kimmins F
3 Padgham D

4139 Storage Department

London Road, Slough
Tel: 0753 34626.

Please note this Department will be relocated at Chatham from July 1989

1 Poulter Dr RG *Head of Department*
2 Prevett P

4140 Chemical Control and Pesticides Analysis

1 Cox J
2 Friendship R
3 Galsworthy D
4 Golob P
5 Halliday D - BSc,CChem,FRSC chemical control and pesticide analysis
6 Harris A
7 Kilminster K
8 Taylor B
9 Walker D

4141 Storage Operations and Management

1 Andrews B
2 Baker AA - BSc storage operations and management
3 Boxall R
4 Gilman GA - BSc,DTA storage operations and management
5 Hindmarsh Mr PS storage operations and management
6 Morley E
7 Tyler P

4142 Pest Biology and Inspection

1 Conway J
2 Dendy J
3 Dick K
4 Dobie P
5 Giles PH - BScmDipAgrSc,DIC,MIBiol pest biology and inspection
6 Haines CP - BSc,PhD,FRES pest biology and inspection
7 Hodges R
8 Locke M
9 McFarlane JA - BSc,DIC,FRES,MIBiol pest biology and inspection
10 Prevett PF - BSc,PhD,ARCS,DIC pest biology and inspection
11 Rees D

4143 Cereals Technology
Culham, Abingdon, OX14 3DA
Tel: 086730-7551.

Please note this section will be relocated to Chatham from Sept. 1989

1 Agranoff J
2 Cecil J
3 Clarke P
4 Dendy DAV - BSc,PhD,CChem,FRSC,AIFST cereals, starches and general food processing
5 James T

4144 Industrial Development Department
Culham, Abingdon, OX14 3DA
Tel: 086730-7551.

Please note this Department will be relocated at Chatham from September 1989

1 Adair D - MSc,CChem,FRSC *Head of Department*

4145 Mechanical Engineering

1 Cox D
2 Hawkes T
3 Nichols W mechanical engineering
4 Pinson G

4146 Chemical Engineering

1 Drew P
2 Kennedy L - BSc,MSc,PhD chemical engineering
3 Timmins W
4 Trim D

4147 Biomass Energy

1 Breag GR - FICE biomass energy
2 Hollingdale Mr AC biomass energy
3 Krishnan R
4 Robinson A
5 Sarwar G
6 Tariq A

4148 Storage Engineering and Packaging
London Road, Slough
Tel: 0753 34626.

Please note this section will be relocated at Chatham from July 1989

1 Bisbrown J
2 Gough M
3 New J
4 O'Dowd T

4149 Pesticides Application and Management Department
Porton Down, Salisbury, Wilts, SP4 0JQ
Tel: 0980-610211.

Please note this Department will be relocated at Chatham from July 1989

1 Lyon DJDEB - BSc,MIBiol,FRES *Head of Department*

4150 Pesticides Application

1 Cooper J
2 Dobson H
3 Flower L
4 Johnstone DR - MSc,FInstP,FIAgrE,MRAeS pesticides application
5 King B
6 Lee C
7 Turner C

4151 Virus and Bioassay

1 Grzywacz D
2 Harris E
3 Jones K
4 McKinley DJ - MSc,MIBiol,FRES virus and bioassay

4152 Pesticide Management

1 Ambridge L
2 Lambert Dr MRK pesticide management

4153 Land Resources
Central Avenue, Chatham Maritime, Chatham, Kent, ME4 4TB
Tel: 0634 880088.

1 Makin MJ - MSc *Head of Department* agriculture and soils

4154 Agriculture

1 Hendy CRC - PhD livestock production
2 Holland MD - BSc agriculture

4155 Agroclimatology

1 Williams JB - PhD agricultural climatology

4156 Forestry

1 Abell TMB - BSc forestry
2 Armstrong GD - BSc forestry

4157 Water Resources

1 Silva GL - BSc,CEng,MICE,MIWE water resource engineering
2 Walker SH - BA hydrology

4158 Remote Sensing Unit

1 King RB - PhD,BSc remote sensing

4159 Training

1 Corker IR - BSc land use planning training
2 Davies K - BSc land use planning training

4160 Tropical Soils Analysis Unit

1 Baker KF - BSc soil chemistry and analysis
2 Baker RM - BSc soil chemistry and analysis

4161 Soil Science

1 Anderson IP - BSc soil science
2 Bennett JG - BSc,ARIC soil science
3 Brinn P
4 Jenkin D
5 Wall JRD - BA,PhD,DipAgrPhoto *Interpretation* soils and geomorphology

4162 Seconded Staff

The ODA also maintains two Liaison Officers within other institutions to provide specialist scientific advice and services to particular disciplines.

1 Riley Miss JR - MSc *Biometrician (Field Crop)*
2 Waller JM - MA,PhD,DIC,DipAgrSc,MIBiol *Plant Pathologist*

GREAT BRITAIN

This section includes those institutions whose remit covers England, Scotland and Wales.

4163 Agricultural and Food Research Council

Headquarters, 160 Great Portland Street, London, W1N 6DT
Tel: 01-580 6655. Telex: 291218 AGRECO G.

1 Stewart Prof. WDP *Secretary*
2 Curtis Prof. RF - CBE *Assistant Secretary (Food Research)*
3 Dickens JMY *Assistant Secretary (Personnel)*
4 Harris RJ - BSc,ARCS *Assistant Secretary (Science Division)*
5 Jamieson Dr BG - BSc,PhD *Director of Central Office*
6 Prideaux RM *Assistant Secretary (Finance)*
7 Goodier J - MSc *Librarian*
8 Goodwin MF *Principal Information Officer (Public Relations)*

4164 AFRC Institute of Animal Physiology and Genetics Research - Edinburgh Research Station
West Mains Road, Edinburgh, EH9 3JQ
Tel: 031-667-6901.

1 Cross Dr BA - CBE,FRS *Director*

4165 Statistics and Computing
West Mains Road, Edinburgh, EH9 3JQ
Tel: 031-667-6901.

1 Simpson SP - BSc MSc PhD Development of statistical and genetic theory and techniques in the context of animal breeding; Genetics of halothane sensitivity

4166 Cattle Blood Typing Service
West Mains Road, Edinburgh, EH9 3JQ
Tel: 031-667-6901.

1 Ross DS - FIMLT Parentage control service for the cattle breeding industry and study of blood type polymorphisms

4167 Research Division
West Mains Road, Edinburgh, EH9 3JQ
Tel: 031-667-6901.

1 Archibald AL - BSc PhD Expression of mammalian genes in vitro to appraise potential for tissue-specific expression in transgenic animals
2 Cameron ND - BSc,MSc Comparison of sheep breeding procedures for genetic improvement in the efficiency of lean meat production; Genetic relationships between growth, carcase quality and appetite in performance-tested pigs
3 Gibson JP - BSc PhD Efficiency of growth and milk production of high and low milk-producing lines of British Friesian and Jersey cattle
4 Glass EJ Bovine MHC class II genes - genetics and function; Class II genes in cell mediated cytotoxicity
5 Haley C - BSc PhD Genetic variation in the components of reproductive rate in the pig, including ovulation rate and embryo survival
6 Land RB - BSc PhD DipAnGen Evaluation of the physiology of male reproductive traits as indicators of genetic merit for female reproduction
7 Lathe R - BSc PhD Expression of mammalian genes in vitro to appraise potential for tissue-specific expression in transgenic animals
8 McNeilly JR Evaluation of the physiology of male reproductive traits as indicators of genetic merit for female reproduction
9 Simons P Artificial transfer of genes into mammalian germline
10 Sinnett-Smith PA - BSc PhD Muscle and adipose tissue metabolism in farm animals and the control of nutrient partition between tissues; Genetic control of hormonal and cellular factors affecting protein and fat deposition; Metabolic variations between genetically high and low yielding dairy cows, particularly energy storage and mobilisation
11 Slee J - BSc PhD DipAnGen Genetic and physiological studies of energy and fat metabolism and survival of lambs in cold environments; Physiological and biochemical characteristics in the calf as indicators of genetic merit for lactation
12 Smith C - BSc MSc DSc PhD Investigations of breeding procedures for improving the efficiency of lean meat production in beef cattle; Comparison of sheep breeding procedures for genetic improvement in the efficiency of lean meat production; Crossbreeding and selection in the improvement of hill sheep
13 Spooner RL - MA PhD BVetMed MRCVS Genetics of humoral immune responses in qualitative responses to complex antigens; Class I genes in cell mediated cytotoxicity
14 Taylor StCS - MA DPhil DipAnGen Multibreed comparisons of feed intake for maintenance, growth and lactation in ruminants

15 Thiessen RB - BScAgrSci MSc PhD Multibreed comparisons of feed intake, growth rate, mature size, body composition and milk yield in cattle
16 Webb AJ - BSc PhD Genetic relationships between growth, carcase quality and appetite in performance-tested pigs; Genetic variation in the components of reproductive rate in the pig, including ovulation rate and embryo survival;Inheritance of porcine stress syndrome and halothane sensitivity and its exploitation in pig production
17 Webb R - BSc PhD Genetic, physiological control of sheep reproduction, specifically hormonal control of ovulation rate and seasonality; Evaluation of male reproductive traits as indicators of genetic merit for female reproduction; Study the control of ovarian follicular maturation, and associated physiological and biochemical changes; Genetic and physiological control of cattle reproduction, utilising hormonal immunological manipulation
18 Wiener G - BSc DSc PhD FIBiol Genetic control of trace element metabolism in ruminants and relationships to health and performance; Between and within breed variation in blood metabolite levels in cattle in relation to milk production
19 Wilmut I - BSc PhD Study of genetic and physiological factors influencing ovulation rate and embryo survival; Studies of factors influencing embryo development and survival; Artificial transfer of genes into mammalian germline
20 Woolliams JA - MA DipMStat Study of genetic variation in inorganic elements (Cu, Se, P) requirements and its consequences for production; Genetic factors influencing growth and the partition of energy into protein and/or fat; Genetic control and manipulation of physiological processes important to inorganic element metabolism; Genetic variation in protein metabolism in relation to underlying physiological processes

4168 Moredun Research Institute
408 Gilmerton Road, Edinburgh, EH17 7JH
Tel: 031-664-3262.

1 Aitken ID - MRCVS BVMS PhD *Director* Host-parasite interactions associated with *Chlamydia psittaci* infections in sheep; *Chlamydia psittaci*; consequences of genital infection in male animals

4169 Microbiology

1 Angus KW - BVMS,DVM,FRCVS Pathogenesis studies of enteric infections. Pathogenesis of and protection against systemic pasteurellosis. Pathogenesis and pathology of sheep pulmonary adenomatsis (SPA)
2 Burrells C - FIMLS Cellular and humoral immunity in the respiratory tract of healthy and infected sheep
3 Donachie W - BSc,PhD Characterisation of antigens of *Pasteurella haemolytica*
4 Gardiner AC - BSc Pathogenesis, aetiology, immunopathology and epidemiology of transplacental pestivirus infections
5 Gilmour JS - BVN&S,FRCVS Pathogenesis, aetiology and immunopathology of transplacental pestivirus infections. Immunopathology parameters which correlate with protection against *P. haemolytica*. Pulmonary defence systems against pathogenic microorganisms. Pathology and pathogenesis of watery mouth
6 Gilmour NJL - MRCVS BVMS PhD DSc The development of vaccines against pasteurellosis in sheep
7 Gray EW - AIMLS Morphological relationships between pestivirus isolates from cattle and sheep. Ultrastructure of SPA tumours and viral morphogenesis
8 Haig D - BSc,MSc,PhD Cellular immune responsiveness associated with transplacental pestivirus infections. Cellular and humoral responses in lung immunity. Immunological analysis of listerial antigens
9 Hay LA - BVMS,MRCVS The experimental induction of watery mouth
10 Herring AJ - BSc PhD Biochemical characterisation of the genome and proteins of ovine chlamydia and cloning the genome; Detection and biochemical characterisation of different types and serotypes of rotavirus; Biochemical characterisation of the genome and proteins of malignant catarrhal fever virus and cloning the viral genome; Biochemical characterisation of bovid herpesviruses; Bovine herpesvirus 1 antigen purification; Detection and biochemical characterisation of the sheep pulmonary adenomatosis associated retrovirus
11 Jenkinson DM - PhD,DSc Host responses and cutaneous barrier mechanisms to microorganisms. Immunology of the skin in response to infection with microorganisms
12 Jones GE - DTVM,MRCVS,BVSc,PhD Transmission and control of *C. psittaci*. Immunity to *P. haemolytica*. Pulmonary defence against microbial infections. Oral bacteriology
13 King TJ - BVMS,MRCVS Watery mouth and its prevention
14 Nettleton PF - MRCVS,BVMS,MSc,PhD Relationships between pestivirus isolates from cattle and sheep; Latency of bovid herpesvirus 1 and related viruses in domestic ruminants; Evaluation of monoclonal antibodies to IBR virus for diagnostic use; The aetiology and pathogenesis of pestivirus-induced diseases of cattle and sheep; The diagnosis of virus diseases of farm ruminants in Scotland
15 Oates CM - BSc,PhD,MIBiol Immunological studies on rotavirus in mice
16 Rae AG - MIBiol,PhD Maintenance of specialist mycoplasma diagnostic service

17 Reid HW - DTVM MRCVS BVM&S PhD Research all aspects of louping-ill virus infection of domestic animals; vaccine quality testing and diagnosis; Characterise the humoral and cellular immune response of sheep to orf virus infection; Coordinate/participate in biochemical, pathological, virological and immunological studies of malignant catarrhal fever

18 Scott FMM - FIMLS,PhD Serological and biochemical studies of rotavirus. Epidemiology and experimental studies on bovine herpes mammillitis

19 Sharp JM - MRCVS BVMS PhD Maintain and operate system for the serological diagnosis of maedi in sheep; Aetiology, immunology and epidemiology of sheep pulmonary adenomatosis

20 Snodgrass DR - MRCVS BVM&S PhD Aetiology, Pathology, Immunology and Vaccination Diarrhoea in Young Animals

21 Sutherland A - BSc Opsonins and bacterial antibodies

22 Sutherland AD - HNC BSc Development and application of methods for measurement of phagocytosis; Analysis of a toxic virulence factor of *Pasteurella haemolytica*; Measurements of phagocytosis and opsonins in the blood and lungs of healthy and infected sheep

23 Williams JT - HNC,LIBiol Anaesthesia and pulmonary lavage by bronchoscopy

4170 Biochemistry

1 Harkins DC - HNC Application of HPLC techniques to the determination of vitamins and related metabolites for diagnosis

2 Hodgson JC - BSc PhD Effects of slow-release anthelmintics on rumen fermentation. Tracer measurements of protein and energy metabolism during infection

3 Jones DG - BSc PhD Interactions between mineral and vitamin deficiencies, immunity and infection in sheep and cattle; Diagnostic relevance of biochemical and immunological changes in ovine gastro-intestinal parasitism; Develop new biochemical diagnostic methods and improve existing ones

4 Knox DP - HNC,MIBiol Isolation and characterisation of nematode secretory products

5 Mellor DJ - BSc,PhD Hormonal effects on host-parasite interactions

6 Slater J - BSc,PhD Separative methods for studying the biochemical responses to infection

7 Slater JS - BSc PhD Improving analytical techniques for the detection of vitamin B12 and E deficiencies in sheep and cattle; Amino acid requirements of the sheep foetus for metabolism and growth

8 Smith BSW - BSc,PhD Effects of dietary calcium, phosphorus, magnesium and vitamin D on calcium homeostasis in ruminants

9 Suttle NF - BSc PhD Sources of variation in availability of dietary copper, selenium and cobalt to sheep and cattle; Improving the diagnosis and treatment of copper, cobalt and selenium disorders in sheep and cattle; Effect of diet and susceptibility of sheep and cattle to infections

10 Wright H - BSc,MIBiol Cloning genes of the ovine major histocompatibility complex (MHC)

4171 Pathology/Immunology

1 Blewett DA - BSc,MSc,PhD Transmission and control of *Toxoplasma gondii*. *In vitro* techniques for studying *Cryptosporidium* and their application to methods of control. Response of sheep to infestations with *Ixodes ricinus*

2 Buxton D - MRCVS BVM&S PhD Pathogenesis, immunology and control of toxoplasma abortion in sheep; Aetiology pathogenesis and epidemiology of malignant catarrhal fever a fatal lymphoproliferative disorder of cattle; Epidemiology, Pathology and Eradication of Scrapie; Pathogenesis immunology and epidemiology of chlamydial abortion in sheep; Neurological disease in farm livestock

3 Dawson A - HNC The role of mucus in protection at mucous surfaces

4 Huntley JF - HNC,MSc,MIBiol Local immune mechanisms in the gastrointestinal mucosa

5 Jenkinson DME - PhD DSc Host responses and cutaneous barrier mechanisms to microorganisms

6 Miller HRP - PhD,BVMS,MRCVS Mechanisms of mucosal resistance; the role of mucus and immediate hypersensitivity reactions. Immunopathogenesis of listerial infections

4172 Parasitology

1 Coop RL - BSc PhD Epidemiology of gastrointestinal nematode infections under alternate grazing management systems; Evaluation of serum pepsinogen concentration as a diagnostic aid in subclinical ostertagiasis

2 Haig DM - BSc,MSc,PhD *In vitro* immune regulatory mechanisms in parasitic gastroenteritis. Cellular immune mechanisms in infectious diseases of ruminants

3 Jackson F - BSc Epidemiology and control of Trichostrongylosis and gastrointestinal nematodosis under a variety of management systems; Immune-Regulatory mechanisms in the control of parasitic gastro-enteritis

4 Smith WD - MRCVS BVMS PhD Investigation of immunological methods for controlling ovine parasitic gastroenteritis

4173 Clinical Research

1 Cuthbertson JC - DVSM DIPBiol MRCVS BVM&S MPVM Collection of sheep disease statistics; Monitor incidence and severity of footrot in naturally infected flocks; Field information on naturally occurring sheep pneumonia

2 Hay L - BVM&S,MRCVS Examination of ram semen quality

3 Henderson DC - MRCVS,BVMS,NCA Evaluation of ram examination by test mating synchronised and naturally-mated flocks; Pattern of Prevalence and Spread of Scrapie in affected flocks

4 McLauchlan M - BSc Computerisation of sheep health schemes

5 Spence JA - BVMS,DTVM,FRCVS The application of research results to sheep health schemes. The pathogenesis, host immune responses and epidemiology of broken mouth. The analysis of survey and diagnostic data and the pathological assessment of disease status and production in ruminants

4174 AFRC Institute for Grassland and Animal Production - Hurley Station

Hurley, Maidenhead, Berkshire, SL6 4LR
Tel: (062882) 3631.

Improving production and utilisation of grass and other forage crops; the nutrition, physiology and production of farm animals.

1 Prescott Prof. JHD - BSc,PhD,FIBiol *Professor*

4175 Soils, Plant and Feed Chemistry

Hurley, Maidenhead, Berkshire, SL6 4LR
Tel: (062882) 3631.

1 Barnes RJ - LRSC Development of methods (near-infrared (NIR)) to predict nutritional quality

2 Braithwaite GD - BSc PhD To follow changes in chemical composition during wilting and storage of crops; Develop methods to predict nutritional quality of silage; Influence of crop, harvest and ensiling conditions on chemical composition of silage; Conservation of moist hays and whole crops with chemicals

3 Hartley RD - BSc PhD CChem FRSC DSc Association of non-carbohydrate constituents of forage cell walls in relation to quality - part with RRI; Changes in cell wall composition and digestibility of poor quality forages induced by chemical and microbial treatments; Chemical/physical limitations to intake, digestion and utilization of conserved and upgraded feeds - chemical aspects; Dietary fibre and the monogastric - physical and chemical aspects

4 Hopper MJ - BSc Factors in the root environment influencing absorption and distribution of nutrients in forage plants; Crop growth and composition in relation to root temperature and form of nitrogen supply; Influence of nutritional and environmental factors on organic and inorganic components of forage plants

5 Jarvis SC - BSc PhD MIBiol Supply, uptake and cycling of nutrients in relation to clover in permanent grassland; NMR studies of forms and distribution of nutrients in forage plants; Influence of nutritional and environmental factors on organic and inorganic components of forage plants; Examine effect of plant nutrient supply on clover performance in the field

6 Mason VC - BSc PhD Influence of crop, harvest and ensiling conditions on chemical composition of silage; Conservation of moist hays and whole crops with chemicals; Factors influencing efficiency of novel methods of upgrading forages; Effects of specific components of fresh, conserved and upgraded forages on ruminant intake, digestion and metabolism

7 Merry RJ - PhD Influence of crop, harvest and ensiling conditions on microbial composition of silage. Conservation of moist hays and whole crops with chemicals

8 Pain BF - BSc PhD Design, operation and performance of farm anaerobic digester systems for waste treatment and biogas production; Characterisation of livestock wastes; influence of management and housing; Optimise the fertiliser equivalent of livestock wastes; minimise adverse environmental effects; Composting of separated slurry solids for horticultural use and heat recovery; Microbial kinetics of rate limiting stages in anaerobic digestion of livestock wastes

9 Pell J - PhD,BSc Nutrient partitioning, amino acid and protein metabolism in growing animals

10 Ryden JC - BSc,MSc,PhD,CBiol,MIBiol Denitrification losses from grassland; factors affecting loss; Volatilisation of gaseous ammonia from swards; deposition/assimilation in crops in the field; Develop methods including models to study transformations and losses of nitrogen as they occur in the field; Inputs of nitrogen to and losses from grassland as affected by soil, climatic and management factors; Inter-relationships between N including fertiliser N, and carbon in soil-plant-animal systems; Strategies to reduce gaseous and leaching losses of nitrogen from grassland; Influence of management on plant nutrients in livestock wastes with special reference to nitrogen; Optimise the fertiliser equivalent of livestock wastes

11 Smith RH - BSc PhD Protein retention in relation to energy metabolism in growing cattle

12 Whitehead DC - BSc PhD CChem MRSC Availability and fractionation of soil nitrogen under grassland; methods for predicting the available supply; Assimilation of gaseous ammonia by the leaves of forage plants; The adequacy of soil and atmospheric sources of sulphur for grassland production; Inter-relationships between N in soil organic matter and that derived from fertiliser and excreta; Chemical characterisation of livestock wastes; Influence of livestock wastes on supply and cycling of plant nutrients; Influence of nutritional and environmental factors on organic and inorganic components of forage plants

13 Williams AP - LRSC,PhD,MIBiol,CBiol To follow changes in chemical composition during wilting and storage of crops; Influence of nutritional and environmental factors on organic components of forage plants; Application of continuous culture systems to predict the nutritional quality of feeds; Influence of crop, harvest and ensiling conditions on chemical composition of silage; Establish the importance of amino acids and peptides for microbial growth in the rumen; Conservation of moist hays and whole crops with chemicals; Protection of dietary protein against microbial degradation in the rumen; To establish the relative importance of amino acids and peptides for rumen motility and microbial growth

4176 Forage Plant Physiology
Hurley, Maidenhead, Berkshire, SL6 4LR
Tel: (062882) 3631.

1 Clements RO - BSc,PhD,FRES Devise and integrate control measures for pests of establishing grass; Investigate factors predisposing grass and clover seedlings to soil-borne disease; develop control measures; Assess damage caused by pests and diseases to temporary grass; devise control measures; Assess losses to grassland legumes caused by pests and diseases

2 Davidson IA - BSc Carbon and nitrogen economies of mixed swards; photosynthesis, nitrate uptake and nitrogen fixation. Carbon economy of simulated swards, growth analysis and data-processing

3 Gordon AJ - BSc PhD Biochemistry and physiology of carbon flux in forage plants; Biochemistry and physiology of carbon supply and utilisation in white clover following defoliation; Biochemistry and physiology of carbon supply to, and utilization in root nodules during nitrogen fixation

4 Lewis GC - HNC Assessment of losses to grassland legumes caused by pests and diseases; devising control methods for diseases of establishing grassland; investigating other seedling losses. The importance and potential of endophytic fungi in grassland in England and Wales

5 Minchin FR - BSc MSc PhD Experimental determination of symbiotic nitrogen fixation and carbohydrate utilisation; Physiological factors, symbiotic nitrogen fixation and carbohydrate utilisation in legumes; Techniques for the measurement of nitrogen fixation by legume crops; Physiology of symbiotic fixation and carbohydrate utilization in genetically manipulated forage legumes;Experimental observations relating to gaseous diffusion in nodules

6 Parsons AJ - BSc PhD Physiology of plant/animal interactions in production and utilisation of grassland: theory, models and practice; Application of the physiology of plant/animal interactions to production and utilisation of grassland

7 Robson MJ - BSc PhD MIBiol Determinants of yield; photosynthesis, respiration and the influence of environment, genotype and management; Carbon and nitrogen economies of mixed swards; effects of environment and management on yield and composition; Application of physiological principles to the grazed sward; carbon input, output use and loss

8 Ryle GJA - BSc PhD DSc MIBiol Carbon and energy fluxes during nitrogen fixation and nitrogen metabolism in forage plants; Defoliation, carbon supply and potential N2 fixation in white clover; Diurnal carbon and nitrogen fluxes through legume root nodules

9 Sheehy JE - BSc MSc PhD Gas exchange of shoots, photosynthesis and respiration; mathematical modelling of gaseous diffusion. Theory of two compartment field enclosures. The physics of techniques

10 Woledge J - BSc Effects of environment on photosynthetic potential of leaves; adaption and readaption, interaction of leaf and canopy; Photosynthetic adaption in grass/clover swards; effects of environment on growth, morphology and composition

4177 Permanent Grassland Division
Hurley, Maidenhead, Berkshire, SL6 5LR
Tel: (062882) 3631.

1 Garwood EA - BSc MSc MIBiol Excess soil water in relation to grassland production and utilization; effects of poaching and drainage; Impact of grazing animals and machinery on soil and sward, and subsequent productivity

2 Mayne CS - BAgrSci PhD Examination of factors influencing the efficiency of milk production from grazing dairy cows; Herbage intake and milk production by cows grazing swards used by sheep in winter; Examine factors determining behaviour and herbage intake by grazing dairy cows; Examination of the use of supplementary feeds with grazing dairy cows

3 Morrison J - BSc,MIBiol,CBiol Influence of management grazing and cutting on the performance of white clover swards; Examine effect of plant nutrient supply on clover performance in the field

4 Newton JE - MA,MSc The development of lamb production systems based on long-term grass/clover swards; To establish optimum procedures for winter grazing by sheep; The effects of sheep grazing in winter on herbage utilization, animal production and sward characteristics; Animal production aspects of auto-immunisation to increase the milk production and lamb growth of lowland sheep

5 Peel S - BSc Herbage production and utilization on dairy farms; Develop optimum strategies for beef production on long term grassland

6 Scholefield D - BSc,PhD The soil/plant/climatic factors involved in grassland poaching. Development of techniques to renovate swards which have been damaged by poaching or machinery

7 Sheldrick RD - BSc MA OBE Study of sward management options (grazing, cutting or spraying) to encourage spread of newly over-sown white clover; To investigate management procedures to raise the clover content of permanent swards with little clover; Integrate grazing management and agro-chemical inputs to ensure long-term persistence of white clover in sheep systems

8 Wilkins RJ - PhD,BSc,CBiol,MIBiol Effects of different N-fertiliser strategies on herbage and animal production. An assessment of the effects of fertiliser on productivity and botanical composition of permanent grassland on the Somerset Levels

4178 Ruminant Nutrition and Metabolism
Hurley, Maidenhead, Berkshire, SL6 5LR
Tel: (062882) 3631.

1 Beever DE - BSc PhD Energy and protein supply and metabolism;Manipulation of protein supply and deposition;Fibre digestion in the rumen; Ammonia uptake by rumen microorganisms; Protein deposition in relation to energy utilization; The effect of turnout to grass on nutrient supply to the dairy cow; Modelling of rumen function, nutrient supply and metabolism; ATP for microbial growth and maintenance; Ammonia metabolism in animals receiving fresh forage; Rare earth markers for studying digesta flow; Manipulation of protein supply and deposition; Energy and protein supply and metabolism

2 Gill EM - BSc PhD Nutrient utilization and oxidation; Mechanism of fibre digestion and ruminal passage rate; Assessment and refinement of empirical intake prediction methods; Mechanistic models of digestion and metabolism for growth, lactation and feed intake; Kinetics of digesta flow in silage-fed animals; Acetate oxidation and incorporation into triglyceride; Energy and protein metabolism

3 McAllan AB - BSc,PhD Fibre digestion in the rumen (dairy cows); dynamics of digestion. Factors influencing microbial synthesis; effect of ammonia on microbial growth

4 Siddons RC - BSc PhD Fibre digestion by rumen microorganisms; Uptake of nitrogen by rumen microorganisms; Effect of diet on rumen fermentation; Particle size and fibre digestion kinetics; Kinetics of ammonia metabolism in the rumen; Kinetics of digesta flow along the ruminant gastro-intestinal tract

5 Smith RW - BSc PhD Acetate utilisation in isolated tissues; Effect of nutrient supply and hormonal control on lipid metabolism; Determination of enzyme activities with respect to acetate utilisation

6 Sutton JD - BA DipAgr PhD Production of volatile fatty acids in relation to milk yield and milk fat; Development of mechanistic models of feed utilization for growth, lactation and the prediction of intake; Effect of type of conserved forage and carbohydrate and protein supplementation on nutrient partition; Relation between body condition at parturition and metabolism and intake post-parturition

7 Theodorou MK - BSc,PhD Growth of microbial communities on solid substrates and identification of rumen microorganisms by monoclonal antibodies; Effect of inhibitory and stimulatory components on fibre digestion in the rumen; Regulation and control of fibre digesting enzymes in the rumen; Factors affecting ammonia metabolism in the rumen; Microbiology of anaerobic digestion of livestock wastes; Chemical/physical limitations to intake, digestion and utilization of conserved and upgraded feeds - microbial aspects; Mechanism and control of the enzymic digestion of fibre; Importance of ammonia in microbial N metabolism

8 Thomson DJ - BSc PhD DipAgr MIBiol Manipulation of physical and metabolic control of food intake, particularly in early lactation; Prediction of volatile fatty acid production; Development of mechanistic models of feed utilization for growth, lactation and the prediction of intake; Effect of type of conserved forage and carbohydrate and protein supplementation on nutrient partition; Relation between body condition at parturition and metabolism and intake post-parturition

4179 Cattle and Sheep Production
Hurley, Maidenhead, Berkshire, SL6 5LR
Tel: (062882) 3631.

1 Baker RD - BSc MSc Measuring herbage and animal responses to nitrogen fertiliser on swards grazed by beef cattle; Optimise the efficiency of forage utilisation by spring calving cows on a whole lactation basis; Establish management guidelines for beef production from grass/clover swards; Improve the efficiency of nutrient use from conserved forages for the growth of lean tissue; Improve utilization efficiency of straw and whole crop cereals for growth by upgrading treatments and supplementation; The influence of diet, animal type and growth promoting agents on lean tissue production; Quantify the residual effects of previous nutrition on the growth of cattle

2 Penning PD - MIBiol,CBiol Develop techniques and measure intake, behaviour and energy expenditure at pasture; Measure effects of supplements on grazing behaviour intake and energy expenditure by sheep; Measure behaviour, intake and performance of sheep in grazing systems

3 Phipps RH - BSc MPhil PhD DipTrAgr The influence of feed characteristics and physiological state of the cow on the substitution of one feed for another; Examine the effect of hormone balance and physiological state on nutrient partitioning; Quantify the response to changes in nutrient supply as influenced by forage type and level and type of supplement; Evaluate systems of feeding and herd management and their effect on intake and performance; Investigate the use of high forage diets for milk production

4 Tetlow RM - MIBiol Improve utilization efficiency of straw and whole crop cereals for growth by upgrading treatments and supplementation; Assess new methods of conservation and their effect on the efficiency of lean meat production

5 Thomas C - BSc PhD The influence of feed characteristics and physiological state of the cow on the substitution of one feed for another; Examine the effect of hormone balance and physiological state on nutrient partitioning; Quantify the response to changes in nutrient supply as influenced by forage type and level and type of supplement; Evaluate systems of feeding and herd management and their effect on intake and performance; Investigate the use of high forage diets for milk production; Conducts studies on the intake of forage and the substitution of one feed for another; Improve the efficiency of nutrient use from conserved forages for the growth of lean tissue

6 Treacher TT - BSc PhD Devise and test long term strategies for supplementation; Devise feeding systems using forages and concentrates for pregnant and lactating ewes; Production aspects of endocrine manipulation; Develop improved grazing systems for lamb production

4180 Pig and Calf Nutrition
Hurley, Maidenhead, Berkshire, SL6 5LR
Tel: (062882) 3631.

1 Close WH - PhD The contribution of VFA to the energy metabolism of the pig; Factors influencing nutrient (feed) utilisation during pregnancy and lactation in the sow; Nutritional and metabolic factors influencing energy and protein metabolism in the pig

2 Heppell LMJ - BSc PhD Immune function and dietary allergenicity during the development of the digestive system and in early life;Effects of dietary fibre upon small intestinal function and development; Allergenicity of diet ingredients and its significance in relation to creep feeding and weaner diet composition; Assessment of hypersensitive responses to dietary antigens

3 Latymer EA - MSc PhD MIBiol Dietary fibre, hindgut function and volatile fatty acid production, absorption and metabolism in pigs

4 Lee PA - BSc PhD Factors influencing nutrient (feed) utilisation during pregnancy and lactation in the sow

5 Low AG - BTech PhD Neural and hormonal modification of gut function as determinants of nutrient absorption; Effects of dietary fibres on endogenous secretions and glucose absorption in stomach and small intestine of pigs; Dietary fibre and nutrient digestibility in the small and large intestines of pigs; Comparison of methods for evaluation of feedstuffs as predictors of nutrient deposition; Dietary fibre and the monogastric - nutritional aspects

6 Sissons JW - LRSC PhD Immune function and dietary allergenicity during the development of the digestive system and in early life; Role and nature of gastrointestinal motility in digestive function, digesta passage and nutrient utilisation; Manipulation of digestive and immune functions in relation to nutrient use

4181 Biomathematics
Hurley, Maidenhead, Berkshire, SL6 5LR
Tel: (062882) 3631.

1 Doyle CJ - BA MSc Grassland animal performance and productivity; Bioeconomic modelling of grassland-animal enterprises and their components

2 France J - BSc MSc DPhil Grazing; animal digestion, metabolism, growth and productivity; Modelling of animal intake, digestion, metabolism, growth and performance; rationing; grazing; Kinetics of digestion in, and passage from, the rumen; Development of prediction equations; Mechanistic modelling of digestion and metabolism in ruminants

3 Thornley JHM - BA MA PhD FInstP Modelling of forage crop growth and productivity: animal physiology and metabolism; Modelling of grassland-animal enterprises and their components

4182 Endocrinology and Animal Physiology
Hurley, Maidenhead, Berkshire, SL6 5LR
Tel: (062882) 3631.

1 Bloomfield GA - BSc PhD Use of hormonal and other biochemical measures to assess the effect of management factors on cattle reproduction; Identify pregnancy specific factors in bovine fluids and tissues by immunological and electrophoretic techniques; Evaluation of new techniques of hormone assay as they become available

2 Buttle HL - BVSC PhD MRCVS Hormonal control of lactation and reproduction in pigs

3 Forsyth IA - BA DPhil MA FIBiol Purification of ruminant placental lactogens, characterization of their physiochemical and biological activity; Hormonal regulation of macromolecular syntheses by mammary tissues *in vivo* and *in vitro*; In vitro studies of pig mammary glands

4 Glencross RG - BSc PhD Effects of immunoneutralization of androgens and inhibin on Gonadotrophin secretion and ovulation rate in dairy cattle

5 Johnsson ID - BRurSc,PhD Growth hormone and growth

6 Jones EA - BSc PhD Biochemistry of mammary gland

7 Mowlem A - FIAT,MIBiol,CBiol Devise improved feeding and management systems for goats

8 Perry BN - BSc PhD Molecular role of hormones in growth and development

9 Tindal JS - BSc PhD DSc Neural mechanisms which control the release of pituitary hormones essential for growth and lactation

4183 AFRC Institute for Animal Health
Pirbright, Woking, Surrey, GU24 0NF
Tel: 0483 232441.

Research on diseases of economic importance, exotic to the UK, which present a major threat to the livestock industry

1 Biggs PM - CBE,FRS *Director*
2 Scott Ms P - BSc,ALA *Information Officer*

4184 Biochemistry
Pirbright, Woking, Surrey, GU24 0NF
Tel: 0483 232441.

1 Black DN - BSc PhD Structure of African swine fever virus DNA; Restriction endonuclease analysis of poxvirus DNA's and cloning of specific fragments

2 Bostock CJ - BSc,PhD Directed reassortment in segmented RNA viral genomes. Mechanism of reassortment and replication of segmented RNA genomes. Structure and function of the African swine fever virus genome. Micro-injection of RNA transcripts of foot-and-mouth disease (FMD) virus produced from cDNA clones into eukaryotic cells.

3 Burroughs JN - BSc Purification and characterisation of bluetongue virus proteins

4 Dixon L - BSc PhD Structure and function of the African swine fever virus genome

5 Gershon P - BSc,PhD Restriction endonuclease mapping of capripoxvirus genome

6 Mertens PPC - BSc DPhil Reassortment, translation, sequencing and purification of bluetongue virus; Cloning and expression of bluetongue virus genome RNA segments

7 Smale CJ - ONC BA Electron microscopic examination of material containing virus

8 Wade-Evans A - BSc Cloning and expression of bluetongue virus genome RNA segments

4185 Genetics
Pirbright, Woking, Surrey, GU24 0NF
Tel: 0483 232441.

1 Belsham G - BSc PhD Expression of cloned foot-and-mouth disease virus genes in eukaryotic cells

2 King AMQ - BA PhD Isoelectric focusing of foot-and-mouth disease virus proteins; Fundamental studies on RNA recombination

3 McCahon D - BSc PhD Cloning and expression of foot-and-mouth disease virus genes; Foot-and-mouth disease virus genetics; Isolation and characterisation of foot-and-mouth disease recombinant viruses

4 Newman JWI Growth and characterisation of foot-and-mouth disease virus recombinants

5 Pedley C - BSc,PhD Cloning and assembly of a complete DNA copy of the FMD virus genome

6 Tiley L - BSc Biochemical studies on complementation in FMD virus

4186 Director's Group
Pirbright, Woking, Surrey, GU24 0NF
Tel: 0483 232441.

1 Barrett T - BSc,MSc,PhD Structure and function of the genomes of morbilliviruses

2 Durrant I - BSc Structure and function of the genome of classical swine fever virus

3 Evans S - BSc Structure of the canine distemper virus genome

4 Mahy BWJ - BSc PhD MA ScD Molecular basis of influenza virus virulence

5 McCauley JW - BSc PhD Structure and function of the influenza virus genome

6 Penn CR - BA MA PhD Genetics and replication of influenza viruses; Structure and function of the genome of classical swine fever virus

4187 Epidemiology
Pirbright, Woking, Surrey, GU24 0NF
Tel: 0483 232441.

1 Denison EP Maintenance of insect colonies for studying exotic virus infections

2 Jennings DM - BSc PhD Role of midges in bluetongue virus transmission

3 Mellor PS - BSc MSc PhD Studies of Ornithodoros spp; Role of midges in bluetongue virus transmission

4 Obi T Epidemiology of peste des petits ruminants; Production of monoclonal antibodies to peste des petits ruminants virus
5 Rowe LW - FIMLT Vaccine manufacturing techniques
6 Taylor WP - BVMS,MRCVS Epidemiology of rinderpest and peste des petits ruminants; Epidemiology and pathogenesis of bluetongue infection

4188 Virus Diagnosis
Pirbright, Woking, Surrey, GU24 ONF
Tel: 0483 232441.

1 Anderson J - HNC LIBIOL Development of monoclonal antibodies for detection and identification of viruses
2 Carpenter WC - ONC HNC Improved immunoassays for exotic viruses
3 Crowther JR - BSc PhD Development of specialised immunoassays for virus antigens
4 Hamblin C - HNC LIBiol MIBiol Development of the ELISA test for identification of vesicular viruses
5 Pedley S - BSc PhD Development of nucleic acid hybridisation methods for detection of exotic viruses

4189 World Reference Laboratory
Pirbright, Woking, Surrey, GU24 ONF
Tel: 0483 232441.

1 Armstrong RM - HNC MIBIOL Typing and subtyping of virus samples received by the World Reference Laboratory; Serological testing of animal sera for foot-and-mouth disease antibodies
2 Barnett ITR - FIMLS Serological testing of animal sera for foot-and-mouth disease antibodies
3 Donaldson AI - MVB MA PhD MRCVS ScD Typing and subtyping of virus samples received by the World Reference Laboratory; Investigation of disease outbreaks: origin of infection and prediction of spread; Airborne spread of vesicular viruses; Initiation of Aujeszky's virus infection in pigs and cattle
4 Ferris NP - BA MSc Typing and subtyping of virus samples received by the World Reference Laboratory
5 Kitching RP - BSc,MRCVS,BVetMed,MSc,PhD Diagnosis, epidemiology and control of sheep and goat pox virus infection
6 Knowles NJ Typing of virus samples received by the World Reference Laboratory; RNA and protein characterisation of field isolates of FMDV
7 Mowat GN - PhD MRCVS Investigation of disease outbreaks: origin of infection and prediction of spread; Establishment of an international foot-and-mouth disease vaccine bank

4190 Experimental Pathology
Pirbright, Woking, Surrey, GU24 ONF
Tel: 0483 232441.

1 Anderson EC - BVMS,PhD,MRCVS The pathophysiology of african swine fever; Pathogenesis of foot-and-mouth disease in farm animals
2 Wilkinson PJ - MVB MRCVS BA PhD The epidemiology of african swine fever; Mechanisms of African swine fever virus persistence and pathogenesis; Persistence and epidemiology of African swine fever

4191 Immunology
Pirbright, Woking, Surrey, GU24 ONF
Tel: 0483 232441.

1 Denyer M - ONC HNC MIBiol Cell-mediated immune responses; Immune response mechanisms in African swine fever
2 Jeggo MH - BVetMed MSc MRCVS DPhil Cell-mediated immunity to bluetongue and African swine fever viruses
3 McCullough KC - BSc PhD Production, characterisation and application of monoclonal antibodies
4 Parkinson D - ONC HNC AIBiol Production and characterisation of monoclonal antibodies
5 Takamatsu H - DVM,MSc,PhD Anti-viral T cell cloning in domestic animals
6 Wardley RC - PhD BVSC Cell-mediated immune mechanisms; Immune response mechanisms in African swine fever; Immune response to equine herpes virus infection

4192 Vaccine Research
Pirbright, Woking, Surrey, GU24 ONF
Tel: 0483 232441.

1 Collen T *In vitro* studies of the immune response to viral antigens
2 Doel TR - BTech PhD Bulk production of virus in microcarrier cultures; Immune response to peptide and whole virus vaccines; Establishment of an international foot-and-mouth disease vaccine bank
3 Staple RF - FIMLS Establishment and maintenance of an international FMD vaccine bank

4193 Disease Security
Pirbright, Woking, Surrey, GU24 ONF
Tel: 0483 232441.

1 Goodridge D - HNC Monitoring filter systems to prevent cross-infection; Study contamination of people and the environment; Investigate effectiveness of various sterilisation procedures
2 Mann JA - BVSC,MRCVS,PhD Monitoring filter systems to prevent cross-infection; Study contamination of people and the environment; Investigate effectiveness of various sterilisation procedures

4194 AFRC Institute of Horticultural Research - East Malling
East Malling, Maidstone, Kent, ME19 6BJ
Tel: 0732 843833.

Research on the productivity, quality and market acceptability of temperate top and soft fruit crops, hops and hardy ornamental nursery stock, including breeding and the control of pests and diseases

1 Swinburne Prof TR *Director*
2 Payne Dr CC *Head of Station*

4195 Zoology
East Malling, Maidstone, Kent, ME19 6BJ
Tel: 0732 843833.

1 Campbell CAM - BSc PhD Evaluate pest management programmes and investigate components for integration into programmes; Devise and develop improved strategies to control damson hop aphid
2 Carter KJ - HNC,CChem,MRSC Gas chromatography-mass spectrometry (GC-MS) and other support work related to metabolic profiling
3 Easterbrook MA - BSc DIC MSc Evaluate pest management programmes and investigate components for integration into programmes; Establish threshold levels for pests for use in pest management strategies; Evaluate new or recently-released pesticides, biotic agents and other orchard chemicals; Identify insect pests and develop chemical, biological and integrated control measures
4 Flegg JJM - BSc PhD ARCS FZS Assess annually bird damage levels in sample apple and pear orchards and determine causative species; Interaction of nematodes and virus and *Verticillium* wilt in hop; Identify nematode pests, assess damage, and develop necessary control measures; Relationship of *Xiphinema* and *Longidorus* to environment and host roots
5 Solomon MG - BSc DIC MSc PhD Evaluate pest management programmes and investigate components for integration into programmes; Establish threshold levels for pests for use in pest management strategies; Evaluate new or recently-released pesticides, biotic agents and other orchard chemicals; Develop metabolic profiling of plants to detect and predict resistance to pests and pathogens

4196 Fruit Storage
East Malling, Maidstone, Kent, ME19 6BJ
Tel: 0732 843833.

1 Dover CJ - BS AIPhys The effects of cooler operations on weight loss in fruit. Developing improved ethylene-removal systems for use in fruit stores
2 Jameson J - HND Study cooler design and weight loss in fruit: develop improved carbon dioxide scrubbing system
3 Johnson DS - BSc The effect of controlled atmosphere storage on bitter pit in apples; Effects of cultural practices on quality and susceptibility to disorders of apples stored under different conditions; Effects of climatic factors on ripening and storage behaviour of pome fruit and prediction of potential storage quality; Use of orchard sprays and post harvest chemicals to control storage disorders of apples
4 Knee M - BSc PhD Investigate control of initiation of ripening in apples and pears by regulation of ethylene biosynthesis; Investigate the relationship between store humidity and apple fruit texture
5 Perring MA - MIBiol The relationship between chemical composition and fruit textural properties; Fruit composition in relation to texture
6 Sharples RO - BSc PhD Advisory and trouble shooting service to commercial fruit growers; Evaluate effects of modern production systems on texture and flavour of stored Cox apples
7 Smith SM - BSc PhD Establish relationships between maturity attributes and fruit quality and develop methods for predicting harvest dates; Techniques for improving the shelf-life of apples; Development of instrumental methods of assessing fruit quality
8 Stow JR - BSc MSc PhD Improvement of storage conditions for apples
9 Topping AJ - BSc Producing standard pre-cooling and storage recommendations for soft fruits

4197 Plant Pathology
East Malling, Maidstone, Kent, ME19 6BJ
Tel: 0732 843833.

1 Adams AN - BSc MSc PhD Develop diagnostic methods; utilise to elucidate virus/mycoplasma epidemiology and produce virus-free stock; Elucidate aetiology/epidemiology of little cherry disease, subject to improvement of diagnostic method; Produce and maintain virus-free clones of new and established hop cultivars; Develop methods to detect and identify strawberry viruses; produce and maintain virus-free stock; Screen plum cultivars for resistance to plum pox virus; produce and maintain healthy stocks; Assess the effects of viruses on hop yield and quality; Appraise urgent problems (viral diseases) and initiate action as required; Appraise urgent problems and initiate action as required
2 Barbara DJ - BSc PhD Develop improved diagnostic methods for cherry diseases; Develop immunoassay procedures; Develop complementary DNA hybridization assays; Virus testing and production of virus free clones; development of broad spectrum tests for viruses; Maintain RBDV diagnostic facility; develop tests for RBDV resistance; produce RBDV-free stock; Appraise urgent problems; monitor quarantine operations etc to assess virus status of plants; Develop ds-RNA detection, and finger-printing

3 Butt DJ - BSc DPP FIBiol Assess the status of post-harvest diseases in untreated fruits; develop alternative strategies of control; Develop disease management strategies utilising epidemiological and meteorological information; Evaluate efficiency and cost-effect of fungicides and application methods for orchard disease management; Evaluate fungicides and develop management strategy for control of apple canker;Elucidate epidemiology of silver leaf in nurseries; evaluate methods of diagnosis and control; Appraise urgent problems (airborne diseases) and initiate action as required

4 Carder JH - HND,MIBiol Improving methods for the rapid detection, quantification and virulence assessment of *Verticillium albo-atrum*. Identifying the metabolic effects on pathogens. Stimulants of appressorial germination and elicitors of phytoalexins. The use of carbohydrates as control agents for fireblight (*Erwinia amylovora*)

5 Chambers DA - BSc *Verticillium* wilt resistance in breeding stocks and new cultivars. Conventional pathogenicity testing with *V. albo-atrum*. Screening chemical and biological agents or agencies for control of hop wilt. Recording plant accessions, quarantine requirements and advisory enquiries

6 Clark MF - BSc MSc PhD Devise and develop diagnostic methods for detecting/identifying viruses and MLOS in woody plants; Develop immunoassay procedures; Develop complementary DNA hybridization assays; Develop ds-RNA detection, and finger-printing

7 Garrett CME - MSc PhD MIBiol Isolate root-surface-inhibiting bacteria, evaluate as biological control agents against soil-borne pathogens; Evaluate bactericides for fireblight control; assess INA as a factor in predisposition to fireblight in pears; Appraise urgent problems (bacterial diseases) and initiate action as required

8 Harris DC - BA MA PhD Elucidate aetiology and epidemiology of crown/collar rot of apple; develop control methods; Evaluate potential biological control agents (bacteria, fungi) against soil-borne pathogens; Elucidate epidemiology of fungal diseases; develop chemical, biological and integrated methods of control; Appraise urgent problems (soil-borne diseases) and initiate action as required; Appraise urgent problems and initiate action as required

9 Sewell GWF - BSc PhD Evaluate the aetiology and epidemiology of apple replant disease; develop methods of control; Elucidate the aetiology of replant diseases in HONS; develop control measures; Assess the significance of replant diseases in strawberry; develop control measures

10 Stickels JE The production, monitoring and maintenance of stocks free from viruses, nematodes and fungal pathogens, with particular reference to virus-free, MLO-free material of cultivars and rootstock for EMLA scheme and virus-free stocks of plum cultivars and rootstocks

4198 Plant Physiology
East Malling, Maidstone, Kent, ME19 6BJ
Tel: 0732 843833.

1 James DJ - BSc PhD Physiological and biochemical mechanisms controlling the *in vitro* regeneration of plants from single cells and tissues; Genetic manipulation of fruit tree rootstocks; Protoplast systems for strawberry somaclone generation; Genetic modification of rootstock

2 Jones HG - MA PhD Investigate water stress and its control; Development of physiological screening tests; Development of immunoassay and immunolocalisation methods for plant hormones, especially ABA; Determine control of water and nutrient especially calcium, movement in fruit trees; Mathematical modelling of plant growth

3 Jones OP - BSc PhD Propagation *in vitro* and regeneration of plants from callus of temperate fruit trees and strawberry; Mechanism of rootstock/scion interaction in relation to control of growth and cropping; Production of somaclonal variants of strawberry cultivars; Genetic manipulation of fruit tree rootstocks; Studies on complex explant systems for strawberry somaclone generation and *in vitro* selection procedures; Genetic modification of rootstock and influence on physiological characteristics

4 MacKenzie KAD - BSc PhD Anatomical and cytological investigations relating to plants produced by *in vitro* techniques

5 Samuelson TJ - BSc PhD Chemical analysis of fruits, leaves and other tissues in relation to uptake and movement studies

6 Venis MA - BSc DPhil Properties of plant hormone receptors; immunochemical assays for plant hormones; Regulation of soft fruit development; Biochemistry of metabolic processes controlling and controlled by cellular calcium

4199 Plant Protective Chemistry
East Malling, Maidstone, Kent, ME19 6BJ
Tel: 0732 843833.

1 Austin DJ - BSc PhD CChem FRSC Chemical analysis of plant tissues and soil in studies of the uptake and metabolism of plant growth regulators; Isolation and identification of natural and synthetic chemicals with root-promoting activity in hardwood cuttings; Chemical analysis of plant tissues after spray application of fungicides in studies of deposition and efficacy; Develop metabolite profiling of plants to detect and predict resistance to pests and pathogens

2 Blake PS - BA Isolation, identification and synthesis of chemicals with root-promoting activity in hardwood cuttings

3 Harrison-Murray RS - BSc,PhD Physiological and environmental studies on rooting and survival of leafy and leafless cuttings

4 Murray RA - BSc PhD Assess application methods and efficacy of chloropicrin and formaldehyde to control replant diseases; Differentiation by electrophoresis of nematodes in strawberry tissue to provide effective assay of pathogenic species; Identify resistance of damson-hop aphid to OP insecticides; Differentiation, by electrophoresis of *Longidorus* spp. nematode species, including known virus vectors

5 Warman TM - BSc Chemical analysis of plant tissues and soil in studies on the uptake and metabolism of plant growth regulators. Chemical analysis of plant tissues after spray application of fungicides in studies on deposition and efficacy. Provision of analytical support for work on novel control agents

4200 Pomology
East Malling, Maidstone, Kent, ME19 6BJ
Tel: 0732 843833.

1 Alston FH - BSc PhD Breed apple and pear scion varieties;Breed pear rootstocks

2 Beech MG - MIBiol Development and evaluation of strawberry production systems; Breeding and genetic manipulation - Fragaria cytology; Production of useful somaclonal variants of strawberry cultivars; Selection criteria and evaluation procedures for strawberry somaclones

3 Browning G - BSc PhD Evaluate new rootstocks for effects on growth and cropping of pear scions; Improvement of pear planting systems by studies on effects of planting density, arrangement and growth retardation; Improvement of fruit setting in pears by study of its hormonal control and new PGR spray technologies; Develop immunological procedures for hormones controlling fruit tree growth and cropping; Hormonal relationships in fruit tree flowering and fruiting; Control of epicormic branching of oak; Genetic manipulation of fruit tree rootstocks; Genetic modification of rootstocks and *in vitro* selection procedures

4 Crisp CM - BA Mycorrhizal infection and nutrition of fruit trees

5 Hamer PJC - BSc PhD Frost protection of blackcurrants; Devise physical methods to combat adverse effects of high early spring temperatures and frosts

6 Holloway RIC - BSc Improvement of cane and bush fruit systems, mechanisation of dessert quality raspberry harvesting

7 Jackson JE - BSc PhD FIBiol Computer models to guide research, orchard design for better fruit quality, improved rooting to reduce staking costs; Investigating environmental effects on fruit set and on year to year variation on yields; Mathematical modelling of crop-environment interactions; Appraise urgent problems and initiate action in consultation with ADAS as required; Temperature effects on apple fruit bud development, fruit set and yield

8 Jones R - LIBiol Breed plum scions and rootstocks

9 Knight VH - BSc MSc Breed *Rubus* - especially primocane types; Completion of transfer of *Ribes* to SCRI

10 Palmer JW - BSc MSc PhD Improve fruit quality through changes in planting density, pruning, plant growth regulators and thinning; Developing and testing new and existing fruit orchard production systems using field trials and computer modelling; Mathematical modelling of fruit tree growth and cropping

11 Quinlan JD - BSc PhD Control tree growth by growth retardants and branching agents and integration of these into orchard systems; Develop plant growth regulator treatments for woody ornamentals

12 Simpson DW - BSc PhD Breeding and genetic manipulation - *Fragaria* gene identification; Breed *Fragaria* for a range of seasons and outlets

13 Sparks TR - HNC Evaluate induced clones of apple and pear scions and rootstocks

14 Tobutt KR - BA Breeding better quality ornamental cultivars; Breed apple rootstocks; Breeding and genetic manipulation - ornamentals, Prunus and Malus; Breeding cherry scions and rootstocks

15 Webster AD - NDH BSc Evaluate induced clones of apple scions and rootstocks; Pomological evaluation of new apple varieties; Evaluate new rootstocks for effects on growth and cropping of apple scion; Develop, select and produce clones with best available or improved horticultural and/or health characteristics; Evaluation of new orchard systems for apples especially using trees raised on their own roots or by other low-cost ways; Control of apple tree growth and cropping using plant growth regulators; Improvement of stone fruit productivity and production systems

16 White GC - BSc Studies of effects of soil and tree management on fruit quality assessed by commercial standards; Provide data on tree performance and economic assessment to determine systems for particular grower/site investment; Soil management systems to combine orchard productivity with maintenance of soil fertility; Soil problems of woody ornamentals and development of soil management systems

4201 Statistics
East Malling, Maidstone, Kent, ME19 6BJ
Tel: 0732 843833.

1 Preece DA - MA,PhD,FIS,FIMA Increasing biometric understanding relating to experimentation on perennials and hardy ornamentals

2 Tamsett J - HNC,AIStat Design and implementation of robust computing systems for control and data handling

3 Thomas CMS - BSc Effects of herbicide and grass/soil management on water use, crop yield and fruit quality. Root growth pattern and nutrient and water uptake. Effects of soil water application and mist irrigation on tree water stress and productivity

4202 Plant Propagation
East Malling, Maidstone, Kent, ME19 6BJ
Tel: 0732 843833.

1 Howard BH - BSc PhD Develop clonal rootstocks from seedling populations and select scionwood sources for improved budtake; Improve vegetative propagation by cuttings and budding; Developmental physiology of vegetative propagation by cuttings and of nursery tree production; Effects of *in vitro* propagation on nursery performance of derived material
2 Marks TR - BSc,PhD Develop methods to micropropagate woody species and increase understanding of physiological processes

4203 AFRC Institute of Food Research - Bristol Laboratory
Langford, Bristol, BS18 7DY
Tel: 0934 852661.

Conversion of meat into fresh and processed products and refrigeration of all foods

1 Bailey Prof. AJ - PhD,ScD,FRSC *Professor*
2 Hutchinson Miss SA - BA,ALA,MIInfSc *Librarian*

4204 Animal Physiology
Langford, Bristol, BS18 7DY
Tel: 0934 852661.

1 Brown AJ - HNC,MInstM Carcass composition and value in cattle and sheep
2 Fisher AV - BSc Techniques for assessment of composition in live animals and carcasses; Investigation of techniques to estimate carcass yields and retail value; Measurement of physico-chemical properties of muscles to develop strategies for fresh meat utilisation
3 Slader RW - HNC Development of computerised data captive systems and software for use in abattoirs
4 Warriss PD - BSc CertEd Identify good preslaughter handling practices to improve meat quality
5 Wood JD - BSc PhD Carcass composition and meat quality in sheep; Carcass composition and meat quality in pigs; Rapid methods of carcass classification

4205 Bioengineering and Development
Langford, Bristol, BS18 7DY
Tel: 0934 852661.

1 Allen RA - BSc,MSc Determining the magnitude of effect of electrical stimulation, conditioning and frozen storage on shelf-life. The effect of mechanically recovered meat on product shelf-life. Mathematical models of product quality to optimise process
2 Burfoot D - BSc PhD Initiate and control research programme on meat process engineering
3 Gigiel AJ - BSc DipEd MSc MCIBS MInstR Developing programme on energy usage and optimisation during production of meat and meat products
4 Grey TC - HNC Factors affecting carcass composition and meat quality in poultry
5 James SJ - BA MInstR Initiate and manage food refrigeration research investigations; Initiate and manage research into meat cooking; Instigating modelling methodology for heat and mass transfer in meat materials
6 Jolley PD - BSc Limiting factors in reforming meat; Processing factors influencing protein solubilisation; Establish biochemical and structural factors affecting adhesion in industrial practice
7 MacDougall DB - ARCST MSc PhD Effects of electrical stimulation, hot-boning, conditioning and frozen storage on meat product quality; Processing properties of meat from different species and mechanically recovered meat products; Relate processing conditions and raw material properties to product quality
8 Malton R - MInstR Survey of industrial performance and technology transfer in the manufacture of meat and meat products
9 Miles CA - BSc PhD MInstP CPhys Investigate physical methods of measuring carcass composition; Improve understanding of the thermophysical properties of meat; Investigate physical properties of meat related to meat quality and composition
10 Newman MJ - BSc,FIFST Programme organization and management in industrial development and technology transfer
11 Newman PB - BSc,MIBiol Development of video image analysis (VIA) to grade carcasses, predict meat yield, assess fat/lean and provide sight for robotic butchery. Relating the functionality of mechanically recovered meat to the recovery process
12 Richardson RI - BSc,PhD Factors affecting functional properties of turkey meat
13 Sheard PR - BSc,PhD Processing factors and restructured meats. Processing factors influencing particle size and aspect ratio
14 Swain MJ - HNC Feasibility study and development of robotic systems for abattoirs and processing plants
15 Taylor AA - BSc SM FIFST Application of electrical stimulation and hot boning for more efficient post-slaughter handling of meat; Improve storage stability and display appearance of chilled and frozen meat; Application of video image analysis in carcase grading and fat:lean assessment in meat and meat products

4206 Microbiology
Langford, Bristol, BS18 7DY
Tel: 0934 852661.

1 Dainty RH - BSc PhD Determine chemical identity of cell components and products; Investigate microbial metabolism during chill storage of meat and its effects on sensory quality
2 Gibson AM - BSc Bacteriology of *Clostridium botulinum* including toxin production and control by preservatives
3 Impey CS - HNC Analysis of intestinal microflora; testing bacterial mixtures for ability to prevent *Salmonella* colonisation in chicks
4 Mackey BM - BTech PhD Effects of physical and chemical damage on bacteria and their recovery and growth
5 Mead GC - BSc,PhD Development of bacterial mixtures from the gut of adult birds which prevent *Salmonella* colonisation in chicks. The incidence, origin and significance of microorganisms associated with the processing of poultry meat
6 Roberts TA - PhD,BSc Access bacteriology of hygiene and develop criteria, standards, specifications; Develop predictive models for microbial growth and toxin production; Food safety by predicting microbial growth
7 Shaw BG - BSc PhD Classification and identification of bacteria on meat and meat products

4207 Muscle Biology
Langford, Bristol, BS18 7DY
Tel: 0934 852661.

1 Barnard K - BSc,PhD Development of HPLC methods and their application to collagen biochemistry
2 Brock CJ - BA,MA,PhD Structure-function relationships of proteins as emulsifiers
3 Cheah KS - BSc MSc PhD Biochemistry of porcine stress syndromes
4 Cheak AM - HNC,MIBiol Biochemistry of porcine stress syndrome
5 Duance VC - BSc,PhD Muscle connective tissue development using recombinant DNA probes. Biochemical and structural studies on cartilage
6 Enser MB - MA DPhil The composition, consistency and cellular development of adipose tissue; Formation and structure of meat protein-stabilised emulsions
7 Etherington DJ - BSc PhD Investigate the mechanisms of proteolytic weakening in the structural elements of meat during conditioning
8 Etherrington DJ - BSc,PhD,FRSC Mode of action of tissue proteinases in the degradation of collagens
9 Jeacocke RE - BA PhD Metabolic changes in muscle post-mortem and their consequences
10 Kent MJC - BSc,DPhil Isolation and structure of glucose-mediated crosslinks in collagen
11 Knight PJ - BSc PhD Formation of muscle filaments during development and growth; X-ray diffraction analysis of myofibril dimensions post-slaughter and during meat processing; Mechanisms of swelling and shrinking of myofibrils and relation to protein solubilization; Mechanism of response of muscle fibres to salt treatments
12 Maciewicz RA - BA,MA,PhD Synthesis and properties of collagen - degrading cathepsins
13 Offer GW - BA PhD Structure and biochemistry of myofibrils; Models for the mechanism of drip and cooking loss; Structure of myosin gels. Structural changes in salt treated meat
14 Purslow PP - BSc,PhD Investigation of the mechanical properties of meat. Basic mechanical techniques for measuring cohesive and adhesive properties of meat particle-gel systems
15 Seymour JVS - BSc PhD Structural basis of meat texture by analysis of endomysium and perimysium by microscopy and image processing
16 Trinick JA - BSc PhD Structure and disposition of cross-bridges and their role in muscle contraction; Structure and composition of protein filaments in myofibrils
17 Wotton SF - BSc,MIBiol Biochemical and structural studies on cartilage

4208 Food Chemistry
Langford, Bristol, BS18 7DY
Tel: 0934 852661.

1 Dransfield E - BSc PhD Develop common sensory scales and instrumental methods for meat and meat products; Reduce variability in eating quality and optimise tenderness
2 Jones SJ - BSc Meat species analysis techniques
3 MacFie HJH - BSc PhD FRSS Define consumer acceptance of meat and meat products
4 Mottram DS - BSc PhD Meat flavour and its precursors
5 Patterson RLS - BSc PhD FRSC Immuno-suppression of boar taint; Processing problems of boars as meat animals; Taints in meat and meat products
6 Salter LJ - BSc,MSc Effect of lipid on flavour-forming reactions between amino acids and sugars

4209 AFRC Institute of Food Research - Norwich Laboratory
Colney Lane, Norwich, NR4 7UA
Tel: 0603 56122.

Quality of food based on plant products and processed primary products

1 Richmond Prof P *Head of Laboratory*
2 Taylor Ms CA - BA *Librarian*

4210 Chemistry and Biochemistry
Colney Lane, Norwich, NR4 7UA
Tel: 0603 56122.

1 Belton PS - BSc PhD Interaction of water with biopolymers; quantitative use of photoacoustic spectroscopy to assess moisture content;Effects of molecular structure and molecular interaction on mechanical properties of polysaccharide and protein gels; Development and implementation of novel nuclear magnetic resonance techniques for foods; Application of nuclear magnetic resonance methods to the non-invasive study of plant tissue

2 Carr HJ - BSc Prediction of secondary structure of legume storage proteins

3 Chan HWS - BA MA PhD Immunoassay of food constituents and contaminants, cells and cell products

4 Coxon DT - BSc PhD Establish oxidation reactions and products of lipids, means of control and action of antioxidants

5 Fenwick GR - BSc PhD FRSC FIST Isolation, quantification and investigation of indole glucosinolates in brassica plants and seeds; Toxicants in foods, feeding stuffs and processed products; Analysis of glucosinolates and degradation products in brassica oilseeds, vegetables and fodder crops; Methods for isolation, characterization and estimation of tannins, saponins and oestrogens in food plants

6 Kemp HA - MSc,PhD Production of monoclonal antibodies as specific immuno-affinity probes; novel immunoassay methods

7 Lambert N - BSc DPhil Structural characterisation and modification of plant proteins for functional purposes

8 Miles MJ - MSc PhD MInstP Small-angle x-ray scattering and wide-angle x-ray diffraction to determine structure-function of food biopolymers; Use of conventional and synchrotron x-ray techniques to study structure-function relationships of food biopolymers

9 Morgan MRA - MSc PhD Immunoassay of food constituents and contaminants, cells and cell products;Development of immunoassays for trichothecene mytotoxins in food; Production of monoclonal antibodies; immuno-affinity techniques; novel immunoassay methodology; Immunoassay for determination of D and B6 group in food

10 Morris VJ - BSc MSc PhD CChem MInstP Determination of the structural, macromolecular and processing factors affecting the rheology of polysaccharide gels; Effects of molecular structure and molecular interaction on mechanical properties of polysaccharide and protein gels; Use physical techniques to determine molecular properties of food proteins and relate to process of gelation

11 Rhodes MJC - BSc PhD Use of plant cell cultures to produce food colours and flavours; development of stable cell lines of high productivity; Develop immobilised plant cell reactors to produce food flavours; study membrane properties related to product release; Selection of plant cell lines with stable high rates of secondary metabolite production and secretion

12 Ring SG - BSc PhD Relate rheological viscoelastic and phase separation properties of pastes and gels of food starches to composition; Effect of other food components, lipids and polymers, on the mechanical and phase behaviour of starch-based composites

13 Robins RJ - MA DPhil Improve the synthesis of useful metabolites in plant cell cultures; develop analytical methods for products; Selection of plant cell lines with stable high rates of secondary metabolite production and secretion

14 Selvendran RR - BSc PhD FRSC Composition of plant cell walls and dietary fibre and their degradation by human colon bacteria; Characterization of cell wall and extracellular polymers of food plants and plant cell cultures; Use of biochemical methods to examine factors influencing the texture of potatoes during cooking and processing; Structural macromolecular and processing factors affecting the rheology of polysaccharide gels; Determination of structure and properties of cereal cell wall polymers and changes during processing and digestion

15 Spinks EA Develop analytical methods for glucosinolates and their products in fresh and processed foods and biological specimens

16 Stevens BJH - BSc Structure of plant cell walls in relation to their degradation by human colonial microflora; Characterisation of extracellular polymers from cell cultures of food plants

17 Tanner SF - MSc PhD Use of nuclear magnetic resonance spectroscopy to study structure-function relationships of food biopolymers; Development and implementation of novel nuclear magnetic resonance techniques for foods;Application of solid state NMR to biological samples

18 Whittam AM - BSc Study molecular interactions between surfactants and alpha 1-4 Glucan chains using biophysical techniques

19 Williamson G - BSc,PhD Isolation of glutathione transferases and investigation of specificity for products of lipid oxidation

20 Wright DJ - BSc PhD Provision of raw material and processed pea fractions for investigation; Physiological and functional properties of vegetable proteins in food; Differential scanning calorimetry of proteins; Develop indigenous crops (peas, beans) for food manufacturing; Cereal protein (Gluten) functionality related to physical properties and structure; effects of commercial processing

21 Wright KM - BSc Digital data restoration using restrained deconvolution methods

4211 Microbiology
Colney Lane, Norwich, NR4 7UA
Tel: 0603 56122.

1 Archer DB - BA MA PhD Understanding the microbial basis of anaerobic digestion to improve methane production from food waste; Use markers, imaging and electrode techniques to observe microbial growth in food and characterize the environment

2 Evans DM - BSc DPhil Development of yeast cloning vectors

3 Hamill JD - BSc DPhil Mutation, selection and genetic analysis of cell lines with improved secondary metabolite production characteristic

4 Harris JE - BSc Develop genetic manipulation of methanogens for investigation of the microbiology of anaerobic digestion

5 Jackman PJH - BSc,PhD Use of markers, imaging and electrode techniques to observe microbial growth in food; Characterisation of the environment

6 Keenan MHJ - BSc PhD Use deacetylating enzymes from selected microbes to upgrade sugarbeet pectin; Use physiological and gene-cloning techniques for production of biotin by yeasts from waste

7 Lund BM - BSc MPharm PhD Obtain predictive data on the effect of conditions in foods on growth and toxin formation by *Clostridium botulinum*; Factors affecting survival and growth of spoilage microorganisms and food-poisoning bacteria in chilled foods

8 Parr AJ - MA DPhil Characterisation of secondary product transport in cell cultures and selection of lines with improved release capacity

9 Peck MW - BSc,MSc Enumeration of methanogens in landfill by indirect assays

10 Powell GE - BSc PhD Mathematical analysis of growth of microbial consortia in anaerobic digesters

11 Robertson J - BSc,PhD The relevance to landfill of known microbiology and biochemistry of methanogenesis from wastes

12 Stratford M - BSc,PhD Selection of yeast strains able to flocculate in logarithmic phase of growth

13 Walton NJ - BA PhD Characterisation of genetically altered plant cell lines in relation to secondary metabolite production

4212 Scientific Services and Development
Colney Lane, Norwich, NR4 7UA
Tel: 0603 56122.

1 Davies AMC - MRSC CChem Application of near infrared reflectance spectroscopy to rapid analysis of food and control of food processing

2 Geeson JD - BSc PhD Optimum conditions for vegetable storage in relation to variety, growth conditions, storage method, and marketing; Plastic films and storage temperatures to optimize shelf-life of packaged fruit and vegetables

3 Hughes JC - BSc,PhD Reducing sugars and nitrogenous constituents of potatoes in relation to fried colour; effects of agronomy and storage; With PBI develop long-dormant potato varieties to avoid sprout growth in storage; study senescent sweetening

4 Mellon FA - BSc PhD Evaluation of thermospray liquid chromatography/mass spectrometry for involatile or polar compounds; Development of mass spectrometry/mass spectrometry as a rapid method of analysis for food constituents

5 Parker ML - BSc,PhD Application of sputter cryo, elemental microanalysis and immuno-labelling techniques in electron microscopy

6 Self R - FIFST Development of combined liquid chromatography-mass spectrometry for analysis of involatile food components; Development of mass spectrometry techniques for direct, rapid analysis of biological material; Evaluation and testing of new mass spectrometric methods with potential for use in AFRC research programmes

4213 Nutrition and Food Quality
Colney Lane, Norwich, NR4 7UA
Tel: 0603 56122.

1 Brownsey GJ - BSc PhD Factors determining the rheological and perceived textural properties of hydrocolloid based food systems

2 Fairweather-Tait SJ - MSc PhD Factors affecting the biological availability of iron and zinc in foods

3 Griffiths NM - BSc MIBiol Methods for sensory testing of food systems and analysis of the results to improve sensitivity and reliability; Commercial contract - details confidential; Sensory evaluation of peas, other legumes and their fractions; Effects of glucosinolates on aroma and taste of fresh and processed vegetables

4 Johnson IT - BSc PhD Biochemical and physiological effects of changes in nutrient intake with special reference to sub-optimal zinc intake; Effect of non-absorbable polysaccharides on nutrient absorption in the rat and assess significance for human nutrition; Develop method to measure availability of folates in human foods using folate depleted rats; Effects of thermal processing techniques on availability of mineral nutrients in human foods; In vivo and in vitro biological assessment of phenolic constituents of plants

5 Livesey G - BSc PhD Provide biochemical data to improve the assessment of the energy values of foods

6 Shepherd R - BA MA PhD Sensory factors as determinants of salt intake through food choice and salt use; relate to physiological factors

7 Southgate DAT - BSc PhD MIBiol Measurement of nutritional and sensory qualities of current potato varieties; effects of storage on these properties; Development of methods for measurement and characterisation of dietary fibre in mixed diets; Vitamin and mineral composition of retail vegetables; Proximate nutritional composition of retail vegetables; Development and operation of computerized nutrient data system

8 Southon S - BSc PhD Biochemical and physiological effects of changes in nutrient intake with special reference to sub-optimal zinc intake

9 Stockley L - MSc Development and evaluation of improved methods for measuring food intake in man; Construction and use of nutrient data bases; develop user files and test accuracy in studies with subjects

4214 Process Physics
Colney Lane, Norwich, NR4 7UA
Tel: 0603 56122.

1 Barker GC - BSc PhD Theoretical analysis of effects of shear and extensional flow on emulsion droplets

2 Chouikhi SM - BSc,PhD Sensor devices: development, applications and problems of usage in the food industry;Methods for measurement of dielectric relaxation of food protein solutions in the frequency range 104–109 kHz

3 Clark DC - BSc DPhil Physical, structural and conformational properties of surface active food proteins in foams and emulsions

4 Grimson MJ - BSc PhD Statistical mechanics of colloidal dispersions; Theoretical analysis of droplet structure in emulsions

5 Howe AM - BA,PhD Mechanics of emulsion breakdown; relation of internal properties and phase diagrams to emulsion stability; Design of equipment to determine effect of bulk composition on food foam stability

6 Javanaud C - BA DPhil Measurement and interpretation of ultrasonic velocity and attenuation in foods and related systems; Interpretation of dielectric relaxation processes in food protein solutions and in emulsions

7 Jug G - MSc PhD Kinetics of crystal growth in food systems; Statistical mechanics of foam stability

8 Mingins J - BSc PhD Interfacial properties of food proteins and associated lipids and relation to form and emulsion behaviour

9 O'Toole C - BSc,MPhil Kinetics and mechanisms of supercritical extraction processes

10 Parker R - BSc,PhD Characterisation of extrusion cooking and its relation to physical properties of soft solid foods

11 Pinder AC - MA,PhD Development of methods for examining structure and dynamics in interfacial layers on solids and liquids

12 Prescott EHA Develop micro-computer based process monitoring and control systems; to establish cooking extruder automation

13 Rahalkar RR - BTech PhD MEng Relate bulk, structural and rheological properties of oil-water emulsions to their stability during processing; Correlate bulk and structural properties of liquid and solid food systems with processing history

14 Richmond P - BSc PhD MInstP DSc Engineering principles of separation and transfer processes using supercritical fluids; Calculation of colloidal interactions, collective structures and physical properties of model food systems

15 Robins MM - MSc PhD Determination of physical mechanisms governing formation, processing and stability of emulsions and liquid foods; Dispersions of crystals in liquids: relationship between processing conditions, structure and properties; Relationship between protein conformation and adhesion at solid-liquid interfaces

16 Smith AC - BSc PhD Physical, rheological and structural properties of solid foods in extrusion cooking; Physical and rheological properties of potato-based food systems during the extrusion cooking process; Rheology of composite materials with reference to die design and modelling of extrusion processes

17 Steytler D - BSc DPhil Fundamental thermodynamics of supercritical extraction processes

18 Toprakcioglu K - BSc,PhD Measurement of colloidal forces in model food systems relating this to the stability of food-foams and emulsions

4215 AFRC Institute of Food Research - Reading Laboratory
Shinfield, Reading, RG2 9AT
Tel: 0734 883103.

Research on quality of milk and its conversion into dairy products

1 Green Dr ML *Acting Head of Laboratory*
2 Briggs NW - BSc,MIInfSc *Librarian*

4216 National Collection of Yeast Cultures

1 Kirsop BE - BSc Yeast culture collection: Development of computer services; cryopreservation; detection, characterisation of plasmids

4217 Food Microbiology
Shinfield, Reading, RG2 9AT
Tel: 0734 883103.

1 Anderson PH - MIBiol Molecular characterisation of bacterial plasmids and construction of cloning vectors in lactic acid bacteria

2 Cliffe AJ - LRIC Develop methods for estimating proteinases of bacteria; Production and properties of bacterial enzymes involved in cheese ripening; Proteinases influencing cheese ripening

3 Cole CB - MIBiol,CIBiol Control gut microflora and diet of experimental animals and measure effect on enzymes concentration in gut; Examine the nutritive value of yoghurt and the claim that it improves gut function

4 Cole WM - IMLT Determine numbers of microorganisms in fermented foods and measure metabolic products

5 Collins MD - BSc PhD Improve classification and identification of bacteria important in dairying and maintain in a national collection

6 Farrow JAE - BSc Determine the content of RNA/DNA and other components of bacterial cells as an aid to classification

7 Fuller R - BSc MSc PhD Study the interaction between diet and gut flora and its influence on the nutrition and health of the consumer; Examine the nutritive value of yoghurt and the claim that it improves gut function

8 Gasson MJ - BSc PhD Study genetics of lactic acid bacteria and develop modified strains for dairy and other fermentation processes; Genetics of lactic acid bacteria, possibility of developing modified strains for dairy and food fermentation

9 Hill SHA - BSc,PhD Genetics of lactic acid bacteria, possibility of developing modified strains for dairy and food fermentation

10 King JS - HNC Metabolism and flavour production of microflora of fermented milks in relation to product quality and consumer health

11 Kirby CJ - BSc,PhD Microencapsulation of enzymes involved in cheese ripening; Microencapsulation of enzymes

12 Kroll RG - BSc PhD MIBiol Improve existing methods for assessing the hygienic quality of milk and milk products; Electron transfer coupling as a rapid method for assessment of bacterial contamination

13 Law BA - BSc PhD Identify ways in which psychrotrophic bacteria affect the quality of milk and milk products and devise control methods; Study the metabolism of microorganisms involved in manufacture and ripening of cheese; Develop ways of accelerating cheese ripening process whilst maintaining good flavour and keeping quality; Microencapsulation of enzymes

14 Marshall VM - BSc PhD Study the metabolism of the microflora of fermented foods in relation to product quality and consumer health

15 Patchett RA - BSc Electron transfer coupling as a rapid method for assessment of bacterial contamination

16 Phillips BA Maintain a culture collection of bacteria and check the identity of the cultures

17 Rodrigues UM - MIBiol Develop methods for the wider exploitation of DEFT for food quality assessment

18 Underwood HM - NDD(T),SDDT Investigate genetic control of resistance to phage in lactic acid bacteria

4218 Food Quality and Human Nutrition
Shinfield, Reading, RG2 9AT
Tel: 0734 883103.

1 Baker KC - BSc The immune response to food antigens

2 Blakeborough P - BSc PhD Influence of dietary components on calcium bioavailability

3 Bryant DTW - LRSC Influence of dietary components on calcium bioavailability

4 Edwards-Webb JD - LRSC The composition and intake of processed fats and foods containing them; Influence of diet on the cellular metabolism of serum lipo-proteins and their lipid and protein components; The effect of isomeric fatty acids formed during processing on requirements for n-6 and n-3 essential fatty acids

5 Farnworth SJ - BSc Nutritional advantages of calcium derived from dairy products

6 Fordham J Interactions between diet and the gut microflora influencing utilization of energy from foods

7 Hewitt D - BSc PhD To determine bioavailability of nutrients from dairy and other foods and the effects of processing on bioavailability; Interactions between diet and the gut microflora influencing utilization of energy from foods

8 Kilshaw PJ - BSc PhD Antigen presentation to the immune system; The allergenicity of diets for maternal and infant feeding

9 Newport MJ - BSc PhD Interactions between proteins, carbohydrates and fats in infant and weaning foods; Effects of lactoferrin in milk on iron absorption and the gut microflora; Interaction between proteins, carbohydrates and fats, micronutrients in infant foods and nutrient absorption

10 Ratcliffe B - BSc PhD MIBiol CBiol Interactions between diet and the gut microflora influencing utilization of energy from foods

11 Rolls BA - MA PhD To determine bioavailability of nutrients from dairy and other foods and the effects of processing on bioavailability; Influence of dietary components on calcium bioavailability; The nutritional advantages of calcium derived from dairy products

12 Sambrook IE - PhD,BTech Influence of diet composition on circulating lipoproteins in man and animals

13 Swindell TE - BSc Evaluation of energy in alternative carbohydrates

14 Welch VA - BSc PhD Influence of diet on the cellular metabolism of serum lipo-proteins and their lipid and protein components; The effect of isomeric fatty acids formed during processing on requirements for n-6 and n-3 essential fatty acids; Influence of diet composition on circulating lipoproteins in man and animals

4219 Food Structure
Shinfield, Reading, RG2 9AT
Tel: 0734 883103.

1 Anderson M - BSc PhD Effects of chemical composition and processing on functional properties of food emulsions, especially cream; Understand compositional factors in milk that affect stability of milk fat; Emulsions based on milk but modified compositionally to improve functional properties; Study individual compounds, especially those derived from milk, as emulsifiers and stabilisers in foods; Chemical composition and processing effects on whippability and whipped stability of emulsions, especially cream; Chemical changes in heated milks, relating heat treatment to composition
2 Andrews AT - MA DPhil DSc Factors affecting activity of proteinases in foods; Relate properties of proteins to their structure, with and without modification by chemical and physical means; Identify the nature and action of heat stable enzymes; Action of proteinases and protein modifying enzymes on milk, food proteins; Modify enzymes by chemical means to improve activity and modify specificity
3 Brooker BE - BSc,FRSTMH Stabilisation of air in multicomponent foods especially those containing fat. The causes of foaming in milks during processing. Interactions between diet and the gut microflora that influence the nutrition and health of the consumer
4 Chaplin LC - CChem,MRSC Action of proteinases and protein modifying enzymes on milk, food proteins; Modify enzymes by chemical means to improve activity and modify specificity
5 Foster CL - BSc,MSc,PhD Application of adhesives science to foods
6 Goodenough PW - BSc PhD Action of proteinases and protein modifying enzymes on milk, food proteins; Modify enzymes by chemical means to improve activity and modify specificity
7 Green ML - MA PhD Structure of milk gels and cheese texture as affected by milk heating, homogenization and preconcentration; Nature of mechanical interactions between heat-setting gels and second phase particulate inclusions
8 Griffin MCA - MA PhD Structure of casein micelle, nature of glycomacropeptide; Relate composition of milk to its physico-chemical properties; Interaction with lipid monolayers of casein with and without sugar groups
9 Langley KR - MSc Relate properties of proteins to their structure, with and without modification by chemical and physical means; Improve methods of measuring, and measure the functional properties of proteins
10 Marshall RJ - HNC MIBiol Increasing yield of cheshire cheese by preliminary heat denaturation of serum proteins in milk
11 Murray EK - BSc PhD Emulsions based on milk but modified compositionally to improve functional properties; Study individual compounds, especially those derived from milk, as emulsifiers and stabilisers in foods
12 Needs EC - HNC Chemical composition and processing effects on whippability and whipped stability of emulsions, especially cream
13 Owen AJ - BSc Relate properties of proteins to their structure, with and without modification by chemical and physical means; Action of proteinases and protein modifying enzymes on milk, food proteins
14 Payne GA - CChem MRSC Development and application of instrumental methods for analysis of flavour-contributing compounds in foods; Off-flavours and taints caused by processing, packaging and storage of milk and dairy products; Chemical changes in heated milks, relating heat treatment to composition
15 Stothart PH - BSc MSc PhD Structure of casein micelle, role of calcium phosphate; Relate composition of milk to its physico-chemical properties; Study of Ca binding sites in calcium-binding proteins
16 Taylor MD Identify the nature and action of heat stable enzymes
17 Wood TJ - LRSC The influence of fat composition and thermal characteristics on the functional properties of food emulsions

4220 Food Process Engineering and Technology
Shinfield, Reading, RG2 9AT
Tel: 0734 883103.

1 Andrews GR - BSc Lactulose levels in heat-treated milks as an indicator of the type of processing and product quality; Production of lactose derivatives by fermentation or chemical modification
2 Belcher JR - HNC Improve methods of milk cooling on farm, reducing costs without increasing risk to milk quality
3 Bines VE - NDD Effect of milk and methods on cheese quality and yield
4 Foxell PR - HNC CEng MIMechE Inactivation of heat resistant enzymes at ultra high temperature
5 Grandison AS - BSc,PhD Reducing deposits in heat exchangers and evaporators to improve efficiency. Development of techniques for recovery, fractionation and utilisation of proteins and sugars from food processing streams. The study and development of reverse osmosis (RO) and ultrafiltration (UF) systems in food processing
6 Hatfield DS - ONC A data collection system for monitoring plant cleaning
7 Heppell NJ - BSc MSc Heat transfer into suspended particles from flowing liquids; residence time distribution in continuous flow sterilizers
8 Perkin AG - BA CEng MIMechE Effectiveness of sterilization processes; Efficient and economic cleaning of food process plant
9 Schroder MJA - BSc PhD Improving the keeping quality of commercial pasteurised milk by modifying plant and operating systems
10 Scurlock PG - HNC,MPhil Study and development of RO and UF systems in food processing

11 Young P - BSc PhD ARCST FIBST DipMICST Effect of milk and methods on cheese quality and yield development of new methods and cheese types; Production of lactose derivatives by fermentation or chemical modification; Study and development of RO and UF systems in food processing

4221 Analytical Services
Shinfield, Reading, RG2 9AT
Tel: 0734 883103.

1 Florence EF - CChem MRSC Develop and evaluate methods to monitor components of interest in foods
2 Scott KJ Analysis and development of analytical methods for vitamins in foods; To determine bioavailability of nutrients from dairy and other foods and the effects of processing on bioavailability; Improve methods for determining the nutritional composition of dairy and other foods; To provide authoritative data on the nutrient composition of dairy and other foods

4222 Food and Beverages
Shinfield, Reading, RG2 9AT
Tel: 0734 883103.

1 Bakker J - BSc PhD Identify changes in port wine pigments due to processing variables
2 Bridle P - LRSC Role of specific components in sensory properties
3 Davenport RR - PhD Identify spoilage yeasts of foods and beverages
4 Goverd KA - BSc Survival of non-acid tolerant bacteria in acid food and beverages
5 Pickard J - LRSC Analytical methods for flavour
6 Preston N - BSc PhD Develop rapid predictive tests for chemical instability of wine components
7 Stead D - BSc,PhD SO(2) and other chemical and natural components of preservatives for foods and beverages
8 Thomas S - BSc,PhD Rapid methods for wine spoilage organisms
9 Tucknott OG - PhD,GRSC,CChem Demonstrate presence of food spoiling organisms by head space analysis
10 Williams AA - BSc,PhD,DSc Flavour of dairy products, fats, oils and confectionery; flavour (taints) from food processing; analytical methods for flavour; methods for authenticating adulteration and origin of foods and beverages. Quantification of factors influencing food acceptability; sensory and analytical components of quality; role of specific components in sensory properties; mathematics of subjective and objective sensory data

4223 AFRC Institute of Horticultural Research - Littlehampton
Worthing Road, Littlehampton, West Sussex, BN17 6LP
Tel: 0903 716123.

Glasshouse crops, mushrooms, bulbs and ornamentals

1 Swinburne Prof TR *Director*
2 Fraser Dr RS *Head of Station*

4224 Plant Physiology and Chemistry
Worthing Road, Littlehampton, West Sussex, BN17 6LP
Tel: 0903 716123.

1 Adams P - HNC MPhil CChem FRSC Calcium nutrition of glasshouse vegetables
2 Grange RI - BA PhD Assimilate compartmentation and reserve remobilization; their effects on photosynthesis and partitioning of assimilates; Water relations techniques and the evaluation of water status in crop growth and nutrition; Physiological mechanisms controlling starch metabolism
3 Ho LC - BSc,PhD,DIPBiol Regulation of sink demand and its effect on partitioning of assimilate
4 Hobson GE - BSc MSc PhD DSc FRSC CChem Biochemistry of ripening of tomatoes especially in relation to quality maintenance
5 Jordan BR - BSc PhD Phytochrome action in photomorphogenetic and photoperiodic responses; Photoregulation of thylakoid membrane composition and organisation
6 Knapp PH - BSc PhD MInstP Instrument design and experimentation for studying light absorption by leaf pigments; Design of light sources, sensors and related equipment and their use for studies on effects of light in plant growth
7 Mitchelson K - BSc,PhD Photoregulation of thylakoid membrane composition and organisation
8 Nichols R - BSc MSc PhD Post harvest physiology of mushrooms in relation to quality, storage and temperature; Post-harvest physiology, quality and control mechanisms in senescence; Chemical treatments for post-harvest effects
9 Partis M - BSc MSc PhD Molecular properties of phytochrome in relation to function; Purification and characterisation of membrane-associated phytochrome
10 Thomas B - BSc PhD Phytochrome action in photomorphogenetic and photoperiodic responses; Characterisation of the effects of light from artificial sources on plant growth and development
11 Vince-Prue D - BSc,MSc,DPhil,DSc,NDH,FIBiol,PhD,CBiol The mechanism of photoperiodic floral induction; Characterisation of the effects of light from artificial sources on plant growth and development

4225 Crop Science
Worthing Road, Littlehampton, West Sussex, BN17 6LP
Tel: 0903 716123.

1 Cockshull KE - BSc PhD Assess the biological consequences of energy-saving measures; Factors controlling flowering and plant development
2 Graves CJ - BSc DPhil Improve methods of growing plants in nutrient film culture; Investigate how plants integrate temperature, and devise computer controlled energy-saving regimes
3 Hand DW - BSc,PhD,MIBiol,MIHort Determination of the optimum level of summer CO_2 enrichment for long-season tomato and cucumber crops; Assessment of the biological consequences of using direct-fired burners for CO_2 enrichment and warm-air heating; Manipulation of environment to either mask or inhibit "bolting" in autumn- or winter-sown celery
4 Hanks GR - BSc MPhil plant growth regulators and bulb growth; Growth retardants for bulb crops; Bulb macro-propagation: morphological, cultural and environmental effects; Vegetative growth of bulbs and corms; Bulb forcing problems and alternative bulb forcing schemes
5 Humphries DJ - BSc,MIHort Effects of aerial environment on mushroom cropping; Materials and methods for mushroom compost production; Systems and methods of mushroom growing; Edible fungi as alternatives to mushrooms
6 Jinks RL - BSc The effects of environmental and cultural factors in container production. Soilless media for container-plant production
7 Langton FA - BSc PhD Bulb micropropagation: evaluation of commercial prospects; Development of growing systems to improve quality in ornamental plants; Use induced mutations and conventional methods to breed resistant cultivars-supervisory role
8 Loach K - BSc CertEd PhD ARCS DIC Investigate factors affecting the rooting of cuttings from representative groups of woody plants; The effect of environmental and cultural factors in container production; Soilless media for container-plant production
9 Menhenett R - BSc MSc PhD Growth retardants for producing bulbs as pot plants; Collaboration with MAFF/ADAS to improve UK yields of dry bulbs and flowers of narcissus and tulip; Modify growth with plant growth regulators to improve plant shape and flower development
10 Rees AR - BSc DSc FIBiol Investigate the role of plant growth regulators in dormancy; Assess factors affecting propagation; Investigate physiological disorders of forced bulbs
11 Slack G - BSc Determination of the optimum level of summer CO_2 enrichment for long-season tomato and cucumber crops; Assessment of the biological consequences of using direct-fired burners for CO_2 enrichment and warm-air heating; Improve systems for tomato pollination
12 Smith JF - MSc Systems and methods of mushroom growing. Effects of aerial environment on mushroom cropping. Materials and methods for mushroom compost production. Edible fungi as alternatives to mushrooms
13 Whalley DN - BSc Investigate factors affecting the rooting of cuttings from representative groups of woody plants; Effects of mineral nutrition on plant growth in the field and on rooting of cuttings; Improvement of plants and their establishment by clonal selection, environmental manipulation and chemical treatment

4226 Biomathematics
Worthing Road, Littlehampton, West Sussex, BN17 6LP
Tel: 0903 716123.

1 Aikman DP - BSc,PhD,DipBiophys Developing computer models relating glasshouse crop growth and energy usage to external conditions and control regimes
2 Fenlon JS - MSc Design and analysis service for bulb experiments. Examination of commodity price records, seasonal trends and relationship to meteorological records

4227 Entomology
Worthing Road, Littlehampton, West Sussex, BN17 6LP
Tel: 0903 716123.

1 Burges HD - BSc PhD DSc *Bacillus thuringiensis*: genetic basis of toxin action; strain improvement; formulation and application methods; Potential of nematodes to control insect pests especially in mushrooms; Diagnosis of pathogens of invertebrate pests; Use of *Bacillus thuringiensis* to control caterpillar pests of glasshouse crops; Development of fungal products for the control of insect pests
2 Chambers RJ - BSc,PhD,ARCS Use of natural enemies, chemical and cultural methods to control aphids and other pests in cereals and other crops. Biological control of aphids on glasshouse crops using syrphids and other natural enemies
3 Crook NE - BSc PhD *Bacillus thuringiensis*: molecular basis of toxin action; Insect virus identification and characterisation; genetic studies on infectivity and host range; Diagnosis of virus infections of invertebrate pests; Identification and quantification of prey in invertebrates; Use of insect viruses to control caterpillar pests of glasshouse crops
4 Gillespie A - BSc PhD Diagnosis of fungal infections of invertebrate pests; The use of fungal insecticides in integrated control programmes; Development of fungal products for the control of insect pests

5 Jarrett P - HNC,CBiol,MIBiol *B. thuringiensis*: genetic basis of toxin action; strain improvement by genetic engineering; formulation and application methods. Diagnosis of bacterial infections of invertebrate pests
6 Payne CC - DPhil,MA,CBiol,MIBiol Studies on insect viruses and their use to control insect pests; Diagnosis of virus infections of invertebrate pests; Use of insect viruses to control caterpillar pests of glasshouse crops
7 Richardson PN - MSc,DIC Potential of nematodes to control insect pests especially in mushrooms. Biological control of insect pests by insect-parasitic nematodes
8 Scopes NEA - BSc DIC PhD Development of biological control methods for arthropod pests of glasshouse crops; Evaluating pesticides for safe use within glasshouses; The biological efficiency of ultra-low volume droplets of insecticide sprays
9 Sunderland KD - BSc PhD Use of natural enemies, chemical and cultural methods to control aphids and other pests in cereals and other crops
10 White PF - HNC,LIBiol The biology and control of mushroom pests, including the effects of pesticides on the crop

4228 Plant Pathology and Microbiology
Worthing Road, Littlehampton, West Sussex, BN17 6LP
Tel: 0903 716123.

1 Atkey PT - HNC,LIBiol,MIBiol,C&G,CBiol Electron microscopy of plant viruses
2 Barton RJ - BSc MSc PhD Characterisation and identification of plant viruses; Purification and characterisation of mushroom and fungal viruses, and development of rapid identification methods
3 Brunt AA - DSc,FIBiol,CBiol Characterisation and identification of plant viruses; Isolation, characterisation, epidemiology and control of viruses infecting glasshouse vegetable crops; Isolation, characterisation, epidemiology and control of viruses infecting glasshouse ornamentals; Testing and multiplication of virus-free glasshouse crops; Isolation, characterisation, epidemiology and control of viruses infecting bulbs and corms; Isolation, characterisation, epidemiology and control of viruses infecting ornamentals
4 Claydon N - BSc DPhil Biochemical and physiological studies of mushrooms; Lignocellulolysis; investigating the factors controlling lignocellulose degradation during mushroom production; Production and role of antibiotic chemicals during microbial interactions; The biochemical basis of virulence in insect-pathogenic fungi
5 Elliott TJ - BSc,CBiol,MIBiol Develop methods for strain improvement of cultivated mushrooms through genetic studies
6 Fermor TR - BSc,PhD,CBiol,MIBiol The microbiology and chemistry of mushroom composts. Isolation, screening and evaluation of potential biocontrol agents against mushroom bacterial and fungal diseases
7 Hand P - BSc,PhD Microbial straw utilisation
8 Lynch JM - BTech,PhD,FRSC,DSc,CChem,CBiol,FIBiol Isolation and screening of potential biocontrol agents against plant diseases; Microbial straw utilisation
9 Magan N - MSc PhD Microbial straw utilisation
10 Thomas BJ - BSc PhD Characterisation and detection of plant viruses; Isolation, characterisation, epidemiology and control of viruses infecting glasshouse vegetable crops; Isolation, characterisation, epidemiology and control of viruses infecting hardy ornamental nursery stock (HONS)
11 Whipps JM - BSc PhD Integrated control: environmental factors affecting disease expression; physiology of antagonist-pathogen interactions
12 Wood DA - BSc,PhD,CBiol,MIBiol Biochemical and physiological studies of mushrooms; Microbiology and chemistry of mushroom composts; investigation of the role of micro-organisms in compost production; Lignocellulolysis; investigating the factors controlling lignocellulose degradation during mushroom production

4229 Macaulay Land Use Research Institute - Bush, Edinburgh
Bush Estate, Penicuik, Midlothian, EH26 0PY
Tel: 031-445 3401.

Research to improve the economic viability of meat production in the hills and uplands of the UK

1 Maxwell Prof. TJ - BSc,PhD *Director* Effects of pre-weaning factors on food intake, growth and carcass composition of lambs; Sheep production efficiency, ewe size, lactation and reproduction responses, upland grazing systems; Deer farming systems, operational research and economic viability; Hill sheep production from grassy hills, ewe and lamb performance, grazing management and profitability; Hill sheep production systems on heather moorland, ewe and lamb performance, grazing management and profitability; Hill sheep production systems, inwintering, increased stocking rates, ewe and lamb performance and their profitability; Upland sheep production systems, ewe and lamb performance, their grazing management and profitability
2 Alcock Ms MM - BA,DipLib,ALA *Information Officer*

4230 Animal Production

Bush Estate, Penicuik, Midlothian, EH26 0PY
Tel: 031-445 3401.

1 Doney JM - BSc PhD Reproductive performance in sheep; pasture and supplementary food intake; body condition, ovulation rate, embryo loss; Reproductive performance in sheep; nutrition during growth and development and during adult productive life; Reproductive performance in sheep; nutrition, body condition, ovulation rate and embryo loss; Reproductive performance in sheep and immunisation against steroid hormones, dosage, body condition, ovulation rate; Measure lactation patterns of various genotypes as affected by herbage intake on different swards; Determine the inter-relationships between milk intake, herbage intake and growth of lambs at pasture

2 Fawcett AR - AIMLS Prophylaxis of cobalt/copper deficiencies in sheep on improved hill pasture; Develop disease preventive programmes based on monitoring; Semen quality in rams in relation to performance in the field

3 Gunn RG - BSc,PhD Reproductive performance in sheep; pasture and supplementary food intake; body condition, ovulation rate, embryo loss; Reproductive performance in sheep; nutrition during growth and development and during adult productive life; Reproductive performance in sheep; nutrition, body condition, ovulation rate and embryo loss; Reproductive performance in sheep and immunisation against steroid hormones, dosage, body condition, ovulation rate; Growth and body composition of lambs on grass and forage crops in relation to intake and growth before weaning; Sheep production efficiency, ewe size, lactation and reproduction responses, upland grazing systems

4 Hamilton WJ - BA NDA NDD CBiol Lifetime performance of red deer and their economic viability; Deer farming systems, their operational management and economic viability; Nutrition and performance of young deer from weaning to slaughter; Performance of grazing deer in pregnancy, lactation and over the reproductive period

5 Lippert M - BSc,PhD Cashmere and meat production in domestic and feral goats; effects of nutrition and genotype on seasonal growth patterns of fibre and on carcass composition. Potential for goat production from hill and upland pastures; grazing management strategies; use of goats in certain sheep production systems and for the purposes of weed control

6 Rhind SM - BSc PhD Reproduction in sheep: effects of nutrition, body condition on endocrine status, ovulation rate and lambing rate; Reproduction in sheep, endocrine manipulation, ovulation, endocrine status and follicle growth; Endocrine manipulation, immunisation against steroid hormones, dosage, body condition and ovulation rate; Lactation physiology in sheep, endocrine status, nutrition, body composition and ewe genotype; Reproductive physiology in beef cows; endocrine manipulation, immunisation against steroids

7 Russel AJF - BSc,MSc,PhD Foetal number determination in sheep by real-time ultrasonic scanning, and management and husbandry benefits; Feed intake and body condition on partition of nutrients between body tissue and milk production in beef cows; Effects of body composition and physiological state on nutrient requirements of beef cows; Effects of herbage mass and sward height on intake and performance of beef cattle; Effects on production from lactating beef cows of interactions between winter feeding and grazing management; Effects of nutrition on production of housed and grazing twin-rearing beef cows; Reproductive physiology in beef cows; endocrine manipulation, immunisation against steroids

8 Ryder ML Cashmere production in domestic and feral goats, skin and hair biology, annual cycle of growth

9 Sibbald AR - HNC Growth and body composition of lambs, carcass evaluation, measurement of herbage and forage crop mass; Sheep production efficiency, grazing management, data analysis; Hill sheep production from grassy hills, animal performance, data analysis and profitability; Hill sheep production systems on heather moorland, animal performance, data analysis and profitability; Hill sheep production systems, animal performance, data analysis and profitability; Upland sheep production systems, animal performance, grazing management and profitability; Beef cattle production systems, their operation, profitability and grazing management; Developing conceptual and quantitative models for sheep grazing systems

10 Wright IA - BSc,PhD Effects of feed intake and body composition on partition of nutrients between body tissue and milk production in beef cows: effects of sward characteristics, winter feeding and grazing management strategies on intake and performance. *In vivo* prediction of body composition in lactating beef cows using deuterium oxide and ultrasonic measurements. Reproductive physiology in beef cows: effects of nutrition and calf suckling on oestrus activity; endocrine manipulation, immunisation against steroids; pregnancy diagnosis and estimation of foetal age

4231 Animal Nutrition

Bush Estate, Penicuik, Midlothian, EH26 0PY
Tel: 031-445 3401.

1 Mayes RW - BSc MSc PhD Nutrition of the grazing sheep during late pregnancy; Nutrition of the lamb from birth to weaning; Development of techniques for quantifying nutrient inputs and metabolism

2 Milne JA - BA BSc(Agr) PhD Nutrition and feeding behaviour of the grazing sheep in mid-pregnancy; Nutrition of the grazing sheep given supplementary feeding in early lactation; Nutrition and supplementation of lambs grazing forage brassicas; Nutritional biochemistry of adipose tissue of the pregnant and lactating sheep; Development of techniques for quantifying nutrient inputs and metabolism; Nutrition over the mating period and reproductive performance of the grazing sheep; Effect of dietary composition on growth and carcass composition of lambs; Nutrition and performance of young deer from weaning to slaughter; Nutrition of grazing deer in lactation, pregnancy and over the reproductive period

4232 Grazing Ecology

Bush Estate, Penicuik, Midlothian, EH26 0PY
Tel: 031-445 3401.

1 Grant SA - BSc MSc Diet selection by grazing animals : effects on plant morphology and growth; Effects of grazing on sward productivity, species composition and morphology; Defoliation and the dynamics of individual plants and plant populations; Defoliation and plant ecology of grass and clover swards; Goat production from hill pastures, grazing effects on botanical composition, herbage productivity and weed control

2 Hodgson J - BSc(Agr) PhD Intake of cut forages by sheep and cattle; Diet selection and herbage intake by grazing animals; patterns of defoliation in grazed swards; Ingestive behaviour, diet selection and herbage intake by grazing animals; patterns of defoliation in grazed swards; Effects of herbage mass and sward height on intake and performance of beef cattle; Effects on production from lactating beef cows of interactions between winter feeding and grazing management

4233 Plants and Soils

Bush Estate, Penicuik, Midlothian, EH26 0PY
Tel: 031-445 3401.

1 Evans CC - HNC Uptake of trace elements in pasture plants and the influence of environmental and management factors

2 Marriott CA - BSc Nitrogen fixation by white clover and its response to environmental and management factors

3 Newbould P - BSc BAgr DPhil Soil chemistry and plant nutrition of hill and upland pastures; Soil chemistry and plant nutrition of hill and upland pastures

4 Thomas RJ - BSc PhD Soil chemistry of nitrogen and its uptake and physiological role in plants; Soil chemistry of nitrogen and its uptake and physiological role in plants

4234 AFRC Institute for Animal Health - Houghton Laboratory

Houghton, Huntingdon, Cambridge, PE17 2DA
Tel: 0480 64101.

Poultry diseases

1 Payne LN - DSc,PhD,BVSc,FRCVS *Director*
2 Szmidt Ms I - BSc,DipLib *Librarian*

4235 Experimental Pathology

Houghton, Huntingdon, Cambridge, PE17 2DA
Tel: 0480 64101.

1 Bumstead NM - MA PhD Genetics of the immune response in commercial flocks; Transfection of env gene; Detection and expression of the genes of the major histocompatibility complex; Immunological detection of gene products; Associations between haplotypes of the major histocompatibility complex and resistance to disease

2 Davison TF - PhD Physiopathology of lymphoid leukosis virus infection in fowl; The physiological responses to putative stressors and pathogenic agents; The effect of stress on susceptibility to and course of infections in poultry

3 Farmer H - PhD,MRCVS Epidemiology and aetiology of stunting syndrome

4 Flack IH - MIST Analysis of DNA in genetic transfer; Analysis of major histocompatibility complex DNA patterns

5 Frazier JA - BSc,PhD Electron microscopy of infectious stunting syndrome

6 Freeman BM - DSc,FIBiol,CBiol Gene transfer to the avian ovum; Factors affecting resistance to disease

7 Greenwood NM - MIBiol,CBiol Vector systems for genetic transfer; Amplification of cloned copies of env gene; Preparation of cosmid libraries and development of DNA probes

8 Howes K Epidemiology of lymphoid leukosis virus infection in fowl

9 Manning ACC - HNC Assessing techniques of gene transfer; Infection and blood group associations

10 Martland MF - MA VetMB MRCVS Pathogenesis of infectious stunting syndrome

11 McDougall JS - MRCVS DipBact Pathogenicity of Marek's disease viruses from vaccinated flocks; Characterisation of viruses causing enteritis in turkeys; Turkey diseases - early poult mortality

12 Messer LI - PhD Vector systems for genetic transfer; Molecular cloning of the env gene of lymphoid leukosis virus; Preparation of cosmid libraries and development of DNA probes

13 Millard BJ - BA,AIMLS Studies on the association between blood groups and resistance to coccidia

14 Orbell S - HNC Pathogenicity studies of MDV isolated from flocks vaccinated against the disease

15 Orbell SJ - HNC Characterisation of viruses causing enteritis in turkeys

16 Palmer C - BSc,PhD Identification of subpopulations of lymphocytes and their role in avian immune system

17 Payne LN - DSc,PhD,BVSc,FRCVS,FRCPath Immunosuppression in Marek's disease; Epidemiology and physiopathology of lymphoid leukosis virus infection in fowl

18 Powell PC - MA PhD MRCVS BVSC Immunology of Marek's disease of poultry; Understand cellular interactions in the avian immune system

19 Rea J Physiopathology of lymphoid leukosis virus infection in fowl; Physiological changes during stress and disease; The effect of stress on susceptibility to and course of infections in poultry

20 Rennie MC - BA Immunology of Marek's disease of poultry; Understand cellular interactions in the avian immune system

4236 Parasitology
Houghton, Huntingdon, Cambridge, PE17 2DA
Tel: 0480 64101.

1 Chapman HD - ARCS DPhil Changes in metabolism and physiology produced by coccidiosis; Resistance to anticoccidial drugs; genetics of *Eimeria*

2 Clarke LE - PhD Cloning and characterisation of genes coding for antigens of *Eimeria*

3 McDonald V - PhD Production and stability of attenuated lines of *Eimeria* spp

4 Rose ME Pathological effects of coccidiosis; immunosuppressive effects of coccidial infections; Mechanisms of protective immunity in coccidiosis; influence of strain on resistance to *Eimeria* infections; Host parasite relationships in coccidiosis; Identification and presentation of coccidial antigens; Intestinal mucosal immunity; Identification of protective antigens in Eimeria spp; Blood groups and resistance to coccidia

5 Shirley MW - PhD MIBiol Effect of coccidial infections on growth; Taxonomy and genetics of *Eimeria*; Introduction of immunity by modulated strains of *Eimeria*

6 Wisher MH - PhD Identification of coccidial antigens; Identification of protective antigens in *Eimeria*

4237 Microbiology
Houghton, Huntingdon, Cambridge, PE17 2DA
Tel: 0480 64101.

1 Avery RJ - BSc,PhD Restriction enzyme analysis of the genomes of vaccine and field strains of ILT virus; Characterisation of the RNAs and proteins of infectious bursal disease virus

2 Barrow PA - PhD Determine the factors that influence the excretion of *Salmonella* organisms by chickens; Determine the basis of the variation in virulence for chickens observed in different strains of *Salmonella typhimurium*

3 Binns MM - BSc,PhD Sequence analysis of IBV cDNA clones; cDNA cloning and clone characterisation; Cloning and characterisation of avipoxvirus DNA

4 Boursnell MEG - MA PhD Sequence analysis of IBV cDNA clones; Cloning and characterisation of avipoxvirus DNA

5 Cavanagh D - BSc PhD Structure and function of IBV glycoproteins; Molecular epidemiology of IBV by peptide mapping of isolates

6 Cook JKA - BSc Study infectious bronchitis cross-protection using an experimental infectious bronchitis and *Escherichia coli* infection; Isolation and serotyping of field isolates of avian infectious bronchitis and role of strain variation in immunity; The association between blood groups and resistance to IBV and *E. coli*. The pathogenesis of rhinotracheitis; characterisation of the aetiological agents

7 Kitchin PA - BSc,PhD Characterization of genes associated with immunity

8 Milne B - FIMLS Determination of genetic structure of MDV and characterisation of genes associated with oncogenicity

9 Mockett APA - MRCVS,BVSc,CBiol,MIBIol Pathogenesis of and immunity to avipoxvirus and avipox/IBV recombinant infection; Mucosal immunity; Study of immunity to avian infectious bronchitis using serological methods including ELISA and monoclonal antibodies

10 Peters RW - FIMLS Characterisation of the RNAs and proteins of infectious bursal disease virus

11 Ross LJN - BSc PhD Genetic structure of Marek's disease virus and characterisation of genes associated with oncogenicity

12 Tomley FM Construction and testing of avipoxvirus/IBV recombinants

4238 AFRC Institute of Horticultural Research - Department of Hop Research
Wye College, Wye, Ashford, Kent, TN25 5AH
Tel: 0223 812179.

Hop breeding and agronomy

1 Gunn RE - BSc,PhD Use of tissue culture techniques in breeding tall and dwarf hops for yield, quality and pest and disease resistance; Effect of harvest date on plant vigour and yield of hops. Development of harvester for dwarf hops; agronomy of dwarf varieties

2 Darby P - BSc,PhD Breeding tall and dwarf hops for yield, quality and pest and disease resistance; Race non-specific resistance to powdery mildew; chemical control of downy mildew and hop canker

3 Green CP - BSc PhD Quality analysis and determination of chemical residues in hops

4 Gunn RE - BSc *Director*

4239 Hannah Research Institute
Ayr, KA6 5HL
Tel: 0292 76013/7.

Animal nutrition and lactation and milk utilisation

1 Peaker Prof. M - BSc,PhD,FIBiol,FRSE *Professor*

2 Wilson PD - BA,DPhil *Information Officer*

4240 Animal Nutrition and Production
Ayr, KA6 5HL
Tel: 0292 76013.

1 Chamberlain DG - BSc PhD Production and utilisation of silages and the effects of enzyme additives on composition and feeding value; Effect of additives on digestion, utilisation and metabolism of silage

2 Martin AK - BSc PhD CChem MRSC Identify, and determine the effects of, flavonoids in forages

3 Martin PA - BSc,MSc Nutrition and metabolism of ruminants with reference to milk production and manipulation of milk composition

4 Morrison IM - BSc PhD CChem FRSC Investigate forage composition and physical properties in relation to ensilage and rumen digestion

5 Reid D - BSc(Agr) PhD MS Study the production of grass, lucerne and white clover crops for ensilage

6 Thomas PC - BSc PhD FIBiol Investigate the production of grass and legume silages and their intake and utilisation by ruminants; Improve the scientific basis for the nutritional formulation of conserved forage diets for dairy cows

7 Thompson GE - BSc,PhD Nutrient metabolism and its regulation in the lactating ruminant

8 Williams AG - BSc,PhD Microbial metabolism and its manipulation in the ruminant

4241 Physical Chemistry
Ayr, KA6 5HL
Tel: 0292 76013/7.

1 Dalgleish DG - BSc PhD Renneting and aggregation of casein micelles; Properties of protein/fat emulsions; Rapid methods for milk protein analysis

2 Davies DT - BSc PhD Partition of casein proteins between soluble and colloidal phases

3 Holt C - MSc PhD Interactions of inorganic ions in milks and simulated milk sera

4 Horne DS - BSc PhD Colloidal stability of casein micelles in different media; Effects of chemical modification on aggregation behaviour of caseins

5 Manson W - BSc PhD Chemical modification of proteins during heating of milk

6 Robson EW - BSc PhD Particle size distribution in homogenised milks

4242 Lactational Physiology and Biochemistry
Ayr, KA6 5HL
Tel: 0292 76013/7.

1 Blatchford DR - HND MIBiol Development of methods for measurement of mammary gland size *in vivo*; Control of secretion of major and minor components of milk under normal and experimental conditions; Limiting factors at different stages of lactation using physiological, pharmacological and nutritional manipulation

2 Calvert DT - BSc PhD Endocrine control of mammogenesis and galactopoiesis in vivo; Control of glucose and other nutrient utilisation by skeletal muscle during lactation; Determine factors limiting milk secretion between sucklings; Control of glucose and other nutrient use by skeletal muscles during lactation; Endocrinology of lactation and reproduction

3 Clegg RA - MA PhD Lipid metabolism in adipose tissue; Acetyl CoA Carboxylase and Lipoprotein Lipase in mammary tissue; Phosphorylation of caseins and other proteins in mammary Golgi vesicles; Composition and metabolism of ruminant lipoproteins

4 Faulkner A - BSc PhD Glucose metabolism in the mammary gland of ruminants *in vivo*; Control of glucose and other nutrient use by skeletal muscles during lactation; Metabolism of glucose and propionate; The mode of transport of metabolites into milk; Glucose metabolism in the mammary gland of ruminants

5 Flint DJ - BSc PhD Cellular metabolism of adipose tissue; Regulation of and processing by insulin receptor sites; Production and efficacy of antibodies to growth hormones; Control of insulin secretory response of Islets of Langerhans during pregnancy and lactation; Control of insulin receptor number and properties in adipose tissue and mammary gland; Endocrinology of adipose tissue and mammary gland during lactation

6 Knight CH - BSc PhD Cell turnover and replacement in lactating mammary gland; mammogenesis and galactopoiesis; Growth responses in the lactating goat mammary gland; Endocrine control of growth of mammary tissue; factors affecting milk yield

7 Vernon RG - BSc PhD Metabolism of adipose tissue; Metabolism of glucose in the mammary gland and skeletal muscles during lactation; Regulation of and processing by insulin receptor sites; Control of glucose and other nutrient use by skeletal muscles during lactation; Control of muscle protein synthesis and degradation during lactation; Contribution of milk protein synthesis to mammary gland and whole body protein synthesis

8 Wilde CJ - BSc PhD Cell turnover and replacement in lactating mammary gland; cell number and activity in the determination of milk yield; Role of mammary tissue factors in metabolic adaptation of adipose tissue during lactation; Different milking regimes and stage of lactation and the hormonal regulation of mammary messenger RNA stability; Chemical inhibitors of milk production and limitation of milk secretion between sucklings; Degradation of newly-synthesized casein during secretion by mammary tissue

9 Zammit VA - MSc DPhil Role of glucagon in control of fatty acid metabolism in peripheral tissues; Regulation of enzymes involved in the supply of substrate by the liver and their utilisation of the mammary gland; Control of metabolism and supply of substrate to the mammary gland; Regulation of cholesterogenesis

4243 Biological Chemistry
Ayr, KA6 5HL
Tel: 0292 76013/7.

1 Annan WD - MA PhD Protein degradation in relation to flavour
2 Christie WW - PhD DSc Biosynthesis structure and activity of lipid related substances in milk; The nature of the flavour compounds and the manner of their biosynthesis
3 Noble RC - BSc PhD DSc MIBiol Metabolism of lipids in broiler chickens
4 Shand JH - BSc PhD The nature of the flavour compounds and the manner of their biosynthesis
5 West DW - BSc PhD To determine the cellular mechanisms for the secretion and control of milk components

4244 Milk Utilization
Ayr, KA6 5HL
Tel: 0292 76013/7.

1 Banks JM - BSc,PhD Relating cheese yield and quality to milk composition
2 Banks W - BSc PhD FRSE Manipulate milk composition by dietary means; Triglyceride composition of milk fat in relation to its physical properties; Factors causing instability in whipped or homogenised milk products
3 Clapperton JL - BSc PhD Manipulate milk composition by dietary means; Elucidate the structures of milk fat triglycerides and how they respond to dietary changes
4 Griffiths MW - BSc PhD Factors affecting the growth of psychrotrophs in milk; Assess the heat-stability characteristics of bacterial spores in milk
5 Muir DD - BSc PhD FIFST Relate cheese yield and quality to milk composition; Understand heat-induced aggregation in skim and concentrated skim milks; Assess the properties of sterile, concentrated milks prepared by ultrafiltration or diafiltration; Factors affecting the growth of psychrotrophs in milk; Factors causing instability in whipped or homogenised milk products; Assess the heat-stability characteristics of bacterial spores in milk
6 Steele W - BAgr PhD Manipulate milk composition by dietary means
7 Sweetsur AWM - PhD Measure heat stability of skim and concentrated skim milks; Measure heat-stability of concentrated milks

4245 Institute of Animal Physiology and Genetics Research - Cambridge Research Station
Babraham, Cambridge, CB2 4AT
Tel: 0223 852312.

Research in physiological sciences to improve animal production and welfare. Research on the safety, quality and processing of milk, meat, vegetables and other foods, and on aspects of human nutrition

1 Cross BA - CBE,MA,PhD,ScS,MRCVS,FIBiol,FRS *Director*

4246 Behavioural Physiology
Babraham, Cambridge, CB2 4AT
Tel: 0223 852312.

1 Baldwin BA - BVetMed PhD Determine the central and peripheral factors controlling food intake in pigs and sheep; Determine the central and peripheral factors controlling fluid intake in pigs and ruminants; Determine the central and peripheral factors controlling thermoregulatory behaviour in pigs and sheep; Determine the central and peripheral factors controlling sexual and aggressive behaviour in pigs and sheep; Quantify behavioural, physiological and neurochemical changes occurring in farm animals exposed to defined stressors; Determine the physiological and behavioural mechanism regulating food and fluid intake in pigs and sheep
2 Booth WD - BSc,PhD,MIBiol Determination of the central and peripheral factors controlling sexual and aggressive behaviour in pigs and sheep. Quantifying the behavioural, physiological and neurochemical changes occurring in farm animals exposed to defined stressors
3 Parrott RF - BSc PhD Determine the central and peripheral factors controlling food intake in pigs and sheep; Determine the central and peripheral factors controlling thermoregulatory behaviour in pigs and sheep; Determine the central and peripheral factors controlling sexual and aggressive behaviour in pigs and sheep; Quantify behavioural, physiological and neurochemical changes occurring in farm animals exposed to defined stressors; Determine the physiological and behavioural mechanism regulating food and fluid intake in pigs and sheep

4 Sharman DF - BSc PhD DSc FIBiol Examine neurochemical changes in the brain in relation to behaviour; Quantify behavioural, physiological and neurochemical changes occurring in farm animals exposed to defined stressors
5 Shillito Walser EE - BSc PhD Evaluate the effects of environmental factors on the behaviour of intensively housed pigs; Determine the sensory factors involved in the maternal/young bond in pigs and sheep
6 Thornton SN - BSc PhD Determine the central and peripheral factors controlling food intake in pigs and sheep; Determine the central and peripheral factors controlling fluid intake in pigs and ruminants; Determine the physiological and behavioural mechanism regulating food and fluid intake in pigs and sheep

4247 Biochemistry
Babraham, Cambridge, CB2 4AT
Tel: 0223 852312.

1 Coleman GS - MA PhD Metabolism of plant components by rumen entodiniomorphid protozoa; Infection of ruminants with entodiniomorphid protozoa
2 Gilmour RS - BSc PhD Investigation of DNA sequence representing the tissue specific expression of beta-globin in transgenic mice
3 Hazlewood GP - BSc PhD Transformation mechanisms in rumen bacteria and role of monoclonal antibodies in digestive microbiology; Immunological methods for quantifying the protein components of ruminant digesta
4 Irvine RF - PhD,BA Mechanisms of activation of phospholipases and their role in regulation of cellular activity. Second messenger metabolism and function
5 Mann SP - PhD,MIBiol Development of cloning systems and vectors for use in bacteria associated with fibre digestion in ruminants
6 Munn EA - BSc PhD DSc FIBiol DipRMS Electron microscopy of rumen bacteriophages; Electron microscopy of rumen phycomycete fungi; Helminth surface membrane antigens and their biosynthesis for use in vaccines
7 Orpin CG - BSc PhD Rumen bacteria: natural plasmids; cellulase gene cloning;isolation, rumen inoculation with highly cellulolytic bacteria; Isolation of lysogenic rumen bacteria, determination of host specificity and search for gene transduction; Metabolism and digestion of plant tissues by rumen phycomycete fungi

4248 Immunology
Babraham, Cambridge, CB2 4AT
Tel: 0223 852312.

1 Beale D - BSc PhD Structural basis of interaction of poly-Ig receptors with IgM
2 Binns RM - MA PhD BVetMed MRCVS Properties and functional behaviour of pig lymphocyte populations and lymphoid tissues
3 Butcher GW - BA PhD Specific T cell functions in the rat; Immunogenetics and functions of the major histocompatibility complex; Immune response gene control of graft rejection
4 Howard JC - MA,PhD Interaction between IgG and CL; Isolate the heavy chain array of rat Ig genes; Mechanism of generating cytotoxic T cells in rat; Specificity of T cell recognition of antigens; Immune response gene control of graft rejection
5 Hughes-Jones NC - MA,DM,PhD,MRCP,FRS Mechanism of activation of CL; Antibody secretion by human lymphoblastoid cell lines; Characterisation of rhesus d antigen of human red cells
6 Roser BJ - MB BS PhD FRCPA Role of antigen and anti-idiotype in transplantation tolerance; soluble histocompatibility antigens
7 Symons DBA - BSc MSc MA PhD Identification and sequence of bovine immunoglobulin genes
8 Taussig MJ - BA PhD Use anti-idiotypes to identify structural family of antisteroid monoclonals; Idiotypic analysis of immune response against progesterone
9 Tregear RT - MSc,MA,ScD Muscle crossbridge structure and mechanism

4249 Neuroendocrinology
Babraham, Cambridge, CB2 4AT
Tel: 0223 852312.

1 Bicknell RJ - BSc PhD Role of opioids in control of pituitary function; Membrane properties of lactotorphs and somatotrophs
2 Dyer RG - BSc MSc PhD Neural control of pituitary gland; Membrane properties of lactotorphs and somatotrophs; Neuropeptides in behaviour
3 Evans MH - BSc PhD Hypothalamic integration of neuroendocrine function with cardiovascular and respiratory systems
4 Leng G - BSc MSc PhD Neural control of oxytocin and vasopressin secretion; Neural circuits for control of drinking behaviour
5 Mason WT - BSc PhD Neural control of oxytocin and vasopressin secretion; Membrane properties of pituitary cells; Neural circuits for control of drinking behaviour
6 Robinson JE - BSc PhD Neural control of luteinising hormone (LH) secretion
7 Sirinathsinghji DJS - BA,MA,BSc,MSc,PhD Brain action of neuropeptides in control of hormone-dependent behavioural responses

4250 Physiology
Babraham, Cambridge, CB2 4AT
Tel: 0223 852312.

1 Brown KD - BSc PhD Synthesis and mechanism of action of polypeptide growth factors; Characterisation of polypeptide growth factors in the mammary gland and its secretions; Growth factor production by the uterus
2 Corps AN - BA PhD Hormonal control of fetal and neonatal growth
3 Fleet IR Galactopietic effects of growth hormone and its cardiovascular effects
4 Flint APF - PhD,DSc,FIBiol Regulation of ovarian function during the oestrus cycle, early pregnancy and post partum period
5 Harrison FA - BSc MVSc DSc MRCVS BVSc Synthesis and mechanism of action of placental hormones; Role of maternal hormones in the onset of parturition; Effects of dietary constituents and local factors on gut function; Role of bile salts in regulating normal hepatic clearance
6 Heap RB - BSc MA PhD ScD FIBiol Regulation of pituitary function during the anoestrus cycle early pregnancy and post partum period; Steroid and growth factor transport across the mammary gland; Cellular and immune responses in early pregnancy and the cause of embryonic mortality
7 Lindley IJD - BSc PhD Regulation of ovarian function during the oestrus cycle, early pregnancy and post partum period
8 Whyte A - BSc,PhD,MIBiol,FRSH Cellular and immune responses in early pregnancy and the cause of embryonic mortality
9 Wooding FBP - MA PhD Regulation of ovarian function during the oestrus cycle, early pregnancy and post partum period; Synthesis and mechanism of action of placental hormones; Cellular and immune responses in early pregnancy and the cause of embryonic mortality

4251 Cell Biology
Babraham, Cambridge, CB2 4AT
Tel: 0223 852312.

1 Dauncey MJ - BSc PhD Effects of energy intake and hormones on metabolism; Cellular mechanisms related to change in metabolic rate induced by nutrition and environment; Spontaneous motor activity metabolic rate and cell structure; Environmental influence on energy regulation
2 Harrison FL - BSc PhD Distribution of endogenous lectins in fibro blast and long term bone marrow cultures
3 Ingram DL - BSc PhD DSc FIBiol Effects of environmental temperature and food intake on metabolism and hormones; Mitochondrial enzymes in relation to energy intake and ambient temperature; Exercise and its influence on metabolic rate and cellular metabolism; Effect of adaptation to environment on thermoregulation and feeding
4 Lawson DEM - BSc PhD cDNA cloning
5 Sepulveda FV - Lic en Ciencias, Dr-es-Sc Ionic and nutrient transport mechanisms across mammalian enterocytes
6 Smith MW - BPharm PhD DSc Enterocyte differentiation in neonatal and adult intestines; Effect of disease and development on intestinal structure and proliferation; Immunological effects on mammalian gut function
7 Tucker EM - MA BSc PhD DSc Correlation of bovine and ovine chromosomes with the presence of bovine and ovine isozymes in heteromyelomas; Bovine and ovine polyclonal and monoclonal antibodies and use in physiological and biochemical studies of membranes; Transport of nutrients and cations across ruminant cells and the role of membrane antigens

4252 Molecular Embryology
Babraham, Cambridge, CB2 4AT
Tel: 0223 852312.

1 Brown CR - MPhil,FIST Molecular mechanisms of sperm/egg recognition and sperm penetration. Structure and function of the egg zona pellucida; interaction with oviduct macromolecules
2 Cran DG - BSc PhD Structural changes during oocyte fertilization; Structural changes during oocyte maturation
3 Gaunt SJ - BSc BVSc PhD Construction, introduction, integration and expression of genes
4 Harrison RAP - MA PhD Interactions of spermatozoa with their environment; Physiology and biochemistry of the acrosome
5 Jones J - BSc PhD Characterisation of sperm surface antigens and DNA
6 Miller JR - BSc,PhD Molecular aspects of genomic imprinting; Construction, introduction, integration and expression of genes
7 Moor RM - BSc PhD Cellular and molecular mechanism in oocyte maturation; Characterization of genes and proteins in early embryos; Construction, introduction, integration and expression of genes; Interactions between somatic cell nuclei and embryonic cytoplasm
8 Polge EJC - BSc,PhD,FRS Storage of spermatozoa *in vitro* and maintenance of fertilizing compounds; *In vitro* fertilization; Production and use of plurotential cell lines prepared from embryos of domestic animals; Production of chimaeras using pluripotential cells; Determine environmental factors governing cryopreservation of pig embryos
9 Powell DJ - BSc,PhD Characterization of genes of the early embryos; Molecular aspects of genomic imprinting; Construction, introduction, integration and expression of genes
10 Surani MAH - BSc PhD Genetic manipulation of eggs and embryos; Construction, introduction, integration and expression of genes

4253 AFRC Institute for Animal Health - Compton Laboratory
Compton, Nr Newbury, Berkshire, RG16 0NN
Tel: 063522 411.

Diseases of farm animals especially cattle and pigs

1 Rutter JM - PhD *Acting Director*

4254 Environmental Sciences
Compton, Nr Newbury, Berkshire, RG16 0NN
Tel: 063522 411.

1 Craven N - BSc,BVSc,PhD,MRCVS Mammary gland immunity. Inflammatory processes and neutrophil activity in the bovine mammary gland. Teat duct defence mechanisms and their effect on bacterial penetration
2 Williams MR - BSc,PhD Interactions between bacteria and neutrophils. Host and pathogen interactions in mastitis pathogenesis

4255 Pathology
Compton, Nr Newbury, Berkshire, RG16 0NN
Tel: 063522 411.

1 Reid IM - BSc,PhD,MNS,MRCPath *(now at AFRC Central Office)* Structure and function of claw horn and relationship to lameness; Footbathing, foot trimming and floor design as means of controlling lameness

4256 Microbiology
Compton, Nr Newbury, Berkshire, RG16 0NN
Tel: 063522 411.

1 Baird GD - MA PhD DPhil Virulence determinants in *Salmonella* serotypes and the interaction of these determinants with the animal host
2 Bridger JC - BSc PhD The primary viral pathogens of calf enteritis. The variations in virulence, antigenicity and cross-protection between bovine rotaviruses; immunity to rotavirus by dam vaccination; viral causes of vaccine breakdown
3 Britton P - BSc,PhD Molecular cloning of TGEV subgenomic RNA: characterisation and expression of clones
4 Chanter N - BSc PhD To characterise the toxin of P
5 Garwes DJ - BSc MSc PhD Monoclonal antibodies against transmissible gastroenteritis virus; Vaccine development against enteric viruses; Investigating the structure and expression of TGEV peplomer antigens
6 Gourlay RN - BSc DipBact DSc MRCVS FRCPath Pathogenicity of P; Immune response in lungs to infection with P; Ability of various antigens of P
7 Jones PW - BSc Induction of resistance to salmonellosis in calves, transfer of immunity and variation in *Salmonella* strain virulence
8 Lax A - BSc PhD Investigation of phase variation in *Bordetella*, and the virulence of *Bordetella* and *Pasteurella*; e of B
9 Lysons RJ - BVSC PhD MRCVS Aetiology and pathogenesis of swine dysentery; Diagnosis of swine dysentery and clinically similar diseases; Vaccines for and the eradication of swine dysentery
10 Morgan JH - MA BSc VetMB MRCVS Determine the bacterial pathogens of calf enteritis
11 Oldham G - BSC,PhD Investigation of active immunity and mechanisms of immune protection. Characterisation of bovine interleukins. Production and characterisation of monoclonal antibodies to surface markers on bovine lymphocytes and accessory cells
12 Pocock DH - HNC MIBiol Molecular basis of rotavirus virulence; Molecular basis for the variation in polypeptides and genome of different BVD virus isolates
13 Rutter JM - BVMS,BSc,PhD,MRCVS Virulence of *Pasteurella* spp. Epidemiology of *P. multocida*
14 Scott EJ - BSc,PhD,MRCPath Production and characterisation of monoclonal antibodies to surface markers on bovine lymphocytes and accessory cells
15 Sellwood R - PhD,FIMLS Genetic basis of virulence of *Treponema hyodysenteriae*. Diagnosis of swine dysentery and clinically similar diseases
16 Stott EJ - BA PhD Diagnosis, epidemiology and pathogenesis of respiratory virus infections; Cell mediated and humoral immunity to Respiratory Syncytial Virus; Development of vaccines and monoclonal antibody therapy against Respiratory Syncytial Virus
17 Thomas LH - MA VetMB MRCVS Assessment of pathological response to microorganisms; Assistance with experimental infections; Design, conduct and assess field trials and preliminary experimental studies with various products

4257 Immunology and Parasitology
Compton, Nr Newbury, Berkshire, RG16 0NN
Tel: 063522 411.

1 Anderson JC - BVMS PhD MRCVS MRCPath The pathology of mastitis and the use of animal models to identify virulence determinants. Methods of increasing the effectiveness of the teat duct barrier in dry cows
2 Bramley AJ - BSc,PhD,CBiol,MIBiol Epidemiology and control of bovine mastitis, particularly coliform and summer mastitis

3 Brownlie J - BVSc PhD MRCVS Determine immunological and immunosuppressive mechanisms in mucosal disease and BVD virus infection
4 Clarke MC - FIMLS,MIBiol Pathogenesis of cytopathic and non-cytopathic isolates of BVD virus
5 Hall GA - BVSC MRCVS PhD MRCPath The nature, severity and extent of intestinal lesions in piglets and calves with enteric disease and repair mechanisms. Characterisation of lesions produced by virulent and avirulent rotaviruses. Pathogenic processes of newly discovered enteropathogens of calves and pathogenesis of mixed infections
6 Hill AW - BSc,PhD Strain typing systems for *Streptococcus uberis* and the epidemiology of infection; pathogenesis of *Str. uberis* infection in lactating and dry cattle
7 Howard CJ - BSc PhD Development and improvement of antigens of *M. dispar*. The immunological and immunosuppressive mechanisms in BVD virus infections. Production and characterisation of monoclonal antibodies to surface markers on bovine lymphocytes and accessory cells
8 McKinnon CH Epidemiology and microbiology of bovine mastitis

4258 AFRC Institute of Plant Science Research - John Innes Institute
Colney Lane, Norwich, NR4 7UH
Tel: 0603 52571.

Research in virus, microbe and higher plant biochemistry and genetics as a foundation for applied work in plant and microbial technology

1 Flavell Dr RB *Director (John Innes Institute)*
2 Woolhouse Prof. HW - PhD,FIBiol,FLS *Director of Research (IPSR)*

4259 Applied Genetics
Colney Lane, Norwich, NR4 7UH
Tel: 0603 52571.

1 Casey R - BSc PhD Isolation and characterisation of seed storage proteins in the pea
2 Criessen G - PhD,BSc Cytological localisation of particular DNA sequences in plant cells
3 Davies DR - BSc PhD Gene mapping using restriction fragment polymorphism and *in situ* hybridisation in the pea
4 Domoney CAM - BSc PhD Analysis of the activity of specific seed storage protein genes
5 Dunwell JM - BA MA PhD Generation of haploid plants from pollen
6 Ellis THN - BSc PhD Genetic mapping using restriction length fragment polymorphisms in the pea
7 Hedley CL - BSc PhD Genetic variation for the cellular development and storage product accumulation in pea embryos
8 Hussey G - BSc PhD Regeneration of plants from cell cultures in peas
9 March J - BSc Isolation and characterisation of seed storage proteins in the pea
10 Matthews P - BSc PhD Disease resistance and pathology of peas
11 Mullineaux PM - BSc,PhD Transformation of plant cells
12 Shannon PRM - BSc,PhD Development of micro-injection techniques for plant cells
13 Thomas CM - BSc Characterisation of mitochondrial genomes of sugar beet
14 Wang TL - BSc PhD Regulation of cellular development and storage product accumulation in pea embryos
15 Young JPW - MA,PhD Genetic variation in natural populations of *Rhizobium*

4260 Genetics
Colney Lane, Norwich, NR4 7UH
Tel: 0603 52571.

1 Asbourn A - BSc,PhD Use of genetics and molecular biology to understand the basis of pathogenicity of *Xanthomonas* to plants
2 Bibb MJ - BSc,PhD Develop methods of genetic manipulation in *Streptomyces* as a background to strain improvement
3 Brewin NJ - BSc,PhD Molecular biology of legume root nodules
4 Carpenter R - HNC,MIBiol Isolation and characterisation, physically and genetically, of transposable genetic elements of plants
5 Chater KF - BSc PhD Develop methods of genetic manipulation in *Streptomyces* as a background to strain improvement
6 Coen ES - BA PhD Isolate and characterise, physically and genetically, transposable genetic elements of plants
7 Collinge DB - BSc PhD Use genetics and molecular biology to understand the basis of pathogenicity of *Xanthomonas* to plants
8 Daniels MJ - BA MA PhD Use genetics and molecular biology to understand the basis of pathogenicity of *Xanthomonas* to plants
9 Davis EO - BSc PhD Structure and function of *Rhizobium* nodulation genes
10 Dow JM - PhD,MA Host-pathogen interactions between turnip and *Xanthomonas*
11 Evans IJ - BSc PhD Analysis of *Rhizobium* nodulation genes
12 Firmin JL - BSc PhD Biochemistry of nitrogen-fixing legume root nodules
13 Hopwood DA - MA PhD DSc FRS Develop methods of genetic manipulation in *Streptomyces* as a background to strain improvement
14 Johnston AWB - BSc PhD Molecular biology of *Rhizobium* symbiotic genes

15 Martin CR - BA,PhD Isolate and characterise, physically and genetically, transposable genetic elements of plants
16 Ridge R - BSc,PhD Cytology of the infection of legumes by *Rhizobium*
17 Robbins T - BSc,PhD Isolation and characterisation, physically and genetically of the delila locus of *Antirrhinum majus*
18 Roberts IN - BSc,PhD Host-pathogen interactions between turnips and *Xanthomonas*
19 Rossen L - BSc,PhD Identification of nodulation genes in legumes

4261 Cell Biology
Colney Lane, Norwich, NR4 7UH
Tel: 0603 52571.

1 Burgess J - MA,PhD Structural basis of plant cell development
2 Dawson PJ - BSc,PhD Molecular characterisation of cytoskeletal proteins
3 Doonan JH - BSc,PhD The cytoskeleton in morphogenesis using mass as a model system
4 Hills G - MIBiol Structural studies in plant development
5 Lloyd CW - BSc,PhD The influence of the cytoskeleton on cell expansion and the plane of division
6 Rawlins D - BSc Computer image processing and plant nuclear structure
7 Roberts K - MA,PhD Structure and biochemistry of plant cell surfaces
8 Shaw P - BA,MA,PhD Computer image processing and plant cell structure
9 Stacey N - BSc Sugar beet protoplast production and regeneration
10 Watts J - PhD Plant protoplast fusion transformation and regeneration

4262 Virus Research
Colney Lane, Norwich, NR4 7UH
Tel: 0603 52571.

1 Ali Hakim A - PhD Development of non-radioactive probes
2 Boulton MI - HNC Exploitation of gemini-viruses as vectors for genetic manipulation of cereals
3 Covey SN - BSc PhD Molecular biology of plant viruses and virus-host interactions
4 Davies JW - PhD Molecular biology of plant viruses
5 Donson J - PhD Replication and expression of DNA plant viruses
6 Hull R - PhD Molecular biology of plant viruses
7 Lomonossoff GP - PhD Molecular biology of RNA plant viruses
8 Markham PG - PhD Transmission of viral plant pathogens by aphids and leafhoppers and the interaction between pathogen and plant host
9 Maule AJ - PhD Molecular biology of plant DNA and RNA viruses and the host response to infection
10 Stanley JR - PhD Replication and expression of DNA plant viruses
11 Townsend R - PhD Replication and expression of DNA plant viruses
12 Wilson TMA - PhD Molecular biology of RNA plant viruses and recombinant RNA packaging

4263 Photosynthesis Group
Colney Lane, Norwich, NR4 7UH
Tel: 0603 52571.

1 Smith AM - BSc,PhD Physiology and biochemistry of photosynthesis, respiration and partitioning of assimilated carbon
2 Woolhouse HW - BSc PhD FInstBiol FLS Photosynthesis and dark respiration in relation to carbon economy of plants

4264 AFRC Institute of Arable Crops Research - Long Ashton Research Station
Long Ashton, Bristol, BS18 9AF
Tel: 0272 392181.

Plant development and the control of weeds, pests and diseases and crop protection practices

1 Treharne Prof. KJ - BSc,PhD,FIBiol *Professor*

4265 Crop Pathology
Long Ashton, Bristol, BS18 9AF
Tel: 0272 392181.

1 Brent KJ - BSc MSc PhD FIBiol Behaviour of fungicide-resistant pathogens; Factors affecting fungicide resistance in crop pathogens
2 Burden RS - BSc,PhD,CChem,MRSC Effects of fungicides on plant metabolism. Metabolism of fungicides in plant tissue
3 Carter GA - BSc,MSc,PhD,MIBiol Bioassay of effects of fungicide regimes on sensitivity of cereal pathogens; Metabolism of fungicides; Behaviour of fungicide-resistant pathogens
4 Deas AHB - MSc,MRSC,CChem Metabolism of fungicides
5 Hollomon DW - BSc PhD Effects of fungicide regimes on sensitivity of cereal pathogens
6 Huband ND - BSc PhD Measurement of weather and crop microclimate relative to plant disease epidemiology
7 Jordan VWL - MSc MIBiol, PhD Forecasting disease epidemics relative to chemical control methods; Effects of fungicide regimes on sensitivity of cereal pathogens; Incidence and effects of fungal diseases in crops, and methods of control; Development of fungal resistance to fungicides; Cultivation and straw disposal relative to cereal pests and diseases; Pests and diseases in grass crops
8 Loeffler RST - BSc,PhD,CChem,MRSC Mechanisms of fungicide action and resistance

9 Royle DJ - BSc PhD MIBiol Disease epidemiology relative to crop growth/environmental conditions; develop forecasting principles and methods

4266 Crop Protection Chemistry
Long Ashton, Bristol, BS18 9AF
Tel: 0272 392181.

1 Anderson NH - BSc PhD Biochemical mechanisms of fungicidal action and selectivity. Metabolism of fungicides by plants and microorganisms
2 Baker EA - MSc,PhD,MIBiol,MRSC,CChem Deposition, redistribution and uptake of agrochemicals on leaves; Determine effects of formulations and environment on cuticular penetration and movement of xenobiotics
3 Holloway PJ - BPharm PhD Behaviour and fate of surfactants applied to leaves and their effects on delivery of agrochemicals

4267 Physiological Plant Pathology
Long Ashton, Bristol, BS18 9AF
Tel: 0272 392181.

1 Bailey JA - BSc,PhD,MIBiol,CBiol Biochemistry, cytology and genetics of plant-pathogen interactions
2 Hargreaves JA - BSc PhD Biochemistry and cytology of infection of plants by fungi. Molecular genetics of plant pathogenic fungi

4268 Spray application
Long Ashton, Bristol, BS18 9AF
Tel: 0272 392181.

1 Cooke BK - BSc,PhD,MIBiol,CBiol Determination of spray deposits and biological efficacy; Examination of farm and orchard soils for accumulation of pesticides; Measurement of deposits of agrochemicals applied by hydraulic spray systems; Chemical assessment of deposits of agrochemicals applied by novel spraying techniques
2 Hislop EC - BSc DPP PhD Required properties of spray deposits and delivery systems relative to efficient and effective agrochemical action; Application of agrochemicals from hydraulic sprayers, deposition and biological efficacy; Application of agrochemicals by new spraying techniques: deposition and biological efficacy

4269 Zoology
Long Ashton, Bristol, BS18 9AF
Tel: 0272 392181.

1 Glen DM - BSc PhD Slug behaviour and control; Cultivation and straw disposal relative to cereal pests and diseases; Pests and diseases in grass crops
2 Kendall DA - BSc PhD Forecasting and control of Barley Yellow Dwarf Virus; Cultivation and straw disposal relative to cereal pests and diseases; Pests and diseases in grass crops
3 Moss S - BSc Effects of tillage and straw disposal systems on weed problems and herbicide activity
4 Smith BD - BSc DIC PhD Forecasting and control of Barley Yellow Dwarf Virus; Cultivation and straw disposal relative to cereal pests and diseases; Pests and diseases in grass crops

4270 Hormone Biochemistry
Long Ashton, Bristol, BS18 9AF
Tel: 0272 392181.

1 Hedden P - BSc PhD Hormonal analysis and metabolism in developmental mutants. Subcellular localisation of gibberellins and their metabolising enzymes. Mode of action of gibberellin in wheat. Biochemical mode of action of plant growth retardants
2 Lenton JR - BSc PhD Growth substance status and responsiveness in plant development. Subcellular distribution of gibberellins in responsive and non-responsive tissues. Gibberellin insensitivity mechanism in wheat. Manipulation of gibberellin and sterol production with plant growth retardants. Study of calcium as a second messenger in plant hormone action
3 Wright M - BSc PhD Effects of gravity on auxin production and metabolism and distribution of auxin and assimilates in grass nodes

4271 Environmental Physiology
Long Ashton, Bristol, BS18 9AF
Tel: 0272 392181.

1 Butler DR - BSc MSc PhD Techniques to measure canopy structure and assess effect of agronomic treatments on canopy structure and plant growth; Study microclimate of cereal crops sprayed with plant growth regulators; Develop and test simulation models for evaporation from wet canopies to predict wetness duration from weather data; Investigate methods of estimating diurnal curves of meteorological variables from synoptic data
2 Porter JR - BSc PhD MIBiol Develop techniques to measure leaf position and leaf area density in a growing canopy; Develop and test predictive crop simulation models of development and growth
3 Rooney JM - BSc Effects of leaf diseases on leaf production, gas exchange and assimilate partitioning in winter wheat
4 Wright MW - BSc,PhD Auxin and assimilate distribution in grass nodes in relation to shoot geotropism

4272 Developmental Physiology
Long Ashton, Bristol, BS18 9AF
Tel: 0272 392181.

1 Abbott AJ - BSc,MSc,PhD,FIBiol Use of tissue-culture techniques to study mechanisms regulating morphogenesis; Effects of environment and growth regulators on the physiology and morphogenesis of plants grown *in vitro*
2 Anderson HM - BA,MIBiol,SDH Effects of applied plant growth regulators on stress tolerance and seed shedding. Use of plant growth regulators to increase seedling survival in oats and barley, and for the control of pod shatter in oilseed rape
3 Barlow PW - BSc,DPhil Analysis of cell growth and division in relation to differentiation and morphogenesis, particularly in roots
4 Barratt DHP - BSc,PhD Hormonal control of gene expression in developing legume cotyledons
5 Child RD - MIBiol Monitor crop response to applied growth regulators under various environmental conditions; Effects of plant growth regulators on canopy structure of cereals in relation to microclimate and disease incidence; Plant growth regulator usage to improve efficiency of arable crop production
6 Fido RJ - MSc,PhD,CBiol Purification of nitrate reductase and determination of its structure and catalytic functions
7 Hooley R - BSc PhD Hormonal control of gene expression in the germinating wild oat. Isolation and characterisation of gibberellin receptor proteins
8 Hucklesby DP - BSc,MSc,PhD Quantifying nitrate and nitrite reductases, nitrate transport and products of assimilated nitrogen in arable crops. Nitrate/nitrite reductase induction and repression *in vivo* and construction of a nitrate reductase cDNA library
9 Jackson MB - BSc,MSc,DPhil Effect of gaseous components of the environment on organogenesis. Effects of aeration and mechanical stress on roots and shoots; involvement of ethylene and other plant hormones. The effect of plant growth regulators on the endogenous concentrations of ethylene and abscisic acid
10 Notton BA - TD,MSc,PhD,CChem,CBiol,MIBiol,FRSC Purification, structural analysis and intracellular localisation of nitrate reductase with mono- and polyclonal antibodies

4273 Plant Science
Long Ashton, Bristol, BS18 9AF
Tel: 0272 392181.

1 Hoad GV - BSc,PhD,MIBiol Structure-activity relations of gibberellins and their derivatives. Phloem translocation of endogenous hormones relative to sieve-tube unloading

4274 Transport Mechanisms
Long Ashton, Bristol, BS18 9AF
Tel: 0272 392181.

1 Butcher DNP - BSc,PhD,MIBiol Plant growth regulators and the physiology of plant development. The transport and fate of applied plant growth regulators in arable crops. Physiological effects of plant growth retardants on whole plants and organ cultures. Translocation of assimilates and plant growth substances. Effect of the gaseous environment on plant development and ethylene production by *in vitro* root cultures
2 Clarkson DT - BSc,MSc,DPhil Physiology and biochemistry of plant cells and membranes; Physiology of ion and water movement in higher plants
3 Drew MC - MA DPhil Effects of root environment on ion and water movement in plants; Effects of root environment on shoot growth and physiology
4 Lee RB - BSc PhD Ion transport and compartmentation using nuclear magnetic resonance (NMR) and flux analysis

4275 Pomology
Long Ashton, Bristol, BS18 9AF
Tel: 0272 392181.

1 Campbell AI - MSc PhD Plum breeding. Clonal selection in hardy ornamentals; EMLA virus tested fruit trees; Assess virus effects on fruit trees; Heat treatment of fruit trees; Mutation breeding of fruit trees; Optimum cropping of apples; Evaluate clonal differences; Evaluate trees raised from tissue culture without rootstocks; Mutant assessment and selection of apples and pears following irradiation
2 Stott KG - BSc,MIFor Assessment of timber willows and comparison of fast-growing trees as source of biomass. Evaluation of plant growth regulators on selected swards, windbreak trees and novel arable crops. Weeds in permanent grass swards; their status and control
3 Williams RR - BSc DHS Pollination studies with self-fertile Cox and cider varieties; Pollination characteristics of ornamental crabs using marker genes; Chemically control biennial bearing in cider orchards; Compare new sward mixtures suitable for cider orchards using heavy harvesting machinery; Compare fruit thinning sprays in cider orchards; Examine growth regulator effects on flowering and fruit set

4276 Herbicide Group
Long Ashton, Bristol, BS18 9AF
Tel: 0272 392181.

1 Breeze VG - BSc PhD Phytotoxicity of herbicides in vapour phase and as small drops in relation to drift damage

2 Clay DV - BSc Evaluation of herbicides for forestry. Develop control measures for problem weeds in fruit based on the weeds' agroecology and sensitivity to herbicides; Develop methods of predicting tolerance of fruit crop to herbicides which avoid the need for extensive field testing
3 Cotterill EG - LRSC,(HNC) Phytotoxicity of herbicides in vapour phase and as small drops in relation to drift damage
4 Hilton JR - BSc,PhD Physiology and biochemistry of weed seed dormancy and germination

4277 Microbiology and Vegetable Management Group
Long Ashton, Bristol, BS18 9AF
Tel: 0272 392181.

1 Barrett PR - BSc Developing environmentally acceptable controls for aquatic weeds emphasising formulation and application techniques
2 Greaves MP - BSc Development of biological weed control methods and integration with chemical methods; Interactions of soil/root region microflora and herbicides as factors in herbicide performance and yield variability
3 Marshall EJ - BA,PhD,MIBiol Vegetation management in field margins and amenity grass by chemical and cultural means

4278 Tropical Weeds Group
Long Ashton, Bristol, BS18 9AF
Tel: 0272 392181.

1 Parker C - MA Physiological studies on the resistance of sorghum, millet and cowpea to parasitic *Striga* species; Selectivity and performance of herbicides on tropical crop and weed species
2 Terry PJ - BTech Selectivity and performance of herbicides on tropical crop and weed species

4279 Annual Crops Group
Long Ashton, Bristol, BS18 9AF
Tel: 0272 392181.

1 Cussans GW - BSc Study of variation in resistance to herbicides in *Alopecurus myosuroides*. Integration of weed control measures into crop management systems. Straw disposal and land cultivation in relation to weed, pest and disease problems in cereals. Population dynamics of arable weeds
2 Lutman PJW - BSc PhD Weed control in cereals, field evaluation of herbicides: effects of formulation mixtures, timing, weather and soil; Research into oil seed rape: the effects of weed on crop growth and yield: crop tolerance and efficacy of herbicides
3 Moss SR - BSc Weed control in cereals
4 Peters NCB - BSc,PhD The study of seed behaviour with special reference to dormancy. Competition with regard to soil variable factors
5 Tottman DR - BSc MIBiol The tolerance of cereals to herbicides: related to growth stage, variety, weather, disease and dose of chemical
6 Wilson BJ - BSc Population dynamics of arable weed: seed production and fate, the influence of cultural factors, species variability; Cereal response to weed competition: effect of crop and weed density, time of weed removal, effect of weed species

4280 Grassland Group
Long Ashton, Bristol, BS18 9AF
Tel: 0272 392181.

1 Standell CJ - BSc Assessment of weed damage in young leys and development of appropriate control measures with special reference to crop safety

4281 Environmental Studies Group
Long Ashton, Bristol, BS18 9AF
Tel: 0272 392181.

1 Blair AM - BSc MPhil MIBiol Effect of weather factors on isoproturon activity on blackgrass, wild oats and wheat; Mechanisms of weather effects on isoproturon on blackgrass, wild oats and wheat
2 Caseley JC - BSc PhD MIBiol Improving performance by understanding the importance of weather factors; Understanding the mechanisms of herbicide/plant environment interactions

4282 Weed Biology and Management
Long Ashton, Bristol, BS18 9AF
Tel: 0272 392181.

1 Cousens RD - BSc PhD Integration of weed control measures into crop management systems. Long-term studies and modelling of weed seed populations. Modelling competition and its economic significance

4283 Macaulay Land Use Research Institute, Aberdeen
Craigiebuckler, Aberdeen, AB9 2QJ
Tel: 0224 318611.

Soil survey and research to advance the understanding of soil science, and to improve the fertility of Scottish soils

1 Dickie Mrs A - ALA,MIInfSc Librarian
2 Maxwell Prof. TJ Director

4284 Mineral Soils
Craigiebuckler, Aberdeen, AB9 2QJ
Tel: 0224-318611.

1 Bain DC - BSc PhD Weathering of minerals in hydrologic sequences of soils to establish relationships with availability of trace elements; Supervision of entry and of requests for retrievals of mineralogical data
2 Bracewell JM - BSc CChem FRSC Organic matter composition, development and degeneration in natural and agricultural soils by analytical pyrolysis
3 Ferrier RC - BSc Role of soils and vegetation in modifying rainwater chemistry
4 MacMahon R Speciation of aluminium in natural waters
5 McHardy WJ - BSc PhD Development of scanning electron microscope techniques for study to root/soil interface. Study of relationship between microstructure and physical properties of Scottish soils
6 Mellor A - BSc,PhD Field and laboratory studies of rates of chemical weathering in different soil types
7 Paterson E - BSc Studies the influence of surface and colloid properties on soil physical and chemical behaviour
8 Smith BFL - BSc MSc CChem MRSC Supervision of systematic chemical and physical analysis of survey samples; Study of release of principal elements from soils by analysis of soil and drainage; Supervision of entry of physical and chemical data and of their retrieval
9 Tait JM - BSc PhD Investigate the crystalline and poorly ordered inorganic constituents of soils. Development of transmission electron microscope techniques for the study of the root/soil interface
10 Walker TAB - BSc Interactions between soil waters and the exchange complexes of soils
11 Wilson MJ - BSc PhD DSc X-ray diffraction studies of crystalline inorganic soil constituents. Investigation of mineral weathering processes in selected catchment sites; Identify and supervise study of soil of variability problems in department of mineral soils

4285 Peat and Forest Soils
Craigiebuckler, Aberdeen, AB9 2QJ
Tel: 0224-318611.

1 Birnie RV - LRSC,MSc Development and application of remote sensing techniques for peat and vegetation surveys; Assessment of snow cover using Landsat data; Assessment of Landsat imagery for bracken survey; Application of radiometers to crop biomass studies
2 Hulme PD - BSC PhD Survey, mapping and classification of peat resources and peatland vegetation
3 Miller DM Mapping of bracken to quantify rates of spread; development of a knowledge-based vegetation mapping system
4 Miller JD - LRSC Design, development and installation of equipment to monitor biogeochemical cycling of nutrients
5 Nisbet AF - BSc,PhD,DIC Changes in the chemistry of rainwater on passage through forest vegetation
6 Nisbet TR - BSc,PhD Changes in the chemistry of rainwater on passage through upland acid soils
7 Proe MF - BSc Modelling the biogeochemical cycling of nutrients in upland ecosystems; Develop simulation models of interaction between nutrient availability and plant growth; Atmospheric "filtering" effect of upland vegetation and consequences for nutrient cycles; Nutrient cycling in mixed stands of coniferous forest
8 Robertson RA - BSc Application of remote sensing techniques for terrain resource surveys; Provide an advisory and application centre for remote sensing within AFRS and facilities for contractual research; Assessment,development and utilisation of Scottish peat resources; Characterisation and assessment of peat and peat products; Effect of soils, vegetation and land use on rainwater chemistry
9 Williams BL - BSc PhD Application of physical and chemical methods for the analysis and characterisation of peat and peat products; To determine the processes controlling nutrient mineralisation and availability in highly organic soils; Nitrogen transformations beneath pure and mixed stands of Sitka spruce
10 Wright GG - BSc Development and application of remote sensing techniques for soil mapping, crop survey and monitoring. Application of radiometers to studies of soil reflectance properties and to crop performance monitoring

4286 Spectrochemistry
Craigiebuckler, Aberdeen, AB9 2QJ
Tel: 0224-318611.

1 Adams MJ - GRSC PhD DIC CChem MRSC Develop instrumentation for multielement spectrochemical methods;Design and testing of portable radiometers
2 Bacon JR - BSc PhD Comprehensive analysis and fingerprinting of soils by spark source mass spectrometry; Development of spark source mass spectrometry: instrumentation, ion-sources detectors and sample preparation
3 Berrow ML - BSc PhD CChem FRSC Establish geochemical and environmental origins, distribution and mobility of trace elements in scottish soils; Investigation of soil aspects of soil/plant trace element; Develop multi-element analytical techniques using dc arc optical emission and spark source mass spectrometry for solids; Suspension of input and retrieval of trace-element data; The use of soil extraction to assess the forms of trace elements in soils

4 Burridge JC - BSc MA Effects of soil conditions on plant trace element uptake and distribution of trace elements in plant parts and species; Development of optical emission spectrometry; instrumentation, sources, spectrometers, chemical concentration methods

5 Goodman BA - BSc PhD Investigation of the form of trace elements in soils and solutions by EPR and Mossbauer spectroscopy; Determine in plants, amounts and forms of elements by EPR and Mossbauer spectroscopy, and free radical contents by EPR; Analysis of organo-metallic complexes by EPR and Mossbauer spectroscopy

6 Linehan DJ - BSc PhD Trace elements in the rhizosphere soil solution and their availability to plants

7 McPhail DB - BSc Investigation of the form of trace elements in soils and solutions by EPR

8 Russell JD - BSc DSc CChem FRSC Soil mineral structures,surface properties and weathering by IR and UV spectroscopic methods; Characterise soil organic matter by infrared and ultraviolet spectroscopy

9 Shand CA - BSc PhD Development of graphite furnace atomic absorption spectrometric methods; Development of chemical and instrumental methods for trace-element speciation in soils and plants; Forms of trace elements in soil and solution by selective extraction and hybrid fractionation/furnace AAS

10 Sharp BL - BSc PhD DIC CChem FRSC Development of instrumentation, computer programming, and analytical methods for ICP emission spectrometry

11 Ure AM - BSc PhD CChem FRSC Establish geochemical and environmental origins, distribution and mobility of trace elements in scottish soils; Effects of soil conditions on plant trace element uptake and distribution of trace elements in plant parts and species; Develop multi-element analytical techniques using dc arc optical emission and spark source mass spectrometry for solids; Develop spectrochemical methods using atomic absorption, inductively coupled plasma emission spectrometry for solutions; Forms of trace elements in soils, their interaction with humic substances and their transport mechanisms to the plant

4287 Soil Organic Chemistry
Craigiebuckler, Aberdeen, AB9 2QJ
Tel: 0224-318611.

1 Anderson HA - BSc PhD Chemical characterisation of organic N components and their transformations in soil; Chemistry and distribution of humic substances in soils and soil solutions

2 Cheshire MV - BSc PhD Origins, nature and behaviour of polysaccharides in soil and effects on soil physical properties; The chemistry and biology of organic matter-trace element reaction in soil

3 MacDonald IR - BSc PhD Seedling growth in relation to physical forces in the environment

4 Vaughn D - BSc PhD Examine the effects of organic constituents of soil on biochemical processes in plants and soils; Devise practical methods to remove pollutants from waste products used

4288 Plant Physiology
Craigiebuckler, Aberdeen, AB9 2QJ
Tel: 0224-318611.

1 Macklon AES - BSc PhD Trace element uptake and transport in plants; Characterisation of ammonium and nitrate ion uptake into root cells, in electrochemical and metabolic terms; Interaction effects on trace element uptake mechanisms in plant; Culture plants in nutrient solutions containing trace elements for speciation studies, interpret results

2 Shand C Development and application of atomic spectroscopic methods for the speciation and determination of trace elements in plant samples

3 Thornton B - BSc,PhD Isolation of trace element complexes from plants for identification by spectroscopic methods; interpretation of results

4289 Microbiology
Craigiebuckler, Aberdeen, AB9 2QJ
Tel: 0224-318611.

1 Chapman SJ - BSc PhD Interrelationships between bacteria and fungi with plant roots, with reference to sulphur and trace elements;Microbial transformation of soil organic matter with reference to sulphur and trace elements

2 Darbyshire JF - MSc PhD DipAgriSc Interrelationships between bacteria and protozoa with plant roots; Transformation of soil organic matter with reference to nitrogen and phosphorus in minerals; Ecology and physiology of soil protozoa

3 Jones D - MSc PhD MIBiol FRMS The interrelationships of fungi with plant roots; The survival of fungi in soil and their transformation into soil organic matter

4290 Soil Fertility
Craigiebuckler, Aberdeen, AB9 2QJ
Tel: 0224-318611.

1 Atkinson D - BSc,PhD Assessing the soil factors influencing selenium availability and uptake. Measurement of plant root system development in relation to nutrient uptake and soil condition. Recording variation in the root system of crop genotypes; assessing their importance. Characterisation of the properties of the root-soil interface; assessing its effects

2 Coutts G Assessing the soil factors influencing selenium availability

3 Duncan C Development of automated methods for the analysis of soil and plant extracts

4 Dyson PW - BSc PhD Designs experiments,interprets and reports results; Growth analysis and nutrient uptake studies in crops

5 Edwards A Assessing the fate of phosphorus in cropped soils and interactions between phosphorus, acidity and aluminium. Development of methods to study trace and other elements in soils

6 Mackie L - BSc PhD The effect of soil structure, texture, macroporosity on root growth, plant seedling and crop yields

7 Millard P - BSc PhD The organic and inorganic constituents of crops in relation to age and yield

8 Robinson D - BSc PhD Estimate mobility of soil native and applied nitrogen relative to water movements and crop uptake

9 Sharp GS - LRSC Measuring the ability of soils and fertilizers to provide the needs of crops from inorganic and organic sources

10 Shepherd H - LRSC Development and application of radioisotope methods

11 Sinclair AH - BSc PhD quantify phosphate levels in soils and phosphate availability; Designs experiments, interprets and reports results; Designs trace element experiments,interprets and reports results

4291 Statistics
Craigiebuckler, Aberdeen, AB9 2QJ
Tel: 0224-318611.

1 Brown KWM - BSc MSc MBCS Production and adaptation of computer programs for applications in soil research; Database design and administration

2 Cooper JM - BSc DiP Stat MIS Experimental design, statistical analysis and computer programming for soil research; Production and revision of computer programs for data processsing and statistical analysis

3 Inkson RHE - BSc FSS FIS Experimental design, statistical analysis, modelling and computer programming for soil research; Multivariate statistical analysis and model-building for crop responses; Planning and management of computing facilities required in soil research; General supervision, management and development of computing resources,both hardware and software

4 Reaves GA - BSc Computer programming for data storage and retrieval

4292 Soil Survey
Craigiebuckler, Aberdeen, AB9 2QJ
Tel: 0224-318611.

1 Bibby JS - BSc Development of interpretation of soil maps

2 Bown CJ - BSc Correlation of map units, classification of soils development and standardization of procedures

3 Dry FT - BSc Soil survey in Eastern Scotland; interpretation of soil maps; plant community recording

4 Futty DW - BSc Development of soil survey publications; cartography

5 Gould JH - BSc,PhD Interpretation of soil maps in Eastern Scotland

6 Heslop REF - BSc(For) Soil survey in Northern Scotland; Interpretations of soil maps in Northern Scotland; plant community recording

7 Hudson G - BSc Soil survey in western Scotland; Interpretation of soil maps in Western Scotland

8 Robertson JS - BSc Develop and co-ordinate the plant community classification; Replacement communities and land use pressure; grazing values

9 Robertson L - BSc Soil survey data management and soil modelling; Supervision of data entry and co-ordination of requests for retrievals relating to soil survey; Study of soil micromorphology using optical microscopy

10 Shipley BM - BSc Design, establish and co-ordinate research on soil moisture

11 Walker AD - BSc Interpretation of soil maps in Northern Scotland

4293 AFRC Institute of Engineering Research
Wrest Park, Silsoe, Bedford, MK45 4HS
Tel: 0525 60000.

Application of engineering to the design, construction and operation of agricultural horticultural and food process machinery

1 Cooper Ms A - BA,ALA Librarian

2 Matthews J - BSc,CEng,FInstP,FIAgrE Director

4294 Crop Engineering
Wrest Park, Silsoe, Bedford, MK45 4HS
Tel: 0525 60000.

1 Audsley E - BA MSc AFIMA Operational research study of economic effects of banning field burning of straw, and alternative uses for straw; Operational research study to determine the consequence of reducing inputs to major agricultural production systems; The development and use of operational research models

2 Frost AR - BSc MPhil MIMechE CEng Design of sprayer boom suspensions and spray application rate controllers

3 Miller PCH - BSc MIAgrE PhD Spray physics, uncharged and charged droplets, their production, movement and deposition; The design of spraying systems

4 Nellist ME - BSc MSc PhD MIAgrE CEng Mathematical modelling of grain drying; quality requirements of the design of high temperature continuous-flow driers; Ambient and low temperature grain driers - design and operational research study

4295 Engineering Design and Development
Wrest Park, Silsoe, Bedford, MK45 4HS
Tel: 0525 60000.

1 Chisholm CJ - BSc PhD MIMechE CEng Formulae for the prediction of static pressures on storage structures. Prediction, testing and improvement of machine fatigue life
2 Collins TS - BSc Prediction, testing and improvement of machine fatigue life
3 Foley AG - BSc PhD MIM CEng Application of wear resistant materials to soil engaging components and bearings
4 Harral BB - BEng MSc Prediction, testing and improvement of machine fatigue life

4296 Farm Buildings
Wrest Park, Silsoe, Bedford, MK45 4HS
Tel: 0525 60000.

1 Billington RS - BSc MSc GradIChemE Factors affecting the use of anaerobic digestion for treating slurries
2 Boon CR - BSc Air mixing in livestock buildings; Engineering for welfare and control of disease
3 Carpenter GA - BSc MIChemE MIAgrE CEng Design and performance of ventilation systems; Air filtration and animal performance
4 Cumby TR - BSc,MIAgrE Factors affecting oxygen transfer to slurry
5 Hoxey RP - PhD,AFIMA Monitoring poultry intake and liveweight gain. Mechanisation of poultry harvesting. Monitoring liveweight of pigs. Milking machines. Wind loading of farm and horticultural buildings; Aerobic deterioration of silage
6 Kettlewell PJ - BSc,GradIAgrE Monitoring poultry intake and liveweight gain. Mechanisation of poultry transport and process engineering. Monitoring liveweight of pigs
7 Lindsay RT - BSc,TEng,MIAgrE Engineering requirements for vermiculture. Milking machines
8 Moran P - BA MIAgrE Wind loading of farm and horticultural buildings
9 Phillips VR - MA PhD GradIChemE Engineering requirements for vermiculture; Factors affecting the use of anaerobic digestion for treating slurries
10 Randall JM - BSc PhD ARAgS FIAgrE Factors affecting fan efficiency
11 Richardson GM - BSc,DipAgEng Wind loading of farm and horticultural buildings
12 Robertson AP - BSc PhD Wind loading of farm and horticultural buildings. Formulae for the prediction of static pressures on storage structures
13 Schofield CP - BSc MIAgrE Livestock waste characterisation
14 Sneath RW - TEng,MIAgrE Minimum aeration regimes to control odour in pig slurries
15 Williams AG - BSc,PhD Minimum aeration regimes to control odour in pig slurries; The control of odour nuisance from livestock buildings in the UK

4297 Instrumentation and Control
Wrest Park, Silsoe, Bedford, MK45 4HS
Tel: 0525 60000.

1 Davis PF - BSc,PhD,MInstP,FIMA Develop statistical time series analysis
2 Lake JR - BSc,MSc,PhD,MIAgrE Development of techniques for detection of mastitis, oestrus and pregnancy
3 Marchant JA - BSc PhD MIEE CEng Development of control algorithms for automatic milking; Specification and design of control systems
4 Stafford JV - BSc,PhD,CEng,MIMechE Instrumentation and control in food processing, storage and distribution. Development of sensors for agriculture and food applications
5 Street MJ - HNC Development of dairy parlour automation systems
6 Weaving GS - MIEE CEng Development of prototype instruments for moisture measurements; Development of prototype instruments for protein analysis

4298 Field Machinery
Wrest Park, Silsoe, Bedford, MK45 4HS
Tel: 0525 60000.

1 Chamen WCT - BSc MIAgrE To determine the effects of field traffic on crop performance; To design a self-propelled gantry system
2 Hale OD - HNC MSc To improve the performance of mowers and conditioners; To develop new secondary crop treatment systems; To develop high-speed pick-up systems for crops
3 Knight AC - BSc MIMechE CEng Development studies of crop slicing mechanisms; To improve mechanical crop conveyors
4 Neale MA - MIAgrE To design equipment for compacting and transporting straw
5 O'Dogherty MJ - PhD FIMechE CEng To research methods of forage cutting
6 Patterson DE - MSc,FIAgrE Development of: seed drilling methods and equipment; and cultivation implements with low draught

4299 Agricultural Vehicles
Wrest Park, Silsoe, Bedford, MK45 4HS
Tel: 0525 60000.

1 Bottoms DJ - TEng,MIAgrE,MEngSoc Ergonomics of tractor steering control
2 Dawson JR - CEng MIMechHE MIAgrE Tyre and track systems

3 Dwyer MJ - BSc MSc PhD CEng MIMechE MIAgrE On-farm evaluation of farm transport vehicle systems
4 Lines JA - BSc,MSc Mathematical modelling of tractor ride vibration
5 McMullan TAG - BSc,MIMechE Improve the tractive efficiency of tractors; Low ground pressure drives
6 Talamo JDC - BA,NDA,MIAgrE Use of changes in noise spectra as driver aids
7 Whyte RT - BSc,MSc Tractor ride vibration; study of seat design for improved ride comfort

4300 Horticultural Engineering
Wrest Park, Silsoe, Bedford, MK45 4HS
Tel: 0525 60000.

1 Bailey BJ - BSc PhD MInsTP The thermal performance of greenhouses; Greenhouse monitoring and control; Monitoring and control of nutrient film culture
2 Bowman GE - BSc Improving the environment in mushroom buildings
3 Brown FR - BSc Design drill to give improved seedling emergence; Instrumentation and control in food processing, storage and distribution;Reducing damage during harvesting of carrots
4 Critten DL - BSc ARCS More efficient use of energy in greenhouses: light transmission
5 Holt JB - BSc MSc MIAgrE MIMechE CEng Improve the efficiency of producing HONS and other pot raised plants; Improve methods and machinery for leaf vegetable harvesting

4301 AFRC Institute of Horticultural Research - Wellesbourne
Wellesbourne, Warwick, CV35 9EF
Tel: 0789 840382.

Studies on all aspects of outdoor vegetable crop breeding and production

1 Bleasdale Prof. JKA - BSc,PhD,FIBiol *Prof.* Professor
2 Woodroffe DA - BSc *Librarian*

4302 Plant Breeding
Wellesbourne, Warwick, CV35 9EF
Tel: 0789 840382.

1 Crisp PC - BSc PhD Breed improved broccoli and cauliflower, screen for quality characters and resistance to stress, pests and diseases
2 Dowker BD - MA Breeding new cultivars of spring-sown bulb onions and develop inbred parent lines for F1 hybrid production; Exploit the genetic potential of autumn-sown bulb onions for improvements in yield and quality; develop F1 hybrids
3 Ockendon DJ - BSc,PhD Development and improvement of techniques for anther culture in brassicas
4 Pink DAC - BSc PhD Search for and study the inheritance of genetic resistance to virus diseases of vegetables, particularly brassica crops
5 Riggs TJ - BSc PhD Combine resistance to downy mildew with resistance to mosaic virus and lettuce root aphid
6 Smith BM - BSc Produce inbred lines of brussels sprouts by single seed descent and assess their field performance
7 Tapsell CR - BSc PhD Variety improvement in winter cauliflowers for South-West England

4303 Soil Science
Wellesbourne, Warwick, CV35 9EF
Tel: 0789 840382.

1 Burns IG - BSc PhD Effects of spatial distribution of nutrients and the continuity of supply on the growth and development of vegetables; Development of quantitative theory of crop performance
2 Costigan PA - BA PhD Site to site variation in vegetable yields caused by differences in soil properties
3 Greenwood DJ - DSc FRSC Development of quantitative theory of crop performance
4 Hunt J - LRSC Develop electronic sensor based field tests for estimating nutrient status of soils and plants; Nitrate concentrations in vegetables as affected by crop type and fertiliser practice
5 McBurney T - BSc Plant water stress; development of a technique for field measurement and determining the dependence on growing conditions
6 Rowse HR - BSc PhD Cultivation systems to extend the number of days suitable for drilling and other management practices
7 Scaife MA - MA PhD DipAgr Optimization of nitrogen fertilizer use on vegetables using quick sap and soil tests; Improve reliability of visual and chemical criteria for diagnosing nutrient deficiency in crops
8 Stone DA - DPhil Develop inexpensive remote sensing techniques for monitoring growth of field crops; Cultivation systems to extend the number of days suitable for drilling and other management practices

4304 Biochemistry
Wellesbourne, Warwick, CV35 9EF
Tel: 0789 840382.

1 Fraser RSS - BSc PhD DSc Biochemistry of resistance to plant viruses, of plant growth substances and development of new methods of virus control

2 Thompson S - BSc PhD Development of novel genes for resistance to plant viruses
3 Whenham RJ - BSc DPhil Biochemical mechanisms of plant resistance to virus disease, and growth regulator metabolism in infected plants

4305 Plant Physiology
Wellesbourne, Warwick, CV35 9EF
Tel: 0789 840382.

1 Benjamin LR - BSc PhD Sources of plant-to-plant variability in weight and size of vegetable crops
2 Biddington NL - BSc PhD Growth regulators and plant growth during plant establishment
3 Brewster JL - DPhil,BSc,MINST,HORT Onions: physiology of bulb development, flowering, growth rate and yield
4 Finch-Savage WE - BA PhD Fluid drilling and other techniques affecting the establishment of vegetable crops from seed; Develop and evaluate techniques for processing and handling germinated seed
5 Gray D - BA PhD Seed source factors affecting germination and establishment of vegetable crops; Physiological factors and seed treatments affecting the establishment of vegetable crops. Agronomic and physiological factors affecting leek production
6 Hardwick RC - BA PhD DipAgSci Vegetable morphogenesis; mechanisms of resource allocation between multiple competing sinks in vegetable crops
7 Hole CC - BSc PhD Comparison of dry matter distribution and development of vegetable root crops in different environments
8 Wurr DCE - MA PhD Environmental and cultural factors affecting growth, development and maturity of crisp lettuce crops; Environmental and cultural factors affecting growth, development and maturity of cauliflower crops

4306 Entomology
Wellesbourne, Warwick, CV35 9EF
Tel: 0789 840382.

1 Cole RA - BSc,PhD,CChem,MRSC Biochemistry of resistance to insect pests in vegetables, including lettuce, carrots and brassicas
2 Ellis PR - BSc MSc PhD Resistance of vegetables to insect pests
3 Finch S - BSc PhD Biology and behaviour of insect pests of vegetables; Interspecific competition during host-plant selection by insect pests of crucifers
4 Jones TH - BSc PhD Interspecific competition during host-plant selection by insect pests
5 Suett DL - MRSC CChem Behaviour and performance of insecticide residues in soils and plants
6 Thompson AR - MA,PhD,MIBiol,CBiol Improve the control of insect pests of vegetables with insecticides, including integration with other control methods

4307 Plant Pathology
Wellesbourne, Warwick, CV35 9EF
Tel: 0789 840382.

1 Burchill RT - BSc PhD Diseases of onions - biology, resistance and control; Diseases of leeks - biology, resistance and control
2 Crute IR - BSc PhD Downy mildew of lettuce - biology, resistance and control; Fungal diseases of brassicas - biology, resistance and control
3 Entwistle AR - BSc PhD Diseases of onions - biology, resistance and control; Diseases of leeks - biology, resistance and control
4 Humpherson-Jones FM - BSc PhD Fungal diseases of brassicas - biology, resistance and control
5 Maude RB - MSc DSc Fungal diseases of brassicas - biology, resistance and control; Diseases of onions - biology, resistance and control; Diseases of leeks - biology, resistance and control; Fungal diseases of umbelliferous crops; Control of leaf blotch of leek
6 Taylor JD - BSc PhD Diseases of onions - biology, resistance and control; Bacterial diseases of *Phaseolus* beans - biology, resistance and control; Nutrition of *Phaseolus vulgaris*; Resistance of beans to halo-blight - a joint NVRS/CIAT study
7 Tomlinson JA - BSc PhD DSc Virus diseases of lettuce - biology, resistance and control; Viral diseases of brassicas - biology, resistance and control; River-borne plant viruses: importance in the epidemiology of vegetable virus diseases
8 Walkey DGA - BSc PhD Virus diseases of lettuce - biology, resistance and control; Viral diseases of brassicas - biology, resistance and control; Virus diseases of *Phaseolus* beans - biology, resistance and control; Cucumber mosaic virus in marrows; Virus diseases of umbelliferous crops; Development of serological methods for study of beet western yellows virus in lettuce; Elimination of viruses from universally infected garlic clones by tissue culture and chemotherapy; Effects of virus infection on seed production and seedling vigour
9 Walsh JA - BSc PhD Detection of lettuce mosaic virus in lettuce seed. Epidemiology and control of virus diseases of oilseed rape
10 White JG - BSc PhD Cavity spot of carrots

4308 Weeds
Wellesbourne, Warwick, CV35 9EF
Tel: 0789 840382.

1 Bond W - BSc Field evaluation of herbicides for weed control in vegetables

2 Walker A - BSc PhD Movement, persistence and activity of herbicides and fungicides in soils

4309 Statistics
Wellesbourne, Warwick, CV35 9EF
Tel: 0789 840382.

1 Sutherland RA - DipStats PhD MA Assist ADAS staff with the design, management, statistical analysis and interpretation of their trials

4310 Vegetable Gene Bank
Wellesbourne, Warwick, CV35 9EF
Tel: 0789 840382.

1 Astley D - BSc,PhD Vegetable Gene Bank: administration of the NVRS Gene Bank and involvement in international activities concerning plant genetic conservation

4311 Overseas Development Administration
Wellesbourne, Warwick, CV35 9EF
Tel: 0789 840382.

1 Teverson DM - BSc Resistance of beans to halo-blight (Joint NVRS/CIAT study)

4312 AFRC Institute of Plant Science Research - Division of Cytogenetics and Plant Breeding
Maris Lane, Trumpington, Cambridge, CB2 2LQ
Tel: 0223 840411.

Production of new and improved varieties of arable crops and research in related scientific disciplines

1 Green Mrs JL - BSc,MIInfSc Information Officer
2 Lane Dr CN Head

4313 Cereals
Maris Lane, Trumpington, Cambridge, CB2 2LQ
Tel: 0223 840411.

1 Bingham J - BSc,FRASE,FRS Breeding winter wheat; accelerated generations; hybrid varieties; take-all resistance
2 Gregory RS - MSc,BSc Breeding winter triticale
3 Hanson PR - BSc Breeding spring barley for improved yield
4 Jenkins G - BSc DipTrAgr Breeding spring wheat for improved bread-making quality; Maintenance and utilisation of genetic resources in cereals
5 Sage GC - DipAgSci MA PhD Breeding winter barley for improved quality; Breeding winter barley for improved yield

4314 Molecular Genetics
Maris Lane, Trumpington, Cambridge, CB2 2LQ
Tel: 0223 840411.

1 Baulcombe DC - BSc PhD Isolation and characterisation of hormonally regulated genes
2 Bevan MW - BSc MSc PhD Transformation of dicotyledonous plants using *Agrobacterium tumefaciens*
3 Bowman CM DNA differences
4 Dyer TA - BSc MSc PhD DIC Biochemical genetics of chloroplasts
5 Flavell RB - BSc PhD Organisation and function of repeated sequence DNA in plant chromosomes; Isolation of rRNA genes and the study of their regulation and activities; Development of rapid screening methods using DNA probes
6 Lonsdale DM - BSc PhD The plant mitochondrial genome and its role in determining plant characters
7 Thompson RD - BSc PhD Cloning and sequencing of grain storage protein genes of wheat and barley

4315 Cytogenetics
Maris Lane, Trumpington, Cambridge, CB2 2LQ
Tel: 0223 840411.

1 Bennett MD - BSc PhD Three dimensional studies of nuclei and chromosome behaviour; Cytochemistry and cell biology of reproductive cells and grain development; Variation in nuclear DNA mass and composition and its biological consequences
2 Forster BP - BSc,PhD Introduction of genes from related species into wheat
3 Gale MD - BSc PhD The contribution of major dwarfing genes to wheat improvement; The genetical description of height differences in barley; Genetic control and mechanisms of hormonal action in cereal development
4 Laurie DA - BSc PhD Wide crosses involving *Zea mays* with special emphasis on maize × sorghum
5 Law CN - BSc PhD Exploitation of chromosome manipulation in genetic analysis of wheat; Cytoplasmic variation in wheat and its use in breeding
6 Mathias RJ - BSc PhD Transformation of cereals using micro-injection techniques
7 Miller TE - Transfer of useful genes from related species into wheat; Genetic control of chromosome pairing in wheat; Chromosome relationships between wheat and related species

8 Sharp PJ - BSc,PhD Genetic markers in wheat using molecular probes
9 Smith JB - BSc,CBiol,MIBiol Reconstruction of plant nuclei from serial thin sections by electron microscopy. Cytochemistry and development of germ-line and grain cells. Variation in nuclear genome size and composition in plants
10 Snape JW - BSc PhD Biometrical studies of metric traits in wheat; The use of doubled haploid lines in barley breeding; Chromosome manipulation and its use in quantitative studies of wheat; Exploit and develop methods of genetic analysis using doubled haploids; Quantitative genetics of cytoplasmic variants
11 Worland AJ - BSc Aneuploids and chromosome substitutions in wheat; Maintenance of aneuploid and substitution lines of wheat; development and analysis of cytoplasmic variants

4316 Pathology and Entomology
Maris Lane, Trumpington, Cambridge, CB2 2LQ
Tel: 0223 840411.

1 Johnson R - BSc MSc PhD Monitor variability in rust pathogens of cereals; Assessment of resistance to rust diseases, its genetic control and exploitation in breeding wheat, barley and triticale
2 Scott PR - MA PhD Search for and study of inherited resistance to *Gaeumannomyces graminis*; Improve resistance to *Septoria* species; resistance mechanisms and inheritance; Improve resistance to eyespot and sharp eyespot; resistance mechanisms and inheritance
3 Summers RW - BSc,MSc UK Cereal Pathogen Virulence Survey to determine changes in variation in *Erysiphe graminis*. Variation and dynamics of populations of the barley and wheat mildew pathogens; strategies for their control. Introduction and analysis of resistance to powdery mildew of barley and wheat
4 Wolfe MS - BSc PhD UK Cereal Pathogen Virulence Survey to determine changes in variation in *Erysiphe graminis*; Variation and dynamics of populations of the barley and wheat mildew pathogens and strategies for their control; Search for and study of inherited resistance to *Gaemannomyces graminis*; Improve resistance to *Septoria* species; resistance mechanisms and inheritance; Improve resistance to eyespot and sharp eyespot; resistance mechanisms and inheritance

4317 Physiology
Maris Lane, Trumpington, Cambridge, CB2 2LQ
Tel: 0223 840411.

1 Austin RB - BSc,DSc Study genetic variation in photosynthesis; Genetic variation in responses of wheat to photoperiod and temperature
2 Bilsborrow PE - BSc Water use efficiency of rice genotypes contrasting in abscisic acid accumulation
3 Clipson NJW Crop characteristics and yield improvement of barley in Mediterranean environments
4 Ford MA - BSc Genetic variation in response of wheat to photoperiod and temperature
5 Payne PI - BSc PhD Biochemistry and genetics of the storage proteins of wheat grains in selection to breadmaking quality
6 Quarrie SA - BA PhD Molecular basis of adaptive responses of wheat to water stress; Molecular basis of genetic variation in photosynthesis in wheat

4318 Chemistry
Maris Lane, Trumpington, Cambridge, CB2 2LQ
Tel: 0223 840411.

1 Smith DB - ONC,HNC,MIBiol,CBiol Investigation of roles of protein subunits in aggregate formation; Devising methods of screening cereal breeders' lines; Biochemical methods for analysing potatoes, oilseed rape

4319 Forage, Oil and Potatoes
Maris Lane, Trumpington, Cambridge, CB2 2LQ
Tel: 0223 840411.

1 Bond DA - BSc PhD Breed field beans
2 Brown SJ - BSc DipAgSci PhD Breed field beans
3 Dale PJ - BSc PhD Tissue culture and micropropagation
4 Hughes WG - ONC,HNC,MSc,BSc Breeding oilseed rape
5 Jellis GJ - BSc,PhD,CBiol,MIBiol Resistance to fungal, viral and bacterial diseases of potatoes; Plant pathology organisation; Pathology of field beans
6 Lacey CND - BSc MSc PhD MIBiol CBiol Breeding potatoes for processing and domestic use; Breeding potatoes for resistance to and tolerance of potato cyst-nematodes; Tissue culture applications for potato breeding; Evaluation of somaclonal variation in field trials
7 Newman PL - BSc,PhD Resistance to oilseed rape diseases
8 Thompson AJ - BSc PhD Breeding potatoes for processing and domestic use; Organisation and logistics of potato breeding; Oilseed rape organisation and logistics; Organisation and logistics of oilseed rape breeding work; Breeding organisation and logistics in field beans; Studies in breeding methods and selection criteria; Somaclonal variation as a plant breeding method
9 Thompson KF - BA MA PhD Breeding oilseed rape as an oil and protein source. Characteristics and manipulation of self incompatibility genes in brassicas

4320 AFRC Institute of Animal Physiology and Genetics Research - Edinburgh Research Station, Roslin
Roslin, Midlothian, EH25 9PS
Tel: 031-440-2726.

Research to improve the efficiency of poultry meat and egg production, with concern for poultry welfare and compositioned aspects of product quality

4321 Ethology
Roslin, Midlothian, EH25 9PS
Tel: 031-440-2726.

1 Clark JSB - BSc MSc Real-time acquisition and analysis of electrical activity in beak nerve fibres
2 Duncan IJH - BAgrSci DipAnHus PhD Identifying factors motivating and inhibiting wing-flapping; Identifying stressful stimuli and the responses chickens make to them; Investigation of the hormonal effects in memory function in the chicken
3 Gentle MJ - BSc PhD The physiological and anatomical consequences of partial beak amputation (Beak Trimming); Sensory composition and brain stem projections of the facial nerve of chickens; Physiological effects of pre-slaughter stunning of poultry
4 Hughes BO - MA VetMB PhD MRCVS The importance of social interactions and environmental constraints in governing the welfare of the domestic fowl; Effect of environment and beak trimming on feather loss in laying stocks
5 Jones RB - BSc PhD Behavioural responses of poultry to blood; Identifying fear inducing stimuli and the responses chickens make to them; Investigation of the hormonal effects in memory function in the chicken
6 Kite VG - BRurSci, PhD Assessing the hen's need to perform nesting behaviour. Identifying the procedures or circumstances that lead to a reduction in welfare before slaughter

4322 Genetics
Roslin, Midlothian, EH25 9PS
Tel: 031-440-2726.

1 Bulfield G - BSc DipAnGen PhD Provision and analysis of novel strains and mutants in poultry; Genetic regulation of gene expression; Investigate the biochemical and molecular genetics of growth in poultry; Molecular basis of selection for egg production;establish genetic regulatory mechanisms of oestrogen-inducible proteins
2 Hanks M - BSc Cloning and analysis of the gene(s) for luteinising hormone-releasing hormone (LH-RH) I and II in chickens
3 McKay JC - BSc PhD Genetic basis of differences in lean tissue growth and efficiency; Provision and analysis of novel strains and mutants in poultry; Semen irradiation as a mechanism for gene transfer in poultry
4 Middleton RJ - BSc PhD Biochemical and molecular genetics of gene regulation
5 Perry MM - BSc The early embryology of the chicken; devising methods for DNA injection into oocytes; growth of embryos in culture
6 Sang HM - BA PhD Analysis of DNA sequences organisation in the chicken genome; Cloning and analysis of genes, construction of novel vectors
7 Scougall RK - HNC BA MIBiol CBiol Develop and use 2-dimensional gel electrophoresis of proteins in analysis of growth

4323 Reproductive Physiology
Roslin, Midlothian, EH25 9PS
Tel: 031-440-2726.

1 Armstrong DG - BSc PhD Investigation of the enzyme activities of the steriodogenic cells of the ovary
2 Gilbert AB - BSc PhD Physiology and endocrinology of ovarian function and follicular maturation in the hen
3 Houston B - BSc PhD Biochemistry and physiology of growth hormone
4 Lake PE - BSc DipAgSci PhD FIBiol FRSE Poultry breeding systems and their efficient application, and germ plasm conservation
5 Sharp PJ - BSc PhD Neuroendocrinology and endocrinology of aging in relation to reproductive performance; Neuroendocrinology and endocrinology of photorefractoriness; Neuroendocrine control of broodiness; Neuroendocrine control of ovulation; Improve lighting patterns for efficient egg production; Production and testing of antibodies to hypothalamic and pituitary hormones; Neuroendocrinology of seasonal breeding; Physiology of muscle growth
6 Talbot R - BSc Chromatography and bioassay of hypothalamic and pituitary hormones
7 Waddington D - BSc MSc Mathematical modelling of ovarian function and follicular growth
8 Wells JW - BSc MSc PhD Investigations of the hormone production potential of the two steroidogenic cells of the ovary; Investigation of the structure and properties of the cuticle protein
9 Wishart GJH - BSc PhD Assessing and applying methods to improve male/female breeding potential by studying avian sperm biology; Semen irradiation and gene transfer; novel methods of transferring DNA into embryos; Survey of egg components of interest to the human diet

4324 AFRC Institute of Grassland and Animal Production - Roslin Research Station (Poultry Department)
Roslin, Midlothian, EH25 9PS
Tel: 031-440-2726.

1 Butterwith SC - BSc PhD Physiology and biochemistry of adipose tissue growth and development in poultry
2 Dewar WA - BSc PhD CChem MRSC Nutritional effects on locomotor disorders in poultry; Growth bioassays for phosphorus; Available phosphorus content of mineral supplements; Mineral requirements of poultry; Effects of dietary minerals on albumen quality
3 Duff SRI - BSc BVSC MVM PhD MRCVS Normal and abnormal skeletal development and morphology in poultry
4 Fisher C - BSc PhD The analysis of response to amino acid supply in laying hens; Development of quantitative growth models in poultry; Nutritional strategies for meat production; Growth bioassays for amino acids; Amino acid composition of poultry feedstuffs; Evaluation of feeds for poultry
5 Griffin HD - BSc PhD Physiology and biochemistry of lipid metabolism in poultry
6 Hocking PM - BSc DipAnGen PhD CBiol MIBiol Examining effects of diet and management systems on the reproductive and economic efficiency in broiler breeders
7 Longstaff M - BSc PhD Digestibility of storage and structural carbohydrates in feedstuffs; Methods for increasing digestibility of carbohydrates
8 MacLeod MG - BSc PhD MIBiol CBiol Assessing metabolic rate in relation to production efficiency; Energy metabolism, thermoregulation and locomotor activity in relation to environment, diet and genotype; The characterisation of the physiological and biochemical mechanisms of heat production in poultry
9 Maxwell MH - FIMLT MSc MIBiol CBiol Soft tissue ultrastructure with particular reference to metabolic disorders in poultry
10 McNab JM - BSc PhD Bioassays for metabolisable energy and amino acid availability; Metabolisable energy values and amino acid contents of poultry feedingstuffs; Effect of physical and chemical treatments on the nutritive value of poultry feedingstuffs
11 Mitchell MA - BSc PhD MIBiol CBiol Physical and physiological measurements of the responses of poultry to their climatic environment; The characterisation of the physiological and biochemical mechanisms of heat production in poultry
12 Saunderson CL - BSc PhD Measurement of protein turnover, muscle growth and leucine oxidation in broiler muscles; To measure and model pathways of methionine metabolism and amino acid oxidation in broiler chicks
13 Sauro H - BSc MSc Construction of a computer model for methionine and betaine utilisation
14 Whitehead CC - BSc PhD CChem FRSC Examining effects of diet and management systems on the reproductive and economic efficiency in turkey breeders; Examining effects of diet and management systems on the reproductive and economic efficiency in broiler breeders; The available biotin content of poultry feedstuffs; Requirements for vitamins and methods for evaluating practical vitamin allowances; Selection of broilers for low and high fat content

4325 AFRC Institute of Arable Crops Research - Rothamsted Experimental Station
Harpenden, Herts, AL5 2JQ
Tel: 05827 63133.

Basic and applied research in plant and soil science and crop production and protection; the emphasis of crop protection studies is on the biology and control of invertebrate pests and soil-borne pathogens

1 Fowden Sir L - PhD,FRS *Director*
2 Palmer Mrs JMP - BSc,ALA *Librarian*

4326 Biochemistry
Harpenden, Herts, AL5 2JQ
Tel: 05827 63133.

1 Antoniw JF - BSc PhD Mechanisms of resistance of plants to virus infections
2 Bright SWJ - BA MA PhD Biochemistry and genetics of pathways of lysine, threonine, methionine, isoleucine and nitrate metabolism; Genetic stocks for protoplast fusion; Genetic analysis of photorespiratory mutants; Investigate transposon mutagenesis as a tool to isolate plant genes
3 Burrell MM - BA PhD Study expression of genes in transformed plants; Alteration of sucrose accumulation during storage of potato tubers for processing
4 Forde BG - BSc PhD Study of the isolation and expression of genes for glutamine synthetase; The molecular basis for organ-specific expression of genes encoding glutamine synthetase and glutamate synthase
5 Gutteridge S - BSc PhD Catalytic mechanisms of ribulose bisphosphate carboxylase; In vitro mutagenesis and expression of genes for ribulose bisphosphate carboxylase; Characterisation of products of vector expression
6 Hammond JBW - BSc,MSc,DPhil Alteration of sucrose accumulation during storage of potato

7 Jones MGK - BSc MA PhD Development of regeneration systems for protoplasts from crop plants; Development of tissue culture techniques for the genetic manipulation of crop plants
8 Karp A - BSc PhD In situ hybridisation and localisation of storage protein genes in chromosomes of barley; Cytogenetical study of plant tissue culture and regeneration and the causes and origins of somaclonal variation
9 Keys AJ - BSc PhD DIC Properties, structure and function of ribulose bisphosphate carboxylase; Characterisation, physiology, biochemistry and use of photorespiratory mutants; Purification of enzymes for C4 photosynthesis and isolation of the corresponding genes
10 Kreis M - PhD Isolation and characterisation of genes for cereal seed proteins; Isolation and characterisation of genes for cereal seed proteins; The control of gene expression during cereal seed development
11 Ooms G - PhD Study of the transfer of genetic material between crop plants; Study of the movement of genetic material amongst crop plants
12 Parry MAJ - BSc MSc DIC Properties and catalytic mechanisms of ribulose bisphosphate carboxylase; Properties of ribulose bisphosphate carboxylase and in vitro mutagenesis
13 Phillips AL - BSc,PhD In vitro mutagenesis and expression of genes for ribulose bisphosphate carboxylase. Isolation of genes for C4 photosynthesis
14 Pierpoint WS - BSc PhD Characterise pathogenesis-related proteins N-R from virus-infected tobacco; identify virus-receptor sites in leaves
15 Shewry PR - BSc PhD Structure, genetics and deposition of cereal seed proteins; Chemical characterisation of wheat proteins important in baking quality; Nutritional control of cereal grain composition and quality
16 Tatham AS - BSc PhD Chemical and physical characterisation of cereal seed proteins; Chemical and physical characterisation of wheat proteins important in baking quality
17 Wallsgrove RM - BSc PhD Studies on nitrogen metabolism, amino acid synthesis and organelle function in normal and mutant plants; Biochemical characterisation and use of photorespiratory mutants

4327 Soils and Plant Nutrition
Harpenden, Herts, AL5 2JQ
Tel: 05827 63133.

1 Addiscott TM - MA PhD Predict solute movement in aggregated soils, including slow-draining and water-logged soils and those with field drains; Model solute leaching in aggregated soils; predicting nitrogen fertiliser use, crop growth and ground water pollution; Modelling solute leaching in aggregated soil, predicting nitrogen use, crop growth and ground water pollution
2 Barraclough PB - PhD Measure and model the processes of diffusion and mass flow of nutrients to roots in the field; Measurement of the form and function of root systems of arable crops; Measuring and modelling root growth of wheat and oil seed rate crops with and without a crop shelter
3 Brookes PC - BSc PhD Soil organic matter and organic farming; The microbial biomass in agricultural soils and how it is influenced by heavy metals and by soil acidification; Biological transformation of fertiliser N in the field; Effect of straw incorporation on soil biomass
4 Brown G - BSc Trace element content of soils and soil fractions; Mineralogy of non-calcareous brown soils; Structure and hydration properties of mixed-layer clays
5 Catt JA - BSc PhD DSc FGS Causes of soil variation on chalky boulder clay, loess and deposits of dry chalk valleys; Remote sensing of short-range variation in cereal growth and yield in relation to soil variability; Surveying sites and describing soils for off-station experiments
6 Christian DG - BSc Design and manage field experiments on straw incorporation; Manage experiments on long-term effects of simplified tillage; Manage field experiments on the interaction of drainage and cultivation
7 Copestake KG - BSc MSc Compare availability to barley and potatoes of phosphorus in different phosphate fertilisers
8 Darby RJ - NCA BSc Relate measurements of mineral N in soil, and fertiliser N applied, to yield; Measure yields of arable crops and grass at different levels of soil P and K; Relate soil mineral nitrogen to wheat and barley growth and yield on several soils; Manage field experiments on straw incorporation on soils of various textures; High-yielding wheat experiments and selection of off-station sites
9 Goss MJ - BSc MSc PhD Mechanisms of reduction in plant growth due to straw; Study soil structure and its effect on root growth and function; Effect of wheel traffic on soil structure and yields of arable crops; Co-ordinate inter-institute research and study soil physical conditions and correlate with root growth
10 Goulding KWT - PhD Equilibria and kinetics of reactions between nutrient ions and colloid surfaces; Amounts and rates of potassium and magnesium released to laboratory extractants from soil constituents; Rates and mechanisms of supply of nutrients and acidity from the atmosphere and their effects on soil leachates
11 Jenkinson DS - BA BSc PhD Turnover of soil organic matter in agricultural soils; Behaviour and properties of the soil microbial biomass; Assess the effects of farming systems on soil nitrogen; Efficiency of use of fertiliser nitrogen; Nitrogen cycling in the crop/soil system; Effect of straw incorporation on nitrogen transformations in soil

12 Johnston AE - BSc Effects of deep loosening & PK enrichment of subsoils on soil strength, crop nutrient availability & uptake; Effects on soil physical and chemical properties of soil organic matter in different farming systems; Relate crop yields to levels of soluble P in soil and model changes in soluble P with time; Availability to crops of soil reserves of potassium and magnesium and their effects on fertiliser inputs; Plan experiments comparing different fertilisers; Test amounts of N on cereals grown continuously and in rotation with and without nematicides and fungicides; Nitrogen balance in long-term field experiments; Planning the cropping of a lysimeter installation; Nitrogen balance in long-term field experiments

13 Leigh RA - BSc PhD Factors controlling nutrient concentrations in crops during growth;Mechanism controlling partition of nutrients between plant cell cytoplasm and vacuole

14 McGrath SP - BSc PhD Effects of trace metal ion activities in soil on root uptake; Relate soil trace element content to crop uptake and response to added trace elements, especially high-yielding crops

15 North PF - BSc PhD MInstP Measure and model effects of cultivation on structure and transport process in soil and on plant establishment; Soil physical measurements and soil monitoring for off-station experiments

16 Penny A - BSc Measure crop response to fertiliser P in field experiments; Effects of urea and urease inhibitors on yields of winter wheat in the field; Relate yields of arable crops and grass to cumulative effects of NPK; Relate soil mineral nitrogen to wheat and barley growth and yield on Batcombe series soils; High-yielding wheat experiments and selection of off-station sites

17 Piper J - BSc PhD Forces between soil particles and their modification by adsorption of ions at the particle surface

18 Pirie NW Leaf protein extraction, extract quality and stability

19 Pope AJ - BSc The mechanism of transport processes at cell membranes

20 Powlson DS - BSc PhD Nuclear magnetic resonance spectra of soil organic matter; Measuring the flux of carbon, nitrogen and phosphorus through the soil microbial biomass; Uptake and loss of nitrogen fertiliser under field conditions; Input, output and carry forward of nitrogen under different systems of land use; Immobilisation and mineralisation of nitrogen in soil as affected by straw incorporation

21 Rayner JH - MA DPhil MInstP Simulation models for the turnover of organic matter in soil;Identification of hydrated and interstratified clay minerals; Model the carry forward of N in the crop/soil system; Models for the behaviour of fertiliser N in arable agriculture; An analysis of the distribution of soil series and of cereal cropping and yields; The interpretation of data from field experiments with isotopically-labelled fertilisers; To improve and apply the crop simulation model of winter wheat and to develop a nitrogen sub-model

22 Rodgers GA - BSc PhD Nitrification inhibitors in soil; Laboratory testing of potentially useful urease inhibitors; Leaching, volatilisation and denitrification of fertiliser and soil N

23 Towner GD - BSc PhD Soil compaction and structure regain; Statics and dynamics of the swelling and shrinkage of clay and their application to soil cracking and drainage

24 Webster C Commissioning a lysimeter installation and measuring the growth and yield of wheat in selected soil cores

25 Webster R - BSc DPhil DSc Measure and model spatial variation in soil, its causes and consequences; Remote sensing of short-range variation in cereal growth and yield and soil

26 Weir AH - BA MA PhD The function of root systems in supplying water and nutrients to arable crops in the field; Water extraction and response to irrigation and droughting with and without a crop shelter on batcombe series soil; Studying root growth and water and nutrient uptake of crops; Selection and collection of soil monoliths for a lysimeter installation; Selection of off-station sites and measurement of soil physical properties; To improve and apply the crop simulation model of winter wheat and to develop a water sub-model

27 Whitmore AP - BSc PhD Model soil and crop nitrogen; factors affecting N-fertiliser requirement, e; Modelling soil and crop nitrogen, mineralization and spatial variation

28 Wood IG - BA MA PhD MInstP Mineralogy of non-calcareous brown soils; Structure and hydration properties of mixed-layer clays

29 Youngs EG - BSc DPhil FInstP DSc Definition and measurement of soil-water properties in heterogeneous soils with particular reference to land drainage; Apply soil physics theory in the modelling of transport processes in soils; Effect of fissures on soil water movement in mole-drained soils

4328 Entomology
Harpenden, Herts, AL5 2JQ
Tel: 05827 63133.

1 Ball BV - BSc AIBiol Study the properties, natural histories and means of control of honeybee pathogens; Study the transmission and distribution of a virus infecting the cereal aphid Rhopalosiphum padi

2 Bent GA - BSc MSc DIC PhD FRES Development of radar network to give instantaneous information on numbers of airborne insects; Radar taxonomy of insect pests

3 Burrows I - PhD Development of plant composts from farm organic wastes processed by earthworms

4 Carter N - BSc PhD To relate aerial samples of cereal aphids to crop infestations, and to model and forecast pest outbreaks; Relating aerial population of cereal aphids to populations of crops

5 Harrington R - BSc ARCS PhD DIC FRES Suction trap monitoring of potato aphids and virus vector relationships; Biology of potato aphids and virus/vector relationships

6 Henderson IF - BSc PhD Development of improved slug control measures; bioassay and field testing of new molluscicides

7 Loxdale HD - BSc DPhil MIBiol Study the spatial and temporal genetic variation (including insecticides resistance) of pest aphid populations

8 Macaulay EDM - BSc MIBiol Improve, maintain and standardise suction traps; Application of pheromones for monitoring and control of lepidopterous pests of brassicas

9 Neale RE - BSc Optimising earthworm processing of organic wastes

10 Powell W - BSc PhD To develop cultural and biological techniques for integrated control of cereal pests, with emphasis on parasitoids; Sampling multidisciplinary wheat experiment for effects of straw incorporation on pests

11 Stafford EA - PhD Comparing heavy metal uptake by *Eisenia* and other worms

12 Tatchell GM - BSc ARCS PhD DIC Interpret current suction trap samples and issue weekly aphid reports to agricultural industry; Interpret current suction trap samples and write weekly aphid and virus reports

13 Wall C - BSc ARCS PhD MIBiol DIC FRES Electrophysiological responses of pea midge; tracking of male pea moth responses in the field;Identification of pea midge sex pheromone and development of monitoring trap; evaluation of pea moth monitoring system

14 Wilding N - BSc ARCS PhD MIBiol To develop methods of using fungi (Entomophthorales) to control aphids and other pests

15 Williams IH - PhD Bioassay of Nasonov and queen honeybee pheromones on behaviour and foraging; Electroantennographic responses of honeybees to pheromones; Pheromonal monitoring and biology of rape pod midge; pollination requirements of rape cultivars

16 Woiwod IP - BSc Computer analysis of Survey data; Computing and statistical analysis to develop forecasting systems for pest insects based on aerial monitoring; Computing statistical analysis and mapping

4329 Insecticides and Fungicides
Harpenden, Herts, AL5 2JQ
Tel: 05827 63133.

1 Blight MM - BSc PhD Isolation and identification of behaviour controlling chemicals

2 Bromilow RH - BSc PhD Behaviour of chemicals in plants and soils; Integrated economic control of cyst and stem nematodes of field crops with nematicides and resistant and tolerant crops

3 Chamberlain K - BSc PhD Phloem mobility of chemicals; synthesis of radio-labelled compounds

4 Dawson GW - BSc Synthesis of behaviour controlling chemicals

5 Denholm AI - BSc PhD Resistance and insensitivity to chemical control agents

6 Devonshire AL - BSc PhD Biochemistry and molecular biology of insecticide resistance

7 Farnham AW - PhD Bioassays of novel compounds against *Musca domestica*, *Phaedon cochleariae* and *Plutella xylostella*; Toxicology and genetics of resistance; Bioassay techniques for assessing activity of pesticides in soil

8 Griffiths DC - BSc MSc MIBiol Studies on insect behaviour and the use of behaviour controlling chemicals to improve crop protection; Biological effectiveness of crop protection agents applied with novel spray equipment

9 Janes NF - BSc PhD Coordinating organic synthesis and bioassay of candidate compounds; spectroscopy to determine structures

10 Khambay BPS - BSc CertEd PhD MRSC Organic synthesis of candidate insecticides; Use of spectroscopic methods (particularly NMR) to determine molecular structure

11 Lewthwaite RJ - BSc MSc Field and laboratory assessment of novel spraying systems; Control of bacterial and fungal diseases of potatoes

12 Mudd A - BSc MPhil PhD Use of spectroscopy and liquid chromatography for identification of behaviour controlling chemicals

13 Nicholls PH - HNC BA Pesticide movement, degradation and biological activity in soil

14 Pickett JA - BSc PhD FRSC CChem Identification, synthesis and development of behaviour controlling chemicals for use in agriculture

15 Pye BJ Factors affecting performance of novel spraying systems

16 Rowland MW - BSc Insecticide resistance in the white-fly

17 Sawicki RM - BSc PhD Resistance and insensitivity to chemical control agents

18 Wadhams LJ - BSc PhD Identification of behaviour controlling chemicals

4330 Nematology
Harpenden, Herts, AL5 2JQ
Tel: 05827 63133.

1 Crump DH - HND BSc Role of soil fungi in population dynamics of sugar beet cyst-nematode

2 Evans KE - BSc PhD Effects of cyst and other nematodes on crop plants, alone and with other organisms

3 Hooper DJ - MIBiol FIBiol Biology and pathogenicity of stem nematode *Ditylenchus dipsaci* on faba beans; Taxonomy of migratory plant-parasitic nematodes

4 Kerry BR - PhD Development of biological control of nematodes by fungi in integrated pest control of arable crops

5 Parrott DM - BSc Host-parasite relations, selection, biology and description of nematode pathotypes and resistance breaking populations; their interaction with resistant potatoes and other crops

6 Perry RM - PhD Studies on compounds affecting hatch of cyst nematodes

7 Perry RN - BSc PhD Biology and physiology of nematodes

8 Whitehead AG - BSc PhD Assess use of tolerant and resistant cultivars in integrated control of nematodes of potatoes and other crops; Integrated, economic control of cyst/stem nematodes of field crops with nematicides and resistant and tolerant crops

4331 Physiology and Environmental Physics
Harpenden, Herts, AL5 2JQ
Tel: 05827 63133.

1 Barnard AJ - NDA Yield variation of maize in relation to agronomic treatments, pests and diseases

2 Chalabi ZS - BSc PhD Modelling physiological processes in heterogeneous field crops

3 Day W - MA,PhD,CPhys Using models in the analysis of crop responses in field experiments; Integration of physiological models to analyse crop responses; Modelling plant and crop responses; Integration of crop physiological principles into a crop simulation of winter wheat

4 Lawlor DW - BSc PhD Carbon and nitrogen metabolism in growing leaves; Effects of water stress on photosynthesis and growth of plants; Physiology and biochemistry of photosynthetic carbon metabolism

5 Leach JE - BSc PhD Yield variation in oilseed rape in relation to light interception and radiation conversion efficiency; Growth of leaves and crop canopies; Effects of water stress on leaf growth and canopy development

6 McCartney HA - BSc PhD AInstP Dispersal and deposition processes important in field crops; The importance of dispersal and deposition in disease epidemiology

7 McEwen J - BSc DipAgricSci DipTrAgr Yield variation of field beans in relation to agronomic treatments and diseases

8 Prew RD - BSc,PhD Yield variation in cereals caused by agronomic treatments, pests and diseases

9 Thorne GN - MA PhD DipAgricSci MIBiol Physiological basis of yield variation caused by agronomic treatments, pests and diseases; Physiological factors determining sink capacity and potential yield in cereals; Physiological basis of yield variation

10 Walklate PJ - BSc MSc PhD Modelling turbulent dispersal and deposition; Theories of dispersal and deposition appropriate for crops

11 Wood DW - BSc PhD Physiological basis of yield variation in cereals caused by agronomic treatments, pests and diseases; Tiller production and survival and effect on grain number

4332 Plant Pathology
Harpenden, Herts, AL5 2JQ
Tel: 05827 63133.

1 Adams MJ - BA MSc DIC PhD The biology of fungal vectors of viruses, especially *Polymyxa graminis* and barley yellow mosaic virus; Forecasting storage diseases; epidemiology and control of gangrene and powdery scab of potatoes; Epidemiology and control of barley yellow mosaic virus and its vector *Polymyxa graminis*

2 Barker I - PhD Characterisation, diagnosis and epidemiology of BYDV in developing countries

3 Bateman GL - BSc PhD The epidemiology and control of take-all, particularly the role of chemicals; The resistance of *Pseudocercosporella herpotrichoides* to fungicides

4 Cockbain AJ - BSc PhD Epidemiology, effects and control of virus diseases of rape; Identification, epidemiology, effects and control of viruses of grain legumes

5 Crook B - PhD Hazards of refuse handling caused by airborne allergenic, infectious and toxic particles

6 Fitt BDL - MA PhD DIC BA Investigate the principles of particle dispersal between and within crop canopies; The development of lesions; and changes in the resistance of *Pseudocercosporella herpotrichoides* to fungicides

7 Gibson RW - BSc PhD Developing and testing chemicals and methods of control for aphid-borne viruses

8 Govier DA - BSc PhD Characterisation, diagnosis and control of plant viruses

9 Hide GA - BSc MSc PhD Forecasting and control of storage diseases; epidemiology and effects of *Rhizoctonia* and other potato diseases

10 Higgins S - PhD Hazards of refuse handling caused by airborne allergenic infections and toxic particles

11 Hornby D - BSc PhD MIBiol The epidemiology and control of graminaceous root diseases, especially take-all

12 Jenkyn JF - BSc PhD The epidemiology and effects of cereal leaf disease; Interactions between husbandry factors, especially straw disposal, and cereal diseases

13 Jones P - BSc FRMS Characterisation, diagnosis and control of plant viruses; Characterisation of viruses, especially barley yellow mosaic virus, transmitted by fungi; To diagnose, characterise and study the epidemiology and control of viruses of important tropical crops

14 Lacey J - BSc PhD Pre- and post-harvest ecology of fungi on rapeseed; The pre- and post harvest ecology of fungi on grain in relation to disease risks in man/animals; Hazards of refuse handling caused by airborne allergenic, infectious and toxic particles; Hay microflora, its enumeration, identification, ecology, control and hazards to health

15 Plumb RT - BSc PhD Epidemiology and control, and the development of forecasting methods, for virus diseases of cereals; Identify and monitor the incidence and severity of virus diseases of grasses

16 Rawlinson CJ - BSc MA PhD Epidemiology, effects and control of fungus diseases of rape and forecasting the need for chemical applications

17 White RF - BSc,PhD Characterisation of beet cryptic virus; Mechanisms of resistance of plants to virus infections

18 Woods RD Characterisation and diagnosis of plant viruses; To diagnose, characterise and study the epidemiology and control of viruses of important tropical crops

4333 Soil Microbiology
Harpenden, Herts, AL5 2JQ
Tel: 05827 63133.

1 Bowen RM - BSc,PhD Microbial and other processes in straw decomposition and their manipulation

2 Day JM - PhD Nutritional constraints on the production of grain legumes

3 Giller KE - BSc Role of BNF in supplying nitrogen to tropical crops

4 Harper SHT - BSc MPhil Effects of additions of nitrogen or other materials on straw decomposition rates in the field; Toxin production by decomposing straw and its alleviation; Microbial processes in straw decomposition and their artificial control

5 Hayman DS - BSc MSc PhD Isolation and screening of VAM endophytes as potential field inoculants; Development of effective inocula for obtaining VAM infection in field crops; Use of VA mycorrhizal fungi to improve growth of clover in hill pastures

6 Hepper CM - BSc MSc PhD Culture of VAM fungi; Biochemical methods for identifying VAM fungi; Factors affecting the susceptibility of roots to VAM infection

7 Hirsch PR - BSc PhD Genetic manipulation of VAM fungi; Manipulate genes that affect the energy cost of nitrogen fixation; Identification of *Rhizobium* genes involved in carbon use for nitrogen fixation by legume nodules; Genetic manipulation of N2-fixing Rhizobium bacteria

8 Koomen I - DRS Interactions between VAM fungi, *Rhizobium* and natural rock fertilizers in acid infertile soils

9 MacDonald RM - BSc PhD The potential of rhizosphere bacteria to affect plant growth; Improved methods for analysis of soil and rhizosphere microorganisms

10 Williams P - PhD Inoculation technology

11 Witty JF - BSc MSc PhD Identification of factors controlling the quantity and efficiency of carbon use for nitrogen fixation by legume nodules; To determine the genetic basis of functions that affect the energy cost of nitrogen fixation; *Rhizobium* genetics and carbon use efficiency

4334 Statistics
Harpenden, Herts, AL5 2JQ
Tel: 05827 63133.

1 Bailey RA - DPhil Design of experiments
2 Gomer JC - BA Applied multivariate analysis and graphical methods
3 Payne RW - BA Analysis of variance and diagnostic keys
4 Ross GJS - BA Non-linear inference

4335 AFRC Institute of Arable Crops Research - Broom's Barn Experimental Station
Bury St. Edmunds, Suffolk, IP28 6NP
Tel: 0284 810363.

1 Armstrong MJ - BSc PhD Analysis and prediction of nutrient and irrigation requirements to maximise yield of extractable sugar; Crop physiology particularly leaf canopy development and plant nutrition

2 Asher MJC - BSc PhD Find suitable non-mercurial seed treatments to control seedling diseases; Study foliar and soil-borne diseases and develop methods of control

3 Cooke DA - BSc ARCS DIC PhD MIBiol Determine the factors affecting seedling pest damage by nematodes and find improved methods of control; Develop control strategies for beet cyst nematode to allow beet to be grown in closer rotations

4 Dewar AM - BSc PhD Determination of factors affecting seedling pest damage by soil arthropods and investigate new methods of control; Improve predictions of the entry and spread of virus yellows and provision of advice on control in individual fields

5 Dunham RJ - BSc DPhil Analysis and prediction of nutrient and irrigation requirements to maximise yield of extractable sugar

6 Durrant MJ - HNC MIBiol MPhil PhD Improvement of seed quality by physical and chemical treatments

7 Gummerson RJ - BSc PhD Effect of soil physical environment on sugar beet seedling establishment

8 Jaggard KW - BSc PhD Analysis and prediction of nutrient and irrigation requirements to maximise yield of extractable sugar

9 Longden PC - BSc MSc PhD Study factors vernalising plants, seed return from bolters and weed beet population dynamics

10 Milford GFJ - BSc MIBiol Effects of environment and nutrition on leaf growth and leaf canopy development

11 Pocock TO - PhD Effects of environment and nutrition on sugar beet storage root development, sugar accumulation and root quality

12 Scott RK - BSc PhD Analysis and prediction of nutrient and irrigation requirements to maximise yield of sugar beet

13 Smith HG - BSc Improve predictions of the entry and spread of virus yellows and provision of advice on control in individual fields

4336 AFRC Institute of Arable Crops Research - Soil Survey and Land Resource Centre
Harpenden, Herts, AL5 2JQ
Tel: 05827 63133.

1 Beard GR - BSc Mapping Soils in the Midlands
2 Bradley RI - BSc MSc Mapping Soils in Northern England
3 Burton RGO - BA MSc The nature and distribution of saline soils, acid sulphate soils and soils of Experimental Husbandry Farms. Soil erosion
4 Carter AD - BSc Evaluation of soils and land for pollution risk
5 Harrod TR - BA,PhD Mapping Soils in South-West England; Soil Survey in Relation to Grassland Classification & Utilisation
6 Hartnup R - BSc Soils of Experimental Husbandry Farms
7 Hartnup RC - BSc Upland soil characterisation and utilisation
8 Hazelden J - MA Mapping Soils in South-East England; Mapping and Characterisation of Saline Soils in Eastern England
9 Hogan DV - BSc Mapping Soils in South-West England
10 Hollis JM - BSc The Development of a Soil Classification System for England and Wales
11 Jones RJA - BSc PhD Models of suitability of soils for specific crops
12 Loveland PJ - BSc PhD Chemical and Physical Characterisation of Soils; Development of Automatic Methods of Soil Analysis; Mineralogical, chemical and pedological studies of soils
13 Palmer RC - BSc Mapping Soils in the Midlands
14 Payton RW - BSc Mapping Soils in Northern England
15 Proctor ME - BSc FBCS Design of Computerised Soil Information System
16 Ragg JM - BSc Mapping Soils in the Midlands; Computerised Methods of Handling Soil Survey Data
17 Reeve MJ - BSc Mapping Soils in the Midlands; Soil Water Regimes
18 Williams AE Chemical and Physical Analysis of Soils
19 Wright PS - BSc MSc Commercial development of soil survey information

4337 Rowett Research Institute
Greenburn Road, Bucksburn, Aberdeen, AB2 9SB
Tel: 0224 712751.

Animal nutrition and production

1 James Prof. WPT - MA,MD,DSc,FRCO,FRCP(E) *Professor*

4338 Biochemistry
Greenburn Road, Bucksburn, Aberdeen, AB2 9SB
Tel: 0224 712751.

1 Aggett PJ - MB ChB MSc MRCP DCH Uptake of trace elements by brush border membrane vesicles, and relation to physiological and disease states; Effects of perinatal trace element status and liquid diets on trace element absorption
2 Arthur JR - BSc PhD Enzyme and metabolic defects in the copper deficient heart; Biochemistry of mechanisms causing tissue damage in selenium/vitamin E deficient animals; Synergism of effects of selenium deficiency; Selenium and copper deficiency and the biochemistry of phagocytic neutrophil function; Biochemistry of the stress susceptibility syndrome in pigs
3 Beckett PR - BSc,MSc Measurement of amino acid catabolism
4 Black AH Specific detection of collagen phenotypes using monoclonal antibodies
5 Black D - PhD,BSc Characterisation of urinary compounds related to collagen crosslinks and their relevance to connective tissue disorders
6 Boyne R - FIMLS,TD Selenium and copper deficiency and the pathology of phagocytic neutrophil function
7 Bremner I - BSc PhD DSc Effects of molybdenum (Mo) and iron (Fe) on Cu availability in calves
8 Cadenhead A - LRSC The optimal balance of dietary amino acids; Responses of growing pigs to protein and energy; Amino acid digestion; Evaluation of feeds for non-ruminants; Measurement of nutrient flux and disposal; Measurement of nitrogen balance, metabolite flux and disposal; Selection and husbandry of pigs given deuterium and oxygen-18 and label exchange in pigs; Measurement of amino acid digestibility
9 Chesters JK - MA,PhD Chemical forms of molybdenum in ruminant digesta; Effects of zinc deficiency on RNA synthesis
10 Coutts AGP - HNC Immune development
11 Davies NT - BSc PhD Oxidative defects and nucleotide translocation during copper deficiency; Prediction of the effects of phytate as an antagonist of dietary zinc
12 Duncan WRH - HNC Structurally modified and saturated fatty acids: influence on membrane phospholipid fatty acid composition; Effect of structurally modified fatty acids on phospholipase activity; Effect of structurally modified fatty acids on eicosanoid synthesis; Dietary trans-fatty acids and the composition of fatty acids and lipids in rat tissues; Dietary supplementation and tissue analysis of trans-fatty acids
13 Duthie GG Biochemistry of the stress susceptibility syndrome in pigs
14 Earl CRA - BSc Cellular membranes from tissues of rats given dietary branched-chain fatty acids; Synthesise methyl-branched fatty acids and corresponding phospholipids; Physical characterisation by fluorescence techniques of cellular structures with reference to phospholipids; Physico-chemical properties of membranes; Synthesise isomeric trans-unsaturated C18 to C22 fatty acids

15 Fuller MF - MA PhD Evaluation of developments in NAA and NMR; The optimal balance of dietary amino acids; Responses of growing pigs to protein and energy; Amino acid digestion; Evaluation of feeds for non-ruminants; Effects of nutrients on protein metabolism; Effects of nutrients on nitrogen balance and hormonal concentrations; Measurement of amino acid digestibility
16 Grant G - HNC The *in vivo* degradation of purified soyabean proteins by rats; Interference of dietary lectin with local immune system in rats; Changes in endogenous nitrogen secretion in gastrointestinal tract of rats in response to ingested soyabean proteins; The *in vivo* responses of rats to dietary soyabean and other legume proteins
17 Hesketh JE - BA PhD Define the importance of zinc in tissue microtubule assembly; The role of cytoskeletal proteins in intracellular transport and secretion
18 Humphries WR Effect of dietary molybdenum and iron on copper status of beef cattle; Therapeutic value of thiomolybdates in ovine copper intoxication
19 Lawrence CB - LRSC Catecholamine metabolism in the copper and selenium deficient heart; Oxidative defects and catecholamine metabolism during copper deficiency
20 Lough AK - BSc PhD DSc FRSE Dietary branched-chain fatty acids and compositional, structural and metabolic changes in animal tissues; Synthesis methyl-branched fatty acids and corresponding phospholipids; Synthesis of structurally modified fatty acids; Synthesis isomeric trans-unsaturated C18 to C22 fatty acids
21 McWilliam R The optimal balance of dietary amino acids. Responses of growing pigs to energy and amino acids. Measurements of amino acid and energy digestion
22 Mills CF - MSc PhD FRSC FRSE Biochemistry of mechanisms causing tissue damage in selenium/vitamin E deficient animals; Synergism of effects of selenium deficiency; Effects of metal sulphides as inhibitors of copper utilisation; Therapeutic value of thiomolybdates in ovine copper intoxication
23 Palmer RM - MIBiol Studies *in vitro* of interactions between hormones, protein turnover and prostaglandins; Relationship of arachidonate metabolites and protein metabolism in muscle
24 Pennie K Responses of body composition of pigs to energy and amino acids
25 Phillipo M - BSc,BVetMed,MRCVS Effect of dietary Mo and Fe on oestrus, ovulation and fertility in cattle
26 Price J - BSc,BAgr,PhD Rumen microbial activity and its modification by diet on forms and availability of Cu; role of transition element sulphides as inhibitors of Cu utilisation
27 Pusztai AJ - MSc PhD The degradation of isolated bean proteins by gut proteases; Effects of lectins on gut IgE response to the diet; Stimulation of endogenous nitrogen secretion in rats by kidney bean lectin and other bean proteins; Antinutrient proteins in legume seeds
28 Quarterman J - BSc,PhD,FRSC,CChem Role of glycocalyx and intestinal mucin in the absorption of heavy metal ions. Uptake of toxic metals by brush border membrane vesicles, and effects of trace element deficiencies. Function of lactoferrin in milk and endogenous secretions on absorption and metabolism of metals
29 Reeds PJ - BSc PhD Studies *in vitro* of interactions between hormones, protein turnover and prostaglandins; Interactions between hormones, prostaglandins and mitochondrial and ribosomal activity; Relationship between protein synthesis, ion transport and cellular energy expenditure; Studies on direct effects of adrenergic agonists on muscle protein turnover and deposition; Measurement of nutrient flux and disposal; Effects on muscle growth and protein turnover of anti-inflammatory drugs; Relationships between metabolite turnover, hormonal status and growth; Protein turnover and storage during drug treatment; Sequential measurements of nitrogen mass by neutron activation analysis in small mammals
30 Robins SP - BSc,PhD Biochemistry of connective tissues. Characterisation of urinary compounds related to connective tissue disorders. Development of collagen degradation indices using intermolecular crosslinking components. Analysis of quantitative changes during growth of the various collagen phenotypes
31 Rucklidge GJ - BSc,PhD Analysis of quantitative changes during growth of the various collagen phenotypes. Immunocytochemical localisation of connective tissues
32 Stewart JC - BSc PhD Isolation and characterisation of products of digestion by gut proteases of isolated bean proteins; Isolation and characterisation of proteins and antinutrients from various bean seeds
33 Wahle KWJ - BSc,PhD Dietary and endocrine factors which influence membrane phospholipid fatty acid composition of heart and pancreas. Determination of phospholipase activity in cell fractions. Effect of trans unsaturated acids on essential fatty acid and eicosanoid metabolism; determination of enzyme activities and eicosanoid production; eicosanoid synthesis from endogenous and exogenous arachidonate; incorporation and release of arachidonic acid in muscle
34 Wang TC - MSc The optimal balance of dietary amino acids
35 Watt WB - LRSC Isolation and characterisation of proteins from soya bean
36 Williams LM - BSc,PhD Role of metallothionein in homeostatic control of Cu and Zn metabolism. Analysis of quantitative changes during growth of various collagen phenotypes. Identification of specific receptor sites to elucidate mode of action of melatonin and functionally related peptides

4339 Nutrition

Greenburn Road, Bucksburn, Aberdeen, AB2 9SB
Tel: 0224 712751.

1   Adam CL - BA PhD Investigate means for the control and manipulation of the reproductive cycle of red deer; Investigate the live-weight gain by calves and carcass composition at 6 months; Investigate diets and feed allowances for deer
2   Chesson A - BSc,MSc,PhD Measurement of carbohydrate digestion. Structure of plant cell walls in relation to the rate and extent of their digestion by gut microorganisms
3   Flint HJ - BSc PhD Celluloysis and digestion of herbage in the rumen
4   Gordon AH - LRSC Structure of plant cell walls in relation to the rate and extent of their digestion by gut microorganisms
5   Hay AJ - HNC Separation by HPLC of the mono and oligosaccharides released by enzyme action
6   Hovell FDDB - BSc(Agric) PhD Protein utilisation during protein depletion and repletion, and during continous and intermittent protein infusion; Effect of adrenergic agonists on metabolism
7   Lomax JA - BSc,PhD The structure and role of cell-surface polysaccharides of rumen bacteria which adhere to fibres
8   MacLeod NA - SDA SDD To obtain the measurements of heat production and general management of the animals; Factors affecting absorption of volatile fatty acids from the rumen
9   Orskov ER - BSc(Agric) MSc PhD DSc The effect of fibre consumption on heat production in animals sustained by intragastric nutrition; The effect of different proportions of volatile fatty acids on absorption from the rumen; The effect of ammonia treated straw on feed intake; levels of treated straw in diets; The effect of level and type of concentrate and supplemented straw diets on cellulolysis in the rumen of cows and sheep; Degradability of dietary proteins and of their outflow rate in the rumen and digestibility in small intestine; Determination of factors affecting rumen microbial protein yield; Effect of protein on fat mobilisation in dairy cows; To manipulate milk protein and milk fat production from knowledge of control mechanisms
10 Reid GW - SDA The effect of urea and ammonia on preservation and utilisation of moist hay; Management of dairy cows given different types and levels of concentrates
11 Richardson AJ - HNC,LIBiol,AMIBiol Celluloysis and digestion of herbage in the rumen Establishing a defined rumen flora capable of supporting the growth of gnotobiotic lambs fed various diets
12 Stewart CS - BSc,PhD Cellulose and digestion of herbage in the rumen. Establishing a defined rumen flora capable of supporting the growth of gnotobiotic lambs fed various diets. The structure and role of cell-surface polysaccharides of rumen bacteria which adhere to fibres
13 Wallace RJ - BSc,PhD The basic mechanisms and consequences of metabolism of nitrogenous compounds by rumen microorganisms. Factors which influence energy conservation and growth yields in rumen bacteria. Establishing a defined rumen flora capable of supporting the growth of gnotobiotic lambs fed various diets
14 Whitelaw FG - BSc(Agric) PhD Factors affecting urea recycling in ruminants; Effect of protein and amino acids on heat production and milk yield in dairy cows
15 Williams PEV - BSc,MSc,PhD Utilisation of low quality roughage and formulation of rations for beef cattle. Energy and protein metabolism of cattle during compensatory growth. Energy metabolism and efficiency of food utilisation in milk fed calves; factors affecting rumen development in pre-weaned calves; growth and balance studies in calves given adrenergic agonists
16 Wood TM - BSc,PhD,CChem,MRSC The structure and mechanism of action of cellulases and xylanases from fungi and rumen anaerobic bacteria. Enzymatic conversion of straw carbohydrates into soluble sugars

4340 Physiology

Greenburn Road, Bucksburn, Aberdeen, AB2 9SB
Tel: 0224 712751.

1   Begbie R - BSc,PhD The degradation of free and cell-bound plant proteins by gastro-intestinal proteases in the pig; changes in mucus production in the gastrointestinal tract in response to ingested plant proteins. Effects of anti-nutritional proteins of legume seeds; intestinal cell receptors for lectins and consequences of lectin binding; the putative interference of dietary lectins with production of secretory IgA
2   Brockway JM - BSc,PhD Measurement of activity and energy expenditure in sheep and cattle. Organ calorimetry of portal chained viscera of sheep
3   Bruce LA - HNC,LRIC Rates of VFA absorption in ruminants
4   Crabtree BF - BA,PhD Futile cycles by which excess acetate is "burnt off" in ruminants. *In vivo* and *in vitro* measurement of intermediary metabolism in hind limb and hepatocytes of sheep
5   Fletcher JM - BSc,DPhil Hormonal control of the abnormal metabolism of the Zucker rat; manipulation of the endocrine status of obese and lean rats. The mode of action of sex-related steroids
6   Fowler VR - BSc,PhD Establishing the principal components of reproductive efficiency using coordinated sow trials. Dietary and social factors which affect feed intake including stock density; substitution of soyabean in pig diets by legume seeds and rapeseeds. Effects of early setbacks and disease on subsequent growth; critical factors affecting adaptation of the gut to novel feeds before and after weaning. Compositional changes in different breeds of pig at different feed intakes close to appetite. The reproductive efficiency and quality of meat from once-bred gilts

7   Garlick PJ - MA,PhD Regulation of muscle protein turnover by food intake, hormones and amino acid concentration; techniques for assessing protein turnover in man. Regulation of protein, fat and carbohydrate storage, in particular the influence of metabolic and nutritional states on protein storage. Role of arachidonic acid metabolites in protein metabolic response to feed intake. Changes in skeletal muscle protein turnover during growth and development; kinetic measurement of proteolysis; functional role of insulin in muscle protein synthesis
8   Gregory PC - BSc PhD Control of gastrointestinal digesta transit in sheep; Mechanics of control of gastrointestinal motility in sheep; Influence of volatile fatty acids on gastrointestinal motility and hormone secretion in sheep; Intestinal motility, digesta propulsion, intestinal secretion, intestinal hormone secretion and absorption of nutrients; Satiety effects of nutrient infusions in the pig gastrointestinal tract; Bypass or resection of the intestine of food intake and hormone secretion; Gastrointestinal and pancreatic hormones and the control of voluntary food intake; Control of gastric emptying and small intestinal digesta transit in relation to food intake in the pig
9   Haggarty P - BSc PhD The use of deuterium and oxygen-18 turnover as an index of energy expenditure particularly in free-living pigs; quantification of energy expenditure in individual tissues *in vivo*; factors controlling energy expenditure *in vitro*
10 Harris CI - BSc PhD 3-Methyl histidine as a measure of muscle protein turnover
11 James WPT - MA,MD,DSc,FRCP,FRCP(E) Assessment of body composition *in vivo*. The hormonal determinants of substrate storage and oxidation
12 Jones AS - BSc,PhD Hormone manipulation in boars. Control of sex hormones and their effect in bulls
13 Kay RNB - MA PhD FRSE BA Establish the mechanisms and inter-relationships of seasonal cycles of growth, appetite, metabolism and reproduction; To see if changes in gut function are associated with seasonal changes in appetite
14 King TP - PhD Immunogold localisation of uncoupling protein in brown adipose tissue mitochondria. Ultrastructure and cytochemistry of neonatal lamb gut; the use of lectin probes to describe gut surface saccharides; the cytopathological effects of dietary lectins
15 Lobley GE - BSc,PhD Nutrient supply and the specific dynamic effect of acetate in ruminants. *In vivo* measurement of intermediary metabolism in the liver and hind limb of sheep. Protein turnover as influenced by repartitioning agents. The mode of action of sex-related steroids
16 MacRae JC - BSc,PhD Amino acid availability and the heat increment of feeding in ruminants. Energy and nitrogen metabolism of sheep given adrenergic agonists. Relationship between efficiency of utilisation of absorbed amino acids and composition of absorbed fatty and amino acids. Aerobic metabolism of portal drained viscera of sheep
17 Maltin CA - BSc,PhD Pathology of muscle damage in Se/Vit E deficient animals. Involvement of proteolytic enzymes in muscle protein degradation. Effects of nutrition and hormones on muscles with different fibre types. Skeletal muscle: changes in skeletal fibre type composition during growth and development; structural changes associated with manipulation of protein degradation; structural role of insulin
18 McNurlan MA - BSc,PhD Regulation of protein turnover in man by food intake, hormones and amino acids; techniques for assessing protein turnover. Regulation of protein, fat and carbohydrate storage, in particular the influence of different metabolic and nutritional states on protein storage. Protein turnover and storage during drug treatment
19 Morgan PJ - BSc,PhD Isolation and characterisation of biologically active peptides from the pineal
20 Partridge GG - BSc,PhD The nutrient and dietary requirements of very early-weaned pigs; critical factors affecting adaptation of the gut to novel feeds before and after weaning
21 Rayner DV - BSc PhD Mechanisms of control of gastrointestinal motility in sheep; Intestinal motility, digesta propulsion, intestinal secretion, intestinal hormone secretion and absorption of nutrients; Metabolic consequences of the batch propulsion of digesta in the pig; The portal absorption of volatile fatty acid in the pig; Satiety effects of nutrient infusions in the pig gastrointestinal tract; Bypass or resection of the intestine of food intake and hormone secretion; Gastrointestinal and pancreatic hormones and the control of voluntary food intake; Control of gastric emptying and small intestinal digesta transit in relation to food intake in the pig
22 Robinson JJ - BAgr,PhD Effects of nutrition, physiological state, and environmental stimuli on the hormone status of ewes; effects of dietary factors on the partitioning of nutrients for reproductive purposes and the subsequent prenatal growth and metabolism of newborn lambs. The use of deuterium and oxygen-18 for measuring energy expenditure in free-living sheep.
23 Scott D - BSc PhD Factors affecting the absorption and excretion of calcium, phosphorus and magnesium in farm animals
24 Smith A - BSc,PhD Measurement of deuterium and oxygen-18 in biological samples
25 Stansfield R - BSc Kinetic analysis of water and bicarbonate turnover
26 Wenham G - DSR(R) SRR Radiographic study of gastrointestinal motility in sheep; Intestinal motility, digesta propulsion, intestinal secretion, intestinal hormone secretion and absorption of nutrients; Foetal growth and development
27 Wood J - BSc,MA Establishing the principal components of reproductive efficiency using coordinated sow trials

4341 Calorimetry
Greenburn Road, Bucksburn, Aberdeen, AB2 9SB
Tel: 0224 712751.

1 Mollison GS Measurement of activity patterns in sheep
2 Smith JS - HNC Measurement of net energy values of ruminant feeds. Determination of maintenance requirements and fasting metabolic rates of growing beef cattle. Calorimetric validation of isotopic measures of energy output

4342 Scottish Crop Research Institute
Invergowrie, Dundee, DD2 5DA
Tel: 0826 7731.

Fruit, cereal, forage and potato breeding, crop production and protection

1 Hillman Prof. J *Director* fruit, cereal, forage and potato breeding, crop production and protection
2 McKean Ms UM - MA,DipLib *Librarian*

4343 Mycology and Bacteriology
Invergowrie, Dundee, DD2 5DA
Tel: 0826 7731.

1 Duncan JM - BSc PhD Autecology and survival of important fungal pathogens of potatoes; Identify root-rotting *Phytophthora* spp of cane fruit and screen fungicides; Compare the stability in the field of population of *Phytophthora infestans* and *P. fragavial*; Inheritance virulence and aggressiveness in Phytophthora infestans and P
2 Fox RA - BSc BAgri MIBiol CBiol Post-planting development of gangrene lesions and contamination of progeny tubers; Autecology and survival of important fungal pathogens of potatoes; Examine the interactions and effects on growth of the soil biota on potato and barley roots
3 Harrison JG - BSc PhD MIBiol Effects of the aerial environment on potato late blight; Epidemiology of chocolate spot disease of field beans and role of pectolytic bacteria
4 Lyon GD - BSc MSc PhD DIC The chemistry of resistance of potato to *Erwinia* spp; Host-pathogen recognition between potato and *Erwinia* spp; Investigation of pollen-stigma interactions in brassicas
5 Malcolmson JF - BSc PhD MIBiol Mechanisms of quantitative resistance to potato blight; Evolution in the field of new *Phytophthora infestans*; Electron microscopy of potato tissue and *Phytophthora infestans*
6 Mowat W - BSc,DipAgr Developing screening methods for virus disease resistance
7 Newton AC - BSc,PhD Epidemiology and control of *Erysiphe graminis* and *Rhynchosporium secalis* in barley
8 Perombelon MCM - BSc MSc Etiology and epidemiology of blackleg in Scotland and overseas; Etiology of bacterial soft rots of stored potatoes; Autecology and survival of *Erwinia* spp pathogenic on potatoes; Detection and control of latent infection of potatoes by *Erwinia* spp; Epidemiology of gall-forming bacteria in *Rubus*; Monitor the species balance in populations of *Erwinia* spp; Genetics of *Erwinia* in relation to pathogenicity and cell wall-degrading enzymes
9 Perry DA - BSc PhD Biology of *Fusarium nivale* and *Typhula incarnata*; action of fungicides; Epidemiology and control of *Erysiphe graminis* and *Rhynchosporium secalis* in autumn sown barley; The origin, biology and pathogenesis of clostridial bacterial soft rots of stored potatoes; Oversee screening for disease resistance
10 Williamson B - BSc MSc PhD Etiology, epidemiology and control of fungal diseases of cane fruit; Inoculation techniques and methods of disease assessment for cane fruit breeding
11 Williamson CJ - BSc PhD The influence of the environment in forage brassica/pathogen interactions and effects on crop yield; Evolution in the field of populations of *Erysiphe cruciferarum* and *Plasmodiophora brassicae*; Inheritance of virulence in, and resistance to, *Erysiphe cruciferarum* and *Plasmodiophora*; Devise methods for resistance breeding and screen material in breeding programs

4344 Zoology
Invergowrie, Dundee, DD2 5DA
Tel: 0826 7731.

1 Alphey TJW - BSc PhD The application and mode of action of nematicides; Integrated control of potato cyst nematodes using nematicides with resistance and partially resistant cultivars
2 Birch ANE - BSc PhD MIBiol Develop and evaluate a screening test for turnip root fly; Biology of turnip root fly on susceptible and resistant swede; Mechanisms of resistance to insects
3 Boag B - BSc PhD Investigate the pathogenicity and control of nematode pests of brassicas; Population dynamics and pathogenicity of virus vector nematodes; Ecology and pathogenicity of nematodes,modelling their population dynamics; Factors determining host status of plants for migratory ecto nematodes
4 Forrest JMS - BSc PhD The behaviour of *Globodera pallida* within and outwith the roots of resistance potatoes; Mechanisms of resistance and hypersensitivity to potato cyst nematodes and effects on their physiology and behaviour; Role of carbohydrates in resistance and host recognition by pathogens
5 Phillips MS - BSc Modelling of population dynamics of cysts nematodes in relation to host resistance; Virulence in potato cyst nematode and identification of pathotypes; Development of screening methods for resistance to potato cyst nematode

6 Robertson WM - HNC Investigate structure and function of virus vectors and mechanisms of virus retention; Role of carbohydrates in virus retention, host recognition and pathogen/prey relations
7 Trudgill DL - BSc PhD Mechanisms of damage thresholds of crop tolerance to damage by root feeding nematodes; Relation between initial density of cyst nematodes, tolerance, nematicides fertiliser and damage to potato; Biology and control of *Pratylenchus penetrans* and virus vector nematodes; Feeding behaviour, specificity and transmission of viruses by *Longidorus* and *Xiphinema*
8 Woodford JAT - MA PhD Ecology of aphid vectors of potato viruses and the factors affecting aphid efficiency in spreading viruses; Devise improved control strategies for pests of cane fruit; Devise methods of controlling virus spread in potato; Modelling aphid population dynamics and virus spread in relation to plant growth; Retention and transmission of stylet-borne viruses and novel means of control; Prevention of virus spread with mineral oils and insecticides

4345 Physiology and Crop Production
Invergowrie, Dundee, DD2 5DA
Tel: 0826 7731.

1 Cormack MR - NDH Dry matter partitioning, and compensation between yield components in cane fruits; Physiological and cultural factors affecting the mechanical harvesting of soft fruits; Selection of cultivars and design of production methods for soft fruit crops
2 Davies HV - BSc PhD Physiological basis of intersprout competition in the potato during the pre-emergence growth phase; Effects of mineral nutrition on growth and development of potatoes; Factors influencing dry matter content and sugars balance of potato tubers
3 Lawson HM - BSc MAgrSc DipAgric Selection of cultivars and design of production methods for soft fruit crops; Prediction and control of weeds and herbicide performance in crop rotations; Ecology and control of weeds and unwanted crop vegetation in agriculture and horticultural crops
4 MacKerron DKL - BSc PhD Variability in size grade distribution in potato crops
5 MacKerron DRL - BSc PhD Physiology of the response of the potato crop to radiation, temperature and water supply
6 Marshall B - BSc ARCS PhD Simulation of the growth and development of arable crops
7 Mason DT - BSc PhD Physiological factors affecting the maturation and quality of raspberry fruits; Dry matter partitioning, and composition between yield components in cane fruits; Physiological and cultural factors affecting the mechanical harvesting of soft fruits
8 Oparka KJ - BSc PhD Physiological processes involved in dry matter partitioning in the potato plant
9 Thompson R - BSc MSc MIBiol Investigate the effects of cultural practices on the growth and development of the potato crop; Investigate the effects of cultural practices, environment, and genotype on growth and development of grain legumes; Devise methods for the production of vegetable crops under northern conditions
10 Waister PD - BSc PhD Sprout number and growth rate following short-term storage of seed. Phasing of inter and intra-plant competition in the potato crop

4346 Chemistry
Invergowrie, Dundee, DD2 5DA
Tel: 0826 7731.

1 Allison MJ - BSc PhD NIR analysis for the rapid estimation of quality attributes of crop plants; Develop rapid tests for the estimation of malting quality in barley; Methods for the estimation of brassica digestibility; Use of Near-Infra red analysis for the rapid estimation of quality factors in potatoes; Develop and use biochemical markers on barley chromosomes
2 Cowe IA - HND NIR spectral structure related to sample composition; Develop near infra-red techniques for the assessment of malting quality

4347 Cereal Breeding
Invergowrie, Dundee, DD2 5DA
Tel: 0826 7731.

1 Ellis RP - BSc PhD Breed improved cultivars of spring barley; Coordinate SCRI barley trials and BAPB spring barley trials; Study the interaction of environment with genotype on plant development and crop yield
2 Powell W - BSc MSc Breed improved cultivars of winter barley; Develop the production and use of inbred lines in barley breeding
3 Thomas WTB - BSc PhD Breed improved cultivars of spring barley with better disease resistance; Develop cross prediction techniques and study the genetic control and relationships of important characters

4348 Forage Brassica Breeding
Invergowrie, Dundee, DD2 5DA
Tel: 0826 7731.

1 Bradshaw JE - MA MSc PhD Breed improved swede cultivars and investigate breeding methods; Breed improved kale cultivars and investigate breeding methods; Breed improved fodder cabbage cultivars and investigate breeding methods; Genetics of heterosis and the genetics of factors affecting root texture of winter-hardy material in the swede
2 Hodgkin JRT - BSc PhD Study genetics of B; Investigate pollen irradiation and pollen selection; Study S-allele incompatibility in brassicas

3 MacFarlane Smith HW - BSc PhD MIBiol FBIM Breed improved rape cultivars and investigate breeding methods; Investigation of variation in synthesised B

4 Wills AB - BSc MS PhD To multiply and stabilise breeders selections and trial selections in collaboration with other organisations; Study genetics of *Brassica oleracea*; Investigate transfer of alien DNA by pollen

4349 Potato Breeding
Invergowrie, Dundee, DD2 5DA
Tel: 0826 7731.

1 Caligari PDS - BSc PhD Breed potatoes for table use and processing in UK; Breed potato cultivars for export as seed from UK; Genetics breeding and trialling methods; Develop screening methods for resistance to pests and diseases of potatoes

2 Carroll CP - BSc Evaluate and develop Group Phureja diploids and their hybrids with tuberosum for use in potato breeding

3 Glendinning DR - BSc Combining Neo-Tuberosum properties to prepare parents for commercial breeding; testing crosses with Tuberosum; Import, maintain, evaluate and distribute true seed of primitive and wild potatoes ;liaise with Dutch-German Collecion

4 Mackay GR - BSc MSc Breed potatoes for table use and processing in UK; Breed potato cultivars for export as seed from UK; Genetics breeding and trialling methods; Develop screening methods for resistance to pests and diseases of potatoes

5 Waistie RL - MA PhD FISP Screening for resistance to fungal and bacterial diseases of potato

4350 Soft Fruit Breeding
Invergowrie, Dundee, DD2 5DA
Tel: 0826 7731.

1 Anderson MM - NDH,SDH,DHE Black currant breeding and related research

2 Jennings DL - BSc PhD Raspberry breeding and related research; Breeding blackberries for earliness, hardiness, spinelessness and fruit quality

3 McNichol RJ - BSc MSc FSS Raspberry breeding and related research; Performance of strawberry selection in the Scottish environment

4351 Virology
Invergowrie, Dundee, DD2 5DA
Tel: 0826 7731.

1 Barker H - BSc PhD Mechanisms and expression in laboratory and field condition of virus resistance in potato

2 Harrison BD - BSc PhD FRSE Virus variation, gene pools and symptom production; Properties and ecology of potato virus and methods for their detection; Infectivity of DNA copies of plant virus genes and reaction to virus of plants transformed with this DNA; Cassava virology

3 Jones AT - BSc PhD Diagnosis, properties, spread and control of Rubus and Ribes viruses; Viruses in Rubus: experimental infection, virus detection, and reactions of genotypes

4 Lennon AM - BSc PhD Detection, diagnosis and characterisation of cassava viruses

5 Massalski PR - BA PhD Preparation, testing and use of monoclonal antibodies (to the potato leafroll virus-PLRV); detection methods, properties, transmission and spread of potato virus

6 Mayo MA - BSc PhD MIBiol Organisation of virus genomes, translation products and variation in nepovirus particle products; Biological and biochemical effects of DNA copies of plant virus genes and their incorporation into plant genomes; Analysis of double-stranded RNA in plant tissue extracts

7 Mowat WP - BSc DipAgrSci Serological and inoculation tests, occurrence and control of forage brassica viruses; Production and maintenence of VT narcissus stocks and assessment of procedures for their propagation; Detection, properties, transmission and control of narcissus viruses; Devise screening methods for virus disease resistance

8 Murant AF - BSc PhD Elucidation of the mechanism of virus transmission by aphids and the roles of virus-coded proteins; Molecular biology of raspberry bushy dwarf virus and its resistance-breaking strain; Analysis of double-stranded RNA in plant virus infections

9 Roberts IM - HNC DipRMS Electron microscopy and electron microscope serology of viral products; Methods for electron microscopy of plant viruses and virus antigens

10 Robinson DJ - MA PhD Virus variation,sequence homologies in virus nucleic acids and subgenomic nucleic acids; Biochemical effects of incorporating DNA copies of plant virus genes into plant genomes; Nucleic acid hybridisation method for virus detection and diagnosis

4352 AFRC Institute of Arable Crops Research - Unit of Insect Neurophysiology and Pharmacology

1 Treharne JE - MA,PhD,ScD *Honorary Director*

4353 Insect Neurophysiology and Pharmacology

1 Berridge MJ - BSc PhD FRS Hormone receptor mechanisms
2 Bridges RG - BSc Choline synthesis in insects: effects on membrane and transmitter properties

3 Buchan PB - ONC HNC Biophysics and physiological instrumentation
4 Dwivedy AK - BSc MSc PhD Insecticidal effects on transmitter uptake and release from insect synaptosomes
5 Evans PD - MA PhD Pharmacology of insect biogenic amines and peptides
6 Harrison JB - ONC HNC LIBiol Electronmicroscopical and immunocytochemical studies of Malpighian tubules
7 Heslop JP - BSc MSc Intracellular signalling mechanisms in insects
8 Howes-Chipchase EA - BA,MA Cell culturing and immunocytochemistry of insect glia
9 Lane NJ - BSc MSc DPhil Structural correlates of secretory function in insects
10 Leech CA - BSc PhD Glial ion channels and receptors
11 Maddrell SHP - MA PhD FRS DSc Membrane properties and permeability of insect Malpighian tubules
12 Moreton RB - BA PhD Biophysical studies on insectidal interaction with ion channels and membranes
13 Sattelle DB - BA PhD Pharmacology of insect cholinergic and GABAergic receptor and ion channels
14 Schofield PK - BSc PhD Pharmacology of insect neuroglia
15 Skaer H - BA MA PhD Pharmacological differentiation of secretory epithelia
16 Swales LS Ultrastructural and immunocytochemical studies

4354 AFRC Institute of Plant Science Research - Division of Nitrogen Fixation Research

1 Smith Dr BE *Head*

4355 Nitrogen Fixation
The Chemical Laboratory, University of Sussex, Brighton, BN1 9RQ
Tel: 0273 678252.

1 Buck M - PhD Site-directed mutagenesis of nitrogen fixation genes
2 Dixon RA - BSc DPhil Molecular biology of nitrogen fixation in *Klebsiella*
3 Eady RR - BSc PhD DSc Physiology of regulation of nitrogenase by oxygen molybdenum nitrogen status; Characterization of nitrogenase in *Azotobacter*; *Azotobacter vinelandii* nif minus mutants properties of alternative nitrogenase functioning under molybdenum deficiency
4 Henderson RA - BSc PhD The mechanisms of activation of dinitrogen bound to transition-metal sites
5 Hill S - BSc,PhD Physiology of nif in facultative anaerobes
6 Hughes DL - BSc PhD X-ray crystal structure analysis of transition metal complexes
7 Kennedy CK - BA PhD Expression of *Klebsiella* nif genes in *Azotobacter* and comparison of gene products; Isolation of genes concerned with regulation of nif from the banks of *Azotobacter*
8 Leigh GJ - BSc PhD DSc Metal complexes of intermediates implicated in nitrogenase function, and their modes of reactivity; Binding, activation and protonation mechanisms of dinitrogen bound to transition metal ions
9 Lowe DJ - MA,PhD Structure and function of nif gene products; Developing computer programs used for nucleic acid sequencing and data analysis; physical techniques; Structure and function of nif gene products
10 Merrick MJ - BSc PhD Molecular biology of regulation of nif in *Klebsiella*
11 Pickett CJ - BSc PhD Electron transfer reactions of prosthetic groups and synthetic analogues of nitrogenase proteins; Electron transfer in nif, energetics intermediates and mechanisms, electrosynthesis of ammonia by cyclic systems
12 Postgate JR - MA,DPhil,DSc,FIBiol,FRS Expression of nif in sulphate-reducing bacteria and genetically engineered *Salmonella*. Expression of *Klebsiella* nif in *Pseudomonas putida*
13 Richards RL - BSc,PhD,DSc Metal complexes of dinitrogen and its analogues as models for the active site of nitrogenase; Reduction of dinitrogen and analogues at transition metal sites-metallic cofactors of molybdo-enzymes
14 Robson RL - BA PhD Cloning and analysis of genes for nif in *Azotobacter*
15 Smith BE - BSc PhD Function and biosynthesis of nif gene products and iron-molybdenum cofactor; Structure, function of nitrogenase, its iron-molybdenum cofactor and other nif gene products
16 Thorneley RNF - BSc PhD Structure, function of nif gene products; oxygen tolerance of nif gene products; Structure, kinetic and mechanism of nitrogenase and related gene products, oxygen tolerance of nitrogenase; Electron transfer and ligand substitution mechanisms with transition metal complexes and metalloproteins
17 Yates MG - BSc, PhD Biochemistry, regulation and genetics of enzymes of hydrogen metabolism in *Azotobacter*; Electron carriers, nitrogenase function, nif and hup minus mutants of aerobic diazotrophs

4356 AFRC Institute of Animal Health - AFRC & MRC Neuropathogenesis Unit
West Mains Road, Edinburgh, EH9 3JF
Tel: 031-667-5204.

1 Kimberlin Dr RH *Acting Director*

Great Britain

4357 Neuropathogenesis
West Mains Road, Edinburgh, EH9 3JF
Tel: 031-667-5204.

1 Dickinson AG - BSc PhD Genetical controls of scrapie pathogenesis in host and agent
2 Fraser H - BVetMed BA MRCVS PhD Pathogenesis and neuropathology of scrapie
3 Kimberlin RH - BSc PhD Pathogenesis of scrapie and nature of the agent
4 Outram GW - BSc PhD Pathogenesis of scrapie and agent strain typing

4358 AFRC Institute for Grassland and Animal Production - Welsh Plant Breeding Station
Plas Gogerddan, Aberystwyth, Dyfed, SY23 3EB
Tel: 0970 828255.

Improvement and agronomy of the main crops, systems of animal production and supporting research into plant science

1 Collins Ms DJ - BSc,DipLib *Librarian*
2 Stoddart Dr JL - PhD,DSc,FIBiol *Director and Head of Division*

4359 Plants and Soils Division
Plas Gogerddan, Aberystwyth, Dyfed, SY23 3EB
Tel: 0970 828255.

1 Stoddart JL - PhD,DSc,FIBiol Regulation of gene expression in meristematic tissues. Low temperature studies on growth and carbon metabolism using contrasting genetic material

4360 Cell Biology
Plas Gogerddan, Aberystwyth, Dyfed, SY23 3EB
Tel: 0970 828255.

1 Chamberlain DA - BSc Production of somatic hybrids from sexually incompatible forage species
2 Dalton SJ - BSc Screening of genotypes and genetic transformation techniques in suspension and protoplast cultures of grasses
3 Webb KJ - BSc PhD Regenerating plants from tissue cell and protoplast cultures; developing techniques for genetic transformation

4361 Developmental Regulation
Plas Gogerddan, Aberystwyth, Dyfed, SY23 3EB
Tel: 0970 828255.

1 Jones TWA - PhD,MSc Regulation of gene expression in meristematic tissues, and during vegetative development and floral initiation in clover. Use of isoenzymes as genetic markers and the adaptive significance of multiple enzyme forms
2 Ougham HJ - PhD,BA Regulation of gene expression in meristematic tissues
3 Thomas HO - BSc PhD Regulation of gene expression in meristematic tissues. Biochemical control of leaf development; duration and performance

4362 Environmental Application
Plas Gogerddan, Aberystwyth, Dyfed, SY23 3EB
Tel: 0970 828255.

1 Eagles CF - BSc,PhD The effect of irradiance and temperature/irradiance interactions on leaf growth and carbon metabolism. Low temperature studies on growth and carbon metabolism using contrasting genetic material; the physiology and biochemistry of cold-hardening and dehardening responses
2 Gay AP - BSc Measurement of photosynthetic parameters in contrasting genotypes developed under varying light and temperature regimes. The physiological basis and use of variation in water-use efficiency in forage species
3 Pollock CF - MA,PhD The effect of irradiance and temperature/irradiance interactions on leaf growth and carbon metabolism
4 Pollock CJ - MA,PhD Low temperature studies on growth and carbon metabolism using contrasting genetic material; the physiology and biochemistry of cold-hardening and dehardening responses. Environmental modulation of carbon partitioning in forage crops and cereals

4363 Pathogen and Pest Resistance
Plas Gogerddan, Aberystwyth, Dyfed, SY23 3EB
Tel: 0970 828255.

1 Carver TLW - BSc,PhD Host-endophyte relationships at the host-plant level. Mechanisms of host resistance to biotrophic pathogens on graminaceous hosts
2 Catherall PL - BSc PhD Characterisation of grass and cereal viruses; the genetics and mechanisms of host responses and an evaluation of resistance. Effects of virus infection on forage grass quality. Epidemic potential of clover viruses; identification and characterisation of host resistances
3 Chamberlain JA - BSc PhD Ultrastructure and serology of resistance to fungal, virus and nematode pathogens of graminaceous hosts; ultrastructure of host-virus interaction and virus serology

4 Clifford BC - BSc,MS,PhD Identifying and characterising host resistances; evaluating the cost:benefits of genetic resistance. Host-pathogen relationships; resistance and pathogen damage in grasses. Epidemic potential of pathogenic fungi in clover; identification and characterisation of host resistances. Temperature sensitive resistance to *Puccinia* spp. Relationship between host resistance and systemic fungicides in the integrated control of cereal diseases. Introduction, characterisation and exploitation of resistant germplasm; coordinating the conservation and supply of resistant germplasm to AFRC user groups
5 Cook R - BSc,PhD,ARCS,DIC Host-nematode relationships; resistance and nematode damage assessment in grasses and forage legumes. Characterisation of resistance to cereal nematodes; pathogen virulence and host-parasite relationships. Mechanisms of resistance to sedentary endoparasitic nematodes
6 Jones ERL - HND Pathogen virulence in oat crown rust and mildew, wheat and barley brown rust, barley net blotch and *Rhynchosporium*. Pathogen variability, host range and epidemic potential on graminaceous hosts
7 Potter LR - HNC,LIBiol Host-pathogen relationships; resistance and pathogen damage in grasses. Effects of viral and fungal infection on quality of forage grasses. Host-endophyte relationships at the plant population level
8 Roderick HW - HNC,BA Introduction, characterisation and exploitation of resistant germplasm; the assembly, conservation and supply of resistant germplasm to AFRC user groups. Identifying and characterising host resistances; evaluating the cost:benefits of genetic resistance. Relationship between host resistance and systemic fungicides in the integrated control of cereal diseases.

4364 Nitrogen Fixation and Nutrient Cycling
Plas Gogerddan, Aberystwyth, Dyfed, SY23 3EB
Tel: 0970 828255.

1 Colbourn P - BSc PhD Measuring nitrogen (N) flows and improving the efficiency of N use in clover based pastures. The mechanisms of denitrification, nitrification and gaseous loss
2 Collison M - HNC The mechanisms of N release, uptake and associated losses
3 Goodman PJ - BSc,PhD Measuring N flows and improving the efficiency of N use in clover based pastures. The mechanisms of N release, uptake and associated losses
4 Hughes DM - MIBiol,MSc Variation in, and ecology of, clover/*Rhizobium* associations; selecting for improved white clover/*Rhizobium* symbiosis; development of inoculation techniques
5 Mytton LR - BSc,PhD Variation in, and ecology of clover/*Rhizobium* associations. Selecting for improved white clover/*Rhizobium* symbiosis; development of inoculation techniques
6 Williams DH - HNC The mechanisms of denitrification, nitrification and gaseous loss

4365 Grass Breeding and Herbage Seed Research
Plas Gogerddan, Aberystwyth, Dyfed, SY23 3EB
Tel: 0970 828255.

1 Hides DH - BSc Investigating species other than *Lolium*; breeding new and improved varieties. Optimising seed yield and quality in forage grasses
2 Hollington PA - BSc,PhD The genetic and management factors affecting white clover seed production in on-farm systems
3 Jones MLC - HNC Producing, developing, assessing and maintaining ryegrass hybrids
4 Marshall A - BSc,PhD Optimising seed yield and quality in forage grasses and forage legumes. Assessing the potential for chemical manipulation of plant development and seed yield in forage legumes
5 Stephens DE - BSc Breed stable ryegrass hybrids for conservation and grazing systems
6 Wilkins PW - BSc PhD Breeding perennial and Italian ryegrass for yield, persistency and disease resistance

4366 Crop Genetics
Plas Gogerddan, Aberystwyth, Dyfed, SY23 3EB
Tel: 0970 828255.

1 Hayward MD - BSc PhD CBiol MIBiol Forages: assessing the genetic control of characters in forages and developing techniques for selection and variety construction; the genetic control of reproductive systems, their exploitation and effect on population structure; the genetic assessment of natural populations of forage in relation to conservation of genetic resources. The influence of polyploidy and hybridization on gene action in the *Lolium*/fescue complex. The internal environment of pollination bags and effects on seed set
2 Thomas HE - BSc PhD ARCS Identification and development of techniques for measuring and selecting for physiological components of drought resistance. Developing techniques and measuring the physiology of conserved germplasm
3 Tyler BF - BSc Collection, characterisation, classification and conservation of the genetic resources of forages and of grass species suitable for amenity purposes. Characterisation of seasonal growth and agronomic performance of conserved germplasm. Documentation of passport and characterisation data of genetic resources. Re-establishment and *in situ* conservation of a population of *Vicia orobus*

4367 Grass and Cereal Cytogenetics

Plas Gogerddan, Aberystwyth, Dyfed, SY23 3EB
Tel: 0970 828255.

1 Humphreys MO - BSc,PhD Development of multi-trait selection procedures. Breeding consistent ryegrasses using a multi-trait approach in order to maximise stress tolerance and seasonal yield. Development of ryegrass fescue hybrids with new growth potentials
2 Leggett JM - BSc Species relationships and gene transfer
3 Lewis EJ - BSc Cytogenetical assessment of species relationships in *Lolium*/fescue; identifying sources with high potential for breeding. Improving meiotic stability in wide hybrids through cytogenetic manipulation
4 Thomas HM - HNC The mechanism and control of the initiation of chromosome pairing in species and interspecific hybrids
5 Thomas HU - PhD,DSc,MIBiol Gene transfer and the use of aneuploids for genetic studies. Genetic relationships between species within the *Lolium/Festuca* group; the control of chromosome pairing in *Lolium/Festuca* hybrids and derivatives

4368 Plant Resource Use

Plas Gogerddan, Aberystwyth, Dyfed, SY23 3EB
Tel: 0970 828255.

1 Davies DA - BSc MSc Herbage growth, quality, intake and animal performance
2 Fothergill M - HND,MIBiol White clover performance on grazed swards
3 Jones D - OND,HNC White clover performance on grazed swards
4 Jones JR - ONC,HND Effect of management on cattle and sheep performance
5 Munro JMM - BSc,MSc,DipAgrSc Effect of management on cattle and sheep performance. Investigating changes in land-use in the uplands

4369 Legume Breeding

Plas Gogerddan, Aberystwyth, Dyfed, SY23 3EB
Tel: 0970 828255.

1 Collins RPA - BSc,PhD Identifying yield related physiological and morphological characters of grass/clover mixtures
2 Evans DR - HNC Breed white clover for yield and persistency under continuous grazing
3 Hill J - BSc PhD The competition between grass and clover
4 Michaelson-Yeates TPT - HNC,LIBiol The inheritance of characters associated with grass and clover yield in mixtures
5 Norris IB - BSc,MSc Identifying the characters related to variation in seed yield potential in white clover; inheritance and response to selection in characters controlling seed yield
6 Rhodes I - BSc PhD Breed white clover for yield, predictability of yield, persistency and seed yield; Breed lucerne for yield, quality and resistance to diseases and pests; Identify characters which affect competitive ability and yielding ability of grasses and clovers in mixed swards

4370 Cereal Breeding

Plas Gogerddan, Aberystwyth, Dyfed, SY23 3EB
Tel: 0970 828255.

1 Clothier RB - HNC Evaluating and exploiting new sources of disease resistance; screening of breeding material
2 Griffiths TER - HNC Field trials of cereal breeding material and innovative varieties
3 Jones JE - HNC Developing spring oats for livestock and human use
4 Valentine J - BSc,PhD Developing winter oats for livestock and human use. Methods of early generation testing in cereals

4371 Variety Assessment

Plas Gogerddan, Aberystwyth, Dyfed, SY23 3EB
Tel: 0970 828255.

1 Evans WB - HND Relating varietal characteristics to animal production. The effects of grass variety on animal production
2 Jones EL - HNC The agronomic and feed quality characteristics of new ryegrass and clover varieties. The effects of grass variety on animal production
3 Walters RJK - BSc Systematic evaluation of new ryegrass and clover varieties. The effects of grass variety on animal production. The performance of new white clover varieties in sheep-grazed swards

4372 Clover Pasture Ecology

Plas Gogerddan, Aberystwyth, Dyfed, SY23 3EB
Tel: 0970 828255.

1 Davies AG - MA,MSc,MIBiol Ecological control mechanisms in white clover growth; development of maintenance guidelines for clover in grazed pastures
2 Haggar RJ - BSc,PhD Impact of agronomic practices on grassland productivity and the environment
3 Williams ED - MSc Ecological control mechanisms in white clover; factors affecting colonisation, establishment and growth

4373 Forage and Soil Chemistry

Plas Gogerddan, Aberystwyth, Dyfed, SY23 3EB
Tel: 0970 828255.

1 Jones DIH - BSc,PhD Chemical composition and ensiling characteristics of forage
2 Thomas TA - HNC Mineral composition of forages

4374 Forage Utilisation

Plas Gogerddan, Aberystwyth, Dyfed, SY23 3EB
Tel: 0970 828255.

1 Moseley G - BSc,PhD Grazing and management strategies for clover-based pastures

4375 Forestry Commission

Forestry Commission Headquarters, 231 Corstorphine Road, Edinburgh, Scotland, EH12 7AP
Tel: 031-334 0303. Telex: 727879 (FORCOM G).

1 Montgomery Sir D *Chairman*
2 Holmes GD - CB,BSc,DSc,FRSE,FICFor,FIWSc *Director-General*

4376 Planning and Surveys

1 Grundy DS - MA,MPhil *Head of Branch*
2 Harper Mrs WCG - BSc,MICFor
3 Oakes RQ - BA recreation studies

4377 Forest Research Station (South)

Alice Holt Lodge, Wrecclesham, Farnham, Surrey
Tel: Bentley (Hants) (0420-4) 22255. Telex: 727879 (FORCOM G).

4378 Research and Development Division

1 Grayson AJ - MA,MLitt,MICFor *Director*
2 Burdekin DA - BA,DipAgSci *Chief Research Officer (South)*

4379 Silviculture

1 Booth TC - BSc,MIFor *Head of Branch*
2 Davies H - BSc broadleaves
3 Davies RJ - BSc,MICFor arboriculture (DOE)
4 Jobling J - BSc vegetative propagation; poplars; urban and industrial sites
5 Patch D - BSc,MSc,MIFor,NDArb,FArborA arboricultural information and advisory service
6 Potter MJ - BSc lowland silviculture
7 Sale JSP - MA,MICFor herbicides; production of planting stock (nurseries)

4380 Site Studies (South)

1 Binns WO - MA,BSc,PhD,FICFor *Head of Branch* 'acid rain'; forest health (air pollution) surveys
2 Anderson MA - BSc effect of trees on sites; ecological and conservation aspects of forestry
3 Carnell R soil physics; water movement in forest soils and effects on quantity and quality of water resources
4 Moffat AJ - BSc,PhD reclamation of surface mineral workings; forest pedology
5 Willson A - BSc,PhD chemical analysis of soil and tree foliage; air pollution effects on trees

4381 Seed Research

1 Gosling PG - BSc,PhD *Head of Branch* testing, storage and seed physiology

4382 Forest Entomology

1 Evans HF - BSc,DPhil,FRES *Head of Branch*
2 Barbour DA - BSc,PhD,FRES nursery pests; *Bupalus* survey
3 Carter CI - MSc,MIBiol,FRES *Elatobium* investigations; host tree susceptibility
4 Wainhouse D - MSc,PhD,FRES *Cryptococcus fagi*

4383 Forest Pathology

1 Gibbs JN - MA,PhD,ScD *Head of Branch* pine stem rust
2 Brasier CM - BSc,PhD,DSc dutch elm disease
3 Lonsdale D - BSc,PhD decay in amenity trees
4 Webber Ms JF - BSc,PhD dutch elm disease

4384 Wildlife & Conservation

1 Ratcliffe PR - BSc,CBiol,MIBiol,MICFor *Head of Branch* conservation, control and impact of birds and mammals in commercial ecosystems

4385 Statistics

1 Howell RS - Head of Branch *design and analysis for experiments and surveys; computer systems for forest research*
2 Boswell RC - BSc,MIS design and analysis of experiments especially for silviculture and site studies
3 Hall GJ senior computer analyst
4 Mobbs ID - MIS statistical service to pathology and work study branches
5 Peace A - BSc design and analysis of experiments for seed and wildlife study
6 Smyth Miss BJ - BSc computer officer

4386 Physiology

1 Harmer R - BSc,PhD *Physiologist*

4387 Forest Surveys

1 Dewar J - BSc,MICFor *Head of Branch*
2 Betts AJA - BSc,MSc field surveys
3 Rollinson TJD - BSc,MICFor forest mensuration

4388 Communications

1 Hibberd BG - MIFor *Head of Branch*
2 Anderson IA - FIIP *Principal Photographer*
3 Harland Mrs E - MA library and technical information
4 Parker J - PhD,MIBiol,MICFor *Publications Officer*

4389 Instrumentation

1 Carnell R electronics and physics

4390 Wood Utilisation

1 Thompson DA - BSc,MIFor

4391 Northern Research Station

Roslin, Midlothian, Scotland
Tel: Fairmilehead (031-445) 2176.

4392 Research and Development Division

1 Neustein SA - BSc *Chief Research Officer (North)*

4393 Silviculture

1 Low AJ - BSc,MScF,PhD,MIC *Head of Branch* silviculture
2 Lines R - BSc,FIFor arboreta; species trials; provenance; hardwoods; air pollution
3 Mason WL - BA,BSc,MICFor vegetative propagation and nursery
4 Miller KF - BSc,MICFor windthrow
5 Tabbush P - BSc,MICFor spacing studies; respacing; forest fertilisation; forest weed control
6 Taylor CMA - BSc,MICFor fertilisation; nutrition

4394 Site Studies (North)

1 Pyatt DG - BSc *Head of Branch* physical and mechanical properties of soils with special reference to cultivation

4395 Forest Genetics

1 Faulkner R - BSc,MICFor *Head of Branch* breeding projects; register of seed sources, seed orchards and seed stands
2 Fletcher AM - BSc,MSc,PhS,MICFor plus tree selection vegetation propagation tree banks; wood investigation; flower induction; progeny tests in the nursery; pollination and pollen
3 Forrest GI - BSc,MSc,PhD variation studies (emphasis on biochemical variation and early test methods); pollen supply and handling phenological records
4 Lee SJ - BSc,MICFor progeny and clonal tests in the field; gene pools
5 Samuel CJA - BSc,PhD quantitative genetics

4396 Physiology

1 Coutts MP - BSc,PhD,MICFor *Head of Branch* physiology of tree root systems
2 John A - BSc
3 Philipson JJ - BSc,PhD root studies
4 Walker C - BSc,PhD

4397 Forest Entomology

1 Stoakley JT - MA,MSc,FIFor,DIC entomological investigations in the north

4398 Forest Pathology

1 Gregory SC - MA,PhD forest damage monitoring

2 Redfern DB - BSc,PhD *Fomes* root and butt rot

4399 Statistics

1 Stewart DH - BSc,MIBiol,FIS design and analysis of experiments for research (north); computer systems; data capture

4400 Nature Conservancy Council

Headquarters, Northminster House, Peterborough, PE1 1UA
Tel: 0733 40345.

1 Langslow DL - PhD *Assistant Chief Scientist*
2 Steele RC - FIBiol,FICFor *Director General*
3 Dair I *Head of Policy and Planning Division*
4 Gay PA - PhD,FIBiol *Head of Data and Information Division*
5 Ratcliffe DA - DSc,FIBiol *Chief Scientist*
6 French CMW - ARICS *Chief Land Agent and Head of Estates Division*
7 Nutt W *Press Officer*
8 Oswald PH *Communications Officer*

4401 Specialist Responsibilities (Biological)

1 Batten LA - PhD ornithological advice
2 Cooke AS - PhD toxicology/herpetology
3 Doody JP - PhD coastal ecology
4 Farrell Miss L lowland heathland and rare plant ecology (temporarily based at Scottish Headquarters)
5 Jefferies DJ - PhD mammalian ecology
6 Lindsay RA peatland ecology
7 Mitchell R - PhD marine and estuarine ecology
8 Newbold C - PhD freshwater ecology
9 Peterken GF - PhD woodland ecology
10 Pienkowski MW - PhD ornithologist
11 Robertson CJ licensing
12 Stubbs AE terrestrial invertebrate zoology
13 Thompson DBA - PhD mountain and moorland ecology based at Scottish Headquarters
14 Vacant monitoring, remote sensing and data-handling
15 Vincent MA - PhD *Species Adviser*
16 Wells DA grassland ecology

4402 Specialist Responsibilities (Geological)

4403 Geology and Physiography Section

1 Duff KL - PhD *Head of Section*

4404 Geological Conservation Review Unit

Pearl House, Bartholomew Street, Newbury, Berkshire
Tel: 0733 40345.

1 Wimbledon WA - PhD *Head of GCRU*

4405 Headquarters for England

Northminster House, Peterborough, PE1 1UA
Tel: 0733 40345.

1 O'Connor FB - PhD,FIBiol *Director*
2 Charman K - PhD *Head of Science Unit*
3 Morris D *Advisory Officer* forestry and agriculture

4406 Headquarters for Scotland

12 Hope Terrace, Edinburgh, EH9 2AS
Tel: 031-447 4784.

1 Francis JM - PhD,DIC *Director*
2 McCarthy J *Deputy Director*

4407 Headquarters for Wales

Plas Penrhos, Ffordd Penrhos, Bangor, Gwynedd, LL57 2LQ
Tel: (0248) 355141.

1 Pritchard T - PhD,JP *Director*
2 Gash MJ *Scientific Advisory Officer*
3 Walters Davies P *Scientific Development Officer*
4 White DA *Scientific Survey Officer*
5 Williams HJ *Deputy Director*

ENGLAND AND WALES

4408 Ministry of Agriculture, Fisheries and Food

Great Westminster House, Horseferry Road, London, SW1P 2AE
Tel: (01) 216 6591.

1 Bell Professor RL - BSc,PhD,CEng,FIM,FInstP,FIAgrE *Chief Scientific Adviser*

4409 Chief Scientists' Group

1 Burgess GHO - BSc,PhD,FZS,FRSE,FIFST *Chief Scientist* agriculture and horticulture
2 Crossett RN - BSc,BAgr,DPhil,FIFST *Chief Scientist* fisheries and food
3 Banbridge A - BSC,PhD *Scientific Liaison Officer* protection and environmental matters
4 Ellis FB - JP,BSc,PhD *Scientific Liaison Officer* agricultural crops and soil science
5 Hall SA - MA,DHMSA,MRCVS *Scientific Liaison Officer* veterinary science, pigs and poultry
6 Hirons JB - BSc,FIFST *Scientific Liaison Officer* food science
7 Hurd RG - BSc,PhD,MiBiol *Scientific Liaison Officer* horticulture
8 Roberts RC - BSc,PhD,DS *Scientific Liaison Officer* animal science (ruminants)
9 Smith DG - BSc,FFS *Scientific Liaison Officer* fisheries
10 White DJ - BSC,PhD,DIC,CEng,FIAgrE,FIMechE *Scientific Liaison Officer* engineering and farm building

4410 Research and Development Requirements Division

Co-ordination and commissioning of the Department's Research and Development Programme

1 Bruce GK *Head of Division*

4411 Food Science Division

Food chemistry, technology and nutrition: advice on food composition, additives, nutrients and contaminants and on use of atomic energy in food and agriculture

1 Knowles ME - B,PhD,CChem,FRSC,FIFST *Head of Division*
2 Coomes TJ - BSc *Head Food Safety and Surveillance*
3 Denner WHB - BSc,PhD *Head Food Composition and Information*
4 McWeeny DJ - BSc,PhD,DSc,FIFST *Head Food Science Laboratory*
5 Meekings GF - BSc,MInstP *Head Atomic Energy Unit*

4412 Agricultural Development Advisory Service

Great Westminster House, Horseferry Road, London, SW1P 2AE
Tel: (01) 216 6311.

1 Ingram P - BSc *Chief Agricultural Officer*
2 Fuller D - BSc *Senior Horticultural Officer*
3 Mansfield GA - NDA *Senior Agricultural Officer* agricultural crops
4 Calvert JA - BSc poultry husbandry
5 Dancey RJ - BSc,MS farm management
6 Finney JB - BSc,DipAgr,NDAgrE *Senior Agricultural Officer* agriculture livestock
7 Hubbard KR - BSc agronomy
8 Kingham HG - NDH *Assistant to Chief Agricultural Officer*
9 MacRae IA - NDD,NDA dairy husbandry
10 Mulholland JR - BSc livestock husbandry

4413 Central Veterinary Laboratory

New Haw, Weybridge, Surrey
Tel: 91-41111.

1 Stevens AJ - MA,BVSc,MRCVS,DipBact *Director*
2 Brinley-Morgan WJ - DSc,PhD,BVSc,MRCVS,DipBact *Deputy Director*
3 Watson WA - PhD,BVSc,FRCVS *Deputy Director*

4414 Bacteriology

1 Shreeve BJ - PhD,BVSc,MRCVS *Head of Department* general bacteriology
2 Boughton E - BVMS,DipBact,MRCVS mycoplasmosis, tuberculosis, anthrax
3 Francis PG - BVSc,MRCVS mastitis
4 Jeffery Mrs JC - MSc,BVMS,MRCVS leptospirosis
5 Pepin GA - BSc mycology
6 Pritchard DG - BSc,BVetMed,MPH,MRCVS respiratory disease, leptospirosis
7 Roeder PL - BVetMed,MSc,PhD,MRCVS biotechnology
8 Shreeve Mrs JE - BSc determinative bacteriology, contagious equine metritis

9 Stuart Miss FA - BVetMed,MSc,MRCVS tuberculosis, anthrax
10 Wray C - BVMS,PhD,MRCVS salmonellosis, colibacillosis, listeriosis

4415 Biochemistry

1 Morris JA - BSc,PhD *Head of Department* environmental contamination toxicology
2 Anderson PH - BSc,BVMed,MRCVS toxicology, chemical pathology
3 Cunningham NF - BSc,PhD,MRSC,CChem
4 Ivins LN - BSc metabolic disorders, chemical pathology
5 Stebbings R STJ - MA,BSc,VetMB,MRCVS metabolic disorders and toxicology
6 Wilson GDA - PhD,BVSc,MRCVS

4416 Biological Products & Standards

1 Davidson I - BSc,MRCVS *Head of Department* national quality control of veterinary immunological products, international standards & specifications
2 Hussainni SN - PhD,BVSc,MRCVS aerobic bacterial products
3 Lee Mrs AMT - PhD,BVMS,MRCVS bacteriological products, FAO/WHO International laboratory for biological standards
4 Luff PR - MSc,BSc,BVetMed,MRCVS viral vaccines
5 Thornton Miss D - PhD,BSc viral vaccines
6 Watson JOS - BVSc,MRCVS viral vaccines

4417 Epidemiology

1 Davies G - BVSc,MRCVS,DipBact *Head of Department* general epidemiology disease information systems
2 Kirby FD - MSc,BVMaS,MRCVS
3 Richards MS - BA,MSc statistics, sample survey techniques
4 Wilesmith JS - BVSc,MRCVS epidemiological investigations
5 Williams Miss PA - MSc,BVetMed,MRCVS advisory unit

4418 Pathology

1 Bradley R - BVetMed,MSc,MRCVS,FRCPath,MIBiol pathology of farmed mammals, myopathology
2 Done SH - BVetMed,PhD,MRCVS respiratory pathology
3 Jeffrey M - BVMS,MRCVS neuropathology
4 Richardson Miss C - BVSc,MSc,MRCVS reproductive and foetal pathology
5 Toole DTO - MVB,PhD,MRCVS cellular pathology, ocular pathology
6 Wells EAH - BVetMed,MRCVS *neuropathology*

4419 Parasitology

1 Cawthorne RJG - BVMS,DipBiol,PhD,MRCVS *Head of Department*
2 Donnelly J - BSc,MIBiol Babesia, ticks
3 Gregory MW - BVSc,DipIEMVT,PhD,MRCVS coccidiosis
4 Ollerenshaw CB - PhD,BSc fascioliasis
5 Sinclair IJB - PhD,BSc serology
6 Tarry DW - BSc,MIBiol entomology
7 Taylor MA - BVMS,MRCVS gastro - intestinal nematodes

4420 Poultry

1 Cullen GA - PhD,BVetMed,FRCVS,DipBact *Head of Department*
2 Alexander DJ - PhD,BTech,MIBiol avian influenza including fowl plague, Newcastle disease, infectious bronchitis
3 Bracewell CD - PhD,BSc,BVSc,MRCVS avian infectious diseases
4 Spackman D - NDB,DipPoul,BVSc,MRCVS avian diseases

4421 Medicines

1 Kidd ARM - MSc,BVSc,MRCVS *Head of Unit*
2 Cameran RS - BSc,MRCVS
3 Gray AK - MSc,BVetMed,MRCVS
4 Hartley EG - MRCVS,MIBiol
5 Watson EEB - BVMS,MRCVS

4422 Virology

1 Roberts DH - BVMS,DipBact,MTech,MRCPath,DVM,MRCVS *Head of Department*
2 Cartwright Miss SF - BSc enteric and infertility viruses of pigs
3 Chasey D - PhD electron microscopy
4 Dawson M - BVetMED,MTech,MRCVS viral diseases of sheep
5 Edwards S - MAVetMed,MSc,MRCVS respiratory viruses in cattle
6 Lucas Miss MH - BSc,BVSc,DipBact bovine leukosis and bovine parvovirus
7 Wood L - BVMS,MRCVS Pestivirus infection of domestic animals

4423 Animal Production

1 Saunders RW - MSc,BVMS,MRCVS *Head of Department*
2 Parker BNJ - MSc,MA,VetMB,MRCVS animal production
3 Wijeratne WVS - BVSc,PhD,FRCVS,DipAnGen breeding and genetics, experimental large animals

4424 Diseases of Breeding

1 Wratnall AE - BVMS,MSc,PhD,MRCVS *Head of Department* productive disease
2 Bell RA - PhD,BVMS,MRCVS,DipBact brucellosis, vibriosis
3 Corbel MJ - PhD,BSc immunology, immunochemistry, microbiology
4 Ellis Mrs B - BVetMed,MSc,MRCVS reproductive diseases
5 Lander KP - PhD,BVMS,MRCVS,DipBact vibriosis, reproductive diseases
6 MacMillan A - BVSc,MRCVS brucellosis diagnosis & experimental testing

4425 Reading Cattle Breeding Centre
Shinfield, Nr Reading, Berkshire
Tel: (0734) 883157/9.

1 Lamont PH - PhD,BSc,MRCVS *Head of Centre* fertility and artificial breeding
2 Foulkes JA - BSc,PhD fertility and artificial breeding, enzyme immunoassay of hormones
3 Glossop CE - PhD,BVetMed,MRCVS fertility and artificial breeding
4 Revell SG - BVMS,MRCVS fertility and artificial breeding

4426 Agricultural Science Service
Great Westminster House, Horseferry Road, London, SW1P 2AE
Tel: 01 216 6311.

1 Bunyan PJ - DSc,PhD,CChem,FRSC *Head of service*
2 Drummond DC - BSc,MS,FIBiol *Senior Agricultural Scientist* research and development
3 Needham P - BSc,MSc,DIC *Senior Agricultural Scientist* advice and promotion
4 Bailey S - CChem,FRSC *Head of Discipline* analytical chemistry
5 Francis GH - BSc *Staff Officer*
6 Hardwick DC - BSc,PhD,CChem *Pollution Scientist*
7 Stubbs WJ - BA,NDA,MD,AgE *Chief Plant Health and Seeds Inspector*
8 Way JM - TD,MSc,PhD *Ecology Specialist*

4427 Central Laboratories , Slough, Tolworth, Worplesdon
50 London Road, Slough, SL3 7HJ
Tel: (02814) 34626.

1 Stanley PI - BSc,PhD *Director*

4428 Storage Pests, Slough

1 Griffiths DA - MSc,PhD *Head of Department*
2 Bell CH - BTech,PhD fumigation
3 Chambers J - BSc,MA,DPhil attractants, syntheses, radiosafety
4 Edwards JP - BTech,PhD hormonal control of insect pests
5 Fishwick FB pesticide chemistry and analysis
6 Halstead DGH - BSc,PhD general systematics of insects and mites
7 Lloyd CJ - LIBiol insecticide susceptibility and synergism
8 Muggleton J - MSc,PhD genetics of resistance and insect populations
9 Pinniger DB - BSc,LIBiol insecticide treatments and insect behaviour
10 Price NR - MSc,PhD insect biochemistry, mode of action of pesticides
11 Rowlands DG pesticide residues and metabolism, resistance mechanisms
12 Wainman HE - LRSC fumigation practice, safety and field studies
13 Wilkin DR - LIBiol control of insects and mites

4429 Microbiology, Slough

1 Buckle AE - Head of Department *feed microbiology, mycotoxins*
2 Clarke JH - BSc,PhD immunology as applied to stored product fungi, mycotoxins

4430 Information and Training, Slough

1 Smith KG - BSc,MIBiol *Head of group*
2 Cullen T - BSc,DipLib,ALA *Librarian*

4431 Mammals and Birds, Tolworth and Worplesdon
4432 Tolworth Laboratory
Hook Rise South, Tolworth, Surbiton, Surrey, KT6 7NF
Tel: 01 337 6611.

1 Greaves JH - BSc,PhD,MIBiol research on rodenticides
2 Redfern R - MA control of moles
3 Rowe FP - BSc,MIBiol mouse biology
4 Townsend MG - BSc,PhD chemical aspects of pest control

4433 Worplesdon Laboratory
Tangley Place, Worplesdon, Guildford, Surrey, GU8 3LQ
Tel: (0483) 232581.

1 Hardy AR - BSc,PhD *Head of Department*
2 Brough T - BSc strike hazard in aviation
3 Feare CJ - BSc,PhD biology and control of birds
4 Gosling LM - BSc,PhD ecology, biology and control of coypu and mink
5 Inglis IR - BSc,PhD development of methods of control of bird damage
6 Lloyd HG - MSc,MIBiol ecology and control of foxes, rabies

7 Rees Miss WA - MSc,PhD rabbits, behavioural control, yield law prediction
8 Ross J - BSc,PhD rabbit disease and control
9 Vaughan JA - MA support services

4434 Wildlife and Storage Biology, Slough

1 Hunter FA - MA *Head of discipline*

4435 Automated Systems Group, Slough

1 Disney RW - BSc *Manager of computer facilities*

4436 Harpenden Laboratory
Hatching Green, Harpenden, Hertfordshire, AL5 2BD
Tel: Harpenden 5241.

1 Gould HJ - BSc *Director*

4437 Entomology and Plant Pathology

1 Gostick KG - BSc,MIBiol *Head of Department*
2 Aitkenhead P - BSc plant health legislation
3 Baker CRB - MA risk potential of alien pests
4 Cotton J - BSc,PhD plant nematology, certification & pest identification
5 Dickens JSW - BSc,PhD,MIBiol seed-borne diseases, mycology
6 Ebbels DL - BSc,PhD certification of planting material
7 Hill SA - BSc,MSc,PhD virus diseases of plants
8 Jones RAC - BSc,PhD virus diseases of plants
9 King JE - BSc,PhD disease assessment
10 Pemberton AW - BSc alien diseases of plants
11 Tarrance Mrs L - BSc,PhD immunology of plant viruses

4438 Pesticide Registration and Surveillance

1 Bates JAR - BSc,CChem,FRSC *Head of Department*
2 Lee DF - BSc pesticides surveillance including pesticide usage surveys
3 Lloyd Mrs GA - BSc technical services and reviews
4 Lloyd GA operator protection
5 Martin AD - PhD toxicology and environment
6 Smart NA - MA,DPhil pesticide residues
7 Stell G - BSc efficacy evaluation
8 Tooby TE - MSc,MIBiol risk evaluation

4439 Regional Centres
Specialist advisers in the Agricultural Science disciplines of analytical chemistry, entomology, microbiology, nutrition chemistry, plant pathology, soil science, wildlife and storage, biology are deployed in the regional centres and subcentres listed below:

4440 Northern Regional Centre
Block 2, Government Buildings, Lawnswood, Leeds, LS16 5PY
Tel: (0532 67) 4411.

1 Wilkinson B - MSc,DipAgrChem *Regional Agricultural Scientist*

4441 Midlands and Western Regional Centre
Woodthorne, Wolverhampton, Staffordshire, WV6 8YQ
Tel: (0902) 754190.

1 Swain RW - BSc *Regional Agricultural Scientist*

4442 Eastern Regional Centre
Block C, Government Buildings, Brooklands Avenue, Cambridge, CB2 2DR
Tel: (0223) 358911.

1 Griffin MJ - BSc,PhD *Regional Agricultural Scientist*

4443 South Eastern Regional Centre
Government Buildings, Coley Park, Reading, Berkshire, RG1 6DT
Tel: (0734) 581222.

1 Benham CL - BSc *Regional Agricultural Scientist*

4444 South Western Regional Centre
Government Buildings, Burghill Road, Westbury-on-Trym, Bristol, BS10 6NJ
Tel: (0272) 500000.

1 Stanks MH - BSc *Regional Agricultural Scientist*

4445 Wales Regional Centre
Welsh Department, Trawsgoed, Aberystwyth, Dyfed, SY23 1NG
Tel: (097) 43301.

1 Hughes AD - BSc,PhD *Regional Agricultural Scientist*

4446 Analytical Chemistry

1 Bishop CK - LRSC
2 Franks CL - LRSC

3 Henderson K - MSc,BSc,LRSC
4 Jane I - BSc
5 Jewell EJ - BSc
6 Jones DI - CChem,MRSC
7 Jones JLO - LRSC
8 Kay JJ - CChem,FRSC

4447 Entomology

1 Alford DV - BSc,PhD,FRES
2 Bassett P - BSc,MI
3 Lane AB - Tech,MIBiol
4 Mathias PL - BSc,MSC,DIC,MPhilCBiol,MBiol
5 Murdoch G - BSc,MIBiol
6 Saynor M - PhD,DIC,MIBiol
7 Winfield AL - BSc,DipAgrZool,MIBiol

4448 Microbiology

1 Cooper Miss AW - BSc,PhD
2 Hacking A - BSc,Grad,RSC,MIBiol
3 Joyce DA - BSc
4 Lyne AR - BSc
5 Ould Mrs AJL
6 Walker GJ

4449 Nutrition Chemistry

1 Adamson AH - BSc
2 Francis GH - BSc
3 Harris CI - BSc
4 Harvard A - MSc
5 Hopkins JR - BSc
6 Jones MGS - BSc

4450 Plant Pathology

1 Cock LJ - BSc,MA,DipAgrSc
2 Cragg IA - BSc
3 Evans EJ - BSc,PhD
4 Fletcher JT - BSc,PhD
5 Gladders P - MA,PhD
6 Jenkins JEE - BSc,MS,MIBiol
7 Yarham DJ - BSc,DipAgriSc

4451 Soil Science

1 Archer JR - BSc,MSc,DIC
2 Davies DB - MA,PhD
3 Farrar K - BSc
4 Hewgill D - BSc
5 Le Grice S - BSc
6 Unwin R - BSc
7 Wadsworth GA - BSc,MSc,DipAgricChem,DIC
8 Waller WM - BSc,MSc,CChem,FRSC

4452 Wildlife and Storage Biology

1 Kennedy R - BScAgric
2 Meyer AN - MSc
3 Shenker AM - BSc,CBiol,MIBiol,MRSH
4 Simpson WJ
5 Symes RG - BSc,MIBiol
6 Wadsworth JT - NDA

4453 Experimental Centres

Great Westminster House, Horseferry Road, London, SW1P 2AE
Tel: 01-216 6053.

1 Wickens R - MA,DipAgric,DTA Director of experimental centres
2 Allington P - BScHort Deputy Director of experimental centres
3 Tas Dr MV - BSc,PhD Head of experimental husbandry farms
4 Thomas Mrs K - BScAgric Head of experimental horticulture stations

4454 Experimental Husbandry Farms (EHF)

4455 Boxworth Experimental Husbandry Farm
Boxworth, Cambridge, CB3 8NN
Tel: (09547)372 and 391.

1 Jarvis RH - MA,DipAgr Director
2 Strickland MJ - MA,DTA Deputy Director

4456 Bridget's Experimental Husbandry Farm
Martyr Worthy, Winchester, Hants, SO21 1AP
Tel: (096278)220 and 330.

1 Mundy EJ - MScAgric Director
2 Francis AL - MA,MIBiol Deputy Director

4457 Drayton Experimental Husbandry Farm
Alcester Road, Stratford-on-Avon, Warwicks, CV37 9RG
Tel: (0789) 293057.

1 Helps MB - NDA,MIBiol Director
2 Yule AH - BSc Deputy Director

4458 Gleadthorpe Experimental Husbandry Farm
Meden Vale, Mansfield, Nottinghamshire, NG20 9PF
Tel: (0623)844331.

1 Martindale JF - NDA,DipFarmMan Director
2 Selman M - BSc,PhD Deputy Director

4459 High Mowthorpe Experimental Husbandry Farm
Duggleby, Malton, N.Yorks, YO17 8BW
Tel: (09443)646 and 647.

1 Perks DA - BSc,DMS,MIBiol Director
2 Bastiman B - BSc Deputy Director

4460 Liscombe Experimental Husbandry Farm
Dulverton, Somerset, TA22 9PZ
Tel: (064385)291.

1 Gurnett LR - BSc,NDA Director
2 Appleton M - BSc Deputy Director

4461 Pwllpeiran Experimental Husbandry Farm
Cwmystwyth, Aberystwyth, Dyfed, SY23 4AB
Tel: (097 422)229 and 261.

1 Powell TL - BSc Director
2 Richards RIWA - BSc,MSc Deputy Director

4462 Arthur Rickwood Experimental Husbandry Farm
Mepal, Ely, Cambridgeshire, CB6 2BA
Tel: (03543)2531.

1 Macleod J - BSc,MS Director
2 Rickard RC Deputy Director

4463 Redesdale Experimental Husbandry Farm
Rochester, Otterburn, Northumberland, NE19 1SB
Tel: (083020)608 and 672.

1 Thompson JR - BSc Director
2 Wilkinson M - BSc Deputy Director

4464 Rosemaund Experimental Husbandry Farm
Preston Wynne, Hereford, HR1 3PG
Tel: (043278)444 and 445.

1 Meadowcroft SC - BSc Director
2 Clare RW - BSc,NDA Deputy Director

4465 Terrington Experimental Husbandry Farm
Terrington St.Clement, Kings Lynn, Norfolk, PE34 4PW
Tel: (0553828)621 & 622.

1 Large JW - NDA,CDA Director
2 Forrest JS - BSc Deputy Director

4466 Trawscoed Experimental Husbandry Farm
Trawscoed, Aberystwyth, Dyfed, SY23 4HT
Tel: (09743)300,307/8/9.

1 Wilcox JC - BSc,MS,MIBiol Director
2 Williams DO - BSc,DipAgric,DTA,MIBiol Deputy Director

4467 Experimental Horticulture Stations (EHS)

4468 Efford Experimental Horticulture Station
Lymington, Hants, SO4 0LZ
Tel: 05907 3341.

1 Clements RF - NDH,MIBiol Director
2 Williams Miss AM Deputy Director

4469 Kirton Experimental Horticulture Station
Kirton, Boston, Lincs, PE20 1EJ
Tel: 0205 722391.

1 Sandwell I - NDH,Hons Director
2 Pinkerton-Hiron Dr RW - BA,Hons,PhD Deputy Director

4470 Lee Valley Experimental Horticulture Station
Ware Road, Hoddesdon, Hertfordshire, EN11 9AQ
Tel: 0992 463623 and 463624.

1 Dyke AJ - NDH Director
2 Farthing JG - NDH Deputy Director

4471 Luddington Experimental Horticulture Station
Luddington, Stratford-on-Avon, Warwicks, CV37 9SJ
Tel: 0789 750601.

1 Shipway Dr MR - BSc,PhD *Director*
2 Parties JT - BSc,MIBiol *Deputy Director*

4472 Rosewarne Experimental Horticulture Station
Camborne, Cornwall, TR14 0AB
Tel: 0209 716673.

1 Pollock MR - CDH *Director*
2 Tompsett AA - NDH *Deputy Director*

4473 Stockbridge House Experimental Horticulture Station
Cawood, Selby, N.Yorks, YO8 0TZ
Tel: 075786 275 & 276.

1 Whitwell JD - NDH,CDH *Director*
2 Bradley MR - BSc,Hort,MPhil *Deputy Director*

4474 National Fruit Trials
Brogdale Farm, Faversham, Kent, ME13 8XZ
Tel: 0795 535462/3/4.

1 Stapleton RR - NDH *Director*
2 Turner Miss EA - *Deputy Director*

4475 Aston University
Aston Triangle, Birmingham, B4 7ET
Tel: 021-359 3611.

4476 Department of Molecular Sciences
1 McWhinnie WR - DSc,CChem,FRSC *Professor and Head of Department*

4477 Biology Division
1 Armstrong RA - PhD plant ecology; lichens
2 Atkins TW - PhD insulin secretion
3 Bailey CJ - PhD carbohydrate metabolism; sex hormones
4 Bromage NR - PhD fish endocrinology, thyroid and pancreas; fish nutrition
5 Hanson PJ - PhD intestinal absorption biochemistry
6 Perris AD erythropoietin, thymus and calcium metabolism
7 Rimmer JJ - PhD comparative immunology
8 Smith CUM - BSc brain physiology; electron microsopy
9 Smith SN - PhD fungal physiology; herbicides
10 Thornburn CC - PhD biochemistry; radiobiology

4478 University of Bath
Claverton Down, Bath, BA2 7AY
Tel: 0225 61244.

4479 School of Biological Sciences
4480 Biochemistry Group
1 Weitzman PDJ - MA,MSc,DPhil,FIBiol *Head of School and Professor and Head of Group* biochemistry
2 Danson MJ - BSc,PhD *Lecturer* metabolism of animal and human parasites
3 Eisenthal RS - AB,PhD *Senior Lecturer* parasitic diseases of domestic cattle
4 Lunt GG - MSc,PhD *Reader* insect neurochemistry
5 Pryke Ms JA - MA,PhD *Lecturer* plant developmental molecular biology
6 Rutherford PP - MA,PhD *Senior Lecturer* biochemistry - plant storage

4481 Animal Physiology & Ecology Group
1 Baker BI - BSc,PhD *Reader and Head of Group* animal physiology (fish)
2 Franks NR - BSc,PhD *Lecturer* insect sociobiology and behavioural ecology
3 Goodchild AJP - BSc,DIC *Lecturer* insect physiology
4 Reynolds SE - MA,PhD *Lecturer* insect physiology

4482 Microbiology Group
1 Rose - DSc,PhD *Professor and Head of Group*
2 Board RG - DSc,PhD,FIFST *Reader* food microbiology
3 Rayner ADM - BA,PhD *Reader* fungal ecology and genetics; forest pathology
4 Wright SJL - BSc,PhD,MIBiol *Senior Lecturer* pesticide microbiology; cyanobacterial ecophysiology; microbial phytotoxins

4483 Plant Biology Group
1 Henshaw GG - BSc,PhD *Professor and Head of Group* plant cell and tissue culture

2 Beeching JR - BA,BSc,PhD *Lecturer* plant molecular genetics
3 Charnley AK - BSc,PhD *Lecturer* insect pathology and physiology
4 Clarkson JM - BSc,PhD *Lecturer* genetics and plant pathogenesis
5 Cooper RM - MSc,PhD,DIC *Lecturer* molecular plant pathogenesis
6 Dodge AD - BSc,PhD,MIBiol *Senior Lecturer* mode of action of herbicides
7 Dourado AM - BSc,PhD *Lecturer* seed physiology
8 Flegmann AW - BSc,MA,PhD *Lecturer* plant cell and tissue culture
9 George RAT - BSc,NDH,MIBiol,MIHort *Senior Lecturer* vegetable seed production
10 Hicks TGT - MSc,PhD *Lecturer* aetiology of virus diseases
11 Mogie M - BSc,PhD *Lecturer* population biology
12 Moore KG - BSc,MA,PhD *Reader* plant physiology (seeds)
13 Robinson LW - BSc,PhD,MIBiol *Lecturer* plant physiology (watercress)
14 Sargent MJ - BSc,PhD,MPhil,NDH *Lecturer* horticultural economics
15 Stephens RJ - BSc,MIBiol,MIHort *Lecturer* crop physiology and weed control
16 Thoday PR - MSc,NDH,MIBiol,MIHort *Senior Lecturer* amenity; land management
17 Wainwright H - BSc,PhD,MIHort *Lecturer* fruit production and micropropagation

4484 University of Birmingham
PO Box 363, Birmingham, B15 2TT
Tel: (021-472) 1301.

4485 Plant Biology
1 Callow JA - BSc,PhD *Mason Professor and Head of Department* molecular plant pathology and biotechnology
2 Butler GM - MA,PhD *Lecturer* host/pathogen interactions, leek rust
3 Ford-Lloyd BV - BSc,PhD *Lecturer* tissue culture and plant breeding, germplasm conservation, biotechnology, cultivated plant biosystematics, cytogenetics and numerical studies
4 Green JR - BSc,PhD *Lecturer* plant cell and molecular biology, biotechnolgy
5 Greig JRA - BA *Research Fellow* archaeology of crop plants
6 Idle DB - BSc,PhD *Senior Lecturer* adaptation to frost
7 Jackson MT - BSc,PhD *Lecturer* systematics of wild and cultivated potatoes, tissue culture and potato breeding, biotechnology
8 Kordan HA - BA,MSc,PhD *Lecturer* physiology of growth and development
9 Lester RN - BSc,PhD *Lecturer* biochemical systematics; solanaceae taxonomoy
10 Lyons IR - BSc,PhD *Research Fellow* plant biotechnology
11 Mumford PM - MPhil,PhD *Lecturer* plant breeding; physiology of seed dormancy
12 Newbury HJ - BSc,PhD *Lecturer* molecular plant pathology and biotechnology
13 Ray T - BSc,PhD *Research Fellow* molecular plant pathology
14 Taylor L - BSc,PhD *Research Fellow* potato breeding
15 Todd G - BSc,MSc *Research Fellow* molecular plant pathology
16 Wilkins DA - BSc,PhD *Senior Lecturer* salt and heavy metal tolerance

4486 University of Bradford
Richmond Road, Bradford, West Yorkshire, BD7 1DP
Tel: (0274) 733466.

1 Harvey-Jones Sir J - KT,MBE *Chancellor*
2 West Professor JC - CBE,DSc,FEng,FIEE *Vice-Chancellor and Principal*

4487 School of Biomedical Sciences
1 Cross T - BSc,PhD *Professor* microbiology
2 Alderson G - BSc,PhD,MIBiol *Lecturer* microbiology
3 Ash R - Bsc,PhD *Lecturer* nutrition, ruminants and fish
4 Wheelock JV - BSc,BAgr,PhD *Reader* food policy

4488 School of Environmental Science
1 Davies BE - BSc,PhD,CChem,FRSC *Professor* soil science, trace metals
2 Seaward MRD - BSc,MSc,PhD,DSc,FIBiol *Reader* ecology, air pollution, heavy metals

4489 School of Pharmacology
1 Costall B - BPharm,PhD,DSc *Reader* pharmacology
2 Foy JM - BPharm,PhD,MPS *Senior Lecturer* pharmacology
3 Hicks R - MPharm,PhD,MIBiol,FRCPath *Reader* pathology and toxicology
4 Naylor RJ - BPharm,PhD,DSc,MPS *Reader* pharmacology
5 Senior J - BPharm,PhD,MPS *Senior Lecturer* pharmacology

4490 School of Science and Society
1 Stonier TT - AB,MS,PhD *Professor* biology of plant cancers

4491 University of Bristol

The Senate House, Tyndall Avenue, Bristol, BS8 1TH
Tel: (0272) 303030.

1 Hodgkin Professor D - OM,MA.PhD,DSc,ScD,DUniv,LLD,HonFRCP,FRS *Chancellor*
2 Kingman Sir J - MA,ScD,DSc,HonFIS,FRS *Vice-Chancellor*

4492 Department of Agriculture and Horticulture
Research Station, Long Ashton, near Bristol
Tel: Long Ashton 392181.

4493 Department of Botany

1 Walsby AE - BSc,PhD *Professor*
2 Campbell RE - BSc,MS,PhD *Lecturer* microbiology
3 Lazarus CM - BSc,PhD *Lecturer* molecular genetics unit
4 Madelin MF - BSc,PhD *Reader* mycology
5 Newman EI - MA,PhD,DSc *Reader* root/microbe interaction

4494 Department of Physical Chemistry

Physical Chemistry of soils and humic acids

1 Everett DH - MBE,MA,DPhil,DSc,FRSE,FRS *Emeritus Professor of Physical Chemistry*
2 Haynes JM - BSc,PhD *Senior Lecturer*
3 Ottewill RH - BSc,MA,PhD,FRS *Leverhulme Professor of Physical Chemistry*

4495 Comparative Animal Respiration, Research Unit

1 Hughes GM - MA,PhD,ScD *Emeritus Professor of Zoology and Head of Research Unit*

4496 Molecular Genetics Unit

1 Beringer JE - BSc,PhD *Professor of Molecular Genetics* plant microbe interactions

4497 Geography

1 Thornes JB - BSc,MSc,PhD *Professor of Geography* soil erosion and land use problems in semi-arid environments
2 Barrett EC - MSc,PhD,DSc *Reader, Director of Remote Sensing Unit* agrometeorological prediction systems
3 Curtis LF - BSc,PhD *HonResFellow* aerial and satellite surveys of land use
4 Osmaston HA - MA,BSc,DPhil,MCIFor *Senior Lecturer* forest ecology and management, remote sensing, subsistence agriculture
5 Ross SM - BSc,PhD,MCIFor *Lecturer* forest and agricultural soils

4498 Veterinary Research

4499 Anatomy

1 Pickering BT - PhD,DSc *Professor of Anatomy* neuroendocrinology, peptides in brain and gonads
2 Porter DG - MSc,BVetMed,PhD,MRCVS *Professor of Pre-Clinical Veterinary Studies* maintenance of pregnancy, initiation of labour
3 Wathes DC - BSc,PhD *Lecturer* maintenance of pregnancy in farm animals

4500 Veterinary Anatomy

1 Butler WF - BVSc,PhD,FRCVS *Lecturer* histochemistry of connective tissues
2 Goodship AE - BVSc,PhD,MRCVS *Lecturer* bone mechanics
3 Smith RN - PhD,DSc,FRCVS *Reader* locomotion in animals
4 Summerlee AJS - BVSc,PhD,MRCVS *Lecturer* neuroendocrinology

4501 Biochemistry

A major interest is metabolic control in mammalian systems

1 Chappell JB - BA,PhD *Professor of Biochemistry*
2 Gutfreund H - PhD *Professor of Physical Biochemistry*
3 Jones OTG - BSc,PhD *Professor of Biochemistry*

4502 Animal Husbandry
Department of Animal Husbandry, Langford House, Langford, Bristol, BS18 7DU
Tel: Churchill 852581.

1 Webster AJF - MA,VetMB,PhD,MRCVS *Professor*
2 Long Miss SE - BVMS,PhD,MRCVS *Lecturer* cytogenetics
3 Manners MJ - MScAgr,PhD,NDA *Lecturer* animal nutrition
4 Nicol Mapes C - BA *Lecturer* ethology, welfare
5 Perry GC - BSc,PhD *Senior Lecturer* applied animal behaviour
6 Wathes CM - BSc,PhD *Lecturer* animal housing

4503 Veterinary Medicine
Department of Veterinary Medicine, Langford House, Langford, Bristol, BS18 7DU
Tel: Churchill 852581.

1 Bourne FJ - BVetMed,PhD,MRCVS *Professor* cattle and pig immunology
2 Holmes JR - MVSc,PhD,MRCVS *Reader* electrocardiography
3 Hinton MH - BVSc,FRCVS *Senior Lecturer (Veterinary Public Health)* anaemia in animals
4 Cripps P - BVSc,MSc,MRCVS *Lecturer* bovine mastitis
5 Harbour D - BSc,PhD *Lecturer* virology, molecular biology
6 Mair T - BVSc,PhD,MRCVS *Lecturer* equine infectious disease
7 Morgan KL - VetMB,PhD,MRCVS *Lecturer* ovine infectious disease
8 O'Brien K - MVB,PhD,MRCVS *Lecturer* equine and bovine internal medicine
9 Stokes CR - BSc,PhD *Research Associate* immunity at mucosal surfaces
10 Taylor FGR - BVSc,PhD,MRCVS *Lecturer* equine and bovine immunology

4504 Pathology
Comparative Pathology Laboratory, Langford House, Langford, Bristol, BS18 7DU
Tel: Churchill 852581.

1 Brown PJ - BVMS,PhD *Lecturer* immunology of parasitic infections
2 Longstaffe JA - BVetMed,PhD,MRCVS *Lecturer* immunology of parasitic disease
3 Lucke Mrs VM - BVSc,PhD,MRCVS *Senior Lecturer* renal disease
4 Pearson GR - BVMS,PhD,MRCVS *Lecturer* neonatal enteritis
5 Silver IA - MA,MRCVS *Professor of Comparative Pathology* wound healing and microcellular environment

4505 Pharmacology

1 Mitchell JF - MA,BSc,PhD,FIBiol *Professor*
2 Keen PM - BVSc,PhD,MRCVS *Reader* central and peripheral nervous systems
3 Livingston A - BSc,BVetMed,PhD,MRCVS *Senior Lecturer (Veterinary Pharmacology)* general anaesthetics and analgesics

4506 Physiology

1 Thomas RC - BSc,PhD *Head of Department*

4507 Veterinary Physiology

1 Carter Miss SB - MSc,PhD *Lecturer* gamete transport in the Fallopian tube; diurnal variations in relation to ovulation time
2 Headley PM - BVSc,BSc,PhD,MRCVS *Lecturer* spinal processing of nociceptive information; spinal actions of analgesics
3 Woolley DM - BVSc,PhD,MRCVS *Senior Lecturer* flagellar movement,in particular the mechanism of motility in spermatozoa

4508 Veterinary Microbiology

1 Linton AH - MSc,PhD,DSc,FRCPath *Professor of Bacteriology* epidemiology of enterobacteriaceae
2 Greenham LW - BSc,BVSc,PhD,MRCVS *Senior Lecturer (Veterinary Bacteriology and Virology)* virus infections of the CNS
3 Hill TJ - BVSc,PhD,MRCVS *Reader (Virology)* latent virus infections of the CNS

4509 Veterinary Surgery
Department of Veterinary Surgery, Langford House, Langford, Bristol, BS18 7DU
Tel: Churchill 852581.

1 Pearson H - BVSc,PhD,FRCVS *Professor*
2 Crispin Miss SM - MA,VetMB,BSc,PhD,MRCVS,DVA,DVOphtal *Lecturer* opthalmology; lipid keratopathy
3 David JSE - BVSc,PhD,MRCVS *Senior Lecturer (Diseases of Reproduction)* oestrus cycle and ovulation in mares; germ cell studies; cryptorchidism in horses; cystic ovaries in cattle
4 Denny HR - MA,VetMB,PhD,FRCVS *Senior Lecturer* orthopaedic surgery
5 Gibbs Miss C - BVSc,PhD,MRCVS,DVR *Senior Lecturer* diagnostic radiology
6 Holt PE - BVMS,MRCVS *Lecturer (Urology)*
7 Lane JG - BVetMed,FRCVS *Lecturer* surgical disorders of the ear, nose and throat; cryosurgery
8 Waterman Mrs AE - BVSc,DVA,PhD,MRCVS *Lecturer* anaesthesia, pharmacokinetics of anaesthetic agents
9 Weaver Miss BMQ - PhD,FRCVS,DVA *Reader* anaesthesia; pharmacokinetics of anaesthetic agents

4510 Zoology

1 Follett BK - BSc,PhD,DSc *Professor of Zoology*
2 Avery RA - BSc,PhD *Senior Lecturer*
3 Goldsmith AR - BSc,PhD *Lecturer*
4 Harris S - BSc,PhD *Research Fellow*
5 Mapes CJ - BSc,PhD *Special Lecturer/Administrator* parasitology
6 Nicholls TJ - BSc,PhD *Lecturer*
7 Rayner JMV - MA,PhD *Royal Society Research Fellow*

8 Soffe SR - BSc,PhD *Research Associate*
9 Strong LD - BSc,PhD *Lecturer* entomology

4511 Brunel University

Kingston Lane, Uxbridge, Middlesex, UB8 3PH
Tel: (0895) 37188.

4512 Biology

1 Dean Professor RT - MA,PhD,DSc biochemistry of animal cell growth
2 Lacey AJ - BSc,PhD environmental studies and weed biology

4513 Cambridge University

4514 Applied Biology
Downing Street, Cambridge, CB2 3DX
Tel: (0223) 358381.

1 Beament Sir J - MA,PhD,ScD,FRS *Drapers Professor of Agriculture*
2 Allen EJ - MA *Director, University Farm*
3 Bache BW - MA,PhD *Lecturer* soil science
4 Brown JC - MA,PhD *Lecturer* mammal ecology
5 Campbell RC - MA,PhD statistics
6 Coaker TH - MA,PhD applied entomology
7 Corbet Miss SA - MA,PhD applied entomology
8 Eltringham SK - MA,PhD vertebrate productivity
9 Flowerdew JR - MA,PhD vertebrate reproductive biology
10 Galwey NW - MA,PhD biometry and computer programming
11 Gilligan CA - MA,PhD applied botany
12 Henderson Miss R - PhD *Demonstrator* animal productivity and nutrition
13 Miller EL - MA,PhD animal nutrition
14 Morgan DG - MA,MSc applied plant physiology
15 Skidmore DI - MA,PhD *Demonstrator* plant breeding
16 Tribe HT - MA,PhD *Assistant Director of Research* soil microbiology

4515 Department of Clinical Veterinary Medicine
Madingley Road, Cambridge
Tel: (0223) 355641.

1 Soulsby EJL - MA,PhD,DVSM,MRCVS *Head of Department and Professor of Animal Pathology* parasitic diseases; immunology of parasitic infections
2 Alexander TJL - PhD,MVSc,BSc,MRCVS *University Lecturer* infectious diseases of pigs; streptococcal meningitis, swine dysentery, medicated early weaning of pigs
3 Blakemore WF - MA,PhD,BVSc,MRCVS *University Lecturer* progressive neurological diseases in young animals; nervous diseases of the dog and cat; chronic compression of spinal cord
4 Connan RM - BVetMed,MA,PhD,MRCVS *University Lecturer* epidemiology and control of helminth parasites of domestic animals
5 Goodwin RFW - MA,PhD,BSc,MRCVS *University Lecturer* pig health control
6 Hall LW - MA,BSc,PhD,DVA,MRCVS *Reader in Comparative Anaesthesia* anaesthesia; diseases of the chest, fluid balance
7 Herrtage ME - MA,BVSc,DVR,MRCVS *University Lecturer* small animal clinical neurology, hormonal disorders and dermatology
8 Houlton JEF - MA,VetMB,MRCVS *University Surgeon* orthopaedics in large and small animals
9 Jackson PGG - BVMS,FRCVS *University Physician* reproduction, infertility, obstetrics and general medicine
10 Jefferies AR - MA,VetMB,MRCVS *University Pathologist* histopathology and pathological diagnosis, neoplastic changes
11 Palmer AC - ScD,MA,PhD,FRCVS *Assistant Director of Research* nervous and muscle diseases of animals
12 Steele-Bodger A - CBE,MA,BSc,FRCVS *Professor of Veterinary Clinical Studies* mineral deficiencies in cattle
13 Wise DR - MA,PhD,VetMB,MRCVS *University Lecturer* skeletal problems in poultry; production and nutrition problems in turkeys and pheasants

4516 University College Cardiff

PO Box 78, Cardiff, CF1 1XL
Tel: (0222) 44211.

4517 Department of Biochemistry

1 Young JE - MA,PhD *Lecturer* research into physical and chemical factors influencing spore germination

4518 Department of Zoology

1 Bellamy Professor D - BSc,DPhil *Head of Department* growth and reproduction of farm livestock on recycled waste nitrogen
2 Bowen ID - BSc,PhD *Reader in Cell Biology* effects of pesticides on molluscs
3 Claridge Professor MF - MA,DPhil *Reader in Entomology* behaviour and ecology of phytophagous and entomophagous insects; pests of rice in S.E. Asia

4 Edington JM - BSc,PhD *Senior Lecturer* application of ecology to land use planning
5 Edington MA - BSc,PhD methods for controlling bacteriological quality of water in developing countries
6 Erasmus Professor DA - BSc,PhD *Reader in Parasitology* reproductive biology of *Schistosoma mansoni*
7 Jefferson GT - BSc,PhD,ARCS *Leverhulme Fellow* heat injury in insects
8 Kidd NAC - BSc,PhD *Lecturer* behaviour and population dynamics of aphids of agricultural and forestry importance
9 Pickard Professor RS - BSc,PhD *Lecturer* biology of honey bees
10 Staddon BW - BSc,PhD *Reader in Entomology* physiology and function of scent glands in Heteroptera
11 Thomas DH - BSc,MSc,PhD *Senior Lecturer* behavioural mechanics in avian thermo and osmo regulation

4519 Cleppa Park Field Research Station

1 Humphreys PN - MSc,MRCVS *Deputy Director* alternative agriculture; organisation of zoological collections

4520 University of Durham

South Road, Durham City, DH1 3LE
Tel: (0385) 64971.

4521 Department of Botany

1 Boulter D - MA,DPhil *Professor and Head of Department* structure and biochemistry of plant proteins and nucleic acids; yield improvement in legumes and rice
2 Croy RRD - BSc,PhD *Senior Experimental Officer* improvement of legumes using genetic engineering; isolation and utilisation of plant genes
3 Gatehouse AMR - BSc,PhD *Senior Research Assistant* resistance of crops to insect pests
4 Gatehouse JA - BA,DPhil *Lecturer* molecular biology of seed development
5 Gates PJ - BA,PhD *Lecturer* yield improvement in legumes and rice using new methods
6 Harris N - BSc,PhD *Senior Experimental Officer* seed ultrastructure
7 Huntley B - MA,PhD *Lecturer* plant community ecology and Quaternary palaeoecology
8 Murphy DJ - BA,DPhil *Lecturer in Biochemistry* membrane organisation and function: plant lipid metabolism
9 Payne JW - BSc,PhD *Reader in Biochemistry* development of nutritional assay methods using microorganisms
10 Pearson JA - BSc,PhD *Lecturer* leaf growth and development in grasses
11 Richardson M - BSc,PhD *Senior Lecturer* structure and function of enzyme inhibitors and lectins
12 Shaw CH - BSc,PhD *Lecturer* plant gene transfer systems
13 Turner J - BSc,MA,PhD *Lecturer* vegetational history and pollen analysis
14 Watson MD - BSc,PhD *Lecturer in Genetics* yeast molecular biology; expression of plant proteins on yeast
15 Whitton BA - MA,DSc *Reader in Botany* nitrogen fixation in deepwater rice-fields
16 Yarwood A - BSc,PhD *Lecturer* structure of toxic/inhibitory plant proteins

4522 Department of Zoology

1 Bowler K - BSc,PhD *Professor and Head of Department* thermal physiology
2 Anstee JH - BSc,PhD *Senior Lecturer* insect physiology, ion and water balance
3 Ashby KR - MA,PhD *Lecturer* mammalian ecology
4 Banks RW - BSc,PhD *Lecturer* vertebrate nerve and muscle
5 Barker D - MA,DPhil,DSc *Professor* vertebrate nerve and muscle
6 Coulson JC - BSc,PhD,DSc *Reader in Animal Ecology* sea-bird ecology
7 Davies L - BSc,PhD *Honorary Research Fellow* entomology; blood-sucking insects
8 Dunstone N - BSc,PhD *Lecturer* behaviour
9 Evans PR - MA,PhD,DPhil *Reader in Estuarine Ecology* shore-bird ecology; bird migration
10 Hemsworth BN - BSc,MSc,PhD *Honorary Research Fellow* carcinogenesis and teratology
11 Horton JD - BSc,PhD *Reader in Immunology* development and comparative immunology
12 Hyde D - BSc,MSc *Lecturer* neurophysiology of vertebrate retina
13 Jones CJ - BSc,MSc,PhD *Cystic Fibrosis Research Fellow* mammalian secretory cell function
14 Manning R - BSc,PhD *Lecturer* biochemistry of development and differentiation
15 Milburn A - BSc,PhD,MRC *Research Fellow* vertebrate nerve and muscle
16 Rowell MJ - BSc,PhD,MIBiol *Lecturer* plant insect interaction and population genetics
17 Scott JJA - BSc,PhD *Addison Wheeler Research Fellow* vertebrate nerve and muscle
18 Stacey MJ - BSc,PhD *Senior Experimental Officer* vertebrate nerve and muscle

19 Wood DW - BSc,PhD *Senior Lecturer* insect physiology and locomotion

4523 University of East Anglia

4524 School of Biological Sciences
University Plain, Norwich, NR4 7TJ
Tel: (0603) 56161. Telex: 975334.

1 Davies Professor DR - BSc,PhD *Dean, John Innes Professor of Applied Genetics* molecular genetics of higher plants
2 Aidley DJ - MA,PhD muscle physiology with especial reference to insects; physiology of African army worm moth
3 Bell DJ - BSc,PhD rabbit social organisation and communication
4 Clarke CH - BSc,PhD ultra-violet and chemical mutagenesis and antimutagenesis in micro-organisms; morphogenesis and photobiology of myxobacteria
5 Coddington A - BSc,PhD biochemical genetics of membrane transport
6 Croghan PC - MA,PhD osmotic relations in animals; membranes
7 Davies Professor DD - BSc,PhD control of metabolism in plants with special reference to carbohydrates and amino and organic acids; protein turnover
8 Davy AJ - BSc,PhD physiological ecology of plants especially in relation to inorganic ion nutrition and photosynthetic productivity
9 Dawson AP - MA,PhD mitochondrial ion transport and energy conservation
10 Dixon Professor AFG - BSc,DPhil life history strategies and population ecology of aphids
11 Duncan G - BSc,PhD vision; lens membranes
12 Gibson I - BSc,PhD gene action of molecular genetics
13 Greenwood Professor C - BSc,PhD fast reaction kinetics; electron transfer; inorganic biochemistry
14 Hewitt GM - BSc,PhD,DSc,FIBiol ecological genetics; chromosomal variation and speciation in Orthoptera and other insects
15 Hopwood Professor DA - FRS,MA,PhD,DSc,FIBiol *John Innes Professor of Genetics* genetics of Actinomycetes, Rhizobium and Agrobacterium
16 James RI - BSc,PhD plasmid-mediated antibiotic resistance
17 James RO - BSc,PhD population ecology of invertebrates
18 Kearn GC - BSc,PhD,DSc biology of monogenean parasites of fishes
19 Lewis BG - BSc,PhD physiological aspects of plant infection: hypersensitive reactions
20 Noble-Nesbitt J - MA,PhD,FRES,MIBiol physiology and ultrastructure of insects, especially Apterygota
21 Oliver RP - BSc,PhD molecular biology of plant-pathogen interactions
22 Ramsay N - BSc,PhD molecular biology of Archaebacteria
23 Selwyn MJ - MA,PhD mitochondrial phosphate metabolism and ion transport
24 Shelton G - MA,PhD comparative physiology of respiratory and circulatory systems; nervous control of breathing
25 Sims AP - BSc,PhD regulatory mechanisms in higher plants and yeast, especially in relation to problems of the control of the uptake and assimilation of inorganic and organic nitrogen compounds
26 Sutherland WJ - BSc,PhD growth patterns of plants
27 Thain J - BSc,PhD ionic and water relations of plant cells and plants
28 Townsend CR - BSc,DPhil ecology and behaviour of freshwater invertebrates and fish
29 Turner JG - BSc,MS,PhD toxigenicity and epidemiology of micro-organisms in relation to crop plant disease
30 Warn RM - MA,DPhil development biology and genetics of *Drosophila*
31 Watkinson AR - BA,PhD plant population dynamics
32 Wildon DC - BScAgr,PhD transport of solutes in plants
33 Woolhouse Professor HW - BSc,PhD *John Innes Professor of Biology* biochemical control of senescence of leaves; adaptive physiology of photosynthesis

4525 University of Exeter

Northcote House, The Queens Drive, Exeter, EX4 4QJ
Tel: (0392) 77911.

4526 Department of Biological Sciences

4527 Hatherly Laboratories
Prince of Wales Road, Exeter, EX4 4PS
Tel: 0392 77911.

1 Webster Professor J - PhD,DSc,FIBiol *Head of Department* ecology and taxonomy of aquatic and aero-aquatic conidial fungi, mechanisms of sporangial discharge in Oomycetes, mechanism of ballistospore discharge, the coprophilous fungus succession
2 Anderson JM - PhD *Reader in Ecology* decomposition processes and nutrient cycling in terrestrial ecosystems, role of invertebrates in soil processes, particularly nitrogen mineralization, impact of management practices on soil biological processes in temperate and tropical regions
3 Harris T *Lecturer* ecology of Ophelia bicornis, littoral ecology and distribution of annelids, ecology of estuarine communities especially those involving polychaetes
4 Ivimey-Cook RB - PhD *Senior Lecturer* taxonomy and ecology of flowering plants, analysis of taxonomic and ecological data
5 Kennedy CR - PhD,DSc *Reader in Zoology* ecology of animal parasites, population and community ecology of parasites of freshwater and migratory fish, co-evolution of host-parasite systems

6 Linn IJ - FLS *Senior Lecturer* distribution, ecology and behaviour of European and African mammals
7 Macnair MR - PhD *Lecturer* experimental population genetics of plants, theoretical modelling of plant and animal populations, mechanisms of metal tolerance in higher plants
8 Nichols Professor D - DPhil,FIBiol reproductive biology and population structure of British echinoderms
9 Proctor MCF - PhD *Reader in Plant Ecology* vegetation ecology and plant autecology, physiological ecology of bryophytes
10 Shuttleworth TJ - PhD *Reader in Animal Physiology* stimulus-response coupling in vertebrate salt-secreting epithelia and the interaction between "second messenger" systems, endocrine control of epithelial ion transport, especially extrarenal salt secretion in vertebrates, physiology of vertebrate salt glands
11 Smirnoff N - PhD *lecturer* physiological plant ecology, stress metabolism and physiology of crop plants: its application to yield improvement, plant response and acclimation to drought and temperate extremes
12 Stradling DJ - PhD *Lecturer* ecology of ants, insect/fungus interactions of leaf-cutting ants, biology of theraphosid spiders, analysis of light-trap data
13 Wooton RJ - PhD *Senior Lecturer* functional morphology and evolution of insect flight systems, particularly of wings, palaeontology, palaeoecology and evolution of insects, especially Hemiptera, Diptera and Lower Pterygota, palaeoecology of aquatic insects

4528 Washington Singer Laboratories
Perry Road, Exeter, EX4 4OC
Tel: (0392) 77911.

1 Aves SJ - PhD *Lecturer* mitotic cell cycle control in yeast, molecular biology of fission yeast
2 Bryant JA - MA,PhD replication and repair of nuclear DNA, particularly in plants, replication of DNA viruses in plants, regulation of gene expression during development in eukaryotic organisms, biochemical responses to environmental stress, especially in higher plants
3 Dinan LN - PhD *Lecturer* insect biochemistry - mode of action of insect steroid hormones (Ecdysteroids), mechanism of uptake of Ecdysteroids into target cells, and identification and titres of Ecdysteroids in insects
4 Pitt D - PhD *Reader in Plant Pathology* physiology of growth and reproductive development in fungi, plant host-parasite physiology, development of the lytic compartment in plants
5 Roddick JG - PhD *Senior Lecturer* physiology and biochemistry of plant steroids: membrane-lytic action of glycoalkaloids and saponins, growth-regulating activity of brassinosteroids
6 Stebbings H - PhD *Lecturer* cell biology, cellular and intracellular movement; cytoskeletons, microtubules and microfilaments
7 Willetts AJ - PhD *Lecturer* biochemistry of aliphatic glycol metabolism by micro-organisms, biodegradation of xenobiotic compounds by micro-organisms, production of aliphatic glycols by immobilized microbial cultures

4529 Department of Economics

4530 Agricultural Economics
University of Exeter, Lafrowda House, St. German's Road, Exeter, EX4 6TL

1 McInerney Professor JP - BSc,Agric,DipAgEcon,PhD *Director of the Unit* agricultural policies in developed and less developed economies; resource economics; economic analysis of technological change in agriculture
2 Dunford WJ - BSc,Econ development of measures to increase the effectiveness of the management and control of farm finance; agricultural training
3 Hessner C - BA,PhD economic evaluation of UK salmon fishery; marketing of agricultural products, particularly the role of advertising and its effectiveness
4 Howe KS - BScEcon,MPhil,PhD economics of animal health; European agricultural policies
5 McCorriston S - BA,MSc economics of international commodity markets; agricultural policy in developed and developing countries; economics of the agricultural input sector
6 Nixon BR - NDA income measures reflecting the economics of the farm business in South West England; computerising analysis of farm accounts
7 Sheldon IM - BSc,PhD rational expectations and agricultural futures markets; economics of the agricultural inputs sector
8 Turner MM - BSc,MA,NDA economics of milk production; agricultural training; economic aspects of farm mechanisation; cost-benefit analysis of agricultural projects
9 Whittaker JM - BA,MPhil economics of rural resources

4531 University of Hull

Cottingham Road, Hull, North Humberside
Tel: (0482) 46311.

4532 Department of Biochemistry

1 Dawes Professor EA - PhD,DSc,FRSC *Head of Department* metabolism and survival of soil bacteria
2 Chipperfield Mrs B - MA,PhD *Lecturer* fish muscle enzymes
3 Large PJ - BSc,DPhil metabolism of amines by bacteria and yeasts

4 Ratledge C - BSc,PhD *Professor* mycobacteria in animals, lipid accumulation by yeasts

4533 Department of Plant Biology and Genetics

1 Friend J - BSc,PhD,CBiol,FIBiol *Professor of Plant Biology* biochemistry of plant diseases
2 Jones DA - MA,DPhil,CBiol,FIBiol *Professor of Genetics* ecological genetics
3 Armstrong W - BSc,PhD *Reader* soil waterlogging and plant aeration
4 Ayling PD - BSc,PhD *Lecturer* bacterial genetics
5 Coley-Smith JR - BSc,PhD,DSc *Professor of Plant Pathology* plant pathology
6 Goulder R - BSc,PhD *Senior Lecturer* aquatic ecology
7 Nicholson GG - BSc *Lecturer* cytogenetics
8 Peel AJ - BSc,PhD *Reader in Plant Physiology* translocation
9 Slusarenko AJ - BSc,PhD *Lecturer* molecular biology of plant-bacteria interactions
10 Threlfall DR - PhD,DSc *Reader in Plant Biochemistry* plant biochemistry

4534 Department of Zoology

1 Donaldson IML - BSc,MA,MBChB,MRCP,FRCP,FZS *Professor of Zoology* mammalian central nervous system
2 Goldspink G - BSc,PhD,ScD *Professor of Molecular and Cellular Biology* aspects of muscle growth
3 Goldsworthy GJ - BSc,PhD *Reader in Comparative Endocrinology* neurosecretory and other hormones in insects with particular reference to intermediary metabolism
4 Jones NV - BSc *Senior Lecturer* freshwater and estuarine ecology
5 Robinson J - BSc,PhD *Lecturer* reproductive endocrinology
6 Sudd JH - MA,PhD *Reader in Entomology* behaviour of ants
7 Uglow RF - BSc,PhD *Senior Lecturer* respiratory behaviour, ionic and osmoregulatory activities of decapod crustaceans
8 Williams IC - BSc,PhD *Senior Lecturer* helminth parasites of wild birds, interactions of digenetic trematode larvae and molluscan hosts; Copepod parasites of Redfish Sebastes marinus L

4535 University of Keele

Keele, Staffordshire, ST5 5BG
Tel: (0782) 621111.

4536 Biological Sciences

1 Arme C - PhD *Professor of Zoology* physiology of animal parasites
2 Badcock Miss RM - MSc *Senior Lecturer* insect ecology, especially aquatic or semi-aquatic habitats
3 Goodway KM - PhD *Senior Lecturer* ecology of land reclamation; ecological genetics
4 Hoole D - PhD *Lecturer* fish parasitology; immunology
5 Polwart A - PhD *Lecturer* stress physiology in plants; grassland energetics; organic waste recycling
6 Richards KS - PhD *Reader (Zoology)* heavy metals and Lumbricid earthworms
7 Ross IS - PhD *lecturer* the toxicity to fungi of fungicides containing heavy metals

4537 University of Kent at Canterbury

Biological Laboratory, The University, Canterbury
Tel: (0227) 66822.

1 Burns RG - PhD microbial degradation of chemicals in soil; survival of exocellular enzymes in soil; cellulose breakdown
2 Gull K - PhD mechanism of action of fungicides and anthelmintics
3 Holt G - PhD *Professor* genetics and molecular biology of filamentous microorganisms
4 Jeffries P - PhD mycorrhizal ecology and technology; ultrastructure of plant-pathogen interactions, mycoparasitism
5 Jenkins N - PhD mechanism of hormone action
6 Knowles CJ - PhD *Professor* cyanide excretion and metabolism by soil microorganisms; toxic waste degradation
7 Salmon I - PhD rumen microbiology; degradation of cellulose materials
8 Stacey K - PhD *Professor* bacterial genetics
9 Tuite MF - PhD genetics and molecular biology

4538 University of Lancaster

4539 Biological Sciences

1 Ayres PG - BA,PhD *Senior Lecturer* physiological plant pathology with special reference to powdery mildew diseases
2 Davies WJ - BSc,PhD *Senior Lecturer* physiological plant ecology, with special reference to plant water relations
3 Heaton FW - BSc,PhD *Reader* nutritional biochemistry; function of metals in animal tissues
4 Lea P - BSc,PhD,DSc *Professor* metabolism of nitrogen in crop plants
5 Mansfield TA - BSc,PhD *Professor* stress physiology of plants with special reference to stomata and effects of atmospheric pollutants
6 Piearce TG - BSc,PhD *Lecturer* ecology and physiology of soil animals

7 Wellburn AR - BSc,PhD *Senior Lecturer* biochemical aspects of chloroplast development and effects of atmospheric pollutants on plants
8 Whittaker JB - BSc,PhD *Senior Lecturer* responses of plants to insect damage

4540 University of Leeds

Leeds, LS2 9JT
Tel: (0532) 431751.

4541 Department of Plant Sciences

1 Leedale GF - PhD,DSc *Professor and Head of Department* microalgal cytology
2 Allen JF - BSc,PhD *Lecturer* photosynthesis
3 Bartley DD - BSc,PhD *Senior Lecturer* ecology
4 Clark SC - BSc,Phd *Lecturer* experimental ecology
5 Elston J - BSc,PhD *Professor* agricultural meteorology, agronomy
6 Evans LV - BSc,PhD,DSc *Reader* seaweed metabolism, ultrastructure, biochemistry
7 Harberd DJ - MSc,DSc *Reader* applied genetics
8 Hodgson DR - BSc,PhD,DipAgr *Senior Lecturer* crop production, cultivation
9 Incoll LD - BAgrSc,PhD *Senior Research Officer* physiology
10 Jewer P - BSc,PhD *Lecturer* plant-water relations
11 Kirkby EA - BSc *Senior Lecturer* nutrition
12 Pilbeam DJ - BSc,PhD *Lecturer* biochemistry of crops, crop protection
13 Preece TH - MSc *Senior Lecturer* pathology
14 Sanders FET - MA,DPhil *Lecturer* plant-soil relations
15 Wesley A - BSc *Senior Lecturer* palaeobotany
16 Whiteley G - BSc,PhD *lecturer* soil physics
17 Wren MJ - BSc,PhD *Lecturer* physiology

4542 Field Station Agronomy Unit

Headley Hall Farm, Tadcaster, N. Yorks
Tel: (0937) 833242.

1 Corry DT *Investigation Officer*

4543 Department of Pure and Applied Zoology

4544 Agricultural Zoology

1 Lee DL - BSc,PhD,FIBiol *Professor of Agricultural Zoology* nematode structure and physiology
2 Atkinson HJ - BSc,PhD *Senior Lecturer* plant-parasitic nematodes, molecular biology
3 Bale JS - BSc,PhD *Lecturer* agricultural entomology
4 Belfield W - MSc,PhD *Honorary Lecturer* soil zoology
5 Du/um/rschner U - DiplIngAgr,PhD *Research Fellow* nematophagous fungi
6 Fox P - BSc,PhD *Research Fellow* plant-parasitic nematodes, biochemistry
7 Futers S - BSc,PhD *Research Fellow* plant-parasitic nematodes, molecular biology
8 Hammond K - BSc,PhD *Research Fellow* plant-parasitic nematodes, molecular biology
9 Isaac RE - BSc,PhD *Lecturer* insect biochemistry and neurochemistry
10 Shannon M - BSc,PhD *Research Fellow* plant-parasitic nematodes, biochemistry
11 Smith JE - BSc,Phd *Lecturer* parasitic protozoa; tissue culture
12 Wright C - BSc *Experimental Officer* agricultural entomology

4545 Zoology

1 Alexander RMCN - MA,PhD,DSc,FIBiol,CBiol *Professor of Zoology and Head of Department* animal mechanics
2 Adams J - BSc,PhD *Lecturer* sexual strategies
3 Baker RA - BSc,MSc,PhD,FIBiol,CBiol *Honorary Lecturer* acarology
4 Brust Miss J - BSc,PhD *Lecturer* embryology
5 Chadwick A - BSc,PhD,DSc,MBiol,CBiol *Reader* endocrine physiology
6 Evennett PJ - BSc,PhD *Lecturer* ultrastructure of endocrine glands
7 Fletcher CR - BSc,MSc,PhD *Lecturer* osmotic and ionic regulation
8 Grahame JW - BSc,PhD *Lecturer* ecological energetics
9 Jennings JB - BSc,PhD,DSc,MBiol,CBiol *Reader* invertebrate digestion
10 Ker RF - MA,BSc,DPhil *Research Fellow* animal mechanics
11 Loxton RG - BSc,PhD *Lecturer* ecology of small mammals; insect physiology
12 Mill PJ - BSc,PhD,DSc *Senior Lecturer* invertebrate neurobiology
13 Shorrocks B - BSc,PhD *Senior Lecturer* insect ecology
14 Sutton L - MA,DPhil *Senior Lecturer* ecology of tropical rainforests; woodlice

4546 Department of Animal Physiology and Nutrition

1 Care AD - MA,Phd,DSc,BVMS,MRCVS *Professor and Head of Department* mineral metabolism
2 Anil MA - DVM,PhD *Research Fellow* appetite control
3 Bennett JA - BSc,PhD *Research Fellow* appetite control
4 Dalley J - BSc *Field Station Director*
5 Driver PM - BSc,PhD *Research Fellow* trace elements
6 Forbes JM - BSc,PhD,DSc *Reader* appetite control
7 Johnson CL - MA,PhD,NDA,DipAgric *Lecturer* dairy cow nutrition

8 Pickard DW - BSc,PhD *Lecturer* mineral metabolism
9 Rodway RG - BSc,PhD *Lecturer* growth and reproduction
10 Smith GH - BA,PhD *Senior Lecturer* biochemistry of lactation
11 Symonds HW - BVetMed,PhD,MRVS *Lecturer* trace elements
12 Telfer SB - BSc,MSc,PhD *Lecturer* animal nutrition
13 Varley MA - BSc,PhD,CBiol,MIBiol *Lecturer* neonatal pigs

4547 Department of Microbiology

1 Lacey RW - MA,MD,PhD,MRCPath,DCH *Professor of Clinical Microbiology and Head of Department*
2 Hayes PR - BSc,MSc,PhD *Lecturer* lipase and proteinase production in milk
3 Killington RA - BSc,PhD *Senior Lecturer* equine herpes virus type 1
4 Knapp JS - BSc,PhD *Lecturer* cereal straw degradation
5 Watson DH - BSc,PhD,ARIC *Professor of General Microbiology*

4548 University of Leicester

University Road, Leicester, LE1 7RH
Tel: (0533) 50000.

4549 Botany Department

1 Cockburn W - BSc,PhD *Senior Lecturer* photosynthesis, CAM and associated carbon metabolism
2 Draper J - BSc,PhD *lecturer* genetic engineering of crop plants
3 Gornall RJ - BSc,PhD *Curator* taxonomy and evolution of currants and gooseberries; *Ribes*
4 Parker PF - BSc,PhD *Lecturer* growth and reproduction at low temperatures; taxonomy of cultivated plants
5 Ratcliffe D - BSc,PhD *Lecturer* lead and other toxic metals in roadside vegetation
6 Smith H - BSc,PhD,DSc *Professor* light control of plant growth and development
7 Stace CA - BSc,PhD *Professor* hybridization and evolution of *Festuca rubra*
8 Whitelam GC - BSc,PhD *Senior Demonstrator* molecular biology of photomorphogenesis

4550 Geography Department

1 Bowler IR - BA,PhD *Lecturer* agriculture in Britain with particular reference to Highland areas
2 King RL - BA,NScEcon,PhD *Lecturer* agriculture in the Mediterranean region; land reform studies
3 Pears NV - BSc,PhD *Senior Lecturer* quaternary forest ecology and tree-lines

4551 Zoology Department

1 Bullock JA - BSc,DIC,PhD *Senior Lecturer* quantitative ecology; insect-plant interaction; ecology of lentic freshwater bodies
2 Harris RR - BSc,PhD *Lecturer* environmental physiology of freshwater species; effects of ions, hypoxia and heavy metals
3 Hart PJB - BSc,PhD *Lecturer* social behaviour in vertebrates with particular reference to foraging
4 Macgregor HC - BSc,PhD *Professor* cytology and biochemistry of gametogenesis in amphibia and insects; evolutionary biology of European newts, genus *Triturus*
5 Shelton PMJ - BSc,PhD *Lecturer* retinal physiology and function of commercially important Crustacea

4552 University of Liverpool

Senate House, Abercromby Square, PO Box 147, Liverpool
Tel: 051-709 6022.

4553 Faculty of Science

4554 Botany Department

1 Bradshaw AD - MA,PhD,FIBiol,FRS *Professor and Head of Department* biological problems in the reclamation of derelict land; processes of evolution in natural plant populations; adaptation of plants to soil containing high concentrations of toxic heavy metals, urban ecology
2 Collin HA - BSc,PhD *Senior Lecturer* physiology of flushing cycle in cocoa, factors affecting secondary product formation in plant tissue cultures; selection for herbicide resistance in plant tissue culture. Plant-fungal protoplast fusion
3 Collins JC - BSc,PhD,DBiophys *Lecturer* uptake of ions by roots and the mechanisms by which these processes are controlled; transport of heavy metals into plant cells; osmoregulation in marine algae
4 Eaton JW - Bsc,PhD *Lecturer* ecophysiology, ecology and conservation of waterplants and algae in rivers and canals; biological aspects of waterweed control
5 Greenhalgh GN - MSc,PhD *Lecturer* ecology, structure and taxonomy of Ascomycetes; lichenised Ascomycetes
6 Hardwick K - BSc,PhD *Senior Lecturer* control of photosynthetic activity during development and senescence; physiology of the flushing cycle in cocoa; physiology of root: shoot interactions

7 Issac S - BSc,PhD *Lecturer* isolation and regeneration of fungal protoplasts; the use of protoplasts for elucidation of the biochemical anatomy of filamentous fungi
8 Jennings DH - MA,DPhil,FIBiol *Professor* ionic and carbohydrate physiology of fungi; structure and physiology of basidiomycete strands, cords and rhizomorphs; biochemistry and physiology of marine fungi
9 Johnson MS - BSc,PhD *Lecturer* polymeric soil conditioners and plant growth; environmental effects of derelict non-ferrous metal mines; bioaccumulation of heavy metals in terrestrial food chains; environmental impact analysis
10 Jones MG - BSc,MSc,PhD *Lecturer* mechanisms of cation transport in fungi
11 McAllisiter HA - BSc,PhD,Hon *Lecturer and Assistant Director of University Botanic Gardens* taxonomy, cytology and evolution in rowan trees (*Sorbus sec. aucuparia*), birches (*Betula* species), *Deschampsia caespitosa*, ivies (*Hedera* spp) and *Ledum*; cultivation and conservation of native alpines
12 McNeilly T - BSc,PhD *Lecturer* genetics and evolution of metal tolerant plant populations; micro-evolutionary changes and neighbour relationships in grazed swards; salinity tolerance in crop plants - selection and genetics
13 Mortimer AM - BSc,PhD *Lecturer* population biology of plants particularly of perennial, arable and grassland weeds; forecasting weed infestation in crops; genetics and ecology of *Phytophthora* species
14 Pearson HW - BSc,PhD,MIBiol *Lecturer* biological nitrogen fixation by free living micro-organisms; effects of environmental stress on the physiology of micro-algae and bacteria; microbiology of sewage treatment in warm climates
15 Putwain PD - BSc,PhD *Lecturer* population dynamics, competition, and evolution in arable weeds; herbicide resistance in weeds and crop plants; reinstatement of natural vegetation on severely disturbed soils
16 Russell G - BSc,PhD,MIBiol,FLS *Senior Lecturer* phytosociology of intertidal marine and estuarine habitats, taxonomy and biosystematics of marine algae
17 Thurman DA - BSc,PhD *Senior Lecturer* mechanisms of heavy metal tolerance in flowering plants and fungi; isolation of plant metallothioneins; trace element analysis by X-ray fluorescence
18 Williams ST - BSc,DSc,PhD *Reader* ecology, taxonomy and genetics of actinomycetes; phage ecology; soil microbiology

4555 Biochemistry Department

4556 Plant Biochemistry

1 Britton G - PhD *Reader* biosynthesis and metabolism of carotenoids, sterols, polyprenols and triterpenoid quinones; photosynthesis
2 Goad LJ - PhD *Reader* biosynthesis and metabolism of carotenoids, sterols, polyprenols and triteropenoid quinones; photosynthesis
3 Jones DS - PhD *Senior Lecturer* nucleic acid structures, gene cloning and expression
4 Pennock JF - PhD *Senior Lecturer* biosynthesis and metabolism of carotenoids, sterols, polyprenols and triterpenoid quinones; photosynthesis
5 Powls R - PhD *Senior Lecturer* biosynthesis and metabolism of carotenoids, sterols, polyprenols and triterpenoid quinones; photosynthesis
6 Rees HH - PhD *Reader* biosynthesis and metabolism of carotenoids, sterols, polyprenols and triterpenoid quinones; photosynthesis
7 Turner PC - PhD *Lecturer* nucleic acid structures, gene cloning and expression

4557 Insect Biochemistry

1 Rees HH - PhD *Reader* insect biochemistry

4558 Animal Biochemistry

1 Glover J - PhD,DSc,MRSC *Professor of Biochemistry, Head of Department* fat-soluble vitamins
2 Barnes M - MSc,PhD *Senior Lecturer* fat-soluble vitamins
3 Pennock JF - PhD *Senior Lecturer* fat-soluble vitamins
4 Pitt GAJP - PhD *Reader* fat-soluble vitamins

4559 Genetics Department

1 Ritchie DA - BSc,PhD,FIBiol,FRSE *Professor and Head of Department* genetics and molecular biology of Streptomycetes including the biosynthesis of novobiocin, interspecies hybrid formation and plasmids; molecular and environmental biology of bacterial heavy metal resistence in aquatic populations
2 Eggleston P - BSc,PhD *Lecturer* genetic analysis of competition in plant and animal systems; structure and function of eukaryotic transposable genetic elements
3 Faulkner BM - BSc,PhD *Lecturer* genetical/biochemical analysis of DNA repair in the fungus *Aspergillus nidulans*
4 Gill JJB - BSc,PhD *Lecturer* the evolutionary significance of duplication of chromosomes and the control of chromosomes pairing; genetics of cats
5 Hill RS - BSc,PhD *Lecturer* studies on the structure and activity of eukaryotic chromatin and chromosomes
6 Strike P - BSc,PhD *Senior Lecturer* error prone repair enhanced by bacterial plasmids; control of DNA repair genes in *E. coli*; repair and mutagenesis in *Streptomycetes* and *Aspergillus nidulans*

7 Tomsett AB - BSc,PhD *Lecturer* genetics and molecular biology of nitrate assimilation and heavy metal tolerance in the fungus *Aspergillus nidulans* and in higher plants

4560 Zoology Department

1 Duncan CJ - BSc,PhD,FIBiol *Professor of Zoology* freshwater pulmonates
2 Begon ME - BSc,PhD *Lecturer* ecology of grasshoppers
3 Cain AJ - MA,DPhil *Professor* ecology, systematics and evolution of gastropods
4 Chubb JC - BSc,PhD,MIBiol *Reader* parasites of freshwater fish
5 Leah RT - BSc,PhD *Lecturer* studies on freshwater fish
6 O'Hara K - BSc,PhD *Lecturer* studies on freshwater fish
7 Parker GA - BSc,PhD *Reader* behaviour of dung flies
8 Pearson RG - MA,PhD *Reader* climatic change
9 Williams TR - MA *Senior Lecturer* ecology of Simulium
10 Young JO - BSc,PhD *Reader* ecology of flatworm and leeches

4561 Faculty of Medicine

4562 Medical Entomology Department

1 Macdonald WW - PhD,DSc,FIBiol *Professor and Head of Department* mosquito ecology and genetics
2 Birley MH - MSc,PhD *Lecturer* computer modelling of vector populations
3 Crampton JM - MSc,PhD *Lecturer* molecular biology
4 Davies JB - MSc,PhD *Senior Lecturer Simulium* ecology and control
5 Post RJ - BSc,PhD *Research Fellow Simulium* cytogenetics
6 Roberts MJ - MSc,PhD *Lecturer* tsetse and *Simulium* ecology and control
7 Service MW - PhD,DSc,FIBiol *Reader* mosquito and *Simulium* ecology
8 Townson H - MSc,PhD,MIBiol *Senior Lecturer* mosquito physiology and genetics, filariasis and *Simulium* biology
9 Ward RD - MSc,PhD,MIBiol *Senior Lecturer* ecology and taxonomy of phlebotomid sandflies

4563 Parasitology Department

1 Nelson GS - MD,DSc,FRCP,FRCPath,DAPE,DTMH *Professor and Head of Department*
2 Ashford RW - PhD *Senior Lecturer* biology and epidemiology of *Leishmania* and coccidia
3 Chance ML - PhD *Senior Lecturer* biochemical systematics of parasitic protozoa; experimental chemotherapy of *Leishmania*
4 Crewe W - PhD *Reader* helminth zoonoses in UK; biology of taeniid tapeworms; blood parasites of birds
5 Hommel M - MD,PhD *Senior Lecturer* immunology of parasites, antigenic variation in malaria
6 Howells RE - PhD *Reader* physiology and chemotherapy of malaria and schistosomes
7 Jewsbury JM - PHD *Senior Lecturer (TC)* schistosomiasis (chemoprophylaxis, snail ecology and control and host-parasite relationship)
8 Marshall I - PhD *Lecturer (TC)* schistosomiasis; reproductive physiology and control
9 Semoff S - PhD *Lecturer* immunocytochemistry and electron microscopy of malaria

4564 Faculty of Veterinary Science

4565 Animal Husbandry Department

1 Anderson RS - BVMS,PhD,MRCVS *Professor of Animal Husbandry* companion animal clinical nutrition
2 Davies DAR - BSc,MA *Lecturer* ovulation rates in highly prolific ewes; development of a multi-teated ewe; nutritional and husbandry factors in the rearing of triplet lambs
3 Lawrence TLJ - MSc,PhD,BScAgr,MIBiol *Reader* utilization of cereals and rape seed meals by the growing pig; nutritional and husbandry factors in gastro-intestinal disease
4 Rowan TG - BVSc,PhD,MRCVS *Lecturer* effects of nutrition and environmental temperature and humidity on growth; immune function and the pathogenesis of respiratory diseases in the young calf; utilization of proteins in ruminants

4566 Veterinary Anatomy Department

1 King AS - BSc,PhD,MRCVS *Professor of Veterinary Anatomy* avian chemo- and baro-receptor pathways
2 Morris R - BSc,PhD *Lecturer* neuronal pathways in dorsal horn
3 Riley Mrs VA - BVSc,MRCVS *Lecturer* radiological anatomy of vertebral columns
4 Skerrit GC - BVSc,MRCVS *Lecturer* clinical neurology, coenuriasis
5 Vaillant C - BSc,PhD *Lecturer* neuropeptides and dysautonomia

4567 Veterinary Clinical Science Department

1 Ford EJH - DVSc,FRCVS,FRCPath *Professor and Head of Department* liver function in health and disease, toxicology, ruminant metabolism, especially glucose and corticol production

2 Clarkson MJ - DVSc,BSc,PhD,MRCVS *Professor of Farm Animal Medicine* ovine abortion; parasitic gastro-enteritis of sheep and cattle; hydatid disease; sheep immune responses
3 Cox JE - BSc,BVetMed,PhD,MRCVS *Senior Lecturer* reproductive endocrinology of male animals, especially stallions, with particular reference to adrenal-testis interaction
4 Curtis PE - BVSc,MRCVS,DipBact *Lecturer* poultry disease, behaviour and husbandry interactions in, and welfare of, non-cage flocks
5 Dobson Mrs H - BSc,PhD *Senior Lecturer* reproductive endocrinology of ruminants; stress
6 Faull WB - BSc,FRCVS *Senior Lecturer* bovine mastitis; sheep health schemes
7 Fitzpatrick RJ - BSc,PhD MRCVS *Professor of Animal Reproduction* endocrine control of labour; clinical pharmacology; ovum transfer
8 Hunt Miss JM - BVSc,PhD,MRCVS *Lecturer* equine internal medicine and gastrointestinal surgery with particular interest in disorders of gut motility
9 Walton GS - BVSc,MRCVS *Senior Lecturer* dermatology; response of the skin to bacteria
10 Walton JR - BVMS,PhS,MRCVS,DipBact *Senior Lecturer* salmonellosis; actions of antibiotics; growth promotion in farm animals
11 Ward WR - BVSc,MRCVS *Senior Lecturer* infertility in ruminants; cattle health schemes and computer recording
12 West Miss HJ - BVSc,MRCVS *Lecturer* liver function in farm animals and horses, (bovine internal medicine and surgery, cardiology)

4568 Veterinary Parasitology Department

1 Beesley WN - MSc,PhD *Reader and Head of Department* biology and control of ectoparasites, particularly myiasis producing insects and ticks; anthelmintics
2 Britt DP - BA,MSc,PhD,FIMLS,CBiol,MIBiol *Research Assistants* chemotherapy and control of ruminant and equine helminths
3 Trees AM - BVM&S,MRCVS,PhD *Lecturer* immunity to parasites especially *Coccidiosis, Onchocerciasis*; tick-borne diseases

4569 Veterinary Pathology Department

1 Kelly DF - MA,PhD,BVSc,MRCVS,FRCPath *Professor and Head of Department* pathology of domesticated animals
2 Allan D - PhD,MRCVS *Honorary Research Fellow* immunology
3 Bailie NC - BVMS,MS,MRCVS *Lecturer* pathology of domesticated animals
4 Baker JR - BVSc,PhD,MRCVS *Senior Lecturer* pathology of non-domesticated animals
5 Batt RM - BVSc,MSc,PhD,MRCVS *Reader* investigation of naturally occurring small intestinal diseases in the dog
6 Bradbury JM - MSc,PhD *Lecturer* avian mycoplasmatales - immunology; diagnostic methods
7 Carter SD - BSc,PhD,FIMLS *Lecturer* applied immunology
8 Dixon JB - MA,VetMB,PhD,MRCVS *Senior Lecturer* studies on histocompatibility and parasite antigens in normal, immunized and tolerant animals
9 Gaskell RM - BVSc,PhD,MRCVS *Honorary Research Fellow* virology
10 Haywood S - BVSc,PhD,MRCVS *Lecturer* mechanisms of toxicity and tolerance to copper poisoning
11 Jones RC - BSc,PhD *Lecturer* avian viruses
12 Jordan FTW - BSc,PhD,DSc,FRCVS *Honorary Research Fellow* avian mycoplasmatales; avian respiratory diseases
13 Lawson G - BVMS,MRCVS *Leverhulme Resident* pathology
14 Newsholme SJ - BSc,BVetMed,MVetMed,DipACVP,MRCVS *Lecturer* pathology of domesticated animals
15 Payne-Johnson CE - BVSc,FRCVS *Lecturer* comparative pathology
16 Thomlinson JR - BSc,PhD,MRCVS *Senior Lecturer* pathogenesis of enteric infections in calves and pigs
17 Varley J - BSc,PhD *Lecturer* pathogenesis of avian mycoplasmas
18 Woldehiwet Z - DVM,PhD,MRCVS *Lecturer* microbiology

4570 Veterinary Physiology & Pharmacology Department

1 Cooke RG - BSc,PhD *Lecturer* hormonal regulation of the myometrium; pharmacokinetics of anticonvulsants
2 Finn CA - PhD,BSc,MRCVS *Professor* implantation of ova
3 Garthwaite J - PhD,BSc *Lecturer* neurotransmission in the central nervous system
4 Knifton A - BVSc,BSc,PhD,MRCVS *Senior Lecturer* hormonal regulation of the myometrium, pharmacokinetics of anticonvulsants
5 Montegomery A - PhD,BSc,MIBiol *Lecturer* patho-physiology of muscle disease
6 Nicholson T - BSc,BVetMed,MRCVS *Lecturer* gastro-intestinal studies in ruminants

4571 London University

4572 King's College

4573 Faculty of Life Sciences

4574 Department of Biology, Half Moon Lane Laboratory
68 Half Moon Lane, Herne Hill, London, SE24 9JF
Tel: 01-733-5666.

1 Bradbeer Professor JW - BSc,PhD,DSc *Professor of Botany* chloroplast development; regulation of the photosynthetic carbon cycle; seed dormancy and germination
2 Charlwood BV - BSc,PhD terpenoid biosynthesis; amino acid biosynthesis; chemotaxonomy
3 Collinson ME - BSc,PhD *Royal Society Research Fellow* tertiary vegetational history
4 Cowan JW - BSc,PhD genetics of fungi
5 Hall DO - Bsc,MS,PhD *Professor of Biology* biological solar energy conversion; iron-sulphur proteins; photosynthesis
6 Moore PD - BSc,PhD *Reader in Ecology* palynology - vegetational history, peatland ecology
7 Wiltshire Mrs PEJ - BSc soil microbiology

4575 Department of Biology, Kensington Laboratory
King's College, Campden Hill Road, London, W8 7AH
Tel: 01-937-5411.

1 Black Professor M - BSC,PhD *Professor of Plant Physiology* physiology, biochemistry and development of seeds; photocontrol of growth
2 Brafield AE - BSc,PhD fish physiology and energetics
3 Chapman JM - BSc,PhD biochemistry of plant development; phyllotaxis
4 Gahan Professor PB - BSc,PhD *Professor of Botany* plant cell biology; differentation of vascular tissue
5 Heale JB - BSc,PhD *Reader in Plant Pathology* plant pathology and biotechnology; biological control
6 Llewellyn M - BSc,PhD *ecology and physiological of aphids; biological control*
7 Peel MP - BSc,PhD cytodifferentiation in algae

4576 Department of Biology, Strand Laboratory
King's College, Strand, London, WC2R 2LS
Tel: 01-836-5454.

1 Sales GD - BSc,PhD *Lecturer in Zoology* rodent behaviour; pest control

4577 Imperial College of Science and Technology
South Kensington, London, SW7 2AZ
Tel: 01-589-5111.

4578 Department of Biochemistry

1 Morris HR - BSc,PhD *Professor, Head of Department* structural studies on molecules of agricultural and medical interest
2 Bradford HF - PhD,DSc,MRCPath *Professor* neurobiology and nervous disorders
3 Cass AEG - BA,DPhil *Lecturer* biotechnology and biosensors as tools in agriculture and medicine
4 Dell A - PhD *Lecturer, Assistant Director* structural studies on molecules of agricultural and medicine interest
5 Dickerson AGF - MSc,PhD,DIC,CBiol,MIBiol *Senior Lecturer, Assistant Director* interrelationships between plants and phytopathogenic fungi
6 Dolly JO - MSc,PhD *Senior Lecturer* neurobiology and neurotoxins
7 Glover DM - PhD *Reader* eukaryotic molecular genetics and cancer
8 Hartley BS - MA,PhD,FRS *Professor* biotechnology in agriculture and medicine
9 Jackson JF - PhD *Lecturer* molecular neurobiology and hormones
10 Mantle PG - PhD,DSc,ARCS,DIC *Reader, Senior Tutor* mycotoxins in agriculture and medicine
11 O'Hare K - PhD *Lecturer* eukaryotic molecular genetics and cancer
12 Smith D - PhD *Lecturer* molecular parasitology
13 Wilkin GP - MSc,PhD *Lecturer* cellular neurobiology

4579 Centre for Biotechnology

1 Hartley BS - MA,PhD,FRS *Professor, Director* biotechnology in medicine and agriculture
2 Banks GT - MSc *Senior Lecturer* microbial fermentations
3 Cass AEG - BA,DPhil *Lecturer* biosensors as tools in medicine and agriculture
4 Leak DJ - MA,MSc,PhD *Lecturer* microbial physiology and biotransformations
5 Lichtenstein CP - BSc,PhD *Lecturer* genetic engineering of plants of agricultural importance
6 Rangarajan M - BSc,PhD *Research Officer* protein chemistry for medicine and agriculture

7 Ribbons DW - PhD,DSc *Professor* biotransformations to specialty chemicals for medicine and agriculture
8 Woodland MP - BSc,PhD *Research Officer* gene-protein relationships of microbial catabolic pathways

4580 Department of Pure and Applied Biology

1 Anderson RM - PhD,DIC,ARCS,CBiol,FIBiol *Professor, Head of Department* parasite ecology and epidemiology
2 Alvin KL - PhD *Senior Lecturer, Senior Tutor* plant anatomy, palaeobotany and taxonomy
3 Archer SA - PhD *Lecturer* genetics, physiology and biochemistry of plant diseases caused by fungi and bacteria
4 Barber J - MSc,PhD,CChem,FRSC,CBiol,FIBiol *Professor* plant physiology and biochemistry
5 Bates JW - BSc,PhD *Lecturer* physiological ecology of bryophytes; oil pollution effects on vegetation
6 Beddington JR - BSc,PhD *Lecturer* mathematical modelling of marine populations
7 Bell JND - PhD *Senior Lecturer* air pollution and plants; uptake of radionuclides into crops
8 Bianco AED - PhD *Research Fellow* immunology of *Onchocerca* infections
9 Brady J - MA,PhD,DSc,CBiol,FIBiol *Reader* tsetse fly behaviour and circadian rhythms
10 Bray RS - PhD,DSc,CBiol,FIBiol *Honorary Lecturer* studies on antigenic fractions of *Leishmania* which stimulate the curative T lymphocytes
11 Brown VK - PhD,ARCS,DIC *Senior Lecturer* role of insect herbivores in plant succession
12 Buck KW - PhD,DSc *Reader* fungal and plant viruses, molecular basis of plant virus pathogenicity
13 Bundy DAP - PhD *Lecturer* parasite epidemiology and ecology
14 Canning EU - PhD,DSc,ARCS,DIC,CBiol,FIBiol *Professor* protozoology
15 Carlile MJ - MA,PhD,ScD *Senior Lecturer* physiology and behaviour of fungi
16 Conway GR - PhD,DipAgrSci,DipTropAgr,CBiol,FIBiol *Professor* environmental technology
17 Coutts RHA - PhD *Lecturer* replication of potato virus X in potato protoplasts
18 Crawley MJ - PhD *Lecturer* dynamics of plant-herbivore interactions
19 Dadd AH - PhD,ARCS *Lecturer* bacteriology
20 Dalby DH - PhD,CBiol,MIBiol *Lecturer* coastal ecology and plant taxonomy
21 David CT - PhD *Research Fellow* orientation mechanisms of flying insects
22 Davies RG - DSc,ARCS,DIC *Emeritus Professor, Senior Research Fellow* entomology and numerical taxonomy
23 Dickinson DJ - PhD,FIWSc *Lecturer* timber technology
24 Djamgoz M - BSc,PhD,ARCS *Lecturer* neurophysiology of vertebrate retina
25 Evans AAF - BAgrSci,PhD *Lecturer* physiology of parasitic nematodes
26 Galley DJ - PhD,ARCS,DIC *Lecturer* movement of systemic insecticides in plants and into aphids
27 Gay JL - MSc,PhD *Reader* biotrophic plant pathogens
28 Goldsworthy A - PhD *Lecturer* electrophysiology of plant tissue culture
29 Goto HE - BSc,CBiol,FIBiol *Senior Lecturer* biology and taxonomy of Collembola
30 Greathead D - PhD,DSc,CBiol,FIBiol *Honorary Lecturer* biological control of agricultural pests
31 Hardie RJ - BTech,PhD *Research Fellow* polymorphism in aphids
32 Hassell MP - MA,DPhil,DSc,CBiol,FIBiol *Professor, Deputy Head of Department* insect ecology, dynamics of population interaction
33 Hominick WM - MSc,PhD *Lecturer* nematodes for biological control of insects
34 Hunter CN - PhD *Lecturer* molecular biology of chlorophyll synthesis, and genetic regulation of photosynthetic membrane assembly
35 Kermack DM - PhD *Lecturer* Middle Jurassic mammals
36 Killick-Kendrick R - MPhil,PhD,DSc,DIC,CBiol,FIBiol *Honorary Lecturer* field biology of phlebotomine sandflies in relation to the transmission of leishmaniasis
37 Lamb BC - PhD *Senior Lecturer* recombination and gene conversion
38 le Patourel GNS - PhD,DipEd *Lecturer* insecticidal action of chemicals on stored product pests
39 Lees AD - MA,PhD,ScD,FRS *Emeritus Professor, Senior Research Fellow* insect daylength clocks especially in relation to aphid polymorphism
40 Levy JF - DSc,ARCS,FIWSc,FIAWS,CBiol,FIBiol *Professor* timber technology
41 Ludlow AR - PhD *Research Fellow* physiological and statistical models of behaviour
42 Maizels RM - PhD *Lecturer* immunology of nematode parasites
43 Matthews G - PhD,ARCS,CBiol,MIBiol *Reader* pesticide application
44 McManus DP - PhD *Lecturer* molecular biology of cestodes
45 McNeill S - PhD,CBiol,MIBiol *Lecturer* plant chemistry and insect herbivory
46 Moorhouse JE - PhD,CBiol,MIBiol *Honorary Lecturer* coordination of behavioural sequences in locusts
47 Morton AJ - PhD,ARCS *Lecturer* moorland and healthland ecology
48 Mueller PW - PhD *Lecturer* plant growth, water relations and pollution
49 Mumford JD - PhD *Lecturer* implementation of integrated pest management
50 Murphy RJ - PhD *Lecturer* timber preservative techniques
51 Norton GA - BScAgr,MScEcon,PhD *Senior Lecturer* information technology and computer models for pest management

52 Palmer JM - BSc,DPhil *Professor* plant biochemistry
53 Price CE - PhD,DIC *Honorary Lecture* weed science
54 Prior JAB - PhD,DipArch *Research Fellow* use of archaeological charcoals in determination of palaeoclimate
55 Ruddock KH - PhD,ARCS *Reader* pharmacological and neurophysiological properties of fish retinal neurones
56 Rutter AJ - PhD,ARCS,CBiol,FIBiol *Emeritus Professor, Senior Research Fellow* plant ecology and air pollution
57 Selkirk MD - PhD *Research Fellow* antigen genes of filarial nematodes
58 Seller TJ - PhD *Lecturer* neurobiology and behaviour of birds
59 Sinden RE - PhD,DSc,DipAnimGenet *Reader* cell biology and immunology of the malaria parasite
60 Smyth JD - PhD,ScD,CBiol,FIBiol *Emeritus Professor, Senior Research Fellow* parasitology; *in vitro* culture of cestodes
61 Waage JK - PhD,DIC *Lecturer* population biology of insect natural enemies
62 Way MJ - MA,DSc,CBiol,FIBiol *Emeritus Professor, Senior Research Fellow* ecology and its application to forecasting insect pest damage
63 Wheeler BEJ - PhD,DSc *Reader* epidemiology of plant diseases
64 Wood RKS - PhD,ARCS,DIC,CBiol,FIBiol *Professor* physiological plant pathology and disease control
65 Wright DJ - PhD,CBiol,MIBiol *Lecturer* mode of action of insecticides and nematicides
66 Young S - BA,PhD *Lecturer* visual responses and behavioural ecology of *Daphnia*

4581 Geology Department

1 Culbard E - BSc,PhD,DIC,FGS *Lecturer* environmental geochemistry
2 Thompson M - BSc,PhD,ARCS,CCHEM,FRSC *Lecturer* geochemistry; soil and plant analysis
3 Thornton I - DSc,PhD,DIC,FGS *Reader* geochemistry applied to crop and animal nutrition

4582 Queen Mary College

Mile End Road, London, E1 4NS
Tel: 01-980-4811.

4583 School of Biological Sciences

1 Betts GF - BSc,PhD mechanisms of enzyme catalyzed reactions; the *in vivo* regulation and organisation of metabolic pathways
2 Bevan Professor EA - BSc,MA,PhD,FIBiol,FLS biochemistry and genetics of yeast viruses; the genetic engineering of brewery yeasts; the synthesis of blood components by genetically engineered yeast cells
3 Bignell DE - BSc,PhD,MSc digestive physiology; symbiotic associations and nutrition of termites, cockroaches and other insects with emphasis on the dynamics, attachment and metabolism of intestinal bacteria
4 Bowmaker JK - BSc,PhD,MIBiol colour vision, colour blindness and visual pigments in man and other primates; the ecology of vision in vertebrates
5 Campbell AC - BSc,DPhil fine structure, physiology and ecology of sea urchins and starfish; coelenterate biology especially reef corals
6 Casselton Mrs LA - BSc,PhD molecular genetics of *Coprinus*; cloning genes which regulate sexual morphogenesis; recombination, gene structure and evolution of the mitochondrial genome
7 Clymo Professor RS - BSc,PhD,MIBiol physiology and ecology of the bog-moss *Sphagnum*; ecology of peat accumulating ecosystems; vegetation survey
8 Courtney Mrs WAM - BSc,PhD distribution of littoral gastropods; polychlorinated biphenyls in marine ecosystems; marine antifouling; macrofauna of the Outer Thames
9 Dalton LT - BSc,PhD hormonal control of ionic transport
10 Denny P - BSc,PhD factors affecting the distribution of submerged waterplants; the vegetation of inland waters in Africa; waterplants, algae and heavy metals in standing waters; the taxonomy of African Potamogetonacae
11 Duckett Professor JG - BA,PhD cytoskeletal microtubule and microfilament systems; ultrastructural cytology of plant and animal spermatozoa; morphogenesis and sex determination in land plants; symbiosis between fungi and liverworts
12 Evans RL - BSc,PhD parasitic and symbiotic associations between fungi and plants; the nature of resistance in leaf spot diseases; the role of ectomycorrhizae in the protection of tree roots against heavy metals
13 Frankland B - BSc,PhD phytochrome and plant morphogenesis; effects of light and gibberellins on seed germination; heavy metals in plants and soils
14 Goodman Mrs LJ - MA,PhD insect neurobiology, especially study of the visual system by neurophysiological and anatomical methods
15 Gray DO - BA,PhD techniques of amine analysis; the isolation, characterisation, metabolism and biological importance of biogenic amines, especially those present in higher plants
16 Green Professor J - BSc,PhD,FIBiol,FLS ecology of tropical lakes and rivers; cyclomorphosis of rotifers
17 Gurnell J - BSc,PhD ecology, behaviour and population dynamics of woodland rodents; genetic composition of populations of mice, voles and squirrels
18 Heathcote P - BSc,PhD biophysics and biochemistry of photosynthetic electron transfer in plants and bacteria
19 Hildrew AG - BSc,PhD freshwater ecology; community structure in streams and rivers; population ecology and production of stream invertebrates; trout growth in acid streams, life history strategies of crustacea and insects from tropical rain pools

20 Hughes RG - BSc,PhD marine ecology, especially benthic and fouling communities; biology of modular marine animals, especially hydrozoa; theoretical community ecology
21 Hunt DM - BSc,PhD mammalian developmental and biochemical genetics; genetic control of trace element metabolism
22 Ketteridge SW - BSc,PhD biochemistry of plant virus multiplication and virus inactivation
23 McCarthy DA - BSc,PhD biochemistry of plant virus structure and replication; immune complexes and polymorph function in rheumatoid arthritis
24 Mitchell Mrs DJ - BSc,PhD molecular biological studies and genetic manipulations of dsRNA viruses in *Saccharomyces cerevisiae* and DNA viruses in *S. lypolytica*
25 Moses Professor V - MA,PhD,DSc,FIBiol organisation and control of metabolism in microorganisms; biochemistry and physiology of the aging process in nematodes; microbial enhancement of oil recovery
26 Newell PF - BSc,PhD biology and ecophysiology of molluscs, with particular reference to terrestrial pulmonates
27 Parker JS - BA,DPhil cytological and genetical studies of natural populations; the control of meiosis
28 Prince SD - BSc,PhD ecophysiology, population and community ecology of higher plants
29 Pye Professor JD - BSc,PhD,FLS bat echolocation by radar snd sound recording techniques; ultrasonic communication in rodents and insects
30 Rainbow PS - BA,PhD the biology of barnacles; the study of the accumulation and detoxification of trace metals such as zinc, cadmium and copper by marine crustaceans and freshwater insect larvae; tropical intertidal ecology
31 Rath Miss EA - BSc,MSc,PhD lipid metabolism in obesity and diabetes, and the effect of diet on milk composition and development of offspring
32 Riegel JA - AB,MA,PhD osmotic and ionic regulation in aquatic animals
33 Robinson JP - BAGrSci,PhD physiology and ecology of anaerobic bacteria in sewage, sediments and animal intestines; microbial enhancement of oil recovery
34 Savage GE - BSc,PhD the neurobiology of memory formation in lower vertebrates
35 Shales SW - BSc,PhD physicohemical properties of cell surfaces, cell adhesion; functional aspects of envelope structure, protein export mechanisms; microbial adaptation to heavy metals and other toxic compounds
36 Smith KE - BsC,PhD microbial hydroxylation of steroids; analysis of ribosomes from drug resistant microorganisms; isolation and characterisation of *Coprinus* mating genes
37 Spencer Ms JE - BSc,MPhil control of gastric acid secretion in mammals employing *in vivo* and *in vitro* models for investigating the physiology and pharmacology of HC1 production
38 Springham DG - MA,DPhil metabolism of Streptomycetes; microbial enhancement of oil recovery; alcoholic fermentation
39 Thake Mrs B - BSc freshwater and marine physiology and ecology of phytoplankton and of seaweeds; eutrophication of freshwaters and seas
40 Thorpe A - BSc,PhD immunocytochemical studies on peptides found in both the gastro-intestinal system and the brain
41 Tinsley RC - BSc,PhD parasitology with special interests in reproductive biology and transmission; ecology and evolution of amphibians in Africa and deserts of North America
42 Walkey M - BSc,PhD ecology and physiology of parasitic helminths, notably those affecting fish
43 Wallwork Professor JA - BSc,PhD,DSc taxonomy of soil mites; role of soil animals as regulators of mineral cycling in deserts; life styles of desert soil animals
44 Williams GB - BSc,PhD ecology of forest animals; population dynamics, trophic relationships and feeding behaviour

4584 Royal Holloway and Bedford New College

Egham Hill, Egham, Surrey, TW20 OEX
Tel: Egham (0784) 34455.

1 Wedderburn Professor D - MA,DLitt *Principal*

4585 Department of Biochemistry

1 Pridham JB - BSc,PhD,DSc *Professor and Head of Department* plant carbohydrate biochemistry with particular reference to storage organ metabolism and fruit ripening
2 Beesley PW - BSc,PhD *Lecturer* synapse development and factors which regulate this; molecular properties and function of neuronal cell surface glycoproteins
3 Bowyer JR - MA,PhD *Lecturer* photosynthetic electron transport in bacteria and chloroplasts and relationships to herbicide mode of action and design
4 Bramley PM - BSc,PhD *Lecturer* carotenoid biosynthesis in microorganisms; mode of action of bleaching herbicides on carotenoid formation
5 Clarke JB - MA,MSc,PhD *Lecturer* kinetics and mechanism of enzyme reactions including fast reaction techniques and NMR spectroscopy
6 Davies DR - BSc,PhD *Lecturer* regulation of key glycolytic and lipogenic enzymes in mammalian tissues notably liver, brain and lymphocytes

7 Dey PM - BSc,PhD *Lecturer* carbohydrates and their enzymic modifications in relation to seed germination, fruit ripening, plant protection and industrial applications

8 Dixon RA - MA,DPhil *Reader* molecular aspects of plant disease resistance and gene expression in plants and plant cell cultures

9 Lacnado JR - MSc,PhD *Lecturer* cytoskeleton-membrane interaction in trypanosomes as potential targets for chemotherapy in the African trypanosomiases

10 Mackenzie A - BSc,PhD *Lecturer* regulatory mechanisms controlling gene expression in microbial systems with special reference to gene structure

11 Owen WJ - BSc,PhD *Lecturer* biochemical aspects of herbicide selectivity and safener mode of action; metabolism of pesticides in plant cell cultures

12 Prebble JN - BSc,PhD *Senior Lecturer* interactions of enzymes and plasma membranes in relation to synthesis and function of isoprenoids

13 Rider CC - BSc,PhD *Lecturer* metabolic specialisation of mammalian cell types, with particular interest in the nervous and immune systems

14 Zagalsky PF - MA,PhD *Senior Lecturer* structural studies on invertebrate carotenoproteins

4586 Department of Botany
Huntersdale, Callow Hill, Virginia Water, Surrey, GU25 4NL
Tel: Egham 35551.

1 Dodge JD - PhD,DSc *Professor and Head of Department* biology of dinoflagellates; plant ultrastructure
2 Brett DW - BSc *Lecturer* plant anatomy; dendrochronology
3 Chaloner WG - PhD,FRS *Professor* palaeobotany; evolution of vascular plants
4 Dixon PA - BSc *Senior Lecturer* microbial ecology; plant pathology
5 Evans JH - PhD *Senior Lecturer* freshwater ecology
6 Ferry BW - PhD *Lecturer* mycorrhizal biology, ecology of shingle; lichen physiology
7 Ford TW - PhD *Lecturer* metabolism of micro-algae
8 Jones BMG - BA,PhD *Senior Lecturer* plant breeding, cytogenetics
9 Lodge E - BSc,PhD *Senior Lecturer* plant ecology; bryophyte physiology
10 McDonough MW - PhD *Lecturer* molecular structure of bacterial flagella
11 Stead AD - PhD *Lecturer* reproductive biology; effects and mode of action of ethylene
12 Stevens WA - PhD *Lecturer* plant virology; mycoplasmas
13 Waters SJP - PhD *Lecturer* plant ecology; quaternary botany

4587 Department of Chemistry

1 Perkins MJ - PhD,DSc *Professor and Head of Department* chemistry of hydroxamic acids; free radical mechanisms, biological antioxidants
2 Beezer AEB - MSc,PhD,DSc *Reader* microcalorimetry; microbial metabolism; drug-cell interactions; particle electrophoresis; thermodynamic parameters for cell and cell-drug interactions
3 Farago ME - PhD,CChem,FRSC *Reader* uptake of metals by plants; metal speciation in plants and soils; environmental chemistry
4 Finch A - PhD,DSc,DIC,ARCS,FRSC *Professor* enzyme - and immobilised-enzyme mediated reactions; use of microcalorimetry as monitor/analytical device with enzymes; thermodynamic studies on collagen
5 Finch P - PhD *Lecturer* structural and synthetic carbohydrate chemistry; chemistry of enzyme-catalysed reactions of carbohydrates; plant-pathogen interactions involving carbohydrates
6 Lewis D - PhD *Senior Lecturer* alditols
7 Weigel H - DipChem,Drrernat,DSc,FRSC *Reader* biochemistry and chemistry of carbohydrates

4588 Department of Geography

1 Cole MM - BSc,PhD *Professor* remote sensing for terrain analysis including plant indentification from multispectral photography; savanna vegetation; use of vegetation in mineral exploration; investigation of the relationship between plant and the mineral content of soils and bedrock; soil and vegetation studies in the South Pennines
2 Eden MJ - MA,MSc *Lecturer* soil and plant-water relations in tropical grasslands, ecology of shifting cultivation in tropics
3 McGregor DFM - PhD *Lecturer* soil erosion and conservation; geomorphological hazards; Latin America and Caribbean

4589 Department of Zoology
Alderhurst, Bakeham Lane, Englefield Green, Surrey, TW20 9TY
Tel: Egham 35553.

1 Lewis CT - MA,PhD *Professor and Head of Department* insect physiology
2 Alibhai SK - BSc,DPhil *Lecturer* population ecology and reproduction of small animals
3 Andrews Miss EB - BSc,PhD *Senior Lecturer* the ultrastructure and function of the excretory systems of prosobranch gastropod molluscs
4 Angus RB - MA,DPhil *Lecturer* entomology
5 Catchpole CK - BSc,PhD *Lecturer* behaviour and ecology of passerine birds, particularly vocalisations; ecology of gravel pit birds
6 Credland PF - BSc,PhD *Lecturer* the growth and physiology of Chironomus larvae, particularly related to the influence of environmental factors

7 Dales RP - BSc,PhD *Professor* polychaete and coelomocytes and defence mechanisms; respiration and metabolism in polychaetes
8 Duncan Miss A - BSc,PhD *Reader in Ecology* limnology
9 Lewis JW - BSc,PhD *Senior Lecturer* parasitology
10 McVean AR - BSc,PhD *Lecturer* anatomical and electrophysiological analysis of the neural mechanisms governing simple behaviour in invertebrates
11 Morris PA - BSc,PhD *Lecturer* mammalogy
12 Pontin AJ - MA,DPhil *Lecturer* ecology, entomology
13 Thorndyke MC - BSc,PhD *Lecturer* comparative endocrinology and ultrastructure of thyroid in lower vertebrates
14 Twigg GI - BSc,PhD *Senior Lecturer* mammalogy

4590 The Royal Veterinary College

4591 Department of Anatomy
Royal College Street, London, NW1 OTU
Tel: 01-387-2898.

1 Batt RAL - BSc,BVSc,PhD,MRCVS *gene expression in the mutant mouse; obese (genotype - ob/ob)*
2 Bee J - PhD,MSc the role of cell communication in the expression and maintenance of the cartilage phenotype
3 Lanyon LE - BVSc,PhD,MRCVS the mechanisms and objectives of functional adaptation in bone
4 O'Shaughnessy P - BSc,PhD development and control of gonadal function in domestic animals
5 Sopher D - BSC,PhD development of the mammalian reproductive system
6 Stickland NC - BSc,PhD factors influencing muscle cellularity in development and growth

4592 Animal Health and Production
Boltons Park, 44 Hawkshead Road, Potters Bar, Herts., EN6 1NB
Tel: (0707)55486.

1 Jones JET - PhD,MRCVS,FRCPath *Coutauld Professor of Animal Health and Production and Head of Department* mastitis in sheep; urogenital tract infection in pigs
2 Goddard PJ - BVetMed,PhD,MRCVS equine and ovine genital tract activity
3 Hill R - DSc,PhD trace elements in animal diets; rapeseed meal in the diets of poultry and pigs; effect of plane of nutrition on infection
4 Madel AJ - BA,VetMB,MRCVS,DipAH,DipTCDHE the assessment of veterinary students
5 Pidduck Ms HG - BSc,PhD animal genetics
6 Stedman A - BSc,PhD,DipNutr use of rapeseed meal in diets of ruminants
7 Vincent Ms IC - BSc,PhD effects on production of feeding rapeseed meal in the diets of ruminants
8 Williams HL - BScAgr,MSc,PhD,NDDH,MIBiol assessment of reproductive activity in sheep: modification of the breeding season of sheep
9 Wilsmore AJ - BVSc,PhD,MRCVS,DipAH ovine perinatal mortality

4593 Medicine
Hawkshead Campus, Hawskshead Lane, N. Mymms, Hatfield, Herts., AL9 7TA
Tel: (0707)55486.

1 Andrews AH - BVetMed,PhD,MRCVS calf diseases, notably enteric diseases
2 Brownlie SE - BVMS,PhD,MRCVS cardiology and biochemical aspects of neurology in dogs and cats
3 Bush BM - BVSc,PhD,FRCVS endocrine and renal disease in dogs and cats
4 Johnston AM - BVMS,MRCVS keratin and lameness in large animals; ante mortem factors affecting carcass quality
5 Kerr MG - BVMS,BSc,PhD,MRCVS clinical biochemistry of equine exertion
6 Mitchell AR - BSc,BVetMed,PhD,MRCVS *Reader in Experimental Veterinary Medicine* fluids and electrolytes; metabolic, renal and enteric disease
7 Stephens DB - BVSc,PhD,FRCVS behavioural aspects of welfare in farm animals
8 Thomsett LR - FRCVS dermatology in farm and companion animals
9 Webbon PM - BVetMed,MRCVS,DVR diagnostic applications of radionuclides in animals
10 White DG - MA,VetMB,MRCVS clinical and immunological aspects of enteric disease and dermatomycosis in ruminants

4594 Pathology
Hawkshead Campus, North Mymms, Hatfield, Herts., AL9 7TA
Tel: 0707 55486.

1 Barlow RM - DSC,DVM&S,MRCVS *Professor of Pathology and Head of Department* neropathology; reproductive wastage in farm animals
2 Middleton DJ - BVSC,PhD pathology, especially of endocrine system
3 Smyth JBA - MVB,MSc,PhD,MRCVS pathology, especially muscle; neoplasia

4595 Pathology
Royal College Street, London, NW1 OTU
Tel: 01-387-2898.

1 Appleby EC - BSc,PhD,MRCVS pathology, especially adrenal glands; neoplasia
2 Humphreys DJ - BSc,PhD,CChem,FRSC toxicology, especially heavy metals and selenium

4596 Microbiology and Parasitology
Royal College Street, London, NW1 OTU
Tel: 01-387-2898.

1 Donnelly HT - BSc,PhD laboratory animal science
2 Eddington N - BVSc,PhD,MRCVS virology
3 Fox MT - PhD,MRCVS parasitology
4 Jacobs DE - BVMS,PhD,MRCVS parasitology
5 Lida J - BVM&S,MRCVS molecular biology
6 Lloyd DH - PhD,MRCVS bacteriology
7 Mackenzie NM - BSc,PhD immunology
8 Rozengurt Mrs N - MSc,MRCVS *Laboratory Animal Science*
9 Russell PH - BVSc,PhD,MRCVS virology
10 Smith IM - MSc,PhD,MRCVS bacteriology

4597 Physiology, Biochemistry and Pharmacology
Royal College Street, London, NW1 OTU
Tel: 01-387-2898.

1 Jukes MGM - BSc,MA,DM *Professor and Head of Department* central nervous and respiratory physiology
2 Dawes CM - BSc,PhD respiration and acid base regulation in embryonic and adult birds
3 Lees P - BPharm,PhD inflammatory processes in the horse
4 Lodge D - BVSc,PhD,MRCVS,DVA interactions of anaesthetic and neurotransmitters
5 Mayes PA - PhD,DSc,MIBiol,CChem,FRSC carbohydrate and lipid metabolism
6 Plummer Miss JM - BSc,PhD fine structural studies of spermatozoa
7 Sedgwick AD - BSc,PhD basic mechanisms of inflammation
8 Tyler DD - MA,PhD biochemistry
9 Watson PF - BVetMed,BSc,PhD,MRCVS structure and function of the spermatozoa

4598 Surgery and Obstetrics
Hawkshead Campus, Hawkshead Lane, North Mymms, Hatfield, Herts.
Tel: (0707) 55486.

1 Vaughan LC - DSc,FRCVS,DVR *Professor of Veterinary Surgery and Head of Department* orthopaedic surgery in small and large animals
2 Allen WE - MVB,PhD,MRCVS animal reproduction
3 Bedford PGC - PhD,MRCVS opthalmology, ENT surgery
4 Clarke Miss KW - MA,VetMB,MRCVS,DVA anaesthesiology
5 Clayton-Jones DG - BVetMed,MRCVS,DVR general and orthopaedic surgery
6 Davies JV - BVetMed,PhD,MRCVS
7 Edwards GB - BVSc,MRCVS general and gastro-intestinal surgery in large animals
8 Frost RC - FRCVS general surgery, urology, orthopaedics in small animals
9 Gerring EEL - BVetMed,PhD,MRCVS the role of ischaemia in equine colic
10 Noakes DE - BVetMed,PhD,MRCVS animal reproduction

4599 University College London

Gower Street, London, WC1E 6BT
Tel: 01-387-7050.

4600 Botany and Microbiology

1 Stewart GR - PhD *Quain Professor of Botany and Head of Department* physiology og plant stress
2 Evans MCW - PhD *Professor of Plant Chemistry* photosynthesis
3 Goldsmith FB - PhD *Lecturer* conservation
4 Hyams JS - PhD *Lecturer* cytoskeleton of yeast; eukaryotic motility
5 Nugent JHA - PhD *Lecturer* photsynthesis and plant biochemistry
6 Phillips R - PhD *Lecturer* plant cell culture
7 Richards H - PhD *Lecturer* genetics and molecular biology of plasmids
8 Roscoe DH - PhD *Lecturer* mammalian cell cycle
9 Rowbury RJ - DPhil *Reader in Microbiology* plasmids and pathogenicity
10 Smith DG - PhD *Senior Lecturer* bacterial behaviour; immobilised bacteria
11 Strange RN - PhD *Lecturer* plant pathology
12 Taylor K - PhD *Senior Lecturer* plant ecology
13 Tristram H - BSc *Senior Lecturer* bacterial membranes

4601 Wye College

Wye, Ashford, Kent, TN25 5AH
Tel: (0233) 812401. Telex: 96118 ANZEEC G.

1 Prescott Prof. JHD - BSc,PhD,FIBiol *Principal*

4602 Agriculture

1 Lever JD *Professor of Agriculture* nutrition of ruminants; land use in food production
2 Campling RC - MSc,PhD *Reader in Ruminant Nutrition and Production*
3 Chatham CJ - BSc,PhD agronomy and crop physiology
4 Curran MK - BSc,PhD *Reader in animal breeding and production*
5 Kempson RE - BSc,MSc,PhD design and analysis of experiments; multivariate analysis
6 Lean IJ - BSc,PhD animal production and behaviour
7 Scarisbrick DH - BSc,MA,PhD crop physiology and agronomy
8 Warboys IB - MSc tillage studies, crop processing
9 Wilkes JM - BSc,MPhil tractor and implement performance, tillage studies

4603 Biochemistry, Physiology and Soil Science

1 Clark NG - BSc,MA,PhD,CChem,FRSC *Senior Lecturer* synthesis and mode of action of organic herbicides
2 Davis RH - BSc,PhD *Lecturer* cyanide formation, metabolism and detoxication in animals and plants
3 Dodds PF - BSc,PhD *Lecturer* biosynthesis of natural and "xenobiotic" glycerolipids
4 Garraway JL - BSc,PhD,AKC,CChem,FRSC *Senior Lecturer* biodegradation of pesticides; studies on *Orobanche crenata*
5 Hodges RRDH - BSc,PhD *Senior Lecturer* ultrastructure and function of the earth digestive tract
6 Ingram Miss JMA - BSc,MSc *Lecturer* methods of stabilizing soil structure
7 Kennedy CD - BSc,PhD *Lecturer* physico-chemical aspects of plant membranes; phytotoxic action of metal ions
8 Moore JH - MSc,PhD,DSc,CChem,FRSC *Professor of Biochemistry in Relation to Agriculture* lipid metabolism
9 Price R - BSc,PhD *Lecturer* aspects of plant carbohydrate metabolism
10 Richards SA - BSc,MSc,PhD *Lecturer* history of physiology and agricultural science
11 Scofield AM - BSc,MSc,PhD *Lecturer* effects of helminth infection on intestinal absorption in rodents
12 Sykes AH - MA,PhD,DipAgric *Reader in Animal Physiology* climatic physiology of poultry
13 Weston EW - MSc,ARCS *Lecturer* chemical changes in plant tissue induced by plant growth regulators

4604 Biological Sciences

1 Baker DA - PhD,FIBiol *Professor of Agricultural Botany* loading and long distance transport of photoassimilates; ion uptake and transport across the root
2 Bailiss KW - PhD legume viruses (particularly lucerne and faba bean); seed transmission; physiology of virus-infected plants; aphid-virus-host interactions
3 Chapman GP - PhD chromosome ultrastructure; new techniques for inducing genetic change
4 Copland MJW - PhD parasitic Hymenoptera structure and physiology: biological control of glasshouse and tropical pests
5 Hill TA - PhD endogenous plant growth substances; ecophysiology of weed species
6 Hodgson CJ - PhD plant host-virus-vector relationships; plant resistance to insect pests in cowpeas and rape
7 Lopez-Real JM - PhD composting of agro-industrial organic wastes; microbiology of biofilters for odour control
8 Madge DS - PhD,DSc,FIBiol *Reader in Applied Zoology* biology of soil invertebrate animals; their importance in grassland and woodland
9 Mansfield JW - PhD mechanisms of disease resistance and microbial pathogenicity in plants
10 Peat WE - PhD physiology and genetics of crop yields

4605 Centre for European Agricultural Studies
Wye College, Ashford, Kent, TN25 5AH
Tel: (0233) 812181. Telex: 96118 ANZEEC G.

Research and consultancy in agricultural economics and marketing

1 Reid IG - BScEcon,DipAgr *Director of Centre For European Agricultural Studies* European agricultural finance and taxation, land ownership and tenure; EC policy
2 Mitchell MK - BSc,MSc *Executive Director, CEAS Consultants* agricultural policy analysis, food security, systems simulation, agricultural trade policy, food aid
3 Young NA - BSc,MSc *Executive Director, CEAS Consultants* European agricultural policy, commodity marketing, agricultural marketing systems

4606 Environmental Studies And Countryside Planning

1 Burnham CP - MA,MSc,PhD *Acting Head of Department* pedology and land evaluation
2 Anderson Miss MA - BSc,PhD environmental policy and planning
3 Edwards Miss A - BSc land use policy
4 Green BH - BSc,PhD *ecology and conservation*
5 McRae SG - PhD land resources and reclamation
6 Potter CA - BSc,PhD agriculture and the environment
7 Redclift MR - BA,DPhil agrarian social development in Latin America
8 Rogers AW - BA,MPhil,PhD rural planning and housing

4607 Hop Research

1 Gunn RE - BSc *Head of Department* plant breeding
2 Darby P - BSc,PhD plant breeding and pathology
3 Green CP - BSc,PhD organic chemistry

4608 Horticulture

1 Moorby J - BSc,PhD *Professor of Horticulture* aspects of crop physiology
2 Blake J - BSc,PhD plant tissue culture in vegetative propagation
3 Branton RL - BSc,PhD tissue culture propagation of palms
4 Buckley GP - BSc,MSc,PhD re-vegetation of industrial spoil and tree establishment
5 Burrage SW - BSc,PhD monitoring and control of glasshouse environments
6 Dodd PB - BSc pomology; physiology of perennial fruit crops and propagation
7 Fordham R - BSc,PhD vegetable physiology
8 Goldwin GK - BSc,PhD improving fruit yields with plant hormones
9 Mantell SH - BSc,MSc,DIC,DTA,PhD application of micropropagation techniques
10 Wright SE - BSc landscape survey and management; vegetation establishment
11 Wright TWJ - BSc,ALI aspects of environmental horticulture and landscape management

4609 Agricultural Economics

4610 Agrarian Development Unit

1 Carruthers ID - BScHort,PhD,DipAgrEcon *Professor of Agrarian Development* planning and management of rural water supplies and irrigation
2 Allanson G - MA,MSc development of livestock industries in UK and overseas
3 Bernstein H - MA,MSc agrarian social structure and change, with special reference to sub-Saharan Africa
4 Kydd JG - BA,MSc,DPhil,PGCE agricultural development planning and policy, especially in sub-Saharan Africa
5 Smith LED - BSc,MSc water management for agriculture in Africa, applications for micro-computers

4611 Agricultural Policy Unit

1 Buckwell AE - BSc,MAEcon *Professor of Agricultural Economics, Head of Department and Policy Unit* structural and technical change in agriculture and the Common Agricultural Policy
2 Burrell Miss A - BA,BScEcon,MScEcon *Lecturer* econometrics applied to agriculture
3 Gasson R - BSc,MSc,MA,PhD *Senior Research Associate* part-time farming, women in agriculture, agriculture and the environment
4 Hill NWFB - BSc,PhD *Lecturer* wealth and income in agriculture; economics of size and tenure
5 Medland JR - BSc *Lecturer* comparative agricultural efficiency in European countries; agricultural policy
6 Ray D - BScEcon,MPhil *Lecturer* economics and marketing of horticultural products

4612 Farm Business Unit

1 Nix Professor JS - BScEcon,MA,CBIM,FRSA,FRAgS *Head of Unit* farm business management and profitability
2 Hill GP - BSc,MSc farm finance, accounting and taxation
3 Nicholson JAH - BScHort,MSc,FBIM management systems and development of new technology
4 Potter Dr CA - BSc,PhD,DipEcon agriculture and the environment
5 Webster Dr JPG - BScAgrSc,DipFBA,PhD,FBIM decision models in agriculture; micro-computers
6 Williams NT - BSc,MAEcon,MBIM micro-computers, natural resource economics

4613 Marketing Unit

1 Hunt AR - BSc,MSc marketing of flowers; the import trade in fresh fruit, vegetables and flowers into N.W. European trade markets; the marketing of horticultural produce from the Channel Islands
2 Newbound P - MA,MIN evolution and operation of input and food marketing systems; agricultural input marketing

4614 Manchester University

Manchester, M13 9PL
Tel: 061 273 3333.

4615 Agricultural Economics

Tel: 061 273 7121.

1 Colman DR - BSc,MS,PhD *Professor and Head of Department* agricultural development, price and supply analysis, agricultural policy
2 Burton MP - BSc,MA *Lecturer* econometric modelling, environmental economics policy
3 Russell NP - MAgrSc,PhD *Lecturer* production economics, management systems

4 Sandiford F - BA,MA,PhD *Research Projects Officer* fisheries and environmental economics agricultural policy
5 Strak J - BSc,MA,PhD international trade marketing
6 Thirtle C - BA,MA,MPhil,PhD *Lecturer* economics of technology and production; agricultural development
7 Young T - BA,MAEcon,PhD *Lecturer* cereals policy, demand analysis, community models

4616 Biochemistry

Tel: 061 273 8241.

1 Grant ME - BScTech,DPhil *Professor* connective tissue biochemistry
2 Anderson JC - MA,PhD *Senior Lecturer* components of connective tissue in health and disease
3 Ayad SR - PhD,DSc *Reader* isolation and characterization of growth factors from mammalian tissues
4 Bardsley WG - PhD,DSc *Reader* enzyme kinetics and statistical analysis of experimental data
5 Bray CM - Bsc,PhD *Lecturer* isolation and characterisation of plant promoter sequences; molecular studies on seed priming; molecular studies on seed viability and vigour
6 Brown BS - BSc,PhD *Lecturer* microbial conversion of waste plastics to protein
7 Burdett K - BSC,PhD *Senior Lecturer* control of dietary carbohydrate adsorption from the gut
8 Dickson AJ - BSc,PhD *Lecturer* control of growth and development of domestic fowls
9 Elliott KRF - MA,PhD *Lecturer* regulation of metabolism in mammalian systems
10 Evans EM - BSc,PhD *Lecturer* integration and control of energy-producing pathways
11 Gregory RPFG - MA,PhD *Senior Lecturer* photosynthesis; organisation of chlorophyll complexes especially photosystem II
12 Itzhaki S - PhD *Lecturer* biochemistry of plant non-protein amino acids
13 Jones MN - PhD,DSc *Reader* studies on the physical properties of humic substances
14 Lowe AG - BSc,PhD *Lecturer* glucose transport in red blood cells
15 Milsom DW - BSc,PhD *Lecturer* control of digestive enzymes in intestinal brush border
16 Robson B - PhD,DSc *Reader* design of synthetic peptide vaccines, neuropeptides and hormones
17 Rosamond J - MA,DPhil *Lecturer* isolation of plant gene regulatory elements
18 Rowsell KV - BSc,PhD *Lecturer* skeletal muscle - amino acid metabolism
19 Shuttleworth CA - BSc,PhD *Senior Lecturer* dentinogenesis and the extracellular matrix
20 Steven FS - PhD,DSc *Reader* probes for early changes in carcinogen treated animals
21 Thompson JS - BSc,PhD *Lecturer* protein engineering of food industry enzymes
22 Tomalin G - BSc,PhD *Lecturer* synthesis of biologically active peptides and peptide analogues
23 Weiss JB - DSc *Reader* new blood vessel formation; connective tissue degrading enzymes
24 Wynn CH - BSc,PhD *Senior Lecturer* molybdenum containing enzymes

4617 Botany

1 Cutter Miss EG - PhD,DSc *Professor of Botany* development morphology
2 Benton RA - BSc,PhD *Lecturer* applied ecology
3 Butler RD - BSc,PhD *Reader* ultrastructure of plant and animal tissues
4 Callow RS - BSc,PhD *Lecturer* cytogenetics
5 Charlton WA - MSc,PhD *Lecturer* developmental studies
6 Earnshaw MJ - BSc,PhD *Lecturer* ion transport photosynthesis
7 Emes MJ - BSc,PhD *Lecturer* nitrogen assimilation, photosynthesis
8 Epton HAS - BSc,PhD,DIC *Senior Lecturer* biochemistry of disease resistance
9 Frost RR - BSc,PhD *Senior Lecturer* virus diseases of economic plants especially roses and celery
10 Hartshorne JN - BSc,PhD *Lecturer* genetics, sex ratio in dioecious plants
11 Lee JA - BSc,PhD *Senior Lecturer* physiological ecology
12 Moore D - BSc,PhD *Senior Lecturer* biochemistry and genetics of fungi
13 Newton P - BSc,PhD *Lecturer* growth analysis of crop plants
14 Peckett RC - BSc,PhD *Senior Lecturer* flower pigmentation, phenolic compounds
15 Shefield Miss E - BSc,PhD *Lecturer* phase changes in ferns
16 Sigee DC - BSc,PhD *Senior Lecturer* cytology and ultrastructure
17 Southern DI - BSc,PhD *Senior Lecturer* cytogenesis of Orthopteran and Diptera insects
18 Tallis JH - BSc,PhD *Senior Lecturer* pollen analysis and ecology of peatlands
19 Trinci APJ - PhD,DSc *Professor of Cryptogamic Botany* fungal physiology
20 Watson Miss J - BSc,PhD *Lecturer* palaeobotany
21 Wood RJ - BSc,PhD *Reader* genetics of tsetse flies

4618 Zoology

Tel: 061 273 3333.

1 Guthrie DM - BSc,PhD *Professor*

2 Askew RR - BSc,DPhil *Reader in Entomology Hymenoptera* biological control

3 Bailey SER - BSc,PhD *Lecturer* ecology and behaviour of land slugs and snails

4 Baker RR - BSc,PhD *Lecturer* evolutionary ecology of migration, navigation and courtship

5 Balment RJ - BSc,PhD *Lecturer* hormonal control of various vertebrate physiological systems

6 Butler RD - BSc,PhD *Reader in Cytology* ultrastructural effects of ionising radiation, ultrastrycture of protozoa

7 Cook LM - BSc,DPhil *Reader* ecology and genetics of populations

8 Dalingwater JE - BSc,PhD *Lecturer* structure and composition of fossil and extant arthropod cuticles

9 Evans MEG - BSc,PhD *Senior Lecturer* entomology of *Coleoptera*

10 Gabbutt PD - BSc,PhD *Senior Lecturer* biology and taxonomy of pseudoscorpions

11 Hartshorne JN - BSc,PhD *Lecturer in Genetics Drosophila* genetics

12 Harwood DW - MA *Lecturer* comparative studies of learning and cognition

13 Hawkins SJ - BSc,PhD *Lecturer* rocky shore community ecology and biology of intertidal prosobrachs and barnacles

14 Kennaugh JH - MSc,PhD *Senior Lecturer* cuticle studies on *Hypodermatidea*

15 Marriott KRM - BSc,PhD *Lecturer* crustacean physiology

16 McCrohan CR - BA,PhD *Lecturer* neural control of behaviour in invertebrates

17 Moore D - BSc,PhD *Senior Lecturer* biochemical aspects of morphogenesis and gene action

18 Sigee DC - MA,PhD *Lecturer in Cytology* histochemistry and autoradiography

19 Southern DE - BSc,PhD *Senior Lecturer in Cytology* tsetse fly control

20 Wood RJ - BSc,PhD *Reader in Genetics* insect genetics

21 Yalden DW - BSc,PhD *Senior Lecturer* vertebrate ecology and anatomy

4619 University of Newcastle upon Tyne

Newcastle Upon Tyne, NE1 7RU
Tel: (0632) 28511.

4620 Department of Agriculture

1 Dickson GR - BSc,PhD *Professor and Head of Department* farm management

2 Ellis M - BSc,PhD *Lecturer* pig production

3 Evans EJ - BSc,PhD *Lecturer* crop agronomy

4 MacRae DJR - BSc *Lecturer* farm management

5 Rowlinson P - BSc,PhD *Lecturer* ruminant livestock

6 Thomas RJ - BVSc,MSc,PhD,MRCVS *Senior Lecturer* parasitology

7 Webster GM - BSc,PhD *Lecturer* sheep production

8 Wilcockson SJ - BSc,PhD *Lecturer* crop agronomy

9 Willis MB - BSc,PhD *Senior Lecturer* animal breeding and genetics

10 Younger A - BSc,PhD *Lecturer* grassland agronomy

4621 Department of Agricultural Biochemistry and Nutrition

1 Armstrong DG - MSc,PhD,DSc,CChemFRS,RIBiol,FRSE *Professor and Head of Department* digestion and metabolism in ruminants

2 Finlayson HJ - BSc,PhD *Lecturer* biochemical microbiology of the rumen, protection of proteins

3 Gilbert HJ - BSc,PhD *Lecturer* recombinant DNA technology with reference to silage inoculants and rumen bacteria

4 Mathers JC - BSc,PhD,DipNutr *Lecturer* role of plant structural carbohydrates in human nutrition

5 Parker DS - BSC,PhD,MNutrS *Lecturer* metabolism of intestinal wall; uptake of nutrients via mesenteric and portal veins in the sheep

6 Rooke JA - BA,PhD *Research Associate* digestion in ruminants with particular reference to nitrogen metabolism

7 Smithard RR - MSc *Lecturer* studies on oil-seed rape; partition of nutrients between mammalian tissues

8 Weekes TEC - BSc,PhD *Senior Lecturer* endocrinology and enzymology of domestic livestock

4622 Department of Agricultural Biology

1 Cain RB - BSc,PhD,DSc,FLS *Professor and Head of Department* biodegradation of recalcitrant and xenobiotic compounds in the environment and the evolutionary aspects of selection pressure by pollution stress; biochemical versatility of soil actinomycetes; exchange of genetic material between micro-organisms under environmental conditions

2 Blackburn F - BSc,MS *Senior Lecturer* physiology of host-parasite interactions; biological nitrogen transformations

3 Kershaw WJS - BSc,PhD *Lecturer* host selection in insects and slugs, pests of brassicas, biology and control of stored product pests

4 Luff ML - BSc,PhD,ARCS,DIC *Senior Lecturer* biology and population dynamics of insects, especially ground beetles; grassland invertebrate ecology

5 Moore K - BSc,fil lic *Lecturer* cytogenesis and breeding of dwarf wheat, potato, tetraploid rye and grasses

6 Ollerenshaw JH - BSc,PhD *Senior Lecturer* ecophysiology of grasses and clovers in relation to hill land production; plant/environment interactions of maize, oilseed rape in northern Britain

7 Pearce RS - BSc,MSc,PhD *Lecturer* biochemical, physiological and ultrastructural studies of plant adaptation of growth at low temperature and to frost: studies of isolated higher plant protoplasts

8 Port GR - BA,PhD,DIC *Lecturer* ecology and behaviour of herbivorous and haematophagous insects and slugs

9 Selman BJ - BSc,PhD *Senior Lecturer* taxonomy and biology of Coleoptera; biological control of insect pests; natural slug parasites

10 Stewart WS - BSc,PhD *Lecturer* low temperature physiology of grasses in upland farming; plant/environment relationships

11 Wilkins RM - BSc,MSc,PhD,MIBiol *Lecturer* use and biochemistry of pest control agents; ecological aspects of pesticides; controlled release formulations; tropical crop protection, especially rice; forest weed control; biomass utilization

4623 Department of Agricultural Economics

1 Ashton JCBE - MA,BLitt,MS *Professor and Head of Department* agricultural policy; international trade and development

2 Dawson PJ - MAEcon,PhD *Lecturer* production economics; resource allocation; labour

3 Ennew CT - BA,PhD *Lecturer (Temporary)* market organisation; agricultural policy; soviet agriculture

4 Hubbard LJ - BSc,MApplSc *Lecturer* C.A.P.; input-output analysis; rural employment

5 Lingard J - BSc,MAEcon *Lecturer* production economics, water resources; overseas development

6 Topham MR - BSc,MSc,PhD *Lecturer* farm management; on-farm computers; rural planning

7 Weightman PWH - BSc,MSc,PhD *Lecturer* farm business; estate and land management

8 Whitby MC - *Reader in Rural Resource Development resource economics; agricultural policy; overseas development*

9 White B - BSc *Demonstrator (Temporary)* farm management; statistical forecasting; potatoes

4624 Department of Agricultural Engineering

1 O'Callaghan JR - MSc *Professor and Head of Department* utilization of agricultural wastes; systems modelling

2 Copland TA - BSc *Lecturer* development of harvesting machinery

3 Gowing JW - MSc *Lecturer* irrigation scheduling and management

4 Hettiaratchi DRP - PhD *Senior Lecturer* soil mechanics; cultivations; mechanics of root growth

5 McCarthy TT - MEngSci *Lecturer* control systems; microprocessor applications

6 Seig DA - PhD *Lecturer* development of cultivating machinery

7 Wills BMD - PhD *Senior Lecturer* design of agricultural machinery; small tractors; implement impact reactions

8 Woods JL - PhD *Lecturer* drying of cereals; cooling of vegetables; heat exchange

4625 Department of Agricultural & Food Marketing

1 Ritson C - BA,MAgrSc *Professor, Head of Department* structure and efficiency in food marketing; European agricultural and food policy

2 Gofton LR - BA,PhD *Lecturer* behavioural science in marketing

3 Ness MR - BAEcon,MAEcon,PhD *Lecturer* time series analysis; econometric modelling; multivariate analysis; managerial economics

4 Oughton EA - BSc,MSc,MLitt *Lecturer* food policy in Britain and Europe; agricultural and food marketing and food policy in developing countries

5 Price DC - BScAgrAgrEcon,AIB,MBIM *Lecturer* financial analysis for agricultural co-operatives

6 Warren RM - BSc *Lecturer* commodity marketing; agricultural and food marketing policy in developing and developed countries

7 Weeks WGR - BSc *Lecturer* processed foods

4626 Department of Soil Science

1 Syers JK - BSc,PhD,DSc,FRSC,FRSNZ,FNZIAS,FNZIC *Professor* soil and fertilizer chemistry

2 Adey MA - MA,PhD *Lecturer* soil physical properties and processes

3 Askew GP - BScAgr,BSc *Senior Lecturer* pedogenesis and soil vegetation relationships

4 Jones MA - BSc *Senior Experimental Assistant* micromorphology of soils

5 King JA - BSC,PhD *Research Associate* land reclamation

6 Montgomery RF - BSc *Senior Lecturer* land use in humid tropics; pedogenesis

7 O'Donnell AG - BSc,PhD *Lecturer* soil microbiology and metabolism

8 Rimmer DL - Bsc,PhD *Lecturer* physicochemical properties of soil

9 Rose DA - BSc,PhD,CPhys,FInstP *Lecturer* soil-plant relations; mathematical modelling

10 Shiel RS - BA,BSc,PhD,FSAScot *Lecturer* land use and soil management

11 Smith RS - BSc,MSc,PhD *Lecturer* wildlife conservation; upland grassland

4627 University College of North Wales

Bangor, Gwynedd, LL57 2DG
Tel: (All Depts.) 0248 351151.

4628 Agriculture Department

1 Owen JB - BSc,MA,PhD,FIBiol *Professor and Head of Department* sheep and cattle breeding and management
2 Alcock MB - BSc,PhD agronomy; grassland management; grazing ecology; agroforestry
3 Bright GA - BSc,MPhil farm management and investment and project appraisal
4 Chamberlain AG - BSc,PhD animal nutrition, pig husbandry, sheep breeding and reproduction
5 Howarth RW - MA,DipEcon agricultural policy
6 Phillips CJC - BSc,PhD cattle nutrition and behaviour
7 Thomas TH - BScEcon,MScEcon *farm management; agricultural marketing; land use evaluation*
8 Wright D - BSc,PhD cereal and root crop physiology and husbandry

4629 Animal Biology

1 Cherrett JM - BSc,PhD,FRES root feeding by pests; biology of tropical leaf-cutting ants
2 Ford JB - BSc,PhD,FRES behaviour of predatory mites and their use in biological control; biology of glasshouse schiarid flies
3 Gatehouse AG - BA,PhD,DIC,FRES migration patterns of the armyworm *Spodoptera*; sublethal effects of insecticides on behaviour of Tsetse flies and other insects
4 Herbert IV - BS,PhD,MS immune responses of animals to parasitic infections; epidemiology of farm livestock infection by animal parasites
5 Lehane MJ - BSC,PhD insect parasites and vectors of importance in tropical medicine; biology and control of the stable fly, *Stomoxys*
6 Morgan Mrs MJ - MSc,FRES taxonomy and distribution of insects in Wales
7 Probert AJ - BSc,PhD physiology of helminth parasites of livestock; mode of action of anti-parasitic drugs; epidemiology of schistosomiasis

4630 Biochemistry & Soil Science

1 Payne JW - BSc,PhD,DSc *Professor and Head of Department* evaluation of nutritional availability of amino acids in foods; membrane transport processes in germinating seeds
2 Axford RFE - MA,VetMB,MRCVS metabolism and nitrogen in the rumen
3 Davies JI - BSc,PhD the role of adipose tissue in the response of domestic animals to changes in their nutritional status
4 Gaunt JK - BSc,PhD biochemical behaviour of phenoxyacetate herbicides on plants of varying sensitivity; regulation of plant growth and development
5 Jenkins DA - BA,PhD chemical and mineralogical analysis of local and arid zone soils
6 Johnson DB - BSc,PhD microbiological changes in soils stored during opencast mining operations; side effects of pesticides on soil microorganisms
7 Kelso WI - Bsc,BAgr microbial activity in field drainage operations; micronutrient and phosphate behaviour in calcareous soils
8 Laidman DL - BSC,PhD biochemical processes during cereal germination
9 Tomos AD - MA,PhD plant water relations at the single cell level
10 Wyn Jones RG - BSc,MSc,DPhil,DSc,FRSC,FIBiol *Professor* plant nutrition; salt and drought tolerance in wheat and related grasses; turgor and osmotic relations in plant cells

4631 Forestry and Wood Science

1 Roche L - MA,MF,PhD *Professor & Head of Department* farm-forestry integration; the role of food and fodder trees in farming systems in the tropics; in situ conservation and management of forest genetic resources
2 Banks WB - DIC,PhD degradation and protection of wood surfaces with special reference to physio-chemical degradation factors; chemical modification of wood with respect to water repellency, surface finishes and adhesion
3 Bolton AJ - BSc,MS,PhD heat and fluid transfer, and rheology, in wood and wood based products during and after processing
4 Cahalan C - BSC,MS,PhD tree genetics and improvement; natural regeneration of Sitka spruce (and other species) in Britain; biomass and energy production from woody species
5 Cooper RJ - BScEcon marketing and economic research for wood processing industries
6 Denne MP - BSc,PhD tree growth and development; environmental control of wood production and structure; urban forestry
7 Hale M - BSc,PhD novel preservative systems for wood protection; micromorphology of wood decay and enzymatic mechanisms of wood decay
8 Hall JB - BSc,PhD tropical agroforestry; tropical catchment forest ecology and conservation; autecology of tropical trees
9 Harding DM - BSc,PhD watershed management; land use and hydrologic processes in mesoscale catchments, with particular reference to river flow variations and sedimentation
10 Hetherington JC - BSc,PhD,FICF site studies; modelling of tree and stand growth; general problems of forest measurement and management

11 Irle MA - BSc adhesion theory and adhesive technology; performance and manufacture of wood based panel products
12 Price C - MA,DPhil natural resource economics; costing of resources; recreation and landscape evaluation; decision-making in land use allocation and forest management; cost-benefit analysis; discount rates, distributional issues, shadow pricing, ripple effects

4632 Plant Biology

1 Sagar GR - MA,DPhil *Professor and Head of School* herbicides; whole plant physiology, experimental horticulture; biology of weeds
2 Bell AD - BSc,PhD plant architecture - branching patterns of rhizomatous plants and their simulation by computer; 3-D plant anatomy, especially of vascular systems
3 Farrar JF - MA,DPhil carbon in plant growth: intracellular compartmentation, assimilate partitioning and the control of dark respiration rate; photosynthesis: physiology of plants infected with biotrophic fungi; physiological ecology
4 Gliddon CJ - BA,DPhil the effects of breeding system on the levels, distribution of genetic variation and neighbourhood size in natural populations; surveys of isoenzyme variation by gel electrophoresis together with mathematical modelling of the system; co-evolution of plant species; plant competition and mixed cropping systems
5 Harper Em. Professor JL - MA,DPhil,FRS plant demography; role of predation in population dynamics of plants; demography of leaves - birth and death rates; resource allocation and reproductive strategies; ecological significance of sex; consequences of interactions between plants
6 Marshall C - BSC,PhD physiology of growth in grasses and cereals - leaf growth, tiller and stolon interrelationships, photosynthesis, translocation of assimilates and minerals, seed production; plant growth regulators
7 Oxley ERB - BSc grassland and plant community ecology; computer simulation of plant systems. plant competition; remote sensing and surveying of vegetation
8 Shattock RC - MSc,PhD,DIC,NDA,MRAC fungal plant pathogens; disease control; inheritance of virulence and fungicide resistance in *Phytophtora infestans*
9 Shaw DS - BSC,PhD genetical analysis of *Phytophthora* spp.; fluorimetric DNA analysis of fungal nuclei
10 Smith AJE - MA,DPhil,DSc taxonomy and biosystematics of bryophytes; bryophyte cytology distribution of bryophytes in Britain; preparation of a liver-wort flora of Great Britain and Ireland
11 Whitbread R - MA,PhD physiology of host-parasite relations, especially of biotrophic associations; plant growth-promoting rhizosphere bacteria; bacterial plant pathogens
12 Wilson JM - BSc,MSc,PhD the physiology of plants at chilling and freezing temperatures; mechanisms of acclimation to low temperatures; changes in photosynthesis, respiration, water relations, and membrane composition and functioning during low temperature stress; chlorophyll fluorescence analysis as a screening test for cold sensitivity
13 Wood CM - BSc,PhD studies of algal ecology, past and present, in Snowdonia and Anglesey lakes; physiological ecology of micro-algae, particularly chlamydomonads and picoplankton; size in diatoms; algal ecology of Australian rice-fields

4633 Centre For Arid Zone Studies

The expertise shown below refers to research activities not detailed under other Departments, but see also Agriculture, Animal Biology, Biochemistry & Soil Science, Forestry & Wood Science and Plant Biology, since many members of the Centre are also members of other Departments

1 Wyn Jones RG - BSC,MSc,DPhil,FIBiol *Professor & Director of Centre* biosaline agriculture
2 Ellis C wide range of horticultural services including landscape rehabilitation
3 Harding DM - BSc,PhD hydrology, with references to run off processes; protected areas - recreation landscape and nature conservation
4 Jenkins DA - BA,PhD soil geochemistry; micromorphology and pedogenesis
5 Kelso WI - BSc,BAgr irrigation and field drainage systems
6 Lehane MJ - BSc,PhD primary health care schemes and parasitological disease
7 MacKay RR - MA economic assessment: regional labour economics
8 Marshall C - BSc,PhD plant growth in arid environment
9 Milner C - BSc,PhD *Inst.Terr. Ecology* environment impact of agricultural practices; desertification
10 Whitbread R - MA,PhD crop protection
11 Wood C - BSc,PhD algal problems in relation to irrigated crops

4634 University of Nottingham

University Park, Nottingham, NG7 2RD
Tel: (0602) 506101. Telex: 37346 UNINOT G.

4635 Department of Biochemistry

1 White DA - BSc,PhD *Senior Lecturer* glycoproteins in the synthesis and secretion of milk products

4636 Department of Botany

1 Cocking Professor ECD - PhD,DSc,FRS *Head of Department* plant genetic manipulations
2 Briatry LG - BSc,PhD *Lecturer* ultrastructure, physiology and biochemistry of seeds and seed reserves; spaceflight effects on growth and development; stereology
3 Crittenden PD - BSc,PhD *Lecturer* nitrogen fixation; ecology of subarctic heaths and forests
4 Davey MR - BSc,PhD *Research Officer* tissue culture; plant cell transformation and somatic hybridization
5 Lucas JA - BSc,PhD *Lecturer* plant pathology; biochemical aspects of host-pathogen interaction; application of tissue culture; host resistance mechanisms
6 Mulligan BJ - BSc,PhD *Lecturer* photosynthesis genes of cyanobacteria and chloroplasts
7 Peberdy Professor JF - BSc,PhD fungal biochemistry and genetics; vector development for fungal genetic engineering; genetics of fungicide resistance; biochemistry and genetics of fungal cellulase
8 Power JB - BSc,PhD *Lecturer* somatic cell genetics; tissue culture and its application to plant breeding; plant cell totipotency
9 Ricketts TR - BSc,PhD *Senior Lecturer* algae as food for marine farming
10 Rieley J - BSc,PhD *Lecturer* heavy metal contamination of arable land; effect of farm wastes on soil quality; plant productivity and groundwater pollution

4637 Department of Economics

1 Rayner Professor AJ - BA,MA,PhD *Professor of Agricultural Economics* quantitative analysis and agricultural policy
2 Hill BE - BSc,PhD *Senior Lecturer* agricultural policy, especially milk and beef, in Europe
3 Hine RC - BA,MSc *Lecturer* European trade and agricultural policies
4 Ingersent KA - BSc,MS *Senior Lecturer* agricultural policies in developing countries

4638 Department of Genetics

1 Clarke Professor BC - MA,DPhil,FRS *Head of Department* the control of gene action in higher organisms; population genetics of polymorphisms
2 Barnes SR - MA,PhD *Lecturer* the organisation and evolution of repeated DNA sequences in related plant genomes
3 Day TH - BSc,PhD *Lecturer* biochemical and population genetics
4 Lloyd RG - BSc,DPhil *Reader* genetic control of DNA repair and recombination in bacteria
5 Morgan GT - BSc,PhD *Lecturer* control of transcription in eukaryotes
6 Parkin DT - BSc,PhD *Senior Lecturer* population genetics of house sparrows; ecology of Canada Geese in farmland

4639 Department of Pharmacy

1 Bycroft Professor BW - BSc,PhD,FRSC *Head of Department* chemistry and biochemistry of invertebrate glutamate receptors and mycotoxins
2 Dewick PM - BSc,PhD *Lecturer* biosynthesis of natural products; especially phytoalexins

4640 Department of Zoology

1 Wakelin Professor D - PhD,DSc *Head of Department* genetic control of eosinophilia; anti-parasite T-cell clones
2 Behnke JM - BSc,PhD *Lecturer* the immunology of chronic nematode infection and immunology rejection of tapeworms in laboratory rodents; the induction of immunosuppression by nematode parasites
3 De Pomerai DI - BSc,PhD *Lecturer* expression of cellular proto-oncogenes and crystallin genes during tissue development *in vivo* and *in vitro*
4 Dowdall MJ - BSc,PhD *Lecturer* biochemical and pharmacological investigations on presynaptic cholinergic membranes using *Torpedo* as a model system
5 Duce IR - BSc,PhD *Lecturer* cell biology of neurotransmitter receptors and mode of action of insecticides
6 Gilbert FS - BA,PhD *Lecturer* individuals, species and communities; the ecology of aphidophagous syrphidae
7 Lowe KC - BSc,MA,PhD *Lecturer* development and physiological assessment of oxygen-carrying blood substitute materials; foetal and placental physiology
8 Usherwood Professor PNR - BSc,PhD,FIBiol,FRSE pharmacological and physiological studies of glutamate receptors on excitable cells

4641 School of Agriculture

Sutton Bonington, Loughborough, Leics., LE12 5RD
Tel: (0602) 506101.

4642 Department of Agriculture and Horticulture

1 Ivins Professor JD - CBE,BSc,MSc,PhD *Head of Department* grassland and crop agronomy
2 Alderson PG - BSc,PhD *Lecturer* plant propagation; tissue culture
3 Almond JA - BSc,PhD *Lecturer* agronomy
4 Atherton JG - BSc,PhD,DipEd *Lecturer* glasshouse crop physiology
5 Cole DJA - BSc,PhD *Reader* animal production and nutrition
6 Garnsworthy PC - BSc,PhD *Lecturer* beef and milk production
7 Haresign W - BSc,PhD *Lecturer* reproduction in sheep

8 Hebblethwaite PD - BSc,MSc,PhD,NDA *Reader* crop agronomy
9 Kerr HWT - MA,MPhil *Senior Lecturer* capital in agriculture
10 Macaskill RA - BA,MSc *Lecturer* planning and control techniques
11 Seabrook MF - BSc,PhD *Lecturer* labour management and work routine simulation
12 Tilston CH - BSc *Lecturer* agricultural marketing
13 Wilton B - MSc,MA *Lecturer* crop mechanization, drying and processing
14 Wiseman J - BSc,PhD *Lecturer* poultry production and nutrition
15 Wright CJ - BSc,PhD *Lecturer* vegetable and fruit crops

4643 Department of Applied Biochemistry and Food Science

1 Lawrie Professor RA - PhD,DSc,ScD *Head of Department* muscle biochemistry; meat quality; protein identification and recovery
2 Bardsley RG - BSc,PhD *Lecturer* enzymology of the contractile process in muscle; conformational changes in muscle proteins, through degradation and its control
3 Blanshard JMVB - MA *Reader* biophysics of meat; polysaccharides and water biopolymer systems
4 Boorman KN - BSc,PhD *Senior Lecturer* avian biochemistry and nutrition
5 Buttery Professor PJ - BSc,PhD,DSc control and growth of protein deposition in farm animals
6 Harding SE - MA,MSc,PhD *Lecturer* macromolecular interactions in foods
7 Ledward DA - BSC,MSc,PhD *Senior Lecturer* meat science, especially connective tissue proteins and haemoproteins
8 Mitchell JE - BSc,PhD *Lecturer* polysaccharide gels; protein surface; chemistry; food rheology
9 Neale RJ - BSc,PhD *Lecturer* amino acid metabolism in protein deficiency; nutritional value of plant proteins
10 Norton G - BSc,PhD *Senior Lecturer* plant biochemistry; minerals and carbohydrates
11 Stewart GSAB - BSc,PhD *Lecturer* genetics of microorganisms important to the food industry
12 Taylor AJ - BSc,PhD *Lecturer* food biochemistry
13 Tucker GA - BSc,PhD *Lecturer* biochemistry of fruit ripening
14 Waites Professor WM - BSc,PhD physiology and growth of microorganisms important to the food industry

4644 Department of Physiology and Environmental Science

4645 Animal Physiology

1 Lamming Professor GE - OBE,BSc,BScAgr,MSc,PhD,DSc *Head of Department* physiology of reproduction
2 Clarke RW - BSc,PhD *Lecturer* neurophysiology of opioids
3 Crighton DB - BScAgr,MSc,PhD,DSc *Senior Lecturer* endocrinology of reproduction
4 Foxcroft GR - BSc,PhD *Lecturer* reproduction; endocrinology
5 Haynes NB - BSc,PhD *Senior Lecturer* endocrinology of reproduction
6 Mepham TB - BSC,PhD *Senior Lecturer* physiology and biochemistry of lactation

4646 Plant Science

1 Whittington Professor WJ - BSc,PhD *agricultural policy*
2 Black CR - BSc,PhD *Lecturer* water relations and carbon assimilation of crop plants
3 Grierson D - BSc,PhD *Reader* RNA synthesis and gene expression
4 Hering TF - MA,PhD *Senior Lecturer* soil-borne plant diseases
5 Lycett GW - BSc,PhD *Lecturer* expression and manipulation of plant genes
6 Roberts JA - BSc,PhD *Lecturer* hormone cell physiology
7 Rossall S - BA,PhD *Lecturer* plant disease resistance
8 Taylor IB - BSc,PhD *Lecturer* hormones; genetics; plant breeding

4647 Environmental Physics

1 Clark JA - BSc,PhD *Senior Lecturer* microclimatology of crops and animals
2 Colls JJ - BSc,PhD *Lecturer* pollution
3 McArthur AJ - BSc,PhD *Lecturer* thermoregulation and microclimatology of animals

4648 Soil Science

1 McGowan M - BSc,MA,PhD *Senior Lecturer* roots and soil water use of crops
2 Young SD - BSc,PhD *Lecturer* soil acidity and pollution

4649 Agricultural Zoology

1 Ritchie JY - BScAgr *Lecturer* invertebrate ecology

4650 Biometry

1 Back HL - BSC,BScSpec *Senior Lecturer* design and analysis of experiments and surveys
2 Gregson K - BTech,PhD *Lecturer* computing and mathematical applications

4651 University of Oxford

4652 Institute of Agriculture Economics
Dartington House, Little Clarendon Street, Oxford, OX1 2HP
Tel: Oxford 52921.

1 Peters GH - MA,MSc *Director of the Institute* economic theory and agricultural history
2 Fennell R - BA,MA,PhD social and institutional aspects of agriculture; C.A.P.
3 Jones GT - MA *mathematical economics and agricultural policy*
4 Maunder AH - MA,BSc,MS farm organisation; production economics
5 Mollett JA - BSc,MSc,PhD developmental economics
6 Palmer-Jones RW - BA,PhD developmental economics
7 Parson ST - BSc,MPhil economics of agriculture
8 Tyler GJ - MA,MSc,PhD microeconomic theory and labour economics
9 Zanias G - BSc,MSc,DPhil econometrics

4653 Department of Plant Science
Agriculture Building, Parks Road, Oxford, OX1 3PF
Tel: Oxford (0865) 57245.

University Field Station, Wytham Tel: (0865) 47368; Farm: (0865) 4735

1 Smith DC - MA,DPhil,FRS *Deputy Head of Department and Sibthorpian Professor of Rural Economy*
2 Barry JM - MA,DPhil animal biochemistry
3 Beckett PHT - MA,DPhil economics of soil survey; waste disposal; heavy metal pollution
4 Hall R - BSc,PhD tropical crops and grassland; plant competition
5 Loughman BC - MA,PhD *plant biochemistry and physiology*
6 Lucas RL - MBE,MA,PhD plant pathology and microbiology
7 Nye PH - MA,DSc *Reader in Soil Science* solute movement in soil and plant uptake

4654 Oxford Forestry Institute
South Parks Road, Oxford, OX1 3RB
Tel: (0865) 511431. Telex: 83147 FOROX.

1 Burley J - MA,MF,PhD *Director* forest genetics and tree improvement; wood structure
2 Dawkins HC - MBE,MA,DPhil statistical method in biology; recurrent inventory of forests; design and analysis of forest experiments
3 Greaves A - BSc,MSc conservation of tropical genetic resources
4 Hardcastle PD - BSc,MBA tropical rural development and training
5 Hughes JF - MA,BSc,DipEd management and utilisation, forest policy
6 Mabberley DJ - MA,DPhil ecology; tropical botany
7 McCready CC - MA,DPhil *plant physiology*
8 Palmer JRP - MA tropical plantations, agroforestry, rural development and training
9 Pannell CM - MA,DPhil tropical botany and ecology
10 Plumptre RA - MA,MSc wood anatomy and quality of fast grown species in the tropics; wood properties in relation to end uses; appropriate technology
11 Reynolds ERC - BSc,PhD water relations of plants; plant root distribution and water absorption; forest hydrology and water balance studies; soil and water conservation
12 Savill PS - BSc,MSc,PhD management of temperate forest; plantation silviculture; British forestry
13 Stewart PJ - BA,MA timber resources; supply and demand trends; prices; marketing; economic aspects of forest policy; land use and economic development
14 Styles BT - MA,DPhil preparation of botanical monographs of important forest trees, particularly *Meliaceae* and *Coniferae*
15 Thompson FB - BSc vegetation hydrology and water balance studies
16 Watkinson SC - MA,DPhil physiology of wood decay fungi
17 White F - MA taxonomy and ecology of tropical trees and shrubs, mainly from Africa, and particularly from Zambia; revisions of the *Ebenaceae* and *Meliaceae*
18 Whitmore TC - MA,DPhil,DSc tropical forest ecology
19 Wood PJ - MA tropical plantation management and silviculture; agroforestry; arid zone forestry
20 Wright HL - MA,MF programming and analysis in forest inventory and experiment; special problems of volume and yield estimation

4655 Department of Zoology
South Parks Road, Oxford, OX1 3PS
Tel: (0865) 56789.

1 Bennet-Clark HC - MA,PhD biophysics of insects
2 Blake CCF - MA,PhD structure and activity of proteins
3 Brunet PCJ - MA,DPhil insect biochemistry
4 Clarke JR - MA,DPhil reproductive biology
5 Coe MJ - MA,PhD tropical vertebrate ecology
6 Crowther WB - MA,PhD zoogeography
7 Dawkins M - MA,DPhil behavioural priorities and welfare of hens
8 Dawkins R - MA animal behaviour
9 Gardner Professor RL - MA,PhD,FRS reproductive physiology
10 Graham Professor CF - MA,PhD,FRS developmental biology
11 Hamilton Professor WD - MA,PhD,FRS natural selection for sexuality
12 Harvey P - MA,DPhil population biology
13 Iles JF - MA,DPhil neurophysiology of vertebrate locomotion

14 Johnson LN - MA,PhD molecular biology
15 Kearsey SE - MA,PhD cell biology
16 Kemp TS - MA,PhD theoretical palaeontology
17 Krebs JR - MA,DPhil,FRS quantitative and experimental behavioural ecology
18 McFarland DJ - MA,DPhil animal behaviour
19 Miller PL - MA,PhD insect neurobiology
20 Phillips Professor Sir D - MA,PhD,FRS molecular biology of proteins
21 Phillipson J - MA,PhD animal ecology
22 Rees AR - MA,DPhil biochemical and biophysical studies of growth control
23 Rogers DJ - MA,DPhil entomology and parasitology
24 Shotton DM - MA,PhD cell biology
25 Smith Professor DS - MA,PhD cell fine structure and function
26 Southwood Professor, Sir R - MA,DSc,FRS *Head of Department* insect/plant relationships
27 Speight MR - MA,DPhil applied entomology
28 Stuart DI - MA,PhD molecular biology

4656 University of Reading
Earley Gate, Reading, RG6 2AT
Tel: Reading (0734) 875123. Telex: 847813.

4657 Agriculture Department

1 Morris Professor TR - BSc,PhD,DSc *Professor of Animal Production and Head of Department* amino acid interrelationships in chick diets; sexual maturity of domestic fowl; photoperiodism in fowls
2 Roberts Professor EH - BSc,PhD,DSc,FIBiol *Professor of Crop Production* seed storage, seed dormancy; weed seed ecology; photo-thermal control of crop development
3 Spedding Professor CRW - BSc,MSc,PhD,DSc,CBiol,FIBiol,FRASE *Professor of Agricultural Systems, Director of the Centre for Agricultural Strategy and Dean of the Faculty of Agriculture and Food* agricultural systems, construction and use of models in the study of biological processes involved in systems of agricultural production; energy use in agriculture; agricultural strategy
4 Ball Dr SL - BSc,PhD *Research Fellow* pathogenic variation of fungi on host plants
5 Betts RDW - BSc,DTA *Lecturer (On secondment to Overseas Relations Committee)* animal production in the tropics
6 Brooke DWI - BA,MIEE,CEng *Lecturer* software and instrumentation for agriculture
7 Bryant Dr MJ - BSc,PhD *Lecturer* ecology of reproduction in cattle and sheep; protein nutrition of cattle and sheep
8 Carruthers Dr SP - BSc,PhD *Research Fellow* biofuels, energy crops; agroforestry
9 Domaniewski Dr JCJ - BSc,PhD,MIFM *Research Scientist* culture and nutrition of cyprinids; aquatic weed control with grass carp
10 Ellis PR - BSc,MPH,MRCVS *Director of the Veterinary Epidemiology and Economics Research Unit* economics of animal health and productivity; development of livestock information systems and services
11 Ellis Dr RH - BSc,PhD *Lecturer* seed storage; genetic conservation; dormancy and germination
12 Errington Dr AJ - BA,MSc,PhD *Lecturer* agricultural and related employment; management of labour on farms
13 Esslemont Dr RJ - BSc,PhD,NDA,CertEd *Lecturer* development of dairy information systems; modelling of livestock production
14 Gibb JAC - OBE,MA,MSc,CEng,FIAgr,FRAGS,MemASAE *Senior Lecturer* agricultural engineering and mechanisation
15 Giles AK - BSc *Professor of Farm Management, Director of the Farm Management Unit* management and managers in agriculture, historical aspects of agricultural economics in the UK
16 Gowen Dr SR - BSc,MSc,PhD *Research Fellow* crop protection, applied tropical nematology
17 Hague Dr NGM - BSc,PhD,DIC *Reader* control of plant parasitic nematodes; nematicides
18 Harris PM - BSc,MSc,DipAgr *Lecturer* agronomy and physiology of intercropping; physiological basis of yield in potato crop
19 Jones Dr JWG - BSc,PhD *Senior Lecturer* agricultural systems; biological models; fish farming
20 Kingwill RG - BSc,MS,MIBiol *Lecturer* dairy husbandry; management; machine milking; bovine mastitis
21 Morgan KE - BSc,NDAgrE,CEng,FIAgrE,MemASAE *Lecturer* recovery of heat and water from water vapour transpired by plants; cryogenic "work" storage and utilisation; hybrid thermal syphons
22 Murdoch AJ - BSc,PhD *Research Fellow* ecology of seeds of annual weeds; microcomputer based environmental monitoring and control
23 Owen Dr E - BSc,PhD,MIBiol *Lecturer* alkali treatment of grass and straw to improve nutritive value for ruminants; comparative digestion of roughages by goats and sheep; beef production
24 Owen JE - MSc,FIAgrE *Lecturer* livestock environment control
25 Rehman Dr T - BSc,MSc,PhD *Lecturer* multiple criteria decision techniques for farm planning and livestock ration formulation; evaluation of the learning methods for pre-experience farm management students
26 Stansfield JM - BSc,DipAgr,FRAgS,FBIM *Deputy Director of Farm Management Unit* organisation of large scale farms and management and feeding practices for dairy cattle
27 Summerfield Dr RJ - BSc,PhD,CBiol,FIBiol *Reader* physiology of grain legume crops

28 Tayler RS - BSc,DipAgr,NDA,MIBiol *Senior Tutor* grain legumes for British conditions

29 Walsingham Dr JM - BSc,PhD,MIBiol *Research Fellow* small-scale farming; conservation; agricultural systems

4658 Agricultural Botany Department

1 Cartwright Miss P - BSc,PhD plant development and grain yield in cereals

2 Dennett MD - BSc,PhD effects of weather on crop growth; agroclimatology of the tropics

3 Drennan DSH - BSc,PhD weed science; seed physiology; crop physiology

4 Froud-Williams RJ - BSc,PhD weed ecology; population dynamics of annual weeds; agroecology of *Galium aperine*; herbicide resistance

5 John P - BSc,PhD plant membranes and crop metabolism; ethylene biosynthesis

6 Jones JK - BSc,PhD interspecific hybridisation and cytogenetics of crop species

7 Pickersgill Miss B - BSc,PhD evolution of *Capsicum* and other cultivated plants; interspecific hybridisation of legumes; archaeobotany

8 Snaydon RW - BSc,PhD ecological genetics of pasture and weed species; crop-weed competition; grass legume interactions; mixed-cropping

4659 Agricultural Economics and Management

1 Marsh JS - MA,PGDip *Professor and Head of Department* European agricultural policy and Food Industry

2 Ansell DJ - BSc economic principles, policy and economic development

3 Beard NF - BSc,MSc agricultural policy with particular reference to the EEC and the US; international agricultural trade issues; the world grain economy

4 Burns JA - BA,MSc food marketing, economics of the food industries

5 Collins EJT - BA,PhD history of agriculture and economic development

6 Giles Professor AK - BScEcon *Professor of Farm Management & Director of the Farm Management Unit* management and managers in agriculture; historical aspects of agricultural economics in the UK

7 Gill AH - BScEcon economics of horticultural production and marketing

8 Hallam D - BA,MSc,PhD quantitative modelling in agricultural economics

9 Harrison A - MA,MSc macroeconomic analysis of change in farming; farm business structure; land ownership and tenure

10 Harvey DR - BSc,MA,PhD *Professor* agricultural policy, technological change and R & D expenditure, rural policy

11 Stansfield JM - BSc,DipAgr,FRAgS,FBIM *Deputy Director of Farm Management Unit* organisation of large-scale farms and management and feeding practices for dairy cattle

12 Swinbank A - BSc,MA,PhD agricultural policy with particular reference to Europe

13 Upton M - BScAgr,MSc economic development; agricultural production economics; technological change and development

14 Ward SB - BA,PhD history of agriculture and the environmental movement

15 Wiggins SL - MA,MAEcon agricultural planning and management in developing countries; nomadic pastoralism

4660 Applied Statistics

1 Curnow RN - BA,DipMathStat,PhD *Professor and Head of Department* data analysis; quantitative and population genetics; planning of experiments; biological modelling

2 Mead R - DipMathStat *Professor* data analysis; response surfaces; quantitative ecology; intercropping experiments, design and analysis

3 Pike DJ - BSc,MSc,PhD *Lecturer* response surfaces; statistical computing; experimental design

4 Stern RD - BSc,MSc,PhD,HonDSc *Lecturer* analysis of climatic data; statistical computing; crop-weather relationships

5 Woods AJ - MA,BSc,PhD *Lecturer* models of epidemics including estimation problems; survey methods for animal health and production studies

4661 Food Science Department

1 Nursten HE - PhD,DSc,CChem,FRSC,FIFST *Professor of Food Science and Head of Department* food chemistry; the substances giving flavour and colour to food; plant phenolics

2 Ames JM - BSc,PhD,AIFST volatile components of foods; non-enzymic browning

3 Apling EC - BSc,CChem,MChemA,FRSC,FIFST,MRSH cereal science; milling and baking; non-digestible carbohydrate; food legislation

4 Bell AE - MSc,PhD,CChem,MRSC food physics: flow and deformation, gelation and emulsification, and textural characteristics

5 Davey MT - BSc,MSc,PhD dairy cream; cheese flavour

6 Dziedzic SZ - BSc,PhD,AIFST carbohydrate chemistry; minor components, such as saponins, fatty acid sterol esters and phenols

7 Gordon MH - MA,DPhil,CChem,MRSC,AIFST physical and chemical properties of lipids; oxidative rancidity; oxidative enzymes

8 Harrigan WF - BA,BSc,PhD,FRSH,FIFST food microbiology; food poisoning organisms and factors influencing their growth and survival

9 Hudson BJF - MA,DPhil,CChem,FRSC all aspects of lipids; utilisation and functional properties of proteins; toxic components

10 Lakin AL - MSc,PhD,CChem,FRSC,AIFST food chemistry and biochemistry; analysis and investigation of proteins using dye-binding

11 Lewis MJ - MSc,PhD,AIFST physical aspects of processing; membrane processes; utilisation of animal and plant protein wastes

12 Lyster RLJ - BA,PhD protein chemistry; protein-metal interactions

13 Macrae R - MA,DPhil,CChem,MRSC,AIFST instrumental methods of analysis; HPLC; coffee

14 Robinson RK - MA,DPhil food microbiology and hygiene; fermentation processes; dairy products

15 Rothwell J - BSc,PhD,FIFST use of milk and milk products as components of foods; physico-chemical studies on cream, ice-cream and cheese

16 Thomson DMH - BSc,PhD sensory assessment and food acceptability

17 Walker AF - MSc,PhD,DipTropAgric,AIFST nutritional implications of food processing; protein quality, toxicity, dietary fibre content, storage and air-classification of grain legumes; energy density of infant foods

18 Wilbey RA - BSc,MSc,AIFST dairy emulsions and their analogues; effects of process variables on the production products; utilisation of alternative milk sources

4662 Food Technology Department

1 Birch GG - BSc,PhD,DSc,FRSC,CChem,FIFST chemistry of rare food sugars and sweeteners

2 Brennan JG - BSC,MSc,AIFST food dehydration and texture studies

3 Cowell ND - BSc,MSc,AInstP heat and mass transfer in foods

4 Davies R - BSc,SM,PhD food poisoning organisms; survival in processed foods

5 Jukes DJ - BSc,MSc food legislation, development and enforcement

6 Kearsley MW - BSc,PhD production and properties of glycose syrups, carbohydrate/iron complex formation

7 King RD - BSc,PhD chemistry and functionality of food proteins

8 Lilly AEV - BSc,CChem,FRSC,FIFST protein hydrolysates, cereals malting, extrusion cooking

9 Owens JD - BSc,PhD conductimetric microbiology assay methods, oriental food fermentations, lactic acid bacteria

10 Pyle DL - BScTech,PhD *Professor of Biotechnology* process biotechnology; design and control of bioreactors and downstream operations

11 Reynolds AJ - BSc,MIBiol food plant hygiene; degradation of lignio-cellulolytic wastes

4663 Horticulture Sub-Department

1 Pegg GF - BSc,MSc,PhD *Professor of Horticulture and Head of Sub-Department* host-parasite physiology, plant disease resistance mechanisms, stress physiology

2 Bisgrove RJ - BSc,MLA *Lecturer* species diversification and management of amenity grassland; development of the English garden 1860-1930

3 Emden Van HF - BSc,PhD,ARCS,DIC *Professor of Applied Entomology* interrelations of insects and plants; plant resistance in pest management systems

4 Fox RTV - BSc,PhD *Lecturer* plant disease diagnosis and control; root pathogens; post-harvest pathology; resistance to fungicides

5 Hadley P - BSc,PhD *Lecturer* plant/crop physiology; vegetable crops; grain legume crops; cocoa; photosynthesis; environmental regulation of growth and development; crop nutrition

6 Harris GP - BSc,PhD *Senior Lecturer* control of flowering in tomatoes, developmental physiology of glasshouse crop plants

7 Wall JK - BSc,PhD *Lecturer* photoperiodic control of flowering; developmental physiology of strawberry; light and *in vitro* plant growth

4664 Physiology & Biochemistry Department

1 Dils RR - BSc,PhD,DSc,CChem,FRSC,CBiol,FIBiol *Professor and Head of Department* mammary gland metabolism and lactation

2 Cunningham FJ - BSc,MSc,PhD *Reader* reproductive endocrinology of farm animals

3 Dawes GS - CBE,BA,MSc,BM,BCh,DM,FRCOG,FRCP,FRS *Professor* fetal physiology

4 Gladwell RT - BSc,PhD *Lecturer* neuroendocrine control of gonadotrophin secretion in chicken

5 Hanson MA - MA,DPhil *Lecturer* fetal and neonatal physiology

6 Hassall KA - BSc,PhD,CChem,FRSC *Senior Lecturer* metabolism of pesticides in animals

7 Holman RB - BA,MS,PhD *Lecturer* behavioural neurochemistry in animals

8 Jeacock Miss MK - MA,PhD *Senior Lecturer* embryonic growth and development in sheep

9 Knight PG - BSc,MSc,PhD *'New Blood' Lecturer* immunological manipulation of mammalian reproduction

10 Lomax MA - BSc,PhD *Lecturer* energy metabolism in sheep

11 Lowry PJ - BSc,PhD *Professor* processing of mammalian neuropeptides and the development of immunoradiatic assay systems

12 Nicholson BH - BSc,PhD *Lecturer* molecular biology; structure of viral proteins

13 Parkes MJ - MA,DPhil *Wellcome Trust Lecturer* fetal and neonatal physiology

14 Savva D - BSc,PhD *'New Blood' Lecturer* molecular biology; gene expression; growth and development in farm animals

15 Shepherd DAL - BVSc,PhD,MRCVS *Senior Lecturer* embryonic growth and development in sheep

16 Skidmore CJ - MA,DPhil *Lecturer* molecular biology; genetics of and gene expression in farm animals; DNA repair

17 Stephens AG - BSc,PhD *Lecturer* nutrition and metabolism of farm animals
18 Walker CH - BSc,MSc *Reader* metabolism of xenobiotics in birds and mammals

4665 Soil Science Department

The University, London Road, Reading, RG1 5AQ
Tel: Reading (0734) 875234. Telex: 847813 RULIB.

1 Wild A - BSc,PhD *Professor of Soil Science and Head of Department* plant nutrients in soils of temperate and tropical regions
2 Brown SC - BSc,PhD root growth and water use in arid environments
3 Campbell DJ - BSc sulphate adsorption and movement in relation to soil acidification; soil solution chemistry
4 Dalrymple JB - BSc,MSc,PhD soil micromorphology; land surface analysis with associated catenary soil sequences; soil modality
5 Gates RPG - BSc,PhD methods of measuring N/sub/2 fixation
6 Gregory PJ - BSc,PhD *effects of soil conditions on root growth and activity; water use by crops*
7 Harris PJ - MSc,PhD ecology of microorganisms in different soil micro-habits; vesicular-arbuscular mycorrhiza
8 Le Mare PH - BSc,PhD chemistry of tropical soils
9 Macduff J - BA,DPhil mineral nutrition of plants
10 Milne Jones AA - BSc,PhD clay mineralogy
11 Mott CJB - MA,DPhil physical chemistry of soils; gypsum in soils; soil acidification
12 Nortcliff S - BA,PhD statistical analysis of soil variability, tropical forest hydrology; soil erosion
13 Opperman MH - BSc decomposition of animal waste applied to soil
14 Payne D - MA,DipAgrSc soil water and soil structure
15 Rowell DL - BSc,DPhil saline and sodic soils; nutrient movement to roots; soil acidity; flooded soils
16 Shepherd KD - BSc,PhD models of water and nitrogen balances of tropical crops and soils
17 Simmonds LP - BSc,PhD root growth; water use by crops; plant water relations
18 Warren GP - BA,MSc,PhD phosphate in tropical soils
19 Wood M - BA,PhD N/sub/2 fixation; selection of *Rhizobium* inoculants; effects of acidity on soil microorganisms

4666 University of Sheffield

Sheffield, SIO 2TN
Tel: (0742) 78555.

4667 Botany Department

1 Willis AJ - DSc,PhD,FLS,FIBiol *Professor and Head of Department* physiological ecology of British plants; coastal vegetation; growth regulators and herbicides in management of roadside vegetation
2 Baker AJM - BSc,PhD,FLS *Lecturer* ecophysiological aspects of heavy metal tolerance and toxicity in plants; pollution studies; geobotanical and biogeochemical prospecting
3 Booth A - BSc,PhD *Senior Lecturer* developmental physiology (growth and differentiation) with particular reference to hormones and the distribution of assimilates
4 Cooke RC - DSc,PhD *Reader* physiology of mycoparasitism; morphogenesis of sclerotia; quantitative fungal ecology
5 Elkington TT - BSc,PhD,FLS *Senior Lecturer* biosystematics, cytogenetics and chemotaxonomy of *Allium, Genista* and *Ulex*
6 Grime JP - BSc,PhD *Deputy Director of the Unit of Comparative Plant Ecology and Professor Associate* experimental ecology with reference to the modelling of physiology and life history strategies of plants in the established and regenerative phases
7 Gupta PL - MSc,PhD *Independent Research Worker* plant and soil chemistry; mineral nutrition
8 Hendry GAF - BSc,PhD *Honorary Lecturer* comparative biochemistry of plants; oxygen radical metabolism; plant cytochrome P450; mechanisms of flood and drought tolerance
9 Hodgson JG - BSc,PhD *Independent Research Worker* plant ecology with special reference to community structure, rare species and phenology
10 Hunt R - BSc,PhD,MIBiol *Honorary Lecturer* comparative physiological ecology; ecological data-banking; modelling growth and resource allocation
11 Jarvis BC - BSc,PhD *Lecturer* physiological and biochemical aspects of adventitious root formation; the role of boron in plant development; seed dormancy and its control
12 Leegood RC - BA,PhD *Lecturer* regulation of carbon assimilation
13 Lewis DH - MA,DPhil *Professor* carbohydrate and nitrogen nutrition of parasitic and mutualistic associations; metabolism of polyhydric alcohols, chemotaxonomy of leafy liverworts
14 Lyon AJE - MA,DPhil *Lecturer* physiology and ecology of litter-decomposing fungi
15 Read DJ - BSc,PhD *Reader* mycorrhizal associations in natural vegetation, including woodlands; water relations of trees; physiological effects of sulphur dioxide
16 Rorison IH - MA,DPhil,FIBiol *Director of the Unit of Comparative Plant Ecology and Professor Associate* experimental ecology with special reference to mineral nutrition, climate and adaptive mechanisms
17 Walker DA - DSc,PhD,FIBiol,FRS *Professor of Biology and Director of the Research Institute for Photosynthesis* photosynthetic control mechanisms in isolated chloroplasts; the role of the limiting envelope; regulation of photosynthesis in intact leaves

18 Wheeler BD - BSc,PhD *Lecturer* factors and processes involved in the composition of wetland vegetation, management and rehabilitation of peatlands

4668 Research Institute for Photosynthesis

Internal factors limiting photosynthesis at the levels of the organelle, protoplast and whole plant

1 Brearley TH - BSc,PhD instrumentational aspects of photosynthesis
2 Foyer Ms CH - BSc,PhD regulation of photosynthesis
3 Furbank RT - BSc,PhD control of photosynthesis by biochemical factors
4 Horton P - BA,DPhil *Reader* regulation of electron transfer and light harvesting in photosynthesis
5 Sivak Ms MN - LicBioSci,DBiolSci photosynthesis and productivity

4669 Genetics Department

1 Anderson RW - BSc,PhD *Lecturer* genetic and biochemical studies of cell differentiation in *Physarum*
2 Burnet B - BSc,PhD *Reader* genetic structure and evolutionary significance of courtship behaviour in *Drosophila*
3 Grindle M - BSc,MSc,PhD *Lecturer* action of fungicides; genetic and biochemical studies of fungicide resistant mutants
4 Hartmann-Goldstein IJ - BSc,PhD *Senior Lecturer* chromosomal and phenotypic effects of rearranged heterochromatin in *Drosophila*
5 Roper JA - BSc,PhD *Professor* studies of spontaneous and induced errors of mitosis in *Aspergillus*
6 Sudbery PE - BSc,PhD *Lecturer* mechanisms controlling cell division in yeast

4670 Wolfson Institute of Biotechnology

1 Fowler MW - BSc,PhD,CBiol,FIBiol *Professor and Head of Department* plant cell culture; enzymology, interactions between primary and secondary metabolism; process systems for cell culture
2 Scragg AH - BSc,PhD *Institute Deputy Director and Senior Lecturer* plant cell culture, in particular mass cultivation; immobilized cells; molecular biology of plant hormone action; plant lipids
3 Allan EJ - BSc,PhD *Scientist* secondary metabolism in plant cell cultures; mass cultivation of plant cells
4 Cresswell RC - BSc,PhD *Scientist* selection in plant cell cultures
5 Grey Ms D - BSc,PhD *Research Assistant* general plant enzymology
6 Kelly SL - BSc,PhD *Lecturer* molecular biology of secondary product synthesis in plants; structure, regulation, and function of cytochrome P450 in antifungal action, xenobiotic metabolism and biotransformations
7 Marison IW - BSc,PhD *Lecturer* applied microbial physiology; fermenters and fermentation; calorimetry
8 Morris P - BSc,PhD *Senior Scientist* plant cell, tissue and organ culture; secondary metabolism and transport
9 Stafford Ms AM - BSc,PhD *Senior Scientist* regulation of secondary product synthesis in plant cell cultures; basis of variability in plant cell cultures
10 Stepan-Sarkissian G - BSc,PhD *Senior Scientist* enzymology of secondary metabolism in plants
11 Warren GS - BSc,PhD *Lecturer* the plant cell membrane and cell selection; interaction of viruses with plant cells

4671 Department of Zoology

1 Calow P - BSc,PhD,DSc,CBiol,FIBiol *Professor, Head of Department* physiological ecology of freshwater invertebrates
2 Bayne BL - BSc,PhD,CBiol,FIBiol *Honorary Professorial Fellow* physiological ecology of marine bivalves
3 Birkhead TR - BSc,DPhil *Lecturer* avian ecology
4 Grenfell BT - BSc,MSc,DPhil *Lecturer* parasite epidemiology
5 Hale PA - BSc,PhD *Lecturer* experimental dermatology
6 Henderson IW - BSc,PhD,DSc,CBiol,FIBiol *Professor* hormones and water and electrolyte balance in vertebrates
7 Hill L - BSc,PhD,FIBiol *Senior Lecturer* insect physiology
8 Kime DE - BSc,PhD *Senior Experimental Officer* endocrinology
9 Lessells CM - BA,DPhil *Lecturer* population and behavioural ecology
10 Messenger JB - MA,PhD *Reader* neurobiology of cephalopods
11 Snart JOH - MSc *Senior Experimental Officer* freshwater ecotoxicology
12 Williams EE - BSc,PhD,MIBiol *Lecturer* heavy metal pollution and biofouling

4672 University of Southampton

Highfield, Southampton, SO9 5NH
Tel: Southampton (0703) 559122.

4673 Department of Biology

1 Allen JA - BSc,PhD *Lecturer* mollusc and bird ecology
2 Bennett CE - BSc,PhD *Lecturer* parasitology, especially fascioliasis
3 Bisby FA - MA,DPhil,FLS *Lecturer Vica fabia* cultivar information
4 Edwards PJ - MA,PhD *Lecturer* effect of grazing on primary production and nutrient turnover
5 Fenner M - BSc,PhD *Lecturer* plant competition; germination water relations
6 Hall JL - BSc,DPhil *Professor* ion and sugar transport in plants; plant development

7 Manners JG - MA,PhD *Reader* cereal plant pathology
8 Morris DA - BSc,PhD *Senior Lecturer* transport of plant hormones; effect of hormones on transport
9 Putman RJ - BSc,DPhil *Lecturer* ecology and behaviour of forest deer
10 Smartt J - BSc,PhD *Senior Lecturer* genetics of beans and ground-nuts
11 Spellerberg IF - BSc,PhD *Lecturer* ecology of animals in forests and heatherlands

4674 Agrochemical Evaluation Unit - Department of Biology

1 Jepson PC - BSc,PhD,ARCS *Lecturer* insect pest management
2 Wratten SD - BSc,MA,PhD *Senior Lecturer* biology and pest status of cereal insects

4675 Plant Improvement Unit - Department of Biology

1 Evans PK - BSc,PhD *Lecturer* plant tissue and cell culture of legumes
2 Haq ANMN - MSc,PhD *Senior Research Fellow* plant breeding

4676 Chemical Entomology Unit - Departments of Biology and Chemistry

1 Howse PE - BSc,PhD *Reader in Biology* entomology - chemical control of insect behaviour
2 Kocienski PJ - PhD,MRSC *Professor of Organic Chemistry* insect chemistry
3 Stevens IDR - BSc,PhD *Senior Lecturer in Chemistry* insect chemistry

4677 Department of Geography

1 Barber KE - BSc,PhD *Lecturer* pollen analysis and history of vegetation
2 Birch BP - MA,PhD *Senior Lecturer* land classification and soil science
3 Milton EJ - BSc,PhD *Lecturer* remote sensing of heathland vegetation
4 Woodruffe BJ - BA,MSc *Lecturer* agriculture geography (New Forest area): rural planning

4678 School of Biochemical and Physiological Sciences

1 Grimble RF - Bsc,PhD *Senior Lecturer* protein turnover in animals; influence of diet, protein and fat during pregnancy; lactation and infection
2 Jackson AA - BA,BCHIR,MB,MD *Professor* intermediatry metabolism on non-essential nitrogen
3 Jordan PM - BSc,PhD *Senior Lecturer* DNA transfer into plant protoplasts
4 Kelly F - BSc,PhD *Lecturer* mechanism of oxygen free radical induced cell damage; temporal development of cellular antioxidants; relationship between oxygen administration and cell damage with special reference to preterm infants
5 Kerkut GA - MA,PhD,ScD,FIBiol *Professor* actions of drugs on the insect nervous system
6 Lee AG - BSc,MA,PhD *Senior Lecturer* biophysics of insecticides
7 Peddie MP - MA,PhD *Lecturer* calcium and gonadal function in birds and mammals
8 Ragan CI - BA,PhD *Senior Lecturer* mode of action of insecticides
9 Walker RJ - BSc,PhD,DSc *Reader* effects of drugs on the nerve muscle system of *Ascaris*
10 Wootton S - BSc,PhD *Lecturer* inter-relationships of diet and exercise in muscle performance
11 York DA - BSc,PhD *Senior Lecturer* energy metabolism in genetically obese animals and nutrient-central nervous system interactions

4679 University of Surrey

Guildford, Surrey
Tel: (0483) 571281.

1 Kelly A - ScD,FEng,FRS *Vice Chancellor*

4680 Department of Biochemistry

1 Parke DVW - PhD,DsC,CChem,FRSC,FIBiol,FRCPath,HonMRCP *Professor and Head of Department* metabolism of food additives; toxicology of pesticides, herbicides and veterinary drugs
2 Ioannides C - BSc,PhD *Senior Experimental Officer* metabolism and mutagenicity of food additives, pesticides, herbicides and veterinary drugs

4681 Biochemistry Division

1 Symons AM - PhD,MIBiol *Senior Lecturer in Biochemistry and Head of Division* hypothalamic-pituitary-gonadal control of reproductive activity in the ewe
2 Goldfarb PSG - BSc,PhD *Lecturer in Molecular Biology* monoclonal antibody production; gene transfer in animal cells
3 Hubbard R - BSc,PhD,CChem,MRSC *Lecturer in Immunology and Biochemistry* monoclonal antibody production against hormones; immunoassay development and physiological mechanisms of hormone action
4 Snell K - BSc,PhD *Reader in Biochemistry* metabolic regulation and enzyme control during neonatal development and in neoplasia

4682 Clinical Biochemistry Division

1 Marks V - MA,DM,FRCP,FRCPath *Professor and Head of Division* insulin, glucagon, gastrointestinal hormones
2 Aherne Mrs GW - BSc,PhD *Experimental Officer* immunoassay; cytotoxic drugs, water analysis
3 Arendt Mrs J - BSc,PhD *Senior Lecturer in Clinical Biochemistry* hypothalamic-pituitary-gonadal control of reproductive activity in the ewe; function of the pineal gland
4 Morris BA - BSc,CChem,FRSC *Senior Research Fellow* immunoassays and vaccines; reproductive physiology; food analysis

4683 Nutrition and Food Science Division

1 Dickerson JWT - PhD,FIBiol,FRSH,FIFST *Professor of Human Nutrition and Head of Division* physiological effects of food additives
2 Clifford MN - BSc,PhD,AIFST *Lecturer in Food Science* coffee phenols in food, food analysis
3 Howell Mrs NK - PhD,AIFST *Lecturer in Food Science* structure and functional properties of proteins in food
4 Walker R - PhD,CChem,NRSC,FIFST *Professor in Food Science* metabolism and toxicology of food additives

4684 Pharmacology and Toxicology Division

1 King LJ - BA,BSc,PhD *Professor in Biochemistry* toxicology of metals and halogenated hydrocarbons
2 Gibson GG - BSc,PhD *Senior Lecturer in Pharmacology and Toxicology* metabolism and toxicology of food additives, insecticides and herbicides
3 Hall DE - BSc,MIBiol,DipRCPath *Lecturer in Comparative Pathology* heavy metal toxicology; spontaneous pathology in animals
4 Howarth Miss J - BSc *Research Officer* comparative pathology; cadmium toxicity
5 Jones RS - PhD,CChem,FRSC *Lecturer in Biochemistry* natural carcinogens; organic anion transport
6 Kentish PA - MIBiol *Research Officer* metabolism of azo dyes

4685 Robens Institute of Industrial and Environmental Health and Safety

1 Bridges JW - PhD *Professor of Toxicology and Director of Institute* mechanisms of toxicity

4686 University of Sussex

Falmer, Brighton, Sussex, BN1 9QO
Tel: (0273) 606755.

4687 School of Biological Sciences

1 Andrew RJ - MA,PhD,ScD *Professor* hormonal effects on behaviour of domestic fowl; mechanisms of memory formation
2 Flowers TJ - BSc,MSc,PhD *Reader* physiology of salt tolerance in crops; development of salt-resistant strains of rice
3 Hutchings MJ - BSc,PhD *Lecturer* plant population biology; competition between crops and weeds; population biology of old pasture species
4 Land MF - BA,PhD,FRS *Professor* ecology of vision in insects
5 Maynard Smith J - BA,BSc,FRS *Professor* ecological modelling
6 Moore A - BSc,PhD *Lecturer* regulation and control of photorespiration; metabolite transport in etiolated photosynthetic and thermogenic tissues
7 Roper TJ - MA,PhD *Lecturer* plant population biology; competition
8 Streeter DT - BSc *Reader* ecology, conservation and land use
9 Thomas JD - BSc,PhD *Reader* freshwater ecology with particular reference to the chemistry, control and ecology of the snail hosts of schistosomiasis and fascioliasis

4688 School of Molecular Sciences

1 Hanson JR - BSc,PhD,DSc *Reader* chemistry and biosynthesis of gibberellins; synthesis of plant growth regulators; structures and biosynthesis of mycotoxins and other fungal metabolites

4689 University College of Swansea

Singleton Park, Swansea, West Glam, SA2 8PP
Tel: Swansea (0792) 205678.

4690 School of Biological Sciences

1 Syrett PJ - MA,DSc,FIBiol *Chairman*

4691 Biochemistry Research Group

1 Brown EG - BSc,PhD,DSc,CChem,FRCS *Professor and Leader of the Biochemistry Research Group* biochemistry of purine and pyrimidine derivatives in plants; the occurrence and metabolism of nucleotides, including cyclic nucleotides
2 Chaplin AE - BSc,PhD *Senior Lecturer* nitrogen fixation by non-heterocystous cyanobacteria; allelochemical metabolism in bryozoa
3 Gallon JR - MA,DPhil *Senior Lecturer* N_2 fixation (acetylene reduction) in unicellular cyanobacterium *Gloeothece* and related organisms; enzymes of alkaloid biosynthesis

4 Newton RP - BSc,PhD,CChem,MRSC,FIAP *Lecturer* cyclic nucleotides and related enzymes in higher plants; tocopherols in higher plants and blue-green algae

5 Smith CJ - BSc,PhD *Lecturer* biochemistry of the plant cell wall; synthesis of carbohydrate polymers; elicitors of phytoalexin accumalation in *Glycine max* L and *Medicago savita* L

6 Walton TJ - BSc,PhD *Lecturer* structure and biochemistry of plant cuticular lipids; glycolipids and terpenoids of cyanobacteria

4692 Biomedical and Physiological Research Group

1 Ratcliffe N - BSc,PhD,DSc *Professor and Leader of the Biomedical and Physiological Research Group* insect immune defences against pathogens and non-self recognition

2 Brain PF - BSc,PhD *Reader* effects of housing on physiology, behaviour and disease resistance of mammals; assessment of behavioural effects of food additives and drugs; studies on horses and injured birds of prey

3 Makings PE - BSc,PhD *Senior Lecturer* oviposition behaviour of Lepidoptera; thermoreception in locusts

4 Parry JM - BSc,PhD *Reader* genetic hazards of environmental chemicals

5 Rowley A - BSc,PhD *Lecturer* fish disease and immunology with particular reference to the inflammatory response

4693 Ecological and Evolutionary Research Group

1 Beardmore JA - BSc,PhD,FIEBiol *Professor and Leader of Research Group* selection of fish and shell fish

2 Kay QON - BA,DPhil *Lecturer* biosystematics and ecology of annual weeds; pollination mechanisms

3 Skibinski DOF - BSc,PhD *Lecturer* breeding and genetics of fish and shellfish

4 Tilney-Bassett RAE - MA,DPhil *Senior Lecturer* genetic factors influencing inheritance, structure and function of chloroplasts; genetics of zonal pelargoniums

4694 Marine Research Group

1 Hayward J - BSc,PhD *Administrative Officer and Senior Lecturer* phytoplankton ecology

2 James BL - BSc,PhD,DSc *Senior Lecturer* biology of *Digenea*

3 King PE - BSc,PhD,DSc *Reader* biology of Chalcidoidea and Coleoptera Adephaga; aspects of the physiology and ultrastructure of insect reproduction; effects of heavy metals on fauna

4695 Plant and Microbial Metabolism Research Group

1 Syrett PJ - MA,DSc,FIBiol *Chairman of the School, Professor and Leader of the Plant and Microbial Metabolism Research Group* algal physiology particularly nitrogen metabolism

2 Halliwell G - MSc,PhD,DSc *Reader* biodegradation of cellulose

3 Hipkin CR - BSc,PhD *Lecturer* nitrate reductase in yeasts

4 Merrett MJ - BSC,PhD,DSc,FIBiol *Professorial Fellow* molecular biology of algal organelles

5 Milton JM - BSc,PhD *Senior Lecturer* cellular recognition systems particularly with *Verticillium* wilt diseases

6 Opik H - BSc,PhD *Senior Lecturer* fine structure of plant cells

7 Trollope DR - MSc *Part-time Lecturer* bacteriology of shell fish

8 Wainwright SJ - BSc,PhD,MIBiol *Lecturer* molecular biology of algal organelles

4696 University College of Wales

4697 Institute of Rural Sciences
Penglais, Aberystwyth, Dyfed
Tel: (0970) 3111.

4698 Department of Agricultural Botany

1 Rees H - PhD,DSc,FRS *Professor and Head* cytology

2 Cannell RQ - BSc,PhD *Professor*

3 Evans GM - BSc,PhD cytology

4 Forster JW - MA,PhD *molecular genetics*

5 Griffiths E - MSc,PhD,DSc *Professor* plant pathology and soil microbiology

6 James DB - MSc agronomy and crop physiology

7 Jones DG - BSc,PhD,DSc *Professor* plant pathology and soil microbiology

8 Jones RN - BSc,PhD cytology

9 Narayan RKJ - BSc,MSc,PhD DNA variation in plants

10 Norrington-Davies J - BSc,PhD agronomy and crop physiology

4699 Department of Agricultural Economics and Marketing

1 Bateman DI - MA *Professor and Head of Department* land use

2 Haines M - BSc,PhD *Professor of Agricultural and Food Marketing* agricultural marketing

3 Delagneau BA - ESC,MA,DScEcon,DRScEconAppl *Lecturer* marketing

4 Dummer MH - BA *Lecturer* agricultural policy in LDCs

5 Green DAG - BSc,MS,PhD *Senior Lecturer* farm management

6 Gwynne DFP - BSc *Lecturer* history of economic thought

7 Jones W - Dyfri,MSc *Senior Lecturer* farm management

8 Lampkin NH - BSc *Lecturer* agriculture environment

9 LeVay C - BA,MSc *Lecturer* cooperation and land use

10 Sroka JR - BACNNA,MBA *Lecturer* marketing strategy and the analysis of competition; retailing and distribution; marketing for the small firm

4700 Department of Agriculture

1 Hayes JD - BSc,MS,PhD,FIBiol *Professor and Head of Department* crop production

2 Beck NFG - BSc,PhD *Lecturer* physiology of animal reproduction

3 Colyer RJ - BSc,PhD,FRhistS,FSA *Lecturer* evolution of the British landscape

4 Jenkins PD - BSc,PhD *Lecturer* agronomy and physiology of arable crops

5 Leitch MH - BSc,PhD *Senior Research Officer* arable crops agronomy

6 Riley JA - BSc,PhD *Lecturer* animal nutrition

7 Wilman D - BSc,MA,PhD,FIBiol *Senior Lecturer* grassland agronomy

4701 Department of Biochemistry & Agricultural Biochemistry

1 Beechey RB - BSc,PhD,CChem,FRSC *Professor and Head of Department* biochemistry of membrane associated processes

2 Davies BH - BSc,PhD,FRSC,CChem *Senior Lecturer* carotenoids

3 Faithfull NT - MSc,PhD,CChem,MRSC *Senior Research Associate*

4 Hopper DJ - BSc,PhD *Senior Lecturer* bacterial metabolism of aromatic compounds

5 Jones M - MSc,MRSC,CChem *Senior Research Officer* microbial transport processes

6 Jones MG - MA,PhD *Lecturer* plant cell walls and growth

7 Kaderbhai MA - BSc,PhD *Lecturer* chloroplast biogenesis

8 Ling JR - BSc,MS,PhD *Lecturer* animal nutrition; rumen microbial metabolism

9 Mercer EI - BSc,PhD,FRSC,MIBiol,CChem *Senior Lecturer* plant sterols, EBI fungicides

10 Rogers LJ - BSc,PhD,CChem,FRSC *Reader* photosynthetic membranes; electron transfer proteins

11 Smith AJ - BSc,DPhil *Senior Lecturer* biochemistry of cyanobacteria

12 Trudgill PW - BSc,PhD *Reader* bacterial metabolism of alicyclic compounds

4702 Soil Science Unit

1 Adams WA - BSc,PhD,CPSS,MIBiol *Senior Lecturer* soil chemistry and plant nutrition

4703 University of Wales Institute of Science and Technology

King Edward VII Avenue, Cardiff, CF1 3NU
Tel: Cardiff (0222) 42588.

4704 Department of Applied Biology

1 Beverton RJH - CBE,FRS *Professor* marine pollution and fish resources

2 Edwards RW - BSc,DSc,FIBiol,FIWPC *Professor* water pollution and herbicidal control of aquatic plants

3 Evans PJ - MA,DPhil *Lecturer* intracellular protein degradation; metabolism; tissue culture

4 Hughes RE - MA,PhD,FIBiol *Reader* vitamin C metabolism and mammalian nutrition

5 Pascoe D - BSc,PhD *Lecturer* physiology of avian trematodes

6 Randerson PF - BSc,MSc *Lecturer* salt marshes - successional development

7 Slater FM - BSc,MSc,PhD *Curator* vegetation of peatlands

4705 University of Warwick

Coventry, CV4 7AI
Tel: (0203) 523523.

4706 Department of Chemistry

1 Wallbridge MGH - BSc,CChem,FRSC *Professor and Chairman of Department* complexation of metals and non-metals in the environment; control and removal of trace elements

2 Cox A - BSc,MA,PhD *Lecturer* photochemical studies of organic compounds

3 Crout DHG - MA,PhD,CChem,FRSC *Professor* biorganic chemistry; biotransformations

4 Dodd GH - BA,DPhil *Lecturer* olfaction, odour/structure relationships flavour

5 Griffiths DE - BSc,MA,PhD *Reader* bioenergetics, cellular conversion mechanistic studies

6 Howarth OW - MA,DPhil *Lecturer* protein structures

7 Hutchinson DW - BSc,PhD *Reader* polynucleotides as interferon inducers, bilirubin and albumin

8 Jennings KR - MA,DPhil *Professor* applications of mass spectrometry to environmental problems

9 Samuel CJ - MA,PhD *Lecturer* mechanistic organic chemistry

10 Swoboda BEP - BSc,PhD *Lecturer* control of eukaryotic cell cycle, control of RNA and DNA synthesis, action of cyclophosphamides
11 Thomas SE - BA,DPhil *Lecturer* synthesis of biologically active compounds using organometallic reagents

4707 Department of Biological Sciences

4708 Animal Molecular Genetics Group

1 Woodland HR - MA,DPhil *Professor* gene expression in early development
2 Brown Ms JE - BSc,PhD *Lecturer* cell-cell interactions
3 Colman A - MA,DPhil *Senior Lecturer* signals in proteins; localisation of messenger RNAs
4 Jones Mrs EA - BSc,DPhil *Lecturer* monoclonal antibodies to early embryos
5 Old RW - BA,PhD *Lecturer* regulation of histone gene expression
6 Vlad Mrs MT - LicBiol,PhD *Lecturer* substances which regulate gene expression

4709 Environmental Microbiology Research Group

1 Kelly DP - PhD,DSc *Professor* inorganic sulphur oxidation by microorganisms
2 Fletcher Mrs MM - BA,PhD bacterial adhesion to surfaces
3 Flint KP - BSc,PhD *Lecturer* detecting pollutants in potential water supplies
4 Norris PR - BSc,PhD *Lecturer* iron and sulphur oxidizing bacteria

4710 Microbiology Group

1 Whittenbury R - MSc,PhD *Professor, Head of Department* general microbiology
2 Carr NG - BSc,PhD *Professor* cyanobacterial biology
3 Dalton H - BSc,DPhil *Professor* oxygenase-catalyzed reactions
4 Dow CS - BSc,PhD *Senior Lecturer* regulation of cellular morphogenesis and differentiation
5 Hodgson DA - BSc,PhD *Lecturer* global regulatory mechanisms in bacteria
6 Mann NH - BSc,PhD *Lecturer* plasmid segregation and plasmid vector development
7 Murrell JC - BSc,PhD *Lecturer* nitrogen fixation and ammonia assimilation by methane-oxidizing bacteria
8 Salmond GPC - BSc,PhD *Lecturer Erwinia* phytopathogenicity and extracellular enzyme export

4711 Plant Biochemistry Group

1 Ellis RJ - BSc,PhD,FRS *Professor* expression of nuclear genes encoding chloroplast proteins
2 Cullimore JV - BSc,PhD *Lecturer* nitrogen metabolism in plants
3 Hartley MR - BSc,PhD *Lecturer* chloroplast ribosomes
4 Lord JM - BSc,PhD *Reader* organelle development in castor bean endosperm
5 Roberts LM - BSc,PhD *Lecturer* genetic manipulation of the plant toxin, ricin
6 Robinson C - BSc,PhD *Lecturer* protein transport in chloroplasts

4712 Virology Group

1 Pringle CR - BSc,PhD *Professor* molecular genetics of RNA viruses
2 Dimmock NJ - BSc,PhD *Reader* interaction of viruses with the host's defence mechanism
3 Easton AJ - BSc,PhD *Lecturer* molecular basis of virus pathogenicity and virulence
4 Evans D - BSc,PhD *Lecturer* use of genetically manipulated plant viruses in the control of plant disease
5 McCrae MA - BSc,PhD *Senior Lecturer* molecular mechanism of viral pathogenesis
6 Morris AG - BSc,PhD *Lecturer* the role of interferons in immune responses

4713 University of York

York, YO1 5DD
Tel: (0904)59861. Telex: 57933 YORKUL.

4714 Department of Biology

1 Chadwick MJ - BSc,PhD,MA *Professor* environmental impacts of energy use
2 Digby J - BSc,PhD *Lecturer* plant development; plant movements
3 Firn RD - BAgrSc,MAgrSc,PhD *Lecturer* plant growth and development
4 Fitter AH - BA,PhD *Lecturer* root architecture; mycorrhizas
5 Law R - BA,PhD *Lecture* plant population biology
6 Lawton JH - BSc,PhD *Professor* entomology; population biology; biological control
7 Leech RM - MA,DPhil *Professor* chloroplasts
8 Robards AW - BSc,PhD,DSc *Reader* plant root ultrastructure
9 Sanders D - BA,DPhil *Lecturer* ion transport
10 Usher MB - BScFor,PhD *Reader* wildlife conservation; soil biology
11 Williamson MH - MA,DPhil *Professor* population biology

4715 Harper Adams Agricultural College

Edgmond, Newport, Shropshire, TF10 8NB
Tel: (0952) 811280. Telex: 57933 YORKUL.

1 Anslow RC - BSc,MSc,MIBiol nutrition of cereals and potatoes; growth patterns of grass
2 Belyavin CG - BSc,NDP,PhD egg quality; nutrition of poultry; computer programming of flock management; egg handling and feeding systems for laying hens
3 Brook PT - BSc,PhD agronomy of triticale
4 Custance PR - BA,PhD market research; consumer attitudes to farming
5 Davies WP - BSc,MSc,PhD,MIBiol diseases of cereals; fungicide and foliar nitrogen on winter cereals
6 Hemingway PF - BSc engineering problems of separating animal wastes
7 Howard MW - BSc,MSc consumer attitudes to farming
8 Kettlewell PS - BSc,MSc growth studies with cereals; foliar nutrition of winter cereals
9 Minter CM - BSc,MIBiol sheep production systems development
10 Parry MA - BSc,PhD,CertEd behaviour of the pregnant sow
11 Parsons ST - BSc,MPhil forecasting animal feed markets; consumer attitudes to farming
12 Sadler AD - BSc,NDA,DipAg nutrition of growing pigs
13 Tempest WM - BSc,PhD sheep and beef production systems development
14 Wells RG - BSc,MSc,NDA,MIBiol feeding of growing pullets and laying hens
15 Williamson P - BSc trace element and mineral content of grass and soils

4716 National Institute of Agricultural Botany

Huntingdon Road, Cambridge, CB3 OLE
Tel: Cambridge (0223) 276381.

1 Milbourn GM - MSc,PhD *Director*
2 Doodson JK - BSc,PhD *Deputy Director, Head of Crops Division*
3 Silvey VS - BSc,DipMathStat *Deputy Director, Head of Seeds and Services Division*

4717 Cereals Branch

Testing varieties for performance and distinctness for statutory and advisory purposes

1 Ramsbottom JE - BSc *Head of Branch*
2 Eade AJ - BScAgric systematic botany of cereals
3 Mann GC - BScAgric cereal trials

4718 Grasses and Herbage Legumes Branch

Testing varieties for performance and distinctness for statutory and advisory purposes

1 Aldrich DTA - BScAgric,DipAgric *Head of Branch*
2 Boyd MM - BScAgric systematic botany of grasses and herbage legumes
3 Ingram J - BScAgric,MPhil grasses and herbage legumes trials

4719 Potato, Sugar Beet, Oilseeds and Fodder Crops Branch

Testing varieties for performance and distinctness for statutory and advisory purposes

1 Richardson DE - MA *Head of Branch*
2 Kerr SP - BSc potatoes
3 Kimber DS - MSc sugar beet, oilseeds and fodder crops

4720 Vegetables and Ornamentals Branch

Testing for performance for advisory purposes and for distinctness for statutory purposes

1 Chowings JW - BSc *Head of Branch*
2 Day MJ - BSc vegetable trials
3 Evans JL - BSc systematic botany of vegetables
4 George AJ - BSc,MSc ornamentals

4721 Chemistry and Quality Assessment Branch

The chemical analysis and quality assessment of crop varieties

1 Morgan AG - BSc,PhD *Head of Branch*
2 Cooke RJ - BSc,PhD research in chemotaxonomy

4722 Pathology Branch

The testing of crop varieties for disease resistance

1 Priestley RH - BSc,PhD *Head of Branch*
2 Bayles RA - BA,PhD cereal pathology
3 Sweet JB - MSc,PhD non-cereal crops pathology

4723 Regional Trials Branch

Growing trials and providing other services for the variety testing work

1 French WM - BTechApplBiol,MIBiol *Head of Branch*

2 Braybrooks J - BSc *Senior Regional Trials Officer* South West Region
3 Broom EW - MIBiol,NDA *Senior Regional Trials Officer* West Central Region
4 Furber MJ - BScAgric,DipAgric *Senior Regional Trials Officer* South Region
5 Kemp MC - BScAgric *Senior Regional Trials Officer* East Central Region
6 Mead RG - BSc *Senior Regional Trials Officer* East Region
7 Pearson N - BSc,MIBiol *Senior Regional Trials Officer* Wales Region
8 Rivett AD - BScAgric *Senior Regional Trials Officer* Northern Region

4724 Statistics and Data Processing Branch

Providing the statistics and data processing services required by all Branches of the Institute

1 Campbell A - MA,MIS *Head of Branch*
2 Hampson AG - MSc *Computer Systems Management*
3 Law JR - MSc *Statistician, DUS Statistics*
4 Tonkin MH - BSc *Statistician, VCU Statistics*

4725 Seed Production Branch

Seed certification, investigation of seed production techniques and weed controls in seed crops

1 Bowring JDC - BScAgric *Head of Branch*
2 Bould A - BSc,BA,MIBiol,PhD *fodder, root and vegetable crops*
3 Evans AW - BScAgric *herbage crops*
4 Wellington RJ - NDA *cereals*

4726 Official Seed Testing Station for England and Wales

Statutory and advisory testing of crop seeds; identification of seeds and seedborne diseases; investigations on aspects of seed quality

1 Draper SR - BSc,PhD *Chief Officer*
2 Tonkin JHB - MA *Assistant Chief Officer*
3 Coster RM - BScAgric *Principal Technical Officer*
4 Hewett PD - BSc,ARCS *Seed Pathologist*

4727 Technical Information Unit

1 Colegate PR - BA,ALA *Librarian and Technical Editor*

4728 Silsoe College - Cranfield Institute of Technology

Silsoe, Bedford, MK45 4DT
Tel: Silsoe (0525) 60428. Telex: 825072 CITECH G.

4729 Faculty of Agricultural Engineering, Food Production & Rural Land Use

1 May BA - BSc,NDAgrE,CEng,MIMechE,FIAgrE *Dean, Head of College, Professor of Environmental Control and Processing*
2 Bascombe MLA - BSc,CEng,MIMechE *Lecturer in Structural Mechanics*
3 Belward AS - BSc,DipAgEng *Lecturer in Remote Sensing for Environmental Management*
4 Carr MKV - BSc,PhD *Senior Lecturer in Irrigation Agronomy*
5 Carter RC - MA,MSc,FGS,AMIG *Lecturer in Water Management*
6 Clarke B - BSc,PhD,DIC,CEng,ARTCS,MIMechE *Lecturer in Agricultural Processing*
7 Cowell PA - BSc,MSc,PhD,CEng,MIMechE,MIAgrE *Reader in Agricultural Engineering*
8 Crawford IM - BA *Lecturer in Marketing*
9 Crossley CP - BSc,MSc,CEng,MIMechE,FIAgrE *Lecturer in Engineering Design*
10 Douglas MP - MSc,MIAgrE,TEngCEI *Lecturer in Farmstead Engineering*
11 Dyson J - BSC,CEng,MIMechE *Senior Lecturer in Engineering Design*
12 Goodwin RJ - BSc,MS,PhD,CEng,MIAgrE,MASE *Professor of Agricultural Machinery Technology*
13 Grant A - MA,DIC,AFIMA *Senior Lecturer in Mathematics*
14 Hann MJ - MSc,BA,CEng,MIAgrE,AICE *Lecturer in Agricultural Engineering*
15 Hill RW - BA,PhD,AIAGrE,MInstM *Professor of Product Management, Head of Department of Marketing and Management*
16 Inns FM - MA,MSc,CEng,MIMechE,FIAgrE *Professor of Agricultural Machinery Engineering*
17 Kay MG - BSc,MSc,DIC,CEng,MICE *Senior Lecturer in Irrigation Engineering*
18 Keech MA - BSc,MSc,FIAgrE *Senior Lecturer in Land Resource Planning*
19 Kilgour J - BSc,MSc,CEng,MIAgrE *Senior Lecturer in Farm Machine Design*
20 Kirkup MH - BA,MSc *Lecturer in Marketing*
21 Knox SD - BSc,PhD,MInstM *Lecturer in Agricultural Marketing and Management*
22 Larkin SBC - BA,DPhil,MIBiol *Lecturer in Agriculture and the Environment*
23 Leeds-Harrison PB - BSc,PhD,MIAgrE *Lecturer in Drainage and Soil Physics*
24 Lewis RT - BSc,MSc,MIAgrE *Lecturer in Agricultural Mechanization*
25 Morgan BA - MLS,ALA *Librarian*
26 Morgan RPC - BA,MA,PhD,FRGS *Professor of Soil Erosion Control*

27 Morris ER - BSc,PhD,CChem,FRSC *Professor of Food Structure and Processing and Head of Department of Food Research and Technology*
28 Murfitt RFA - BSc,DTA,MIBiol *Lecturer in Agricultural Science*
29 Noble DH - BSc,MA,FSS *Senior Lecturer in Operational Research*
30 Radley RW - BSc,PhD *Professor of Agricultural Production Technology*
31 Spoor G - BSc,MSc,CEng,MIAgrE *Professor of Applied Soil Physics*
32 Stenning BC - BSc,CIAgrE *Lecturer in Environmental Instrumentation*
33 Stone GT - DipTech,AMIEE *Lecturer in Electronics and Control*
34 Strang IG - BSc,CEng,MIMechE,MCIBS,MRAeS *Senior Lecturer in Agricultural Thermodynamics*
35 Taylor JC - ANCAE,MS,PhD,MIAgrE *Lecturer in Agricultural Engineering*
36 Tindall HD - MBE,BSc,MSc,FLS,FIBiol,CIAgrE,NDH *Professor of Tropical Agronomy*
37 Watt CD - MSc,BSc,CEng,MIMechE,MIAgrE *Lecturer in Machinery Engineering*
38 Weatherhead EK - BA,MA,CEng,MICE *Lecturer in Soil and Water*

4730 Norfolk Agricultural Station

Morley St Botolph, Wymondham, Norfolk, NR18 9DB
Tel: Wymondham (0953) 605511.

1 McClean SP - BSc *Director*
2 Hilton JG - BSc *sugar beet herbicides*
3 Madge WER *cereal fungicides and general agronomy; oilseed rape fungicides, herbicides and agronomy*
4 May MJ - HNC *sugar beet herbicides and general agronomy*
5 Nuttall M - BSc *sugar beet cultivations; cereal agronomy; beef cattle*
6 Palmer GM - BSc *cereal fungicides and general agronomy; oilseed rape fungicides; herbicides and agronomy*

4731 Seale-Hayne College

Newton Abbot, Devon
Tel: Newton Abbot 2323.

4732 Animal and Crop Production Farm Management

1 Dowrick GJ - BSc,PhD,MIBiol *Principal*
2 Halley RJ - MA,MSc *Deputy Principal*

4733 Agriculture

1 Brooks PH - BSc,PhD *animal production; crop production; agricultural engineering*
2 Camm BM - MSc,MA *agricultural management*

SCOTLAND

4734 Department of Agriculture and Fisheries for Scotland

Agricultural Advisory Service - See Agricultural Colleges

4735 Agricultural Economics
Chesser House, 500 Gorgie Road, Edinburgh, EH11 3AW
Tel: 031-4434020.

1 Dunn JM - MA,BLitt,DPhil *Chief Agricultural Economist*
2 Greig DJ - BSc *Economic Advisor*
3 Martin PC - BSc,DipAgr *Economic Adviser*
4 McIntosh F - BScAgr *Economic Adviser*
5 Oakeshott WEF - BLitt,MA *Economic Adviser*

4736 Agricultural Scientific Services
East Craigs, Edinburgh, EH12 8NJ
Tel: 031-339 2355. Telex: 727348 (DAF ASS G).

1 Graham DC - DSc,PhD,CChem,FRSC,FIBiol,FRSA,FRSE *Director*

4737 Plant Varieties and Seeds Division

1 Seaton RD - BSc *Head of Division and Deputy Director*

4738 Official Seed Testing Station

1 Cooper SR - MA,MSc,DipAgrSt *Head of Section* testing methods; seed physiology and taxonomy
2 Rennie WJ - HNC,MIBiol seed pathology

4739 Cereals

1 Keppie JL - BScAgr *Head of Section*
2 Chadwick GR - BA,DipAgr varietal taxonomy
3 Davidson JC - BScAgr seed certification

4740 Herbage and Vegetable Crops

1 Sutton WG - BSc *Head of Section* varietal taxonomy

4741 Potato and Plant Health Division

1 Richardson MJ - BSc,MSc *Head of Division*

4742 Potato

1 Hall TD - BScAgr,NDA *Head of Section* production and development of nuclear stock
2 Carnegie SF - BSc,PhD fungal disease epidemiology; varietal susceptibility
3 Jeffries CJ - BSc,PhD production of nuclear stock; associated pathology
4 MacDonald DM - BSc varietal taxonomy physiology

4743 Plant Health

1 Howell PJ - BSc,PhD,DIC,ARCS,CBiol,MIBiol *Head of Section* statutory aspects of plant health: non-indigenous diseases
2 Haris PS - BScAgr non-indigenous potato diseases

4744 Nematology

1 Mabbott TW - BScAgr *Head of Section* potato cyst eelworms and related species

4745 Potato Pathology

1 Quinn CE - HNC,MIBiol *Head of Section*
2 Jeffrey IG - HNC,MIBiol potato pathology
3 Laidlaw WMR *potato pathology*

4746 Pest Control and Pesticides Division

1 Cutler JR - BSc *Head of Division*

4747 Pesticide Usage and Crop Loss Assessment

1 Hosie G - BSc,FRES *Head of Section* usage surveys
2 Bowen HM - HNC usage surveys
3 Davis RP - MIBiol biology and taxonomy of mites

4748 Chemistry

1 Hamilton GA - CChem,FRSC *Head of Section* pesticide residues
2 Findlay Miss E - HNC pesticide residue analyses
3 Griffiths CJ - BSc,PhD pesticide residue analyses
4 Hunter K - BSc,MSc,PhD pesticide residue analyses

5 Ruthven AD - ONC,LRSC pesticide residue analyses

4749 Crop Entomology

1 Turl Mrs LAD - BSc,FRES *Head of Section* aphid biology; non-indigenous pests
2 Anderson Miss IB - BSc wheat bulb fly

4750 Infestation Control (Stored products and associated premises)

1 Edwards DA - BA *Head of Section*
2 Lupton AM - BSc infestation by insects and mites

4751 Wildlife Investigation

1 Cuthbert JH - BScAgr *Head of Section*
2 Brodie J - BSc,MIBiol fat ecology
3 Kolb HH - BA,PhD urban fox biology

4752 Scientific Advisers Unit
Chesser House, 500 Gorgie Road, Edinburgh, EH11 3AW
Tel: (031-443) 4020.

1 Raven AM - BSc,PhD,CChem,FRSC,MIBiol *Scientific Adviser*
2 Dowdell RJ - BSc,PhD,MIBiol grassland and dairying
3 Hegarty TW - BSc,PhD,DipAgrSc plants and soils
4 Thornton DD - BSc,MSc,MS,DipAgr animals

4753 Royal Botanic Garden
Inverleith Row, Edinburgh, EH3 5LR
Tel: 031-522 7171.

1 Henderson DM - CBE,BSc,FRSE,FLS *Professor, Regius Keeper*
2 Anderson G horticultural training
3 Argent GCG - BSc,PhD tropical botany; Ericaceae
4 Bennell AP - BA mycology: palynology
5 Chamberlain DF - MA,PhD rhododendron; bryophyta
6 Coppins BJ - BSc,PhD mycology; lichenology
7 Cullen J - BSc,PhD *Assistant Keeper*
8 Edwards ID - BSc,PhD education/PR
9 Eudall R photography
10 Gregory NM mycology
11 Grierson AJC - BSc flora of Bhutan; compositae
12 Hedge IC - BSc *Curator of Herbarium* SW Asian flora
13 King R - BSc Arabian flora
14 Long DM - BSc flora of Bhutan
15 Mathew MV - BA,MA,DLSc,ALA *Librarian*
16 McKean DR British plants
17 Miller AG - BSc,MSc Arabian flora
18 Page CN - BSc,PhD gymnospermae; pteridophyta
19 Rae DAH - BSc,MSc horticultural training
20 Ratter JA - BScBot,PhD,MIBiol cytology, Brasilian flora
21 Shaw RL - NDH *Curator of Gardens*
22 Smith Miss RM Zingiberaceae
23 Watling R - BSc,PhD,MIBiol mycology
24 Woods J Umbelliferae
25 Woods PJ Orchidaceae

4754 University of Aberdeen

Regent Walk, Aberdeen, AB9 1FX
Tel: (0224) 480241.

4755 Agriculture

The School of Agriculture, 581 King Street, Aberdeen AB9 1UD: Tel (0244) 480291 is formed by the joint institutes of the North of Scotland College of Agriculture and the Department of Agriculture, University of Aberdeen

1 Martin A - BScAgr,MSFRAgrSc *Professor of Agriculture and Deputy Principal of College* agricultural extension
2 Edwards IE - BScAgr,PhD *Lecturer in Agriculture* ruminant nutrition; beef and sheep production; agronomy of micropropagated potatoes

4756 Animal Production and Health

1 Greenhalgh JFD - BA,MS,MAP,PhD *Professor of Animal Production and Health*
2 Bampton PR - BSc,PhD animal breeding
3 English PR - BScAgr,NDA,PhD animal production
4 Hutchinson JSM - BSc,PhD,DipAgrSc,MIBiol endocrinology of reproduction

4757 Animal Husbandry

1 Kay M - BSc,PhD *Head* beef production
2 Broadbent PJ - BScAgr,PhD beef production
3 Keeling Miss BJ - BA,PhD sheep production
4 Thomas S - BScAgr extension work, sheep production
5 Vipond JE - BScAgr extension work, sheep production

4758 Farm Buildings

1 Robinson TW - BA,MSc *Head* general farm buildings design
2 McHattie KL - ARIBA general farm buildings design

4759 Poultry Husbandry

1 Michie W - SDA,NDD,NDP *Head* egg production
2 Rose SP - BSc,PhD environment interactions

4760 Veterinary Hygiene and Physiology

1 Macdonald DC - MSc,MRCVS *Head* diagnosis, control and prevention of production diseases in beef and dairy herds, sheep flock production and health programmes
2 Downie JG - BSc,BVMS,PhD,MRCVS health and production programmes for beef herds

4761 Veterinary Investigation

1 Halliday GJ - BVMS,MRCVS *Acting Head* diseases general
2 Gunn GJ - BVMS,MRCVS studies of infertility in the bovine and ovine; mastitis in the dairy herd with particular reference to pseudomonas infection
3 Hollands RD - BVMS,MRCVS sheep and poultry diseases
4 Inglis DM - BVMS,MRCVS metabolic disorders
5 Johnston WS - MRCVS sheep diseases
6 MacLachlan GK - BScVeySc,MRCVS diseases general
7 Roberts L - BVMS,MRCVS reproductive diseases
8 Ross HM - BVMS,MRCVS diseases general
9 Smith WJ - BVMS,MRCVS pig diseases

4762 Horticulture

1 Dixon GR - BSc,PhD *Head* biological constraints on horticultural production
2 Berridge JW - BSc field vegetable production
3 Blain AQM - BSc ornamental nursery stock production
4 Sutherland JP - SDH,NDA,SHM soft fruit production
5 Sutton MW - BSc ornamental bulb and protected crop production

4763 Agricultural Zoology

1 Shaw MW - BScAgr,MSc,PhD *Head* aphids; *Plutella*; turnip root fly
2 Blasdale P - BA invertebrate faunas of grassland
3 Mackintosh GM - MSc,MS,PhD plant nematology
4 Mobus BOE - NDB beekeeping

4764 Grassland Husbandry

1 Copeman GJF - BScAgr *Head* animal grass relationships
2 Heath SB - BSc,PhD agronomy of grassland and crops
3 Mackie CK - BScAgr animal grass relationships
4 Weddell JR - BScAgr grass production and utilisation
5 Younie D - BScAgr,MAgrSc hills and uplands; herbage legumes

4765 Crop Husbandry

1 Paterson WGW - BScAgr,MSc *Head* crop production - general
2 Addison WG - BScAgr,DipAgr,DTA crop production - potatoes
3 Blackett GA - BScAgr,MSc crop production - cereals
4 McMartin I - BScAgr crop production - general
5 Morrice LAF - BScAgr crop production - National List Trials
6 Taylor BR - BSc,PhD crop production - general
7 Walker KC - BScAgr crop production - general

4766 Agricultural Botany

1 Matthews S - BSc,PhD,DSc *Head* seed physiology; growth regulators
2 Munro JM - BSc,PhD plant pathology - wheat; rape; potatoes
3 Naylor REL - BSc,PhD grass and cereal ecology and physiology
4 Powell AA - BSc,PhD seed physiology and pathology
5 Scragg EB - BScAgr weed competition and control
6 Wale SJ - BSc,PhD plant pathology - barley; potatoes; vegetables
7 Watson RD - BScAgrBot,SDH,DIC mycotoxins; respiratory allergies

4767 Agricultural Chemistry & Biochemistry

1 Topps JH - BSc,PhD,DSc,FRSC,CChem *Head* ruminant nutrition
2 Farr E - BSc,DPhil,DipSoilSc soil science
3 Galbraith H - BSc,PhD endocrinology of growth in ruminants
4 Gelman AL - BSc,PhD,DipAgrSc spectrochemistry
5 Murray I - BSc,PhD evaluation of foods near infrared reflectance spectroscopy
6 Scaife JR - BSc,PhD propionate metabolism and induced changes in fat metabolism in ruminants
7 Stephen NH - BSc,PhD chemistry of crop protection and chemical pollution
8 Stuchbury T - BSc,PhD metabolism of the cytokinin growth regulators
9 Thompson JK - BSc,BAgr,MS,PhD mineral nutrition of ruminants
10 Walker HF - BScAgr,PhD metabolism of vitamins in ruminants

4768 Agricultural Bacteriology

1 Hunter AC - BScAgr *Head* dairy microbiology
2 Fenlon DR - BA,MSc,PhD transmission of *Listeria*
3 Jeffrey DC - BScAgr,MSc dairy microbiology
4 Jones Miss SM - BScAgr,PhD L-form bacteria in plants
5 Paton AM - BSc,PhD L-form bacteria; fluorescence microscopy
6 Robinson K - BSc,PhD,MIBiol identification of *Erwinia* spp. and control of potato blackleg disease
7 Seddon B - BSc,PhD antibiotic production of spore-forming organisms

4769 Engineering

1 Shiach JG - JP,BScAgr,BScEng,FIAgrE *Head* general mechanisation
2 Bates CM - BScMechEng,AMIAgrE soil machinery; general mechanisation
3 Feilden NEH - MA,CEng,MIChemE farm wastes; energy studies
4 Hendry JC - EngTech,AMIAgrE general mechanisation
5 Jamieson JE - BScAgrE drainage
6 McGovern RE - BScAgrEng,AMIAgrE general mechanisation
7 Morris RJF - BSc drainage
8 Pringle RT - BScEng,MSc,MIAgrE ventilation; costings and general mechanisations
9 Robertson TW - BScAgr,MScAgrEng,TEng,MIAgrE materials handling
10 Scott DG - BScAgrEng,TEng,MIAgrE general mechanisation
11 Shepherd HM - SDA,NDAgrE,TEng,MIAgrE dairy engineering
12 Tupper J - HNDAgrEng,MSc,AMIAgrE spraying and grain drying

4770 Economics

1 Dalton GE - BScAgrEcon,MScAgrEconPhD *Head of Division* farm management; simulation; decision making; regional development
2 Thomson ITJ - MA,MSc,MS *Professor of Agricultural Economics* agricultural policy especially in a European context; rural and regional development planning; cereal marketing
3 Bone JS - BScAgrMSc farm business management; land use
4 Burgess G - BAMAgrSc agricultural economics; land use
5 Chalmers N - BScAgr farm business management; computer applications in farm business management
6 Cook P - BScAgr farm business management
7 Crabtree JR - BScAgr,PhD,MPhil business management; cereals marketing finance; forestry economics
8 Critchley R - BScAgrEcon farm management and agricultural extension
9 Daw ME - BScAgr,MScAgrEcon,MAgrFmECon,NDA farm economics; marketing; foreign agricultural development; project appraisal
10 Entwistle G - BSc agricultural marketing; futures trading
11 Hartley N - BScAgrEcon farm business management
12 Haughs Miss MA - BScAgr farm business management; farm debt
13 Isaacs RJ - BAgrSc,DipFMBA farm business organisation and management; profitability of farming; taxation
14 Johns P - PhD,BA sheep meat marketing and policy; input - output modelling
15 Leat PMK european community farm structures policy; input - output modelling and rural development; agricultural marketing and support policies
16 MacLaren D - BScAgrMs,PhD agricultural trade policy; applied econometrics
17 Revell BJ - BA,MAEcon policy analysis; marketing; quantitative methods
18 Sutherland RM - BScAgrEcon risk and uncertainty in agriculture; economics of beef production; farm business management

4771 Scottish Farm Building Investigation Unit

1 Bruce JM - MSc,PhD,CEng,MIMechE,MBIM *Head* building climatology
2 Baxter MR - BSc animal behaviour and welfare
3 Burnett GA - BA livestock and crop environment
4 Higgins KP - BE structures
5 MacCormack JAD - BSc,MIInfSc,TEng,MIAgrE information
6 Petchey AM - BSc,PhD,MIBiol livestock buildings
7 Robertson AM - BScAgr livestock and crop environment
8 Welch AR - MSc,BSc,BA,NDA,MRAC,MIBiol environment and health

4772 Forestry

St Machar Drive, Aberdeen, AB9 2UU

4773 Forest Biogeochemistry and Tree Nutrition

1 Miller HG - BSc,PhD,DSc,FICFor,FRSE *Professor and Head of Department* nutrient cycling in forests
2 Armstrong JA - BSc,MSc *ion transfer at leaf surfaces*
3 Kelly JM use of sewage sludge in forests
4 Rees RM - BSc,PhD acid production in forest soils

4774 Forest Biomass

1 Gervais PJ - BSc harvesting systems for short rotation coppice
2 Mitchell CP - BSc,PhD forest biomass as an energy feedstock
3 Wightman AD - BSc management of short rotation energy forests

4775 Forest Entomology

1 Parry WH - BSc,PhD,FRES insects pests on forest trees

4776 Forest Genetics and Tree Biotechnology

1 Blackburn P - BSc breeding of birch, alder and cherry
2 Brown IR - BSc,MS,PhD,MICFor breeding of forest trees; cytogenetics of birch species
3 Gordon AM - BSc protoplast fusion in birch
4 Pirrie P - BSc,PhD micropropagation and protoplast fusion in hardwood species

4777 Forest Management and Policy

1 Holmes WD - BA economic survey of private forestry in Scotland
2 Phillip MS - MBE,MA,BSc,FICFor *Reader in Forestry* forest management systems; econometric modelling of the forest sector
3 Todd JD - BSc economic survey of private forestry in Scotland

4778 Forest Mycology and Pathology

1 Millar CS - BSc,PhD needle fungi and litter-decomposing fungi
2 Pawsey RG - BSc,MSc,PhD,PCE tree pathology; ash decline

4779 Tree Physiology

1 Thompson S - BSc,PhD,MICFor physiology of early growth of conifers

4780 Wood Science and Timber Management

1 Henderson J - BSc,MICFor timber harvesting and extraction
2 Petty JA - MA,PhD,DSc,MInstP,MICFor fluid movement in wood; physics of windblow

4781 Plant Science (Botany)
St Machar Drive, Aberdeen, AB9 2UD

1 Gimingham CH - BA,ScD,PhD,FRSE,FIBiol *Professor and Head of Department* vegetation management in the uplands; ecology of land use and conservation
2 Alexander IJ - BScFor,PhD ectomycorrhizas, nutrient cycling in coniferous forests
3 Kenworthy JB - BSc,PhD restitution of vegetation on disturbed ground; phenology and age structure of birch woods
4 Sabnis DD - BSc,MSc,PhD tissue culture and micropropagation
5 Swaine MDS - BSc,PhD tropical forest dynamics and recovery after disturbance
6 Wilcock CC - BSc,MSc,PhD,FLS plant taxonomy and horticulture

4782 Soil Science
Old Aberdeen, Aberdeen, AB9 2UE

1 Parsons JW - BSc,PhD,FIBiol *Professor and Head of Department* soil chemistry; soil organic matter
2 Batey T - BSc,PhD,FIBiol soil management
3 Court MN - BSc,PhD soil physics and physical chemistry
4 Cresser MS - BSc,PhD,DIC,ARCS,FRSC,CChem soil and environmental chemical analysis
5 FitzPatrick EA - PhD pedology and micromorphology
6 Killham KS - BSc,PhD,DipALC,MIBiol soil microbiology; soil nitrogen and sulphur
7 Ley GJ - MSc plant root growth and hard setting soils
8 Livesey NT - BSc,PhD,CChem,FRSC soil mineralogy and pedogenesis
9 Mullins CE - BA,MSc,PhD soil physical conditions and plant root growth
10 Peirson-Smith TJ - BSc acid deposition and upland soils
11 Skiba UM - BSc,PhD acid deposition and upland soils
12 Whalley WR - BSc relation of soil clays to other soil properties
13 Young IM - BSc soil strength and root growth

4783 Zoology
University of Aberdeen, Department of Zoology, Tillydrone Avenue, Aberdeen, AB9 2TN

1 Mordue WM - BSc,PhD,DSc *Professor and Head of Department* insect physiology
2 Chappell LH - BSc,PhD animal parasitology
3 Dunnet GM - BSc,PhD,DSc *Regius Professor of Natural History* birds and mammals in agricultural land
4 Marshall AG - BScAgr,PhD,MS pollination and seed dispersal of tropical rain forest trees
5 Mordue (Luntz) AJ - BSc,MSc,PhD insect physiology; toxic effects of plant secondary compounds
6 Patterson IJ - BSc,DPhil bird damage to crops
7 Pike AW - BSc,PhD helminth parasites in farmed and sport fish (salmonoids)
8 Priede IG - BSc,PhD fish physiology and aquaculture
9 Secombes CJ - BSc,PhD fish immunology and reproduction and diseases
10 Young MR - BSc,PhD analysis of insect communities in variously managed stands of commercial Scots Pine forest

4784 University of Dundee
Dundee, DD1 4HN
Tel: (0382) 23181.

4785 Biological Sciences

1 Stewart Professor WDP - PhD,DSc,FRS biological nitrogen fixation; agricultural fertilizers; eutrophication and water quality
2 Daft MJ - BSc,PhD,DSc,FRSE mycorrhizal plants in semi arid areas, and pathogens of procaryotes
3 Hastie AC - BSc,PhD genetics of plant-pathogenic fungi
4 Ingram HAP - BA,PhD peatland ecohydrology, soil physics, drainage, management
5 Sprent JI - BSc,ARCS,PhD symbiotic biological nitrogen fixation

4786 University of Edinburgh

4787 The Edinburgh School of Agriculture
West Mains Road, Edinburgh, EH1 3JG
Tel: 031 667 1041. Telex: 727617.

The East of Scotland College of Agriculture, including the Scottish Agricultural Advisory Service, is shown attached to the University's School of Agriculture

1 Wilson PN - BSc,MSc,PhD,DipAnGen,FIBiol *Head of School and of ESCA, Professor of Agriculture and Rural Economy* animal nutrition

4788 Animal Division

1 Whittemore CT - NDA,BSc,PhD *Professor of Animal Production* growth, reproduction, lactation and bio-economic modelling in farm animals

4789 Agricultural Biochemistry

1 Duffus CM - BSc,MS,PhD,DIC,DSc *Acting Head* biochemistry and physiology of seed development
2 D'Mello JPF - BSc,PhD amino acid metabolism
3 Henderson AR - BSc,PhD silage biochemistry
4 McLellan AR - BSc,PhD open circuit calorimetry

4790 Animal Nutrition

1 Donaldson E - BSc,ARIC,PhD silage fermentation and ruminant nutrition
2 Edwards RA - MSc,PhD *Head* nutrition of ruminants
3 Lewis M - BSc,PhD nutrition of ruminants
4 Morgan CA - BSc,PhD nutrition of non-ruminants

4791 Animal Production (Teaching)

1 Whittemore CT - NDA,BSc,PhD *Head* growth, reproduction, lactation and bioeconomic modelling in farm animals
2 Appleby M - BSc,PhD animal behaviour
3 Black WJM - BSc,PhD intensive sheep production
4 Feilding D - BSc,MSc,PhD exploitation of equines in the tropics
5 Guild WJ - BSc,PhD thermoregulation in mammals
6 Hinks CE - BSc,PhD nutrition of calves and beef cattle; carcase evaluation
7 Illius AW - BSc,PhD physiology and behaviour of grazing ruminants
8 Jessop NS - BSc,PhD energy metabolism in ruminants
9 Manson JM - BSc,MS,PhD poultry husbandry and genetics
10 Wood-Gush DGM - BSc,PhD,DipAnGen,FRSA,FRSE animal behaviour
11 Wyllie D - BSc,MS,PhD agricultural education, tropical and temperate animal production

4792 Animal Production (Advisory and Development)

1 Oldham JD - BSc,PhD *Head* ruminant nutrition; dairy cow production; simulation of animal performance
2 Bell EM - NDD,SDPH systems of large scale dairying
3 Dingwall WS - BSc,PhD nutrition and reproduction in sheep
4 Emmans GC - BSc description of growth in farm animals; simulation of animal performance
5 Hillyer GM - BSc production systems and nutrition of pigs
6 Hodgson-Jones LS - BSc intensive dairy herd management; farm waste disposal
7 Lawrence AB - BSc,PhD behaviour of farm animals
8 Lloyd MD - BSc growth and carcase quality in sheep
9 Lowman BG - BSc,PhD nutrition and management of beef cattle; intensive beef production systems
10 Neilson DR - HND,MIBiol,MSc dairy cow production; intensive beef production systems
11 Simm G - BSc,PhD breeding and genetic improvement of farm animals; assessment of live animal carcase composition

4793 Crop Division

1 Holmes JC - BSc,MS,PhD *Professor of Crop Production* crop production, crop nutrition, root/soil interaction

4794 Applied Plant Science

1 Lennard JH - BSc,PhD *Head* foliar pathogens in cereals; genetic variation and resistance; seed pathology
2 Gerard BM - BSc,ARCS,PhD integrated pest management
3 Habeshaw D - BSc,PhD plant physiology
4 Smith ML - BSc,PhD,MIBiol seed physiology; plant growth regulators
5 Spoor W - BSc,PhD plant tissue culture; genetics of host parasite interactions
6 Turner MR - BSc,PhD,MIBiol seed technology

4795 Crop Production and Physiology (Teaching)

1 Thow RF - MSc,PhD *Head* growth and development of the potato crop
2 Hughes G - BA,PhD environmental physiology of tropical and sub-tropical grain legumes
3 Russell G - BSc,PhD cereal and oilseed physiology

4796 Crop Production (Advisory and Development)

1 Fisher NM - BA,PhD *Head* cereal agronomy
2 Berndt GW - BSc,PhD,CChem,MRSC trace element nutrition in relation to soils, plants and animals
3 Crooks P - BSc,PhD chemical methods of assessing nutrient availability in soil
4 Davies DHK - BSc,PhD weed science
5 Gill WD - BSc brassica production
6 Lang RW - SDA,NDA,PhD potato production
7 Lockhart DAS - TD,MA,MSc,PhD cereal production
8 Morrison MW - DHE,NDH herbage variety testing
9 Purves D - BSc,PhD,FRSC trace element deficiency and toxicity problems
10 Richards MC - BSc,DCP weed science; cereal variety testing
11 Swift G - BSc,MSc grassland production

4797 Crop Protection

1 Gilmour J - BSc,PhD,MIBiol *Head* cereal diseases; fungicide insensitivity
2 Brokenshire T - BSc,PhD brassica clubroot; raspberry diseases
3 McKinlay RG - BSc,PhD potato aphids, cabbage root fly; pesticide application technology
4 Oxley SJP - BSc,PhD bacterial and storage diseases of potatoes
5 Spaull AM - BSc,PhD potato cyst and cereal cyst nematode control

4798 Horticulture

1 Fordyce W - SDH,NDH *Head* soft fruit and field vegetable production
2 Halford M - BSc soft fruit crop production
3 Horgan AR - DHE,BSc vegetable crop production
4 McGhee JGJ - SDH soft fruit crop production
5 Mitchell DM - BSc vegetable crop production; plant tissue culture

4799 Soil Science

1 Smith KA - BSc,PhD,FRSC *Head* soil nitrogen studies, including denitrification and N-15 tracer element availability
2 Frost CA - BSc soil physical conditions
3 Naysmith DB - BSc,PhD,MInstP nitrogen uptake by crops; soil water movement
4 Speirs RB - AH-WC soil erosion

4800 Extension Division

1 Runcie KV - BSc,MS,FRAgS *Director of Extension Services* extension methods

4801 Agricultural Engineering and Mechanisation

1 Boyd JEL - BSc,MIAgrE forage conservation
2 Jack DA - MSc,NDA,NDAgrE,MIAgrE seed technology
3 Jeffrey WA - BSc crop spraying
4 Langley A - MSc,NDAgrE,MIAgrE soil tillage and crop establishment
5 Oskoui KE - DipSci,BSc,PhD,MIAgrE machinery management
6 Parkes ME - BSc,MSc,PhD,MIAgrE soil and water conservation
7 Witney BD - MSc,NDA,NDA,MIMechE,CEng,FIAgrE *Head* soil machine mechanisms; potato crop mechanisation; machinery management

4802 Agricultural Resource Management (Advisory and Development)

1 Lilwall NB - BSc,MA,PhD *Head* marketing and rural development
2 Anderson JL - BSc accounts; potato economics
3 Beattie IC - BSc,MAgrSc,DipAnHusb cereal economics; EEC policy
4 Blight CM - BL,MSc non-price competition; agricultural law
5 Chadwick LD - BSc agricultural inputs
6 McLean J - BSc farm, business management
7 Mordaunt FD - BA,DipAgEcon farm business management
8 Morgan OW - BSc systems analysis
9 Scanlan WS - BA,MSc beef economics
10 Tait SE - HND,DipFBM,MSc farm business management
11 Volans JKS - BSc,MSc sheep economics; marketing and information technology
12 Wright EM - BSc,NDA horticultural economics

4803 Agricultural Resource Management (Teaching)

1 Fawcett RH - BSc alternative land use strategies; risk management
2 Heney JA - BSc,DipAgEcon,MA overseas development
3 Stott AW - BSc,PhD economic efficiency versus biological efficiency as a means of selection in dairy cows
4 Wagstaff HR - BA,MSc lower input systems in agriculture
5 Wilson RM - BSc,MSc,PhD farm capital structure and debt; lamb marketing strategy

4804 Scottish Centre for Agricultural Engineering
Bush Estate, Penicuik, Midlothian, EH26 0PH
Tel: 031-445-2147.

Mechanisation of crops, particularly the potato; soil mechanics and engineering for the Uplands

1 Blight DP - BSc,MSc(AgrEng),PhD,CEng,FIMechE,FIAgrE *Director*
2 Fraser Mrs HM - MSc *Librarian*

4805 Agriculture
Bush Estate, Penicuik, Midlothian, EH26 0PH
Tel: 031-445-2147.

1 Ball BC - BSc PhD Establish the effects of several levels of seedbed traffic; Compare reduced and zero-cultivation with ploughing for cereals; Develop new and existing methods of measuring soil pore characteristics and gas movement
2 Campbell DJ - BSc MPhil Establish the effects of several levels of seedbed traffic; Examine normal and zero traffic systems for cereals, grass and potatoes; Compare reduced and zero-cultivation with ploughing for cereals; Measure, compare and predict (using a model) compaction by different wheel systems; Develop new and existing methods of measuring soil physical properties; Examine soil and crop responses to subsoiling and other drainage operations; Testing forest soils for mechanical properties in relation to windthrow problems (Forestry Commission)
3 McRae DC - BSc MSc(AgrEng) CEng FIAgrE Test potato planter for high speed precision planting of chitted seed; Improve shares, webs and other components of potato harvesters; The factors affecting and methods of measuring the mechanical damage of potatoes; Develop mechanisms for conveying and handling of potatoes with low damage; Develop aids for quality inspection and grading using electronic and optical techniques; Design, construct, assemble and test semi-automatic separator for apples (Normark); Develop potato harvesting attachment and prepare drawings for manufacture (Elbar)
4 Pascal JA - BA NDA TEng(CEI) MIAgrE Apply rotary principle to a ridger/bed shaper and to an implement for straw incorporation; Improve depth control and general performance of shallow ploughs; Improve and test a coulter for drilling cereals in the presence of trash; Develop and test equipment for improving pastures on hill land
5 Smith DLO - BSc,MS,PhD Predicting (using a model), the compaction resulting from different wheel systems on different soils. Developing new and refining existing methods of measuring soil strength characteristics with triaxial equipment. Testing forest soils for mechanical properties in relation to windthrow problems

4806 Engineering
Bush Estate, Penicuik, Midlothian, EH26 0PH
Tel: 031-445-2147.

1 Duncan EJ - BSc(Maths),BSc(Comp.Sci) Swath drying of grass in the laboratory and air flow modelling
2 Haughey DP - BE ME PhD CEng FIChemE Data collection and analysis on the energy requirements of various systems of crop production; Strategy and interfaces necessary for renewable energy sources; Mathematical modelling of alternative energy systems
3 Hunter AGM - MA CEng MIMechE Forage mechanisation on slopes; Tyre parameters affecting braking and side force performance; Safe working on slopes; Slope monitor evaluation (HSE contract)
4 Lamond WJ - HNC MIAgrE Swath drying of grass in the field; Optimum design of forage equipment for Scottish conditions
5 Lewis C - BSc,PhD,MIBiol Data collection and analysis on the energy requirements of various crop production systems. Strategy and interfaces necessary for renewable energy sources. Mathematical modelling of alternative energy systems
6 Mathers MT - BSc,FTC,AMIMI Data collection and analysis on the energy requirements of various systems of crop production
7 McGechan MB - BSc PhD CEng MIMechE Field and operational research studies of forage systems
8 Moore AB - HNC Loads on tined implements while in work; Development of techniques for grain yield mapping within a field
9 Owen GM - HNC CEng MIMechE MIAgrE Forage mechanisation on slopes and braking and traction; Safe working on slopes; Slope monitor evaluation (HSE contract)
10 Palmer J - BSc MSc(AgrEng) MA CEng FIAgrE Loads on tined implements while in work; Development of techniques for grain yield mapping within a field
11 Quigley AD - BSc,MSc Tyre parameters that affect braking and side force performance
12 Spencer HB - HND CEng MSc FIAgrE Swath drying of grass in the field; Swath drying of grass in the laboratory; Optimum design of forage equipment for Scottish conditions; Solar-assisted barn drying of hay and grain

4807 Instrumentation
Bush Estate, Penicuik, Midlothian, EH26 OPH
Tel: 031-445-2147.

1 Muir AY - MPhil,MInstP,CPhys Instrumentation for detecting diseases and other defects in plants. Relationship between mechanical abuse of crops and the resulting mechanical damage. Multichannel optical scanning systems
2 Porteous RL - BSc Instrumentation for detecting diseases and other defects in plants. Relationship between mechanical abuse of crops and the resulting mechanical damage. Multichannel optical scanning system

4808 Veterinary Division

1 Mathieson AO - BVM&S,MSc,MRCVS *Regional Veterinary Investigation Officer* gastro intestinal parasitism in sheep
2 Appleyard WT - BVMS,MRCVS diagnostic methods in sheep diseases
3 Gould PW - BVetMed,PhD,MSc,MRCVS sheep pneumonia
4 Greig A - BVM&S,FRCVS sheep infertility
5 Hosie BD - BVM&S,MSc,MRCVS diagnostic methods in animal diseases
6 Linklater KA - BVM&S,PhD,MRCVS border disease; enzootic abortion of sheep; respiratory disease in sheep
7 Low JC - BVSc,MRCVS diseases of poultry
8 MacLachlan I - DVSM,MRCVS diseases of poultry
9 Miller JK - MVSc,MRCVS premature tooth losses in sheep; genetic susceptibility to diseases in dairy cattle
10 Sutherland HK - BSc epidemiological studies in animal diseases
11 Synge BA - BVM&S,MPhil,MRCVS enteric disease in cattle and sheep
12 Thomson JR - BVSc,PhD,MRCVS respiratory disease in cattle

4809 Microbiology Division

1 Wilkinson JF - MA,PhD *Professor of Microbiology* methane utilising micro-organisms
2 Chard JM - BSc,PhD biological control of plant pathogens
3 Dawes IW - BSc,DPhil regulation of morphogenesis
4 Deacon JW - BSc,PhD biological control of plant pathogenic fungi; successions of ectomycorrhizas; cereal root senescence and rhizosphere biology
5 Lowe JF - BSc,PhD microbiology of silage; bacterial plant diseases
6 McLarty RM - BSc,SDD general agricultural microbiology particularly with water/environment
7 Moseley BEB - BSc,PhD DNA repair and mutogenesis; genetics of silage bacteria; genetics of heat resistance in bacteria
8 Reid GA - BSc,PhD biochemistry and molecular biology of protein targeting
9 Robb J - NDD,MSc biodeterioration of hay and grain; mycotoxins in animal feedingstuffs
10 Sutherland IW - PhD,DSc HPLC of carbohydrates; polysaccharide-degrading enzymes
11 Ward FB - MA,PhD microbial energy metabolism, bacterial proteases

4810 Department of Botany
The King's Buildings, Mayfield Road, Edinburgh, EH9 3JH
Tel: (031 667) 1081.

1 Yeoman MM - PhD *Regius Professor and Head of Department* biosynthetic potential of cultured plant cells; physiology of grafting
2 Dale JE - PhD *Professor* leaf growth, and development of cereals
3 Dixon ROD - PhD *Senior Lecturer* nitrogen fixation in legume root nodules
4 Dyer AF - DPhil *Senior Lecturer* biology of ferns
5 Fry SC - PhD *Lecturer* cell wall biochemistry and metabolism
6 Jeffree CE - BSc *Director of EM Unit* plant surface structure; plant cellular differentation
7 Kemp RFO - DPhil *Lecturer* genetics, incompatibility and speciation of the genus *Coprinus*
8 Leaver CJ - PhD *Reader* mitochondrial genome organisation and expression with particular reference to cytoplasmic male sterility
9 Lyndon RF - PhD *Senior Lecturer* morphogenesis and physiology of meristems
10 Mann DG - PhD *Lecturer* biology of diatoms
11 Read ND - PhD *Lecturer* fungal development
12 Smith JAC - PhD *Lecturer* bioenergetics, membrane transport and plant water relations
13 Smith PM - PhD *Senior Lecturer* evolution and taxonomy of the Gramineae
14 Smith SM - PhD *Lecturer* genetic manipulation
15 Trewavas AJ - PhD *Reader* biochemistry of plant hormones and regulation of plant development by calcium
16 Tulett AJ - BSc *Lecturer* electron microscopy of plant cells

4811 Department of Forestry & Natural Resources
Darwin Building, King's Buildings, Mayfield Road, Edinburgh, EH9 3JU
Tel: 031 667 1081.

1 Mutch WES - OBE,PhD *Head of Department* resource economics; farm-forest management
2 Blyth JF - PhD *Lecturer* forest management; irregular woodland and agro-forestry systems
3 Ennos RA - PhD *Lecturer* forest genetics; genetics of forest pathogens

4 Grace J - PhD *Reader* environmental physiology of plants; tropical forest ecology
5 Hayes AJ - PhD *Senior Lecturer* forest entomology and pathology
6 Jarvis PG - Professor *environmental physiology; assimilation and transpiration; production modelling*
7 Jones PJ - DPhil *Lecturer* wildlife ecology; tropical ornithology; agricultural bird pests
8 Ledger DC - PhD *Lecturer* water shed hydrology
9 Legg CJ - PhD *Lecturer* heathland ecology and vegetation dynamics
10 Malcolm DC - PhD *Senior Lecturer* silviculture; tree-soil interactions
11 Mills DH - PhD *Senior Lecturer* fisheries management; salmonoid ecology
12 Moncrieff JB - PhD *Lecturer* micro-meteorology; mass and energy exchange from atmosphere to forest
13 Muetzelfeldt RI - PhD *Lecturer* computer modelling
14 Taylor IR - PhD *Lecturer* wildlife ecology; conservation and farming

4812 Department of Genetics
University of Edinburgh, West Mains Road, Edinburgh, EH9 3JN
Tel: (031) 667 1081.

1 Truman DES - PhD *Head of Department* cellular differentiation
2 Auerbach Professor EC - DSc,FRSE,FRS *Professor Emeritus* mutation and mutagens
3 Beale Professor GH - MBE,FRS,FRSE genetics of protozoa, especially parasitic
4 Bishop JO - PhD,FRSE cloning genes from animals
5 Bond DJ - PhD fungal morphogenesis and development
6 Campbell JC - JP,PhD tissue regeneration
7 Clayton Mrs RM - MA,MIIE animal development, especially lens and retina
8 Falconer Professor EDS - ScD,FRSE,FRS population and quantitative genetics
9 Hill Professor WG - DSc,FRSE,FRS *Professor of Animal Genetics* animal breeding, population genetics, theory of selection
10 Jacob J - PhD ultrastructure of cellular development
11 Jones KW - PhD DNA sequences, especially in animal viruses and sex chromosomes
12 Jurand A - PhD,DSc,FRSE teratogens and teratogenesis
13 Kascer H - PhD control of metabolism
14 Kilbey BJ - PhD,DSc mutation and mutagens
15 Leigh Brown AJ - PhD evolutionary genetics
16 Mackay Miss TFC - PhD population and ecological genetics
17 Reeve ECR - DPhil,FRSE *Editor - 'Genetical Research'*
18 Robertson Professor A - OBE,DSc,FRS animal breeding, population genetics, theory of selection
19 Selman GG - PhD mechanisms of morphogenesis
20 Tait A - PhD genetics of protozoa, especially parasitic
21 Walliker D - DAP&E,PhD genetics of malarial parasites

4813 Royal (Dick) School of Veterinary Studies
University of Edinburgh, Summerhall, Edinburgh, EH9 1QH
Tel: 031 667 1011. Cables: Veter.

4814 Department of Pre-Clinical Veterinary Sciences

1 House CR - BSc,MSc,PhD,FRSE *Professor and Head of Debt.* mammalian fertilization
2 Anderson L - BSc,ARCS,PhD *Research Fellow* cardiovascular and respiratory mechanisms
3 Bartlet AL - BPharm,PhD *Senior Lecturer* avian pharmacology
4 Bland KP - BSc,PhD *Lecturer* reproductive endocrinology
5 Brown AG - BSc,MB,ChB,PhD,FRSE *Professor* structure and function in the mammalian spinal cord
6 Carlyle SS - MA,PhD,VetMB,MRCVS *Lecturer* calcium metabolism
7 Copland AN - BSc,MSc,BVMS,MRCVS *Lecturer* red-foot
8 Cottrell DF - BVSc,MSc,PhD,MRCVS *Lecturer* visceral sensory mechanisms
9 Fleetwood-Walker SM - BSc,MSc,PhD *Lecturer* endogenous analgesic mechanisms
10 Georgiou P - BSc,PhD *Research Fellow* mammalian fertilization
11 Haigh AL - BSc,PhD,MIBiol *Senior Lecturer* cardiovascular and respiratory mechanisms
12 Hope PJ - BSc,PhD *Research Fellow* endogenous analgesic mechanisms
13 Houchin RJ - BSc,PhD *Research Fellow* structure and function in the mammalian spinal cord
14 Iggo A - BSc,MAgricSc,PhD,DSc,FRSE,FRS *Professor* cutaneous sensory mechanisms
15 Kai Kai Miss MA - BSc,MA,PhD *Lecturer* histochemistry of sensory ganglia
16 Kempson Miss SA - BSc,PhD *Lecturer* oral mucosa/surgical implant material
17 Leaver JL - BSc,PhD *Senior Lecturer* bacterial respiration
18 Macdonald AA - BSc,PhD,DVM *Lecturer* fetal physiology, growth and development
19 Martin RJ - PhD,BVSc,MRCVS *Lecturer* neuropharmacology
20 Maxwell DJ - BSc,PhD *Research Fellow* structure and function in the mammalian spinal cord
21 McLelland J - BVMS,MVSc,PhD *Reader* avian anatomy
22 Molony V - BVSc,MSc,PhD,MRCVS *Lecturer* avian respiration and pain mechanisms
23 Noble R - BSc,PhD,MIBiol *Research Fellow* structure and function in the mammalian spinal cord

24 Pettigrew GW - BSc,PhD *Lecturer* bacterial respiration

25 Short AD - MA,BSc,BM,BCh *Lecturer* vascular endothelial cells

26 Stafford WL - BSc,PhD *Lecturer* nutritional aspects of zinc

27 Steedman WM - BSc,PhD *Research Fellow* sensory mechanisms in the spinal cord

4815 Department of Veterinary Clinical Studies

1 Imlah P - PhD,MSc,MRCVS *Acting Head of Department and Senior Lecturer* lymphosarcoma in pigs, porcine stress syndrome

2 Beck R - MRCVS *Lecturer* origins of native ponies

3 Borthwick R - MRCVS *Senior Lecturer* cryosurgery, oncology

4 Camburn MA - BVSc,MRCVS *Lecturer* anaesthesia, fluidtherapy, orthopaedics

5 Cockram M - BVetMed,MRCVS *Lecturer* animal welfare, environmental studies

6 Cuddeford D - BSc,MSc,PhD *Lecturer* urolithiasis, animal nutrition, alternative horse feeds

7 Darke PGG - BVSc,PhD,MRCVS,DVR *Director of Small Animal Clinic* small animal cardiology and respiratory disorders

8 Doxey DL - BVM&S,PhD,MRCVS *Senior Lecturer* isoenzyme studies in sheep and horses

9 Fraser JA - BSc,FRCVS *Senior Lecturer* orthopaedics

10 Kelly JM - BVMS,MRCVS *Senior Lecturer* dairy herd health scheme, equine nutrition

11 Lawson JB - BVMS,DVSM,MRCVS *Senior Lecturer* epidemiology of Orf

12 McPherson EA - MSc,MRCVS *Senior Lecturer* chronic respiratory disease in horses

13 Milne EM - BVM&S,PhD,MRCVS *Demonstrator* equine muscular problems

14 Munro CD - BVM&S,PhD,MRCVS *Lecturer* obstetrics and reproduction

15 Robertson IS - BSc,PhD,MIBiol *University Fellow* beef production, immunological castration of cattle

16 Simpson JW - SDA,BVM&S,MPhil,MRCVS *Lecturer* equine digestive disturbances

17 Smith Miss EJ - BSc *Research Associate* dairy herd health scheme

18 Thoday KL - BVetMed,MRCVS *Lecturer* dermatology and feline thyroid dysfunction

19 Thrusfield MV - MSc,BVMS,DTVM,MIBiol,MRCVS *Lecturer* computerised data storage and retrieval systems, urinary incontinence in bitches, canine heart disease

20 Van den Broek AHM - BVSc,FRCVS,DVR *Lecturer* small animal dermatology

21 Whitaker D - MA,VetMB,MVScSyd,MRCVS *Research Fellow* dairy herd health scheme

22 Wilson JC - BSc,MRCVS *Senior Lecturer* obstetrics, pregnancy diagnosis and infertility

4816 Veterinary Pathology

1 McConnell I - BVMS,MA,PhD,MRCPath,MRCVS *Professor and Head of Dept.* immunology, immunopathology and clinical veterinary immunology

2 Bingham RW - MA,PhD *Lecturer* molecular virology

3 Bujdoso R - BSc,PhD *Research Fellow* cellular immunology

4 Dutia B - BSc,PhD *Research Fellow* immunochemistry/virology

5 Else RW - BVSc,PhD,MRCVS *Lecturer* cancer studies

6 Fraser G - BSc,PhD,MRCVS *Senior Lecturer* virology

7 Harkiss GD - BSc,PhD *New Blood Lecturer* molecular immunology

8 Head KW - BSc,MRCVS *Senior Lecturer* cancer studies

9 Hopkins J - BSc,PhD,MIBiol *Lecturer* cellular immunology

10 Lawson GHK - BSc,PhD,BVM&S *Reader* bacteriology and intestinal disease

11 Phillips JE - DVMS,BSc,MRCVS *Senior Lecturer* bacteriology

12 Ritchie JSD - MRCVS *Lecturer* diagnostic pathology

13 Rowland AC - BSc,MRCVS *Lecturer* intestinal disease studies

14 Sargan D - BSc,PhD *New Blood Lecturer* molecular biology

15 Watt NJ - BVM&S,PhD,MRCVS *Lecturer* molecular pathology

16 Young PM - BSc *Research Associate* cellular immunology

4817 Tropical Animal Health

1 Brocklesby DW - DrMedVet,FRCPath,FRCVS *Professor and Head of Dept.* tropical veterinary diseases

4818 Animal Husbandry Section

1 Fielding D - BSc,MSc,PhD *Lecturer* tropical animal husbandry

2 Lawrence PR - BSc,PhD *Research Fellow* draught animals

3 Matthewman RW - BSc,MAgSc *Lecturer* tropical animal husbandry

4 Pearson RA - BSc,PhD *Research Fellow* draught animals

5 Smith AJ - BSc,MSc,NDA,PhD *Senior Lecturer* tropical animal husbandry

4819 Helminthology Section

1 Hammond JA - PhD,MRCVS,DVSM,DTVM,DAP&E *Senior Lecturer* helminthology

2 Harrison LJS - BSc,PhD *Research Fellow* cestode zoonoses

3 Joshua GWP - BSc,MSc,PhD *Research Fellow* cestode zoonoses

4 Sewell MMH - MA,PhD,VetMB,MRCVS *Reader* helminthology

4820 Microbiology Section

1 Edelsten RM - BVM&S,DVSM,MSc,MRCVS *Lecturer* tropical bacteriology

2 Morrow AN - BA,MVB,MRCVS,PhD *Research Fellow* streptothricosis

3 Scott GR - BSc,MSc,PhD,MRCVS *Reader* tropical microbiology

4821 Protozoology Section

1 Brown CGD - BVM&S,MRCVS *Senior Research Fellow* tropical protozoology (theileriosis)

2 Bell LJ - BSc,MPhil *Research Associate* tick tissue culture

3 Boid R - BSc,PhD *Research Fellow* trypanosomiasis

4 Frame IA - BSc *Research Associate* trypanosomiasis

5 Hunter AG - BVM&S,MRCVS,DTVM *Lecturer* tropical protozoology

6 Jones TW - BSc,PhD *Research Fellow* trypanosomiasis

7 Luckins AG - BSc,PhD *Senior Research Fellow* trypanosomiasis

8 Prain CJ - BSc,PhD *Research Fellow* trypanosomiasis

9 Ross CA - BSc,PhD *Research Fellow* trypanosomiasis

10 Walker AR - BSc,MIBiol,PhD *Research Fellow* entomology

4822 Zoology

West Mains Road, Edinburgh, EH9 3JT
Tel: (031-667) 1081.

1 Mitchison Professor JM - ScD,FRS *Professor of Zoology* the cell cycle in fission yeast; synchronous cultures; control of synthesis and division

2 Ansell JD - BSc,PhD normal and neoplastic differentiation of haemopoietic cells; immunobiology of pregnancy

3 Ashmole NP - PhD feeding behaviour and ecology of birds; distribution patterns and ecology of spiders; cave biology

4 Cosens DJ - BSc,PhD insect vision; a study of formicoidea indigenous to Scotland, and their role in conservation

5 Deag JM - PhD quantitative analysis of sociodynamics; interaction between ecology and social behaviour; evolution of social behaviour; material behaviour of domestic cats; ecology and ethology of the Barbary macaque

6 Ewing AW - PhD *Drosophila* behaviour and neurophysiology

7 Fantes P - MA,PhD genetics and molecular genetics of the cell cycle in *Schizosaccharomyces pombe*; periodicity of transcription, cell structural elements and the control of division

8 French V - BSc,DPhil the formation and regeneration of spatial patterns in developing animal tissues, particularly insect epidermis, developing and evolution

9 Inchley CJ - PhD regulation of immune responses; factors influencing the behaviour and distribution of lymphocytes

10 Loecing UE - PhD concogenic viruses, cell transformation and RNA (ribonucleic acid) metabolism

11 Maddy AH - MSc,PhD role of proteins in membranes, senescene of erythrocytes

12 Manning Professor AWG - DPhil *Professor of Natural History* behaviour genetics of Drosophila and mice; behavioural development in domestic rats

13 Matthews BE - BSc,PhD physiology and behaviour of helminths particularly invasion and eclosion mechanisms

14 Micklem HS - BA,MA,DPhil aging and the immune system: immunology of malaria and haematopoietic stem cells

15 Partridge L - MA,DPhil sexual selection in Drosophila, food preferences in rodents and habitat preferences in titmice

16 Preston PM - BSc,PhD,DAP&E immunity in infections with parasitic protozoa

17 Rogers ME - PhD vertebrate functional morphology - research for a book

18 Saunders DS - PhD insect circadian rhythms and photoperiodism; control of diapause and pupal eclosion

19 Tuft PH - PhD the state of water and the mechanisms of water regulation in amphibian and mammalian embryos; the effect of morphogenic inhibitors and haploidy on the water regulating mechanism

20 Wilson PAG - BSc,PhD roundworms - (a) growth; (b) larval metabolism and physiology adaption; (c) larval interaction with the mammal microvasculature

4823 MRC Mammalian Genome Unit

1 Bird AP - PhD methylation of the DNA in higher organisms; the distribution of methylated bases on well-characterized DNA fractions in relation to regulation of gene behaviour

2 Cooke H - PhD organisation of human dihydropholate reductase

4824 University of Glasgow

Glasgow, G12
Tel: 041 339 8855.

4825 Department of Animal Husbandry

1 Hemingway RG - MSc,PhD *Professor and Head of Department* ruminant nutrition

2 Fishwick G - BSc,PhD *Lecturer* ruminant nutrition

3 Parkins JJ - BSc,PhD *Senior Lecturer* ruminant nutrition

4 Pepper TA - BVMS,MRCVS *Lecturer* pig management

5 Ritchie NS - BSc,PhD *Senior Lecturer* ruminant nutrition

6 Stewart MF - MRCVS,DVS *Lecturer* companion animals in society

4826 Department of Botany

1 Berrie AMM - BSc,PhD *Reader in Agricultural Botany, Head of Department* control mechanisms in seed germination and bud dormancy
2 Boney AD - PhD,DSc,FRSE,FLS,FIBiol *Titular Professor* biology and ecology of marine algae
3 Bowes BG - BSc,PhD *Senior Lecturer* structural aspects of morphogenesis
4 Brett CT - BA,PhD *Lecturer* biochemistry of plant cell walls
5 Clarke DD - BSc,PhD *Lecturer* host-parasite interactions
6 Cogdell RG - BSc,PhD *Lecturer* primary reactions of bacterial photosynthesis
7 Crozier A - BSc,PhD *Reader* physiology and biochemistry of plant hormones
8 Dickson JH - BSc,MA,PhD *Senior Lecturer* palaeoecology; vegetational history of Scotland; archaeobotany
9 Elliott CG - MSc,PhD *Reader in Agricultural Botany* reproduction in oomycete fungi
10 Hipkins MF - BSc,PhD *Lecturer* primary processes of photosynthesis in chloroplasts, intact leaves and unicellular algae
11 Jenkins GI - BSc,PhD *Lecturer* plant molecular biology, photosynthesis and photomorphogenesis
12 Milner JJ - BA,DPhil *Lecturer* plant virology
13 Murphy KJ - BSc,PhD *Lecturer* community ecology of freshwater macrophytes and ecology of terrestrial weeds
14 Wheeler CT - BSc,PhD *Lecturer* symbiotic nitrogen fixation
15 Wilkins MB - PhD,DSc,FRSE *Regius Professor* circadian rhythms in plant metabolism and the sensory physiology of plants

4827 Department of Agricultural Chemistry

1 Duncan HJ - BSc,PhD *Reader* agrochemicals and plant nutrients - their analysis, fate and behaviour in soil plant systems and crop storage
2 Flowers TH - BSc,PhD *Lecturer* behaviour of nitrogen in soil, both inorganic forms and organic wastes
3 Jarvis MC - BSc,PhD *Lecturer* chemistry of carbohydrates - their relevance to fungal disease control, growth regulators, digestibility and fruit processing
4 Pulford ID - BSc,PhD *Lecturer* derelict land reclamation and soil trace metal chemistry

4828 Department of Veterinary Parasitology

1 Urquhart GM - PhD,MRCVS *Professor and Head of Department*
2 Armour J - PhD,VMD,Drhc,MRCVS *Titular Professor (Dean)*
3 Duncan JL - BVMS,PhD,MRCVS *Reader*
4 Dunn AM - PhD,MRCVS,DVM *Senior Lecturer*
5 Henderson JJW - BSc *Research Technologist*
6 Jennings FW - BSc,PhD,MAgr *Reader*

4829 Department of Veterinary Pathology

1 Jarrett WFH - PhD,MRCVS,FRCPath,FRSE,FRS *Professor and Head of Department* viral leukaemia and neoplasia
2 Cornwall HJC - PhD,BVMS,MRCVS *Senior Lecturer* virology of distemper and herpes
3 Dawson Miss CO - PhD,BSc *Lecturer* mycosis
4 Hay D - FIMLS *Research Technologist*
5 Jarrett JO - PhD,BVMS,MRCVS *Titular Professor* leukaemia viruses
6 Laird Miss HM - PhD,MA *Lecturer* electron microscopy and viral leukaemia
7 McCandlish Miss IAP - BVMS,PhD,MRCVS *Lecturer* infectious diseases of respiratory and alimentary tract
8 McNeil Mrs PE - BVMS,PhD,MRCVS *Lecturer* alimentary disease
9 Onions DE - BVSc,PhD,MRCVS *Reader* leukaemia virus
10 Pirie HM - PhD,BVMS,MRCVS,MRCPath *Reader* respiratory diseases in cattle and dogs
11 Rycroft AN - BSc,PhD *Lecturer* microbial pathogenicity
12 Taylor DJ - PhD,MAVet,MRCVS *Lecturer* swine dysentery
13 Thompson H - PhD,BVMS,MRCVS *Senior Lecturer* respiratory diseases of dogs
14 Toth Mrs SR - PhD,DVM *Lecturer* haematology

4830 Department of Veterinary Pharmacology

1 Bogan JA - MSc,PhD,FRSC *Titular Professor and Head of Department* absorption and metabolism of drugs in ruminant and equine species; pharmacological mediators of helminth infections; pesticide residue analysis
2 Marriner SE - BVMS,PhD,MRCVS *Lecturer* pharmacokinetics of drugs of veterinary importance in the horse and ruminant; biochemical pharmacology and anthelmintics
3 McKellar QA - BVMS,PhD,MRCVS *Lecturer* mode of action of anthelmintic drugs

4831 Department of Veterinary Physiology

Glasgow, G12
Tel: 041 339 8855.

1 Holmes PH - BVMS,PhD,MRCVS *Titular Professor and Head of Department* pathophysiology and immunology of parasitic infections principally using radioisotopic techniques

2 Hosie KF - BSc,PhD,MRCVS *Senior Lecturer* cardiovascular and renal physiology
3 Jeffcoate IA - BSc,PhD *Lecturer* reproductive physiology
4 Maclean JM - MIBiol,MSc *Research Technologist* pathophysiology of helminth infections principally using radioisotopic techniques
5 Mulligan W - MSc,PhD,FRSE *Professor* the physiology of the host helminth relationship in domestic animals

4832 Department of Veterinary Surgery

1 Ayliffe TR - BSc,BVetMed,PhD,MRCVS *Lecturer* urine environment in bovine
2 Boyd HW - VMD,MRCVS *Senior Lecturer* herd fertility in cattle
3 Carmichael S - BVMS,MRCVS *Lecturer* small animal orthopaedics
4 Griffiths IR - BVMS,PhD,FRCVS *Reader* canine neurology
5 Harvey MJA - BVMS,PhD,MRCVS *Lecturer* chromosomal abnormalities; chromosomal content of all species
6 Lawson DD - BSc,DVR,MRCVS *Professor* radiology, neurology and opthalmology
7 Lee R - BVSc,PhD,MRCVS,DVR *Reader* veterinary radiology
8 Reid J - BVMS,MRCVS *Lecturer* anaesthesia
9 Renton JP - PhD,MRCVS *Senior Lecturer* control of breeding in equine both male and female; freezing of semen and use of frozen semen in canine
10 Robins RJ - BVSc,FRCVS *Lecturer* equine orthopaedics
11 Sullivan M - BVMS,MRCVS *Lecturer* soft tissue surgery

4833 Department of Zoology

1 Phillips RS - BSc,PhD,FIBiol *Professor and Head of Department* immune responses to protozoan infections
2 Bruce RG - BScAgr,PhD *Lecturer* cytology, immunology and epidemiology of helminths
3 Coombs GH - BSc,PhD *Lecturer* biochemistry and chemotherapy of parasitic protozoa
4 Crompton DWT - MA,PhD,ScD,FIBiol *John Graham Kerr Professor* parasitic helminths and nutrition
5 Dobson RM - BA,MA,PhD *Senior Lecturer* carophilus taxonomy; cranefly bionomics; pathogenic acari
6 Ede DA - BSc,MS,PhD *Reader* analysis of morphogenesis in normal and mutant vertebrate embryos at the cellular level
7 Furness RW - BSc,PhD *Lecturer* applied ornithology
8 Houston DC - BSc,DPhil *Lecturer* bird and mammal ecology
9 Kennedy MW - BSc,PhD *Lecturer* immune responses to helminth infections
10 Lackie AM - MA,PhD *Lecturer* invertebrate immune mechanisms
11 Lockey KH - BSc,PhD *Lecturer* insect cuticular lipids
12 Monaghan P - BSc,PhD *Lecturer* applied ornithology
13 Vickerman K - BSc,PhD,DSc,FRSE,FRS *Regius Professor* trypanosomiasis, parasitic protozoa

4834 Department of Biochemistry

1 Coggins JR - PhD *Senior Lecturer* biosynthesis of amino acids and herbicide resistance in plants
2 Fewson CA - PhD,FRSE *Professor* regulation of enzyme activity in plants and micro-organisms; metabolism of aromatic compounds by soil micro-organisms
3 Fixter LM - DPhil *Lecturer* lipids and membranes of *Erwinia* in relation to plant disease
4 Hamilton ID - PhD *Lecturer* physiology and biochemistry of *Erwinia* as a plant pathogen
5 Holms WD - PhD *Senior Lecturer* regulation of microbial metabolism
6 Jenkins GI - PhD *Lecturer* regulation of transcription of plant nuclear genes by light
7 Knowler JT - PhD *Senior Lecturer* wound healing in crop plants
8 Nimmo HG - PhD *Lecturer* purification and characterisation of plant and microbial enzymes

4835 Department of Veterinary Medicine

1 Murray M - BVMS,MRCVS,PhD,DVM,FRCPath,FRSE *Professor and Head of Department* genetic resistance in ruminants; molecular parasitology; trypanosomiasis
2 Fisher EW - PhD,BSc,DVM,MRCVS *Titular Professor* cardiovascular diseases of domestic animals; calf diarrhoea; body fluids; rabies
3 Gibbs HA - BVMS,PhD,MRCVS *Lecturer* cattle respiratory disease
4 McEwan NA - BVMS,MVM,MRACV *Lecturer* dermatology; endocrine diseases in dog and cat
5 Nash AS - BVMS,PhD,MRCVS *Senior Lecturer* renal disease; internal medicine dog and cat
6 Selman IE - BVMS,PhD,MRCVS *Titular Professor* respiratory diseases of cattle; calf diarrhoea

4836 Heriot-Watt University

Chambers Street, Edinburgh, EH1 1HX
Tel: (031-225) 8432.

4837 Department of Brewing and Biological Sciences

4838 General Biochemistry

1 Manners Professor DJ - DSc,ScD *Head of Department* carbohydrate biochemistry
2 Fleming M - PhD structure and metabolism of polysaccharides
3 Park MV - PhD enzymology
4 Stark JR - PhD starch metabolism; digestive enzymes in marine species
5 Sturgeon RJ - PhD biochemistry of microbial cell walls

4839 Microbiology

1 Campbell J - PhD taxonomy and continuous cultures of yeast
2 Catley BJ - PhD carbohydrate metabolism in *Candida albicans*
3 Flanningan B - PhD *microflora of cereal grains*
4 Stewart DJ - PhD physiology, ecology and taxonomy of Pseudomonas

4840 Plant Biochemistry

1 Duffus JH - PhD cell division, the cell cycle and differentiation in plants
2 Palmer GH - PhD barley germination and seedling growth
3 Slaughter JC - PhD amino acid metabolism in plants
4 Wilkinson M - PhD algal physiology

4841 Aquaculture

1 Brown Professor CM - DSc microbiology and development in aquaculture
2 Poxton MG - PhD aquaculture engineering; fish biology and food chain studies

4842 University of St Andrews

College Gate, St Andrews
Tel: (0334) 76161.

4843 Department of Plant Biology

1 Crawford Professor RMM - BSc,DSc adaptations of roots to depth, cold and flooding
2 Weeks Dr DC - BSc,MA,PhD depth sensing in roots

4844 Department of Biochemistry and Microbiology

1 Russell WC - BSc,PhD *Professor, Chairman of Department* molecular analysis of ovine mycoplasmas
2 Wray JL - BSc,PhD molecular genetics of nitrate assimilation in higher plants

4845 University of Stirling

Stirling
Tel: Stirling (0786) 73171.

4846 Department of Biological Science

1 Sargent Professor JR - BSc,PhD,FIBiol lipid biochemistry
2 Berry AJ - BSc,PhD *Senior Lecturer* littoral animal biology
3 Brophy PJ - BSc,PhD *Lecturer* neurobiochemistry
4 Bryant DM - BSc,PhD avian energetics
5 Dix NJ - BSc,PhD fungal ecology
6 Edwards BA - BSc,PhD endocrinology
7 McLusky DS - BSc,PhD estuarine biology
8 Newbery DM - BSc,PhD forest ecology
9 North MJ - MA,PhD microbial biochemistry
10 Phillips RD - BSc,PhD photosynthesis
11 Price NC - MA,DPhil protein biochemistry
12 Proctor J - MA,DPhil soil science and plant nutrition
13 Reid JSG - BSc,PhD plant biochemistry
14 Sexton R - BSc,DPhil developmental plant physiology
15 Stevens L - MA,DPhil microbial biochemistry
16 Tytler P - BSc,PhD fish physiology
17 Wales W - BSc,PhD invertebrate neurophysiology
18 Willmer CM - BSc,PhD stomatal and plant metabolism

4847 Scottish Marine Biological Association, Oban

1 Ansell AD - BSc,PhD *Honorary Lecturer* invertebrate physiology
2 Blaxter JHS - MA,DSc,FRSE *Honorary Reader* fish physiology
3 Currie RI - BSc,FRSE *Honorary Research Lecturer* oceanography
4 Matthews JBL - BSc,PhD *Honorary Professor* biological oceanography
5 Stanley SO - BSc,PhD *Honorary Lecturer* microbiology

4848 DAFS Freshwater Fisheries Laboratory, Pitlochry

1 Thorpe JE - BA,PhD *Honorary Lecturer* behaviour and ecology of freshwater fishes

4849 Institute of Aquaculture

1 Adams S - BSc,PhD immunobacteriology
2 Beveridge MCM - BSc,PhD cage fish culture in tropical systems; environmental impact of aquaculture
3 Brown JH - BSc,PhD prawn culture; biology and culture of giant clams
4 Cattan A - BSc,PhD immunology
5 Clifton-Hadley RS - BScVetMB,MRCVS proliferative kidney disease in salmonoids
6 Frerichs GN - BVMS,PhD,MRCVS biology of bacteria and viruses associated with fish and shellfish; tissue culture of fish cells
7 Hartley SE - BSc,PhD salmonoid cytogenetics
8 Horne MT - BA,PhD,FIMLS,MBIM fish vaccines and their production
9 Jauncey K - BSc,PhD fish and prawn nutrition
10 Macintosh DJ - BSc,PhD biology of mangrove ecosystems; induced breeding of fish, hatchery techniques
11 McAndrew BJ - BSc,PhD genetics and stock improvement of *Tilapia* and other tropical fish, salmonoids and prawns; cryopreservation of fish gametes and embryos
12 Muir JF - BSc,PhD aquaculture engineering system design and waste management; salmonoid ranching
13 Phillips MJ - BSc,PhD environmental impact of aquaculture; toxic effects of algae and pesticides to fish
14 Polglase JL - BSc,PhD invertebrate pathology, particularly mycoses
15 Richards RH - MA,VetMB,PhD,MRCVS pathology of temperate and tropical fish, specialising in mycoses nutritional and eye pathology
16 Roberts Professor RJ - BVMS,PhD,FRCPath,MRCVS,FIBiol,FRSE pathology of temperate and tropical fish
17 Robertson DA - BSc,PhD use of triploid in the production of sterile salmonoids for farm production
18 Ross B - BSc,PhD fish nutrition, particularly replacement of fish meal in standard fish diets; novel components for fish diets
19 Ross GL - BSc,PhD fish energetics and anaesthesiology
20 Sommerville C - BA,PhD pathology of parasites of economic importance in fish
21 Stirling HP - MA,PhD,MIBiol shellfish & finfish fisheries; water quality parameters for aquaculture
22 Tatner MF - BSc,PhD fish immunology and its relation to vaccine production
23 Webster J - BSc,PhD fish physiology and nutrition particularly replacement of fish meal in fish diets

4850 Department of Management Science

1 Donnelly M - BA,MSc modelling of deer farming; applications of mathematical programming

4851 Department of Environmental Science

1 Grieve IC - BSc,PhD soil chemistry
2 Harrison SJ - BSc,PhD spring warning of upland soils; Ochil Hills
3 Stott TA - BSc monitoring of suspended sediment and bedload sources and outputs from stream basins

4852 University of Strathclyde

McCance Building, 16 Richmond Street, Glasgow, G1 1XQ
Tel: 041 552 4400.

4853 Department of Bioscience and Biotechnology

1 Marshall RD - BSc,MSc,PhD,CChem,FRSC,FRSA,FRSE *Professor and Chairman*

4854 Biology Division

1 Burdon RH - BSc,PhD,FRSA,FRSE *Professor and Head of Division* molecular biology of stress; response in plant and animal cells, DNA methylation and gene regulation; toxoplasma genes
2 Burge MN - BSc,PhD *Lecturer* pest control - bracken control with fungal pathogens
3 Gardner Miss IC - BSc,PhD *Senior Lecturer* plant fungal symbiosis in land reclamation and forestry
4 Hull P - BSc,PhD,DipGen *Lecturer* population genetics of plant and animal pests
5 Hutchinson WM - BSc,PhD,DSc,FLS,FIBiol,FRSE *Personal Professor* parasite control - toxoplasma gondii
6 Kirkwood RC - BSc,PhD,FRSE *Reader* pest control; herbicide mode of action
7 Scott A - BSc,PhD *Senior Lecturer* toxicity testing

4855 Applied Microbiology Division

1 Smith JE - BSc,MSc,PhD,DSc,FIBiol,FRSE *Professor and Head of Division* biochemistry of differentiation in filamentous fungi; mycotoxins; fungal biotechnology
2 Allan MC - BSc,DipMic *Lecturer* food microbiology; bacterial ecology in relation to fish farming
3 Anderson JG - BSc,PhD *Senior Lecturer* bacterial flora in estuarine silt in relation to predators and pollution
4 Berry DR - MA,MSc,PhD *Reader* biochemistry of differentiation in yeasts; microbial utilisation of agricultural wastes; genetic engineering

5 Halling PJ - BA,PhD *Lecturer* enzyme technology

6 Johnston JR - BSc,PhD,FIBiol *Reader* yeast genetics; genetic engineering

7 Senior E - BSc,PhD *Lecturer* anaerobic technology; landfill biotechnology

8 Wood BJB - BSc,PhD,CChem,MRSC *Reader* primary production in aquatic systems; fermented food stuffs

4856 Immunology Division

1 Stimson WH - BSc,PhD,CBiol,FIBiol,FRSE *Professor and Head of Division* immunodiagnostics, immunopharmacology, production of monoclonal antibodies

2 Alexander J - BSc,PhD *Lecturer* immunoparasitology

3 MacDonald CM - BSc,PhD *Lecturer* immunoglobulin genetics, immortalisation of animal cells using oncogenes and recombinant DNA techniques

4 McCruden AB - BSc,PhD *Lecturer* neuronal culture and physiology, steroid immunology

4857 Food Science Division

1 Morrison WR - BSc,PhD,DSc,ARCST,FIFST,FRSE *Professor and Head of Division* cereal starches and lipids; lipid-protein and lipid-starch interactions

2 Duckworth RB - BSc,PhD,FIFST *Reader (part-time)* properties of water in foods

3 Karkalas J - BSc,MSc,FIFST *Lecturer* enzymic analysis of starch and other polysaccharides

4 Paterson A - BSc,PhD *Lecturer* new enzymes for modifying food qualities

5 Piggot JR - BA,MSc,PhD *Lecturer* sensory analysis of foods and beverages and food acceptability

6 Scholefield J - BSc,MSc,PhD,FIFST,MBiol *Senior Lecturer* rapid and automated systems for screening micro-organisms and cells in foods; fermentation liqueur; effluents and medical specimens

7 Smith J - BSc,MSc,PhD,AIFST *Lecturer* factors influencing the quality of farmed rainbow trout

4858 Biochemistry Division

1 Marshall RD - BSc,MSc,PhD,CChem,FRSC,FRSA,FRSE *Professor and Head of Division* structure, function and metabolism of glycoproteins, especially with regard to renal function and tumour cells

2 Brzeski H - BSc,PhD *Lecturer* nucleic acid biochemistry; molecular biology

3 Green AL - BSc,PhD *Reader* amine oxidase inhibitors

4 Green B - BSc,PhD,CChem,MRSC *Senior Lecturer* biochemistry of reproduction, hormone action

5 Kibby MR - MA,PhD *Senior Lecturer* pharmacokinetics, computer control of fermenter reactions

6 Mattey M - BSc,PhD *Lecturer* enzymes and enzyme technology, microbial synthesis

7 O'Grady JE - BSc *Lecturer* biochemistry of reproduction

8 Patterson JDE - BSc,PhD *Senior Lecturer* enzymes and enzyme technology, protein fractionation

9 Robb DA - BSc,PhD *Lecturer* plant enzymes, mechanisms of enzyme action

10 Watson J - BSc,PhD,ARCST *Reader* ovarian metabolism and prostaglandin function

4859 The West of Scotland Agricultural College

Auchincruive, Ayr, KA6 5HW
Tel: (0292) 520331.

Diploma and university level education, research and developmental and advisory work in agriculture, horticulture and food technology, poultry production and agricultural engineering

1 Cunningham Professor JMM - CBE,BScAgr,PhD,FRAGS,FIBiol,FRSE *Principal*

2 Martin DJ - BSc,MS,PhD,CBiol,MIBiol,MIHort *Deputy Principal*

3 Waterson HA - MSc *Deputy Principal*

4860 Education

1 Martin Dr D *Head of Division*

4861 Agriculture

1 Buckett M - BSc,MIBiol,ARAgrS *Head of Department* farm management

2 Evans RG - BScAgr,PhD agronomy

3 Gill A - NDA,SDA,LIBiol *Farm Management*

4 Gladstone Miss JE - BScAgr *Farm Management*

5 Gooding RF - SDA,HNC,MIBiol agronomy

6 Griffith DR - BScAgr animal production

7 Layhe WH - BScAgr,MS bovine reproduction

8 Leggate AT - BSc,SCMPTC animal production

9 McLean AF - BScAgr,PhD animal production

10 Norris FG - BAgr farm management

4862 Agricultural Engineering

1 Morrison RR - BScAgrE,MIAgrE *Head of Department*

2 Davidson W - BScAgr,MSc,MIAgrE horticultural engineering; aerogeneration

3 Hair IG - BScAgrE tractor monitoring

4 Howet D - SDA,NDAgrE,MIAgrE field engineering

5 Jones A - BScAgrE,OND livestock engineering, environmental control

6 Parish S - BEng,CertEd,MSc tractor power utilizing

7 Pickering J - NDAgrE,CertEd machinery assessment

8 Walker AEL - BA,NDA,NDAgrE,HNC,TEng,MIAgrE field engineering

9 Wallington RS - BScAgr,MSc,FIAgrE control systems

4863 Dairy Technology

1 Crawford RJM - BScAgr,PhD,NDA,NDD,FIFST *Head of Department*

2 Davies G - BScAgr,AIFST product chemistry; membrane processing

3 Fyfe J - NDA,NDD milking equipment

4 Galloway Miss JH - NDD,CDD,AIFST cheese production; dairy microbiology

5 Hamilton Miss MP - NDD,SDDH butter and ice-cream; food additives

6 Howie Mrs LJM - BSc *Lecturer*

7 Rutter Miss MA - BSc,ONDFT canning food and dairy products

8 Tamine AY - MSc,PhD,SDDH,SDDT fermented dairy products

9 Wade VN - BSc,PhD,AIFST food engineering

10 Wiggins AL - BScAgr,MSc,FIFST,MIBiol,FRSH milk processing

4864 Horticulture and Beekeeping

1 Gooding Professor HJ - BSc,PhD,MIBiol,FLS,MIHort *Head of Department*

2 Dougall ICG - NDH,SDH amenity horticulture

3 Dudney PJ - BScHort,MSc fruit and turf culture

4 Hardy FS - NDH,MIHort amenity horticulture

5 Maxwell IC - NDB,SDH glasshouse crops; bee management

6 McDonald PT - MSc,MIBiol,MIHort propagation, nursery practice and garden centres

7 Taylor CW - BSc,MPhil plant and crop physiology

8 Watson GD - MSc,NDH,MIHort intensive vegetable and field crops

4865 Poultry Husbandry

1 Dunn P - NDA,NDP,MIBiol *Head of Department* poultry nutrition

2 Davies DE - CDP,NDP egg and poultry marketing

3 Lindsay Miss EH - NDD,NDP poultry management

4 McDonald R - SDA,SDP,MIBiol poultry housing and management

5 Smith Miss EFH - NDD,NDP poultry management

6 Smith WK - BAgS,DPH,PhD *Senior Lecturer/Adviser* poultry breeding and nutrition

7 Tullet SG - BSc,PhD embryology and reproduction

8 Wright D - SDA advanced husbandry and management

4866 Agricultural Economics

1 Clark J - BScAgr,NDA *Head of Division*

2 Ashworth SW - BSc,MBA farm management

3 Godfrey D - BScAgr accounting and finance

4 Groves CR - BA,MSc agricultural policy, marketing

5 Horsburgh AS - BScAgr horticultural economics and research

6 Hunt RWT - BCom,MSc marketing

7 Lindsay LC - BScAgr work organisation

8 Macpherson JF - BScAgr low ground sheep development

9 Mainland DD - BSc,PhD econometrics

10 McGregor IM - OND,MIBiol horticultural econometrics

11 McKinnon Miss SG - BScAgr work organisation

12 Munro RF - BScAgr dairy farming development work

13 Smith PG - MA,DipAgr farm management

14 Tweddle JM - BSc,NDA,SDA farm management

15 Twigg Miss JA - BScAgr marketing

4867 Brickrow Farm Unit

1 Laird R - BScAgr,MS,DipAgr,NDA,NDD,MIBiol *Head of Department* genetic improvement and dairying

4868 Glasshouse Investigational Unit for Scotland

1 Szmidt RAK - BSc,HortSci *Head of Department*

2 Hitchow GM - BSc,MSc,MIBiol glasshouse crop protection

3 Smith JHF - NDB glasshouse crop production

4869 Biological Sciences

4870 Botany and Plant Pathology

1 Hag RKM - BSc,MSc,PhD,MIBiol *Head of Department* plant and crop physiology

2 Alcorn JWS - BSc,MSc,MIBiol rhizobium

3 Gwynne DC - BSc,MIBiol ecology and use of hill pasture

4 Holmes SJI - BSc,PhD plant virus epidemiology in crops

5 Marshall G - HNDA,PhD,MIBiol weed science

6 Svoboda Mrs KP - DiplIng,MIBiol culinary and medicinal herbs

7 Walker AJ - BSc,DPhil crop physiology

8 Walters DR - BSc,PhD,MIBiol physiological responses of plants to pathogen infection

9 Williams GH - BSc,PhD,MIBiol weed science

4871 Microbiology

1 Baines S - BSc,PhD *Head of Department*
2 Bruce J - BSc,PhD food and dairy microbiology
3 Deans SG - BSc,PhD food microbiology
4 Evans MR - BSc,PhD animal wastes
5 Fallowfield HJ - BSc,PhD algal biomass and products
6 Martin NJ - BSc,PhD soils and microbiology
7 Svoboda IF - DipIIng,AMIWPC animal wastes
8 Thacker FE - BSc animal wastes

4872 Zoology

1 Newbold JW - BSc,DIC,FRES *Head of Department*
2 Foster GN - MA,PhD,MIBiol,FRES integrated control of pests
3 Stewart RM - BSc potato cyst eelworm
4 Titchener RN - BScAgr,MSc,PhD,CertEd,MIBiol animal ectoparasites
5 West DI - BSc,MIBiol mammalian physiology and history

4873 Agricultural Chemistry

1 Barber GD *Head of Division* animal nutrition
2 Alexander RH - NDA,CDA sampling and analytical techniques
3 Barclay Miss MNI - BSc,PhD,AIFST food science
4 Dixon J - HNC,ARIC,PhD trace elements
5 Ferns HR - BScAgr,MIBiol milk composition
6 Golightly RD - BSc,NDA,SDA soils and plant nutrition
7 Hall DA - BSc,PhD,ARIC,MIBiol,CertEd soils and plant nutrition
8 Johnson AH - BSc,ARIC inorganic and physical chemistry
9 Klessa DA - BSc soil science
10 MacPherson A - BSc,PhD trace element nutrition
11 McGowan Miss M - BScAgr,SDD sampling and analytical techniques
12 Merrilees DW - BScAgr soils and drainage
13 Munro P - MAgrSc dairy chemistry
14 Offer NW - BSc,PhD animal nutrition and biochemistry
15 Weir J - BSc animal nutrition and biochemistry
16 Wilson GCS - CDA,CDD horticultural chemistry and plant nutrition

4874 Veterinary Investigation

1 Wright CL - MSc,BVSc,MRCVS,DTVM *Head of Division*

4875 Auchincruive VI Centre

1 Barber DML - MA,VetMB,MRCVS calf rearing
2 Gray D - BVMS,MSc trace element deficiencies in cattle & sheep
3 Legue DN - FRCVS,BVM&S,PhD infertility & reproduction of sheep and cattle
4 MacLaren APC - BVMS,MRCVS infertility; poultry diseases
5 Mitchell GBB - BVMS,PhD helminthology; tick ecology

4876 Oban VI Centre

1 Brodie TA - BVMS,MVM tick borne fever, parasitology

4877 Dumfries VI Centre

1 MacLeod NSM - BVMS,MRCVS diagnosis and pathology cattle and sheep diseases

4878 Advisory and Development

1 Waterson HA - MSc *Director*

4879 Agronomy

1 Frame J - BSc,MAgrSc,DipAgr,PhD,NDA,NDD,SDDH,FIBiol,CBiol *Head of Department*
2 Boyd AG - BScAgr grassland agronomy
3 Harkess RD - BScAgr,MS,PhD,NDA,CBiol,MIBiol grassland agronomy
4 Heppel Miss VAF - BSc,PhD crop agronomy
5 Pridgeon Miss I - BScAgr cereal agronomy
6 Tiley GED - BSc,PhD hill and upland agronomy
7 Whytock Miss GP - BSc crop agronomy

4880 Crichton Royal Farm

1 Leaver JD - BSc,PhD,MIBiol *Farm Director*
2 Box JE - HNDA,MIBiol dairy young stock and bull beef
3 Kelly Miss EF - BVMS,PhD,MRCVS calf rearing, health and fertility of dairy cows
4 Roberts DJ - BSc,PhD fodder beet, buffer feeding of dairy cows

4881 Farm Buildings

1 Kelly M - BSc,MSc,PhD *Head of Department*
2 Johnston CA - AHWC,MICE farm buildings structures
3 Shepherd CS - MSATT,MRSH livestock housing and handling facilities
4 Stevenson DE - BSc,BArch,RIBA livestock housing and handling facilities

4882 Farm Planning Unit

1 Sargent ED - BSc *Head of Department* farm management
2 Burr MP - BSc socio-economics, countryside and wildlife conservation
3 Douglas DH - BSc,DipFBA farm management and computer applications
4 Ritchie JR - BSc farm management

4883 Kirkton Farm

1 McClelland TH - BAgr,MIBiol *Head of Department*
2 Dickson IA - BSc,MAgr,NDD,MIBiol *Sheep Specialist*
3 Waterhouse A - BSc,PhD *Beef and Sheep Specialist*

4884 Argyll Area

1 MacLeod A - NDA,NDD *Head of Department*

4885 Central Area

1 Lyons W - BScAgr,DipFBA *Head of Department*

4886 Clyde Area

1 Ross SA - NDA,NDD,DipFBA,ARAgrS,MIBiol *Head of Department*

4887 Southern Area

1 Thorburn J - BScAgr,NDA *Head of Department*

4888 South Western Area

1 Campbell A - BScAgr,MS,NDA,MIBiol *Head of Department*

NORTHERN IRELAND

4889 Department of Agriculture

Dundonald House, Belfast, BT4 3SB
Tel: Belfast (0232) 650111.

1 Wright CE - Dr BSc,BAgr,PhD,FIBiol *Chief Scientific Officer*
2 Larmour TA - BAgr,MIBiol *Chief Agricultural Officer*

4890 Agricultural and Food Bacteriology Division

Agriculture and Food Science Centre, Newforge Lane, Belfast, BT9 5PX
Tel: Belfast (0232) 661166.

1 Holding Professor AJ - BSc,MS,PhD,FIFST *Head of Division* soil and water
2 Collins MA - BSc,PhD food microbial technology
3 Cooper JE - BA,DPhil soil and water ecology
4 Damoglou AP - BA,PhD plant products
5 Gilmour A - BSc,PhD milk and milk products
6 Madden RH - BSc,PhD meat and meat products
7 McKay AM - BSc,PhD food microbial technology
8 Rowe MT - BSc milk and milk products

4891 Agricultural Economics and Statistics Division

Dundonald House, Belfast, BT4 3SB
Tel: Belfast (0232) 650111.

1 Furness GW - DFM,MSc *Chief Agricultural Economist*

4892 Agricultural Zoology Research Division

Agriculture and Food Science Centre, Newforge Lane, Belfast, BT9 5PX
Tel: Belfast (0232) 661166.

1 Marks RJ - MSc,DIC,PhD,MIBiol *Head of Division*
2 Bell AC - BSc,PhD aphids; rabbits
3 Blackshaw RP - BSc,PhD pests of grassland
4 Mowat DJ - MSc,MIBiol,PhD pests of grassland

4893 Nematology Laboratories

Felden, Mill Road, Newtownabbey, Co Antrim
Tel: Whitabbey (0231) 51361.

1 Clayden IJ - BSc,PhD nematology
2 Fleming CC - BSc,PhD nematology
3 Turner SJ - MSc,PhD nematology

4894 Crop and Animal Husbandry Division

Agricultural Research Institute, Hillsborough, Co Down
Tel: Hillsborough (0846) 682484.

1 Murdoch JC - OBE,BSc,PhD *Head of Division, Director of Institute* grass production and utilisation
2 Chestnutt DMB - MAgr,PhD beef and sheep production
3 Easson DL - BSc,PhD grass and crop production
4 Frost JP - BAgr,MSc farm machinery
5 Gordon FJ - BAgr,DFM milk production
6 Patterson DC - MAgr,FRags,PhD pig production
7 Steen RWJ - BAgr,PhD beef production
8 Walker N - BSc,PhD pig production

4895 Northern Ireland Plant Breeding Station

Loughgall, Armagh, BT61 8JB
Tel: (076289) 436.

1 Costelloe BE - BSc,CBiol,MIBiol *Officer in Charge* plant breeding; potatoes
2 Evans NE - BSc,PhD,CBiol,MIBiol plant breeding; tissue culture; potatoes
3 Faulkner JS - MA,DPhil,CBiol,MIBiol plant breeding - grasses, flax, oats
4 Lee HC - MSc,PhD plant breeding; potatoes

4896 Plant Testing Station

Crossnacreevy, Castlereagh
Tel: Castlereagh (023123) 8121.

1 Camlin MS - BSc,BAgr,PhD plant variety rights; grass cultivar competition
2 Gilliland TJ - BSc,BAgr,PhD herbage cultivars
3 White Miss EM - BSc,PhD cereal and potato cultivars

4897 Fisheries Research Laboratory

38 Castleroe Road, Coleraine, Co Londonderry, BT51 3RL
Tel: Coleraine (0265) 4521/4594.

1 Boyd RJ - BSc,PhD *Head of Laboratory* marine demersal fish studies
2 Briggs RP - BSc,PhD,MIBiol marine shellfish and crustacean studies
3 Kennedy GJA - BSc,PhD salmonid ecology and behaviour

4898 Fresh Water Biological Investigation Unit

Greenmount Road, Muckamore, Antrim
Tel: Antrim (08494) 62660.

1 Smith RV - BSc,PhD *Head of Unit* physiology of blue-green algae; eutrophication of fresh water
2 Gibson CE - BSc,PhD ecology of fresh water

4899 Plant Pathology Division

Agriculture and Food Science Centre, Newforge Lane, Belfast, BT9 5PX
Tel: Belfast (0232) 661166.

1 Blackeman Professor JP - MSc,PhD *Head of Division* biological control of plant diseases
2 Brown AE - BSc,PhD fungicides; potato diseases
3 Cooke LR - BSc,PhD fungicides; potato diseases
4 Copeland RB - BSc,BAgr,PhD plant virology; potato diseases
5 McCracken A - BSc,PhD,MIBiol,MIHort diseases of horticultural crops
6 Mercer PC - BSc,PhD,DIC cereal and other arable crop diseases
7 Mills PR - BSc,PhD,DIC plant virology
8 Seaby DA - BSc,BAgr forest pathology; mushroom biology

4900 Veterinary Research Division

Stormont, Belfast, BT4 3SD
Tel: Belfast (0232) 760011.

1 Dow Professor C - BVMS,PhD,MRCVS,MRCPath *Chief Veterinary Research Officer and Head of Division* neuropathology of animals
2 Adair B - McC,BSc adenoviruses
3 Armstrong JD - MVB,MRCVS reproductive physiology
4 Ball HJ - MSc mycoplasmosis
5 Bryson TD - BVMS,MRCVS pathology of pneumonia
6 Ellis WA - BVMS,MRCVS infectious reproductive diseases
7 Kennedy S - MVB,MRCVS pathology
8 Logan EF - BVMS,PhD,FRCVS immunology
9 Mackie DP - MVB,MRCVS mastitis
10 McCaughey WJ - MA,MS,MVB,PhD,MRCVS reproductive physiology
11 McCracken RM - BVMS,PhD,MRCVS pathology
12 McFerran JB - BSc,MRCVS,PhD *Deputy Head* virology
13 McLoughlin Mrs MF - MVB,MRCVS pathology
14 Rice DA - MVB,MRCVS biochemistry

4901 Biometrics Division

Castle Grounds, Stormont Estate, Belfast, BT4 3SD
Tel: Belfast (0232) 63939.

1 Weatherup STC - MSc,PhD *Head of Division* design and analysis of experiments and multivariate methods
2 McCallion T - MSc,PhD computer techniques in the design and analysis of censuses and surveys

4902 Agricultural and Food Science Centre

Newforge Lane, Belfast, BT9 5PX
Tel: Belfast (0232) 661166.

1 Goodall EA - MSc,AFIMA computer techniques in the design and analysis of censuses and surveys
2 Kilpatrick DJ - MSc,PhD design and analysis of experiments and sampling methods
3 Stewart DA - BSc,AFRAeS,CEng,MBCS model building and simulation studies
4 Watson Miss S - BSc,MSc design and analysis experiments

4903 Agricultural and Food Chemistry Division

Agriculture and Food Science Centre, Newforge Lane, Belfast, BT9 5PX
Tel: Belfast (0232) 661166.

1 McMurray Professor CH - BSc,BAgr,PhD,FRSC *Head of Division* mineral metabolism
2 Anderson R - BSc,PhD chemistry of herbage conservation
3 Christie P - BA,PhD soil chemistry; plant nutrition
4 Dickson DA - BSc,PhD forest soils peats and forest tree nutrition
5 Fearon Mrs AM - BSc,AIFST food chemistry; dairy products
6 Garrett MK - BSc,PhD plant biochemistry; farm waste utilization
7 Gault NFS - BSc,PhD food chemistry; meat products
8 Harper DB - BSc,PhD,CChem,FRSC plant and microbial biochemistry; pesticides residues
9 Jackson N - BAgr,MA,PhD,DSc,CChem,FRSC,FIBiol poultry nutrition; herbage conservation
10 Johnston DE - BSc,PhD,CChem,FRSC,AIFST food chemistry; food structure/property relationships
11 King Miss K - BSc,PhD food chemistry; fruit and vegetable products
12 McCracken KJ - BSc,BAgr,PhD,CChem,FRSC pig nutrition
13 Moss BW - BSc,PhD,MSc,AIFST food chemistry; muscle biochemistry; preslaughter stress
14 Pearce J - BSc,PhD avian and animal biochemistry and carcase lipid quality
15 Stevens RJ - BSc,BAgr,PhD,FRSE soil chemistry; plant nutrition
16 Stevenson Mrs MH - BSc,BAgr,MSc,PhD poultry nutrition
17 Unsworth EF - MSc,PhD voluntary feed intake and utilisation of roughages

18 Watson CJ - BSc,PhD soil chemistry; plant nutrition
19 Wylie ARG - BSc,PhD ruminant nutrition

4904 Agricultural Botany Division

Agriculture and Food Science Centre, Newforge Lane, Belfast, BT9 5PX
Tel: Belfast (0232) 661166.

1 McWha Professor JA - BSc,BAgr,PhD *Head of Division*
2 Courtney AD - BSc,PhD weed biology and control
3 Harvey Ms BMR - BSc,PhD plant physiology
4 Laidlaw AS - BSc,PhD,MIBiol agronomy of herbage plants
5 McAdam JH - BSc,MAgr hill land agronomy
6 McBratney JLM - BSc,CBiol,MIBiol grass and legume investigations
7 McCullough SJ - BVSC,MRCVS virology
8 McNulty MS - MVB,PhD,MRCVS virology
9 Neill SD - BSc bacteriology
10 O'Brien JJ - BA,MVB,PhD,DipBact,MRCVS pig diseases
11 Selby C - BSc tissue culturing of forest trees
12 Smyth Mrs JA - MVB,MRCVS pathology
13 Taylor S - BVMS,MRCVS parasites of ruminants

4905 Forest Service

Dundonald House, Belfast, BT4 3SB
Tel: Belfast (0232) 650111.

1 Schaible RC *Research and Inventory Officer*

4906 The Queen's University of Belfast

Belfast, BT7 1NN
Tel: (0232) 661111. Telex: 74487.

4907 Department of Biochemistry

1 Martin SJ - BSc,MSc,PhD *Professor: Director N.I. Centre for Genetic Engineering*
2 Cosby SL - BSc,PhD *Lecturer* canine distemper virus
3 Hoey EM - BSc,PhD *Senior Research Officer/Lecturer* bovine enterovirus; DNA sequencing
4 Rima BK - PhD *Lecturer* canine distemper virus; measles virus

4908 Department of Botany

1 Clifford PE - BSc,PhD *Lecturer* whole plant physiology - assimilate distribution; water stress; tropisms
2 Dring MJ - BSc,PhD *Reader* photobiology of marine algae
3 Govier RN - BSc,PhD *Lecturer* foliar uptake; biology of angiosperm hemiparasites
4 Mackender RO - BSc,DPhil *Lecturer* composition and function of plastid envelope membranes; chloroplast development
5 Pilcher JR - BSc,PhD *Senior Lecturer* pollen analysis and dendrochronology; C-14 dating; post-glacial vegetational history of Ireland
6 Quinn JP - BSc,PhD *Lecturer* manipulation of plant and viral genomes; gene expression and regulation in lower plants
7 Robinson PM - BSc,PhD *Reader* fungal physiology and morphogenesis; chemotropism in fungi; continuous culture of fungi; physiology of fern gametophytes

4909 Sub-Department of Microbiology

1 Dent VE - BSc,PhD *Lecturer* chemotaxonomy of oral gram-positive rods of the genera *Actinomyces* and *Lactobacillus*
2 Larkin MJ - BSc,PhD *Lecturer* mutation and selection in bacterial populations

4910 Department of Genetics

1 Hickey GI - BSc,PhD *Head of Department, Lecturer* mutation in tissue cultures; cell fusion studies of tumorigenecity
2 Butcher AC - BSc,PhD *Lecturer* population and biometrical genetics
3 Fletcher HL - BSc,PhD *Lecturer* meiosis; chromosome pairing; recombination, both meiotic and in the repair of DNA double-strand breaks
4 Hart MGR - BSc,PhD *Lecturer* mechanisms of bacterial mutagenesis

4911 Department of Zoology

1 Fairweather I - BSc,PhD *Lecturer* physiology of trematodes
2 Halton DW - BSc,PhD,MIBiol *Professor* trematoda
3 Hanna REB - BSc,PhD *Lecturer* physiology of trematodes
4 Roberts D - BSc,PhD *Senior Lecturer* marine pollution

4912 Enniskillen Agricultural College

Castle Archdale Experimental Husbandry Farm, Irvinstown, Co Fermanagh
Tel: (03656) 21245.

1 O'Neill DG - BAgr,FRAgS *Principal*

2 McLaughlan W - BAgr,PhD drainage, grass production, conservation and utilisation; beef production

4913 Greenmount Agricultural and Horticultural College

Antrim, Northern Ireland, BT41 4PU
Tel: Antrim 62114/6.

1 Fulton RB - MAgr,MSAgr,CBiol,MIBiol *Principal*

4914 Research and Development Division

1 Stewart TA - MA,MAgr,CBiol,MIBiol *Head of Division*

4915 Experimental Husbandry Department

Antrim, BT41 4PU
Tel: Antrim 62114/6.

1 Gracey HI - BAgr,MSc,PhD *Head of Department* conservation and grassland fertilisation
2 Kennedy SJ - BSc,BAgr,PhD,DipAgrComm,CBiol,MIBiol grass production and silage
3 Saunders AR - BSc,MSc,PGCE,CBiol,MIBiol potatoes and other arable crops

4916 Hill Farm Department

Glenwherry, Co Antrim, BT42 4RF
Tel: Glenwherry 8183.

1 Wilson JK - BSc,DipAgrComm *Head of Department* hill sheep and suckler cattle

4917 Livestock Husbandry Farm

Manor House, Loughgall, Co Armagh, BT61 8JB
Tel: Loughgall 206.

1 McGaughey SJ - BScAgr *Head of Department* lowland sheep and beef

4918 Loughry College of Agriculture and Food Technology

Cookstown, Co Tyrone, BT80 9AA
Tel: Cookstown (06487) 62491.

1 Stevenson RC - BAgr,MS,PhD *Principal*

4919 Agricultural Communication and Rural Development Division

1 Gould M - BAgr,DipAgrComm interpersonal skills
2 McGurnaghan P - BAgr,DipAgrComm computing; information technology; educational methods
3 Michail D - BSc,PhD research and survey methods; educational studies

4920 Agriculture Division

1 Haycock RE - BAgr *Head of Division* clover
2 Armour DM - NDA,SDA,NDP,SDP goose production
3 Barr HW - BSc,MIAgE tractor power measurement
4 Bayne BEF - MSc,NDP,NDR,DipPoult,DMA,MBIM intensive pheasant production
5 Brodison Mrs JA - BSc,MAgr goats
6 Knox LM - BSc slurry spreading
7 Mayes FJ - BSc,PhD,MIBiol poultry processing
8 McLain HD - BSc,CEng,MIMechE,MIAgrE *Head of Section* biomass mechanisation
9 Park RJD - BSc farm buildings

4921 Food Technology Division

1 Lang GIA - BSc,MS,NDA,NDD,FIFST *Head of Division: Vice Principal*
2 Crawford JH - BSc,BAgr,PhD,MIBiol biology; microbiology
3 Evans IR - BSc food engineering
4 Hamilton Mrs MH - BSc,AIFST oils and fats; cereals
5 Haydock Mrs CJL - BSc fruit and vegetables
6 Kinghan S - BSc fish processing
7 Kormos J - MAgr,PhD food chemistry
8 Loan Miss BK - BSc butter; concentrated and dried milks
9 McEvoy MW - BSc,AIFST pigmeat; processed meat products
10 Mullan WMA - BSc,MSc,PhD *Head of Section* starter and cheese technology; liquid milk quality; ice cream
11 Rooney Miss CT - BSc,AIFST meats; cooked meats and offals
12 Steele HW - BSc,BAgr eggs; egg products
13 Weatherup W - BA,HND microbiology; quality control; fermented milk products
14 Wootton AE - BSc,PhD,CChem,FRSC,FIFST *Head of Section* meat and meat products

4922 Agricultural Research Institute of Northern Ireland

Hillsborough, Co Down
Tel: Hillsborough 0846 682484.

1 Murdoch Professor JC *Director* see Crop and Animal Husbandry Division, Department of Agriculture for Northern Ireland, page

ZAMBIA

4923 Department of Agriculture

4924 Head Office
P.O. Box 50291, Mulungushi House, Lusaka
Tel: 213551/212342. Cables: DEPAGRIC TELEX.

1 Mutelo JB - MSc *Acting Director*
2 Patel BK - PhD *Acting Assistant Director (Research)*

4925 Research Division

4926 Mount Makulu Central Research Station
Private Bag 7, Chilanga
Tel: 278655/278008. Cables: Research.

1 Kaliangile I - BAgricSc,DipSeedTech,MSc *Acting Chief Agricultural Research Officer*
2 Lumande E - BA,MLS,AIS *Librarian*
3 Mwampole R - MSc *Research Extension Liaison Officer*

4927 Agricultural Chemistry
1 Malama C - BSc *Chemist*
2 McPhillips JK *Soils Crop Advisory Officer*

4928 Germplasm Bank
1 Nkhoma C - BSc,MSc

4929 ARPT
1 Edwards R - MSc *Farming Systems Agronomist*
2 Masi C - BSc,MSc *Farming Systems Agronomist*
3 Singongo LP - BSc,MSc *ARPT National Coordinator*
4 Sutherland A - PhD *Rural Sociologist*

4930 Plant Protection
1 Chakupurakal J - PhD *Entomologist*
2 Kabuswe M - PhD *Entomologist*
3 Raemaekers R - MSc *Plant Pathologist*

4931 Soil Science
1 Stapleton A *Soil Physics*

4932 Food Conservation and Storage Unit
1 Siandwa M - BSc *Research/Extension Training Officer*

4933 National Oilseeds Development Programme
1 Chimbwe B - BSc *Agronomist, Sunflower*
2 Eylands V - PhD *Agronomist, Sunflower*
3 Javaheri A - BSc *Agronomist, Soybeans*
4 Lepoint P - MSc *Plant Pathologist, Sunflower*
5 Lubozya BH - MSc *Plant Breeder, Sunflower*
6 Mwala MS - PhD *Plant Breeder, Sunflower*

4934 Tree and Plantation Crops Research
1 Rajendram NS - BSc *Agronomist, Tea*

4935 Soil Survey Unit
1 Chileshe - MSc *Acting Senior Soil Surveyor*
2 Chirwa B - BSc *Soil Surveyor*
3 Damaseke I - BSc *Soil Chemist*
4 Kalima C - MSc *Land Evaluation*
5 Mkanda N - MSc *Soil Correlation*
6 Tia Tiang W - MSc *Soil Surveyor*
7 Veldkamp W - PhD *Soil Correlator*

4936 Cereals Research Team
1 Chumvwa M - BSc *Maize Pathologist*
2 Gumbo M *Maize Agronomist*
3 Hultgren S - MSc *Sorghum Agronomist*
4 Kasalu H - BSc *Agrc. Sorghum*
5 Kaula G - MSc *Sorghum Entomologist*
6 Little R - MSc *Wheat Breeder*
7 Mulenga G - MSc *Maize Pathologist*
8 Muuka FP - BSc *Sorghum Agronomist*
9 Mwale W - MSc *Maize Breeder*
10 Mwambula C - MSc *Maize Breeder*
11 Myers O - PhD *Maize Breeder*
12 Rao N - PhD *Maize Pathologist*
13 Ristanovic D - PhD *Maize Breeder*
14 Verma BN - PhD *Sorghum Breeder*

4937 Seed Control and Certification Institute
1 Moonga JM - BSc *Seed Analyst. Acting Head*
2 Fredrickson P - MSc *Seed Agronomist*
3 Mwanza M - BSc *Seed Agronomist*
4 Phiri BK - BSc *Seed Agronomist*
5 Silwimba W - BSc *Seed Agronomist*
6 Wiberg H - PhD *Seed Technology Expert*
7 Zulu E - BSc *Seed Agronomist (Potato)*

4938 Mochipapa Regional Research Station
P.O. Box 90, Choma
Tel: 20078. Cables: Research Choma.

1 Kaluba EM - BSc,MSc *Animal Husbandry Research Officer, Officer-in-charge*
2 Chitambaika C - BSc *Provincial Soil Surveyor*
3 Lubinda S - BSc *SOil Surveyor*
4 Simakando KG - BSc *Animal Husbandry Research Officer*

4939 Kabwe Regional Research Station
P.O. Box 80908, Kabwe
Tel: 224211. Cables: Research Kabwe.

1 Kale JA - BSc *Officer-in-charge*
2 Bezune M - PhD *Farming Systems Economist*
3 Olsen F - PhD *Agronomist*
4 Shamaambo M - BA *Farming Systems Economist*
5 Shingalili E - MSc *Reaearch Extension Liaison Officer*
6 Siddiqui IJ - MSc *Agronomist*
7 Swallow G - MSc *Research Extension Liaison Officer*
8 Zakaria IM - MSc *Agronomist*

4940 Mansa Regional Research Station
P.O. Box 129710, Mansa
Tel: 821617/821618. Cables: Mansa Research.

1 Allen JMS - BA,Dipl,MSc *Farming Systems Economist*
2 Banda D - BSc *Soil Surveyor*
3 Chewe CMM - BSc *Rice Agronomist*
4 Daka J - BSc *Soil Surveyor*
5 Dougnac M - MSc *Farming Systems Agronomist*
6 Kunnert C - MSc *Root and Tuber Agronomist*
7 Msoni R - BSc *Soil Surveyor*
8 Mutonga B - BSc *Farming Systems Agronomist*
9 Zimba B - BA,BSc *Farming Systems Economist*

4941 Mongu Regional Research Station
P.O. Box 910064, Mongu
Tel: 221446. Cables: Mongu Research.

1 Nambayo GS - MSc *Animal Husbandry Research Officer, Officer-in-charge*
2 Broekhuis JF *Soil Surveyor*
3 De Boer W - MSc *Farming Systems Economist*
4 Jacob N *FAO Team Leader-Cashew Project*
5 Latis T *Agronomist*
6 Penninkhoff EP - MSc *Agronomist Adaptive Research Planning Team Coordinator*

4942 Msekera Regional Research Station
P.O. Box 510089, Chipata
Tel: 217225. Cables: Chipata Research.

1 Musanya JC - BSc *Groundnut Agronomist, Officer-in-charge*
2 Chindinda M - BSc *Soil Surveyor*
3 Hachiwa C - MSc *Plant Pathologist with Grain Legume Research Team*
4 Irving NS - *Entomologist*
5 Makungu PM - BA,MSc *Farming Systems Economist*
6 Mbewe MN - BSc *Plant Breeder with Grain Legume Research Team*
7 Mulila MJ - BSc,MSc,PhD *Plant Breeder (Groundnuts)*
8 Munyenyembe F - BSc *Soil Surveyor*
9 Sandhu D - PhD *Groundnut Breeder*
10 Sohati PH - BSc *Entomologist with Grain Legume Research Team*
11 Waterworth J - BSc *Farming Systems Agronomist*

4943 Misamfu Regional Research Station
P.O. Box 410055, Kasama
Tel: 221215. Cables: Kasama Research.

1 N'gambi W - BA *Officer-in charge*
2 Bolt R - MSc *Farming Systems Economist*
3 Mapiki A - MSc

4 Mulekwa C - MSc
5 Reid P - MSc *Farming Systems Agronomist*
6 Sikana P - BA *Rural Sociologist*
7 Silavwe M - BA *Farming Systems Economist*
8 Sokotela SB - BSc *Soil Surveyor*
9 Uprichard J - MSc *Research Extension Liaison Officer*

4944 Mutanda Regional Research Station

1 Chuzu P - BA *Farming Systems Economist*
2 Hopkinson - BSc *Farming Systems Agronomist*
3 Opeke R - PhD *Research Extension Liaison Officer*
4 Phiri - MSc *Research Extension Liaison Officer*
5 Rukandema M - PhD *Farming Systems Economist*

4945 National Irrigation Research Station

Private Bag, S.3,, Mazabuka
Tel: 30405. Cables: Mazabuka Research.

1 Siakantu JB - MSc *Irrigation Engineer, Officer-in-charge*
2 Mingochi DS - MSc *Horticulturist/Plant Protection*
3 Mnzava NA - PhD *Vegetable Research Officer*
4 Msikita W - BSc *Horticulturist*
5 Qasem MA - MSc *FAO Project Manager*
6 Stoutjesdijk JA - MSc *Irrigation Engineer*
7 Syankwilimba ISK - MSc *Rice Agronomist*

4946 Magoye Regional Research Station

P.O. Box 11, Magoye
Tel: 30377. Cables: Magoye Research.

1 Chitah WT - MSc *Officer-in-charge*
2 Bernard M *Cotton Entomologist*
3 Bruno B - MSc *Cotton Agronomist*
4 Chola M - BSc *Crop Husbandry Officer*
5 Dadson D - PhD *Soybean Breeder*
6 Lombe F - MSc *Agricultural Engineer*
7 Marc L *Cotton Breeder*
8 Mpata FE - MSc *Cotton Breeder*

4947 Copperbelt Regional Research Station

Private Bag 8, Mufulira
Tel: 410477. Cables: Mufulira Research.

1 Mwila GP - BSc *Agronomist, Officer-in-charge*

4948 Ministry of Lands, Agriculture and Rural Resettlement

4949 Department of Research and Specialist Services

PO Box 8108, Causeway, Harare
Tel: 704531.

1 Fenner RJ - BSc *Director*
2 Gata Mrs NR - BSc,MSc *Assistant Director Research Services Division*
3 Mufandaedza OT - BAgricSc *Assistant Director Livestock and Pastures Division*
4 Whingwiri EE - BAgricSc,BSc,MPhil,PhD *Assistant Director Crop Research Division*

4950 Biometrics Bureau
PO Box 8108, Causeway, Harare
Tel: 704531.

1 McLaren CG - MSc,PhD *Chief Biometrician*
2 Chakanyuka Miss B - BSc *Assistant Research Officer* biometrician
3 Gumbie C - BSc *Biometrician*
4 Kangai Mrs J - BA *Biometrician*
5 Makore J - BSc,MSc *Biometrician*
6 Zinyandu Mrs F - BSc *Biometrician*

4951 Chemistry and Soil Research Institute
PO Box 8100, Causeway, Harare
Tel: 704531.

1 Madziva TJT - BSc,MSc *Head of Institute*

4952 General Analytical Laboratory

1 Muguto Mrs AC - BAgricSc *Assistant Research Officer-in-Charge* general analysis
2 Dube JS - BSc,MSc *Assistant Research Officer* general analysis
3 Samufonda Mrs MBA - BSc *Research Officer* general analysis

4953 Special Analysis Laboratory

1 Mushambi CF - BSc,MSc *Senior Research Officer-in-Charge* special analysis
2 Mhlanga Mrs A - BSc,MSc *Senior Research Officer* pesticide residues
3 Nziramasanga Mrs N - NHDip *Asst. Research Officer* pesticide residues
4 Siwela AH - BSc *Research Officer* biochemistry

4954 Crop Nutrition Section

1 Tagwira F - BSc *Senior Research Officer-in-Charge* crop nutrition
2 Dangarembga G - BSc,MSc *Research Officer* advisory services
3 Nleya G - BSc,MSc *Research Officer* advisory services

4955 Pedology and Soil Survey Section

1 Garikayi AOB - BA,MSc *Research Officer* pedology
2 Kanyanda CW - BSc *Senior Research Officer-in-Charge*
3 Moyo M - MSc *Research Officer* soil survey
4 Mughodi FK - BSc *Research Officer* soil survey
5 Mushiri S - BSc,MSc *Research Officer* soil survey and pedology
6 Nyamugafata PE - BSc *Research Officer* soil survey
7 Nyamwanza B - MSc *Research Officer* soil survey
8 Spurway JK - BSc,MSc *Research Officer* soil mineralogy
9 Wuta M - BSc *Research Officer* soil survey

4956 Soil Productivity Research Laboratory

1 Mandiringana OT - BSc,MSc *Research Officer-in-Charge*
2 Butai PC - BSc *Research Officer* soil chemistry
3 Mugwira LM - BSc,MSc,PhD *Principal Research Officer* soil chemistry
4 Mukurumbira L - BSc,MSc *Research Officer* soil chemistry
5 Nyika M - BAgrSc *Assistant Research Officer* microbiology
6 Ryder Mrs MR - BSC *Assistant Research Officer* microbiology

4957 Cotton Research Institute
P.M.B. 765, Kadoma
Tel: 3927/3928.

1 Rabey GG - BSc,MSc *Head of Institute*

4958 Agronomy, Physiology

1 Gopoza W *Assistant Research Officer*

2 Mashavira T - BSc *Research Officer*

4959 Pest Research

1 Brennan Miss MJ - BSc *Assistant Research Officer (At Chiredzi Research Station)*
2 Brettell JH - BSc,PhD *Chief Research Officer, Team-Leader*
3 Jowah P - BSc,MPhil *Research Officer*

4960 Pathology

1 Chinodya R - BAgrSc *Senior Research Officer*

4961 Crop Breeding Institute
P.O. Box 8100, Causeway, Harare
Tel: 704531.

1 Olver RC - BSc *Head of Institute and Leader of Maize Team*

4962 Maize Team
P.O. Box 8100, Causeway, Harare
Tel: 704531.

4963 Small Grain Team
P.O. Box 8100, Causeway, Harare
Tel: 704531.

1 Mushonga JN - BSc,MSc *Senior Research Officer and Team Leader* sorghum breeding
2 Muza FR - BSc *Research Officer* millet breeding

4964 Winter Cereals Team
P.O. Box 8100, Causeway, Harare
Tel: 704531.

1 Mashiringwani AN - BSc *Research Officer/Team Leader*

4965 Oilseeds Team
P.O. Box 8100, Causeway, Harare
Tel: 704531.

1 Tichagwa JS - BA *Senior Res Officer* soyabeans
2 Chiteka AZ - BSc,MSc *Research Officer* groundnut breeding
3 Venge Miss O - BSc,MPhil *Research Officer* field beans

4966 Agronomy Institute
P.O. Box 8100, Causeway, Harare
Tel: 704531.

1 Hikwa Mrs D - MSc *Acting Head of Institute*
2 Gono L - BAgrSc *Asst. Res Officer* sorghum & millet agronomy
3 Machado S - BAgrSc,MSc *Res Officer* wheat & barley agronomy
4 Mafongoya P - BAgrSc *Asst Res Officer* new crops programme
5 Mataruka D - BAgrSc *Res Officer* maize agronomy
6 Niaya Mrs T - BAgrSc *Asst. Res Officer* field beans & cowpeas
7 Shamudzarira Z - BSc *Asst. Res Officer* groundnuts programme
8 Zharare G - BAgrSc,MSc *Res Officer* groundnuts programme

4967 Dairy Services
P.O. Box 8100, Causeway, Harare
Tel: 704531.

1 Borland Mrs PA - BAgrSc *Chief Dairy Officer*
2 Chademana Miss I - BAgrSc *Research Officer* dairy
3 Jones R - BAgricSc *Research Officer* dairy
4 Kasaka L - BSc *Research Officer* dairy
5 Lakey Miss G - BAgrSc,MSc *Senior Research Officer* dairy
6 Mpofu Miss N - BSc,MSc *Research Officer* milk recording
7 Mupunga EG - BAgrSc,MSc *Senior Research Officer* dairy
8 Mutsau E - BSc,MSc *Research Officer* dairy
9 Ngwarwone Miss F - BSc *Asst. Research Officer* dairy

4968 Farming Systems Research Unit
P.O. Box 8100, Causeway, Harare
Tel: 704531.

1 Dhliwayo DKC - BSc *Research Officer*
2 Makombe G - BSc *Asst. Research Officer*
3 Mombeshora B - BAgrSc,MSc *Senior Research Officer and Team Leader*
4 Shumba E - BAgrSc,MSc *Research Officer*

4969 Grasslands Research Station
P.B. 3701, Marondera
Tel: 3527.

1 Taonezvi HPR - BSc,MSc,PhD *Head of Station*

4970 Livestock Unit
P.B. 3701, Marondera
Tel: 3527.

1 Manyuchi B - BSc *Asst. Research Officer* livestock
2 Mazorodza Miss CR - BSc *Asst Res Officer* livestock

4971 Pasture
P.B. 3701, Marondera
Tel: 3527.

1 Clatworthy JN - BAgrSc,DPhil *Head of Veld & Pastures*
2 Mariyau GJ - BSc *Asst. Res Officer* pastures
3 Musvoto Miss C - BSc *Asst. Res Officer* pastures

4972 Legume Innoculants
P.B. 3701, Marondera
Tel: 3527.

1 Marufu L - BSc *Asst Res Officer*

4973 Henderson Research Station
P.B. 2004, Mazowe
Tel: Mazowe 481/482.

4974 Livestock Research Section
P.B. 2004, Mazowe
Tel: Mazowe 481/482.

1 Manyenga Miss A - BAgrSc *Res. Officer In-Charge* livestock
2 Murinola SE - MSc *Research Officer* laboratory
3 Rigaba OM - BSc *Research Officer* laboratory

4975 Pasture Research Section
P.B. 2004, Mazowe
Tel: Mazowe 481/482.

1 Mills Sir PFL - BAgrSc *Chief Res. Officer In-Charge*
2 Madakadze C - BAgrSc *Research Officer* pastures
3 Nyamadzawo E - BSc *Asst. Research Officer* pastures

4976 Dairy Services Research Section
P.B. 2004, Mazowe
Tel: Mazowe 481/482.

1 Hamudikuwanda H - BSc *Research Officer In-Charge*
2 Mutsuanga T - BSc *Asst Res Officer* dairy

4977 Poultry Unit
P.B. 2004, Mazowe
Tel: Mazowe 481/482.

1 Kulube K - BAgrSc,MSc *Research Officer* poultry

4978 Weed Research Programme
P.B. 2004, Mazowe
Tel: Mazowe 481/482.

1 Rambakudzibga A - BAgrSc *Asst. Research Officer* weed research
2 Mabasa S - BAgrSc *Asst. Research Officer* weed research
3 Takavarasha Miss B - BAgrSc *Asst. Research Officer* weed research

4979 Horticulture Research Centre
P.B. 3748, Marondera
Tel: 4122.

1 Vogel ML - BSc *Research Officer In-Charge*

4980 Information Services
P.O. Box 8108, Causeway, Harare
Tel: 704531.

1 Mandiveyi MT - BAdmin *Research Officer A.I.C.*

4981 Lowveld Research Stations Headquarters Chiredzi
P.O. Box 97, Chiredzi
Tel: 387/398.

1 Mharapara IM - BSc *Head of Station*

4982 Agronomy (Chiredzi)
P.O. Box 97, Chiredzi
Tel: 387/398.

1 Jones E - BSc,MSc *Research Officer*
2 Nyamudaza P - BSc *Research Officer*

4983 Horticulture (Chiredzi)
P.O. Box 97, Chiredzi
Tel: 387/398.

1 Chipara MB - BSc *Asst. Res. Officer*
2 Manzungu E - BSc *Research Officer*
3 Nzima MDS - BSc,MSc *Head of Horticulture and Coffee Res. Inst.*

4984 Makoholi Experiment Station
P. Bag 9182, Masvingo
Tel: 3255/271624.

1 Chikomo J - MSc *Asst. Res. Officer* crop production
2 Khombe CT - BSc *Res. Officer* livestock

4985 Matopos Research Station
P.B. K5137, Bulawayo
Tel: Matopos 2.

1 Calvert GM - BSc *Acting Head of Station*

4986 Range Livestock Nutrition
P.B. K5137, Bulawayo
Tel: Matopos 2.

1 Dube JS - BSc,MSc *Asst. Res. Officer* agricultural chemistry
2 Hatendi PR - BAgrSc *Res. Officer*
3 Ncube S - BSc,MSc *Research Officer*
4 Nyoni VS - MSc *Research Officer*

4987 Livestock Breeding
P.B. K5137, Bulawayo
Tel: Matopos 2.

1 Baffa ML - BAgrSc *Research Officer*
2 Dlodlo Miss S - MSc *Research Officer*

4988 Veld Management and Ecology
P.B. K5137, Bulawayo
Tel: Matopos 2.

1 Gambiza J - BSc *Research Officer* range management
2 Motazedian I - PhD *Res. Officer* range management

4989 Small Ruminant Production
P.B. K5137, Bulawayo
Tel: Matopos 2.

1 Sibanda R - BAgrSc *Asst. Res. Officer*
2 Sikosana J - MAgrSc *Research Officer*

4990 National Herbarium & Botanic Garden
P.O. Box 8100, Causeway, Harare
Tel: 725313/702236.

1 Drummond RB - BSc *Chief Res. Officer: Keeper of National Herbarium*
2 Best EB *Honorary Botanist*
3 Kativu S - BSc *Asst Res Officer* botany
4 Muller TH - BSc *Head of NHBG*
5 Nobanda Miss N - BSc *Asst. Res. Officer* ecology

4991 Plant Protection Research Institute
P.O. Box 8100, Causeway, Harare
Tel: 704531.

1 Mlambo SS - BSc,MSc,PhD *Head of Institute*

4992 Entomology Section

1 Chikwenhara GP - MSc *Research Officer* entomology
2 Mubvuta D - BAgrSc *Research Officer* entomology
3 Sithole SZ - BSc,MSc *Research Officer* entomology
4 Tanyongana Miss R - BSc,MSc *Research Officer* entomology
5 Tshuma J - BSc *Research Officer* entomology
6 Zitsanza E - BAgrSc,MSc *Research Officer* entomology

4993 Pathology Section

1 Mguni C - MSc *Research Officer-in-Charge*
2 Gubba A - BSc *Research Officer* pathology
3 Mahuku GS - BSc *Research Officer* pathology
4 Mtisi Mrs E - BSc,MSc *Senior Research Officer* pathology
5 Ramachela K - BSc *Research Officer* pathology
6 Zvoutata D - BSc *Research Officer* pathology

4994 Nematology Section

1 Muchena P - BSc,MSc *Senior Research Officer-in-Charge*
2 Dube B - BSc,MSc *Research Officer* nematology
3 Ndlovu Miss Z - BSc *Research Officer* nematology

4995 Rhodes Inyanga Experiment Station
P.M.B. 8044, Rusape
Tel: 28125.

1 Payne CB - BScHort *Chief Research Officer-in-Charge*

4996 Seed Services
PO Box 8100, Causeway, Harare
Tel: 720370.

1 Mtindi Mrs K - BSc,MSc *Head of Seed Services*
2 Chigogora Mrs JL - BSc *Research Officer* seeds
3 Maramba P - BSc,MSc,PhD *Principal Research Officer* seeds
4 Munyaradzi Mrs B - BSc *Research Officer* seeds
5 Simbi F - BSc *Research Officer* seeds

4997 Department of Agricultural Technical and Extension Services
PO Box 8117, Causeway, Harare
Tel: 707311/794601.

1 Matanyaira CM - BSc *Director*
2 Denga E - BSc *Assistant Director: Technical*
3 Pazvakavambwa S - BA,MScAgrEng *Assistant Director: Irrigation*
4 Vaughan-Evans RH - BAgrSc *Assistant Director: Field*

4998 Monitoring and Evaluation Unit
PO Box 8117, Causeway, Harare
Tel: 707311/794601.

1 Chipika S - BSc *Extension Specialist: Research*
2 Mashonjowa L - BSc *Extension Specialist: Research*

4999 Agricultural Engineering Institute
PO Box BW 330, Borrowdale, Harare
Tel: 725936/883119.

1 Elliot KM - BScEng,MScEcol,MIAgrE,MZwelE *Assistant Chief Engineer*
2 Elwell HA - MScEng,CEng,MICE,MZwelE soil and water conservation; field and simulator studies
3 Karima A - BScAgrEng crop production
4 Moyo N - BScAgrEng mechanisation
5 Nazare RM - BScAgrEng testing farm implements
6 Norton AJ - BScAgrEng crop production; tillage
7 Small RB - BArch farm buildings; brick vaulted roofing

5000 Department of Veterinary Services
P.O. Box 8012, Causeway, Harare
Tel: 791355.

1 Thomson JW - BVMS,DTVM,DVSM,MRCVS *Director*
2 Hargreaves SK - BVSc *Deputy Director:Field*
3 Madzima WN - BAgrSc,BMedVet *Deputy Director : Technical*
4 Stewart GHG - MRCVS,FRSH *Asst. Director : Research*

5001 Veterinary Research Laboratory
P.O. Box 8101, Causeway, Harare
Tel: 705885.

1 Munatswa FC - BSc,DVM *Chief Vet. Res. Officer* protozoology
2 Foggin CM - BVSc,MRCVS *Senior Vet. Res. Officer* histopathology and rabies
3 Gwaze G - DVM,MVSc *Veterinary Res. Officer* poultry pathologist
4 Minarik P *Veterinary Res. Officer* poultry pathology

5002 Tick Ecology
P.O. Box 8101, Causeway, Harare
Tel: 705885.

1 Colborne JRA ticks
2 Mazhou GWA ticks
3 Muvazarivwa P virology
4 Tabala N protozoology

5003 Library
P.O. Box 8101, Causeway, Harare
Tel: 705885.

1 Mufunda G - BSc *Librarian*

5004 Tsetse and Trypanosomiasis Control Branch
P.O. Box 8283, Causeway, Harare
Tel: 707381.

1 Mtrindurwa A - BSc target technology

2 Sherani W - BSc,MSc,DIC Tsetse bait technology
3 Vale GA - BA,PhD *Assistant Director* odour attractants & traps

5005 Tobacco Research Board
Kutsaga Research Station, P.O. Box 1909, Harare
Tel: 50411.

1 Cousins LTV - BAgrSc,MPhil *Director*

5006 Research Division
Kutsaga Research Station, P.O. Box 1909, Harare
Tel: 50411.

1 Cole JS - BSc,PhD,DSc,DIC,CBiol,FIBiol *Asst. Director* phytopathology; biological control

5007 Agronomy Department
Kutsaga Research Station, P.O. Box 1909, Harare
Tel: 50411.

1 Lapham J - BAgrSc,MSc,PhD *Senior Agronomist* agronomy; weed biology; herbicides
2 Flower KC - BAgrSc agronomy
3 Ushamba Miss J - BAgrSc agronomy; growth regulation

5008 Analytical Chemistry Department
Kutsaga Research Station, PO Box 1909, Harare
Tel: 50411.

1 Toet L - DipEng *Senior Analytical Chemist* analytical chemistry
2 Mudungwa Miss L - BSc Chemistry and Biology

5009 Entomology Department
Kutsaga Research Station, P.O. Box 1909, Harare
Tel: 50411.

1 Blair BW - BSc,PhD,CBiol,MBiol *Senior Entomologist* entomology, mites
2 Kunjeku Miss E - BAgrSc,MScApplEnt,DIC,FRES entomology, insect viruses

5010 Liaison Department
Kutsaga Research Station, P.O. Box 1909, Harare
Tel: 50411.

1 Stocks JL - BAgrSc *Senior Liaison Officer* tobacco extension

5011 Nematology Department
Kutsaga Research Station, P.O. Box 1909, Harare
Tel: 50411.

1 Shepherd JA - MA,DipAgr,DIC,MPhil *Senior Nematologist* nematology; nematicides
2 Way Miss J - BSc,MPhil nematology, biology of resistance

5012 Plant Breeding Department
Kutsaga Research Station, P.O. Box 1909, Harare
Tel: 50411.

1 Ternouth RAF - BAgrSc *Senior Plant Breeder* tobacco breeding
2 Matibiri EA - BAgrSc,MPhil plant breeding; tissue culture
3 Woodand JJ - BSc,MSc,PhD Plant Breeding

5013 Plant Pathology Department
Kutsaga Research Station, P.O. Box 1909, Harare
Tel: 50411.

1 Fisher CR - BAgrSc *Senior Plant Pathologist* phytopathology; fungal diseases

5014 Plant Physiology Department
Kutsaga Research Station, P.O. Box 1909, Harare
Tel: 50411.

1 Garvin RT - CAE,BAgrSc,MScAgrExtn,DPhil *Senior Plant Pathologist* crop physiology; tobacco
2 Dix DL - BSc crop physiology; irrigation
3 Norton AG - BAgrSc,MScAgr crop physiology; tobacco

5015 Productivity Department
Kutsaga Research Station, P.O. Box 1909, Harare
Tel: 50411.

1 Mason CM - BAgrSc agronomy; extension

5016 Soil Chemistry Department
Kutsaga Research Station, P.O. Box 1909, Harare
Tel: 50411.

1 Ryding WW - BSc,MSc,PhD,CChem,MRSc *Senior Soil Chemist* soil chemistry; fertilizers

5017 Pig Industry Board

Experiment Farm, P.O. Box HG 297, Highlands, Harare
Tel: 25724.

1 Mandisodza KT - DipAgr,BSc,PhD *Director*
2 Masvaura S - NDipAgr,BSc,MSc *Research Officer* pig nutrition

5018 Forestry Commission

PO Box 8111, Causeway, Harare
Tel: 706216/706217/706218.

1 Banks PF - BScFor *Director*

5019 Forest Research Centre
PO Box HG 595, Highlands, Harare
Tel: 46878/46879.

1 Bleakley Miss SE - BA eucalypt provenance studies & breeding
2 Gwaze DP - BSc multipurpose species screening; provenance studies; nitrogen fixing
3 Mills WR - BScDipAgrSc biometrics

5020 Agricultural Research Trust

PO Box MP 84, Mount Pleasant, Harare
Tel: 726061/726062/726063.

1 Winkfield RA - DipAgr,BEd *Director*

5021 Research Farm
PO Box MP 84, Mount Pleasant, Harare
Tel: 726061/726062/726063.

5022 Crop Research

1 MacRobert JF - BAgrSc *Research Manager*

5023 Farm Management

1 Pilbrough S - BAgrSc *Economist*

5024 University of Zimbabwe

5025 Faculty of Agriculture
PO Box MP 167, Mount Pleasant, Harare
Tel: 303211.

1 Ngongoni NT - DipAgr,BAgrSc,MAgrSc,PhD *Deputy Dean* animal nutrition, farming systems
2 Rukuni M - DipAgr,BSc,MSc,PhD *Doctor, Dean* project appraisal, sector planning, irrigation analysis

5026 Animal Science

1 Chesworth J - BSc,PhD *Associate Professor* animal nutrition
2 Mutisi C - BSc,PhD *Lecturer* animal production systems
3 Ndlovu L - BSc,MSc *Lecturer* animal nutrition
4 Titterton Mrs M - BAgrSc,MPhil *Lecturer* animal production systems

5027 Crop Science

1 Schweppenhauser MA - MSc,DSc *Professor and Chairman of Department* genetics and plant breeding
2 Canhao Sister J - BSc,PCE,MA *Lecturer* biometrics
3 Chivinga OA - BSc,MSc *Lecturer* weed science
4 Cole Mrs DL - BSc,MSc,DPhil *Senior Lecturer* plant pathology and disease
5 Giga DP - BSc,MSc,PhD *Lecturer* entomology
6 Maasdorp Mrs BV - BAgrSc,MPhil *Lecturer* grassland science
7 Mariga IK - BSc,MSc,PhD *Lecturer* agronomy
8 Mashingaidza K - BSc,MPhil,PhD *Lecturer* genetics/agronomy
9 McLaurin AR - BSc,MSc *Lecturer* grassland science and legume introduction
10 Northcraft PD - BSc,MSc,PhD *Lecturer* agronomy
11 Rice RP - BSA,MS,PhD *Lecturer* horticulture
12 Robertson AI - BSc,DipEd,PhD *Lecturer* crop physiology
13 Tongoona P - BSc,MPhil,PhD *Lecturer* genetics

5028 Soil Science and Agricultural Engineering

1 Nyamapfene KW - BAGradCE,MSc,DipSoilSurv,PhD,MIBiol land use; pedology
2 Ascough WJ - BAgrSc,DipAgrEng,MIAE *Senior Lecturer* agricultural engineering
3 Hungwe AP - BAgrSc *Research Fellow* pedology/soil chemistry
4 Hussein Mrs J - BAgrSc,MSc *Lecturer* soil physics
5 M'peperaki S - MSc *Lecturer* soil microbiology
6 Piha M - PhD *Lecturer* soil fertility

7 Senzanja A - MSc *Lecturer* irrigation engineering
8 Tambo S - MSc *Lecturer* agricultural engineering

5029 Agricultural Economics and Extension

1 Barnstan R - BA,MSc,PhD *Lecturer* data collection, farming systems research
2 Chimadza Mrs R - BAEcon,MAEcon,DipAgrEcon,MScAgrEcon *Lecturer* development, rural planning, women in agriculture
3 Chopak C - BS,MS *Lecturer* agricultural marketing
4 Mbwanda C - BScEcon,MScAgrEcon *Lecturer* project planning & analysis, micro & macro economics
5 Mudimu G - BScAgr,BScEcon,MScAgrEcon *Lecturer* farm management, production economics, natural resource economics
6 Muir-Lerasche K - BAEcon,PhD *Lecturer* policy analysis, marketing, natural resources economics
7 Stanning Ms J - BA,MA,MScAgrEcon *Lecturer* production economics, research methods

SUBJECT INDEX

The references in the subject entries of this index give first a country code (see table below), followed by the institution number in bold type and person number in small type. Thus "AUS **13** 50" refers to the fiftieth person at institute 13 (Australia).

Aylemore LAG *AUS* **876** 3
Ayliffe TR *UK* **4832** 1
Ayling PD *UK* **4533** 4
Aylward JH *AUS* **26** 2
Ayo C *UG* **3989** 4
Ayoade SK *NGA* **3685** 2
Ayob b. Sukra *MY* **3176** 25
Ayodele EA *NGA* **3563** 2
Ayorinde KL *NGA* **3668** 3
Ayre GL *CAN* **1113** 2
Ayres JF *AUS* **241** 2
Ayres PA *AUS* **228** 1
Ayres PG *UK* **4539** 1
Ayyagari V *IN* **2097** 2
Ayyamperumal A *IN* **2575** 2
Ayyanna T *IN* **2246** 2
Ayyappan S *IN* **2112** 2
Ayyar SP *IN* **2115** 2
Ayyasamy MK *IN* **2530** 2
Ayyavoo R *IN* **2548** 2
Azad Dr AA *AUS* **31** 1
Azad Thakur NS *IN* **2007** 3
Azam KM *IN* **2246** 3
Azeez Basha A *IN* **2583** 2
Azhakiamanavalan RS *IN* **2564** 3
Aziz IH *CY* **1601** 2
Aziz SA *IN* **2246** 4
Aziz SA *IN* **2634** 3
Aziz-Demova Mrs M *CY* **1588** 2
Azizah bte Hashim *MY* **3240** 8
Azizah bte Osman *MY* **3234** 4
Azizuddin M *IN* **2701** 1
Azubuike AA *NGA* **3717** 3
Azuolas J *AUS* **612** 3

B

Baafi-Yeboah S *GH* **1696** 7
Baah SO *GH* **1693** 3
Baars JA *NZ* **3329** 5
Babaleye SOT *NGA* **3717** 4
Babalola ML *NGA* **3642** 2
Babarinsa FA *NGA* **3658** 3
Babatunde OO *NGA* **3630** 1
Babb JC *CAN* **1221** 2
Babidge P *AUS* **499** 2
Babiuk LA *CAN* **1384** 2
Babiuk LA *CAN* **1479** 3
Babji AS *MY* **3217** 3
Babu AGC *LK* **3828** 2
Babu CR *IN* **2639** 2
Babu RS *IN* **2244** 1
Babu TS *IN* **2263** 2
Bacchus Dr S *UK* **4124** 1
Bach L *CAN* **1456** 3
Bach L *CAN* **1460** 3
Bache BW *UK* **4514** 3
Bachelard EP *AUS* **754** 2
Bacher GJ *AUS* **666** 4
Bachew S *TT* **3973** 2
Bachraz DY *MU* **3245** 2
Bacic A *AUS* **862** 4
Back HL *UK* **4650** 1
Back Miss JF *AUS* **38** 4
Backholer JR *AUS* **690** 1
Backhouse MH *AUS* **195** 1
Bacon AJ *MY* **3171** 2
Bacon G *AUS* **364** 1
Bacon JR *UK* **4286** 2
Bacon P *JAM* **2891** 2
Bacon PE *AUS* **287** 3
Bacon PE *AUS* **292** 2
Badanur VP *IN* **2404** 2
Badaru K *NGA* **3561** 2
Badawy NS *AUS* **658** 2
Badcock Miss RM *UK* **4536** 2
Badejo SOO *NGA* **3584** 3
Badger MR *AUS* **736** 3
Badmus GA *NGA* **3630** 2
Badu-Araku B *GH* **1641** 5
Baffa ML *ZW* **4987** 1
Baffoe-Bonnie E *GH* **1713** 6
Baffour-Senkyire JK *GH* **1710** 3
Bafokuzara DN *UG* **3989** 5
Bafor Mrs ME *NGA* **3631** 3
Bag SC *IN* **1876** 2
Bagavandoss *IN* **2627** 2
Bagchi MM *IN* **2114** 2
Bagga AK *IN* **1785** 2
Bagga RK *IN* **2349** 1
Baghel AS *IN* **2454** 1
Baghel SS *IN* **2432** 5

Bagine RKN *KE* **3049** 2
Bagle BG *IN* **1940** 1
Bagnall DJ *AUS* **66** 4
Bagnall RH *CAN* **1039** 2
Bagust TJ *AUS* **17** 1
Bahadur P *IN* **1792** 2
Baharuddin b Abd Ghani *MY* **3234** 5
Baharudin b. Baharom *MY* **3182** 1
Bahga HS *IN* **2826** 1
Bahl DK *IN* **1761** 3
Bahl Mrs N *IN* **1768** 2
Bahuguna GN *IN* **1800** 3
Bahule Miss J *UG* **3989** 6
Baidu J *GH* **1693** 4
Baier W *CAN* **985** 1
Baig M *IN* **2692** 1
Baigent DR *AUS* **241** 3
Baigent DR *AUS* **292** 3
Bailey Prof. AJ *UK* **4203** 1
Bailey AW *CAN* **1461** 2
Bailey BJ *UK* **4300** 1
Bailey CBM *CAN* **1152** 2
Bailey CJ *UK* **4477** 3
Bailey DRC *CAN* **1152** 1
Bailey J *PNG* **3718** 2
Bailey JA *UK* **4267** 1
Bailey JV *CAN* **1476** 2
Bailey KL *CAN* **1126** 2
Bailey LD *CAN* **1104** 1
Bailey PJ *AUS* **594** 3
Bailey PT *AUS* **505** 5
Bailey RA *UK* **4334** 1
Bailey S *UK* **4426** 4
Bailey SER *UK* **4618** 3
Bailey WJ *AUS* **877** 2
Bailie NC *UK* **4569** 3
Bailiss KW *UK* **4604** 2
Baillargeon G *CAN* **998** 5
Bain DC *UK* **4284** 1
Bain J *NZ* **3356** 1
Bain JF *CAN* **1549** 2
Bain Dr R *AUS* **169** 1
Bainbridge MH *AUS* **340** 2
Baines JT *NZ* **3495** 4
Baines S *UK* **4871** 1
Bainton J *UK* **4104** 1
Baird DB *NZ* **3289** 2
Baird GD *UK* **4256** 1
Baird J *NZ* **3494** 2
Baird JD *CAN* **1515** 4
Bajpai LD *IN* **2426** 3
Bajpai MR *IN* **2500** 1
Bajpai RP *IN* **2434** 2
Bakanwangiraki B *UG* **3989** 7
Bakar MA *BGD* **932** 1
Bakare BO *NGA* **3583** 2
Bakau B *PNG* **3728** 1
Baker AA *AUS* **850** 3
Baker AA *UK* **4141** 2
Baker AJM *UK* **4667** 2
Baker Dr AT *AUS* **31** 2
Baker Prof. BE *CAN* **1543** 5
Baker BI *UK* **4481** 1
Baker CRB *UK* **4437** 3
Baker D *BWA* **976** 2
Baker DA *UK* **4604** 1
Baker Mr DM *UK* **4108** 1
Baker EA *UK* **4266** 2
Baker GH *AUS* **50** 1
Baker GJ *AUS* **505** 6
Baker GL *AUS* **293** 2
Baker IW *AUS* **342** 1
Baker J *CAN* **1275** 2
Baker JM *UK* **4036** 1
Baker Dr JR *UK* **4066** 2
Baker JR *UK* **4569** 4
Baker JT *AUS* **216** 1
Baker KC *UK* **4218** 2
Baker KF *UK* **4160** 2
Baker LBB *CAN* **1544** 3
Baker MF *CAN* **1192** 2
Baker MJ *NZ* **3422** 2
Baker NR *AUS* **229** 1
Baker P *UK* **4010** 1
Baker RA *UK* **4545** 3
Baker RD *UK* **4179** 1
Baker RJ *AUS* **430** 2
Baker RJ *CAN* **1470** 3
Baker RJ *JAM* **2877** 2
Baker RL *NZ* **3322** 3
Baker RM *UK* **4160** 1
Baker RR *UK* **4618** 4
Baker RT *NZ* **3281** 1

Baker TE *CAN* **1355** 1
Baker TG *AUS* **659** 2
Bakker J *UK* **4222** 1
Baksh N *TT* **3943** 1
Baktavatsalam G *IN* **1811** 3
Bal AR *IN* **1974** 2
Bala G *TT* **3943** 2
Bala Nambisan Mrs *IN* **2176** 9
Bala Ravi S *IN* **1795** 5
Balachandran R *IN* **2041** 1
Balachandran S *IN* **2612** 1
Balagopalan C *IN* **2176** 1
Balaguru T *IN* **2228** 2
Balaine DS *IN* **2356** 1
Balakrishna VK *IN* **2579** 2
Balakrishna Rao K *IN* **2404** 3
Balakrishnamurthy PB *IN* **2670** 1
Balakrishnan CR *IN* **2029** 1
Balakrishnan M *IN* **2050** 3
Balakrishnan N *CAN* **1436** 1
Balakrishnan R *IN* **2565** 1
Balakrishnan V *IN* **2543** 1
Balami DK *IN* **2354** 2
Balan V *IN* **2136** 5
Balaraj CTA *IN* **2627** 3
Balasooriya I *LK* **3898** 1
Balasubramaniam A *IN* **1757** 13
Balasubramaniam A *IN* **2530** 3
Balasubramaniam EM *LK* **3797** 2
Balasubramaniam Prof S *LK* **3887** 4
Balasubramanian A *IN* **2559** 1
Balasubramanian G *IN* **2564** 1
Balasubramanian KA *IN* **2245** 1
Balasubramanian M *IN* **2520** 1
Balasubramanian M *IN* **2523** 1
Balasubramanian M *IN* **2530** 4
Balasubramanian N *IN* **2534** 1
Balasubramanian R *IN* **2627** 4
Balasubramanian S *IN* **2587** 3
Balasubramanian TN *IN* **2569** 2
Balasubramanian V *IN* **1988** 1
Balasubramanian V *NGA* **3709** 6
Balasubramanya RH *IN* **1853** 2
Balasundaram VR *IN* **1775** 3
Balasundaram S *IN* **2601** 1
Balasunderam N *IN* **1835** 4
Balasuriya BRU *LK* **3823** 1
Balasuriya G *LK* **3811** 4
Balasuriya I *LK* **3790** 1
Balatinecz Prof. JJ *CAN* **1523** 5
Balderson J *AUS* **46** 4
Baldev B *IN* **1785** 3
Baldock RC *AUS* **403** 2
Baldwin AJ *NZ* **3534** 6
Baldwin BA *UK* **4246** 1
Baldwin CS *CAN* **1420** 3
Baldwin GE *AUS* **8** 1
Bale JO *NGA* **3600** 2
Bale JS *UK* **4544** 3
Balingasa EN *TZ* **3906** 1
Ball BC *UK* **4805** 1
Ball BV *UK* **4328** 1
Ball DF *UK* **4064** 1
Ball EE *AUS* **738** 2
Ball HJ *UK* **4900** 4
Ball Ms KM *AUS* **186** 1
Ball PR *NZ* **3428** 2
Ball RO *CAN* **1503** 3
Ball Dr SL *UK* **4657** 4
Ball WJ *CAN* **1286** 2
Ballah-Conteh Mrs A *UK* **3994** 4
Ballal AL *IN* **2467** 2
Ballance GM *CAN* **1489** 2
Ballantyne Dr BJ *AUS* **286** 4
Ballard R *NZ* **3353** 1
Ballard TM *CAN* **1442** 2
Ballard TM *CAN* **1451** 2
Ballinger DJ *AUS* **621** 2
Balment RJ *UK* **4618** 5
Balnave D *AUS* **804** 1
Balneaves JN *NZ* **3380** 2
Balogun JK *NGA* **3666** 2
Balogun MD *NGA* **3571** 2
Balu A *IN* **2593** 1
Bamford S *UK* **4024** 3
Bampton PR *UK* **4756** 2
Banbridge A *UK* **4409** 3
Bancroft Prof. JB *CAN* **1520** 1
Banda D *ZM* **4940** 2
Banda JW *MW* **3147** 3
Banda MHP *MW* **3111** 2
Banda P *MW* **3109** 2

Gupta GP *IN* **1765** 16
Gupta HS *IN* **2007** 17
Gupta IC *CAN* **1473** 5
Gupta JN *IN* **2044** 2
Gupta JP *IN* **1786** 14
Gupta JP *IN* **1979** 1
Gupta JP *IN* **1980** 4
Gupta JP *IN* **2229** 5
Gupta KC *IN* **1800** 13
Gupta KP *IN* **2431** 6
Gupta KR *IN* **2335** 13
Gupta Mrs M *IN* **1765** 17
Gupta MC *IN* **2820** 4
Gupta Miss MD *IN* **1770** 7
Gupta MG *IN* **2015** 5
Gupta MK *IN* **2514** 2
Gupta MN *IN* **2845** 16
Gupta MN *IN* **2845** 17
Gupta MP *IN* **2387** 5
Gupta N *IN* **1778** 3
Gupta NP *IN* **1782** 4
Gupta NP *IN* **2079** 2
Gupta OP *IN* **2334** 4
Gupta OP *IN* **2436** 1
Gupta OP *IN* **2500** 7
Gupta PC *IN* **2006** 1
Gupta PC *IN* **2334** 5
Gupta PC *IN* **2337** 7
Gupta PC *IN* **2339** 3
Gupta PC *IN* **2351** 3
Gupta PC *IN* **2500** 8
Gupta PC *IN* **2812** 3
Gupta PD *IN* **1987** 4
Gupta PK *IN* **2067** 1
Gupta PL *UK* **4667** 7
Gupta PN *IN* **2007** 18
Gupta R *IN* **2182** 2
Gupta RA *IN* **2113** 2
Gupta RD *IN* **2379** 1
Gupta RD *IN* **2398** 3
Gupta RK *IN* **1961** 6
Gupta RK *IN* **1969** 7
Gupta RK *IN* **2018** 1
Gupta RK *IN* **2427** 5
Gupta RP *IN* **1763** 6
Gupta RP *IN* **2360** 4
Gupta RS *IN* **2398** 4
Gupta RSR *IN* **2345** 1
Gupta S *IN* **2655** 5
Gupta Mrs S *IN* **2677** 4
Gupta SC *IN* **2007** 19
Gupta SC *IN* **2639** 7
Gupta SC *IN* **2728** 1
Gupta SC *IN* **2752** 3
Gupta SD *IN* **2112** 10
Gupta SK *IN* **1972** 6
Gupta SK *IN* **2015** 6
Gupta SK *IN* **2040** 1
Gupta SK *IN* **2056** 2
Gupta SK *IN* **2335** 14
Gupta SK *IN* **2429** 1
Gupta SL *IN* **1765** 18
Gupta SL *IN* **2358** 4
Gupta SR *IN* **2015** 7
Gupta SS *IN* **1800** 14
Gupta UC *CAN* **1046** 6
Gupta US *IN* **2429** 2
Gupta VK *IN* **2024** 3
Gupta VK *IN* **2425** 3
Gupta VP *IN* **2368** 2
Gupta VS *CAN* **1481** 6
Gupta YP *IN* **1764** 5
Gurchan Singh a/l Harnam Singh *MY* **3239** 5
Gurib A *MU* **3271** 4
Gurnah AM *PNG* **3737** 4
Gurnani M *IN* **2028** 1
Gurnell J *UK* **4583** 17
Gurnett LR *UK* **4460** 1
Gurnsey MP *NZ* **3389** 19
Gurr CG *AUS* **74** 18
Gurusinghe P DE A *LK* **3841** 1
Guruswamy M *IN* **2520** 2
Guruswamy Raja VD *IN* **2518** 1
Gusta LV *CAN* **1470** 9
Gutfreund H *UK* **4501** 2
Guthrie DM *UK* **4618** 1
Gutteridge RC *AUS* **841** 16
Gutteridge S *UK* **4326** 5
Guutazi AS *TZ* **3910** 8
Guy GL *AUS* **632** 8
Guy PL *AUS* **575** 2
Gwaiseuk W *PNG* **3729** 4
Gwal HB *IN* **2430** 6

Gwayumba WE *KE* **2968** 1
Gwaze DP *ZW* **5019** 2
Gwaze G *ZW* **5001** 3
Gwynne DC *UK* **4870** 3
Gwynne DFP *UK* **4699** 6
Gwyther D *AUS* **667** 12
Gyan Miss N *TT* **3954** 1
Gyang FN *GH* **1704** 6
Gyawu P *GH* **1712** 6
Gyening KO *GH* **1679** 1
Gyimah Mrs A *GH* **1690** 4
Gyles Prof. CL *CAN* **1518** 3
Gyles OA *AUS* **627** 2

H

Haavisto VF *CAN* **1267** 3
Haber SM *CAN* **1111** 3
Haberakade Miss RR *LK* **3779** 4
Habeshaw D *UK* **4794** 3
Habib AF *IN* **2405** 9
Habibuddin b. Hashim *MY* **3178** 6
Habibullah AKM *BGD* **923** 2
Habsah Abdul Kadir *MY* **3228** 6
Hachiwa C *ZM* **4942** 3
Hacker Dr JB *AUS* **97** 5
Hacker RB *AUS* **718** 4
Hacker RR *CAN* **1503** 12
Hackett AJ *CAN* **991** 3
Hackett C *AUS* **127** 6
Hacking A *UK* **4448** 2
Hacking CJ *AUS* **897** 2
Hackman RH *AUS* **46** 36
Haddad KS *AUS* **291** 9
Hadfield W *PNG* **3718** 5
Hadimani AS *IN* **2404** 6
Hadjichristodoulou A *CY* **1598** 1
Hadjichristophorou Miss M *CY* **1607** 3
Hadjidemetriou D *CY* **1600** 1
Hadjipanayiotou M *CY* **1599** 2
Hadjistephanou N *CY* **1607** 5
Hadley P *UK* **4663** 5
Hadziyev D *CAN* **1459** 2
Haegi LAR *AUS* **543** 2
Hafeezuddin N *IN* **2258** 1
Hafi AAB *LK* **3811** 6
Hag RKM *UK* **4870** 1
Hagarty JJ *CAN* **1395** 1
Hagen Prof. DW *CAN* **1579** 8
Haggar RJ *UK* **4372** 2
Haggarty P *UK* **4340** 9
Haggerstone B *CAN* **1450** 4
Haggett TOR *NZ* **3534** 31
Haggis G *CAN* **1035** 1
Haggis M *UK* **4129** 2
Hagley EAC *CAN* **1097** 3
Hague NGM *UK* **4016** 2
Hague Dr NGM *UK* **4657** 17
Hagyard CJ *NZ* **3540** 7
Hahn ND *NGA* **3709** 13
Hahn SK *NGA* **3708** 1
Haig D *UK* **4169** 8
Haig DM *UK* **4172** 2
Haig RA *CAN* **1259** 2
Haigh AL *UK* **4814** 11
Haigh J *CAN* **1478** 8
Hails MR *UK* **4024** 11
Hailstone TS *AUS* **852** 7
Haines CP *UK* **4142** 6
Haines DM *CAN* **1479** 6
Haines M *UK* **4699** 2
Haines MW *AUS* **354** 1
Haines PJ *AUS* **635** 5
Haines SR *NZ* **3303** 3
Hair IG *UK* **4862** 3
Haizel KA *GH* **1718** 1
Haji Ahmad b. Hj. Ab. Wahab *MY* **3176** 36
Haji Mustapha Mamat *MY* **3233** 8
Hajra A *IN* **2217** 1
Haki JM *TZ* **3908** 1
Hakim KL *IN* **2230** 3
Hakiza Mrs G *UG* **3989** 11
Halchuk C *CAN* **986** 6
Halder DD *IN* **2107** 2
Hale CJ *AUS* **615** 6
Hale CN *NZ* **3466** 1
Hale DJ *UK* **485** 1
Hale KB *AUS* **398** 1
Hale M *UK* **4631** 7
Hale OD *UK* **4298** 2
Hale PA *UK* **4671** 5
Haleem R *MY* **3222** 3

Hales DF *AUS* **767** 11
Hales JRS *AUS* **21** 24
Haley C *UK* **4167** 5
Haley D *CAN* **1442** 7
Haley Prof. LE *CAN* **1561** 12
Halford M *UK* **4798** 2
Haliburton TH *CAN* **1568** 5
Halijah Ibrahim *MY* **3226** 6
Halimah AS *MY* **3215** 4
Halimah binti Haji Ahmad *MY* **3238** 3
Hall AM *UK* **4029** 4
Hall Prof. BK *CAN* **1561** 13
Hall DA *UK* **4873** 7
Hall DE *UK* **4684** 3
Hall DG *AUS* **276** 4
Hall DI *UK* **4033** 6
Hall DJM *AUS* **268** 7
Hall DJM *AUS* **292** 11
Hall DO *UK* **4574** 1
Hall DR *UK* **4115** 5
Hall G *UK* **4013** 7
Hall GA *UK* **4257** 5
Hall GJ *UK* **4385** 3
Hall GMJ *NZ* **3379** 5
Hall HK *NZ* **3412** 3
Hall I *CAN* **1004** 3
Hall IR *NZ* **3298** 3
Hall IR *NZ* **3507** 2
Hall IV *CAN* **1050** 4
Hall JB *UK* **4631** 8
Hall JC *CAN* **1506** 12
Hall JL *UK* **4673** 6
Hall JP *CAN* **1236** 2
Hall JW *CAN* **1172** 2
Hall LW *UK* **4515** 6
Hall MJ *AUS* **311** 1
Hall NH *AUS* **195** 4
Hall R *CAN* **1506** 13
Hall R *UK* **4653** 4
Hall RJ *CAN* **1275** 4
Hall SA *UK* **4409** 5
Hall TD *UK* **4742** 1
Hall TJ *AUS* **451** 2
Hall WA *CAN* **1357** 3
Hallam D *UK* **4659** 8
Hallegraeff GM *AUS* **110** 3
Hallett IC *NZ* **3468** 3
Halley RJ *UK* **4732** 2
Halliday D *UK* **4140** 5
Halliday GJ *UK* **4761** 1
Halliday RB *AUS* **46** 37
Halliday WJ *AUS* **846** 2
Halligan G *NZ* **3536** 4
Hallim MK *TT* **3939** 1
Halling PJ *UK* **4855** 5
Halliwell G *UK* **4695** 2
Halloran GM *AUS* **864** 12
Halm AT *GH* **1664** 1
Halm AT *GH* **1669** 1
Halm BJ *GH* **1686** 3
Halm Miss M *GH* **1650** 1
Halnan Dr CRE *AUS* **812** 3
Halsall DM *AUS* **65** 48
Halsall Miss JL *UK* **4024** 12
Halse NJ *AUS* **705** 1
Halse SA *AUS* **726** 5
Halstead DGH *UK* **4428** 6
Halton DW *UK* **4911** 2
Haluschak P *CAN* **1492** 1
Halyal KG *IN* **2298** 1
Ham GJ *AUS* **359** 1
Hamblin AP *AUS* **67** 1
Hamblin C *UK* **4188** 4
Hamdan bin Jol *MY* **3240** 9
Hamdorf IJ *AUS* **343** 1
Hamdzah b. A. Rahman *MY* **3176** 37
Hamer G *AUS* **228** 3
Hamer PJC *UK* **4200** 5
Hamid L *MY* **3216** 2
Hamid MA *BGD* **921** 9
Hamid MA *BGD* **965** 3
Hamill AS *CAN* **1095** 1
Hamill JD *UK* **4211** 3
Hamilton BA *AUS* **241** 8
Hamilton CJ *UK* **3998** 1
Hamilton CP *AUS* **424** 2
Hamilton D *AUS* **636** 3
Hamilton D *AUS* **848** 7
Hamilton DJ *AUS* **422** 3
Hamilton DL *CAN* **1481** 7
Hamilton E *CAN* **1346** 21
Hamilton E *CAN* **1355** 16
Hamilton GA *UK* **4748** 1

Jeffree CE *UK* **4810** 6
Jeffrey DC *UK* **4768** 3
Jeffrey IG *UK* **4745** 2
Jeffrey M *UK* **4418** 3
Jeffrey SR *CAN* **1484** 11
Jeffrey SW *AUS* **110** 1
Jeffrey WA *UK* **4801** 3
Jeffries CJ *UK* **4742** 3
Jeffries P *UK* **4537** 4
Jeganathan M *LK* **3847** 5
Jegatheesan G *IN* **2586** 1
Jegede JO *NGA* **3603** 3
Jeger Dr MJ *UK* **4110** 4
Jeggo MH *UK* **4191** 2
Jeglum JK *CAN* **1260** 2
Jehangeer MI *MU* **3256** 4
Jelen P *CAN* **1459** 4
Jell PA *AUS* **479** 3
Jellis GJ *UK* **4319** 5
Jellis Mrs S *UK* **4030** 5
Jellyman DJ *NZ* **3384** 4
Jena BC *IN* **2498** 5
Jena D *IN* **2497** 5
Jena S *IN* **2112** 13
Jenkin D *UK* **4161** 4
Jenkin JJ *AUS* **685** 1
Jenkins CLD *AUS* **65** 63
Jenkins DA *UK* **4630** 5
Jenkins DA *UK* **4633** 4
Jenkins E *AUS* **183** 3
Jenkins G *UK* **4313** 4
Jenkins GI *UK* **4826** 11
Jenkins GI *UK* **4834** 6
Jenkins J *AUS* **184** 2
Jenkins JEE *UK* **4450** 6
Jenkins KJ *CAN* **992** 5
Jenkins N *UK* **4537** 5
Jenkins PD *UK* **4700** 4
Jenkins TC *AUS* **215** 2
Jenkinson DM *UK* **4169** 11
Jenkinson DME *UK* **4171** 3
Jenkinson DS *UK* **4327** 11
Jenkinson GM *CAN* **1501** 3
Jenkinson GM *CAN* **1503** 14
Jenkyn JF *UK* **4332** 12
Jenner CF *AUS* **787** 3
Jennings AC *AUS* **782** 3
Jennings DH *UK* **4554** 8
Jennings DL *UK* **4350** 2
Jennings DM *UK* **4187** 2
Jennings FW *UK* **4828** 6
Jennings JB *UK* **4545** 9
Jennings KR *UK* **4706** 8
Jennings PA *AUS* **43** 8
Jennings PG *JAM* **2881** 1
Jensen DEN *CAN* **1499** 9
Jensen KIN *CAN* **1051** 1
Jensen LCW *NZ* **3479** 6
Jensen RC *AUS* **844** 2
Jepson PC *UK* **4674** 1
Jeremiah LE *CAN* **1144** 4
Jericho KWF *CAN* **1211** 3
Jerie PH *AUS* **628** 1
Jerie RH *AUS* **626** 1
Jermyn WA *NZ* **3408** 1
Jernakoff P *AUS* **117** 5
Jerrett AR *NZ* **3441** 3
Jessop JP *AUS* **542** 1
Jessop NS *UK* **4791** 8
Jessop RS *AUS* **822** 7
Jeswani MD *IN* **1953** 5
Jetuah FK *GH* **1690** 1
Jewell EJ *UK* **4446** 5
Jewer P *UK* **4541** 10
Jewers K *UK* **4105** 4
Jewett TJ *CAN* **1094** 2
Jewsbury JM *UK* **4563** 7
Jeyandran Mrs G *GUY* **1735** 3
Jeyaraj S *IN* **2546** 1
Jeyarajan R *IN* **2547** 1
Jeyarajan S *IN* **2627** 10
Jha GJ *IN* **2287** 2
Jha KP *IN* **1802** 4
Jha MP *IN* **1799** 1
Jhala VM *IN* **2302** 1
Jhingran AG *IN* **2105** 1
Jhingran AG *IN* **2215** 1
Jhoree H *MU* **3249** 9
Jhundoo CR *MU* **3244** 6
Jibogun Mrs AC *NGA* **3574** 2
Jinadasa DM *LK* **3778** 4
Jinap bte Selamat *MY* **3234** 8
Jindal JK *IN* **1771** 2

Jindal PC *IN* **1772** 3
Jindal SK *IN* **1977** 9
Jindal SK *IN* **2074** 1
Jinks RL *UK* **4225** 6
Jjemba PK *UG* **3989** 13
Joachim R *SLB* **3760** 1
Joanides C *AUS* **619** 3
Joannides G *CY* **1582** 2
Job TV *IN* **2528** 2
Jobin L *CAN* **1249** 1
Jobity A *TT* **3957** 1
Joblin ADH *NZ* **3287** 1
Joblin KN *NZ* **3389** 22
Jobling J *UK* **4379** 4
Jocumsen GL *AUS* **890** 2
Jodha NS *IN* **2740** 2
Jofriet JC *CAN* **1505** 8
Johal SS *IN* **2645** 1
Johannes RE *AUS* **114** 1
Johannson E *CAN* **1182** 2
Johansen C *IN* **2726** 4
Johanson A *UK* **4110** 5
Johar KS *IN* **2428** 4
Johari DC *IN* **2100** 1
Johari RP *IN* **1764** 6
John A *NZ* **3389** 23
John A *UK* **4396** 2
John B *AUS* **739** 1
John D *IN* **2600** 2
John P *UK* **4658** 5
John PCL *AUS* **735** 2
John VT *NGA* **3707** 7
Johns BW *CAN* **1338** 4
Johns G *AUS* **244** 3
Johns P *UK* **4770** 14
Johns PM *NZ* **3485** 8
Johns R *AUS* **639** 2
Johns RJ *PNG* **3738** 1
Johnson AD *AUS* **81** 2
Johnson AH *UK* **4873** 8
Johnson AL *MW* **3141** 5
Johnson AO *NGA* **3602** 7
Johnson AW *UK* **4024** 13
Johnson B *AUS* **614** 7
Johnson BC *CAN* **1316** 5
Johnson BY *AUS* **40** 10
Johnson C *CAN* **1346** 25
Johnson CB *AUS* **204** 1
Johnson CB *NZ* **3389** 24
Johnson CD *UK* **4024** 14
Johnson CL *UK* **4546** 7
Johnson DA *CAN* **1532** 9
Johnson DB *UK* **4630** 6
Johnson DE *AUS* **572** 4
Johnson DJ *CAN* **1485** 6
Johnson DL *CAN* **1153** 5
Johnson DL *NZ* **3322** 2
Johnson DS *UK* **4196** 3
Johnson DW *NZ* **3285** 4
Johnson ER *AUS* **849** 2
Johnson GR *NZ* **3358** 4
Johnson HT *AUS* **302** 5
Johnson IG *AUS* **310** 15
Johnson IR *AUS* **822** 8
Johnson IT *UK* **4213** 4
Johnson J *JAM* **2876** 2
Johnson JD *CAN* **1263** 2
Johnson JD *CAN* **1279** 3
Johnson KA *AUS* **813** 11
Johnson KG *AUS* **777** 10
Johnson LN *UK* **4655** 14
Johnson MS *AUS* **877** 8
Johnson MS *UK* **4554** 9
Johnson PL *CAN* **1414** 9
Johnson PN *NZ* **3403** 2
Johnson PNT *GH* **1650** 5
Johnson PW *CAN* **1088** 1
Johnson R *UK* **4316** 1
Johnson RC *CAN* **1166** 1
Johnson RJ *AUS* **602** 4
Johnson RL *AUS* **38** 27
Johnson RP *CAN* **1518** 4
Johnson RW *AUS* **456** 1
Johnson T *AUS* **707** 16
Johnson W *GUY* **1741** 4
Johnson WH *CAN* **1517** 6
Johnson WMP *AUS* **44** 4
Johnsson ID *UK* **4182** 5
Johnston AE *UK* **4327** 12
Johnston AM *UK* **4593** 4
Johnston AWB *UK* **4260** 14
Johnston BG *AUS* **191** 2
Johnston BR *AUS* **334** 2

Johnston CA *UK* **4881** 2
Johnston DC *NZ* **3482** 4
Johnston DE *UK* **4903** 10
Johnston HW *CAN* **1046** 7
Johnston JH *AUS* **229** 10
Johnston JH *CAN* **1244** 1
Johnston JR *UK* **4855** 6
Johnston KG *AUS* **813** 12
Johnston LAY *AUS* **26** 8
Johnston PG *AUS* **767** 15
Johnston PR *NZ* **3468** 4
Johnston RP *AUS* **438** 2
Johnston RW *CAN* **1420** 8
Johnston WB *NZ* **3482** 1
Johnston WS *UK* **4761** 5
Johnstone DR *UK* **4150** 4
Johnstone GR *AUS* **575** 3
Johnstone IB *CAN* **1514** 19
Johnstone IP *AUS* **842** 10
Johnstone PD *NZ* **3305** 1
Johnstone PK *AUS* **658** 3
Johnstone RS *AUS* **310** 16
Johnstone WD *CAN* **1355** 21
Johri BN *IN* **2844** 1
Johri BN *IN* **2845** 18
Johri NK *IN* **2778** 1
Johri TS *IN* **2101** 2
Jolaosho AA *NGA* **3692** 5
Jolette J *CAN* **1407** 1
Jolley PD *UK* **4205** 6
Jolliffe PA *CAN* **1450** 7
Jolly LO *AUS* **202** 3
Jolly MS *IN* **2691** 1
Jonas WE *NZ* **3340** 1
Jones A *GMB* **1634** 2
Jones A *UK* **4017** 3
Jones A *UK* **4862** 5
Jones AR *AUS* **322** 3
Jones ARC *CAN* **1550** 10
Jones AS *UK* **4340** 12
Jones AT *UK* **4351** 3
Jones B *UK* **4105** 5
Jones BMG *UK* **4586** 8
Jones CJ *UK* **4522** 13
Jones D *NZ* **3509** 1
Jones D *UK* **4289** 3
Jones D *UK* **4368** 3
Jones DA *UK* **4533** 2
Jones DB *AUS* **66** 20
Jones DG *UK* **4170** 3
Jones DG *UK* **4698** 3
Jones DI *UK* **4446** 6
Jones DIH *UK* **4373** 1
Jones DS *UK* **4556** 3
Jones DT *AUS* **295** 13
Jones DT *MY* **3226** 7
Jones E *TT* **3961** 1
Jones E *ZW* **4982** 1
Jones EA *UK* **4182** 6
Jones Mrs EA *UK* **4708** 4
Jones EL *AUS* **287** 13
Jones EL *UK* **4371** 2
Jones ERL *UK* **4363** 6
Jones G *GUY* **1735** 4
Jones GA *CAN* **1469** 1
Jones GE *UK* **4169** 12
Jones GT *UK* **4652** 3
Jones HG *UK* **4198** 2
Jones J *UK* **4252** 5
Jones JD *CAN* **1080** 2
Jones JE *UK* **4370** 3
Jones JET *UK* **4592** 1
Jones JK *UK* **4658** 6
Jones JLO *UK* **4446** 7
Jones JR *UK* **4368** 4
Jones Dr JWG *UK* **4657** 19
Jones K *UK* **4151** 3
Jones KM *AUS* **569** 4
Jones KW *UK* **4812** 11
Jones LD *AUS* **704** 6
Jones LP *AUS* **597** 5
Jones M *TT* **3943** 9
Jones M *UK* **4701** 5
Jones MA *UK* **4626** 4
Jones ME *CAN* **1346** 26
Jones MG *UK* **4554** 10
Jones MG *UK* **4701** 6
Jones MGK *UK* **4326** 7
Jones MGS *UK* **4449** 6
Jones MLC *UK* **4365** 3
Jones MN *UK* **4616** 13
Jones MR *AUS* **21** 30
Jones NO *AUS* **883** 3

Q

R

Rabey GG *ZW* **4957** 1
Rabindra RJ *IN* **2548** 4
Rabuor JO *KE* **2962** 19
Rachem K *BWA* **976** 25
Racz Prof. GJ *CAN* **1490** 1
Radadia NS *IN* **2325** 20
Radburn LR *CAN* **984** 3
Radcliffe JC *AUS* **488** 2
Radcliffe JE *NZ* **3294** 3
Radder GD *IN* **2403** 17
Rade VM *IN* **2849** 10
Radford BJ *AUS* **420** 2
Radford GL *UK* **4065** 2
Radford HM *AUS* **21** 49
Radford JB *NZ* **3534** 74
Radha Mrs NV *IN* **2546** 2
Radha Krishna T *IN* **2520** 3
Radhakrishanan AN *IN* **2641** 2
Radhakrishna K *IN* **2144** 3
Radhakrishna Pillay PN *IN* **2706** 6
Radhakrishnamurthy CH *IN* **2641** 3
Radhakrishnan EV *IN* **2138** 10
Radhakrishnan R *IN* **2521** 4
Radhakrishnan S *IN* **2126** 4
Radhakrishnan T *IN* **2607** 2
Radiro MPO *KE* **2959** 8
Radke CD *CAN* **1356** 5
Radley RW *UK* **4729** 30
Radokrishnan R *LK* **3781** 5
Radostits OM *CAN* **1477** 1
Radziah bte Othman *MY* **3240** 17
Rae AG *UK* **4169** 16
Rae DAH *UK* **4753** 19
Rae JR *AUS* **185** 2
Raemaekers R *ZM* **4930** 3
Raeside Prof. JI *CAN* **1514** 22
Raff JW *AUS* **623** 7
Ragab MTH *CAN* **1051** 10
Ragan CI *UK* **4678** 8
Ragan P *CAN* **1363** 5
Ragetli HWJ *CAN* **1175** 1
Ragg JM *UK* **4336** 16
Raghava SPS *IN* **1916** 6
Raghavaiah CV *IN* **1881** 3
Raghavan GSV *CAN* **1545** 9
Raghavan GV *IN* **2271** 5
Raghavan MRV *IN* **2639** 9
Raghavan N *IN* **2604** 2
Raghavan RS *IN* **2264** 4
Raghavendra AS *IN* **2641** 4
Raghavendra M *IN* **2697** 2
Raghavendra Rao NN *IN* **1919** 5
Raghavendra Rao V *IN* **2228** 5
Raghavulu P *IN* **2242** 11
Raghu JS *IN* **2433** 4
Raghunathan Nair N *IN* **2707** 1
Raghupathy A *IN* **2548** 5
Raghuveer P *IN* **1785** 11
Raghwanshi RKS *IN* **2427** 10
Ragobur D *MU* **3254** 2
Ragoonath Miss J *TT* **3942** 14
Ragwa LR *KE* **2979** 1
Rahal MS *IN* **2845** 28
Rahaley RS *AUS* **614** 15
Rahalkar RR *UK* **4214** 13
Rahamathulla Khan GA *IN* **2592** 4
Rahat D *GUY* **1735** 10
Rahe JE *CAN* **1453** 12
Rahim A *BGD* **919** 16
Rahmah M *MY* **3215** 7
Rahmah Ms WR *MY* **3211** 8
Rahman A *BGD* **959** 1
Rahman A *IN* **1865** 2
Rahman A *NZ* **3329** 4
Rahman ABMF *BGD* **920** 2
Rahman L *BGD* **923** 4
Rahman L *BGD* **957** 2
Rahman MA *BGD* **924** 1
Rahman MA *BGD* **936** 10
Rahman MA *BGD* **954** 1
Rahman ML *BGD* **939** 1
Rahman MM *BGD* **920** 1
Rahman MM *BGD* **968** 1
Rahman MS *BGD* **950** 2
Rahman MS *IN* **2760** 2
Rahman b. Syed Hj. Abd. Rashid SA *MY* **3179** 3
Rahming G *BHS* **905** 1
Rahnefeld GW *CAN* **1103** 1
Rahnefeld GW *CAN* **1486** 14
Rai A *IN* **2065** 12
Rai AK *IN* **2087** 1

Rai KN *IN* **2721** 2
Rai M *IN* **2007** 32
Rai M *IN* **2240** 5
Rai MM *IN* **2429** 7
Rai MP *IN* **2771** 3
Rai P *IN* **2023** 3
Rai P *IN* **2631** 8
Rai RD *IN* **2226** 1
Rai RN *IN* **2007** 33
Rai RP *IN* **1954** 6
Rai S *IN* **1765** 36
Rai SC *IN* **1800** 26
Rai SN *IN* **2032** 6
Rai SP *IN* **2127** 1
Raikhelkar SV *IN* **2459** 1
Raina BL *IN* **2053** 6
Raina BL *IN* **2677** 10
Raina HS *IN* **2135** 2
Raina SK *IN* **1778** 9
Rainbow AF *NZ* **3281** 10
Rainbow Prof. AJ *CAN* **1499** 20
Rainbow PS *UK* **4583** 30
Raine J *CAN* **1173** 3
Rainer SF *AUS* **117** 8
Raintree JB *KE* **3080** 21
Raison JK *AUS* **39** 5
Raison JM *AUS* **225** 14
Raison RJ *AUS* **59** 29
Raitra KC *IN* **2246** 9
Raiu B *IN* **2337** 11
Raj AS *IN* **2247** 5
Raj BT *IN* **1953** 8
Raj J *IN* **2411** 3
Raj RB *IN* **2245** 6
Raj Y *CAN* **1431** 3
Raj Kumar D *TT* **3986** 7
Raja Mrs IM *IN* **2563** 4
Raja a/l Amarthalingam *MY* **3232** 23
Raja Muhammad bin Raja Harun *MY* **3232** 24
Rajagopal CK *IN* **2530** 26
Rajagopal D *IN* **2406** 8
Rajagopal KV *IN* **2415** 1
Rajagopalan G *IN* **2626** 3
Rajagopalan M *IN* **2138** 11
Rajagopalan M *IN* **2228** 6
Rajagopalan MS *IN* **2136** 33
Rajagopalan V *IN* **2516** 1
Rajaguru ASB *LK* **3878** 1
Rajakrishnamoorthy V *IN* **2585** 9
Rajamani M *IN* **2140** 8
Rajamani S *IN* **1807** 8
Rajamanickam G *IN* **2581** 6
Rajamannar A *IN* **2575** 1
Rajamma Mrs P *IN* **2176** 21
Rajan AV *IN* **2530** 27
Rajan KN *IN* **2136** 34
Rajan MSS *IN* **2242** 12
Rajan SSS *NZ* **3328** 16
Rajan TSS *IN* **2593** 4
Rajanauth G *TT* **3943** 14
Rajanna L *IN* **2694** 1
Rajapakse NC *LK* **3877** 6
Rajapakse Mrs S *LK* **3877** 7
Rajapakse TBJ *LK* **3849** 8
Rajapandian ME *IN* **2140** 9
Rajapaske D *LK* **3842** 3
Rajapaske R *LK* **3894** 9
Rajapaske RMT *LK* **3803** 10
Rajappan K *IN* **2776** 3
Rajasekaran J *IN* **2613** 2
Rajasekaran P *IN* **2527** 4
Rajasekaran R *IN* **2584** 1
Rajasekaran S *IN* **2552** 4
Rajashekaraiah S *IN* **2405** 20
Rajasingham CC *LK* **3856** 1
Rajavelu G *IN* **2601** 6
Rajbanshi VK *IN* **2501** 1
Raje Miss CR *IN* **1852** 3
Raje SR *IN* **2425** 6
Rajendram GF *LK* **3890** 3
Rajendram NS *ZM* **4934** 1
Rajendran *IN* **2205** 3
Rajendran K *IN* **2520** 4
Rajendran P *IN* **2564** 7
Rajendran PG *IN* **2176** 22
Raji JA *NGA* **3679** 6
Rajkhowa C *IN* **2007** 34
Rajkomar B *MU* **3250** 7
Rajkomar I *MU* **3245** 1
Rajmane KD *IN* **2463** 3
Rajor RB *IN* **2041** 8
Rajorhia GS *IN* **2041** 9
Rajpal *IN* **2331** 9

Rajput DS *IN* **1992** 1
Rajput GS *IN* **2427** 11
Rajput MS *IN* **1943** 7
Rajput RK *IN* **1971** 1
Rajput TBS *IN* **1777** 5
Raju AS *IN* **2245** 7
Raju AS *IN* **2248** 5
Raju CA *IN* **1887** 2
Raju DG *IN* **2245** 8
Raju KS *IN* **2245** 9
Raju KV *IN* **2248** 6
Raju MS *IN* **2242** 13
Raju NBT *IN* **2246** 10
Raju VT *IN* **2250** 5
Rajukannu K *IN* **2548** 6
Rajyalakshmi Mrs T *IN* **2132** 2
Rakaiyappan P *IN* **1826** 2
Rakow GFW *CAN* **1125** 8
Rakowski P *CAN* **1340** 2
Ralph BJ *AUS* **645** 6
Ralph IG *AUS* **707** 24
Ram B *IN* **2808** 1
Ram G *IN* **1954** 7
Ram H *IN* **2819** 8
Ram KJ *IN* **2112** 19
Ram M *IN* **2007** 35
Ram MHY *IN* **2639** 10
Ram RD *IN* **1793** 3
Ram RS *IN* **1821** 6
Ram S *IN* **1803** 1
Ram S *IN* **2498** 11
Ram S *IN* **2818** 1
Ram Babu *IN* **1961** 12
Ram Balak *IN* **1976** 5
Ram Kumar *IN* **2046** 4
Ram Nath *IN* **2185** 1
Ram Phal *IN* **1757** 40
Ram Ratan *IN* **2080** 2
Rama Mohan Rao MS *IN* **1963** 1
Rama Rao K *IN* **2230** 13
Ramachandaran TK *IN* **2587** 15
Ramachander PR *IN* **1924** 1
Ramachandra Prasad TV *IN* **2403** 18
Ramachandran K *IN* **2564** 8
Ramachandran M *IN* **2240** 6
Ramachandran M *IN* **2704** 1
Ramachandran Mrs P *IN* **1770** 12
Ramachandran PV *IN* **2616** 1
Ramachandran S *IN* **2591** 2
Ramachela K *ZW* **4993** 5
Ramadasan Krishnan *MY* **3176** 79
Ramadoss K *IN* **2137** 5
Ramadoss N *IN* **2571** 2
Ramahotar K *MU* **3244** 8
Ramaiah B *IN* **2247** 6
Ramaiah M *IN* **2579** 5
Ramaiah PK *IN* **2700** 2
Ramakrishna *IN* **1979** 4
Ramakrishna KV *IN* **2126** 5
Ramakrishna O *IN* **2265** 1
Ramakrishna Rao G *IN* **2459** 16
Ramakrishnan G *IN* **2576** 2
Ramakrishnan MS *IN* **2575** 4
Ramakrishnan N *IN* **1765** 37
Ramakrishnan S *IN* **2573** 4
Ramakrishnan Mrs U *IN* **1765** 38
Ramakrishnan Nayar TV *IN* **2176** 23
Ramakrishnayya BV *IN* **1893** 1
Ramakrishnayya G *IN* **1812** 4
Ramakrishniah M *IN* **2120** 2
Ramalingam C *IN* **2552** 5
Ramalingam SR *IN* **2572** 7
Ramamirtham CP *IN* **2136** 35
Ramamurthi R *IN* **2626** 1
Ramamurthi TG *IN* **2587** 16
Ramamurthy K *IN* **2473** 1
Ramamurthy MK *IN* **2050** 13
Ramamurthy S *IN* **2138** 12
Ramamurthy VV *IN* **1765** 39
Ramamurty PS *IN* **2641** 5
Raman K *IN* **2126** 6
Raman KV *IN* **2228** 7
Raman Mrs S *IN* **2248** 7
Raman S *IN* **2325** 21
Raman TS *IN* **1786** 22
Ramana Rao BV *IN* **1990** 6
Ramana Rao TC *IN* **1830** 5
Ramanagowda P *IN* **2403** 19
Ramanaiah NV *IN* **2242** 14
Ramananda RD *IN* **2167** 1
Ramanatha Chetty CK *IN* **1989** 1
Ramanatha Rao V *IN* **2742** 5
Ramanathan G *IN* **2521** 5

INDEX OF INSTITUTIONS

Magoye Regional Research Station *ZM* **4946**
Maharashtra Association for the Cultivation of Science *IN* **2664**
Mahatma Phule Agricultural University *IN* **2846**
Makoholi Experiment Station *ZW* **4984**
Makoka Agricultural Research Station *MW* **3111**
Malawi Tobacco Research Authority *MW* **3137**
Malaysian Agricultural Research and Development Institute *MY* **3175**
Mallee Research Station *AUS* **643**
Manchester University *UK* **4614**
Manipur Agricultural College *IN* **2649**
Manitoba Natural Resources *CAN* **1385**
Mansa Regional Research Station *ZM* **4940**
Marathwada Agricultural University *IN* **2458**
Marcus Oldham Farm Management College *AUS* **896**
Mariakani Animal Husbandry Research Station *KE* **2907**
Marine Biological Association of the United Kingdom *UK* **4088**
Market Analysis Directorate, Agriculture Canada *CAN* **1179**
Market Outlook and Analysis Division, Agriculture Canada *CAN* **1180**
Matopos Research Station *ZW* **4985**
Mauritius Sugar Industry Research Institute *MU* **3257**
McGill University *CAN* **1541**
McMaster Laboratory *AUS* **19**
McMaster University *CAN* **1498**
Meat and Wool Division, Department of Primary Industry *AUS* **184**
Meat Industry Research Institute of New Zealand (Inc.) *NZ* **3539**
Meat Industry Research Institute of New Zealand (Inc.) *NZ* **3539**
Mechery Sheep Research Station, Potteneri *IN* **2619**
Medical Research Council of New Zealand Virus Research Unit *NZ* **3510**
Melbourne and Metropolitan Board of Works *AUS* **691**
Merlewood Research Station - ITE *UK* **4075**
Meru Sub-Research Station *MW* **3116**
Migratory Birds and Threatened Species Conservation Division *CAN* **1328**
Migratory Birds Management Division *CAN* **1331**
Minerals Unit *AUS* **24**
Ministry of Agriculture *BWA* **973**
Ministry of Agriculture *GH* **1670**
Ministry of Agriculture *GMB* **1630**
Ministry of Agriculture *GUY* **1727**
Ministry of Agriculture *JAM* **2872**
Ministry of Agriculture *MW* **3097**
Ministry of Agriculture and Fisheries *NZ* **3278**
Ministry of Agriculture and Forestry *UG* **3987**
Ministry of Agriculture and Lands *SLB* **3742**
Ministry of Agriculture and Livestock Development *KE* **2897**
Ministry of Agriculture and Livestock Development *TZ* **3902**
Ministry of Agriculture and Natural Resources *CY* **1580**
Ministry of Agriculture, Fisheries and Food *UK* **4408**
Ministry of Agriculture, Fisheries and Natural Resources *MU* **3243**
Ministry of Agriculture, Lands, & Food Production *TT* **3936**
Ministry of Agriculture, Trade and Industry *BHS* **901**
Ministry of Agriculture, Trade, Lands & Housing *MSR* **3276**
Ministry of Forestry *NZ* **3353**
Ministry of Forestry and Natural Resources *MW* **3131**
Ministry of Lands, Agriculture and Rural Resettlement *ZW* **4948**
Ministry of Natural Resources *SLB* **3759**
Ministry of Primary Industries *FJ* **1610**
Ministry of Tourism and Wildlife *KE* **3037**
Minnipa Research Centre *AUS* **540**
Misamfu Regional Research Station *ZM* **4943**
Mochipapa Regional Research Station *ZM* **4938**
Moist Forest Research Station, Sapoba *NGA* **3597**
Monash University *AUS* **771**
Mongu Regional Research Station *ZM* **4941**
Monks Wood Experimental Station - ITE *UK* **4052**
Monogastric Research Centre *PNG* **3728**
Montpelier Agricultural Research Station *JAM* **2884**
Moredun Research Institute *UK* **4168**
Mount Allison University *CAN* **1572**
Mount Makulu Central Research Station *ZM* **4926**
MRC Mammalian Genome Unit *UK* **4823**
Msabaha Agricultural Research Station *KE* **2921**
Msekera Regional Research Station *ZM* **4942**
Muck Research Station *CAN* **1400**
Murdoch University *AUS* **773**
Museum of Victoria *AUS* **695**

Mutanda Regional Research Station *ZM* **4944**
Mwea Irrigation Research Station *KE* **3056**

N

Nakawa Forestry Research Station *UG* **3992**
Namulonge Research Station *UG* **3991**
National Academy of Agricultural Research Management *IN* **2228**
National Agricultural Research Centre, Muguga *KE* **2934**
National Agricultural Research Institute *GUY* **1733**
National Agricultural Research Laboratories *KE* **2957**
National Agricultural Research Station *KE* **2964**
National Agricultural Training Institute *SLB* **3772**
National Animal Disease Centre *TT* **3960**
National Animal Husbandry Station *KE* **2971**
National Animal Production Research Institute, Shika *NGA* **3598**
National Bureau of Animal Genetic Resources and National Institute of Animal Genetics *IN* **2210**
National Bureau of Fish Genetic Resources *IN* **2215**
National Bureau of Plant Genetic Resources (IARI Campus) *IN* **2178**
National Bureau of Soil Survey and Land Use Planning *IN* **2197**
National Centre for Mushroom Research and Training *IN* **2226**
National Coconut Development Project *TZ* **3906**
National Dairy Research Institute *IN* **2027**
National Dryland Farming Research Station Katumani (NDFRS) *KE* **2976**
National Herbarium & Botanic Garden *ZW* **4990**
National Horticultural Research Institute, Ibadan *NGA* **3608**
National Institute of Agricultural Botany *UK* **4716**
National Institute of Plantation Management *LK* **3812**
National Irrigation Board Operational Research and Training Project *KE* **3051**
National Irrigation Research Station *ZM* **4945**
National Museums of Kenya *KE* **3045**
National Padi and Rice Grading Centre *GUY* **1742**
National Potato Research Station *KE* **2984**
National Pulses Research Centre, Pudukkottai *IN* **2567**
National Pyrethrum and Horticultural Research Station *KE* **2991**
National Range Research Station, Kiboko *KE* **2997**
National Research Centre for Groundnut *IN* **2218**
National Research Centre on Camel *IN* **2227**
National Research Centre on Equines *IN* **2214**
National Root Crops Research Institute *NGA* **3616**
National Seed Quality Control Service *KE* **3000**
National Sugar Research Station, Kibos *KE* **3005**
National Veterinary Laboratory, Kila Kila *PNG* **3727**
National Veterinary Research Centre, Mugaga *KE* **3015**
Natural Environment Research Council *UK* **4039**
Natural Resources Conservation Department *JAM* **2885**
Natural Resources Energy & Science Authority *LK* **3845**
Nature Conservancy Council *UK* **4400**
New Liskeard College of Agricultural Technology *CAN* **1415**
New Zealand Dairy Board *NZ* **3527**
New Zealand Dairy Research Institute *NZ* **3534**
New Zealand Leather and Shoe Research Association *NZ* **3536**
New Zealand Logging Industry Research Association Inc. *NZ* **3535**
New Zealand Meat & Wool Boards' Economic Service *NZ* **3537**
New Zealand Meteorological Service *NZ* **3519**
Newfoundland Forestry Centre *CAN* **1234**
Ngabu Agricultural Research Station *MW* **3112**
Nigerian Institute For Oil Palm Research *NGA* **3627**
Nigerian Institute for Trypanosomiasis, Kaduna *NGA* **3640**
Nigerian Stored Products Research Institute, Lagos *NGA* **3655**
Nimbkar Agricultural Research Institute *IN* **2667**
Norfolk Agricultural Station *UK* **4730**
North and Northeast Regional Laboratory, CSIRO, Division of Fisheries *AUS* **116**
North Coast Agricultural Institute *AUS* **246**
Northern Forestry Centre *CAN* **1271**
Northern Ireland Plant Breeding Station *UK* **4895**
Northern South Island Region *NZ* **3287**
Northfield Research Laboratories *AUS* **505**
Northfield Research Laboratories *AUS* **512**
Nova Scotia Agricultural College *CAN* **1562**

Nova Scotia Research Foundation Corporation *CAN* **1438**
NSW Department of Agriculture Region 2, New England, Hunter & Metropolitan *AUS* **251**
NSW Department of Agriculture Region 4, Central West, South East and Illawarra *AUS* **271**
NSW Department of Agriculture Region 5, Murray and Riverina *AUS* **281**
NSW Department of Agriculture Region I, North Coast *AUS* **240**
Nuclear Research Laboratory *IN* **1786**
Nyandarua Agricultural Research Station *KE* **2922**
Nyanza Agricultural Research Station *KE* **2925**

O

Oban VI Centre *UK* **4876**
Official Seed Testing Station for England and Wales *UK* **4726**
Oil Palm Research Association *PNG* **3725**
Oil Palm Research Centre *GH* **1663**
Ontario Agricultural College *CAN* **1501**
Ontario Agricultural Museum *CAN* **1396**
Ontario Ministry of Agriculture and Food *CAN* **1392**
Ontario Ministry of Natural Resources *CAN* **1426**
Ontario Tree Improvement and Forest Biomass Institute *CAN* **1428**
Ontario Tree Improvement and Forest Biomass Institute Field Units *CAN* **1434**
Ontario Veterinary College *CAN* **1513**
Oonoonba Animal Health Station *AUS* **404**
Orange Agricultural College *AUS* **899**
Orange Agricultural Research and Veterinary Centre *AUS* **277**
Orissa University of Agriculture and Technology *IN* **2473**
Overseas Development Administration *UK* **4090**
Overseas Development Natural Resources Institute (ODNRI) *UK* **4098**

P

Pacific Forestry Centre *CAN* **1290**
Paddy Research Station, Ambasamudram *IN* **2587**
Pastoral Research Centre, Erap *PNG* **3729**
Perkerra Agricultural Research Station *KE* **2932**
Petawawa National Forestry Institute *CAN* **1223**
Pig Industry Board *ZW* **5017**
Plant Diseases Division *NZ* **3465**
Plant Health and Plant Products Directorate *CAN* **1213**
Plant Health Division *CAN* **1214**
Plant Introduction and Quarantine Station *PNG* **3724**
Plant Physiology Division, DSIR *NZ* **3472**
Plant Protection Research Institute *ZW* **4991**
Plant Research Centre *CAN* **1024**
Plant Testing Station *UK* **4896**
Plantation Forest Research Centre *AUS* **62**
Policy Branch, Agriculture Canada *CAN* **1178**
Polytechnic of Sokoto State, Birnin Kebbi *NGA* **3699**
Polytechnic, Ibadan *NGA* **3698**
Potato Research Station, Toolangi *AUS* **644**
Poultry Research and Development Centre, Madurai *IN* **2610**
Poultry Research and Development Centre, Namakkal *IN* **2620**
Poultry Research and Development Centre, Pudukkottai *IN* **2621**
Poultry Research and Development Centre, Trichy *IN* **2628**
Poultry Research Station, Nandanam *IN* **2622**
Prairie Agricultural Machinery Institute *CAN* **1381**
Prairie Migratory Birds Research Centre *CAN* **1336**
Program Coordination Directorate, Agriculture Canada *CAN* **985**
Pulp and Paper Research Organisation of New Zealand *NZ* **3373**
Punjabi University *IN* **2645**
Pwllpeiran Experimental Husbandry Farm *UK* **4461**

Q

Queen Mary College *UK* **4582**
Queen Victoria Museum and Art Gallery *AUS* **589**
Queen's University of Belfast *UK* **4906**
Queensland Animal Research Institute, Queensland *AUS* **401**